Register Now to Improve Your Grade!

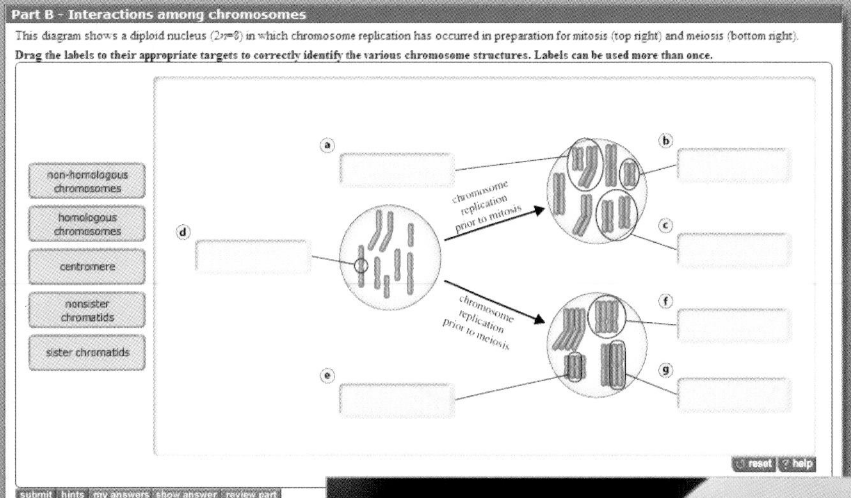

- Access your textbook online 24/7
- Take practice quizzes to prepare for tests
- Grasp difficult concepts with 3-D BioFlix® animations, engaging activities, and more

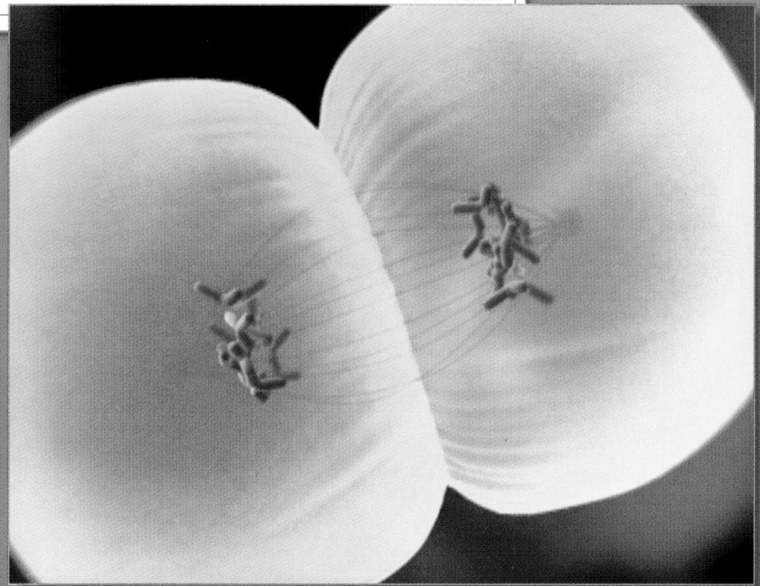

www.masteringbiology.com

STUDENTS

To Register Using the Student Access Kit

Your textbook may have been packaged with a **Mastering Student Access Kit**. This kit contains your access code to this valuable website.

To Purchase Access Online

If your textbook was not packaged with a **Mastering Student Access Kit**, you can purchase access online using a major credit card or PayPal account.

INSTRUCTORS

To Request Access Online

To request access online, go to www.masteringbiology.com and click New Instructors. Please contact your sales representative for more information.

Find important information in this book by focusing on the Gold Thread.

These black swan chicks are being introduced to the world by their mother, who trails close behind them. Likewise, this chapter introduces the key ideas that launched biological science as a discipline.

Biology and the Tree of Life 1

In essence, biological science is a search for ideas and observations that unify our understanding of the diversity of life, from bacteria living in rocks a mile underground to hedgehogs and humans. Chapter 1 is an introduction to this search.

The goals of this chapter are to introduce the nature of life and explore how biologists go about studying it. The chapter also introduces themes that will resonate throughout this book: (1) analyzing how organisms work at the molecular level, (2) understanding organisms in terms of their evolutionary history, and (3) helping you learn to think like a biologist.

Let's begin with what may be the most fundamental question of all: What is life?

KEY CONCEPTS

- Organisms obtain and use energy, are made up of cells, process information, replicate, and as populations evolve.
- The cell theory proposes that all organisms are made of cells and that all cells come from preexisting cells.
- The theory of evolution by natural selection maintains that species change through time because individuals with certain heritable traits produce more offspring than other individuals do.
- A phylogenetic tree is a graphical representation of the evolutionary relationships between species. These relationships can be estimated by analyzing similarities and differences in traits. Species that share distinctive traits are closely related and are placed close to each other on the tree of life.
- Biologists ask questions, generate hypotheses to answer them, and design experiments that test the predictions made by competing hypotheses.

1.1 What Does It Mean to Say That Something Is Alive?

An **organism** is a life-form—a living entity made up of one or more cells. Although there is no simple definition of life that is endorsed by all biologists, most agree that organisms share a suite of five fundamental characteristics.

- *Energy* To stay alive and reproduce, organisms have to acquire and use energy. To give just two examples: plants absorb sunlight; animals ingest food.
- *Cells* Organisms are made up of membrane-bound units called cells. A cell's membrane regulates the passage of materials between exterior and interior spaces.
- *Information* Organisms process hereditary or genetic information, encoded in units called genes, along with information they acquire from the environment. Right now cells throughout your body are using genetic information to make the molecules that keep you alive; your eyes and brain are decoding information on this page that will help you learn some biology.

✔ When you see this checkmark, stop and test yourself. Answers are available in Appendix B.

1

Key Concepts

Start with Key Concepts on the first page of every chapter. Read these gold key points first to familiarize yourself with the chapter's big ideas.

MORE! Bulleted Lists

Take note of bulleted lists that "chunk" information and ideas. This will help you manage the information that you are learning in the course.

Gold Highlighting

Watch for important information highlighted in gold. Gold highlighting is always a signal to slow down and pay special attention.

Gold Key 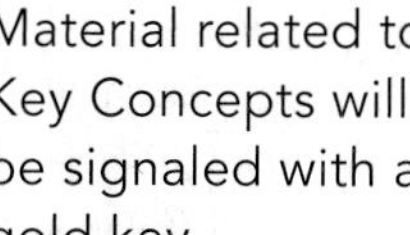

Material related to Key Concepts will be signaled with a gold key.

As you read the text and view the figures, practice with the Blue Thread.

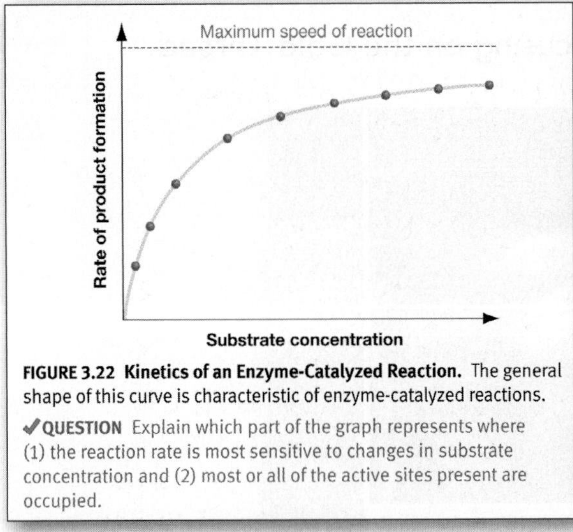

FIGURE 3.22 Kinetics of an Enzyme-Catalyzed Reaction. The general shape of this curve is characteristic of enzyme-catalyzed reactions.

✔**QUESTION** Explain which part of the graph represents where (1) the reaction rate is most sensitive to changes in substrate concentration and (2) most or all of the active sites present are occupied.

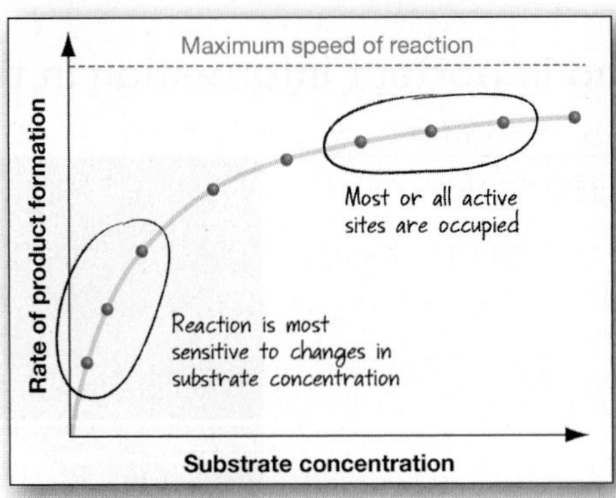

Blue Thread Questions

Many figures include Blue Thread Questions or Exercises to help you check your understanding of the material they present.

Drawing Exercises

Some Blue Thread Questions contain artwork from the textbook that you will be asked to draw on or modify.

NEW! Suggested Answers

Suggested answers for the Blue Thread Questions and Exercises are provided in Appendix B.

"You Should be Able To" Exercises

Text passages flagged with blue type and the words "you should be able to" offer exercises on concepts that professors and students have identified as most difficult. These are the topics most students struggle with on exams.

To read this graph, put your finger on the *x*-axis at time 0. Then read up the *y*-axis, and note that kernels averaged about 11 percent protein at the start of the experiment. Now read the graph to the right. Each dot is a data point, representing the average kernel protein concentration in a particular generation. (A generation in maize is one year.) The lines on this graph simply connect the dots, to make the pattern in the data easier to see. At the end of the graph, after 100 generations of selection, average kernel protein content is about 29 percent. (For more help with reading graphs, see **BioSkills 2** in Appendix A.)

This sort of change in the characteristics of a population, over time, is evolution. Humans have been practicing artificial selection for thousands of years, and biologists have now documented evolution by *natural* selection—where humans don't do the selecting—occurring in thousands of different populations, including humans.

To practice applying the principles of artificial selection, go to the online study area at *www.masteringbiology.com.*

(MB) **Web Activity** Artificial Selection

Evolution occurs when heritable variation leads to differential success in reproduction. ✔If you understand this concept, you should be able to describe how protein content in maize kernels changed over time, using the same *x*-axis and *y*-axis as in Figure 1.3, when researchers selected individuals with *lowest* kernel protein content to be the parents of the next generation. (This experiment was actually done, starting with the same population at the same time as selection for high protein content.)

FITNESS AND ADAPTATION Darwin also introduced some new terminology to identify what is happening during natural selection.

CHECK YOUR UNDERSTANDING

If you understand that . . .

- Natural selection occurs when heritable variation in certain traits leads to improved success in reproduction. Because individuals with these traits produce many offspring with the same traits, the traits increase in frequency and evolution occurs.
- Evolution is a change in the characteristics of a population over time.

✔ **You should be able to . . .**

On the graph you just analyzed, describe the average kernel protein content over time in a maize population where *no* selection occurred.

Answers are available in Appendix B.

1.4 The Tree of Life

Section 1.3 focused on how individual populations change through time in response to natural selection. But over the past several decades, biologists have also documented dozens of cases in which natural selection has caused populations of one species to diverge and form new species. This divergence process is called **speciation**.

Research on speciation has two important implications: All species come from preexisting species, and all species, past and present, trace their ancestry back to a single common ancestor.

The theory of evolution by natural selection predicts that biologists should be able to reconstruct a **tree of life**—a family tree

CHAPTER 1 Biology and the Tree of Life 5

Check Your Understanding

The blue half of the Check Your Understanding boxes asks you to do something with the information in the top half. If you can't complete these exercises, go back and re-read that section of the chapter.

MasteringBIOLOGY®

Make Learning Part of the Grade®

It's possible that your professor will include these Blue Thread Questions in a graded assignment at www.masteringbiology.com.

NEW! Bulleted Summary of Key Concepts

The succinct Summary of Key Concepts reviews important concepts in short, manageable bullet points.

Blue Thread Exercises

End-of-chapter Blue Thread Exercises help you review the major themes of the chapter and synthesize information.

CHAPTER 6 REVIEW *For media, go to the study area at www.masteringbiology.com*

Summary of Key Concepts

Phospholipids are amphipathic molecules—they have a hydrophilic region and a hydrophobic region. In solution, phospholipids spontaneously form bilayers that are selectively permeable—meaning that only certain substances cross them readily.

- The plasma membrane forms a physical barrier between the internal and external environment—often between life and nonlife.
- The basic structure of plasma membranes is created by a phospholipid bilayer.

solutes to the region of lower water concentration and higher solute concentration.

- Osmosis is a passive process driven by an increase in entropy.

✔You should be able to imagine a beaker with solutions separated by a plasma membrane, and then predict what will happen after addition of a solute to one side if the solute (1) crosses the membrane readily or (2) is incapable of crossing the membrane.

Web Activity Diffusion and Osmosis

Make Learning Part of the Grade®

Visit www.masteringbiology.com for practice quizzes, 3-D animations, the eText, and more.

BioFlix™

BioFlix™ 3-D animations are included in MasteringBiology's Study Area and are available as automatically graded assignments.

Analyze: Can I recognize underlying patterns and structure?

Evaluate: Can I make judgements on the relative value of ideas and information?

Synthesize: Can I put ideas and information together to create something new?

Apply: Can I use these ideas in a new situation?

Explain: Can I explain this concept in my own words?

Remember: Can I recall the key terms and ideas?

Bloom's Taxonomy

Bloom's Taxonomy categorizes six levels of learning competency. The Blue Thread Questions and Exercises in the textbook test on the higher levels of the scale—Explain, Apply, Analyze, Evaluate, and Synthesize—to help you develop critical thinking skills and prepare you for exams.

Steps to Understanding

End of Chapter questions are scaled along Bloom's Taxonomy.

✔TEST YOUR KNOWLEDGE

Begin by testing your knowledge of new facts.

✔TEST YOUR UNDERSTANDING

Once you're confident in your knowledge of the material, demonstrate your understanding by answering the Test Your Understanding questions.

✔APPLYING CONCEPTS TO NEW SITUATIONS

Challenge yourself even further by applying your understanding of the concepts to new situations.

A unique emphasis on the process of scientific discovery and experimental design teaches you how to think like a scientist as you learn fundamental biology concepts.

Experiment Boxes

Study Experiment Boxes to help you understand how experiments are designed and give you practice interpreting data.

www.masteringbiology.com

NEW! Experimental Inquiry Tutorials

Experimental Inquiry Tutorials based on some of biology's most seminal experiments can be found on **www.masteringbiology.com**. Your instructor may assign these. They will give you practice analyzing the experimental design and data, and help you understand reasoning that led scientists from the data they collected to their conclusions.

Some of the topics include:

- The Process of Science
- Engelmann's Photosynthesis and Wavelengths of Light
- Morgan's Cross with White-Eyed Males
- Meselson-Stahl's Semiconservative Replication
- Steinhardt et al and Hafner et al's Polyspermy
- Grant's Changes in Finch Beak Size
- Went's Phototropism and Auxin Distribution
- Coleman's Obesity Gene
- Connell's Competition in Barnacles
- Bormann, Likens et al's Nutrient Cycling in Hubbard Brook Forest

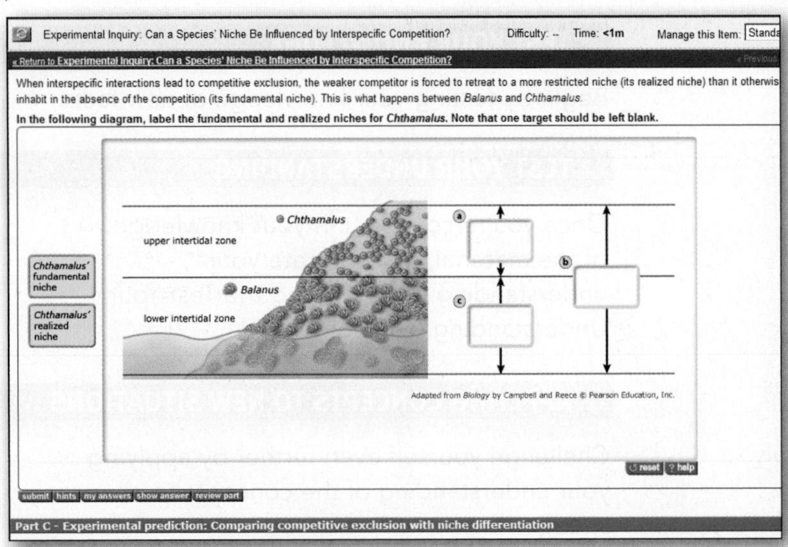

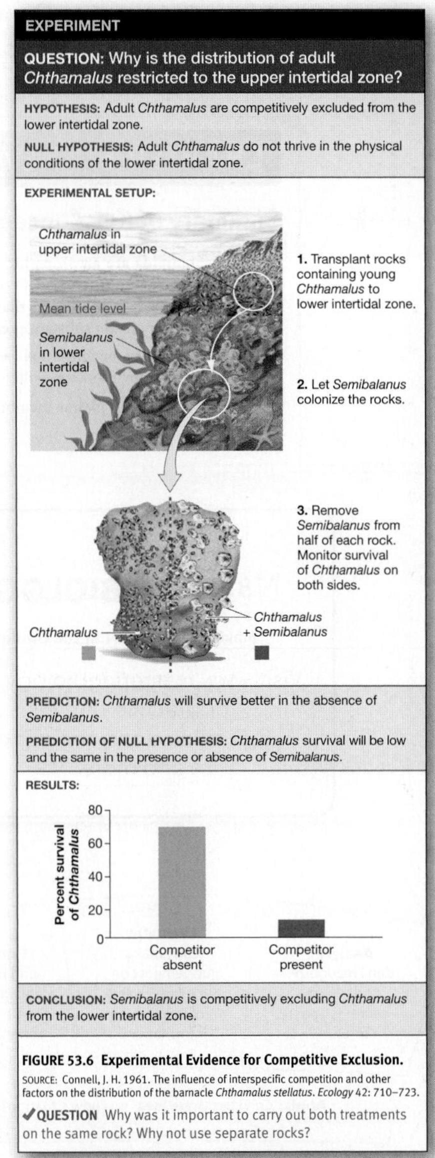

FIGURE 53.6 Experimental Evidence for Competitive Exclusion.

SOURCE: Connell, J. H. 1961. The influence of interspecific competition and other factors on the distribution of the barnacle *Chthamalus stellatus*. *Ecology* 42: 710–723.

✔QUESTION Why was it important to carry out both treatments on the same rock? Why not use separate rocks?

NEW! Source Citations

Each Experiment Box now cites the original research paper, encouraging you to extend your learning by exploring the primary literature.

NEW! Experiment Box Questions

Each Experiment Box now includes a question that asks students to analyze the design of the experiment.

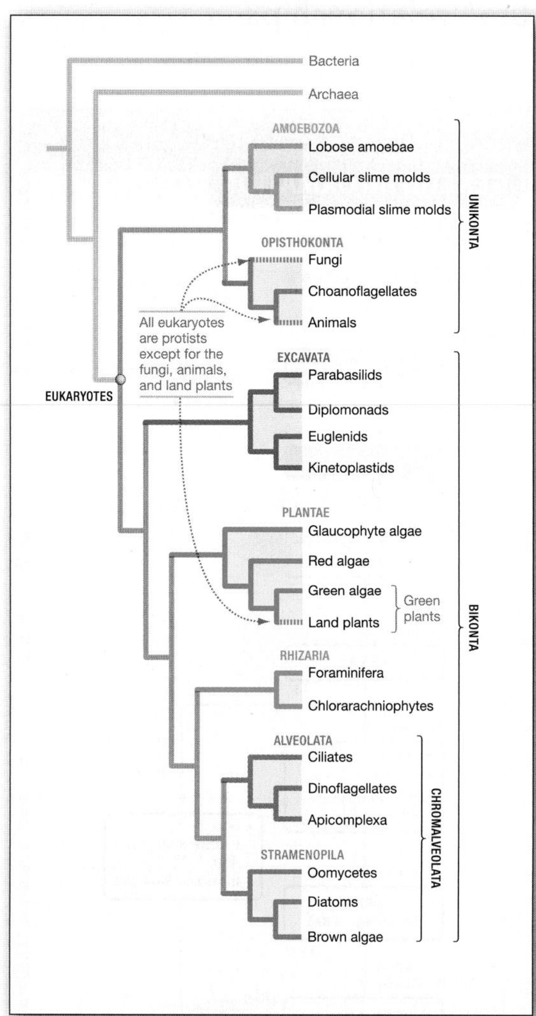

NEW! Redesigned Phylogenetic Trees

Practice "tree thinking" using these newly redesigned phylogenetic trees. Their U-shaped, top-to-bottom format is consistent with the way such trees are most commonly depicted in the scientific literature.

Expanded BioSkills Appendix

BIOSKILLS

Build skills that will be important to your success in future courses. At relevant points in the text, you'll find references to the expanded BioSkills Appendix that will help you learn and practice the following foundational skills:

- NEW! The Metric System
- Reading Graphs
- Reading a Phylogenetic Tree
- NEW! Some Common Latin and Greek Roots Used in Biology
- Using Statistical Tests and Interpreting Standard Error Bars
- Reading Chemical Structures
- Using Logarithms
- Making Concept Maps
- Separating and Visualizing Molecules
- Biological Imaging: Microscopy and X-Ray Crystallography
- NEW! Separating Cell Components by Centrifugation
- NEW! Cell Culture Methods
- Combining Probabilities
- NEW! Model Organisms

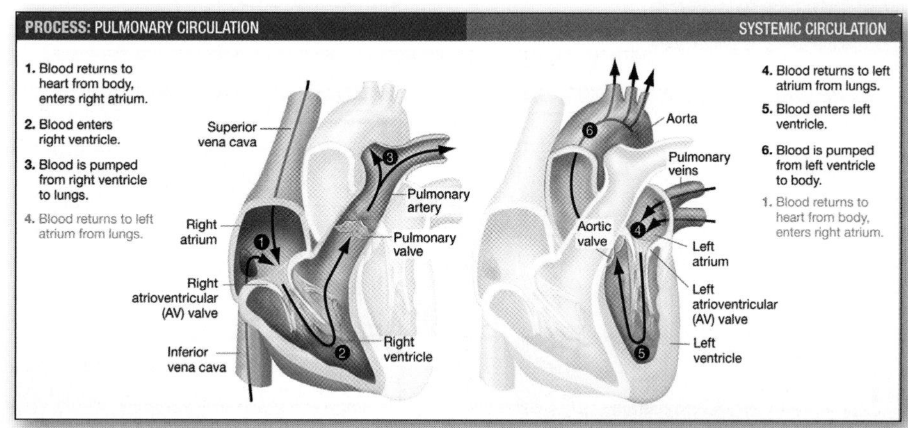

Informative Figures

Think through complex biological processes with figures that clearly define concepts.

Concept maps help you to keep sight of "big picture" relationships among biological concepts.

NEW! Big Picture Concept Maps

Four remarkable Big Picture concept maps help you synthesize information across the chapters on energy, genetics, evolution, and ecology.

Check Your Understanding

Check your understanding of these big picture relationships by answering the Blue Thread Questions.

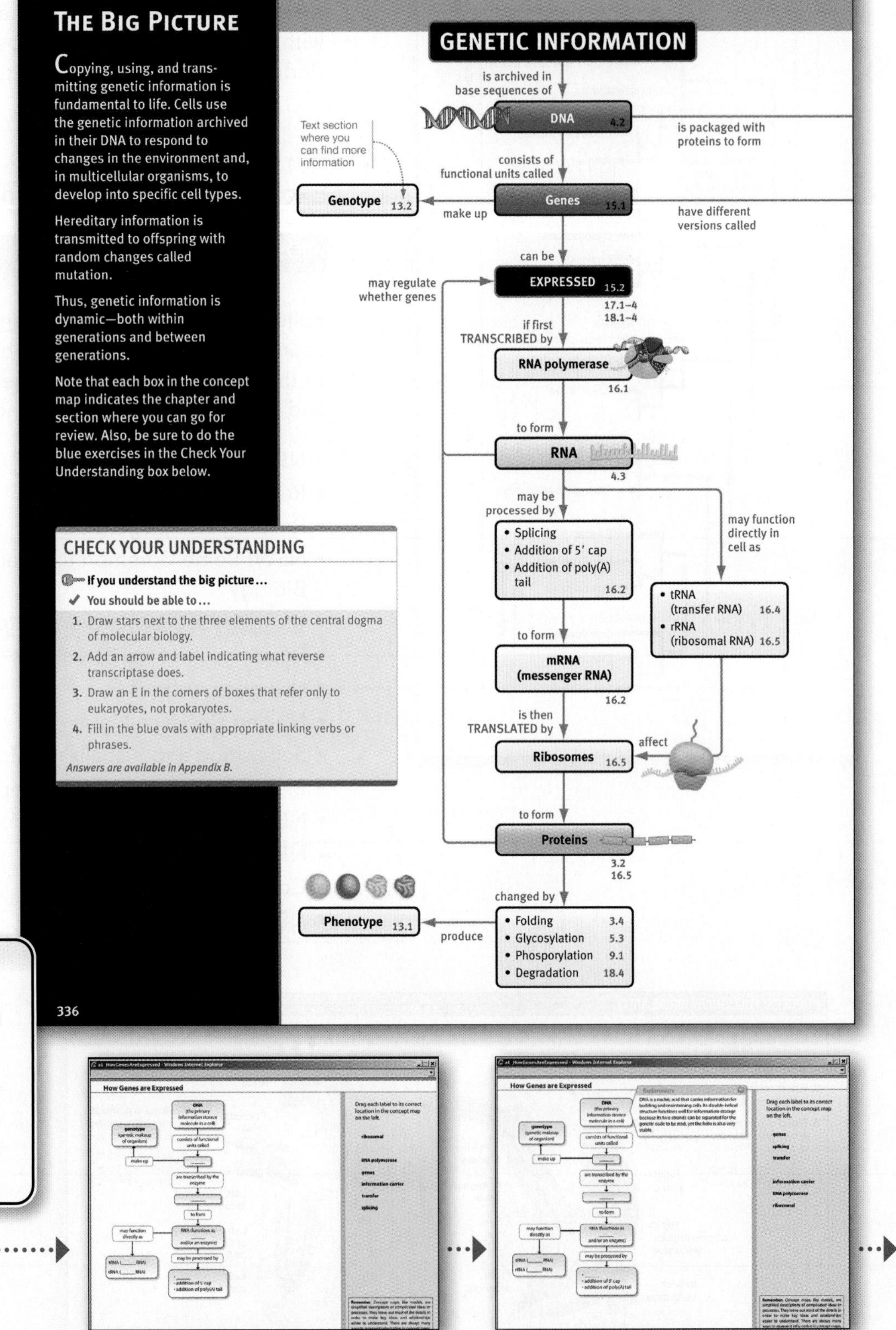

THE BIG PICTURE

Copying, using, and transmitting genetic information is fundamental to life. Cells use the genetic information archived in their DNA to respond to changes in the environment and, in multicellular organisms, to develop into specific cell types.

Hereditary information is transmitted to offspring with random changes called mutation.

Thus, genetic information is dynamic—both within generations and between generations.

Note that each box in the concept map indicates the chapter and section where you can go for review. Also, be sure to do the blue exercises in the Check Your Understanding box below.

CHECK YOUR UNDERSTANDING

If you understand the big picture...

✓ You should be able to...

1. Draw stars next to the three elements of the central dogma of molecular biology.
2. Add an arrow and label indicating what reverse transcriptase does.
3. Draw an E in the corners of boxes that refer only to eukaryotes, not prokaryotes.
4. Fill in the blue ovals with appropriate linking verbs or phrases.

Answers are available in Appendix B.

336

MasteringBIOLOGY®

Make Learning Part of the Grade®

Your professor may assign interactive Big Picture concept map exercises at www.masteringbiology.com.

More Big Picture activities are available in the study area at www.masteringbiology.com

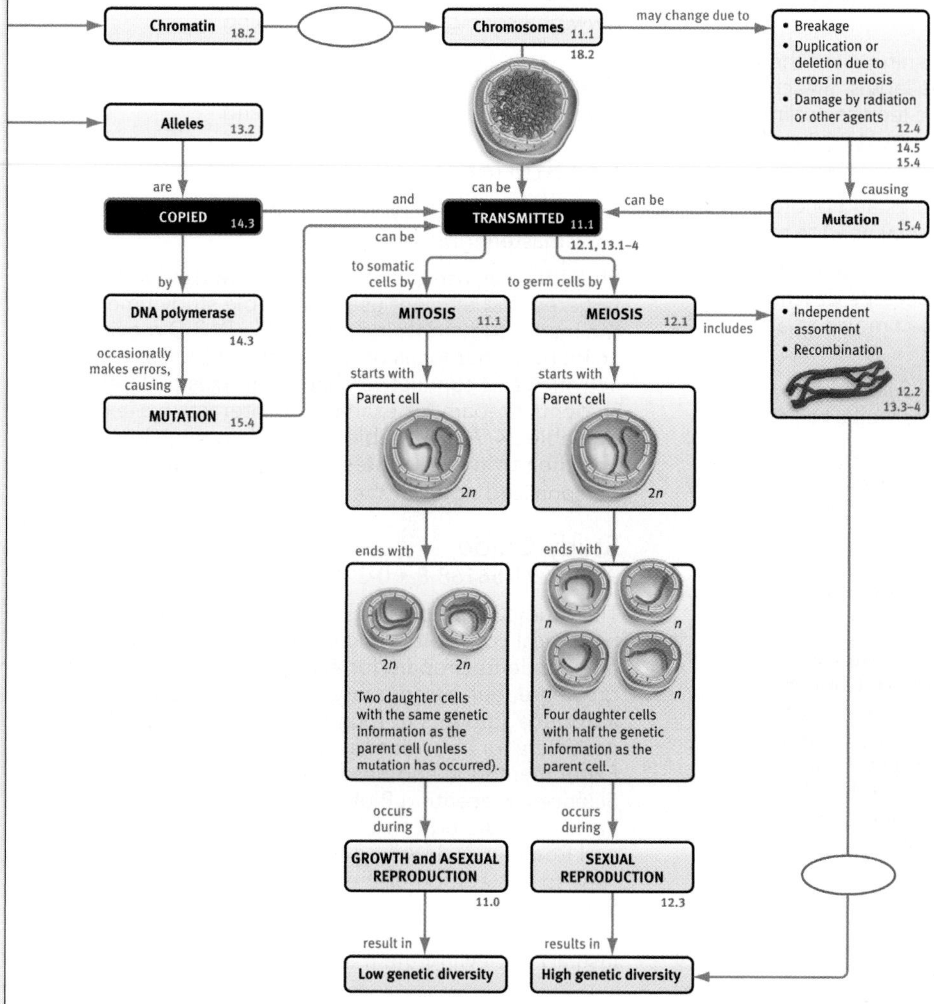

Chromatin 18.2 → ( ) → Chromosomes 11.1, 18.2

Chromosomes may change due to:
- Breakage
- Duplication or deletion due to errors in meiosis
- Damage by radiation or other agents

12.4, 14.5, 15.4

causing **Mutation** 15.4, which can be TRANSMITTED

Alleles 13.2 are **COPIED** 14.3 and TRANSMITTED 11.1

COPIED by **DNA polymerase** 14.3, which occasionally makes errors, causing **MUTATION** 15.4, which can be TRANSMITTED

Chromosomes can be TRANSMITTED 11.1 (12.1, 13.1–4)

to somatic cells by **MITOSIS** 11.1

to germ cells by **MEIOSIS** 12.1

MEIOSIS includes:
- Independent assortment
- Recombination

12.2, 13.3–4

MITOSIS starts with Parent cell (2n); ends with 2n, 2n: Two daughter cells with the same genetic information as the parent cell (unless mutation has occurred).

MITOSIS occurs during **GROWTH and ASEXUAL REPRODUCTION** 11.0, which result in **Low genetic diversity**

MEIOSIS starts with Parent cell (2n); ends with n, n, n, n: Four daughter cells with half the genetic information as the parent cell.

MEIOSIS occurs during **SEXUAL REPRODUCTION** 12.3, which results in **High genetic diversity**

Independent assortment and Recombination → ( ) → High genetic diversity

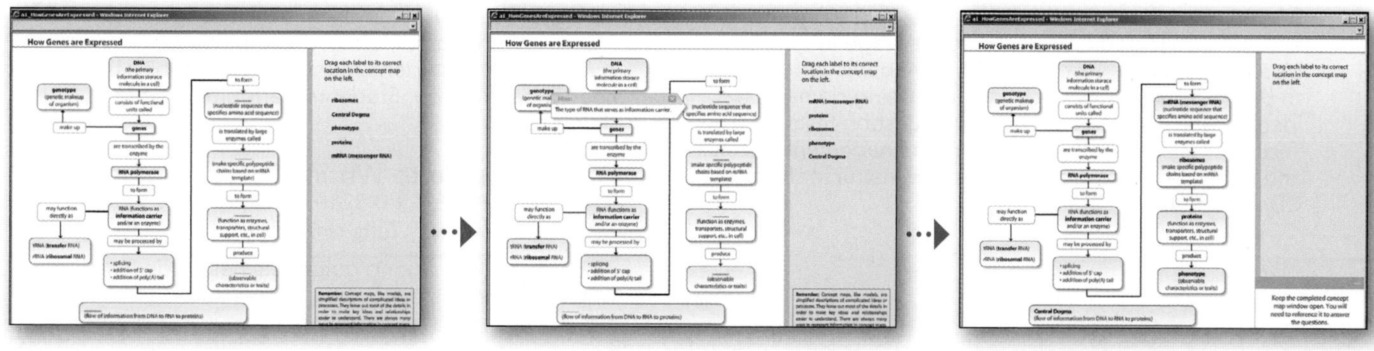

For Instructors

Instructor Resource CD/DVD-ROM

978-0-321-61351-6 • 0-321-61351-1

Everything instructors need for lectures is in one place, including video segments that demonstrate how to incorporate active-learning techniques into your own classroom. The Instructor Resource CD/DVD-ROM includes:

- PowerPoint® Lecture Tools containing all of the figures and photos, which have editable labels; five clicker questions per chapter; pre-made lecture outlines containing select images from the text with embedded animations
- JPEG images of all textbook figures and photos including printer-ready transparency acetate masters
- Over 300 animations and videos that accurately depict complex topics and dynamic processes described in the book; five new BioFlix™ 3-D movie-quality animations
- Instructor's Guides for *Biological Science* and *Practicing Biology* are available as well as a list of all primary literature citations
- The full test bank for *Biological Science* and the *Active Learning Workshop DVD* also come included with the IR-DVD

MasteringBiology® with Pearson eText

www.masteringbiology.com

Assign dynamic homework into your course with automatic grading and adaptive tutoring. Choose from a wide variety of stimulating activities, including visually stunning and scientifically accurate tutorials, ranking questions, 3-D animations, and test bank questions. The powerful gradebook compiles all your favorite teaching diagnostics—the hardest concept, class grade distribution, which students are spending the most or the least time on homework—with the click of a button. Instructors are empowered to customize the eText for themselves and students, including highlighting key text, annotating with comments, adding weblinks, and hiding chapters.

TestGen®

978-0-321-60533-7 • 0-321-60533-0

All of the exam questions in the test bank have been rigorously peer reviewed and revised using the metadata collected from real student usage in MasteringBiology. Test questions have been correlated to Bloom's Taxonomy of cognitive learning domains to identify which level of learning the question tests on. The Test Bank is also available in course management systems and in Microsoft® Word format on the Instructor Resources CD/DVD-ROM.

Course Management Options

CourseCompass™

www.pearsonhighered.com/elearning

This course management system contains preloaded content such as testing and assessment question pools.

WebCT

www.pearsonhighered.com/elearning

Blackboard

www.pearsonhighered.com/elearning

For Students

MasteringBiology with Pearson eText

www.masteringbiology.com

Students may use the Study Area in MasteringBiology for targeted and efficient use of valuable study time. Some of the many study tools include BioFlix™ 3-D movie-quality animations that focus on the toughest topics, engaging activities and cumulative chapter quizzes that help students prepare for exams. The interactive eText is available 24/7 and enables students to highlight text, add their own study notes, and review their Instructor's personalized notes at their convenience.

Study Guide

978-0-321-56168-8 • 0-321-56168-6

The Study Guide presents a breakdown of key biological concepts, difficult topics, and quizzes to help students prepare for exams. Unique to this study guide are four introductory, stand-alone chapters that introduce students to foundational ideas and skills necessary for classroom success: Introduction to Experimentation and Research in the Biological Sciences, Presenting Biological Data, Understanding Patterns in Biology and Improving Study Techniques, and Reading and Writing to Understand Biology. New to this edition of the Study Guide are "Looking Forward" and "Looking Back" sections that help students make connections across the chapters instead of viewing them as discrete entities.

Practicing Biology: A Student Workbook

978-0-321-61264-9 • 0-321-61264-7

This workbook focuses on key ideas, principles, and concepts that are fundamental to understanding biology. A variety of hands-on activities such as mapping and modeling suit different learning styles and help students discover which topics they need more help on. Students learn biology by doing biology.

A Short Guide to Writing About Biology

978-0-321-66838-7 • 0-321-66838-3

by Jan A. Pechenik, Tufts University

This best-selling writing guide teaches students to write and think as biologists.

BIOLOGICAL SCIENCE

VOLUME 3 How Plants & Animals Work

Black Swan, *Cygnus atratus*
Male and female black swans have identical coloration—an all-black body with white flight feathers at the tips of the wings. Although the significance of their orange-red beak coloration is unknown, experiments with other species have shown that individuals with particularly bright beaks or feathers are in exceptionally good health and are attractive to potential mates. To explore how you might test this hypothesis in black swans, see Chapter 25.

BIOLOGICAL SCIENCE

FOURTH EDITION

SCOTT FREEMAN

University of Washington

Benjamin Cummings

Boston Columbus Indianapolis New York San Francisco Upper Saddle River
Amsterdam Cape Town Dubai London Madrid Milan Munich Paris Montréal Toronto
Delhi Mexico City São Paulo Sydney Hong Kong Seoul Singapore Taipei Tokyo

VP, Editor-in-Chief, Biology: Beth Wilbur
Acquisitions Editor: Becky Ruden
Executive Director of Development, Biology: Deborah Gale
Editorial Project Manager: Sonia DiVittorio
Development Editors: Alice Fugate, Moira Lerner-Nelson, William O'Neal, Susan Teahan
Art Editor: Kelly Murphy
Assistant Editor: Brady Golden
Editorial Assistant: Leslie Allen
Senior Media Producer: Laura Tommasi
Director of Editorial Content, Mastering Biology: Tania Mlawer
Developmental Editor, Mastering Biology: Sarah Jensen
Director of Marketing: Christy Lawrence
Executive Marketing Manager: Lauren Harp
Director of Production, Science: Erin Gregg
Managing Editor, Biology: Michael Early
Production Supervisor: Lori Newman
Media Production Supervisor: James Bruce
Supplements Production Supervisor: Jane Brundage
Production Management and Composition: S4Carlisle Publishing Services
Design Manager and Interior Designer: Marilyn Perry
Cover Designer: Riezebos Holzbaur Design Group
Illustrators: Kim Quillin, Imagineering Media Services
Photo Researcher: Maureen Spuhler
Manufacturing Buyer: Michael Penne
Cover Printer: Phoenix Color Corp.
Printer and Binder: Courier, Kendallville

Cover Photo Credits: Black Swan—*Cygnus atratus,* © Eric Isselée/Fotolia (front cover); Black swan among white swans, Hokkaido, Japan, North-East Asia © Keren Su/Getty Images, Inc. (back cover)

Credits and acknowledgments borrowed from other sources and reproduced with permission in this textbook appear on the appropriate page within the text or beginning on page C:1 of the backmatter.

Library of Congress Cataloging-in-Publication Data

Freeman, Scott
Biological science / Scott Freeman.—4th ed.
p. cm.
Includes index.
ISBN 978-0-32-159820-2 (student ed.)—ISBN 978-0-32-159819-6 (professional copy)—
ISBN 978-0-32-161347-9 (v. 1 : the cell, genetics, and development—ISBN 978-0-32-160530-6
(v. 2 : evolution, diversity, and ecology)—ISBN 978-0-32-157676-7 (v. 3 : how plants and animals work)
1. Biology—Textbooks. I. Title.

QH308. 2. F73 2011
570—dc22 2009047825

ISBN 10: 0-32-159820-2; ISBN 13: 978-0-32-159820-2 (Student edition)
ISBN 10: 0-32-159819-9; ISBN 13: 978-0-32-159819-6 (Professional copy)
ISBN 10: 0-32-161347-3; ISBN 13: 978-0-32-161347-9 (Volume 1)
ISBN 10: 0-32-160530-6; ISBN 13: 978-0-32-160530-6 (Volume 2)
ISBN 10: 0-32-157676-4; ISBN 13: 978-0-32-157676-7 (Volume 3)

2 3 4 5 6 7 8 9 10—CRK—14 13 12 11 10

Benjamin Cummings
is an imprint of

www.pearsonhighered.com

Brief Contents: Volume 3

1 Biology and the Tree of Life 1

UNIT 1 THE MOLECULES OF LIFE 15

2 Water and Carbon: The Chemical Basis of Life 15
3 Protein Structure and Function 38
4 Nucleic Acids and the RNA World 59
5 An Introduction to Carbohydrates 71
6 Lipids, Membranes, and the First Cells 82

UNIT 2 CELL STRUCTURE AND FUNCTION 102

7 Inside the Cell 102
8 Cell-Cell Interactions 131
9 Cellular Respiration and Fermentation 148
10 Photosynthesis 172
11 The Cell Cycle 194

UNIT 3 GENE STRUCTURE AND EXPRESSION 211

12 Meiosis 211
13 Mendel and the Gene 230
14 DNA and the Gene: Synthesis and Repair 258
15 How Genes Work 276
16 Transcription, RNA Processing, and Translation 289
17 Control of Gene Expression in Bacteria 307
18 Control of Gene Expression in Eukaryotes 319
19 Analyzing and Engineering Genes 338
20 Genomics 359

UNIT 4 DEVELOPMENTAL BIOLOGY 374

21 Principles of Development 374
22 An Introduction to Animal Development 388
23 An Introduction to Plant Development 401

UNIT 5 EVOLUTIONARY PROCESSES AND PATTERNS 414

24 Evolution by Natural Selection 414
25 Evolutionary Processes 435
26 Speciation 458
27 Phylogenies and the History of Life 474

UNIT 6 THE DIVERSIFICATION OF LIFE 496

28 Bacteria and Archaea 496
29 Protists 519
30 Green Algae and Land Plants 546
31 Fungi 579
32 An Introduction to Animals 601
33 Protostome Animals 623
34 Deuterostome Animals 646
35 Viruses 675

UNIT 7 HOW PLANTS WORK 695

36 Plant Form and Function 695
37 Water and Sugar Transport in Plants 717
38 Plant Nutrition 737
39 Plant Sensory Systems, Signals, and Responses 755
40 Plant Reproduction 783

UNIT 8 HOW ANIMALS WORK 803

41 Animal Form and Function 803
42 Water and Electrolyte Balance in Animals 822
43 Animal Nutrition 841
44 Gas Exchange and Circulation 861
45 Electrical Signals in Animals 885
46 Animal Sensory Systems and Movement 907
47 Chemical Signals in Animals 929
48 Animal Reproduction 950
49 The Immune System in Animals 973

UNIT 9 ECOLOGY 993

50 An Introduction to Ecology 993
51 Behavioral Ecology 1019
52 Population Ecology 1037
53 Community Ecology 1058
54 Ecosystems 1083
55 Biodiversity and Conservation Biology 1105

Detailed Contents

1 Biology and the Tree of Life 1

1.1 **What Does It Mean to Say That Something Is Alive?** 1

1.2 **The Cell Theory** 2
Are *All* Organisms Made of Cells? 2
Where Do Cells Come From? 2

1.3 **The Theory of Evolution by Natural Selection** 4
What Is Evolution? 4
What Is Natural Selection? 4

1.4 **The Tree of Life** 5
Using Molecules to Understand the Tree of Life 6
How Should We Name Branches on the Tree of Life? 7

1.5 **Doing Biology** 8
The Nature of Science 8
Why Do Giraffes Have Long Necks? An Introduction to Hypothesis Testing 9
How Do Ants Navigate? An Introduction to Experimental Design 10

CHAPTER REVIEW 13

UNIT 1 THE MOLECULES OF LIFE 15

2 Water and Carbon: The Chemical Basis of Life 15

2.1 **Atoms, Ions, and Molecules: The Building Blocks of Chemical Evolution** 16
Basic Atomic Structure 16
How Does Covalent Bonding Hold Molecules Together? 17
Ionic Bonding, Ions, and the Electron-Sharing Continuum 18
Some Simple Molecules Formed from C, H, N, and O 19
The Geometry of Simple Molecules 20
Representing Molecules 20
Basic Concepts in Chemical Reactions 21

2.2 **The Early Oceans and the Properties of Water** 22
Why Is Water Such an Efficient Solvent? 22
How Does Water's Structure Correlate with Its Properties? 22
Acid–Base Reactions Involve a Transfer of Protons 25

2.3 **Chemical Reactions, Chemical Evolution, and Chemical Energy** 27
How Do Chemical Reactions Happen? 27
What Is Energy? 27
Chemical Evolution: A Model System 29
How Did Chemical Energy Change during Chemical Evolution? 33

2.4 **The Importance of Carbon** 33
Linking Carbon Atoms Together 34
Functional Groups 34

CHAPTER REVIEW 36

3 Protein Structure and Function 38

3.1 **Early Origin-of-Life Experiments** 39

3.2 **Amino Acids and Polymerization** 40
The Structure of Amino Acids 40
The Nature of Side Chains 40
How Do Amino Acids Link to Form Proteins? 42

3.3 **Proteins Are the Most Versatile Large Molecules in Cells** 45

3.4 **What Do Proteins Look Like?** 45
Primary Structure 46
Secondary Structure 46
Tertiary Structure 47
Quaternary Structure 48
Folding and Function 50

3.5 **Enzymes: An Introduction to Catalysis** 51
Enzymes Help Reactions Clear Two Hurdles 51
How Do Enzymes Work? 53
Was the First Living Entity a Protein Catalyst? 56

CHAPTER REVIEW 57

4 Nucleic Acids and the RNA World 59

4.1 **What Is a Nucleic Acid?** 59
Could Chemical Evolution Result in the Production of Nucleotides? 60
How Do Nucleotides Polymerize to Form Nucleic Acids? 61

4.2 **DNA Structure and Function** 62
What Is the Nature of DNA's Secondary Structure? 62
DNA Functions as an Information-Containing Molecule 65
Is DNA a Catalytic Molecule? 65

4.3 **RNA Structure and Function** 66
Structurally, RNA Differs from DNA 66
RNA's Structure Makes It an Extraordinarily Versatile Molecule 67
RNA Is an Information-Containing Molecule 67
RNA Can Function as a Catalytic Molecule 68

4.4 **The First Life-Form** 68

CHAPTER REVIEW 69

5 An Introduction to Carbohydrates 71

5.1 **Sugars as Monomers** 71
How Monosaccharides Differ 72
Monosaccharides and Chemical Evolution 73

5.2 **The Structure of Polysaccharides** 73
Starch: A Storage Polysaccharide in Plants 74
Glycogen: A Highly Branched Storage Polysaccharide in Animals 74
Cellulose: A Structural Polysaccharide in Plants 76

Chitin: A Structural Polysaccharide in Fungi and Animals 76
Peptidoglycan: A Structural Polysaccharide in Bacteria 76
Polysaccharides and Chemical Evolution 76

5.3 What Do Carbohydrates Do? 77
The Role of Carbohydrates as Structural Molecules 77
The Role of Carbohydrates in Cell Identity 77
The Role of Carbohydrates in Energy Storage 78

CHAPTER REVIEW 80

6 Lipids, Membranes, and the First Cells 82

6.1 Lipids 83
A Look at Three Types of Lipids Found in Cells 83
The Structures of Membrane Lipids 84

6.2 Phospholipid Bilayers 85
Artificial Membranes as an Experimental System 85
Selective Permeability of Lipid Bilayers 86
How Does Lipid Structure Affect Membrane Properties? 87
How Does Temperature Affect the Fluidity and Permeability of Membranes? 88

6.3 Why Molecules Move across Lipid Bilayers: Diffusion and Osmosis 89
Diffusion 89
Osmosis 90

6.4 Membrane Proteins 92
Evolution of the Fluid-Mosaic Model 92
Systems for Studying Membrane Proteins 94
Protein Transport I: Facilitated Diffusion via Channel Proteins 94
Protein Transport II: Facilitated Diffusion via Carrier Proteins 96
Protein Transport III: Active Transport by Pumps 97
Plasma Membranes and the Intracellular Environment 98

CHAPTER REVIEW 100

UNIT 2 CELL STRUCTURE AND FUNCTION 102

7 Inside the Cell 102

7.1 Bacterial and Archaeal Cell Structures and Their Functions 102
A Revolutionary New View 103
Prokaryotic Cell Structures: A Parts List 103

7.2 Eukaryotic Cell Structures and Their Functions 105
The Benefits of Organelles 107
Eukaryotic Cell Structures: A Parts List 107

7.3 Putting the Parts into a Whole 115
Structure and Function at the Whole-Cell Level 115
The Dynamic Cell 116

7.4 Cell Systems I: Nuclear Transport 116
Structure and Function of the Nuclear Envelope 116
How Are Molecules Imported into the Nucleus? 117

7.5 Cell Systems II: The Endomembrane System Manufactures and Ships Proteins 118
Studying the Pathway through the Endomembrane System 119
Entering the Endomembrane System: The Signal Hypothesis 120
Moving from the ER to the Golgi 122
What Happens inside the Golgi Apparatus? 122
How Do Proteins Reach Their Destinations? 122

7.6 Cell Systems III: The Dynamic Cytoskeleton 123
Actin Filaments 123
Intermediate Filaments 125
Microtubules 125
Flagella and Cilia: Moving the Entire Cell 127

CHAPTER REVIEW 129

8 Cell-Cell Interactions 131

8.1 The Cell Surface 132
The Structure and Function of an Extracellular Layer 132
The Cell Wall in Plants 132
The Extracellular Matrix in Animals 133

8.2 How Do Adjacent Cells Connect and Communicate? 134
Cell-Cell Attachments in Eukaryotes 135
Cells Communicate via Cell-Cell Gaps 138

8.3 How Do Distant Cells Communicate? 139
Cell-Cell Signaling in Multicellular Organisms 139
Signal Reception 140
Signal Processing 140
Signal Response 144
Signal Deactivation 144
Cross-Talk: Synthesizing Input from Many Signals 145
Quorum Sensing in Bacteria 145

CHAPTER REVIEW 146

9 Cellular Respiration and Fermentation 148

9.1 The Nature of Chemical Energy and Redox Reactions 149
The Structure and Function of ATP 149
What Is a Redox Reaction? 151

9.2 An Overview of Cellular Respiration 153

9.3 Glycolysis: Processing Glucose to Pyruvate 155
Glycolysis Is a Sequence of 10 Reactions 155
How Is Glycolysis Regulated? 156

9.4 Processing Pyruvate to Acetyl CoA 156

9.5 The Citric Acid Cycle: Oxidizing Acetyl CoA to CO_2 158
How Is the Citric Acid Cycle Regulated? 158
What Happens to the NADH and $FADH_2$? 160

9.6 Electron Transport and Chemiosmosis: Building a Proton Gradient to Produce ATP 161
Components of the Electron Transport Chain 161
The Chemiosmosis Hypothesis 162
How Is the Electron Transport Chain Organized? 163
The Discovery of ATP Synthase 164
Organisms Use a Diversity of Electron Acceptors 165

9.7 Fermentation 166

9.8 **How Does Cellular Respiration Interact with Other Metabolic Pathways?** 168
Catabolic Pathways Break Down Molecules as Fuel 168
Anabolic Pathways Synthesize Key Molecules 169
CHAPTER REVIEW 169

10 Photosynthesis 172

10.1 **Photosynthesis Harnesses Sunlight to Make Carbohydrate** 172
Photosynthesis: Two Linked Sets of Reactions 173
Photosynthesis Occurs in Chloroplasts 174

10.2 **How Does Chlorophyll Capture Light Energy?** 174
Photosynthetic Pigments Absorb Light 175
When Light Is Absorbed, Electrons Enter an Excited State 177

10.3 **The Discovery of Photosystems I and II** 179
How Does Photosystem II Work? 180
How Does Photosystem I Work? 182
The Z Scheme: Photosystems II and I Work Together 182

10.4 **How Is Carbon Dioxide Reduced to Produce Glucose?** 184
The Calvin Cycle Fixes Carbon 185
The Discovery of Rubisco 186
Carbon Dioxide Enters Leaves through Stomata 187
Mechanisms for Increasing CO_2 Concentration 187
How Is Photosynthesis Regulated? 189
What Happens to the Sugar That Is Produced by Photosynthesis? 189
CHAPTER REVIEW 190
The Big Picture: Energy for Life 192

11 The Cell Cycle 194

11.1 **Mitosis and the Cell Cycle** 195
What Is a Chromosome? 195
Cells Alternate between M Phase and Interphase 196
The Discovery of S Phase 196
The Discovery of the Gap Phases 196
The Cell Cycle 196

11.2 **How Does Mitosis Take Place?** 197
Events in Mitosis 197
Cytokinesis Results in Two Daughter Cells 200
How Do Chromosomes Move during Mitosis? 201

11.3 **Control of the Cell Cycle** 202
The Discovery of Cell-Cycle Regulatory Molecules 203
Cell-Cycle Checkpoints Can Arrest the Cell Cycle 204

11.4 **Cancer: Out-of-Control Cell Division** 206
Properties of Cancer Cells 206
Cancer Involves Loss of Cell-Cycle Control 207
CHAPTER REVIEW 209

UNIT 3 GENE STRUCTURE AND EXPRESSION 211

12 Meiosis 211

12.1 **How Does Meiosis Occur?** 212
Chromosomes Come in Distinct Types 212
The Concept of Ploidy 212
An Overview of Meiosis 213
The Phases of Meiosis I 216
The Phases of Meiosis II 218
A Closer Look at Prophase I 219

12.2 **The Consequences of Meiosis** 220
Chromosomes and Heredity 221
Independent Assortment Produces Genetic Variation 221
The Role of Crossing Over 222
How Does Fertilization Affect Genetic Variation? 222

12.3 **Why Does Meiosis Exist?** 223
The Paradox of Sex 223
The Purifying Selection Hypothesis 224
The Changing-Environment Hypothesis 224

12.4 **Mistakes in Meiosis** 225
How Do Mistakes Occur? 225
Why Do Mistakes Occur? 226
CHAPTER REVIEW 227

13 Mendel and the Gene 230

13.1 **Mendel's Experimental System** 230
What Questions Was Mendel Trying to Answer? 231
Garden Peas Served as the First Model Organism in Genetics 231

13.2 **Mendel's Experiments with a Single Trait** 232
The Monohybrid Cross 232
Particulate Inheritance 234

13.3 **Mendel's Experiments with Two Traits** 236
The Dihybrid Cross 236
Using a Testcross to Confirm Predictions 238

13.4 **The Chromosome Theory of Inheritance** 239
Meiosis Explains Mendel's Principles 240
Testing the Chromosome Theory 241

13.5 **Extending Mendel's Rules** 243
Linkage: What Happens When Genes Are Located on the Same Chromosome? 243
Do Heterozygotes Always Have a Dominant or Recessive Phenotype? 245
BOX 13.1 QUANTITATIVE METHODS: Linkage 245
How Many Alleles and Phenotypes Exist? 247
Does Each Gene Affect Just One Trait? 247
Are Phenotypes Determined by Genes? 247
What About Traits Like Human Height and Intelligence? 248

13.6 **Applying Mendel's Rules to Humans** 250
Identifying Human Alleles as Recessive or Dominant 250
Identifying Human Traits as Autosomal or Sex-Linked 251
CHAPTER REVIEW 253

14 DNA and the Gene: Synthesis and Repair 258

14.1 **What Are Genes Made Of?** 259
The Hershey-Chase Experiment 259
The Secondary Structure of DNA 260

14.2 **Testing Early Hypotheses about DNA Synthesis: The Meselson-Stahl Experiment** 261

14.3 A Comprehensive Model for DNA Synthesis 263
How Does Replication Get Started? 264
How Is the Helix Opened and Stabilized? 264
How Is the Leading Strand Synthesized? 265
How Is the Lagging Strand Synthesized? 266

14.4 Replicating the Ends of Linear Chromosomes 269

14.5 Repairing Mistakes and Damage 271
Correcting Mistakes in DNA Synthesis 271
Repairing Damaged DNA 272
Xeroderma Pigmentosum: A Case Study 272

CHAPTER REVIEW 274

15 How Genes Work 276

15.1 What Do Genes Do? 277
The One-Gene, One-Enzyme Hypothesis 277
An Experimental Test of the Hypothesis 277

15.2 The Central Dogma of Molecular Biology 279
The Genetic Code Hypothesis 279
RNA as the Intermediary between Genes and Proteins 279
Dissecting the Central Dogma 280

15.3 The Genetic Code 282
How Long Is a Word in the Genetic Code? 282
How Did Researchers Crack the Code? 283

15.4 What Is the Molecular Basis of Mutation? 285
Point Mutation 285
Chromosome-Level Mutations 286

CHAPTER REVIEW 287

16 Transcription, RNA Processing, and Translation 289

16.1 An Overview of Transcription 289
Characteristics of RNA Polymerase 290
Initiation: How Does Transcription Begin? 291
Elongation and Termination 292

16.2 RNA Processing in Eukaryotes 293
The Startling Discovery of Eukaryotic Genes in Pieces 293
RNA Splicing 294
Adding Caps and Tails to Transcripts 295

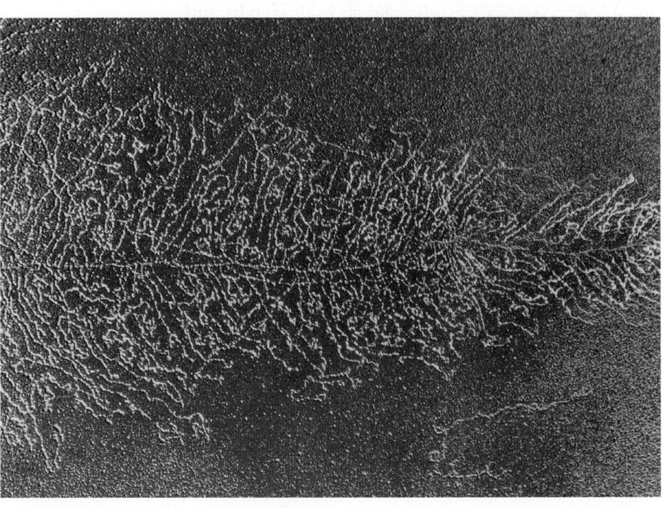

16.3 An Introduction to Translation 295
Ribosomes Are the Site of Protein Synthesis 295
Comparing Translation in Bacteria and Eukaryotes 296
How Does an mRNA Triplet Specify an Amino Acid? 297

16.4 The Structure and Function of Transfer RNA 297
What Do tRNAs Look Like? 299
How Many tRNAs Are There? 299

16.5 The Structure and Function of Ribosomes 300
Initiating Translation 301
Elongation: Extending the Polypeptide 301
Terminating Translation 302
Post-Translational Modifications 304

CHAPTER REVIEW 304

17 Control of Gene Expression in Bacteria 307

17.1 Gene Regulation and Information Flow 307
Mechanisms of Regulation—An Overview 308
Metabolizing Lactose—A Model System 309

17.2 Identifying Genes under Regulatory Control 310
Replica Plating to Find Mutant Genes 310
Different Classes of Lactose Metabolism Mutants 311
Several Genes Are Involved in Lactose Metabolism 312

17.3 Mechanisms of Negative Control: Discovery of the Repressor 312
The *lac* Operon 313
Why Has the *lac* Operon Model Been So Important? 314

17.4 Mechanisms of Positive Control: Catabolite Repression 314
The CAP Protein and Binding Site 315
How Does Glucose Influence Formation of the CAP–cAMP Complex? 316

CHAPTER REVIEW 317

18 Control of Gene Expression in Eukaryotes 319

18.1 Mechanisms of Gene Regulation in Eukaryotes—An Overview 320

18.2 Chromatin Remodeling 320
What Is Chromatin's Basic Structure? 320
Evidence That Chromatin Structure Is Altered in Active Genes 321
How Is Chromatin Altered? 322
Chromatin Modifications Can Be Inherited 323

18.3 Initiating Transcription: Regulatory Sequences and Regulatory Proteins 323
Some Regulatory Sequences Are Near the Promoter 323
Some Regulatory Sequences Are Far from the Promoter 324
The Role of Regulatory Proteins in Differential Gene Expression 326
The Initiation Complex 326

18.4 Post-Transcriptional Control 328
Alternative Splicing of mRNAs 328
mRNA Stability and RNA Interference 329

How Is Translation Controlled? 330
Post-Translational Control 330

18.5 How Does Gene Expression in Bacteria Compare with That in Eukaryotes? 331

18.6 Linking Cancer with Defects in Gene Regulation 332
Causes of Uncontrolled Cell Growth 332
p53: A Case Study 332

CHAPTER REVIEW 333

The Big Picture: Genetic Information 336

19 Analyzing and Engineering Genes 338

19.1 Case 1—The Effort to Cure Pituitary Dwarfism: Basic Recombinant DNA Technologies 338
Why Did Early Efforts to Treat the Disease Fail? 339
Steps in Engineering a Safe Supply of Growth Hormone 339
Ethical Concerns over Recombinant Growth Hormone 343

19.2 Case 2—Amplification of Fossil DNA: The Polymerase Chain Reaction 344
Requirements of PCR 344
PCR in Action 345

19.3 Case 3—Sanger's Breakthrough Innovation: Dideoxy DNA Sequencing 346
The Logic of Dideoxy Sequencing 346
"Next Generation" Sequencing 347

19.4 Case 4—The Huntington's Disease Story: Finding Genes by Mapping 348
How Was the Huntington's Disease Gene Found? 348
What Are the Benefits of Finding a Disease Gene? 350
Ethical Concerns over Genetic Testing 350

19.5 Case 5—Severe Immune Disorders: The Potential of Gene Therapy 351
How Can Novel Alleles Be Introduced into Human Cells? 351
Using Gene Therapy to Treat X-Linked Immune Deficiency 352
Ethical Concerns over Gene Therapy 354

19.6 Case 6—The Development of Golden Rice: Biotechnology in Agriculture 354
Rice as a Target Crop 355
Synthesizing β-Carotene in Rice 355
The *Agrobacterium* Transformation System 355
Using the Ti Plasmid to Produce Golden Rice 356

CHAPTER REVIEW 356

20 Genomics 359

20.1 Whole-Genome Sequencing 359
How Are Complete Genomes Sequenced? 360
Which Genomes Are Being Sequenced, and Why? 361
Which Sequences Are Genes? 362

20.2 Bacterial and Archaeal Genomes 363
The Natural History of Prokaryotic Genomes 363
Lateral Gene Transfer 364
Environmental Sequencing 364

20.3 Eukaryotic Genomes 365
Parasitic and Repeated Sequences 365
Gene Families 367
Insights from the Human Genome Project 368

20.4 Functional Genomics and Proteomics 370
What Is Functional Genomics? 370
What Is Proteomics? 371
Applied Genomics in Action: Understanding Cancer 371

CHAPTER REVIEW 372

UNIT 4 DEVELOPMENTAL BIOLOGY 374

21 Principles of Development 374

21.1 Shared Developmental Processes 375
Cell Proliferation 375
Programmed Cell Death 376
Cell Movement or Cell Growth 376
Cell Differentiation 377
Cell-Cell Interactions 377

21.2 The Role of Differential Gene Expression in Development 377
Evidence That Differentiated Plant Cells Are Genetically Equivalent 377
Evidence That Differentiated Animal Cells Are Genetically Equivalent 377
How Does Differential Gene Expression Occur? 378

21.3 Cell-Cell Signals Trigger Differential Gene Expression 379
Master Regulators Set Up the Major Body Axes 379
Regulatory Genes Provide Increasingly Specific Positional Information 381
Cell-Cell Signals and Regulatory Genes Are Evolutionarily Conserved 383
Common Signaling Pathways Are Active in Many Contexts 383

21.4 Changes in Developmental Pathways Underlie Evolutionary Change 384

CHAPTER REVIEW 385

22 An Introduction to Animal Development 388

22.1 Gamete Structure and Function 389
Sperm Structure and Function 389
Egg Structure and Function 390

22.2 Fertilization 390
How Do Gametes from the Same Species Recognize Each Other? 391
Why Does Only One Sperm Enter the Egg? 391

22.3 Cleavage 392
Partitioning Cytoplasmic Determinants 393
Cleavage in Mammals 393

22.4 Gastrulation 394
Formation of Germ Layers 394
Definition of Body Axes 395

22.5 Organogenesis 396
Organizing Mesoderm into Somites: Precursors of Muscle, Skeleton, and Skin 396
Differentiation of Muscle Cells 398

CHAPTER REVIEW 399

23 An Introduction to Plant Development 401

23.1 Gametogenesis, Pollination, and Fertilization 402
How Are Sperm and Egg Produced? 402
Pollen–Stigma Interactions 402
Double Fertilization 403

23.2 Embryogenesis 404
What Happens during Plant Embryogenesis? 404
Which Genes and Proteins Set Up Body Axes? 406

23.3 Vegetative Development 407
Meristems Provide Lifelong Growth and Development 407
Which Genes and Proteins Determine Leaf Shape? 408

23.4 Reproductive Development 409
The Floral Meristem and the Flower 409
The Genetic Control of Flower Structures 409

CHAPTER REVIEW 412

UNIT 5 EVOLUTIONARY PROCESSES AND PATTERNS 414

24 Evolution by Natural Selection 414

24.1 The Evolution of Evolutionary Thought 415
Plato and Typological Thinking 415
Aristotle and the Great Chain of Being 415
Lamarck and the Idea of Evolution as Change through Time 415
Darwin and Wallace and Evolution by Natural Selection 415

24.2 The Pattern of Evolution: Have Species Changed through Time? 416
Evidence for Change through Time 416
Evidence of Descent from a Common Ancestor 418
Evolution's "Internal Consistency"—the Importance of Independent Datasets 422

24.3 The Process of Evolution: How Does Natural Selection Work? 422
Darwin's Four Postulates 423
The Biological Definitions of Fitness and Adaptation 424

24.4 Evolution in Action: Recent Research on Natural Selection 424
Case Study 1: How Did *Mycobacterium tuberculosis* Become Resistant to Antibiotics? 424
Case Study 2: Why Are Beak Size, Beak Shape, and Body Size Changing in Galápagos Finches? 426

24.5 Common Misconceptions about Natural Selection and Adaptation 429
Selection Acts on Individuals, but Evolutionary Change Occurs in Populations 429
Evolution Is Not Goal Directed 430
Organisms Do Not Act for the Good of the Species 430
Limitations of Natural Selection 431

CHAPTER REVIEW 432

25 Evolutionary Processes 435

25.1 Analyzing Change in Allele Frequencies: The Hardy-Weinberg Principle 436
The Gene Pool Concept 436
Deriving the Hardy-Weinberg Principle 436
The Hardy-Weinberg Model Makes Important Assumptions 437
How Does the Hardy-Weinberg Principle Serve as a Null Hypothesis? 438

25.2 Types of Natural Selection 440
Directional Selection 440
Stabilizing Selection 441
Disruptive Selection 442
Balancing Selection 442

25.3 Genetic Drift 443
Simulation Studies of Genetic Drift 443
Experimental Studies of Genetic Drift 445
What Causes Genetic Drift in Natural Populations? 445

25.4 Gene Flow 447
Gene Flow in Natural Populations 447
How Does Gene Flow Affect Fitness? 448

25.5 Mutation 448
Mutation as an Evolutionary Mechanism 448
Experimental Studies of Mutation 449

25.6 Nonrandom Mating 450
Inbreeding 450
Sexual Selection 452

CHAPTER REVIEW 456

26 Speciation 458

26.1 How Are Species Defined and Identified? 458
The Biological Species Concept 459
The Morphospecies Concept 460
The Phylogenetic Species Concept 460
Species Definitions in Action: The Case of the Dusky Seaside Sparrow 461

26.2 Isolation and Divergence in Allopatry 462
Dispersal and Colonization Isolate Populations 463
Vicariance Isolates Populations 464

26.3 Isolation and Divergence in Sympatry 464
Can Natural Selection Cause Speciation Even When Gene Flow Is Possible? 465
How Can Polyploidy Lead to Speciation? 465

26.4 What Happens When Isolated Populations Come into Contact? 468
Reinforcement 468
Hybrid Zones 468
New Species through Hybridization 470

CHAPTER REVIEW 472

27 Phylogenies and the History of Life 474

27.1 Tools for Studying History: Phylogenetic Trees 474
How Do Researchers Estimate Phylogenies? 475
How Can Biologists Distinguish Homology from Homoplasy? 475
Whale Evolution: A Case History 477

27.2 Tools for Studying History: The Fossil Record 479
How Do Fossils Form? 479
Limitations of the Fossil Record 480
Life's Time Line 481

27.3 Adaptive Radiation 484
Why Do Adaptive Radiations Occur? 484
The Cambrian Explosion 486

27.4 Mass Extinction 488
How Do Mass Extinctions Differ From Background Extinctions? 489
The End-Permian Extinction 489
What Killed the Dinosaurs? 490

CHAPTER REVIEW 492

The Big Picture: Evolution 494

UNIT 6 THE DIVERSIFICATION OF LIFE 496

28 Bacteria and Archaea 496

28.1 Why Do Biologists Study Bacteria and Archaea? 497
Biological Impact 497
Medical Importance 498
Role in Bioremediation 500
Extremophiles 501

28.2 How Do Biologists Study Bacteria and Archaea? 501
Using Enrichment Cultures 501
Using Direct Sequencing 502
Evaluating Molecular Phylogenies 503

28.3 What Themes Occur in the Diversification of Bacteria and Archaea? 504
Morphological Diversity 504
Metabolic Diversity 506
Ecological Diversity and Global Change 509

28.4 Key Lineages of Bacteria and Archaea 512
Bacteria 512
Archaea 512
- Bacteria > Firmicutes 513
- Bacteria > Spirochaetes (Spirochetes) 513
- Bacteria > Actinobacteria 514
- Bacteria > Chlamydiae 514
- Bacteria > Cyanobacteria 515
- Bacteria > Proteobacteria 515
- Archaea > Crenarchaeota 516
- Archaea > Euryarchaeota 516

CHAPTER REVIEW 517

29 Protists 519

29.1 Why Do Biologists Study Protists? 520
Impacts on Human Health and Welfare 520
Ecological Importance of Protists 522

29.2 How Do Biologists Study Protists? 524
Microscopy: Studying Cell Structure 524
Evaluating Molecular Phylogenies 525
Discovering New Lineages via Direct Sequencing 525

29.3 What Themes Occur in the Diversification of Protists? 526
What Morphological Innovations Evolved in Protists? 526
How Do Protists Obtain Food? 529
How Do Protists Move? 532
How Do Protists Reproduce? 533
Life Cycles—Haploid- versus Diploid-Dominated 533

29.4 Key Lineages of Protists 536
Amoebozoa 536
Excavata 536
Plantae 536
Rhizaria 536
Alveolata 537
Stramenopila (Heterokonta) 537
- Amoebozoa > Myxogastrida (Plasmodial Slime Molds) 537
- Excavata > Parabasalida 538
- Excavata > Diplomonadida 538
- Excavata > Euglenida 539
- Plantae > Rhodophyta (Red Algae) 539
- Rhizaria > Foraminifera 540
- Alveolata > Ciliata 540
- Alveolata > Dinoflagellata 541
- Alveolata > Apicomplexa 541
- Stramenopila > Oomycota (Water Molds) 542
- Stramenopila > Diatoms 542
- Stramenopila > Phaeophyta (Brown Algae) 543

CHAPTER REVIEW 543

30 Green Algae and Land Plants 546

30.1 Why Do Biologists Study the Green Algae and Land Plants? 546
Plants Provide Ecosystem Services 547
Plants Provide Humans with Food, Fuel, Fiber, Building Materials, and Medicines 548

30.2 How Do Biologists Study Green Algae and Land Plants? 549
Analyzing Morphological Traits 549
Using the Fossil Record 550
Evaluating Molecular Phylogenies 551

30.3 What Themes Occur in the Diversification of Land Plants? 553
The Transition to Land, I: How Did Plants Adapt to Dry Conditions? 553
Mapping Evolutionary Changes on the Phylogenetic Tree 555
The Transition to Land, II: How Do Plants Reproduce in Dry Conditions? 556
The Angiosperm Radiation 564

30.4 Key Lineages of Green Algae and Land Plants 566
Green Algae 566
Non-Vascular Plants ("Bryophytes") 567
Seedless Vascular Plants 567
Seed Plants 567
- Green Algae > Ulvophyceae (Ulvophytes) 568
- Green Algae > Coleochaetophyceae (Coleochaetes) 568
- Green Algae > Charophyceae (Stoneworts) 569
- Non-Vascular Plants > Hepaticophyta (Liverworts) 569
- Non-Vascular Plants > Bryophyta (Mosses) 570
- Non-Vascular Plants > Anthocerophyta (Hornworts) 571
- Seedless Vascular Plants > Lycophyta (Lycophytes, or Club Mosses) 571
- Seedless Vascular Plants > Psilotophyta (Whisk Ferns) 572
- Seedless Vascular Plants > Equisetophyta (or Sphenophyta) (Horsetails) 572
- Seedless Vascular Plants > Pteridophyta (Ferns) 573
- Seed Plants > Gymnosperms > Cycadophyta (Cycads) 574
- Seed Plants > Gymnosperms > Ginkgophyta (Ginkgos) 574
- Seed Plants > Gymnosperms > Redwood group (Redwoods, Junipers, Yews) 575
- Seed Plants > Gymnosperms > Pinophyta (Pines, Spruces, Firs) 575
- Seed Plants > Gymnosperms > Gnetophyta (Gnetophytes) 576
- Seed Plants > Anthophyta (Angiosperms) 576

CHAPTER REVIEW 577

31 Fungi 579

31.1 Why Do Biologists Study Fungi? 580
Fungi Provide Nutrients for Land Plants 580
Fungi Speed the Carbon Cycle on Land 580
Fungi Have Important Economic Impacts 581

31.2 How Do Biologists Study Fungi? 582
Analyzing Morphological Traits 582
Evaluating Molecular Phylogenies 584
Experimental Studies of Mutualism 586

31.3 What Themes Occur in the Diversification of Fungi? 586
Fungi Participate in Several Types of Mutualisms 586
What Adaptations Make Fungi Such Effective Decomposers? 589
Variation in Reproduction 590
Four Major Types of Life Cycles 591

31.4 Key Lineages of Fungi 594
- Fungi > Microsporidia 594
- Fungi > Chytrids 595
- Fungi > Zygomycetes 595
- Fungi > Glomeromycota 596
- Fungi > Basidiomycota (Club Fungi) 596
- Fungi > Ascomycota > Lichen-Formers 597
- Fungi > Ascomycota > Non-Lichen-Formers 598

CHAPTER REVIEW 599

32 An Introduction to Animals 601

32.1 Why Do Biologists Study Animals? 602
Biological Importance 602
Role in Human Health and Welfare 602

32.2 How Do Biologists Study Animals? 603
Analyzing Comparative Morphology 603
Evaluating Molecular Phylogenies 607

32.3 What Themes Occur in the Diversification of Animals? 610
Sensory Organs 610
Feeding 610
Movement 613
Reproduction 615
Life Cycles 615

32.4 Key Lineages of Animals: Non-Bilaterian Groups 617
- Porifera (Sponges) 618
- Cnidaria (Jellyfish, Corals, Anemones, Hydroids) 619
- Ctenophora (Comb Jellies) 620
- Acoelomorpha (Acoels) 620

CHAPTER REVIEW 621

33 Protostome Animals 623

33.1 An Overview of Protostome Evolution 624
What Is a Lophotrochozoan? 624
What Is an Ecdysozoan? 625

33.2 Themes in the Diversification of Protostomes 625
How Do Body Plans Vary among Phyla? 626
The Water-to-Land Transition 627
Adaptations for Feeding 628
Adaptations for Moving 628
Adaptations in Reproduction 630

33.3 Key Lineages: Lophotrochozoans 630
- Lophotrochozoans > Rotifera (Rotifers) 631
- Lophotrochozoans > Platyhelminthes (Flatworms) 631
- Lophotrochozoans > Annelida (Segmented Worms) 633
- Lophotrochozoans > Mollusca > Bivalvia (Clams, Mussels, Scallops, Oysters) 634
- Lophotrochozoans > Mollusca > Gastropoda (Snails, Slugs, Nudibranchs) 635
- Lophotrochozoans > Mollusca > Polyplacophora (Chitons) 636
- Lophotrochozoans > Mollusca > Cephalopoda (Nautilus, Cuttlefish, Squid, Octopuses) 636

33.4 Key Lineages: Ecdysozoans 637
- Ecdysozoans > Nematoda (Roundworms) 638
- Ecdysozoans > Arthropoda > Myriapods (Millipedes, Centipedes) 639
- Ecdysozoans > Arthropoda > Insecta (Insects) 639
- Ecdysozoans > Arthropoda > Chelicerata (Spiders, Ticks, Mites, Horseshoe Crabs, Daddy Longlegs, Scorpions) 642
- Ecdysozoans > Arthropoda > Crustaceans (Shrimp, Lobster, Crabs, Barnacles, Isopods, Copepods) 643

CHAPTER REVIEW 644

34 Deuterostome Animals 646

34.1 What Is an Echinoderm? 647
The Echinoderm Body Plan 647
How Do Echinoderms Feed? 648
Key Lineages 649
- Echinodermata > Asteroidea (Sea Stars) 649
- Echinodermata > Echinoidea (Sea Urchins and Sand Dollars) 650

34.2 What Is a Chordate? 650
Three "Subphyla" 651
Key Lineages: The Invertebrate Chordates 651
- Chordata > Cephalochordata (Lancelets) 652
- Chordata > Urochordata (Tunicates) 652

34.3 What Is a Vertebrate? 653
An Overview of Vertebrate Evolution 653
Key Innovations 655
Key Lineages 660
- Chordata > Vertebrata > Myxinoidea (Hagfish) and Petromyzontoidea (Lampreys) 661
- Chordata > Vertebrata > Chondrichthyes (Sharks, Rays, Skates) 662
- Chordata > Vertebrata > Actinopterygii (Ray-Finned Fishes) 662
- Chordata > Vertebrata > Actinistia (Coelacanths) and Dipnoi (Lungfish) 663
- Chordata > Vertebrata > Amphibia (Frogs, Salamanders, Caecilians) 664
- Chordata > Vertebrata > Mammalia > Monotremata (Platypuses, Echidnas) 665
- Chordata > Vertebrata > Mammalia > Marsupiala (Marsupials) 665
- Chordata > Vertebrata > Mammalia > Eutheria (Placental Mammals) 666
- Chordata > Vertebrata > Reptilia > Lepidosauria (Lizards, Snakes) 666
- Chordata > Vertebrata > Reptilia > Testudinia (Turtles) 667
- Chordata > Vertebrata > Reptilia > Crocodilia (Crocodiles, Alligators) 667
- Chordata > Vertebrata > Reptilia > Aves (Birds) 668

34.4 The Primates and Hominins 668
The Primates 668
Fossil Humans 670
The Out-of-Africa Hypothesis 672

CHAPTER REVIEW 673

35 Viruses 675

35.1 Why Do Biologists Study Viruses? 676
Recent Viral Epidemics in Humans 676
Current Viral Epidemics in Humans: HIV 677

35.2 How Do Biologists Study Viruses? 678
Analyzing Morphological Traits 679
Analyzing Variation in Growth Cycles: Replicative and Latent Growth 679
Analyzing the Phases of the Replicative Cycle 681

35.3 What Themes Occur in the Diversification of Viruses? 686
The Nature of the Viral Genetic Material 686
Where Did Viruses Come From? 686
Emerging Viruses, Emerging Diseases 688

35.4 Key Lineages of Viruses 689
- Double-Stranded DNA (dsDNA) Viruses 690
- RNA Reverse-Transcribing Viruses (Retroviruses) 691
- Double-Stranded RNA (dsRNA) Viruses 691
- Negative-Sense Single-Stranded RNA ([−]ssRNA) Viruses 692
- Positive-Sense Single-Stranded RNA ([+]ssRNA) Viruses 692

CHAPTER REVIEW 693

UNIT 7 HOW PLANTS WORK 695

36 Plant Form and Function 695

36.1 Plant Form: Themes with Many Variations 696
The Importance of Surface Area/Volume Relationships 696
The Root System 697
The Shoot System 699
The Leaf 701

36.2 Primary Growth Extends the Plant Body 704
How Do Apical Meristems Produce the Primary Plant Body? 704
How Is the Primary Root System Organized? 705
How Is the Primary Shoot System Organized? 706

36.3 Cells and Tissues of the Primary Plant Body 706
The Dermal Tissue System 707
The Ground Tissue System 708
The Vascular Tissue System 710

36.4 Secondary Growth Widens Shoots and Roots 712
What Is a Cambium? 712
What Does Vascular Cambium Produce? 713
What Does Cork Cambium Produce? 713
The Structure of a Tree Trunk 714

CHAPTER REVIEW 715

37 Water and Sugar Transport in Plants 717

37.1 Water Potential and Water Movement 717
What Is Water Potential? 718
What Factors Affect Water Potential? 718
Calculating Water Potential 719
Water Potentials in Soils, Plants, and the Atmosphere 720

37.2 How Does Water Move from Roots to Shoots? 721
Movement of Water and Solutes into the Root 722
Water Movement via Root Pressure 723
Water Movement via Capillary Action 723
The Cohesion-Tension Theory 724

37.3 Water Absorption and Water Loss 727
Limiting Water Loss 727
Obtaining Carbon Dioxide under Water Stress 728

37.4 Translocation 728
Tracing Connections between Sources and Sinks 728
The Anatomy of Phloem 729
The Pressure-Flow Hypothesis 730
Phloem Loading 731
Phloem Unloading 733

CHAPTER REVIEW 735

38 Plant Nutrition 737

38.1 Nutritional Requirements of Plants 738
Which Nutrients Are Essential? 738
What Happens When Key Nutrients Are in Short Supply? 740

38.2 Soil: A Dynamic Mixture of Living and Nonliving Components 741
The Importance of Soil Conservation 742
What Factors Affect Nutrient Availability? 742

38.3 Nutrient Uptake 744
Mechanisms of Nutrient Uptake 744
Mechanisms of Ion Exclusion 746

38.4 Nitrogen Fixation 748
The Role of Symbiotic Bacteria 749
How Do Nitrogen-Fixing Bacteria Colonize Plant Roots? 749

38.5 Nutritional Adaptations of Plants 750
Epiphytic Plants 750
Parasitic Plants 751
Carnivorous Plants 751

CHAPTER REVIEW 752

39 Plant Sensory Systems, Signals, and Responses 755

39.1 Information Processing in Plants 756
How Do Cells Receive and Transduce an External Signal? 756
How Are Cell-Cell Signals Transmitted? 756
How Do Cells Respond to Cell-Cell Signals? 757

39.2 Blue Light: The Phototropic Response 758
Phototropins as Blue-Light Receptors 758
Auxin as the Phototropic Hormone 759

39.3 Red and Far-Red Light: Germination and Stem Elongation 762
The Red/Far-Red "Switch" 763
Phytochromes as Red/Far-Red Receptors 763
How Were Phytochromes Isolated? 763

39.4 Gravity: The Gravitropic Response 764
The Statolith Hypothesis 764
Auxin as the Gravitropic Signal 765

39.5 How Do Plants Respond to Wind and Touch? 766
Changes in Growth Patterns 766
Movement Responses 766

39.6 Youth, Maturity, and Aging: The Growth Responses 767
Auxin and Apical Dominance 767
Cytokinins and Cell Division 768
Gibberellins and ABA: Growth and Dormancy 769
Brassinosteroids and Body Size 773
Ethylene and Senescence 773
An Overview of Plant Growth Regulators 774

39.7 Pathogens and Herbivores: The Defense Responses 776
How Do Plants Sense and Respond to Pathogens? 776
How Do Plants Sense and Respond to Herbivore Attack? 778

CHAPTER REVIEW 781

40 Plant Reproduction 783

40.1 An Introduction to Plant Reproduction 784
Sexual Reproduction 784
The Land Plant Life Cycle 784
Asexual Reproduction 786

40.2 Reproductive Structures 786
When Does Flowering Occur? 787
The General Structure of the Flower 788
How Are Female Gametophytes Produced? 790
How Are Male Gametophytes Produced? 790

40.3 Pollination and Fertilization 792
Pollination 792
Fertilization 794

40.4 The Seed 795
Embryogenesis 796
The Role of Drying in Seed Maturation 797
Fruit Development and Seed Dispersal 797
Seed Dormancy 798
Seed Germination 799

CHAPTER REVIEW 800

UNIT 8 HOW ANIMALS WORK 803

41 Animal Form and Function 803

41.1 Form, Function, and Adaptation 804
The Role of Fitness Trade-Offs 804
Adaptation and Acclimatization 804

41.2 Tissues, Organs, and Systems: How Does Structure Correlate with Function? 806
Structure-Function Relationships at the Molecular and Cellular Levels 806
Tissues Are Groups of Similar Cells That Function as a Unit 806
Organs and Organ Systems 810

41.3 How Does Body Size Affect Animal Physiology? 811
Surface Area/Volume Relationships: Theory 811
Surface Area/Volume Relationships: Data 812
Adaptations That Increase Surface Area 814

41.4 Homeostasis 814
Homeostasis: General Principles 814
The Role of Regulation and Feedback 815

41.5 **How Do Animals Regulate Body Temperature?** 816
Mechanisms of Heat Exchange 816
Variation in Thermoregulation 816
Endothermy and Ectothermy: A Closer Look 817
Temperature Homeostasis in Endotherms 817
Countercurrent Heat Exchangers 818
CHAPTER REVIEW 820

42 Water and Electrolyte Balance in Animals 822

42.1 **Osmoregulation and Osmotic Stress** 823
What Is Osmotic Stress? 823
Osmotic Stress in Seawater 824
Osmotic Stress in Freshwater 824
Osmotic Stress on Land 825
How Do Cells Move Electrolytes and Water? 825

42.2 **Water and Electrolyte Balance in Aquatic Environments** 826
How Do Sharks Excrete Salt? 826
How Do Freshwater Fish Osmoregulate? 827

42.3 **Water and Electrolyte Balance in Terrestrial Insects** 828
How Do Insects Minimize Water Loss from the Body Surface? 828
Types of Nitrogenous Wastes: Impact on Water Balance 829
Maintaining Homeostasis: The Excretory System 830

42.4 **Water and Electrolyte Balance in Terrestrial Vertebrates** 832
The Structure of the Kidney 832
The Function of the Kidney: An Overview 832
Filtration: The Renal Corpuscle 832
Reabsorption: The Proximal Tubule 834
Creating an Osmotic Gradient: The Loop of Henle 835
Regulating Water and Electrolyte Balance: The Distal Tubule and Collecting Duct 837
CHAPTER REVIEW 839

43 Animal Nutrition 841

43.1 **Nutritional Requirements** 842
Meeting Basic Needs in Humans 842
Studying Nutrient Requirements 842

43.2 **Capturing Food: The Structure and Function of Mouthparts** 843
Mouthparts as Adaptations 843
A Case Study: The Cichlid Jaw 844

43.3 **How Are Nutrients Digested and Absorbed?** 845
An Introduction to the Digestive Tract 845
An Overview of Digestive Processes 846
The Mouth and Esophagus 847
The Stomach 848
The Small Intestine 851
The Cecum and Appendix 854
The Large Intestine 854

43.4 **Nutritional Homeostasis—Glucose as a Case Study** 856
The Discovery of Insulin 856
Insulin's Role in Homeostasis 856
Diabetes Can Take Several Forms 856
The Type 2 Diabetes Mellitus Epidemic 857
CHAPTER REVIEW 858

44 Gas Exchange and Circulation 861

44.1 **The Respiratory and Circulatory Systems** 861

44.2 **Air and Water as Respiratory Media** 862
How Do Oxygen and Carbon Dioxide Behave in Air? 862
How Do Oxygen and Carbon Dioxide Behave in Water? 863

44.3 **Organs of Gas Exchange** 864
Physical Parameters: The Law of Diffusion 864
How Do Fish Gills Work? 865
How Do Insect Tracheae Work? 866
How Do Vertebrate Lungs Work? 867
Homeostatic Control of Ventilation 870

44.4 **How Are Oxygen and Carbon Dioxide Transported in Blood?** 870
Structure and Function of Hemoglobin 871
CO_2 Transport and the Buffering of Blood pH 873

44.5 **The Circulatory System** 874
What Is an Open Circulatory System? 875
What Is a Closed Circulatory System? 875
How Does the Heart Work? 877
Patterns in Blood Pressure and Blood Flow 882
CHAPTER REVIEW 883

45 Electrical Signals in Animals 885

45.1 **Principles of Electrical Signaling** 885
Types of Neurons in the Nervous System 886
The Anatomy of a Neuron 886
An Introduction to Membrane Potentials 887
BOX 45.1 QUANTITATIVE METHODS: Using the Nernst Equation to Calculate Equilibrium Potentials 888
How Is the Resting Potential Maintained? 888
Using Microelectrodes to Measure Membrane Potentials 890
What Is an Action Potential? 890

45.2 **Dissecting the Action Potential** 891
Distinct Ion Currents Are Responsible for Depolarization and Repolarization 891
How Do Voltage-Gated Channels Work? 891
How Is the Action Potential Propagated? 893

45.3 **The Synapse** 895
Synapse Structure and Neurotransmitter Release 895
What Do Neurotransmitters Do? 896
Postsynaptic Potentials 897

45.4 **The Vertebrate Nervous System** 899
What Does the Peripheral Nervous System Do? 899
Functional Anatomy of the CNS 900
How Does Memory Work? 902

CHAPTER REVIEW 904

46 Animal Sensory Systems and Movement 907

46.1 **How Do Sensory Organs Convey Information to the Brain?** 908
Sensory Transduction 908
Transmitting Information to the Brain 909

46.2 **Hearing** 909
How Do Sensory Cells Respond to Sound Waves and Other Forms of Pressure? 909
The Mammalian Ear 910
Sensory Worlds: What Do Other Animals Hear? 912

46.3 **Vision** 913
The Insect Eye 913
The Vertebrate Eye 914
Sensory Worlds: Do Other Animals See Color? 917

46.4 **Taste and Smell** 918
Taste: Detecting Molecules in the Mouth 918
Olfaction: Detecting Molecules in the Air 919

46.5 **Movement** 920
Skeletons 920
Muscle Types 921
How Do Muscles Contract? 922

CHAPTER REVIEW 926

47 Chemical Signals in Animals 929

47.1 **Cell-to-Cell Signaling: An Overview** 929
Major Categories of Chemical Signals 930
Hormone Signaling Pathways 931
What Makes Up the Endocrine System? 932
Chemical Characteristics of Hormones 933
How Do Researchers Identify a Hormone? 934

47.2 **What Do Hormones Do?** 935
How Do Hormones Direct Developmental Processes? 935
How Do Hormones Coordinate Responses to Environmental Change? 937
How Are Hormones Involved in Homeostasis? 938

47.3 **How Is the Production of Hormones Regulated?** 940
The Hypothalamus and Pituitary Gland 940
Control of Epinephrine by Sympathetic Nerves 943

47.4 **How Do Hormones Act on Target Cells?** 943
Steroid Hormones Bind to Intracellular Receptors 943
Hormones That Bind to Cell-Surface Receptors 945
Why Do Different Target Cells Respond in Different Ways? 947

CHAPTER REVIEW 948

48 Animal Reproduction 950

48.1 **Asexual and Sexual Reproduction** 950
How Does Asexual Reproduction Occur? 951
Switching Reproductive Modes: A Case History 951
Mechanisms of Sexual Reproduction: Gametogenesis 952

48.2 **Fertilization and Egg Development** 954
External Fertilization 954
Internal Fertilization 954
Unusual Aspects of Mating 955
Why Do Some Females Lay Eggs while Others Give Birth? 956

48.3 **Reproductive Structures and Their Functions** 957
The Male Reproductive System 957
The Female Reproductive System 959

48.4 **The Role of Sex Hormones in Mammalian Reproduction** 960
Which Hormones Control Puberty in Mammals? 961
Which Hormones Control the Menstrual Cycle in Mammals? 962

48.5 **Pregnancy and Birth in Mammals** 967
Gestation and Early Development in Marsupials 967
Major Events during Human Pregnancy 967
How Does the Mother Nourish the Fetus? 968
Birth 970

CHAPTER REVIEW 971

49 The Immune System in Animals 973

49.1 **Innate Immunity** 974
Barriers to Entry 974
The Innate Immune Response 975

49.2 **The Adaptive Immune Response: Recognition** 977
An Introduction to Lymphocytes 978
The Discovery of B Cells and T Cells 979
The Clonal-Selection Theory 979
How Does the Immune System Distinguish Self from Nonself? 983

49.3 **The Adaptive Immune Response: Activation** 984
T-Cell Activation 984
B-Cell Activation and Antibody Secretion 986

49.4 **The Adaptive Immune Response: Culmination** 987
How Are Bacteria and Other Foreign Cells Killed? 987
How Are Viruses Destroyed? 987
Why Does the Immune System Reject Foreign Tissues and Organs? 988
Responding to Future Infections: Immunological Memory 989

49.5 **What Happens When the Immune System *Doesn't* Work Correctly?** 990
Immunodeficiency Diseases 990
Allergies 990

CHAPTER REVIEW 991

UNIT 9 ECOLOGY 993

50 An Introduction to Ecology 993

50.1 Areas of Ecological Study 993
Organismal Ecology 994
Population Ecology 994
Community Ecology 994
Ecosystem Ecology 995
How Do Ecology and Conservation Efforts Interact? 995

50.2 Types of Aquatic Ecosystems 995
Nutrient Availability 995
Water Flow 996
Water Depth 996
- Freshwater Environments > Lakes and Ponds 997
- Freshwater Environments > Wetlands 998
- Freshwater Environments > Streams 999
- Freshwater/Marine Environments > Estuaries 1000
- Marine Environments > The Ocean 1000

50.3 Types of Terrestrial Ecosystems 1001
- Terrestrial Biomes > Tropical Wet Forest 1003
- Terrestrial Biomes > Subtropical Deserts 1004
- Terrestrial Biomes > Temperate Grasslands 1005
- Terrestrial Biomes > Temperate Forests 1006
- Terrestrial Biomes > Boreal Forests 1007
- Terrestrial Biomes > Arctic Tundra 1008

50.4 The Role of Climate and the Consequences of Climate Change 1008
Global Patterns in Climate 1009
How Will Global Climate Change Affect Ecosystems? 1011

50.5 Biogeography: Why Are Organisms Found Where They Are? 1013
Abiotic Factors 1013
The Role of History 1013
Biotic Factors 1015
Biotic and Abiotic Factors Interact 1015

CHAPTER REVIEW 1017

51 Behavioral Ecology 1019

51.1 An Introduction to Behavioral Biology 1019
Proximate and Ultimate Causation 1020
Conditional Strategies and Decision Making 1020
Five Questions in Behavioral Ecology 1021

51.2 What Should I Eat? 1021
Foraging Alleles in *Drosophila melanogaster* 1021
Optimal Foraging in White-Fronted Bee-Eaters 1022

51.3 Who Should I Mate With? 1022
Sexual Activity in *Anolis* Lizards 1023
How Do Female Barn Swallows Choose Mates? 1024

51.4 Where Should I Live? 1026
How Do Animals Find Their Way on Migration? 1026
Why Do Animals Move with a Change of Seasons? 1027

51.5 How Should I Communicate? 1027
Honeybee Language 1028
Modes of Communication 1029
When Is Communication Honest or Deceitful? 1029

51.6 When Should I Cooperate? 1031
Kin Selection 1031
BOX 51.1 QUANTITATIVE METHODS: Calculating the Coefficient of Relatedness 1032
Reciprocal Altruism 1033
An Extreme Case: Abuse of Non-Kin in Humans 1034

CHAPTER REVIEW 1035

52 Population Ecology 1037

52.1 Demography 1037
Life Tables 1038
The Role of Life History 1039
BOX 52.1 QUANTITATIVE METHODS: Using Life Tables to Calculate Population Growth Rates 1040

52.2 Population Growth 1041
Quantifying the Growth Rate 1041
Exponential Growth 1042
Logistic Growth 1042
BOX 52.2 QUANTITATIVE METHODS: Developing and Applying Population Growth Equations 1043
What Limits Growth Rates and Population Sizes? 1044

52.3 Population Dynamics 1046
How Do Metapopulations Change through Time? 1046
Why Do Some Populations Cycle? 1047
BOX 52.3 QUANTITATIVE METHODS: Mark-Recapture Studies 1048
How Does Age Structure Affect Population Growth? 1050
Analyzing Change in the Growth Rate of Human Populations 1052

52.4 How Can Population Ecology Help Endangered Species? 1053
Using Life-Table Data 1054
Preserving Metapopulations 1055

CHAPTER REVIEW 1056

53 Community Ecology 1058

53.1 Species Interactions 1058
Three Themes 1059
Competition 1059
Consumption 1063
Mutualism 1068

53.2 Community Structure 1070
How Predictable Are Communities? 1070
How Do Keystone Species Structure Communities? 1072

53.3 Community Dynamics 1073
Disturbance and Change in Ecological Communities 1073
Succession: The Development of Communities after Disturbance 1074

53.4 Species Richness in Ecological Communities 1077
Predicting Species Richness: The Theory of Island Biogeography 1077
Global Patterns in Species Richness 1078
BOX 53.1 QUANTITATIVE METHODS: Measuring Species Diversity 1079

CHAPTER REVIEW 1080

54 Ecosystems 1083

54.1 How Does Energy Flow through Ecosystems? 1083
Why Is NPP So Important? 1084
Solar Power: Transforming Incoming Energy to Biomass 1084
Trophic Structure 1085
Energy Transfer between Trophic Levels 1086
Trophic Cascades and Top-Down Control 1087
Biomagnification 1088
Global Patterns in Productivity 1089
What Limits Productivity? 1090

54.2 How Do Nutrients Cycle through Ecosystems? 1092
Nutrient Cycling within Ecosystems 1092
Global Biogeochemical Cycles 1094

54.3 Global Warming 1098
Understanding the Problem 1098
Positive and Negative Feedback 1099
Impact on Organisms 1099
Productivity Changes 1101

CHAPTER REVIEW 1103

55 Biodiversity and Conservation Biology 1105

55.1 What Is Biodiversity? 1106
Biodiversity Can Be Measured and Analyzed at Several Levels 1106
How Many Species Are Living Today? 1107
BOX 55.1 QUANTITATIVE METHODS: Extrapolation Techniques 1108

55.2 Where Is Biodiversity Highest? 1109
Hotspots of Biodiversity and Endemism 1109
Conservation Hotspots 1110

55.3 Threats to Biodiversity 1110
Changes in the Nature of the Problem 1110
How Can Biologists Predict Future Extinction Rates? 1114
BOX 55.2 QUANTITATIVE METHODS: Population Viability Analysis 1115

55.4 Why Is Biodiversity Important? 1117
Economic Benefits of Biodiversity 1117
Biological Benefits of Biodiversity 1117
An Ethical Dimension? 1120

55.5 Preserving Biodiversity 1120
Designing Effective Protected Areas 1120
Beyond Protected Areas: A Comprehensive Approach 1121

CHAPTER REVIEW 1123

The Big Picture: Ecology 1126

APPENDIX A: **BioSkills** B:1
1 The Metric System B:1
2 Reading Graphs B:2
3 Reading a Phylogenetic Tree B:4
4 Some Common Latin and Greek Roots Used in Biology B:6
5 Using Statistical Tests and Interpreting Standard Error Bars B:7
6 Reading Chemical Structures B:8
7 Using Logarithms B:9
8 Making Concept Maps B:10
9 Separating and Visualizing Molecules B:11
10 Biological Imaging: Microscopy and X-Ray Crystallography B:13
11 Separating Cell Components by Centrifugation B:16
12 Cell and Tissue Culture Methods B:17
13 Combining Probabilities B:19
14 Model Organisms B:19

APPENDIX B: **Answers** A:1

Glossary G:1

Credits C:1

Index I:1

About the Author

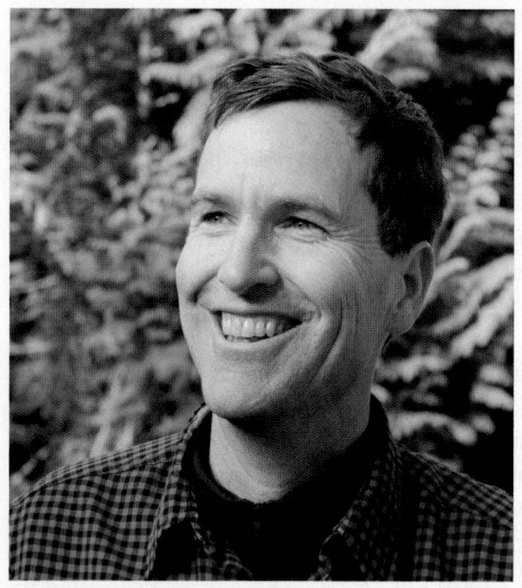

SCOTT FREEMAN received his Ph.D. in Zoology from the University of Washington and was subsequently awarded an Alfred P. Sloan Postdoctoral Fellowship in Molecular Evolution at Princeton University. His current research focuses on the scholarship of teaching and learning—specifically **(1)** how active learning and peer teaching techniques increase student learning and improve performance in introductory biology, and **(2)** how the levels of exam questions vary among introductory biology courses, standardized postgraduate entrance exams, and professional school courses. He has also done research in evolutionary biology on topics ranging from nest parasitism to the molecular systematics of the blackbird family. Scott teaches introductory biology for majors at the University of Washington and is coauthor, with Jon Herron, of the standard-setting undergraduate text *Evolutionary Analysis.*

Unit Advisors

Twelve cherished colleagues guided the revision process by synthesizing reviews, providing citations for recent high-impact publications, and drawing on their extensive teaching experience and subject-matter expertise to advise Scott on hundreds of questions, ranging from what to include to which analogies might communicate best to students. It is hard to overstate just how critical these people were to this revision. Through their own teaching and research and their work on this book, they are having a profound effect on how biology is taught.

Jason Flores, University of North Carolina, Charlotte (Unit 1)

Lisa Elfring, University of Arizona, Tucson (Unit 1)

Suzanne Simon-Westendorf, Ohio University (Unit 2)

Gregory Podgorski, Utah State University (Unit 3)

Kathleen Marrs, Indiana University–Purdue University, Indianapolis (Units 4, 6, and 7)

Jon Monroe, James Madison University (Units 4, 6, and 7)

Warren Burggren, University of North Texas (Units 4 and 8)

Joan Sharp, Simon Fraser University (Units 5 and 6)

Michael Black, California Polytechnic State University, San Luis Obispo (Units 6 and 8)

Kathleen Hunt, University of Portland (Units 6 and 9)

Emily Taylor, California Polytechnic State University, San Luis Obispo (Unit 8)

Fred Wasserman, Boston University (Unit 9)

Illustrator

KIM QUILLIN combines expertise in biology and information design to create lucid visual representations of biological principles. She received her B.A. in Biology at Oberlin College and her Ph.D. in Integrative Biology from the University of California, Berkeley (as a National Science Foundation Graduate Fellow), and taught undergraduate biology at both schools. Students and instructors alike have praised Kim's illustration programs for *Biological Science,* as well as *Biology: A Guide to the Natural World* by David Krogh and *Biology: Science for Life* by Colleen Belk and Virginia Borden, for their success in the visual communication of biology. Kim is a lecturer in the Department of Biological Sciences at Salisbury University.

Preface to Instructors

This book is a response to calls—from the National Academy of Sciences, the Howard Hughes Medical Institute and American Association of Medical Colleges, and the National Science Foundation—for changes in the way introductory biology is taught. Reports like *Biology 2010, Scientific Foundations for Future Physicians,* and *Vision and Change* are asking that introductory students not only learn the language of biology and understand fundamental concepts, but begin to apply those concepts in new situations, analyze experimental design, synthesize results, and evaluate hypotheses and data.

I wrote this book for instructors who embrace this challenge—who want to help their students learn how to think like a biologist. The essence of higher education is to promote higher-order thinking. Our job is to help students understand biological science at all six levels of Bloom's taxonomy of learning.

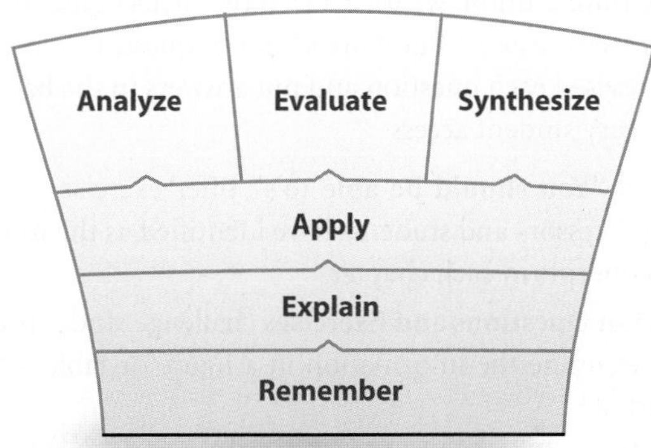

Bloom's Taxonomy. An annotated version of this graphic can be found in "To the Student: How to Use This Book" at the front of this book.

The Evolution of a Textbook

Evolution can be extremely fast in populations with short generation times and high mutation rates. Biology textbooks are no exception. Generation times have to be short because the pace of research in biology and student learning is so fast. This book, in particular, evolves quickly because it incorporates so many new ideas with each edition. Some of these "alleles" are novel mutations, but most arrive via lateral transfer—from advisors, reviewers, friends, students, and the literature.

In comparing the first edition of *Biological Science* with the fourth, the thing that jumps out is what educational researchers call "scaffolding"—tools that help professors and students teach and learn at the level demanded by the NAS, HHMI, AMC, and NSF. It's essential to have high expectations of our students, but we also need to provide the help and practice they need to meet those expectations.

What's New in This Edition

This revision was about making the book a better teaching and learning tool. To help students manage the mass of information and ideas that is contemporary biology, I broke long paragraphs into shorter paragraphs, made liberal use of numbered lists and bulleted lists to "chunk" information and ideas, and broke out dozens of new sections and subsections. I could almost hear my mother's voice: "Take small bites, and chew them well." I also made the book over 100 pages shorter by following Thoreau's dictum to "Simplify, simplify, simplify"—focusing on the most critical concepts that an introductory student needs to master.

In addition, the book team and I came up with a long list of new or expanded features.

- **The Big Picture** These new, two-page spreads are meant to help students see the forest for the trees. They are concept maps that focus on particularly critical areas—Energy, Genetic Information, Evolution, and Ecology. Each synthesizes content and concepts from an array of chapters and includes exercises for students to complete. You'll recognize these pages readily—their edges are colored black (for example, see The Big Picture: Energy on pages 192–193). In addition, the book's MasteringBiology® website has ten new concept map activities based on Big Picture content that will allow you to explore the concepts and their connections with your students during lecture.
- **BioSkills** *Biology 2010, Scientific Foundations,* and *Vision and Change* all place a premium on skills—the ability to read a graph, interpret an equation, understand the bands on a gel. The third edition of *Biological Science* introduced a series of appendixes focused on key skills for introductory biology students. Instructors and students found them extraordinarily helpful. New in this edition are BioSkills on using the metric system, common Latin and Greek roots, techniques for isolating and visualizing cell components, cell and tissue culture methods, and model organisms. BioSkills are located in Appendix A.
- **Answer Key** New to the Fourth Edition are suggested answers to all questions and exercises in the textbook. Students asked us to make this important change between editions to make the book a more complete study tool. The answer key will allow them to self-check their understanding while reading, and when reviewing for exams. Answers are in Appendix B.
- **Experiment Boxes** This text's hallmark has always been its emphasis on experimental evidence—on teaching how we know what we know. In the second edition, key experiments were converted to a boxed format so students could easily

navigate through the logic of the question, hypothesis, and test. In this edition, I added a new question to every experiment box to encourage students to analyze some aspect of the experiment's design.

- **Art Program** Recent research shows that students are more likely to interpret phylogenetic trees correctly if the trees are designed with U-shaped branches instead of Y-shaped branches. We responded by redesigning every phylogenetic tree in the text. To make other subject areas more accessible to visual learners, we enlarged figures, replaced hundreds of photos with clearer images, and strove to streamline labels and graphics across the board. (More on improvements to the art program below.)
- **MasteringBiology Quizzes** MasteringBiology gives students round-the-clock access to quizzes. We developed 550 new assignable questions based on the book's "Blue Thread" questions (more on the "Blue Thread" and its evolution below). We also developed 1100 new quiz questions, along with a cumulative practice test to simulate what a real exam might be like. To help students keep up with their reading, we created 55 new reading quizzes—one for each chapter—that you can assign through Mastering Biology.
- **MasteringBiology Experimental Inquiry Tutorials** The call to teach students about the process of science has never been louder. In response, a team lead by Tom Owens of Cornell University developed 10 new interactive tutorials on classic scientific experiments—ranging from Meselson–Stahl on DNA replication to the Grants' work on Galápagos finches and Connell's work on competition. Students who use these interactive tutorials should be better prepared to think critically about experimental design and evaluate the wider implications of the data—preparing them to do the work of real scientists in the future.
- **MasteringBiology BioFlix Animations and Tutorials** BioFlix™ are movie-quality, 3-D animations available on MasteringBiology. They focus on the most difficult core topics and are accompanied by in-depth, online tutorials that provide hints and feedback to help guide student learning. Thirteen BioFlix were available with the third edition of *Biological Science.* Five new BioFlix 3-D animations and tutorials have been developed for this edition—on mechanisms of evolution, homeostasis, gas exchange, population ecology, and the carbon cycle.

Changes to Gold Thread Scaffolding

The third edition introduced a dramatically expanded set of tools designed to help with a chronic problem for novice learners: picking out important information. Novices highlight every line in the text and try to memorize everything mentioned in lecture; experts instinctively home in on the key unifying ideas.

For students to make the novice-to-expert transition, we have to help them with features like:

1. **Key concepts** that are declared at the start of each chapter, highlighted with a key icon within the chapter, and reviewed at the end of the chapter.
2. **In-text highlighting**, in gold, that directs their attention to particularly important ideas.
3. **Check Your Understanding boxes**, at the end of key sections, with a bulleted list of key points.
4. **Summary tables** that pull information together in a compact format that is easy to review and synthesize.

The book team and I scrubbed every aspect of the gold thread—reviewing, revising, and re-revising. The third edition was a coming-out party for the gold thread; in the Fourth Edition we wanted it to dance.

Changes to Blue Thread Scaffolding

Each edition of this text has added tools to help students with metacognition—understanding what they do and don't understand. Novices like to receive information passively, and easily persuade themselves that they know what's going on. Experts are skeptical—they want to solve some problems before they're convinced that they know and understand an idea.

In the third edition, we formalized the metacognitive tools in *Biological Science* as a "Blue Thread" set of questions; in this edition, we revised each question and put answers in the back of the book for easy student access.

1. **In-text "You should be able to's"** offer exercises on topics that professors and students have identified as the most difficult concepts in each chapter.
2. **Caption Questions and Exercises** challenge students to critically examine the information in a figure or table—not just absorb it.
3. **Check Your Understanding boxes** present two to three tasks that students should be able to complete in order to demonstrate a mastery of summarized key ideas.
4. **Chapter Summaries** include "You should be able to" problems or exercises related to each of the key concepts declared in the gold thread.
5. **End-of-Chapter Questions** are organized around Bloom's taxonomy of learning, so students can test their understanding at the knowledge, comprehension, and application levels.

The fundamental idea is that if students really understand a piece of information or a concept, they should be able to do something with it. How do you get to Carnegie Hall? Practice.

As students mature as biologists-in-training and start taking upper-division courses, most or all of this scaffolding can disappear. By the time our students are juniors and seniors, they should have enough expertise to construct a high-level understanding on their own. But if a well-designed scaffold isn't there to get them started in their first and second years, when they are novices, most will flounder. We have to help them learn how to become good students.

Supporting Visual Learners

Figures can help students, especially visual learners, at all levels of Bloom's taxonomy—not only to understand and remember the material, but also to exercise higher levels of critical thinking. The overall goal of the Fourth Edition art revision was to hone the figures for accessibility to help novice learners recognize and engage with important visual information. In addition to re-designing the previously mentioned phylogenetic trees, Kim Quillin led the effort to enhance virtually every other aspect of the visual-teaching program.

- **Art and Photos** Kim enlarged art and photographs in figures throughout the book to increase clarity by making details physically easier to see. She also reduced the amount of detail in labels and graphics to simplify, simplify, simplify.
- **Color Use** Kim continues to use color strategically to draw attention to important parts of the figures. In this revision, she boosted color contrast in many figures to make the art more vibrant and the details easier to see.
- **Molecular Icons** Kim redesigned many molecular icons to simplify their shapes. The overall contours are based on molecular coordinates, when available, to accurately represent size and geometry, but she smoothed the textures for a simpler appearance—one that is more memorable and pleasing.
- **Molecular Models** New molecular models have been introduced to help students visualize structure-function relationships. In Chapter 5, for example, redesigned 2-D line drawings of sugars are now paired with 3-D ball-and-stick models.
- **"Pointers"** The Fourth Edition figures still use pointer annotations as a "whisper in the ear" to guide students in interpreting figures, but Kim has replaced the hand with an arrow to be more precise.

Serving a Community of Teachers

I love students, but I love teachers even more. There is nothing I like better than to sit in a workshop with a bunch of biology instructors, steal some good ideas, and come home to try them out—in some cases in a framework where I can collect data and test the hypothesis that the new approaches are improving student learning.

Research on biology education is gathering momentum, trying to catch up on the trail blazed by physics education researchers, bringing the same level of rigor to our classrooms that we bring to our lab benches and field sites. I try to bring the spirit and practice of evidence-based teaching into this textbook, and welcome your comments, suggestions, and questions.

Thank you for considering this text, and for your work on behalf of your students. We have the best jobs in the world.

SCOTT FREEMAN
University of Washington

Content Highlights of the Fourth Edition

As discussed in the preface, a major focus of this revision is to enhance the pedagogical utility of *Biological Science.* Another major goal is to ensure that the content reflects the current state of science and is accurate. In addition, every chapter has been rigorously evaluated for discussions that, in the previous edition, may have been too complex or overly detailed. As a result of this scrutiny, certain sections in every chapter have been simplified, content has been pruned judiciously, and the approach to certain topics has been re-envisioned to enhance student comprehension. In this section, some of the key content improvements to the textbook are highlighted.

Unit 1 The Molecules of Life

Chapter 1 A new experiment on ant navigation and discussions of tree-based naming systems and artificial selection in maize has been added. Coverage is expanded on the definition of life and the nature of science and religion.

Chapter 2 The descriptions of bond angles and the geometry of simple molecules are simplified. Added is a discussion on the hot-start hypothesis as well as a new Key Concept on the nature of chemical energy.

Chapter 3 This chapter has been streamlined by eliminating discussion of optical isomers/chirality and reducing coverage of enzyme kinetics and reaction rates.

Chapter 4 The discussion of RNA is expanded to include recently discovered roles for RNAs in cells. Also added is a new summary table (**Table 4.1**) comparing DNA and RNA structure.

Chapter 5 A stronger emphasis on the link between electronegativity of atoms and potential energy in C–C, C–H, and C–O bonds is developed. New ball-and-stick models are added to clarify the differences in location and orientation of functional groups.

Chapter 6 Coverage of secondary active transport has been expanded. Also included in this chapter is current research on the "first cell" and a discussion of non-random distribution of membrane proteins and phospholipids.

Unit 2 Cell Structure and Function

Chapter 7 New research on bacterial cell structure has been included, and a more explicit connection between lysosomes and the endomembrane system is emphasized. Centrifugation is moved to **BioSkills 11** in Appendix A.

Chapter 8 New sections on quorum sensing in bacteria and cross-talk among signal-transduction pathways have been added.

Chapter 9 The discussions of mitochondrial structure, ATP yield from glucose oxidation, and the role of GDP in the citric acid cycle have been updated. The introductory section on cellular respiration has been simplified.

Chapter 10 A new section on regulation (inhibition) has been added. The sections on C_4 and CAM photosynthesis now emphasize the role of these pathways in increasing CO_2 concentrations versus water conservation.

Chapter 11 Micrographs have been added to the phases of mitosis figure (**Figure 11.5**). The discussion on the role of activated MPF has been updated to include the triggering M phase of the cell cycle. Animal-cell culture methods are moved to **BioSkills 12** in Appendix A.

Unit 3 Gene Structure and Expression

Chapter 12 The discussions of recombination rates and aneuploidy rates in humans are updated. New micrographs have been added to the phases of meiosis figure (**Figure 12.7**).

Chapter 13 The linkage discussion and notation in fly crosses have been simplified. Sex-linkage is moved to the Mendelian section (Section 13.4 The Chromosome Theory of Inheritance), and mapping is now covered in Box 13.1 Quantitative Methods: Linkage. A new summary table (**Table 13.3**) presenting basic vocabulary used in Mendelian genetics has been added.

Chapter 14 A new space-filling model of DNA has been added to **Figure 14.4**.

Chapter 15 Discussions on mutation in the melanocortin receptor (link to mouse-coat-color camouflage) and karyotypes of cancerous cells have been added.

Chapter 16 The sections on transcription in bacteria and eukaryotes are now combined. The structure of the translation initiation complex in bacteria has been updated to reflect current science; snRNAs have been added to the discussion of RNA splicing.

Chapter 17 The chapter was streamlined with the removal of discussions of DNA fingerprinting and the structure of the operator and DNA-binding proteins. Treatment of catabolite repression/positive control has been trimmed.

Chapter 18 Included in this chapter is a new summary table (**Table 18.1**) comparing control of gene expression in bacteria and eukaryotes. Also added are discussions on ubiquitination and protein degradation, the importance of epigenetic inheritance (chromosome structure), and the histone code hypothesis.

Chapter 19 Southern/Northern/Western blots have moved to **BioSkills 9** in Appendix A. The discussions on golden rice, the impact of GM crops, and SNP association studies for human diseases have been updated with the most recent research. Notes on "next generation" sequencing technologies have been included.

Chapter 20 Human health applications now emphasize the use of genomics and microarrays to study cancer. Several datasets are updated, including sequencing database totals. New notes on miRNA genes, metagenomics, and the definition of the gene have been added.

Unit 4 Developmental Biology

Chapter 21 The discussions of *bicoid* and regulatory gene cascades are simplified. New material on auxin as a master regulator in early development and the importance of apoptosis has been added.

Chapter 22 The discussion about sea urchin fertilization and variation has been streamlined.

Chapter 23 A new section introducing basic concepts in angiosperm gametogenesis is added.

Unit 5 Evolutionary Processes and Patterns

Chapter 24 A section on the internal consistency of diverse data as evidence for evolution, including a new phylogeny and timeline of whale evolution, has been added. **Figure 24.6**, depicting the evolution of the Galápagos mockingbird, and long-term data on ground finches (**Figure 24.17**) are updated to reflect the most current science. There is a new graph on the evolution of drug resistance in pathogenic bacteria (**Figure 24.14**).

Chapter 25 The genetic drift example has changed from breeding in a small population on Pitcairn Island to coin flips simulating mating in a single couple (using data from the author's classroom). The prairie lupine gene flow example is replaced by recent work on an island population of *Parus major*. Notes on balancing selection and interactions among evolutionary forces have been included.

Chapter 26 The speciation-by-vicariance example has been changed from ratites to snapping shrimp, and the sympatric speciation example featuring soapberry bugs has been changed to apple/hawthorn flies.

Chapter 27 The sections on adaptive radiation and mass extinction have been completely reorganized. A new hypothesis for the cause of the Cambrian explosion is included, and detail on the "new genes, new bodies" hypothesis has been removed. Presentation of "Life's Timeline" has been significantly overhauled (see **Figures 27.8, 27.9,** and **27.10**).

Unit 6 The Diversification of Life

The model organisms have been moved to **BioSkills 14** in Appendix A. Phylogenetic trees have been redrawn to reflect a horizontal orientation with U-shaped branches for easier comprehension.

Chapter 28 New information on mechanisms of pathogenicity is added. Extensive updates include new notes on archaeon-eukaryote polymerases, the discovery of extensive biomass in the marine subfloor, an archaeon associated with a human disease, discovery of N-fixation and nitrification in archaea, and bacteriorhodopsin's role in phototrophy.

Chapter 29 A stronger emphasis on endosymbiosis as a theme in protist diversification has been threaded throughout this chapter.

Chapter 30 New content on green algae as a grade and on convergence in vascular tissue in mosses-vascular plants and gnetophytes-angiosperms has been added.

Chapter 31 The dynamic nature of mycelia, the importance of glomalin in soil, the role of mating types, and the discovery of "multigenomic" asexual glomales all have new supporting material.

Chapter 32 The treatment of embryonic tissues, developmental patterns, the coelom, and body symmetry have been updated to reflect the latest scientific thinking. A shift in emphasis to the origin of the neuron and cephalization has been implemented.

Chapter 33 New commentary on the independent transitions to land as well as a clarified discussion on the nature of the ecdysozoan-lophotrochozoan split are included. The discussion of annelids is updated to reflect recent results.

Chapter 34 The coverage of the echinoderm endoskeleton has been expanded and a phylogeny of early tetrapods has been added to the fin-to-limb transition figure (**Figure 34.16**). New data have been incorporated in the evolution-of-fishes timeline (**Figure 34.11**). The treatment of hagfish-lamprey, evolution-of-the-jaw (**Figure 34.14**) and *H. sapiens* migration (**Figure 34.40**) also include the most recent data available. The emphasis on the adaptive significance of the amniotic egg has changed from watertightness to increased size and support. Emphasis in the discussion of viviparity has changed to the adaptive advantage of embryo portability and temperature control. The recent analysis of *Ardipithecus ramidus* as the first hominin, with data on estimated body mass and braincase volume, has been included.

Chapter 35 The material on HIV phylogeny has been moved to the section on emerging viruses.

Unit 7 How Plants Work

Chapter 36 Surface area-to-volume ratios have been added as a theme in root and shoot systems. New information on contractile roots in *Ficus* and bulbs is incorporated into this chapter.

Chapter 37 New content on aquaporins and the transmembrane route to root xylem has been added, and coverage of why air has such low water pressure potential has been expanded.

Chapter 38 The description of nitrogen fixation has been clarified.

Chapter 39 **Figure 39.8** on the acid-growth hypothesis has been redesigned, and the discussion of polar auxin transport is simplified. New commentary on the role of brassinosteroids in growth regulation and on "talking trees" is included. The coverage of the receptors for GA, auxin, ABA, and brassinosteroids, and MeSA's role in the SAR has been updated with the most current research. Plant-tissue culture methods have been moved to **BioSkills 12** in Appendix A.

Chapter 40 Comments on day-length sensing and on pollination syndromes are new to this chapter.

Unit 8 How Animals Work

Chapter 41 New details on tissue types (especially connective tissue) have been incorporated. The discussion of thermoregulation has been completely reorganized for a more logical flow.

Chapter 42 The sections on the shark rectal gland and the mammalian loop of Henle have been revised to improve focus.

Chapter 43 A description of incomplete digestive systems is now included, and coverage of comparative aspects of digestive tract structure and function has been expanded.

Chapter 44 Information on the types of circulatory systems and types of blood vessels has been consolidated. Details on surface tension and lung elasticity have been removed while new content on countercurrent exchange in fish gills has been added.

Chapter 45 The chapter and section introductions have been rewritten to introduce a comparative context and to make the neuron-to-systems chapter organization more transparent. New content on interspecific variation in nervous systems has been added.

Chapter 46 The chapter has been shortened and its focus sharpened by the removal of nonessential information.

Chapter 47 New material on EPO abuse in athletes has been included.

Chapter 48 The section on sperm competition includes new data from experiments on seed beetles.

Chapter 49 The discussion of the V regions of BCRs and antibodies and recombination in BCR/TCR genes has been simplified. New content on autoimmune disorders and diseases associated with immunosuppression, allergies, and immunodeficiency diseases has been added. The discussion of vaccination has been expanded.

Unit 9 Ecology

Chapter 50 New information on the importance of nutrient availability in aquatic ecosystems, with details on lake turnover and ocean upwelling, is included. A new section on the Wallace line has also been added.

Chapter 51 The content in this chapter has been completely reorganized to increase cohesiveness. It is presented as a series of questions in behavioral ecology, with each question addressed at the proximate and ultimate levels with separate case studies. Material on modes of learning, innate behavior, bat-moth interactions, sex change in wrasses, and acoustic and visual signaling in red-winged blackbirds has been trimmed or dropped. New content on child abuse in humans is added.

Chapter 52 Discussion of the hare–lynx-cycle field experiment has been reorganized for clarity, with new supporting "Results" data added to accompanying **Figure 52.12**.

Chapter 53 New content has been added on species richness and resistance of communities to invasion, the use of predators or parasites as biocontrol agents, and character displacement in finches. The discussion of succession in Glacier Bay is reorganized and simplified. The discussion of alternative hypotheses to explain the latitudinal gradient in species richness has been expanded and clarified.

Chapter 54 The chapter was rewritten and reorganized to sharpen its focus on human impacts. Sections on trophic cascades and biomagnification have been added, as have recent data on human appropriation of NPP, sources of nutrient gain and loss, and the impact of ocean acidification on coral growth.

Chapter 55 New content on the impact of global climate change and a new section on ways to preserve biodiversity are now included. Two new boxes on quantitative methods have been added: one on estimating species numbers and species losses and the other on population viability analysis.

Acknowledgments

Reviewers

The peer review system is the key to quality and clarity in science publishing. In addition to providing a filter, the investment that respected individuals make in vetting the material—catching errors or inconsistencies and making suggestions to improve the presentation—gives authors, editors, and readers confidence that what they are publishing and reading meets rigorous professional standards.

Peer review plays the same role in textbook publishing. The time and care that this book's reviewers have invested is a tribute to their professional integrity, their scholarship, and their concern for the quality of teaching. Virtually every paragraph in this edition has been revised and improved based on insights from the following individuals.

Ann Aguanno, *Marymount Manhattan College*
Adrienne Alaie, *City University of New York, Hunter College*
John Alcock, *Arizona State University*
Sylvester Allred, *Northern Arizona University*
Suzanne Alonzo, *Yale University*
Dan Ardia, *Franklin and Marshall College*
Peter Armbruster, *Georgetown University*
David Asch, *University of Pennsylvania*
Andrea Aspbury, *Texas State University, San Marcos*
Nicanor Austriaco, *Providence College*
Mitchell Balish, *Miami University*
Elizabeth Balko, *State University of New York, Oswego*
Ralston Bartholomew, *Warren County Community College*
Christine Barton, *Centre College*
Robert Bauman, *Amarillo College*
Wayne Becker, *University of Wisconsin, Madison*
Peter Berget, *Carnegie Mellon University*
Ethan Bier, *University of California, San Diego*
Michael Black, *California Polytechnic State University, San Luis Obispo*
Anthony Bledsoe, *University of Pittsburgh*
James Bottesch, *Brevard Community College*
Scott Bowling, *Auburn University*
Robert Boyd, *Auburn University*
Ronald Breaker, *Yale University*
Diane Bridge, *Elizabethtown College*
Andrew Brower, *Middle Tennessee State University*
Rebecca Brown, *College of Marin*
Mark Browning, *Purdue University*
Arthur Buikema, *Virginia Polytechnic Institute and State University*
Warren Burggren, *University of North Texas*
Jennifer Carbrey, *Duke University*
Dale Casamatta, *University of North Florida*
Gregory Chandler, *University of North Carolina, Wilmington*
Deborah Chapman, *University of Pittsburgh*
Curt Coffman, *Vincennes University*
Patricia Colberg, *University of Wyoming*
Kathleen Cornely, *Providence College*
Elizabeth Cowles, *Eastern Connecticut State University*
Jason Curtis, *Purdue University, North Central*
Karen Curto, *University of Pittsburgh*
Farahad Dastoor, *University of Maine*
Robin Lee Davies, *Sweet Briar College*
Charles Delwiche, *University of Maryland, College Park*
Jean DeSaix, *University of North Carolina, Chapel Hill*
Hudson DeYoe, *University of Texas, Pan American*
Sunethra Dharmasari, *Texas State University, San Marcos*
Lisa Elfring, *University of Arizona, Tucson*
Eric Engstrom, *College of William and Mary*
Jean Everett, *College of Charleston*
Brent Ewers, *University of Wyoming*
Andrew Fabich, *Tennessee Temple University*
Gary Firestone, *University of California, Berkeley*
Ryan Fisher, *Salem State College*
David Fitch, *New York University*
Jason Flores, *University of North Carolina, Charlotte*
Andrew Forbes, *University of Notre Dame*
Edward Freeman, *St. John Fisher College*
Caitlin Gabor, *Texas State University, San Marcos*
George Gilchrist, *College of William and Mary*
Lynda Goff, *University of California, Santa Cruz*
Elliot Goldstein, *Arizona State University*
Andrew Goliszek, *North Carolina Agricultural & Technical State University*
Margaret Goodman, *Wittenberg University*
Joyce Gordon, *University of British Columbia*
Steve Gorsich, *Central Michigan University*
Susan Michele Green, *Texas State University, San Marcos*
Paul Greenwood, *Colby College*
Stanley Guffey, *University of Tennessee, Knoxville*
Wendy Hanna-Rose, *Pennsylvania State University*
Kiki Harbitz, *Gustavus Adolphus College*
Jeffrey Hardin, *University of Wisconsin, Madison*
Jana Henson, *Georgetown College*
Helen Hess, *College of the Atlantic*
Tracey Hickox, *University of Illinois, Urbana-Champaign*
Sara Hoot, *University of Wisconsin, Milwaukee*
Laurie Host, *Harford Community College*
Kelly Howe, *University of New Mexico, Valencia*
Kathleen Hunt, *University of Portland*
Christine Janis, *Brown University*
Eric Jellen, *Brigham Young University*
Warren Johnson, *University of Wisconsin, Green Bay*
Cindy Johnson-Groh, *Gustavus Adolphus College*
Greg M. Kelly, *University of Western Ontario*
Paul King, *Massasoit Community College*
Joel Kingsolver, *University of North Carolina, Chapel Hill*
Roger Koeppe, *Oklahoma State University*
David Kooyman, *Brigham Young University*

Jocelyn Krebs, *University of Alaska, Anchorage*
John Krenetsky, *Metro State College of Denver*
Patrick Krug, *California State University, Los Angeles*
Holly Kupfer, *Central Piedmont Community College*
Mary Rose Lamb, *University of Puget Sound*
John Lammert, *Gustavus Adolphus College*
Hans Landel, *Edmonds Community College*
Dominic Lannutti, *El Paso Community College*
Janet Lanza, *University of Arkansas, Little Rock*
Georgia Lind, *Kingsborough Community College*
Debra Linton, *Central Michigan University*
Curtis Loer, *University of California, San Diego*
Frank Logiudice, *University of Central Florida*
Barbara Lom, *Davidson College*
James Maller, *University of Colorado, Denver*
James Manser, *Harvey Mudd College (retired)*
Kathleen Marrs, *Indiana University–Purdue University, Indianapolis*
Jennifer Martin, *University of Colorado, Boulder*
Kathy Martin-Troy, *Central Connecticut State University*
John H. McDonald, *University of Delaware*
Robert McLean, *Texas State University, San Marcos*
Michael Meighan, *University of California, Berkeley*
Mariana Melo, *North Essex Community College*
Philip Meneely, *Haverford College*
Dennis Minchella, *Purdue University*
Alan Molumby, *University of Illinois, Chicago*
Vertigo Moody, *Santa Fe Community College*
Elizabeth Morgan, *Lonestar College Kingswood*
Nancy Morvillo, *Florida Southern College*
Mike Muller, *University of Illinois, Chicago*
Dennis Nyberg, *University of Illinois, Chicago*
Amanda N. Orenstein, *Centenary College of New Jersey*
Stephanie Scher Pandolfi, *Michigan State University*
Lisa Parks, *North Carolina State University*
Robert Paul, *Kennesaw State University*
Andrew Pease, *Stevenson University*
Andrew Pekosz, *Johns Hopkins University*
Nancy Pelaez, *Purdue University*
Roger Persell, *City University of New York, Hunter College*
John Peters, *College of Charleston*
Debra Pires, *University of California, Los Angeles*
Gregory Podgorski, *Utah State University*
Robert Podolsky, *College of Charleston*
Therese Poole, *Georgia State University*
Harvey Pough, *Rochester Institute of Technology*
Vanessa Quinn, *Purdue University North Central*
Stephanie Randell, *McLennan Community College*
Clifford Ross, *University of North Florida*
Michael Rutledge, *Middle Tennessee State University*
James Ryan, *Hobart and William Smith Colleges*
Margaret Saha, *College of William and Mary*
Mark Sandheinrich, *University of Wisconsin, La Crosse*
Terry Saropoulos, *Vanier College*
Thomas Sasek, *University of Louisiana, Monroe*
Jon Scales, *Midwestern State University*
Gregory Schmaltz, *Great Basin College*
Oswald Schmitz, *Yale University*
Joan Sharp, *Simon Fraser University*
Michele Shuster, *New Mexico State University*
Suzanne Simon-Westendorf, *Ohio State University*
David Skelly, *Yale University*
Meredith Somerville-Norris, *University of North Carolina, Charlotte*
Sally Sommers-Smith, *Boston University*
Eric Stavney, *Pierce College*
Scott Steinmaus, *California Polytechnic State University, San Luis Obispo*
Janet Steven, *Sweet Briar College*
Kirk Stowe, *University of South Carolina*
Christine Strand, *California Polytechnic State University, San Luis Obispo*
Cynthia Surmacz, *Bloomsburg University*
Jackie Swanik, *North Carolina Central University*
Jerilyn Swann, *Maryville College*
Brad Swanson, *Central Michigan University*
Emily Taylor, *California Polytechnic State University, San Luis Obispo*
Eric Thobaben, *Carroll University*
Ken Thomas, *North Essex Community College*
Briana Timmerman, *University of South Carolina*
Alexandru M. F. Tomescu, *Humboldt State University*
James Traniello, *Boston University*
William Velhagen, *New York University*
Sara Via, *University of Maryland, College Park*
Susan Waaland, *University of Washington*
Frederick Wasserman, *Boston University*
Elizabeth Weiss-Kuziel, *University of Texas, Austin*
Jason Wiles, *Syracuse University*
Kelly P. Williams, *Virginia Polytechnic Institute and State University*
Elizabeth Willott, *University of Arizona, Tucson*
Clifford Wilson, *Kennedy King College*
Charles Wimpee, *University of Wisconsin, Milwaukee*
Leslie Wooten-Blanks, *College of Charleston*
Todd Yetter, *University of the Cumberlands*

Correspondents

One of the most enjoyable interactions I have as a textbook author is correspondence or conversations with researchers and teachers who take the time and trouble to contact me to discuss an issue with the book, or who respond to my queries about a particular data set or study. I'm always amazed and heartened by the generosity of these individuals. They care, deeply.

Julie Aires, *Florida Community College, Jacksonville*
Göran Arnqvist, *Uppsala University*
Terry Bidleman, *Environment Canada*
Brian Buchwitz, *University of Washington*
Helaine Burstein, *Ohio University*
Curt Coffman, *Vincennes University*
Mark Cooper, *University of Washington*
Scott Creel, *Montana State University*
Laura DiCaprio, *Ohio University*
Mary Durant, *Lone Star College*
Victoria Finnerty, *Emory University*
Kathleen Foltz, *University of California, Santa Barbara*
Kathy Gillen, *Kenyon College*
Peter Grant, *Princeton University*
Rosemary Grant, *Princeton University*
Takato Imaizumi, *University of Washington*
Kathleen Janech, *College of Charleston*
Paul King, *Massosoit Community College*

John Lammert, *Gustavus Adolphus College*
Curtis Loer, *University of San Diego*
Scott Meissner, *Cornell University*
Philip Meneely, *Haverford College*
Diarmaid Ó Foighil, *University of Michigan*
John Roth, *University of California, Davis*
Brian Spohn, *Florida Community College, Jacksonville*
Fayla Schwartz, *Everett Community College*
Elizabeth Willott, *University of Arizona, Tucson*
Dan Wulff, *University at Albany, State University of New York*
Glenn Yasuda, *Seattle University*

Supplements Contributors

Instructors depend on an impressive array of support materials—in print and online—to design and deliver their courses. The student experience would be much weaker without the study guide, test bank, activities, animations, quizzes, and tutorials written by the following individuals.

Brian Bagatto, *University of Akron*
Michael Black, *California Polytechnic State University, San Luis Obispo*
Jay L. Brewster, *Pepperdine University*
Warren Burggren, *University of North Texas*
Patricia Colberg, *University of Wyoming*
Clarissa Dirks, *Evergreen State College*
Lisa Elfring, *University of Arizona, Tucson*
Brent Ewers, *University of Wyoming*
Miriam Ferzli, *North Carolina State University*
Cheryl Frederick, *University of Washington*
Cindee Giffen, *University of Wisconsin, Madison*
Kathy M. Gillen, *Kenyon College*
Mary Catherine Hager
Christopher Harendza, *Montgomery County Community College*
Laurel Hester, *University of South Carolina*
Jean Heitz, *University of Wisconsin, Madison*
Tracey Hickox, *University of Illinois, Urbana-Champaign*
Kathleen Hunt, *University of Portland*
Barbara Lom, *Davidson College*
Jennifer Nauen, *University of Delaware*
Chris Pagliarulo, *University of Arizona, Tucson*
Stephanie Scher Pandolfi, *Michigan State University*
Debra Pires, *University of California, Los Angeles*
Gregory Podgorski, *Utah State University*
Carol Pollock, *University of British Columbia*
Jessica Poulin, *University at Buffalo, the State University of New York*
Vanessa Quinn, *Purdue University North Central*
Eric Ribbens, *Western Illinois University*
Joan Sharp, *Simon Fraser University*
Suzanne Simon-Westendorf, *Ohio University*
Emily Taylor, *California Polytechnic State University, San Luis Obispo*
Fred Wasserman, *Boston University*
Cindy White, *University of Northern Colorado*

Book Team

People who watch film credits won't be surprised to learn how many talented people are involved in making a textbook happen. Ruth Steyn provided incisive comments on the earliest drafts of the revised manuscript, which were then implemented by Development Editors Alice Fugate, Moira Lerner-Nelson, Bill O'Neal, and Susan Teahan; the final version of the text was copyedited by Michael Rossa and proofread by Pete Shanks. This editorial team was directed by Executive Director of Development Deborah Gale.

As in the first three editions, the book's figures were designed by Dr. Kim Quillin. Kim's artistic sensibilities, scientific training, and teaching talent are what make the figures in this book sing. If imitation really is the sincerest form of flattery, then Kim should be blushing—we've started seeing her style copied in textbook figures and scientific illustrations by other illustrator/designers. Kim's designs were rendered by Imagineering Media Services; the final art manuscript was vetted by Art Editor Kelly Murphy and rendered art expertly proofread by Frank Purcell. Maureen Spuhler did a thorough review of the photography program and researched the hundreds of images new to the Fourth Edition.

The book's clean, elegant design is the brainchild of Marilyn Perry; the text and art were set in Marilyn's design by S4Carlisle Publishing Services. The book's production was supervised by Lori Newman and Mike Early.

The extensive supplements program was managed by Assistant Editor Brady Golden. All of the individuals I've mentioned—and more—were supported by Editorial Assistant Leslie Allen.

Creating MasteringBiology® tutorials and activities requires a talented team of people playing many different roles. Media content development was overseen by Tania Mlawer and Sarah Jensen who benefited from the program expertise of Caroline Power and Karen Sheh. Julia Henderson and Laura Tommasi worked together as Media Producers with the leadership of Senior Media Producer Deb Greco. Lauren Fogel (VP, Director, Media Development), Stacy Treco (VP, Director, Media Product Strategy), and Laura ensured that the complete media program that accompanies the Fourth Edition, including MasteringBiology, will meet the needs of the students and professors who use our offerings.

Pearson's talented Sales Reps, who listen to professors, advise the editorial staff, and get the book in students' hands, are supported by tireless Executive Marketing Manager Lauren Harp and Director of Marketing Christy Lawrence. The marketing materials that support the outreach effort were produced by Lillian Carr and her colleagues in Pearson's MarCom group.

The vision and resources required to run this entire enterprise are the responsibility of Vice President and Editor-in-Chief Beth Wilbur, Senior Vice President and Editorial Director Frank Ruggirello, President of Pearson Science Paul Corey, and President of Pearson Science & Math Linda Davis.

Finally, the driving forces behind this edition were Project Manager Sonia DiVittorio and Acquisitions Editor Becky Ruden. Sonia's razor-sharp mind, depth of heart, and dedication to excellence shine in every page. Becky's energy and belief in this book surmounted obstacles that would have felled any other editor. I thank the two of them from the bottom of my heart.

Media Guide

Students who purchase a new copy of the text receive free access to MasteringBiology® *(www.masteringbiology.com)*, which contains valuable videos, animations, and practice quizzes to help students learn and prepare for exams.

THE BIG PICTURE New to the Fourth Edition, The Big Pictures are interactive concept maps based on four overarching topics in biology that help students synthesize information across broad concepts and not get lost in the details.

Energy (Chapters 9 and 10)

- How Photosynthesis Yields Sugar
- How Cellular Respiration Yields ATP
- How Photosynthesis Relates to Cellular Respiration

Genetic Information (Chapters 12–18)

- How Genes Are Expressed
- How Genetic Information Is Copied and Transmitted
- How Genetic Information Changes

Evolution (Chapters 24–27)

- How Species Evolve
- How Species Form the Tree of Life

Ecology (Chapters 50–55)

- How Organisms Interact in Their Environment
- How Energy and Nutrients Flow through Ecosystems

BIOFLIX™ BioFlix are 3-D movie-quality animations with carefully constructed student tutorials, labeled slide shows, study sheets, and quizzes which bring biology to life.

WEB ACTIVITIES Web Activities help students learn biological concepts via simple, cartoon-style animations and contain pre-quizzes and post-quizzes to test student's understanding of biology's dynamic processes and concepts.

DISCOVERY VIDEOS Brief videos from the Discovery Channel on 29 different biology topics are available for student viewing along with a corresponding video quiz.

VIDEOS Additional molecular and microscopy videos provide vivid images of processes of the cell.

BIOSKILLS BioSkills (in Appendix A) provide background on key skills and techniques for introductory biology students. New to the Fourth Edition are online questions that give students practice building their skill set.

GRAPHIT! Graphing tutorials show students how to plot, interpret, and critically evaluate real data.

Chapter 1

- An Introduction to Graphing

Chapter 50

- Animal Food Production Efficiency and Food Policy
- Atmospheric CO_2 and Temperature Changes

Chapter 52

- Age Pyramids and Population Growth

Chapter 53

- Species Area Effect and Island Biogeography

Chapter 55

- Forestation Change
- Global Fisheries and Overfishing
- Municipal Solid Waste Trends in the U.S.
- Global Freshwater Resources
- Prospects for Renewable Energy
- Global Soil Degradation

WORD STUDY TOOLS New to the Fourth Edition are Latin and Greek root word flash cards to help students practice the language of biology. In addition, an audio glossary provides correct pronunciation to help students learn key terms introduced in the book.

CUMULATIVE TEST Every chapter offers 20 Practice Test questions that students can pool from different chapters into a Cumulative Test to simulate a practice exam.

RSS FEEDS Real Simple Syndication directly links breaking news from four important sources: National Public Radio, *Scientific American*, *Science Daily News*, and *BioScience*. Current articles reinforce the dynamic nature of science in our daily lives.

eTEXT The eText of *Biological Science*, Fourth Edition, is available online 24/7 for students' convenience. New annotation, highlighting, and bookmarking tools allow students to personalize the material for efficient review.

MEDIA AT A GLANCE

	BIOFLIX	WEB ACTIVITIES	DISCOVERY VIDEOS	VIDEOS	BIOSKILLS
1 Biology and the Tree of Life		Artificial Selection; Introduction to Experimental Design	Cells; Charles Darwin; Early Life; Bacteria		The Metric System; Reading Graphs; Reading a Phylogenetic Tree; Some Common Latin and Greek Roots Used in Biology
Unit 1 The Molecules of Life					
2 Water and Carbon: The Chemical Basis of Life		The Properties of Water			Reading Chemical Structures; Using Logarithms; Making Concept Maps; Reading Graphs
3 Protein Structure and Function		Condensation and Hydrolysis Reactions; Activation Energy and Enzymes		An Idealized Alpha Helix (A); An Idealized Alpha Helix (B); An Idealized Beta-Pleated Sheet (A); An Idealized Beta-Pleated Sheet (B)	
4 Nucleic Acids and the RNA World		Structure of RNA and DNA		Stick Model of DNA; Surface Model of DNA	Separating and Visualizing Molecules; Biological Imaging: Microscopy and X-Ray Crystallography
5 An Introduction to Carbohydrates		Carbohydrate Structure and Function			
6 Lipids, Membranes, and the First Cells	Membrane Transport	Diffusion and Osmosis; Membrane Transport Proteins		Space-filling Model of Cholesterol; Stick Model of Cholesterol; Space-filling Model of Phosphatidylcholine; Stick Model of a Phosphatidylcholine	Biological Imaging: Microscopy and X-Ray Crystallography; Separating and Visualizing Molecules
Unit 2 Cell Structure and Function					
7 Inside the Cell	Tour of an Animal Cell; Tour of a Plant Cell	Transport into the Nucleus; A Pulse-Chase Experiment	Bacteria; Cells	Confocal vs. Standard Fluorescence Microscopy; Cytoplasmic Streaming; Crawling Amoeba	Separating Cell Components by Centrifugation; Biological Imaging: Microscopy and X-Ray Crystallography; Separating and Visualizing Molecules
8 Cell-Cell Interactions			Cells	Connexon Structure	Separating and Visualizing Molecules
9 Cellular Respiration and Fermentation	Cellular Respiration	Redox Reactions; Glucose Metabolism		Space-filling Model of ATP (adenosine triphosphate); Stick Model of ATP (adenosine triphosphate)	
10 Photosynthesis	Photosynthesis	Chemiosmosis; Photosynthesis; Strategies for Carbon Fixation	Space Plants	Space-filling Model of Chlorophyll	
11 The Cell Cycle	Mitosis	The Phases of Mitosis; Four Phases of the Cell Cycle	Fighting Cancer	Mitosis	Separating and Visualizing Molecules; Cell and Tissue Culture Methods

MEDIA AT A GLANCE *(continued)*

	BIOFLIX	WEB ACTIVITIES	DISCOVERY VIDEOS	VIDEOS	BIOSKILLS
Unit 3 Gene Structure and Expression					
12 Meiosis	Meiosis	Meiosis; Mistakes in Meiosis			Combining Probabilities; Using Statistical Tests and Interpreting Standard Error Bars
13 Mendel and the Gene		Mendel's Experiments; The Principle of Independent Assortment			Model Organisms; Combining Probabilities; Reading Graphs
14 DNA and the Gene: Synthesis and Repair	DNA Replication	DNA Synthesis			Separating Cell Components by Centrifugation; Cell and Tissue Culture Methods; Using Logarithms; Reading Graphs
15 How Genes Work		The One-Gene One-Enzyme Hypothesis; The Triplet Nature of the Genetic Code			
16 Transcription, RNA Processing, and Translation	Protein Synthesis	RNA Synthesis; Synthesizing Proteins		A Stick-and-Ribbon Rendering of a tRNA	
17 Control of Gene Expression in Bacteria		The *lac* Operon		Cartoon Model of the *lac* Repressor from *E. coli*	
18 Control of Gene Expression in Eukaryotes		Transcription Initiation in Eukaryotes		Cartoon Model of the DNA-Binding Portion of TATA-Binding Protein Interacting with DNA; Cartoon Model of the GAL4 Transcription Factor from the yeast *S. cerevisiae*	Biological Imaging: Microscopy and X-Ray Crystallography; Separating and Visualizing Molecules
19 Analyzing and Engineering Genes		Producing Human Growth Hormone; The Polymerase Chain Reaction	Transgenics	Cartoon Model of the BamH1a Endonuclease	Separating and Visualizing Molecules
20 Genomics		Human Genome Sequencing Strategies	DNA Forensics		Model Organisms; Using Logarithms
Unit 4 Developmental Biology					
21 Principles of Development		Early Pattern Formation in *Drosophila*	Cloning	A Cartoon and Stick Model of the Homeodomain of the *Engrailed* Protein from *Drosophila* Interacting with DNA	Model Organisms; Cell and Tissue Culture Methods
22 An Introduction to Animal Development		Early Stages of Animal Development			
23 An Introduction to Plant Development					Model Organisms
Unit 5 Evolutionary Processes and Patterns					
24 Evolution by Natural Selection		Natural Selection for Antibiotic Resistance	Charles Darwin; Antibiotics		Reading a Phylogenetic Tree; Model Organisms; Reading Graphs

	BIOFLIX	WEB ACTIVITIES	DISCOVERY VIDEOS	VIDEOS	BIOSKILLS
25 Evolutionary Processes	Mechanisms of Evolution	The Hardy-Weinberg Principle; Three Modes of Natural Selection			Combining Probabilities; Using Statistical Tests and Interpreting Standard Error Bars; Reading Graphs
26 Speciation		Allopatric Speciation; Speciation by Changes in Ploidy			Reading a Phylogenetic Tree
27 Phylogenies and the History of Life		Adaptive Radiation	Mass Extinctions		Reading a Phylogenetic Tree
Unit 6 The Diversification of Life					
28 Bacteria and Archaea		The Tree of Life	Bacteria; Antibiotics; Molds; Tasty Bacteria; Early Life		Reading a Phylogenetic Tree; Model Organisms
29 Protists		Alternation of Generations in a Protist		A Crawling Amoeba	Biological Imaging: Microscopy and X-Ray Crystallography; Model Organisms
30 Green Algae and Land Plants		Plant Evolution and the PhylogeneticTree	Colored Cotton; Space Plants; Plant Pollination		
31 Fungi		Life Cycle of a Mushroom	Leafcutter Ants; Molds		
32 An Introduction to Animals		The Architecture of Animals	Leafcutter Ants; Invertebrates		
33 Protostome Animals		Protostome Diversity			Model Organisms
34 Deuterostome Animals		Deuterostome Diversity	Invertebrates		
35 Viruses		The HIV Replicative Cycle	Vaccines; Emerging Diseases		Biological Imaging: Microscopy and X-Ray Crystallography; Separating and Visualizing Molecules
Unit 7 How Plants Work					
36 Plant Form and Function		Plant Growth			
37 Water and Sugar Transport in Plants	Water Transport in Plants	Solute Transport in Plants		Plasmolysis of Plant Cells	
38 Plant Nutrition		Soil Formation and Nutrient Uptake			
39 Plant Sensory Systems, Signals, and Responses		Sensing Light; Plant Hormones; Plant Defenses			Cell and Tissue Culture Methods
40 Plant Reproduction		Reproduction in Flowering Plants; Fruit Structure and Development	Colored Cotton; Plant Pollination		
Unit 8 How Animals Work					
41 Animal Form and Function		Surface Area/Volume Relationships; Homeostasis	Blood; Human Body		Using Logarithms
42 Water and Electrolyte Balance in Animals		The Mammalian Kidney			

MEDIA AT A GLANCE *(continued)*

	BIOFLIX	WEB ACTIVITIES	DISCOVERY VIDEOS	VIDEOS	BIOSKILLS
43 Animal Nutrition	Homeostasis: Regulating Blood Sugar	The Digestion and Absorption of Food; Understanding Diabetes Mellitus	Nutrition		Biological Imaging: Microscopy and X-Ray Crystallography; Separating and Visualizing Molecules
44 Gas Exchange and Circulation	Gas Exchange	Gas Exchange in the Lungs and Tissues; The Human Heart	Blood		
45 Electrical Signals in Animals	How Neurons Work; How Synapses Work	Membrane Potentials; Action Potentials	Novelty Gene; Teen Brains	The Acetylcholine Receptor	Using Logarithms
46 Animal Sensory Systems and Movement	Muscle Contraction	The Vertebrate Eye; Structure and Contraction of Muscle Fibers	Muscles & Bones	The Acetylcholine Receptor	
47 Chemical Signals in Animals		Endocrine System Anatomy; Hormone Actions on Target Cells	Endocrine System	Cartoon Model of the DNA Binding Motif of a Zinc Finger Transcription Factor Binding to DNA	Separating Cell Components by Centrifugation
48 Animal Reproduction		Human Gametogenesis; Human Reproduction			Using Logarithms; Reading a Phylogenetic Tree
49 The Immune System in Animals		The Inflammatory Response; The Adaptive Immune Response	Emerging Diseases; Vaccines	Chemotaxis of a Neutrophil	
Unit 9 Ecology					
50 An Introduction to Ecology		Tropical Atmospheric Circulation	Rain Forests; Trees; Introduced Species		
51 Behavioral Ecology		Homing Behavior in Digger Wasps	Novelty Gene		
52 Population Ecology	Population Ecology	Modeling Population Growth; Human Population Growth and Regulation			
53 Community Ecology		Life Cycle of a Malaria Parasite; Succession	Leafcutter Ants		
54 Ecosystems	The Carbon Cycle	The Global Carbon Cycle			
55 Biodiversity and Conservation Biology		Habitat Fragmentation	Rain Forests		Using Logarithms

These black swan chicks are being introduced to the world by their mother, who trails close behind them. Likewise, this chapter introduces the key ideas that launched biological science as a discipline.

Biology and the Tree of Life 1

In essence, biological science is a search for ideas and observations that unify our understanding of the diversity of life, from bacteria living in rocks a mile underground to hedgehogs and humans. Chapter 1 is an introduction to this search.

The goals of this chapter are to introduce the nature of life and explore how biologists go about studying it. The chapter also introduces themes that will resonate throughout this book: (**1**) analyzing how organisms work at the molecular level, (**2**) understanding organisms in terms of their evolutionary history, and (**3**) helping you learn to think like a biologist.

Let's begin with what may be the most fundamental question of all: What is life?

KEY CONCEPTS

- Organisms obtain and use energy, are made up of cells, process information, replicate, and as populations evolve.
- The cell theory proposes that all organisms are made of cells and that all cells come from preexisting cells.
- The theory of evolution by natural selection maintains that species change through time because individuals with certain heritable traits produce more offspring than other individuals do.
- A phylogenetic tree is a graphical representation of the evolutionary relationships between species. These relationships can be estimated by analyzing similarities and differences in traits. Species that share distinctive traits are closely related and are placed close to each other on the tree of life.
- Biologists ask questions, generate hypotheses to answer them, and design experiments that test the predictions made by competing hypotheses.

1.1 What Does It Mean to Say That Something Is Alive?

An **organism** is a life-form—a living entity made up of one or more cells. Although there is no simple definition of life that is endorsed by all biologists, most agree that organisms share a suite of five fundamental characteristics.

- *Energy* To stay alive and reproduce, organisms have to acquire and use energy. To give just two examples: plants absorb sunlight; animals ingest food.
- *Cells* Organisms are made up of membrane-bound units called cells. A cell's membrane regulates the passage of materials between exterior and interior spaces.
- *Information* Organisms process hereditary or genetic information, encoded in units called genes, along with information they acquire from the environment. Right now cells throughout your body are using genetic information to make the molecules that keep you alive; your eyes and brain are decoding information on this page that will help you learn some biology.

✔ When you see this checkmark, stop and test yourself. Answers are available in Appendix B.

- *Replication* One of the great biologists of the twentieth century, François Jacob, said that the "dream of a bacterium is to become two bacteria." Almost everything an organism does contributes to one goal: replicating itself.
- *Evolution* Organisms are the product of evolution, and their populations continue to evolve.

You can think of this text as one long exploration of these five traits. Here's to life!

1.2 The Cell Theory

Two of the greatest unifying ideas in all of science laid the groundwork for modern biology: the cell theory and the theory of evolution by natural selection. Formally, scientists define a **theory** as an explanation for a very general class of phenomena or observations. The cell theory and theory of evolution address fundamental questions: What are organisms made of? Where do they come from?

When these concepts emerged in the mid-1800s, they revolutionized the way biologists think about the world. They established two of the five attributes of life: Organisms are cellular, and their populations change over time.

Neither insight came easily, however. The cell theory, for example, emerged after some 200 years of work. In 1665 Robert Hooke used a crude microscope to examine the structure of cork (a bark tissue) from an oak tree. The instrument magnified objects to just 30× (30 times) their normal size, but it allowed Hooke to see something extraordinary. In the cork he observed small, pore-like compartments that were invisible to the naked eye. These structures came to be called cells.

Soon after Hooke published his results, Anton van Leeuwenhoek succeeded in developing much more powerful microscopes, some capable of magnifications up to 300×. With these instruments, Leeuwenhoek inspected samples of pond water and made the first observations of human blood cells, and of sperm cells, shown in **Figure 1.1**.

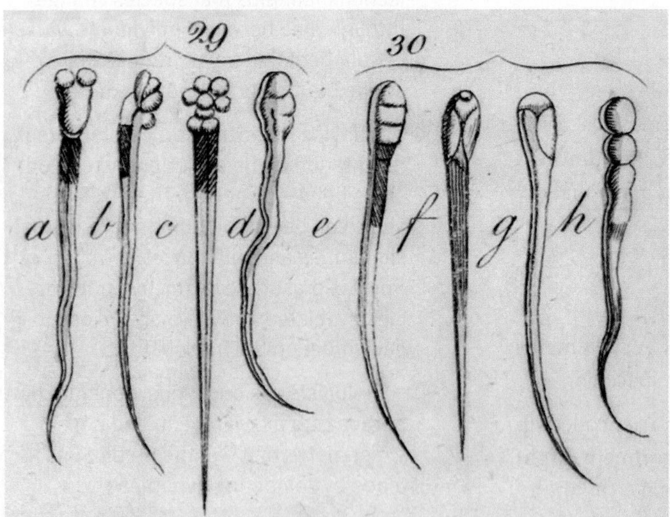

FIGURE 1.1 First Images of Cells. This drawing, by Anton van Leeuwenhoek, shows sperm cells.

In the 1670s a researcher who was studying the leaves and stems of plants with a microscope concluded that these large, complex structures are composed of many individual cells. By the early 1800s enough data had accumulated for a biologist to claim that *all* organisms consist of cells.

Are *All* Organisms Made of Cells?

The smallest organisms known today are bacteria that are barely 80 nanometers wide, or 80 *billionths* of a meter. (See **BioSkills 1** in Appendix A to review the metric system and its prefixes.[1]) It would take 12,500 of these organisms lined up end to end to span a millimeter. This is the distance between the smallest hash marks on a metric ruler.

In contrast, sequoia trees can be over 100 meters tall. This is the equivalent of a 20-story building. Bacteria and sequoias are composed of the same fundamental building block, however—the cell. Bacteria consist of a single cell; sequoias are made up of many cells.

Today, a **cell** is defined as a highly organized compartment that is bounded by a thin, flexible structure called a plasma membrane and that contains concentrated chemicals in an aqueous (watery) solution. The chemical reactions that sustain life take place inside cells. Most cells are also capable of reproducing by dividing—in effect, by making a copy of themselves.

The realization that all organisms are made of cells was fundamentally important, but it formed only the first part of the cell theory. In addition to understanding what organisms are made of, scientists wanted to understand how cells come to be.

Where Do Cells Come From?

Most scientific theories have two components: The first describes a pattern in the natural world; the second identifies a mechanism or process that is responsible for creating that pattern. Hooke and his fellow scientists articulated the pattern component of the cell theory. In 1858 Rudolph Virchow added the process component by stating that all cells arise from preexisting cells.

The complete **cell theory** can be stated as follows: All organisms are made of cells, and all cells come from preexisting cells.

TWO HYPOTHESES The cell theory was a direct challenge to the prevailing explanation of where cells come from, called spontaneous generation. At the time, most biologists believed that organisms arise spontaneously under certain conditions. For example, the bacteria and fungi that spoil foods such as milk and wine were thought to appear in these nutrient-rich media of their own accord—springing to life from nonliving materials. Spontaneous generation was a **hypothesis**: a proposed explanation.

The all-cells-from-cells hypothesis, in contrast, maintained that cells do not spring to life spontaneously but are produced only when preexisting cells grow and divide.

Biologists usually use the word theory to refer to proposed explanations for broad patterns in nature and prefer hypothesis to refer to explanations for more tightly focused questions.

[1]BioSkills are located in the first appendix at the back of the book. They focus on general skills that you'll use throughout this course. More than a few students have found them to be a life-saver. Please use them!

AN EXPERIMENT TO SETTLE THE QUESTION Soon after the all-cells-from-cells hypothesis appeared in print, Louis Pasteur set out to test its predictions experimentally. A **prediction** is something that can be measured and that must be correct if a hypothesis is valid.

Pasteur wanted to determine whether microorganisms could arise spontaneously in a nutrient broth or whether they appear only when a broth is exposed to a source of preexisting cells. To address the question, he created two treatment groups: a broth that was not exposed to a source of preexisting cells and a broth that was.

The spontaneous generation hypothesis predicted that cells would appear in both treatment groups. The all-cells-from-cells hypothesis predicted that cells would appear only in the treatment exposed to a source of preexisting cells.

Figure 1.2 shows Pasteur's experimental setup. Note that the two treatments are identical in every respect but one. Both used glass flasks filled with the same amount of the same nutrient broth. Both were boiled for the same amount of time to kill any existing organisms such as bacteria or fungi. But because the flask pictured in Figure 1.2a had a straight neck, it was exposed to preexisting cells after sterilization by the heat treatment. These preexisting cells are the bacteria and fungi that cling to dust particles in the air. They could drop into the nutrient broth because the neck of the flask was straight.

In contrast, the flask drawn in Figure 1.2b had a long swan neck. Pasteur knew that water would condense in the crook of the swan neck after the boiling treatment and that this pool of water

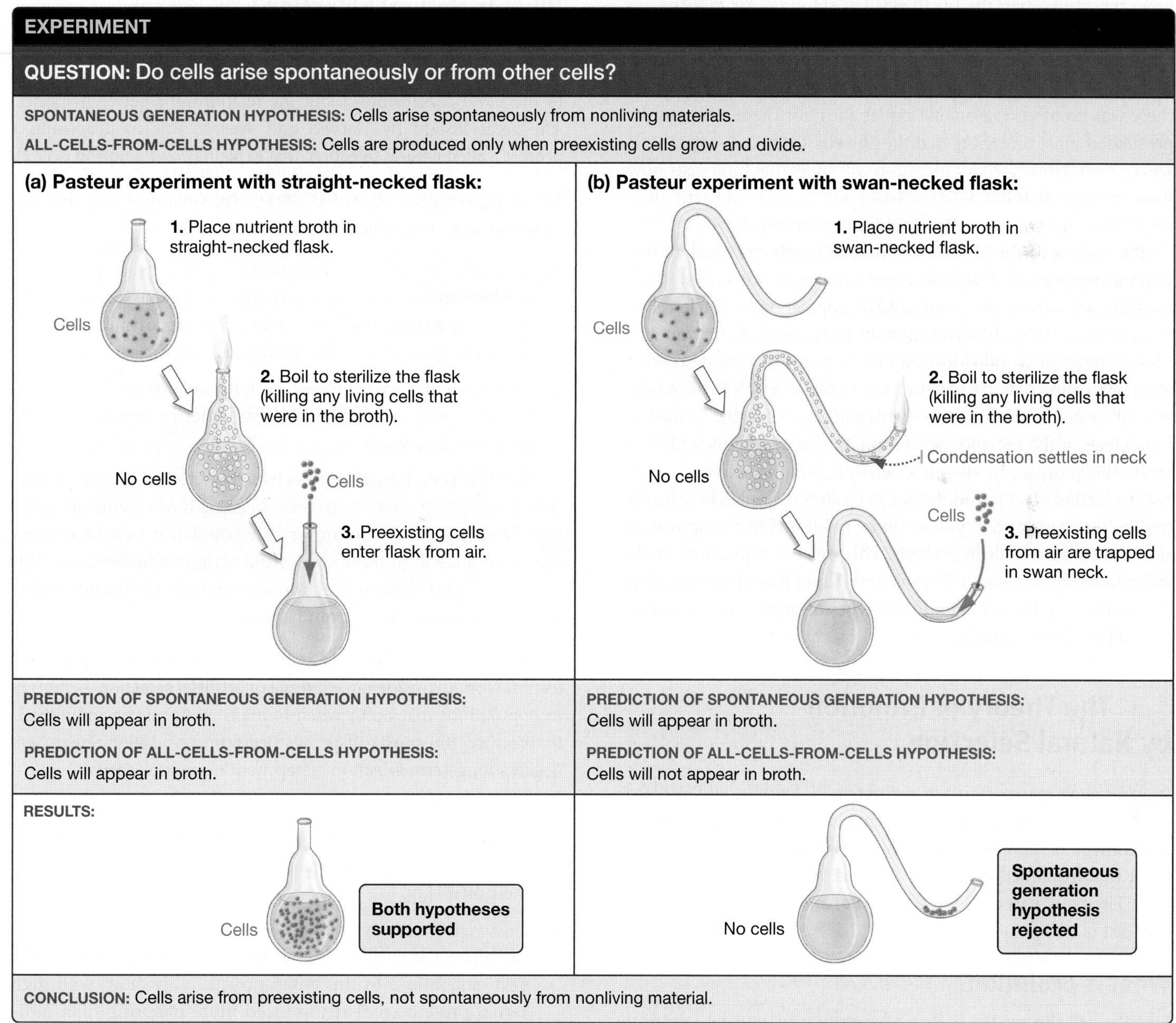

FIGURE 1.2 The Spontaneous Generation and All-Cells-from-Cells Hypotheses Were Tested Experimentally.

✔**QUESTION** What problem would arise if Pasteur had (1) put different types of broth in the two treatments, (2) heated them for different lengths of time, or (3) used a ceramic flask for one treatment and a glass flask for the other?

would trap any bacteria or fungi that entered on dust particles. Thus, the contents of the swan-necked flask was isolated from any source of preexisting cells even though still open to the air.

Pasteur's experimental setup was effective because there was only one difference between the two treatments and because that difference was the factor being tested—in this case, a broth's exposure to preexisting cells.

ONE HYPOTHESIS SUPPORTED And Pasteur's results? As Figure 1.2 shows, the treatment exposed to preexisting cells quickly filled with bacteria and fungi. This observation was important because it showed that the heat sterilization step had not altered the nutrient broth's capacity to support growth.

The treatment in the swan-necked flask remained sterile, however. Even when the broth was left standing for months, no organisms appeared in it. This result was inconsistent with the hypothesis of spontaneous generation.

Because Pasteur's data were so conclusive—meaning that there was no other reasonable explanation for them—the results persuaded most biologists that the all-cells-from-cells hypothesis was correct. However, Chapters 2–6 will show that biologists now have evidence that life did arise from nonlife early in Earth's history, through a process called chemical evolution.

The success of the cell theory's process component had an important implication: If all cells come from preexisting cells, it follows that all individuals in an isolated population of single-celled organisms are related by common ancestry. Similarly, in you and other multicellular individuals, all the cells present are descended from preexisting cells, tracing back to a fertilized egg. A fertilized egg is a cell created by the fusion of sperm and egg—cells that formed in individuals of the previous generation. In this way, all the cells in a multicellular organism are connected by common ancestry.

The second great founding idea in biology is similar, in spirit, to the cell theory. It also happened to be published the same year as the all-cells-from-cells hypothesis. This was the realization, made independently by Charles Darwin and Alfred Russel Wallace, that all species—all distinct, identifiable types of organisms—are connected by common ancestry.

1.3 The Theory of Evolution by Natural Selection

In 1858 short papers written separately by Darwin and Wallace were read to a small group of scientists attending a meeting of the Linnean Society of London. A year later, Darwin published a book that expanded on the idea summarized in those brief papers. The book was called *The Origin of Species.* The first edition sold out in a day.

What Is Evolution?

Like the cell theory, the theory of evolution by natural selection has a pattern and a process component. Darwin and Wallace's theory made two important claims concerning patterns that exist in the natural world.

- Species are related by common ancestry. This contrasted with the prevailing view in science at the time, which was that species represent independent entities created separately by a divine being.
- In contrast to the accepted view that species remain unchanged through time, Darwin and Wallace proposed that the characteristics of species can be modified from generation to generation. Darwin called this process "descent with modification."

Evolution is a change in the characteristics of a population over time. It means that species are not independent and unchanging entities, but are related to one another and can change through time.

What Is Natural Selection?

This pattern component of the theory of evolution was actually not original to Darwin and Wallace. Several scientists had already come to the same conclusions about the relationships between species. The great insight by Darwin and Wallace was in proposing a process, called **natural selection**, that explains *how* evolution occurs.

TWO CONDITIONS OF NATURAL SELECTION Natural selection occurs whenever two conditions are met.

1. Individuals within a population vary in characteristics that are **heritable**—meaning, traits that can be passed on to offspring. A **population** is defined as a group of individuals of the same species living in the same area at the same time.
2. In a particular environment, certain versions of these heritable traits help individuals survive better or reproduce more than do other versions.

If certain heritable traits lead to increased success in producing offspring, then those traits become more common in the population over time. In this way, the population's characteristics change as a result of natural selection acting on individuals. This is a key insight: Natural selection acts on individuals, but evolutionary change occurs in populations.

SELECTION ON MAIZE AS AN EXAMPLE To clarify how natural selection works, consider an example of **artificial selection**—changes in populations that occur when *humans* select certain individuals to produce the most offspring. Beginning in 1896, researchers began a long-term selection experiment on maize (corn).

1. In the original population, the percentage of protein in maize kernels was variable among individuals. Kernel protein content is a heritable trait—parents tend to pass the trait on to their offspring.
2. Each year for many years, researchers chose individuals with the highest kernel protein content to be the parents of the next generation. In this environment, individuals with high kernel protein content produced more offspring than individuals with low kernel protein content.

Figure 1.3 shows the results. Note that this graph plots generation number on the *x*-axis, starting from the first generation

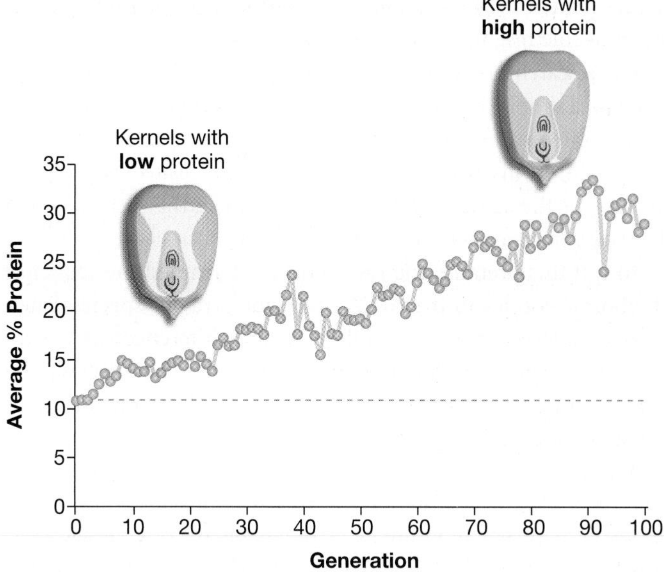

FIGURE 1.3 Response to Selection for High Kernel Protein Content in Maize. The average protein content goes down in a few years because of poor growing conditions or chance changes in how the many genes responsible for this trait interact.

("0") and continuing for 100 generations. The average percentage of protein in a kernel among individuals in this population is plotted on the *y*-axis.

To read this graph, put your finger on the *x*-axis at time 0. Then read up the *y*-axis, and note that kernels averaged about 11 percent protein at the start of the experiment. Now read the graph to the right. Each dot is a data point, representing the average kernel protein concentration in a particular generation. (A generation in maize is one year.) The lines on this graph simply connect the dots, to make the pattern in the data easier to see. At the end of the graph, after 100 generations of selection, average kernel protein content is about 29 percent. (For more help with reading graphs, see **BioSkills 2** in Appendix A.)

This sort of change in the characteristics of a population, over time, is evolution. Humans have been practicing artificial selection for thousands of years, and biologists have now documented evolution by *natural* selection—where humans don't do the selecting—occurring in thousands of different populations, including humans.

To practice applying the principles of artificial selection, go to the online study area at *www.masteringbiology.com.*

(MB) **Web Activity** Artificial Selection

Evolution occurs when heritable variation leads to differential success in reproduction. ✔If you understand this concept, you should be able to describe how protein content in maize kernels changed over time, using the same *x*-axis and *y*-axis as in Figure 1.3, when researchers selected individuals with *lowest* kernel protein content to be the parents of the next generation. (This experiment was actually done, starting with the same population at the same time as selection for high protein content.)

FITNESS AND ADAPTATION Darwin also introduced some new terminology to identify what is happening during natural selection.

- In everyday English, fitness means health and well-being. But in biology, **fitness** means the ability of an individual to produce offspring. Individuals with high fitness produce many surviving offspring.
- In everyday English, adaptation means that an individual is adjusting and changing to function in new circumstances. But in biology, an **adaptation** is a trait that increases the fitness of an individual in a particular environment.

Once again, consider kernel protein content in maize: In the environment of the experiment graphed in Figure 1.3, individuals with high kernel protein content produced more offspring and had higher fitness than individuals with lower kernel protein content. In this population and this environment, high kernel protein content was an adaptation that allowed certain individuals to thrive.

Note that during this process, the amount of protein in the kernels of any individual maize plant did not change within its lifetime—the change occurred in the characteristics of the population over time.

Together, the cell theory and the theory of evolution provided the young science of biology with two central, unifying ideas:

1. The cell is the fundamental structural unit in all organisms.
2. All species are related by common ancestry and have changed over time in response to natural selection.

CHECK YOUR UNDERSTANDING

If you understand that . . .

- Natural selection occurs when heritable variation in certain traits leads to improved success in reproduction. Because individuals with these traits produce many offspring with the same traits, the traits increase in frequency and evolution occurs.
- Evolution is a change in the characteristics of a population over time.

✔ **You should be able to . . .**

On the graph you just analyzed, describe the average kernel protein content over time in a maize population where *no* selection occurred.

Answers are available in Appendix B.

1.4 The Tree of Life

Section 1.3 focused on how individual populations change through time in response to natural selection. But over the past several decades, biologists have also documented dozens of cases in which natural selection has caused populations of one species to diverge and form new species. This divergence process is called **speciation.**

Research on speciation has two important implications: All species come from preexisting species, and all species, past and present, trace their ancestry back to a single common ancestor.

The theory of evolution by natural selection predicts that biologists should be able to reconstruct a **tree of life**—a family tree of organisms. If life on Earth arose just once, then such a diagram would describe the genealogical relationships between species with a single, ancestral species at its base.

Has this task been accomplished? If the tree of life exists, what does it look like?

Using Molecules to Understand the Tree of Life

One of the great breakthroughs in research on the tree of life occurred when Carl Woese (pronounced *woes*) and colleagues began analyzing the chemical components of organisms, as a way to understand their evolutionary relationships. Their goal was to understand the **phylogeny** of all organisms—their actual genealogical relationships. Translated literally, phylogeny means "tribe-source."

To understand which organisms are closely versus distantly related, Woese and co-workers needed to study a molecule that is found in all organisms. The molecule they selected is called small subunit ribosomal RNA (rRNA). It is an essential part of the machinery that all cells use to grow and reproduce.

Although rRNA is a large and complex molecule, its underlying structure is simple. The rRNA molecule is made up of sequences of four smaller chemical components called ribonucleotides. These ribonucleotides are symbolized by the letters A, U, C, and G. In rRNA, ribonucleotides are connected to one another linearly, like boxcars of a freight train (**Figure 1.4**).

ANALYZING rRNA Why might rRNA be useful for understanding the relationships between organisms? The answer is that the ribonucleotide sequence in rRNA is a trait that can change during the course of evolution. Although rRNA performs the same function in all organisms, the sequence of ribonucleotide building blocks in this molecule is not identical among species.

In land plants, for example, the molecule might start with the sequence A-U-A-U-C-G-A-G. In green algae, which are closely related to land plants, the same section of the molecule might contain A-U-A-U-G-G-A-G. But in brown algae, which are not closely related to green algae or to land plants, the same part of the molecule might consist of A-A-A-U-G-G-A-C.

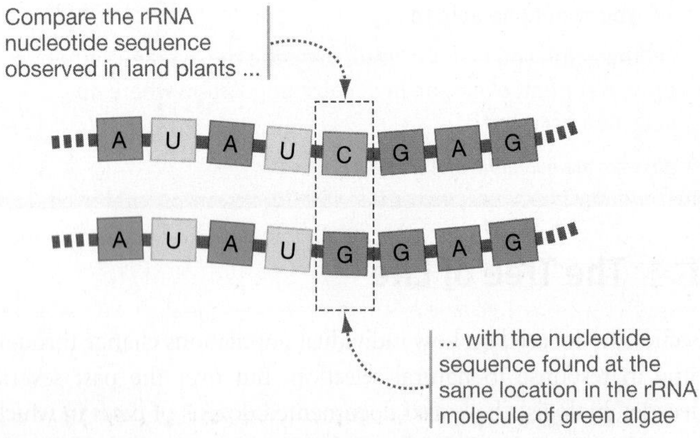

FIGURE 1.4 RNA Molecules Are Made Up of Smaller Molecules. The complete small subunit rRNA molecule contains about 2000 ribonucleotides; just 8 are shown in this comparison.

✔**QUESTION** Suppose that in the same portion of rRNA, molds and other fungi have the sequence A-U-A-U-G-G-A-C. According to these data, are fungi more closely related to green algae or to land plants? Explain your logic.

The research program that Woese and co-workers pursued was based on a simple premise: If the theory of evolution is correct, then rRNA sequences should be very similar in closely related organisms but less similar in organisms that are less closely related. Species that are part of the same evolutionary lineage, like the plants, should share certain changes in rRNA that no other species have.

To test this premise, the researchers determined the sequence of ribonucleotides in the rRNA of a wide array of species. Then they considered what the similarities and differences in the sequences implied about relationships between the species. The goal was to produce a diagram that described the phylogeny of the organisms in the study.

A diagram that depicts evolutionary history in this way is called a phylogenetic tree. Just as a family tree shows relationships between individuals, a phylogenetic tree shows relationships between species. On a phylogenetic tree, branches that share a recent common ancestor represent species that are closely related; branches that don't share recent common ancestors represent species that are more distantly related.

THE TREE OF LIFE ESTIMATED FROM AN ARRAY OF GENES To construct a phylogenetic tree, researchers use a computer to find the arrangement of branches that is most consistent with the similarities and differences observed in the data.

Although the initial work was based only on the sequences of ribonucleotides observed in rRNA, biologists now use data sets that include sequences from a wide array of genes. **Figure 1.5** shows a recent tree produced by comparing these sequences. Because this tree includes such a diverse array of species, it is often called the universal tree, or the tree of life. (For help in learning how to read a phylogenetic tree, see **BioSkills 3** in Appendix A.)

The tree of life implied by rRNA and other genetic data established that there are three fundamental groups or lineages of organisms: (**1**) the Bacteria, (**2**) the Archaea, and (**3**) the Eukarya. In all **eukaryotes**, cells have a prominent component called the nucleus (**Figure 1.6a**). Translated literally, the word eukaryotes means "true kernel." Because the vast majority of bacterial and archaeal cells lack a nucleus, they are referred to as **prokaryotes** (literally, "before kernel"; see **Figure 1.6b**). The vast majority of bacteria and archaea are unicellular ("one-celled"); many eukaryotes are multicellular ("many-celled").

When these results were first published, biologists were astonished. For example:

- Prior to Woese's work and follow-up studies, biologists thought that the most fundamental division among organisms was between prokaryotes and eukaryotes. The Archaea were virtually unknown—much less recognized as a major and highly distinctive branch on the tree of life.
- Fungi were thought to be closely related to plants. Instead, they are actually much more closely related to animals.
- Traditional approaches for classifying organisms—including the system of five kingdoms divided into various classes, or-

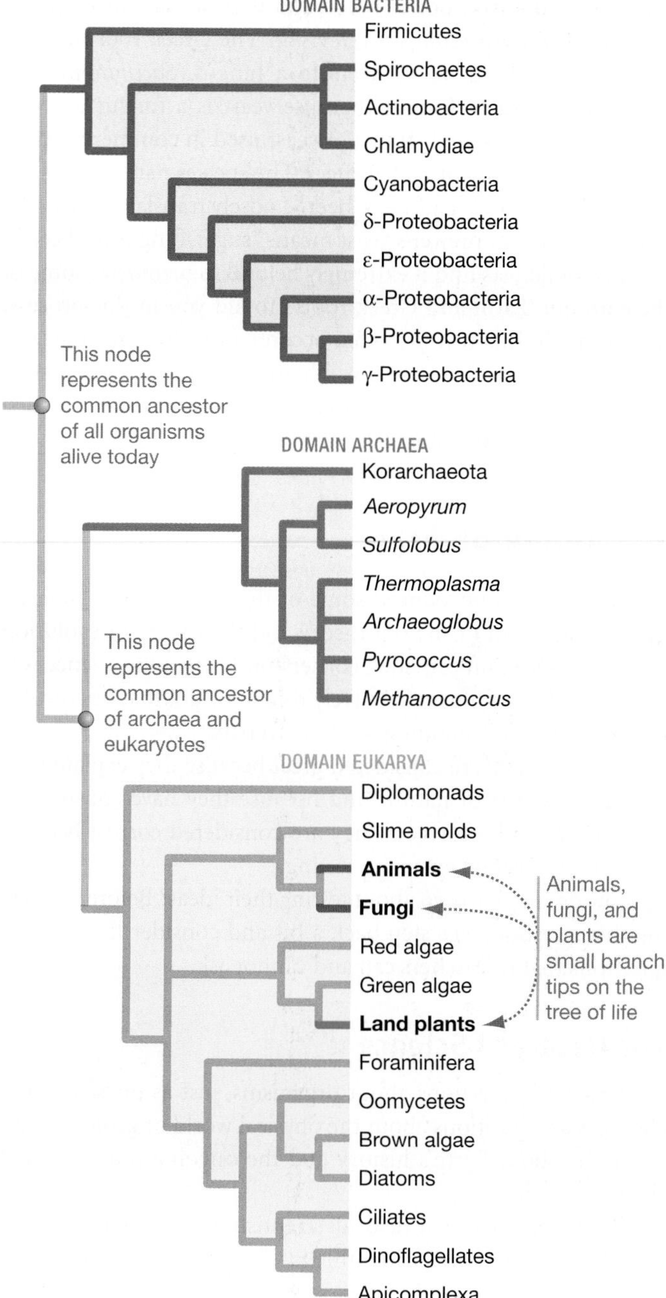

FIGURE 1.5 The Tree of Life. "Universal tree" estimated from a large amount of gene sequence data. The three domains of life revealed by the analysis are labeled. Common names are given for most lineages in the domains Bacteria and Eukarya. Genus names are given for members of the domain Archaea, because most of these organisms have no common names.

ders, and families that you may have learned in high school—are inaccurate in many cases, because they do not reflect the actual evolutionary history of the organisms involved.

THE TREE OF LIFE IS A WORK IN PROGRESS Just as researching your family tree can help you understand who you are and where you came from, so the tree of life helps biologists understand the relationships between organisms and the history of species. The discovery of the Archaea and the accurate placement of lineages such as the fungi qualify as exciting breakthroughs in our understanding of evolutionary history and life's diversity.

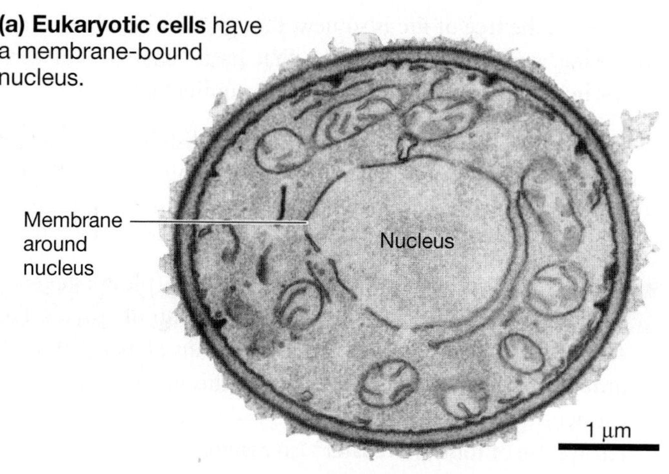

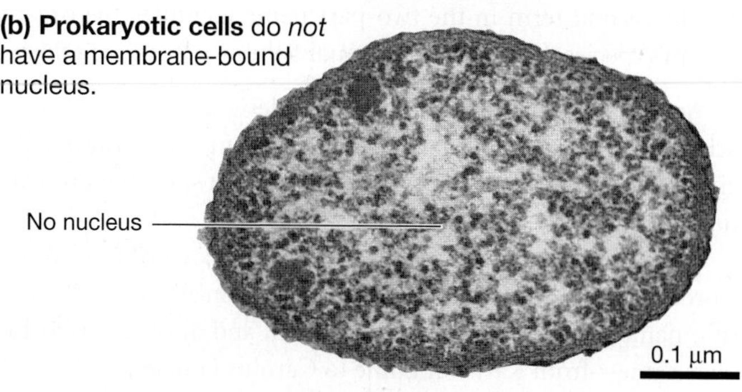

FIGURE 1.6 Eukaryotes and Prokaryotes.

✔**QUESTION** How many times larger is the eukaryotic cell in this figure than the prokaryotic cell? (Hint: study the scale bars.)

Work on the tree of life continues at a furious pace, however, and the location of certain branches on the tree is hotly debated. As databases expand and as techniques for analyzing data improve, the shape of the tree of life presented in Figure 1.5 will undoubtedly change. Our understanding of the tree of life, like our understanding of every other topic in biological science, is dynamic.

How Should We Name Branches on the Tree of Life?

In science, the effort to name and classify organisms is called **taxonomy**. Any named group is called a **taxon** (plural: **taxa**). Currently, biologists are working to create a taxonomy, or naming system, that accurately reflects the phylogeny of organisms.

For example, Woese proposed a new taxonomic category called the **domain**. The three domains of life are the Bacteria, Archaea, and Eukarya.

Biologists often use the term **phylum** (plural: **phyla**) to refer to major lineages within each domain. Although the designation is somewhat arbitrary, each phylum is considered a major branch on the tree of life. For example, within the lineage called animals, biologists name about 35 phyla—each of which is distinguished by distinctive aspects of its body structure as well as by distinctive gene sequences. The mollusks (clams, squid, octopuses) constitute a phylum, as do chordates (the vertebrates and their close relatives).

Because the tree of life is so new, though, naming systems are still being worked out. One thing that hasn't changed for centuries, however, is the naming system for individual species.

SCIENTIFIC (SPECIES) NAMES In 1735, a Swedish botanist named Carolus Linnaeus established a system for naming species that is still in use today. Linnaeus created a two-part name unique to each type of organism.

- The first part indicates the organism's **genus** (plural: **genera**). A genus is made up of a closely related group of species. For example, Linnaeus put humans in the genus *Homo*. Although humans are the only living species in this genus, at least five extinct organisms, all of which walked upright and made extensive use of tools, were later also assigned to *Homo*.
- The second term in the two-part name identifies the organism's species. Linnaeus gave humans the species name *sapiens*.

An organism's genus and species designation is called its **scientific name** or Latin name. Scientific names are always italicized. Genus names are always capitalized, but species names are not—for instance, *Homo sapiens*.

Scientific names are based on Latin or Greek word roots or on words "Latinized" from other languages. Linnaeus gave a scientific name to every species then known, and also Latinized his own name—from Karl von Linné to Carolus Linnaeus.

Linnaeus maintained that different types of organisms should not be given the same genus and species names. Other species may be assigned to the genus *Homo*, and members of other genera may be named *sapiens*, but only humans are named *Homo sapiens*. Each scientific name is unique.

SCIENTIFIC NAMES ARE OFTEN DESCRIPTIVE Scientific names and terms are often based on Latin or Greek word roots that are descriptive. For example, *Homo sapiens* is derived from the Latin *homo* for "man" and *sapiens* for "wise" or "knowing." The yeast that bakers use to produce bread and that brewers use to brew beer is called *Saccharomyces cerevisiae*. The Greek root *saccharo* means "sugar," and *myces* refers to a fungus. *Saccharomyces* is aptly named "sugar fungus" because yeast is a fungus and because the domesticated strains of yeast used in commercial baking and brewing are often fed sugar. The species name of this organism, *cerevisiae*, is Latin for "beer." Loosely translated, then, the scientific name of brewer's yeast means "sugar-fungus for beer."

Most biologists find it extremely helpful to memorize some of the common Latin and Greek roots. To aid you in this process, new terms in this text are often accompanied by a reference to their Latin or Greek word roots in parentheses, and a glossary of common root words with translations and examples is provided in **BioSkills 4** in Appendix A.

CHECK YOUR UNDERSTANDING

If you understand that . . .

- A phylogenetic tree shows the evolutionary relationships between species.
- To infer where species belong on a phylogenetic tree, biologists examine the characteristics of the species involved. Closely related species should have similar characteristics, while less closely related species should be less similar.

✓ You should be able to . . .

Examine the following sequences and draw a phylogenetic tree showing the relationships between species A, B, and C that these data imply:

Species A: A A C T A G C G C G A T
Species B: A A C T A G C G C C A T
Species C: T T C T A G C G G T A T

Answers are available in Appendix B.

1.5 Doing Biology

This chapter has introduced some of the great ideas in biology. The development of the cell theory and the theory of evolution by natural selection provided cornerstones when the science was young; the tree of life is a relatively recent insight that has revolutionized our understanding of life's diversity.

These theories are considered great because they explain fundamental aspects of nature, and because they have consistently been shown to be correct. They are considered correct because they have withstood extensive testing.

How do biologists go about testing their ideas? Before answering this question, let's step back a bit and consider the types of questions that researchers can and cannot ask.

The Nature of Science

Biologists ask questions about organisms, just as physicists and chemists ask questions about the physical world or geologists ask questions about Earth's history and the ongoing processes that shape landforms.

No matter what their field, all scientists ask questions that can be answered by measuring things—by collecting data. Conversely, scientists cannot address questions that can't be answered by measuring things.

This distinction is important. It is at the root of continuing controversies about teaching evolution in publicly funded schools. In the United States and in Turkey, in particular, some Christian and Islamic leaders have been particularly successful in pushing their claim that evolution and religious faith are in conflict. Even though the theory of evolution is considered one of the most successful and best-substantiated ideas in the history of science, they object to teaching it.

The vast majority of biologists and religious leaders reject this claim; they see no conflict between evolution and religious faith. Their view is that science and religion are compatible because they address different types of questions.

- Science is about formulating hypotheses and finding evidence that supports or conflicts with those hypotheses.

- Religious faith addresses questions that cannot be answered by data. The questions addressed by the world's great religions focus on why we exist and how we should live.

Both types of questions are seen as legitimate and important.

So how do biologists go about answering questions? Let's consider two issues currently being addressed by researchers.

Why Do Giraffes Have Long Necks? An Introduction to Hypothesis Testing

If you were asked why giraffes have long necks, you might say that long necks enable giraffes to reach food that is unavailable to other mammals. This hypothesis is expressed in African folktales and has traditionally been accepted by many biologists. The food competition hypothesis is so plausible, in fact, that for decades no one thought to test it.

In the mid-1990s, however, Robert Simmons and Lue Scheepers assembled data suggesting that the food competition hypothesis is only part of the story. Their analysis supports an alternative hypothesis—that long necks allow giraffes to use their heads as effective weapons for battering their opponents.

How did biologists test the food competition hypothesis? What data support their alternative explanation? Before attempting to answer these questions, it's important to recognize that hypothesis testing is a two-step process:

1. State the hypothesis as precisely as possible and list the predictions it makes.
2. Design an observational or experimental study that is capable of testing those predictions.

If the predictions are accurate, the hypothesis is supported. If the predictions are not met, then researchers do further tests, modify the original hypothesis, or search for alternative explanations.

THE FOOD COMPETITION HYPOTHESIS: PREDICTIONS AND TESTS

The food competition hypothesis claims that giraffes compete for food with other species of mammals. When food is scarce, as it is during the dry season, giraffes with longer necks can reach food that is unavailable to other species and to giraffes with shorter necks. As a result, the longest-necked individuals in a giraffe population survive better and produce more young than do shorter-necked individuals, and average neck length of the population increases with each generation.

To use the terms introduced earlier, long necks are adaptations that increase the fitness of individual giraffes during competition for food. This type of natural selection has gone on so long that the population has become extremely long necked.

The food competition hypothesis makes several explicit predictions. For example, the food competition hypothesis predicts that

1. neck length is variable among giraffes;
2. neck length in giraffes is heritable; and
3. giraffes feed high in trees, especially during the dry season, when food is scarce and the threat of starvation is high.

The first prediction is correct. Studies in zoos and natural populations confirm that neck length is variable among individuals.

The researchers were unable to test the second prediction, however, because they studied giraffes in a natural population and were unable to do breeding experiments. As a result, they simply had to accept this prediction as an assumption. In general, though, biologists prefer to test every assumption behind a hypothesis.

What about the prediction regarding feeding high in trees? According to Simmons and Scheepers, this is where the food competition hypothesis breaks down.

Consider, for example, data collected by a different research team on the amount of time that giraffes spend feeding in vegetation of different heights. **Figure 1.7a** plots the height of vegetation versus the percentage of bites taken by a giraffe, for males and for females from the same population in Kenya. The dashed line on each graph indicates the average height of a male or female in this population.

Note that the average height of a giraffe in this population is much greater than the height where most feeding takes place. In

(a) Most feeding is done at about shoulder height.

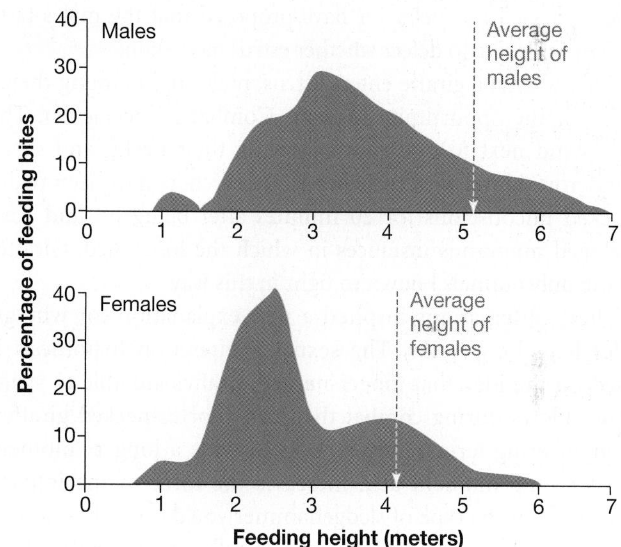

(b) Typical feeding posture in giraffes

FIGURE 1.7 Giraffes Do Not Usually Extend Their Necks to Feed.

this population, both male and female giraffes spend most of their feeding time eating vegetation that averages just 60 percent of their full height. Studies on other populations of giraffes, during both the wet and dry seasons, are consistent with these data. Giraffes usually feed with their necks bent (**Figure 1.7b**).

These data cast doubt on the food competition hypothesis, because one of its predictions does not appear to hold. Biologists have not abandoned this hypothesis completely, though, because feeding high in trees may be particularly valuable during extreme droughts, when a giraffe's ability to reach leaves far above the ground could mean the difference between life and death. Still, Simmons and Scheepers have offered an alternative explanation for why giraffes have long necks. The new hypothesis is based on the mating system of giraffes.

THE SEXUAL COMPETITION HYPOTHESIS: PREDICTIONS AND TESTS

Giraffes have an unusual mating system. Breeding occurs year round rather than seasonally. To determine when females are coming into estrus or "heat" and are thus receptive to mating, the males nuzzle the rumps of females. In response, the females urinate into the males' mouths. The males then tip their heads back and pull their lips to and fro, as if tasting the liquid. Biologists who have witnessed this behavior have proposed that the males taste the females' urine to detect whether estrus has begun.

Once a female giraffe enters estrus, males fight among themselves for the opportunity to mate. Combat is spectacular. The bulls stand next to one another, swing their necks, and strike thunderous blows with their heads. Researchers have seen males knocked unconscious for 20 minutes after being hit and have cataloged numerous instances in which the loser died. Giraffes are the only animals known to fight in this way.

These observations inspired a new explanation for why giraffes have long necks. The sexual competition hypothesis is based on the idea that longer-necked giraffes are able to strike harder blows during combat than can shorter-necked giraffes. In engineering terms, longer necks provide a longer "moment arm." A long moment arm increases the force of the impact. (Think about the type of sledgehammer you'd use to bash down a concrete wall—one with a short handle or one with a long handle?)

The idea here is that longer-necked males should win more fights and, as a result, father more offspring than shorter-necked males do. If neck length in giraffes is inherited, then the average neck length in the population should increase over time. Under the sexual competition hypothesis, long necks are adaptations that increase the fitness of males during competition for females.

Although several studies have shown that long-necked males are more successful in fighting and that the winners of fights gain access to estrous females, the question of why giraffes have long necks is not closed. With the data collected to date, most biologists would probably concede that the food competition hypothesis needs further testing and refinement and that the sexual selection hypothesis appears promising. It could also be true that both hypotheses are correct. For our purposes, the important take-home message is that all hypotheses must be tested rigorously.

In many cases in biological science, testing hypotheses rigorously involves experimentation. Experimenting on giraffes is difficult. But in the case study considered next, biologists were able to test an interesting hypothesis experimentally.

How Do Ants Navigate? An Introduction to Experimental Design

Experiments are a powerful scientific tool because they allow researchers to test the effect of a single, well-defined factor on a particular phenomenon. Because experiments testing the effect of neck length on food and sexual competition in giraffes haven't been done yet, let's consider a different question: When ants leave their nest to search for food, how do they find their way back?

The Saharan desert ant lives in colonies and makes a living by scavenging the dead carcasses of insects. Individuals leave the burrow and wander about searching for food at midday, when temperatures at the surface can reach 60°C (140°F) and predators are hiding from the heat.

Foraging trips can take the ants hundreds of meters—an impressive distance when you consider that these animals are only about a centimeter long. But when an ant returns, it doesn't follow the same long, wandering route it took on its way away from the nest. Instead, individuals return in a straight line. How do they do this?

THE PEDOMETER HYPOTHESIS Early work on navigation in desert ants showed that they use the Sun's position as a compass—meaning that they always know the approximate direction of the nest relative to the Sun. But how do they know how far to go?

After experiments had shown that the ants do not use landmarks to navigate, Matthias Wittlinger and co-workers set out to test a novel idea. The biologists proposed that Saharan desert ants know how far they are from the nest by integrating information from leg movements.

According to this pedometer hypothesis, the ants always know how far they are from the nest because they track the number of steps they have taken and their stride length. The idea is that they can make a beeline back to the burrow because they integrate information on the angles they have traveled *and* the distance they have gone—based on step number and stride length.

TESTING THE HYPOTHESIS To test their idea, Wittlinger's group allowed ants to walk from a nest to a feeder through a channel—a distance of 10 m. Then they caught ants at the feeder and created three test groups, each with 25 individuals:

- *Stumps* By cutting the lower legs of some individuals off, they created ants with shorter-than-normal legs.
- *Normal* Some individuals were left alone, meaning that they had normal leg length.
- *Stilts* By gluing pig bristles onto each leg, the biologists created ants with longer-than-normal legs.

Next they put the ants in a different channel and recorded how far they traveled before starting the characteristic set of 180° turns that ants make when they are looking for the nest hole. To see the data they collected, look at the graph on the left side of the "Results" section in **Figure 1.8**.

- *Stumps* The ants with stumps stopped short, by about 5 m, before starting to look for the nest opening.
- *Normal* The normal ants walked the correct distance—about 10 m.
- *Stilts* The ants with stilts walked about 5 m too far before starting to look for the nest opening.

To check the validity of this result, they put the test ants back in the nest and recaptured them one to several days later, when they had walked to the feeder on their stumps, normal legs, or stilts. Now when the ants were put into the other channel to "walk back," they all traveled the correct distance—10 m—before starting to look for the nest (see the graph on the right side of the "Results" section in Figure 1.8).

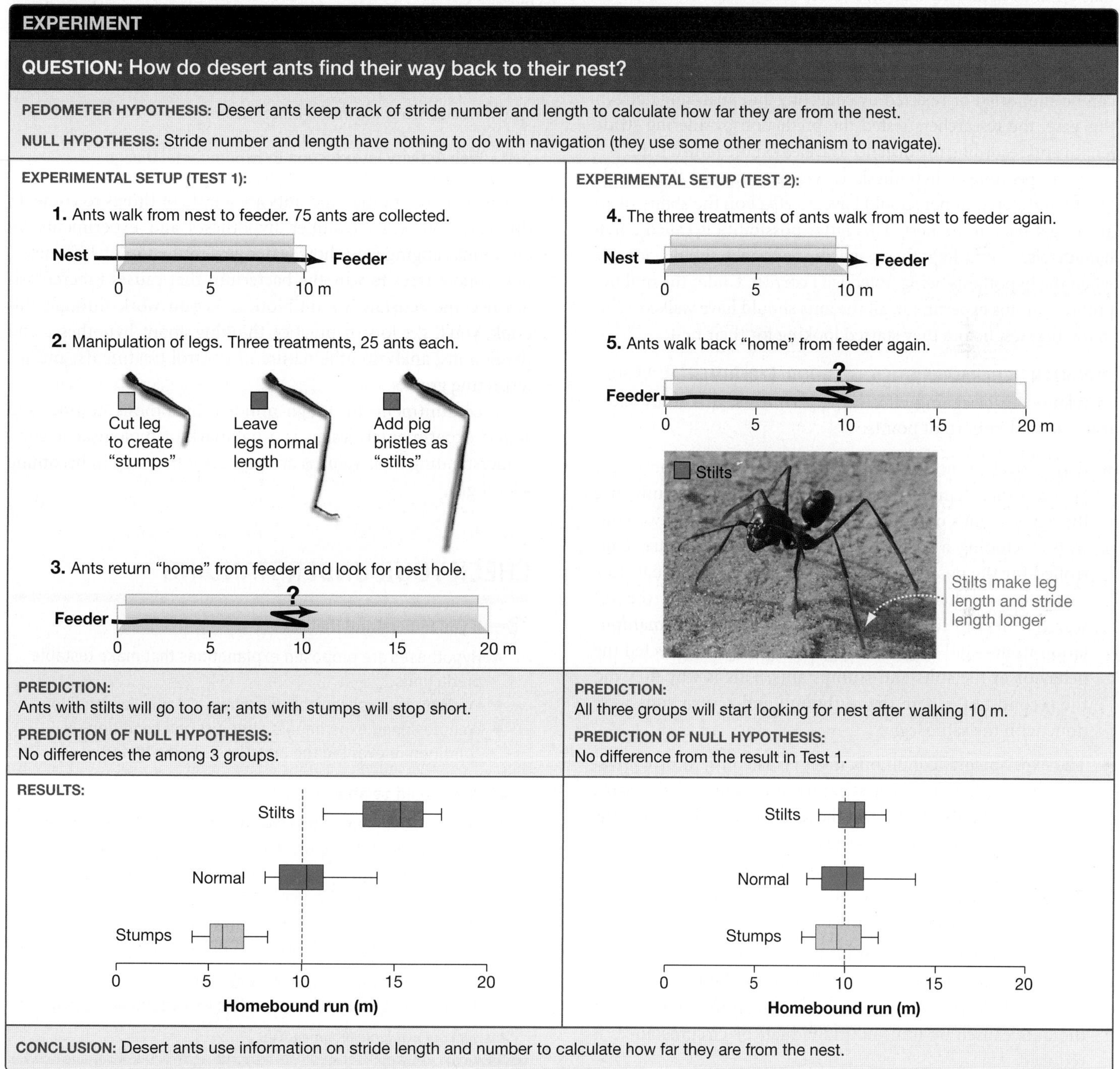

FIGURE 1.8 An Experimental Test: Do Desert Ants Use a "Pedometer"?

SOURCE: Wittlinger, M., R. Wehner, and H. Wolf. 2006. The ant odometer: Stepping on stilts and stumps. *Science* 312: 1965–1967.

✔**QUESTION** How would you interpret the experiment if the researchers had used just one ant in each group instead of 25?

The graphs in the "Results" display "box-and-whisker" plots. Each box indicates the range of distances where 50 percent of the ants stopped to search for the nest; the whiskers indicate the range where 95 percent of the ants stopped to search. The vertical line inside each box indicates the median—meaning that half the ants stopped above this distance and half below. For more details on how biologists report averages and indicate the variability and uncertainty in data, see **BioSkills 5** in Appendix A.

INTERPRETING THE RESULTS The pedometer hypothesis predicts that an ant's ability to walk home depends on the number and length of steps taken on its outbound trip. Recall that a prediction specifies what we should observe if a hypothesis is correct. Good scientific hypotheses make testable predictions—predictions that can be supported or rejected by collecting and analyzing data. In this case, the researchers tested the prediction by altering stride length and recording the distance traveled on the return trip.

If the pedometer hypothesis is wrong, however, then stride length and step number should have no effect on the ability of an ant to get back to its nest. This latter possibility is called a **null hypothesis**. A null hypothesis specifies what we should observe when the hypothesis being tested isn't correct. Under the null hypothesis in this experiment, all the ants should have walked 10 m in the first test before they started looking for their nest.

IMPORTANT CHARACTERISTICS OF GOOD EXPERIMENTAL DESIGN In relation to designing effective experiments, this study illustrates several important points:

- It is critical to include **control** groups. A control checks for factors, other than the one being tested, that might influence the experiment's outcome. In this case, there were two controls. Including a normal, unmanipulated individual controlled for the possibility that switching the individuals to a new channel altered their behavior. In addition, the researchers had to control for the possibility that the manipulation itself—and not the change in leg length—affected the behavior of the stilts and stumps ants. This is why they did the second test, where the outbound and return runs were done with the same legs.
- The experimental conditions must be as constant or equivalent as possible. The investigators used ants of the same species, from the same nest, at the same time of day, under the same humidity and temperature conditions, at the same feeders, in the same channels. Controlling all the variables except one—leg length in this case—is crucial because it eliminates alternative explanations for the results.
- Repeating the test is essential. It is almost universally true that larger sample sizes in experiments are better. By testing many individuals, the amount of distortion or "noise" in the data caused by unusual individuals or circumstances is reduced.

✔**You should be able to explain: (1) What issue would arise if in the first test, the normal individual had not walked 10 m on the return trip before looking for the nest; and (2) What you would conclude if the stilts and stumps ants had not navigated normally during the second test.**

From the outcomes of these experiments, the researchers concluded that desert ants use stride length and number to measure how far they are from the nest. They interpreted their results as strong support for the pedometer hypothesis.

Biologists practice evidence-based decision-making. They ask questions about how organisms work, pose hypotheses to answer those questions, and use experimental or observational evidence to decide which hypotheses are correct.

To review the principles of experimental design, go to the study area at *www.masteringbiology.com*.

(MB) **Web Activity** Introduction to Experimental Design

The data on giraffes and ants are a taste of things to come. In this text you will encounter hypotheses and experiments on questions ranging from how water gets to the top of 100-meter-tall sequoia trees to why the bacterium that causes tuberculosis has become resistant to antibiotics. As you work through this book, you'll get lots of practice thinking about hypotheses and predictions, analyzing the nature of control treatments, and interpreting graphs.

A commitment to tough-minded hypothesis testing and sound experimental design is a hallmark of biological science. Understanding their value is an important first step in becoming a biologist.

CHECK YOUR UNDERSTANDING

If you understand that . . .

- Hypotheses are proposed explanations that make testable predictions.
- Predictions are observable outcomes of particular conditions.
- Well-designed experiments alter just one condition—a condition relevant to the hypothesis being tested.

✔ **You should be able to . . .**

Design an experiment to test the hypothesis that desert ants feed during the hottest part of the day because it allows them to avoid being eaten by lizards. Then answer the following questions about your experimental design:

1. How does the presence of a control group in your experiment allow you to test the null hypothesis?
2. How are experimental conditions controlled or standardized in a way that precludes alternative explanations of the data?

Answers are available in Appendix B.

CHAPTER 1 REVIEW

For media, go to the study area at www.masteringbiology.com 

Summary of Key Concepts

Organisms obtain and use energy, are made up of cells, process information, replicate, and as populations evolve.

- There is no single, well-accepted definition of life. Instead, biologists point to five characteristics that organisms share.

  ✔You should be able to explain why the cells in a dead organism are different from the cells in a live organism.

The cell theory proposes that all organisms are made of cells and that all cells come from preexisting cells.

- The cell theory identified the fundamental structural unit common to all life.

  ✔You should be able to describe the evidence that supported the pattern and the process components of the cell theory.

The theory of evolution by natural selection maintains that species change through time because individuals with certain heritable traits produce more offspring than other individuals do.

- The theory of evolution states that all organisms are related by common ancestry.
- Natural selection is a well-tested explanation for why species change through time and why they are so well adapted to their habitats.

  ✔You should be able to explain why the average protein content of seeds in a natural population of a grass species would increase over time, if seeds with higher protein content survive better and grow into individuals that produce many seeds with high protein content when they mature.

  MB **Web Activity** Artificial Selection

A phylogenetic tree is a graphical representation of the evolutionary relationships between species. These relationships can be estimated by analyzing similarities and differences in traits. Species that share distinctive traits are closely related and are placed close to each other on the tree of life.

- The cell theory and the theory of evolution predict that all organisms are part of a genealogy of species, and that all species trace their ancestry back to a single common ancestor.
- To reconstruct this phylogeny, biologists have analyzed the sequence of components in rRNA and other molecules found in all cells.
- A tree of life, based on similarities and differences in these molecules has three major lineages: the Bacteria, Archaea, and Eukarya.

  ✔You should be able to explain how biologists can determine whether newly discovered species are members of the Bacteria, Archaea, or Eukarya by analyzing their rRNA or other molecules.

Biologists ask questions, generate hypotheses to answer them, and design experiments that test the predictions made by competing hypotheses.

- Biology is a hypothesis-driven, experimental science.

  ✔You should be able to explain (1) the relationship between a hypothesis and a prediction and (2) why experiments are convincing ways to test predictions.

  MB **Web Activity** Introduction to Experimental Design

Questions

✔TEST YOUR KNOWLEDGE

Answers are available in Appendix B

1. Anton van Leeuwenhoek made an important contribution to the development of the cell theory. How?
   - **a.** He articulated the pattern component of the theory—that all organisms are made of cells.
   - **b.** He articulated the process component of the theory—that all cells come from preexisting cells.
   - **c.** He invented the first microscope and saw the first cell.
   - **d.** He invented more powerful microscopes and was the first to describe the diversity of cells.
2. What does it mean to say that experimental conditions are controlled?
   - **a.** The test groups consist of the same individuals.
   - **b.** The null hypothesis is correct.
   - **c.** There is no difference in outcome between the control and experimental treatment.
   - **d.** Physical conditions are identical for all groups tested.
3. What does the term *evolution* mean?
   - **a.** The strongest individuals produce the most offspring.
   - **b.** The characteristics of an individual change through the course of its life, in response to natural selection.
   - **c.** The characteristics of populations change through time.
   - **d.** The characteristics of species become more complex over time.
4. What does it mean to say that a characteristic of an organism is heritable?
   - **a.** The characteristic evolves.
   - **b.** The characteristic can be passed on to offspring.
   - **c.** The characteristic is advantageous to the organism.
   - **d.** The characteristic does not vary in the population.
5. In biology, to what does the term *fitness* refer?
   - **a.** The degree of training and muscle mass an individual has, relative to others in the same population
   - **b.** An individual's slimness, relative to others in the same population
   - **c.** The longevity of a particular individual
   - **d.** An individual's ability to survive and reproduce
6. Could *both* the food competition hypothesis and the sexual selection hypothesis explain why giraffes have long necks? Why or why not?
   - **a.** No. In science, only one hypothesis can be correct.
   - **b.** No. Observations have shown that the food competition hypothesis cannot be correct.
   - **c.** Yes. Long necks could be advantageous for more than one reason.
   - **d.** Yes. All giraffes have been shown to feed at the highest possible height and fight for mates.

✓TEST YOUR UNDERSTANDING

Answers are available in Appendix B

1. What would researchers have to demonstrate to convince you that they had discovered life on another planet?

2. It was once thought that the deepest split between life-forms was between two groups: prokaryotes and eukaryotes. Draw and label a phylogenetic tree that represents this hypothesis. Then draw and label a phylogenetic tree that shows the actual relationships between the three domains of organisms.

3. Why was it important for Linnaeus to establish the rule that only one type of organism can have a particular genus and species name?

4. What does it mean to say that a species is adapted to a particular habitat?

5. Explain how selection occurs during natural selection. What is selected, and why?

6. The following two statements explain the logic behind the use of molecular sequence data to estimate evolutionary relationships:

   "If the theory of evolution is true, then rRNA sequences should be very similar in closely related organisms but less similar in organisms that are less closely related."

   "On a phylogenetic tree, branches that share a recent common ancestor represent species that are closely related; branches that don't share recent common ancestors represent species that are more distantly related."

   Is the logic of these statements sound? Why or why not?

✓APPLYING CONCEPTS TO NEW SITUATIONS

Answers are available in Appendix B

1. A scientific theory is a set of propositions that defines and explains some aspect of the world. This definition contrasts sharply with the everyday usage of the word theory, which often carries meanings such as "speculation" or "guess." Explain the difference between the two definitions, using the cell theory and the theory of evolution by natural selection as examples.

2. Turn back to the tree of life shown in Figure 1.5. Note that Bacteria and Archaea are prokaryotes, while Eukarya are eukaryotes. On the simplified tree below, draw an arrow that points to the branch where the structure called the nucleus originated. Explain your reasoning.

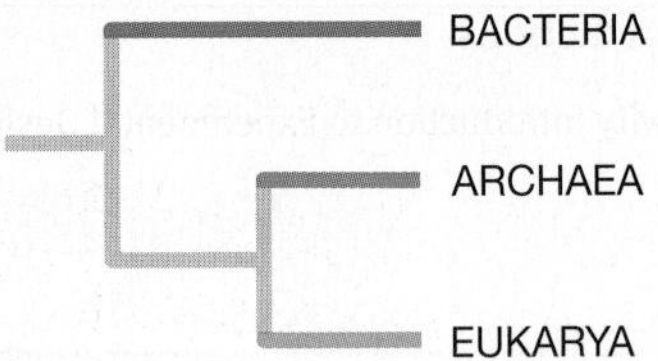

3. The proponents of the cell theory could not "prove" that it was correct in the sense of providing incontrovertible evidence that all organisms are made up of cells. They could state only that all organisms examined to date were made of cells. Why was it reasonable for them to conclude that the theory was valid?

4. Some humans have heritable traits that make them resistant to infection by HIV. In areas of the world where HIV infection rates are high, are human populations evolving? Explain your logic.

All plants are able to harvest diffuse resources and concentrate them in cells and tissues, but their forms and strategies are diverse. These baobab trees in Madagascar may live to be hundreds of years old despite drought conditions, in part by storing water in their enormous trunks.

Plant Form and Function 36

Photosynthetic plants do the most remarkable chemistry of any terrestrial organism. Using the energy in sunlight and the simplest of starting materials—carbon dioxide, water, and ions containing nitrogen, phosphorus, potassium, and other key atoms—plants synthesize thousands of different carbohydrates, proteins, nucleic acids, and lipids. They use these compounds to build bodies that may live for thousands of years.

This feat is even more impressive when you consider that the simple starting materials that plants need to grow are tiny and diffuse—carbon dioxide molecules, water molecules, ammonium ions, and other resources are usually found at low concentrations over a large area. To gather the raw materials required for their sophisticated biosynthetic machinery, a plant's roots and shoots grow outward, extending the individual into the soil and atmosphere.

In essence, a plant's body harvests diffuse resources and concentrates them in cells and tissues. The structure of its body is dynamic, because most plants exhibit **indeterminate growth**, that is, they grow throughout their lives. A 4750-year-old bristlecone pine has roots and shoots that are still growing. In response to favorable conditions, a plant sends roots and shoots in the most promising directions, seeking light and the simple compounds it requires.

The contrast between the plant and animal way of life is striking. Most animals move around and eat concentrated sources of food. But plants stay in one place, extend their roots and shoots to harvest diffuse resources, and make their own food.

This chapter focuses on three fundamental questions:

1. How is the plant body organized?
2. Why are plants so diverse in size and shape?
3. How do plants grow throughout their lives?

KEY CONCEPTS

- The vascular plant body consists of (1) a root system that anchors the individual and absorbs water and key ions, and (2) a shoot system that absorbs carbon dioxide and sunlight. Both systems are dynamic—they grow and change throughout life.
- Variation in body size and shape allows different species to harvest water, light, and other resources in unique ways.
- Primary growth occurs when cells located at the tips of each root and shoot divide and enlarge. Primary growth lengthens roots and shoots and gives rise to three primary tissue systems that are specialized for protection, food production and storage, and transport.
- In some species, secondary growth occurs when cells near the perimeter of a root or shoot divide and enlarge, widening the structure. Secondary growth adds transport tissue and provides structural support.

✔ When you see this checkmark, stop and test yourself. Answers are available in Appendix B.

Instead of surveying the entire catalog of land plants, though, the focus here is on the angiosperms. Recall from Chapter 30 that angiosperms, the flowering plants, are the most recent major group of plants to appear in the fossil record and the most abundant, species-rich, and geographically widespread on Earth today. It is also difficult to overstate their economic and medical importance to humans. Most of the food we eat and the drugs we use are derived from angiosperms.

If you look outside or down the produce aisle of a grocery store, you will see a wide diversity of plants and plant products. By the time you finish this chapter, you'll understand how these plant bodies are put together and how they grow. Exploring questions about the anatomy of flowering plants is vital to understanding the world at large as well as the other chapters in this unit.

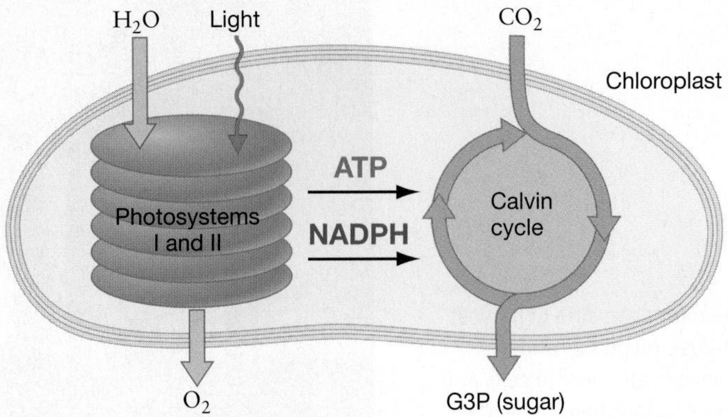

FIGURE 36.1 Plants Need Resources to Perform Photosynthesis. Plants perform some of the most remarkable biochemistry of any organism—manufacturing sugars from water, carbon dioxide, and sunlight. In the process, plants absorb diffuse resources and concentrate them in the form of complex organic compounds.

36.1 Plant Form: Themes with Many Variations

Chapter 10 detailed how plants—along with algae, cyanobacteria, and a variety of protists—obtain the energy and carbon they need to grow and reproduce. Plants use light energy (photons) to synthesize carbohydrates using carbon dioxide from the air and water from the soil.

For photosynthesis to occur, plants need large amounts of light and carbon dioxide, along with water as an electron source (**Figure 36.1**). Plants also need large amounts of water to fill their cells and maintain them at normal volume and pressure.

To synthesize nucleic acids, enzymes, phospholipids, and the other macromolecules needed to build and run cells, plants must obtain nitrogen (N), phosphorus (P), potassium (K), magnesium (Mg), and a host of other nutrients. Most of these key elements exist in nature as ions that dissolve in water found in soil.

Figure 36.2 labels the major structures in the two basic systems that plants use to acquire the resources they need for photosynthesis. A belowground portion called the **root system** anchors the plant and takes in water and nutrients from the soil; an aboveground portion called the **shoot system** harvests light and carbon dioxide from the atmosphere to produce sugars. Both systems grow throughout the life of the individual, allowing the plant to increase in size, outcompete other individuals, and acquire resources.

In most plants, vascular tissue connects the root and shoot systems. Water is transported from roots to shoots through vascular tissue; sugars and other nutrients are transported in both directions.

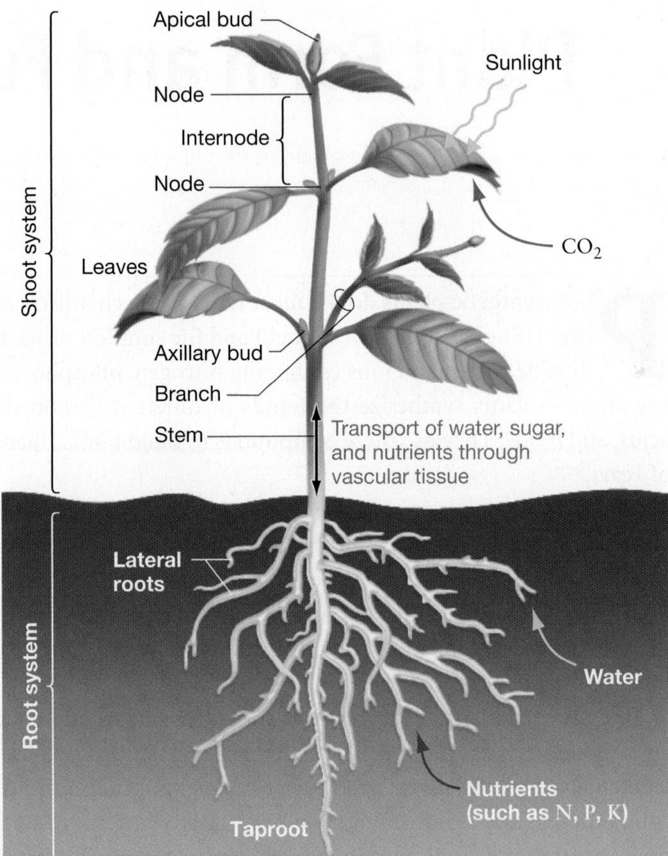

FIGURE 36.2 Root and Shoot Systems Acquire and Transport Resources. Most root systems have the same general structures, and most shoot systems have the same general structure. Shoot systems are specialized for harvesting light and CO_2. Root systems absorb water and key nutrients such as nitrogen (N), phosphorus (P), and potassium (K).

✔**QUESTION** Suppose that this plant's growth was limited by access to light and nutrients, and that there was a new deposit of nutrient-rich soil to the left and much more sunlight was suddenly available to the right. Explain what you would expect this individual to look like in one month.

The Importance of Surface Area/Volume Relationships

Before exploring the nature of root and shoot systems in more detail, it's important to recognize a key physical relationship that connects their structure to their function.

Root and shoot systems both function in absorption—of light or key ions and molecules. Absorption takes place across a surface. But

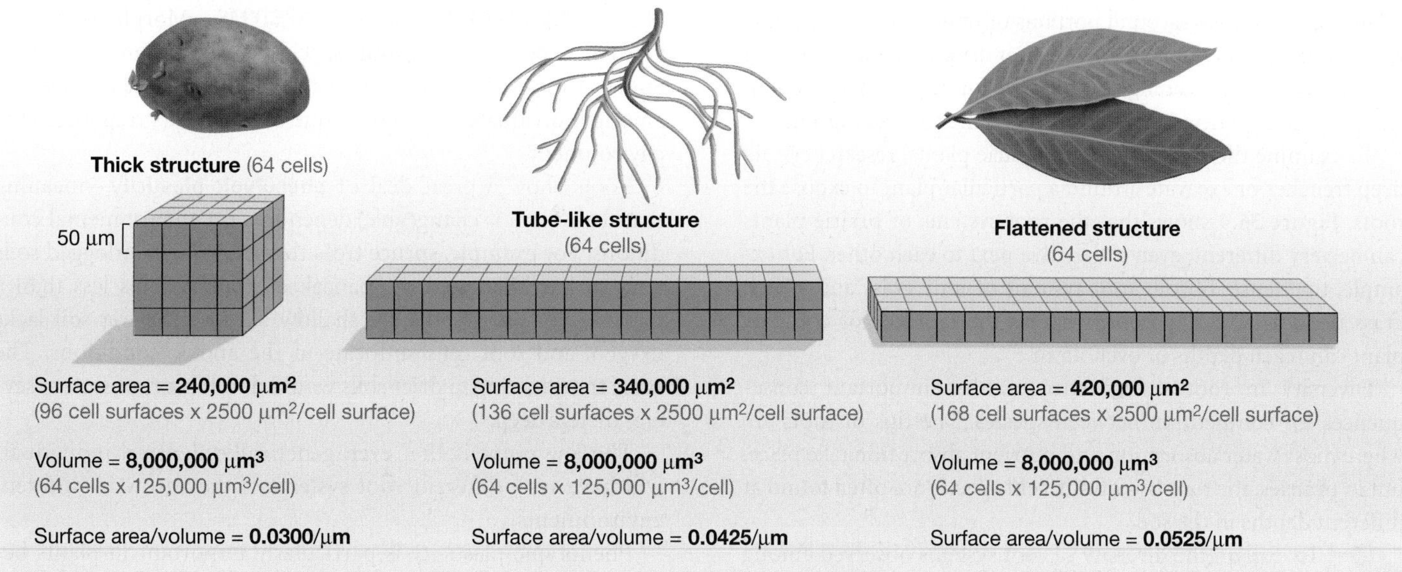

FIGURE 36.3 The Morphology of Roots and Shoots Gives Them a High Surface-Area-to-Volume Ratio. In this example, the "thick structure" represents a tree trunk or potato-like storage organ; the "tube-like structure" represents a root; the "flattened structure" represents a leaf. Note that each schematic structure has the same number of cells and the same total volume—but a very different amount of surface area.

the cells that use the absorbed light and molecules occupy a volume. Thus, a plant body is more efficient as an absorbance-and-synthesis machine when it has a large surface area relative to its volume.

Figure 36.3 illustrates this point. In this example, the cells in a plant are represented by cubes; the side of each cell is 50 μm long. Thus, each face of a cell has a surface area of $50 \times 50 = 2500$ μm^2; each cell has a volume of $50 \times 50 \times 50 = 125{,}000$ μm^3. Follow the calculations in the figure and note that:

- If 64 cells are arranged in a thick block, the surface area/volume relationship is 0.0300/μm.
- If 64 cells are arranged in a long tube, the surface area/volume relationship is 0.0425/μm.
- If 64 cells are arranged in a flat sheet, the surface area/volume relationship is 0.0525/μm.

This simple exercise has an important punchline: tubes and sheets have much more surface area relative to their volume than squares. It's no surprise, then, that the absorptive regions of a root system are tubelike, and the absorptive regions of a shoot system are the flattened structures called leaves. Storage tissues such as tubers and seeds have a low surface-area-to-volume ratio because they are not involved in absorption.

The Root System

Many root systems have a vertical **taproot**, as well as numerous **lateral roots** that run more or less horizontally. The root system anchors the plant in soil, absorbs water and ions from the soil, conducts water and selected ions to the shoot, and stores material produced in the shoot for later use.

Root systems can be impressive in extent. For example, a researcher grew a winter rye plant in a container full of soil for four months, then unearthed the plant and meticulously measured its roots. The root system of this single individual contained more than 13 million identifiable structures with a combined length of over 11,000 km—almost one-third of Earth's circumference! A root system like this contains an enormous surface area for absorbing diffuse resources located underground.

Other studies have shown that **(1)** the roots of trees routinely extend wider than their aboveground canopy, and **(2)** it is not unusual for a plant's root system to represent over 80 percent of its total mass. Many plants devote a great deal of energy and resources to the growth of their root systems.

Although most root systems contain the same general structures, the root systems observed in different species are diverse, as well. This diversity can be analyzed on three levels:

1. morphological diversity among species;
2. phenotypic plasticity, or changes in the structure of an individual's root system over time; and
3. modified roots that are specialized for unusual functions.

Let's consider each level in turn.

MORPHOLOGICAL DIVERSITY IN ROOT SYSTEMS As an example of the range of morphological diversity observed in the root systems of angiosperms, consider prairie plants.

Prairies are grassland ecosystems found in areas of the world such as central North America, the Serengeti Plain of East Africa, the Pampas region of Argentina, and the steppes of central Asia. Rain is abundant enough in these areas to support a lush growth of **herbaceous plants**—meaning, seed plants that lack woody tissue. Rain is scarce enough to exclude trees and most shrubs, however. The growth of woody species is also discouraged by fires that regularly sweep through these ecosystems.

Although the aboveground portions of prairie plants burn during fires and die back during the winter or dry season, their root systems are **perennial**, meaning that they live for many years. The root system sends up a new shoot system after a fire and each spring.

To examine the root systems of prairie plants, researchers dig deep trenches or excavate around a particular plant to expose the roots. **Figure 36.4** shows that the root systems of prairie plants can be very different, even if they live next to each other. For example, the dense, fibrous root systems of june grass and switch grass do not have a taproot, whereas the taproots of compass plant can reach depths of over 4.5 m.

Diversity in root system structure has important consequences for competition between species. The tips of roots are where most water absorption and nutrient absorption take place, but in prairies, the root tips of different species are often found at different depths in the soil.

To explain the diversity of root systems observed among species that grow in the same habitat, biologists suggest that natural selection has favored structures that minimize competition for water and nutrients.

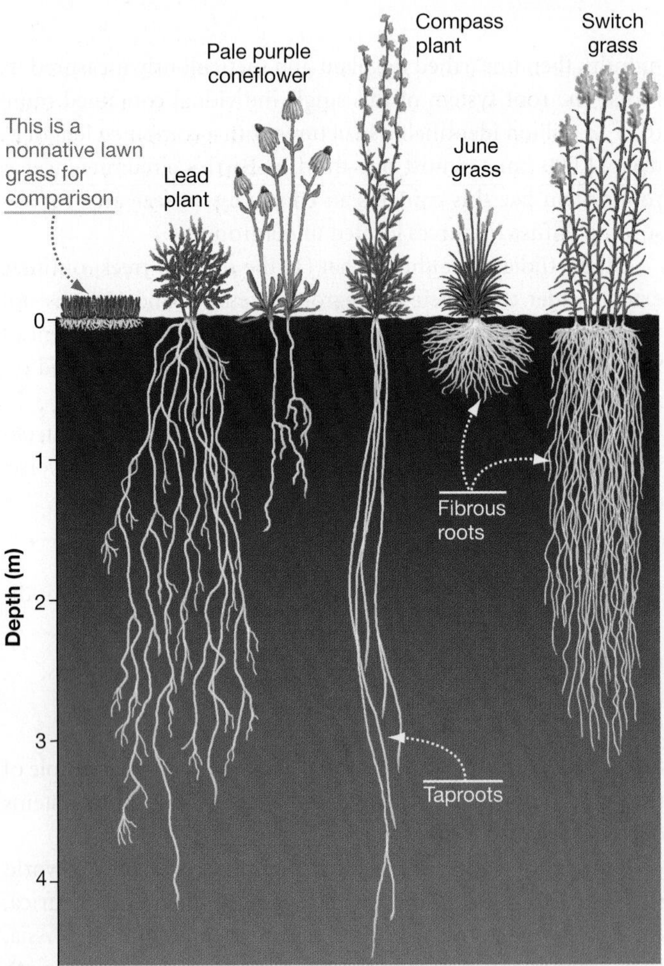

FIGURE 36.4 Plants Have Diverse Root Systems. The roots of prairie plants that live side by side can be very different.

✔**QUESTION** Why are lawns not drought tolerant?

PHENOTYPIC PLASTICITY IN ROOT SYSTEMS Morphological diversity in roots occurs within species as well as among species. Some of the within-species variation is due to genetic diversity among individuals, but some is due to how roots respond to the environment.

Roots show a great deal of **phenotypic plasticity**—meaning that their form is changeable, depending on environmental conditions. For example, spruce trees that grow in waterlogged soils tend to have flattened or "pancaked" root systems less than a meter deep. Their roots are shallow because the wet soil lacks oxygen, and root cells suffocate in the anoxic conditions. The same tree growing in drier soils would develop a root system several meters deep.

The key point is that even genetically identical individuals will have very different root systems if they grow in different environments.

Phenotypic plasticity is particularly important in plants because they grow throughout their lives. Thanks to this mechanism, plants can respond when environmental conditions change over the course of their lifetime. Root systems that grow into nutrient-rich septic fields or sewer pipes that leak human waste are a prime example. Roots actively grow into areas of soil where resources are abundant; roots stop growing or die back in areas where resources are used up or lacking.

MODIFIED ROOTS The taproots and fibrous roots illustrated earlier do not begin to exhaust the types of roots found among plants. For example, some roots are **adventitious**, meaning they develop from the shoot system instead of the root system.

- In ivy, adventitious roots that grow from nodes in the shoot system help individuals cling to brick walls or other structures.
- The prop roots of corn are adventitious roots that help brace individuals in windy weather (**Figure 36.5a**).

Even roots that are part of the root system can have specialized functions—meaning that they do things other than absorbing water and nutrients and anchoring the shoot system.

- The pneumatophores of mangroves in the genus *Avicennia* are specialized lateral roots that function in gas exchange (**Figure 36.5b**). These mangroves grow in habitats where fine silt is deposited, cutting off oxygen from their roots. Their root cells do not suffocate, however, because oxygen from the atmosphere can diffuse into the root system through the pneumatophores. These roots grow upward—not downward—in response to gravity.
- The roots of some plants are contractile, meaning that their cells can shorten much the way an animal muscle cell does. Some species of *Ficus* that grow as vines have adventitious roots that grow from the shoot system down into the ground, and then contract to form a tight supporting structure. In the greenhouse, contractile roots from *Ficus bengalensis* have been known to grow into a bucket of soil and lift it off the ground as they contract. In many cases, the

(a) Prop roots support.

(b) Pneumatophores function in gas exchange.

FIGURE 36.5 Modified Roots Have Unusual Structures or Functions. (a) The prop roots of corn plants that help stabilize the stem are adventitious—they develop from the shoot system. **(b)** The pneumatophores of mangrove trees allow gas exchange to occur between root tissues and the atmosphere.

✔**QUESTION** Why do root cells need oxygen?

roots of plants with bulbs contract, pulling the bulb deeper into the soil over time—in effect, planting themselves.

The Shoot System

As Figure 36.2 indicated, the shoot system has an array of important anatomical features.

- The shoot system consists of one or more **stems**, which are vertical aboveground structures.
- A stem consists of **nodes**, where leaves are attached, and **internodes**, or segments between nodes.
- A **leaf** is an appendage that projects from a stem laterally. Leaves usually function as photosynthetic organs.
- The nodes where leaves attach to the stem are also the site of **axillary** (or **lateral**) **buds**, which form just above the leaf.
- If conditions are appropriate, an axillary bud may grow into a **branch**—a lateral extension of the shoot system.
- The tip of each stem and branch contains an **apical bud**, where growth occurs that extends the length of the stem or branch.
- If conditions are appropriate, apical or axillary buds may develop into flowers or other reproductive structures.

In essence, the shoot system is a repeating series of nodes, internodes, leaves, and apical and axillary buds. As plants grow, the number of nodes, internodes, and leaves increases.

After an initial period of growth, however, an internode does not increase much in size over time. The shoot system of a plant grows by adding more parts rather than by increasing the size of each part.

As with root systems, diversity in shoots can be analyzed on three levels: morphological diversity among species, phenotypic plasticity within individuals, and modified shoots with specialized functions.

MORPHOLOGICAL DIVERSITY IN SHOOT SYSTEMS The shoot systems of land plants range in size from species like the tiny (<5-mm diameter) duckweed that you may have seen growing on the surface of stagnant ponds to redwood trees that reach heights of over 100 m (300 ft) and giant sequoia trunks that weigh 2.6 million kg (over 5.7 million lbs)—about the same as 10 diesel locomotives.

The shape of the shoot system also varies a great deal among species. For example, the manner in which new branches are added as the shoot system grows affects the shape of the individual and its ability to compete for light. As **Figure 36.6** shows, a plant growing with wide branching angles and short internodes has a very different shape than that of a plant with narrow branching angles and long internodes.

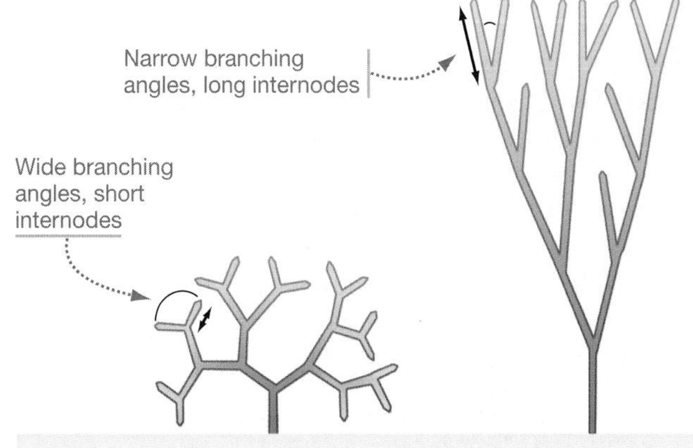

FIGURE 36.6 Plant Form Can Vary as Function of Branch Angle and Internode Length.

Variation in the size and shape of the shoot system is important: it allows plants of different species to harvest light at different locations and thus minimize competition. It also allows them to thrive in a wide array of habitats.

As an example of how the shape of a shoot system varies among species in different environments, consider the silversword plants native to Hawaii. You might recall from Chapter 27 that all of the silverswords are descended from the same ancestor—a species of tarweed that arrived in Hawaii from the west coast of North America, about 5 million years ago.

Silverswords represent an adaptive radiation: a lineage that rapidly split into many species occupying a wide array of habitats. Their shoot systems are particularly diverse in size, shape, and growth habit (see Figure 27.11). Some silverswords grow low to the ground in dense mats; some form bunched rosettes of leaves; some are vines; others are woody shrubs or even small-to-medium-sized trees.

Biologists interpret this diversity of shoot systems as a suite of adaptations for harvesting light and carbon dioxide in different environments. In lush habitats, where competition for light is intense, woody individuals grow tall and are favored by natural selection. But in dry, windblown habitats, individuals with short stems or rosettes thrive because they require less water than taller individuals do, and they don't blow over. The adaptive radiation of silverswords has been based in part on diversification in shoot systems.

PHENOTYPIC PLASTICITY IN SHOOT SYSTEMS The size and shape of an individual's shoot system can vary dramatically based on variation in growing conditions: temperature, exposure to wind, and availability of water, nutrients, and light.

This conclusion was driven home in an experiment conducted by Jens Clausen and colleagues in the late 1930s. These biologists transplanted several species of herbaceous plants between sites along an elevational gradient: from sea level to alpine habitats. In each case, the transplanted individuals were propagated from cuttings—meaning that they were genetically identical to individuals growing at the other locations. As the "Results" section in **Figure 36.7** shows, the overall size and shape of the shoot system varied markedly among locations.

Because an individual's shoot system continues to grow over the course of its lifetime, it can respond to changes in environmental conditions just as the root system can. Experiments highlighted in Chapter 39, for example, established that shoot systems can bend toward light if an individual is shaded on one side. Plants also undergo differential growth, producing more branches and leaves in regions of the body that are exposed to the highest light levels. A plant's shoot system grows in directions that maximize its chances of capturing light.

MODIFIED SHOOTS Even though they are a single lineage, silverswords illustrate many of the modified shoot systems found among land plants. But still other variations occur. Not all stems grow vertically, and not all stems acquire carbon dioxide and photons.

EXPERIMENT

QUESTION: How much does a plant's growth form depend on its environment?

HYPOTHESIS: (No explicit hypothesis—the goal of this experiment was to explore the interaction between genetic makeup versus environmental influence on size and shape.)

EXPERIMENTAL SETUP:

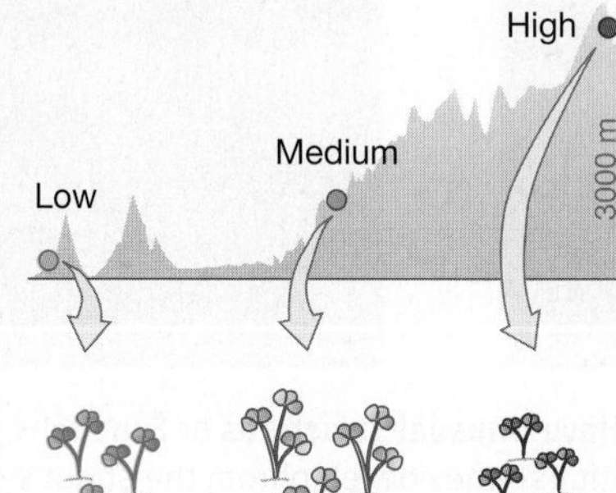

1. Take cuttings from individuals of *Potentilla glandulosa* growing at low, medium, and high elevation habitats in the Sierra Nevada mountains.

2. Propagate genetically indentical individuals.

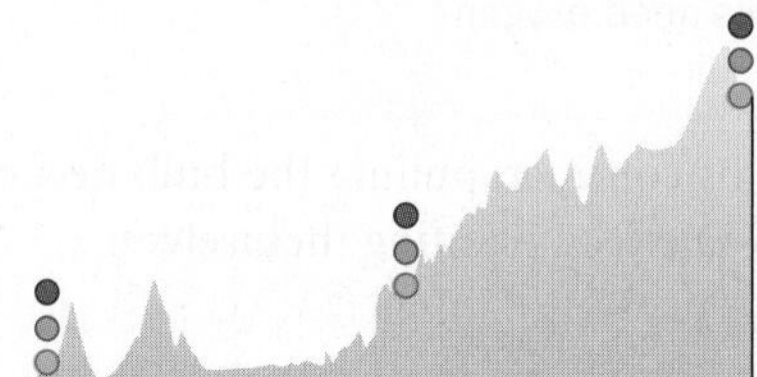

3. Transplant individuals from each source population into each habitat (low, medium, and high elevation). Allow to grow and observe mature plants.

PREDICTION: (No explicit predictions.)

RESULTS:

Examples of mature plants observed:

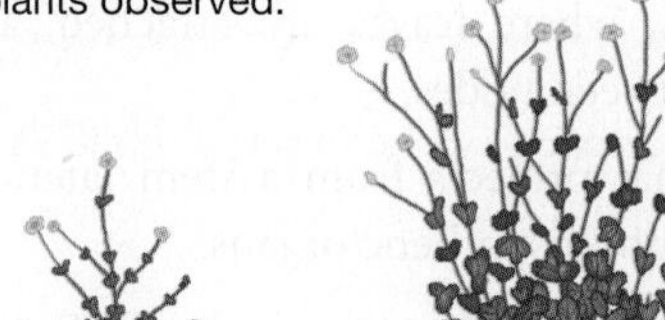

CONCLUSION: Environmental conditions have a profound influence on body size and shape (genetically identical plants look different at each site). BUT, genetic makeup also has a large influence on plant morphology (plants from each source population look different, even when grown in the same habitat).

FIGURE 36.7 Experimental Evidence for Phenotypic Plasticity in Shoot Systems.

SOURCE: Clausen, J., D. D. Keck, and W. M. Hiesey. 1945. Experimental studies on the nature of species. II. Plant evolution through amphiploidy and autoploidy, with examples from the Madiinae. Washington D.C., Carnegie Institution of Washington.

✔**QUESTION** Why was it important for the researchers to propagate the individuals from cuttings?

(a) Cactus stems (shown here in cross section) store water.

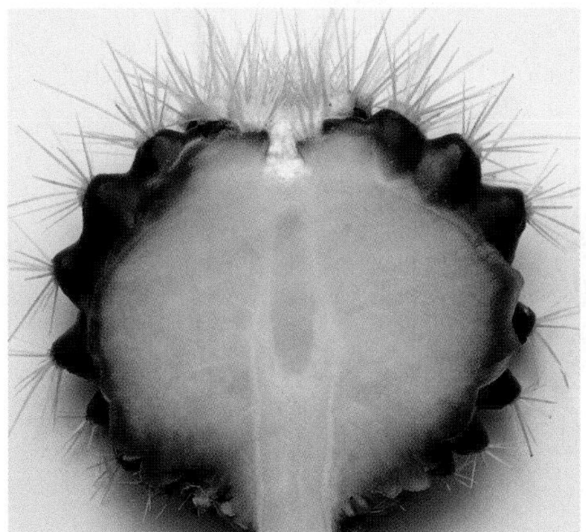

(b) Stolons produce new individuals at nodes aboveground.

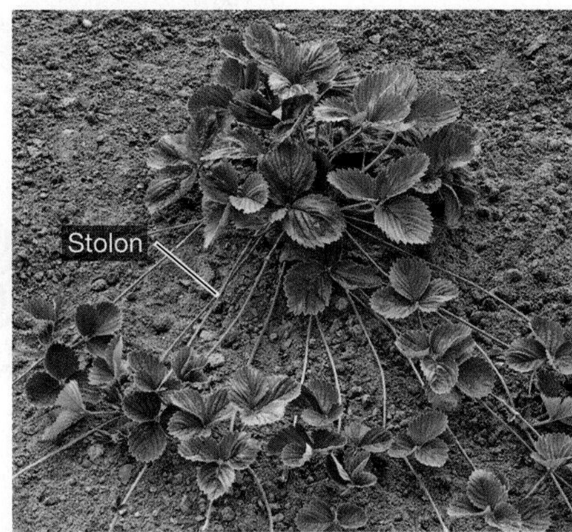

(c) Rhizomes produce new individuals at nodes belowground.

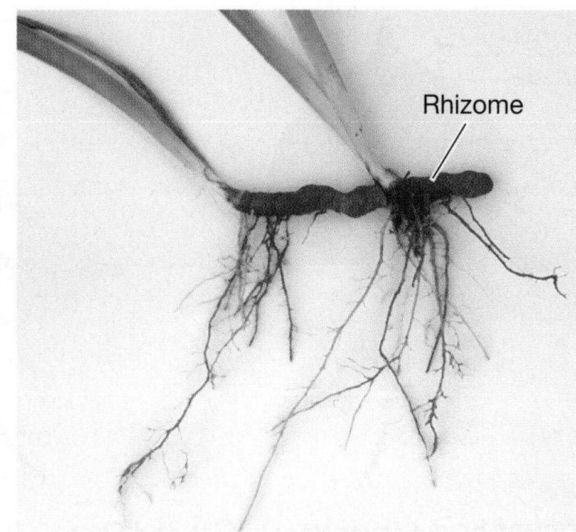

(d) Tubers store carbohydrates.

(e) Thorns provide protection.

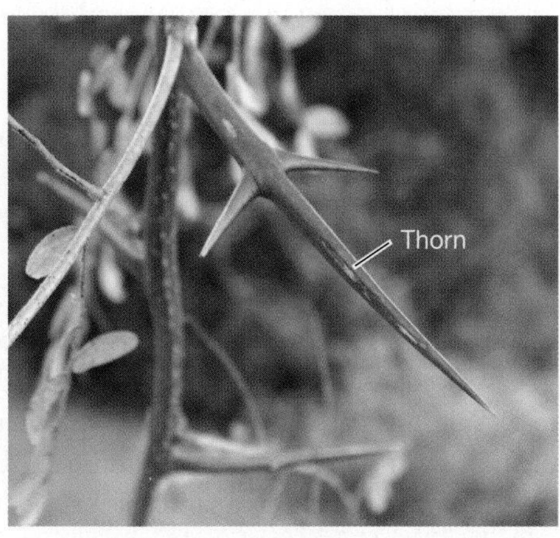

FIGURE 36.8 Modified Stems Have Unusual Structures or Functions. Instead of providing support, the stems of some species have been modified to function as water- or starch-storage tanks, a means of asexual reproduction, or a defense weapon.

- Many desert cacti have highly modified stems. Instead of functioning primarily to support leaves, cactus stems often enlarge into water-storage organs (**Figure 36.8a**). Water accounts for up to 98 percent of the weight of a cactus stem. A cactus stem also contains the plant's photosynthetic tissue. Instead of being the main food-producing organ, its leaves are modified into protective structures called **spines**.
- **Stolons** are modified stems that grow over the soil surface, producing adventitious roots and leaves at each node (**Figure 36.8b**). Because new plants form at these nodes, stolons function in asexual reproduction (see Chapter 12).
- Like stolons, **rhizomes** are stems that grow horizontally instead of vertically. They produce new plants at nodes and thus participate in asexual reproduction. But while stolons grow aboveground, rhizomes spread belowground (**Figure 36.8c**). Rhizomes also store starch.
- **Tubers** are underground, swollen ends of rhizomes that function as carbohydrate-storage organs (**Figure 36.8d**). The eyes of a potato—a typical tuber—are nodes in the stem where new branches may arise.
- **Thorns** are modified stems that help protect the plant from attacks by large **herbivores**, or plant-eaters, such as deer, giraffe, or cattle (**Figure 36.8e**).

The Leaf

In most plant species, the vast majority of photosynthesis occurs in leaves. The total area of leaf produced by a single plant can be enormous—a single tree can have hundreds of thousands of leaves with a total leaf surface area equivalent to that of a football field. All of this area is available for absorbing photons and supporting photosynthesis.

A simple leaf (**Figure 36.9a** on page 702) is composed of just two major structures: an expanded portion called the **blade** and a stalk called the **petiole**. But leaves exhibit many variations on the central theme of a flattened structure specialized for performing photosynthesis.

MORPHOLOGICAL DIVERSITY IN LEAVES Glance outside or stroll through a garden, and you'll find many types of simple leaves with an easily recognizable blade and petiole. (Grass leaves will stump you, though, because they lack petioles entirely.) You

(a) Simple leaves have a petiole and a single blade.

Blade

Petiole

(b) Compound leaves have blades divided into leaflets.

(c) Doubly compound leaves are large yet rarely damaged by wind or rain.

(d) Species from very cold or hot climates have needlelike leaves.

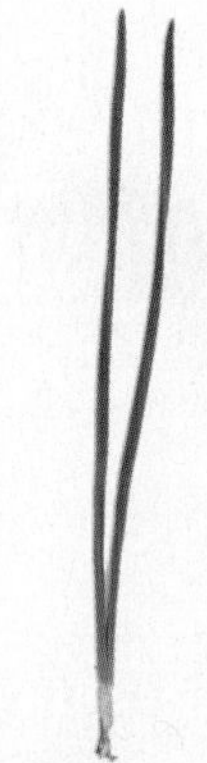

FIGURE 36.9 Leaves Vary in Size and Shape. The structures in parts **(a)** through **(c)** represent single leaves; part **(d)** shows two leaves.

✔**QUESTION** For capturing photons, what is the advantage of having a leaf with a large surface area? In terms of wind damage and water loss, what is the disadvantage of having a leaf with a large surface area?

will also find compound leaves that have blades divided into a series of leaflets (**Figure 36.9b**). You may even encounter doubly compound leaves, which have leaflets that are again divided (**Figure 36.9c**).

Not all leaf blades are thin with a large surface area, however. For example, plants that thrive in deserts and in cold, dry habitats tend to have needle-shaped leaves (**Figure 36.9d**). The leading hypothesis to explain this pattern is based on two observations: **(1)** Water is often in short supply in these environments because it is absent in deserts or frozen and thus unavailable in cold habitats, and **(2)** leaves with large surface areas lose large amounts of water through an evaporative process called **transpiration** (discussed in Chapter 37).

Thus, needlelike leaves are interpreted as adaptations that minimize transpiration in water-scarce habitats. Small, narrow leaves are also much less susceptible to wind damage than are large, broad leaves.

The arrangement of leaves on a stem can vary as much as leaf shape. For example, leaves can be:

- paired opposite each other on the stem (**Figure 36.10a**);
- arranged in a whorl (**Figure 36.10b**);
- arranged to alternate on either side of the stem (**Figure 36.10c**);
- found in a compact basal arrangement where internodes are extremely short—leading to the rosette growth form (**Figure 36.10d**).

PHENOTYPIC PLASTICITY IN LEAVES Even though leaves do not grow continuously, they exhibit phenotypic plasticity just as root and shoot systems do.

Leaves from the same individual that grow in sun versus shade are a prominent example of phenotypic plasticity in leaf morphology. As the oak tree leaves in **Figure 36.11** show:

(a) Opposite leaves

(b) Whorled leaves

(c) Alternate leaves

(d) Rosette

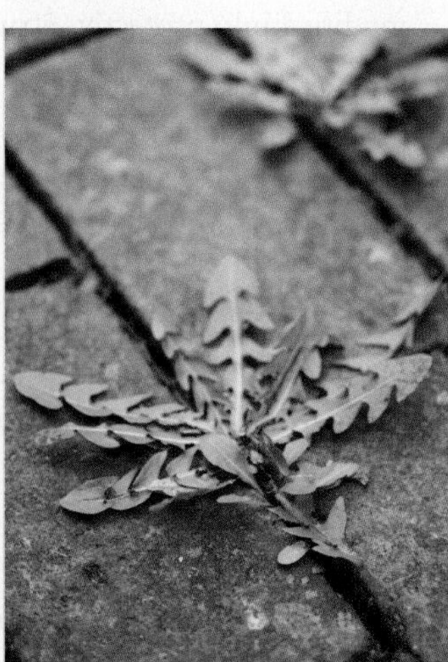

FIGURE 36.10 The Arrangement of Leaves on Stems Varies.

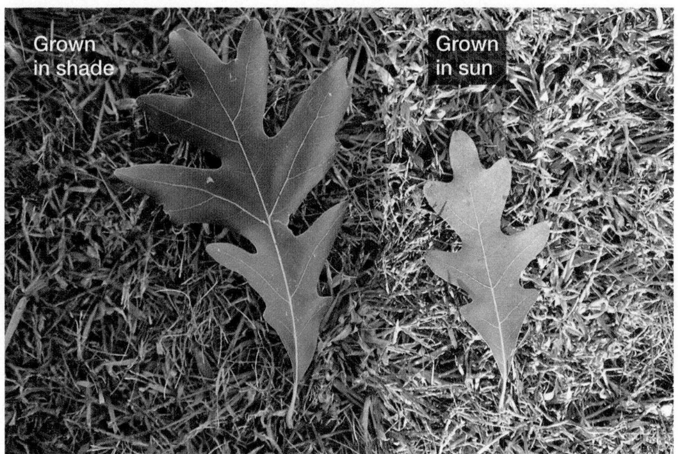

FIGURE 36.11 Phenotypic Plasticity in Leaves. These leaves came from the same tree.

- *Sun leaves* have a relatively small surface area, which reduces water loss in areas of the body where light is abundant.
- *Shade leaves* are relatively large and broad, providing a high surface area that maximizes absorption of rare photons.

Water loss is less of a problem for shade leaves, because temperatures are cooler in shade than in bright sun.

MODIFIED LEAVES Not all leaves function primarily in photosynthesis; some perform other roles.

- Cactus spines are modified leaves that protect the stem (see Figure 36.8a).
- Onion bulbs consist of thickened leaf bases, separated by highly condensed internodes, that store nutrients (**Figure 36.12a**).
- The thick leaves of species called succulents, such as aloe vera, store water (**Figure 36.12b**).
- The tendrils that enable garden peas and other vines to climb are modified leaflets or leaves (**Figure 36.12c**).
- The bright red leaves of poinsettias attract pollinators to the tiny yellow flowers that they surround (**Figure 36.12d**).
- The tube-like leaves of the pitcher plant trap insects (**Figure 36.12e**). When insects enter, they are discouraged from flying out by the dark "hood" that covers the opening. As they feed on the plant's nectar, they appear to become dizzy. Eventually they fall into the bottom of the tube and drown in water that has accumulated. The nutrients are digested by enzymes secreted by the plant and taken up by epidermal cells.

The variability of plant root systems, shoot systems, and leaves is impressive. Diversity, plasticity, and dynamism are recurring themes in the study of plant anatomy.

(a) Onion leaves store food.

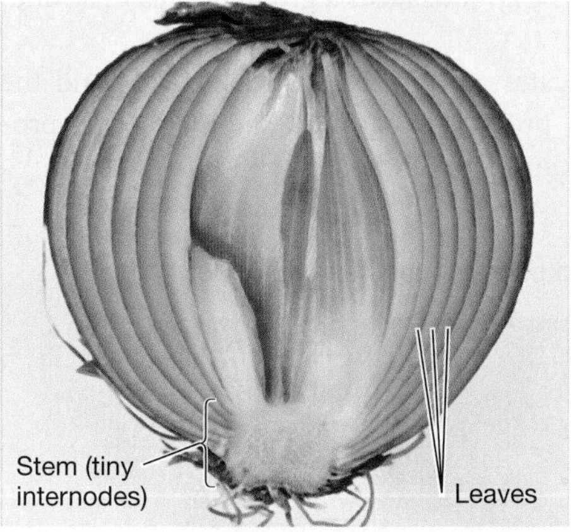

(b) Aloe vera leaves store water.

(c) Pea tendrils aid in climbing.

(d) Poinsettia leaves attract pollinators.

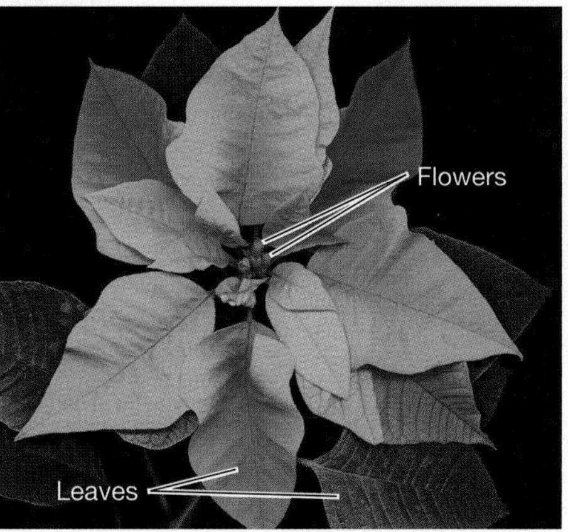

(e) Pitcher plant leaves trap insects.

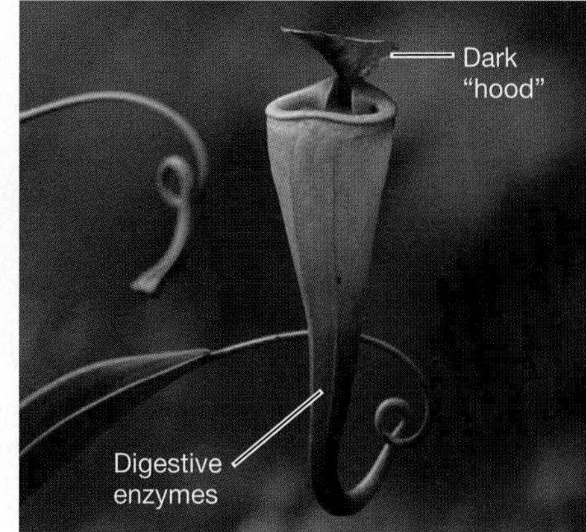

FIGURE 36.12 Modified Leaves Have Unusual Structures or Functions. Instead of functioning primarily as the site of photosynthesis, the leaves of some species have been modified to function as water- or starch-storage tanks or in climbing, sexual reproduction, or nutrient acquisition.

CHECK YOUR UNDERSTANDING

If you understand that . . .

- The plant body is organized into a root system and a shoot system.
- Roots and shoots explore the environment via continuous growth and efficiently absorb diffuse resources like water, ions, carbon dioxide, and sunlight.
- Roots and shoots may also function to anchor the plant, store water, produce offspring asexually, provide protection, or store carbohydrates.
- Leaves vary among species and within individuals, and may be modified to store food or water, capture insects, or attract pollinators.

✔ **You should be able to . . .**

1. Diagram a generalized version of the angiosperm body, labeling each major part.
2. Provide two examples each of root systems, shoot systems, and leaves that differ from the generalized body shown in Figure 36.2, in structure and/or function.

Answers are available in Appendix B.

36.2 Primary Growth Extends the Plant Body

Plants grow continuously because they have **meristems**—populations of undifferentiated cells that retain the ability to undergo mitosis and produce new cells. When meristematic cells divide, some of the daughter cells remain in the meristem, allowing the meristem to persist. Other cells, though, undergo differentiation. You might recall from Chapter 21 that differentiation is a developmental process that produces a specialized cell—one that expresses only certain genes and has a distinctive structure and function.

Apical meristems are located at the tip of each root and shoot. As cells in apical meristems divide, enlarge, and differentiate, root and shoot tips extend the plant body outward, allowing it to explore new space.

This process is **primary growth**. The major consequence of primary growth is to increase the length of the root and shoot systems. Cells that are derived from apical meristems form the primary plant body.

To understand how primary growth occurs, let's look at the overall organization of the primary plant body and then delve into a detailed look at its tissues and cells.

How Do Apical Meristems Produce the Primary Plant Body?

Whether located in the root or the shoot, apical meristems give rise to three distinct populations of cells: protoderm, ground meristem, and procambium. These cells are partially differentiated but retain the character of meristematic cells because they keep dividing.

The three types of primary meristematic cells are important because they give rise to three major tissue systems that extend throughout the plant body. A **tissue** is a group of cells that functions as a unit.

Figure 36.13 indicates where the apical meristems and the primary meristems—protoderm, ground meristem, and procambium—are found in shoots and roots.

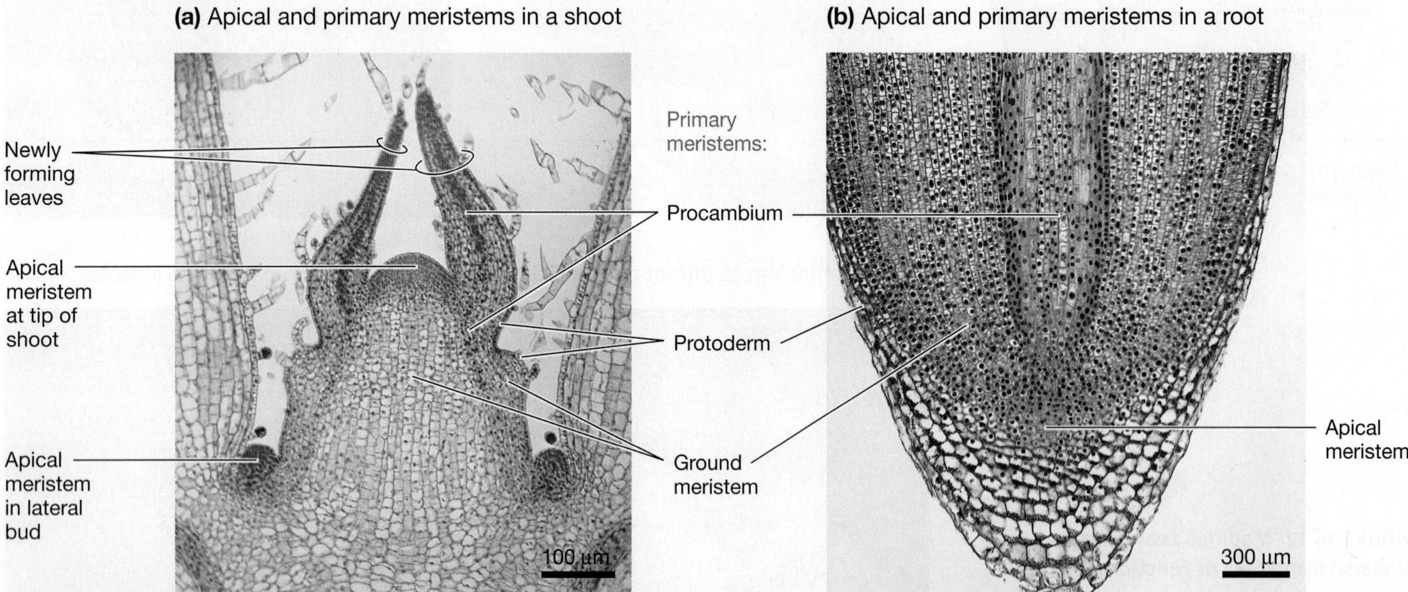

FIGURE 36.13 The Structure of Apical Meristems in the Shoot and Root. Apical meristems consist of small, similar-looking cells that divide when water and nutrients are plentiful. Three types of cells—protoderm, ground meristem, and procambium—are derived from the apical meristem and consist of partially differentiated cells that can still divide.

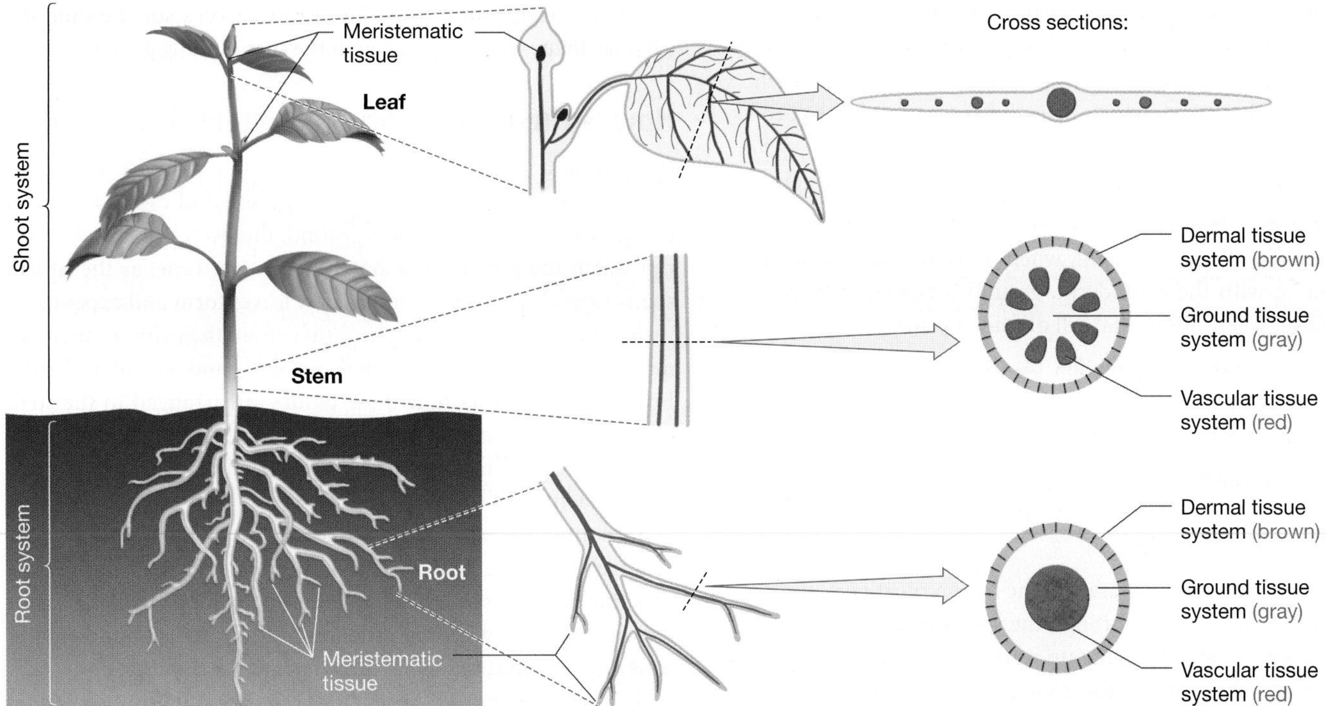

FIGURE 36.14 The Primary Plant Body Comprises the Dermal, Ground, and Vascular Tissue Systems. The dermal, ground, and vascular tissue systems arise from the protoderm, ground meristem, and procambium illustrated in Figure 36.13.

- **Protoderm** gives rise to the **dermal** (literally, "skin") **tissue system.** The dermal tissue system, or **epidermis**, is a single layer of cells that covers the plant body and protects it.
- **Ground meristem** gives rise to the **ground tissue system**, which makes up the bulk of the plant body and is responsible for photosynthesis and storage.
- **Procambium** gives rise to the **vascular tissue system**, which provides support and transports water, nutrients, and photosynthetic products between the root system and shoot system. Vascular tissue runs through ground tissue, so the cells that make up ground tissue are adjacent to cells that conduct the water and nutrients they need.

Figure 36.14 shows how the dermal, ground, and vascular tissues are distributed in the plant body. In contrast to these tissues, meristematic cells are highly localized at the tips of shoots and roots.

The key point to remember is that the dermal, ground, and vascular tissue systems are originally derived from cells in apical meristems. Thus, they represent the primary plant body.

How Is the Primary Root System Organized?

Roots have several features that allow them to grow into new regions of the soil, so they can furnish cells throughout the body with water and key nutrients.

As **Figure 36.15** shows, a group of cells called the **root cap** protects the root apical meristem. Cells produced by the meristem constantly replenish the cap, which regularly loses cells.

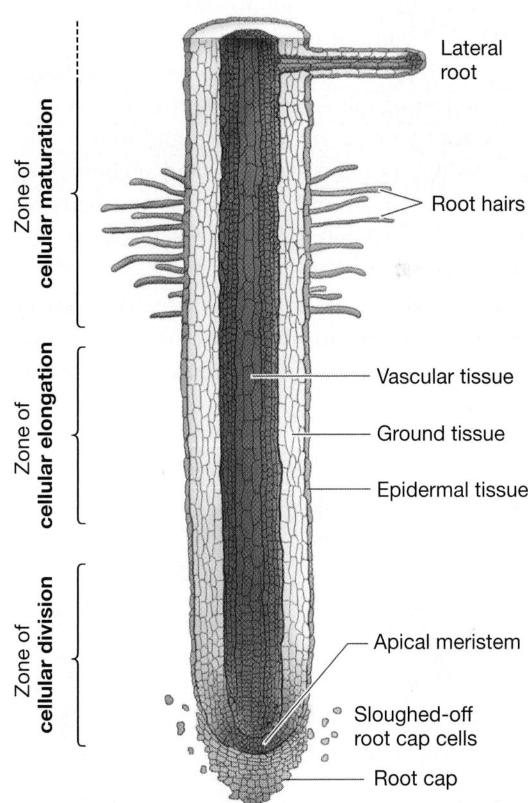

FIGURE 36.15 Roots Extend into the Soil via Growth of Apical Meristems and Cell Elongation. This is a longitudinal (lengthwise) section. The zone of cellular maturation is actually much larger than can be shown here. Most absorption of water and nutrients occurs at root hairs.

In addition to protecting the root tip, root cap cells are important in sensing gravity and determining the direction of growth. They also synthesize and secrete a slimy, polysaccharide-rich substance called **mucigel**, which helps lubricate the root tip, reducing friction and making movement more efficient.

Three distinct populations of cells exist behind the root cap:

1. The **zone of cellular division** (0.5–1.5 mm behind the root tip) contains the apical meristem, where cells are actively dividing, along with the protoderm, ground meristem, and procambium, where additional cell division occurs.
2. The **zone of cellular elongation** (4–10 mm behind the root tip) is made up of cells that are recently derived from the primary meristematic tissues and actively increasing in length.
3. The **zone of cellular maturation** (1–5 cm from the root tip) is where older cells complete their differentiation into dermal, vascular, and ground tissues.

The zone of cellular elongation is the region most responsible for the movement of roots through the soil. The cells in this region increase in length by taking up water. Their expansion provides the force that pushes the root cap and apical meristem through the soil. When conditions are good, roots can extend by as much as 4 centimeters per day.

The zone of cellular maturation is the most important root segment in terms of water and nutrient absorption. In this region, epidermal cells produce outgrowths called **root hairs**, which greatly increase the surface area of the dermal tissue.

Root hairs furnish the actual sites of water and nutrient absorption. The rest of the root system provides structural support for the root hairs, conducts water and ions to the shoot, stores the products of photosynthesis, and anchors the plant in the soil. Uptake of water and nutrients in root hairs is vital to plants; portions of Chapters 37 and 38 focus on how these processes occur.

The zone of cellular maturation is also where lateral roots begin to grow. In contrast to lateral branches in the shoot, which arise from apical meristems in axillary buds (see Figure 36.2), lateral roots arise from cells within a ring of cells surrounding the vascular tissue, and erupt through the surrounding ground tissue.

How Is the Primary Shoot System Organized?

If you visit a garden regularly, you can only imagine the movement of root tips as they penetrate the soil and expand to form complex networks deep underground. But even a casual observer can watch the growth of shoot systems over time, as the tips of stems extend and branch and as new leaves form and expand.

Just behind each shoot apical meristem, the primary meristematic cells give rise to dermal, ground, and vascular tissues. **Figure 36.16a** shows how these tissues are arranged in the stem of a sunflower plant when it matures. Note that the vascular tissues are grouped into **vascular bundles**, which form strands running the length of the stem.

In sunflowers and other eudicots, the vascular bundles are arranged in a ring near the stem's perimeter. The ground tissue that the vascular tissue runs through is divided into two major regions: **pith**, the ground tissue inside the vascular bundles, and **cortex**, the ground tissue outside the vascular bundles.

The arrangement of the vascular bundles and ground tissue is dramatically different in the stems of monocots, however. As **Figure 36.16b** shows, vascular bundles tend to be scattered throughout the ground tissue of monocot stems.

Now let's drill down a bit deeper, and look at the composition of the dermal, ground, and vascular tissue systems. Each of these tissue systems is made up of an array of distinct cell and tissue types. What are they?

36.3 Cells and Tissues of the Primary Plant Body

Recall from Chapter 7 that plant cells and animal cells share most of their key characteristics: both have chromosomes enclosed in a nuclear envelope, a plasma membrane studded with proteins

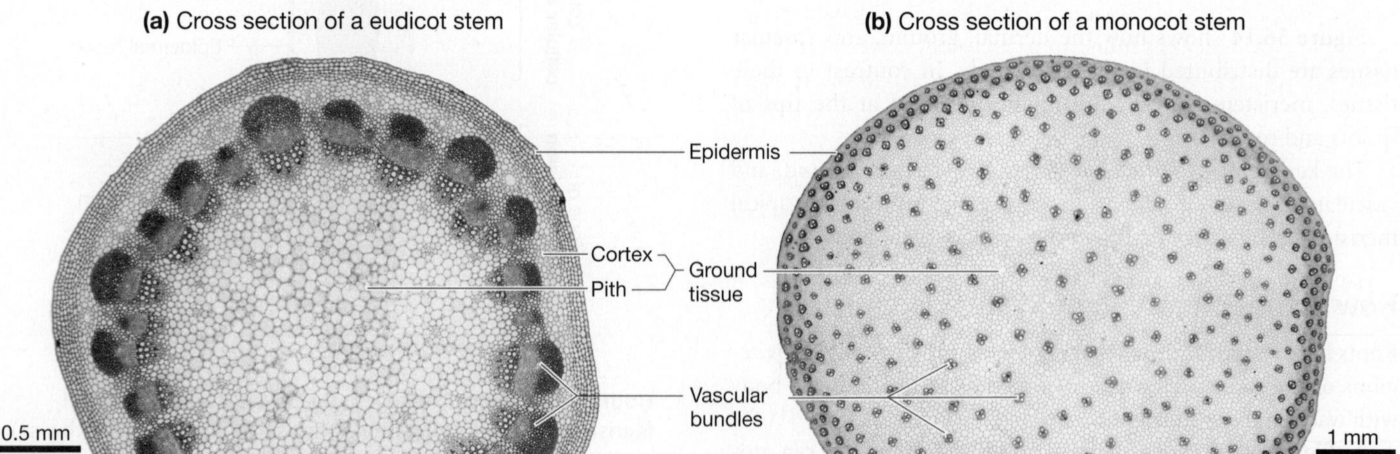

FIGURE 36.16 Stems Contain a Variety of Cell and Tissue Types. As these cross sections show, vascular bundles are **(a)** arranged in a ring near the perimeter of eudicot stems but **(b)** scattered throughout the pith in monocots.

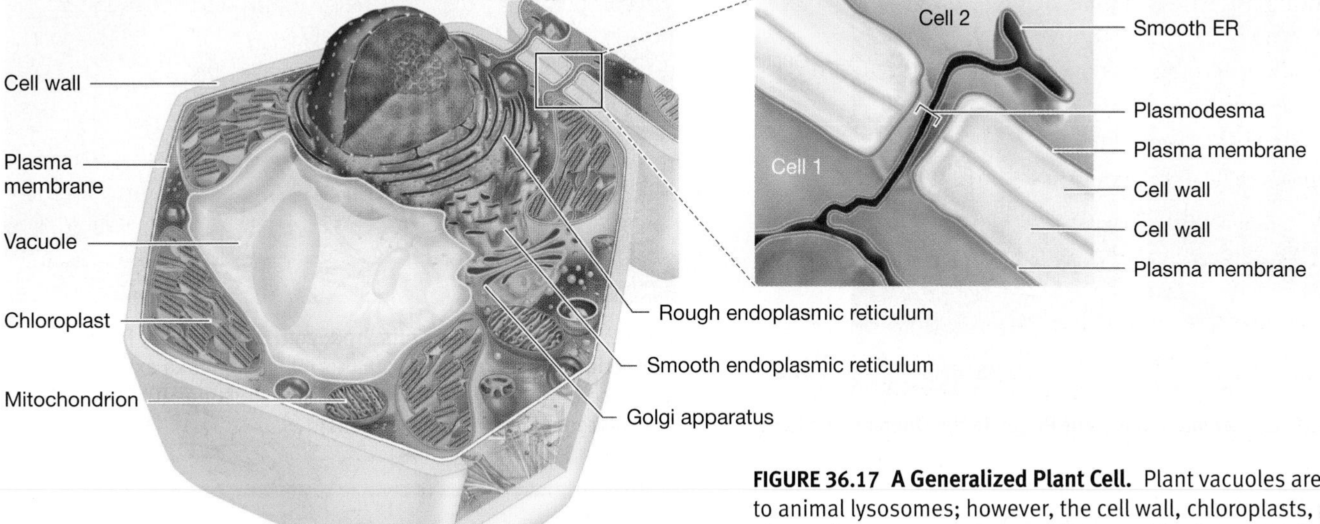

FIGURE 36.17 A Generalized Plant Cell. Plant vacuoles are similar to animal lysosomes; however, the cell wall, chloroplasts, and plasmodesmata are unique to plants. Unlike the extracellular matrix of animals, the plant cell wall is rigid.

that regulate the passage of materials in and out, mitochondria that produce ATP by oxidizing sugars, and an array of other organelles that synthesize or degrade key molecules.

In addition, plant cells have several features that are absent in animal cells (**Figure 36.17a**):

1. All plant cells are surrounded by a stiff, cellulose-rich **cell wall** that supports the cell and defines its shape.
2. The cytoplasm of adjacent plant cells is connected via **plasmodesmata** (singular: **plasmodesma**; see Chapter 8). Plasmodesmata consist of cytoplasm and segments of smooth endoplasmic reticulum (smooth ER) that run through tiny, membrane-lined gaps in the cell wall (**Figure 36.17b**).
3. Plant cells often contain several types of organelles that are not found in animals—specifically chloroplasts and a large, membrane-bound organelle called a vacuole, which fills most of the cell's volume.

Chloroplasts are the site of photosynthesis (see Chapter 10). Non-photosynthetic cells found in roots, seeds, flower petals, and other locations may have organelles that are related to chloroplasts but are specialized for storing pigments, starch, oils, or protein.

Vacuoles, which contain an aqueous solution called **cell sap**, store wastes and in some cases also digest wastes, as do animal lysosomes. In addition, plant vacuoles store water and nutrients. They may also hold pigments that provide color or poisons that deter herbivores.

Another important distinction between plant cells and animal cells is that plant cells do not change position once they form. Some animal cells change positions either early in the development of an individual or as mature cells.

During both plant and animal development, a cell's location in the body determines which tissue system it contributes to and what type of cell it becomes. Let's examine the cells and tissues of the three primary tissue systems in turn.

The Dermal Tissue System

Dermal tissue is the interface between the individual and the external environment. Its primary function is to protect the plant body—from water loss, disease-causing agents, and herbivores.

Most tissues are made up of several different cell types, each of which has a distinct structure and function. Let's consider the cell types found in dermal tissue.

EPIDERMAL CELLS PROTECT THE SURFACE Most of the cells in the dermal tissue system are epidermal cells, which are flattened and usually lack chloroplasts. Epidermal cells in the root are responsible for absorbing water and nutrients. In addition, epidermal cells in both the root and shoot systems play a key role in protecting the plant.

Epidermal cells in the shoot system fulfill their protective role in part by secreting the **cuticle**: a waxy layer that forms a continuous sheet on the surface (see Chapter 30). Waxes are lipids and are thus highly hydrophobic. As a result, the presence of cuticle on stems and leaves drastically reduces the amount of water that is lost by evaporation.

From a human perspective, the water-repellent properties of cuticle make it a valuable ingredient in polishes and lipsticks. The carnauba wax used in car and floor polishes, for example, is secreted by epidermal cells in the leaves of carnauba palms native to Brazil.

Besides minimizing water loss, cuticle forms a barrier to protect the plant from viruses, bacteria, and the spores or growing filaments of parasitic fungi. In this way, the plant epidermis forms the first line of defense against disease-causing agents, or **pathogens**. For pathogens to enter the plant body and initiate an infection, they must either secrete enzymes that digest the cuticle or enter via a wound where the cuticle has been torn away.

The waxes found in cuticle can also be detrimental to the plant, however, by reducing gas exchange. This can be a serious

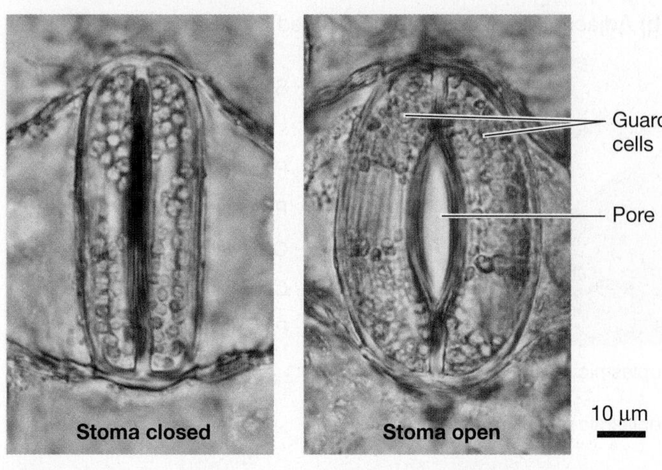

FIGURE 36.18 In a Stoma, Guard Cells Regulate the Opening of a Pore.

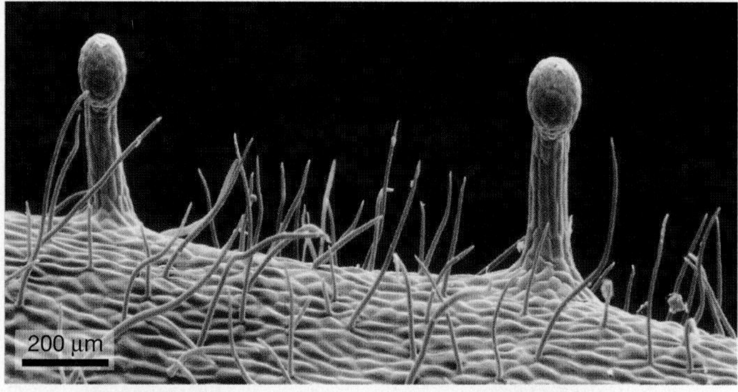

FIGURE 36.19 Epidermal Cells Produce Trichomes That Provide Protection. Some trichomes on this leaf are hairlike extensions of a single cell, while others are multicellular structures (the bulbs that are colored orange here hold toxins).

problem because photosynthesis depends on the free flow of carbon dioxide to photosynthetic cells. The problem is solved by specialized structures in dermal tissue called stomata.

STOMATA REGULATE GAS EXCHANGE AND WATER LOSS Most land plants have structures called **stomata** (singular: **stoma**) that allow carbon dioxide to enter photosynthetically active tissues. A stoma consists of two specialized **guard cells**, which change shape to open or close an opening in the epidermis known as a **pore** (**Figure 36.18**).

When stomata are open, CO_2, O_2, water vapor, and other gases can move between the atmosphere and the interior of the plant by diffusion. Stomata open when CO_2 is needed. They close when CO_2 is not needed or when large amounts of water are lost by transpiration. Chapter 39 explores the molecular mechanisms responsible for stomatal opening and closing.

TRICHOMES PERFORM AN ARRAY OF FUNCTIONS In addition to minimizing water loss and regulating gas exchange, cells in dermal tissue may protect the individual from the damaging effects of intense sunlight and attacks by herbivores.

Trichomes are protective, hairlike appendages made up of specialized epidermal cells. They are found in shoot systems, and come in a wide variety of shapes, sizes, and abundance.

Depending on the species, trichomes may **(1)** keep the leaf surface cool by reflecting sunlight, **(2)** reduce water loss by forming a dense mat that limits transpiration, **(3)** provide barbs, or store toxic compounds that thwart herbivores (**Figure 36.19**), or even **(4)** trap and digest insects.

The Ground Tissue System

Most photosynthesis, as well as most carbohydrate storage, takes place in ground tissue. Cells in ground tissue are also responsible for most of the synthesis and storage of specialized products such as colorful pigments, the chemical signals called hormones, and toxins required for defense.

If the primary business of the dermal tissue system is protection, the ground tissue is all about producing and storing valuable molecules.

Ground tissue is made up of three distinct tissue types: parenchyma (pronounced *pa-REN-ki-ma*), collenchyma (*ko-LEN-ki-ma*), and sclerenchyma (*skle-REN-ki-ma*).

PARENCHYMA ARE "WORKHORSE" CELLS **Parenchyma cells** have relatively thin primary cell walls and are the most abundant and versatile plant cells. Groups of parenchyma cells form parenchyma tissue.

The parenchyma tissue in leaves consists of parenchyma cells filled with chloroplasts, and is the primary site of photosynthesis (**Figure 36.20a**). But in other organs, parenchyma cells store starch granules (**Figure 36.20b**). When you eat a salad, a potato, or an apple, you are ingesting primarily parenchyma cells in ground tissue.

Many parenchyma cells are **totipotent**, meaning they retain the capacity to divide and develop into a complete, mature plant. The totipotency of parenchyma cells is important in healing wounds and in reproducing asexually via stolons or rhizomes. In each case, parenchyma cells may begin to divide, grow, and differentiate to form new roots and shoots.

The totipotency of parenchyma cells also allows gardeners to clone plants by making cuttings. For example, if you cut a piece of coleus stem and place it in water, parenchyma cells will divide to produce a mass of undifferentiated cells called a **callus**. Roots develop from the callus (**Figure 36.21**), and the new individual can be planted in soil.

Bananas, seedless grapes, and several other commercially important strains cannot undergo sexual reproduction to produce seeds. Instead, they are propagated entirely by cuttings.

COLLENCHYMA CELLS FUNCTION PRIMARILY IN SHOOT SUPPORT **Collenchyma cells** have primary cell walls that are thicker in some areas than others, and their overall shape is longer and thinner than that of parenchyma cells. Groups of collenchyma cells form collenchyma tissue that supports the plant body.

Even when collenchyma cells are mature, their cell walls retain the ability to stretch and elongate. As a result, collenchyma cells can continue to lengthen as they provide structural support to the growing regions of shoots.

(a) In leaves: photosynthesis and gas exchange

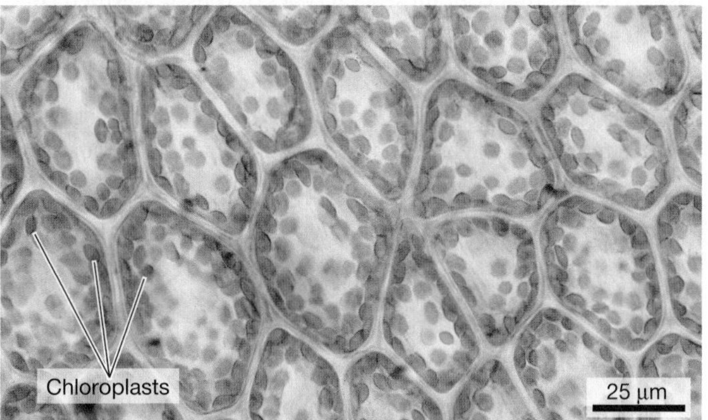

(b) In roots: carbohydrate storage

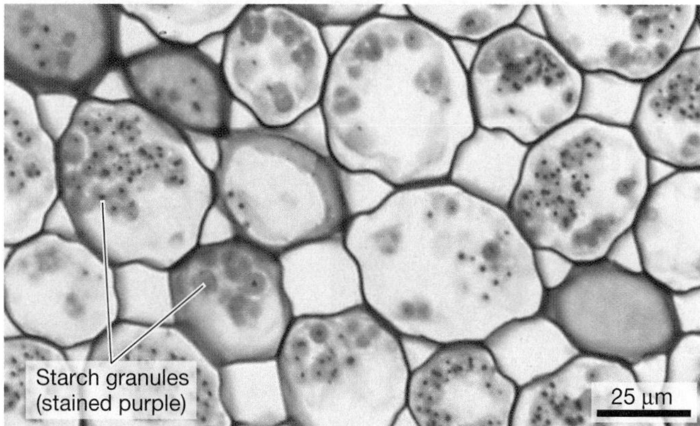

FIGURE 36.20 Parenchyma Cells Perform a Wide Array of Tasks.

✔**EXERCISE** Give an example of a gene that is likely to be expressed in these leaf cells, but not in the root cells.

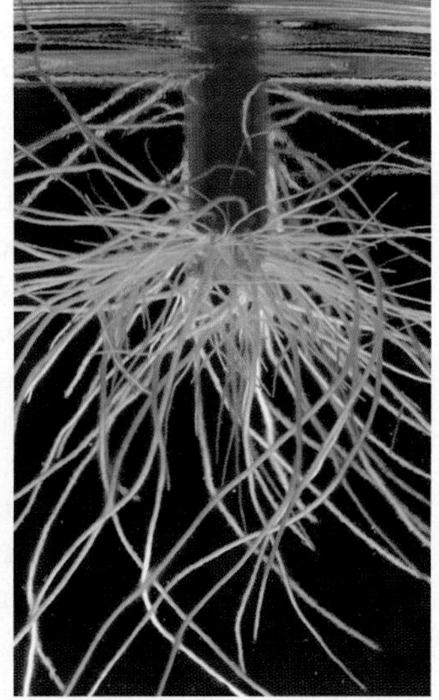

FIGURE 36.21 Parenchyma Cells in Cut Stems Can Form Adventitious Roots. Parenchyma cells in a cut coleus stem (left) divide to form a mass of undifferentiated cells called a callus, which then sprouts roots (right).

Because they can elongate, collenchyma cells are particularly abundant in growing stems and in the stalk portions of leaves. The "strings" you may have peeled from a stalk of celery or rhubarb—which is actually the petiole of a celery or rhubarb leaf—include many strands of collenchyma cells (**Figure 36.22**).

SCLERENCHYMA: TWO TYPES OF SPECIALIZED SUPPORT CELLS The cells that are classified as **sclerenchyma** produce a thick **secondary cell wall** in addition to the relatively thin primary cell wall found in all cells. Unlike a primary cell wall, the secondary cell wall contains the tough, rigid compound **lignin** in addition to cellulose (see Chapter 30).

Collenchyma cells can support actively growing parts of the plant because they have an expandable primary cell wall. In contrast, the nonexpandable secondary cell wall of sclerenchyma cells specializes them for supporting stems and other structures after active growth has ceased. Another key difference between collenchyma and sclerenchyma is that the cells in sclerenchyma tissue are usually dead at maturity—meaning they contain no cytoplasm.

Ground tissue typically contains two types of sclerenchyma cells: fibers and sclereids.

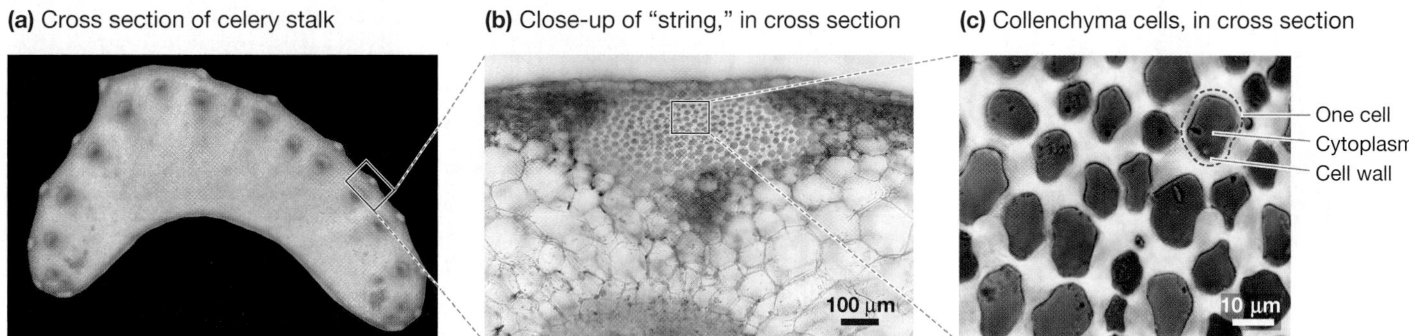

FIGURE 36.22 Collenchyma Cells Support Growing Tissues. A celery stalk is actually a petiole; the strands you can peel from it are columns of collenchyma cells.

(a) Fibers

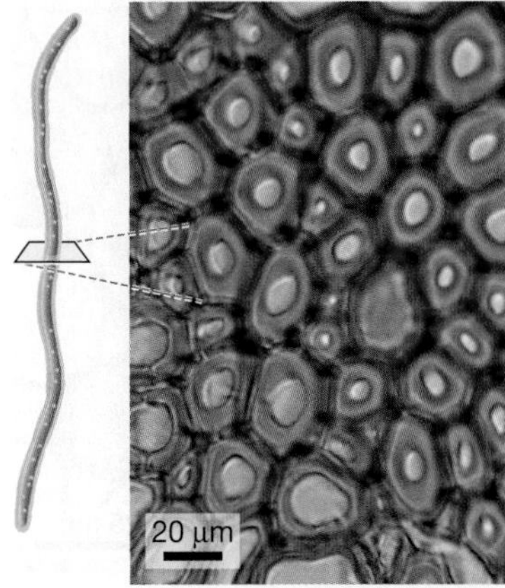

(b) Sclereids

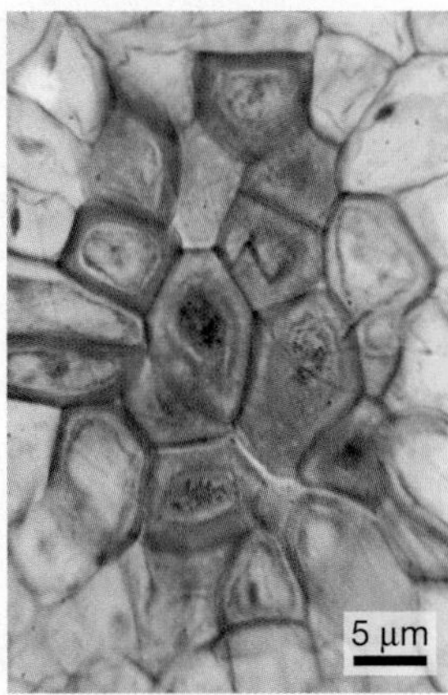

FIGURE 36.23 Sclerenchyma Cells Support Mature Tissues. **(a)** Fibers and **(b)** sclereids have thickened secondary cell walls. These cells provide support for tissues that are no longer growing.

- **Fibers** are extremely elongated. The fiber cells from ramie plants, for example, can be over half a meter long. Fiber cells are important in the manufacture of paper, hemp or jute ropes, or linen and other fabrics (**Figure 36.23a**).
- **Sclereids** are relatively short, have variable shapes, and often function in protection. The tough coats of seeds and the thick shells of nuts are composed of sclereids; these cells are also responsible for the gritty texture of pears (**Figure 36.23b**).

The Vascular Tissue System

The vascular tissue system functions in support and long-distance transport of water and dissolved nutrients. It moves the products made and stored in ground tissue.

Plant tissues that consist of a single cell type are called simple tissues; tissues that contain several types of cells are termed complex tissues. The vascular tissue system is made up of two complex tissues, xylem and phloem.

- **Xylem** (pronounced *ZYE-lem*) conducts water and dissolved ions in one direction: from the root system to the shoot system.
- **Phloem** (*FLO-em*) conducts sugar, amino acids, chemical signals, and other substances in two directions: from roots to shoots and from shoots to roots.

XYLEM STRUCTURE The most important cell types in xylem tissue are tracheids and vessel elements.

- In all vascular plants, xylem contains water-conducting cells called **tracheids** (*TRAY-kee-ids*).
- In angiosperms and species in the group Gnetophyta, xylem also contains conducting cells called **vessel elements**.

Tracheids and vessel elements have thick, lignin-containing secondary cell walls that are often deposited in ringlike or spiral patterns. Both tracheids and vessel elements are dead at maturity. As a result, they are filled with the fluids that they conduct instead of with cytoplasm.

Tracheids are long, slender cells with tapered ends (**Figure 36.24a**). The sides and ends of tracheids have **pits**, which are gaps in the secondary cell wall where only the primary cell wall is present. Because the cell is dead, pits have no plasma membrane spanning the opening. When water is moving up a plant through tracheids, it moves from cell to cell both vertically and laterally through pits, because that is where resistance to flow is lowest.

Vessel elements, in contrast, are shorter and wider than tracheids (**Figure 36.24b**). In addition to having pits, vessel elements have **perforations**—openings that lack both primary and secondary cell walls. In some species, the ends of vessel elements lack any cell

(a) Tracheids are spindle shaped and have pits.

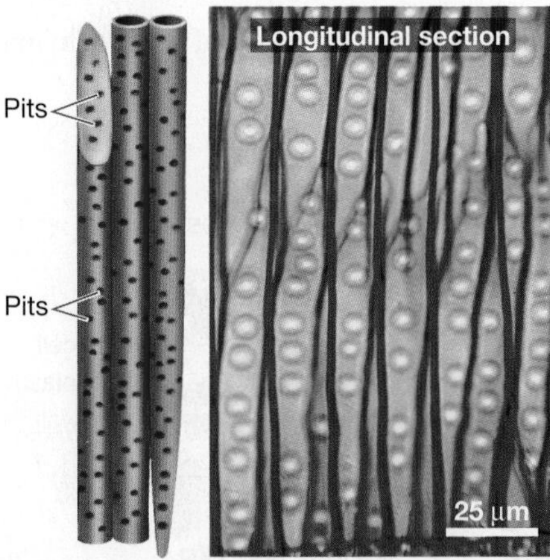

(b) Vessel elements are short and wide and have perforations as well as pits.

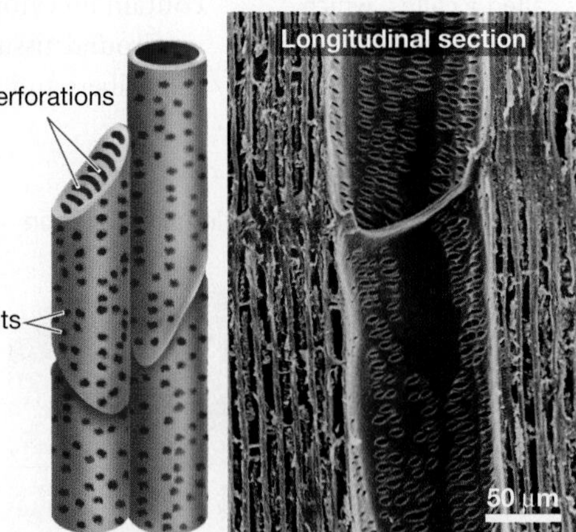

(c) Tracheids and vessel elements together in vascular tissue

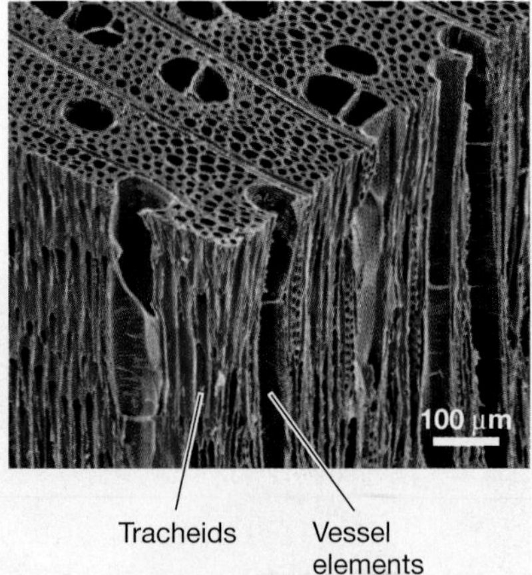

FIGURE 36.24 Xylem May Contain Two Types of Water-Conducting Cells. **(a)** Tracheids are long and thin compared to **(b)** vessel elements, which are much shorter and wider. **(c)** Both types of water-conducting cells are found in the vascular tissue of angiosperms. The image in (a) is a light micrograph of stained tissue; the images in (b) and (c) are colorized scanning electron micrographs.

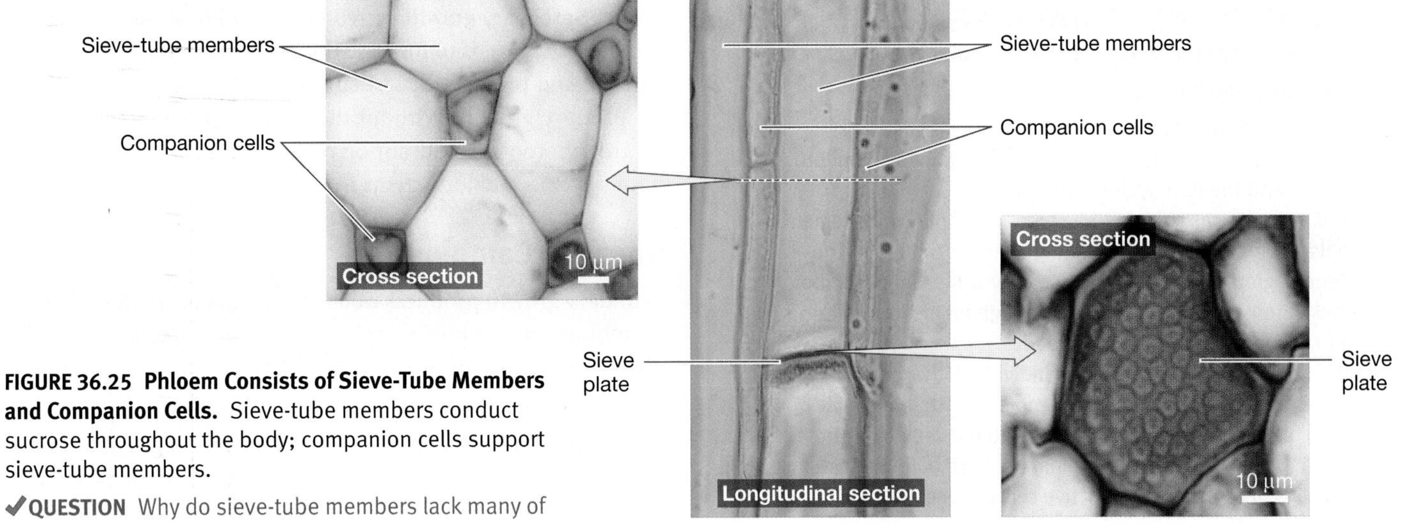

FIGURE 36.25 Phloem Consists of Sieve-Tube Members and Companion Cells. Sieve-tube members conduct sucrose throughout the body; companion cells support sieve-tube members.

✔**QUESTION** Why do sieve-tube members lack many of the organelles found in companion cells?

wall at all, and stacked cells form open pipes called vessels. Vessel elements conduct water much more efficiently than do tracheids because their width and perforations offer less resistance to flow.

In angiosperms, tracheids and vessel elements are found adjacent to each other (**Figure 36.24c**). Xylem also contains some parenchyma cells that transport materials laterally in the stem—not vertically.

PHLOEM STRUCTURE Phloem is made up primarily of two specialized types of parenchyma cells: sieve-tube members and companion cells. Both sieve-tube members and companion cells are alive at maturity, lack lignified secondary cell walls, and arise from division of a common precursor cell.

- **Sieve-tube members** (also called sieve-tube elements) are long, thin cells that have perforated ends called **sieve plates** (**Figure 36.25**). They are responsible for transporting sugars and other nutrients.
- **Companion cells** are not conducting cells, but instead provide materials to maintain the cytoplasm and plasma membrane of sieve-tube members.

Sieve-tube members lack nuclei and most other major organelles, but are directly connected to adjacent companion cells by means of numerous plasmodesmata. Companion cells contain all the organelles normally found in a plant cell.

As Chapter 37 will show, companion cells are also involved in loading and unloading carbohydrates and other nutrients from the solution inside sieve-tube members. Phloem also contains fibers and sclereids, which provide support.

Table 36.1 summarizes the major tissue and cell types found in the dermal, ground, and vascular tissue systems. Once you've mastered the structure and function of the primary plant body, you're ready to consider the next level of complexity: secondary growth.

SUMMARY TABLE 36.1 **Components of the Primary Plant Body**

Tissues Present	Description of Tissue	Function
Dermal Tissue System (arises from protoderm)		
Epidermal	Complex tissue consisting of epidermal cells, guard cells, and trichome cells	Shoots: Protection, gas exchange Roots: Protection, water and nutrient absorption
Ground Tissue System (arises from ground meristem)		
Parenchyma	Simple tissue consisting of parenchyma cells	Synthesis and storage of sugars and other compounds
Collenchyma	Simple tissue consisting of collenchyma cells	Support (expandable in size)
Sclerenchyma	Simple tissues consisting of sclereids or fibers	Support (fixed in size)
Vascular Tissue System (arises from procambium)		
Xylem	Complex tissue consisting of tracheids, vessels, and parenchyma cells and sclerenchyma cells (fibers)	Transport of water and ions; support
Phloem	Complex tissue consisting of parenchyma cells (sieve-tube members, companion cells) and sclerenchyma cells (fibers, sclereids)	Transport of sugars, amino acids, hormones, etc.; support

CHECK YOUR UNDERSTANDING

If you understand that . . .

- Primary growth results from cell division in apical meristems. Its function is to extend the shoot system into the air and the root system into the soil.
- Apical meristems contain three types of primary meristematic cells: protoderm, ground meristem, and procambium. The dermal, ground, and vascular tissue systems that arise from these meristematic cells extend throughout the individual and make up the primary plant body.
- The dermal system protects the individual; the ground system produces and stores the molecules that make life possible; the vascular system moves those molecules from place to place and holds the plant up. Each of these systems consists of an array of distinctive cell and tissue types.

✔ You should be able to . . .

1. Explain the relationship between an apical meristem and the three primary meristematic tissues.
2. Describe the structure and function of epidermal cells, parenchyma cells, tracheids, and sieve-tube elements.

Answers are available in Appendix B.

36.4 Secondary Growth Widens Shoots and Roots

Primary growth increases the length of roots and shoots; its major function is to extend the reach of the root and shoot system and thus increase a plant's ability to absorb photons and acquire carbon dioxide, water, and ions.

Secondary growth increases the width of the plant body. Its major function is to increase the amount of conducting tissue available and provide the structural support required for extensive primary growth. Without the support provided by secondary growth, roots would not be massive enough to anchor large shoot systems, and long stems would fall over or break.

What Is a Cambium?

Secondary growth produces **wood** and occurs in species that have a cambium in addition to apical meristems. A **cambium** is also called a secondary meristem or **lateral meristem.** A cambium differs from an apical meristem in two ways:

1. A cambium forms a cylinder that runs the length of a root or stem and is made up of a single layer of meristematic cells. In contrast, apical meristems are localized at root tips and shoot tips and are dome shaped.
2. In a cambium, cells divide in a way that increases the width of roots and shoots (**Figure 36.26a**). Cells in an apical meristem divide in a way that extends the root and shoot tips.

As **Figure 36.26b** shows, there are two distinct types of cambium in plants that undergo secondary growth.

- A ring of meristematic cells called the **vascular cambium** is located between the secondary xylem and phloem, inside the stem.
- A second ring of meristematic cells called the **cork cambium** is located near the perimeter of the stem.

One other observation is critical to understanding how lateral meristems work: The cork cambium produces new cells primarily to the outside. Vascular cambium, in contrast, generates new layers of cells both to the inside and outside. The new cells

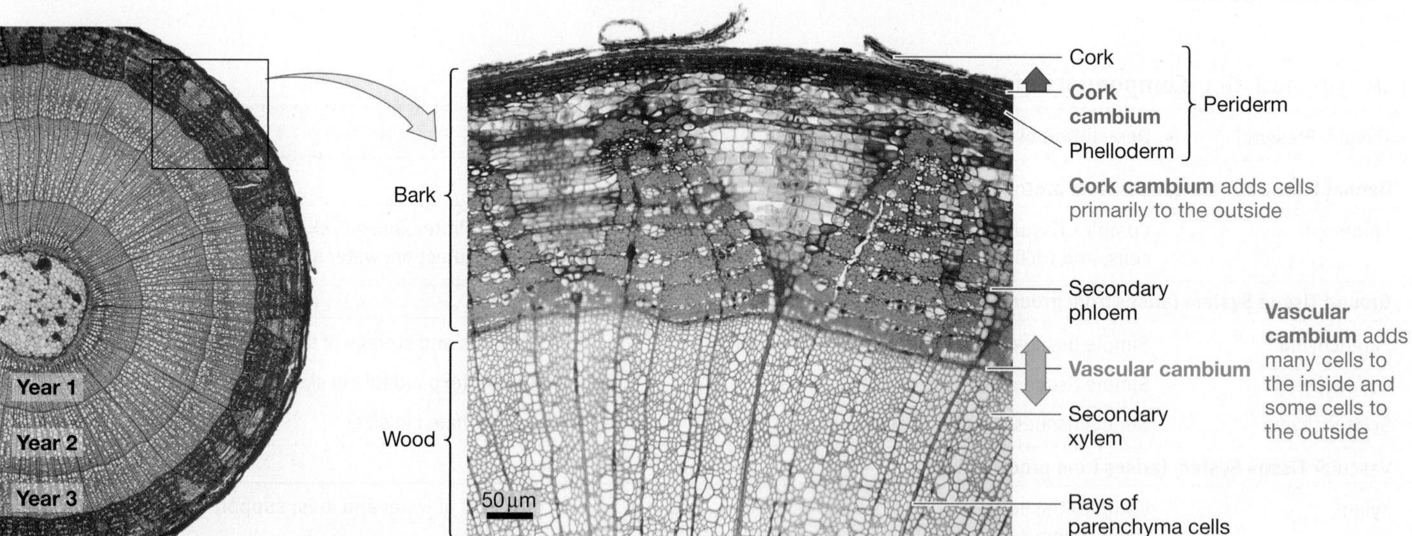

FIGURE 36.26 Vascular Cambium Is Responsible for Secondary Growth. Tree trunks contain two types of lateral meristems: cork cambium and vascular cambium. Bark consists primarily of the cork cells produced by the cork cambium. Wood consists of the secondary xylem and the parenchyma cells produced by vascular cambium.

formed to the inside push all of the other cells toward the outside, causing an increase in girth.

✔If you understand this concept, you should be able to draw a cross section of a stem, draw rings representing the vascular cambium and the cork cambium, and add arrows showing the direction of growth in each meristem.

Now let's consider how each of these meristems works.

What Does Vascular Cambium Produce?

Vascular cambium produces both phloem and xylem (see Figure 36.26b). New cells that are produced to the outside of the meristem differentiate into phloem; new cells produced to the inside differentiate into xylem.

More specifically, cells produced by vascular cambium develop into secondary phloem and secondary xylem. (In contrast, the procambium at each apical meristem produces primary phloem and primary xylem.) Secondary phloem and secondary xylem cells are not always produced simultaneously. In most cases, the vascular cambium produces many more secondary xylem cells than secondary phloem cells.

✔If you understand this concept, you should be able to add "Secondary phloem" and "Secondary xylem" labels to the arrows on your stem cross-section diagram. (Make one arrow fatter than the other to reflect the relative amount of cell division).

Primary phloem and xylem are found throughout the roots and shoots of all vascular plants, but secondary phloem and xylem are found only in some lycophytes, the gymnosperms, and certain angiosperms.

Structurally, primary and secondary phloem and primary and secondary xylem are complex tissues, made up of more than one cell type. Functionally, primary and secondary phloem are similar; primary and secondary xylem are also similar.

- Secondary phloem functions in sugar transport. In combination with cork cambium tissues, it forms bark.
- Secondary xylem functions in water transport and structural support, forming the structural material called wood.

Besides producing conducting cells such as sieve-tube members, tracheids, and vessel elements, the vascular cambium produces fibers for additional strength, along with parenchyma cells. The parenchyma cells radiate laterally across the xylem and form structures called **rays** (see Figure 36.26b), which transport water and nutrients laterally across the stem.

Secondary growth in roots is similar to secondary growth in stems. In both portions of the plant, width increases as cells produced on the inside of the vascular cambium form secondary xylem, and cells produced on the outside form secondary phloem.

It's important to realize, though, that the results of cell division in lateral meristems are highly asymmetrical. As the vascular cambium grows, all of the secondary xylem is retained and accumulates but the primary xylem eventually dies and may rot away. In addition, the outermost secondary phloem and cork layers are sloughed off as the stem increases in diameter. As a result, mature woody roots and stems are dominated by secondary xylem, or wood.

Table 36.2 summarizes the major tissue types and cell types involved in secondary growth. You can also go to the study area at *www.masteringbiology.com* and review both primary and secondary growth.

 **Web Activity** Plant Growth

What Does Cork Cambium Produce?

The cork cambium produces **cork cells** to the outside and a smaller layer of cells called the **phelloderm** ("cork-skin") to the inside (Figure 36.26b). Together, the cork cambium, cork cells, and phelloderm make up the tissue called **bark**.

✔If you understand the structure of bark, you should be able to add labels to your stem cross-section diagram that read "Cork

SUMMARY TABLE 36.2 **Components of Secondary Growth**

Region	Tissue Type	Cell Composition	Function
Bark (arises from and includes cork cambium)	Cork	Cork cells	Protection
	Cork cambium	Meristematic cells	Production of cork and phelloderm
	Phelloderm	Parenchyma cells	Synthesis and storage
Secondary phloem (arises from vascular cambium)	Phloem	Parenchyma cells (sieve-tube members, companion cells) and sclerenchyma cells (fibers, sclereids)	Transport of sugars, amino acids, hormones, etc.; support
Secondary xylem* (arises from vascular cambium)	Xylem	Tracheids, vessels, parenchyma cells (arranged in rays) and sclerenchyma cells (fibers)	Transport of water and ions; support

*Secondary xylem is also called wood.

cells" and "Phelloderm." You should also be able to make one arrow from the cork cambium fatter than the other to reflect the relative amount of cell division that occurs.

Bark is important because it protects the woody stem as it increases in girth. As a woody stem or root matures, the epidermal tissue produced by the apical meristem during primary growth is replaced by the bark, which takes over the role of preventing water loss and protecting the stem and root from pathogens and herbivores. In some species, exceptionally thick bark can even protect the shoot system from fire damage. Redwood trees, for example, which grow in fire-prone habitats, can have bark that is 20 cm (12 in.) thick.

Bark provides a particularly tough barrier in species where cork cells secrete a strong secondary cell wall containing lignin. Bark also helps prevent water loss because cork cells produce a layer of wax and other molecules inside their cell walls, making them impermeable to water and gases. Gas exchange can still occur between the atmosphere and living tissues inside the stem, though—through small, spongy segments of the bark called **lenticels**.

Cork cells die when they mature. As a stem continues to widen, the cork layer often cracks and flakes.

✔If you understand the structure of the cork cambium and the cells it produces, you should be able to add a label to your stem cross-section diagram that reads "Bark." You should also be able to label the areas of the stem that function in structural support, transport of water and nutrients, and protection.

The Structure of a Tree Trunk

Trees are perennial plants that live for many years. As a tree matures and grows in width, the innermost xylem layers stop transporting water—only the xylem from the most recent years actually transports fluid.

HEARTWOOD AND SAPWOOD Xylem that no longer transports begins accumulating protective compounds secreted by other tissues. These compounds form resins, gums, and other complex mixtures. The deposition of these molecules causes the oldest portions of secondary xylem to become darker than the younger portions.

The darker-colored, inner xylem region is called **heartwood** while the lighter-colored, outer xylem is called **sapwood** (**Figure 36.27a**). If you look closely at wood furniture or flooring, you may see a color difference between sapwood and heartwood in some boards.

ANNUAL GROWTH RINGS Another important phenomenon occurs in environments where the vascular cambium stops growing for a portion of each year. This period of **dormancy** occurs during the winter in cold climates and during the dry season in tropical habitats.

When the vascular cambium resumes growth in the spring or at the start of the rainy season, it produces large, relatively thin-walled cells. As the growing season nears its end, conditions tend to dry out or become cooler; the secondary xylem cells that are produced at this time tend to be smaller, thicker walled, and

(a) Heartwood and sapwood have different functions.

(b) Growth rings result from variation in cell size.

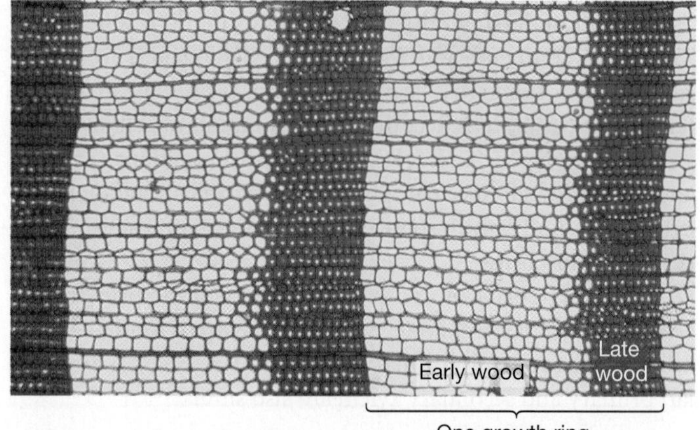

(c) Patterns in growth rings can tell a tree's history.

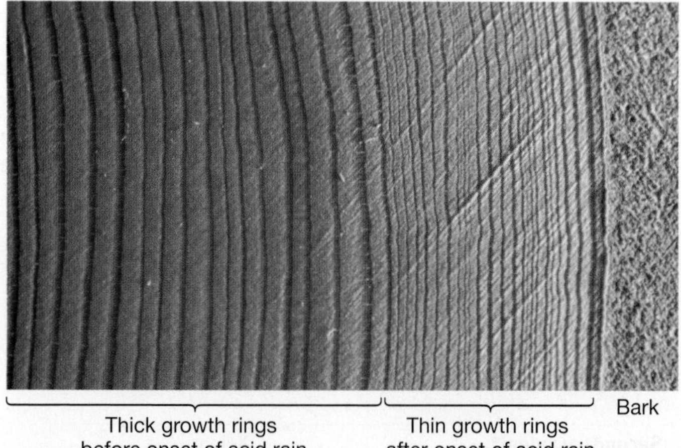

FIGURE 36.27 Anatomy of a Tree Trunk. (a) Unstained section of wood. **(b)** Section of wood, stained to show individual cells. **(c)** Unstained section through a fir tree from Germany's Black Forest.

✔**EXERCISE** In part (c), pick a thick growth ring and label the early wood and late wood.

darker. Thus when growth is seasonal, regions of large, thin-walled cells alternate with layers of small, thick-walled cells, resulting in annual growth rings (**Figure 36.27b**).

Analyzing patterns in tree growth rings is an important field of study in biology. Because trees grow faster when moisture and nutrients are plentiful, wide tree rings are reliable indicators of wet years. In contrast, narrow rings signal drought years—or in the case of the fir tree shown in **Figure 36.27c**, years when abundant acid rain, due to air pollution, reduced growth.

By studying the growth rings in fossil trees and extremely old living trees, biologists can often assemble a continuous record that dates back thousands of years. In doing so, they gain a better understanding of climate changes that occurred in the past. With continued research, researchers also hope to predict how forests might respond to the global warming that is currently under way.

CHECK YOUR UNDERSTANDING

If you understand that . . .

- Secondary growth occurs in species with lateral meristems and results in a broadening of the shoot and root systems.
- Secondary growth results from cell division in the vascular cambium and the cork cambium. Vascular cambium gives rise to secondary vascular tissues. Cork cambium gives rise to the protective tissue called bark.

✔ **You should be able to . . .**

1. Explain how secondary growth relates to primary growth.
2. Describe what the rings in a cross section of a tree trunk would look like if the individual is heavily shaded on one side but exposed to sunlight on the other side.

Answers are available in Appendix B.

CHAPTER 36 REVIEW

For media, go to the study area at www.masteringbiology.com

Summary of Key Concepts

The vascular plant body consists of (1) a root system that anchors the individual and absorbs water and key ions, and (2) a shoot system that absorbs carbon dioxide and sunlight. Both systems are dynamic—they grow and change throughout life.

- The root and shoot systems of plants are specialized for harvesting the light, water, and nutrients required for performing photosynthesis. Structures involved in absorption have a high surface-area-to-volume ratio.
- Roots extract water and nutrients such as nitrogen, phosphorus, and potassium from the soil.
- Shoots capture light and carbon dioxide from the atmosphere.
- Leaves, the major organ for performing photosynthesis, usually consist of a flattened blade that extends from a petiole.
- Roots, stems, and leaves may be modified to perform a variety of other functions, however, including nutrient storage, water storage, protection, and asexual reproduction.
- Because roots and shoots grow throughout life, a plant is able to respond appropriately to changes in environmental conditions.

✔ You should be able to explain why phenotypic plasticity in roots, shoots, and leaves is expected to be more important (1) in environments where conditions are variable versus stable, and (2) in long-lived versus short-lived species.

Variation in body size and shape allows different species to harvest water, light, and other resources in unique ways.

- The overall morphology of root and shoot systems varies widely among plant species.
- In prairie plants, root systems range from long, linear taproots to shallow, dense mats.
- Among the silverswords of Hawaii, shoot systems vary from low mats or rosettes to woody, highly branched tree trunks.
- Among species, variation in plant size and shape allows individuals to reduce competition for resources and thrive in a particular habitat.

✔ You should be able to describe a habitat where (1) plants would be expected to have relatively large root systems and small shoot systems, and (2) plants would be expected to have relatively small root systems and large shoot systems.

Primary growth occurs when cells located at the tips of each root and shoot divide and enlarge. Primary growth lengthens roots and shoots and gives rise to three primary tissue systems that are specialized for protection, food production and storage, and transport.

- Each apical meristem gives rise to three primary meristematic tissues: protoderm, ground meristem, and procambium.
- The primary meristematic tissues give rise to the dermal, ground, and vascular tissue systems, which extend throughout the plant body.
- The dermal tissue system is usually one cell layer thick and plays a role in protection, water conservation, and water absorption.
- The ground tissue system performs photosynthesis and stores carbohydrates and other compounds.
- The vascular tissue system transports materials throughout the plant. Within the vascular system, xylem tissue transports water and dissolved ions up the plant; phloem tissue transports sugars up and down.
- Each tissue system is made up of simple tissues that contain a single cell type, complex tissues that contain two or more cell types, or a combination of simple and complex tissues.

- Ground tissue contains (1) parenchyma cells, which function in material synthesis and storage, (2) collenchyma cells, which provide structural support for growing regions, and (3) the fiber and sclereid cells of sclerenchyma tissue, which strengthen regions of the body that have stopped growing.

✔You should be able to predict the results of an experiment where a drug was used to (1) poison an apical meristem in the shoot and an apical meristem in the root, and (2) selectively poison protoderm cells in a shoot apical meristem.

In some species, secondary growth occurs when cells near the perimeter of a root or shoot divide and enlarge, widening the structure. Secondary growth adds transport tissue and provides structural support.

- In some plant species, shoots and roots are widened by lateral meristems that produce secondary xylem, secondary phloem, and bark.
- Wood consists of secondary xylem, while bark consists of all tissue outside of the vascular cambium.

✔You should be able to predict the results of an experiment where a drug was used to (1) slow the growth of the vascular cambium but not the cork cambium, and (2) slow the growth of both the vascular and cork cambia on one side of a tree trunk.

MB **Web Activity** Plant Growth

Questions

✔TEST YOUR KNOWLEDGE

Answers are available in Appendix B

1. Which of the following functions is not performed by parenchyma cells?
   a. synthesizing key molecules
   b. transporting water
   c. performing photosynthesis
   d. storing nutrients
2. How do tracheids differ from vessel elements, in addition to their overall shape?
   a. Tracheids are stacked end to end to form continuous, open columns.
   b. In tracheids, water flows from cell to cell primarily through gaps in the secondary cell wall called pits.
   c. Tracheids are dead at maturity.
   d. Tracheids have secondary cell walls reinforced with lignin.
3. What is a sieve-tube member?
   a. the sugar-conducting cell found in phloem
   b. the widened, perforation-containing, water-conducting cell found only in angiosperms
   c. the nutrient- and water-absorbing cell found in root hairs
   d. the nucleated and organelle-rich support cell found in phloem
4. What is an adventitious root?
   a. one that performs a function other than anchoring the plant or absorbing water and ions
   b. one that erupts through the root cortex and spreads laterally
   c. a long, filamentous extension that increases the surface area of the root system, for efficient absorption
   d. one that arises from the shoot system
5. Which statement best characterizes primary growth?
   a. It does not occur in roots, only in shoots.
   b. It leads to the development of cork.
   c. It produces the dermal, ground, and vascular tissues.
   d. It produces rings of xylem and phloem tissue as well as rings of cork tissue.
6. Which statement best characterizes secondary growth?
   a. It results from divisions of the vascular cambium cells.
   b. It increases the length of the plant stem.
   c. It results from divisions in the apical meristem cells.
   d. It often produces phloem cells to the inside and xylem cells to the outside of the vascular cambium.

✔TEST YOUR UNDERSTANDING

Answers are available in Appendix B

1. Describe the general function of the shoot system and the general function of the root system. Which tissues are continuous throughout these two systems? Suggest a hypothesis to explain why the shoot and root systems of different species are so variable in size and shape.
2. Explain why continuous growth enhances the phenomenon known as phenotypic plasticity.
3. To illustrate the concept that a plant structure can be modified for many different functions, give an example of a leaf and a stem that protect individuals against large herbivores.
4. What does cuticle do? What do stomata do? Predict how the thickness of cuticle and the number of stomata differ in plants from wet habitats versus dry habitats.
5. Compare and contrast the roles of parenchyma cells in the ground tissue system versus the vascular tissue system.
6. Describe how the vascular cambium produces secondary xylem and secondary phloem.

✔APPLYING CONCEPTS TO NEW SITUATIONS

Answers are available in Appendix B

1. The shoot systems of domesticated broccoli family plants (broccoli, cauliflower, cabbage, kohlrabi, and others) changed in response to artificial selection from the wild mustard that was their ancestor. Would you predict that leafy populations such as cabbage, kale, and savoy have more or fewer sclerenchyma cells in their stems and leaves than the wild population has? Explain your logic.
2. Identify the structure you are consuming when you eat the following vegetables: asparagus, Brussels sprouts, celery, spinach, carrots, potato.
3. Why do trees that grow in tropical rain forests lack growth rings?
4. Trees can be killed by girdling—meaning the removal of bark and vascular cambium in a ring all the way around the tree. Explain why.

This chapter explores how plants move water from their roots to their leaves and how they transport sugars to all of their tissues—sometimes over great distances.

Water and Sugar Transport in Plants

O n a hot summer day, a large deciduous tree can lose enough water to fill three 55-gallon drums. To understand why, recall from Chapter 36 that the surfaces of leaves are dotted with stomata, which open during the day so gas exchange can occur between the atmosphere and the cells inside the leaf. This exchange is crucial. For photosynthesis and food production to continue, leaf cells must acquire carbon dioxide (CO_2).

There's a catch, however. While stomata are open, the moist interior of the leaf is exposed to the dry atmosphere. As a result, large quantities of water evaporate from the leaf.

In essence, then, water loss is an inevitable consequence of a plant's need to obtain carbon and release oxygen. Water loss is a side effect of photosynthesis.

Evaporation from leaves can actually be beneficial under some conditions, because it cools the plant, just as sweating cools your body. Heavy rates of evaporation can lower leaf temperatures by as much as 10–15°C.

If the lost water is not replaced, however, plant cells will dry out and die. In the case of a redwood tree, the leaves that lose water may be 100 m from the root hairs that absorb water.

How do plants transport water against the force of gravity—in some cases, the length of a football field? And how do plants move the sugar they produce from active photosynthetic sites to storage sites in roots? These questions are the heart and soul of this chapter. Answering them is a fundamental part of understanding how plants work.

KEY CONCEPTS

- Water moves from areas of high water potential to areas of low water potential. Water's potential energy in plants is a combination of (1) its tendency to move in response to differences in solute concentration and (2) the pressure exerted on it.
- Plants lose water to transpiration when stomata are open and photosynthesis is occurring, but they do not expend energy to replace it. Instead, water moves from soil and roots to leaves along a water-potential gradient. Evaporation of water from leaves, driven by the Sun, creates a negative pressure (tension) that pulls water up.
- In phloem, sugars are transported from "sources"—tissues that release sugars for use elsewhere—to "sinks"—tissues in which sugars are being used or stored. Movement occurs because large amounts of sucrose move into phloem cells near source tissues. Water follows by osmosis, creating a pressure gradient that favors the movement of water and sucrose to sinks.

37.1 Water Potential and Water Movement

Loss of water via evaporation from the aerial parts of a plant is called **transpiration**. Transpiration occurs whenever two conditions are met: (1) Stomata are open, and (2) the air surrounding leaves is drier than the air inside leaves.

The first condition is usually met during the day, when photosynthesis takes place. The second condition occurs whenever atmospheric humidity is less than 100 percent.

Plants replace water that is lost from leaves with water that is absorbed by roots. One of the most astonishing observations in biology is that water moves from roots to leaves passively—that is, with no expenditure of ATP. Plants do not need a heart muscle to pump water from roots to shoots. Even in 100-m-tall redwood trees, water flows passively from the root system to the shoot system. This movement occurs because of differences in the potential energy of water.

What Is Water Potential?

Recall from Chapter 2 that potential energy is stored energy. Changes in potential energy are associated with changes in position, such as the position of a molecule or an electron.

Biologists use the term **water potential** to indicate the potential energy that water has in a particular environment compared with the potential energy of pure water at room temperature and atmospheric pressure. Under these conditions, pure water has a water potential of 0.

Water potential is symbolized by the Greek letter ψ (psi, pronounced *sigh*). Differences in water potential determine the direction that water moves. Water always flows from areas of high water potential to areas of lower water potential.

What Factors Affect Water Potential?

To understand how water moves from cell to cell in a plant, consider the cell in the beaker on the left in **Figure 37.1a**. Notice that it is sitting in a **solution**—a homogenous, liquid mixture containing several substances. In this case, the solution consists of water and dissolved substances, or **solutes**.

In the beaker on the left, the solute concentrations in the cell and in the surrounding solution are the same. Recall from Chapter 6 that such a solution is **isotonic** to the cell. When two solutions are isotonic, there is no net movement of water between them.

THE ROLE OF SOLUTE POTENTIAL What happens when the cell is transferred to the beaker on the right in Figure 37.1a? This beaker contains pure water, which has no solutes. As a result, the solution surrounding the cell is strongly **hypotonic** relative to the cell.

The concentration of the solution is important, because when solutions are separated by a selectively permeable membrane, such as the plasma membrane of a cell, water passes through the membrane from regions of low solute concentration to regions of high solute concentration. This movement of water across membranes, in response to differences in water potential, is called **osmosis** (see Chapter 6).

The tendency for water to move by osmosis, in response to differences in solute concentrations, is called the **solute potential** (**ψ_S**) or **osmotic potential.**

The solute potential of a solution is defined by its solute concentration relative to pure water. If water contains a high concentration of solutes, then it has a low solute potential compared with pure water.

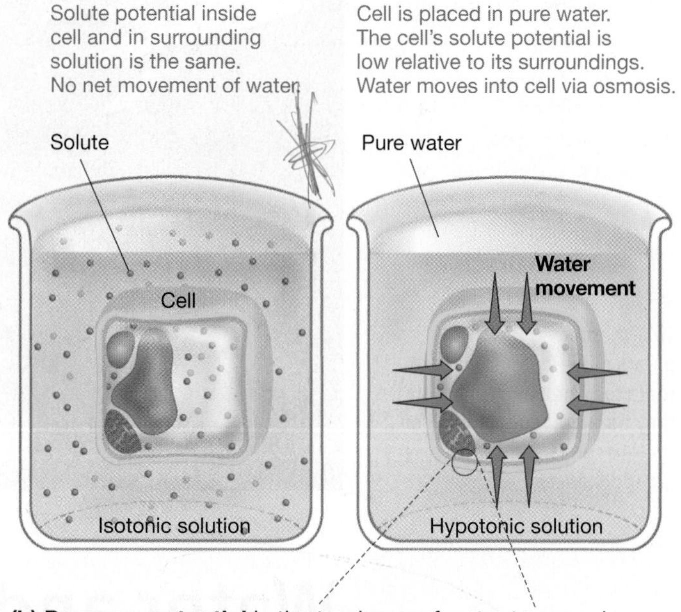

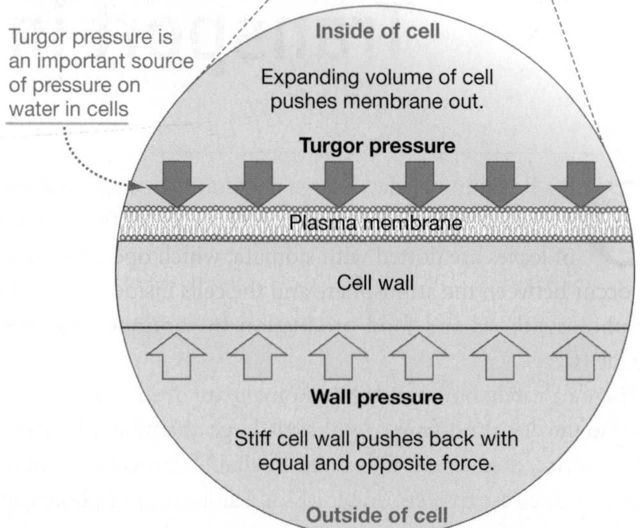

FIGURE 37.1 Water Potential Has Two Major Components: Solute Potential and Pressure Potential.

THE ROLE OF PRESSURE POTENTIAL When an animal cell is placed in a hypotonic solution and water enters the cell via osmosis, the volume of the cell increases until the cell bursts. This does not happen to plant cells, however.

If a plant cell swells in response to incoming water, its plasma membrane pushes against the relatively rigid cell wall. The cell wall resists expansion of the cell volume by pushing back, much as the walls of a basketball push back when the ball is inflated.

The force exerted by the wall is called **wall pressure** (**Figure 37.1b**). As water moves into the cell, the pressure inside the cell, known as **turgor pressure**, increases until wall pressure is induced. Cells that are firm and that experience wall pressure are said to be **turgid**.

Turgor pressure is important, because it counteracts the movement of water due to osmosis. In the right-hand example in Figure 37.1a, the solute potential favors water moving into the cell. However, the rigid cell walls limit the amount of water that can enter the cell.

Pressure potential (ψ_P) refers to any kind of physical pressure on water. Inside a cell, the pressure potential consists of turgor pressure.

When osmosis and pressure affect a cell at the same time, how do biologists determine the direction of water movement?

Calculating Water Potential

As a form of potential energy, water potential (ψ) can be thought of as water's stored energy or its tendency to move to a new position. Thus, a water potential summarizes the stored energy that will tend to make water move—in response to the combined effects of a pressure potential and a solute potential.

When selectively permeable membranes are present, water tends to move by osmosis from areas of high solute potential to areas of low solute potential. When no membranes are present to stop it, water moves from areas of high pressure potential to areas of low pressure potential.

If we ignore the effects of gravity, water potential is defined algebraically as

$$\psi = \psi_P + \psi_S$$

In words, the potential energy of water in a particular location is the sum of the pressure potential and the solute potential that it experiences.

ASSIGNING UNITS OF PRESSURE AND SIGNS Water potential is measured in units called **megapascals** (**MPa**, 10^6 Pa). A **pascal** (**Pa**) is a unit of measurement commonly applied to pressures—force per unit area. A car tire is inflated to about 0.2 MPa, and the water pressure in home plumbing is usually 0.2 to 0.3 MPa.

Solute potentials (ψ_S) are always negative, because they are measured relative to the solute potential of pure water, which is 0 Mpa because it contains no dissolved substances. And because there are always some solutes inside a cell, the water inside always has a solute potential lower than that of pure water. Therefore pure water will tend to move *into* the cell. Increasing the concentration of solutes in a cell lowers its water potential even more—making it more negative.

In contrast to the solute potential in cells, the pressure potential (ψ_P) from turgor pressure is positive inside cells. It increases the potential energy of the water inside by exerting pressure on the water, making it more likely to move *out of* the cell. But pressure potential can also be negative.

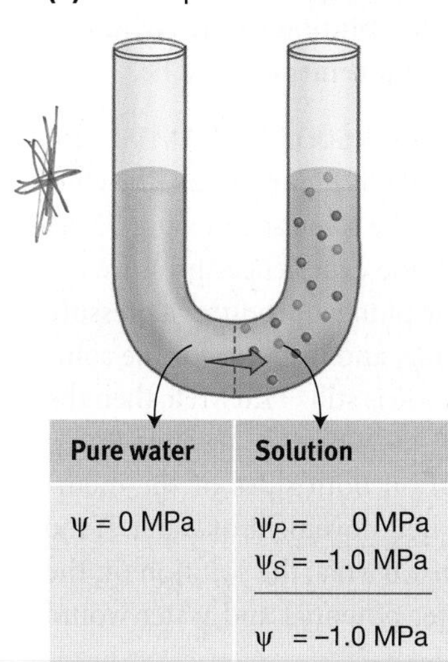

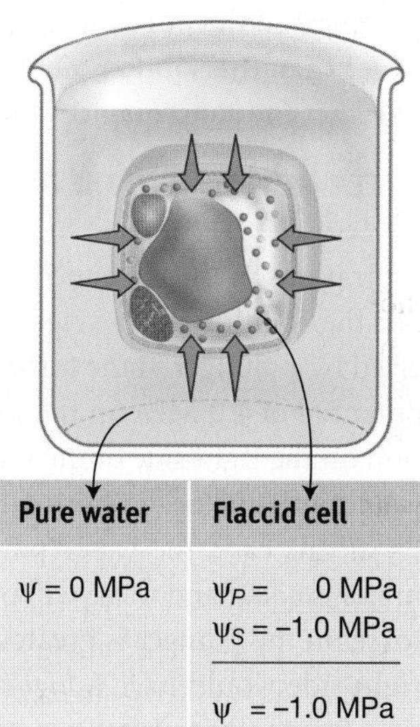

Pure water	Solution	Pure water	Flaccid cell
ψ = 0 MPa	ψ_P = 0 MPa ψ_S = −1.0 MPa ψ = −1.0 MPa	ψ = 0 MPa	ψ_P = 0 MPa ψ_S = −1.0 MPa ψ = −1.0 MPa

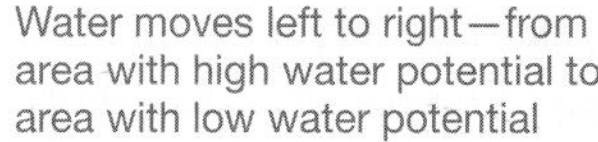

Water moves into cell—from area with high water potential to area with low water potential

(b) Solute and pressure potentials differ.

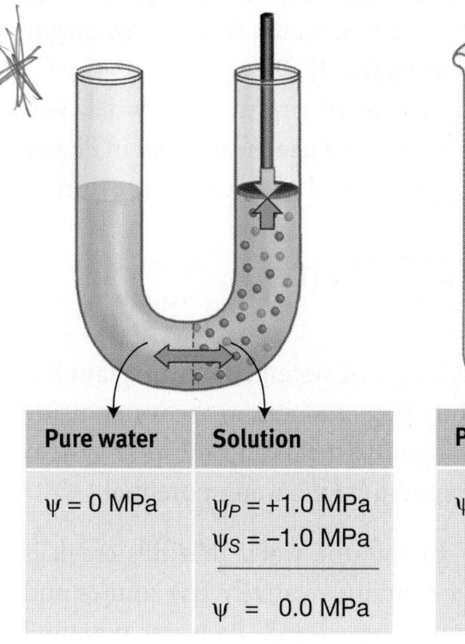

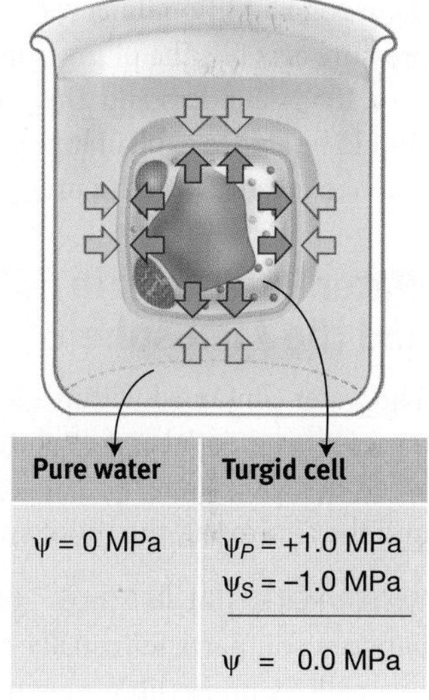

Pure water	Solution	Pure water	Turgid cell
ψ = 0 MPa	ψ_P = +1.0 MPa ψ_S = −1.0 MPa ψ = 0.0 MPa	ψ = 0 MPa	ψ_P = +1.0 MPa ψ_S = −1.0 MPa ψ = 0.0 MPa

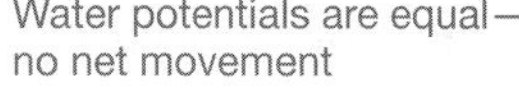

Water potentials are equal—no net movement

FIGURE 37.2 Solute Potential and Pressure Potential Interact.

✔**QUESTION** In the left side of part (a), does the solute potential on the right side of the tube increase, decrease, or stay the same when water flows from left to right by osmosis?

WATER MOVEMENT IN THE ABSENCE OF PRESSURE In the U-shaped tube on the left side of **Figure 37.2a**, two solutions are separated by a selectively permeable membrane. The system is open to the atmosphere and thus is not under additional pressure, meaning $\psi_P = 0$ MPa.

Note that the left side of the tube contains pure water, which has a ψ_S of 0 MPa. The ψ_S for the solution on the right side of the membrane is −1.0 MPa. Because water potential is higher on the left side of the tube than on the right side, water moves from left to right.

The right side of Figure 37.2a models the same situation with a cell that is **flaccid**—meaning it has no turgor pressure and thus a pressure potential of 0. Note that the cell has been placed in a solution of pure water. Because the cell has low solute potential (−1.0 MPa) and the pure water has a higher water potential than the cell, water enters the cell via osmosis.

✔**If you understand the concept of solute potential, you should be able to explain (1) how you would change the solute** potential of the pure water on the left of Figure 37.2a to make it lower than the solute potential in the solution to the right, and (2) what the consequences for water movement would be.

WATER MOVEMENT IN THE PRESENCE OF A SOLUTE POTENTIAL AND PRESSURE POTENTIAL On the left side of **Figure 37.2b**, the concentrations in the U-shaped tube are the same as in Figure 37.2a, but the solution on the right-hand side experiences pressure exerted by a plunger. If the force on the plunger produces a pressure potential of 1.0 MPa on the right side, and if the ψ_S for the solution on the right side of the membrane is still -1.0 MPa, then the water potential of the right side is $-1.0 \text{ MPa} + 1.0 \text{ MPa} = 0$.

In this case, the water potential on both sides of the membrane is equal and there will be no net movement of water. If the force on the plunger is greater than 1.0 MPa, the solution on the right side would have a *higher* water potential and water would flow from right to left.

The right side of Figure 37.2b models this situation in a cell. In this case, the incoming water creates turgor pressure. When the positive turgor pressure (+1.0 MPa) plus the cell's negative solute potential (−1.0 MPa) equals 0 MPa—the water potential of pure water—the system reaches equilibrium. At equilibrium, there is no additional net movement of water. In this way, turgor pressure acts like the plunger in Figure 37.2b.

✔If you understand the concept of pressure potential, you should be able to explain how you would use the plunger in Figure 37.2b to create a negative pressure instead of a positive pressure.

Water Potentials in Soils, Plants, and the Atmosphere

The water contained within a leaf, root system, or entire plant has a pressure potential and a solute potential, just as the water inside a cell does. Likewise, both the soil surrounding the root system and the air around the shoot system have a water potential.

WATER POTENTIAL IN SOILS In soil, the water that fills crevices between soil particles usually contains relatively few solutes and normally is under little pressure. As a result, its water potential tends to be high relative to the water potential found in a plant's roots, which is high in solutes.

There are important exceptions to this rule, however.

- *Salty soils* The soils near ocean coastlines and in deserts may have water potentials as low as −4 MPa or less due to high solute concentrations. This is much lower than the water potential typically found inside plant roots.
- *Dry soils* When soils dry, water no longer floats freely in the spaces between soil particles. All of the remaining water clings tightly to soil particles, creating a negative pressure that lowers the pressure potential of soil water.

When the water potential in soil drops, water is less likely to move from soil into plants. If soil water potential is low enough, water may even move from plants to the soil.

This is an enormously important issue for world agriculture. When soils are irrigated to boost crop yields, much of the water evaporates. The solutes in the irrigation water are left behind and tend to collect in the first few inches below the surface. Over time, then, irrigated soils tend to become salty. In some parts of the world, formerly productive soils have become so salty that they are now abandoned as cropland.

HOW DO PLANTS THAT ARE ADAPTED TO SALTY OR DRY HABITATS COPE? Salt-adapted species respond to low water potentials in soil by lowering the solute potential of their root cells. These plants have enzymes and transport proteins that increase the concentration of sugars and other organic molecules in the cytoplasm. As a result, they can keep the water potential of their tissues even lower than the water potential of salty soils.

Species that are adapted to dry sites cope by tolerating low solute potentials. **Figure 37.3**, for example, plots changes in solute potential that were recorded in tissue from a shrub species called ninebark. In this graph, each data point represents the solute potential recorded on a particular day; the lines between the data points are drawn simply to make the trend clear.

Ninebarks thrive on dry sites in the Rocky Mountains of western North America. Notice that the solute potential of tissue is relatively high in June, at the start of the growing season. In the Rockies, June tends to be rainy and cool. July and August are progressively hotter and drier, however. As the graph indicates, the solute potential in tissue drops dramatically during this period.

This is a key observation. As the summer progresses and water potential in soil drops, ninebark shrubs are able to keep acquiring water and grow because the solute potentials of their tissues can drop to match a decline in soil water potential.

Plants that are adapted to wetter sites cannot tolerate such low solute potentials in their tissues. When conditions get hot and dry, they have to close their stomata and stop photosynthesis—meaning that growth will stop or slow down dramatically compared to dry-adapted plants.

If plants keep their stomata open and leaf cells lose water faster than the water is replaced, plasma membranes contract. If the cells do not regain turgor soon, they are at risk of dehydra-

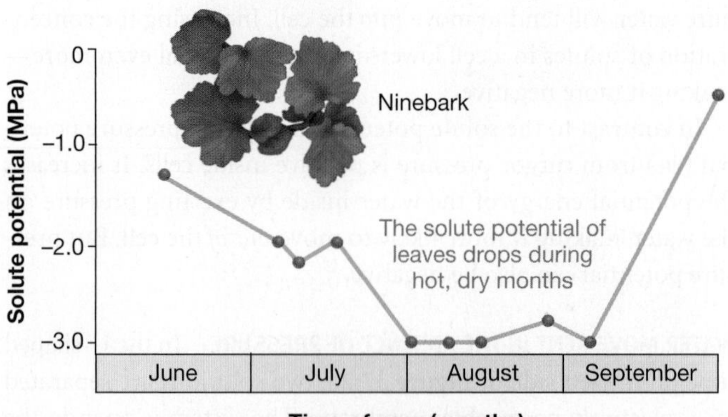

FIGURE 37.3 Plants with Low Solute Potentials Can Grow in Dry Soils. As soils dry, their water potential declines. But if a plant's solute potential also drops, it can maintain a water-potential gradient that continues to bring water into the plant.

tion and death as cells shrivel like grapes drying into raisins. When an entire tissue loses turgor, it will **wilt** (**Figure 37.4**).

Turgor pressure is required for growth to occur; without it, cells cannot expand once cell division is complete. In addition, turgor pressure provides structural support—which is why wilted plants droop. Unless corrected, extensive wilting may lead to the death of the tissue and, eventually, the plant.

FIGURE 37.4 Wilting Occurs when Water Loss Leads to Loss of Turgor Pressure. Wilting is a life-threatening condition in plants, analogous to severe dehydration in humans.

FIGURE 37.5 A Water-Potential Gradient Exists between Soil, Plants, and Atmosphere. Water moves from regions of high water potential to regions of low water potential.

WATER POTENTIAL IN AIR In the atmosphere, water exists as a vapor with a solute potential of 0 MPa. The pressure exerted by water vapor in the atmosphere can be low or high, depending on conditions.

- When air is dry, there are few water molecules present and the pressure they exert is low.
- When air is warm, water molecules move farther apart and exert lower pressure.

Warm, dry air has an extremely low water potential. When it is rainy or foggy, however, the water potential of the atmosphere may be equal to the water potential inside a leaf. But otherwise, the water potential of the atmosphere is much lower than the water potential inside a leaf.

In most cases, water potential is high in soil and roots but low in leaves and the atmosphere. This situation sets up a **water-potential gradient** between roots and shoots.

To move *up* a plant, water moves *down* the water-potential gradient that exists between the soil, its tissues, and the atmosphere. When it does so, it replaces the water lost to transpiration (**Figure 37.5**).

CHECK YOUR UNDERSTANDING

If you understand that . . .

- Water moves along a water-potential gradient, from areas of high water potential to low water potential.
- In areas separated by a selectively permeable membrane, part of water's potential energy is made up of its solute potential—its tendency to move via osmosis.
- Water also has a pressure potential. In plant cells, for example, the cell wall can exert pressure on water and affect its pressure potential.
- Soils, plant tissues, and the atmosphere all have a water potential made up of a solute potential and a pressure potential.

✓ **You should be able to . . .**

Compare and contrast:

1. The pressure potential of wet soils versus dry soils.
2. The solute potential of salty soils versus more-typical soils.

Answers are available in Appendix B.

37.2 How Does Water Move from Roots to Shoots?

Suppose that you are caring for the wilted plant in Figure 37.4. If you add water to the soil, the water potential of the soil increases and water will move into the plant along a water-potential gradient. Biologists have tested three major hypotheses for how the water could be transported:

1. Root pressure—a pressure potential that develops in roots—could drive water up against the force of gravity.

2. Capillary action could draw water up the cells of xylem.
3. Cohesion-tension, a force generated in leaves, could pull water up from roots.

Note that all three hypotheses could be correct—they are not mutually exclusive.

Let's begin by considering how water and solutes move from the soil across the root and into the vascular tissue. Then we can analyze each of the three hypotheses for how water and solutes finish the trip up the xylem to the shoot system.

Movement of Water and Solutes into the Root

To understand how water enters a root, consider the cross section through a young buttercup root shown in **Figure 37.6**. Starting at the outside of the root and working inward, notice that several distinct tissues are present:

- The **epidermis** (literally, "outside skin") is a single layer of cells. In addition to protecting the root, some epidermal cells produce **root hairs**, which greatly increase the total surface area of the root (see Chapter 36).
- The **cortex** consists of ground tissue—usually parenchyma cells—and stores carbohydrates.
- The **endodermis** ("inside skin") is a cylindrical layer of cells that forms a boundary between the cortex and the vascular tissue. The function of the endodermis is to control ion uptake.
- The **pericycle** ("around-circle") is a layer of cells that can become meristematic and produce lateral roots.
- Conducting cells function in transport and are located in the center of roots in buttercups and other eudicots. Notice that, in these species, phloem is situated between each of four arms formed by xylem, which is arranged in a cross-shaped pattern.

THREE ROUTES THROUGH ROOT CORTEX TO XYLEM When water enters a root along a water-potential gradient, it does so through root hairs. As water is absorbed, it moves through the root cortex toward the xylem through three distinct routes (**Figure 37.7a**).

1. The **transmembrane route** is based on flow through aquaporin proteins—water channels located in the plasma membranes of root cells (see Chapter 6), as well as some direct diffusion across plasma membranes.
2. The **apoplastic** ("away-from-particle") pathway is outside the plasma membrane. It consists of cell walls, which are porous, and the spaces that exist between cells. The name apoplastic is apt because movement takes place outside the plasma membrane and cell interior.
3. The **symplastic** ("with-particle") pathway consists of the continuous connection through cells that exists via plasmodesmata (see Chapter 8).

In essence, water can flow through cells via aquaporins, around cells, or through cells via plasmodesmata.

THE ROLE OF THE CASPARIAN STRIP The situation changes when water reaches the endodermis, however. Endodermal cells are tightly packed and secrete a narrow band of wax called the **Casparian strip**. This layer is composed primarily of a compound called **suberin**, which forms a water-repellent cylinder at the endodermis.

The Casparian strip blocks the apoplastic pathway by preventing water from moving through the walls of endodermal cells and proceeding to the vascular tissue (**Figure 37.7b**). The Casparian strip does not affect water that is moving through the symplastic pathway.

The Casparian strip is important because it means that for water and solutes to reach vascular tissue, they have to move into the cytoplasm of an endodermal cell. Endodermal cells, in turn, act as filters.

Endodermal cells allow ions such as potassium (K^+) that are needed by the plant to pass through to the vascular tissue. In contrast, these cells can prevent the passage of ions such as sodium (Na^+) that are not needed or may be harmful.

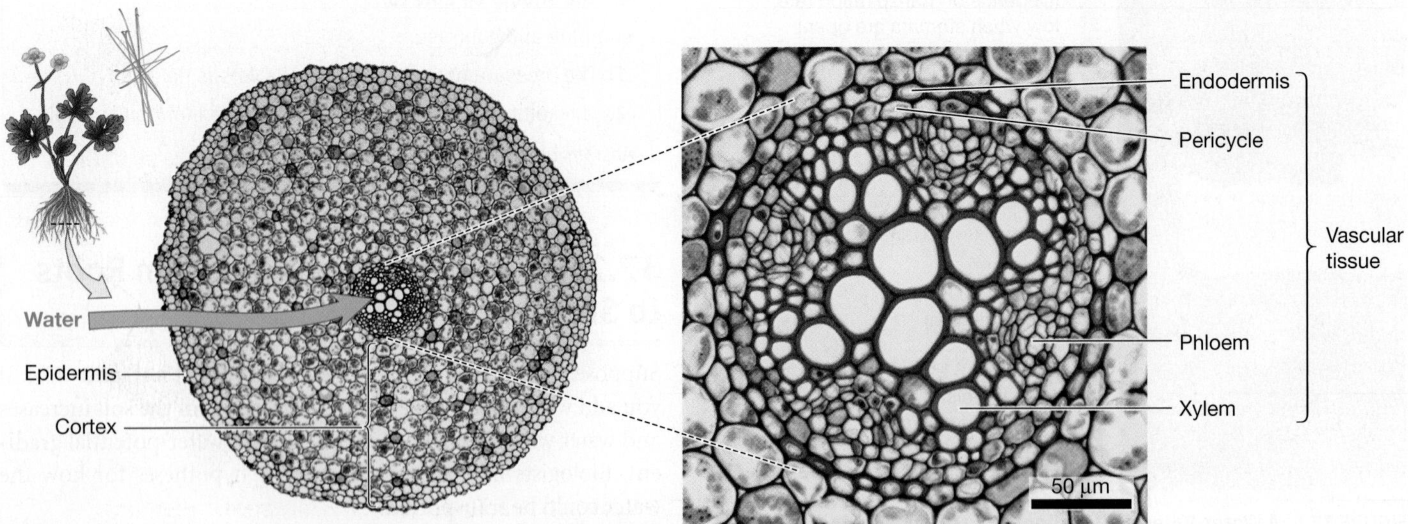

FIGURE 37.6 In Roots, Water Has to Travel through Several Tissue Layers to Reach Vascular Tissue.
Cross section through a buttercup root, showing the anatomy that is typical of roots in eudicots.

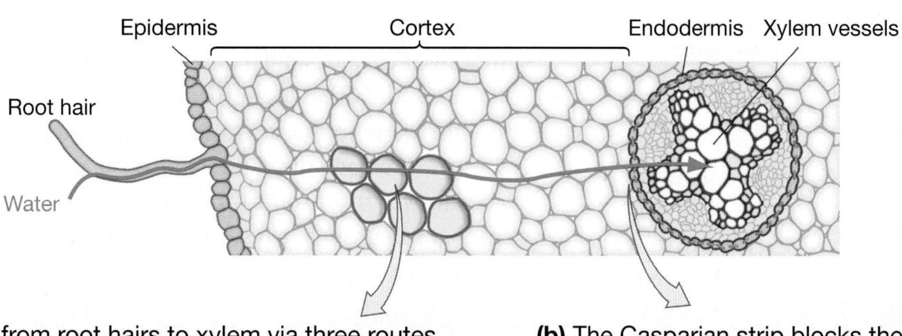

(a) Water travels from root hairs to xylem via three routes.

(b) The Casparian strip blocks the apoplastic route at the endodermis.

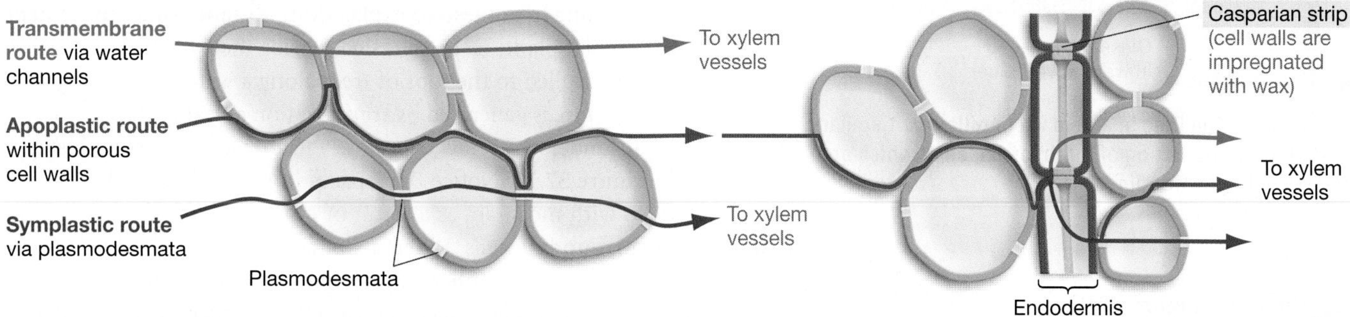

FIGURE 37.7 How Water Travels from Root Hair to Xylem. Water in the apoplastic pathway *must* pass into the cytoplasm of endodermal cells before entering xylem.

✔**EXERCISE** Add dots representing aquaporins in the transmembrane route.

✔If you understand this concept, you should be able to predict what would happen if a plant had a mutation that prevented synthesis of suberin and formation of the Casparian strip.

Water Movement via Root Pressure

Movement of ions and water into the root cortex is responsible for the phenomenon known as **root pressure**. Recall that root pressure is one of three hypothesized mechanisms for moving water up xylem, from root to shoot.

Stomata normally close during the night, when photosynthesis is not occurring and CO_2 is not needed. Their closure minimizes water loss and slows the movement of water into roots. But roots often continue to accumulate ions that their epidermal cells acquire from the soil as nutrients. These nutrients move into xylem. The influx of ions lowers the water potential of xylem below the water potential in the surrounding cells.

As water flows into xylem from other root cells in response, a positive pressure is generated that forces fluid up the xylem. More water moves up xylem and into leaves than is being transpired from the leaves.

In low-growing plants, enough water can move to force water droplets out of the leaves, a phenomenon known as **guttation**. If you are up early in the morning, you may observe water drops on leaf edges formed by guttation (**Figure 37.8**).

At one time, positive root pressure was a leading hypothesis to explain how water moves from roots to leaves in trees. However, research showed that over long distances, such as a tree trunk, the force of root pressure is not enough to overcome the force of gravity on the water inside xylem.

In addition, researchers demonstrated that cut stems, which have no contact with the root system, are still able to transport water to leaves. Biologists concluded that there must be some other mechanism involved in the long-distance transport of water. What is it?

Water Movement via Capillary Action

Researchers have also evaluated a hypothesis based on the phenomenon of **capillarity**, or movement of water up a narrow tube. When a thin glass tube is placed upright in a pan of water, water

FIGURE 37.8 Root Pressure Causes Guttation. When ions accumulate in the xylem of roots at night, enough water may enter xylem via osmosis to force water up and out of low-growing leaves.

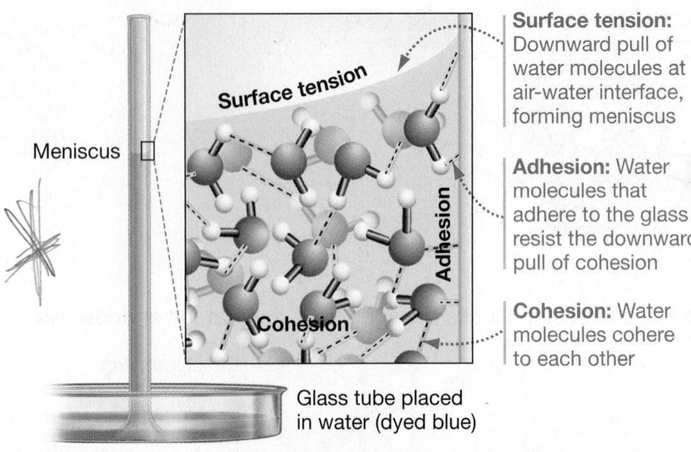

FIGURE 37.9 Water Can Rise in Xylem via Capillarity. Capillarity occurs through a combination of three forces, all of which are generated by hydrogen bonding.

creeps up the tube (**Figure 37.9**). The movement occurs in response to three forces: (**1**) surface tension, (**2**) adhesion, and (**3**) cohesion. Let's consider each force in turn.

Surface tension is a downward pull that exists on water molecules at an air-water interface. In the body of a solution, all the water molecules present are surrounded by other water molecules and form hydrogen bonds in all directions. The water molecules at a surface, however, can form hydrogen bonds in one direction only—with the water molecules below them. As a result, the topmost layer of water molecules is pulled downward by the bonds. Because pulling forces create tension, the pulling force on the water molecules at the air-water interface is called surface tension.

Adhesion is a molecular attraction among unlike molecules. In this case, water interacts with a solid substrate—such as the glass walls of a capillary tube or the cell walls of tracheids or vessel elements—through hydrogen bonding. As water molecules bond to each other and adhere to the side of the tube, they are pulled upward.

Cohesion is a molecular attraction among like molecules, such as the hydrogen bonding that occurs among molecules in water. Because water molecules cohere, the tension at the surface of a thin tube is transmitted downward through the water column. Water molecules at the surface are pulled down; the water molecules below them are pulled up.

Surface tension creates a pull at the surface; adhesion creates a pull at the water-container surface; cohesion transmits both forces to the water below. The result is capillarity.

Capillarity is resisted by the force of gravity and by surface tension. The interaction between cohesion and surface tension is also responsible for the formation of a concave boundary layer called a **meniscus** (plural: **menisci**). A meniscus forms at most air-water interfaces—including those found in narrow tubes. Menisci form because cohesion pulls water molecules up along the tube-water surface, while surface tension pulls the surface down in the middle.

Like root pressure, capillarity can transport only a limited amount of water. Capillary action moves water along the surfaces of mosses and other low-growing, non-vascular plants; but it can raise the water in the xylem of a vertical stem only about 1 m.

Thus, root pressure and capillary action cannot explain how water moves from soil to the top of a redwood tree—an organism that can grow 5 to 6 stories higher than the Statue of Liberty. How does it happen?

The Cohesion-Tension Theory

The leading hypothesis to explain long-distance water movement in vascular plants is the **cohesion-tension theory**, which states that water is pulled to the tops of trees along a water-potential gradient, via forces generated by transpiration at leaf surfaces.

To understand how cohesion-tension works, start with step 1 in **Figure 37.10**. Notice that spaces in the middle of the leaf are filled with moist air, as a result of evaporation from the surfaces of surrounding cells. When a stoma opens, this humid air is exposed to the atmosphere, which in most cases is much drier.

Open stomata usually create steep concentration gradients between the leaf interior and its surroundings. The steeper the gradient, the faster water vapor diffuses out through the stomata.

Step 2 shows that as water is lost from the leaf to the atmosphere, the humidity of the gas-filled space inside the leaf drops. In response, more water evaporates from the walls of the parenchyma cells, where menisci exist at the air-water interface.

If water molecules leave menisci rapidly due to a steep water-potential gradient with the atmosphere, then fewer water molecules are available at the surface than before. The water molecules that remain at the surface are pulled downward more strongly, the meniscus deepens, and the water molecules below the surface are pulled up more strongly.

In 1894 Henry Dixon and John Joly hypothesized that the formation of steep menisci produces a force capable of pulling water up from the roots, dozens or hundreds of meters into the air. Is this really possible?

THE ROLE OF SURFACE TENSION IN WATER TRANSPORT The key concept in the cohesion-tension theory is that the negative force or pull (tension) generated at the air-water interface is transmitted through the water outside of leaf cells (step 3 in Figure 37.10), to the water in xylem (step 4), to the water in the vascular tissue of roots (step 5), and finally to the water in the soil (step 6).

The continuous transmission of pulling force from the leaf surface to the root is possible because (**1**) there is water throughout the plant, and (**2**) all of the water molecules present hydrogen bonds to one another in a continuous fashion (cohesion).

Note that the plant does not expend energy to create the pulling force. The force is generated by energy from the Sun, which drives evaporation from the leaf surface. Water transport is solar-powered.

In effect, the cohesion-tension theory of water movement states that, because of the hydrogen bonding between water molecules, water is pulled up through xylem in continuous columns.

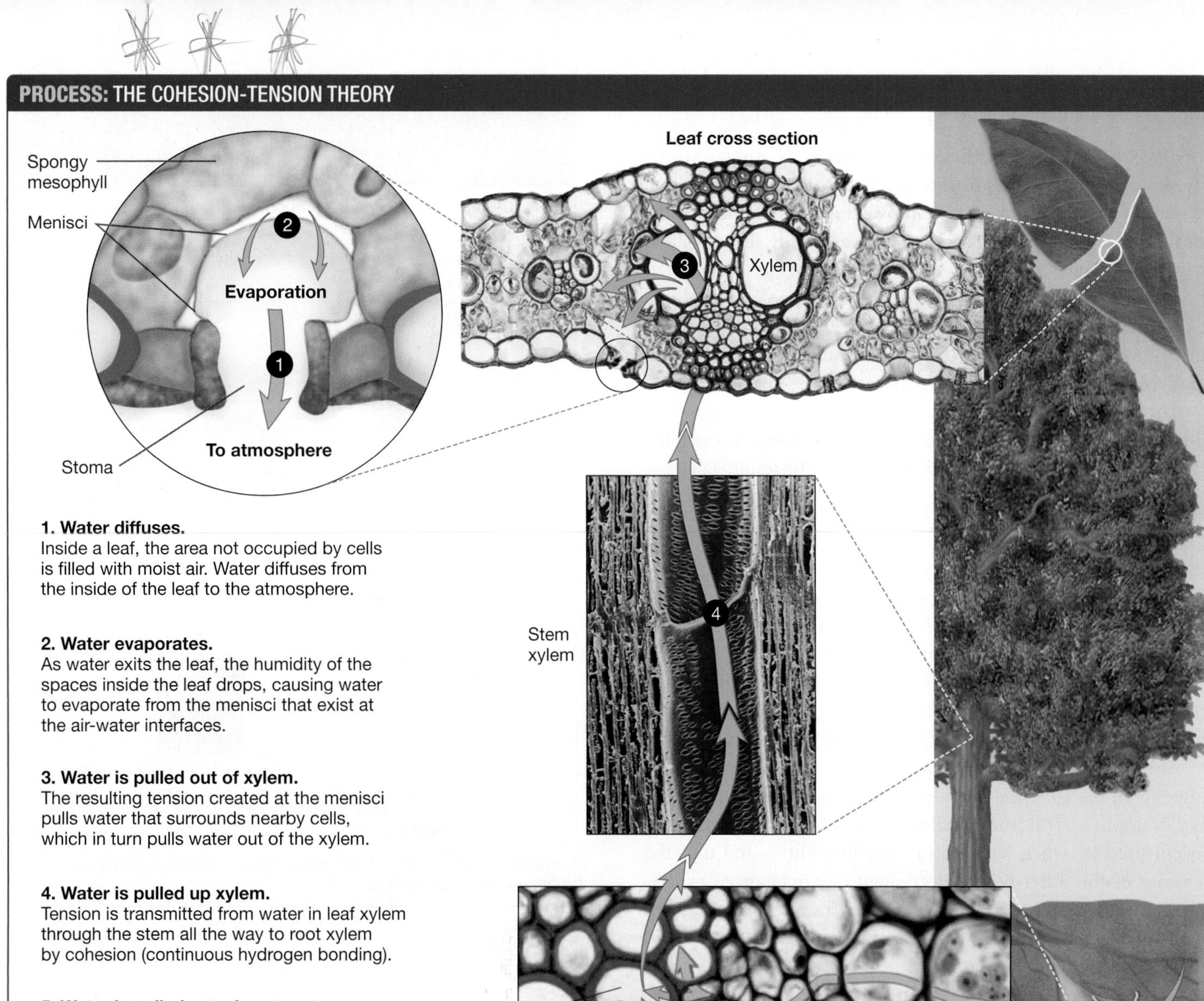

FIGURE 37.10 Transpiration Creates Tension That Is Transmitted from Leaves to Roots.

✔**EXERCISE** Using the data in Figure 37.5, label typical values for the water potential present in soil, roots, leaves, and the atmosphere.

CREATING A WATER-POTENTIAL GRADIENT You can also think about the cohesion-tension theory in terms of water potentials. Note that because tracheids and vessels are dead at maturity, the water in xylem does not cross plasma membranes. As a result, water does not move between cells by osmosis. In xylem, water movement is driven entirely by differences in pressure potential.

The pulling force generated at menisci lowers the pressure potential of water in leaves. Even though the tension created at each meniscus is relatively small, there are millions or billions of menisci in the leaves of the entire plant. The tension created by summing many small pulling forces is remarkable. It creates a water-potential gradient between leaves and roots that is steep enough to overcome the force of gravity and pull water up long distances.

To appreciate just how great the forces involved are, think of the vessel elements or tracheids in xylem as groups of straws. When you use a straw, the vacuum that you create causes liquid to rise. Sucking on a straw creates a pressure difference of about −0.1 MPa, which can draw water up a maximum of about 10 m. In contrast, the negative pressure exerted by the menisci in leaves can be as high as −2.0 MPa. The tension in xylem tissue may be ten times the amount of pressure on a fully inflated car tire. (But note that the pressure on a car tire is positive, not negative.) The force is enough to draw water up 100 m.

THE IMPORTANCE OF SECONDARY CELL WALLS If you suck on a straw hard enough, the pressure gradient between the inside of the straw and the atmosphere can overcome the stiffness of the straw and cause it to collapse. How can vascular tissue withstand negative pressures as large as −2.0 MPa without collapsing?

The answer is found in a key adaptation: The secondary thickenings characteristic of the cell walls in tracheids and vessel elements. As Chapter 36 noted, the cells in vascular tissue have walls that are reinforced with tough lignin molecules.

The evolution of lignified secondary cell walls was an important event in the evolution of land plants, because it allowed vascular tissue to withstand extremely negative pressures. The result? Tall trees.

WHAT EVIDENCE DO BIOLOGISTS HAVE FOR THE COHESION-TENSION THEORY? If the cohesion-tension theory is correct, the water present in xylem should experience a strong pulling force. A simple experiment supports this prediction.

If you find a leaf that is actively transpiring and cut its petiole, the watery fluid in the xylem, or **xylem sap**, withdraws from the edge toward the inside of the leaf (**Figure 37.11**). According to the cohesion-tension theory, this observation is due to a transpirational pull at the air-water interface in parenchyma cells of leaves.

Although the observation that xylem sap is under tension was important, the cohesion-tension theory remained controversial. For example, several studies failed to document rapid changes in xylem pressure when temperature and humidity—and thus the severity of the water-potential gradient—were changed experimentally. Advocates of the theory blamed the negative results on instruments that could not document small rapid changes in pressure potential. Who was right?

Chunfang Wei, Melvin Tyree, and Ernst Steudle answered the question in the late 1990s with an instrument called a xylem pressure probe (**Figure 37.12**). A xylem pressure probe includes an oil-

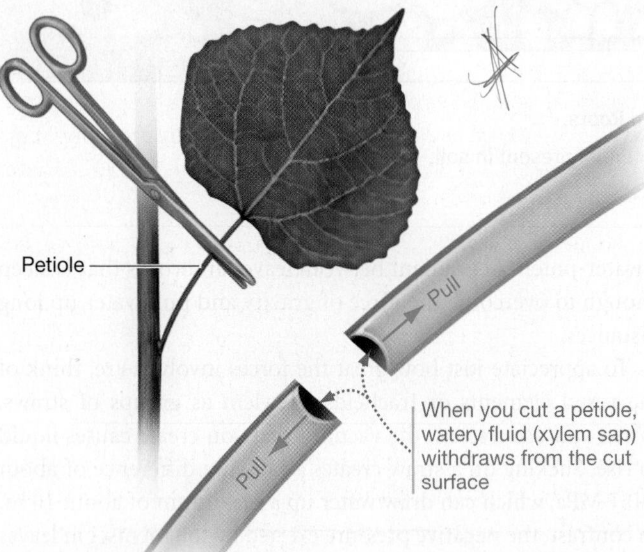

FIGURE 37.11 Xylem Sap in Cut Stems "Snaps Back."

EXPERIMENT

QUESTION: Do direct measurements of pressure in xylem tissue support the cohesion-tension theory?

HYPOTHESIS: Increasing transpiration by raising light intensity will lower xylem pressure in leaves.

NULL HYPOTHESIS: Increasing light intensity will not affect xylem pressure in leaves.

EXPERIMENTAL SETUP:

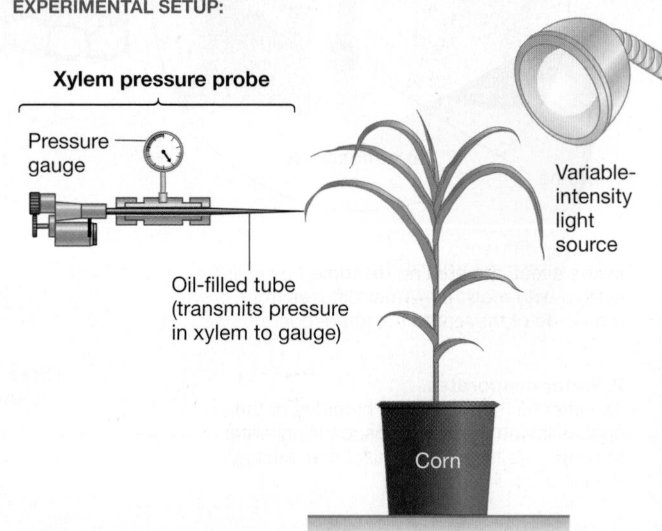

PREDICTION: Xylem pressure will decrease as transpiration increases with higher light levels.

PREDICTION OF NULL HYPOTHESIS: Xylem pressure will not be affected by light intensity.

RESULTS:

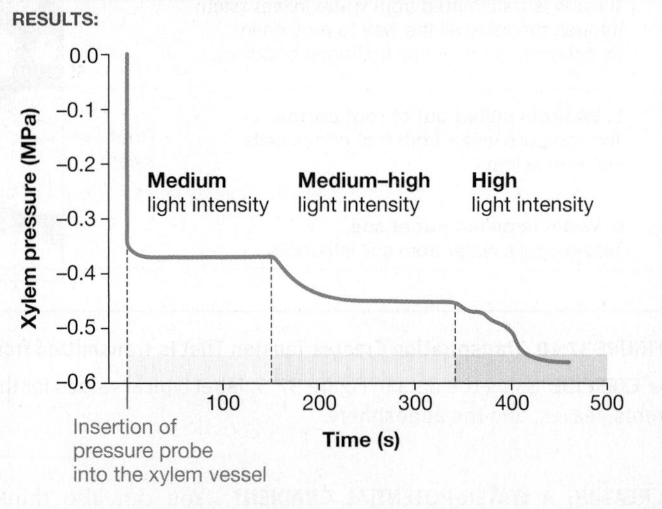

CONCLUSION: Xylem pressure decreases when light intensity increases. The data support the cohesion-tension theory.

FIGURE 37.12 Measuring Changes in Pressure inside Xylem.

SOURCE: Wei, C., M. T. Tyree, and E. Steudle. 1999. Direct measurement of xylem pressure in leaves of intact maize plants. A test of the cohesion-tension theory taking hydraulic architecture into consideration. *Plant Physiology* 121: 1191–1205.

✔**QUESTION** Suppose the researchers had chosen to plot changes in the water-potential gradient between the xylem in roots and leaves on the *y*-axis, during the same experiment. What would the graph look like?

filled glass tube that can be inserted directly into the xylem of a leaf. The oil transmits changes in pressure within the xylem to a gauge, allowing researchers to record changes in xylem pressure instantly and directly.

To test the cohesion-tension theory, the researchers altered xylem pressure by raising light levels to alter transpiration rates in the leaves of corn plants. Their results?

- The graph in Figure 37.12 plots how xylem pressure changed as the researchers changed light levels, over a 7-minute period. Note that as light intensity increased, the xylem pressure probe documented increased tension, or pull—negative pressure.
- In addition, higher light levels reduced the weight of the entire plant—suggesting that higher transpiration rates caused water loss.

Both observations are consistent with predictions that follow from the cohesion-tension theory.

These experiments convinced most biologists that increased transpiration leads to increased tension on xylem sap. Rising tension, in turn, lowers the water potential of leaves and exerts a pull on water in the roots and soil, where the water potential is high. On the basis of these and other results, most biologists now accept the cohesion-tension theory.

To review how water moves from soil to root cells and from root to shoots, go to the study area at *www.masteringbiology.com.*

BioFlix™ Water Transport in Plants

CHECK YOUR UNDERSTANDING

If you understand that . . .

- Water can move a short distance in xylem via root pressure or capillarity.
- Long-distance transport of water depends on movement along a steep water-potential gradient. This gradient is created primarily by the negative pressure potential of water in leaves, due to surface tension that develops in response to transpiration.

✓ You should be able to . . .

1. Explain how asymmetrical hydrogen bonding and transpiration at the surface of a meniscus near a stoma creates a pull on the water in leaves and xylem.
2. Predict what happens to the meniscus when each of the following occurs: a nearby stoma closes, a rain shower starts, and weather changes and dry air blows in.

Answers are available in Appendix B.

37.3 Water Absorption and Water Loss

One of the most important features of cohesion-tension is that it does not require plants to expend energy. Instead, the Sun furnishes the energy required to pull water from roots to shoots—not ATP supplied by the plant.

1. Energy from the Sun heats water molecules at the air-water interface inside leaves enough to break the hydrogen bonds between them and cause transpiration.
2. Rapid transpiration creates deep menisci in the walls of leaf cells, causing tension that lowers the water potential of leaves.
3. Hydrogen bonding between water molecules transmits this tension down to water molecules in the root and soil.

Xylem acts as a passive conduit—a set of pipes that allows water to move from a region of high water potential (the soil) to a region of low water potential (the leaves). Water flows from roots to shoots as long as the water-potential gradient—from soil to root to leaf to atmosphere—is intact.

When soils begin to dry, however, it becomes difficult for plants to replace water being lost via transpiration. If water is not replaced fast enough, the solute potentials of leaves drop and leaves and branches begin to wilt. In response, stomata may close down partially or completely to reduce transpiration rates and conserve water. However, closing stomata affect the ability of plants to carry on photosynthesis, because CO_2 acquisition slows or stops.

The balance between conserving water and maximizing photosynthesis is termed the photosynthesis-transpiration compromise. This compromise is particularly delicate for species that grow on dry sites. How do they cope?

Limiting Water Loss

Plants that thrive in dry sites have several adaptations that help them slow transpiration and limit water loss. Consider the oleander plant, which is native to the dry shrub-grassland habitats of southern Eurasia. The micrograph in **Figure 37.13** on page 728 shows a cross section through an oleander leaf, which has the following special features:

- A particularly thick cuticle covers the upper surface of oleander leaves. This waxy layer minimizes water loss from cells that are directly exposed to sunlight. In general, species that are adapted to dry soils have much thicker cuticles than do species adapted to wet soils.
- The epidermis of most plants is a single cell layer thick, but the epidermis in oleanders is several cell layers deep. The thick epidermis of species found in dry environments is thought to reduce water loss from parenchyma cells in the leaf, where most photosynthesis takes place.
- The stomata of oleanders are located on the undersides of their leaves, inside deep pits in the epidermis. Hairlike extensions of epidermal cells called trichomes (see Section 36.3), shield these pits from the atmosphere. The leading hypothesis to explain these traits is that they slow the loss of water vapor from stomata to the dry air surrounding the leaf.

Other species have other adaptations for limiting water loss. For example, recall from Chapter 36 that many species adapted to water-short habitats—either cold environments where water is often frozen or deserts where rainfall is rare—have needlelike leaves. Long, thin leaf shapes minimize the surface area exposed

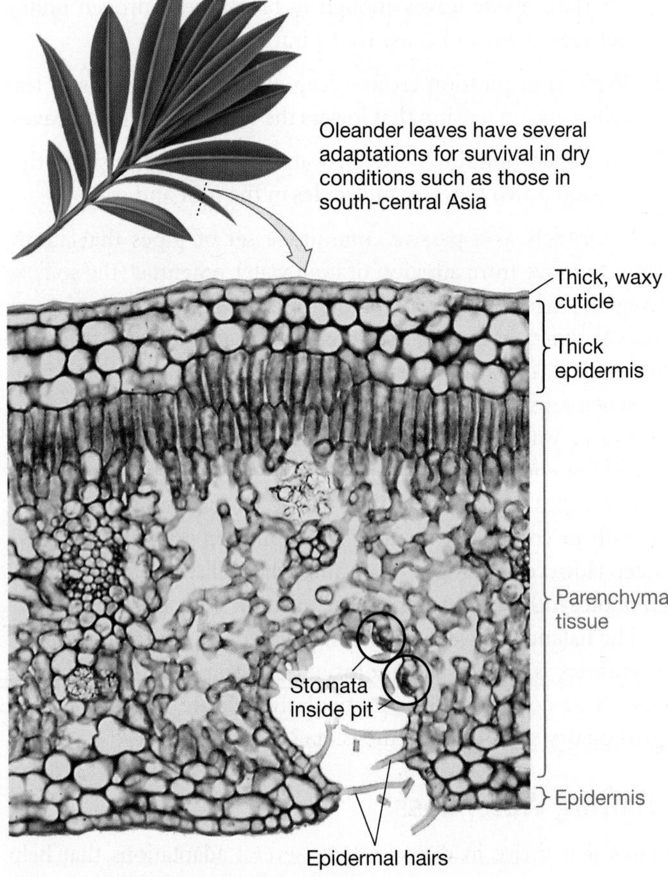

FIGURE 37.13 Species That Are Adapted to Dry Habitats Have Modified Leaf Structures.

to sunlight and thus minimize transpiration. Plants adapted to wetter habitats, in contrast, tend to have broad leaves with a large surface area.

Obtaining Carbon Dioxide under Water Stress

Many of the species that thrive in deserts and other hot, dry habitats can continue photosynthesizing even when soil moisture content is low. Recall from Chapter 10 that two novel biochemical pathways, **crassulacean acid metabolism** (**CAM**) and **C_4 photosynthesis**, allow plants to increase CO_2 concentrations in their leaves and conserve water.

CAM plants open their stomata at night and store the CO_2 that diffuses into their tissues by adding the carbon dioxide molecules to organic acids. When sunlight is available during the day and photosynthesis begins, the CO_2 molecules are released from the organic acids and transferred to **rubisco**—the enzyme that initiates the Calvin cycle. In this way, CAM plants can photosynthesize and grow even with their stomata closed during the day.

C_4 plants minimize the extent to which their stomata open because they use CO_2 so efficiently. Mesophyll cells in C_4 plants take up CO_2 and add it to organic acids. The CO_2 is then transferred to specialized cells called **bundle-sheath cells** (see Chapter 10), where rubisco is abundant. In effect, the C_4 pathway is a mechanism for concentrating carbon dioxide in cells deep inside the leaf, so stomata do not have to be wide open continuously.

Like the cuticle and stomata-containing pits of oleanders, CAM and C_4 photosynthesis are adaptations that help plants conserve water by limiting transpiration.

37.4 Translocation

Translocation is the movement of sugars throughout a plant—specifically, from sources to sinks. In vascular plants, a **source** is a tissue where sugar enters the phloem; a **sink** is a tissue where sugar exits the phloem. Sources contain a high concentration of sugar; sinks have a low concentration of sugar.

Where do sources and sinks occur in a plant? The answer often depends on the time of year.

- *During the growing season* Mature leaves and stems that are actively photosynthesizing produce sugar in excess of their own needs. These tissues act as sources. Sugar moves from leaves and stems to a variety of sinks, where sugar use is high and production is low. Apical meristems, lateral meristems, developing leaves, flowers, developing seeds and fruits, and storage cells in roots all act as sinks (**Figure 37.14**).
- *Early in the growing season* When a plant resumes growth after the winter or the dry season, sugars move from storage areas to growing areas. Storage cells in roots act as sources; developing leaves act as sinks.

Tracing Connections between Sources and Sinks

To explore the relationship between sources and sinks in more detail, consider research on sugar-beet plants that were exposed to carbon dioxide molecules containing the radioactive isotope ^{14}C. The goal was to track where carbon atoms moved after they were incorporated into sugars via photosynthesis.

The location of ^{14}C atoms inside a plant can be documented in two ways: **(1)** by measuring the number of radioactive emis-

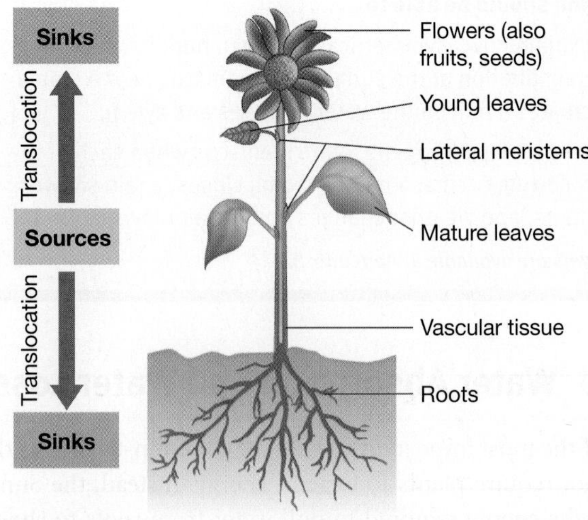

FIGURE 37.14 Sugars Move from Sources to Sinks.

sions emanating from different tissues, or (2) by laying plant parts on X-ray film and allowing the radioactivity to expose and blacken the film.

Researchers have enclosed individual leaves of intact plants in a bag and introduced a fixed amount of radioactive CO_2 for a fixed amount of time. One typical experiment documented that mature leaves retained just over 9 percent of the labeled carbon. In contrast, growing leaves retained 67 percent. These data are consistent with the prediction that fully expanded leaves act as sources of sugar, while actively growing leaves and roots act as sinks.

In similar experiments, researchers have exposed all the leaves on a growing plant to labeled carbon. One experiment like this found that over 16 percent of the total carbon was translocated to root tissue within 3 hours. This result is consistent with the prediction that, during the growing season, roots also act as sinks.

Similar experiments support two generalizations.

1. Sugars can be translocated rapidly—typically 50–100 cm/hr.
2. There is a strong correspondence between the physical locations of sources and sinks.

The second point is particularly interesting. For example, mature leaves that act as sources send sugar to tissues on the same side of the plant (**Figure 37.15a**). In addition, experiments with tall herbaceous plants show that leaves on the upper part of the stem send sugar to apical meristems, but leaves on the lower part of the plant send sugar to the roots (**Figure 37.15b**).

Why would leaves send sugar to tissues on a certain side or part of the body? The answer hinges on understanding the structure of phloem.

The Anatomy of Phloem

Chapter 36 introduced the two specialized parenchyma cell types that make up phloem: **sieve-tube members** and **companion cells**. Unlike the tracheids and vessel elements that make up most of the xylem, sieve-tube members and companion cells are alive at maturity.

Recall that, in most plants, sieve-tube members lack nuclei and many major organelles. They are connected to one another, end to end, by perforated **sieve plates** (**Figure 37.16**). The pores

(a) Source leaves send sugar to the same side of the plant.

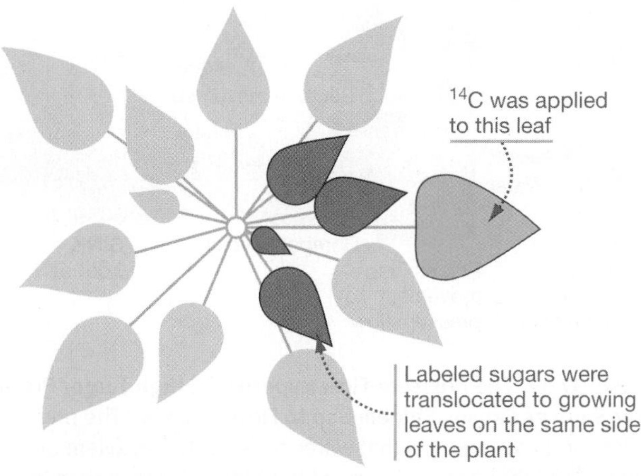

(b) Source leaves send sugar to tissues on the same end of the plant.

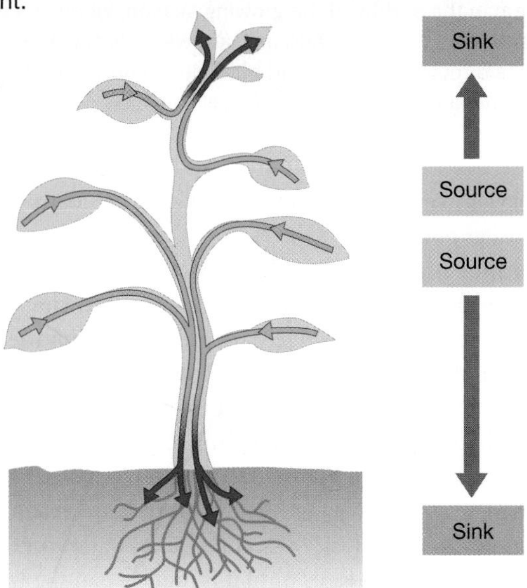

FIGURE 37.15 Sources Supply Sinks on the Same Side and Same End of the Body.

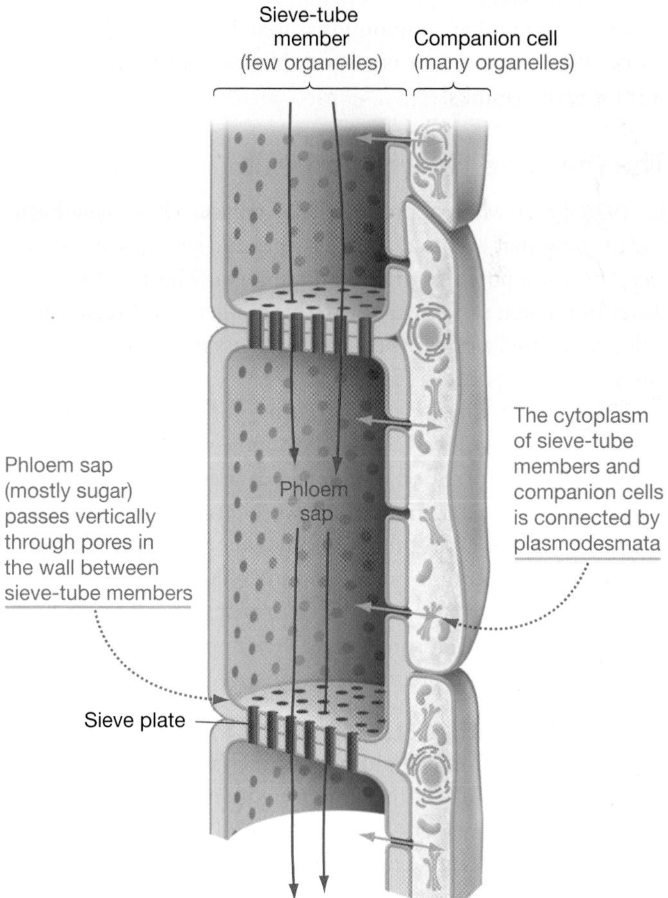

FIGURE 37.16 Sieve-Tube Members Are Connected by Pores.

create a direct connection between the cytoplasms of adjacent cells. Companion cells, in contrast, have nuclei and a rich assortment of ribosomes, mitochondria, and other organelles. Companion cells function as "support staff" for sieve-tube members.

You might also recall from Chapter 36 that secondary phloem is part of a tree's bark. In addition, it's important to recognize that the phloem in secondary vascular tissue is continuous throughout the plant—meaning that there is a direct anatomical connection to the phloem in trunks, stems, branches, and roots. The sieve-tube members in bark phloem represent a continuous system for transporting sugar throughout the plant body.

As Chapter 36 noted, however, phloem is restricted to discrete vascular bundles in tissues that do not form wood. Each vascular bundle runs the length of stems and roots, and certain bundles extend into specific branches, leaves, and lateral roots. In primary vascular tissue, phloem sap does not move from one vascular bundle to another—instead, each bundle is independent.

Based on these results, the physical relationships observed between sources and sinks in herbaceous plants are logical. For example, the phloem in the leaves on one side of an herbaceous plant connects directly with the phloem of branches, stems, and roots on the same side of the individual, through a specific set of vascular bundles.

The phloem sap that flows through vascular tissue is often dominated by the disaccharide sucrose—table sugar. Phloem sap can contain small amounts of minerals, amino acids, mRNAs, hormones, and other compounds as well. How does this solution move? What mechanism is responsible for translocating sugars from sources to sinks?

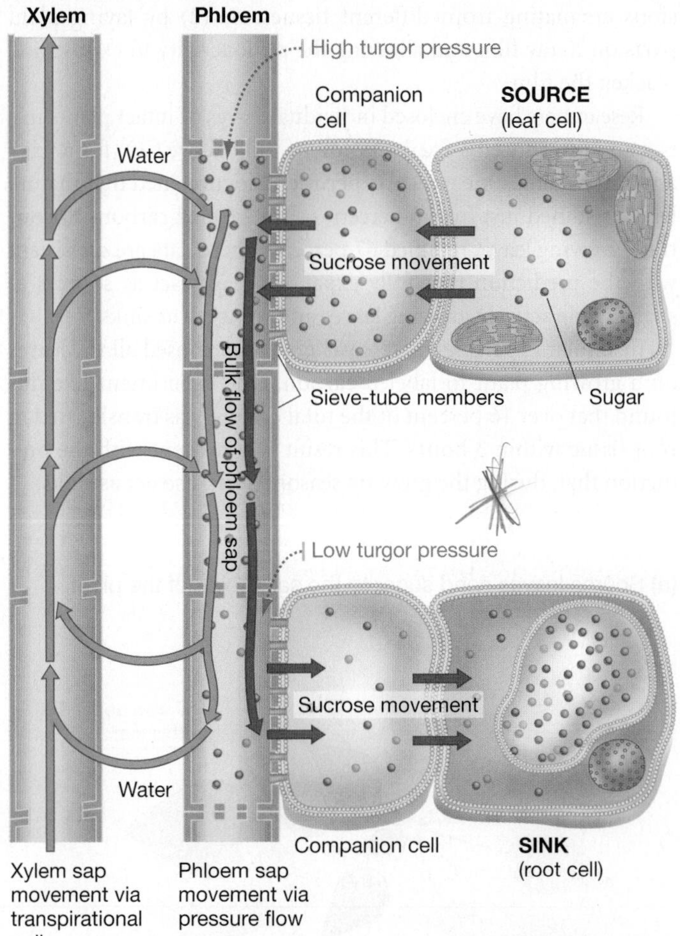

FIGURE 37.17 The Pressure-Flow Hypothesis: High Turgor Pressure Near Sources Causes Phloem Sap to Flow to Sinks. The pressure-flow hypothesis predicts that water cycles between xylem and phloem and that water movement in phloem is a response to a gradient in pressure potential.

✔**EXERCISE** This diagram shows how water moves between xylem and phloem in the middle of the growing season, when leaves are sources and roots are sinks. Add new arrows, in new colors, to indicate the direction of water and phloem sap flow in spring, when roots act as sources and leaves act as sinks.

The Pressure-Flow Hypothesis

In 1926 Ernst Münch proposed the **pressure-flow hypothesis**, which states that events at source tissues and sink tissues create a steep pressure potential gradient in phloem (**Figure 37.17**). The water in phloem sap moves down this gradient, and sugar molecules are carried along by **bulk flow**—a mass movement of molecules along a pressure gradient.

Like the cohesion-tension theory for water transport, the pressure-flow hypothesis is based on movement along a water-potential gradient created by changes in pressure potential. Unlike the cohesion-tension model, however, transpiration does not provide the driving force to move phloem sap. Instead, large differences between turgor pressure in the phloem near source tissues and turgor pressure in the phloem near sink tissues generate the necessary force. In some cases, creating these differences in turgor pressure requires an expenditure of ATP.

CREATING HIGH PRESSURE NEAR SOURCES AND LOW PRESSURE NEAR SINKS To understand how Münch's model works, start with the source cell at the upper right in Figure 37.17. The small red arrows reflect Münch's proposal that sucrose moves from source cells into companion cells and from there into sieve-tube members.

Because of this phloem loading, the phloem sap near the source has a high concentration of sucrose. Compared to the water in the adjacent xylem cells in a vascular bundle, the phloem sap has a very low water potential.

As the blue arrows in the upper part of the diagram show, water moves along a water-potential gradient—flowing passively from xylem across the selectively permeable plasma membrane of sieve-tube members. In response, pressure begins to build in the sieve-tube members nearest the source region.

What is happening at the sink? Münch proposed that cells in the sink (bottom right in Figure 37.17) remove sucrose from the phloem sap by passive or active transport. As a result of this phloem unloading—a loss of solutes—the water potential in sieve-tube members increases until it is higher than the water potential in adjacent xylem cells. As the blue arrows at the bottom of the figure show, water flows across the selectively permeable membranes of sieve-tube members into xylem along a water-potential gradient. In response, turgor pressure in the sieve-tube members near the sink drops.

The net result of these events is high turgor pressure in phloem near the source and low turgor pressure in phloem near the sink, created by the loading and unloading of sugars. This difference in pressure potential drives phloem sap from source to sink via bulk flow. There is a one-way flow of sucrose and a continuous loop of water movement, with water being supplied to and from the xylem.

TESTING THE PRESSURE-FLOW MODEL The pressure-flow hypothesis is logical, given the anatomy of vascular tissue and the principles that govern water movement. But has any experimental work supported the theory? Some of the best tests have relied on aphids—small insects that make their living ingesting phloem sap.

Aphids insert a syringe-like mouthpart, called a stylet, into sieve-tube members. The pressure on the fluid in these cells forces it through the stylet, into the aphid's digestive tract, and out its anus as droplets of "honeydew" (**Figure 37.18**).

If the aphids are then severed from their stylets, sap continues to flow out through the stylets. This phenomenon allows researchers to collect phloem sap efficiently for analysis. It also confirms that the aphids do not actively suck the fluid. As predicted by the pressure-flow model, phloem is indeed under pressure.

This observation supports one of the fundamental predictions of the pressure-flow hypothesis. Now the question is, How does sucrose enter and leave phloem in a way that sets up the water-potential gradient?

Phloem Loading

In contrast to the cohesion-tension model of water movement in xylem, pressure flow often requires that plants expend energy to set up a water-potential gradient in phloem. To establish a high pressure potential in sieve-tube members near source cells, large amounts of sugar have to be transported into the phloem sap—enough to raise the solute concentration of sieve-tube members. This requirement is illustrated in Figure 37.17, top right.

In some cases, loading is active—it requires an expenditure of ATP and some sort of membrane transport system. But when sucrose concentrations in source cells are extremely high, movement into sieve-tube members can also occur via passive diffusion through plasmodesmata.

Conversely, sugar must sometimes be unloaded against its concentration gradient at sinks. When active unloading occurs, it requires an expenditure of ATP and a second membrane transport mechanism. How do phloem loading and unloading occur? What specific membrane proteins are involved?

To answer these questions, let's start by reviewing how transport proteins make it possible for sugars and other large or charged substances to cross a phosholipid bilayer.

HOW ARE SUCROSE AND OTHER SOLUTES TRANSPORTED ACROSS MEMBRANES? **Passive transport** (**Figure 37.19**) occurs when ions or molecules move across a plasma membrane by diffusion—that is, with their electrochemical gradient. The adjective passive is appropriate because no expenditure of energy is required for the movement to occur.

Recall from Chapter 6 that small, nonpolar molecules diffuse across phospholipid bilayers rapidly (Figure 37.19, left). But ions and many large molecules diffuse across phospholipid bilayers slowly if at all, even when their movement is favored by a strong electrochemical gradient. To diffuse rapidly, they must avoid direct contact with the phospholipid bilayer by passing through a membrane protein.

As Chapter 6 explained, two types of membrane protein—channels and carriers—permit the passive diffusion of specific ions or molecules.

- **Channels** form pores that selectively admit certain ions (Figure 37.19, middle).
- **Carriers** work like enzymes, undergoing a conformational change that transports a bound substrate across the lipid bilayer (Figure 37.19, right).

Channels and carriers are responsible for **facilitated diffusion**.

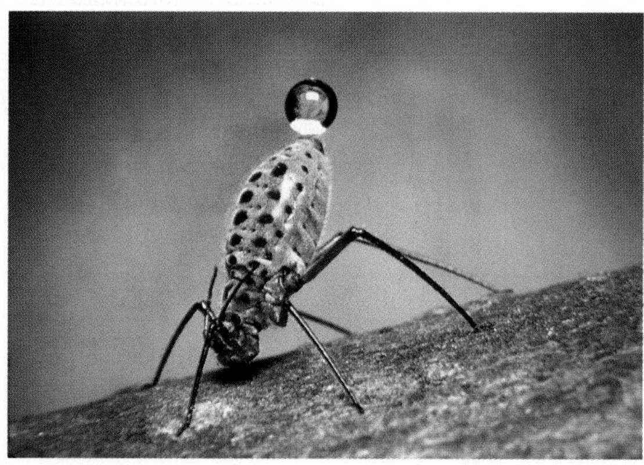

FIGURE 37.18 Aphids Feed on Phloem Sap. The tip of this aphid's mouthpart (the stylet) is in a sieve-tube member within the plant stem. The droplet emerging from the aphid's anus is honeydew, which consists of sugary phloem sap.

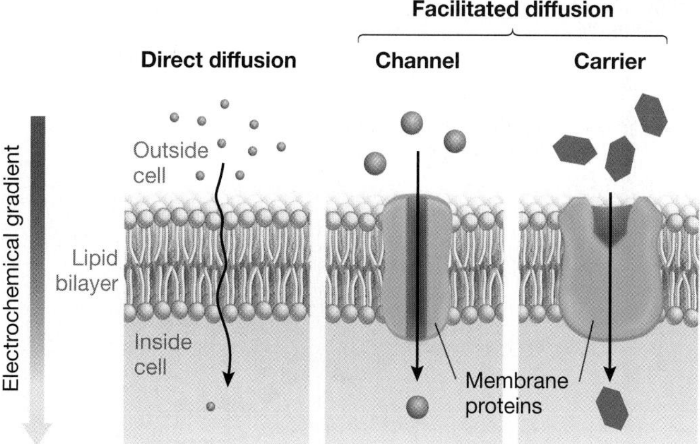

FIGURE 37.19 Passive Transport Is Based on Diffusion. In passive transport, ions or molecules diffuse across membranes—meaning they follow their electrochemical gradient. The movement can occur directly through the phospholipid bilayer or be facilitated by a channel or carrier protein.

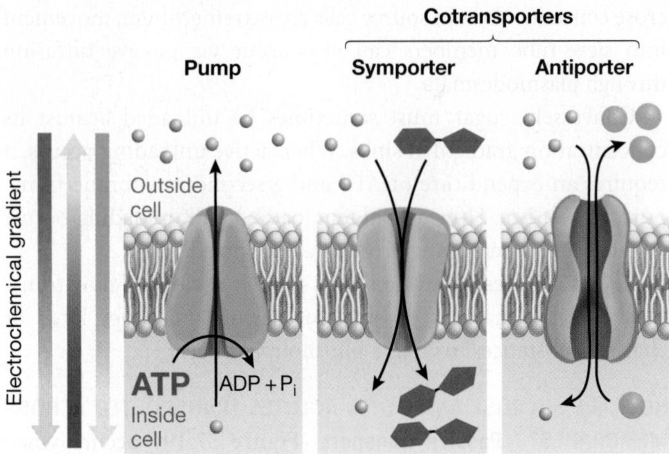

FIGURE 37.20 Active Transport Moves Ions or Molecules against an Electrochemical Gradient. All forms of active transport require an expenditure of ATP. Pumps use ATP directly; cotransporters use ATP indirectly. Cotransport depends on a previous expenditure of ATP by a pump.

Active transport (**Figure 37.20**) occurs when ions or molecules move across a plasma membrane against their electrochemical gradient. The adjective active is appropriate because cells must expend energy in the form of ATP to move solutes in an energetically unfavorable direction.

Active transport always involves membrane proteins. **Pumps** are proteins that change shape when they bind ATP or a phosphate group from ATP. As they move, pumps transport ions or molecules against an electrochemical gradient.

Pumps can establish an electrochemical gradient that favors movement of an ion or molecule across the plasma membrane. For example, the pump on the left side of Figure 37.19b has established an electrochemical gradient for bringing the ions or molecules symbolized by the light gray balls into the cell.

In many cases, the electrochemical gradients established by pumps are used to transport other molecules or ions by two types of membrane proteins called **cotransporters.**

- **Symporters** transport solutes *against* a concentration gradient, using the energy released when a different solute moves in the same direction *down* its electrochemical gradient. The red molecules in the middle of Figure 37.20 are moving through a symporter.
- **Antiporters** work in a similar way, except that the solute being transported against its concentration gradient moves in the direction *opposite* that of the solute moving down its concentration gradient. The orange ions in Figure 37.20, right, are moving through an antiporter.

When solutes or ions move through cotransporters, **secondary active transport** occurs (see Chapter 6).

Active transport, secondary active transport, and passive transport are all involved in moving sugars around plants. Let's look first at events at source tissues, where the active transport of sucrose into sieve-tube members results in a high pressure potential and high water potential. The chapter concludes with a look at how the same molecules are unloaded to maintain low water potentials at sinks.

HOW ARE SUGARS CONCENTRATED IN SIEVE-TUBE MEMBERS AT SOURCES? Because sucrose may be more highly concentrated in companion cells than in photosynthetic cells where it is produced, researchers hypothesized that sucrose transport from source cells into companion cells may be active. Another key observation—that strong pH differences exist between the interior and exterior of phloem cells—suggested that sucrose might enter companion cells with protons.

Figure 37.21 explains the logic behind this hypothesis. Note a key claim: that a membrane protein in companion cells hy-

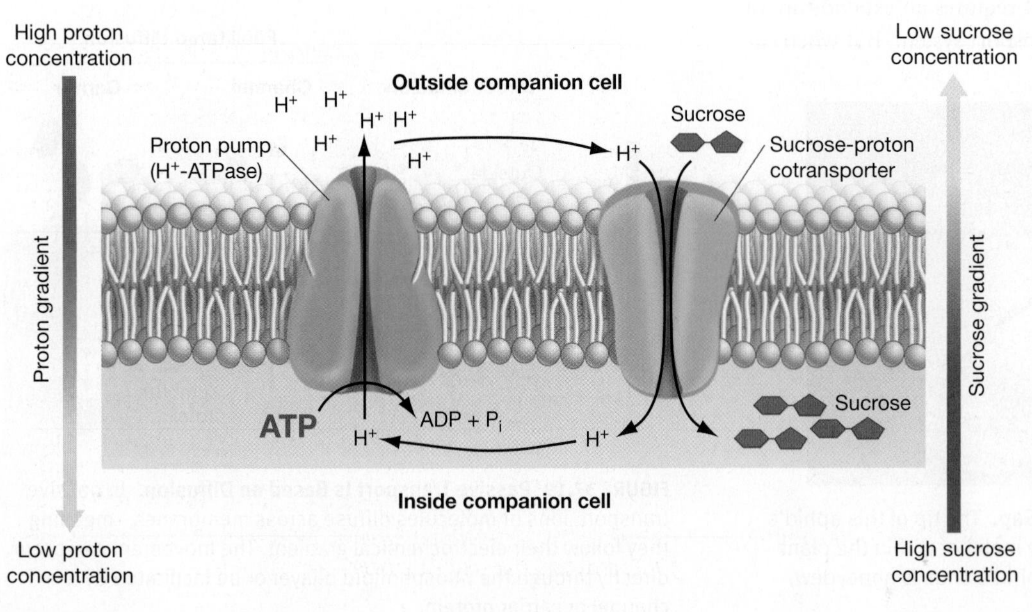

FIGURE 37.21 A Model for Cotransport of Protons and Sucrose. According to the model of cotransport, a proton pump hydrolyzes ATP to move hydrogen ions to the exterior of the cell. The resulting high concentration of H$^+$ establishes an electrochemical gradient that allows the transport of sucrose into the cell against its concentration gradient.

drolyzes ATP and uses the energy that is released to transport protons (H^+) across the membrane to the exterior of the cell. Proteins like these are called **proton pumps**, or more formally, **H^+-ATPases.**

Proton pumps establish a large difference in charge and in hydrogen ion concentration on the two sides of the membrane. The resulting electrochemical gradient favors the entry of protons into the cell.

The right side of the figure shows the second key claim: A membrane protein called a cotransporter acts as a conduit for protons and sucrose to enter the cell together. Protons move along their electrochemical gradient; sucrose moves against its concentration gradient.

If phloem loading depends on the activity of a proton pump, researchers should be able to find and characterize the pump proteins.

WHERE ARE H^+-ATPases LOCATED? Proton pumps are found in the plasma membranes of a wide variety of organisms, including bacteria, fungi, and animals. To analyze the proton pumps in plants, researchers from several laboratories focused on a small member of the mustard family called *Arabidopsis thaliana.*

Researchers determined the amino acid sequence of a proton pump purified from the plasma membranes of *Arabidopsis* and used these data to infer the DNA sequence of the corresponding gene. They found that the *Arabidopsis* genome actually codes for 10 different proton pump proteins. One of these genes, *AHA3*, appeared to be expressed primarily in vascular tissues.

These observations led Natalie DeWitt and Michael Sussman to hypothesize that *AHA3* encodes the proton pump responsible for phloem loading. To test this hypothesis, they produced antibodies to the AHA3 protein. You might recall from earlier chapters that an antibody is a protein that binds to a specific location on a molecule—typically, a specific protein.

To track the antibody, the investigators attached gold particles to it. When viewed with the electron microscope, a gold particle looks like a black dot. DeWitt and Sussman's goal was to treat *Arabidopsis* leaves with the AHA3 antibody, examine treated leaves under the electron microscope, and determine where the pump proteins are located.

As the bar graphs in **Figure 37.22** show, the proton pumps responsible for phloem loading were found almost exclusively in the plasma membranes of companion cells. This result supported the following model for phloem loading:

1. Proton pumps in the membranes of companion cells create a strong gradient that favors a flow of protons into companion cells.
2. A cotransporter protein in the membranes of companion cells uses the proton gradient to bring sucrose into companion cells from the surrounding cell walls and intercellular spaces around source cells.
3. Once inside companion cells, sucrose travels into sieve-tube members via plasmodesmata.

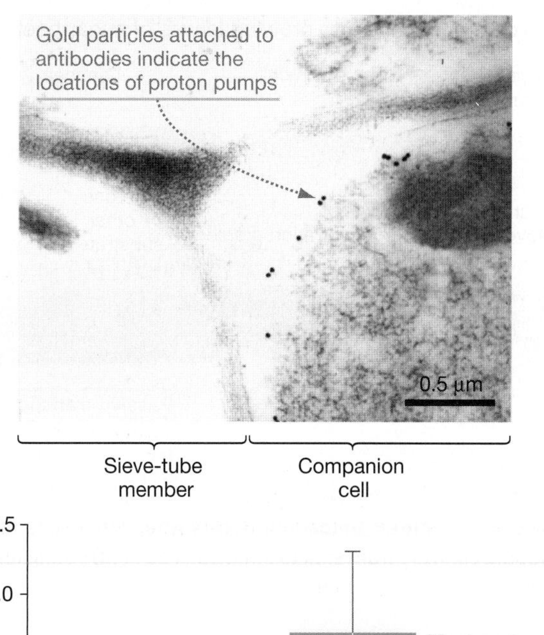

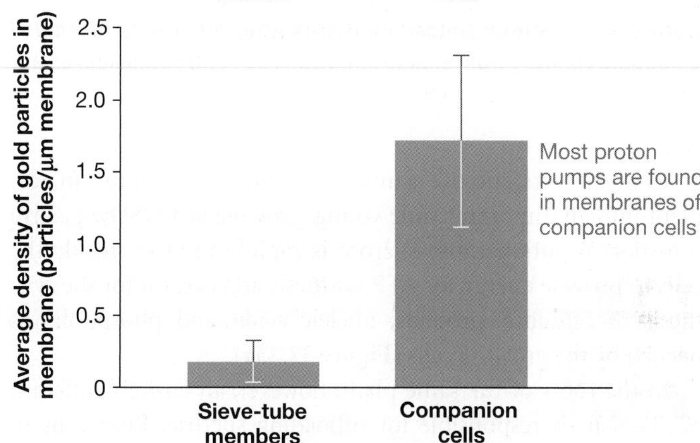

FIGURE 37.22 The Membranes of Companion Cells Contain H^+-ATPases. The micrograph shows that the proton pumps responsible for phloem loading are located in the plasma membranes of companion cells. The histogram compares the average density of proton pumps in the membranes of companion cells versus those of sieve-tube members. These pumps create a proton gradient that allows sucrose to be transported into companion cells against a concentration gradient.

Although work on the mechanism of phloem loading in *Arabidopsis* and other species continues, most researchers are convinced that proton pumps and proton-sucrose cotransporters play a key role. Now, once sucrose has been loaded into sieve-tube members near sources and follows a water-potential gradient to sinks, how is it unloaded?

Phloem Unloading

The membrane proteins that are involved in transporting sucrose molecules, and the mechanism of movement, vary among different types of sinks within the same plant. Mechanisms for phloem unloading also vary among different species.

To appreciate this diversity, consider how sucrose is unloaded in the phloem of sugar beets—a crop grown for the storage tissues in its root, which are a major source of the granulated and powdered sucrose sold in grocery stores.

(a) Phloem unloading into growing leaves of sugar beets

Sieve-tube member
Companion cell
Leaf cell
Sucrose
Movement along concentration gradient
Sucrose + other substrate + ATP
Proteins and nucleic acids
Passive transport across membrane, then use of ATP within cell indirectly

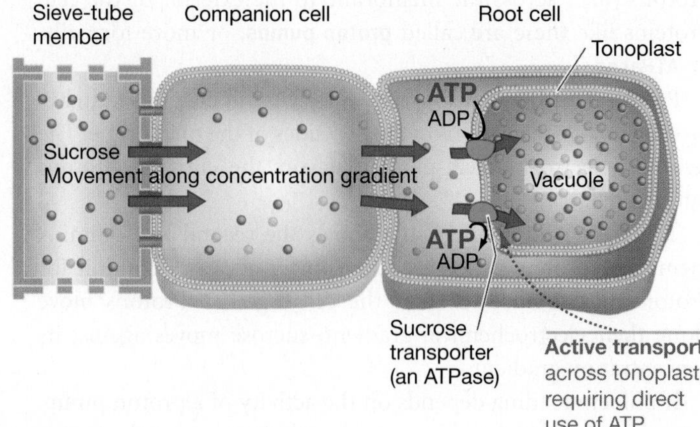

FIGURE 37.23 Phloem Unloading Occurs when Sucrose Is Taken Up by Cells. The mechanism of phloem unloading can vary from sink to sink, such as in **(a)** young leaves and **(b)** roots of the same plant.

In sugar beets, sucrose is unloaded along a concentration gradient into an important sink: young, growing leaves. The passive transport occurs because sucrose is rapidly used up inside the cells to provide energy for ATP synthesis and carbon for the synthesis of cellulose, proteins, nucleic acids, and phospholipids needed by the growing cells (**Figure 37.23a**).

In the roots of the same plant, however, an entirely different mechanism is responsible for offloading sucrose. Root cells in this species have a large vacuole that stores sucrose. The membrane surrounding this organelle is called the **tonoplast**. It contains a protein that hydrolyzes ATP and uses the energy released to transport sucrose into the vacuole, against its concentration gradient (**Figure 37.23b**). The active transport of sucrose into the vacuole allows sucrose to move passively from phloem into the storage cells, keeping the water potential of the phloem sap near the sink low—as required by the Münch pressure-flow model.

To summarize, more than seven decades of research provide convincing evidence that the pressure-flow hypothesis is fundamentally correct. To see the pressure-flow mechanism in action and review the cohesion-tension theory for water movement in xylem, go to the study area at *www.masteringbiology.com*.

(MB) **Web Activity** Solute Transport in Plants

The cohesion-tension theory for water movement and the pressure-flow model for phloem sap movement represent major advances in our understanding of how plants work.

CHECK YOUR UNDERSTANDING

If you understand that . . .

- Phloem sap moves from areas of high water potential to areas of low water potential.
- In phloem, high water potential is due to the high turgor pressure observed in the sieve-tube members near source cells. This pressure is created by pumps that actively load sucrose into companion cells against a concentration gradient. Water follows by osmosis, creating high turgor pressure inside sieve tubes.
- At sinks, turgor pressure is much lower than it is at sources because storage cells or growing tissues remove sucrose from phloem sap. The decreased concentration of solutes in phloem causes water to leave phloem and enter xylem.

✔ **You should be able to . . .**

Predict the water potential of phloem sap near leaves and near roots in a deciduous tree under the following conditions:

1. at the start of growth in spring
2. at midday when growing conditions are ideal in summer

Answers are available in Appendix B.

CHAPTER 37 REVIEW

For media, go to the study area at www.masteringbiology.com 

Summary of Key Concepts

Water moves from areas of high water potential to areas of low water potential. Water's potential energy in plants is a combination of (1) its tendency to move in response to differences in solute concentration and (2) the pressure exerted on it.

- Plants lose water as an inevitable side effect of exchanging gases with the atmosphere. The flow of water from soil to air via plant tissues follows a water-potential gradient.
- Water potential (ψ) is a measure of the tendency of water to move down its potential energy gradient.
- In plants, water potential has two components: (1) a solute potential, formed by the concentration of solutes in a cell or tissue; and (2) a pressure potential, provided by the cell wall and other factors. The water potential of a cell, tissue, or plant is the sum of its solute potential and pressure potential.
- When selectively permeable membranes are present, water moves by osmosis from areas of high potential to areas of low potential.
- When no membranes are present, water moves by bulk flow from areas of high pressure to areas of low pressure, independently of differences in solute potential.

✔ You should be able to explain why water in the xylem of a root moves up through the shoot in response only to a pressure gradient—not a solute gradient.

Plants lose water to transpiration when stomata are open and photosynthesis is occurring, but they do not expend energy to replace it. Instead, water moves from soil and roots to leaves along a water-potential gradient. Evaporation of water from leaves, driven by the Sun, creates a negative pressure (tension) that pulls water up.

- According to the cohesion-tension theory, water is pulled in one continuous column from the soil to roots to shoots, against the force of gravity, by the surface tension caused by transpiration from leaves.
- Surface tension occurs at menisci that form as water evaporates from the walls of leaf cells, and is transmitted downward via hydrogen bonding between water molecules. In this way, the energy in sunlight is responsible for the movement of water from roots to shoots.
- Plants that occupy dry habitats have traits that limit the amount of water they lose to transpiration. In some species, stomata are located in pits on the undersides of their leaves, where humidity is higher. The CAM and C_4 photosynthetic pathways are adaptations that limit water loss and photorespiration in dry habitats.

✔ You should be able to explain how the pressure potential of water in a leaf, and thus the rate of water movement up a stem, changes when the Sun goes behind a cloud.

MB **BioFlix™** Water Transport in Plants

In phloem, sugars are transported from "sources"—tissues that release sugars for use elsewhere—to "sinks"—tissues in which sugars are being used or stored. Movement occurs because large amounts of sucrose move into phloem cells near source tissues. Water follows by osmosis, creating a pressure gradient that favors the movement of water and sucrose to sinks.

- Translocation is the movement of sucrose and other products through the plant.
- According to the Münch pressure-flow model, sugars move from sources to sinks via bulk flow along a pressure gradient that develops in phloem.
- A pressure gradient is generated by the transport of sugars into sieve-tube members in source tissues, coupled with the transport of sucrose out of sieve-tube members at sink tissues. Water moves from xylem into sieve-tube members near sources and cycles back to xylem near sinks.

✔ You should be able to explain why the pressure-flow model would not work if xylem and phloem were not bundled together.

MB **Web Activity** Solute Transport in Plants

Questions

✔ TEST YOUR KNOWLEDGE

Answers are available in Appendix B

1. Under what conditions does the rate of transpiration increase?
   a. in species in which stomata are located in pits on the bottom of leaves
   b. when stomata close at night
   c. during rainstorms, when atmospheric pressure is low
   d. when the weather changes and air becomes drier
2. Which of the following does *not* affect the pressure potential of water?
   a. High sucrose concentrations in companion cells create turgor pressure.
   b. Menisci around spongy mesophyll cells create tension when transpiration rates are high.
   c. Few ions or other solutes are present in soil water.
   d. Water tends to adhere tightly to soil particles when soils dry.
3. The cells of a certain plant species can tolerate extremely high solute potentials. Which of the following statements is correct?
   a. The plant's transpiration rates will tend to be extremely low.
   b. The plant can compete for water effectively and live in dry soils.
   c. The plant will grow most effectively in soils that are saturated with water year round.
   d. The plant's leaves will wilt easily.

4. What forces are responsible for capillarity?
   a. adhesion of water molecules to the sides of xylem cells, cohesion of water molecules to each other, and surface tension
   b. surface tension created by transpiration and cohesion of water molecules in a continuous flow from leaf to root
   c. high solute potentials created by the entry of ions during the night, when transpiration rates are low, followed by an influx of water
   d. gravity and wall pressure (from the sides of xylem cells)

5. What is a proton pump?
   a. a membrane protein that transports sucrose against a concentration gradient
   b. a membrane protein that transports protons against an electrochemical gradient
   c. a membrane protein that transports protons *with* an electrochemical gradient and sucrose *against* a concentration gradient
   d. any membrane protein that acts as a channel—meaning it does not consume ATP

6. Why is the transport of phloem sap considered an active process?
   a. The manufacture of sucrose via photosynthesis is driven by the energy in sunlight.
   b. Transpiration is driven by the energy in sunlight.
   c. ATP is used to transport sucrose into companion cells near sources against a concentration gradient.
   d. In spring, phloem sap moves against the force of gravity.

TEST YOUR UNDERSTANDING

Answers are available in Appendix B

1. Draw a plant cell in pure water. Add dots to indicate solutes inside the cell. Now add dots to indicate an increase in solute potential inside the cell. Add an arrow showing the direction of water movement in response. Add arrows showing the direction of wall pressure and turgor pressure in response to water movement. Repeat the same exercise, but this time add solutes to the solution outside the cell.
2. Compare and contrast the forces involved in transporting water in xylem via root pressure, capillarity, and transpiration. Which of these mechanisms are passive?
3. Why are "cohesion-tension" and "pressure-flow" sensible names for the hypotheses analyzed in this chapter?
4. Suppose Aphid A and Aphid B are sitting on the same plant. A inserts her stylet into the phloem at the base of a large, mature leaf, while B prefers to probe the phloem of the young growing tissue near the shoot's apical meristem. Which aphid is attacking a source, and which is preying on a sink? Which aphid is getting a higher concentration of sugar, A or B? Where would you expect aphids to be found on a plant growing in the wild?
5. How does cotransport result in phloem loading?
6. A seed is a sink when it is forming inside the parent plant. When is it a source?

APPLYING CONCEPTS TO NEW SITUATIONS

Answers are available in Appendix B

1. The text claims that water loss is an inevitable side effect of gas exchange. What data or observations support or challenge this claim? Would the same statement be true in terrestrial animals?
2. Suppose that plants over 1 meter tall had to expend energy to transport water from their roots to their leaves. What would be the consequences in terms of growth rates and overall height?
3. When young trees are transplanted to a new site, it takes several weeks or months for their root systems to grow and establish a high capacity to take up water. If a heat wave occurs during this period, the trees are likely to die—but not of starvation or loss of turgor. What kills them?
4. A recent paper indicates that the aquaporins in plasma membranes of cells throughout a plant close in response to drought stress. How does the closing of these channels help plant cells maintain turgor?

In most plants, roots obtain the water and key elements required for individuals to survive and thrive.

Plant Nutrition 38

The most urgent tasks facing any organism are to acquire (**1**) carbon-containing molecules that will be used as cellular building blocks and (**2**) the chemical energy required to make ATP. Plants acquire both by producing sugar through the process of photosynthesis.

Yet plants cannot live on sugar alone. Besides making the carbohydrates they need, plants synthesize all of their own nucleic acids, amino acids, enzymes, chlorophylls, enzyme cofactors, and other molecules. Plants do some of the world's most impressive synthetic organic chemistry.

A plant's ability to perform sophisticated reactions depends on its capacity to harvest a wide variety of simple ions and elements as raw materials. In addition to carbon dioxide and water, plants have to obtain nitrogen, phosphorus, potassium, sulfur, magnesium, and other elements. Soil provides most of these nutrients, the majority existing as ions that are dissolved in soil water at low—sometimes extremely low—concentrations. Once these ions are inside root cells, the one-way flow of water up xylem carries the nutrients throughout the plant body.

The plant body is an efficient machine for harvesting these diffuse resources and concentrating them in cells and tissues. Chapter 36 introduced how the organization and growth of the plant root and shoot systems make resource acquisition possible; Chapter 37 focused on how water and nutrients are transported throughout the plant body. This chapter concentrates on how plants take up simple ions and elements from the soil, so that these nutrients can be transported to the cells that need them.

Questions about nutrition are fundamental to understanding how plants work, increasing agricultural productivity, and maintaining the productivity of forests that supply lumber and fuel—as well as mitigate global warming (see Chapter 54) by transforming atmospheric CO_2 into wood. Let's begin by analyzing the basic nutritional needs of plants—the equivalent of the minimum daily requirements in humans.

KEY CONCEPTS

- In addition to needing carbon dioxide and water, plants require an array of essential nutrients to support growth. These nutrients are available as ions dissolved in soil water and are taken up by roots.
- Nutrient absorption occurs via specialized proteins in the plasma membranes of root cells. Most plants also obtain nitrogen or phosphorus from fungi associated with their roots. Toxins that enter roots are either excluded or actively transported into cell vacuoles and stored.
- Some species of plants have specialized methods of obtaining nutrients, including associations with nitrogen-fixing bacteria, parasitism, and carnivory.

✔ When you see this checkmark, stop and test yourself. Answers are available in Appendix B.

38.1 Nutritional Requirements of Plants

What do plants need to live? In the early 1600s, Jean-Baptiste van Helmont performed a classic experiment designed to answer this question. Van Helmont wanted to know where the mass of a growing plant comes from, and he used a willow tree as a study organism.

As **Figure 38.1** shows, he began by placing 200 pounds of soil in a pot with a 5-pound willow sapling. He allowed the plant to grow for five years, adding only water. At the end of the experiment, he weighed the willow and the soil. The willow weighed 169 lb, 3 ounces; the soil weighed 199 lb, 14 ounces.

Where had the additional 164 pounds, 3 ounces, of tree come from? Because he was not aware that gases have mass, van Helmont hypothesized that the new plant material came from water. He also ignored the loss of 2 ounces in the soil, chalking it up to measurement error.

As it turned out, van Helmont's measurements were not the problem—his conclusions were. Most of the mass of the tree came from carbon dioxide in the atmosphere. The 2 ounces removed from the soil contained vital ions and elements—the nutrients that are the focus of this chapter. What are they?

About half the elements in the periodic table—more than 60—can be found in the tissues of one or more plant species. The question that biologists, farmers, and foresters ask is, Which of these elements are essential for growth and reproduction in most species, and in what quantities?

Which Nutrients Are Essential?

Biologists define an **essential nutrient** as an element or compound that fulfills two criteria:

1. It is required for normal growth and reproduction—meaning that the plant cannot complete its life cycle without this nutrient.
2. It is required for a specific structure or metabolic function.

Researchers test whether a nutrient is essential by denying a specific element to plants and documenting the effects or lack of effect. For most vascular plants, 17 elements are essential. Just three of these—carbon, hydrogen, and oxygen—typically make up about 96 percent of the dry weight of a plant. The remaining 14 elements are sometimes called mineral nutrients, because they originate in soil.

Although different classification schemes for the essential elements have been proposed, the most common is based on distinguishing nutrients that are obtained from water or carbon dioxide versus soil, and then dividing soil nutrients into macronutrients and micronutrients (**Table 38.1**).

MACRONUTRIENTS Certain elements in the soil are required during the synthesis of nucleic acids, proteins, carbohydrates, phospholipids, and other key molecules. Because they are required in relatively large quantities, these elements are called **macronutrients**.

Among the macronutrients, nitrogen (N), phosphorus (P), and potassium (K) are particularly important because they often act as **limiting nutrients**, meaning their availability limits plant growth. If N, P, and/or K are added to soil as fertilizer, plant growth usually increases. This observation explains why the leading ingredients in virtually every commercial fertilizer are N, P, and K—usually listed in that order on the container.

EXPERIMENT

QUESTION: Where does the mass of a growing plant come from?

HYPOTHESIS: The mass of a growing plant comes from soil.

NULL HYPOTHESIS: The mass of a growing plant does not come from soil.

EXPERIMENTAL SETUP:

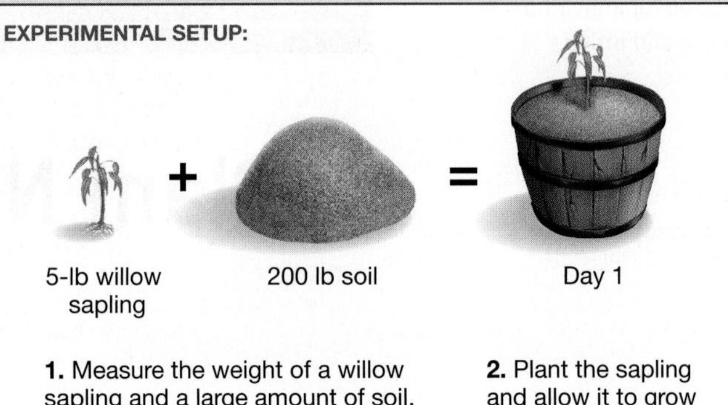

PREDICTION: After 5 years, the soil mass will decrease by the same amount that the plant mass increased.

PREDICTION OF NULL HYPOTHESIS: The soil mass will not decrease.

RESULTS:

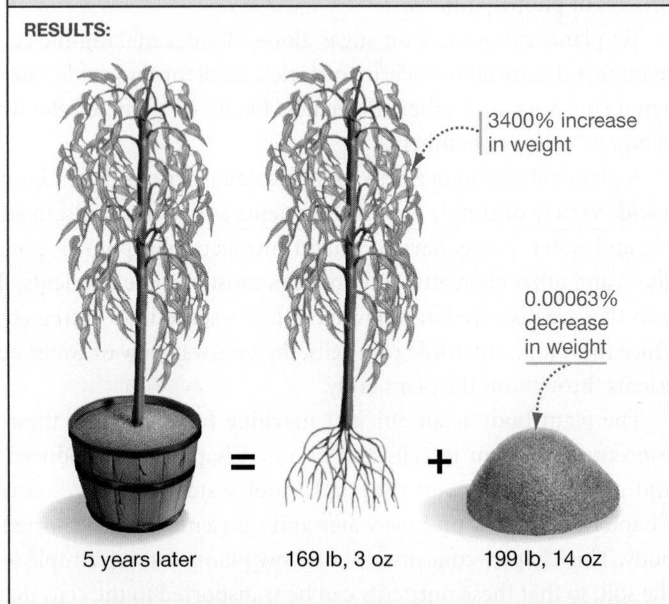

CONCLUSION: The mass of a growing plant does not come from soil.

FIGURE 38.1 An Early Experiment on the Role of Soil in Plant Nutrition.

✔**QUESTION** Many nonbiologists think that most of a plant's mass comes from soil or water. Describe an experiment that would convince someone that most of a plant's mass comes from CO_2.

TABLE 38.1 **Essential Nutrients**

Element	Form Available to Plants	Functions	Average % Dry Weight*	Deficiency Symptoms
Obtained from H_2O or CO_2				
Oxygen	O_2, H_2O	Electron acceptor in cellular respiration; major component of organic compounds	45	Usually affects roots—cells suffocate, leading to root rot and wilting
Carbon	CO_2	Substrate for photosynthesis; major component of organic compounds	45	Slow growth (starvation)
Hydrogen	H_2O	Major component of organic compounds; electrical balance and establishment of electrochemical gradients	6	Slow growth due to cell death (desiccation)
Obtained from Soil: Macronutrients				
Nitrogen	NO_3^- (nitrate) NH_4^+ (ammonium ion)	Component of nucleic acids, ATP, chlorophyll, proteins, hormones, and coenzymes	1.5	Failure to thrive; chlorosis (yellowing of older leaves)
Potassium	K^+	Cofactor for many enzymes; necessary for osmotic adjustment in cells; required for synthesis of organic molecules	1.0	Chlorosis at margins of leaves or in mottled pattern; weak stems; short internodes
Calcium	Ca^{2+}	Regulatory functions; role in cell wall structure; stabilizes membranes; second messenger in signal transduction	0.5	Necrosis (small spots of dead cells) in meristems; deformation of young leaves; stunted, highly branched root system
Magnesium	Mg^{2+}	Chlorophyll component; activates many enzymes	0.2	Chlorosis between leaf veins; premature leaf drop
Phosphorus	$H_2PO_4^-$ (dihydrogen phosphate ion) HPO_4^{2-} (hydrogen phosphate ion)	Component of ATP nucleic acids, phospholipids, and several coenzymes	0.2	Stunted growth in young plants; dark green leaves with necrosis
Sulfur	SO_4^{2-} (sulfate ion)	Component of protein and coenzymes	0.1	Stunted growth; chlorosis
Obtained from Soil: Micronutrients				
Chlorine	Cl^- (chloride ion)	Needed for water-splitting step of photosynthesis; functions in water balance and electrical balance	0.01	Wilting at leaf tips; general chlorosis and necrosis of leaves or development of bronze color
Iron	Fe^{3-} (ferric ion) Fe^{2-} (ferrous ion)	Necessary for chlorophyll synthesis; component of cytochromes and ferredoxin; enzyme cofactor	0.01	Chlorosis between veins of young leaves
Manganese	Mn^{2+}	Involved in photosynthetic O_2 evolution; enzyme activator; important in electron transfer	0.005	Chlorosis between leaf veins and small necrotic spots
Zinc	Zn^{2+}	Involved in synthesis of the plant hormone auxin; maintenance of ribosome structure; enzyme activation	0.002	Small internodes; stunted and distorted ("puckered") leaves
Boron	$H_2BO_3^-$ (borate ion)	Strengthens cell walls; required for pollen tube growth and normal membrane function	0.002	Black necrosis in young leaves and buds
Copper	Cu^+ (cuprous ion) Cu^{2+} (cupric ion)	Cofactor of some enzymes; present in lignin of xylem	0.0006	Light-green leaves with necrotic spots; twisted and malformed leaves
Nickel	Ni^{2+}	Cofactor for enzyme functioning in nitrogen metabolism	[no data]	Necrosis at leaf tips
Molybdenum	MoO_4^{2-} (molybdate ion)	Cofactor in nitrogen reduction; essential for nitrogen fixation	0.00001	Chlorosis between veins; necrosis of older leaves

*These percentages were obtained by drying vascular plants and documenting what proportion of the waterless mass consists of various elements.

MICRONUTRIENTS In contrast to macronutrients, **micronutrients** are required in small quantities. When plant tissues are dried and analyzed, micronutrients are typically present in 1–100 parts per trillion. Instead of acting as components of macromolecules, micronutrients usually function as cofactors for specific enzymes—substances that are required for normal enzyme function (Chapter 3).

It's important not to underestimate the importance of micronutrients, even though only tiny amounts are needed. For example, a typical plant contains just one molybdenum atom for every 60 million hydrogen atoms in its body, not including water. Yet plants die without molybdenum, because it functions as a cofactor for several enzymes involved in nitrogen processing. What happens to plants when other essential nutrients are missing?

What Happens When Key Nutrients Are in Short Supply?

In some cases, biologists can examine a plant that is growing poorly and diagnose a nutrient deficiency (**Figure 38.2**). If older leaves are in poor condition, the problem is probably due to lack of N, P, K, or magnesium. These elements are mobile—meaning they are readily transferred from older leaves to newer leaves when they are in short supply—so older leaves deteriorate first when these atoms are scarce. Immobile nutrients like iron or calcium, in contrast, stay tied up in older leaves. When they are in short supply, newer leaves are the first to show symptoms.

In large part, the ability to diagnose nutrient deficiencies is based on studies involving hydroponic growth systems. **Hydroponic growth** takes place in liquid cultures, without soil, so researchers can precisely control the availability of nutrients.

Consider an experiment on copper deficiency in tomatoes (**Figure 38.3**). Researchers grew seedlings in two types of treatments. One treatment consisted of flasks containing water and all the essential nutrients in the relative concentrations that are optimal for tomato growth. The second treatment was identical, except that the nutrient solution lacked copper.

EXPERIMENT

QUESTION: How does copper deficiency affect plants?

HYPOTHESIS: Plants denied copper will grow poorly.

NULL HYPOTHESIS: Plants denied copper will grow normally.

EXPERIMENTAL SETUP:

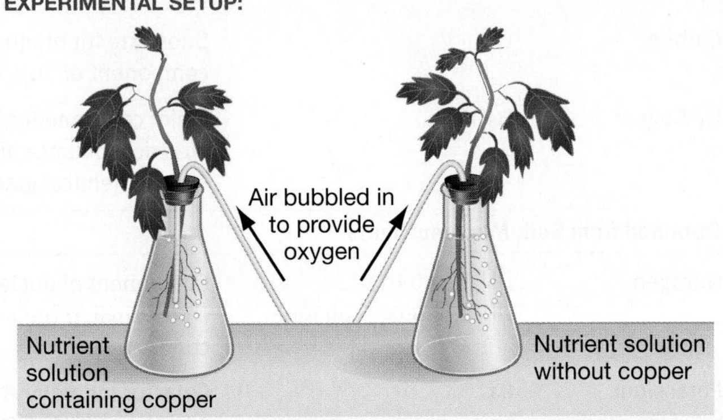

PREDICTION: The copper-deficient plant will grow less than the normal plant.

PREDICTION OF NULL HYPOTHESIS: Both plants will grow the same.

RESULTS:

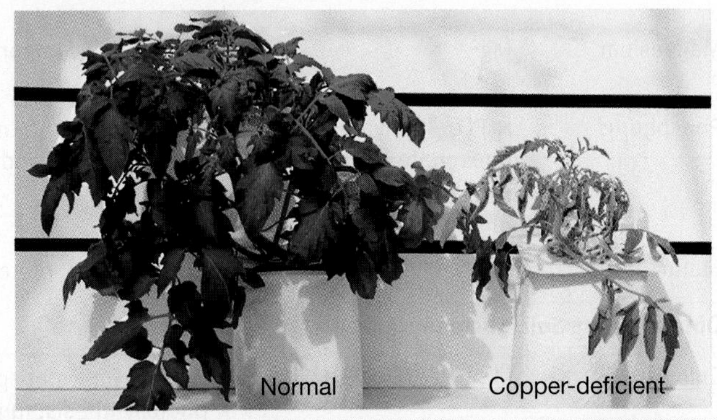

CONCLUSION: Copper deficiency leads to poor growth. All tissues appear to be affected adversely.

FIGURE 38.3 Hydroponics Is Used to Study Nutrient Deficiencies.

SOURCE: Arnon, D. I. and P. R. Stout. 1939. The essentiality of certain elements in minute quantity for plants with special reference to copper. *Plant Physiology* 14: 371–375.

✔**QUESTION** What problem would arise if this experiment had been done in soil with and without added copper?

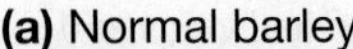
(a) Normal barley

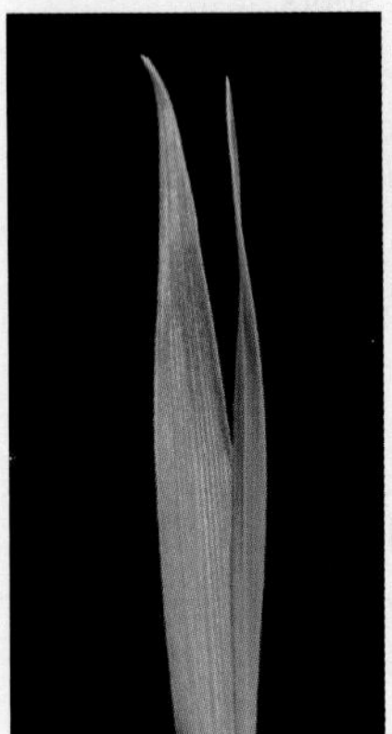

(b) N deficiency

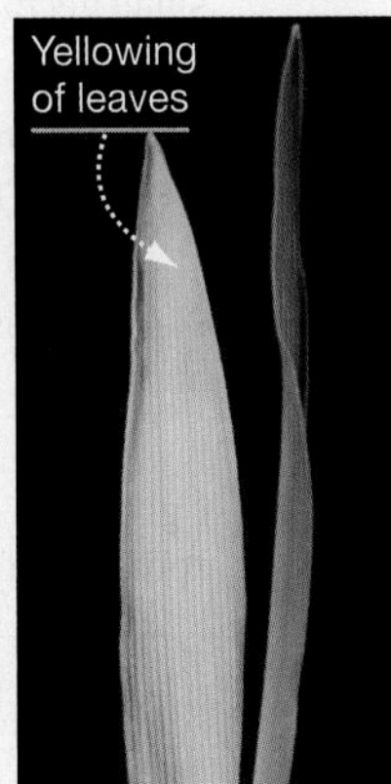

(c) P deficiency

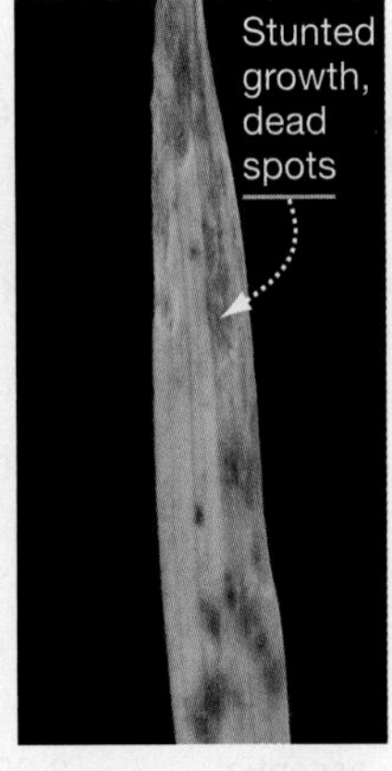

FIGURE 38.2 Nutrient Deficiencies Can Have Distinctive Symptoms.

✔**QUESTION** When nitrogen is in short supply, which molecules or processes are affected?

As Figure 38.3 shows, copper-deprived individuals have stunted shoots, unnaturally light foliage, and curled leaves. Given copper's role as a cofactor or component of several enzymes involved in redox reactions required for ATP production, it is understandable that all tissues in the plant were severely affected. And because copper is a micronutrient, it is reasonable to expect that a relatively small amount would cure the deficiency. In line with this prediction, the researchers found that the symptoms were prevented if the plants were cultured in a solution containing just 0.002 mg/L of copper. Analogous studies have been done on the other essential nutrients.

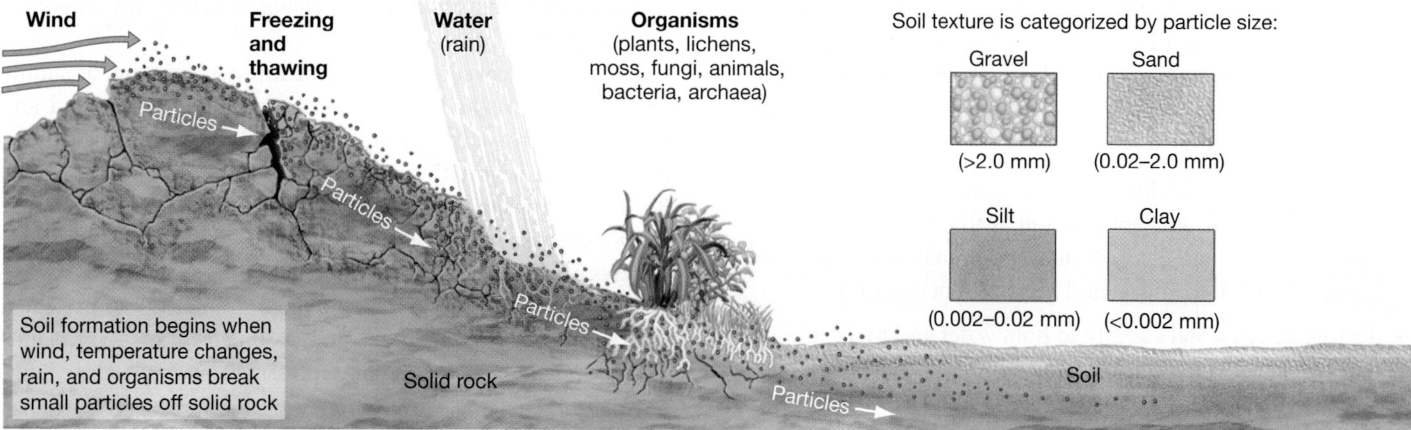

FIGURE 38.4 Soil Formation Begins with Erosion of Rock.

For farmers, foresters, and plant ecologists, understanding which nutrients are essential, and why, is basic to understanding why certain plants thrive and others fail. Now, where do these nutrients come from? The answer—soil—is simple. But soil itself is astonishingly complex.

38.2 Soil: A Dynamic Mixture of Living and Nonliving Components

The process of soil building begins with solid rock. As **Figure 38.4** shows, **weathering**—the forces applied by rain, running water, and wind—continually breaks tiny pieces off large rocks. The weathering process is accelerated if small cracks develop in the rock. If plant roots grow into the crack, they expand as they grow, widen the crack, and break off small flakes or pebbles. A similar effect occurs in high latitudes or at high elevations when water enters the cracks, freezes in winter, expands, and breaks off pieces.

Depending on their size and composition, the particles resulting from these processes are called gravel, sand, silt, or clay. These rock fragments are the first ingredient in soil. As organisms occupy the substrate, they add dead cells and tissues and feces. This decaying organic matter is called **humus** (pronounced *HEW-muss*).

With time, soil eventually becomes a complex and dynamic mixture of inorganic particles, organic particles, and living organisms. It is commonplace to find thousands of species living in the top few centimeters of a single square meter of soil. In addition to plants, soil-dwelling organisms include a variety of fungi and animals, along with vast numbers of bacteria, archaea, and microscopic protists (**Figure 38.5**).

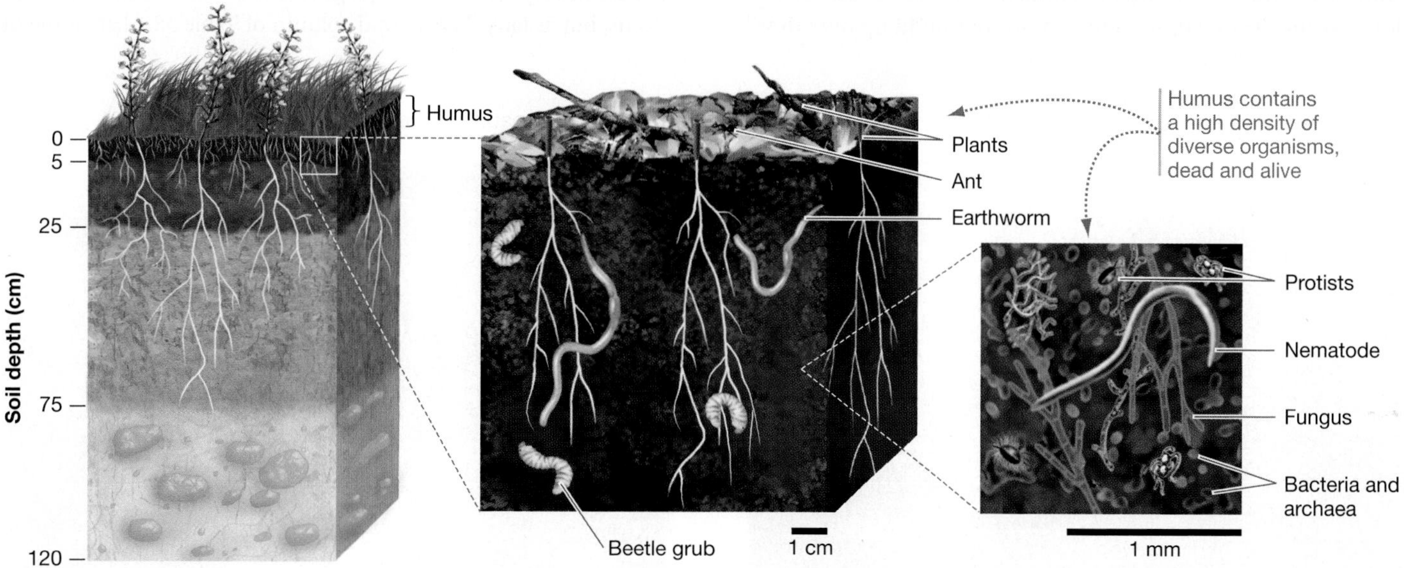

FIGURE 38.5 Mature Soils Are a Complex Mixture of Organic and Inorganic Components. In addition to mineral particles, soil contains humus—organic material derived from dead organisms—and a wide array of living organisms.

Both the parent rock that contributes inorganic soil components and the organisms and organic matter that occupy soils vary from one site to another. **Texture**—the proportions of gravel, sand, silt, and clay—and other qualities vary as well. Soil texture is important for several reasons:

- Texture affects the ability of roots to penetrate more deeply to obtain water and nutrients, as well as to anchor and support the body. For example, soil that is dominated by clay-sized particles tends to compact and resist root penetration.
- Texture affects a soil's ability to hold water and make it available to plants. Water tends to adhere to clay and silt particles but runs through sand and gravel.
- A soil's texture and water content dictate the availability of oxygen. Like other eukaryotes, plants have to take in oxygen to use as an electron acceptor during cellular respiration. The oxygen used by plant root cells is found in air pockets among soil particles. This explains why overwatering a plant is just as detrimental as underwatering it: Overwatering drowns a plant's roots.

The best soils, called loams, contain roughly equal amounts of sand, silt, and clay along with a high proportion of humus.

Loams and other topsoils that have good texture and large amounts of organic matter can take thousands of years to develop through the weathering of rocks and the continual addition of humus. Unfortunately, it can take just a few years of abuse by humans for them to blow or wash away.

The Importance of Soil Conservation

Soil erosion occurs when soil is carried away from a site by wind or water. Soil erosion occurs naturally, such as when rivers cut away at their banks and carry material downstream. In most natural environments, though, the rate of soil formation exceeds the rate of soil erosion, so soils build up over time. Unfortunately, the situation can change dramatically when humans exploit an area.

When plant cover is removed for forestry, farming, or suburbanization, stems and leaves can't lessen the force of wind and rain, and plant roots no longer hold soil particles in place. The results can be devastating. At some locations in the United States, 8–10 cm of topsoil were blown away during the Dust Bowl of the 1930s, when drought and poor farming practices left thousands of acres of soil unprotected (**Figure 38.6a**).

U.S. soil erosion rates have declined dramatically since then. Still, researchers estimate that almost 30 percent of all croplands in the United States are eroding too fast to maintain their long-term productivity. Deforestation is also exposing forest soils, contributing to disasters such as the mudslides and flooding that occurred in the Dominican Republic and Haiti in 2004, killing close to 5000 people. **Figure 38.6b** provides an aerial view of devastating slides that occurred in the U.S. state of Washington, after recent deforestation. Worldwide, it is estimated that 36 billion tons of soil are lost to erosion every year.

If soil is managed carefully, however, it can be a renewable resource. Techniques that maintain long-term soil quality and productivity are the basis of **sustainable agriculture** and sustainable forestry. Farmers can reduce soil loss dramatically by:

- planting rows of trees as windbreaks;
- using techniques that minimize the amount of plowing and tilling needed to control weeds; and
- planting crops in strips that follow the contour of hillsides.

They can also maintain soil quality by adding organic material in the form of manures and planting cover crops that are plowed in and allowed to decompose.

What Factors Affect Nutrient Availability?

The elements required for plant growth are found in the soil not as atoms, but as **ions**. The second column of Table 38.1 lists some of

(a) Wind erosion in the United States, 1930s

(b) Mudslides caused by deforestation

FIGURE 38.6 Soil Erosion Can Have Devastating Consequences.

the ions in soil that contain essential nutrients. Notice that some of these nutrients are available as elemental ions, such as K^+ or Cl^-, while others exist as molecular ions, such as HPO_4^{2-} or NO_3^-.

ANIONS AND CATIONS BEHAVE DIFFERENTLY The ions present in soil tend to behave in one of two ways, depending on their charge (**Figure 38.7**). Anions—ions with negative charges—usually dissolve in soil water, because they interact with water molecules via hydrogen bonding. (Phosphate ions, an exception to this rule, tend to form insoluble complexes with iron, aluminum, calcium, or other positively charged cations.)

Because they exist as solutes, negatively charged anions are readily available to plants for absorption. They are also easily washed out of the soil by rain, however. The loss of nutrients via the movement of water through soil is called **leaching**.

Cations—ions with positive charges—dissolve in soil water but are not as immediately available as anions. In solution, cations interact with the negative charges found on two types of soil particles: (**1**) organic matter that is rich in negatively charged organic acids; and (**2**) the surfaces of the tiny, sheetlike particles called clay, which are rich in mineral anions (see Figure 38.7).

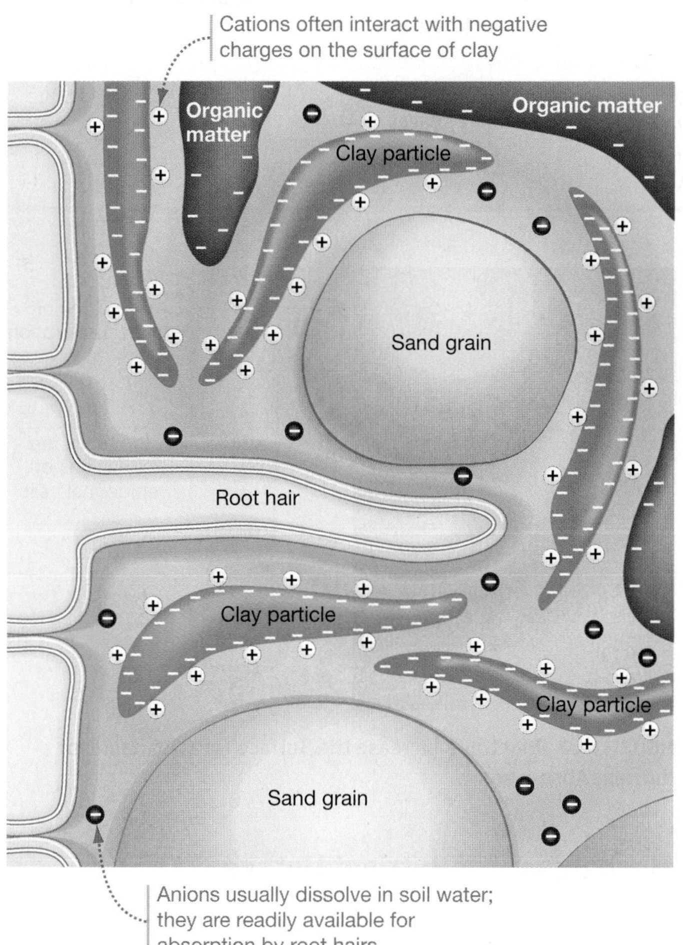

FIGURE 38.7 Cations Tend to Bind to Soil Particles; Anions Stay in Solution. Cations bind to organic matter in soil as well as to clay particles. Anions, in contrast, tend to go into solution.

Organic soils that contain clay tend to retain nutrients, because few positively charged cations leach away and because these soils hold water (and thus anions) better than sandy soils do. The presence of clay makes cations more difficult for plants to extract and use, however, because the cations are tightly bound.

THE ROLE OF SOIL PH In addition to soil texture, other factors can influence the availability of essential elements. Perhaps the most important of these is soil pH.

Recall from Chapter 2 that the pH scale indicates the relative concentration of hydrogen ions in a solution, and that it ranges from 0 to 14. Soils with low pH have a relatively high concentration of hydrogen ions and are considered acidic. Soils with a high pH contain relatively few hydrogen ions and are termed basic or alkaline.

Acidic soils are found in regions such as conifer forests, where the decomposition of organic matter produces carbonic acid, phosphoric acid, or nitric acid. Alkaline soils, in contrast, are common in regions where limestone ($CaCO_3$) is abundant. When limestone reacts with water, the calcium ions that are released take the place of protons that cling to soil particles. The protons then react with CO_3^{2-} to form bicarbonate ions (HCO_3^-), lowering the hydrogen ion concentration of the soil and raising its pH.

Most plants prefer soils with a relatively neutral pH—between 6 and 7. But some plants, such as blueberries, grow best in acidic soils (pH around 3 to 5); others require an alkaline soil (pH above 8). Certain species have enzymes that are adapted to work well in low-pH or high-pH environments.

CATION EXCHANGE Soil pH affects the availability of plant nutrients in a number of ways. For example, the presence of protons in soil water can cause the release of cations that are bound to soil particles. The process responsible is called **cation exchange**. Cation exchange occurs when protons or other cations bind to negative charges on soil particles and cause bound cations, such as magnesium or calcium, to be released (**Figure 38.8a** on page 744). Nutrients that are released by cation exchange become available for uptake by nearby plant roots (**Figure 38.8b**).

Plants influence cation exchange because root cells release CO_2 as a by-product of cellular respiration (see Chapter 9). The CO_2 reacts with H_2O to form carbonic acid, which releases protons.

If soil is too acidic, however, rain may wash cations away before the roots can take up nutrients (**Figure 38.8c**). In wet environments, acidic soils tend to be nutrient-poor.

To summarize, anions stay in solution in soil water. They are readily available to plants but may wash away easily. Positive ions, in contrast, tend to bind to soil particles but can be released by cation exchange.

Table 38.2 on page 744 details how the presence of sand, clay, and organic matter affects ion availability and other soil properties, including water availability.

Nutrients are found at extremely low concentrations in soil but at high concentrations in plant cells. How are plants able to bring N, P, K, and other key elements into their bodies against a concentration gradient?

(a) Cation exchange releases nutrients...

H+
Clay or organic matter
Mg^{2+}
Ca^{2+}
H+
H+
Protons in soil water
Released nutrients
H+

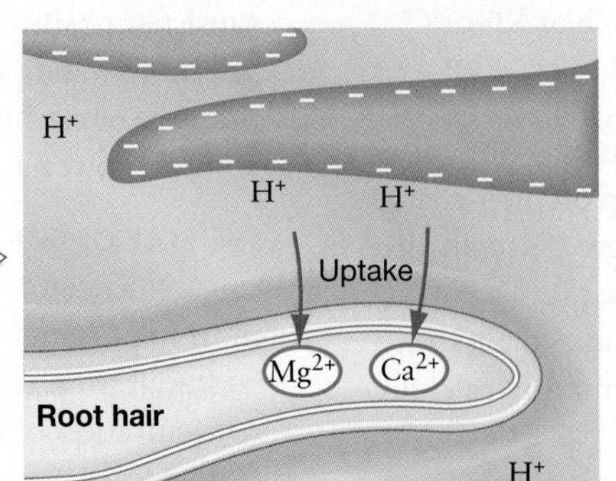

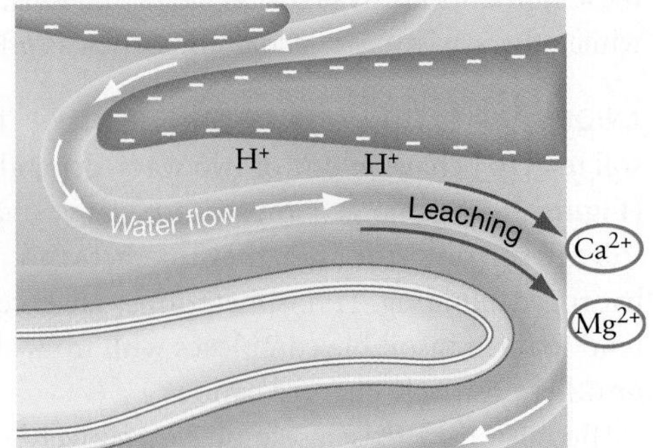

FIGURE 38.8 Cation Exchange Releases Nutrients Bound to Soil Particles. When cation exchange occurs, a proton binds to negative charges on clay or organic matter, releasing bound cations.

✔**QUESTION** Why is cation exchange an appropriate term?

TABLE 38.2 **Effects of Soil Composition on Soil Properties**

Composition	Water Availability	Nutrient Availability	Oxygen Availability	Root Penetration Ability
Sand	***Low:*** water drains through	***Low:*** poor capacity for cation exchange; anions leave in solution	***High:*** many air-containing spaces	***High:*** does not pack tight
Clay	***High:*** water clings to charged surface	***High:*** large capacity for cation exchange; anions remain in solution	***Low:*** few air-containing spaces	***Low:*** packs tight
Organic matter	***High:*** water clings to charged surface	***High:*** source of nutrients; large capacity for cation exchange; anions remain in solution	***High:*** many air-containing spaces	***High:*** does not pack tight

38.3 Nutrient Uptake

In most species of plants, the root system is the site of nutrient uptake. Chapter 36 introduced the general features of root anatomy, and Chapter 37 analyzed the role of roots in water uptake. Now we need to explore the detailed anatomy of roots and analyze events that occur in the plasma membranes of their epidermal cells.

Nutrient uptake occurs just above the growing root tip, in the region called the **zone of maturation**. Recall from Chapter 36 that epidermal cells in this part of the root have extensions called **root hairs** (**Figure 38.9**). Root hairs dramatically increase the surface area available for nutrient and water absorption. For example, the root system of a single annual rye plant can have a total surface area the size of a basketball court. Most of this area—60 percent—is found in an estimated 10^{10} root hairs.

Root hairs are so numerous and so efficient at absorbing nutrients from soil that, over time, they create a "zone of nutrient depletion" in soil immediately surrounding them. The creation of this mined-out region is why continued root growth is vital to a plant's health. Because roots continue to grow throughout a plant's life, the zone of maturation is continually entering new and potentially nutrient-rich areas of soil.

If an epidermal cell encounters soil where nutrients are available, what happens? How do ions enter?

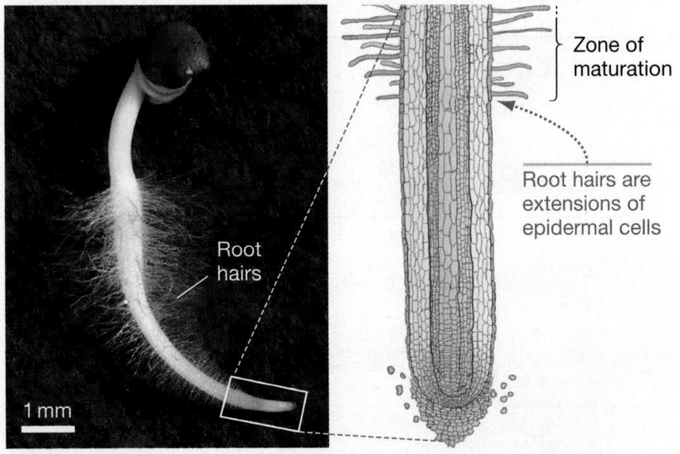

FIGURE 38.9 Root Hairs Increase the Surface Area Available for Nutrient Absorption.

Mechanisms of Nutrient Uptake

Ions, small molecules, and even large molecules can pass through the plant cell wall freely. The plasma membrane, in contrast, is highly selective. Recall from Chapter 6 that the plasma membrane is a fluid, sheetlike structure consisting of a phospholipid bilayer studded with proteins. Because the interior of the phospholipid bilayer is nonpolar and hydrophobic, it resists the passage of ions—even ions that are required by the plant. Some membrane proteins span the bilayer, however, and allow specific ions to cross the membrane.

Because root hairs have such a large surface area, they have large numbers of membrane proteins that bring nutrients into the cortex of the root. These membrane proteins cannot import the ions that the plant needs completely on their own, however.

ESTABLISHING A PROTON GRADIENT Plants harvest diffuse nutrients and concentrate them in their tissues. In an epidermal cell, ions such as potassium (K^+), hydrogen phosphate (HPO_4^{2-}), and nitrate (NO_3^-) are many times more concentrated than they are in the soil water outside the cell. For these and other ions to enter the cell, they have to cross the plasma membrane against a strong concentration gradient. How is this possible?

As **Figure 38.10a** shows, the answer hinges on **proton pumps**, or H^+-ATPases, in the plasma membranes of epidermal cells. These H^+-ATPases are similar to the pumps that make it possible for companion cells to load sucrose into phloem against a strong concentration gradient. Recall from Chapter 37 that when a phosphate group from ATP binds to the pumps, they change conformation in a way that allows them to transport protons to the exterior of the cell.

The activity of proton pumps leads to a strong excess of protons on the exterior of the plasma membrane relative to the interior. In the case of root hairs, this differential results in a strong concentration gradient favoring the movement of protons into the epidermal cell. In addition, the outside of the epidermal membrane becomes positively charged relative to the inside. Stated another way, there is a separation of charge—a **voltage**—across the membrane.

Proton pumps are found in all types of plant cells, and all plant cell membranes carry a voltage. Because voltage is a form of potential energy, the charges that are separated by the membrane create a **membrane potential**, or difference in electrical charge across a cell membrane. By convention, membrane potentials are expressed as inside-cell relative to outside-cell. Typically, the membrane potential of an active plant cell is about -200 mV (millivolts).

USING A PROTON GRADIENT TO IMPORT CATIONS In the membranes of root hairs, the electrical gradient established by proton pumps, which favors the entry of positive ions, is strong enough to overcome the concentration gradient, which opposes the entry of these cations. In essence, plant cells are batteries that are charged up to attract nutritionally necessary cations.

To drive this point home, consider potassium cations. Potassium is an essential nutrient in plants because it is required as a cofactor by over 40 enzymes. In addition, it is found at relatively high concentrations inside plant cells and plays a key role in bringing water into cells via osmosis and maintaining normal turgor pressure.

Researchers who added radioactive potassium ions to the solution outside a root cell and followed their movement found that K^+ does indeed flow into cells along an electrochemical gradient (**Figure 38.10b**). In follow-up experiments, biologists were able to isolate the membrane protein responsible for the uptake of K^+ and sequence the gene that encodes the protein.

✔If you understand this concept, you should be able to explain what happens to cation uptake if (1) H^+-ATPases fail, and (2) a molecule blocks a cation channel.

Cations like K^+ enter root hairs through membrane proteins along an electrochemical gradient, but how is it possible for anions to enter? The negatively charged interior of the cell should repel these ions.

USING A PROTON GRADIENT TO IMPORT ANIONS **Figure 38.10c** shows why anions such as NO_3^- are able to enter root hairs against their electrochemical gradient. The key is that anions enter through membrane transport proteins called cotransporters (see Chapter 37), which transport two solutes at once. In this case,

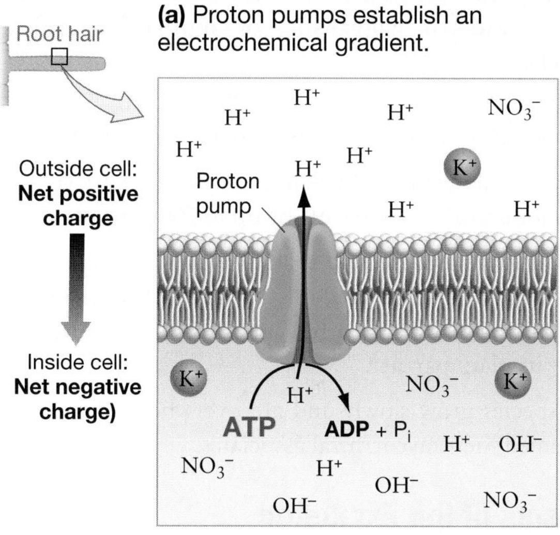

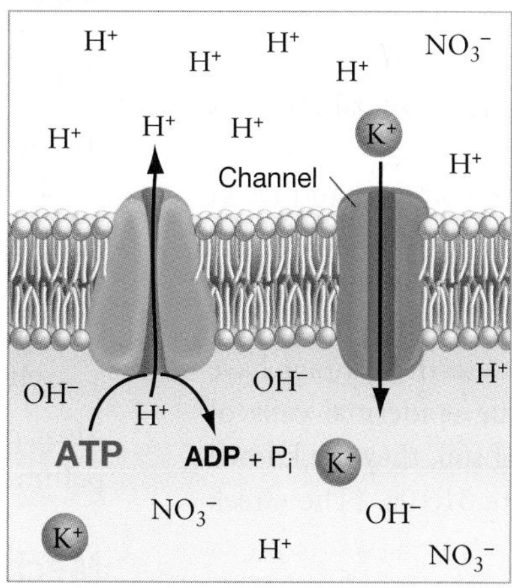

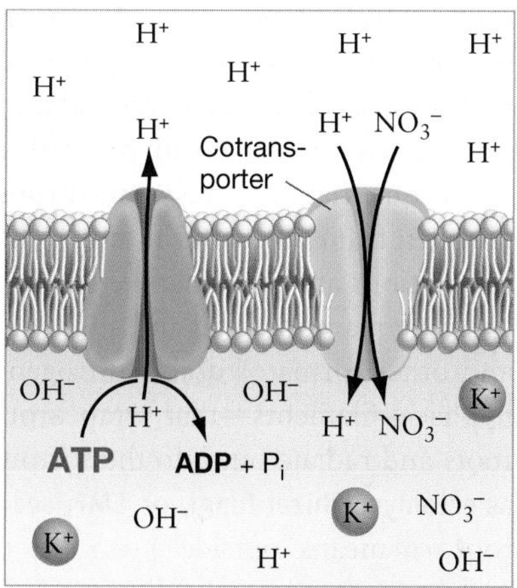

FIGURE 38.10 Ions Enter Roots along Electrochemical Gradients Created by Proton Pumps.

✔**EXERCISE** In part (b), add an arrow labeled "Proton gradient." In part (c), add an arrow labeled "Electrical gradient" and indicate its positive-negative polarity.

the cotransporter is a symporter: It brings NO_3^- into the cell along with a proton, and both solutes move in the same direction.

Here's the key observation: So much energy is released when a proton enters the cell along its electrochemical gradient that nitrate, phosphate ions, or other anions can be cotransported *against* their electrochemical gradients.

Symporters are also involved in bringing an array of cations into root cells. Ions that are missed by root hairs and that move into the root through cell walls can be taken into cortical cells or endodermal cells in the same fashion.

To summarize, the electrochemical gradient set up by proton pumps makes it possible for plant roots to absorb key cations and anions via ion channels and symporters. ✔If you understand this concept, you should be able to explain what happens to anion uptake if (1) H^+-ATPases fail, and (2) a molecule blocks an anion symporter.

To review how soils form and how nutrients enter via passive and active transport, go to the study area at *www.masteringbiology.com*.

(MB) **Web Activity** Soil Formation and Nutrient Uptake

Once ions have entered the root, they are transported across the root cortex via the symplastic, transmembrane, or apoplastic pathways introduced in Chapter 37. To enter xylem, however, the presence of the Casparian strip (see Chapter 37) means that nutrients moving through the apoplastic pathway have to pass through the plasma membranes of endodermal cells. From there they diffuse into the xylem tubes. Once they become part of xylem sap, the ions are transported passively to tissues throughout the plant.

NUTRIENT TRANSFER VIA MYCORRHIZAL FUNGI To synthesize proteins and nucleic acids, plants need to extract large quantities of nitrogen and phosphorus from the soil. The vast majority of plants take up nutrients by establishing a proton gradient across the membranes of their root hairs. But most can only absorb enough N and P to satisfy their nutritional needs with help from fungi that live in close association with their roots.

You might recall from Chapter 31 that fungi that live in association with plant roots are called **mycorrhizae** (literally, "fungus-root"). Mycorrhizae and plants are **symbiotic** ("living together"), meaning that they live in physical contact with each other. Biologists estimate that more than 80 percent of all vascular plant species associate with mycorrhizae.

In some habitats, trees and shrubs receive large quantities of nitrogen from mycorrhizal fungi. The fungal symbionts in these associations are particularly efficient at digesting amino acids in decaying plant material and absorbing ammonium ions (NH_4^+) and other forms of usable nitrogen. Because these fungi have hyphae—filaments—that wrap around the epidermal cells of roots and radiate out into the surrounding soil, they are known as **ectomycorrhizal fungi**, or **EMF**; see Figure 31.10a. (The Greek root *ecto* means "outside.")

In contrast, plants that live in grasslands and tropical forests receive much of the phosphorus and nitrogen they need from species of fungi whose hyphae can actually penetrate the walls of plant root cells. Because some of these fungal hyphae form branched clusters that look like little trees when viewed under a microscope, these symbionts are called **arbuscular mycorrhizal fungi**, or **AMF**; see Figure 31.10b. (The Latin root *arbor* means "tree.")

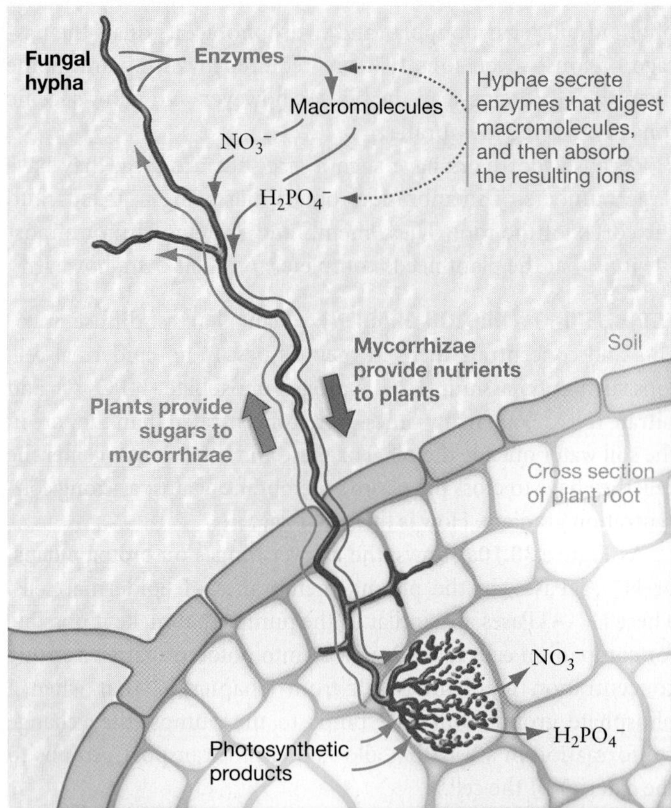

FIGURE 38.11 Most Mycorrhizae and Plants Are Mutualists. The AMF illustrated here provides nutrients to its host plant in exchange for photosynthetic products.

Chapter 31 presented experimental evidence that EMF and AMF transfer nitrogen and phosphorus from soil to plant roots. In exchange, the plants that they associate with transfer sugars and other photosynthetic products to the fungi. Because there is a reciprocal exchange of nutrients, the symbiotic relationship is considered **mutualistic**, or mutually beneficial (**Figure 38.11**).

Chapter 31 also emphasized two reasons why fungi are particularly efficient at acquiring the nutrients required by plants:

1. Networks of filamentous hyphae increase the surface area available for absorbing nutrients by up to 700 percent.
2. Fungi synthesize and secrete digestive enzymes that break down dead plant material and other sources of nutrients, releasing ions that can be absorbed and used by the fungus itself or by its plant partner.

Most plant species grow slowly and are overwhelmed by competitors if denied their mycorrhizal associates.

Mechanisms of Ion Exclusion

Plants have sophisticated systems for absorbing nutrients via the proton gradients they generate and mutualistic relationships with fungi. Not all ion uptake is beneficial, however.

Certain types of natural soils—as well as soils that have been contaminated by waste products from mining or smelting operations—contain enough cadmium, zinc, nickel, lead, or other metals to poison enzymes in most plants. Sodium is also detrimental at high concentrations: Too much Na^+ inside cells can disrupt enzyme function; and too much sodium in extracellular spaces can create a solute potential that pulls enough water out of cells to result in a loss of turgor.

Sodium poisoning is a key issue in an array of environments, including (**1**) ocean coastlines, (**2**) habitats near roads that are treated with salt to melt ice and snow, and (**3**) irrigated farmlands. When soils are irrigated, solutes are left behind when water evaporates from the surface—leading to salt buildup over time.

How do plants exclude ions that are detrimental?

PASSIVE EXCLUSION Many ions do not enter the transmembrane or symplastic routes through roots (see Chapter 37), because the epidermal and cortical cells lack the requisite membrane channels. And ions that enter roots via the apoplastic pathway may never make it to xylem. To understand why, recall from Chapter 37 that the Casparian strip forces all solutes that travel through the root apoplast to cross the plasma membrane of endodermal cells, before moving further into the root. Ions that cannot enter endodermal cells in this way are excluded from entering the rest of the plant.

In the apoplastic pathway, ion exclusion occurs at endodermal cells because their plasma membranes contain only certain types of transporters and channels. The presence of the Casparian strip allows endodermal cells to act as a selective filter—preventing some ions from entering the symplast and reaching the xylem. Because the excluded ions are not actively pumped out of the root cortex, the Casparian strip qualifies as a mechanism for the passive exclusion of metals, sodium, or other ions (**Figure 38.12**).

A similar type of passive exclusion also occurs in root hairs. If root hairs lack the membrane protein required for a certain ion to enter the cell, the ion won't enter. For example, salt-tolerant strains of corn transport much less salt into their roots than salt-intolerant strains—possibly because of genetic variation in the number of sodium channels found in root hairs. Among species, rice is notoriously sensitive to salt buildup. Barley, in contrast, is relatively tolerant to high salt concentrations in soil.

ACTIVE EXCLUSION BY METALLOTHIONEINS Plants also have mechanisms for coping with toxins once they are inside their cells. This is important, because all nutrients are toxic at high enough concentrations.

Copper channels, for example, admit the small amounts of copper ions required for normal cell function. But plants that grow on soils near copper-mining operations experience large concentration gradients that favor an influx of this nutrient, so a surplus is likely to build up inside the plant body. How do plants neutralize copper and other types of excess nutrients before they poison key enzymes?

One mechanism for coping with toxic concentrations of metals involves small proteins called **metallothioneins**. Metallothioneins bind to metal ions and prevent them from acting as a poison. Producing metallothionein proteins requires an expenditure of energy and represents a form of active exclusion.

Genes for metallothioneins have been found in a wide variety of organisms—bacteria, fungi, and animals as well as plants. Recent research on the mustard-family species *Arabidopsis thaliana* has shown that individuals from populations with a high tolerance for copper produce many more metallothionein proteins than do individuals from populations with a low tolerance for copper.

ACTIVE EXCLUSION BY ANTIPORTERS A second mechanism for actively neutralizing specific toxins involves transport proteins located in the **tonoplast**—the membrane surrounding the vacuole. Proteins in the tonoplast membrane allow plants to actively remove toxic substances from the cytosol and store them in the vacuole.

Perhaps the best-studied example involves proteins that move sodium ions from the cytosol into vacuoles, where they cannot

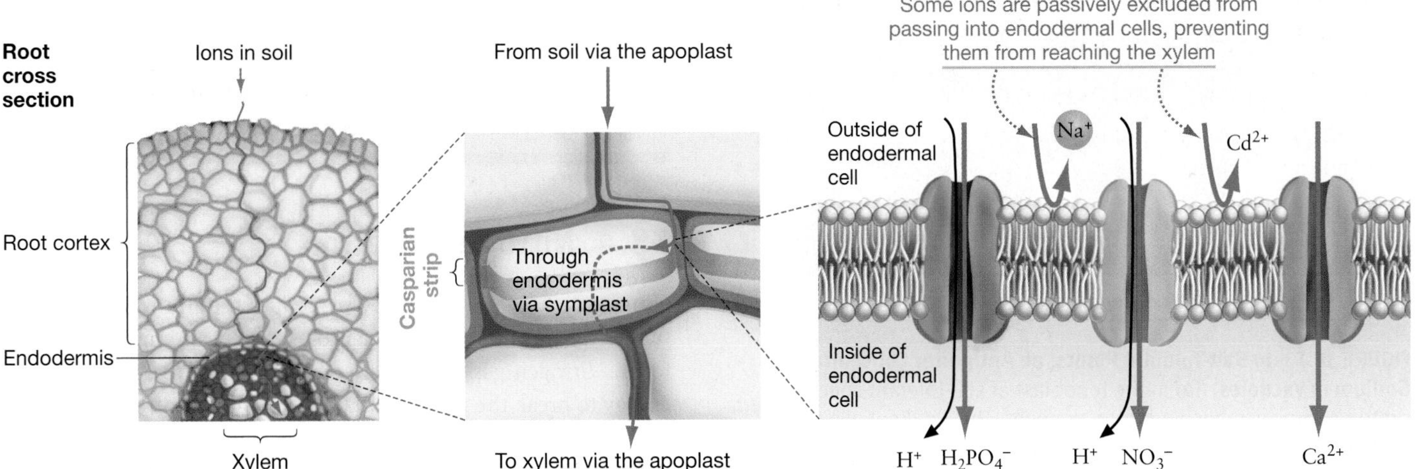

FIGURE 38.12 Passive Exclusion Occurs in Endodermal Cells. Certain ions are excluded from endodermal cells because it is difficult for them to cross the plasma membrane.

(a) In the tonoplast, antiporters send H^+ out and Na^+ in.

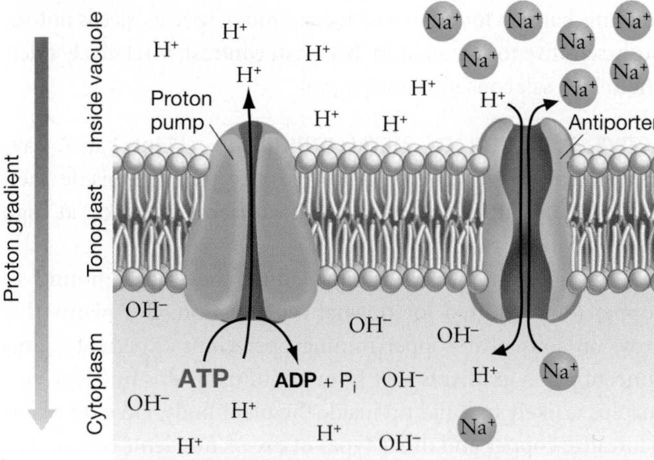

(b) Plants with H^+/Na^+ antiporters tolerate salt.

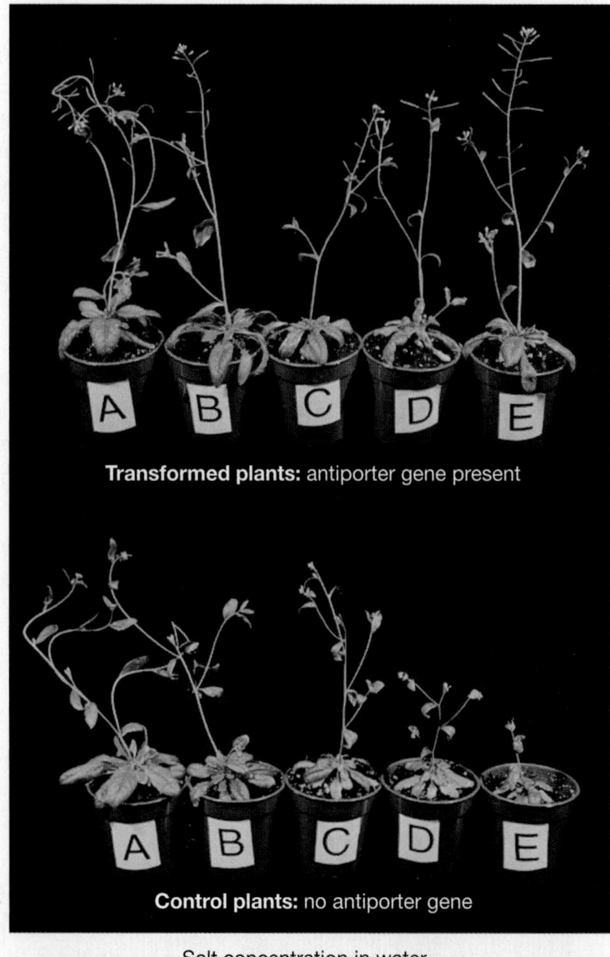

FIGURE 38.13 In Salt-Tolerant Plants, an Antiporter Concentrates Sodium in Vacuoles. (a) In the tonoplast of salt-tolerant species, proton pumps establish a proton gradient. This gradient allows antiporters to concentrate sodium ions inside the vacuole, where they cannot poison enzymes in the cytoplasm. **(b)** *Arabidopsis* plants that were transformed with the gene for the antiporter protein grow well even when watered with salty water.

poison enzymes (**Figure 38.13a**). H^+-ATPases in the tonoplast move protons into the vacuole, creating an electrochemical gradient that favors the movement of H^+ out of the vacuole. A transport protein that functions as an **antiporter** then uses this gradient to conduct protons *out* of the vacuole, along their electrochemical gradient, and bring sodium ions *into* the vacuole—against the sodium concentration gradient.

Recall from Chapter 37 that an antiporter is a cotransporter where the solutes move in opposite directions instead of the same direction—as they do through symporters.

This H^+/Na^+ antiporter has created a great deal of excitement among biologists, because it offers a way to genetically engineer crop plants that may grow well in salty soils created by poor irrigation practices. As an example, consider an experiment in which normal *Arabidopsis* plants and individuals transformed with copies of the gene for the H^+/Na^+ antiporter were exposed to different levels of NaCl. As **Figure 38.13b** shows, the genetically transformed individuals were able to grow efficiently even when exposed to high concentrations of salt.

CHECK YOUR UNDERSTANDING

If you understand that . . .

- Plants absorb most of the nutrients they need via membrane proteins in root hairs.
- Proteins in the plasma membranes of root-hair cells pump protons out of the cells, creating an electrochemical gradient that favors the entry of selected ions via ion channels or cotransporters.
- If no membrane proteins in the epidermal, cortex, or endodermal cells of a root admit a certain type of ion, that ion is excluded from entering the root.
- In many species, mycorrhizal fungi are important for bringing ions that contain nitrogen or phosphorus atoms into the root.

✔ You should be able to . . .

1. Explain how plants generate and use electrical power to import cations and anions.
2. Make a diagram that traces a cadmium ion (Cd^{2+}) from the soil into the plant, showing how it is excluded at different locations by passive and active mechanisms of ion exclusion.

Answers are available in Appendix B.

38.4 Nitrogen Fixation

Nitrogen gas (N_2) makes up 80 percent of the atmosphere. Unfortunately, plants and other eukaryotes cannot use nitrogen in this form. Nitrogen gas is unreactive because it takes a great deal of energy to break the triple bond between the two nitrogen atoms.

To synthesize amino acids, nucleic acids, and other nitrogen-containing compounds, plants have to absorb nitrogen in forms such as ammonium or nitrate ions. But these ions are in short supply in many soils, meaning that plant growth is often restricted by the availability of usable nitrogen. Most plants grow much faster when they receive a nitrogen-containing fertilizer.

Nitrogen-based fertilizers have drawbacks, however. Fertilizer production is extremely energy-intensive; and in many parts of the world, ammonia and other nitrogen-based fertilizers are too expensive for farmers. In more affluent regions, these fertilizers are used so extensively that they are causing serious pollution problems (see Chapter 28). For these and other reasons, there is intense interest in understanding the molecular basis of a phenomenon called biological nitrogen fixation.

The Role of Symbiotic Bacteria

Among all the organisms on the tree of life, only a few species of bacteria and archaea are able to absorb N_2 from the atmosphere and convert it to ammonia, nitrites, or nitrates. This process is called **nitrogen fixation**.

Nitrogen fixation requires a series of specialized enzymes and cofactors, including a large multi-enzyme complex called nitrogenase. The process is extremely energy demanding. An expenditure of 16 ATP molecules is required for nitrogenase to reduce one molecule of N_2 to two molecules of NH_3. The process can be summarized as follows:

$$N_2 + 8\,e^- + 8\,H^+ + 16\,ATP \longrightarrow 2\,NH_3 + H_2 + 16\,ADP + P_i$$

In many cases, bacterial cells that are capable of nitrogen fixation take up residence *inside* plant root cells. Although several different bacteria and plant hosts can be involved, the best-studied nitrogen-fixing bacteria are members of the genus *Rhizobium* that associate with plants in the pea family. Members of the genus *Rhizobium* and closely related species are often called **rhizobia**; pea family plants are often called **legumes**.

As **Figure 38.14** shows, the root cells of legumes form distinctive structures called **nodules**, where nitrogen-fixing rhizobia are found. The nodules are pink because they contain an iron-containing molecule called leghemoglobin (short for legume hemoglobin). **Leghemoglobin** is related to the hemoglobin that carries oxygen in your blood. Like hemoglobin, leghemoglobin binds oxygen.

Leghemoglobin is important because nitrogenase—the enzyme complex responsible for nitrogen fixation—is poisoned by the presence of oxygen. Inside root nodules, oxygen molecules bind to leghemoglobin instead of binding to nitrogenase. In this way, leghemoglobin keeps levels of free oxygen low enough to keep nitrogenase functioning, while maintaining a source of oxygen for electron transport in cortex cells, where intense cellular respiration and ATP synthesis are occurring.

Root cortex cells begin producing leghemoglobin when they are colonized by rhizobia. How does this colonization process start?

How Do Nitrogen-Fixing Bacteria Colonize Plant Roots?

Like mycorrhizal fungi and their host plants, legumes and rhizobia have a mutualistic relationship. The nitrogen-fixing bacteria provide the plant with ammonia, while the legume provides the bacteria with carbohydrates and protection. The association is costly for the host plant, however. For the bacteria to synthesize the enormous amounts of ATP required to manufacture ammonia, plants have to supply them with large quantities of sugar.

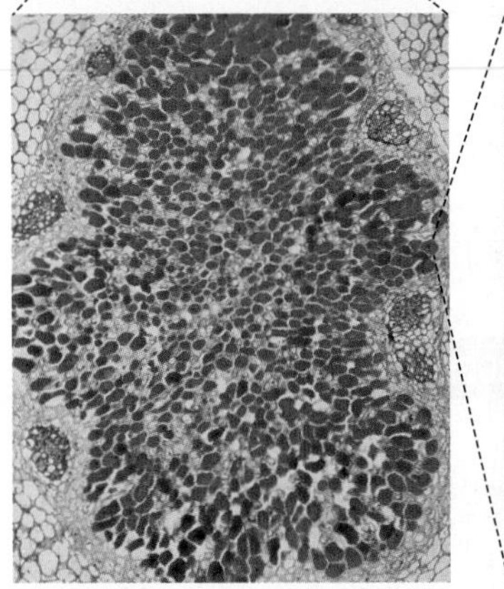
Nodule in cross section (red dots are infected plant cells)

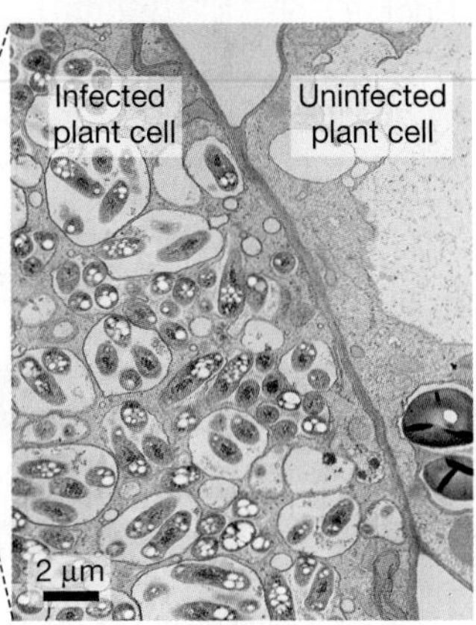

Close-up of rhizobia inside vesicles, inside infected plant cell

FIGURE 38.14 In Some Plants, Roots Form Nodules where Nitrogen-Fixing Bacteria Live.

Given the cost of the plant-bacterium interaction and its importance to both species, it's not surprising that the colonization process is carefully regulated.

When a pea seed germinates, its roots do not contain a population of rhizobia. Instead, the root makes contact with bacterial cells existing in the soil, the rhizobia colonize root cells, and the root cells grow into a nodule.

The first event in colonization is a recognition step that occurs between a pea family plant and its symbiotic bacterium. First, young roots release compounds called flavonoids. When rhizobia contact the flavonoids, the bacteria respond by producing sugar-containing molecules called **Nod factors** (for *nod*ule-formation). Nod factors, in turn, bind to proteins on the membrane surface of root hairs.

The recognition step is specific to the species involved. Each legume species produces a different flavonoid that acts as a recognition signal, and each rhizobium species responds with one or more unique Nod factors. When researchers have switched recognition signals or Nod factors between species, the recognition step fails.

When Nod factors bind to the root-hair surface, they set off a chain of events that leads to dramatic morphological changes in the host legume, as **Figure 38.15** shows.

1. The bacterial cells contact the root-hair surface.
2. Rhizobia begin to multiply and move, through an invagination of the root-hair membrane called an **infection thread**.
3. The infection thread extends toward the root, invading the cortex.

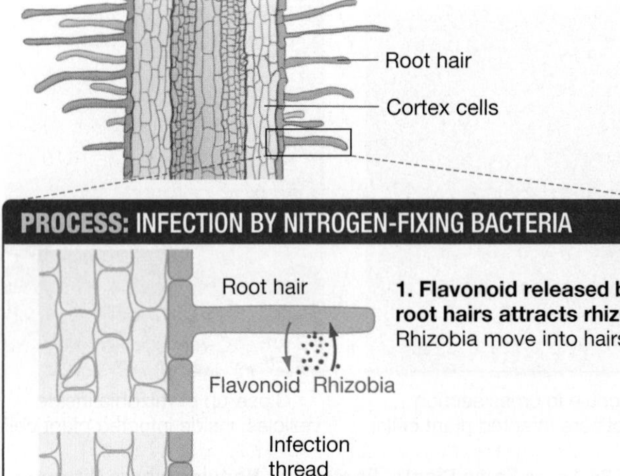

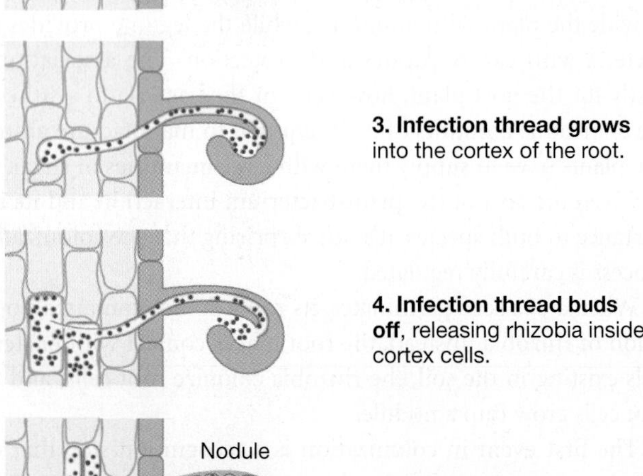

FIGURE 38.15 Infection by Nitrogen-Fixing Bacteria Is a Multistep Process. After rhizobia bind to a root hair, they enter the cytoplasm and travel down an infection thread into the root cortex, where they enter cortex cells.

4. Portions of the infection thread bud off, forming membrane-bound clusters within cortex cells, each filled with rhizobia.
5. The infected cortex cells divide rapidly, forming root nodules.

The interaction between rhizobia and legumes is one of the most complex and best-studied of all mutualisms.

CHECK YOUR UNDERSTANDING

If you understand that. . .

- Certain species of bacteria infect plant root cells and fix nitrogen, which is then made available to the plant.

✓ You should be able to. . .

Generate hypotheses addressing two additional questions about nitrogen fixation:

1. Why don't all plants have symbiotic bacteria that fix nitrogen?
2. Why is the interaction that establishes the infection by nitrogen-fixing bacteria so complex?

Answers are available in Appendix B.

38.5 Nutritional Adaptations of Plants

Based on the data available so far, over 95 percent of vascular plants import nutrients from soil. And over 80 percent of all plants supplement their "diet" with nutrients acquired from mycorrhizal fungi, while a small but significant fraction associate with nitrogen-fixing bacteria. Perhaps 99 percent of all living plant species make their own sugar through the process of photosynthesis.

What about the small number of plant species that don't follow these rules? Some appear to live on air, some parasitize other plants, and others catch insects and digest them.

Epiphytic Plants

Species from a diverse array of plant lineages do not absorb nutrients from soil. In fact, these species never even make contact with soil. As **Figure 38.16a** shows, they often grow on the leaves or branches of trees. For this reason they are called **epiphytes** ("upon-plants").

In northern forests, it is common for mosses and ferns to grow epiphytically on tree trunks. The so-called Spanish moss that hangs from oak trees in the southern United States is another familiar epiphyte—although this species is actually not a moss but a bromeliad, a relative of the pineapple. In the tropics and subtropics, one-third of all ferns grow as epiphytes; and there are thousands of species of epiphytic orchids, bromeliads, and lycophytes.

Epiphytes absorb most of the water and nutrients they need from rainwater, dust, and particles that collect in their tissues or in the crevices of bark. As **Figure 38.16b** shows, some epiphytes have leaves that grow in rosettes and form "tanks" that collect

(a) Epiphytes grow on trees.

(b) Water-holding "tanks" formed by leaves of an epiphyte

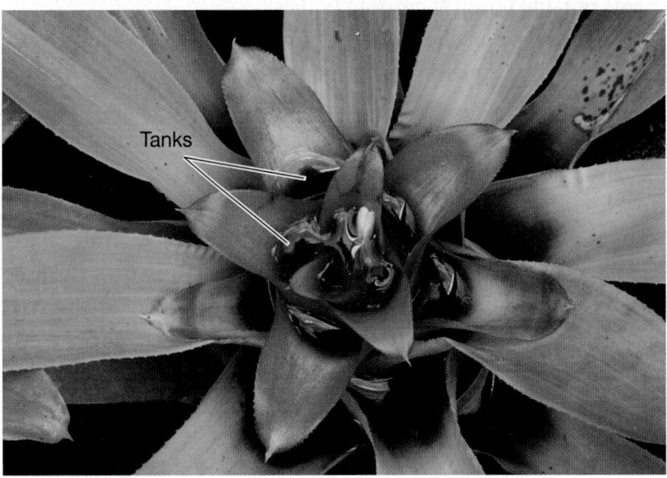

FIGURE 38.16 Epiphytes Are Adapted to Grow in the Absence of Soil. The environments where many epiphytes grow—tree trunks and branches—are dry and nutrient poor.

water and organic debris. In such cases, nutrients are actually absorbed through the leaves themselves.

Parasitic Plants

Parasites are organisms that live in close physical contact with individuals from another species and that lower the fitness of those individuals, usually by obtaining water or nutrients from them. Biologists estimate that about 3000 species of angiosperms are parasitic—less than 1 percent of the plant species that have been studied and named to date.

Some plant parasites are non-photosynthetic and obtain all of their nutrition by tapping into the vascular tissue of the host individual. But most parasitic plants make their own sugars through photosynthesis and tap the xylem of other species for water and essential nutrients.

For example, the mistletoe (**Figure 38.17**) is green and photosynthetic but has structures called haustoria that penetrate a host's xylem and extract water and ions that the mistletoe uses as nutrients. In the forests of western North America, mistletoe infection is a serious cause of economic loss.

Carnivorous Plants

Carnivorous plants trap insects and other animals, kill them, and absorb the prey's nutrients. Carnivorous species make their own carbohydrates via photosynthesis but use carnivory to supplement the nitrogen available in the environment. Most are found in bogs or other habitats where nitrogen is scarce or unavailable.

MODIFIED LEAVES FORM AN ARRAY OF TRAPPING MECHANISMS Carnivorous plants use modified leaves or roots to trap insects. Chapter 36 highlighted the leaves of pitcher plants, which form tubes (see Figure 36.12e). Prey are enticed to the tube by an attractive odor, then have a difficult time climbing back out. Eventually they fall into the pool of water below, where they drown. Enzymes

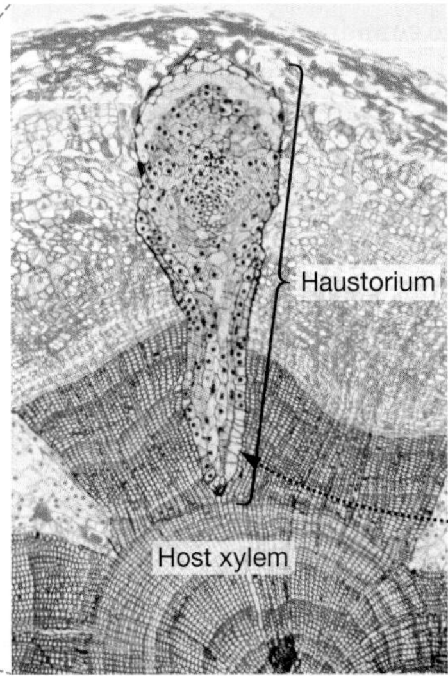

FIGURE 38.17 Some Plant Parasites Tap into the Vascular Tissue of Their Hosts.

✔**QUESTION** The bromeliads in Figure 38.16 and mistletoe are epiphytes. How do they differ in their relationship to the trees on which they grow?

released by the plant digest the prey, and the plant absorbs the nutrients obtained from their dead bodies.

Sundews have an alternative hunting strategy: They have modified leaves that function like flypaper. The leaf surfaces develop hairs that exude a sticky substance and trap insects (**Figure 38.18**). After the insects die, glands near the sundew's trap release enzymes that slowly digest the prey. The leaf then absorbs the nutrients that are released.

In the Venus flytrap, modified leaves trap insects mechanically. Sensory hairs protrude from the epidermis of each leaf. When an insect lands on the trap and bumps two or three of these hairs, the hair cells produce an electrical signal and the leaf responds by snapping shut (see Chapter 39). If the prey continues to stimulate hairs, the leaf secretes digestive enzymes that slowly digest the prey.

COSTS AND BENEFITS OF CARNIVORY Initially, research on carnivorous plants focused on documenting that meat-eating was adaptive. Early experiments in natural environments supported this hypothesis by showing that, compared to individuals of the same species that are not fed fruit flies, carnivorous plants that were fed fruit flies grew faster and flowered more often.

Follow-up work explored the costs as well as the benefits of carnivory. For example, several teams of researchers showed that when carnivorous plants are provided with nitrogen-based fertilizers, they produce fewer of their specialized insect-trapping leaves and a higher proportion of leaves that function primarily in photosynthesis. These results suggest carnivory is a trait that shows phenotypic plasticity (see Chapter 36): Plants increase investment in prey-capture devices when nitrogen is rare, but they decrease investment in prey-capturing structures when nitrogen is readily available.

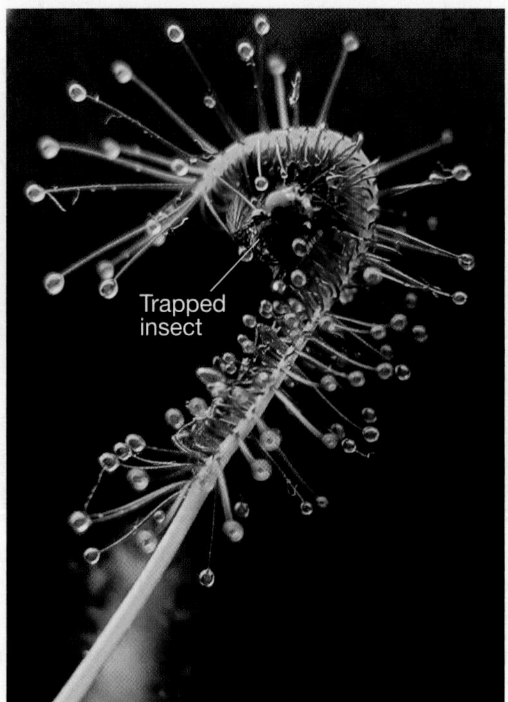

FIGURE 38.18 Sundews Have Modified Leaves with Sticky Surfaces That Catch Insects.

Currently, some research groups are doing "evo-devo" research (see Chapter 21) on carnivory. They are working to identify the novel alleles that made the evolution of carnivory possible in certain groups. Meat-eating may be relatively rare in plants, but it has inspired interesting new directions for research.

CHAPTER 38 REVIEW

For media, go to the study area at www.masteringbiology.com

Summary of Key Concepts

In addition to needing carbon dioxide and water, plants require an array of essential nutrients to support growth. These nutrients are available as ions dissolved in soil water and are taken up by roots.

- About 96 percent of the dry weight of a typical plant consists of carbon, hydrogen, and oxygen. Plants obtain these elements by absorbing carbon dioxide from the atmosphere and water from the soil.
- The other 4 percent of the plant body consists of a complex suite of elements, acquired from soil, that are required for normal growth and reproduction. These essential elements are usually absorbed in the form of ions like nitrate and phosphate.
- Soil is a complex and dynamic mixture of inorganic particles such as clay and sand, organic matter, and organisms. Soil provides plants with oxygen, water, and nutrients as well as a physical substrate for anchoring and supporting the body.

✔You should be able to design a field experiment that would determine whether nitrogen is a limiting nutrient in a habitat near your campus.

Nutrient absorption occurs via specialized proteins in root membranes. Most plants also obtain nitrogen or phosphorus from fungi associated with their roots. Toxins that enter roots are either excluded by root cells or are actively transported into cell vacuoles and stored.

- Nutrients are present at low concentrations in the soil surrounding roots. Plants import nutrients against a concentration gradient by pumping protons into the extracellular space.
- A large excess of protons outside the root hair creates a strongly negative membrane potential and thus an electrochemical gradient favoring the entry of positively charged ions through membrane channels or symporters.
- The proton gradient in root hair membranes also allows negatively charged ions to cross the plasma membrane via symporters.
- In addition to acquiring nutrients by establishing a membrane potential and a proton gradient, many plants obtain nitrogen or phosphorus from mycorrhizal fungi.

- The relationship between plants and the fungi is mutually beneficial. The host plant provides the fungi with carbohydrates; the fungi harvest N or P from the surrounding soil and transport it to the root system.
- Passive and active systems also exist for excluding certain ions. This is important because all nutrients are toxic in high concentrations.
- Passive exclusion occurs when ions cannot pass through the Casparian strip.
- Active exclusion often results from the production of specialized proteins that bind a particular ion, or the active transport of toxic ions into vacuoles.

✔You should be able to predict the consequences of mutations that allow root cells to establish a membrane voltage that is much more negative than normal, without a large increase in energy expenditure.

 **Web Activity** Soil Formation and Nutrient Uptake

Some species of plants have specialized methods of obtaining nutrients, including associations with nitrogen-fixing bacteria, parasitism, and carnivory.

- Certain plants, including those in the pea family, are capable of symbiosis with nitrogen-fixing bacteria. The bacteria colonize the interior of root cells after exchanging species-specific signals.
- Nitrogen-fixing bacteria receive protection and sugar from the host plant, in exchange for producing ammonia from nitrogen gas.
- Some parasitic plants produce their own carbohydrates through photosynthesis but steal water and nutrients by infecting the xylem of host plants.
- Carnivorous plants have evolved mechanisms for trapping and digesting insects and other animals. The primary benefit of carnivory is to obtain nitrogen in low-nitrogen environments.

✔You should be able to design an experiment, using radioactive carbon and the heavy isotope of nitrogen (^{15}N), that would test whether the rhizobia–pea plant interaction is mutualistic.

Questions

✔TEST YOUR KNOWLEDGE

Answers are available in Appendix B

1. Which of the following characteristics defines an element as essential for a particular species?
   a. It has to be added as fertilizer to achieve maximum seed production.
   b. If it is missing, a plant cannot grow or reproduce normally.
   c. If it is present in high concentration, plant growth increases.
   d. If it is absent, other nutrients may be substituted for it.
2. Why is the presence of clay particles important in soil?
   a. They provide macronutrients—particularly nitrogen, phosphorus, and potassium.
   b. They bind metal ions, which would be toxic if absorbed by plants.
   c. They allow water to percolate through the soil, making oxygen-rich air pockets available.
   d. The negative charges on clay bind to positively charged ions and prevent them from leaching.
3. Where does most nutrient uptake occur in roots?
   a. at the root cap, where root tissue first encounters soil away from the zone of nutrient depletion
   b. at the Casparian strip, where ions must enter the symplast prior to entering xylem cells
   c. in the symplastic and apoplastic pathways
   d. in root hairs, in the zone of maturation
4. Why is the activity of proton pumps in root hair membranes important?
   a. It creates a strongly positive membrane potential (voltage).
   b. It allows toxins to be concentrated in vacuoles, so the toxins do not poison enzymes in the cytoplasm.
   c. It sets up an electrochemical gradient that makes it possible to absorb cations and anions.
   d. It sets up the membrane voltage required for action potentials to occur.
5. Why is the relationship between most mycorrhizal fungi and their host plants considered mutualistic?
   a. They live in close physical association.
   b. Both species benefit from the association.
   c. The host plant cannot live without the mycorrhizae.
   d. The mycorrhizae cannot live without the host plant.
6. How do most epiphytes obtain nutrients?
   a. by trapping and digesting insects
   b. from their host plant
   c. via symbiotic bacteria called rhizobia
   d. from dust or other airborne particles

✔TEST YOUR UNDERSTANDING

Answers are available in Appendix B

1. A farmer is concerned that her corn crop is suffering from iron deficiency. Design an experiment that uses hydroponic cultures to identify (1) the symptoms of iron deficiency in corn and (2) the minimum amount of iron required to support normal growth rates.
2. Explain why carnivorous and parasitic plants are most common in nutrient-poor habitats.
3. Would you expect plant productivity to be higher in sandy soils or in soils containing both sand and clay? Why?
4. Sketch the plasma membrane of a root hair. Include proton pumps, ion channels, and cotransporters. Now add ATP to the system, then protons in their correct relative concentration, and ions that move across the membrane via channels or cotransporters. Explain why plant cells have a negative membrane potential.
5. Why is it important for plants to exclude certain ions? Summarize the difference between active and passive exclusion mechanisms.
6. Why is it logical for biologists to (1) distinguish nutrients that are obtained from the atmosphere or water (C, H, and O) from mineral nutrients obtained from soil, and (2) distinguish macronutrients from micronutrients?

✔ APPLYING CONCEPTS TO NEW SITUATIONS

Answers are available in Appendix B

1. There is a conflict between van Helmont's data on willow tree growth and the data on essential nutrients listed in Table 38.1. According to the table, nutrients other than C, H, and O should make up about 4 percent of a willow tree's weight. Most or all of these nutrients should come from soil. But van Helmont claimed that the soil in his experiment lost just 2 ounces, while the tree gained 2627 ounces. If so, then soil contributed just 0.08 percent of the added weight instead of 4 percent. State at least one hypothesis to explain the conflict. How would you test this hypothesis?
2. Acid rain occurs when sulfur oxides and nitrous oxides released by cars or factories react with water vapor, forming sulfuric and nitric acids, respectively. In areas affected by acid rain, cations quickly leach out of the soil. Explain why.
3. Design an experiment using the drug vanadate, which poisons proton pumps, to test the hypothesis that phosphate ions enter cells via an H^+/HPO_4^{2-} cotransporter.
4. Alder trees have nitrogen-fixing bacteria associated with their roots and have a much higher proportion of nitrogen in their tissues than spruce trees that grow in the same habitat. Which species decays faster after death: alder or spruce? Explain your logic.

Plants have sophisticated information processing systems. These radish seedlings sense the presence of light and are bending toward it.

Plant Sensory Systems, Signals, and Responses 39

Imagine standing in place for several hundred years, like an oak tree. Each spring you produce flowers. All summer you absorb light from the Sun, carbon dioxide from the atmosphere, and water and nutrients from the soil. Water, ions, and sugars flow up and down your vascular tissue. For six months or more your body grows upward, outward, and downward. In fall, thousands of your offspring drop to the ground as acorns. Then as winter approaches, your metabolism slows and you stop growing. Like a hibernating animal, you spend the long, hard months of winter in a state of suspended animation, before awakening the following spring.

To stay alive, an oak needs to gather information about its environment. It has to sense the season of the year, the time of day, the pull of gravity, the force of wind, and attacks by enemies. It needs to sense when its leaves are being shaded and which leaves are receiving the wavelengths of light that are required to support photosynthesis.

Plants may not have eyes or ears, but they can sense light, gravity, pressure, and wounds. They have the equivalent of a sense of smell, because they can perceive certain airborne molecules. It could even be argued that they have a sense of taste, because their roots sense the presence of nutrients in the soil.

In addition to gathering information about the conditions around them, plants have to respond in an appropriate way. They do not jump or swim or run, but their shoot systems grow toward light or end up shorter and stockier in response to wind. In response to gravity, shoots grow up and roots grow down. In response to touch, the modified leaves of a Venus flytrap shut fast enough to catch flying insects. If a plant is being attacked or if it senses that a neighboring individual is under attack, it may lace its tissues with toxic compounds or mobilize other defenses.

The message of this chapter is simple: Plants have sophisticated systems for collecting information about their environment and responding in ways that maximize their chances of surviving, thriving, and producing offspring. The ability to gather information and respond to it is one of the five fundamental attributes of life (see Chapter 1).

KEY CONCEPTS

- Plants are selective about the information they process. They perceive a wide variety of environmental stimuli that affect their ability to grow and reproduce.
- When sensory cells receive a stimulus, they transduce the signal and respond by producing hormones that carry information to target cells elsewhere in the body.
- Target cells respond to hormonal stimulation in ways that increase the ability of the plant to survive and reproduce.
- Hormones are also responsible for regulating how plants grow throughout their lives—especially in response to changes in environmental conditions. Each type of plant growth regulator plays a general role in the life of a plant, and growth responses are usually affected by interactions among several different hormones.

✓ When you see this checkmark, stop and test yourself. Answers are available in Appendix B.

39.1 Information Processing in Plants

Every environment is full of information. But, like other organisms, plants monitor only aspects of the environment that matter to them—that affect their ability to stay alive and produce offspring.

Figure 39.1 summarizes the three-step process by which plants gather, process, and respond to the information they monitor:

1. Receptor cells receive an external signal and change it to an intracellular signal.
2. Receptor cells sends a signal to cells in other parts of the body that can respond to the information.
3. Responder cells receive this signal, transduce it to an intracellular signal, and change their activity in a way that produces an appropriate response.

Let's briefly consider how each of these steps works, then delve into the details of how plants sense and respond to light, gravity, and other information.

How Do Cells Receive and Transduce an External Signal?

When you text a friend, the signal travels from you to the receiver via airwaves. When the message arrives, the receiving cell phone changes the information in the airwaves into electrical signals and then into words that your friend can understand.

Plants work in much the same way. When light strikes a sensory cell, for example, the information that it carries has to be changed into a form that is meaningful to that cell. Signals from the environment are usually received by a protein specialized for that function. Receptor proteins change shape in response to an environmental stimulus, such as being struck by a particular wavelength of light, having pressure applied, or binding to a particular type of molecule.

When a receptor changes shape in response to a stimulus, the information changes form—from an external signal to an intracellular signal. This process is called **signal transduction** (**Figure 39.2**); the verb transduce means "to convert energy from one form to another." Once information has been transduced to an intracellular form, it travels down what biologists call a signal transduction pathway.

You might recall from Chapter 8 that there are two basic types of signal transduction pathways: phosphorylation cascades and second messengers. Both begin with a receptor protein in the plasma membrane.

- **Phosphorylation cascades** are triggered when the change in the receptor protein's shape leads to the addition of a phosphate group (PO_4^{3-}) from ATP to the receptor or an associated protein. Phosphorylation activates proteins involved in signal transduction cascades, causing them to phosphorylate and activate a different set of proteins, which in turn catalyze the phosphorylation and activation of still other proteins, and so on.
- **Second messengers** are produced when hormone binding results in the production of intracellular signals or their release from storage areas. Calcium ions (Ca^{2+}) stored in the vacuole, ER, or cell wall are one of several ions or molecules that function as the second messenger in plants.

In some cases, phosphorylation cascades and second messengers interact.

Signal transduction primes the receptor cell for action. In many cases, though, the cells that receive information from the environment are located in a part of the body that is distant from the cells that need to respond to the information. How does information from an activated receptor cell get to responder cells?

In most cases, the answer is a **hormone**—an organic compound that is produced in small amounts in one part of a plant and transported to target cells, where it causes a physiological response. Signal transduction in a receptor cell often results in the release of a hormone that carries information to responder cells.

How Are Cell-Cell Signals Transmitted?

Because plants perceive such a wide array of stimuli, they have a wide array of hormones coursing through their bodies. These signals may be transmitted from cell to cell by specialized transport proteins in cell membranes, in xylem sap or phloem sap, or by simple diffusion from the originating cell. In most cases the signals act on target tissues throughout the body.

Plant cells routinely receive information from several different hormones at the same time, so it is common for different types of hormones to interact with each other and modulate the cell's response.

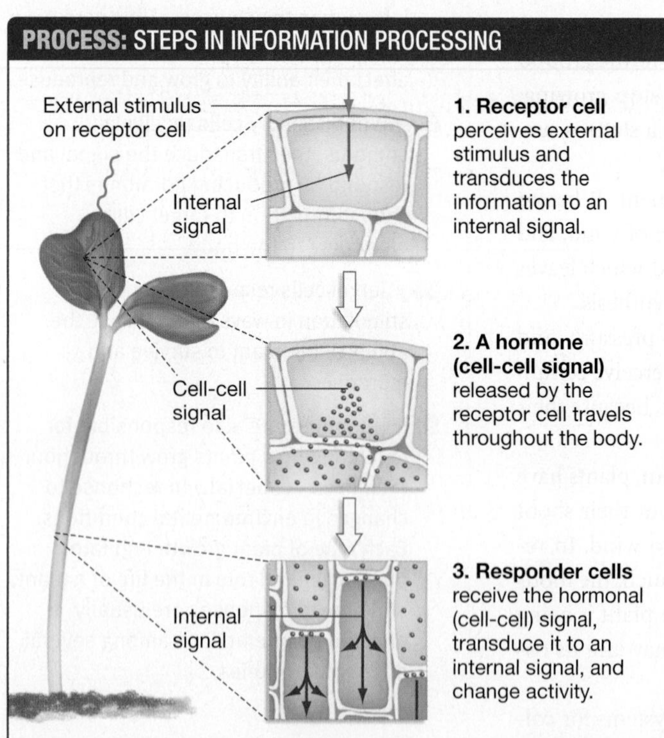

FIGURE 39.1 In Many Cases, Hormones Link a Stimulus and a Response.

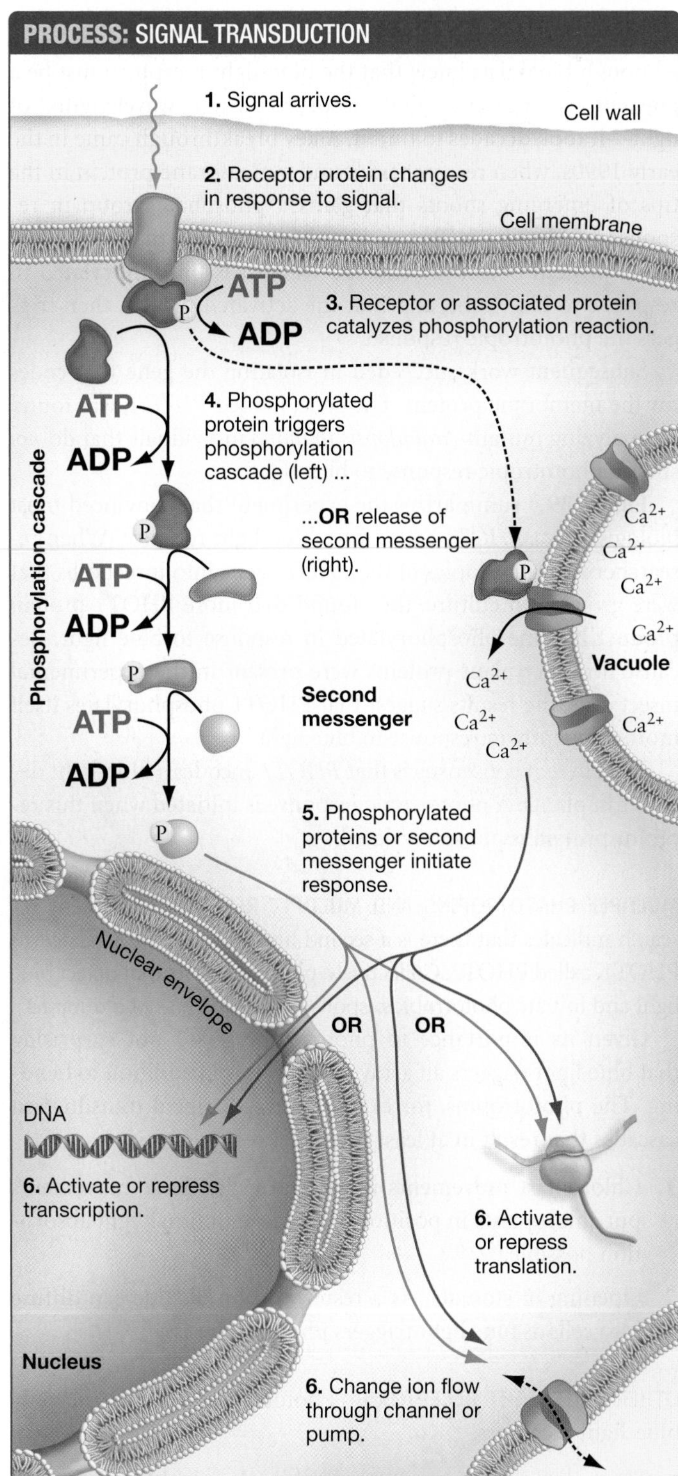

FIGURE 39.2 Signal Transduction Changes an External Signal to an Internal Signal. After an environmental or cell-cell signal is transduced to an intracellular signal, it triggers a change in the cell's activity.

How Do Cells Respond to Cell-Cell Signals?

Cells are exposed to a constant stream of hormones. Many of these signals have little to no effect on what is happening inside. The reason is simple: Hormones can elicit a response only if a cell has an appropriate receptor.

But if a receptor exists on or in the cell, an arriving hormone binds to it. When the receptor changes shape in response, the effect is like a sharp rap on the door of your room. The signal—often received at the cell's periphery—is rapidly transduced to a series of phosphorylated proteins or the release of a second messenger inside the cell. The result may be a dramatic change in the cell's activity.

To understand why these binding events can have such significant effects on cell activity, look again at Figure 39.2. Activation of a signal transduction cascade results in the production of many phosphorylated proteins, or the release of many second messengers. These events amplify the original signal many times. In this way, low concentrations of plant hormones can have a large impact on target cells. Hormones are tiny molecules in tiny concentrations, but because their signal is amplified during signal transduction, they produce big results.

As Figure 39.2 shows, the response to hormone binding can include:

- activation of membrane transport proteins (Chapter 6), which produces a change in the membrane's electrical potential or the cell wall's pH; or
- changes in gene expression—via alterations in transcription activators or repressors (Chapter 18) or the translation machinery—that result in new suites of proteins or RNAs in the cell.

The response to the signal is important. When cells respond to a hormone, the change in their activity helps the plant cope with the environmental change sensed by the receptor cell.

CHECK YOUR UNDERSTANDING

If you understand that . . .

- Plants can respond to changes in their environment because sensory cells receive information from the environment, sensory cells produce hormones or other cell-cell signals, and target cells respond to hormones.
- Information processing in sensory cells and the cells that respond to cell-cell signals involves three steps:
  Step 1 A receptor molecule changes in response to the signal.
  Step 2 A signal transduction pathway transforms the hormonal signal into an intracellular signal.
  Step 3 The intracellular signal triggers a response—transcription of target genes, activation of specific enzymes or transport proteins, or other changes in cell activity.

✓ **You should be able to . . .**

1. Explain why only certain cells in the body—not all—respond to an environmental signal, and why only certain cells respond to a hormone.
2. Explain how cell-cell signals are amplified inside a target cell.

Answers are available in Appendix B.

39.2 Blue Light: The Phototropic Response

Most of the general principles of plant communication emerged from studies of how plants respond to light. For example, consider the claim that plants are highly selective about the information they process. Light is made up of a wide array of wavelengths (see Chapter 10), but plants sense and respond to only a few.

This conclusion traces back to experiments that Charles Darwin and his son Francis did in 1881 with coleoptiles of a plant called reed canary grass. A **coleoptile** is a modified leaf that forms a sheath protecting the emerging shoots of young grasses.

The Darwins germinated seeds in the dark, placed the young, straight shoots next to a light source, and noted that the shoots grew toward the light—a response, due to differential cell elongation, called bending. You have probably seen the same response in houseplants that are near a window (see the photo at the start of the chapter). Directed movement in response to light is called **phototropism** (literally, "light-turn").

When the Darwins exposed shoots to light filtered through a solution of potassium dichromate, however, the shoots did not bend toward the light. Potassium dichromate solutions filter out wavelengths in the blue part of the visible spectrum. The photo in **Figure 39.3** shows the results of a follow-up experiment, with coleoptiles that have been exposed to blue, yellow, green, orange, and red light. Bending occurs only toward light that contains blue wavelengths.

To understand the specificity of the response is important, recall from Chapter 10 that (1) chlorophylls *a* and *b* are the primary photosynthetic pigments, and (2) these pigments absorb strongly in the blue and red parts of the spectrum. Plants exhibit a phototropic response if blue wavelengths are available, but show no response if blue wavelengths are not present. Plants move toward blue light because it is important in photosynthesis.

Let's look at what happens when blue light strikes a receptor cell, and follow through to the cells that undergo differential growth in response.

FIGURE 39.3 Experimental Evidence That Plants Sense Specific Wavelengths of Light. Each of these shoots was exposed to a different wavelength of light. In each case, the light source was on the left as the shoots are shown here; only the shoot exposed to blue light bent toward the source.

Phototropins as Blue-Light Receptors

Although biologists knew that the blue-light receptor must be a **pigment**—a molecule that absorbs certain wavelengths of light—it took decades to find it. A key breakthrough came in the early 1990s, when researchers found a membrane protein in the tips of emerging shoots that gains a phosphate group in response to blue light. Researchers hypothesized that the membrane protein becomes activated when it is phosphorylated in response to blue light, and that the activated protein then triggers the phototropic response.

Subsequent work succeeded in isolating the gene that codes for the membrane protein. The gene, named *PHOT1*, was found by analyzing mutant *Arabidopsis thaliana* individuals that do *not* show a phototropic response to blue light.

Figure 39.4 summarizes the experiment that convinced most biologists that *PHOT1* codes for a blue-light receptor. When researchers inserted copies of the *PHOT1* gene into insect cells that were growing in culture, they found that more PHOT1 protein product became phosphorylated in response to blue light. Because no other plant proteins were present in the experimental insect cells, the results suggest that PHOT1 phosphorylates itself more frequently in response to blue light.

The current consensus is that *PHOT1* encodes a blue-light detector in plants. A phototropic response is initiated when this receptor protein is phosphorylated.

MULTIPLE PHOTOTROPINS AND MULTIPLE RESPONSES Recent research indicates that there is a second blue-light receptor related to PHOT1, called PHOT2. Collectively, photoreceptors that detect blue light and initiate phototropic responses are known as **phototropins**.

Given its importance to photosynthesis, it's not surprising that blue light triggers an array of responses in addition to bending. The phototropins, for example, trigger signal transduction cascades that result in at least two other responses.

1. Chloroplast movements inside leaf cells. These movements put chloroplasts in positions that make optimal light absorption possible.
2. Opening of stomata. As a result, carbon dioxide can diffuse into cells as blue light triggers photosynthesis.

OTHER BLUE-LIGHT RECEPTORS Phototropins are not the only blue-light receptors.

- The carotenoid pigment zeaxanthin absorbs blue light and, along with phototropins, is involved in the opening of stomata in response to blue light.
- A group of photoreceptors called cryptochromes (literally, "hidden-colors") are blue-light photoreceptors involved in stem growth under shady conditions and the developmental change that leads to flower production.

Phototropism, however, ranks as the best-studied of all blue-light responses. How is the signal from PHOT proteins transmitted to the cells that cause bending?

EXPERIMENT

QUESTION: Does *PHOT1* encode a blue-light receptor?

HYPOTHESIS: The *PHOT1* gene codes for a blue-light receptor.

NULL HYPOTHESIS: The *PHOT1* gene does not code for a blue-light receptor.

EXPERIMENTAL SETUP:

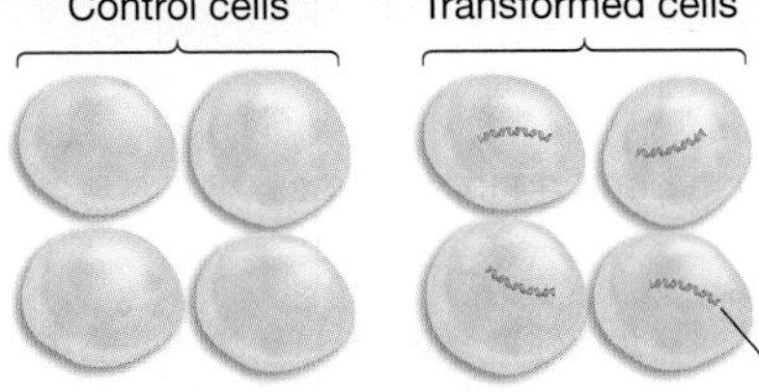

1. Transform cells. Insert *PHOT1* genes into half of insect cells in culture. These cells should produce PHOT1 protein. The other half serve as the control.

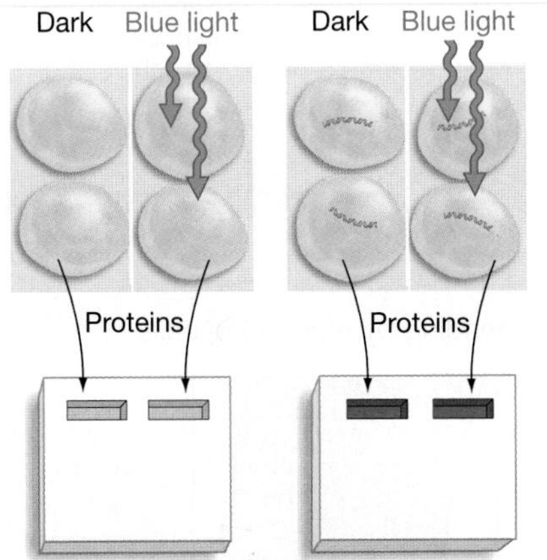

2. Expose to light. Grow cells in a medium containing radioactive phosphorus and subject cells to darkness or to blue light.

3. Isolate proteins and separate via electrophoresis. See BioSkills 9 for an introduction to this technique.

PREDICTION: Insect cells containing the *PHOT1* gene will have a radiolabeled band when treated with blue light.

PREDICTION OF NULL HYPOTHESIS: Insect cells containing the *PHOT1* gene, like control cells, will not have a radiolabeled band.

RESULTS:

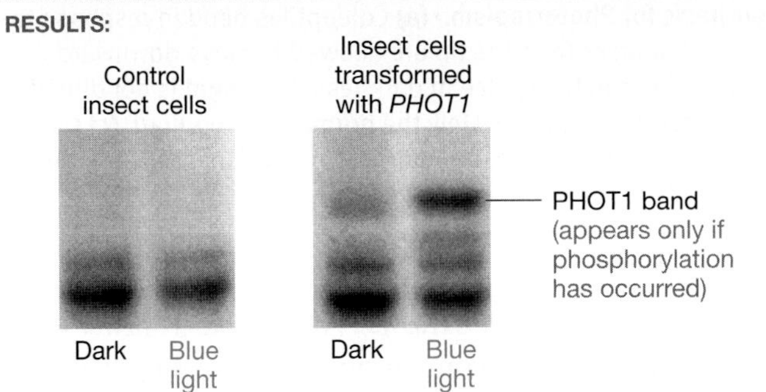

CONCLUSION: Much more PHOT1 is phosphorylated in response to blue light than in darkness. Because no other plant proteins were present in the experimental cells, PHOT1 must phosphorylate itself.

FIGURE 39.4 Experimental Evidence That PHOT1 Is a Blue-Light Receptor That Phosphorylates Itself.

SOURCE: Christie, J. M., P. Reymond, G. K. Powell, P. Bernasconi, A. A. Raibekas, E. Liscum, and W. R. Briggs. 1998. *Arabidopsis* NPH1: A flavoprotein with the properties of a photoreceptor for phototropism. *Science* 282: 1698–1701.

✔**QUESTION** Why did the researchers bother to analyze cells that were not transformed with the *PHOT1* gene and that were not exposed to blue light?

Auxin as the Phototropic Hormone

Long before the phototropins were identified, biologists knew that receptor cells responded to blue light by releasing a hormone. The Darwins established this result when they followed up on their initial experiments. **Figure 39.5** shows their experimental setup and results.

- If they removed the tips of coleoptiles, they found that the decapitated seedlings stopped bending toward the light.
- If they covered the tips of coleoptiles with opaque material, the seedlings did not bend toward light.

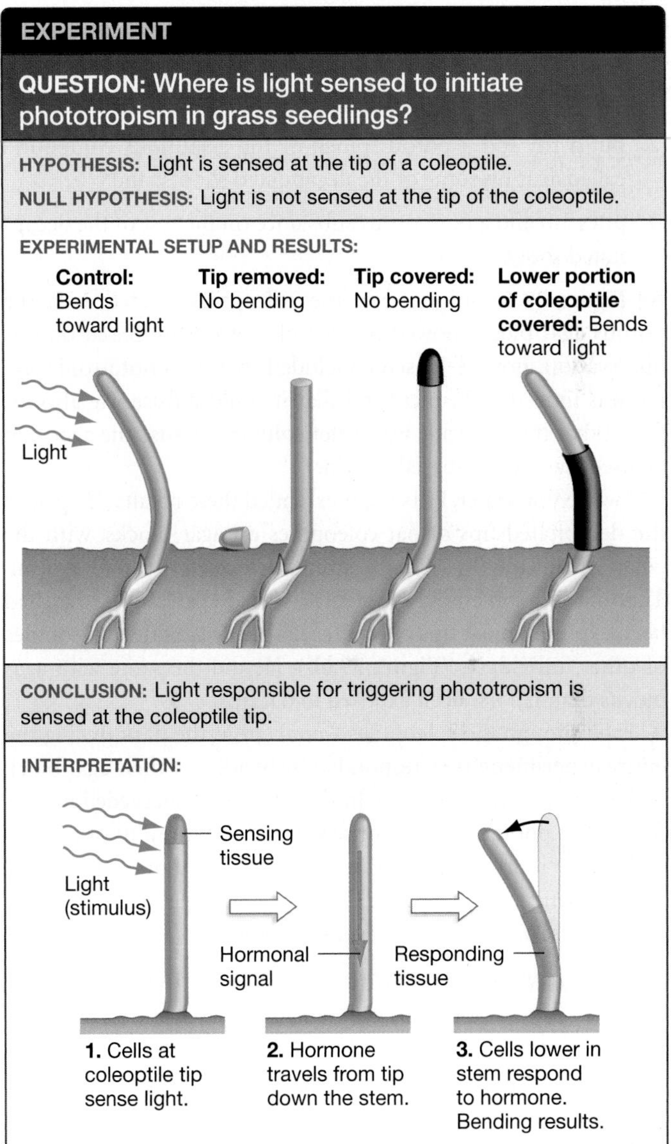

FIGURE 39.5 The Sensory and Response Cells Involved in Phototropism Are Not the Same.

SOURCE: Darwin, C. and F. Darwin. 1897. *The Power of Movement in Plants* (D. Appleton & Co. New York).

✔**QUESTION** A critic could argue that this experiment lacked appropriate controls for the treatments labeled "Tip removed" and "Tip covered." Suggest better controls for these treatments than the unmanipulated individual at the far left.

- If they put opaque collars below the tips, in the area where bending occurs, the seedlings bent toward light normally.

These data provided convincing evidence that the blue-light sensors were located in the tips of the coleoptiles.

How did the sensory cells in the tip communicate with the cells that actually elongate? The Darwins proposed that phototropism depends on "some matter in the upper part which is acted on by light, and which transmits its effects to the lower part." Their hypothesis was that a substance produced at the tip of the coleoptile acts as a signal and is transported to the area of bending. This was the first explicit hypothesis stating that hormones—signaling molecules that can act at a distance—must exist.

The hormone hypothesis was not tested rigorously until 1913, when Peter Boysen-Jensen:

1. cut the tips off young oat shoots;
2. put a tip and a porous block of the gelatinous compound called agar on some of the decapitated shoots; and
3. put a tip and a nonporous substance on the rest of the decapitated shoots.

As **Figure 39.6a** shows, only the coleoptiles treated with the porous agar block showed normal phototropism. Based on this observation, Boysen-Jensen concluded that the phototropic signal was indeed a chemical and that it could diffuse. He also inferred that the molecule was water soluble, because the agar that he used was a water-based gelatin.

Twelve years later, Frits Went extended these results. He placed the decapitated tips of oat coleoptiles on agar blocks, with the goal of collecting the hypothesized hormone for phototropism. Then he did something clever: He placed agar blocks that had been exposed to oat tips off-center on the decapitated coleoptiles of other individuals (**Figure 39.6b**). He did the same with agar blocks that had *not* been exposed to oat tips.

Even though the coleoptiles were kept in the dark during the entire experiment, they responded by bending if their agar block had been exposed to oat tips. In this way, Went succeeded in producing the phototropic response without the stimulus of light.

Because it promotes cell elongation in the shoot, Went named the hormone **auxin** (from the Greek *auxein*, "to increase"). Auxin was the first plant hormone ever discovered.

ISOLATING AND CHARACTERIZING AUXIN After years of effort, researchers in two laboratories independently succeeded in isolating and characterizing auxin. The hormone turned out to be indole acetic acid, or IAA. It was hard to find because it is rare. Depending on the species and tissue involved, IAA concentrations range from 3 to 500 nanograms per gram of tissue. (The prefix *nano–* refers to billionths.)

Like the other plant hormones introduced in this chapter, auxin is a small molecule with a relatively simple structure. It is present in quantities so small that its concentration is difficult to measure. Yet its impact is huge. Auxin can bend stems—producing, in some cases, tree trunks that are permanently bowed.

(a) The phototropic signal is a chemical.

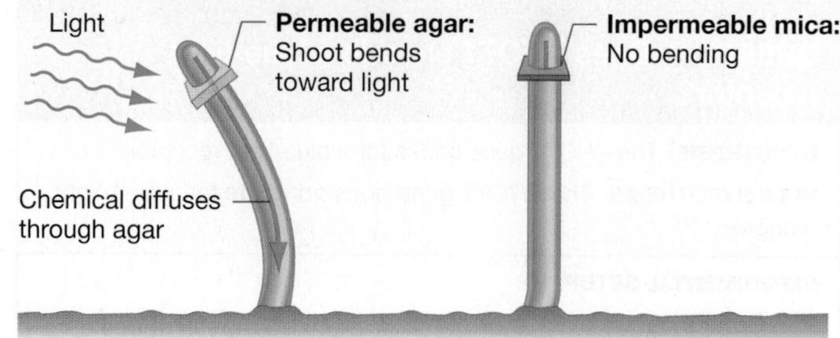

(b) The hormone can cause bending in darkness.

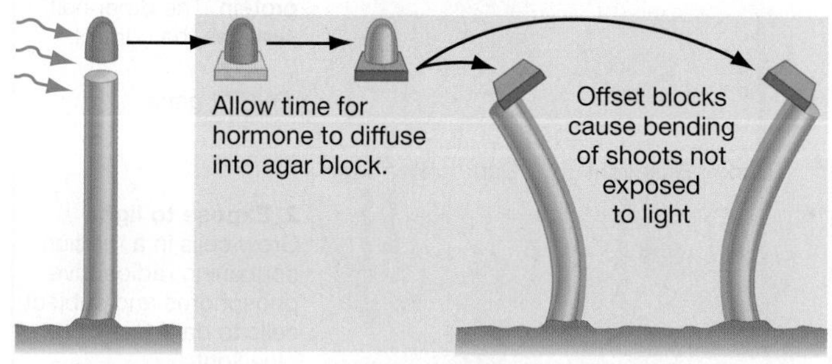

(c) The hormone causes bending by elongating cells.

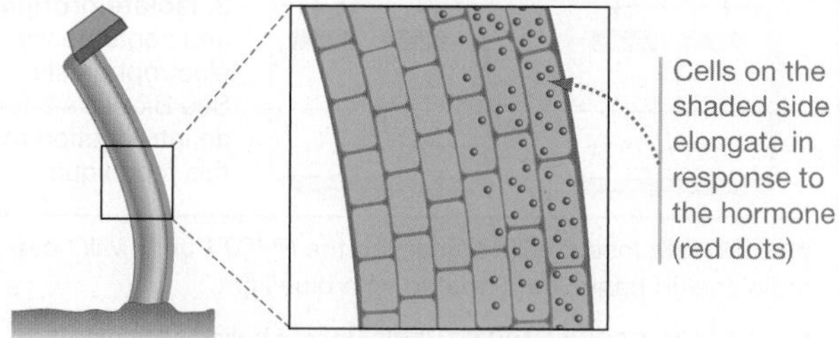

FIGURE 39.6 Experimental Evidence Supports the Hormone Hypothesis for Phototropism. (a) Coleoptiles bend in response to light if substances from the tip are allowed to move downward. **(b)** If bending can take place in darkness, then light is not directly required for the response. Only the hormone is required. **(c)** During the phototropic response, bending occurs because cells on the shaded side of the shoot elongate.

THE CHOLODNY-WENT HYPOTHESIS Went's experiments were a breakthrough in research on information processing: They confirmed the hormone hypothesis and led to the discovery and characterization of IAA. But Went's experiments also inspired an important hypothesis for *how* the hormone produces the bending response. Working independently, both N. O. Cholodny and Went proposed that phototropism results from an asymmetric distribution of auxin.

The Cholodny-Went hypothesis contends that:

1. Auxin produced in the tips of coleoptiles is shunted from one side of the tip to the other in response to light.
2. Auxin is then transported straight down one side of the shoot.

3. The asymmetric distribution of auxin causes cells on the shaded side of the coleoptile to elongate more than cells on the illuminated side (**Figure 39.6c**).

Bending results. In essence, bending results from differential elongation.

To test the Cholodny-Went model, Winslow Briggs divided coleoptile tips completely or partially in half with a thin piece of mica, then exposed one side to sunlight (**Figure 39.7**). Mica is impermeable to dissolved molecules. Briggs's idea was that the movement of auxin would be stopped in the completely divided tips and agar blocks, but it would not be stopped in the partially divided tips and agar blocks.

Next, Briggs placed the resulting blocks on one side of decapitated shoots and recorded the bending response.

- If the tip was completely divided, there was no difference in bending induced by the sunlit or shaded side of the tip.
- If the tip was partially divided, the bending responses differed in the sunlit and shaded sides. The side away from light induced much more bending, indicating that auxin had been transported from one side of the tip to the other.

The Cholodny-Went asymmetric distribution model was correct. It explained how auxin leads to asymmetric cell elongation, and thus the bending response called phototropism.

THE CELL-ELONGATION RESPONSE How do cells in the stem respond to auxin? Experiments in corn plants and *Arabidopsis* suggest that proteins called TIR1 and auxin-binding protein 1, or ABP1, are auxin receptors found in stem and leaf cells. Researchers proposed that once TIR1 or ABP1 has bound to auxin, the signal transduction cascade that follows increases the number of membrane H^+-ATPases, or proton pumps, in the plasma membrane.

Recall from Chapter 38 that **proton pumps** use the energy in ATP to drive protons out of the cell against an electrochemical gradient. Because the pH of the cell wall decreases when H^+-ATPases are active, the idea that these pumps are responsible for cell elongation became known as the **acid-growth hypothesis**.

To understand the rationale behind the acid-growth hypothesis, it's important to realize that two things have to happen for a plant cell to get larger:

1. The cell wall has to expand to create a larger volume.
2. Water has to enter the cell and generate turgor pressure on the cell wall to make an increase in volume possible.

EXPERIMENT

QUESTION: How does an asymmetric distribution of auxin in shoot tips, which causes bending, develop?

HYPOTHESIS: Auxin moves from the sunny side to the shady side of shoot tip, resulting in an asymmetric distribution.

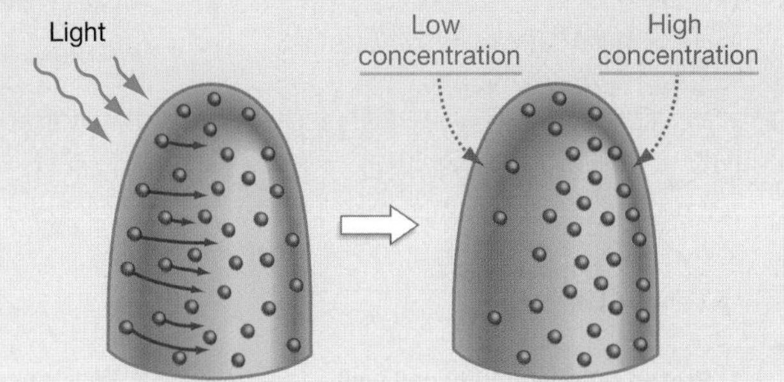

NULL HYPOTHESIS: Auxin does not move from the sunny side to the shady side of the shoot tip.

EXPERIMENTAL SETUP:

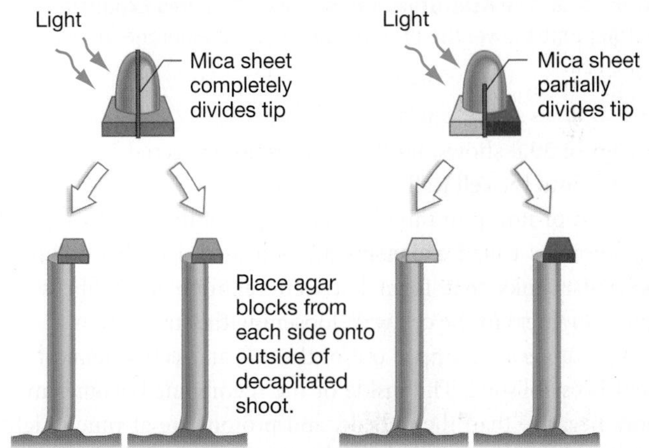

PREDICTION: In completely divided tip, both agar blocks will elicit the same degree of bending. In partially divided tip, block receiving the most light will elicit less bending.

PREDICTION OF NULL HYPOTHESIS: The same degree of bending will occur in all treatments.

RESULTS:

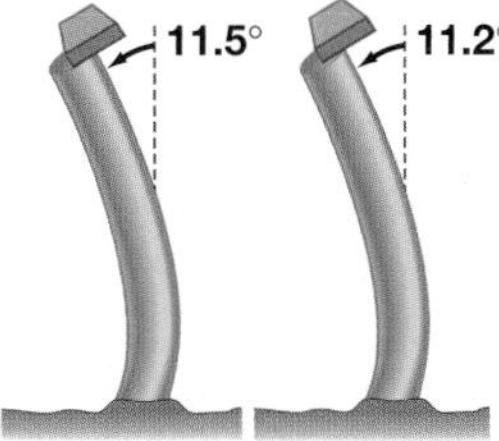

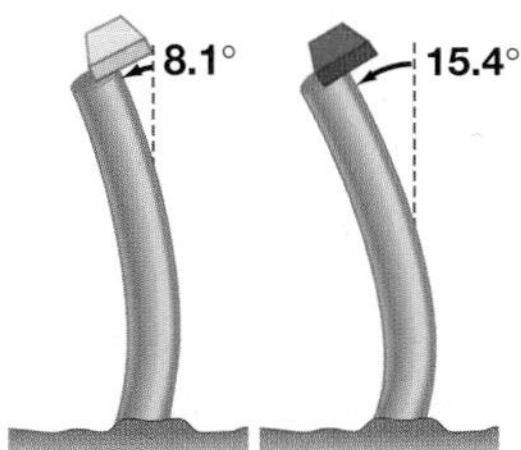

Interpretation: Mica prevented flow of auxin to shaded side, and shoots bent the same amount.

Interpretation: Auxin was redistributed to shaded side, which bent more than sunny side.

CONCLUSION: Asymmetric distribution of auxin results from lateral redistribution in tip.

FIGURE 39.7 Testing the Auxin Redistribution Hypothesis.

SOURCE: Baskin, T. I., M. Iino, P. G. Green, and W. R. Briggs. 1985. High-resolution measurements of growth during first positive phototropism in maize. *Plant, Cell & Environment* 8: 595–603.

✔**QUESTION** Consider the hypothesis that cutting coleoptile tips and inserting mica sheets disrupts normal cell activity. What prediction does this hypothesis make?

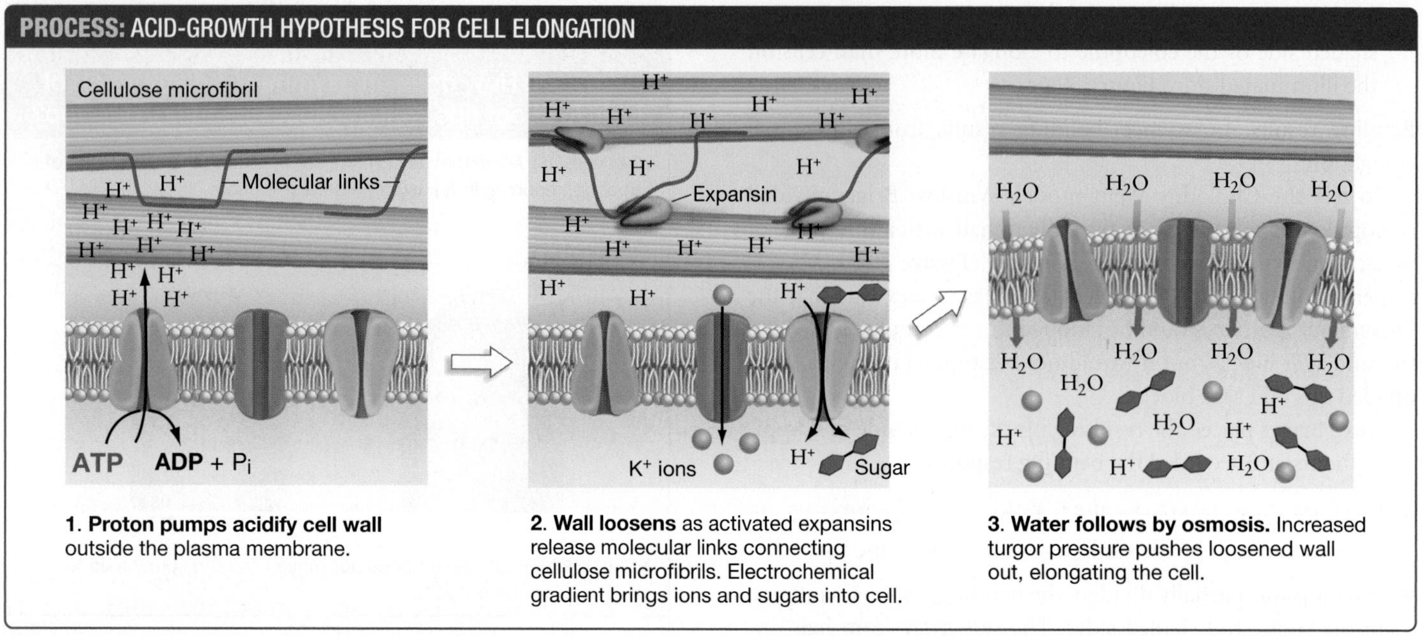

PROCESS: ACID-GROWTH HYPOTHESIS FOR CELL ELONGATION

Cellulose microfibril — Molecular links — ATP — ADP + Pi

Expansin — K+ ions — Sugar

1. Proton pumps acidify cell wall outside the plasma membrane.

2. **Wall loosens** as activated expansins release molecular links connecting cellulose microfibrils. Electrochemical gradient brings ions and sugars into cell.

3. **Water follows by osmosis.** Increased turgor pressure pushes loosened wall out, elongating the cell.

FIGURE 39.8 The Acid-Growth Hypothesis Requires Expansion of Cell Wall and Intake of Water. When activated proton pumps lower the pH outside the cell membrane in response to auxin, a series of events leads to elongation of the cell.

As **Figure 39.8** shows, both processes are triggered by pumping protons into the cell wall.

When proton pumping lowers the pH of the wall to 4.5, cell-wall proteins called **expansins** are activated. Expansins "unzip" molecular links that form between cellulose microfibrils and other polymers in the cell wall, loosening the structure.

As protons are pumped out of the cell, an electrochemical gradient is established. The inside of the membrane becomes much more negative than the outside, and protons are at much higher concentrations outside than inside. The gradient favors the entry of potassium (K^+) or other positively charged ions, as well as sugars that enter via proton cotransporters. As the concentration of solutes increases inside the cell, water follows via osmosis.

This is a key point: Cells don't move water directly. Instead, they create an osmotic gradient that favors water movement. Pumping protons *out* of a cell is a way to bring water *into* the cell.

The incoming water increases turgor pressure, which pushes out the loosened cell wall. The cell gets bigger.

The upshot? When cell walls on one side of a stem are acidified in response to a signal from auxin, bending results. Auxin's role in phototropism may qualify as the best-understood example of information processing in plants.

CHECK YOUR UNDERSTANDING

If you understand that . . .

The chain of events involved in phototropism can be summarized as follows:

- When phototropins in shoot-tip cells absorb blue light, auxin is redistributed to the shaded side of the tip.
- Auxin is transported down the shoot and binds to receptors in target cells.
- These cells elongate when activated receptors lead to the activation of proton pumps, acidification of the cell wall, and activation of expansin proteins.

✔ **You should be able to . . .**

Assuming that you have a large supply of purified auxin, state how you would manipulate a large bed of roses so their stems bend toward the east.

Answers are available in Appendix B.

39.3 Red and Far-Red Light: Germination and Stem Elongation

Plants are sensitive to wavelengths in the red and far-red portions of the visible spectrum, as well as to blue light. This sensitivity is interesting, because red wavelengths (about 660 to 700 nm) and far-red wavelengths (over 710 nm) signal very different things to a plant.

- Red light drives photosynthesis, just as blue light does.
- Far-red wavelengths are not absorbed strongly by photosynthetic pigments, so they tend to pass through leaves. As a result, far-red wavelengths are prominent in light that is filtered through tree leaves before it reaches the forest floor. Far-red light indicates shade.

To review experimental work on the differences between red and far-red light, go to the study area at *www.masteringbiology.com.*

Web Activity Sensing Light

The Red/Far-Red "Switch"

The first hint that plants monitor red and far-red light emerged from studies on how lettuce seeds germinate. By exposing lettuce seeds to various wavelengths of light and plotting their frequency of germination, researchers discovered that germination rates peak when seeds receive red light (about 660 nm).

This observation made sense, because lettuce thrives best when it grows in bright sunlight. But the stimulatory effect of red light disappeared if seeds were later exposed to far-red light.

This observation also made sense, because far-red light indicates that the seeds are shaded. Wavelengths near 735 nm inhibit germination the most effectively.

Follow-up experiments showed that red and far-red light act like an on-off switch for lettuce seed germination (**Table 39.1**). Red light promotes lettuce germination; far red inhibits it.

The key observation, though, is that the last wavelength sensed by the seed determines whether germination occurs at a high rate. This result implies that plants sense red and far-red light *together*. How could this happen?

Phytochromes as Red/Far-Red Receptors

To interpret the red/far-red switch in seed germination, biologists hypothesized that the same pigment absorbs both wavelengths. Further, they suggested that the pigment exists in two shapes, or conformations: one shape absorbs red light, and one shape absorbs far-red light.

The idea was that switching behavior, or **photoreversibility**, occurs because light absorption makes the photoreceptor pigment change shape, like a light switch moving up or down in response to touch. Each conformation would be responsible for a different response.

TABLE 39.1 **How Do Red Light and Far-Red Light Affect the Germination of Lettuce Seeds?**

Biologists exposed moistened lettuce seeds to flashes of light containing one of two wavelengths: red or far-red (FR). After exposure to light, the seeds were held in the dark for several days.

Light Exposure	Germination (%)
None (control)	9
Red	98
Red → FR	54
Red → FR → Red	100
Red → FR → Red → FR	43
Red → FR → Red → FR → Red	99
Red → FR → Red → FR → Red → FR	54
Red → FR → Red → FR → Red → FR → Red	98

SOURCE: H. A. Borthwick et al. 1952. A reversible photoreaction controlling seed germination. *PNAS* 38: 662–666, Table 1.

✔**QUESTION** According to the data above, what is the average germination rate of lettuce seeds that were last exposed to red light? To far-red light? How do these values compare with the germination rate of seeds that are buried underground and receive no light at all?

Biologists called the hypothesized pigment **phytochrome** ("plant-color"). Phytochrome was thought to be a specialized light receptor, different from any of the pigments involved in absorbing light during photosynthesis.

Figure 39.9 illustrates the photoreversibility hypothesis. One conformation of phytochrome called P_r (phytochrome red) absorbs red light. Another conformation of the same molecule, called P_{fr} (phytochrome far-red), absorbs far-red light. According to the photoreversibility hypothesis, each conformation switches to the other when it absorbs its preferred wavelength.

How Were Phytochromes Isolated?

Young corn and bean plants lengthen their stems in response to light deprivation or exposure to excessive far-red light. If you have grown these seeds indoors, you have seen this response to far-red light firsthand. The shoots get long and spindly.

Corn and beans normally grow in open sunlight. If they are grown under incandescent light indoors, the plants react as though they are being shaded—they attempt to grow high enough to reach full sunlight. This behavior suggested that young corn and bean plants have a receptor protein for far-red light.

To follow up on this observation, researchers purified proteins from corn coleoptiles and succeeded in isolating one that was photoreversible. Specifically, when the protein was placed in solution and exposed to alternating red and far-red light, the color of the solution switched from blue to blue-green and back. The color switches supported the hypothesis that the same protein absorbed red as well as far-red light. Phytochrome was found.

Follow-up work showed that a region of the phytochrome molecule changes shape in response to red and far-red light. The

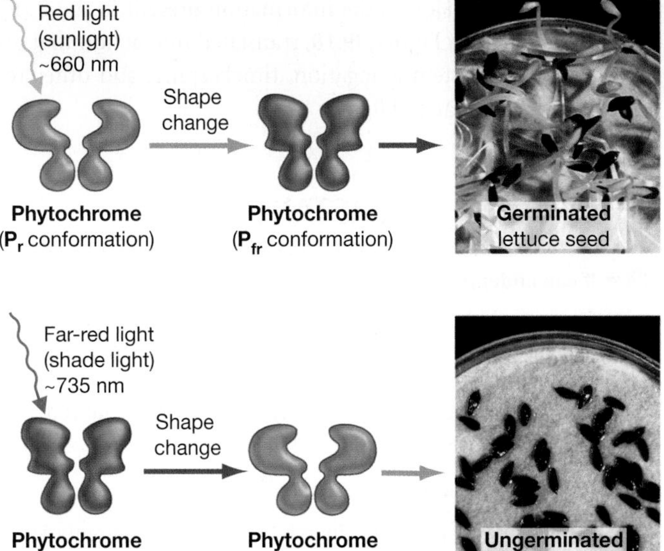

FIGURE 39.9 The Photoreversibility Hypothesis for Phytochrome Behavior. According to the photoreversibility hypothesis, phytochrome switches between the P_r conformation and the P_{fr} conformation when it absorbs red or far-red light respectively. Each form of the protein has a different effect on germination.

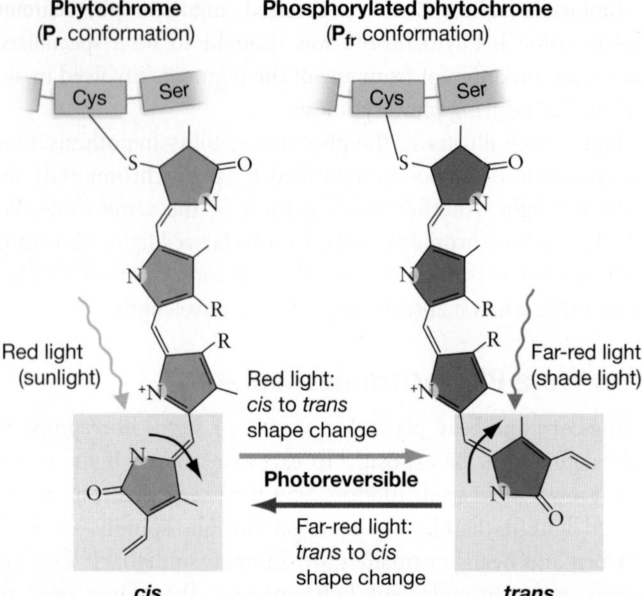

FIGURE 39.10 Phytochrome Changes Shape after Absorbing Red and Far-Red Light. This diagram shows how a subunit of phytochrome changes shape in response to absorbing red or far-red light. The shape changes correspond to phytochrome's P_r and P_{fr} conformations.

✔**QUESTION** How does this shape change relate to signal transduction?

shape changes cause phytochrome to take on or lose a phosphate group, just as the shape changes triggered by blue light result in phosphorylation of the phototropins. Conformational changes in proteins are frequently associated with a change in phosphorylation.

As Chapter 40 will show, phytochrome also plays a key role in timekeeping—the ability of plants to track changes in the length of nights and days. How is the information present in the P_r/P_{fr} switch, illustrated in **Figure 39.10**, translated into action that affects germination, stem elongation, timekeeping, and other responses to red and far-red light?

CHECK YOUR UNDERSTANDING

If you understand that . . .

- Plants respond to red and far-red light via phytochromes, which change shape and activity when they absorb red or far-red light.
- Red light acts as a sunlight or day indicator, while far-red light acts as a shade or night indicator.

✔ **You should be able to . . .**

1. Explain why it is adaptive for red light to trigger germination in lettuce seeds while far-red light inhibits it.
2. Explain why it is adaptive for plants that normally grow in sunny habitats to elongate their stems in response to far-red light.

Answers are available in Appendix B.

This question brings us to the forefront of research on phytochromes. The answer appears to be complex; one study suggests that almost one-third of all *Arabidopsis* genes are involved in the elongated growth characteristic of shade avoidance. Work on linking the phytochrome switch to the elongation and germination responses continues.

39.4 Gravity: The Gravitropic Response

The wavelengths, quantity, and direction of light that a plant receives change with the season, weather, time of day, and shading by other plants. However, gravity is constant and unidirectional. Light means food; gravity provides information about how the plant should orient itself in space.

Shoots usually respond to gravity by growing in an upward direction; roots usually respond by growing downward or laterally. How do plants sense gravity, so that it can be used as a signal to orient the body?

In 1881 Charles and Francis Darwin published one of the first experimental results on **gravitropism** ("gravity-turn")—the ability to move in response to gravity. Recall from Chapter 36 that the ends of root tips are covered by a protective collection of cells called the **root cap**. The Darwins found that roots stop responding to gravity if the root caps are removed. This observation suggested that gravity sensing occurs somewhere inside the root cap.

Recently biologists demonstrated precisely which cells are involved in gravity sensing in *Arabidopsis* roots. By killing tiny blocks of cells with laser beams, researchers showed that the cells illustrated in **Figure 39.11**, at the center of the root cap, are the most important for regulating the gravitropic response. Cells in the tips of roots respond to gravity and initiate gravitropism. How do root cap cells sense the force of gravity?

The Statolith Hypothesis

The leading explanation for how plants sense gravity—the **statolith hypothesis**—is based on two interconnected ideas:

(a) Root tips have a protective cap.

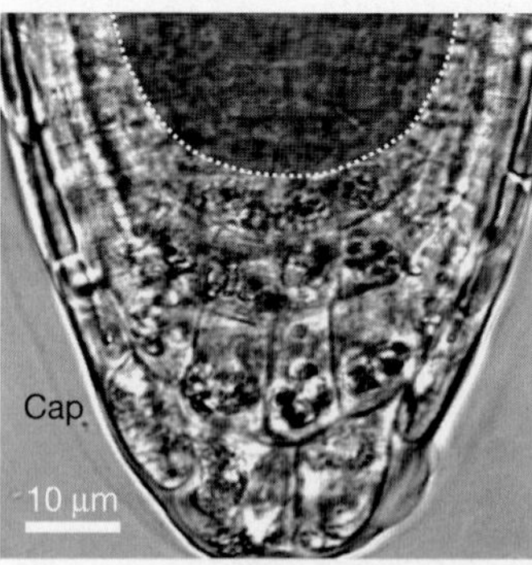

(b) Gravity-sensing cells are in the center of the cap.

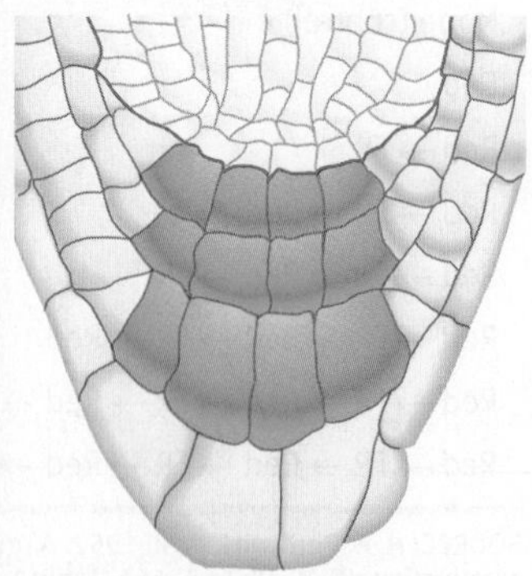

FIGURE 39.11 Gravity-Sensing Occurs in the Root Cap. The root cap is a protective structure. When the cells marked in orange are killed experimentally, the gravitational response in roots is dramatically reduced.

1. Dense, starch-storing organelles called **amyloplasts** respond to gravity by being pulled to the bottom of root cap cells (**Figure 39.12**).

2. The weight of the amyloplasts activates sensory proteins located in the plasma membrane. These sensory proteins initiate the gravitropic response.

The statolith hypothesis was inspired by animals that use dense particles to sense gravity. Lobsters, for example, take up grains of sand that become positioned in specialized gravity-sensing organs in their antennae. The grains of sand are called **statoliths** ("place-stones"). When the animal tilts or flips over, the statolith moves in response to gravity. Inside the organ, the sand grain ends up pushing against a sensory cell. When this cell is activated, it indicates that the animal is no longer upright.

According to the statolith hypothesis, the same thing happens in root cap cells. If the wind tips a plant over, for example, the amyloplasts settle onto the new "lower" cell walls. The weight activates a new set of receptors, which signal that the root no longer faces in the correct direction.

Although recent experiments strongly support the statolith hypothesis, the search for the gravity receptor itself continues. In contrast, the second and third steps of information processing in response to gravity—the production of a cell-cell signal and the response of target cells—is much better understood.

Auxin as the Gravitropic Signal

Root cap cells that sense changes in the direction of gravitational pull respond by changing the distribution of auxin in the root tip. **Figure 39.13** illustrates this chain of events.

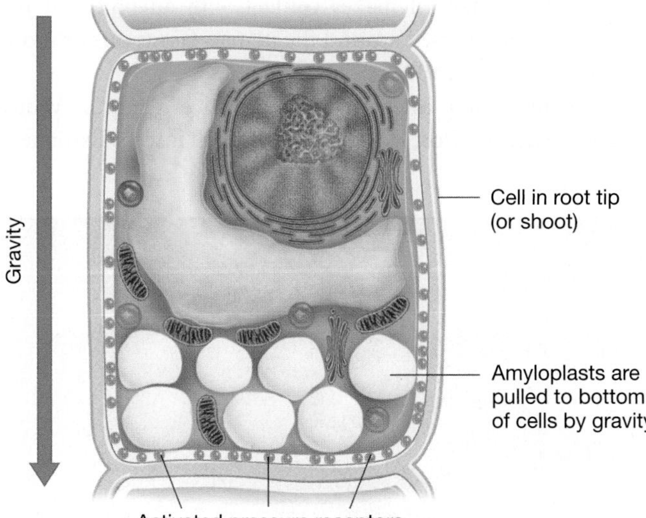

FIGURE 39.12 The Statolith Hypothesis States That Amyloplasts Stimulate Sensory Cells. Amyloplasts are filled with starch. They are dense, so they sink in response to gravity. The statolith hypothesis predicts that pressure receptors in the plasma membrane become activated as a result.

✔**EXERCISE** Assuming that the receptor is a protein, suggest a hypothesis to explain how signal transduction occurs in response to pressure from a statolith.

Step 1 Under normal conditions, auxin flows down the middle of the root and then toward the perimeter and away from the root cap.

Step 2 If the root is tipped, sensory receptors trigger changes in transport proteins that redistribute auxin.

Step 3 Auxin is redistributed: The lower portion of the root receives increased concentrations of auxin; the upper portion receives lower concentrations. The redistribution is mediated by changes in the location of auxin-binding PIN proteins.

Step 4 In response to the differences in auxin concentration, cells in the lower portion of the root grow more slowly and cells in the upper portion grow more quickly. In roots, high auxin concentrations inhibit growth. The result is bending.

Note that the way root cells respond to auxin redistribution during the gravitropic response is opposite to the way that cells in the stem respond during phototropism. In stems, high concentrations of auxin lead to *increased* cell elongation and bending. In roots, high concentrations of auxin lead to *decreased* cell division and elongation. Roots bend as cells on the other side of the zones of cellular division and elongation

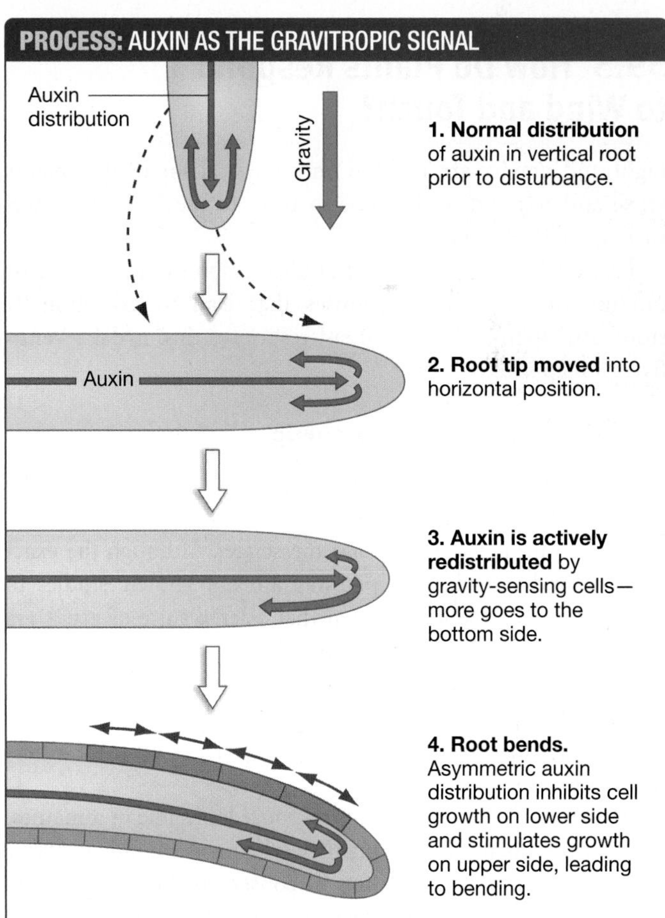

FIGURE 39.13 The Auxin Redistribution Hypothesis for Gravitropism. This sequence of events might begin when a growing root tip hits a rock and is displaced horizontally or when a plant is tipped in a windstorm and partially uprooted.

(see Chapter 36) continue to grow, in response to low auxin concentrations. The mechanism responsible for this difference is still unknown.

To review how auxin is redistributed during phototropism and gravitropism, go to the study area at *www.masteringbiology.com.*

Web Activity Plant Hormones

CHECK YOUR UNDERSTANDING

If you understand that . . .

- Cells in root caps sense gravity via pressure that amyloplasts exert on receptors.
- Changes in gravity sensing result in a redistribution of auxin and changes in the growth rate of root tips.

✓ You should be able to . . .

1. Explain the parallel that exists between auxin redistribution in shoots and roots and the responses called phototropism and gravitropism.
2. Explain why auxin could be considered a gravitropic hormone.

Answers are available in Appendix B.

FIGURE 39.14 Plant Growth Changes in Response to Wind or Touch. The tomato plants shown here were touched lightly 0, 10, or 20 times per day each day for 10 consecutive days.

✓QUESTION What is the adaptive significance of this response? Give an example of when it would occur in nature.

39.5 How Do Plants Respond to Wind and Touch?

Light and gravity are not the only physical forces that plants sense and respond to. Plants also react to mechanical stresses such as wind and touch.

Let's analyze two of the best-studied responses to pushing or pulling forces: growth responses that lead to exceptionally stout, stiff stems, and movement responses that make a Venus flytrap snap shut.

Changes in Growth Patterns

When plants are buffeted by wind, receptor cells transduce the mechanical force into an internal signal in the form of phosphorylated proteins or a second messenger. Although the exact receptor and transduction pathway are not known, studies in *Arabidopsis thaliana* have shown that a large suite of genes are transcribed in response to touch or other mechanical stimuli that mimic the effect of wind. The protein products of some of these genes act to stiffen cell walls, resulting in plants that are shorter and stockier than plants that do not experience repeated vibrations or touching.

Figure 39.14 shows what this response looks like in tomatoes. In this experiment, tomato plants were touched lightly 0, 10, or 20 times per day, each day for 10 consecutive days. The more plants were touched in the experiment, the slower they grew and the stockier they became.

In response to wind, then, plants change their growth patterns in ways that make them more likely to withstand the force, stay upright, and live long enough to produce flowers and fruit.

Movement Responses

In some cases, plants respond to touch by moving. This response, **thigmotropism** ("touch-bending"), can be fast. For example, species that grow by climbing objects or other plants may have modified leaves or stems that form long, thin structures called tendrils. When a tendril makes contact with an object, it responds by wrapping itself around the item as fast as one or more times per hour (**Figure 39.15**).

Movement is even faster in "touch-sensitive" plants. A Venus flytrap, for example, closes fast enough to catch insects. Extremely

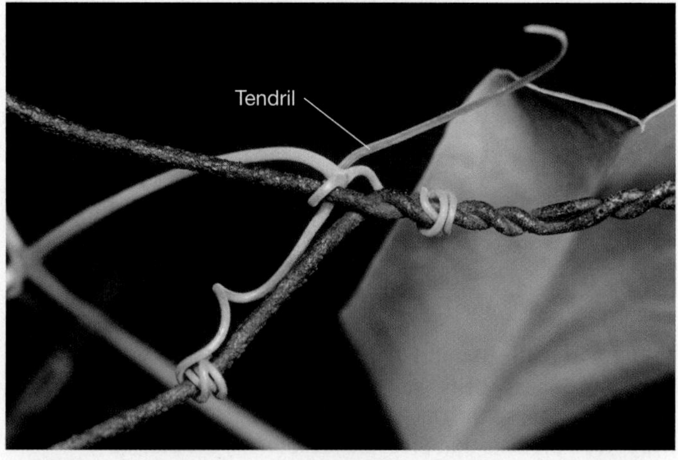

FIGURE 39.15 Thigmotropism Is Movement in Response to Touch. Portions of pea plants called tendrils wind around support structures after contacting them. The attachment provided by the tendrils allows peas to climb.

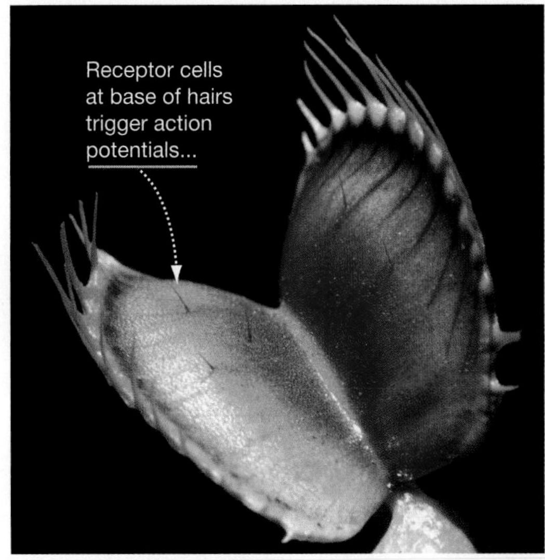

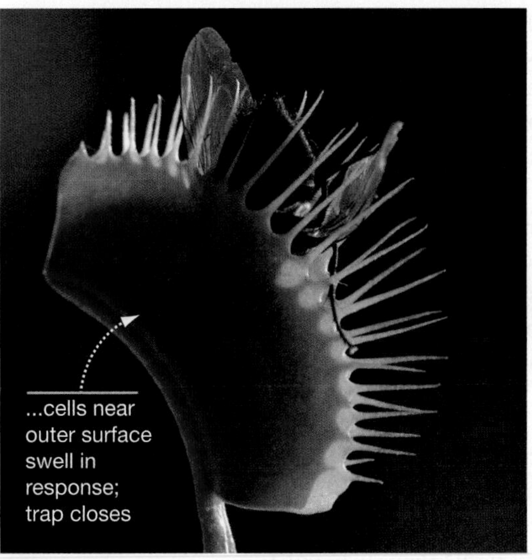

FIGURE 39.16 Venus Flytraps Close in Response to Action Potentials Generated by Sensory Hairs.

rapid movements like this are possible when a touch-receptor cell transduces the mechanical signal to an electrical signal.

To understand how plants use electrical signaling, recall from Chapter 38 that proton pumps in the plasma membrane of most plant cells give them a negative charge relative to the exterior environment. The separation of charges creates a membrane voltage, or **membrane potential**. If a receptor protein responds to touch by allowing ions to flow across the membrane—which changes the amount of charge on either side—then the membrane potential changes. In this way, the mechanical signal (touch) can be transduced to an electrical signal.

To travel from a sensory cell to a response cell, electrical signals are propagated in a characteristic form called an action potential. (Chapter 45 analyzes the action potentials that occur in animal nerve cells in detail.) In a Venus flytrap, action potentials race across the leaf at a rate of about 10 cm/sec.

When the action potentials reach cells on the outer surface of the trap, the cells change shape and push the trap shut (**Figure 39.16**). The response is rapid enough to resemble the way an animal's muscle contracts in response to an action potential.

Although biologists have been fascinated by the speed of the trap closing for decades, the molecular mechanism involved remains controversial. Rapid water loss from cells along the inside of the hinge is the most probable cause, but how water moves in response to the change in membrane voltage remains a mystery.

39.6 Youth, Maturity, and Aging: The Growth Responses

Plants grow throughout their lives, from the time they germinate until the time they die. As the plant body grows, it matures into an efficient machine for absorbing sunlight, water, nutrients, and other diffuse resources. But growth is not constant. It speeds up when water, light, and nutrients are abundant and slows or stops when conditions are poor, or when it is time for leaves to drop or fruits to ripen.

Controlling growth in response to changes in age or environmental conditions is one of the most important aspects of information processing in plants. Hormones play a key role in regulating growth.

To explore how plants grow in response to changing conditions, let's consider five of the best-studied hormones involved in growth responses. We'll start with auxin—the molecule responsible for phototropism and gravitropism.

Auxin and Apical Dominance

When **apical dominance** occurs, growth is restricted to the main stems, and the lateral buds in the axils of each leaf remain dormant. But if the apical bud dies, the dormancy of the lateral buds is broken and lateral branches begin to grow. **Figure 39.17** shows this phenomenon in action.

(a) Apical meristem intact

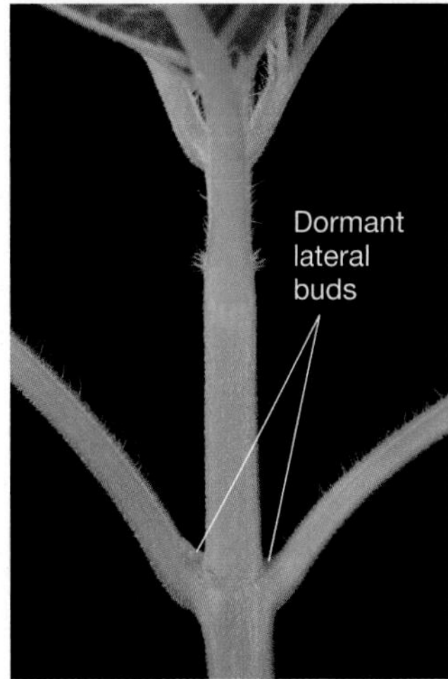

(b) Apical meristem cut off

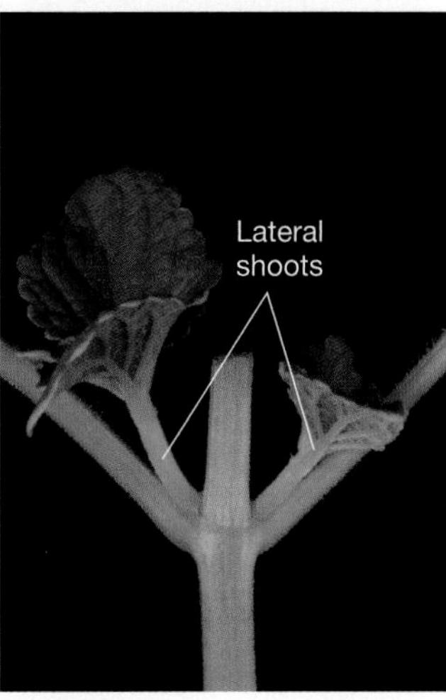

FIGURE 39.17 When Apical Dominance Occurs, Growth of Lateral Buds Is Suppressed. (a) The stem of a coleus plant is shown still intact. **(b)** The same plant, several weeks after the stem was cut. The lateral shoots will orient themselves vertically.

What caused the change? Because auxin is produced in shoot tips, researchers suspected that it might have a role in apical dominance as well as phototropism. This hypothesis was confirmed when it was shown that apical dominance could be sustained by adding auxin to a shoot's cut surface after its tip had been removed.

Auxin's role in apical dominance suggests that tip cells send a constant stream of information down to other organs and tissues. If the signal stops, it means that apical growth has been interrupted. In response, lateral branches sprout and begin to take over for the main shoot. Now the question is, How does this signal move?

POLAR TRANSPORT OF AUXIN Auxin transport is **polar**, or unidirectional. If radioactively labeled auxin is added to the top of a cut stem, the hormone is transported toward the base. But if labeled auxin is added to the base of a cut stem, it is not transported toward the apex. Auxin is the only plant hormone known to be transported in one direction only.

Studies with labeled auxin have also shown that the hormone is transported all the way down the stem through the root, via parenchyma cells in the ground tissue and vascular tissue. Auxin enters the apical end of cells via a specialized membrane protein, diffuses to the other end of the cell, and then is transported out by carrier proteins located in the basal portion of the plasma membrane. Labeled auxin moves from cell to cell at about 10 cm/hr—approximately 10 times more slowly than substances traveling in the phloem or xylem.

Because enzymes destroy some auxin molecules as they travel down the long axis of the plant, polar transport sets up a strong gradient in auxin concentration. Auxin concentrations are much higher in shoots than they are in roots.

WHAT IS AUXIN'S OVERALL ROLE? Auxin clearly plays a key role in controlling growth via apical dominance, phototropism, and gravitropism. But this chemical messenger has other important effects as well:

- Auxin produced by seeds within the fruit influences fruit development.
- Falling auxin concentrations are involved in the **abscission**, or shedding of leaves and fruits, associated with the genetically programmed aging process called senescence.
- The presence of auxin in growing roots and shoots is essential not only for the proper differentiation of xylem and phloem cells in vascular tissue but also for the development of vascular cambium.
- Auxin stimulates the development of adventitious roots in tissue cultures and cuttings.

Auxin has so many different effects on plants that it has been difficult for biologists to understand its overall role. Recently, several investigators have proposed that auxin's overall function is to signal where cells are in space. The idea is that auxin concentration identifies where a cell is located relative to the long axis of the plant body—the axis that runs from shoot to root.

If conditions relating to the long axis change—for instance, a windstorm tips the plant or a deer eats the shoot apex—changes in auxin concentration effectively signal how the individual's tissues should respond. Phototropism, gravitropism, apical dominance, and the production of adventitious roots are all ways of coping with changes in the long axis of the plant.

Cytokinins and Cell Division

Cytokinins are a group of plant hormones that promote cell division. (*Cyto* is the Greek root for cell; *kinin* refers to kinesis, meaning "movement" or "division.")

THE DISCOVERY OF CYTOKININS When biologists were first attempting to grow plant cells and embryos in culture, they found that coconut milk, which stores nutrients used by growing coconut embryos, promoted cell division. This was the first hint that certain molecules can promote cell division in plants. Later experiments showed that molecules derived from the nitrogenous base adenine also stimulate the growth of cells in culture.

Eventually, naturally occurring adenine derivatives that stimulate growth were discovered in corn and apples, and were named cytokinins. Zeatin, the cytokinin that has been found in the most species, is derived from adenine.

Cytokinins are synthesized in root tips, young fruits, seeds, growing buds, and other developing organs. But most of the zeatin and other cytokinins that are active in plants are synthesized in the apical meristems of roots and transported up into the shoot system via the xylem. Biologists still add cytokinins to plant cells growing in culture, to stimulate cell division (see **BioSkills 12** in Appendix A).

HOW DO CYTOKININS PROMOTE CELL DIVISION? After years of searching, a group of closely related proteins that act as cytokinin receptors has now been isolated and characterized. When cytokinins bind to these receptors in the plasma membranes of target cells, the receptors activate genes that regulate cell division.

Recent research on cytokinins has explored whether they affect molecules that regulate the cell cycle. Chapter 11 introduced some of these cell-cycle regulators, including the cyclins and the cyclin-dependent kinases (Cdks). Recall that activated cyclins and Cdks allow cells to progress through checkpoints in the cell cycle and continue dividing.

To assess whether cytokinins affect cell-cycle genes, researchers grew *Arabidopsis* cells in culture so the nutrients and other molecules available could be carefully controlled. The biologists starved the cells of cytokinins for a day, then added the hormones again to half of the cells. When they assessed the level of mRNA from a cyclin gene called *CycD3*, they documented significant increases in the cells that were exposed to cytokinins again compared with the level in cells that were not reexposed to cytokinins.

This is strong evidence that cytokinins regulate growth by activating genes that keep the cell cycle going. In the absence of cytokinins, cells arrest at the G_2 checkpoint in the cell cycle and stop dividing (**Figure 39.18**).

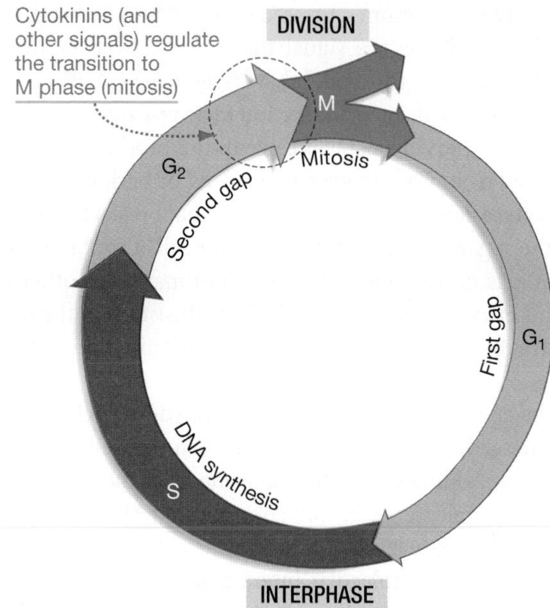

FIGURE 39.18 Cytokinins Affect the Cell Cycle.

Gibberellins and ABA: Growth and Dormancy

In high latitudes and at high elevations, most seeds and mature plants start growing in spring. Conditions for growth are good at that time of year, because temperatures are warming and soil moisture levels are usually at their peak. Seedlings and mature plants continue to grow throughout the summer and early fall if moisture and nutrients are still available.

During drought conditions, however, growth stops. Growth also stops in the embryos inside seeds. Embryos begin to develop as seeds mature, but cease this initial growth and remain dormant throughout the cold winter months. **Dormancy** is a temporary state of reduced metabolic activity or no metabolic activity.

Which signals initiate growth in response to changing environmental conditions, and which signals stop it? Two hormones provide the answer. **Gibberellins**, a large family of closely related compounds, stimulate growth in plants. **Abscisic acid**, commonly abbreviated **ABA**, inhibits growth. In at least some cases, the two hormones interact like start and stop signals.

THE DISCOVERY OF GIBBERELLINS Over 100 years ago, Japanese farmers noticed that some of their rice seedlings grew exceptionally quickly but fell over before they could be harvested. Biologists found that the diseased plants were infected with the fungus *Gibberella fujikuroi.*

Researchers confirmed a causal connection between *Gibberella* infection and rapid stem elongation when they treated rice seedlings with an extract from the fungus. As predicted, the treated seedlings produced abnormally long shoots.

The active component in the extract was eventually isolated and named gibberellic acid (GA), which is a gibberellin. Follow-up research showed that rice plants produce their own gibberellin but respond to applications of additional hormone by elongating their stems. In effect, the infected rice seedlings were suffering from a gibberellin overdose.

Gibberellins are found in a wide array of fungi and plants. Most plant species produce several different gibberellins that are active as hormones.

Even though gibberellins have dramatic effects on growth, they are present in vanishingly small concentrations. In growing stems and leaves, active forms of gibberellin may be present in concentrations of about 10 nanograms per gram of tissue.

DEFECTIVE GIBBERELLIN GENES CAUSE DWARFING To find the genes that are responsible for producing gibberellins, biologists analyzed mutant plants with abnormal stem length (**Figure 39.19**). Recall from Chapter 13 that Gregor Mendel analyzed the transmission of two alleles at a single gene that affected stem height in garden peas. One allele was associated with tall stems; the other was associated with dwarfed growth. The tall allele was dominant to the dwarf allele.

The gene responsible for the stem-length differences in garden peas is known as *Le* (for *le*ngth). Early work on dwarf mutants showed that they attain normal height if they are treated with the gibberellin called GA_1. This observation suggested that dwarf peas can respond to gibberellins normally—meaning that the problem is not with a hormone receptor.

Follow-up experiments treated dwarf peas with a radioactively labeled molecule used in the synthesis of GA_1. These plants did not produce radioactively labeled GA_1, even though plants

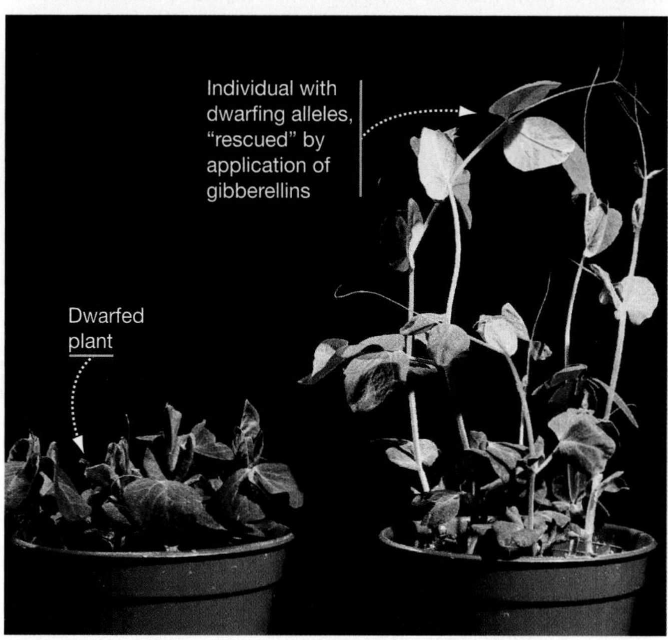

FIGURE 39.19 Dwarfed Individuals May Have Mutations That Affect Gibberellins. Dwarfed plants have much shorter stems than do normal individuals of the same age. By analyzing dwarfed individuals, biologists were able to identify genes involved in gibberellin synthesis.

✔**QUESTION** If dwarfed individuals receive the same amount of sun and thus perform as much photosynthesis as taller individuals, they often produce more flowers and seeds. Why?

with the normal allele did. Based on these results, researchers became convinced that the *Le* locus encodes an enzyme involved in GA synthesis.

Investigators recently provided support for this hypothesis by finding a gene in peas that encodes an enzyme called 3β-hydrolase. This enzyme adds a hydroxyl group (—OH) to a gibberellin called GA_{20}, producing the biologically active molecule GA_1. Researchers who compared the DNA sequences of this gene from normal and dwarf pea plants found an important difference: In a part of the enzyme near the active site, the mutant DNA sequence codes for the amino acid threonine instead of alanine. Follow-up tests showed that the mutant enzymes are unable to convert GA_{20} to GA_1.

These experiments provided strong evidence that the *Le* gene encodes the enzyme GA 3β-hydroxylase, and that a single amino acid change renders the enzyme largely ineffective and causes dwarfing. Over 100 years after Mendel did his experiments, biologists finally understood the molecular basis of the dwarfing phenotype he studied.

In stems, gibberellins appear to promote both cell elongation and rates of cell division. But it is well established that auxin also promotes cell elongation, and that cytokinins also promote cell division. Research continues on how GAs, cytokinins, and auxin interact on the molecular level to control plant growth and development.

GIBBERELLINS AND ABA INTERACT DURING SEED DORMANCY AND GERMINATION Many plants produce seeds that have to undergo a period of drying or a period of cold, wet conditions before they are able to germinate in response to warm, wet conditions. A requirement for drying ensures that mature seeds will not sprout on the parent plant; an obligatory cold period prevents seeds from germinating just before the onset of winter. Some seeds also have to receive a dose of red light, which indicates that they are in a sunny location.

In essence, then, seeds have an "off" setting that discourages germination and an "on" setting that initiates growth. The appropriate state for the on/off switch is determined by environmental cues such as temperature, moisture, and light.

By applying hormones to seeds, researchers learned that in many plants, ABA is the signal that inhibits seed germination and gibberellins are the signal that triggers germination. To understand how these messengers interact, researchers have concentrated on studying a specific event: the production of an enzyme called α-amylase in germinating oat or barley seeds.

α-Amylase acts as a digestive enzyme that breaks the bonds between the sugar subunits of starch. (Your saliva contains an amylase that acts on the starch in food. In humans and other mammals, this enzyme initiates carbohydrate digestion in the mouth.)

Figure 39.20 shows that, during the germination of a barley seedling, α-amylase is released from a tissue called the aleurone layer. The enzyme diffuses into the carbohydrate-rich storage tissue in the seed and releases sugars that can be used by the growing embryo. Adding GA to the aleurone layer increases the production and release of α-amylase; adding ABA to that layer decreases α-amylase levels.

Research on the molecular interaction between GA and ABA carries several important messages:

1. A cell's response to a hormone often occurs because specific genes are turned on or off.
2. Hormones don't act on genes directly. Instead, a receptor on the surface of a cell or in the cytosol receives the message and responds by initiating a signal transduction cascade, which activates specific gene regulatory proteins—the transcription activators and repressors introduced in Chapter 18.
3. Different hormones interact at the molecular level because they induce different gene regulatory proteins, which increase or decrease expression of key genes.

The logic runs as follows: Different hormones trigger the production of different regulatory transcription factors. Hormone concentration affects the amount of each transcription factor produced. Changes in transcription factors are responsible for changes in gene expression.

GA's role in seed germination has important commercial applications. Brewers, for example, routinely use gibberellins in the

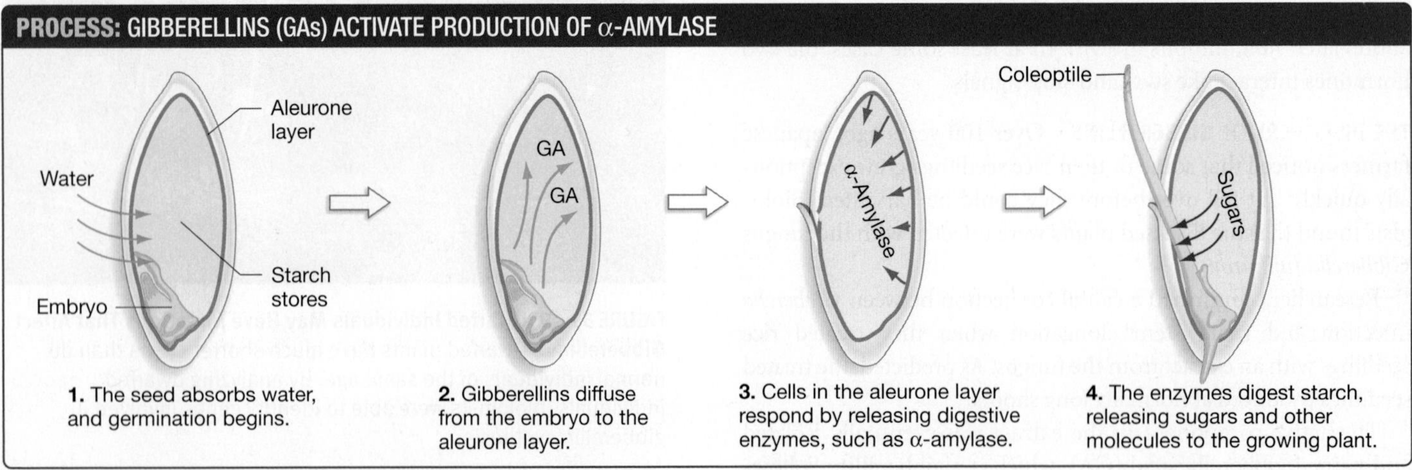

FIGURE 39.20 The Molecular Mechanism of Gibberellin Action.

malting process—the conversion of starches stored in barley seeds to sugars. The sugars that are released in response to GA treatment support fermentation by yeast and provide flavor in the finished beer.

ABA CLOSES GUARD CELLS IN STOMATA Chapter 37 introduced one of the major problems faced by land plants: replacing water that is lost to the atmosphere when **stomata** are open. Because stomata open in response to blue light, they allow gas exchange to occur while the plant is receiving the wavelengths of light used in photosynthesis.

However, if plant roots are unable to obtain enough water to replace the fluid being lost at the leaves, stomata close. Closing stomata is an adaptive response when roots cannot find adequate water, because continued transpiration would lead to wilting and potential tissue damage.

Early work on the mechanism of stomatal closing suggested that ABA is involved. For example, applying ABA to the exterior of stomata causes them to close.

To explore the hypothesis that a hormone regulates stomatal closing, researchers performed the experiments summarized in **Figure 39.21**. The fundamental idea was to grow plants whose roots had been divided. Only one side of the experimental plants was watered, while both sides of the control plants were watered. During this treatment, investigators documented that the water potential of the leaves remained the same in both control and experimental plants. Yet the stomata of experimental plants began to close. This result suggested that roots from the dry side of the pot were signaling drought stress, even though the leaves were not actually experiencing a water shortage.

Follow-up experiments have supported two important predictions: ABA concentrations in roots on the dry side of the pot are extraordinarily high relative to the watered side, and ABA concentrations in the leaves of experimental plants are much higher than in the leaves of control plants.

These results suggest that ABA from roots is transported to leaves and that it serves as an early warning of drought stress. In doing so, ABA overrides the signal from the blue-light photoreceptors introduced earlier in this chapter.

THE MOLECULAR MECHANISM OF GUARD CELL CLOSURE To understand how stomata open and close, recall from Chapter 36 that a stoma consists of two **guard cells**. When the vacuoles of guard cells are filled with water, the cells are turgid. The shape of turgid guard cells results in an open pore, which allows gas exchange between the atmosphere and the interior of the leaf. But when vacuoles lose water and guard cells become flaccid and lack turgor, cell shape changes in a way that closes the pore and stops both gas exchange and loss of water via transpiration.

Based on these observations, the question of how stomata open and close is the same as asking how guard cells become turgid or flaccid—meaning, how water flows into or out of vacuoles. Activation of PHOT by blue light leads to water entry and

EXPERIMENT

QUESTION: Can roots communicate with shoots?

HYPOTHESIS: Roots that are dry can signal shoots to close stomata.

NULL HYPOTHESIS: Roots cannot communicate with the shoot.

EXPERIMENTAL SETUP:

1. Divide roots of many plants into two sides.

2. In experimental group, water one side.

3. In control group, water both sides.

4. In both groups, measure water potential of leaves and observe stomata.

PREDICTION: Stomata in experimental plants will close; stomata in control plants will stay open.

PREDICTION OF NULL HYPOTHESIS: Stomata in both experimental and control plants will stay open.

RESULTS: No difference between experimental and control plants in water potential of leaves.

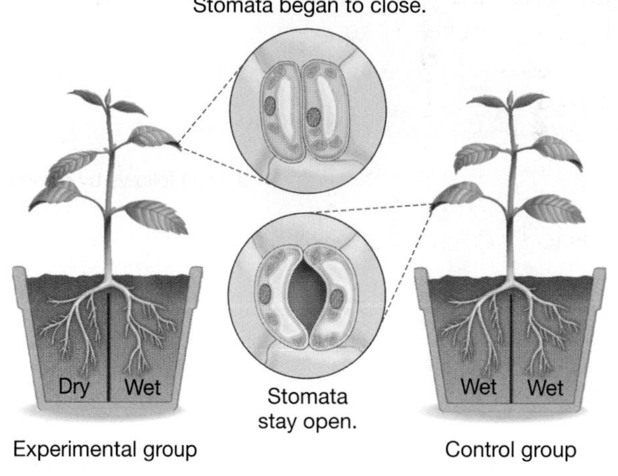

CONCLUSION: Roots can communicate with shoots. Dry roots signal the shoot and cause stomata to close, even though leaves are receiving sufficient water (from roots on the wet side of the plant).

FIGURE 39.21 Experimental Evidence That Roots Produce an "It's Too Dry" Signal.

SOURCE: Blackman, P. G. and W. J. Davies. 1985. Root to shoot communication in maize plants of the effects of soil drying. *Journal of Experimental Botany* 36: 39–48.

✔**QUESTION** Why was it important to show that the water potential of leaves was the same in the two treatments?

stomatal opening; activation of ABA receptors leads to water exit and stomatal closing.

Remember that cells do not transport water directly. Instead, they change ion concentrations, creating osmotic gradients that result in water movement.

Based on this observation, it shouldn't be surprising to learn that guard cell opening and closing is based on changes in the activity of H^+-ATPases in the plasma membrane.

When PHOTs or cryptochromes are stimulated by blue light, large numbers of protons are pumped out of each guard cell. As **Figure 39.22a** shows, increased H^+-ATPase activity creates a strong electrochemical gradient that brings potassium and chloride ions into the interior of the guard cells. Water follows the incoming ions via osmosis. The upshot? The cells swell and pores open.

But as **Figure 39.22b** shows, guard cells respond to ABA in a very different way. When ABA reaches the guard cells, two things happen:

1. Channels that allow chloride and other anions to leave along their electrochemical gradients are opened.
2. H^+-ATPases and inward-directed potassium channels are inhibited.

When the anions leave guard cells, the change in membrane potential causes outward-directed potassium channels to open. Large amounts of K^+ leave the cells, with water following by osmosis. The result is a loss of turgor and closing of the pore.

(a) PROCESS: STOMATA OPEN IN RESPONSE TO BLUE LIGHT

Blue light strikes photoreceptor.

H^+ H^+ H^+ H^+

1. Pumping by H^+-ATPases increases. Protons leave guard cells.

K^+ Cl^- H^+ K^+ Cl^- H^+

2. K^+ and Cl^- enter cells along electrochemical gradients via inward-directed K^+ channels and H^+/Cl^- cotransporter.

H_2O H_2O H_2O H_2O

3. H_2O follows by osmosis.

4. Cells swell. Pore opens.

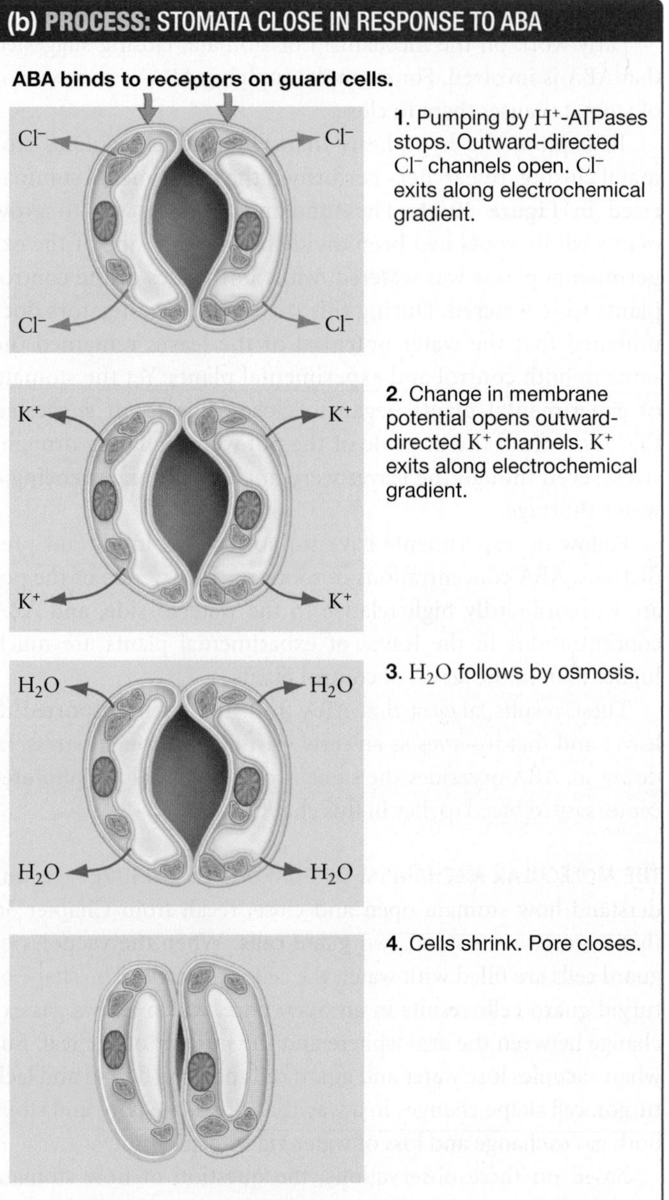

FIGURE 39.22 Changes in Ion Flows Are Responsible for Opening and Closing Stomata. (a) Activated blue-light receptors trigger ion flows into guard cells. The cells swell when water follows by osmosis. **(b)** Activated ABA receptors trigger ion flows out of guard cells. The cells shrink when water follows by osmosis.

Whether it acts on guard cells or seeds, ABA fulfills a general role in plants as a dormancy or "no-growth" signal. In many cases, its action depends on input from other hormones and photoreceptors. To survive and reproduce successfully, plants have to integrate information from a variety of sources.

Brassinosteroids and Body Size

When you went through puberty, you underwent a growth spurt triggered by surges in steroid hormones called testosterone and estradiol. In plants, growth spurts are triggered by surges in steroid hormones called **brassinosteroids**.

The name brassinosteroid was inspired by two observations:

1. The hormones were initially discovered in *Brassica napus*—a crop plant that is the source of the canola oil you may use in cooking.
2. They are steroids—part of a family of lipid-soluble compounds introduced in Chapter 6.

Brassinosteroids promote growth and are a key regulator of overall body size in plants. In the model organism *Arabidopsis thaliana*, for example, mutant individuals that cannot synthesize brassinosteroids are extremely dwarfed.

Recent research has highlighted a fascinating difference between the brassinosteroids and the steroid hormones found in animals. As Chapter 47 will show, steroid signals in animals act by entering cells, binding to receptors in the cytosol, and forming a hormone-receptor complex that enters the nucleus, binds to DNA, and directly changes gene expression. Because steroids are lipid-soluble, researchers were not surprised to find that testosterone and estradiol cross the plasma membrane before binding to a receptor.

Brassinosteroids, in contrast, never enter the cell. Instead, they bind to receptors on the plasma membrane and activate signal transduction events—probably phosphorylation cascades—that lead to changes in gene expression.

The genes for the synthesis of brassinosteroids appear to be homologous with the genes required for steroid hormone in animals, meaning that they are derived from a similar gene in the common ancestor of plants and animals. So why do they have such different modes of action? And how do brassinosteroids interact with auxin, cytokinins, and gibberellins to regulate growth and body size? These are questions for future research.

Ethylene and Senescence

Senescence is a regulated process of aging and eventual death of an entire organism or organs such as fruits and leaves. Like most aspects of plant growth and development, senescence is regulated by complex interactions between several hormones in response to changes in temperature, light, and other factors.

The hormone most strongly associated with senescence is **ethylene**. Like other plant hormones, ethylene is simple in structure and active at small concentrations. Unlike other plant hormones, however, ethylene is a gas at normal temperatures.

Ethylene is synthesized from the amino acid methionine and is strongly involved in three aspects of senescence in plants:

1. fruit ripening, which eventually leads to the aging and rotting of fruit;
2. flowers fading; and
3. leaf abscission—meaning their detachment and fall.

In addition, ethylene influences plant growth, and is a stress hormone induced by drought and other conditions. Ethylene regulates a surprisingly large range of physiological responses.

THE DISCOVERY OF ETHYLENE Ethylene was initially discovered in ancient China, when fruit growers noticed that burning incense in closed rooms made pears ripen faster. Westerners made a similar observation in the late 1800s, when gas street lamps came into wide use in cities and plants growing near leaky gas lines dropped their leaves prematurely.

Researchers showed that ethylene in lamp gas was the molecule responsible for the leaf loss; ethylene is also present in incense smoke. In the 1930s ethylene was found in the gases that are released by ripening apples.

Subsequently, biologists documented sharp spikes in ethylene production during fruit ripening in tomatoes, bananas, and certain other species in addition to apples. Follow-up research on these species showed that ethylene induces (1) the production of some of the enzymes required for the ripening process, and (2) an increase in cellular respiration, which furnishes ATP.

ETHYLENE AND FRUIT RIPENING During ripening, stored starch is converted to sugar, enhancing sweetness; protective toxins are removed or destroyed; cell walls are degraded, softening the fruit; chlorophyll is broken down; and pigments and aromas that signal ripeness are produced. Biologists interpret fruit ripening as an adaptation that enhances its attractiveness to birds, mammals, and other animals that disperse seeds to new locations.

Today, fruit growers manipulate ethylene levels to control fruit ripening. For example, they treat green bananas with ethylene after the bananas have been shipped to encourage ripening (**Figure 39.23**). Conversely, apples are stored in warehouses with

FIGURE 39.23 Ethylene Speeds Ripening and Other Aspects of Senescence. These bananas are identical, except that the bunch on the right was exposed to the plant growth regulator ethylene.

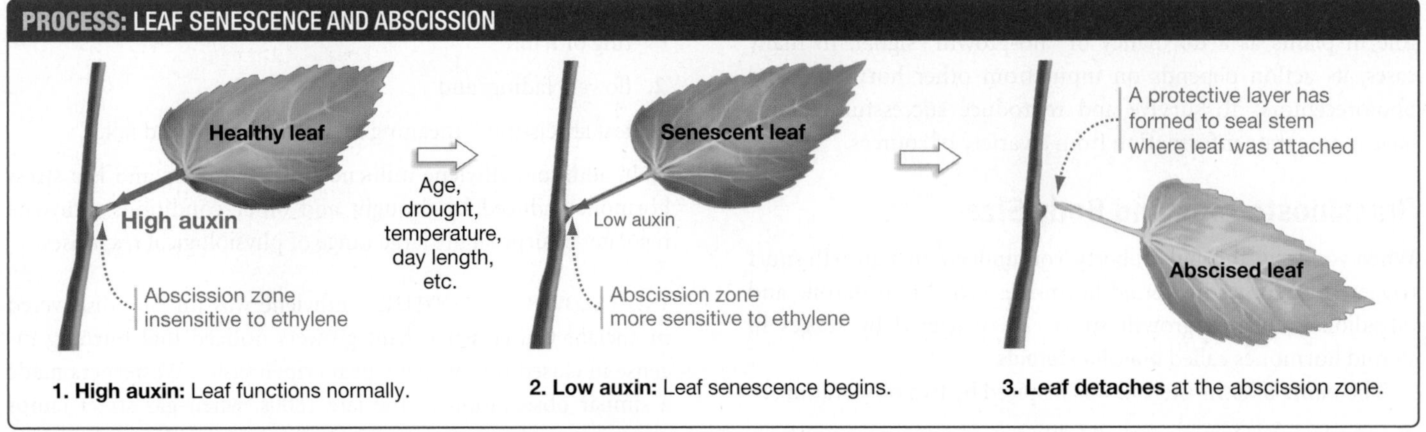

FIGURE 39.24 Leaves Drop in Response to Signals from Auxin and Ethylene. Young leaves produce much larger amounts of auxin than old leaves do. The combination of low auxin and high ethylene concentrations triggers leaf senescence and abscission.

✔**QUESTION** When ethylene levels in leaves are high relative to auxin, (1) nutrients are transported from leaves to the stem and (2) chlorophyll synthesis stops (this is why leaves change color in the fall). What is the adaptive significance of these two events?

high concentrations of CO_2 and low concentrations of O_2, which inhibits ethylene production in the fruit. Apples stored under these conditions can be sold long after their original harvest date, when untreated fruits have rotted.

ETHYLENE AND LEAF ABSCISSION Ethylene's effects on leaf senescence and leaf abscission involve complex interactions with auxin and cytokinins. In addition to being sequestered in apical meristems, auxin is synthesized in healthy leaves. It is then transported from the leaf to the stem through the petiole. In response to age or to changes in ambient temperature or day length, leaves produce less auxin (**Figure 39.24**). As a result, cells in a region of the leaf petiole called the **abscission zone** become more sensitive to ethylene in the tissue.

Increased ethylene sensitivity activates enzymes that weaken the cell walls of cells near the base of the petiole. At the same time, chlorophyll in the leaf degrades, and nutrients are withdrawn and stored in parenchyma cells in the stem. Eventually the cell walls at the base of the petiole degrade enough that the leaf falls.

Applications of cytokinins, in contrast, reverse these effects and dramatically extend the life span of leaves. As a result, ethylene and cytokinins are thought to have opposite effects on at least some of the processes involved in senescence. With two hormones involved, the process can be sped up or slowed down.

An Overview of Plant Growth Regulators

Understanding how different signals interact is an exciting frontier in research on plant growth regulators. And work continues on characterizing the genes that are directly regulated by auxin, cytokinins, ABA, GAs, brassinosteroids, and ethylene. Although progress has been rapid, a great deal remains to be learned about the response step in information processing during growth responses.

Table 39.2 provides notes on the structure and function of hormones discussed in this section. As you study this table, two key observations should emerge:

1. It is common for a single hormone to affect many different target tissues. This means that there can be an array of responses to the same cell-cell signal. To interpret this pattern, biologists point out that hormones may carry a common message to a variety of tissues and organs. Auxin can define the long axis of the body; gibberellins trigger stem growth; cytokinins promote cell division; ABA slows or prevents growth; ethylene signals senescence; brassinosteroids increase overall mass.
2. In most cases, several hormones affect the same response. Stated another way, hormones do not work independently—they interact with each other. To make sense of this pattern, biologists point out that individual hormones tend to be produced by an environmental cue at a certain location, such as water availability at root tips. Many environmental cues may be changing at the same time, however. For plants to respond appropriately, they need to integrate information from various environmental cues perceived at various locations in the body.

Chapter 8 introduced the concept of **cross-talk**: interactions between the signal transduction cascades triggered by different hormones. The key insight is that signaling systems form communication networks. Cross-talk is the molecular mechanism responsible for integrating information from many sensory cells and signals.

The complex interactions among hormones involved in the growth response have a purpose: allowing individuals to survive and thrive long enough to reproduce. The same can be said for the hormones involved in protecting plants from danger.

SUMMARY TABLE 39.2 **Plant Growth Regulators**

Hormone	Function(s)	Notes	Chemical Structure
Auxin	• Helps define long axis of body (phototropism and gravitropism responses) • Involved in cell elongation and apical dominance • Promotes cell division • Induces ethylene production • Development of adventitious roots and secondary growth • Differentiation of xylem and phloem	• First plant hormone ever characterized and isolated • Produced in shoot apical meristems and young leaves • Receptor discovered 2005; receptor structure solved 2007	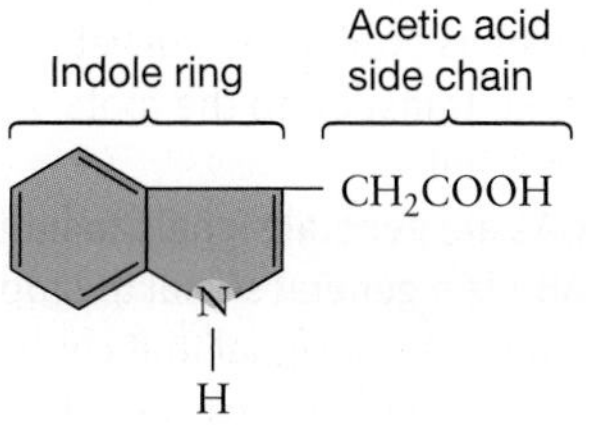 Indole ring; Acetic acid side chain; CH_2COOH; N; H
Cytokinins	• Promote cell division in the presence of auxin • Promote chloroplast development and break lateral bud dormancy • Delay senescence (aging)	• New data indicate they may act on cell cycle regulators • Produced in root apical meristems, many other tissues • 3 distinct receptors have been identified; each found in a different location in cells	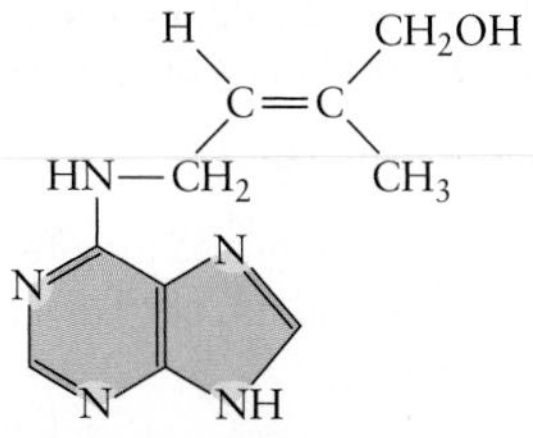 H; CH_2OH; C=C; $HN-CH_2$; CH_3; N; N; N; NH
Gibberellins (GAs)	• Promote stem growth via both cell elongation and division • Encourage seed germination • Involved in flowering	• Fungi that produce gibberellins infect rice plants and induce hyper-elongated stems • Analysis of these fungi led to discovery of gibberellins • Produced in apical meristems, immature seeds, anthers (pollen-producing organs) • Receptor discovered 2005	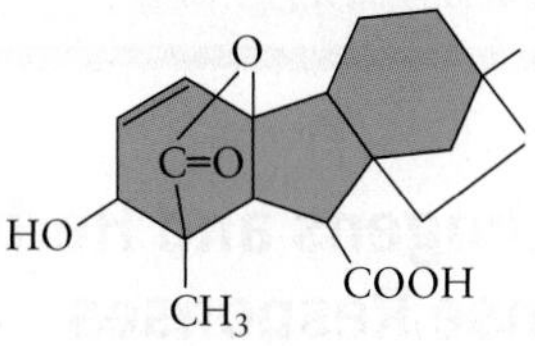 O; C=O; HO; CH_3; COOH
Abscisic Acid (ABA)	• Inhibits bud growth and seed germination • Induces closure of stomata in response to water stress	• Acts as a stress hormone analogous to cortisol in humans • Produced in almost all cells	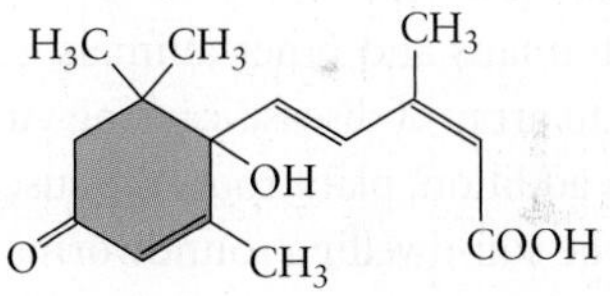 H_3C; CH_3; CH_3; OH; O; CH_3; COOH
Brassinosteroids	• Promote cell elongation in stems and leaves (mutants that lack these hormones or their receptors are dwarfed)	• First steroid hormones discovered in plants • Structurally related to steroid hormones in animals • Produced in almost all tissues • Act on receptor at cell surface	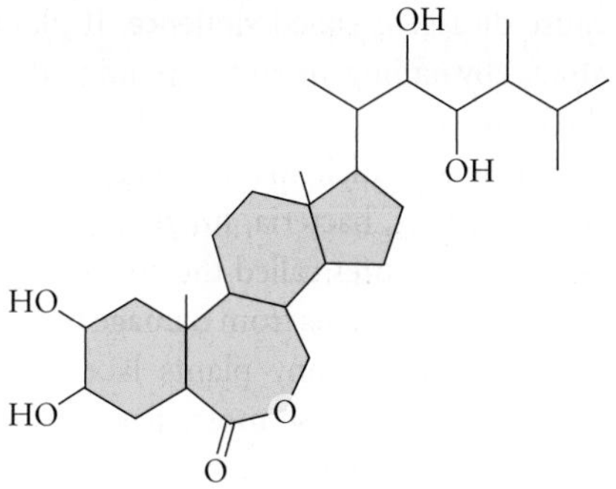 OH; OH; HO; HO; O; O
Ethylene	• Involved in fruit ripening • Induces senescence of fruits, flowers, and leaves • Produced when plants are under stress	• A gas; first identified through unusual morphology of plants growing near gas lines for illuminating streets • Produced in all organs but highest in aging tissues and fruits	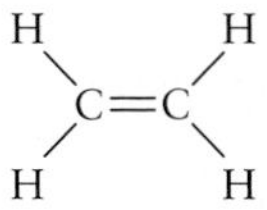 H; H; C=C; H; H

CHECK YOUR UNDERSTANDING

If you understand that . . .

- Auxin's primary role is to signal the position of cells along the long axis of the plant body. This is possible because auxin is produced in developing leaves and undergoes polar transport to the roots, forming a concentration gradient.
- GAs are general signals to initiate or continue growth.
- ABA is a general signal to stop growth or remain dormant.
- Ethylene is a signal that controls senescence.
- Brassinosteroids promote large body size.
- GAs and ABA interact at the molecular level to control seed germination and dormancy.
- Signals from blue light and ABA interact to control the opening and closing of stomata.

✓ **You should be able to . . .**

Predict the short-term and long-term effects of watering a large number of daisy plants, each in a pot in a greenhouse, with water that contains:

1. ABA, or
2. GAs.

Answers are available in Appendix B.

39.7 Pathogens and Herbivores: The Defense Responses

Plants cannot run away from danger. Instead, they have to stand and fight.

Like humans and other animals, plants are constantly threatened by an array of disease-causing viruses, bacteria, and parasitic fungi. In addition, plant roots are susceptible to attacks by nematodes—the soil-dwelling roundworms introduced in Chapter 33.

Disease-causing agents are termed **pathogens**; the ability to cause disease is called **virulence**. If plants were not able to sense attacks by pathogens and respond to them quickly and effectively, the landscape would be littered with dead and dying vegetation.

The waxy cuticle that covers epidermal cells is an effective barrier to viruses, bacteria, fungi, and other disease-causing agents, and the structures called thorns, spines, and trichomes help protect leaves and stems from damage by herbivores (see Chapter 36).

In addition, many plants lace their tissues with **secondary metabolites**—defense molecules that are closely related to compounds in key synthetic pathways. Some secondary metabolites poison herbivores.

- The flavorful oils in peppermint, lemon, basil, and sage have insect repellent properties.
- The pitch that oozes from pines and firs contains a molecule called pinene, which is toxic to bark beetles.
- The pyrethroids produced by *Chrysanthemum* plants are a common ingredient in commercial insecticides.
- Molecules called tannins are found in a wide array of plant species; when they are ingested by animals, tannins bind to digestive enzymes and make the herbivore sick.
- Compounds like opium, caffeine, cocaine, nicotine, and tetrahydrocannabinol (THC) disrupt the nervous systems of plant-eating insects and vertebrates.

Although these defenses are effective, they are also expensive to produce in terms of the ATP and materials invested. It is not surprising that plants may produce defenses or increase their existing defenses only in direct response to attacks by pathogens or herbivores.

Responses to attacks are called **inducible defenses**, because they are induced by the presence of a threat. Let's first consider how plants sense and respond to viruses and other pathogens, then explore what they do when attacked by insects and other herbivores.

How Do Plants Sense and Respond to Pathogens?

If a virus, bacterium, or fungus is able to get inside a plant, the cells at the infection site respond by committing suicide. The rapid and localized death of one or a few infected cells is called the **hypersensitive response (HR)**.

In several respects, the hypersensitive response in plants is similar to the cell-mediated immune response in mammals, which leads to the death of infected cells (see Chapter 49). The HR is also extremely effective; when individuals mount a hypersensitive response, they rarely succumb to disease. How do cells sense the presence of pathogens, so that the HR can kill them?

THE GENE-FOR-GENE HYPOTHESIS In the early twentieth century, crop breeders established that plants have disease-resistance genes, known as **resistance (*R*) genes**, that are inherited according to Mendel's rules. Follow-up research showed that many of the *R* genes are responsible for sensing the presence of pathogens and triggering the HR.

The fungi that cause disease in wheat, flax, barley, and other crops have alleles that make individuals either virulent or avirulent (not virulent) in certain strains or varieties of these crops. The genes associated with virulence in pathogens are known as **avirulence (*avr*) genes**.

In 1956 H. H. Flor published data demonstrating a one-to-one correspondence between the resistance alleles found in host plants and the avirulence alleles found in pathogens. Each *R* allele in the host corresponded to an *avr* allele in the pathogen.

Flor's **gene-for-gene hypothesis** expanded on this observation by proposing that *R* and *avr* gene products interact in a specific way. What molecular mechanism could be responsible? Researchers who followed up on Flor's work suggested that the HR begins when proteins produced by host plants bind to proteins or other molecules produced by the pathogen (**Figure 39.25**). More specifically, the hypothesis was that *R* genes

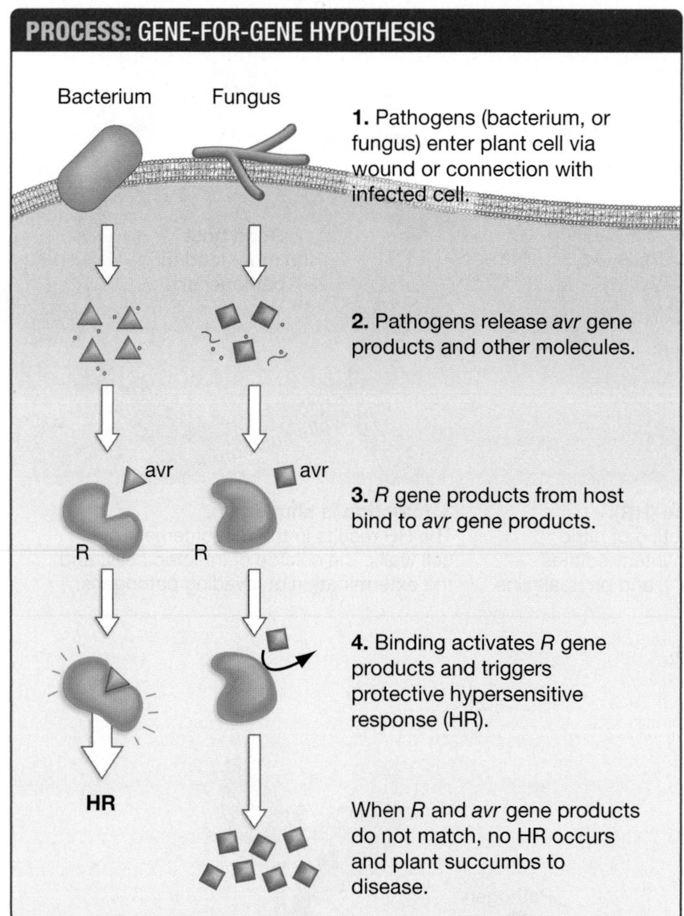

FIGURE 39.25 The Hypersensitive Response Begins when *R* Gene Products Bind to *avr* Gene Products. The gene-for-gene hypothesis predicts that *R* gene products act as receptors for *avr* gene products and initiate a defense response.

produce receptors and *avr* genes produce **ligands**—molecules that bind to receptors.

The first breakthrough in testing the gene-for-gene hypothesis occurred when researchers cloned and sequenced a series of *R* genes from crop plants and *avr* genes from bacterial and fungal pathogens. These results confirmed that *R* genes and *avr* genes exist and that they code for products that could interact at the start of an infection.

A second major advance, published in 1996, supported the gene-for-gene hypothesis by showing that *R* products and *avr* products actually do bind to one another. *R* gene products function as "pathogen receptors."

WHY IS THE EXISTENCE OF SO MANY RESISTANCE GENES AND ALLELES SIGNIFICANT? A large number of *R* genes have been identified in *Arabidopsis*, tomato, flax, tobacco, and other plants. Two general patterns are emerging as data on these genes accumulate:

1. *R* genes that are similar in sequence and structure tend to cluster together on the same chromosome.
2. Within a population of plants, there are usually many different alleles at each *R* locus. To use a term introduced earlier in the text, *R* genes are highly polymorphic.

These observations are important because they provide hints about the history and function of these genes. For example, clusters of similar genes, or what biologists call **gene families**, are thought to originate through errors in recombination that result in **gene duplication** events (see Chapter 20). Mutations in duplicated genes can result in novel gene products.

The idea here is that duplicated *R* genes change over time due to mutation, and begin to produce new proteins that give individuals the ability to recognize and respond to novel *avr* products and thus new pathogens.

Why is it significant that many different alleles exist at each *R* locus? The hypothesis here is that different alleles allow plants to recognize different proteins from the same pathogen. Plants are diploid or polyploid, so they have at least two copies of each *R* gene. If many different alleles exist in a population, each individual is likely to have at least two different alleles of each gene. The different alleles allow the host to recognize different *avr* products. This is important because new *avr* products constantly arise in pathogen populations via mutation.

Plants with different alleles for each of many *R* genes should be able to sense a wide variety of disease-causing agents and respond by triggering the HR. ✔If you understand this concept, you should be able to explain why bananas—which are propagated asexually and are thus genetically identical—are considered particularly vulnerable to disease epidemics.

THE HYPERSENSITIVE RESPONSE Once an *R* gene product is activated by binding to a ligand from a pathogen, what happens? Recent work has established that the HR consists of two key events (**Figure 39.26** on page 778):

- Production of nitric oxide (NO) and **reactive oxygen intermediates (ROIs)** such as hydrogen peroxide (H_2O_2) and superoxide ions (O_2–). ROIs (**1**) trigger reactions that help reinforce cell walls, and (**2**) kill host cells at the point of infection by disrupting their enzymes. Immune system cells in your body also use a lethal combination of NO and ROIs to kill cells that have become infected.
- Production of antibacterial and antifungal compounds that are collectively known as **phytoalexins**.

As a result, the HR leads to the walling-off of the infected area, the direct killing of pathogens by phytoalexins, and the starvation of pathogens as nearby cells commit suicide. Although the signal transduction events responsible for these events are not well understood, they are currently the focus of intense research.

AN ALARM HORMONE EXTENDS THE HR Once the HR is under way in a localized area of infection, a hormone produced at the infection site travels throughout the body and triggers a slower and more widespread set of events called **systemic acquired resistance (SAR)**. Over the course of several days, SAR primes cells throughout the root or shoot system for resistance to assault by a pathogen—even cells that have not been directly exposed to the disease-causing agent.

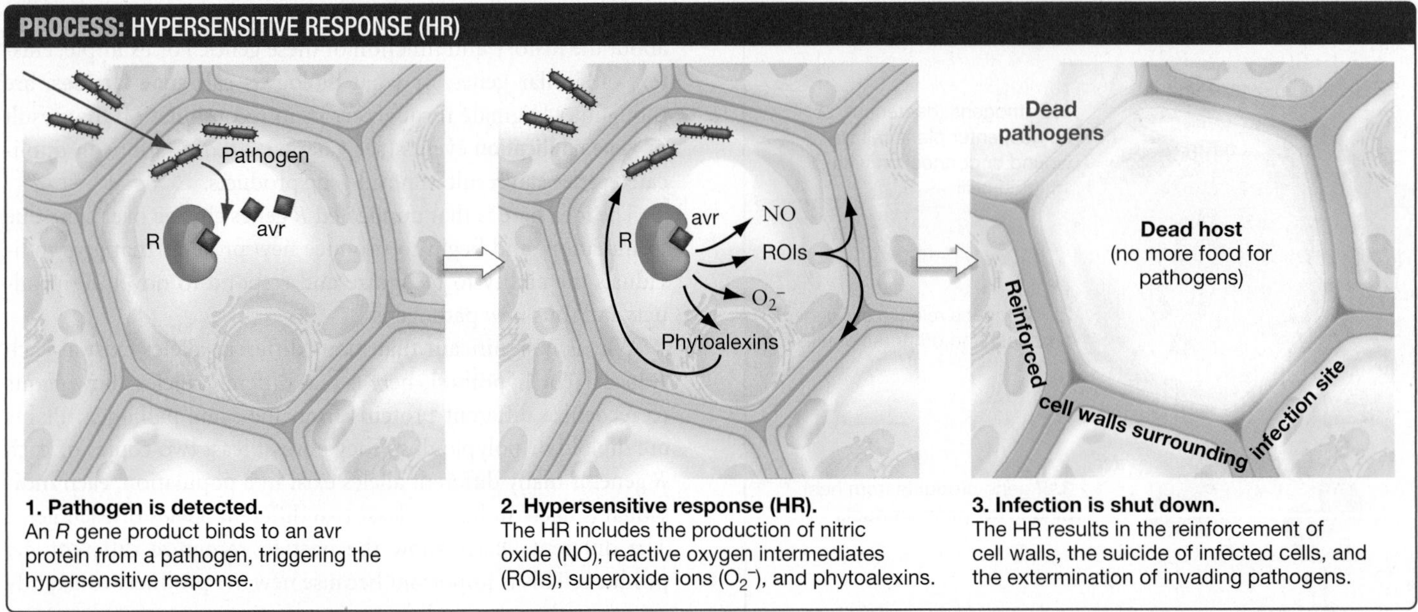

FIGURE 39.26 The Hypersensitive Response Protects Plants from Pathogens.

Figure 39.27 illustrates how the HR and SAR are thought to work together. In addition to triggering the HR, interaction between the *R* and *avr* gene products releases a hormone that initiates SAR. This signal acts globally as well as locally—that is, at the point of infection—and results in the expression of a large suite of genes called the pathogenesis-related (*PR*) genes.

When biologists set out to locate the hormone responsible for SAR, they found that levels of **methyl salicylate (MeSA)**—a molecule derived from salicylic acid—increase dramatically after tissues are infected with a pathogen. Follow-up work showed that:

- Phloem sap leaving infected sites has elevated levels of MeSA.
- Treatments that reduce MeSA reduce or abolish SAR.
- Adding MeSA to the lower leaves of tobacco plants leads to SAR in the upper, untreated leaves.

MeSA is a short-lived molecule, however, and it is still not clear that MeSA is transported throughout the plant. Researchers continue to look for the defense hormone that travels from the site of infection, through phloem sap, to the rest of the body—where it sounds a Paul Revere-like warning to prepare for attack.

How Do Plants Sense and Respond to Herbivore Attack?

Over a million species of insects have been discovered and named so far. Most of them make their living by eating leaves, stems, phloem sap, seeds, roots, or pollen. Plants have effective induced defenses in response to pathogens like viruses, bacteria, and fungi. But can they ramp up defenses in response to insect attack?

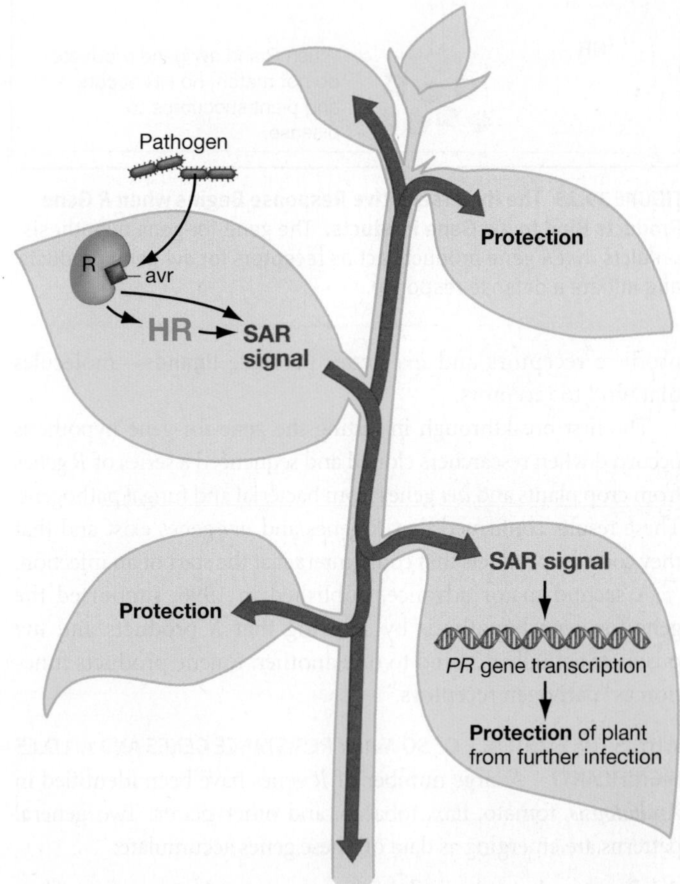

FIGURE 39.27 The Hypersensitive Response Produces a Signal That Induces Systemic Acquired Resistance. This diagram summarizes the current consensus on how the HR and SAR interact.

THE ROLE OF PROTEINASE INHIBITORS When researchers started studying why some plant tissues are more palatable and digestible than others, biochemists discovered that many seeds and some storage organs, such as potato tubers, contain proteins called **proteinase inhibitors**.

Proteinase inhibitors block the enzymes—found in the mouths and stomachs of animals—that are responsible for digesting proteins. When an insect or a mammalian herbivore ingests a large dose of a proteinase inhibitor, the herbivore gets sick. As a result, herbivores learn to detect proteinase inhibitors by taste and avoid plant tissues containing high concentrations of these molecules.

Although many plant tissues contain proteinase inhibitors in low concentrations, biologists wanted to test the hypothesis that these proteins might also be part of an induced defense by the plant. To evaluate this idea, researchers allowed herbivorous beetles to attack one leaf on each of several potato plants.

- In liquid extracted from the other leaves on the attacked plants, proteinase inhibitor concentrations averaged 336 μg per mL.
- In leaves of control plants, where no insect damage had occurred, proteinase inhibitor levels averaged just 103 μg per mL of leaf juice.

This result supported the hypothesis that a hormone produced by wounded cells travels to undamaged tissues and induces the production of proteinase inhibitors.

THE DISCOVERY OF SYSTEMIN Biologists isolated the wound-response hormone by purifying the compounds found in tomato leaves and testing them for the ability to induce proteinase inhibitor production. The hormone that is active in tomato plants turned out to be **systemin**, a polypeptide just 18 amino acids long. Systemin was the first peptide hormone ever described in plants.

Researchers who labeled copies of systemin with a radioactive carbon atom, injected the hormone into plants, and then monitored its location confirmed that systemin moves from damaged tissues to undamaged tissues.

Currently, work on the production of systemin and proteinase inhibitors focuses on determining each step in the signal transduction pathway that alerts undamaged cells to danger (**Figure 39.28**). The data indicate that:

1. Systemin is released from damaged cells.
2. Systemin travels through the body via phloem and binds to membrane receptors on target cells.
3. The activated receptor triggers a long series of chemical reactions that eventually synthesize a molecule called jasmonic acid.
4. Jasmonic acid activates the production of at least 15 new gene products, including proteinase inhibitors.

In this way, plants build potent concentrations of insecticides in tissues that are in imminent danger of attack.

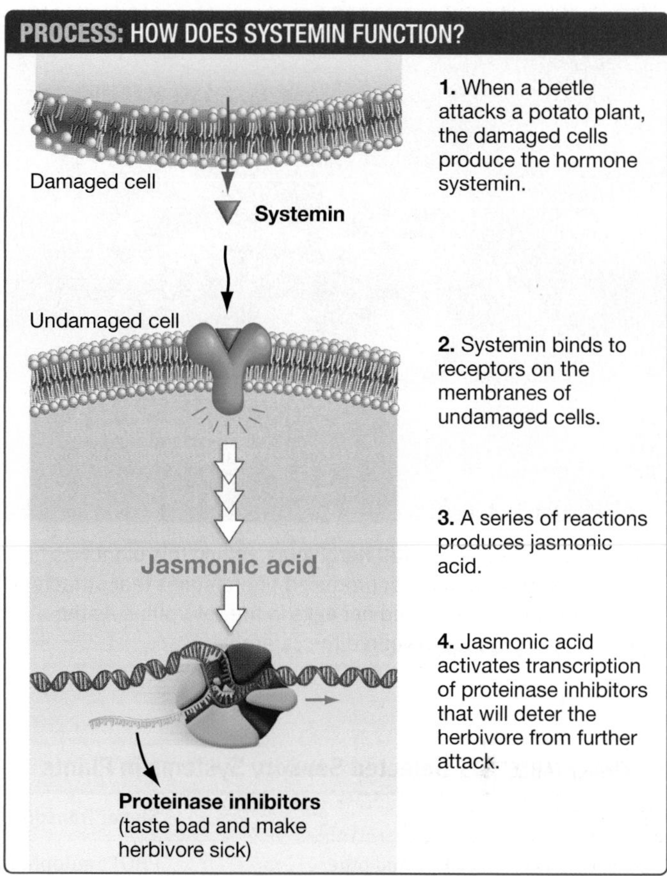

FIGURE 39.28 Signals from Insect-Damaged Cells Prepare Other Cells for Attack. Systemin, a hormone produced by herbivore-damaged cells, initiates a protective response in undamaged cells.

✔**QUESTION** Why is "systemin" an appropriate name for this hormone?

To review SAR and the systemin-based systems for plant defense, go to the study area at *www.masteringbiology.com*.

MB **Web Activity** Plant Defenses

"TALKING TREES": RESPONSES FROM NEARBY PLANTS When an insect starts munching on a leaf, volatile compounds evaporate from the surface and travel through the air. In the 1980s, researchers began to suspect that individuals growing near plants under herbivore attack "eavesdrop" on these volatiles. In response, they increase their own defenses—even though they've yet to be attacked.

The "talking trees" hypothesis has received extensive support, primarily through experiments based on exposing plants to volatiles in the absence of any insects. Even if the volatiles come from an entirely different species, research has shown that plants can sense the presence of these chemicals and respond by increasing the production of proteinase inhibitors and other defense compounds.

In some cases, plants that are under attack even call for help from other organisms.

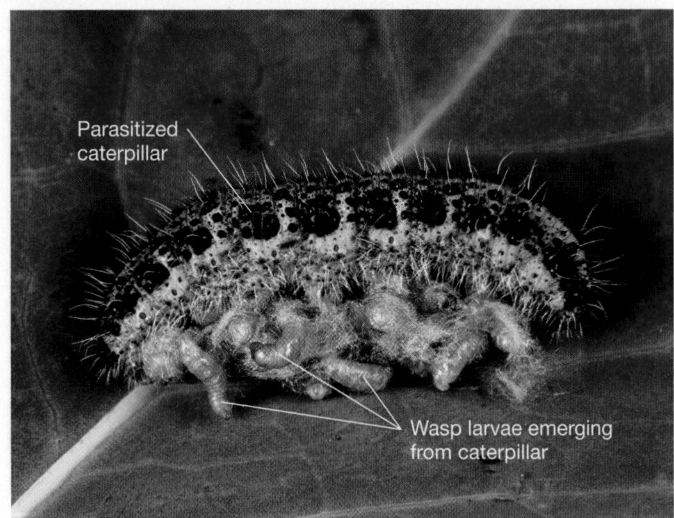

FIGURE 39.29 Parasitoids Kill Herbivores. When this plant was first attacked by this caterpillar, it produced pheromones that attracted a female wasp. The wasp laid her eggs in the caterpillar. As the wasp larvae grew, they devoured the caterpillar.

PHEROMONES RELEASED FROM PLANT WOUNDS RECRUIT HELP FROM WASPS Caterpillars and other herbivorous insects have enemies of their own—often wasps that lay their eggs in the insects' bodies. When wasp eggs hatch inside a caterpillar, the wasp larvae begin eating the host from the inside out. An organism that is free living as an adult but parasitic as a larva, such as these wasps, is called a **parasitoid** (**Figure 39.29**).

Biologists observed that parasitoids are particularly common when insect outbreaks occur in croplands. They wondered whether wounded plants release compounds that actively recruit parasitoids. More specifically, they hypothesized that plants produce **pheromones**—chemical messengers that are synthesized by an individual, released into the environment, and elicit a response from a different individual. Hormones act on cells inside an individual; pheromones act on another individual.

To explore this idea, researchers analyzed compounds that were released from corn seedlings during attacks by caterpillars. The insect-damaged leaves produced 11 molecules that were not produced by undamaged leaves. These compounds were not

SUMMARY TABLE 39.3 **Selected Sensory Systems in Plants**

Stimulus	Receptor	Signal Transduction	Response	Adaptive Significance
Blue light	PHOT1 and other phototropins in stem and leaves	PHOT1 autophosphorylates; remainder of signal transduction systems unknown	Phototropism occurs; also involved in stomatal opening	Stems grow toward light with wavelengths needed for photosynthesis
Blue light	Cryptochromes in stems and leaves	Details under investigation	Inhibits stem elongation; regulates daily rhythms and flowering response to day length	Keeps stems short if light is abundant; allows plant to time events
Red light	Phytochrome in seeds and elsewhere	Phytochrome changes to P_{fr} form and activates responses	Seed germinates	Sunlight triggers germination
Far-red light	Phytochrome in stem and elsewhere	P_r moves into nuclei and induces genetic responses	Stems lengthen	Species that require full sunlight attempt to escape shade from leaves
Gravity	Proteins located in plasma membrane?	Details unknown	Cells on opposite side of root or shoot elongate; tissue curves	Roots grow down; shoots grow up
Touch or wind	Stretch receptors; location unknown	Details unknown, but result is transcription activation in target genes	Stems grow shorter and thicker; in some cases vining (twining) growth	Individual is more resistant to damage; plants grow toward light
Touch	Receptor hair cell in Venus flytrap	Electrical changes in receptor cell's plasma membrane trigger action potentials	Target cells change shape; trap shuts	Plant can capture prey
Pathogens	*R* gene products	Details unknown	Hypersensitive response (HR); death of infected cells	Pathogens starve, so infection is slowed or stopped
Herbivores	Unknown; activated in response to molecule from herbivore	Details unknown	Insecticide production; signals to parasitoids	Herbivores are sickened or killed

released by leaves that had been cut with scissors or crushed with a tool; only insect damage triggered their production.

To follow up on this result, the investigators put female wasps in an arena that contained leaves damaged by insects or leaves that had suffered mechanical damage. In more than two-thirds of the tests that were performed, the wasps preferred to fly toward the insect-damaged leaves.

These results support the hypothesis that plants produce wasp attractants in response to attack by caterpillars. Biologists are increasingly convinced that plants can produce pheromones that recruit help in the form of egg-laden wasps.

From the research that has been done on plant sensory systems, it is abundantly clear that plants don't just sit there. These organisms may be stationary, but they constantly monitor and respond to a wide array of information about their environment (**Table 39.3**). Gaining a better understanding of phototropism, gravitropism, response to disease, and other aspects of plant behavior is an exciting frontier in biology.

CHECK YOUR UNDERSTANDING

If you understand that . . .

- The presence of *R* gene products allows plants to recognize attacks by parasites such as viruses, bacteria, and fungi.
- When the protein product of an *R* allele binds to a specific *avr* gene product from a pathogen, a signal transduction sequence results in the hypersensitive response.
- When plants are attacked by insect herbivores, damaged tissues release signals that trigger the production of toxins in undamaged leaves and that recruit enemies of the insects.

✔ **You should be able to . . .**

1. Explain the fitness advantage of having a wide array of *R* alleles in a single individual.
2. Explain why plants do not maintain high proteinase inhibitor concentrations in their tissues at all times.

Answers are available in Appendix B.

CHAPTER 39 REVIEW

For media, go to the study area at www.masteringbiology.com

Summary of Key Concepts

Plants are selective about the information they process. They perceive a wide variety of environmental stimuli that affect their ability to grow and reproduce.

- In most cases, information processing starts when a receptor protein changes shape in response to a stimulus. For example, a portion of the phytochrome protein changes shape when it absorbs red light. The same protein changes to an alternate shape when it absorbs far-red light.
- Plants monitor information about the nature and amount of light they receive, the direction of gravity, mechanical forces, and attacks by pathogens and herbivores.

✔ You should be able to state a hypothesis to explain why animals do not have phytochromes (a red/far-red light switch).

MB **Web Activity** Sensing Light

When sensory cells receive a stimulus, they transduce the signal and respond by producing hormones that carry information to target cells elsewhere in the body.

- When signal transduction occurs, an external signal is changed into an internal signal.
- In receptor cells, signal transduction culminates in the production of hormones that are transported throughout the plant body.

✔ You should be able to explain the analogy between how a plant sensory cell works and the following events: A person sees a barn on fire and calls the rural fire department; the dispatcher rings a siren; in response, members of the volunteer fire department race to the station to get the pumper truck.

Target cells respond to hormonal stimulation in ways that increase the ability of the plant to survive and reproduce.

- If a hormone binds to a receptor on a target cell, signal transduction occurs and culminates in changes in gene expression, altered translation rates, or changes in the activity of specific membrane pumps, channels, or ion carriers.
- Cells near the tips of shoots sense changes in blue light and respond by altering the distribution of the hormone auxin. Cells on one side of the shoot elongate in response to auxin much more than do cells on the other side of the stem. In this way, plants bend toward sunlight.

✔ You should be able to predict whether the phototropic response differs in plants that require high-light conditions versus plants that thrive best in low-light conditions.

Hormones are also responsible for regulating how plants grow throughout their lives—especially in response to changes in environmental conditions. Each type of plant growth regulator plays a general role in the life of a plant, and growth responses are usually affected by interactions among several different hormones.

- Auxin establishes and maintains the long axis of the plant body, playing a key role in phototropism and gravitropism and maintaining apical dominance. All of these responses rely on the polar transport of auxin, which establishes a gradient in auxin from the plant's apex to its roots.
- Gibberellins signal that conditions for growth are good and promote the initiation or continuation of growth and development.
- Abscisic acid (ABA) signals that environmental conditions are bad by suppressing growth and enforcing dormancy.

- Regulation of dormancy and growth by ABA and by gibberellins (GAs) are examples of how hormones interact—allowing plants to integrate information from several different stimuli and respond appropriately.

✔ You should be able to suggest a hypothesis explaining why some hormones are not transported in a single direction.

(MB) **Web Activity** Plant Hormones, **Web Activity** Plant Defenses

Questions

✔ TEST YOUR KNOWLEDGE

Answers are available in Appendix B

1. Which of the following statements about phytochrome is *not* correct?
   a. It is photoreversible.
   b. Its function was understood long before the protein itself was isolated.
   c. The P_{fr} form activates the responses to red light.
   d. It is involved in guard-cell opening.
2. Why was it logical to predict that amyloplasts function as statoliths?
   a. They are dense and settle to the bottom of gravity-sensing cells.
   b. They are only present in gravity-sensing cells.
   c. They make a direct physical connection with membrane proteins called integrins, which have been shown to be the gravity receptor molecule.
   d. Their density changes in response to gravity.
3. If a plant is touched repeatedly over the course of many days or if it experiences long-term exposure to wind, what happens?
   a. Growth of the root system is accelerated, making the plant more stable.
   b. Large-scale changes in gene expression occur, resulting in shoots that are short and stout.
   c. Electrical signals cause leaves to fold up, avoiding damage.
   d. Continued mechanical stimulation indicates that the individual is threatened with destruction, so it initiates flowering in an attempt to reproduce before it dies.
4. Which of the following statements about hormones is *not* correct?
   a. They tend to be small molecules.
   b. They exert their effects only on the same cells that produce them.
   c. They can exert strong effects even when they are present in extremely low concentrations.
   d. They trigger a response by binding to receptors in target cells.
5. In order for auxin to stimulate cell elongation, what two things have to happen?
   a. Transport proteins on the apical and basal ends of the cell must be activated.
   b. Proteins in the aleurone layer must be activated and α-amylase transcription increased.
   c. Water must flow into the cell, and the cell wall must become more extensible.
   d. Water must flow out of the cell, and the cell wall must become more extensible.
6. What evidence suggests that ABA from roots can signal guard cells to close?
   a. If roots are given sufficient water, guard cells close anyway.
   b. If roots are dry, guard cells begin to close—even though leaves are not experiencing water stress.
   c. Applying ABA on guard cells directly causes them to close.
   d. If roots are dry, ABA concentrations in leaf cells drop dramatically.

✔ TEST YOUR UNDERSTANDING

Answers are available in Appendix B

1. Phytochromes can be considered "shade detectors," while phototropins such as PHOT1 can be considered "sunlight detectors." Explain why these characterizations are valid.
2. In the experiment that confirmed the self-phosphorylation hypothesis for the blue-light receptor, researchers inserted the *PHOT1* gene into insect cells. Why was it important for the gene to be expressed in an organism other than a plant?
3. What does *transduce* mean? Give an example of a signal transduction event in plant sensory systems.
4. A plant's response to a given hormone depends on the cells or tissues that receive the signal, the plant's developmental stage or age, the concentration of the hormone, and the concentration of other plant hormones that are present. Provide examples that support each of these claims.
5. How do changes in hormonal signals allow plants to respond to changes in their physical environment over the course of their lives? Answer this question using phototropism as an example.
6. Discuss the general role that ethylene serves in plants. Provide evidence that supports your claim.

✔ APPLYING CONCEPTS TO NEW SITUATIONS

Answers are available in Appendix B

1. In general, small seeds that have few food reserves must be exposed to red light before they will germinate. (Lettuce is an example.) In contrast, large seeds that have substantial food reserves typically do not depend on red light as a stimulus to trigger germination. State a hypothesis to explain these observations.
2. To explore how hormones function, researchers have begun to transform plants with particular genes. In one experiment, a gene involved in cytokinin synthesis was introduced into tobacco plants. When the recombinant individuals matured, they produced more than the usual number of lateral branches. How does this experiment inform our understanding of how cytokinins function?
3. In many species native to tropical wet forests, seeds do not undergo a period of dormancy. Instead, they germinate immediately. Make a prediction about the role of ABA in these seeds. How would you test your predictions?
4. Researchers have shown that stomata do not close if ABA is injected into guard cells. Stomata do close, however, if ABA is applied to the surface of guard cells. Based on these results, researchers claim that the ABA receptor must be on the surface of guard cells, not in the interior. Do you agree? Why or why not?

This chapter focuses on the structure and function of plant reproductive structures, such as those of this flower.

Plant Reproduction 40

It would be difficult to overemphasize the importance of the reproductive organs and processes that are analyzed in this chapter—for plants, for biologists, and for you.

- For plants, every structure in the body and every physiological process—from water transport to photosynthesis—exist for one reason: to maximize the chances that the individual will produce offspring. As Chapter 1 pointed out, reproduction is the unconscious goal of everything that an organism does.
- For biologists, plant reproduction is not only fundamental to understanding how plants work but also the basis for major industries. Agriculture, horticulture, forestry, biotechnology, and ecological restoration draw extensively on what biologists know about plant reproduction.
- For you, plant reproduction means food. Human diets are based on consuming plant reproductive structures—primarily the seeds and fruits derived from flowers. A **flower** is a reproductive structure that produces gametes, attracts gametes from other individuals, nourishes embryos, and develops seeds and fruits. **Seeds** consist of an embryo and nutrient stores surrounded by a protective coat. **Fruits** develop from the flower's seed-producing organ and contain seeds.

This chapter's analysis of plant reproduction focuses on angiosperms, for three reasons: (**1**) Angiosperms represent over 85 percent of the land plants described to date; (**2**) virtually every important domesticated plant is an angiosperm; and (**3**) Chapter 30 introduced aspects of reproduction in other land plant lineages. By the end of this chapter, you'll appreciate the practical aspects of flowers as well as their beauty.

KEY CONCEPTS

- Plants undergo alternation of generations, in which a diploid sporophyte phase alternates with a haploid gametophyte phase. Sporophytes produce spores by meiosis. Gametophytes produce gametes by mitosis.
- In angiosperms, male and female gametophytes are microscopic and are produced inside flowers. Male gametophytes (pollen grains) are portable. Female gametophytes are encased in an ovary and are retained in the flower. When pollen grains land on a flower, they deliver sperm cells that fertilize the egg produced by the female gametophyte.
- Seeds contain an embryo and a food supply surrounded by a coat. In angiosperms, the walls of the ovary develop into a fruit that encloses the seed or seeds. In many cases, fruits function in seed dispersal.

✔ When you see this checkmark, stop and test yourself. Answers are available in Appendix B.

40.1 An Introduction to Plant Reproduction

Plant reproductive structures and processes vary among species. Consider just one aspect of reproductive organs—size. Flowers vary from microscopic to the size of a small child; seeds and fruits range from dustlike particles to coconuts.

Fortunately for students of plant biology, several basic principles unify this diversity of reproductive systems. Let's begin by defining sex.

Sexual Reproduction

Most plants reproduce sexually. As Chapter 12 pointed out, **sexual reproduction** is based on meiosis and fertilization, and results in offspring that are genetically unlike each other and unlike their parents. In plants, meiosis and fertilization occur in alternate phases of a life cycle.

To review briefly, **meiosis** is a type of nuclear division that results in four daughter cells, each of which has half the number of chromosomes present in the parent cell. **Fertilization** is the fusion of haploid cells termed **gametes**. The result of fertilization is the production of a single, diploid cell called the **zygote**, which will develop into a multicellular individual. Male gametes, or **sperm**, are small cells that contribute genetic information in the form of DNA but few or no nutrients to the offspring. Female gametes, or **eggs**, also contain DNA. But the female parent contributes nutrient stores as well. The important point is that although both sperm and egg contribute DNA, the male and female parent contribute vastly different amounts of other resources to the offspring.

Meiosis and fertilization result in offspring that are genetically unlike the parents. Sexual reproduction produces genetically diverse offspring that may be much more successful at warding off attacks from viruses, bacteria, and other pathogens than their parents (see Chapter 12).

The advantages of sexual reproduction are common to all eukaryotes that undergo meiosis. When and where meiosis occurs, however, is highly variable. Let's take a closer look.

The Land Plant Life Cycle

In most animals, meiosis leads directly to the formation of gametes. In plants, the situation is much different.

Land plants are characterized by a life cycle with two distinct multicellular forms—one diploid and one haploid. An individual in the diploid phase of the life cycle is called a **sporophyte**, while an individual in the haploid phase of the life cycle is called a **gametophyte**.

WHAT IS ALTERNATION OF GENERATIONS? This type of life cycle, **alternation of generations**, was introduced in Chapters 29 and 30 and is reviewed in **Figure 40.1**. Figures in those chapters detail how alternation of generations works in various protists and land plant groups. It is a life cycle that has evolved several times independently during the history of life.

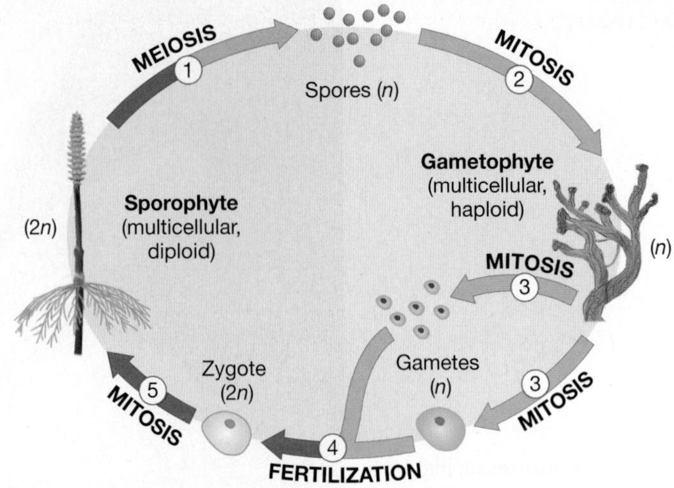

FIGURE 40.1 All Plants Undergo Alternation of Generations.

When alternation of generations occurs, meiosis does not lead directly to the formation of gametes as it does in humans and other animals. Instead it leads to the production of haploid cells called spores. A **spore** is a cell that grows directly into an adult individual. Several structures and processes are common to all land plant life cycles:

1. Meiosis occurs in sporophytes and results in the production of haploid spores. Unlike zygotes, spores are not produced by the fusion of two cells. Unlike gametes, spores produce an adult without fusing with another cell. Meiosis and spore production occur inside structures called **sporangia**.
2. Spores divide by mitosis to form multicellular, haploid gametophytes.
3. Gametophytes produce gametes by mitosis.
4. Fertilization occurs when two gametes fuse to form a diploid zygote.
5. The zygote grows by mitosis to form the sporophyte.

A good way to keep these terms straight is to remember that sporophyte means "spore-plant," while gametophyte means "gamete-plant." Sporophytes produce spores by meiosis. Gametophytes produce gametes by mitosis.

VARIATION IN LAND PLANT LIFE CYCLES Chapter 30 pointed out that land plant species vary a great deal in how large and long lived the gametophyte and sporophyte are relative to each other. The examples in **Figure 40.2** represent the extremes of the variation in life cycles observed within land plants.

- Figure 40.2a diagrams the life cycle of a liverwort. Notice that the largest stage in the liverwort life cycle is the gametophyte. The sporophyte is dependent on the gametophyte for nutrition.
- Figure 40.2b diagrams the life cycle of an angiosperm. Notice that the male and female gametophytes are microscopic, physically separate, and completely dependent on the sporophyte for nutrition.

(a) Liverworts: Gametophytes are large and long lived; the sporophyte is small and short lived.

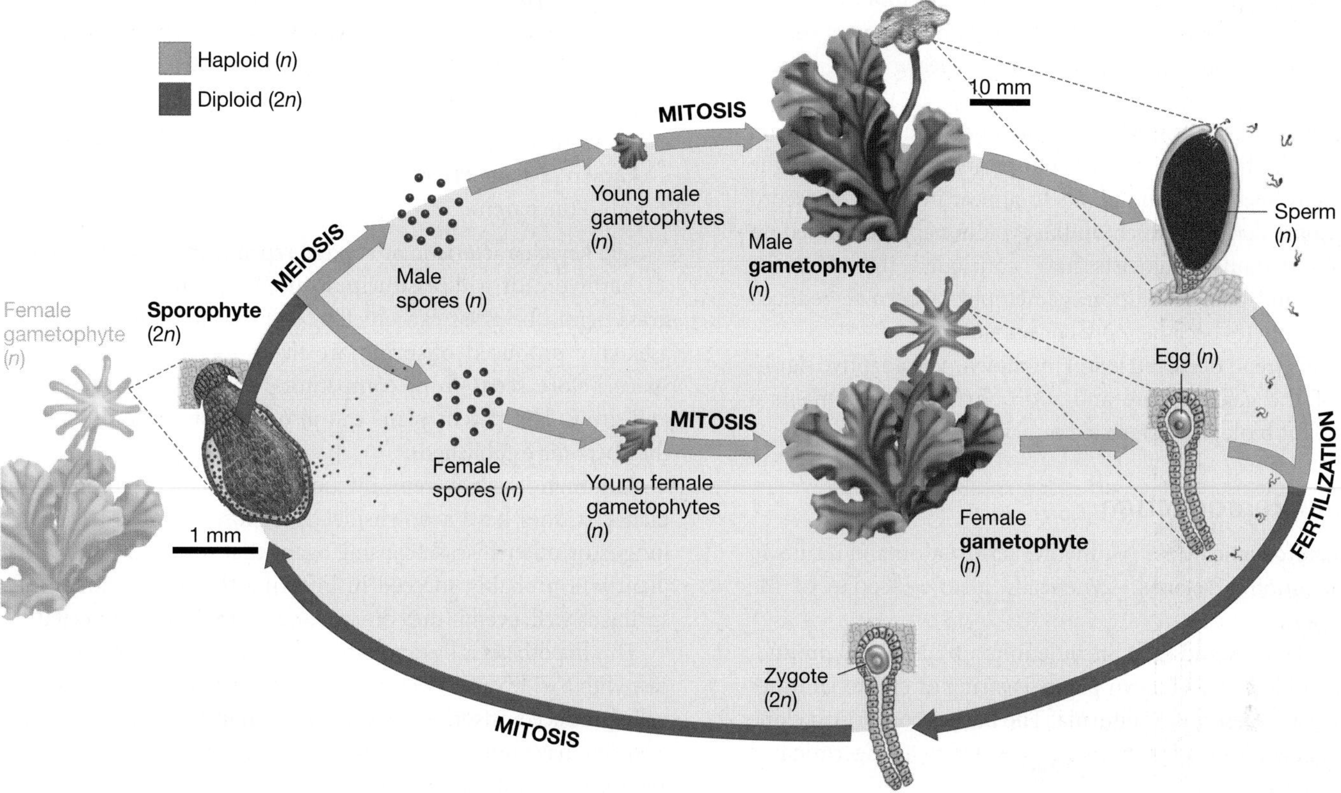

(b) Angiosperms: Sporophyte is large and long lived; gametophytes are small (microscopic) and short lived.

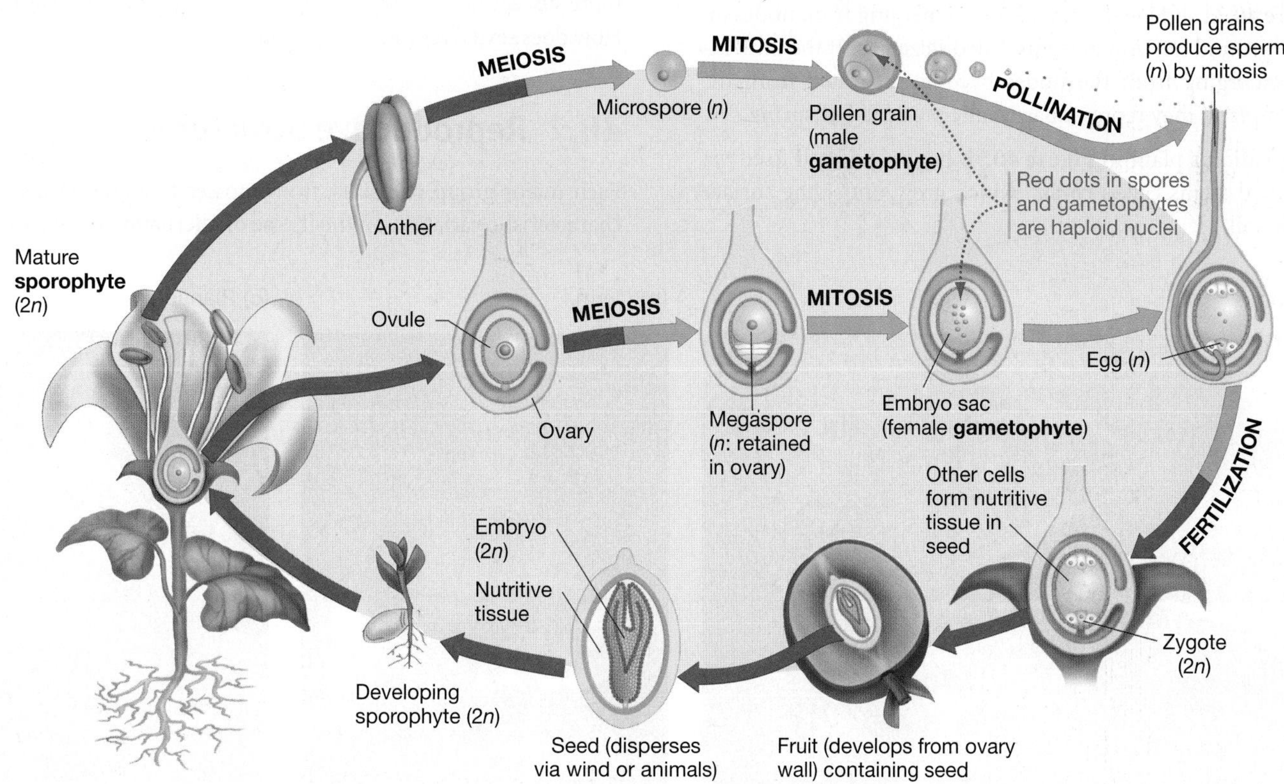

FIGURE 40.2 There Is Wide Variation in Plant Life Cycles. **(a)** In liverworts and other basal groups of land plants, the sporophyte depends on the gametophyte for nutrition. **(b)** In angiosperms and other more recent groups of land plants, the gametophytes depend on the sporophyte for nutrition.

✔If you understand the angiosperm life cycle, you should be able to (1) identify the male spore and female spore in Figure 40.2b, and (2) provide evidence to support the statement that in angiosperms, female gametophytes never leave their parent plant.

Liverworts are the sister group to all other land plants, and are found early in the fossil record. Angiosperms, in contrast, are the most recent lineage of land plants to appear in the fossil record and the most derived branch on the evolutionary tree. Based on these observations, biologists have concluded that over the course of land plant evolution, gametophytes became reduced while sporophytes became more conspicuous.

Why this trend occurred is still unknown. The adaptive significance of the gametophyte and sporophyte phases remains a challenge for biologists interested in plant reproduction.

Asexual Reproduction

Asexual reproduction does not involve fertilization and results in the production of **clones**—genetically identical copies of the parent plant.

Some plants extend their life indefinitely by asexual reproduction. The oldest of all known plants is a ring of creosote bushes in the Mojave Desert of California. The bushes comprise a clone that originated from a parent plant estimated to have germinated 12,000 years ago.

Although all asexual reproduction is based on mitosis, a wide array of mechanisms is involved.

- **Figure 40.3a** shows shoots and roots emerging from nodes on underground horizontal stems called **rhizomes**. If the individuals emerging from the nodes become separated from the parent plant, they represent asexually produced offspring.
- The gladiolus plant in **Figure 40.3b** has propagated itself via modified stems called **corms**, which grow under the surface of the soil.
- The kalanchoe in **Figure 40.3c** produces **plantlets**, which form from meristematic tissue located along the margins of its leaves. When the plantlets mature, they drop off the parent plant and grow into independent individuals.
- In dandelion and certain other species, mature seeds can form without fertilization occurring. This phenomenon, known as **apomixis**, results in seeds that are genetically identical to the parent.

The key characteristic of asexual reproduction is efficiency. If an herbivore or a disease wipes out the plants surrounding a grass plant, the grass can quickly send out horizontal stems. Its asexually produced offspring are likely to fill the unoccupied space before seeds from competitors can establish themselves and grow. The parent plant can also nourish these progeny as they become established.

Although asexual reproduction is extremely common in plants, it does have a downside. The most important is that a fungus or other disease-causing agent that infects an individual plant will probably succeed in infecting the plant's cloned offspring as well, even if they are no longer physically connected.

This hypothesis is based on the observation that plants fight disease with a wide variety of molecules (see Chapter 39). Because sexually produced offspring are genetically unlike their parents, these offspring have unique combinations of disease-fighting molecules and may be able to resist infections that devastate their parents.

This is an important point in agriculture and horticulture because asexually propagated apples, bananas, and other crops are more susceptible to epidemics than are sexually propagated species. How does sexual reproduction occur?

40.2 Reproductive Structures

Each major group of plants, from mosses to angiosperms, has a characteristic variation on the theme of alternation of generations,

(a) Rhizome

(b) Corm

(c) Plantlets

FIGURE 40.3 The Mechanisms of Asexual Reproduction Are Diverse. These are just three of many mechanisms of asexual reproduction in plants.

as well as characteristic male and female reproductive structures (see Chapter 30). Here, though, the focus is on the flower.

Let's start with a simple question: When do plants produce them?

When Does Flowering Occur?

Anatomically, a flower is a modified shoot that develops from a compressed stem and highly modified leaves. In essence, flower formation begins when an apical meristem stops making energy-harvesting stems and leaves and begins to produce the modified stems and leaves that make up flowers. Instead of making more food through photosynthesis, a sporophyte commits to investing energy in sexual reproduction—by producing gametophytes that will produce gametes.

When does this happen? Early experiments on the environmental signals that promote flowering focused on the number of hours of light and dark during a day.

RESPONDING TO CHANGES IN PHOTOPERIOD **Photoperiodism** is any response by an organism that is based on photoperiod—the relative lengths of day and night. In plants, the ability to measure photoperiod is important because it allows individuals to respond to seasonal changes in climate—for example, to flower when pollinators are available and when resources for producing seeds are abundant.

Experiments on photoperiodism in plants have shown that, with respect to flowering, plants fall into three main categories:

1. **Long-day plants** bloom in midsummer, when days are longest and nights shortest. Radishes, lettuce, spinach, corn, irises, and other long-day plants flower only when days are longer than a certain length—usually between 10 and 16 hours, depending on the species.
2. **Short-day plants** bloom in spring, late summer, or fall. Asters, chrysanthemums, poinsettias, and other short-day plants flower only if days are shorter than a certain species-specific length.
3. **Day-neutral plants** flower without regard to photoperiod. Day length has no effect on flowering in roses, snapdragons, dandelions, tomatoes, cucumbers, and many **weeds**—plants that are adapted to grow in soils that have been disturbed enough to remove or damage existing vegetation.

How do plants sense changes in day length and initiate the chain of events that leads to flowering? Early experiments suggested that phytochrome—the photoreversible pigment introduced in Chapter 39—is involved.

Figure 40.4 summarizes some of the key experiments. Researchers showed that interrupting the night period with a light flash changed the flowering response. In short-day plants, for example, flowering was inhibited if the dark period was interrupted by red light—wavelengths around 660 nm. But a subsequent flash of far-red light—wavelengths around 735 nm—erased the effect. In some way, a red/far-red switch linked changes in day length to flowering.

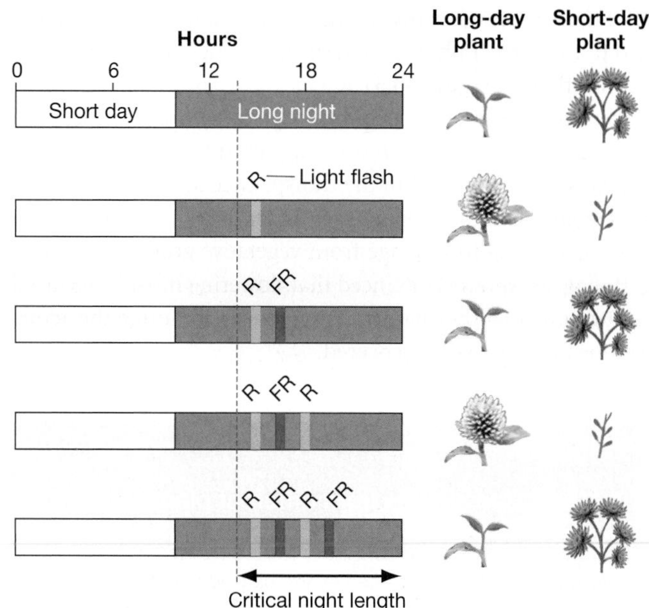

FIGURE 40.4 Flashes of Red Light and Far-Red Light Switch the Photoperiod Response On and Off. If flashes of red light (R) and far-red light (FR) are alternated during the night, the plant's flowering response correlates with the last light it experienced. A flash of red light turns a long night into a short night; a flash of far-red light restores the perception of a long night.

More recent research has shown that phytochrome's effect is closely tied to the molecular mechanisms of timekeeping in plants. Plants have a clock that is reset each morning. Clock proteins rise during the day and trigger expression of a gene called *CONSTANS* (*CO*). The CO protein is a transcription factor that affects the production of a flowering hormone. When phytochrome is activated by light, it stabilizes CO—so that CO accumulates in cells.

- In long-day plants, high levels of CO stimulate production of the flowering hormone.
- In short-day plants, high levels of CO *inhibit* production of the flowering hormone.

Many questions remain, however, and research is continuing. In the meantime, it is clear that there is a complex interaction between clock proteins, phytochrome activation, and the production of the signal that initiates flowering. What is that signal?

THE DISCOVERY OF THE FLOWERING HORMONE Since the 1930s, biologists have known that exposing even one leaf on a plant to the conditions necessary to induce flowering may result in the whole plant flowering. This result suggests that the signal to flower comes from leaves and travels to the apical meristem.

Grafting experiments, in which an organ from one individual is physically attached to a different individual, support this result. If you provide an experimental plant with the appropriate flower-triggering night length and then cut off a leaf or stem and graft it onto a second, experimental plant that has never been exposed to the correct photoperiod for flowering, the experimental plant will flower (**Figure 40.5**).

Like the experiments on phototropism with agar blocks reviewed in Chapter 39, this result supported the hypothesis that some substance produced in the transplanted leaf must travel up the recipient plant and trip a developmental switch in the apical meristem, causing the change from vegetative growth to flowering. Biologists were so convinced that flowering must be induced by a hormone that they named it **florigen**, even though the actual hormone had not been discovered.

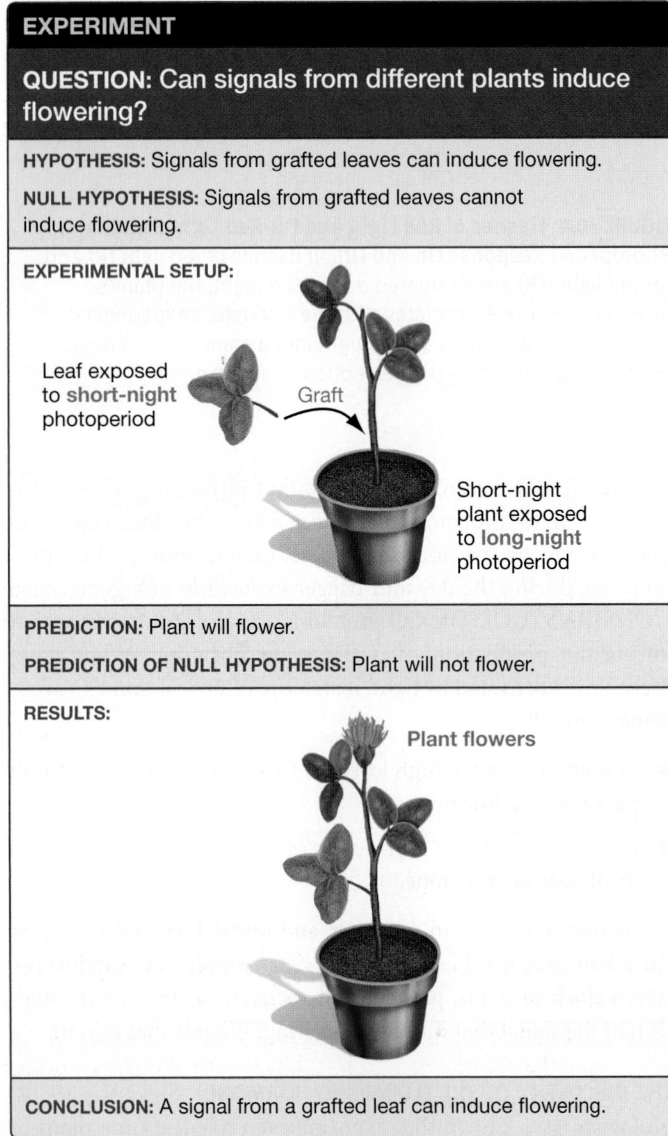

FIGURE 40.5 Experimental Support for the Hypothesis That a Hormone for Flowering Exists.

SOURCE: Lang, A., M. K. Chailakhyan, and I. A. Frolova. 1977. Promotion and inhibition of flower formation in a day-neutral plant in grafts with a short-day plant and a long-day plant. *Proceedings of the National Academy of Sciences, USA* 74: 2412–2416.

✔**QUESTION** For this experiment to support the claim that a flowering hormone exists, a control treatment needs to be done. What is this control treatment? If the hypothesis is correct, what is the predicted outcome of the control treatment?

Almost 80 years later, researchers finally found the florigen molecule. This work began by focusing on the *FLOWERING LOCUS T (FT)* gene in *Arabidopsis thaliana*, which is known to promote flowering when activated. When researchers exposed leaves to short nights, they found that the gene was expressed in the leaf vascular tissue. Subsequently, some research indicates that the protein product of the *FT* gene is transported from leaves to the shoot apical meristem, and that the protein's presence triggers the activation of genes required for converting the stem to a flower.

Now the question is, What does the flower itself look like?

The General Structure of the Flower

Structurally, all flowers are variations on a theme. They are made up of four basic organs that are essentially modified leaves: (1) sepals, (2) petals, (3) stamens, and (4) one or more carpels. These organs are attached to a compressed portion of stem called the receptacle (**Figure 40.6a**).

Not all four organs are present in all flowers, however; and as **Figure 40.6b** shows, the coloration, size, and shape of these four components are fabulously diverse. Let's consider each of the four parts, in turn.

SEPALS FORM AN OUTER, PROTECTIVE WHORL **Sepals** are leaflike structures that make up the outermost parts of a flower. Sepals are usually green and photosynthetic, and they are relatively thick compared with other parts of the flower.

Because they attach to the receptacle in a circle or whorled arrangement, sepals enclose the flower bud as it develops and grows—protecting young buds from damage by insects or disease-causing agents. The entire group of sepals in the flower is called the **calyx**.

PETALS FURNISH A VISUAL ADVERTISEMENT Like sepals, **petals** are arranged around the receptacle in a whorl. Often brightly colored and scented, petals function to advertise the flower to bees, flies, hummingbirds, and other pollinators.

In some cases, the color of the petals correlates with the visual abilities of particular animals. Bees, for example, respond strongly to wavelengths in the blue and purple regions of the light spectrum, as well as yellow (they don't see red well). Flowers that attract bees, in turn, often have yellow, blue, or purple petals with ultraviolet patches.

The ultraviolet sections of petals in "bee flowers" frequently highlight the center of the flower (**Figure 40.7**), where the stamens and carpels are located. Why? In these flowers, the base of the petals contains a gland called a **nectary**. The nectary produces the sugar-rich fluid **nectar**, which is harvested by many of the animals that visit flowers, along with pollen. In the process of collecting pollen or nectar, the visiting animal usually deposits pollen from a different plant on the stamens—accomplishing pollination.

The entire group of petals in a flower is called the **corolla**. In some species, the petals within the corolla vary in size, shape, and function:

(a) Basic parts of a flower

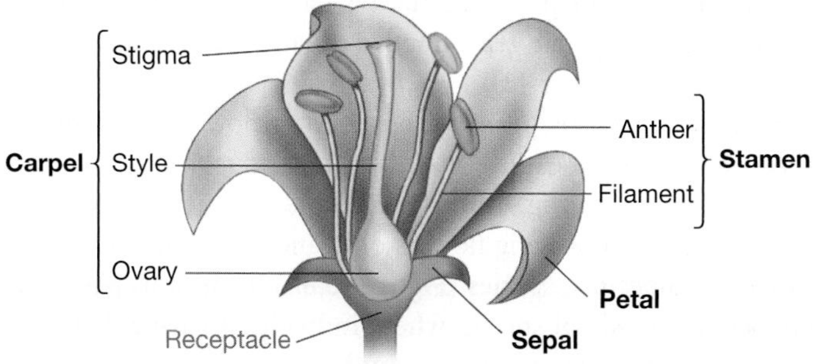

(b) Examples of flower diversity

FIGURE 40.6 The Basic Structures in Flowers Are Highly Variable. **(a)** Flowers comprise sepals, petals, stamens, and carpels. **(b)** The four parts vary among species.

(a) What you and a bee see

(b) What a bee sees in addition

FIGURE 40.7 Insects See in the Ultraviolet Range. **(a)** The inflorescence (flower cluster) of a black-eyed Susan, seen by the unaided human eye. **(b)** The same structure, photographed with a camera that records ultraviolet wavelengths that are visible to bees but invisible to humans.

- Flattened petals may provide a landing pad for flying insects.
- Elongated, tubelike petals frequently have a nectary at their base that can be reached only by animals with a long beak or tongue-like proboscis.
- Some petals protect the reproductive organs located inside the corolla.
- Specialized cells in some petals synthesize and release molecules that provide a scent attractive to certain species of pollinating insects.

In contrast, wind-pollinated angiosperms such as oaks, birches, pecans, and grasses have flowers that have small petals or no petals at all, and that lack nectaries. These species do not invest in structures that aren't required for pollination.

STAMENS PRODUCE POLLEN **Stamens** are reproductive structures that produce male gametophytes—also known as pollen grains. The male gametophytes, in turn, produce sperm.

Each stamen consists of two components:

1. a slender stalk termed the **filament**, and
2. the pollen-producing organs called **anthers** (see Figure 40.6a).

The anther is the business end of the stamen—where meiosis and pollen formation take place. The function of the filament is to hold the stamen in a place where wind, insects, hummingbirds, or other agents can make contact with the pollen grains produced in the anther.

CARPELS PRODUCE OVULES The fourth reproductive structure is the **carpel**, which produces female gametophytes. A carpel consists of three regions:

1. The **stigma** is a moist tip that receives pollen.
2. The **style** is a slender stalk.
3. The **ovary** is an enlarged structure at the base of the carpel (see Figure 40.6a).

Inside the ovary, female gametophytes are produced in structures called **ovules**. An ovary may contain more than one ovule. When the female gametophytes that are produced inside ovules mature, they produce eggs.

THE "SEX" OF FLOWERS VARIES In most angiosperm species, stamens and carpels are produced on the same individual. Flowers that contain both stamens and carpels are referred to as **perfect**.

Flowers can also be **imperfect**, however, meaning they contain either stamens *or* carpels, but not both. Imperfect flowers that contain only stamens can be considered "male" flowers. Similarly, imperfect flowers that contain only carpels can be considered "female" flowers.[1]

In some cases, separate stamen- or carpel-producing flowers occur on the same individual. Species like these, including the corn plants illustrated in **Figure 40.8a**, are **monoecious** (literally, "one-house"). In corn, the tassel is a collection of stamen-producing "male" flowers, and the ear contains a group of carpel-producing "female" flowers.

[1]Technically, flowers are not referred to as male and female. Instead, they are staminate or carpellate. Staminate flowers produce stamens, which produce pollen grains, which produce male gametes (sperm). Carpellate flowers produce carpels, which contain ovaries. Female gametophytes develop inside ovaries and produce female gametes (eggs). For convenience, though, the text will sometimes refer to male and female flowers and reproductive structures.

(a) Corn is monoecious.

(b) *Cannabis* is dioecious.

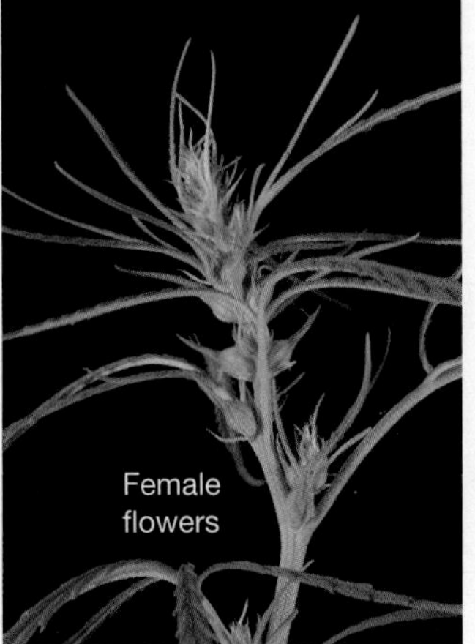

FIGURE 40.8 Male and Female Flowers Can Occur on the Same Individual or on Different Individuals. (a) The tassels of corn are male flowers; ears are female flowers. **(b)** In *Cannabis sativa*, male and female flowers are found on different individuals.

Species that are monoecious or that have perfect flowers are analogous to hermaphroditic species of animals, which contain both male and female reproductive organs and produce sperm and eggs (see Chapter 48).

In contrast, some species with imperfect flowers are **dioecious** ("two-houses")—meaning that each individual plant produces either stamen-bearing flowers only, and could be considered "male," or carpel-bearing flowers only and is "female." *Cannabis sativa* is a dioecious species (**Figure 40.8b**). Dioecious plants are analogous to animal species where individuals have either male or female reproductive organs—not both.

How Are Female Gametophytes Produced?

What purposes do the three parts of the carpel serve? The function of the stigma and style will become clear in Section 40.3; for now let's concentrate on what happens inside the ovary.

Figure 40.9 is a longitudinal section showing the inside of a typical angiosperm ovary. Notice that it contains one or more ovules. Each ovule contains a structure called the megasporangium, which contains a cell called the megasporocyte. (The use of "mega" is appropriate, because these structures are much larger than their counterparts in the stamen.) The megasporangium is comparable to spore-producing organs found in other plants, such as the sporangia found on the back of fern leaves.

When you study Figure 40.9, note four important points:

1. The megasporocyte divides by meiosis.
2. Four haploid nuclei called **megaspores** result from meiosis, but three degenerate. No one is sure how or why this happens.
3. The surviving megaspore divides by mitosis to produce a structure with haploid nuclei. This is the female gametophyte—usually known as the **embryo sac**.
4. The haploid nuclei segregate to different positions in the embryo sac, and cell walls form around them.

In the carpel, then, a diploid megasporocyte divides by meiosis to form a megaspore, which then divides by mitosis to form the female gametophyte. Female gametophytes are encased in an ovary, are retained in the flower, and produce an egg.

In many angiosperms the embryo sac contains eight haploid nuclei and seven cells. Typically, two **polar nuclei** stay together within one central cell—the largest cell in the ovule. The number of polar nuclei varies among species, however.

The egg cell is located at one end of the female gametophyte, near an opening in the ovule called the **micropyle** ("little-gate"). The micropyle is where a sperm nucleus will enter the ovule, if fertilization occurs.

How Are Male Gametophytes Produced?

Figure 40.10 provides a detailed look at the stamen and the steps that occur in the production of male gametophytes. Recall that a stamen consists of two major parts: an anther and a filament. Inside the anther, structures known as microsporangia contain diploid cells called microsporocytes.

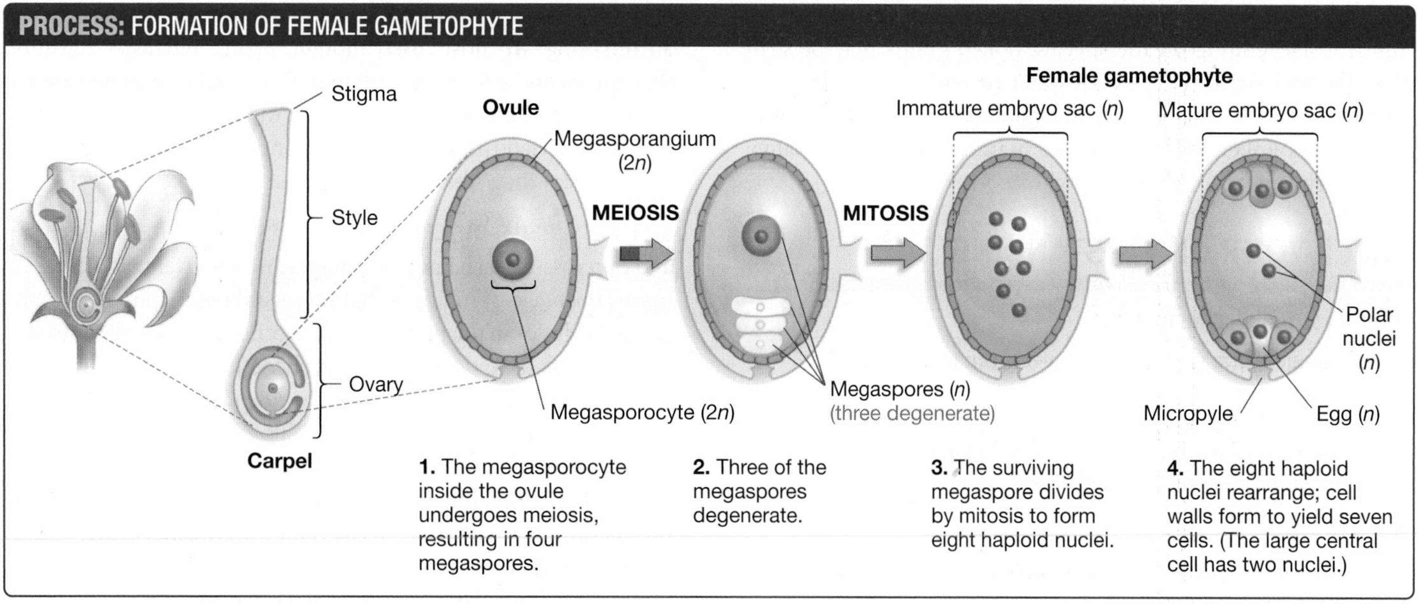

FIGURE 40.9 In Angiosperms, Megaspores Produce Female Gametophytes.

✔**QUESTION** Define a gametophyte. Why does the embryo sac conform to this definition?

When you study Figure 40.10, note three important points:

1. Microsporocytes undergo meiosis.
2. Each haploid cell that results is a **microspore**. Normally, all of the microspores survive. Microspores divide by mitosis.
3. The two nuclei that result from mitotic division in a microspore form a haploid, immature male gametophyte, also known as the **pollen grain**.

In the anther, then, a diploid microsporocyte divides by meiosis to form microspores, which then divide by mitosis to form male gametophytes. Male gametophytes are dispersed from the flower, and eventually produce sperm.

At the immature stage—before it has produced sperm—the male gametophyte consists of two cells: a small generative cell enclosed within a larger tube cell. The male gametophyte is considered mature when the haploid generative cell produces two sperm cells via mitosis.

In some species, this maturation step occurs while pollen is still in the anther. In other species, maturation and sperm production don't occur until after the pollen grain lands on a stigma and begins to grow.

The wall of a pollen grain develops a tough outer coat that includes the watertight compound called sporopollenin, introduced in Chapter 30. This coat protects the male gametophyte

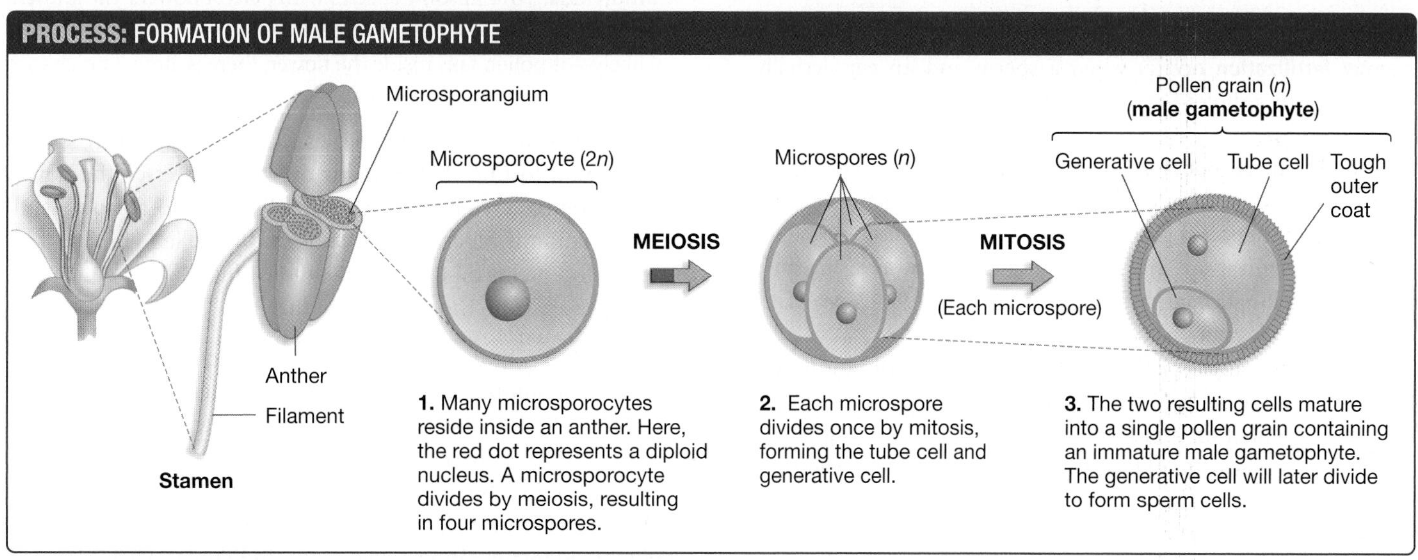

FIGURE 40.10 In Angiosperms, Microspores Produce Male Gametophytes.

✔**QUESTION** Why do pollen grains conform to the definition of a gametophyte?

when the pollen is released from the parent plant into the environment. Depending on the species, pollen grains may be dispersed by an animal, the wind, or water currents.

And now we're finally ready to explore the moment of truth: How does a pollen grain get to the mature carpel of the same species, where an egg cell is waiting?

CHECK YOUR UNDERSTANDING

If you understand that . . .

- In angiosperm sporophytes, flowers produce spores that develop into female and male gametophytes.
- The female reproductive structures called carpels contain ovaries. Ovaries enclose structures called ovules. Female gametophytes are produced inside ovules.
- Formation of a female gametophyte begins when a diploid megasporocyte inside an ovule undergoes meiosis. The product of meiosis is a haploid megaspore. The megaspore divides by mitosis to form the female gametophyte—including the egg and polar nuclei.
- Male gametophytes are produced inside reproductive structures called anthers.
- Formation of the male gametophyte begins when a diploid microsporocyte undergoes meiosis to form haploid microspores. Microspores divide by mitosis to form the male gametophyte—including a generative cell that will divide by mitosis to form two sperm cells.

✓ **You should be able to . . .**

1. Compare and contrast a megasporocyte and a microsporocyte.
2. Compare and contrast a female gametophyte and a male gametophyte.

Answers are available in Appendix B.

40.3 Pollination and Fertilization

Pollination is the transfer of pollen grains from an anther to a stigma; **fertilization** occurs when a sperm and an egg actually unite to form a diploid zygote. The two events are separated in space and time.

Pollination is not restricted to angiosperms. The gymnosperms introduced in Chapter 30 also package their male gametophytes into pollen grains. This section will focus on pollination and fertilization in flowering plants, however, because managing pollination and fertilization in angiosperms is a critical challenge for fruit growers and plant breeders.

In addition, angiosperms' pollination and fertilization systems are thought to be key to their evolutionary success. What aspects of pollination and fertilization allowed flowering plants to become so successful in terms of their numbers of species?

Pollination

Pollen can fall on the stigma of the same individual or the stigma of a different individual. **Self-fertilization**, or **selfing**, occurs when a sperm and an egg from the same individual combine to produce an offspring. In most cases, though, plants **outcross**—meaning that sperm and eggs from different individuals combine to form an offspring. Outcrossing is the result of **cross-pollination**—when pollen is carried from the anther of one individual to the stigma of a different individual.

SELFING VERSUS OUTCROSSING: COSTS AND BENEFITS Selfing and outcrossing each have advantages and disadvantages. The primary advantage of selfing is that successful pollination is virtually assured—it doesn't depend on agents other than the plant itself.

Biologists have documented the benefit of pollination-assurance by hand-pollinating plants that normally outcross. In most cases, the hand-pollinated individuals produce far more seed than do individuals that are pollinated naturally.

Other things being equal, self-pollination should result in the production of many more seeds than outcrossing. Other things are not equal, however. Selfing has a distinct disadvantage: Even though it still involves meiosis, selfed offspring are usually much less diverse genetically than outcrossed offspring are (see Chapter 12). In some cases, selfed offspring may also suffer from inbreeding depression (see Chapter 25).

Although outcrossing is riskier in terms of the chances that pollination will occur, it results in genetically diverse offspring that may be much more successful at warding off attacks from viruses, bacteria, and other pathogens (see Chapter 12).

Outcrossing is much more common than selfing, and in many cases plants have elaborate mechanisms to prevent selfing. These mechanisms include:

- *Temporal avoidance* In some species that have perfect flowers, male and female gametophytes mature at different times. Thus, selfing cannot occur.
- *Spatial avoidance* Selfing isn't possible in dioecious species and may be rare in monoecious species, unless pollinators transfer pollen between different-"sexed" flowers on the same individual. And in some species with perfect flowers, the anthers and stigma are so far apart that self-pollination is extremely unlikely—if pollen falls inside the flower, there is almost no chance that it will land on the stigma.
- *Molecular matching* In species that normally outcross, it is common to find that proteins on the surfaces of the pollen and stigma have to interact for pollination to be successful. Pollination is blocked if proteins on the pollen grain surface match proteins on the stigma—indicating that the pollen and stigma are from the same individual.

POLLINATION SYNDROMES Cross-pollination can be accomplished in a number of ways: Pollen can be carried from flower to flower by physical agents such as wind or water, or by organisms such as insects, birds, or bats.

Animals visit flowers to eat pollen grains, harvest nectar, or both. As an animal feeds from a flower, pollen grains adhere to its body incidentally. When the same individual visits another flower of the same species to feed, some of these grains are deposited on a stigma of the second flower.

In most cases, animal pollination is an example of **mutualism**: a mutually beneficial relationship between two species. Pollinators usually benefit by receiving food; flowering plants gain by having their male gametophytes transferred to a different individual so that outcrossing takes place.

Chapter 30 introduced pollination syndromes: close correlations between the structure of a flower and the size, shape, and behavior of its pollinators. Most moths and bats, for example, are active at night. If they feed on nectar or pollen, the flowers that they visit tend to be white—and thus more visible in low light—have a strong scent, and open up at night.

One of the most famous examples of a pollination syndrome involves a species called Darwin's orchid (**Figure 40.11**), which is native to Madagascar. The orchid attracted a great deal of attention when it was first discovered by western scientists because it has a "spur" that can be as much as 28 cm (11 in) long, with a nectary at its base. Charles Darwin hypothesized that it must be pollinated by a moth with a tongue-like proboscis as long as the spur. Although the idea seemed preposterous at the time, 40 years later a hawkmoth species with a proboscis that averages 24 cm in length was discovered pollinating the orchid in the wild.

Structures associated with pollination syndromes are thought to be adaptations: traits that increase the fitness of individuals in a particular environment. In this case, flowers and pollinators have adaptations that increase pollination frequency and feeding efficiency, respectively. To capture this point, biologists say that **coevolution** has occurred—a topic that is taken up in more detail in Chapter 53.

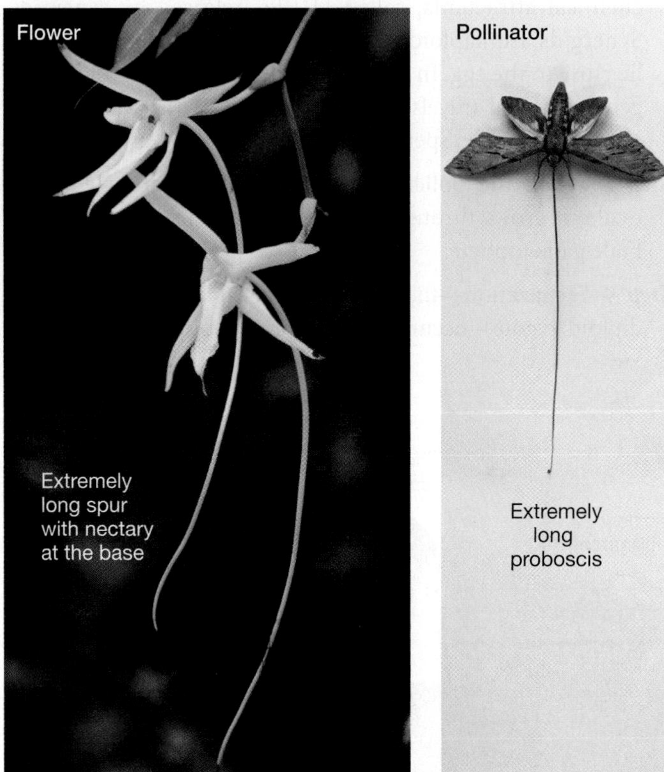

FIGURE 40.11 Pollination Syndrome Can Be Inferred from Flower Structure and Color. Charles Darwin hypothesized that this white orchid flower with an extremely long spur must be pollinated by a moth with a tongue-like proboscis that was long to reach the nectar at the end. He was right.

WHY DID POLLINATION EVOLVE? In mosses, ferns, and other groups that do not form pollen, sperm have flagella and swim to the egg through droplets of water, or are transferred on water droplets that cling to the legs of tiny insects called springtails. In conifers and most other gymnosperms, wind transmits pollen from male to female. In some of the other groups that produce pollen, such as the cycads, gnetophytes, and angiosperms, many species are pollinated by animals—particularly by insects.

When these observations are mapped onto the phylogenetic tree shown in **Figure 40.12**, two important patterns emerge.

1. Pollination evolved late in land plant evolution. Mosses and other groups that do not form pollen appear first in the fossil record of land plants. Conifers and other groups that are strictly or primarily wind pollinated evolved later but before angiosperms.
2. Seed plants do not need water for sexual reproduction to occur. As a result, the evolution of pollen has allowed these species to be much less dependent on wet habitats. Along with the evolution of the seed—highlighted in Section 40.4—pollen paved the way for the colonization of drier environments.

In addition, it's critical to realize that pollination became a much more precise process when plants began to recruit animals

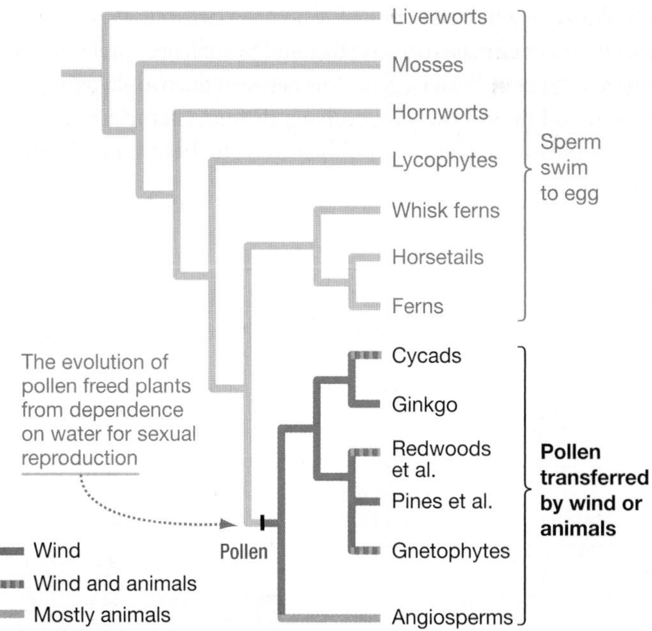

FIGURE 40.12 Pollen Is a Relatively Recent Innovation in Land Plant Evolution. Evolutionary relationships among the major groups of plants. In the lineages colored gray, sperm have flagella and swim to the egg. In the lineages colored blue or orange, sperm are produced by pollen grains.

to act as pollinators. Wind-borne pollen grains have a low probability of landing successfully on a flower stigma. Insect-borne pollen, in contrast, is much more likely to be successfully transferred to flowers of the same species.

Wind-pollinated species invest in making large numbers of pollen grains; animal-pollinated species make fewer pollen grains but invest in structures that attract and reward animals.

In effect, plants "pay" nectar- and pollen-eating insects to work for them. Wind is free, but insects are more precise. Insect pollination is an important adaptation because it makes sexual reproduction much more efficient.

DOES POLLINATION BY ANIMALS ENCOURAGE SPECIATION? In addition to affecting the fitness of individual plants, does pollination by animals make the formation of entirely new species more likely?

To answer this question, consider the situation shown in **Figure 40.13**. A biologist has documented that two populations of a mountain-dwelling species called the alpine skypilot have flowers with different characteristics.

- Alpine skypilots that grow in forested habitats at or below timberline have small flowers with short stalks and an aroma described as "skunky."
- Individuals that grow in the tundra habitats above timberline have large flowers with long stalks and smell sweet.

These differences are interesting, because different insects pollinate the two populations. Small flies are abundant at slightly lower elevations and pollinate the timberline individuals; large bumblebees are abundant at higher elevations and pollinate the tundra flowers. Experiments have shown that bumblebees prefer to pollinate big flowers—probably because larger flowers can support their larger mass. Because flies and bumblebees prefer to visit different types of flowers, gene flow between the two skypilot populations is low and they are evolving distinct characteristics. The two populations may be on their way to becoming different species.

The message here is that evolutionary changes in the size or food-finding habits of a pollinator affect the angiosperm populations they pollinate. In return, changes in flower size and shape affect the insects pollinating that population. If a small population of Darwin's orchids evolved longer spurs, for example, the hawkmoth pollinators that lived in that area would be under intense selection that favored the evolution of a longer proboscis.

Because mutation continuously introduces variations in traits, insect and angiosperm populations frequently change, diverge, and form new species. Changes in pollination syndromes can trigger the evolution of new species. It is no surprise that insects and angiosperms are exceptionally species-rich groups.

It is clear that pollination was a crucial innovation during plant evolution. Now let's get down to mechanics. What happens once a pollen grain is deposited on a stigma?

Fertilization

Figure 40.14 walks you through the steps in fertilization.

Step 1 After landing on the stigma of a mature flower from the same species, a pollen grain absorbs water and germinates. **Germination** is a resumption of growth.

Step 2 When the male gametophyte germinates, a long filament called a **pollen tube** grows through the stigma and down the length of the style. This growth is directed by chemical attractants, called LUREs, released by synergids. Synergids are haploid cells in the female gametophyte that lie close to the egg. In the species illustrated in the figure, the generative cell travels down the length of the tube and divides to form two sperm.

Step 3 When the pollen tube reaches the micropyle of the ovule, it grows through it and enters the interior of the female gametophyte.

Step 4 Fertilization—the fusion of sperm and egg to form a diploid zygote—occurs. In most plant groups, fertilization is

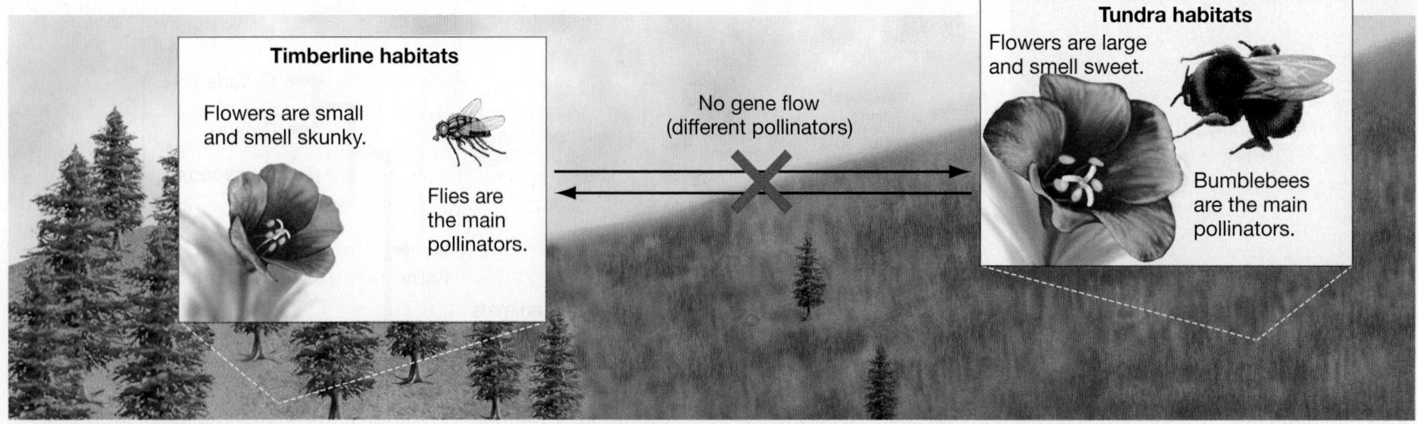

FIGURE 40.13 Interactions between Angiosperms and Animal Pollinators Affect Gene Flow. Flowers in alpine skypilot populations from timberline and tundra habitats look and smell different, and attract different pollinators that are common in timberline versus tundra habitats. As a result, interbreeding is rare.

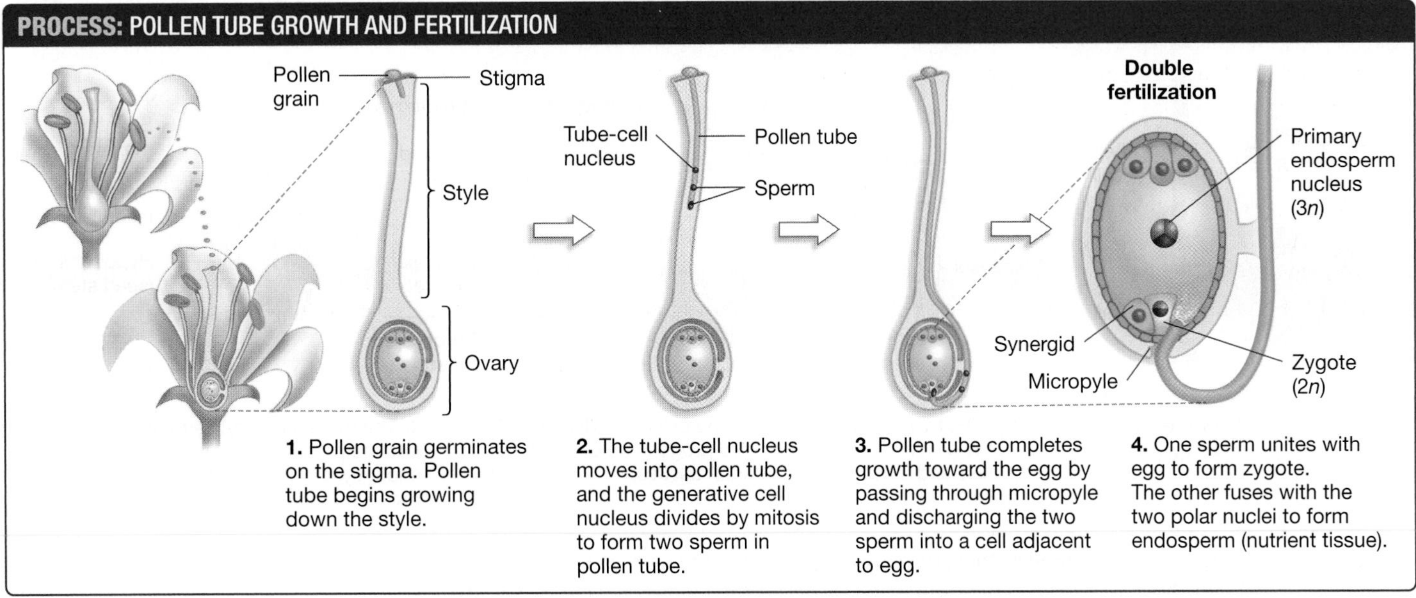

FIGURE 40.14 Double Fertilization Produces a Zygote and an Endosperm Nucleus. When the pollen tube reaches the female gametophyte, one sperm nucleus fertilizes the egg while the second fuses with the polar nuclei.

straightforward—sperm and egg simply combine, and a diploid nucleus is formed. In angiosperms, however, an unusual event called **double fertilization** takes place. One sperm nucleus unites with the egg nucleus to form the zygote. The other sperm nucleus moves through the female gametophyte and fuses with the polar nuclei in the central cell. In most cases, two polar nuclei are present and a large triploid ($3n$) cell forms.

✔If you understand double fertilization, you should be able to identify cells in a female gametophyte, immediately after fertilization occurs, that are haploid, diploid, and triploid.

CHECK YOUR UNDERSTANDING

If you understand that . . .

- Wind, water, or animals carry pollen grains from one plant to another.
- When a pollen grain lands on a stigma, the grain germinates. A pollen tube forms and grows until it reaches the ovule.
- Sperm cells produced by the male gametophyte fertilize the egg and the polar nuclei, forming a diploid zygote and in most cases a triploid endosperm.

✔ **You should be able to . . .**

1. Explain why insects increase their fitness by visiting flowers and why flowers increase their fitness by rewarding insects.
2. Describe the function of the cells produced by double fertilization.

Answers are available in Appendix B.

The triploid nucleus resulting from this second fertilization undergoes mitosis and cytokinesis to form the **endosperm** ("inside-seed") tissue. In most species, endosperm is triploid and functions to store nutrients. Endosperm cells are loaded with starch or oils (lipids) plus proteins and other nutrients that the embryo will need after it germinates. But before germination can occur, seeds must develop and be dispersed.

40.4 The Seed

Fertilization triggers the development of a young sporophyte. In angiosperms, the first stage in the sporophyte's life is the maturation of the seed.

As a seed matures, the embryo and endosperm develop inside the ovule and become surrounded by a covering called a **seed coat**. At the same time, the ovary around the ovule develops into a fruit, which encloses and protects the seed (or seeds, if a single ovary contains multiple ovules). In addition to providing protection, fruits often aid in dispersing seeds away from the parent plant. The mature seed consists of an embryo, a food supply—originating with endosperm—and a seed coat. In most cases, mature seeds leave the parent plant encased in a fruit.

Along with pollen, the evolution of the seed was a crucial innovation as land plants diversified. Because seeds contain stored nutrients, they allow offspring to be much more successful in colonizing habitats that are crowded with competitors than are offspring produced from spores, which are single cells. As a young plant emerges from the seed, it can subsist on stored nutrients until it is well enough established to absorb water from the soil and feed itself via photosynthesis.

Let's analyze how seeds work, beginning with a closer look at how the embryo develops.

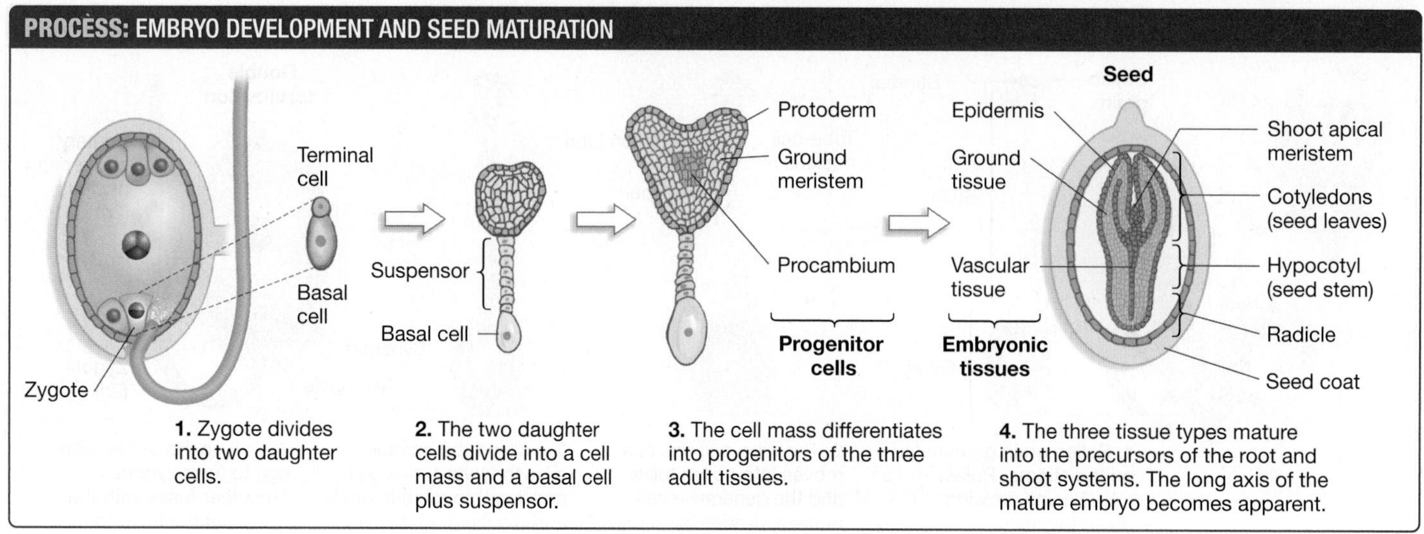

FIGURE 40.15 Embryonic Tissues and Structures Develop inside Seeds. The embryo inside a seed has the beginnings of root and shoot systems and the plant's first leaves, or cotyledons. The embryonic epidermis and the ground and vascular tissues are organized in distinct layers.

Embryogenesis

Recall from Chapter 23 that **embryogenesis** is the process by which a single-celled zygote becomes a multicellular embryo. As **Figure 40.15** shows, embryogenesis in angiosperms can be analyzed as a four-step process.

Step 1 The zygote divides to form the two daughter cells.

Step 2 The lower daughter cell, or basal cell, divides to form a cell that forms part of the root tip and a cell that produces a row of cells. This row of cells, called the suspensor, provides a route for nutrient transfer from the parent plant to the developing embryo. The upper daughter cell, or terminal cell, is the parent of almost all the cells in the embryo. It divides to form a mass of cells.

Step 3 Cells within the mass differentiate into groups conforming to one of the three adult tissue types introduced in Chapter 36. The exterior layer of embryonic cells, the **protoderm**, is the progenitor of the adult dermal tissue, or epidermis. The **ground meristem**—the cells just within the protoderm—gives rise to the ground tissue found in adults. And the **procambium** is a group of cells in the core of the embryo that becomes the vascular tissue.

Step 4 As the embryo continues to develop, the long axis of the plant begins to emerge and several important structures take shape: (1) the **cotyledons**, or seed leaves, (2) the **hypocotyl** ("under-cotyledon"), or seed stem, which is the embryonic stem, and (3) the **radicle**, or embryonic root. Some embryos also have an **epicotyl** ("above-cotyledon"), which is a portion of the embryonic stem that extends above the cotyledons.

✔If you understand the basic steps in embryonic development, you should be able to explain the relationship between the populations of embryonic cells labeled in step 3 of Figure 40.15 and the tissues labeled in step 4.

Recall from Chapter 30 that one prominent lineage of angiosperms—the monocotyledons, or monocots—has just one seed leaf, whereas eudicotyledons, or eudicots, have two. In most eudicots, the cotyledons take up the nutrients in the endosperm and store them. In these species, there is no endosperm left by the time the seed matures—instead the cotyledons function as the nutrient storage organ. **Figure 40.16** compares the seed structure in beans and corn—a representative eudicot and monocot, respectively.

By the time a seed matures, then, the three major embryonic tissue types have developed. The precursors of the root and shoot systems, along with the seed-leaves, have formed. Once these events are accomplished, the seed tissues dry and the embryo becomes quiescent—meaning it stops growing.

To review gametogenesis, pollination, and embryogenesis in angiosperms, go to the study area at *www.masteringbiology.com.*

(MB) **Web Activity** Reproduction in Flowering Plants

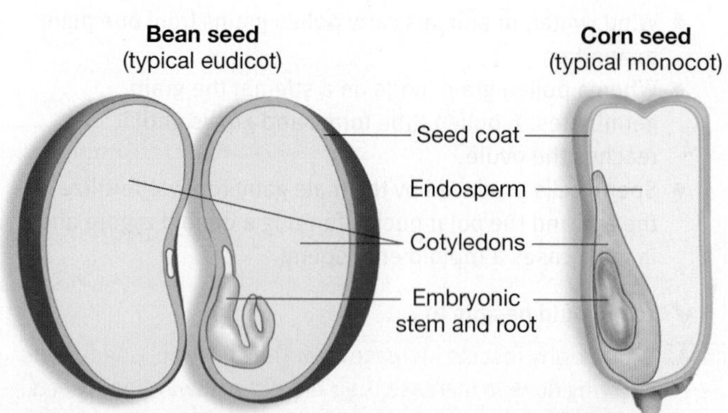

FIGURE 40.16 Seeds Contain an Embryo and a Food Supply Surrounded by a Tough Coat. In beans (left), the nutrients in the endosperm are absorbed by the cotyledons and stored. In corn (right), the endosperm is intact.

The Role of Drying in Seed Maturation

The seeds of many species dry out as they mature. Water makes up 90 percent of normal plant cells, but dried seeds contain just 5–20 percent water.

Loss of water is interpreted as an adaptation that prevents seeds from germinating on the parent plant, where they would compete with the parent for resources. In addition, the dry condition of seeds ensures that once they have dispersed from the parent plant, they will not germinate until water is available. This is logical, because water is crucial to the survival of germinated seedlings. Dry seeds are less susceptible than wet seeds to damage from freezing, and are lighter and more easily transported.

How do the plasma membranes and proteins in the embryo and endosperm survive the drying process? When researchers reduce the amount of water surrounding isolated plasma membranes or isolated proteins to the levels observed in extremely dry seeds, some of the membranes and proteins disintegrate. Clearly, something is happening at the molecular level in seeds to keep these cell components intact.

Researchers have recently established that one of these "somethings" involves sugars. As water leaves the seed during drying, sugars replace it and maintain the integrity of plasma membranes and proteins. If drying is extreme, the sugars form an extremely viscous liquid that contains little if any water. Substances such as this are considered vitrified, or glass-like (glass is a liquid solution with the viscosity of a solid). Biologists propose that this glassy, sugar-coated state helps maintain the integrity of plasma membranes and proteins in seeds that experience extremely dry conditions. When seeds imbibe water, the glassy sugars dissolve and germination proceeds.

Drying is only one part of the seed maturation process, however. Equally important is the development of tissues surrounding the seed itself. In many cases, these tissues are required for the seed to be dispersed from the parent plant.

Fruit Development and Seed Dispersal

Fertilization initiates not only the development of the seed and embryo in angiosperms, but also the development of the fruit.

FRUIT STRUCTURE Fruits come in three basic types (**Figure 40.17**).

- **Simple fruits** like the apricot develop from a single flower that contains a single carpel or several carpels that are fused together.
- **Aggregate fruits** like the raspberry also develop from a single flower, but one that contains many separate carpels.
- **Multiple fruits** like the pineapple develop from many flowers and thus many carpels.

As a fruit matures, the walls of the ovary thicken to form the **pericarp,** the part of the fruit that surrounds and protects the seed or seeds (**Figure 40.18** on page 798). Fruits can be dry when they are mature, as in nuts, or fleshy, as in cherries and tomatoes.

Note that fruits are formed from cells derived from the parent plant—not the embryo. They are tissue from the mother, encasing the offspring.

(a) Simple fruit (e.g., cherry):
Develops from a single flower with one carpel or fused carpels

(b) Aggregate fruit (e.g., blackberry):
Develops from a single flower with many separate carpels

(c) Multiple fruit (e.g., pineapple):
Develops from many flowers with many carpels

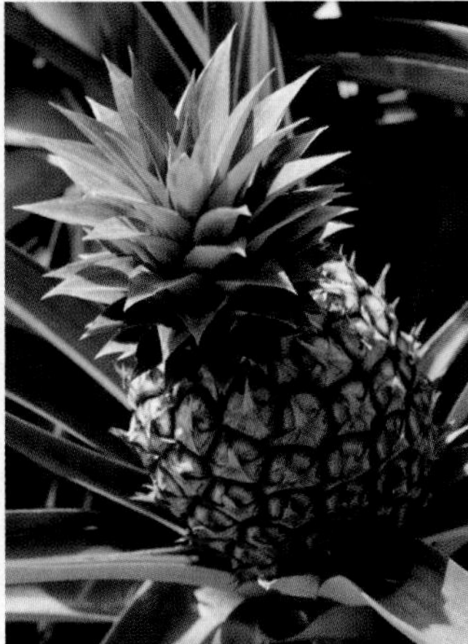

FIGURE 40.17 Three Major Types of Fruits. The structure of a fruit depends on the number of carpels found in each flower and whether or not the carpels are fused.

✔**QUESTION** When fruits ripen, their color changes in a way that makes them more conspicuous to fruit eaters. State a hypothesis to explain why this color change might increase the fitness of an individual.

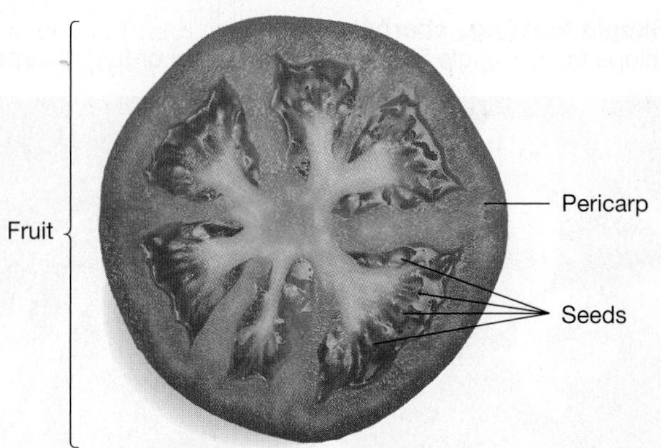

FIGURE 40.18 As Fruits Mature, the Ovary Wall Develops into a Pericarp That Surrounds the Seed or Seeds. A fruit consists of a pericarp and the enclosed seeds.

To learn more about the structure of fruits and the relationship between seeds and fruits, go to the study area at *www.masteringbiology.com.*

 **Web Activity** Fruit Structure and Development

FRUIT FUNCTION Fruits have two functions: They protect seeds from physical damage and seed predators, and they frequently aid in seed dispersal. Dispersal is important to the fitness of the young sporophyte. This is especially true in long-lived species, in which the offspring may compete with the parent plant for light, water, and nutrients if there is no dispersal.

Fruits sometimes split open and release seeds to be dispersed directly. In many cases, however, seeds are dispersed to new locations while they are still enclosed in the fruit.

Dry fruits simply fall to the ground or are dispersed by wind, propulsion, or animals (**Figure 40.19**). Some dry fruits have hooks or barbs that adhere to passing animals, while nuts are dispersed by seed predators. Fruits that are dispersed by wind often have external structures to catch the breeze and extend the distance they travel; the fruits of dandelions and maple trees are familiar examples. Fruits that float can disperse seeds in water.

Some plants actually disperse dry fruits via propulsion. The sandbox tree, for example, produces a seed pod that shrinks as it dries. Eventually the pod splits apart violently, spraying seeds in all directions with so much force that the plant is sometimes called the dynamite tree. The bursting seed pod sounds like a pistol shot, and seeds can be scattered as much as 40 m away from the parent plant. Similarly, the dwarf mistletoe fruit fills with sugars as it matures. Enough water follows via osmosis to make the fruit explode and shoot seeds as far as 5 m.

Animals are the most common dispersal agent for fleshy fruits. In animal-dispersed fruits, the seed coat has to be tough enough to resist the mechanical forces and chemical conditions in the animal's mouth and digestive tract, so that seeds can emerge in the feces unscathed. In cases like this, seed dispersal is an example of mutualism. The plant provides a fruit rich in sugars and other nutrients; in return, the animal carries the fruit to a new location and excretes the seeds along with a supply of fertilizer. In fact, some seeds won't germinate unless their hard coats are exposed to stomach acids and abrasion as they pass through an animal's digestive tract.

Seed Dormancy

Once they have dispersed from the parent plant, seeds may not germinate for a period of time. This condition is known as **dormancy**.

Dormancy is usually a feature of seeds from species that inhabit seasonal environments, where for extended periods of time conditions may be too cold or dry for seedlings to thrive. Based on this observation, dormancy is interpreted as an adaptation that allows seeds to remain viable until conditions improve.

Consistent with this hypothesis, dormancy is rare or nonexistent in seeds produced by plants that inhabit tropical wet forests or other areas where conditions are suitable for germination year-round.

What molecular mechanisms are responsible for dormancy? And how does dormancy cease so that germination can begin?

FIGURE 40.19 Seeds Are Often Contained in Structures That Allow Them to Be Dispersed.

WHAT ROLE DOES ABA PLAY IN DORMANCY? Chapter 39 introduced the hormone abscisic acid (ABA) and described its role in preventing germination. In some species, seeds that enter dormancy have a high concentration of this hormone. The seeds of desert plants, for example, have high levels of ABA in their seed coats. When these seeds are exposed to large amounts of water during rare or seasonal rains, the hormone literally washes out of the seed's outer tissues and germination proceeds.

There is not a strict correlation between ABA concentration and degree of seed dormancy, however. In peas and many other species, seeds routinely contain high levels of ABA and yet are not dormant. In *Arabidopsis*, ABA concentrations rise as seeds mature and appear to impose dormancy. ABA levels eventually fall, however, so dormant, mature seeds contain only trace amounts of the hormone.

Researchers have concluded that there is no single, universal mechanism for initiating and maintaining seed dormancy. In some cases, changing ABA levels or the ratio of ABA to gibberellin present control the dormant state. In other cases, changes in sensitivity to ABA, rather than the sheer amount of hormone present, appear to be important. It is also likely that novel mechanisms for initiating and maintaining dormancy are still to be discovered.

HOW IS DORMANCY BROKEN? The coats of some seeds are thick enough to prevent water and oxygen from physically reaching the embryo. For germination to occur, these seed coats must be disrupted, or **scarified**.

Crop seeds that require scarification are placed in large, revolving drums with pieces of sandpaper that abrade and scarify the seeds. In nature, seed coats can be disrupted by a fire, by the passage of the seed through an animal's digestive tract, or by abrasion against soil particles. The basic principle is that the seed coat has to be broken for water to enter the seed.

Other seeds must experience particular environmental conditions in addition to exposure to water. Species native to high latitudes or high elevations often produce seeds that must undergo cool, wet conditions before they will germinate. No one knows how these seeds perceive cold or what molecular mechanisms are involved in breaking dormancy when the cool, wet period ends.

Because small seeds have few nutrient reserves in their cotyledons or endosperm, many small-seeded species need to germinate near the soil surface, where individuals are exposed to light and can feed themselves via photosynthesis. As Chapter 39 indicated, lettuce seeds and other small seeds must be exposed to red light before they will break dormancy and germinate. Red light is an important environmental cue, because wavelengths in the red portion of the light spectrum support photosynthesis. Red light and blue light indicate that sunlight is abundant.

Finally, many of the seeds produced by species native to habitats where wildfires are frequent, such as the California chaparral and South African fynbos, have an unusual chemical requirement to break dormancy: They must be exposed to fire or smoke before they will germinate. In fact, the commercial food product "liquid smoke" induces germination in these seeds as well as actual smoke does. In fire-prone habitats, it is advantageous for seeds to germinate after fire has cleared away existing vegetation.

The message here is that dormancy can be broken in response to a wide variety of environmental cues. In general, the cue that triggers germination is a reliable signal that conditions for seedling growth are favorable for a particular species in a particular environment.

Seed Germination

Even if specific environmental signals are required to break dormancy, seeds do not germinate without water. Water uptake is the first event in germination. Once the seed coat allows water penetration, water enters by a steep water-potential gradient, because the seed is so dry.

The graph in **Figure 40.20** plots changes in water uptake in seeds through time. If you trace the graph from the start of seed germination onward, you should see that water uptake in a typical angiosperm seed has three distinct phases:

Phase 1 Germination begins with a rapid influx of water. Oxygen consumption and protein synthesis in the seed increase dramatically, but no new messenger RNAs are transcribed. Based on these observations, biologists have concluded that some of the key early events in germination are driven by mRNAs that are stored in the seed prior to maturation.

Phase 2 The second phase is an extended period during which water uptake stops. Newly transcribed mRNAs appear and are translated into protein products. Mitochondria also begin to multiply. In effect, seeds take up enough water in Phase 1 to hydrate their existing proteins and membranes, and then begin to manufacture the proteins and mitochondria needed to support growth.

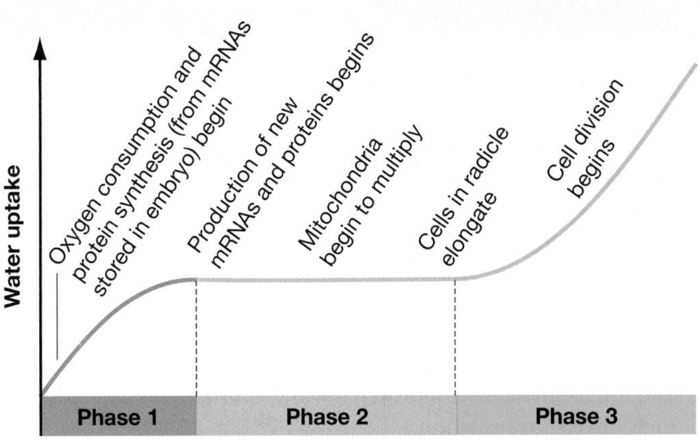

FIGURE 40.20 Germination Begins with Three Distinct Phases. This graph plots the rate of water uptake through time as a typical angiosperm seed germinates. (The graph is conceptual, meaning it represents a general pattern observed in data from many species, so the axes have no units.)

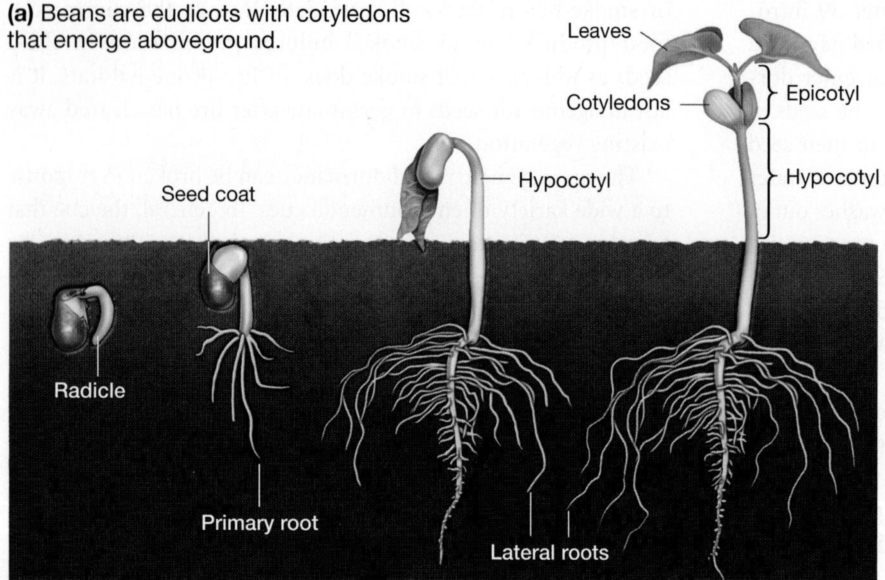

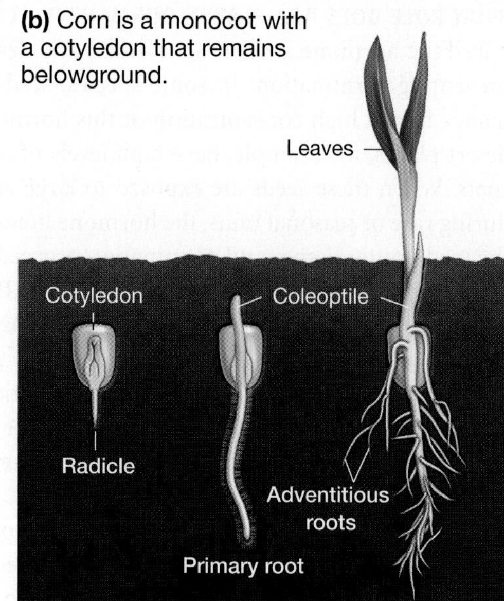

FIGURE 40.21 The Germination Sequence Varies among Species.

✔**QUESTION** In which of these species are cotyledons photosynthetic?

Phase 3 Water uptake resumes as growth begins. This renewed phase of water uptake enables cells to develop enough turgor pressure to enlarge. Eventually, the seedling bursts from the seed coat.

Figure 40.21 shows what happens as eudicot and monocot embryos emerge from the seed. First the radicle emerges, and subsequently develops into the mature root system. In eudicots, the shoot system with its cotyledons usually emerges shortly after the radicle appears. In corn, the radicle and the coleoptile, which covers the young shoot, emerge at the same time. Note that in eudicots, the emerging stem has a hook shape. Like the coleoptile of monocots, this trait is thought to protect the apical meristem from damage as the shoot works its way upward through rough soil particles.

The next major event in the seedling's life occurs when either the cotyledons or the earliest leaves produced by the growing seedling commence photosynthesis. The seedling is said to be established when the young plant no longer relies on food reserves in its endosperm or cotyledons; instead, it receives all of its nourishment from its own photosynthetic products. With this, a new generation is under way.

CHAPTER 40 REVIEW

For media, go to the study area at www.masteringbiology.com 

Summary of Key Concepts

Plants undergo alternation of generations, in which a diploid sporophyte phase alternates with a haploid gametophyte phase. Sporophytes produce spores by meiosis. Gametophytes produce gametes by mitosis.

- The relative size and life span of the gametophyte and sporophyte phases vary a great deal among plant groups.
- In the most basal groups of land plants, gametophytes are larger and longer-lived than sporophytes, and sporophytes depend on gametophytes for nutrition.
- In angiosperms, or flowering plants, sporophytes are the large and long-lived phase where photosynthesis takes place; gametophytes consist of just a few cells.

✔ You should be able to state the ploidy of a sporophyte, gametophyte, spore, gamete, and zygote.

In angiosperms, male and female gametophytes are microscopic and are produced inside flowers. Male gametophytes (pollen grains) are portable. Female gametophytes are encased in an ovary and are retained in the flower. When pollen grains land on a flower, they deliver sperm cells that fertilize the egg produced by the female gametophyte.

- Angiosperms initiate flowering and sexual reproduction in response to external cues from the environment as well as internal cues based on the individual's condition. Frequently, these cues allow individuals to flower when environmental conditions are favorable.
- Flowers are made up of sepals, petals, stamens, and one or more carpels.

- The lower part of the carpel, the ovary, contains one to many ovules. Within the ovule, a megasporocyte undergoes meiosis, producing a megaspore that develops into the female gametophyte.
- In the anthers of stamens, microsporocytes undergo meiosis. The resulting microspores develop into male gametophytes, which are enclosed in pollen grains.
- Pollination occurs when pollen grains are transported to the stigma of the carpel. In most cases, the structure of a flower correlates with the morphology and behavior of its pollinator.
- If allowed to germinate on the stigma, a pollen grain sends a long pollen tube down the style. Two sperm travel down the pollen tube and enter the female gametophyte.
- In double fertilization, one sperm fuses with the egg to form a zygote, while the other fuses with polar nuclei within the female gametophyte. The fusion of sperm and polar nuclei produces endosperm—nutritive tissue that in most species is triploid.

✔ You should be able to describe what would happen if all meiotic products in megasporangia went on to form female gametophytes.

Seeds contain an embryo and a food supply surrounded by a coat. In angiosperms, the walls of the ovary develop into a fruit that encloses the seed or seeds. In many cases, fruits function in seed dispersal.

- The development of an angiosperm embryo includes the formation of dermal tissue (epidermis), ground tissue, and vascular tissue layers and the development of the radicle, hypocotyl, and cotyledons.
- As an embryo develops, endosperm cells divide to form a nutrient-rich tissue. In addition, cells along the outside of the ovules form a protective seed coat, and the ovary develops into a fruit.
- In many cases, the mature fruit contains structures that help disperse the mature seed via wind, water, propulsion, or animals.
- Many seeds do not germinate immediately but instead experience a period of dormancy.
- A wide variety of conditions, ranging from scarification to exposure to red light, may break seed dormancy. In many cases, the event that triggers germination ensures that the seed germinates when environmental conditions are favorable.
- Germination begins when the seed takes up water and mRNAs already present in the seed are translated. It ends when the radicle breaks the seed coat and begins to penetrate the soil.

✔ You should be able to explain the relationships among carpels, ovaries, ovules, fruits, and seeds.

MB **Web Activity** Reproduction in Flowering Plants, **Web Activity** Fruit Structure and Development

Questions

✔ TEST YOUR KNOWLEDGE

Answers are available in Appendix B

1. What is the major evolutionary trend in land plant life cycles?
   - **a.** Instead of being approximately the same size and shape, gametophytes and sporophytes began to look different.
   - **b.** Sporophytes became larger and long lived while gametophytes became drastically reduced.
   - **c.** In lineages that evolved more recently, such as angiosperms, spores are no longer produced.
   - **d.** Sporophytes began to rely on gametophytes for all of their nutritional needs.
2. What happens when double fertilization occurs?
   - **a.** Two zygotes are formed, but only one survives.
   - **b.** Two sperm fertilize the egg, forming a triploid zygote.
   - **c.** One sperm fertilizes the egg, while another sperm fuses with the polar nuclei.
   - **d.** One sperm fertilizes the egg, while two other sperm fuse with a polar nucleus.
3. What is a fruit?
   - **a.** a structure formed from the ovary wall that contains a seed or seeds
   - **b.** a structure consisting of an embryo and a food supply surrounded by a tough coat
   - **c.** a female gametophyte
   - **d.** a male gametophyte
4. Which of the following is a key event during embryogenesis?
   - **a.** The seed coat takes on water so that germination can begin.
   - **b.** Starches are hydrolyzed, providing sugars that fuel the early stages of germination.
   - **c.** The megasporocyte divides by mitosis, forming the cells that will become the embryonic female gametophyte.
   - **d.** Distinct groups of cells form that will become dermal, ground, and vascular tissues.
5. Why is the interaction between angiosperms and pollinators considered mutualistic?
   - **a.** New species can form if mutant flowers attract new types of pollinators.
   - **b.** Flowers may have an array of traits, including corolla shape, color, scent, and the presence of nectar, to attract a specific type of pollinator.
   - **c.** Wind pollination is much "cheaper," but animal pollination is much more precise.
   - **d.** Angiosperms get their pollen dispersed, while pollinators get food.
6. What happens when outcrossing occurs?
   - **a.** Inbred offspring are produced.
   - **b.** The same flower can be visited by many different types of pollinators—not a single specialist species.
   - **c.** Gametes from different individuals fuse to form a zygote.
   - **d.** Gametes from the same individual fuse.

✔TEST YOUR UNDERSTANDING

Answers are available in Appendix B

1. In terms of maximizing reproductive success, what is the advantage of asexual reproduction? What is the disadvantage?
2. In the angiosperm life cycle, which cells undergo meiosis? Which cells are spores? Which structures are gametophytes?
3. How do the structure and function of sepals and petals differ? How would you expect these structures to differ in species that are pollinated by wind versus bumblebees?
4. What are the advantages and disadvantages of self-fertilization versus those of outcrossing?
5. Consider a carpel, an ovary, and an ovule. Which is responsible for producing the female gametophyte, and which produces the pericarp of a fruit? Are these structures part of the sporophyte, the gametophyte, or a combination of the two?
6. What is the relationship between the endosperm of corn and the cotyledons of beans?

✔APPLYING CONCEPTS TO NEW SITUATIONS

Answers are available in Appendix B

1. Suppose you discovered an angiosperm that was new to science. The population grows on an island near the equator. The island is dry 10 months of the year but experiences a 2-month period of frequent rains. Predict what cues trigger flowering and seed germination in the newly discovered species.
2. Some flowering plants "cheat" their pollinators because they offer no food reward. Likewise, certain pollinators cheat plants by removing nectar from flowers without picking up pollen. (In some cases, they do so by chewing through the petals that hold the store of nectar.) Speculate on the types of mutations that might modify insect behavior and/or plant structure in a way that limits cheating and enforces mutualism.
3. Pollinators frequently deposit pollen from more than one individual on a stigma. When they do, pollen grains from different males compete to fertilize the egg. Design an experiment to test the hypothesis that pollen tubes grow faster when pollen grains germinate in the presence of pollen from a different individual.
4. Consider the following fruits: an acorn, a cherry, a burr, and a milkweed seed. Based on the structure of each of these fruits, predict how the seed is dispersed. Design a study that would estimate the average distance that each type of seed is dispersed from the parent plant.

Oryx are adapted to desert life. They have an exceptional ability to withstand heat, and can acquire all of their water from the food they eat.

Animal Form and Function 41

The Sahara and Arabian deserts are extreme environments. In some parts of the Sahara, several years can pass between rainfalls. In a single day, temperatures in these deserts can fluctuate between −0.5°C and 37.5°C; midday temperatures in summer can exceed 50°C. Yet few places in the Sahara or Arabian deserts are devoid of life. Even large animals, such as the oryx, thrive in both regions.

How do the animals native to these areas cope? Small animals avoid the midday heat by retreating to a cool burrow deep underground or the shade of a small shrub. Large mammals have a harder time hiding from the Sun but possess traits that allow them to keep cool and conserve water.

The Arabian oryx, for example, never drinks. It gets 86 percent of the water it needs from the vegetation it eats and the other 14 percent from water synthesized as a by-product of cellular respiration—what biologists call **metabolic water**. To conserve what water they have, Arabian oryx produce extremely concentrated urine and exceptionally dry fecal pellets.

In addition, oryx do not sweat to cool off. Instead, their body temperature rises from a normal 37°C to just over 40°C as temperatures increase during the day. The excess body heat is released during the cool desert nights, when their body temperature returns to normal.

In contrast to oryx, humans die of dehydration when denied water for just three days. And for a human, a body temperature of 40°C is a life-threatening situation—equivalent to running a fever of 104°F. How do oryx do it? What aspects of their anatomy and physiology allow them to thrive in such an extreme environment?

Anatomy is the study of an organism's physical structure. **Physiology** is the study of how the physical structures in an organism function. The anatomy and physiology of an oryx are clearly different from those of a human—or those of a shark, frog, tuna, fruit fly, or crab, for that matter. This chapter is an introduction to the study of anatomy and physiology in animals.

KEY CONCEPTS

- In biology, structure has a profound influence on function. Biologists analyze the structure and function of animals at a variety of levels: molecules, cells, tissues, organs, and organ systems.
- Body size has a strong influence on how animals work, in large part because a body's volume increases faster than its surface area as body size increases.
- Animals use an array of methods to maintain a relatively constant environment inside their bodies. They have systems that sense changes in internal conditions and trigger responses that return conditions to normal.
- Some animals have sophisticated systems for generating and conserving heat and regulating body temperature.

✔ When you see this checkmark, stop and test yourself. Answers are available in Appendix B.

41.1 Form, Function, and Adaptation

Biologists who study animal anatomy and physiology are studying **adaptations**—heritable traits that allow individuals to survive and reproduce in a certain environment better than individuals that lack those traits (see Chapter 24).

Recall from Chapter 24 that adaptation results from evolution by natural selection. Natural selection, in turn, occurs whenever individuals with certain alleles leave more offspring than do individuals with different alleles. Because of this difference in reproductive success, the frequency of the selected alleles increases from one generation to the next.

Oryx with alleles that allow them to extract more water from their feces survive better and produce more offspring than do oryx with alleles that allow water to be lost in feces. The ability to produce extremely dry fecal pellets is an adaptation that helps oryx thrive in water-short environments.

The Role of Fitness Trade-Offs

Adaptations increase fitness—the ability to produce offspring. But no adaptation is "perfect." Instead, adaptations are limited by which alleles are present in a population and by the nature of the traits that already exist—because all adaptations derive from preexisting traits.

The human spine, for example, is a highly modified form of the vertebral column in ancestors that walked on all fours (see Chapter 34). The modifications in the human spine can be considered adaptations to support our upright posture, but they are far from perfect—85 percent of U.S. adults under the age of 50 experience back pain. The evolution of the human spine has been constrained by the nature of the ancestral trait and by a lack of alleles that would improve its structure and function.

The most important constraint on adaptation, though, may be **trade-offs**—inescapable compromises between traits. For example, every female animal has a finite amount of time and energy available for producing offspring. In species that do not care for their young, a female's entire reproductive investment consists of the eggs she lays. Given that the total amount of energy available for egg production is limited, there should be a trade-off between the number of eggs a female produces and the quality of those eggs. Egg quality is determined by egg size—specifically by the amount of yolk, or nutrient-rich cytoplasm, in the egg.

How do biologists study trade-offs in animal anatomy and physiology? Let's consider experimental work on egg size and egg number in the side-blotched lizard, a reptile that lives in the deserts of western North America (**Figure 41.1**). Like many animals, side-blotched lizards lay eggs in groups called clutches. Theory predicts that there has to be a trade-off between the number and size of eggs in a clutch.

To document this trade-off in nature, biologists introduced a large amount of variation in egg size and egg number.

- They induced the production of small eggs by catching females, surgically removing yolk from their eggs early in development, replacing the eggs, and releasing the mothers back into the wild.
- To form clutches with small numbers of eggs, biologists caught females and removed all but two or three of their eggs early in development.
- As an experimental control, the researchers did "sham operations." They caught a large number of females and performed surgery to expose their eggs, but they left the eggs alone.

These manipulations gave them a study population with a large variation in egg size and number. Did fitness trade-offs occur?

The graph labeled Results 1 in Figure 41.1 plots the average mass of eggs that were laid against clutch size, meaning the number of eggs laid by each female. The data show that as egg size increases, clutch size decreases. The pattern confirms a trade-off between egg size and egg number. It is not possible for a female in this population to produce large numbers of large eggs.

What about the prediction that large offspring are higher in quality, meaning that they survive better? To test this idea, the researchers marked 1668 newly hatched lizards in the experimental population, released them, and recaptured those that survived one month later. The graph labeled Results 2 in Figure 41.1 indicates that survival increases as egg mass increases. These data support the prediction that larger offspring survive much better than do smaller offspring.

Given this trade-off between offspring quantity and quality, does natural selection favor mothers that produce large numbers of small offspring or those that produce small numbers of large offspring? The graph labeled Results 3 in Figure 41.1 shows the combined effect of egg number and offspring survival by plotting "Mother's fitness"—the number of surviving offspring that she produces—on the y-axis. In this population, mothers that produced an intermediate number of offspring of intermediate size generated the highest total number of surviving offspring.

Trade-offs, such as the inescapable compromise between egg size and egg number, are pervasive in nature. Desert animals that sweat to cool off are threatened with dehydration. An eagle's beak is superbly adapted for tearing meat but not for weaving nesting materials together. In studying animal anatomy and physiology, biologists study compromise and constraint as well as adaptation.

Adaptation and Acclimatization

In everyday English, the word adaptation describes short-term, reversible responses to environmental fluctuations. In biology, physiological and biochemical changes like these are referred to as **acclimatization**, or **acclimation**. Acclimatization is a phenotypic change in an individual in response to short-term changes in the environment. Adaptation refers only to a genetic change in a population in response to natural selection exerted by the environment.

If you moved to Tibet, your body would acclimatize to high elevation by making more of the oxygen-carrying pigment hemoglobin and more hemoglobin-carrying red blood cells. But populations that have lived in Tibet for many generations are adapted to this environment through genetic changes. Among native Tibetans, for example, an allele that increases the ability of

EXPERIMENT

QUESTION: Is there a trade-off between the quality and quantity of offspring that a female can produce?

HYPOTHESIS 1: Females can produce many small eggs or a few large eggs. **NULL HYPOTHESIS 1:** There is no relationship between egg size and egg number.	**HYPOTHESIS 2:** Offspring quality increases with increasing egg size. **NULL HYPOTHESIS 2:** There is no relationship between offspring quality and egg size.	**HYPOTHESIS 3:** There is an optimal clutch size based on a trade-off between the quality and quantity of offspring a female can produce. **NULL HYPOTHESIS 3:** No optimal clutch size exists.
EXPERIMENTAL SETUP 1: Eggs early in development: Reduced yolk Reduced number Left alone Mothers Vary egg size and egg number by catching females and removing yolk from eggs or removing all but 2–3 eggs. Also do sham operations, with eggs left alone. Record size and number of eggs laid.	**EXPERIMENTAL SETUP 2:** Offspring Catch and mark large number of newly hatched offspring from experiment 1. Re-catch survivors one month later.	**STUDY DESIGN 3:** Number of surviving offspring? Number of surviving offspring? Number of surviving offspring? Mothers Calculate number of surviving offspring per female, based on results of experiments 1 and 2.
PREDICTION 1: Females with small eggs produce a large number; females with few eggs produce large eggs. **PREDICTION OF NULL HYPOTHESIS 1:** As average egg size increases, average clutch size stays the same.	**PREDICTION 2:** Larger offspring survive better than smaller offspring. **PREDICTION OF NULL HYPOTHESIS 2:** No difference in survival of large versus small offspring.	**PREDICTION 3:** The number of surviving offspring is maximum at intermediate egg size and intermediate clutch size. **PREDICTION OF NULL HYPOTHESIS 3:** As egg size increases, number of surviving offspring does not change.
RESULTS 1: Females can produce many small eggs or a few large eggs Clutch size (0, 2, 4, 6) Egg mass (g) (0.2, 0.3, 0.4, 0.5, 0.6, 0.7)	**RESULTS 2:** Large offspring survive best (larger eggs produce higher-quality offspring) Probability of survival (0, 0.1, 0.2, 0.3) Egg mass (g) (0.2, 0.3, 0.4, 0.5, 0.6, 0.7)	**RESULTS 3:** Mothers that produce intermediate numbers of mid-sized offspring have highest fitness Mother's fitness (0, 0.1, 0.2, 0.3, 0.4, 0.5) Egg mass (g) (0.2, 0.3, 0.4, 0.5, 0.6, 0.7)

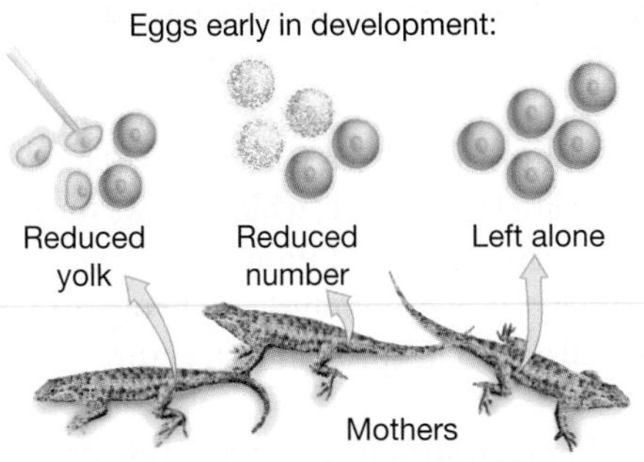

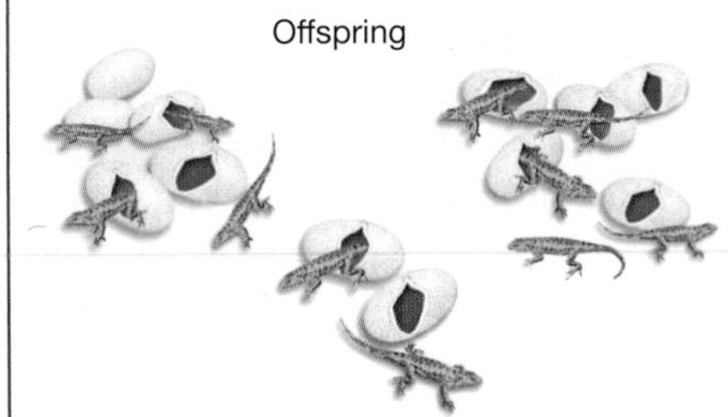

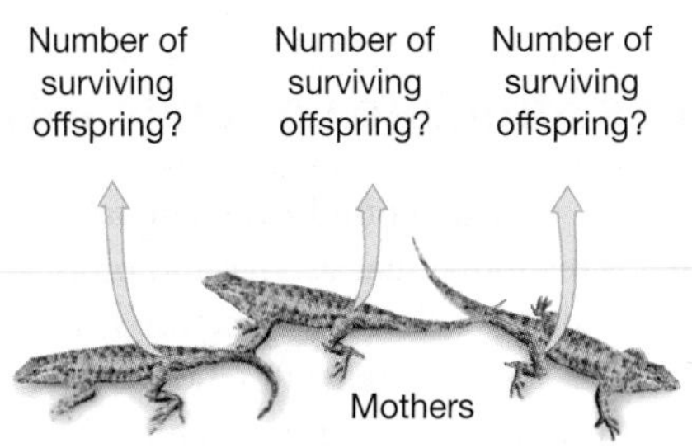

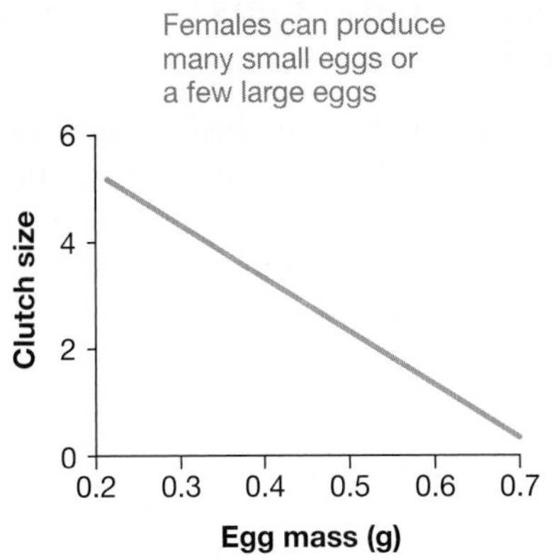

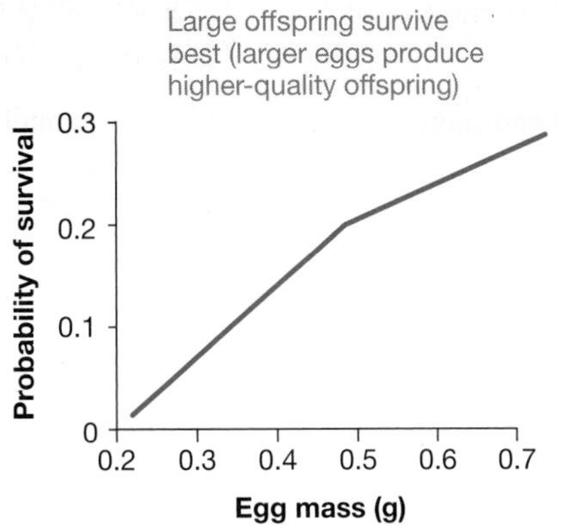

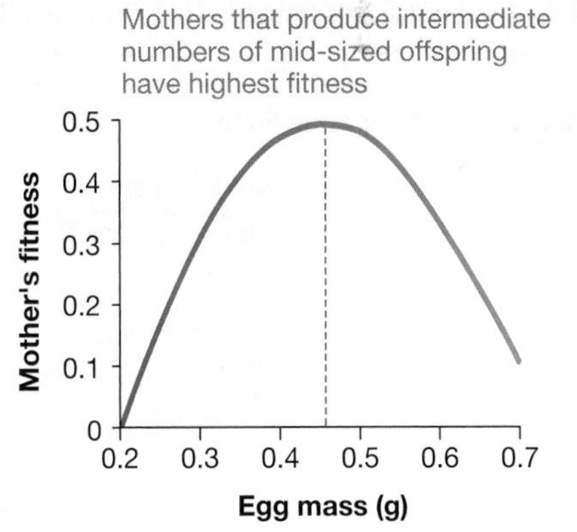

CONCLUSION: There is a trade-off between offspring quality (egg size) and quantity (egg number).

FIGURE 41.1 Fitness Trade-Offs between Clutch Size and Egg Size in Lizards. The egg size and clutch size manipulation here accomplished two key goals: (1) increasing the variation among individuals, so that fitness differences are easier to detect, and (2) ensuring that—because manipulated individuals were chosen at random—the only thing that differed among females in the study was their egg size or clutch size.

SOURCE: Sinervo, B., P. Doughty, R. B. Huey, and K. Zamudio. 1992. Allometric engineering: A causal analysis of natural selection on offspring size. *Science* 258: 1927–1930.

✔**QUESTION** Describe the results predicted by the null hypothesis in all three of these experiments, and explain what they would look like on the graphs.

hemoglobin to hold oxygen has increased to high frequency. In populations that do not live at high elevations, this allele is rare or nonexistent.

The ability to acclimatize is itself an adaptation. Light-skinned humans, for example, vary in the ability to tan in response to sunlight. Some individuals tan easily—they have alleles that allow them to acclimate efficiently to environments with intense sun—while others do not. In this and many other cases, the ability to acclimatize is a genetically variable trait that can respond to natural selection.

41.2 Tissues, Organs, and Systems: How Does Structure Correlate with Function?

If a structure found in an animal is adaptive—meaning that it helps the individual survive and produce offspring—it is common to observe that the structure's size, shape, or composition correlates closely with its function.

For example, recall from Chapter 24 that biologists have documented extensive changes in beak size and shape in medium ground finches from the Galápagos Islands. Such changes are due to natural selection. Individuals with deep beaks are better able to crack the large fruits that predominate during drought years, while individuals with small beaks are better able to harvest the small seeds that predominate during wet years.

As **Figure 41.2** shows, a strong correlation between diet and beak structure is also found *among* species of Galápagos finch. Species with large, cone-shaped beaks eat large seeds; species with small, cone-shaped beaks eat small seeds. Species with long, tweezer-like beaks pick insects off tree trunks or other surfaces.

The mechanism responsible for these structure-function correlations is straightforward: If a mutant allele alters the size or shape of a structure in a way that makes it function more efficiently, individuals who have that allele will produce more offspring than will other individuals. As a result, the allele will increase in frequency in the population over time.

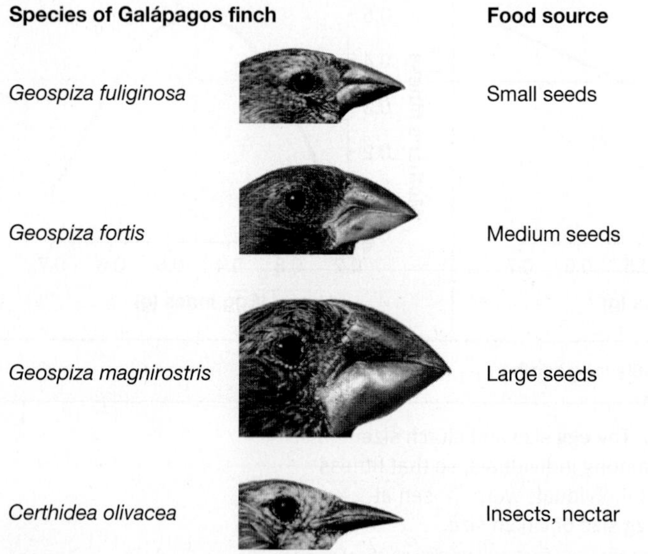

Species of Galápagos finch	Food source
Geospiza fuliginosa	Small seeds
Geospiza fortis	Medium seeds
Geospiza magnirostris	Large seeds
Certhidea olivacea	Insects, nectar

FIGURE 41.2 In Animal Anatomy and Physiology, Form Often Correlates with Function.

Structure-Function Relationships at the Molecular and Cellular Levels

Correlations between form and function start at the molecular level. For example, earlier chapters emphasized that the shape of proteins correlates with their role as enzymes, structural components of the cell, or transporters. The membrane proteins called channels form pores that allow specific ions or molecules to pass in or out of cells (see Chapter 6). The ends and interior of a channel are hydrophilic, which allows the protein to interact with the surrounding solution or the interior of the cell, while the perimeter is hydrophobic—allowing it to interact with the lipid bilayer. The protein's structure fits its function.

Similar correlations between structure and function occur at the level of the cell. Cells that manufacture and secrete hormones or digestive enzymes are packed with rough ER and Golgi; cells that store energy are dominated by large fat droplets; cells that ingest and destroy invading bacteria have numerous lysosomes.

The overall shape of a cell can also correlate with its function. For example, cells that are responsible for transporting materials into or out of the body often have extremely large areas of plasma membrane. As a result, they have room to accommodate the thousands of membrane channels, transporters, and pumps required for extensive transport.

Tissues Are Groups of Similar Cells That Function as a Unit

Animals are **multicellular**, meaning that their bodies contain distinct types of cells that are specialized for different functions. Frequently, animal cells that are similar in structure and function are physically attached to each other and form a tissue. A **tissue** is a group of similar cells that function as a unit.

The embryonic tissues called ectoderm, mesoderm, and endoderm are found in most animals and were introduced in Chapter 22 and Chapter 32. As an individual develops, the embryonic tissues give rise to four adult tissue types: (**1**) connective tissue, (**2**) nervous tissue, (**3**) muscle tissue, and (**4**) epithelial tissue. In each case, the structure of the tissue correlates closely with its function. Let's consider each in turn.

CONNECTIVE TISSUE **Connective tissue** consists of cells that are loosely arranged in a liquid, jellylike, or solid matrix. The matrix comprises extracellular fibers and other materials, and is secreted by the connective tissue cells themselves. Each type of connective tissue secretes a distinct type of extracellular matrix. The nature of the matrix determines the nature of the connective tissue.

(a) Loose connective tissue: soft extracellular matrix; provides padding

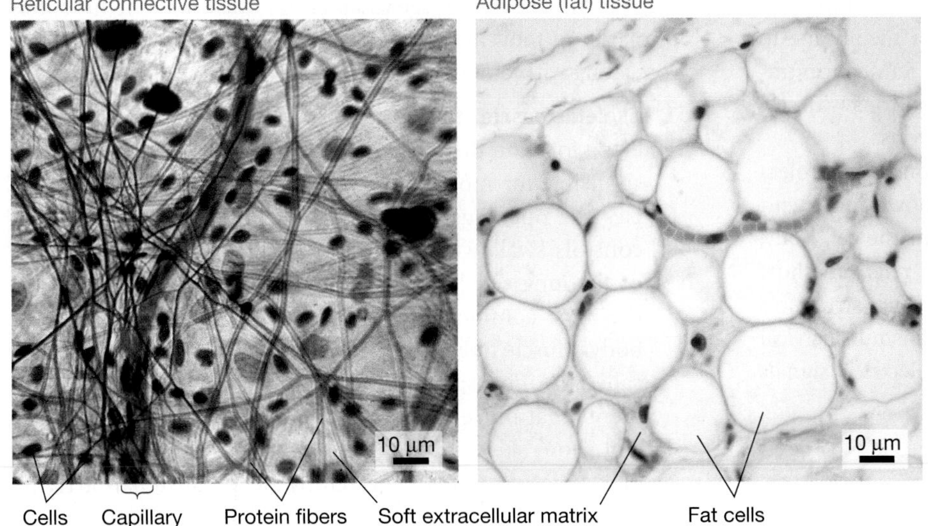

(b) Dense connective tissue: fibrous extracellular matrix; provides connections

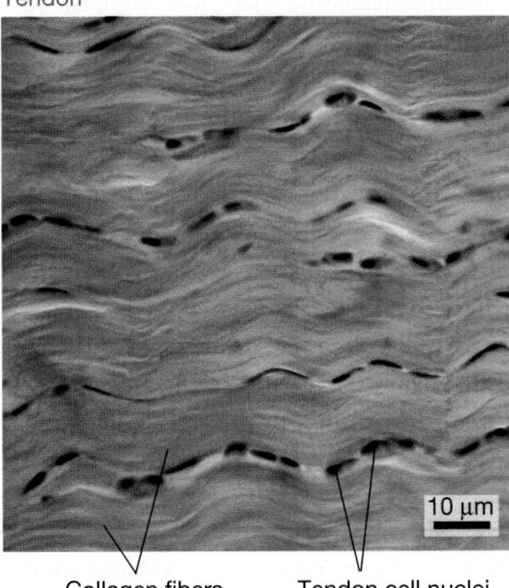

(c) Supporting connective tissue: firm extracellular matrix; functions in structural support and protection

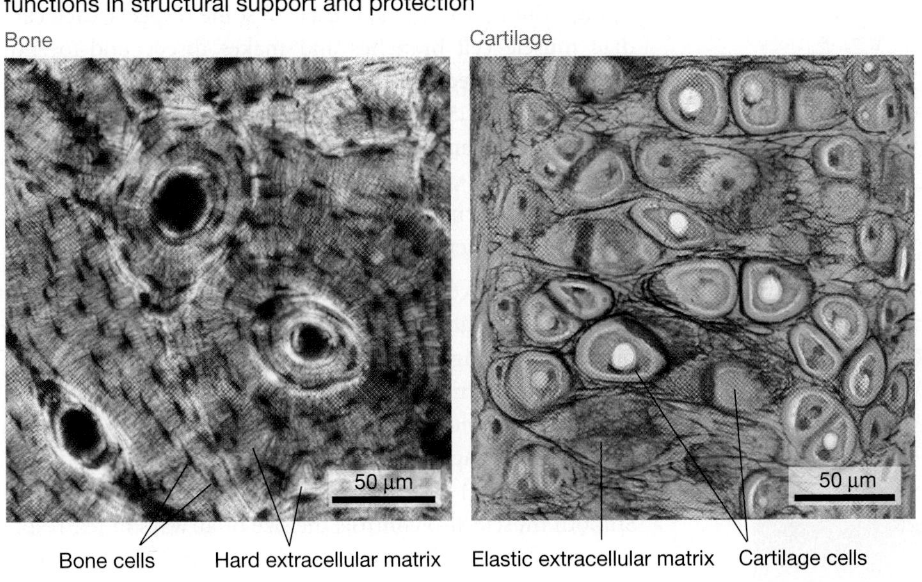

(d) Fluid connective tissue: liquid extracellular matrix; functions in transport

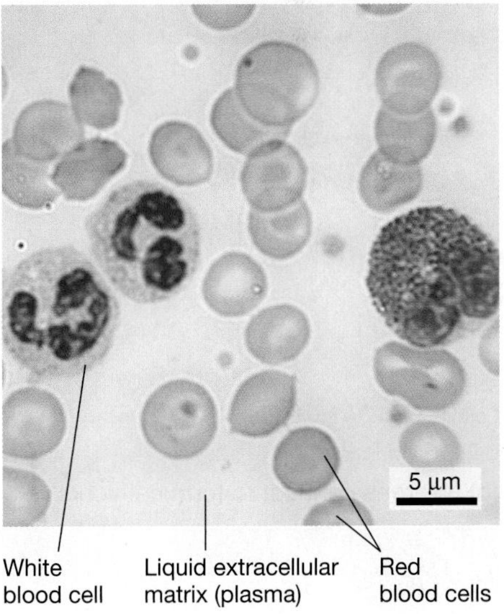

FIGURE 41.3 Connective Tissues Consist of Cells Surrounded by an Extracellular Matrix.

✔**QUESTION** How does the stiffness of the extracellular matrix in a connective tissue correlate with that tissue's function?

- **Loose connective tissue** contains an array of fibrous proteins in a soft matrix; it serves as a packing material between organs or padding under the skin. The **adipose tissue**, or fat tissue, illustrated in **Figure 41.3a** is a loose connective tissue made up of cells that are dominated by fat droplets and a loose matrix of fibers and fluid.
- **Dense connective tissue** is found in the tendons and ligaments that connect muscles, bones, and organs. As **Figure 41.3b** shows, the matrix in tendons and ligaments is dominated by the tough collagen fibers introduced in Chapter 8.
- **Supporting connective tissue** has a firm extracellular matrix. **Bone** and **cartilage** (**Figure 41.3c**) are connective tissues that provide structural support for the vertebrate body, as well as protective enclosures for the brain and other components of the nervous system.
- **Fluid connective tissue** consists of cells surrounded by a liquid extracellular matrix. **Blood**, which transports materials throughout the vertebrate body (**Figure 41.3d**), contains a variety of cell types and has a specialized extracellular matrix called plasma (see Chapter 44).

NERVOUS TISSUE **Nervous tissue** consists of nerve cells, which are also called **neurons**, and several types of supporting cells. Neurons transmit electrical signals, which are produced by changes in the permeability of the cell's plasma membrane to ions (see Chapter 45). Supporting cells regulate ion concentrations in the space surrounding neurons, supply neurons with nutrients, or serve as scaffolding or support for neurons.

Although they vary widely in shape, all neurons have projections that contact other cells. As **Figure 41.4a** shows, most neurons have two distinct types of projections from the cell body, where the nucleus is located: **(1)** highly branched, relatively short processes called **dendrites**, and **(2)** relatively long structures called **axons**. Dendrites contact other cells and transmit electrical signals from them to the cell body; axons carry electrical signals from the cell body to other cells (**Figure 41.4b**).

(a) In a neuron, information is transmitted from dendrites to the cell body to the axon.

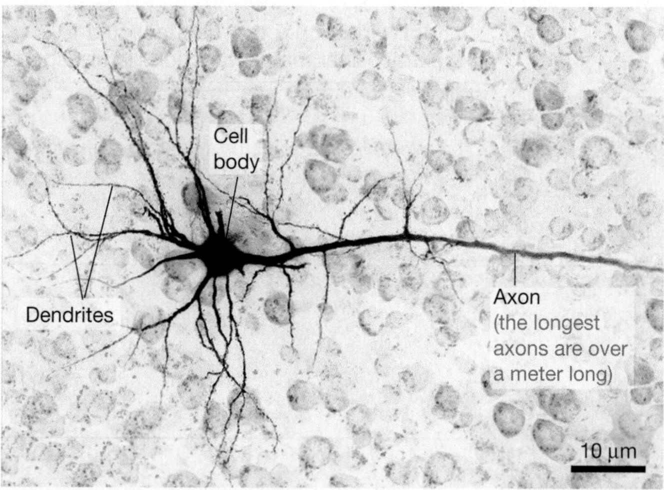

(b) Neurons connect to form networks.

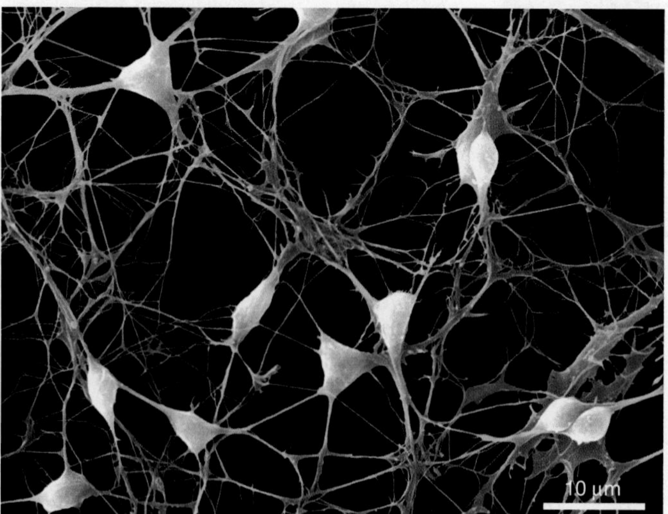

FIGURE 41.4 Neurons Transmit Electrical Signals.

✔**QUESTION** How does the presence of projections on nerve cells support their function in electrical signaling?

MUSCLE TISSUE **Muscle tissue** was a key innovation in the evolution of animals—like nervous tissue, it appears in no other lineage on the tree of life (see Chapter 32). Muscle tissue functions primarily in movement. There are three types of muscle tissue; you, along with other vertebrates, have all three.

1. **Skeletal muscle** attaches to the bones of the skeleton and exerts a force on them when it contracts. Skeletal muscle is responsible for most body movements. In addition, it encircles the openings of the digestive and urinary tracts and controls swallowing, defecation, and urination. It consists of the long cells called **muscle fibers** (**Figure 41.5a**) and is the most common type of muscle tissue in the vertebrate body. Muscle fibers are packed with long tubelike structures called myofibrils; each myofibril is packed with protein filaments that move by sliding past each other. As Chapter 46 will show, skeletal muscle contracts in response to a complex series of events triggered by electrical signals arriving from nerve cells.
2. **Cardiac muscle** makes up the walls of the heart and is responsible for pumping blood throughout the body. Although similar to skeletal muscle in some respects, each cardiac muscle cell branches and makes direct, end-to-end, physical and electrical contact with other cardiac muscle cells (**Figure 41.5b**). These connections help transmit signals from one cardiac muscle cell to another during a heartbeat. In this way, the contraction of cardiac muscle cells is coordinated during the series of events known as the cardiac cycle (see Chapter 44).
3. **Smooth muscle** cells, which are tapered at each end, form a muscle tissue that lines the walls of the digestive tract and the blood vessels (**Figure 41.5c**). Different types of neurons control the contraction of smooth muscle cells versus striated muscle cells. Smooth muscle is responsible for movements such as the passage of food down the digestive tract or the dilation (opening) of arteries near the skin in hot weather. Smooth muscle also controls the size of airways in the respiratory system and is responsible for expelling the fetus during birth (see Chapter 48).

In addition, biologists make three other key distinctions about muscle cells and tissues:

- Tissues that can contract in response to conscious thought are said to be **voluntary muscle**; tissues that contract only in response to unconscious electrical activity are said to be **involuntary muscle**. Skeletal muscle is voluntary; cardiac and smooth muscle are involuntary.
- Muscle cells may have one or many nuclei. Skeletal muscle cells and some cardiac muscle cells are multinucleate; most cardiac and all smooth muscle cells are uninucleate.
- In some muscle cells, the protein filaments responsible for contractions are organized into repeating structures that give the cells and tissues a banded or **striated** appearance. Skeletal and cardiac muscle are striated; smooth muscle is unstriated.

(a) Skeletal muscle: voluntary, multinucleate, striated

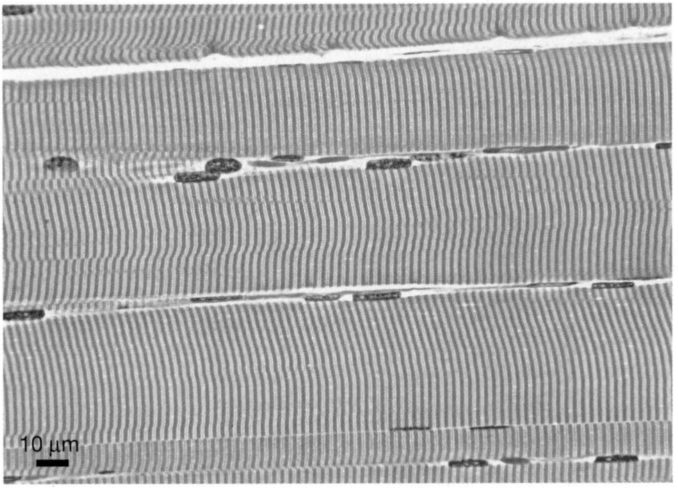

Long cells (muscle fibers)

(b) Cardiac muscle: involuntary, usually uninucleate, striated

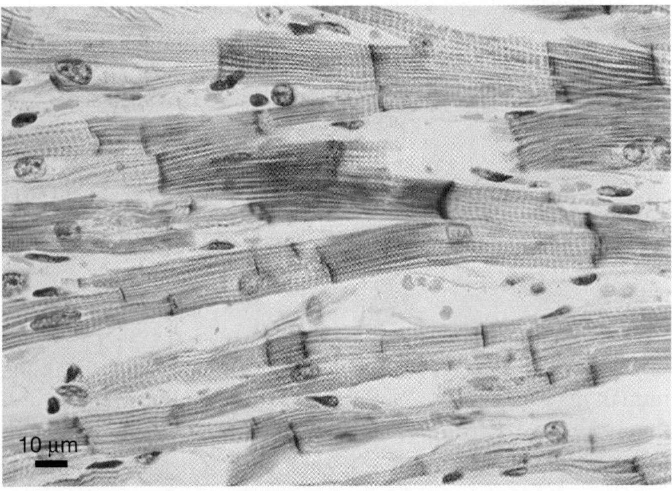

Branched cells

(c) Smooth muscle: involuntary, uninucleate, unstriated

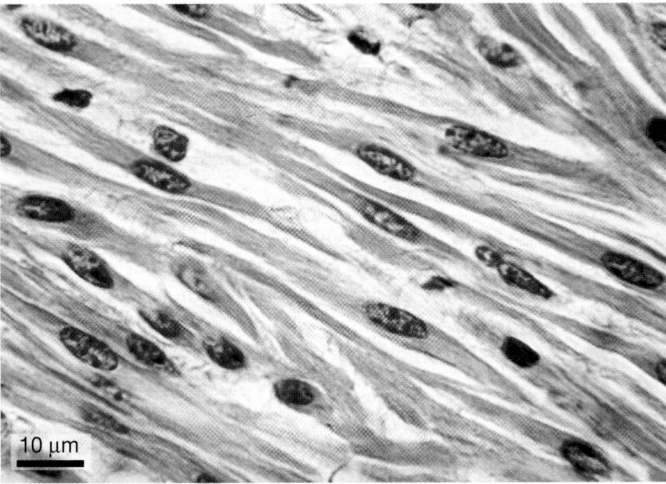

Tapered cells

FIGURE 41.5 Muscle Tissues Comprise Cells That Contract. The three types of muscle tissue have distinctive structures and functions.

EPITHELIAL TISSUES **Epithelial tissues** are also called **epithelia** (singular: **epithelium**). Epithelium covers the outside of the body, lines the surfaces of organs, and forms glands. An **organ** is a structure that serves a specialized function and consists of several tissues; a **gland** is a group of cells that secrete specific molecules or solutions.

Epithelia form the interface between the interior of an organ or body and the exterior. In addition to providing protection, epithelial tissues are gatekeepers. Epithelia regulate the transfer of heat between the interior and exterior of structures, as well as the transfer of water, nutrients, and other substances.

Because the primary function of epithelium is to act as a barrier and protective layer, it's not surprising to observe that epithelial cells typically form layers of closely packed cells (**Figure 41.6**). In many cases, adjacent epithelial cells are joined by structures that hold them tightly together, such as tight junctions and desmosomes (introduced in Chapter 8).

Epithelial tissue has polarity, or sidedness. An epithelium has an **apical** side, which faces away from other tissues and toward the environment, and a **basolateral** side, which faces the interior of the animal and connects to connective tissues. This connection is made by a layer of fibers called the **basal lamina**.

The apical and basolateral sides of an epithelium have distinct structures and functions. Epithelial cells, for example, line the surface of your trachea, or windpipe. The apical side of these cells secretes mucus and is covered with cilia that help sweep away dust, bacteria, and viruses. The basolateral side lacks these features, but is cemented to the basal lamina.

Epithelial cells have short life spans. The cells that line your esophagus—the tube connecting your mouth and stomach—live for 2 to 3 days, while the cells that line your large intestine live for a maximum of 6 days. Muscle cells and neurons, in contrast,

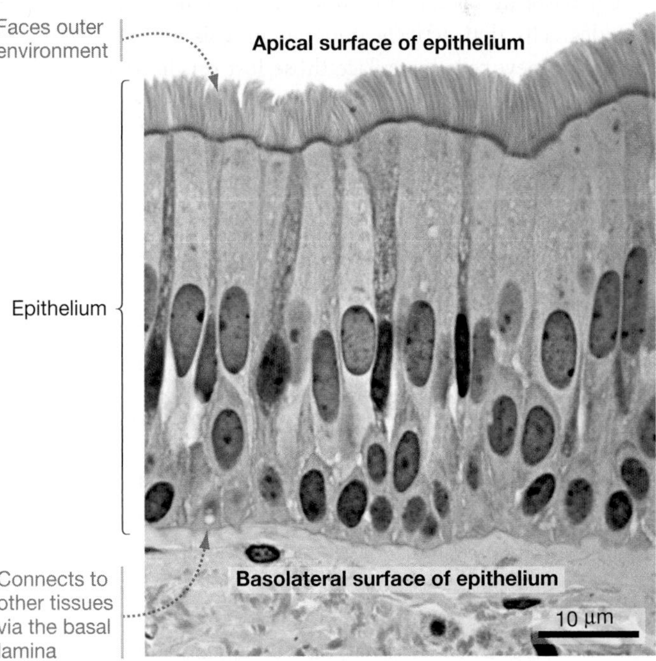

FIGURE 41.6 Epithelial Cells Provide Protection and Regulate Which Materials Pass across Body Surfaces.

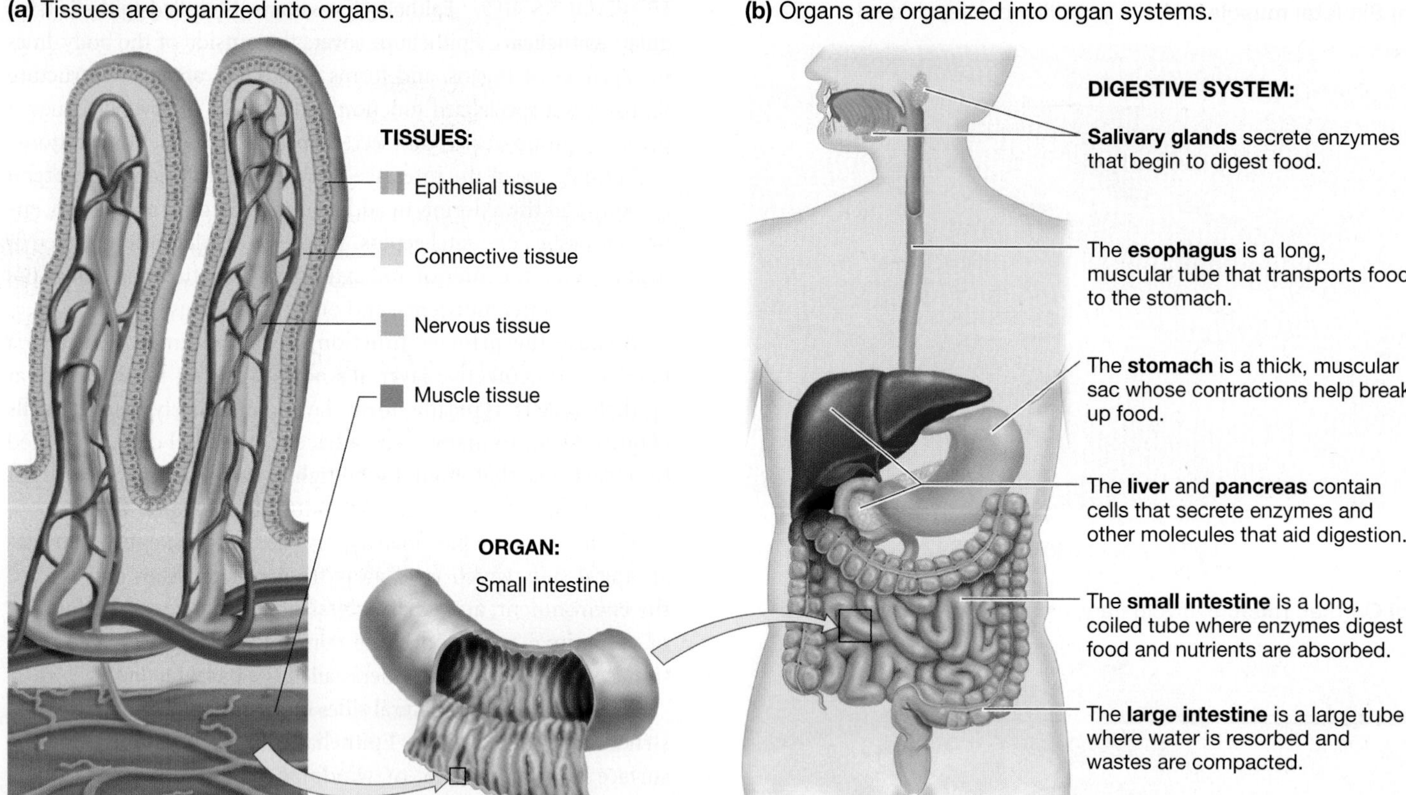

FIGURE 41.7 Organs Are Composed of Tissues; Organ Systems Are Made Up of Organs. **(a)** The human small intestine is an organ composed of all four major tissue types. **(b)** The human digestive system is essentially one long tube divided into chambers where food is processed and nutrients are absorbed. The salivary glands, liver, and pancreas are organs that secrete specific enzymes or compounds into the tube.

normally live as long as the individual does. Epithelial cells are short-lived because they are exposed to harsh environments, where they are likely to be killed or scraped away.

The tissue as a whole does not wear away, however, because it includes cells that actively undergo mitosis and cytokinesis—producing new cells to replace those lost on the side that faces the environment.

Organs and Organ Systems

Cells with similar functions are organized into tissues, and tissues are organized into specialized structures called organs. Recall that an organ is a structure that serves a specialized function and consists of several types of tissues. The small intestine, for example, consists of muscle, nervous, connective, and epithelial tissues (**Figure 41.7a**).

An **organ system** consists of groups of tissues and organs that work together to perform one or more functions. Using the digestive system as an example, **Figure 41.7b** illustrates how the structure of organs correlates with their function and how the components of an organ system work together in an integrated fashion.

Because an animal's body contains molecules, cells, tissues, organs, and organ systems, biologists who study animal anatomy and physiology must work at various levels of organization to understand how that body operates.

Figure 41.8 illustrates these levels of organization, using the human nervous system as an example. Because the structure and function of each component in the body are integrated with other components, and because each level of organization is integrated with other levels of organization, the organism as a whole is greater than the sum of its parts. In other words, an organism is more than just a collection of individual systems, and each system is more than just a collection of individual cells or tissues or even organs.

CHECK YOUR UNDERSTANDING

If you understand that . . .

- Biologists study structure and function at the molecular, cellular, tissue, organ, and organ system levels.
- Events at each level of organization in an individual interact to form an integrated whole that responds to the environment in appropriate ways.

✔ You should be able to . . .

Describe the structure and function of the four major types of animal tissues.

Answers are available in Appendix B.

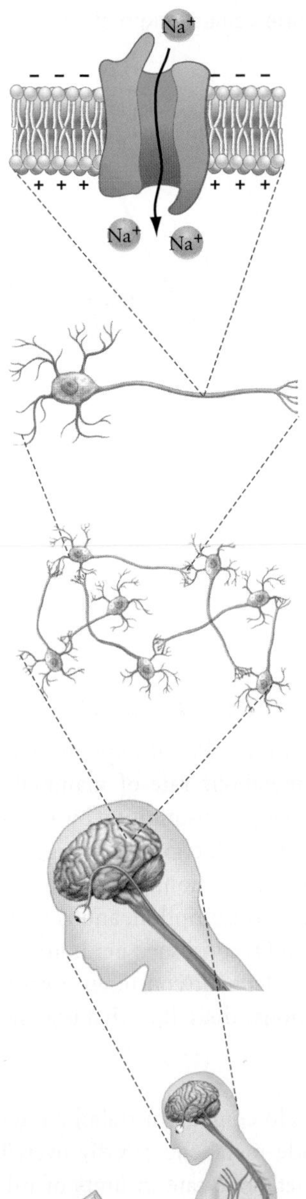

FIGURE 41.8 Biologists Study Anatomy and Physiology at Many Levels. The levels of organization within an organism are not independent of each other. Instead, they are tightly integrated.

In effect, each subsequent chapter in this unit focuses on a different organ system found in animals, beginning with the excretory system and ending with the immune system. Each of these systems can be interpreted as a suite of adaptations and trade-offs. Each system accomplishes a specific task required for survival and reproduction, and each works in conjunction with other systems.

Before delving into the various systems, however, it's essential to examine general phenomena that affect all systems in animals. Let's start by looking at how body size affects animal physiology.

41.3 How Does Body Size Affect Animal Physiology?

Animals are living machines, made up of molecules, cells, tissues, organs, and organ systems that have changed over time in response to natural selection.

The laws of physics affect the anatomy and physiology of a living machine. The force of gravity, for example, limits how large an animal can be and still move efficiently. Or consider the forces exerted by the medium in which animals live. Because water is much denser than air, it is harder for animals to move through water. As a result, fish and aquatic mammals have much more streamlined bodies than terrestrial animals do.

Physical laws clearly affect body size. Just as clearly, body size has pervasive effects on how animals function. Large animals need more food than small animals do. Large animals also produce more waste, take longer to mature, reproduce more slowly, and live longer. Conversely, small animals are more susceptible to damage from cold and dehydration than large animals are, because they lose heat and water faster. Juveniles and adults of the same species face different challenges simply because their body sizes are different.

Why is body size such an important factor in how animals work? How do biologists study the consequences of size? Let's consider each question in turn.

Surface Area/Volume Relationships: Theory

From microscopic roundworms to gigantic blue whales, animals span an incredible range of body masses—a total of twelve orders of magnitude. Many of the challenges posed by increasing size are based on the relationship between surface area and volume.

To understand why surface area is important, recall from Chapter 6 that diffusion takes place across the surface of the plasma membrane. Oxygen and nutrients such as glucose must diffuse into the cell, and waste products such as urea and carbon dioxide must diffuse out. The rate at which these and other molecules and ions diffuse depends in part on the amount of surface area available for diffusion. In contrast, the rate at which nutrients are used and heat and waste products are produced depends on the volume of the cell.

The contrast between processes that depend on surface area and those that depend on volume is important for a simple reason. As a cell gets larger, its volume increases much faster than its surface area does.

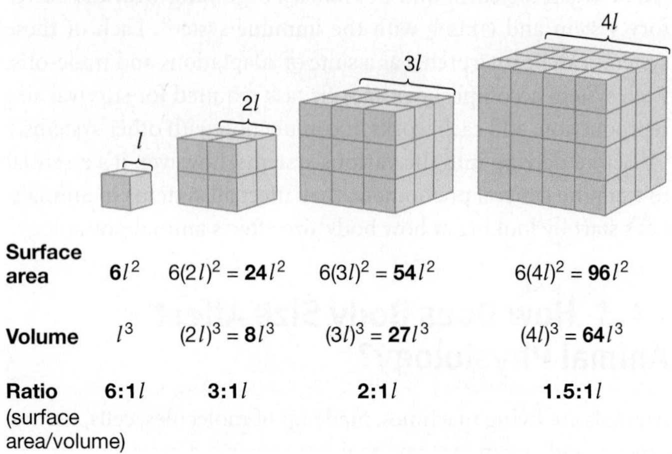

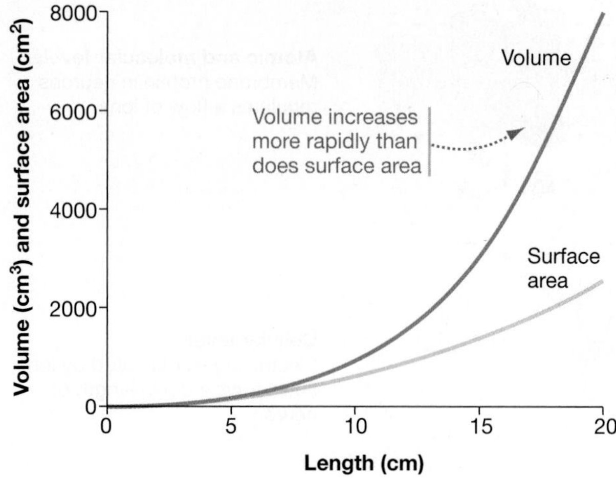

FIGURE 41.9 Surface Area and Volume Change as a Function of Overall Size. **(a)** The surface area of an object increases as the square of the length (l). The volume increases as the cube of that linear dimension. **(b)** Volume increases much more rapidly than does surface area as linear dimensions increase.

Reviewing a little basic geometry will convince you why this is so. As **Figure 41.9a** shows:

- The surface area of a cube increases as a function of its linear dimension *squared*. Because a cube has six sides, the surface area of a cube of length l is $6l^2$ (six times the area of any one side).
- The volume of the same structure increases as a function of its linear dimension *cubed*. Hence, the volume (or mass) of a cube of length l is l^3.

Area has two dimensions; volume has three. In general:

$$\text{Surface area} \propto (\text{length})^2$$

$$\text{Volume (or mass)} \propto (\text{length})^3$$

$$\text{Surface area} \propto (\text{volume})^{2/3}$$

(The symbol $\propto$ means "is proportional to.")

Figure 41.9b graphs the consequences of these relationships. The x-axis plots the length of a side in a cube; the y-axis plots the cube's volume (orange line) or surface area (yellow line). As a cube gets bigger, its surface area increases much more slowly than does its volume (or mass).

The same general relationship holds for cells, tissues, organs, and systems. Quantities that are based on volume, such as body mass, increase disproportionately fast with increases in linear dimensions.

✔**If you understand the relationship between surface area and volume, you should be able to predict which of the following has the higher surface area/volume ratio: a newborn or an adult human.** You should also go to the study area at *www.masteringbiology.com* to review how body size affects surface area and volume.

(MB) **Web Activity** Surface Area/Volume Relationships

How does the relationship between surface area and volume affect animal form and function?

Surface Area/Volume Relationships: Data

As an example of how surface area/volume relationships affect an animal's physiology, consider the metabolic rate of mammals. **Metabolic rate** is the overall rate of energy consumption by an individual. Because consumption and production of energy in mammals depend largely on aerobic respiration, metabolic rate is often measured in terms of oxygen consumption, and is typically reported in units of milliliters of O_2 consumed per hour.

Because it is so much larger, an elephant consumes a great deal more oxygen per hour than a mouse does. But what is going on at the levels of cells and tissues?

COMPARING MICE AND ELEPHANTS To compare metabolic rates in different species, biologists divide metabolic rate by overall mass and report a mass-specific metabolic rate in units of mL O_2/gram/hour. This mass-specific metabolic rate gives the rate of oxygen consumption per gram of tissue.

Because an individual's metabolic rate varies dramatically with its activity, the accepted convention is to report the **basal metabolic rate (BMR)**—the rate at which an animal consumes oxygen while at rest, with an empty stomach, under normal temperature and moisture conditions.

Figure 41.10 plots per-gram or "mass-specific" BMR as a function of average body mass. Note that the x-axis on the graph is logarithmic, to make it easier to compare very small with very large species. For help with logarithms, see **BioSkills 7** in Appendix A.

The graph's take-home message? On a per-gram basis, small animals have higher BMRs than do large animals. An elephant has more mass than a mouse, but a gram of elephant tissue consumes much less energy than a gram of mouse tissue does.

The leading hypothesis to explain this pattern is based on surface area/volume ratios. Many aspects of metabolism—including oxygen consumption, food digestion, delivery of nutrients to tissues, and removal of wastes and excess heat—depend on ex-

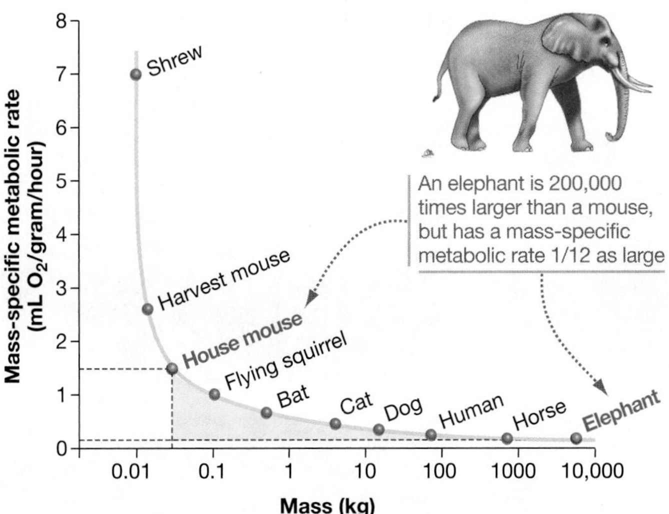

FIGURE 41.10 Small Animals Have Higher Metabolic Rates than Large Animals Do. Overall body mass, plotted on a logarithmic scale, versus metabolic rate per gram of tissue.

✔**QUESTION** Which mammal has to eat more to support each gram of its tissue: a dog or a human?

change across surfaces. As an organism's size increases, its mass-specific metabolic rate must decrease. Otherwise the surface area available for exchange of materials would fail to keep up with the metabolic demands generated by the enzymes in the organism.

Small animals can "live fast" because they have enough surface area to support rapid metabolism. Large animals don't have enough surface area to keep up and have to "live slow." There is a trade-off between size and metabolic rate.

CHANGES DURING DEVELOPMENT A king salmon weighs a few milligrams or less at hatching but grows into an adult weighing 50 kg or more—a millionfold increase in body mass. To explore the consequences of this change, biologists have studied how gas exchange—uptake of oxygen and removal of carbon dioxide—occurs in newly hatched Atlantic salmon.

Like most fish species, young salmon have rudimentary gills but also exchange gases across their skin. In aquatic animals, **gills** are organs that allow the exchange of gases and dissolved substances between the animals' blood and the surrounding water.

To document the amount of gas exchange that occurs in the gills versus the general body surface, researchers inserted the heads of individual salmon through a pinhole in a soft rubber membrane, then recorded the rate of oxygen uptake on either side of the membrane. As the "Experimental Setup" section of **Figure 41.11** indicates, the gills were responsible for oxygen uptake on one side of the membrane; skin was responsible for oxygen uptake on the other side of the membrane.

The graph in the figure's "Results" section plots the percentage of total oxygen uptake that took place across the skin (green line) versus gills (purple line), as a function of body mass. Each data point represents the recordings from an individual fish.

Note that newly hatched larvae take up most of the oxygen they need by diffusion across the body surface. As an individual grows, however, its skin surface area decreases in relation to its volume. To avoid suffocation, individuals switch from skin-breathing to gill-breathing at about 0.1 grams. What makes gills so effective in oxygen uptake?

EXPERIMENT

QUESTION: Newly hatched salmon can breathe through their skin and through their gills. Which predominates?

HYPOTHESIS: No explicit hypothesis is being tested. Experiment is exploratory in nature.

EXPERIMENTAL SETUP:

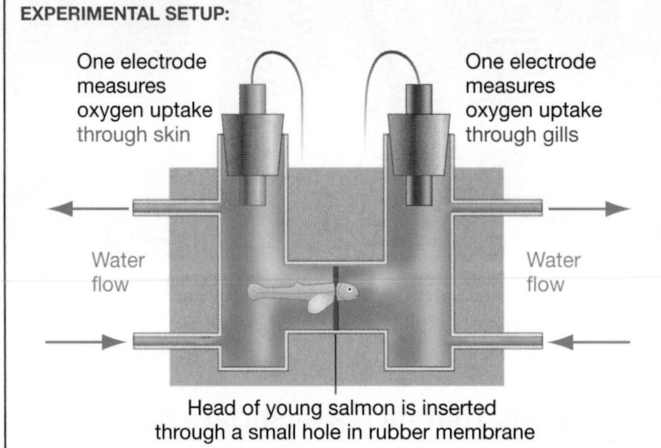

PREDICTION: No explicit predictions.

RESULTS:

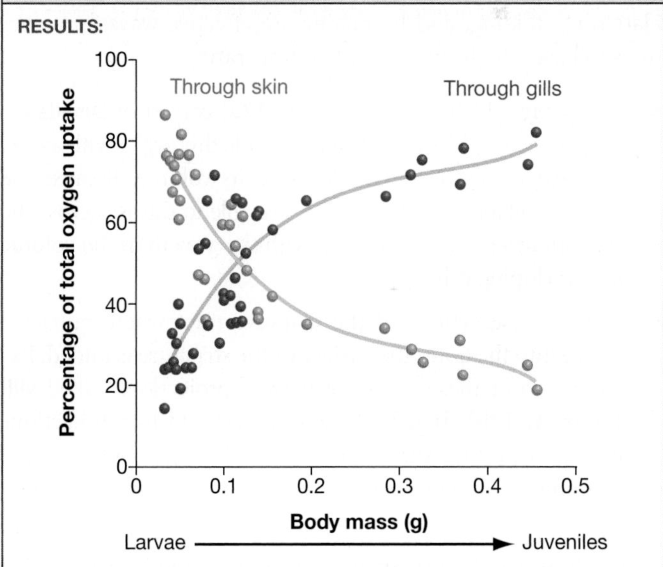

CONCLUSION: Breathing changes from skin to gills as larvae grow. Interpretation: Gills provide larger surface area relative to increasing volume of body.

FIGURE 41.11 How Do Young Salmon Breathe?

SOURCE: Wells, P. R. and A. W. Pinder. 1996. The respiratory development of Atlantic salmon. *Journal of Experimental Biology* 199: 2737–2744.

✔**QUESTION** Suppose the experimenters had measured oxygen uptake on either side of the apparatus in the absence of a fish. What would the results be?

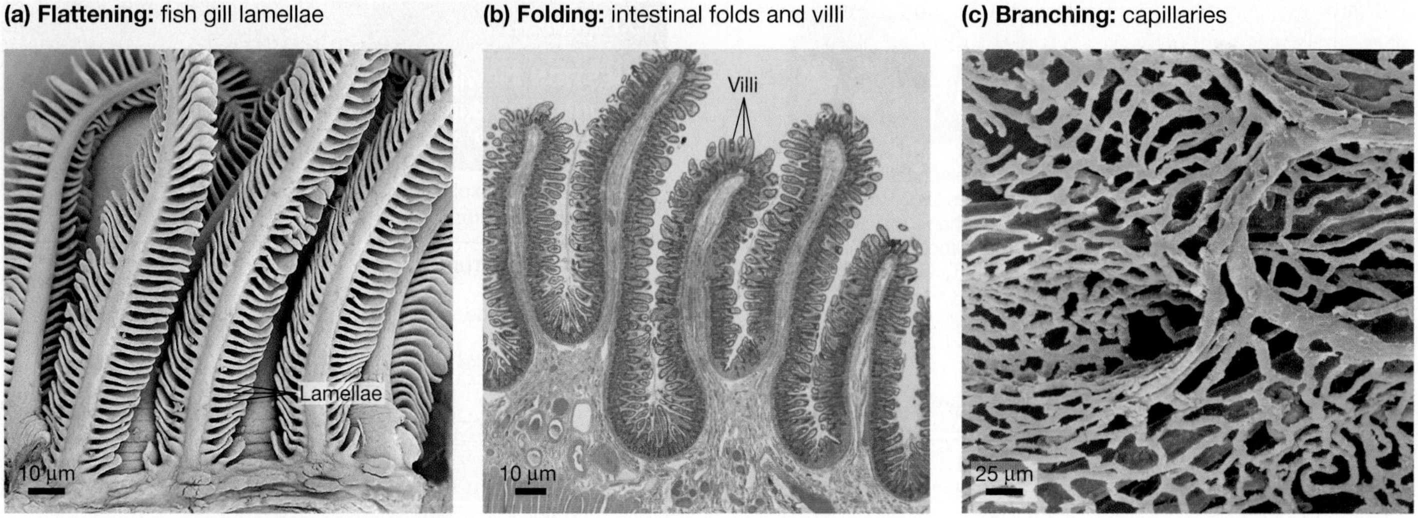

(a) **Flattening:** fish gill lamellae

(b) **Folding:** intestinal folds and villi

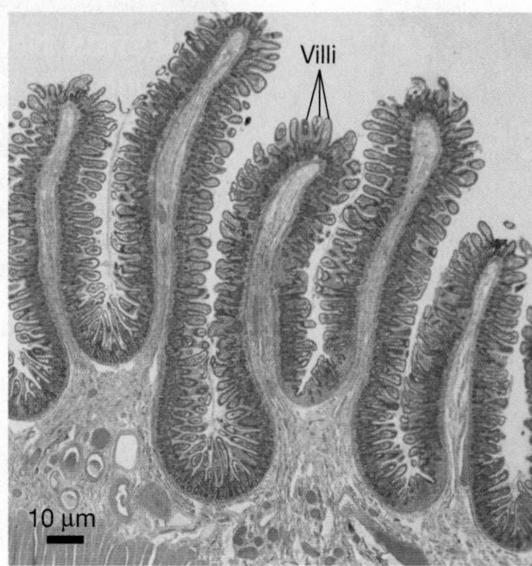

(c) **Branching:** capillaries

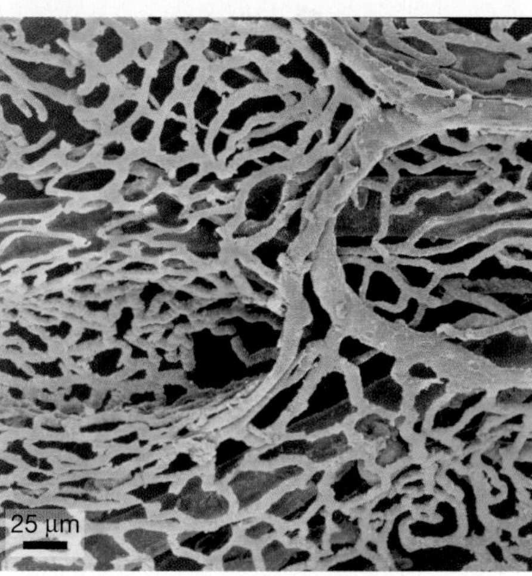

FIGURE 41.12 Certain Structures Have High Surface Area/Volume Ratios. The micrograph in part (c) has been colorized to highlight capillaries, colored pink.

Adaptations That Increase Surface Area

If the function of a cell or tissue depends on diffusion, it usually has a shape that increases its surface area relative to its volume. Flattening, folding, and branching are effective ways for structures to have a high surface area/volume ratio:

- *Flattening* Fish gills (**Figure 41.12a**) consist of **lamellae**—thin sheets of epithelial cells that provide this organ with an extremely high surface area relative to its volume. Because the surface available is so large, gases are able to diffuse across the gills rapidly enough to keep up with the growth in the volume of a developing fish.
- *Folding* In portions of the digestive tract where nutrients diffuse into the body, the surface of the structure is folded. Extending from these folds are narrow projections called **villi** (**Figure 41.12b**). Together, the folds and fold-like projections make an extensive surface area available. Folded surfaces are common in diffusion-dependent organs.
- *Branching* The highly branched network shown in **Figure 41.12c** is a system of small, thin-walled blood vessels called **capillaries.** Capillaries have a high surface area available for gases, nutrients, and waste products to diffuse into and out of blood; branching increases the surface available in each square centimeter of tissue. In general, highly branched structures increase the surface area available for diffusion.

The amount of surface area created by flattening, folding, and branching can be impressive. The highly branched capillaries in a human have a total surface area of up to 1000 m^2; extensive folding gives a total surface area of about 140 m^2 in your lungs and 250 m^2 in your small intestine. For comparison, a doubles tennis court has a surface area of 261 m^2.

Surface area/volume relationships have a pervasive influence on the structure and function of animals. They will be an issue in almost every chapter in this unit.

CHECK YOUR UNDERSTANDING

If you understand that . . .

- An animal's overall size is important in part because body mass is affected by an array of physical forces.
- The amount of heat and waste that an animal produces and the amount of food and oxygen that it requires are proportional to its mass.
- The amount of surface area available relative to that mass is critical, because heat exchange and other important processes take place across surfaces.

✓ **You should be able to . . .**

1. Explain why large animals have a relatively small surface area/volume ratio.
2. Explain the relationship between body mass and mass-specific basal metabolic rate.

Answers are available in Appendix B.

41.4 Homeostasis

Adaptation and surface area/volume ratios are important themes in the analysis of animal form and function. So is homeostasis.

Homeostasis (literally, "alike-standing") is defined as stability in the chemical and physical conditions within an animal's cells, tissues, and organs. Although conditions may vary as an animal's environment changes, internal chemical and physical states are kept within a tolerable range.

Homeostasis: General Principles

Many of the structures and processes observed in animals can be interpreted as mechanisms for maintaining homeostasis with respect to some quantity, such as pH or calcium ion concentration.

Let's review some important general ideas about homeostasis, then analyze how homeostasis can be maintained in the face of environmental fluctuations.

TWO APPROACHES TO ACHIEVING HOMEOSTASIS Constancy of physiological state can be achieved by two processes: (1) conformation or (2) regulation.

- *Conformation* The body temperature of Antarctic rock cod closely matches that of the surrounding seawater, which is typically −1.9°C (seawater is still liquid at this temperature because of the high concentration of solutes). The rock cod does not actively regulate its body temperature to match that of seawater. Instead, its body temperature remains constant because it conforms to the temperature of its surroundings.
- *Regulation* Regulatory homeostasis is based on mechanisms that adjust the internal state to keep it within limits that can be tolerated, no matter what the external conditions. For example, a dog maintains a body temperature of about 38°C whether it's cold or hot outside. If its body temperature rises, it might pant to cool off and maintain homeostasis. If its body temperature falls, it might shiver to bring its temperature back up to the target value.

THE ROLE OF EPITHELIUM Because epithelium is the interface between the internal and external environments, it plays a key role in achieving homeostasis. Epithelium is responsible for forming an internal environment that can be dramatically different from the external environment, and for maintaining physical and chemical conditions inside an animal that are relatively constant.

As subsequent chapters will show, many epithelial cells are studded with membrane proteins that regulate the transport of ions, water, nutrients, and wastes. No molecule can enter or leave the body without crossing an epithelium. Homeostasis is possible because epithelia control this exchange.

WHY IS HOMEOSTASIS IMPORTANT? Much of the answer to this question is based on enzyme function. Recall from Chapter 3 that enzymes are proteins that catalyze chemical reactions within cells. Temperature, pH, and other physical and chemical conditions have a dramatic effect on the structure and function of enzymes. Most enzymes function best under a fairly narrow range of conditions.

Other processes depend on homeostasis, too. Temperature changes affect membrane permeability and how quickly solutes diffuse. To take an extreme case, the expansion of water as it freezes can rip cells apart if tissues are allowed to drop much below 0°C. Conversely, extremely high temperatures can denature proteins—meaning that they lose their tertiary structure and cease to function.

When homeostasis occurs, conditions inside the body allow molecules, cells, tissues, organs, and organ systems to function at an optimal level.

The Role of Regulation and Feedback

To achieve homeostasis, most animals have regulatory systems that constantly monitor internal conditions such as temperature, blood pressure, blood pH, and blood glucose. If one of these variables changes, a homeostatic system acts quickly to modify it. Like the thermostat in a home heating system, each of these systems has a **set point**—a normal or target value for the controlled variable.

Animals have a set point for blood pH, blood oxygen concentration, nutrient availability, and other parameters. In most mammals, the set point for body temperature is somewhere between 35°C and 39°C. How does an individual maintain its tissues at the set point despite changes in activity and the environment?

A homeostatic system is based on three general components: a sensor, an integrator, and an effector. **Figure 41.13** shows how these components interact:

1. A **sensor** is a structure that senses some aspect of the external or internal environment.
2. An **integrator** evaluates the incoming sensory information and "decides" whether a response is necessary to achieve homeostasis. (The word decides is in quotation marks because the decision is not a conscious one.)
3. An **effector** is any structure that helps restore the desired internal condition.

Without these three elements, homeostatic control systems can't maintain a desired set point; homeostasis is impossible.

Homeostatic systems are based on negative feedback. When **negative feedback** occurs, effectors reduce or oppose the change in internal conditions. For example, a rise in blood pH triggers effectors that act to reduce that rise. Blood pH returns to the set

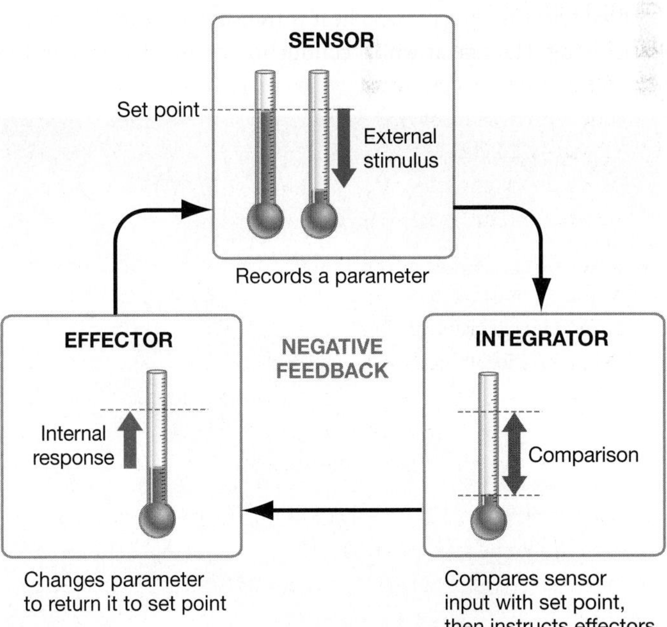

FIGURE 41.13 Animals Achieve Homeostasis through Negative Feedback. Many animals use homeostatic systems similar to this one to maintain a preferred range of hydration (water concentration), blood pH, blood pressure, calcium ion concentration, body temperature, and so on.

point in response to this negative feedback. For more detail on how negative feedback works in homeostasis, go to the study area at *www.masteringbiology.com.*

MB **Web Activity** Homeostasis

Homeostatic systems are a key aspect of one of the five attributes of life introduced in Chapter 1: acquiring information from the environment and responding to it. Subsequent chapters in this unit explore how animals use sensor-integrator-effector systems to achieve homeostasis with respect to the solute concentrations of their cells and tissues, their oxygen supply, and nutrient availability. In this chapter, let's focus on how different animals achieve homeostasis with respect to body temperature.

41.5 How Do Animals Regulate Body Temperature?

All animals exchange heat with their environment. Heat flows "downhill," from regions of higher temperature to regions of lower temperature. If an individual is warmer than its surroundings, it will lose heat; if it is cooler than its environment, it will gain heat.

How does heat exchange occur?

Mechanisms of Heat Exchange

As **Figure 41.14** shows, animals exchange heat with the environment in four ways: conduction, convection, radiation, and evaporation.

- **Conduction** is the direct transfer of heat between two physical bodies that are in contact with each other. For instance, when a turtle sits on a warm rock, heat is transferred from the rock to its body. The rate at which conduction occurs depends on the surface area of transfer, the steepness of the temperature difference between the two bodies, and how well each body conducts heat. Water, for example, conducts heat much better than air does. As a result, a person immersed in 15°C water loses heat much faster than does a person who is exposed to air at the same temperature.
- **Convection** is a special case of conduction. During conduction, heat is transferred between two solids; but during convection, heat is exchanged between a solid and a liquid or gas. For example, the heat loss that occurs when wind blows on your skin is due to convection. As the speed of the air or water flow increases, so does the rate of heat transfer.
- **Radiation** is the transfer of heat between two bodies that are not in direct physical contact. All objects, including animals, radiate energy as a function of their temperature. The Sun radiates heat; so does your body, but to a much lesser degree.
- **Evaporation** is the phase change that occurs when liquid water becomes a gas. Conduction, convection, and radiation can cause heat gain or loss, but evaporation leads only to heat loss. The turtle in the photograph is losing heat as water evaporates off its shell and skin. Because of the extensive hydrogen bonding in liquid water, a large amount of energy is needed to heat water and produce evaporation (see Chapter 2). If you get overheated on a summer day, splashing water on your skin and sweating will absorb a large amount of heat and cool you off. Conversely, getting wet on a cold day can be deadly. The water on your skin absorbs so much heat from your body that your temperature may drop dangerously.

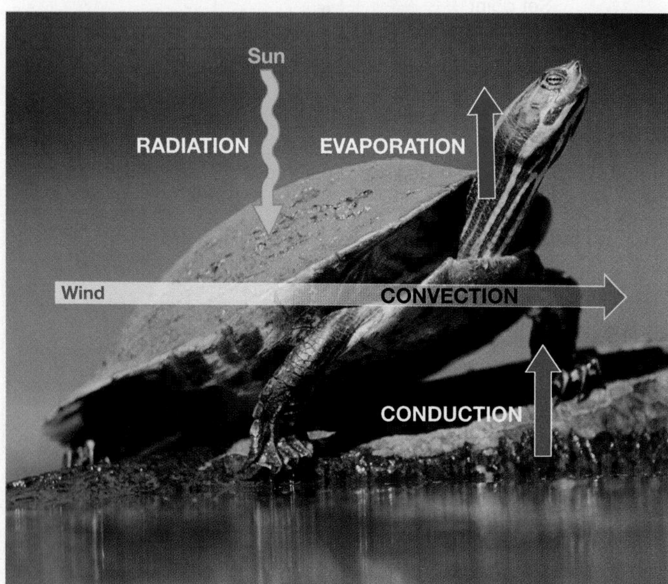

FIGURE 41.14 Four Methods of Heat Exchange. The arrows indicate the direction of heat exchange from the warmer body to the cooler body.

Heat exchange is critical in animal physiology because individuals that get too hot or too cold may die. Overheating can cause enzymes and other proteins to denature and cease functioning. It may also lead to excessive water loss and dehydration. A sharp drop in body temperature, in contrast, can slow enzyme function and energy production. In humans, both heat stroke and hypothermia ("under-heating") are life-threatening conditions.

Although most organisms cannot regulate their body temperature to keep it in an optimal range, some animals can. Let's take a closer look.

Variation in Thermoregulation

The ability of animals to **thermoregulate**, or control body temperature, varies widely. One way to organize this variation is by examining (1) how animals obtain heat, and (2) whether body temperature is held constant.

An **endotherm** ("inner-heat") produces adequate heat to warm its own tissues, while an **ectotherm** ("outer-heat") relies principally on heat gained from the environment. The resulting difference in body temperature can be dramatic, as illustrated in the thermogram—an image taken with a camera that is sensitive to infrared radiation—in **Figure 41.15**.

Endotherms and ectotherms represent two opposite extremes along a continuum of heat sources. Many animals are partially endothermic and partially ectothermic.

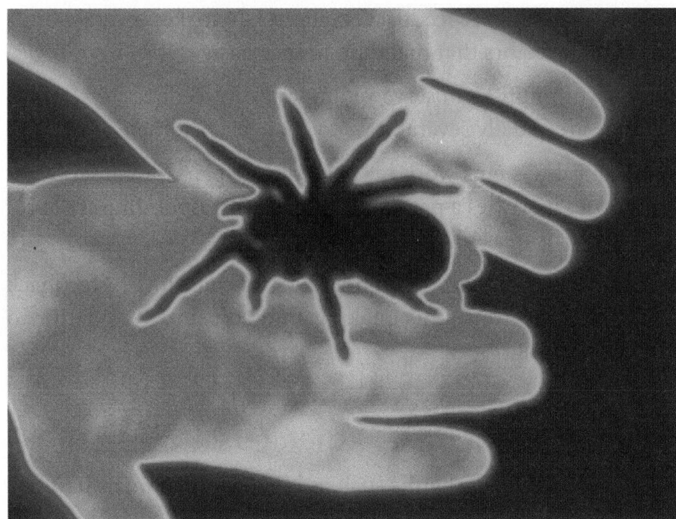

FIGURE 41.15 Endotherms Produce Heat, Ectotherms Do Not. The temperatures recorded in this thermogram range from 35.9°C (light red) to room temperature at 22.8°C (dark blue). Note the dramatic difference in surface temperature between a human (produces heat) and a tarantula (relies on heat from environment).

There are also two extremes on a continuum describing whether animals hold their body temperature constant: **Homeotherms** ("alike-heat") keep their body temperature constant, while **heterotherms** ("different-heat") allow their body temperature to rise or fall depending on environmental conditions.

Humans and other mammals, along with most birds, are strictly endothermic homeotherms. These species produce their own heat and maintain a constant body temperature. In contrast, mosquito larvae and other freshwater invertebrates are ectothermic heterotherms. But many animal species lie somewhere between these extremes:

- Some desert-adapted mammals, such as the oryx featured in the introduction to this chapter, allow their body temperature to rise during the hotter part of the day—meaning they are somewhat heterothermic.
- Small mammals that inhabit cold climates lose heat rapidly because their surface area is large relative to their volume. To survive when temperatures are cold, species such as ground squirrels reduce their metabolic rate and allow their body temperature to drop. This condition is called **torpor**. If torpor persists for weeks or months, it is called **hibernation**.
- Naked mole rats are mammals but lack insulation, because they have no fur. They live in underground tunnels and allow their body temperature to rise and fall with burrow temperatures. They are heterothermic and intermediate between ectotherms and endotherms.
- On cold mornings, bumblebees "shiver" by contracting their flight muscles together. (During flight, these same muscles contract in an alternating pattern to beat the wings.) Shivering generates enough heat to raise the individual's body temperature in preparation for flight.

Even in a homeothermic endotherm such as a mammal or bird, body temperature can vary widely in different body regions. When a Canada goose is standing on ice, its feet may be at a temperature of just 9°C, even though its body core is at 35°C. Similar variations exist in tuna and mackerel. These fish are ectotherms but generate heat to warm certain sections of their bodies, such as their eyes or swimming muscles.

Endothermy and Ectothermy: A Closer Look

Endotherms can warm themselves because their basal metabolic rates are extremely high—the heat given off by the high rate of chemical reactions is enough to warm the body. Mammals and birds retain this heat because they have elaborate insulating structures such as feathers or fur.

Ectotherms can also generate heat as a by-product of metabolism. The amount of heat they generate is small compared with the amount generated by endotherms, however, because ectotherms have relatively low metabolic rates. The most important sources of heat gain in ectotherms are radiation and conduction: they bask in sunlight or lie on warm rocks or soil.

Endothermy and ectothermy are best understood as contrasting adaptive strategies. Because endotherms maintain a high body temperature at all times, they can be active in winter and at night. Their high metabolic rates also allow them to sustain high levels of aerobic activities, such as running or flying.

These abilities come at a cost, however: To fuel their high metabolic rates, endotherms have to obtain large quantities of energy-rich food. The energy used to produce heat is then unavailable for other energy-demanding processes, such as reproduction and growth.

In contrast, ectotherms are able to thrive with much lower intakes of food. And because they are not oxidizing food to provide heat, they can use a greater proportion of their total energy intake to support reproduction.

What's the downside of ectothermy? Chemical reaction rates are temperature dependent, so muscle activity and digestion slow dramatically as the body temperature of an ectotherm drops. As a result, ectotherms are more vulnerable to predation in cold weather and in general are less successful than endotherms at inhabiting cold environments or remaining active in cool nighttime temperatures.

In short, each suite of adaptations has advantages and disadvantages. Like all adaptations, endothermy and ectothermy involve trade-offs.

Temperature Homeostasis in Endotherms

Thermoregulation is an important aspect of homeostasis in some animals. **Figure 41.16** on page 818 illustrates how the sensor-integrator-effector components of mammals function in thermoregulation.

Temperature receptors located throughout the body constantly monitor information about body temperature. For example, temperature receptors in the skin sense cooling or heating, and respond by altering the pattern of electrical signals that they send to adjacent neurons. Receptors in the brain region called the anterior **hypothalamus** respond in a similar fashion to changes in blood temperature.

The electrical signals that originate with temperature receptors are transmitted to an integrator located in the brain. Current evidence indicates that separate centers in the hypothalamus of the brain integrate and respond to increases and decreases in body temperature.

If a mammal is cold, cells in the posterior hypothalamus send signals to effectors that return body temperature to the set point. Signals from the posterior hypothalamus might induce shivering to generate warmth, and fluffing of fur or feathers to improve insulation and retain heat. Signals from the same or nearby cells can also result in the release of blood-borne chemical signals that increase the rate at which cellular respiration takes place throughout the body—generating more body heat.

But if the same individual is too hot, an integrator in the anterior hypothalamus sends signals that initiate sweating or panting—responses that cool the body. Other signals can induce behavioral changes that slow heat production, such as seeking shade or a cool burrow and then resting. Within cells, temperature spikes that are dramatic enough to denature proteins may activate the **heat-shock proteins** introduced in Chapter 3. Heat-shock proteins speed the refolding of proteins—a key step in the recovery process.

In response to either cooling or heating, behavioral and physiological responses move the body temperature back toward the set point via negative feedback. Figure 41.16 makes several points about the effectors that maintain homeostasis:

- It is common to observe redundancy in feedback systems—there are usually several ways to change a parameter.
- Feedback systems work in "antagonistic pairs": One set of responses increases a parameter while a corresponding set of responses decreases it.
- Input from sensors and integrators is constant, so feedback systems are constantly making fine adjustments relative to the set point.

Countercurrent Heat Exchangers

Homeothermic endotherms such as birds and mammals have sophisticated systems for thermoregulation. One of their most impressive adaptations allows them to minimize heat loss from limbs.

Heat loss is a particularly important problem for mammals that live in aquatic environments. If you've ever gone swimming in cold water, you can appreciate the problem faced by seals, otters, and whales. Water is such an effective conductor of heat that aquatic organisms lose metabolic heat rapidly. To conserve heat, otters have dense, water-repellent fur that maintains a layer of trapped air next to the skin. Seals and whales are insulated by thick layers of fatty blubber.

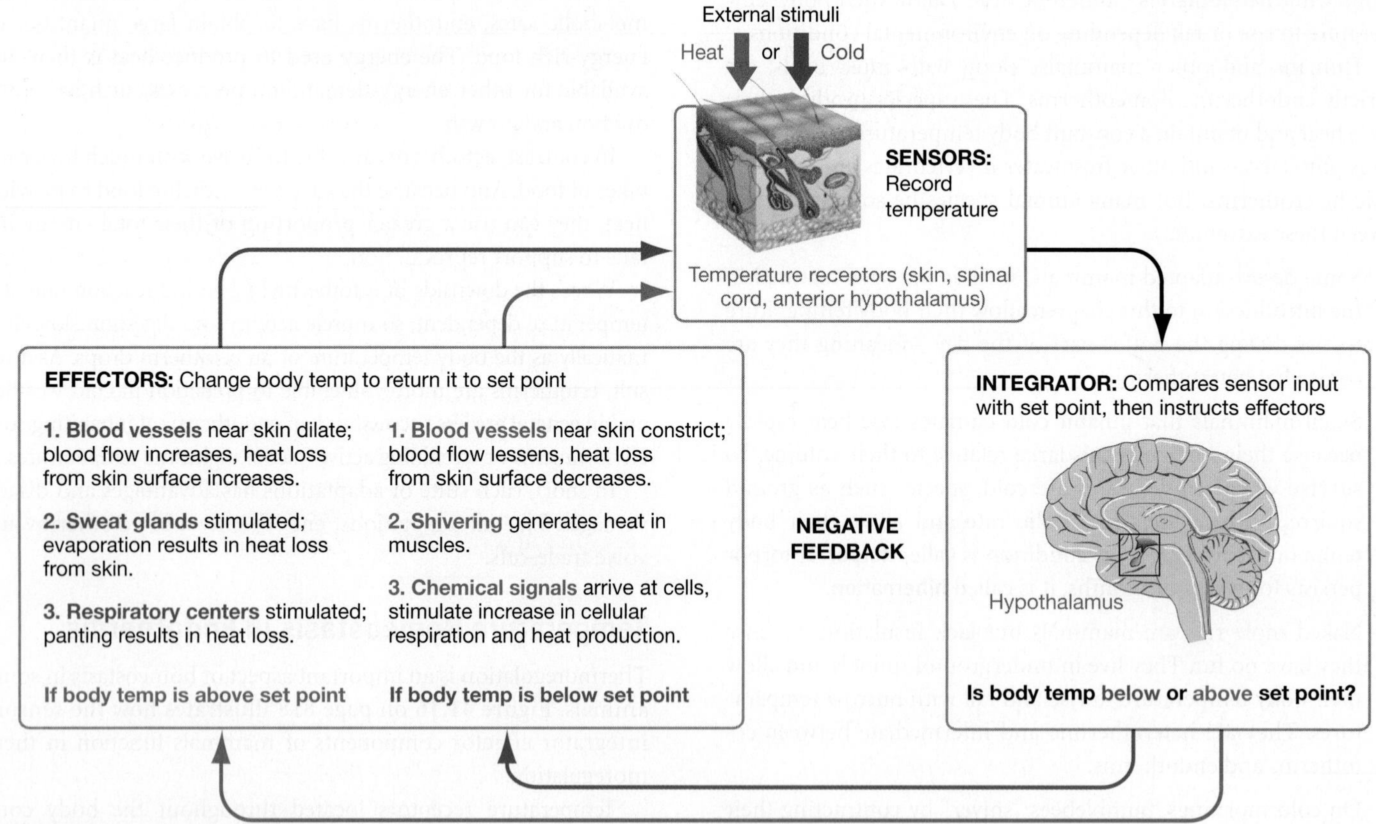

FIGURE 41.16 Mammals Regulate Temperature through Negative Feedback. In mammals, a set point for temperature is maintained by a complex negative feedback system that includes integrators in the anterior and posterior hypothalamus and sensors located throughout the body. The set point varies among species, from 30°C in monotremes to over 39°C in rabbits.

AN EXAMPLE: GRAY WHALE TONGUES Gray whales have a feature that minimizes heat loss from their tongue, which is exposed to cold water during feeding. As **Figure 41.17a** shows, the tongue contains bundles of arteries and veins. Each bundle has an artery that carries warm, oxygenated blood from the body core. The artery is encircled by smaller veins, which transport cool blood from the tongue surface back toward the body core.

The key is that the two types of blood vessel are arranged in an antiparallel fashion. Warm blood traveling out to the tongue is in close contact with cool blood traveling back to the body.

This type of arrangement, with fluids flowing through adjacent pipes in opposite directions, is called a **countercurrent exchanger**. The "exchanger" part of the name is apt because in a case like the whale's tongue, heat is exchanged between the warm blood in the artery and the cool blood in the veins.

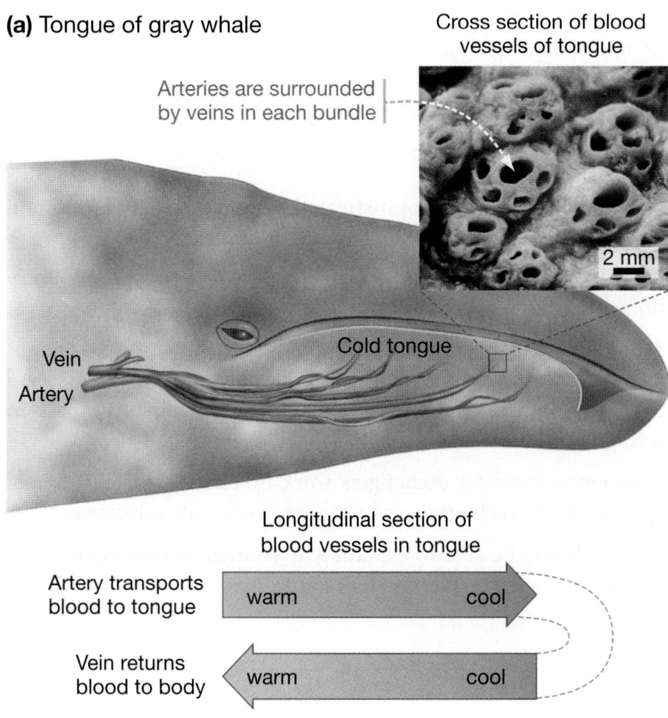

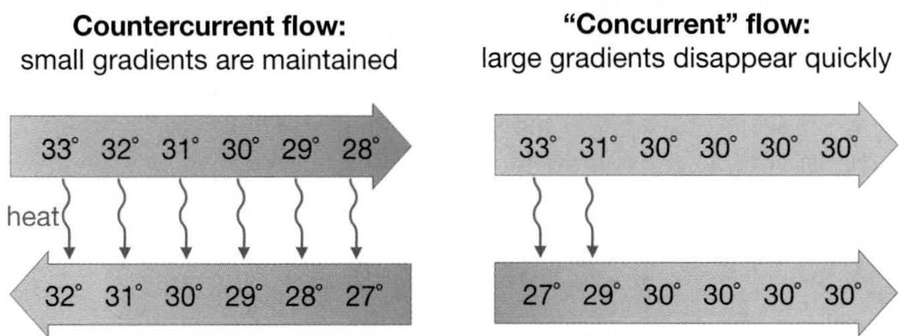

(b) Contrasting countercurrent with "concurrent" heat exchange

Countercurrent flow: small gradients are maintained

33° 32° 31° 30° 29° 28°

heat

32° 31° 30° 29° 28° 27°

"Concurrent" flow: large gradients disappear quickly

33° 31° 30° 30° 30° 30°

27° 29° 30° 30° 30° 30°

FIGURE 41.17 Countercurrent Exchangers Conserve Heat.
(a) Bundles of arteries and veins in whale tongues form heat exchangers that minimize heat loss from the tongue to the cold ocean water during feeding. **(b)** Countercurrent arrangements are much more efficient than "concurrent" arrangements. The data given here are hypothetical.

To see how the countercurrent exchange system works, study the longitudinal section in **Figure 41.17b**. Note that the fluid that enters the countercurrent heat exchanger is initially warm but steadily transfers heat to the adjacent, cooler fluid flowing in the opposite direction. There is a warmer-to-cooler gradient between the two currents at every point along the length of the countercurrent exchanger.

If the two solutions ran in the same direction, as in the right side of Figure 41.17b, the gradient between the two solutions would disappear quickly as the source current cooled and the recipient current heated. Countercurrent exchangers are efficient because they maintain a gradient between the two fluids along their entire length.

A MULTIPLIER EFFECT A second key point about countercurrent heat exchangers concerns the temperature differential between the solutions. Although the differential is small at any point along the pipes, there is a large temperature differential from one end of each pipe to the other.

In effect, small differences in heat along the length of the exchanger sum up to create a large overall temperature gradient from beginning to end. The longer the system, the greater the overall differential will be. To highlight this property, the systems are sometimes called countercurrent multipliers.

Similar heat-conserving arrangements of arteries and veins are found in the flippers of whales and dolphins, and in the legs of many mammals and birds that live in cold terrestrial environments.

Countercurrent exchangers are just one of many sophisticated adaptations you'll encounter in this unit—structures that allow animals to thrive in a wide array of environments.

CHECK YOUR UNDERSTANDING

If you understand that . . .

- Two major aspects of temperature regulation vary among animal species: the amount of heat generated by the animal's own tissues and the degree to which body temperature varies over time.
- Negative feedback allows some endotherms to maintain homeostasis with respect to body temperature.
- Countercurrent heat exchangers are efficient ways to minimize heat loss from extremities.

✓ **You should be able to . . .**

1. Discuss the advantages and disadvantages of endothermy and ectothermy.
2. Diagram a countercurrent system that exchanges sodium ions (Na^+) via diffusion. Your diagram should include (1) labels indicating areas of high versus low Na^+ concentration, (2) arrows that indicate the direction that solutions are moving in the exchanger, and (3) arrows that indicate the direction of Na^+ diffusion.

Answers are available in Appendix B.

CHAPTER 41 REVIEW

For media, go to the study area at www.masteringbiology.com 

Summary of Key Concepts

In biology, structure has a profound influence on function. Biologists analyze the structure and function of animals at a variety of levels: molecules, cells, tissues, organs, and organ systems.

- Cells with a similar structure and common function are grouped together into four general types of tissue: connective tissue, nervous tissue, muscle tissue, and epithelial tissue.
- Epithelium is a particularly important type of tissue because it defines the interface between the animal's external and internal environments. Epithelial cells and tissues have a distinct polarity.
- Organs are structures that are composed of two or more tissues and that perform specific tasks.
- Organ systems are comprised of organs that work together in an integrated fashion to perform a function.

✔ You should be able to predict the effect of a drug that loosens the tight junctions between epithelial cells.

Body size has a strong influence on how animals work, in large part because a body's volume increases faster than its surface area as body size increases.

- Large animals have low metabolic rates, because they have a relatively small surface area for exchanging the oxygen and nutrients required to support metabolism.
- The relatively high surface area of small animals means that they lose heat extremely rapidly. As a result, there are no tiny endothermic animals.

✔ You should be able to explain why it would be impossible for a gorilla the size of King Kong to have fur. (In answering this question, explain how the surface-area-to-volume ratio of a normal size gorilla would compare to Kong's, then relate this to role of surface area and volume in heat generation and transfer and the function of fur.)

MB **Web Activity** Surface Area/Volume Relationships

Animals use an array of methods to maintain a relatively constant environment inside their bodies. They have systems that sense changes in internal conditions and trigger responses that return conditions to normal.

- Homeostasis refers to relatively constant physical and chemical conditions inside the body.
- Animals have a set point, or target value, for blood pH, tissue oxygen concentration, nutrient availability, and other parameters.
- Negative feedback occurs when a condition in the body is not at its set point.
- Responses to negative feedback return conditions to the set point and result in homeostasis.

✔ You should be able to explain how, due to acclimatization as well as adaptation, the homeostatic system for body temperature in a species of mammal might change as global temperatures rise.

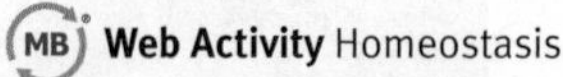

Some animals have sophisticated systems for generating and conserving heat and regulating body temperature.

- Animals vary from endothermic to ectothermic, and from homeothermic to heterothermic.
- Most mammals have a set point for body temperature of about 37°C. If an individual overheats, it will pant or sweat and seek a cool environment; if an individual is cold, it will shiver, bask in sunlight, or fluff its fur.
- Countercurrent heat exchangers work by placing warm and cool liquids next to each other and running in opposite directions.

✔ You should be able to explain why countercurrent systems are sometimes called countercurrent multipliers.

Questions

✔ TEST YOUR KNOWLEDGE

Answers are available in Appendix B

1. How do biologists measure an animal's metabolic rate?
   a. by taking its temperature
   b. by measuring how rapidly it uses oxygen
   c. by measuring how rapidly it uses glucose
   d. by measuring how rapidly it produces wastes
2. How is the structure of a connective tissue most closely correlated with its function?
   a. The density of cells in the tissue correlates with the tissue's function.
   b. The surface area of the tissue correlates with the tissue's function.
   c. The origin of the tissue (from endoderm, mesoderm, or ectoderm) correlates with the tissue's function.
   d. The nature of the extracellular matrix correlates with the tissue's function.
3. As an animal gets larger, which of the following occurs?
   a. Its surface area grows more rapidly than its volume.
   b. Its volume grows more rapidly than its surface area.
   c. Its volume and surface area increase in perfect proportion to each other.
   d. Its volume increases, but its total surface area decreases.
4. Which of the following best describes the set point in a homeostatic system?
   a. the cells that collect and transmit information about the state of the system
   b. the cells that receive information about the state of the system and that direct changes to the system

c. the various components that produce appropriate changes in the system
d. the target or "normal" value of the parameter in question

5. What does it mean to say that an animal is a heterothermic endotherm?
   a. Its body temperature can vary, but it produces heat from its own tissues.
   b. Its body temperature varies because it gains most of its heat from sources outside its body.
   c. Its body temperature does not vary, because it produces heat from its own tissues.
   d. Its body temperature does not vary, even though it gains most of its heat from sources outside its body.

6. Which of the following is an advantage that ectotherms have over endotherms of the same size?
   a. They require much less food.
   b. They can save energy in cold weather by hibernating (entering torpor for long periods).
   c. They can remain active in cold weather or at nighttime—when temperatures cool.
   d. They have higher metabolic rates and grow much more quickly.

TEST YOUR UNDERSTANDING

Answers are available in Appendix B

1. Why is epithelium a particularly important tissue in achieving homeostasis with respect to temperature?
2. The metabolic rate of a frog in summer (at 35°C) is about eight times higher than in winter (at 5°C). Compare and contrast the individual's ability to move, exchange gases, and digest food at the two temperatures. During which season will the frog require more food energy, and why?
3. Consider the following:
   - Absorptive sections of digestive tract
   - Capillaries
   - Beaks of Galápagos finches
   - Fish gills

   In each case, how does the structure relate to the function?
4. Why is the surface area/volume ratio different in a small sphere versus a large sphere? If materials diffuse into and out of each sphere, in which case will diffusion occur more efficiently relative to the volume? Explain your answer.
5. Consider a day in which daytime temperatures reach 30°C and nighttime temperatures drop to 18°C. Analyze how an ant might gain and lose heat by conduction, convection, radiation, and evaporation to avoid overheating during the day and escape cold-induced lethargy in the early morning and evening.
6. Why is "negative" an appropriate adjective to describe the feedback that occurs in homeostatic systems?

APPLYING CONCEPTS TO NEW SITUATIONS

Answers are available in Appendix B

1. When food is scarce, bigger *Geospiza fortis* (medium ground finches) win contests over seeds. Biologists have documented that there is strong natural selection in favor of large body size under these conditions, and that the finch population evolves in response. If so, why aren't *G. fortis* a lot bigger than they are?
2. What data would you need to collect in order to document that adaptation is occurring in a population of lizards, as global warming intensifies?
3. An engineer has to design a system for dissipating heat from a new type of car engine that runs particularly hot. Recall that heat is gained and lost as a function of surface area. Suggest ideas to consider that are inspired by biological structures with exceptionally high surface area/volume ratios.
4. Suppose a friend of yours is trying to decide whether to buy a pet turtle or a pet mouse. In making the decision, all he cares about is how much it will cost to feed the animal. Your friend says he can't decide because the two animals weigh the same and their food costs the same per pound. As a biologist, what's your advice?

Terrestrial animals lose water every time they breathe and urinate. For many animals, drinking is an important way to gain water and achieve homeostasis. This chapter explores how terrestrial and aquatic animals maintain water balance.

42 Water and Electrolyte Balance in Animals

KEY CONCEPTS

- Freshwater, marine, and terrestrial habitats pose different challenges to animals with regard to maintaining water and electrolyte balance.
- In marine animals, specialized epithelial cells have membrane proteins that remove excess salt (NaCl) from the body so that it can be excreted. The same types of cells are found in the kidneys of mammals.
- In terrestrial insects, the hindgut and Malpighian tubules are responsible for excreting water-soluble waste products and achieving homeostasis with respect to water and electrolyte concentrations.
- In terrestrial vertebrates, the kidney is responsible for excreting water-soluble waste products and achieving homeostasis with respect to water and electrolyte concentrations.

The chemical reactions that make life possible occur in an aqueous solution. If the balance of water and dissolved substances in the solution is disturbed, those chemical reactions—and life itself—may stop. Humans can stay alive for weeks without eating but survive just three days without drinking water. If hurricanes introduce enough freshwater to the ocean shore to disrupt normal salt concentrations, marine animals die. Maintaining water balance is a matter of life or death.

An animal achieves water balance when its intake of water equals its loss of water. Water balance is an important element in homeostasis—the ability to keep cells and tissues in constant and favorable conditions. Both water intake and water loss must be carefully controlled if the individual is to function at optimum levels. In humans, even relatively minor disruptions in water balance can lead to dehydration and symptoms such as headache, dizziness, muscle weakness, and fatigue (sleepiness).

Water balance is intimately associated with sustaining a balanced concentration of electrolytes throughout the body. An **electrolyte** is a compound that dissociates into ions when dissolved in water. Electrolytes got their name because they conduct electrical current.

In many animals, the most abundant electrolytes are sodium (Na^+), chloride (Cl^-), potassium (K^+), and calcium (Ca^{2+}). Cells require precise concentrations of these ions to function normally. In humans, electrolyte imbalances can lead to muscle spasms, confusion, irregular heart rhythms, fatigue, paralysis, or even death.

This chapter is focused on a single question: How do animals maintain water and electrolyte balance? Answering it will introduce you to some of the most complex and important homeostatic systems known.

✔ When you see this checkmark, stop and test yourself. Answers are available in Appendix B.

42.1 Osmoregulation and Osmotic Stress

In organisms, electrolytes and water move by diffusion and osmosis—two processes introduced in Chapter 6.

- **Diffusion** is the movement of substances from regions of higher concentration to regions of lower concentration.
- **Osmosis** is a special case of diffusion. It is the movement of water from regions of higher water concentration to regions of lower water concentration, across a selectively permeable membrane.

A **selectively permeable membrane**, such as a phospholipid bilayer, is a membrane that some solutes can cross more easily than other solutes can. When either diffusion or osmosis occur, ions and molecules move along their **concentration gradient**.

Figure 42.1a illustrates how dissolved substances, or **solutes**, move down their concentration gradients via diffusion across a selectively permeable membrane. When the solutes are randomly distributed throughout the solutions on both sides of the membrane, an equilibrium is established. Molecules continue to move back and forth across the membrane at equilibrium, but at equal rates.

As **Figure 42.1b** shows, water can also move down its concentration gradient. The concentration of dissolved substances in a solution, measured in moles per liter, is the solution's **osmolarity**. When dissolved substances are separated by a selectively permeable membrane and the solutes cannot cross that membrane, water moves from areas of lower osmolarity—that is, solute concentrations are lower—to areas of higher osmolarity—that is, solute concentrations are higher.

Diffusion and osmosis affect animals differently in marine, freshwater, and terrestrial habitats. As a result, these environments pose different challenges to animals in maintaining water and electrolyte balance.

What Is Osmotic Stress?

Osmotic stress occurs when the concentration of dissolved substances in a cell or tissue is abnormal. It means that water and solute concentrations are different from their set point.

Organisms respond to osmotic stress by osmoregulating, just as they respond to heat or cold stress by thermoregulating (see Chapter 41). **Osmoregulation** is the process by which living organisms control the concentration of water and salt in their bodies.

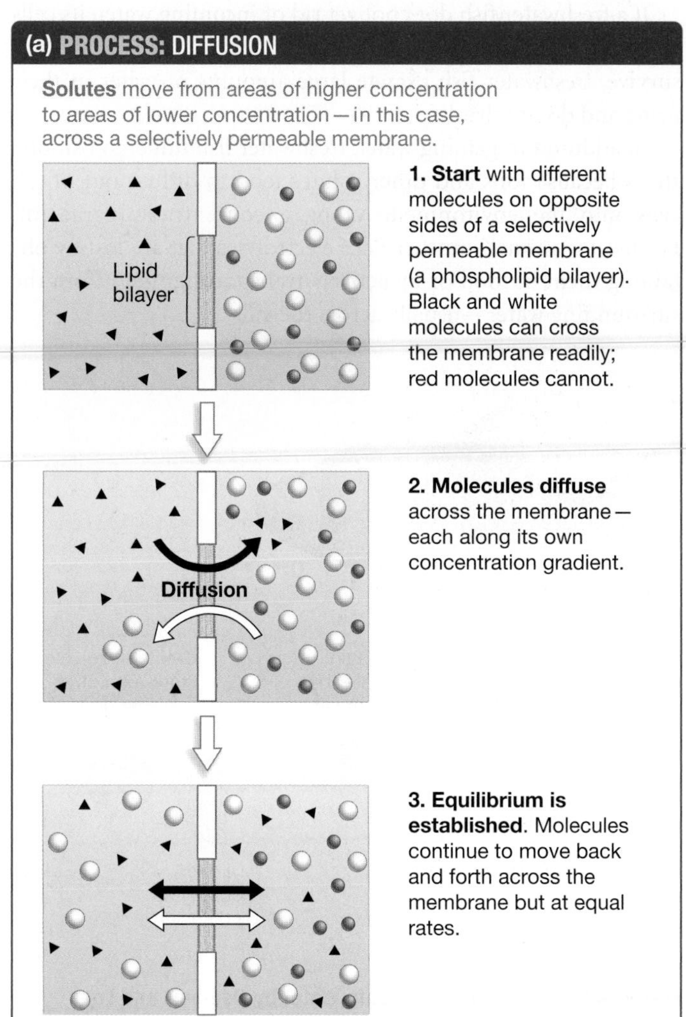

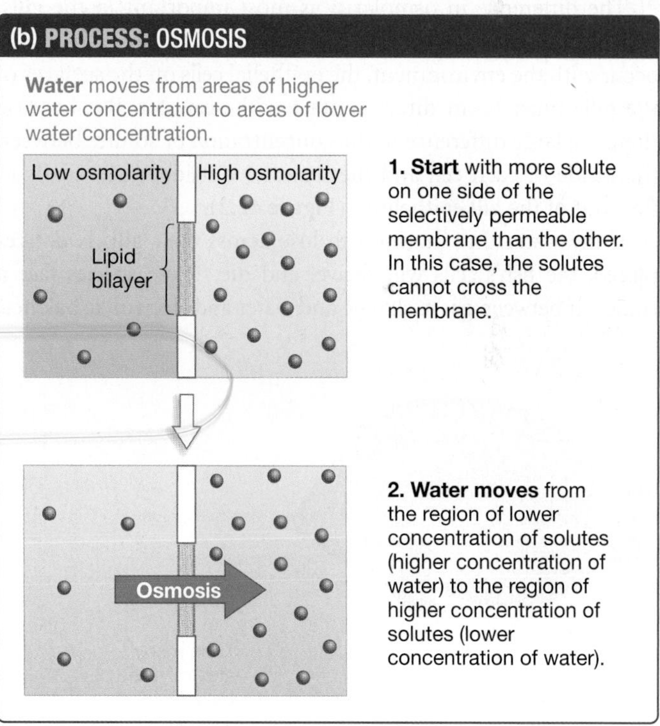

FIGURE 42.1 Solutes Move Down a Concentration Gradient via Diffusion; Water Moves Down a Concentration Gradient via Osmosis. (a) Diffusion occurs any time a solute is at higher concentration in one location than another. **(b)** Osmosis is a special case of diffusion involving the movement of water across a selectively permeable membrane.

✔**QUESTION** Why doesn't the presence of red molecules on the right side of part (a) affect the movement of the black and white molecules?

Not all animals osmoregulate, however, because not all encounter osmotic stress. For marine species such as sponges, jellyfish, and flatworms, achieving homeostasis with respect to water and electrolyte balance is straightforward. Seawater is a fairly constant ionic and osmotic environment, and it nearly matches the normal electrolyte concentrations found within these animals.

To use vocabulary introduced in Chapter 6, tissues in these species are **isotonic** with respect to seawater. Stated another way, solute concentrations inside and outside these animals are equal. As a result, diffusion and osmosis don't alter water and electrolyte balance and induce osmotic stress.

These species are **osmoconformers**. Their set point for water and electrolyte concentration closely matches their environment.

Osmotic Stress in Seawater

In contrast to most marine invertebrates, most marine fish are **osmoregulators**. Fish actively regulate osmolarity inside their bodies to achieve homeostasis.

Marine fish have to osmoregulate because their tissues are **hypotonic** relative to salt water—the solution inside the body contains fewer solutes than does the solution outside.

The difference in osmolarity is most important in the gills, which are organs involved in gas exchange. For gas exchange to occur with the environment, the epithelial cells on the surfaces of the gills must be in direct contact with seawater. But because there is a large difference in the concentration of solutes between the inside of each cell and the seawater outside, water tends to flow out of the gill epithelium (**Figure 42.2**).

If the water that marine fish lose across their gills is not replaced, the fish's cells will shrivel and die. These species face a trade-off between gas exchange and water and electrolyte balance.

Marine fish replace the lost water by drinking large quantities of seawater. Drinking brings in excess electrolytes, however. Electrolyte balance is thrown even further out of whack because ions and other solutes diffuse into the fish across the gills, following a concentration gradient from seawater to tissues.

To rid themselves of these excess electrolytes, marine fish have to actively pump ions out of their bodies and back into seawater, using membrane proteins found in the gill epithelium.

Osmotic Stress in Freshwater

Freshwater animals osmoregulate in a dramatically different environment than the ocean. Marine fish are under osmotic stress because they lose water and gain salt; freshwater animals are under osmotic stress because they gain water and lose salt.

Why? In the gills of freshwater fish, epithelial cells contain more solutes than the solution outside. To use the vocabulary introduced in Chapter 6, the tissue is **hypertonic** relative to the surrounding water. As a result, the epithelial cells gain water via osmosis (**Figure 42.3**). When this water moves from the epithelium into adjacent tissues via osmosis, it puts the rest of the body under osmotic stress. Just as in marine fish, there is a trade-off between gas exchange and osmoregulation.

If a freshwater fish does not get rid of incoming water, its cells will burst and the individual will die. To achieve homeostasis and survive, freshwater fish excrete large amounts of water in their urine and do not drink.

In addition to gaining water, freshwater fish undergo osmotic stress because ions and other solutes tend to diffuse out of gill cells into the environment, along a concentration gradient. Freshwater animals must replace electrolytes that are lost by obtaining them in food or by actively transporting them from the surrounding water—usually across the gills.

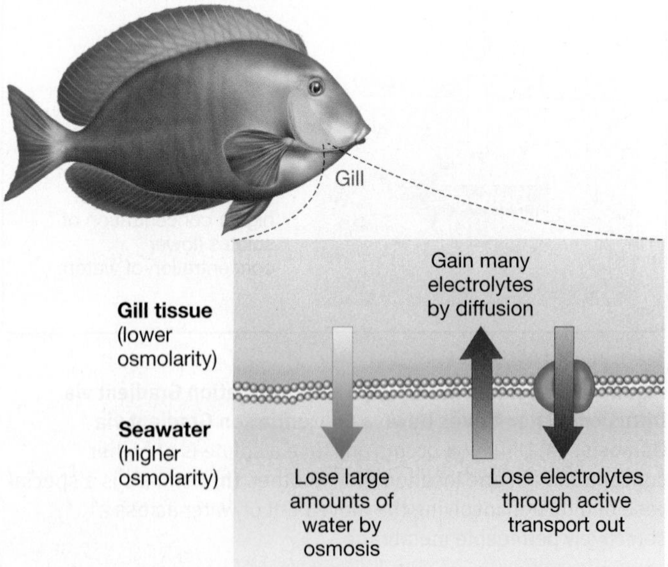

Gill

Gill tissue (lower osmolarity)

Seawater (higher osmolarity)

Gain many electrolytes by diffusion

Lose large amounts of water by osmosis

Lose electrolytes through active transport out

FIGURE 42.2 Marine Fish Lose Water by Osmosis and Gain Electrolytes by Diffusion.

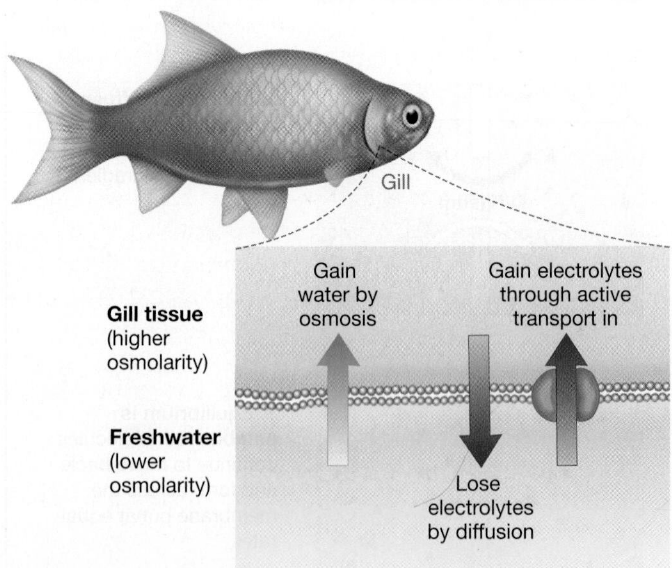

FIGURE 42.3 Freshwater Fish Gain Water by Osmosis and Lose Electrolytes by Diffusion.

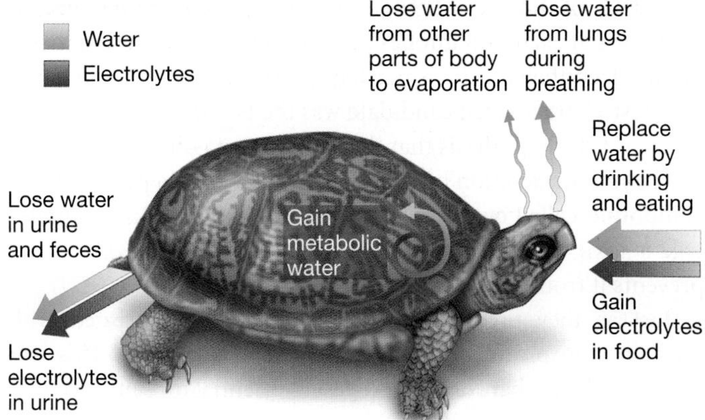

FIGURE 42.4 Maintaining Water and Electrolyte Balance in Terrestrial Environments Is a Challenge. In terrestrial environments, animals lose water by evaporation from the body surface and as water vapor during breathing. Electrolytes are lost primarily in the urine and feces.

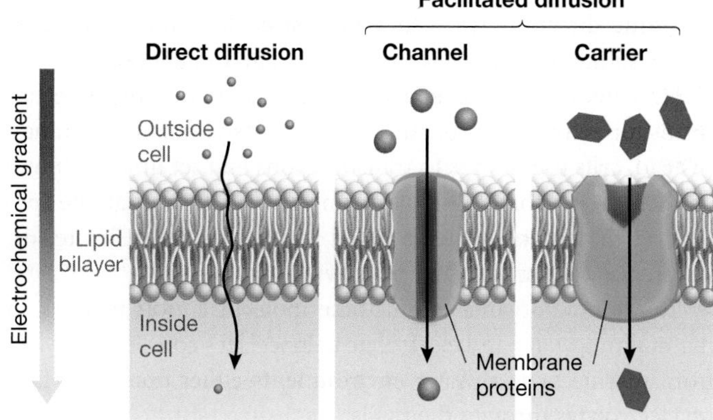

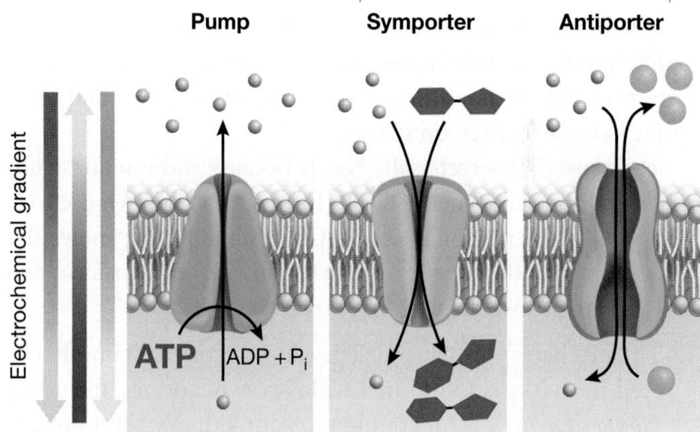

FIGURE 42.5 Mechanisms of Passive and Active Transport. **(a)** In passive transport, solutes diffuse along their electrochemical gradient either directly or through a membrane protein. **(b)** In active transport, a pump uses ATP to move solutes against their electrochemical gradients.

Osmotic Stress on Land

What about land animals? In terms of water balance, terrestrial environments are similar to the ocean. Land animals constantly lose water to the environment, just as many marine animals do. In this case, however, the process involved is not osmosis but evaporation (see Chapter 41).

The epithelial cells that line a turtle's lung and a fruit fly's gas exchange structures have a moist surface, in order to protect the integrity of their plasma membranes. Because the atmosphere is almost always drier than the wet gas exchange surface, terrestrial animals lose water by evaporation. Once again, there is a trade-off between breathing and water and electrolyte balance.

Water balance is further complicated because all terrestrial animals lose water in the form of urine, and some species lose additional water when they sweat or pant to lower their body temperature (**Figure 42.4**). The lost water has to be replaced by drinking, ingesting water in food, or gaining metabolic water—the H_2O produced during cellular respiration (see Chapter 9).

How Do Cells Move Electrolytes and Water?

What molecular mechanisms allow animals to cope with the diverse challenges they face in maintaining water and electrolyte balance? You might recall from earlier chapters that solutes move across membranes by passive or active transport. **Passive transport** is driven by diffusion along an electrochemical gradient and does not require an expenditure of energy in the form of ATP. **Active transport**, in contrast, occurs when ATP powers the movement of a solute against its electrochemical gradient.

Because ions and large molecules such as glucose do not cross phospholipid bilayers readily, most passive transport and all active transport takes place via membrane proteins.

- In many cases, passive transport occurs through **channels**—proteins that form a pore, or opening, that selectively admits a specific ion or ions (**Figure 42.5a**).
- Passive transport also occurs via **carriers**, which are transmembrane proteins that bind a specific ion or molecule and transport it across the membrane by undergoing a conformational change.
- When solutes move via channels or carriers, **facilitated diffusion** is said to occur.

Active transport is based on membrane proteins called pumps, which change conformation when they bind ATP or are phosphorylated (**Figure 42.5b**). This energy-demanding change in shape allows them to transport ions or molecules against their concentration gradient. Chapter 6 introduced the **sodium-potassium pump**, or Na^+/K^+-ATPase, which is the most important type of pump in animals.

Once a pump has established an electrochemical gradient, **secondary active transport** can occur. Specifically, a **cotransporter** can use the energy released when an ion is transported *along* that electrochemical gradient to transport a different solute *against* its electrochemical gradient. A cotransporter that moves solutes in the same direction is called a **symporter**; a cotransporter that moves solutes in opposite directions is called an **antiporter**.

How does water move? To date, there are no known mechanisms for actively transporting water across plasma membranes. Instead, cells use pumps to transport ions and set up an osmotic gradient; water then follows by osmosis—often through the specialized membrane proteins called aquaporins (see Chapter 6). In essence, cells move water by moving salt.

To see how organisms put these molecular tools to work in maintaining water and electrolyte balance, let's consider how fish from marine or freshwater environments either excrete or take up electrolytes through their gills.

42.2 Water and Electrolyte Balance in Aquatic Environments

Whether they live in seawater or freshwater, virtually all of the 26,000 species of fish living today experience osmotic stress. Early research on how these species maintain water and electrolyte balance focused on sharks.

Sharks need to secrete salt (NaCl) because sodium and chloride ions diffuse into their gill cells from seawater, down the concentration gradients for the ions. Understanding the molecular mechanism of salt excretion in sharks turned out to have two wide-ranging consequences:

1. The mechanism is general. The salt-secreting system discovered in sharks is found in a wide array of species, including *Homo sapiens*. It is functioning in your kidneys, right now.
2. The research revealed a critically important concept in physiology. Plant and animal cells use active transport to set up a strong electrochemical gradient for one ion—typically Na^+ in animals and H^+ in plants. The sodium or proton gradient is then used to transport a wide array of other substances, without further expenditure of energy.

This is fundamental stuff. Let's dig in.

How Do Sharks Excrete Salt?

Research on salt excretion focused on an organ called the **rectal gland**, which secretes a concentrated salt solution. To determine how this gland works, researchers studied it in vitro—meaning outside the shark's body. The basic approach was to dissect rectal glands, immerse them in a solution with a defined composition and osmolarity, and analyze the fluid that the rectal gland produced in response.

Early experiments showed that normal salt excretion only occurred if the solution in the rectal gland contained ATP. This result supported the hypothesis that salt excretion is an energy-demanding activity. Ions can be concentrated only if they are actively transported against a concentration gradient. The question was, How does concentration occur?

THE ROLE OF NA^+/K^+-ATPASE An energy-demanding mechanism for salt excretion implies that a protein in the plasma membrane of epithelial cells is actively pumping Na^+, Cl^- or both. The best-characterized candidate was the Na^+/K^+-ATPase.

To test the hypothesis that the sodium-potassium pump is involved in salt excretion by sharks, biologists used a plant defense compound called **ouabain** (pronounced *WAA-bane*). This molecule is toxic to animals because it binds to Na^+/K^+-ATPase and prevents it from functioning.

Just as they predicted, rectal glands that are treated with ouabain stop producing a concentrated salt solution. This was strong evidence that Na^+/K^+-ATPase is essential for salt excretion.

A MOLECULAR MODEL FOR SALT EXCRETION Subsequent work has shown that salt excretion is a multistep process, summarized in **Figure 42.6**.

1. Na^+/K^+-ATPase pumps sodium ions out of epithelial cells across the basolateral surface, into the extracellular fluid. The pump creates an electrochemical gradient favoring the diffusion of Na^+ into the cell. This "master gradient" allows the cell to transport other ions without an additional expenditure of energy.
2. Na^+, Cl^-, and K^+ all enter the cell, powered by the Na^+ "master gradient."
3. As Cl^- builds up inside the cell, a chloride channel located in the apical membrane allows Cl^- to diffuse down its concentration gradient into the lumen of the gland.
4. Following their charge and concentration gradients, sodium ions diffuse into the lumen of the gland through spaces between the cells.

A COMMON MOLECULAR MECHANISM UNDERLIES MANY INSTANCES OF SALT EXCRETION In many animals, epithelial cells that transport sodium and chloride ions contain the same combination of membrane proteins found in the shark rectal gland. These species include:

- marine birds and reptiles that drink salt water and excrete NaCl via glands in their nostrils;
- marine fish that excrete salt from their gills (see Figure 42.2);
- mammals that transport salt in their kidneys.

Research on the shark rectal gland also had an unforeseen benefit for biomedical research. Several years after the shark chloride channel was characterized, investigators identified a human protein called cystic fibrosis transmembrane regulator (CFTR). As Chapter 6 pointed out, cystic fibrosis is the most common genetic disease in populations of northern European extraction. Although the disease was known to be associated with defects in the CFTR protein, no one knew what the molecule did.

When investigators realized that the amino acid sequence of CFTR is 80 percent identical to that of the shark chloride chan-

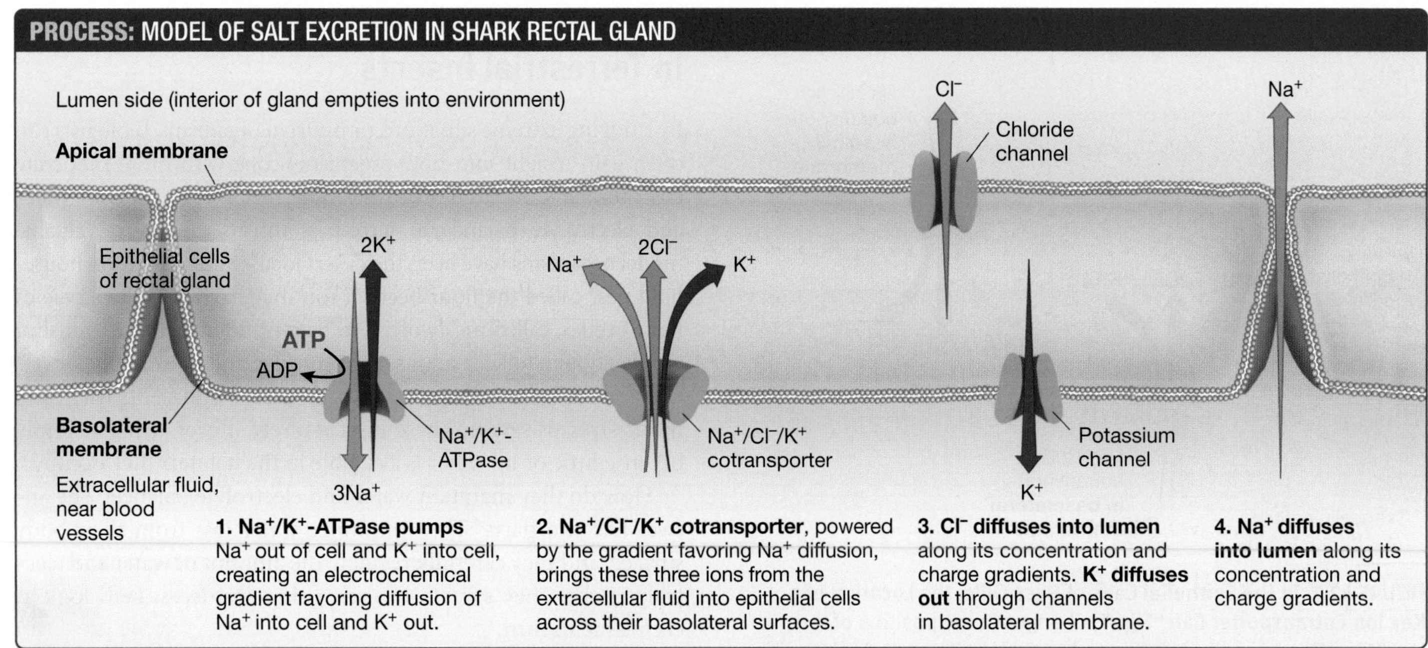

FIGURE 42.6 The Shark Rectal Gland Rids the Body of Excess Salt.

✔**QUESTION** Which of these membrane proteins are involved in (1) active transport, (2) secondary active transport, or (3) passive transport?

nel, it was their first hint that CFTR is involved in Cl^- transport. Subsequent studies supported the hypothesis that cystic fibrosis results from a defect in a chloride channel. In this way, studies on water and electrolyte balance in sharks shed light on an important human disease.

How Do Freshwater Fish Osmoregulate?

The molecular mechanisms of salt balance in marine animals are now well known. How freshwater fish achieve homeostasis with respect to electrolytes is still unclear, however.

Recall from Figure 42.3 that freshwater fish have to cope with an osmotic stress that is opposite the challenge facing marine fish. Freshwater fish lose electrolytes across their gill epithelium by diffusion across a concentration gradient. To maintain homeostasis, they have to actively transport ions back into the body across the gill epithelium. How do they do this?

SALMON AND SEA BASS AS MODEL SYSTEMS To understand how freshwater fish gain electrolytes, researchers have focused on sea bass and several species of salmon. Over the course of a lifetime, individuals of these species move between salt water and freshwater. Thus, they move between environments with dramatically different osmotic stresses.

In marine fish, specialized cells in the gill epithelium, called chloride cells, move salt using the combination of membrane proteins illustrated in Figure 42.6. When sea bass and salmon are in salt water, these cells are abundant and active. What happens to these cells when individuals move into freshwater? Do the changes that occur provide any insight into how they acclimatize to the new environment and avoid dying of osmotic stress?

A FRESHWATER CHLORIDE CELL? Although research is continuing at a brisk pace, recent results support the hypothesis that there is a freshwater version of the classical chloride cell: one that moves ions in the opposite direction of the saltwater version. Instead of excreting salt, these cells import it.

Three lines of evidence have accumulated to date:

1. ***Osmoregulatory cells may be in different locations.*** Young salmon taken from freshwater versus salt water have cells with Na^+/K^+-ATPase in different locations on the gills. Adult salmon from freshwater versus salt water show the same pattern, and similar changes have been observed in other fish species that switch between freshwater and saltwater habitats. The pattern suggests that when the nature of osmotic stress changes, the nature of the gill epithelium changes. Specifically, active pumping of ions takes place in a different population of cells in seawater versus freshwater.
2. ***Different forms of Na^+/K^+-ATPase may be activated.*** The salmon genome contains genes for several different forms of the Na^+/K^+-ATPase. There is now strong evidence that different forms are activated when individuals are in salt water versus freshwater.
3. ***The orientation of key transport proteins "flips."*** In sea bass, researchers have been able to stain epithelial cells to determine the location of the cotransporter illustrated in Figure 42.6—the one that brings Na^+, Cl^-, and K^+ into the

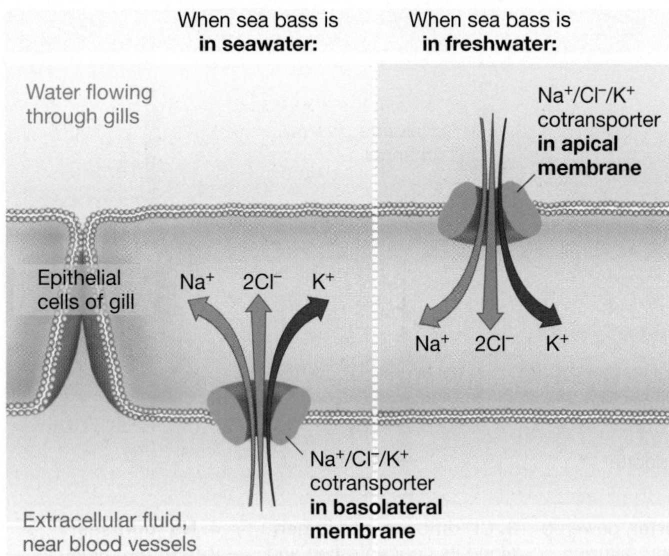

FIGURE 42.7 In the Epithelial Cells of Bass Gills, the Location of a Key Ion Cotransporter Can "Flip." Changes in the position of the $Na^+/Cl^-/K^+$ cotransporter help sea bass deal with osmotic stress in both seawater and freshwater environments.

cell. When individuals are in seawater, the protein is located in the basolateral side of chloride cells. But when individuals are in freshwater, the protein is located in the apical side (**Figure 42.7**).

Taken together, the data suggest that freshwater fish have a freshwater version of the chloride cell, with pumps and transporters arranged in the opposite orientation to that found in marine fish. The evidence is far from conclusive, however, and research continues. Identifying the mechanisms of electrolyte uptake in freshwater fish is an important challenge for researchers who want to know how aquatic organisms cope with osmotic stress.

CHECK YOUR UNDERSTANDING

If you understand that . . .

- Marine fish lose water by osmosis. To replace it, they have to drink salt water.
- Marine fish have to rid themselves of salt. They gain salt when they drink salt water or when sodium and chloride ions diffuse into their cells along a concentration gradient.
- Freshwater fish gain water by osmosis. They have to rid themselves of excess water by urinating.
- Freshwater fish lose electrolytes to the surrounding water by diffusion. They gain electrolytes in their food and by active transport from the surrounding water.

✔ **You should be able to . . .**

Predict what happens when epithelial cells in the gills of a freshwater fish are treated with ouabain—a molecule that poisons Na^+/K^+-ATPase.

Answers are available in Appendix B.

42.3 Water and Electrolyte Balance in Terrestrial Insects

By studying extreme situations or unusual organisms, biologists can often gain insight into how organisms cope with more moderate environments. In studies on the molecular mechanisms of water and electrolyte balance in terrestrial insects, the most valuable model organisms have been the desert locust and a common household pest called the flour beetle. (You may have seen the larvae of flour beetles, called mealworms, in bags of flour that were not shut tightly enough to keep adults from entering and breeding.)

Desert locusts and flour beetles live in environments where osmotic stress is severe. These insects rarely, if ever, drink—simply because little or no water is available in the habitats they occupy.

How do they maintain water and electrolyte balance? The answer has two parts: They minimize water loss from their body surface, and they carefully regulate the amount of water and electrolytes that they excrete in their urine and feces. Let's look at each issue in turn.

How Do Insects Minimize Water Loss from the Body Surface?

As Chapter 44 will show, terrestrial animals breathe by exposing an extremely thin layer of epithelium to the atmosphere. Oxygen diffuses into this epithelium, and carbon dioxide diffuses out. But water constantly leaks across the thin respiratory surface and is lost to the atmosphere via evaporation.

Evaporation from the body surface itself is another threat—a particular challenge to insects, because they are small. As Chapter 41 emphasized, small organisms have a high surface area/volume ratio. Insects have a relatively large surface area from which to lose water but a small volume in which to retain it.

How do desert locusts, flour beetles, and other insects minimize water loss during gas exchange? In these species, gas exchange occurs across the membranes of epithelial cells that line the **tracheae** (pronounced *TRAY-kee-ee*), an extensive system of tubes. The insect tracheal system connects with the atmosphere at openings called **spiracles** (**Figure 42.8a**). Muscles just inside each spiracle open or close the pore, much as guard cells open or close the pores in plant leaves and stems.

When investigators manipulated insects called *Rhodnius* so that their spiracles stayed open and placed the animals in a dry environment, they died within three days. These data support the hypothesis that the ability to close spiracles is an important adaptation for minimizing water loss during respiration. If an insect is under osmotic stress, it may be able to close its spiracles and wait until conditions improve before resuming activity.

Figure 42.8b shows how insects minimize evaporation from the surface of their bodies. This diagram is a cross-sectional view of the exoskeleton of an insect, which consists of a tough, nitrogen-containing polysaccharide called chitin and layers of protein. This combination of chitin and protein is known as **cuticle**.

As the figure shows, cuticle includes a layer of wax on its surface. Recall from Chapter 6 that waxes, a type of lipid, are highly

(a) Spiracles can be closed to minimize water loss from tracheae.

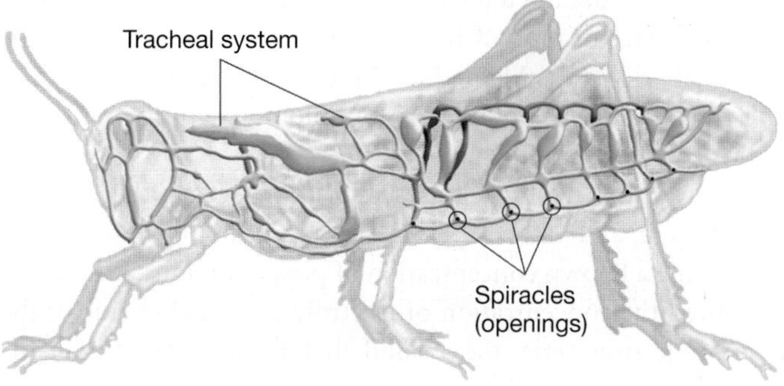

(b) Except at spiracles, the insect body is covered with wax.

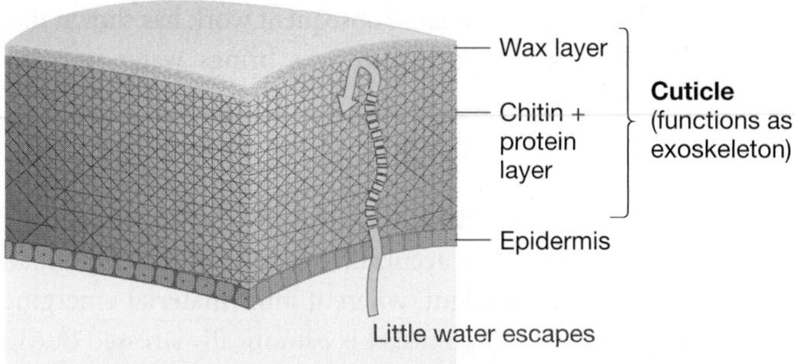

FIGURE 42.8 In Desert Locusts, Adaptations Limit Water Loss during Respiration and from the Body Surface.

✔**QUESTION** In desert grasshoppers, spiracles are located in a row along the underside of the abdomen. Why would this location help minimize water loss?

hydrophobic and thus highly impermeable to water. Researchers who removed the wax from insect exoskeletons have confirmed that the rate of water loss from the body surface increases sharply. Based on this observation, the wax layer is interpreted as an adaptation that minimizes evaporative water loss.

Types of Nitrogenous Wastes: Impact on Water Balance

Animal cells contain amino acids and nucleic acids that are used to synthesize proteins, RNA, and DNA. Both amino acids and nucleic acids are nitrogenous (nitrogen-containing) monomers.

If amino acids and nucleic acids are present in excess of a cell's needs, they are broken down in catabolic reactions that result in the production of **ammonia** (NH_3). Ammonia, a strong base, readily gains a proton to form an ammonium ion (NH_4^+). Ammonia is toxic to cells, because at high concentrations it raises the pH of intracellular and extracellular fluids enough to poison enzymes.

FORMS OF NITROGENOUS WASTE VARY AMONG SPECIES How do animals get rid of ammonia safely and efficiently? Different species solve the problem in different ways (see **Table 42.1**).

- In freshwater fish, ammonia is diluted to low concentration and excreted in a watery urine.
- In freshwater and saltwater fish, ammonia diffuses across the gills into the surrounding water along a concentration gradient.
- In humans, enzyme-catalyzed reactions convert ammonia to a much less toxic compound called urea, which is excreted in urine.
- In birds, reptiles, and terrestrial arthropods, reactions convert ammonia to **uric acid**, the white, paste-like substance that you have probably seen in bird feces.

Uric acid is a particularly interesting form of nitrogenous waste. Compared with urea and ammonia, uric acid is extremely insoluble in water—which explains why it is so difficult to wash bird droppings off a car. As a result, uric acid provides a mechanism for birds, snakes, lizards, and terrestrial arthropods to rid themselves of excess nitrogen while losing a minimum amount of water. Many birds and some insects, in fact, do not produce any urine at all.

SUMMARY TABLE 42.1 **Attributes of Nitrogenous Wastes Produced by Animals**

Attribute	Ammonia	Urea	Uric Acid
Solubility in water (moles/liter)	high	medium	very low
Water loss (amount required for excretion of waste)	high	medium	very low
Energy cost (amount of ATP required)	low	high	high
Toxicity	high	medium	low
Groups where it is the primary waste	fish, aquatic invertebrates	mammals,* sharks	birds† and other reptiles, most terrestrial insects and spiders
Method of synthesis	breakdown of amino acids and nucleic acids	synthesized in liver, starting with amino groups from amino acids	synthesis starts with amino acids and nucleic acids
Method of excretion	in urine and diffuses across gills	in urine (mammals); diffuses across gills (sharks)	in feces (in birds, uric acid is derived from the urine but excreted with the feces)

*Mammals also excrete a small amount of uric acid, synthesized from excess nucleic acids.
†Birds also excrete a small amount of ammonia.

WHY DO NITROGENOUS WASTES VARY AMONG SPECIES? The type of nitrogenous waste produced by an animal correlates with its lineage—its evolutionary history. For example, mammals excrete urea while reptiles (including birds) and insects excrete uric acid (Table 42.1).

Evolutionary history is not the entire story, however. Waste production also correlates with the habitat that a species occupies, and thus the amount of osmotic stress it endures.

- Terrestrial birds conserve water by excreting about 90 percent of their nitrogenous waste as uric acid and only 3–4 percent as NH_3, but ducks and other birds with ready access to water excrete just 50 percent of their excess nitrogen as uric acid and 30 percent as NH_3.
- Tadpoles are aquatic and excrete ammonia, but adult frogs are terrestrial and excrete urea.
- Production of urea and uric acid is particularly common in animals that live in dry habitats.

To make sense of these observations, biologists point out that there is a fitness trade-off between the energetic cost of excreting urea or uric acid and the benefit of conserving water. Ammonia excretion requires a large water loss but little energy expenditure, because the molecule isn't processed by enzymes. Uric acid excretion, in contrast, requires almost no loss of water but a sophisticated series of enzyme-catalyzed, energy-demanding reactions. Different trade-offs are favored in different environments.

Maintaining Homeostasis: The Excretory System

For insects, minimizing water loss is only half the battle in avoiding osmotic stress. To maintain homeostasis, insects must also carefully regulate the composition of a blood-like fluid called **hemolymph**. Hemolymph is pumped by the heart and transports electrolytes, nutrients, oxygen, and waste products.

How do insects regulate the composition of the hemolymph? This question is important for three reasons: **(1)** Nitrogenous wastes have to be removed before they build up to toxic concentrations, **(2)** excess electrolytes must be excreted before they lead to osmotic stress, and **(3)** water balance must be regulated constantly.

To maintain water and electrolyte balance, insects rely on **Malpighian tubules**, which are an excretory organ, and on their hindgut—the posterior portion of their digestive tract (**Figure 42.9a**).

FILTRATE FORMS IN THE MALPIGHIAN TUBULES As the enlarged section of Figure 42.9a shows, Malpighian tubules have a large surface area, are in direct contact with the hemolymph, and empty into the hindgut. The Malpighian tubules are responsible for forming a **filtrate** from the hemolymph. This "pre-urine" then passes into the hindgut, where it is processed and modified prior to excretion.

To explore how the Malpighian tubules work, biologists collected fluid from the lumen of Malpighian tubules in mealworms and compared the composition of the filtrate with that of hemolymph from the same individuals. The two solutions were roughly isotonic, but not identical.

How do they differ? Compared to hemolymph, the solution inside the tubules has a lower concentration of sodium ions and a higher concentration of K^+. This observation suggests that the epithelial cells in the Malpighian tubules are relatively impermeable to sodium ions but contain a pump that actively transports potassium ions into the tubules.

To test this hypothesis, researchers dissected the tubules, rinsed them, and bathed their interior and exterior in a solution containing a known concentration of potassium ions. When they measured the concentration of electrolytes on either side of the membrane over time, they found that K^+ accumulated in the tubule lumen, against its concentration gradient.

This result supported the hypothesis that cells in the membranes of Malpighian tubules contain a pump that transports potassium ions into the lumen of the organ. Subsequent work has shown that a high concentration of potassium ions brings water into the tubules by osmosis. Other electrolytes and nitrogenous wastes then diffuse into the tubules along their concentration gradients.

THE HINDGUT: SELECTIVE REABSORPTION OF ELECTROLYTES AND WATER The solution that accumulates inside the Malpighian tubules flows into the hindgut, where it joins material emerging from the digestive tract. If an insect is osmotically stressed due to a shortage of electrolytes and water, electrolytes and water from the filtrate are reabsorbed in the hindgut and returned to the hemolymph. Reabsorption results in formation of a hypertonic final urine, conservation of water, and efficient elimination of nitrogenous wastes.

In desert locusts, flour beetles, and other species in extremely dry environments, 80 to 95 percent of the water in the filtrate is recovered and kept inside the body. The ability to recover this water allows these insects to live in dry habitats such as deserts and flour bins. How does reabsorption happen?

The mechanism involves a series of specific membrane pumps and channels, not unlike the system found in the chloride cells of fish. To study it, researchers positioned the rectal epithelium from a desert locust as a sheet dividing two solutions. They manipulated electrolyte concentrations on either side of the rectal wall and measured changes in the solutions over time.

For example, when investigators removed K^+ and Na^+ from the solution on the lumen side of the organ, water reabsorption stopped. These data established that the hindgut's ability to recover water from urine depends on ion movement: the epithelial cells in the hindgut transport ions out of the filtrate and into the hemolymph. Water follows by osmosis, forming a concentrated urine.

How do the ions move? By poisoning the experimental membranes with ouabain, biologists confirmed that Na^+/K^+-ATPase is involved in moving ions out of the lumen and into the hemolymph. Ouabain-treated membranes continued to transport Cl^-, however. Later experiments confirmed that the rectal epithelium transports Cl^- against electrical and concentration gradients. These results support the hypothesis that insects' hindguts have two active pumps: a chloride pump and Na^+/K^+-ATPase.

Years of experiments on the locust hindgut resulted in the model in **Figure 42.9b**:

(a) Malpighian tubules produce an isotonic pre-urine.

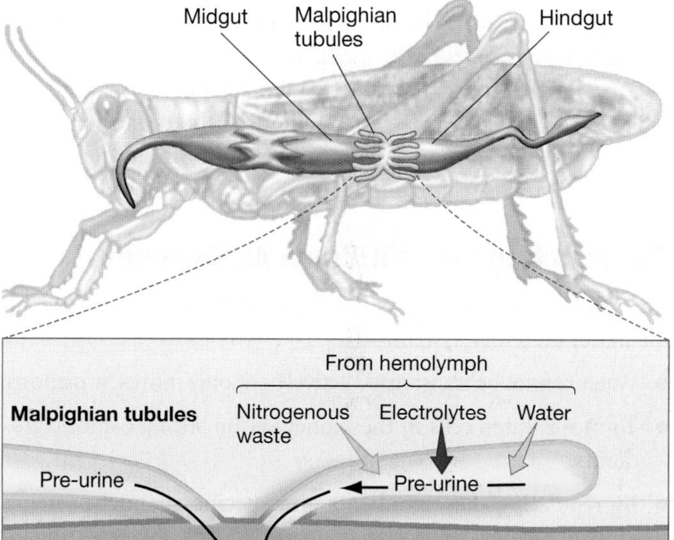

(b) Under osmotic stress, the hindgut reabsorbs electrolytes and water to form a hypertonic urine.

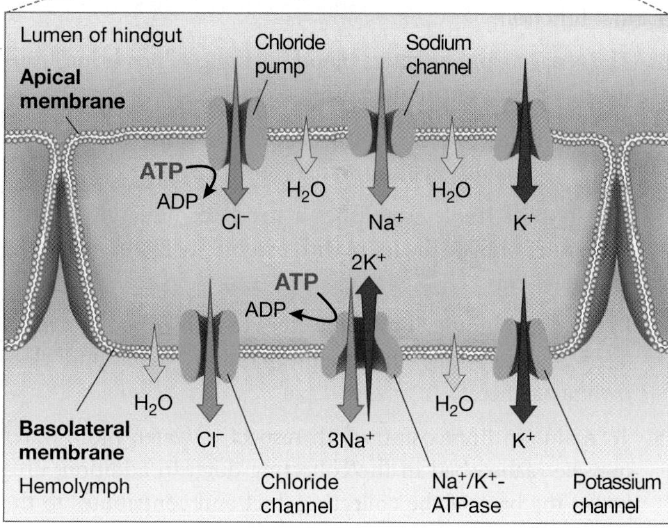

FIGURE 42.9 In Insects, Urine Forms in the Malpighian Tubules and Hindgut. (a) The isotonic filtrate that forms in the Malpighian tubules empties into the hindgut. **(b)** In the hindgut, the primary driving forces for reabsorption of electrolytes and water are chloride pumps in the apical membrane and Na^+/K^+-ATPases in the basolateral membrane.

- Cl^- is pumped into cells from the hindgut lumen, with K^+ following through potassium channels along an electrochemical gradient and water following via osmosis.
- In the basolateral membrane, Na^+/K^+-ATPase sets up electrochemical and osmotic gradients that favor movement of Cl^-, K^+, and H_2O into the hemolymph.

There is also strong experimental evidence for the existence of other channels, pumps, and cotransporters in the epithelium of the insect hindgut. These membrane proteins transport protons, ammonia, amino acids, and other molecules across the tissue and help insects maintain water and electrolyte balance.

REGULATING WATER AND ELECTROLYTE BALANCE: AN OVERVIEW Several general principles that have emerged from studies of insect excretion turn out to be relevant to vertebrate systems as well:

- Water moves only by osmosis—it is not pumped directly. Water moves between cells or body compartments via osmotic gradients that are set up by the active transport of ions.
- The formation of the filtrate is not particularly selective. Most of the molecules present in the hemolymph are also present in the Malpighian tubules.
- In contrast to filtrate formation, reabsorption is highly selective. The protein pumps and channels involved in reabsorption are highly specific for certain ions and molecules. Waste products do not pass through the rectal membrane. Instead, they remain in the urine and feces and are eliminated from the body. Only valuable ions and molecules are reabsorbed.
- In contrast to filtrate formation, reabsorption is tightly regulated. The membrane pumps and channels involved in reabsorption are activated and deactivated in response to osmotic stress. If an insect is dehydrated, virtually all of the water in the filtrate is reabsorbed. But if it has plenty to drink, reabsorption does not occur and the urine is watery and hypotonic to the individual's hemolymph. The system is dynamic and allows precise control over water and electrolyte balance.

Given the success of insects in terms of the numbers of species and individuals and the array of habitats they occupy, it is clear that their systems for maintaining water and electrolyte balance are remarkably effective.

CHECK YOUR UNDERSTANDING

If you understand that . . .

- Terrestrial insects are prone to dehydration, primarily via evaporation from their respiratory surfaces.
- Terrestrial insects have a cuticle, respiratory system, and excretory system that are designed to conserve water.
- In terrestrial insects, urine formation begins with formation of a solution that is isotonic with hemolymph, followed by selective and tightly regulated reabsorption of ions, nutrients, and water.

✔ **You should be able to . . .**

Explain how the following traits are involved in water retention:

1. Excretion of ammonia in the form of uric acid.
2. Selective reabsorption of electrolytes in the hindgut.

Answers are available in Appendix B.

42.4 Water and Electrolyte Balance in Terrestrial Vertebrates

With respect to water loss, terrestrial vertebrates face the same hazards that terrestrial insects do. Crocodiles, turtles, lizards, frogs, birds, and mammals lose water from their body surfaces and from the surface of their lungs every time they breathe. Electrolytes are lost in urine and in some species, in sweat.

To replace the water they lose, most terrestrial vertebrates drink. They also ingest electrolytes in food.

In land-dwelling vertebrates, osmoregulation occurs primarily through events that take place in the **kidney.** The kidney is responsible for water and electrolyte balance, as well as the excretion of nitrogenous wastes. In terms of function, it is analogous to the Malpighian tubules and hindgut of insects.

The Structure of the Kidney

Kidneys occur in pairs and tend to be bean shaped (**Figure 42.10a**). A large blood vessel called the renal artery brings blood that contains nitrogenous wastes into the organ; the renal vein is a large blood vessel that carries "clean" blood away.

The urine that forms in the kidney is transported via a long tube called the **ureter** to a storage organ, the **bladder**. From the bladder, urine is transported to the body surface through the **urethra**, then excreted. In most vertebrates, the kidneys are located near the dorsal (back) side of the body.

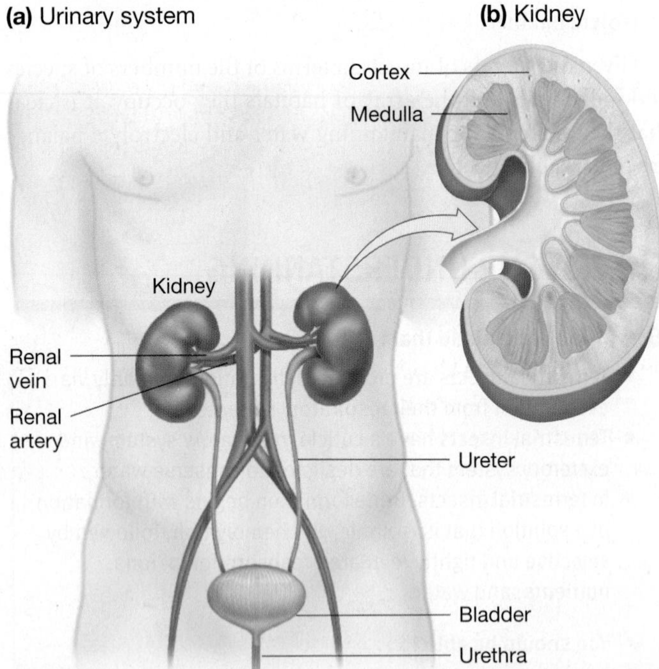

FIGURE 42.10 Anatomy of the Human Urinary System and Kidney. **(a)** In mammals, kidneys are paired and are located near the spinal column. **(b)** The kidney has an outer region called the cortex and an inner area called the medulla.

Most of the kidney's mass is made up of small structures called nephrons. The nephron is the basic functional unit of the kidney. The work involved in maintaining water and electrolyte balance occurs in the nephron.

Most of the approximately 1 million nephrons in a human kidney are located in the outer region of the organ, or **cortex** (**Figure 42.10b**). But some nephrons extend from the cortex into the kidney's inner region, or **medulla**.

The Function of the Kidney: An Overview

The nephron shares important functional characteristics with the insect excretory system:

- Water cannot be transported actively—it only moves by osmosis.
- To move water, cells in the kidney set up strong osmotic gradients.
- By regulating these gradients and specific channel proteins, kidney cells exert precise control over loss or retention of water and electrolytes.

Figure 42.11a provides a detailed view of the nephron. Note that it has four major regions, and is closely associated with a tube called the collecting duct. **Figure 42.11b** illustrates a second key point about the anatomy of the nephron: It is served by blood vessels that wrap around each of its four regions.

The four major regions and the collecting duct each have a distinct function:

1. The renal corpuscle filters blood, forming a "pre-urine" consisting of ions, nutrients, wastes, and water.
2. In the proximal tubule, epithelial cells reabsorb nutrients, vitamins, valuable ions, and water.
3. The loop of Henle establishes a strong osmotic gradient in the tissues outside the loop, with osmolarity increasing as the loop descends.
4. In the distal tubule, ions and water are reabsorbed in a regulated manner—one that helps maintain water and electrolyte balance.
5. To maintain homeostasis with respect to water, more water may be reabsorbed in the collecting duct. In addition, urea leaves the base of the collecting duct and contributes to the osmotic gradient set up by the loop of Henle.

The blood vessels that are juxtaposed with the nephron play a key role as well: they bring "dirty" blood into the nephron and then take away the molecules and ions that are reabsorbed from the initial filtrate.

Now let's delve into the details. The sections that follow trace the flow of material through each component of the nephron and out of the collecting duct.

Filtration: The Renal Corpuscle

In terrestrial vertebrates, urine formation begins in the **renal corpuscle** (literally, "kidney-little-body"). Notice in Figure 42.11a

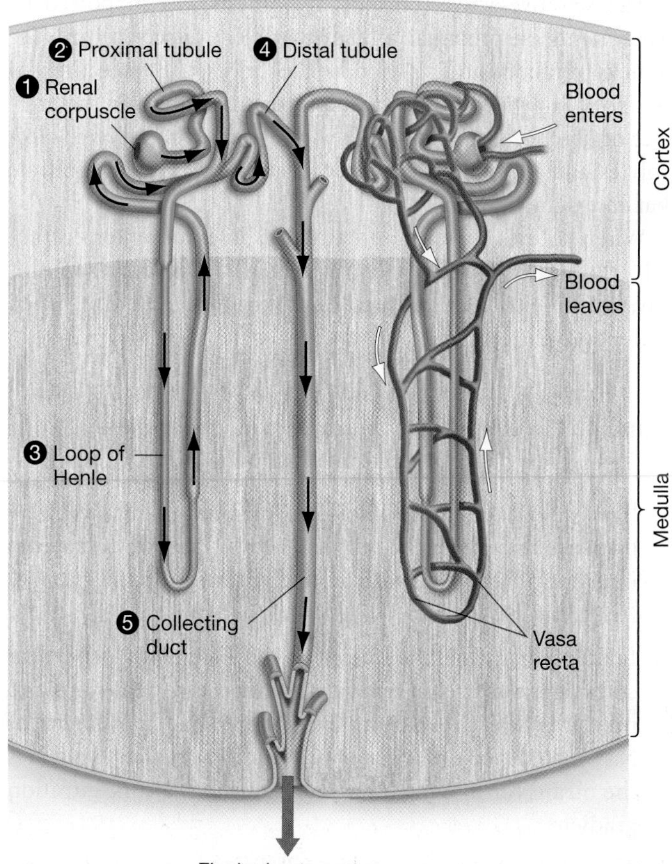

FIGURE 42.11 A Nephron Has Four Major Parts, Empties into a Collecting Duct, and Is Served by Blood Vessels. Urine formation begins in the renal corpuscle and ends in the collecting duct.

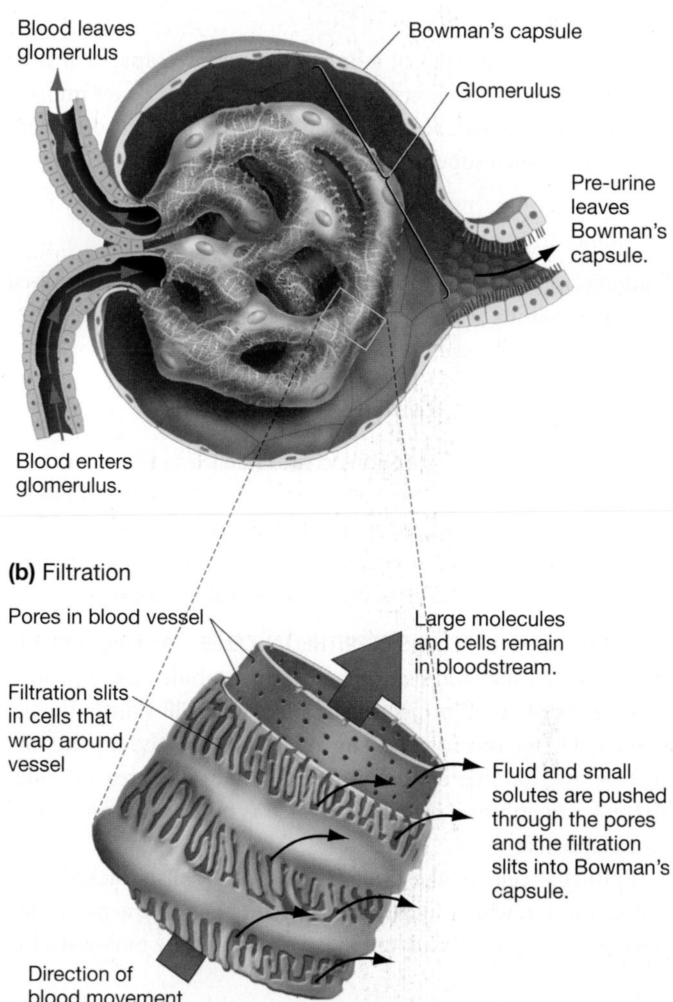

FIGURE 42.12 Urine Formation Begins When Blood Is Filtered in the Renal Corpuscle. (a) The renal corpuscle consists of Bowman's capsule and the glomerulus. **(b)** The capillaries in the glomerulus have pores and are surrounded by cells that have filtration slits. Blood pressure forces water and small molecules out of the capillaries, through the slits, and into Bowman's capsule.

that the nephron is a tube that is closed at one end and open at the other. The closed end is the beginning of the nephron; the open end is the terminus of the collecting duct.

As **Figure 42.12a** shows, the closed end of the nephron forms a capsule that encloses a cluster of tiny blood vessels, or capillaries. These vessels bring blood to the nephron from the renal artery. Collectively, the cluster of capillaries is called the **glomerulus** ("ball of yarn"). The region of the nephron that surrounds the glomerulus is named **Bowman's capsule**. Together, the glomerulus and Bowman's capsule make up the renal corpuscle.

Figure 42.12b illustrates a key feature of the glomerular capillaries: They have large pores, or openings. In addition, they are surrounded by unusual cells whose membranes fold into a series of slits and ridges.

The structure of the renal corpuscle allows it to function as a **filtration** device. Water and small solutes from the blood pass through the pores and slits into the nephron. Filtration is based on size: Proteins, cells, and other large components of blood do not fit through the pores and do not enter the nephron. They remain in the blood instead.

Stated another way, urine formation starts with a size-selective filtration step—with blood pressure supplying the force required to perform filtration. In vertebrates, blood is under higher pressure than the surrounding tissues because it is pumped by the heart through a closed system of vessels. This pressure is enough to force water and small solutes through the pores in the glomerulus, so the renal corpuscle strains large volumes of fluid without expending energy in the form of ATP.

To summarize, the renal glomerulus filters the blood to create a filtrate comprising water, electrolytes, and other small substances. During the formation of this filtrate, up to 25 percent of the water and solutes present in the blood is removed.

✔If you understand this concept, you should be able to describe the contents of blood on one side of the filter and the contents of the filtrate on the other side.

It is critical to note two additional facts about the filtration step in urine formation:

1. The renal corpuscles of a human kidney are capable of producing about 180 liters of filtrate per day. This is an impressive volume—think of 180 one-liter bottles of soft drink arranged on a supermarket shelf.
2. About 99 percent of the filtrate is recycled—only a tiny fraction of the original volume is actually excreted.

Filtering large volumes from the blood allows wastes to be removed effectively; pairing this process with reabsorption allows waste excretion to occur with a minimum of water and nutrient loss.

Reabsorption: The Proximal Tubule

Where does filtrate reabsorption occur? Fluid leaves Bowman's capsule and enters a convoluted structure called the **proximal tubule**. The fluid inside this tubule contains water and small solutes such as urea, glucose, amino acids, vitamins, and electrolytes. Some of these molecules are waste products; others are valuable nutrients.

ACTIVE TRANSPORT OCCURS IN EPITHELIAL CELLS As **Figure 42.13a** shows, the epithelial cells of the proximal tubule have a prominent series of small projections, called **microvilli** ("little shaggy hairs"), facing the lumen. The microvilli greatly expand the surface area of this epithelium. A large surface area provides space for membrane proteins that act as pumps, channels, and cotransporters.

Epithelial cells in the proximal tubule are also packed with mitochondria, which suggests that ATP-demanding active transport is occurring. Based on these observations, biologists hypothesized that the proximal tubule functions in the active transport of selected-molecules out of the filtrate.

The selective-active-transport hypothesis was supported by experiments on proximal tubules that were dissected from the kidneys of rabbits and rats and isolated in vitro. By injecting solutions of known composition into proximal tubules in the presence or absence of ATP, researchers confirmed that selected electrolytes and nutrients are actively reabsorbed from the filtrate that enters the tubules.

When solutes leave the proximal tubule and enter epithelial cells, water follows along the osmotic gradient. In this way, valuable solutes and water are reabsorbed and returned to the body.

ION AND WATER MOVEMENT IS DRIVEN BY A "MASTER GRADIENT" **Figure 42.13b** summarizes the current model of the molecular mechanisms involved in selective reabsorption:

1. Na^+/K^+-ATPase in the basolateral membranes removes Na^+ from the interior of the cell. The active transport of sodium ions out of the cell creates a gradient favoring the entry of Na^+ from the lumen.
2. In the apical membrane adjacent to the lumen, Na^+-dependent cotransporters use this gradient to remove valuable ions and nutrients selectively from the filtrate. The movement of Na^+ into the cell, *with* its concentration gradient, provides the means for moving other solutes *against* a concentration gradient.
3. The solutes that move into the cell diffuse across the basolateral membrane into nearby blood vessels.

(a) Microvilli expand surface area of lumen of proximal tubule.

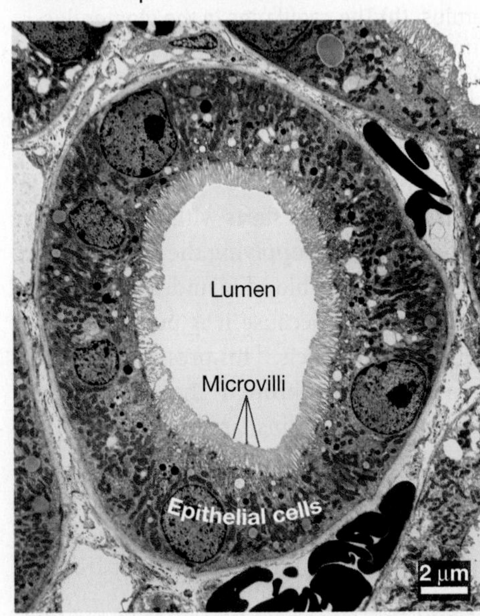

(b) Model of selective reabsorption in proximal tubules

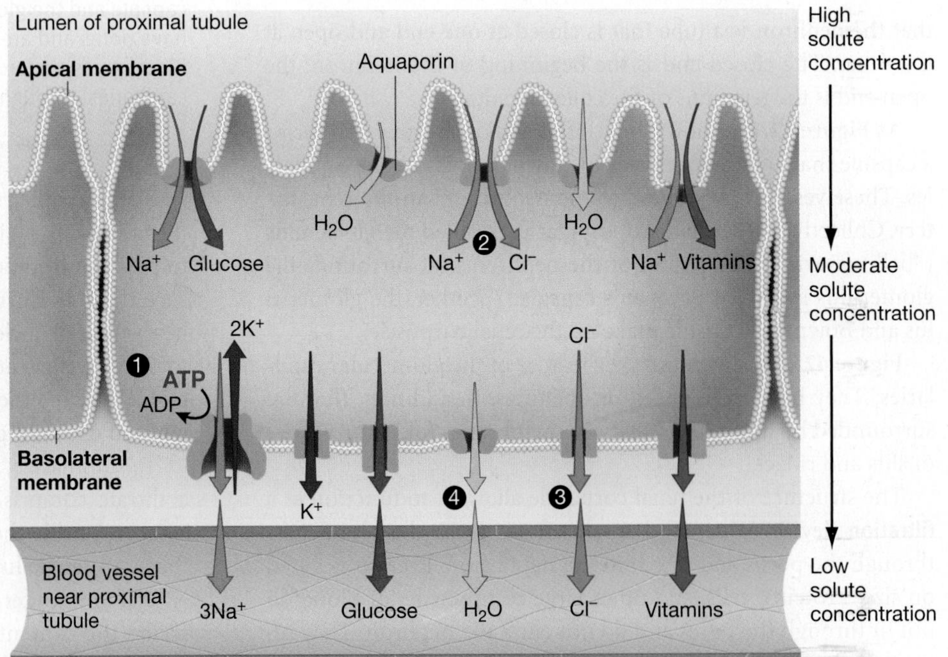

FIGURE 42.13 Water and Electrolytes Are Reabsorbed in the Proximal Tubule.

4. Water follows the movement of ions from the proximal tubule into the cell, and then out of the cell and into blood vessels.

Recent work has shown that water moves across the membranes of these epithelial cells through specialized membrane proteins called **aquaporins** ("water-pores"; see Figure 42.13b). Aquaporins are water channels: When one is open, 3 billion water molecules pass through it per second.

✔If you understand this concept, you should be able to explain why caffeine increases urine output, based on recent data indicating that caffeine may inhibit sodium ion reabsorption in the proximal tubule.

About two-thirds of the NaCl and water that is originally filtered by the renal corpuscle is reabsorbed in the proximal tubule. The osmolarity of the tubular fluid is unchanged despite this huge change in volume, however, because water reabsorption is proportional to solute reabsorption.

In effect, then, the cells that line the proximal tubule act as a recycling center. The filtration step in the renal corpuscle is based on size; the reabsorption step in the proximal tubule selectively retrieves small substances that are valuable. The pumps and cotransporters in the proximal tubule recover water, nutrients, and electrolytes but leave wastes. As the filtrate flows into the loop of Henle, it has a relatively high concentration of waste molecules and a relatively low concentration of nutrients.

Creating an Osmotic Gradient: The Loop of Henle

In mammals, the fluid that emerges from the proximal tubule enters the **loop of Henle**—named for Jacob Henle, who described it in the early 1860s. In most nephrons, the loop is short and does not leave the cortex. But in about 20 percent of the nephrons present in a human kidney, the loop is long and plunges from the cortex of the kidney deep into the medulla.

In 1942 Werner Kuhn offered a hypothesis, inspired by countercurrent heat exchangers, to explain what the loop of Henle does. Recall from Chapter 41 that a countercurrent heat exchanger is a system in which two adjacent fluids flow through pipes in opposite directions. Kuhn proposed that the loop of Henle is a countercurrent exchanger and multiplier. It doesn't exchange heat, however. Instead, it sets up an osmotic gradient.

Specifically, Kuhn proposed that the osmolarity of the fluid inside the loop of Henle is low in the cortex and high in the medulla. Further, Kuhn maintained that the osmolarity in tissues surrounding the loop mirrors the gradient inside the loop. This is a key point.

TESTING KUHN'S HYPOTHESIS A series of papers published during the 1950s supplied important experimental support for the countercurrent exchange model. **Figure 42.14** reproduces two particularly important data sets, obtained by comparing the osmolarity of kidney tissue slices. In both graphs, the *x*-axis shows the location in the kidney, from cortex to medulla. The *y*-axis indicates osmolarity, either measured as the percentage of the maximum observed or as solute concentration.

- Part (a) of the figure shows data on the osmolarity of fluid inside the loop of Henle. The vertical lines represent the range of values observed at a particular location. As predicted by Kuhn's model, a strong gradient in osmolarity exists from the cortex to the medulla.
- The data in part (b) show that outside the loop of Henle, the concentrations of Na^+, Cl^-, and urea also increase sharply from the cortex to the medulla.

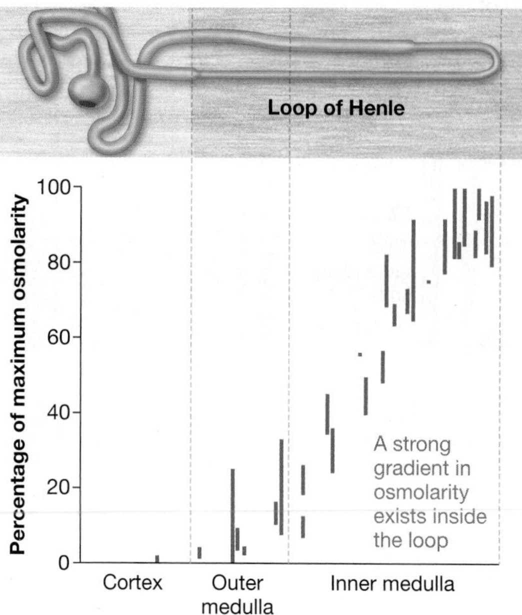

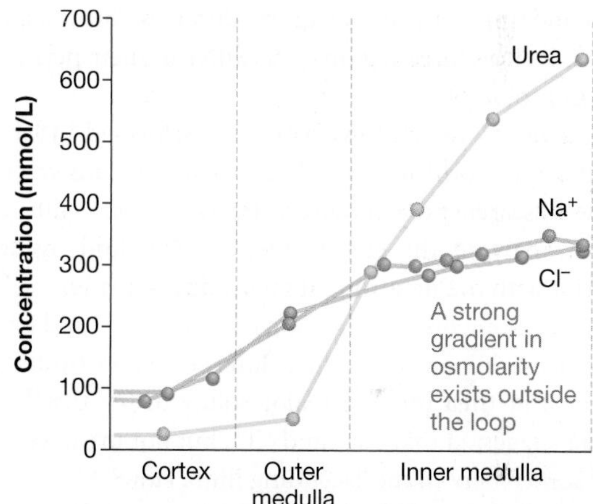

FIGURE 42.14 Data Confirm the Existence of a Strong Osmotic Gradient Both inside and outside the Loop of Henle. As the nephron plunges into the inner medulla, the concentration of dissolved solutes increases both inside **(a)** and outside **(b)** the loop of Henle.

The second observation was particularly important, because it suggested that the solutes responsible for the gradient outside the loop are Na^+, Cl^-, and urea. The change in concentration of urea turned out to be particularly important.

HOW IS THE OSMOTIC GRADIENT ESTABLISHED? The loop of Henle has three distinct regions: the descending limb, the thin

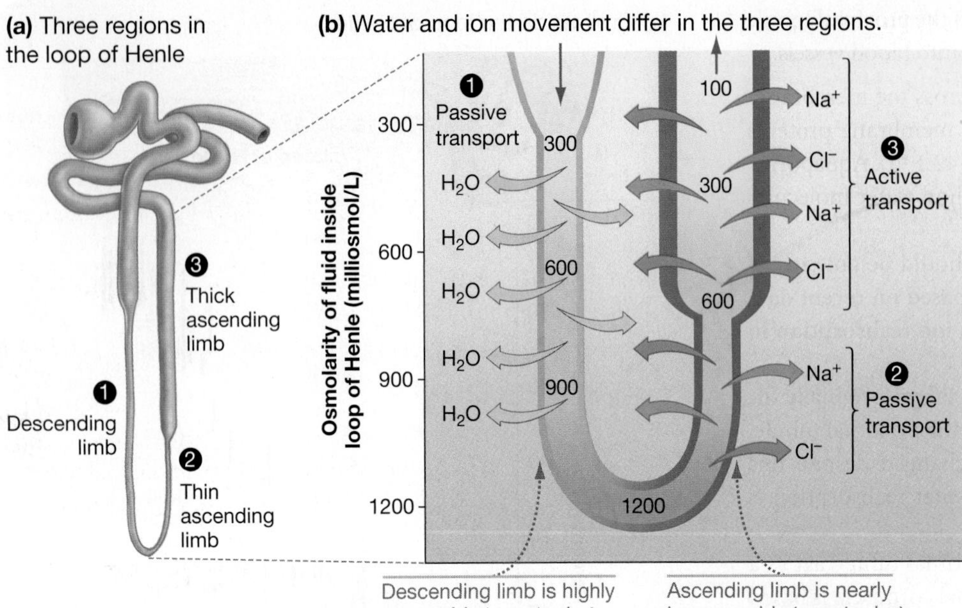

FIGURE 42.15 The Loop of Henle Maintains an Osmotic Gradient because Water Leaves the Descending Limb and Salt Leaves the Ascending Limb. The numbers inside the nephron in part (b) represent the osmolarity of the filtrate.

ascending limb, and the thick ascending limb (**Figure 42.15a**). The thin and thick ascending limbs differ in the thickness of their walls. Do the three regions also differ in their permeability to water and solutes?

It took over 15 years of experiments performed in laboratories around the world to formulate a definitive answer to that question. Researchers punctured Henle's loop with a micropipette, analyzed the composition of the fluid inside, and compared it with the nephron's final product—urine.

In the ascending limb of the loop of Henle, Na^+ and Cl^- constituted at least 60 percent of the solutes; urea constituted about 10 percent. But urea was the major solute in the distal tubule. These data suggested that Na^+ and Cl^-, but not urea, were being removed somewhere in the ascending limb. How?

Na^+ and Cl^- were also present at high concentrations in the tissue surrounding the thick ascending limb, so researchers hypothesized that sodium might be actively pumped out of this portion of the nephron. The hypothesis was that the active transport of Na^+ out of the thick ascending limb would create an electrical gradient that would also favor the loss of Cl^-.

Follow-up experiments using ouabain and other poisons supported the hypothesis that sodium ions are actively transported out of the solution inside the thick ascending limb, with chloride ions following along an electrochemical gradient. The epithelial cells responsible for salt excretion are configured almost exactly like the epithelium of the shark rectal gland (see Figure 42.6).

What is happening in the descending limb and the thin ascending limb of the loop of Henle? By injecting solutions of known concentration into the nephrons of rabbits, biologists documented that the descending limb is highly permeable to water but almost completely impermeable to solutes. The thin ascending limb of the loop, in contrast, is highly permeable to Na^+ and Cl^-, moderately permeable to urea, and almost completely impermeable to water.

A COMPREHENSIVE VIEW OF THE LOOP OF HENLE All of the observations just summarized came together in 1972 when two papers, published independently, proposed the same comprehensive model for how the loop of Henle works. To understand this model, follow the events in **Figure 42.15b**:

1. As fluid flows down the descending limb, the fluid inside the loop loses water to the tissue surrounding the nephron. This movement of water is passive—it does not require an expenditure of ATP. The water follows an osmotic gradient created by the ascending limb.

At the bottom of the loop—in the inner medulla—the fluids inside and outside the nephron have high osmolarity. The filtrate does not continue to lose water, though, because the membrane in the ascending limb is nearly impermeable to water.

2. The fluid inside the nephron loses Na^+ and Cl^- in the thin ascending limb. The ions move passively, along their concentration gradients.
3. Toward the cortex, the osmolarity of the surrounding solution is low. Additional Na^+ and Cl^- ions are actively transported out of the nephron in the thick ascending limb.

The countercurrent flow of fluid, combined with changes in permeability to water and in the types of channels and pumps that are active in the epithelium of the nephron, creates a self-reinforcing system. The presence of an osmotic gradient stimulates water and ion flows that in turn maintain an osmotic gradient.

Here's how it works: The movement of NaCl from the ascending limb into the surrounding tissue increases the osmotic concentration outside the descending limb, which results in a flow of water out of the water-permeable walls of the descending limb, via osmosis. This loss of water in the descending limb increases the osmolarity in the fluid entering the ascending limb.

The high concentration of salt in the fluid at the base of the ascending limb triggers a passive flow of ions out—reinforcing the osmotic gradient.

✔If you understand this concept, you should be able to predict what happens to the osmotic gradient when the drug furosemide inhibits membrane proteins that pump sodium and chloride ions out of the thick ascending limb. Specifically, how does this drug affect (**1**) water loss in the descending limb, (**2**) the osmolarity of the solution inside the bottom of the nephron, and (**3**) loss of salt in the thin ascending limb?

THE VASA RECTA REMOVES WATER AND SOLUTES THAT LEAVE THE LOOP OF HENLE What happens to the water and salt that move out of the loop? They quickly diffuse into the **vasa recta**, a network of blood vessels that runs along the loop. As a result, the water and electrolytes are returned to the body (**Figure 42.16**).

The removal of water that leaves the descending limb is particularly important. If it were not drawn off into the bloodstream, it would dilute the concentrated fluid outside the loop of Henle and quickly destroy the osmotic gradient.

THE COLLECTING DUCT LEAKS UREA The loop of Henle maintains an osmotic gradient, but a key feature of the nephron's final portions—the **distal tubule** and the **collecting duct**—helps establish it.

Urea is the solute that is most responsible for the steep osmotic gradient in the space surrounding the nephron. Urea is at high concentration in the inner medulla and low concentration in the outer medulla. This gradient exists because the innermost section of the collecting duct is permeable to urea.

Although the system created by the nephron, vasa recta, and collecting duct may seem complex, its outcome is simple: the creation and maintenance of a strong osmotic gradient with the minimum possible expenditure of energy.

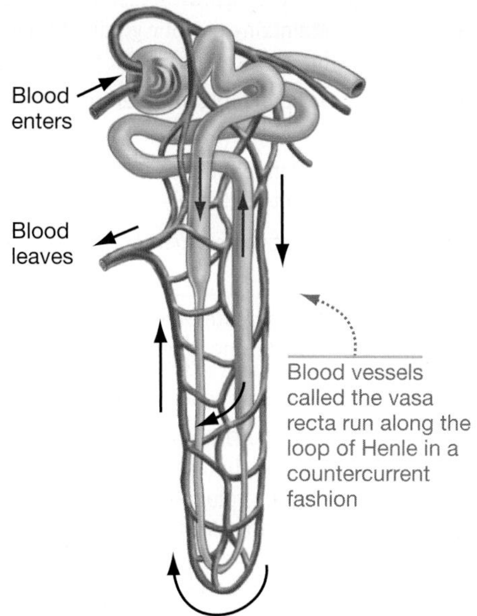

FIGURE 42.16 Blood Supply to the Loop of Henle. Water and solutes from the loop of Henle move into blood vessels (vasa recta).

Regulating Water and Electrolyte Balance: The Distal Tubule and Collecting Duct

The first three steps in urine formation—filtration, reabsorption, and establishment of an osmotic gradient—result in a fluid that is slightly hypotonic to blood. Once the filtrate has passed through the loop of Henle, the major solutes that it contains are urea and other wastes.

The fluid that enters the distal tubule is relatively constant in composition over time. In contrast, the urine that leaves the collecting duct is highly variable in osmolarity and in Na^+ and Cl^- concentration. How is this possible?

URINE FORMATION IS UNDER HORMONAL CONTROL The answer is based on two observations about the distal tubule and collecting duct: Their activity is (**1**) highly regulated, and (**2**) altered in response to osmotic stress. The amount of Na^+, Cl^-, and water that is reabsorbed in the distal tubule and in the collecting duct varies with the animal's condition.

Changes in the distal tubule and collecting duct are controlled by **hormones**—signaling molecules that were introduced in Chapter 8 and are explored further in Chapter 47. Specifically:

- If Na^+ levels in the blood are low, the adrenal glands release the hormone **aldosterone**, which leads to activation of sodium pumps and reabsorption of Na^+ in the distal tubule. Aldosterone saves sodium.
- If an individual is dehydrated, the brain releases **antidiuretic hormone (ADH)**. (The term *diuresis* refers to increased urine production, so *antidiuresis* means inhibition of urine production. ADH is also referred to as vasopressin or arginine vasopressin.) ADH saves water.

HOW DOES ADH WORK? ADH has two important effects on epithelial cells in the collecting duct:

1. ADH triggers the insertion of aquaporins into the apical membrane. As a result, cells become much more permeable to water and large amounts of water are reabsorbed.
2. ADH increases permeability to urea, which increases the osmolarity of the surrounding fluid and thus water loss from the filtrate.

As **Figure 42.17a** on page 838 shows, water leaves the collecting duct passively—following the concentration gradient maintained by the loop of Henle. When ADH is present, water is conserved and urine is strongly hypertonic relative to blood.

When ADH is absent, however, few aquaporins are found in the epithelium of the collecting duct, and the structure is relatively impermeable to water (**Figure 42.17b**). In this case, a hypotonic urine is produced.

People with defective forms of ADH or aquaporins produce copious amounts of urine—up to 30 liters per day. They suffer from the condition **diabetes insipidus**. (The word diabetes means "to run through"; insipidus means "tasteless." You may also have heard of the disease diabetes mellitus, which is due to problems

(a) ADH present: Collecting duct is highly permeable to water.

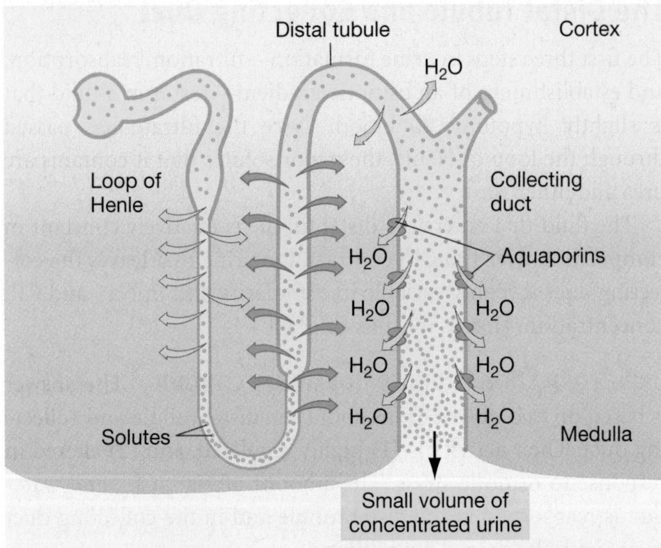

(b) No ADH present: Collecting duct is not permeable to water.

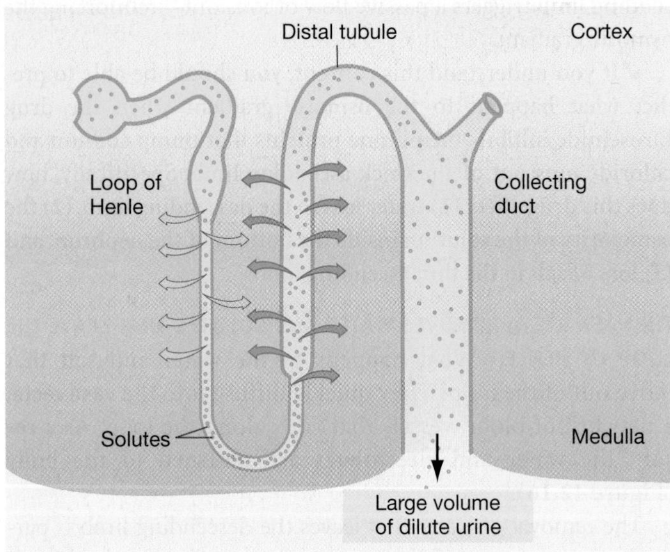

FIGURE 42.17 ADH Regulates Water Reabsorption by the Collecting Duct.

with regulating glucose concentrations in blood, rather than water retention in the kidneys—see Chapter 43.)

✔If you understand ADH's effect on the collecting duct, you should be able to predict how urine formation is affected by ingestion of ethanol, which inhibits ADH release, versus nicotine, which stimulates ADH release.

Table 42.2 reviews the functions of the four major regions of the nephron and the collecting duct. You can also review how the kidney works by going to the study area at *www.masteringbiology.com*.

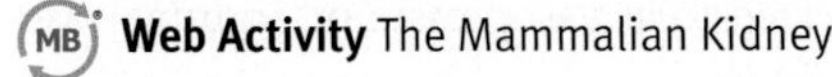
Web Activity The Mammalian Kidney

To summarize, the first three segments of the nephron—the renal corpuscle, proximal tubule, and loop of Henle—concentrate nitrogenous wastes, and create the possibility for Na^+, Cl^-, and water to be either excreted or reabsorbed by the distal tubule and collecting duct. The nephron is a remarkably effective mechanism for regulating water and electrolyte balance and achieving homeostasis.

CHECK YOUR UNDERSTANDING

If you understand that . . .

- The vertebrate kidney is specialized for the production of hypertonic urine.
- The loop of Henle is a countercurrent system that maintains a strong osmotic gradient in the medulla.
- The characteristics of urine change in response to hormonal signals. The signals trigger changes in the nephron that either save or eliminate water and solutes. The result is negative feedback and homeostasis.

✔ **You should be able to . . .**

1. Explain how aldosterone regulates the characteristics of urine.
2. Predict how the following events would affect urine production: drinking massive amounts of water, eating large amounts of salt, and refraining from drinking for 48 hours.

Answers are available in Appendix B.

SUMMARY TABLE 42.2 **Structure and Function of the Nephron and Collecting Duct**

Structure	Function
Renal corpuscle (Bowman's capsule and glomerulus)	Size-selective filtration: forms filtrate from blood (water and other small substances enter nephron)
Proximal tubule	Reabsorbs electrolytes, nutrients, water (active transport)
Loop of Henle	Maintains osmotic gradient from outer to inner medulla
• Descending limb	• Permeable to water (passive transport out of filtrate)
• Thin ascending limb	• Permeable to Na^+, Cl^- (passive transport out of filtrate)
• Thick ascending limb	• Active transport of Na^+, Cl^- out of filtrate
Distal tubule	With aldosterone: reabsorbs Na^+. Without aldosterone: does not reabsorb Na^+
Collecting duct	Regulates water retention and loss
• Main portion	• With ADH: water leaves filtrate; produces urine that is hypertonic to blood (reduced urine volume) Without ADH: water stays in filtrate; produces urine that is hypotonic to blood (increased urine volume)
• Innermost portion	• Urea leaks out to establish/maintain high osmolarity of inner medulla

Summary of Key Concepts

Freshwater, marine, and terrestrial habitats pose different challenges to animals with regard to maintaining water and electrolyte balance.

- The mechanisms involved in regulating water and electrolyte balance vary widely among animal groups, because different habitats present different types of osmotic stress.
- Freshwater fish are strongly hypertonic to their environment and tend to gain water and lose electrolytes.
- Marine fish are strongly hypotonic in relation to seawater and tend to lose water and gain electrolytes.
- Terrestrial animals lose water every time they breathe and from their body surfaces by evaporation.
- In most animals, epithelial cells that selectively transport water and electrolytes are responsible for homeostasis.

✔ You should be able to describe the water and electrolyte balance challenges faced by river otters that live near the ocean and move from freshwater to saltwater habitats and back on a daily or weekly basis.

In marine animals, specialized epithelial cells have membrane proteins that remove excess salt (NaCl) from the body so it can be excreted. The same types of cells are found in the kidneys of mammals.

- Epithelial cells in the shark rectal gland excrete excess salt, based on a "master gradient" established by sodium-potassium pumps.
- Salt-excreting cells similar to those found in the shark rectal gland occur in the gills of saltwater fish, the salt glands of marine birds and reptiles, and the kidneys of mammals.
- The leading hypothesis to explain how freshwater fish import salt is that they have cells in their gill epithelium that are similar to the salt-excreting cells found in marine fish, but oriented in the opposite direction.

✔ You should be able to explain how an electrochemical gradient for sodium ions makes it possible for water and ions to move across a membrane passively.

In terrestrial insects, the hindgut and Malpighian tubules are responsible for excreting water-soluble waste products and achieving homeostasis with respect to water and electrolyte concentrations.

- A waxy coating on the insect exoskeleton limits evaporative water loss. The openings to insect respiratory organs close when osmotic stress is severe.
- Insects can form a hypertonic urine that minimizes water loss during the excretion of nitrogenous wastes.
- The Malpighian tubules of insects form a filtrate that is isotonic with the hemolymph. If pumps in the epithelium of the hindgut are activated, then electrolytes and water are reabsorbed from the filtrate and returned to the hemolymph.

✔ You should be able to predict whether desert locusts or grasshoppers that live near ponds expend more energy during urine formation, and explain your reasoning.

In terrestrial vertebrates, the kidney is responsible for excreting water-soluble waste products and achieving homeostasis with respect to water and electrolyte concentrations.

- Nephrons in the mammalian kidney form a filtrate in the renal corpuscle and then reabsorb valuable nutrients, electrolytes, and water in the proximal tubule.
- A solution containing urea and electrolytes flows through the loop of Henle, where changes in the permeability of epithelial cells to water and salt—along with active transport of salt—create a steep osmotic gradient.
- If ADH increases the water permeability of the collecting duct, water is reabsorbed along the osmotic gradient and a hypertonic urine is produced.
- In animals, maintaining water and electrolyte balance is an active, energy-demanding process based on the action of membrane proteins. Water and electrolyte excretion and reabsorption are carefully controlled to achieve homeostasis.

✔ You should be able to explain why desert-dwelling kangaroo rats, which are under extreme osmotic stress due to lack of drinking water, have extremely long loops of Henle relative to their body size.

MB **Web Activity** The Mammalian Kidney

Questions

✔ TEST YOUR KNOWLEDGE

Answers are available in Appendix B

1. Which one of the following statements is true of fish that live in freshwater?
   a. Their tissues are isotonic with respect to their environment. As a result, they do not require a specialized organ to maintain water and electrolyte balance.
   b. They lose water to their environment primarily through the gills. They replace this water by drinking.
   c. Water enters epithelial cells in their gills via osmosis. Electrolytes leave the same cells via diffusion.
   d. They have specialized cells that actively pump Na^+ and Cl^- from blood into epithelial cells, so the ions can be excreted.
2. Na^+/K^+-ATPase is found in epithelial cells that produce urine or other excretions that are hypertonic relative to tissues. Why?
   a. To form a hypertonic solution, ions or molecules must be actively transported across their concentration gradient.
   b. Cells similar to those found in the shark rectal gland occur in many other species that need to excrete salt (NaCl).

c. The sodium-potassium pump is directly responsible for moving sodium ions from blood or body tissues into the urine for salt excretion, against a concentration gradient.
d. They are usually located in the basolateral membrane of those epithelial cells, along with other channels and cotransporters.

3. Regarding urine formation, the function of the insect hindgut is closest to which of the following structures in the vertebrate kidney?
   a. renal corpuscle
   b. proximal tubule
   c. loop of Henle
   d. distal tubules and collecting duct

4. Which of the following gives the correct sequence of the regions of the nephron with respect to fluid flow?
   a. distal tubule → ascending limb of Henle → descending limb of Henle → proximal tubule → collecting duct
   b. ascending limb of Henle → descending limb of Henle → proximal tubule → distal tubule → collecting duct
   c. proximal tubule → ascending limb of Henle → descending limb of Henle → distal tubule → collecting duct
   d. proximal tubule → descending limb of Henle → ascending limb of Henle → distal tubule → collecting duct

5. Which of the following statements about kidney function is *not* correct?
   a. The loop of Henle acts as a countercurrent exchanger, so no energy is required to maintain the osmotic gradient.
   b. The descending limb of the loop of Henle is highly permeable to water.
   c. The thin ascending limb of the loop of Henle is highly permeable to salt.
   d. Reabsorption of water and solutes takes place primarily in the proximal tubule.

6. What effect does antidiuretic hormone (ADH) have on the nephron?
   a. It increases water permeability of the descending limb of the loop of Henle.
   b. It decreases water permeability of the descending limb of the loop of Henle.
   c. It increases water permeability of the collecting duct.
   d. It decreases water permeability of the collecting duct.

✓TEST YOUR UNDERSTANDING

Answers are available in Appendix B

1. Explain the changes that need to occur in water and electrolyte balance as a salmon moves from freshwater to salt water and back. Specifically, state when the animal should drink or not drink, when cells in the gill epithelium should excrete or import electrolytes, and why each change occurs.
2. The chloride cells of fish gills are sometimes called mitochondria-rich cells due to their high density of mitochondria. How does high mitochondrial density relate to the functional role of chloride cells? Would you expect other epithelial cells involved in ion transport to contain large numbers of mitochondria? Explain.
3. Why is it significant that cells involved in transport processes often have microvilli?
4. This chapter introduced a number of features that help terrestrial animals reduce water loss. These traits include the layer of wax found on insect exoskeletons, the ability of insects to close the openings to their respiratory passages, the excretion of nitrogenous wastes as insoluble uric acid, and long loops of Henle in the mammalian kidney. Predict how each of these traits differs in animals that live in very humid versus very dry habitats. How would you test your predictions?
5. In insects, active transport of electrolytes into the Malpighian tubules leads to the formation of a filtrate that is isotonic with respect to the hemolymph. In mammals, urine formation begins with blood pressure in the renal corpuscle that leads to the formation of a filtered filtrate that is isotonic with respect to the blood. In insects, the filtrate is processed in the hindgut; in mammals, the filtrate is processed in the remainder of the nephron. How are the processing steps in insects and mammals similar? How are they different?
6. Compare and contrast the types of nitrogenous wastes observed in animals. Identify which compound can be excreted with a minimum of water, which is most toxic, and which waste product is found in fish, mammals, and insects. Which type of nitrogenous waste would you expect to find produced by embryos inside eggs laid on land?

✓APPLYING CONCEPTS TO NEW SITUATIONS

Answers are available in Appendix B

1. Examine Figure 42.16 again, and notice that the network of blood vessels that runs along the mammalian nephron is arranged so that blood flows in a countercurrent arrangement relative to the flow of fluid in the loop of Henle. The water and electrolytes that leave the loop of Henle and collecting duct diffuse into these blood vessels and are returned to the body. Why is the countercurrent arrangement for blood flow important?
2. You have isolated a segment of a rat nephron and introduced a solution of known composition. When you compare the fluid collected at the end of the segment with the test solution introduced at the beginning, you find that the volume has decreased by 30 percent and the Na^+ concentration has decreased by 30 percent, but the urea concentration has increased by 50 percent. What processes in this segment might account for these changes?
3. In some areas of the world, highway maintenance crews use salt extensively to keep roads free of ice in winter. Biologists are concerned about the effect of this salt on freshwater organisms. Why?
4. Biologists have recently been able to produce mice that lack functioning genes for aquaporins. How does their urine compare to that of individuals with normal aquaporins?

A young crocodile has just caught a frog. Animals obtain nutrients by ingesting food.

Animal Nutrition 43

Animals get the two basic requirements for life—the chemical energy required to synthesize ATP and the carbon-containing compounds required to synthesize complex macromolecules—from other organisms. They are heterotrophs. They eat to live.

The types of food that are available to different animals vary widely, and food is often in dangerously short supply. Based on these observations, it is logical to predict that animals have a variety of means for obtaining food and they are under intense natural selection to make efficient use of the food they have.

Acquiring and processing energy is a fundamental attribute of life (see Chapter 1). How do animals get their food, and how do they process it? Which substances in food are used as nutrients, and how do humans and other animals maintain appropriate levels of key nutrients in their bodies?

If you're like most people, you've probably given a fair amount of thought to the food you eat but little thought to what happens to that food once it's inside you. As a meal moves through a digestive system, its chemical composition and physical characteristics change dramatically. Large packets of food that enter the mouth are reduced to monomers that can be absorbed by cells and enter the bloodstream.

Ingesting food is the first of four processes needed to obtain energy. For you or any other animal to stay alive, ingestion must be followed by digestion, absorption, and elimination of wastes (**Figure 43.1** on page 842). The digestive system that accomplishes these tasks is analogous to a lumber mill, where bulky, complex raw materials are processed into smaller and more usable products.

Research on feeding and digestion is fundamental to understanding basic aspects of animal anatomy and physiology. But research on animal nutrition has important practical applications as well. For example, this chapter addresses questions about why ulcers can develop and why type 2 diabetes and other nutrition-related diseases are on the rise in many human populations.

KEY CONCEPTS

- Animals require an array of nutrients to stay healthy, including specific amino acids, vitamins, and elements, as well as organic compounds that act as building blocks in chemical synthesis or have high potential energy.
- There is usually a close correspondence between the structure of an animal's mouthparts and the function of those mouthparts in capturing and processing food.
- Digestion occurs in the digestive tract, which is compartmentalized into organs that have specialized functions in the ingestion and digestion of food, absorption of nutrients and water, or excretion of wastes.
- Lack of homeostasis with respect to nutrients such as glucose can cause disease.

✔ When you see this checkmark, stop and test yourself. Answers are available in Appendix B.

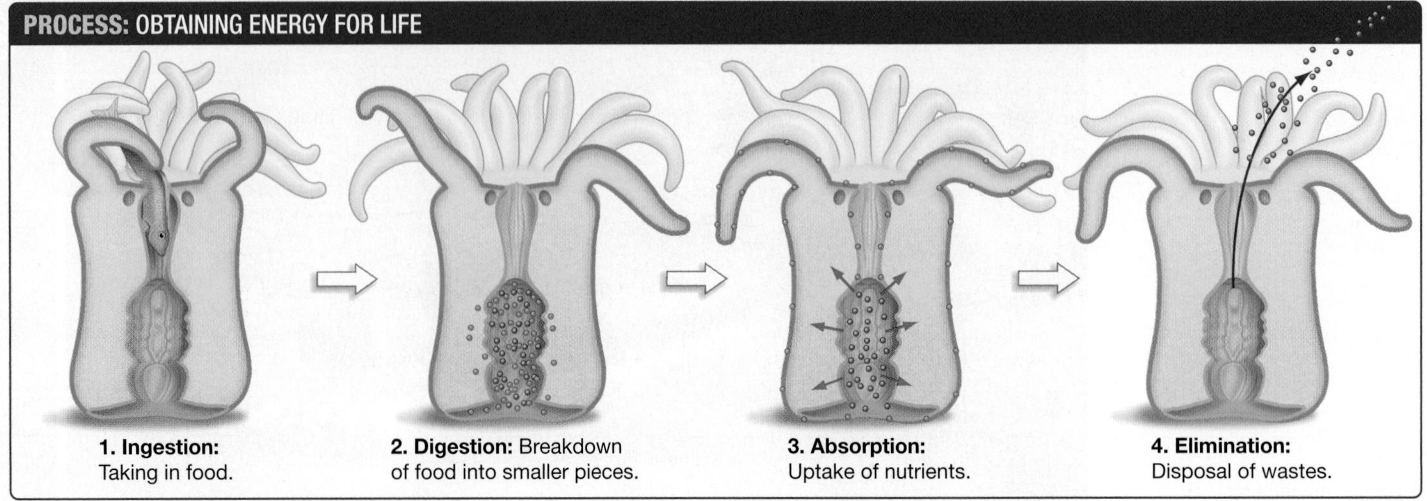

FIGURE 43.1 Four Steps in Obtaining Nutrients.

Just as plant nutrition is fundamental to the productivity of agricultural and natural ecosystems, nutrition is a basic component of animal health and welfare. Let's begin with a look at what animals must eat to live.

43.1 Nutritional Requirements

Animals get the chemical energy and carbon-containing building blocks they need from carbohydrates and fats, which are carbon compounds with high potential energy. Because fats and other lipids contain more C–H bonds than do carbohydrates such as starch and sugars, lipids provide over twice the energy per gram. Chapters 5 and 6 analyzed the structures of carbohydrates and fats; Chapter 9 detailed how these compounds are used to synthesize ATP and key macromolecules.

A carbohydrate is an example of a **nutrient**: a substance that an organism needs to remain alive. **Food** is any material that contains nutrients.

Understanding which nutrients an individual needs, and in what amounts, are basic issues in research on animal nutrition. To illustrate how biologists go about the task, let's consider what humans need to maintain good health.

Meeting Basic Needs in Humans

In 1943 the Food and Nutrition Board of the U.S. National Academy of Sciences[1] published the first Recommended Dietary Allowances (RDAs). The goal of the RDAs was to specify the amount of each essential nutrient that an individual must ingest to meet the needs of most healthy people.

In addition to obtaining "building blocks" for chemical synthesis and chemical energy from carbohydrates and other compounds, humans require several other **essential nutrients**—meaning, nutrients that cannot be synthesized and must be obtained in the diet:

- Proteins provide amino acids that are used to synthesize polypeptides; amino acids may also be oxidized to provide energy. Of the 20 amino acids required to manufacture most proteins, humans can synthesize 12. The others are called **essential amino acids** because they must be obtained from food.
- **Vitamins** are organic compounds that are vital for health but are required in only minute amounts. They have a variety of roles; several function as coenzymes in critical reactions. **Table 43.1** lists a few of the vitamins for which RDAs have been established, notes their functions, and indicates the problems that develop if they are missing in the diet.
- **Electrolytes** are inorganic ions that influence osmotic balance and are required for normal membrane function. Sodium (Na^+), potassium (K^+), and chloride (Cl^-) are the major ions in the body.
- Inorganic substances fulfill a wide variety of functions not performed by electrolytes—often because they are important components of enzyme cofactors or structural materials (see **Table 43.2**). Some, such as calcium and phosphorus, are needed in relatively large quantities. Others, such as iron and magnesium, are required in small or trace amounts.

Studying Nutrient Requirements

Research on the basic nutritional requirements of nonhuman animals can sometimes be based on experiments that vary dietary intake of specific ions or molecules under controlled conditions, and compare indices of health. Based on experiments like these and analyses of diets eaten by healthy individuals, researchers have been able to publish the equivalent of RDAs for a wide array of domestic livestock and animals commonly maintained in zoos and aquariums.

Studying nutrient requirements in humans is more difficult. It is unethical to assign different, carefully controlled diets to sev-

[1]The National Academy of Sciences is a group of scientists and engineers that advises the U.S. Congress on scientific and technical matters. Its Food and Nutrition Board is made up of biologists who specialize in animal nutrition.

TABLE 43.1 **Some Important Vitamins (Required by Humans)**

	Source in Diet	Function	Symptoms if Deficient
Vitamin B_1 (thiamine)	legumes, whole grains, potatoes, peanuts	formation of coenzyme in citric acid cycle	beriberi (fatigue, nerve disorders, anemia)
Vitamin B_{12}	red meat, eggs, dairy products; also synthesized by bacteria in intestine	coenzyme in synthesis of proteins and nucleic acids; formation of red blood cells	anemia (fatigue and weakness due to low hemoglobin content in blood)
Niacin	meat, whole grains	component of coenzymes NAD^+ and $NADP^+$	pellagra (digestive problems, skin lesions, nerve disorders)
Folate	green vegetables, oranges, nuts, legumes, whole grains; also synthesized by bacteria in intestine	coenzyme in nucleic acid and amino acid metabolism	anemia
Vitamin C (ascorbic acid)	citrus fruits, tomatoes, broccoli, cabbage, green peppers	used in collagen synthesis, prevents oxidation of cell components, improves absorption of iron	scurvy (degeneration of teeth and gums)
Vitamin D	fortified milk, egg yolk; also synthesized in skin exposed to sunlight	aids absorption of calcium and phosphorus in small intestine	rickets (bone deformities) in children; bone softening in adults

TABLE 43.2 **Essential Elements (Required by Humans)**

	Source in Diet	Function	Symptoms if Deficient
Calcium (Ca)	dairy products, green vegetables, legumes	bone and tooth formation, nerve signaling, muscle response	loss of bone mass, slow growth
Fluorine (F)	fluoridated water, seafood	maintenance of tooth structure	higher frequency of tooth decay
Iodine (I)	iodized salt, algae, seafood	component of the thyroid hormones thyroxin and T_3	goiter (enlarged thyroid gland)
Iron (Fe)	meat, eggs, whole grains, green leafy vegetables, legumes	enzyme cofactor; synthesis of hemoglobin and electron carriers	anemia, weakness
Magnesium (Mg)	whole grains, green leafy vegetables	enzyme cofactor	nerve disorders
Phosphorus (P)	dairy products, meat, grains	bone and tooth formation, synthesis of nucleotides and ATP	weakness, loss of bone
Sulfur (S)	any source of protein	amino acid synthesis	swollen tissues, degeneration of liver, mental retardation

eral groups of subjects if there is a chance that one of the diets might lead to illness. As a result, researchers must rely on observational studies that document the health of people with known intake levels of specific nutrients.

43.2 Capturing Food: The Structure and Function of Mouthparts

Instead of making their own food as plants do, animals obtain the energy and nutrients they need from other organisms. The question is, How?

As Chapter 32 noted, biologists assign animal food-getting techniques to one of four strategies:

- **Suspension feeders**, such as sponges and tubeworms, filter small organisms or bits of organic debris from water, by means of cilia, mucous-lined "nets," or other structures.
- **Deposit feeders**, including earthworms and sea cucumbers, swallow organic-rich sediments and other types of deposited material.
- **Fluid feeders** suck or lap up fluids—blood, nectar, or sap, for example.
- **Mass feeders** are the majority of animals. They seize and manipulate chunks of food by using mouthparts such as jaws and teeth, beaks, or special toxin-injecting organs.

Mouthparts as Adaptations

The types of food that animals harvest range from soupy solutions in decaying carcasses to hard nuts. Solutions have to be lapped up; nuts have to be cracked. Given the diversity of food sources that animals exploit, it is not surprising that a wide variety of mouthpart structures have evolved to facilitate capturing and processing food.

Natural selection has tightly matched the structure of animal mouthparts to their function in obtaining food. For example:

- Mammals are the only animals that chew their food and swallow distinct packets or boluses. The extinct mammal species shown in **Figure 43.2a** illustrates one example of the diversity in tooth shapes that evolved from the relatively simple and similar teeth that occurred in the common ancestor of all mammals. Diversification in tooth shape has allowed mammals to exploit a wide range of foods.
- Complex, multipart skull and jawbones, with associated musculature, have evolved in snakes. The highly movable structure that results allows snakes to ingest and swallow large prey whole (**Figure 43.2b**).

The reason that there is such a close correlation between the structure and function of mouthparts is simple: Natural selection is particularly strong when it comes to food capture, because obtaining nutrients is so fundamental to fitness—the ability to produce offspring.

It's important to remember some key points about natural selection and adaptation, however (see Chapter 24):

- *Evolution is not progressive.* Animal mouthparts do not get "better" over time in the sense of being more complex. Most nematomorphs (hair worms) do not capture food and do not have a mouth—even though they evolved from ancestors that had mouths. Instead, nematomorphs absorb nutrients directly across their body wall.
- *Adaptation is not perfect.* Humans are often cited as an example of an omnivore with teeth adapted for both tearing meat and mashing plant leaves or stalks. But most dentists would also claim that our "wisdom teeth" are maladaptive—meaning, a trait that actually lowers fitness on average.

Let's pursue the correlation between mouthparts and food sources further, by analyzing the structure and function of jaws and teeth in what may be the most diverse lineage among vertebrates: the cichlid fishes of Africa.

(a) The large canines of saber-toothed cats stabbed and sliced prey.

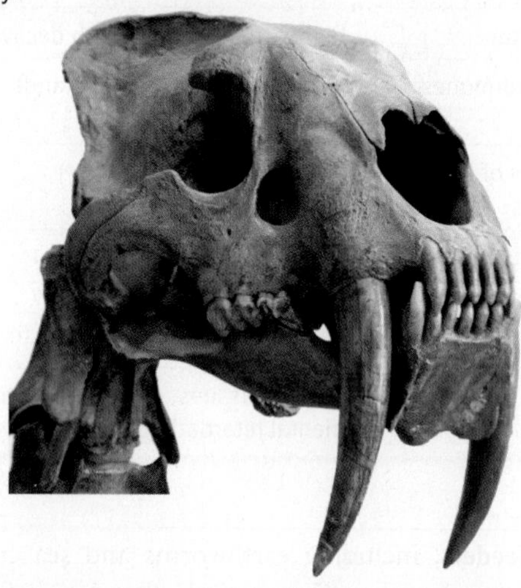

(b) A highly movable skull and jaw allows snakes to swallow large prey whole.

FIGURE 43.2 Mouthpart Structure Correlates with Function.

A Case Study: The Cichlid Jaw

The cichlids that inhabit the Rift Lakes of East Africa are a spectacular example of **adaptive radiation**—the diversification of a single ancestral lineage into many species, each of which lives in a different habitat or employs a distinct feeding method (see Chapter 27). Lake Victoria, for example, is home to 300 **endemic** cichlids—meaning species that live nowhere else.

Most of the Lake Victoria cichlids feed on a specific item, but as a group they exploit almost every food source in the lake: planktonic organisms, crust-forming algae, leaflike algae, eggs, fish scales, fish fins, whole fish, plants, insects, and snails.

How can a group of closely related species exploit so many different food sources? To answer this question, biologists point to a structure called the pharyngeal jaw.

Many fish species have pharyngeal (throat) jaws located well behind the normal oral (mouth) jaws. Non-cichlids use their pharyngeal jaws to move food down their throats, but cichlids can also use theirs to bite. This is possible because the upper pharyngeal jaw attaches to the skull in cichlids, and because the muscles of their lower pharyngeal jaw allow it to move against the upper jaw (**Figure 43.3**).

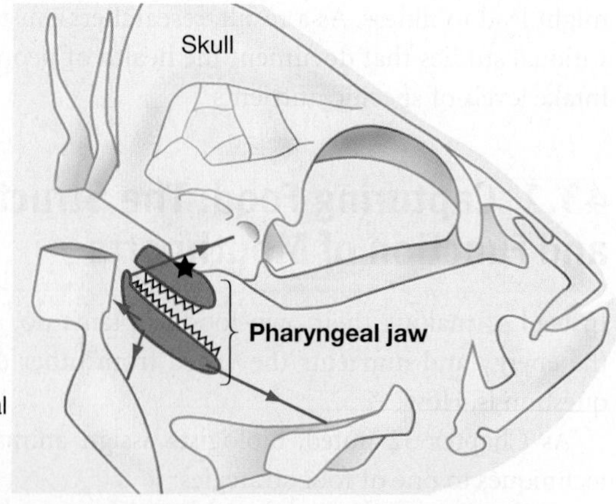

FIGURE 43.3 Rift Lake Cichlids Have Two Sets of Biting Jaws. Oral jaws capture food; pharyngeal jaws process it.

✔**EXERCISE** Label the oral jaw.

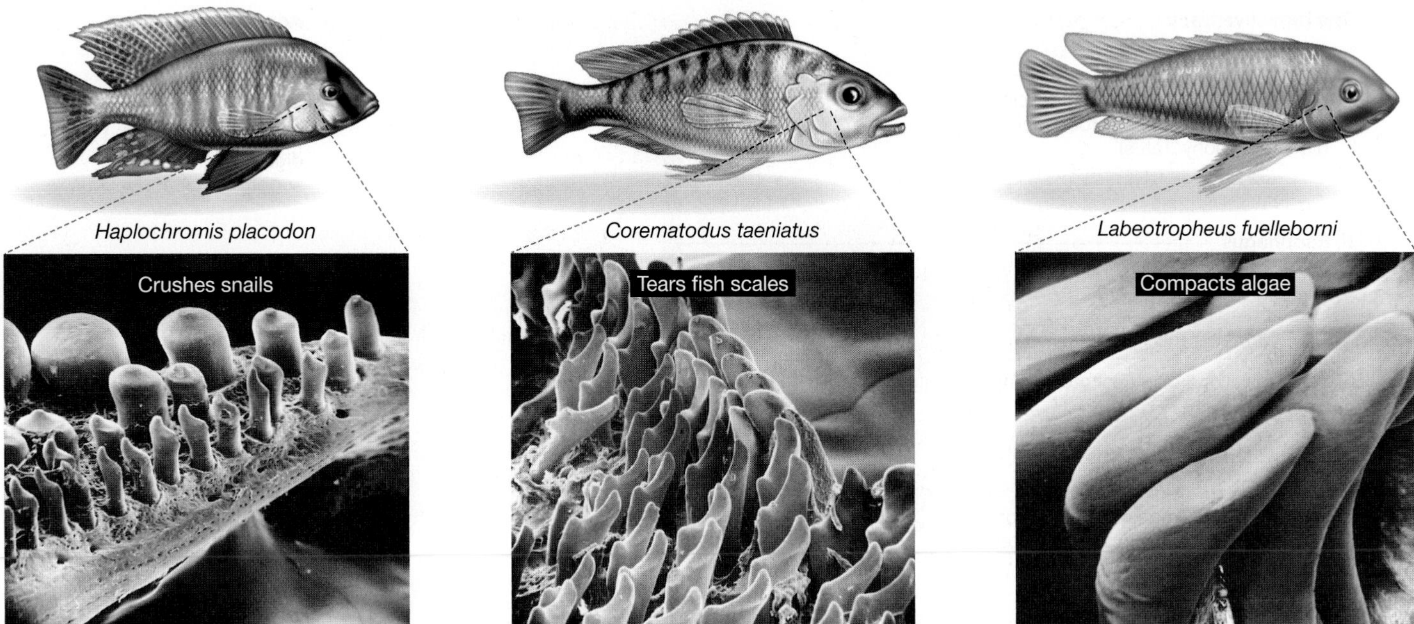

FIGURE 43.4 In Cichlids, the Structure of the Pharyngeal Jaw Correlates with the Type of Food Ingested.

In addition to furnishing a second set of biting jaws that make food processing more efficient, cichlid pharyngeal jaws provide an extra set of toothlike structures. These protuberances vary in size and shape, correlating with their function such as crushing snail shells, tearing fish scales, or compacting algae (**Figure 43.4**).

These observations add to a large body of evidence supporting a general pattern in animal evolution: Mouthparts have diversified in response to natural selection for exploiting a diversity of food sources. The structure of jaws, teeth, and other mouthparts correlates with their function in harvesting and processing food.

43.3 How Are Nutrients Digested and Absorbed?

Animals ingest just about every type of food conceivable. Ingestion is only the first of four processes needed to obtain energy, however. For an animal to stay alive, ingestion must be followed by digestion, absorption, and elimination of wastes.

Digestion is the breakdown of food into small enough pieces to allow for **absorption**—the uptake of specific ions and molecules that act as nutrients. Digestion is a key process in animals because, unlike plants, unicellular organisms, and certain parasites, animals do not acquire nutrients as individual molecules. Instead, they take in packets of food that must be broken down into small pieces. Nutrients must be extracted from the small pieces and waste materials must be eliminated. How and where does this processing occur?

An Introduction to the Digestive Tract

In many animals, digestion takes place in a tube called the **digestive tract**—also known as the alimentary (literally, "nourishment") canal or gastrointestinal (GI) tract. Digestive tracts come in two general designs:

- **Incomplete digestive tracts** have a single opening that doubles as the location where food is ingested and wastes are eliminated. The mouth opens into a chamber, called a gastrovascular cavity, where digestion takes place (**Figure 43.5**).
- **Complete digestive tracts** have two openings—they start at the mouth and end at the anus. The interior of this tube communicates directly with the external environment via these openings (**Figure 43.6** on page 846). In embryos, the gut derives from the hollow tube that forms as cells invaginate during gastrulation (see Chapter 22).

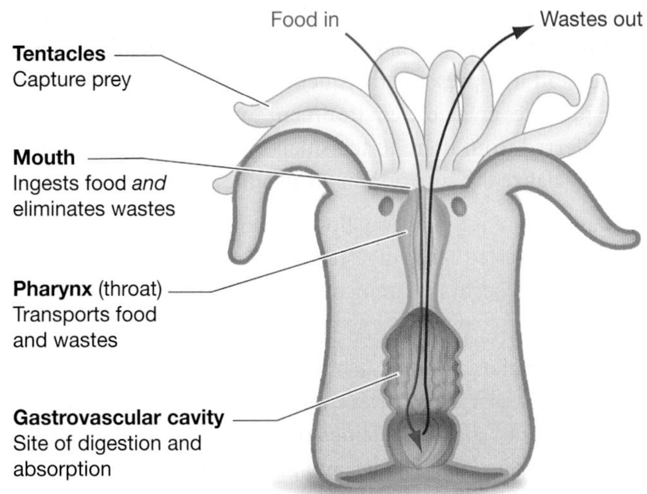

FIGURE 43.5 An Incomplete Digestive Tract. Anemones use stinging cells located on their tentacles to capture small fish, crustaceans, and other prey. Prey are taken into the mouth and digested in the gastrovascular cavity.

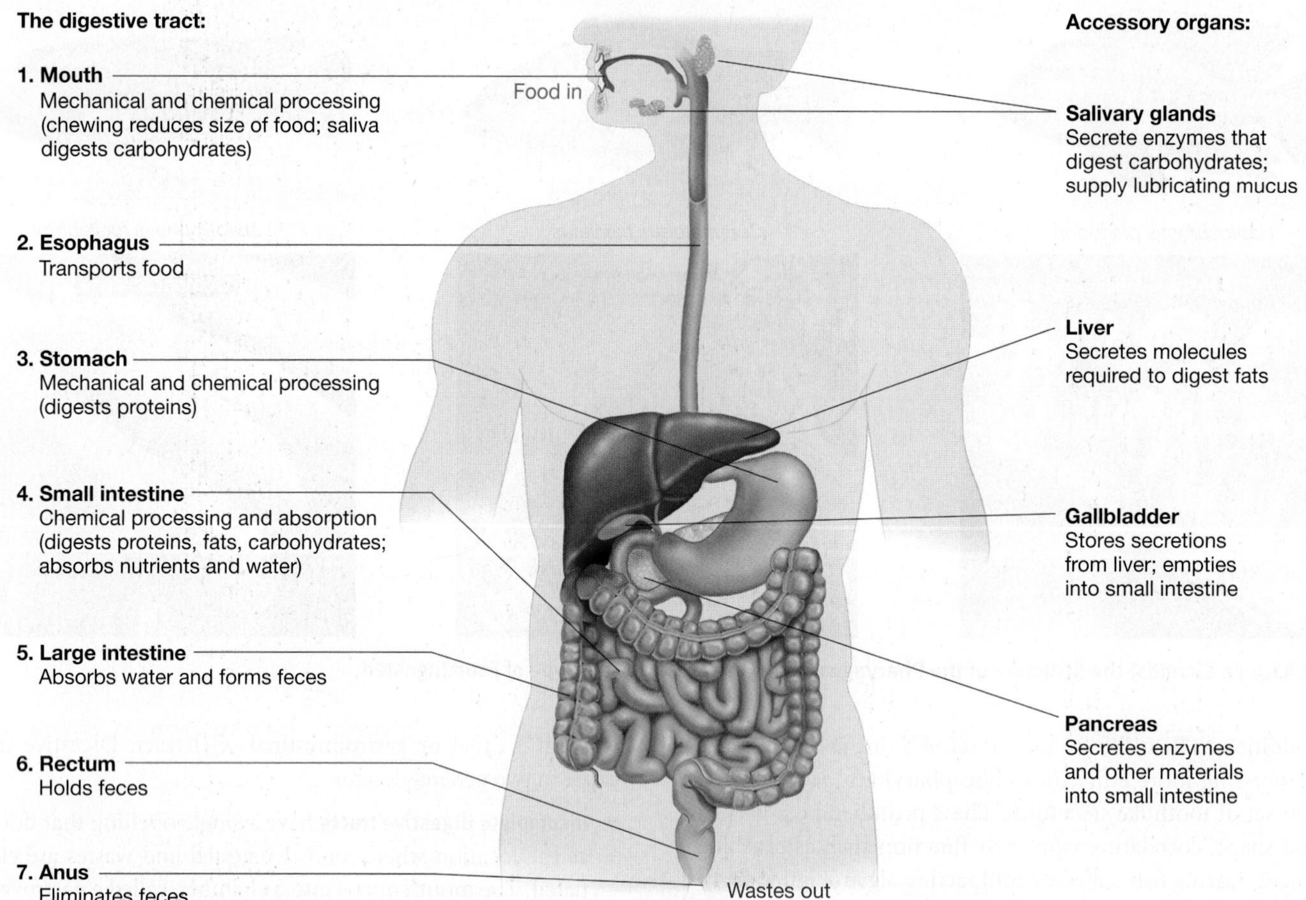

FIGURE 43.6 A Complete Digestive Tract. In humans, the digestive tract is a tube that runs from the mouth to the anus. The salivary glands, liver, gallbladder, and pancreas are not part of the tract itself. Instead, they secrete material into the tract at specific points.

A tubelike digestive tract has three advantages:

1. It allows animals to feed on large pieces of food—expanding the types of food sources that can be ingested.
2. Different chemical and physical processes can be separated within the canal, so that they occur independently of each other and in a prescribed sequence. The stomach, for example, provides an acidic environment for digestion. Material is transferred from there to the small intestine, where enzymes are specialized to function in an alkaline environment.
3. Because there is a one-way flow of food and wastes, material can be ingested and digested continuously instead of in batches—as occurs in an incomplete digestive tract.

The digestive tract is only one part of the digestive system, however. Several vital organs and glands are connected to the digestive tract. These accessory structures contribute digestive enzymes and other products to specific portions of the tract. They include the salivary glands, liver, gallbladder, and pancreas (see Figure 43.6).

An Overview of Digestive Processes

Before analyzing the function of each component of the digestive system in detail, let's consider the general changes that happen to food on its way through the digestive tract. In this brief overview and in the detailed discussion that follows, humans will serve as a model species—simply because so much is known about human digestion.

In mammals, digestion begins with the tearing and crushing activity of teeth. Chewing reduces the size of food particles and softens them. Humans augment the mechanical breakdown of food by their use of knives and cooking. In fact, the invention of cutting tools and cooking, which make food easier to chew, is the leading hypothesis to explain why average tooth size has declined steadily over the past several million years of human evolution.

Distinct chemical changes occur as food moves through each compartment in the digestive tract (**Figure 43.7**):

- The chemical breakdown of carbohydrates begins in the mouth, through enzymes in saliva.
- Protein digestion begins in the acidic environment of the stomach.
- Chemical processing of the three major types of macromolecules—carbohydrates, lipids, and proteins—is completed in the small intestine. The small molecules that result from the digestion of these macromolecules are absorbed in the small intestine, along with water, vitamins, and ions.

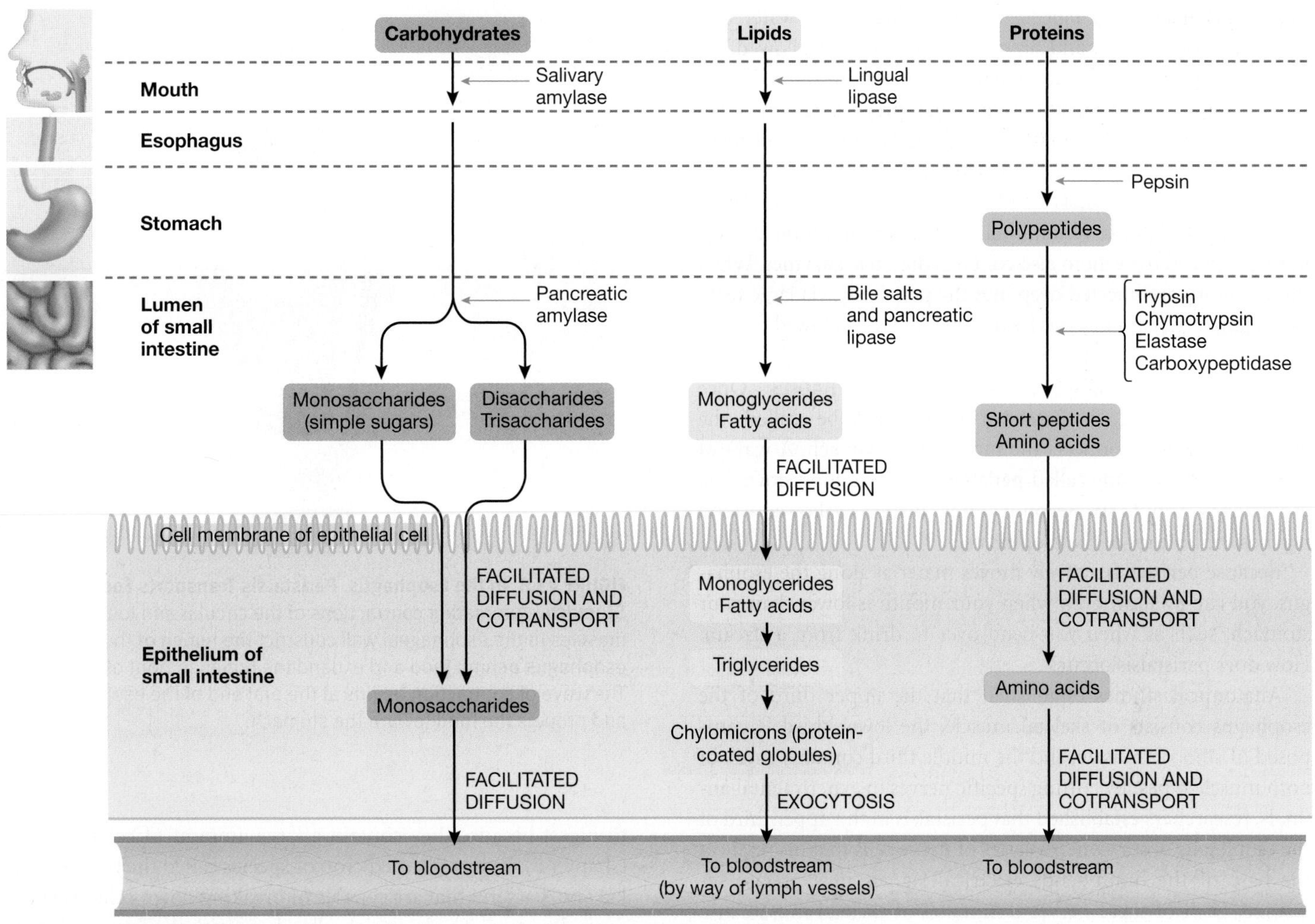

FIGURE 43.7 Carbohydrates, Lipids, and Proteins Are Processed in a Series of Steps. Three key types of macromolecules enter the digestive system (top of diagram). As they proceed through the digestive tract, they are broken apart by a variety of enzymes. Simple sugars, fatty acids, and amino acids then enter epithelial cells in the small intestine and are transported to the bloodstream.

- In the large intestine, or colon, more water is absorbed. The result is **feces**, which are eventually excreted.

For a human to stay healthy, each step in this process must be completed correctly. Problems in the digestive tract can lead to heartburn, ulcers, nausea, constipation, and other maladies. By far the most serious of these is diarrhea, which kills more than 4 million children each year.

Because digestion is so important to understanding how animals work, let's analyze each step in more detail. As the following sections track food as it moves from the mouth to the anus in humans—a particularly well-strudied organism—watch for notes highlighting the diversity of structures found in the digestive tracts of nonhuman animals.

The Mouth and Esophagus

If you hold a cracker in your mouth long enough, it will start to taste sweet. The sensation occurs because an enzyme in your saliva hydrolyzes the starch molecules in the cracker to maltose. Maltose is a disaccharide that is split in the small intestine to form two glucose monomers.

Starch breakdown was actually the first enzyme-catalyzed reaction ever discovered. In the early 1800s several researchers found that a component of certain plant extracts digested starch; in 1831 the same activity was discovered in human saliva.

DIGESTION STARTS IN THE MOUTH **Amylase**, the enzyme responsible for starch digestion in the mouth, ranks as one of the best-studied enzymes. Amylase cleaves bonds that release maltose dimers from starch, glycogen, and other glucose polymers. In doing so, it initiates the digestion of carbohydrates.

Cells in the tongue synthesize and secrete another important salivary enzyme, **lipase**, which begins the digestion of lipids. More specifically, the lingual lipase (lingual refers to the tongue) produced in the mouth begins breaking triglycerides into fatty acids and monoglycerides.

Salivary glands in the mouth also release water and glycoproteins called mucins. When mucins contact water, they form the slimy substance called **mucus**. The combination of water and mucus makes food soft and slippery enough to be swallowed.

In snakes, however, the statement "digestion starts in the mouth" is not correct. Snakes swallow their prey whole—meaning that the function of the mouth is to take in large packets of food, rather than begin digestion by chewing and secretion of enzymes. And some venomous snakes add a twist: Digestion begins *before* food is ingested. Snake venoms kill or otherwise immobilize prey; in some species the venom also contains digestive enzymes. When these venoms are injected deep into the prey body, via large teeth called fangs, digestion begins before the prey is swallowed.

PERISTALSIS MOVES MATERIAL DOWN THE ESOPHAGUS Once food is swallowed, it enters a muscular tube called the **esophagus**, which connects the mouth and stomach. A wave of muscular contractions called **peristalsis** propels food down the esophagus. About six seconds after being swallowed, food reaches the bottom of the esophagus.

Because peristalsis actively moves material along the esophagus, you can swallow even when your mouth is lower than your stomach, such as when you bend over to drink from a stream. How does peristalsis occur?

Anatomical studies established that the upper third of the esophagus consists of skeletal muscle, the lower third is composed of smooth muscle, and the middle third contains a mix of both muscle types. By cutting specific nerves in experimental animals, researchers established that peristalsis in the upper third of the esophagus occurs when a series of nerve cells that originate at the base of the brain sends electrical signals in a precise sequence. Each of these nerves terminates at skeletal muscle at a different location along the esophagus. Some of these muscles wrap around the esophagus; others run along its length.

In response to nerve signals, the muscle contracts or relaxes in a coordinated fashion (**Figure 43.8**). In this way, a wave of muscle contractions propagates down the tube, propelling the food mass ahead of it. The action of these nerves is not the result of a conscious choice. The system is a **reflex**—an automatic reaction to a stimulus—that is stimulated by the act of swallowing.

Even after decades of experiments, though, the mechanism of peristalsis in the lower third of the esophagus is still not well understood. Research is continuing.

A MODIFIED ESOPHAGUS: THE BIRD CROP Almost all animals have a tubelike esophagus that connects the mouth and the stomach. But in an array of bird species, the esophagus has a prominent, widened segment called the **crop** where food can be stored and, in some cases, processed.

The structure and function of the crop varies among the bird species that have one. In many groups, the crop is a simple sac that holds food and regulates its flow into the stomach. In these species, the crop is interpreted as an adaptation that allows individuals to eat a large amount in a short time, then retreat to a safe location while digestion occurs.

The crop has evolved into a digestive organ, though, in two leaf-eating species that are not closely related. Leaves are difficult

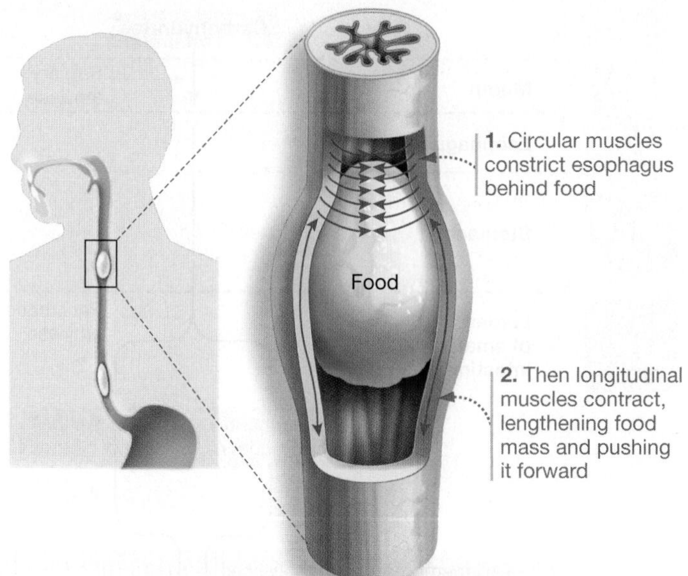

FIGURE 43.8 In the Esophagus, Peristalsis Transports Food to the Stomach. Alternating contractions of the circular and longitudinal muscles in the esophageal wall constrict the lumen of the esophagus behind food and expand the lumen in front of food. The wave of contraction begins at the oral end of the esophagus and propels the food toward the stomach.

to digest because they contain a large amount of cellulose (see Chapter 5). In the enlarged crop of species called the hoatzin and kakapo, bacteria that are capable of breaking down cellulose perform digestion. The bacterial cells, along with the fatty acids that result from bacterial metabolism, leak out of the crop and are used as food by the bird.

The Stomach

Although little if any digestion occurs in the esophagus of most animals, the situation changes dramatically when food reaches the stomach. The vertebrate **stomach** is a tough, muscular pouch in the digestive tract, bracketed on both ends by ringlike muscles called **sphincters**, which control movement of material through the gut (**Figure 43.9**).

When a meal fills the stomach, muscular contractions result in churning that mixes the contents and reduces food to a uniform consistency and solute concentration, or osmolarity; a certain amount of mechanical breakdown of food also occurs in response to this churning. The other main function of the stomach is the partial digestion of proteins.

Compared with the mouth or esophagus (or virtually any other tissue), the lumen of the stomach is highly acidic. Early researchers documented this fact by analyzing vomit or the contents of sponges that were tied to strings, swallowed, and pulled back up; chemists confirmed that the predominant acid in the stomach is hydrochloric acid (HCl).

Not long after, a physician named William Beaumont established that digestion takes place in the stomach. He reached this

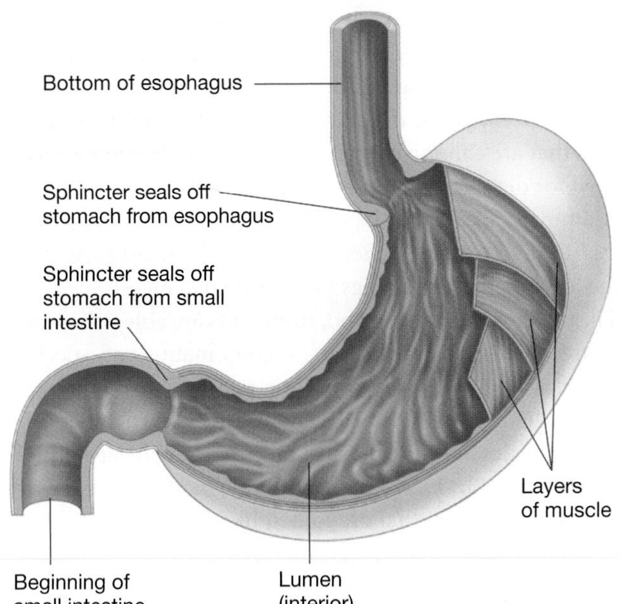

FIGURE 43.9 The Stomach Is a Muscular Outpocketing of the Digestive Tract. The stomach encloses an acidic environment. Muscular contractions mix food and break it into smaller pieces.

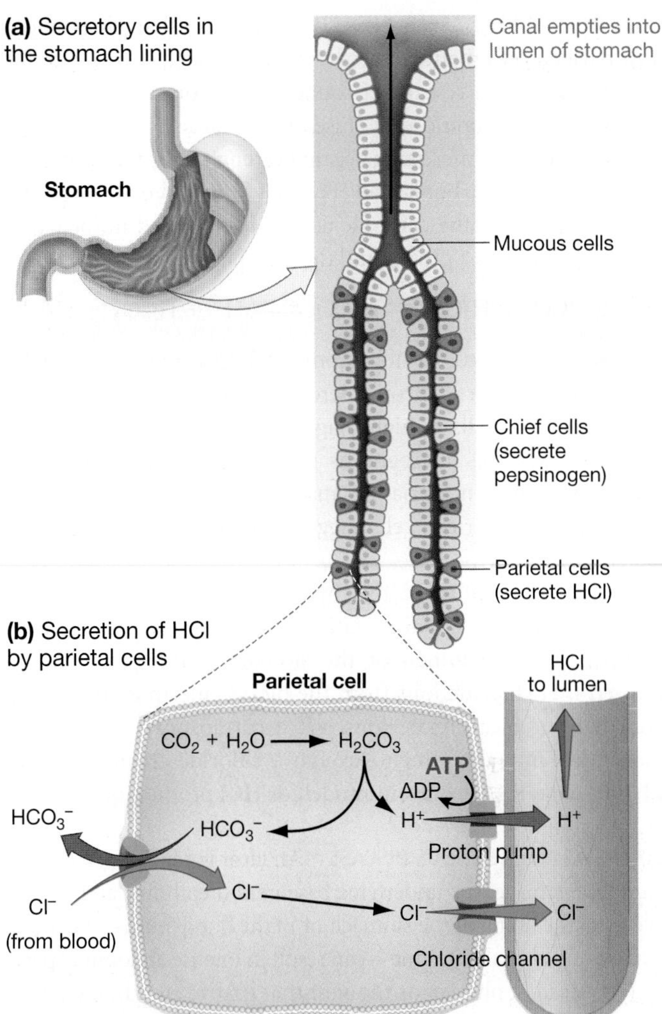

FIGURE 43.10 Cells in the Stomach Lining Secrete Mucus, Pepsinogen, and Hydrochloric Acid.

conclusion through an extraordinary series of experiments on a young man named Alexis St. Martin.

THE STOMACH AS A SITE OF PROTEIN DIGESTION In 1822, when St. Martin was 19 years old, a shotgun accidentally discharged into his abdomen and created a series of wounds. Despite repeated attempts, Beaumont was unable to close a hole in St. Martin's stomach. Eventually Beaumont inserted a small tube through the opening; the tube remained in St. Martin's body for the rest of his life. (Today, biologists insert tubes into various parts of the digestive tract of cows or sheep to study how these animals digest different types of feed.)

With the tube in place, Beaumont was able to tie a string onto small pieces of meat or vegetables, insert the food directly into St. Martin's stomach, and draw it out after various intervals. Beaumont also removed liquid from inside the stomach and observed how this gastric (stomach) juice acted on food in vitro. His experiments showed that gastric juice digests food—particularly meat.

Theodor Schwann later purified the enzyme that is responsible for digesting proteins in the stomach and named it **pepsin**. Because it destroys proteins, biologists hypothesized that pepsin must be synthesized and stored in cells while it is in an inactive form—otherwise it would kill the cells that make it.

In 1870 a biologist established through careful microscopy that granules occur in specialized stomach cells called chief cells. These granules were hypothesized to be a pepsin precursor. Follow-up work showed that this hypothesis was correct. The precursor compound found in chief cells, which came to be called **pepsinogen**, is converted to active pepsin by contact with the acidic environment of the stomach.

Secretion of a protein-digesting enzyme in inactive form is important: It prevents destruction of proteins in the cells where the enzyme is synthesized.

WHICH CELLS PRODUCE STOMACH ACID? The acidic environment of the human stomach denatures (unfolds) proteins and makes it possible for pepsin to work efficiently. But where does the acid come from?

Researchers who were studying the anatomy of the stomach wall noticed clusters of distinctive **parietal cells** located in pits that communicate with the lumen of the stomach (**Figure 43.10a**). An investigator also documented that the shape and activity of these cells appeared to vary as the digestion of a meal proceeded. On the basis of these observations, he inferred that parietal cells are the source of the HCl in gastric juice, which may have a pH as low as 1.5.

Earlier microscopists had shown that another type of cell, called a **mucous cell**, secretes the mucus that is found in gastric juice. Mucus lines the gastric epithelium and protects the stomach from damage by HCl. To summarize, these anatomical studies showed that the epithelium of the stomach contains several types of secretory cells, each of which is specialized for a particular function.

HOW DO PARIETAL CELLS SECRETE HCl? The first clues about how parietal cells manufacture hydrochloric acid emerged in the late 1930s, when a researcher found a high concentration of an enzyme called carbonic anhydrase in parietal cells.

This result was interesting because **carbonic anhydrase** catalyzes the formation of carbonic acid (H_2CO_3) from carbon dioxide and water. In solution, the carbonic acid that is formed immediately dissociates to form a proton and the bicarbonate ion (HCO_3^-):

$$CO_2 + H_2O \rightleftharpoons H_2CO_3 \rightleftharpoons H^+ + HCO_3^-$$

A second clue to the formation of HCl came in the 1950s, when transmission electron microscopes allowed researchers to analyze parietal cells at high magnification (see **BioSkills 10** in Appendix A). The micrographs showed that parietal cells are packed with mitochondria. Because mitochondria produce ATP, the structure of parietal cells suggested that they might function in active transport.

Later work confirmed this hypothesis by showing that the protons formed by the dissociation of carbonic acid are actively pumped into the lumen of the stomach. Subsequent studies showed that chloride ions from the blood enter parietal cells in exchange for bicarbonate ions, via a cotransport protein, and then move into the lumen through a chloride channel. **Figure 43.10b** diagrams the current model for HCl production.

ULCERS AS AN INFECTIOUS DISEASE An **ulcer** is a hole in an epithelium that damages the underlying basement membrane and tissues. Ulcers in the lining of the stomach or in the duodenum—the initial section of the small intestine—can result in intense abdominal pain.

For decades, physicians thought that gastric and duodenal ulcers resulted from the production of excess acid in the stomach. They treated ulcers by prescribing alkaline compounds (bases—see Chapter 2) that neutralized hydrochloric acid in the stomach.

In 1983, however, Robin Warren and Barry Marshall published data indicating that ulcers were associated with infections from a bacterium called *Helicobacter pylori.* Instead of being caused by environmental influences like acid-rich diets and psychological factors like anxiety, Warren and Marshall hypothesized that ulcers were an infectious disease.

This hypothesis met with intense skepticism, for two reasons: **(1)** the low acidity of the stomach was thought to sterilize the environment, and **(2)** treating ulcers as a bacterial rather than environmental disease required a radical change in thinking. In science, radical changes in thinking require exceptionally high standards of evidence.

To help provide this evidence, Marshall performed an experiment on himself. After a colleague cultured *H. pylori* in his lab, Marshall drank fluid from a petri plate where the bacteria were growing. As predicted, he developed gastritis—an inflammation of the stomach that precedes development of ulcers.

This experiment fulfilled several requirements of Koch's postulates for linking an organism to a specific disease (see Chapter 28). Subsequent work, in labs throughout the world, helped cement the link between *H. pylori* infection and ulcers. Physicians now routinely prescribe antibiotics to relieve ulcers.

THE RUMINANT STOMACH It is common for animals to have a stomach or stomach-like organ. The structure and function of this organ can vary, however, depending on the nature of the diet. In cattle, sheep, goats, deer, antelope, giraffe, and pronghorn—species that are collectively called **ruminants**—the stomach is specialized for digesting cellulose—not proteins.

Mammals, like birds, lack the enzymes called cellulases required to digest cellulose. Yet cellulose is the main carbohydrate in the leaves, stems, and twigs that ruminants ingest.

Like the hoatzin and kakapo, ruminants are able to harvest energy from cellulose thanks to a combination of specialized anatomical structures and symbiotic relationships with bacteria and unicellular protists. When **symbiosis** occurs, members of two different species live in close physical contact with each other.

As **Figure 43.11** shows, ruminant mammals have four-chambered stomachs that are folded in complex ways.

1. Food initially enters the largest chamber, the rumen, which serves as a fermentation vat. The rumen is packed with symbiotic bacteria and protists. These organisms have enzymes capable of breaking apart the chemical bonds in cellulose, yielding glucose. The rumen is an oxygen-free environment, and the symbiotic organisms produce ATP from this glucose via fermentation, releasing fatty acids as a by-product (see Chapter 9).
2. The chamber adjacent to the rumen, called the reticulum, is similar in function. After plant material has been partially digested in the rumen and the reticulum, the animal regurgitates portions of that material into its mouth, forming a cud. The ruminant chews that regurgitated material further to enhance mechanical breakdown, then re-swallows it.
3. Processed food that moves out of the rumen enters the third chamber, or omasum, where water is removed.
4. The final chamber is the abomasum, which contains the ruminant's own digestive enzymes and corresponds to a true stomach.

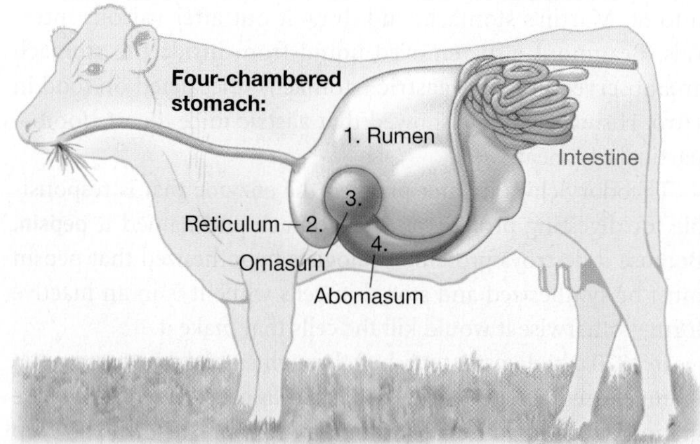

FIGURE 43.11 Ruminant Stomachs Facilitate the Digestion of Cellulose by Symbiotic Organisms. Ruminants obtain many of their nutrients from symbiotic bacteria and protists that live in the rumen and reticulum chambers of the stomach.

Most of a ruminant's food consists of (1) fatty acids and other compounds that are released as waste products of fermentation reactions in symbiotic organisms, and (2) the symbiotic cells themselves.

THE AVIAN GIZZARD The avian gizzard is another prominent type of modified stomach. Birds do not have teeth and cannot chew food into small pieces. Instead, most species swallow sand and small stones that lodge in the gizzard. As this muscular sac contracts, food is pulverized by the grit.

The gizzard is particularly large and strong in bird species that eat coarse foods such as seeds and nuts. The gizzard of a wild turkey, for example, can crack large walnuts.

Like the crop, the gizzard is interpreted as an adaptation that allows birds to ingest food quickly—by avoiding the need to chew—and digest it later. Biologists invoke the same hypothesis to explain why ruminants chew the cud. The ability to regurgitate material and finish chewing, while hiding in a place safe from predators, is thought to increase fitness.

The Small Intestine

In humans, the stomach is responsible for mixing the contents of a meal into a homogenous slurry, mechanically breaking up food material, and providing the acid and enzymes required to partially digest proteins. Peristalsis in the stomach wall then moves small amounts of material through the valve created by a sphincter muscle at the base of the stomach and into the small intestine.

The **small intestine** is a long tube that is folded into a compact space between the stomach and the last major section of the digestive tract—the large intestine. In the small intestine, partially digested food mixes with secretions from the pancreas and the liver and begins a journey of about 6 m (20 ft). When passage through this structure is complete, digestion is finished and most nutrients—along with large quantities of water—have been absorbed.

FOLDING AND PROJECTIONS INCREASE SURFACE AREA The surface area available for nutrient and water absorption in the small intestine is nothing short of remarkable. As **Figure 43.12** shows, the organ's epithelial tissue is folded and covered with fingerlike projections called **villi** (singular: **villus**). In turn, the cells that line the surface of villi have tiny projections on their apical surfaces called **microvilli** (singular: **microvillus**). Microvilli project into the lumen of the digestive tract.

If the small intestine lacked folds, villi, and microvilli, it would have a surface area of about 3300 cm^2 (3.6 ft^2). Instead, the epithelium covers about 2 million cm^2 (over 2200 ft^2)—an area about the size of a tennis court.

The enormous surface area of the small intestine increases the efficiency of nutrient absorption. And because each villus contains blood vessels and a lymphatic vessel called a **lacteal**, nutrients pass quickly from epithelial cells into the body's transport systems. (The circulatory system and lymphatic system are analyzed in detail in Chapters 44 and 49, respectively.)

✔If you understand the importance of surface area in the small intestine, you should be able to explain why surface area is so much higher in this structure than it is in the stomach or esophagus.

To understand how digestion is completed and absorption occurs, let's explore what happens to proteins, lipids, and carbohydrates as they move through this section of the digestive tract—again using humans as a model organism.

PROTEIN PROCESSING BY PANCREATIC ENZYMES The acidic environment of the stomach denatures proteins, destroying their secondary and tertiary structures. In addition, pepsin cleaves the peptide bonds next to certain amino acids, reducing long polypeptides to relatively small chains of amino acids. In the small intestine, protein digestion is completed so that individual amino acids can enter the bloodstream and be transported to cells throughout the body.

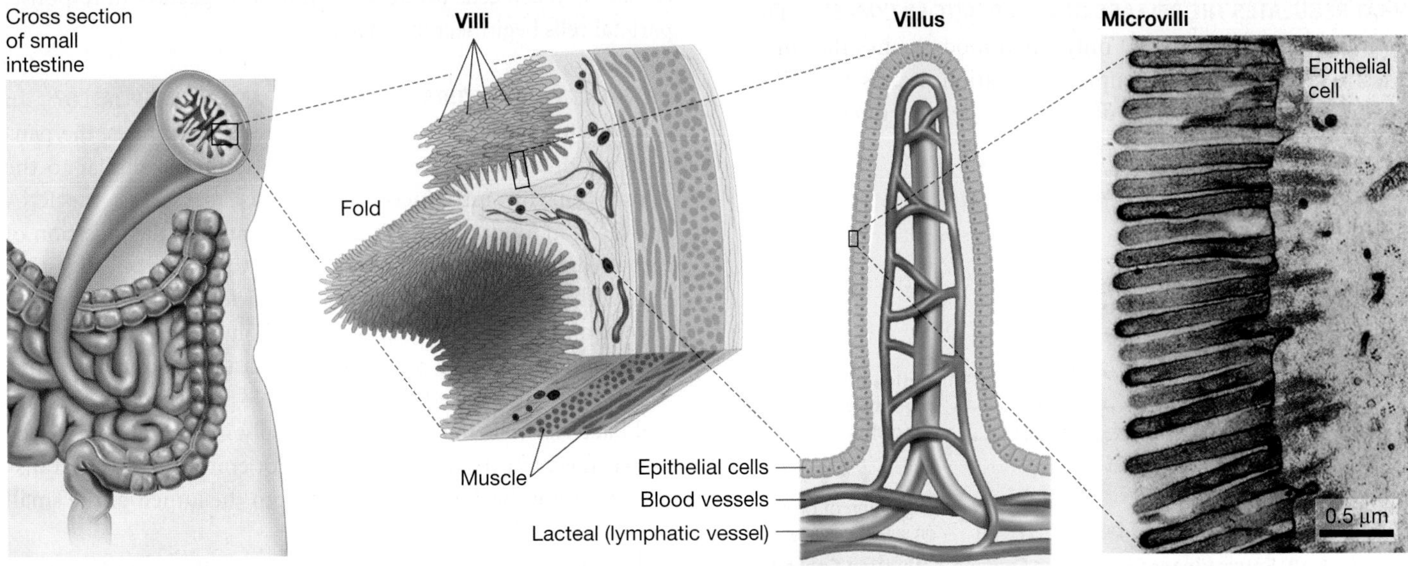

FIGURE 43.12 The Small Intestine Has an Extremely Large Surface Area. The villi that project from folds in the small intestine are covered with microvilli (colorized brown in the micrograph at the far right).

How does protein digestion occur? By the end of the nineteenth century, it had been established that enzymes in the small intestine digest polypeptides to monomers. Later work showed that each of these protein-digesting enzymes, or **proteases**, is specific to certain types or configurations of amino acids in a polypeptide chain. Thus, a suite of proteases is required to completely digest polypeptides to amino acid monomers.

In addition, by 1900 biologists had determined that proteases are synthesized in an inactive form in the **pancreas**, which is connected to the small intestine by the pancreatic duct. Like the production of inactive pepsinogen by chief cells in the stomach, the production of digestive enzymes in an inactive conformation prevents pancreatic cells from digesting themselves.

It took decades of work to understand how pancreatic enzymes are activated in the small intestine. In 1900 a researcher showed that contact with juice from the upper part of the small intestine activates pancreatic enzymes. Because activation did not occur when he heated the intestinal juice, and because heat denatures proteins, he hypothesized that the agent responsible for activating the pancreatic enzymes was also an enzyme. He called the unknown enzyme enterokinase.

Decades later, a researcher succeeded in purifying a pancreatic enzyme called **trypsinogen** and demonstrated that enterokinase activates it in vitro. Enterokinase activates trypsinogen by phosphorylating it, resulting in the active enzyme **trypsin**. Trypsin then triggers the activation of other protein-digesting enzymes, such as chymotrypsin, elastase, and carboxypeptidase. These enzymes are also synthesized by the pancreas and secreted in an inactive form.

Figure 43.13 summarizes the sequential activation of digestive enzymes in the small intestine. Enterokinase triggers the activation of trypsinogen to trypsin, which in turn triggers the activation of the three other protein-digesting enzymes. Once these enzymes are activated in the upper reaches of the small intestine, each begins cleaving specific peptide bonds. Eventually polypeptides are broken up into amino acid monomers.

WHAT REGULATES THE RELEASE OF PANCREATIC ENZYMES? Digestive enzymes are needed only when food reaches the small intestine. Based on this simple observation, it was logical to predict that their release would be carefully controlled.

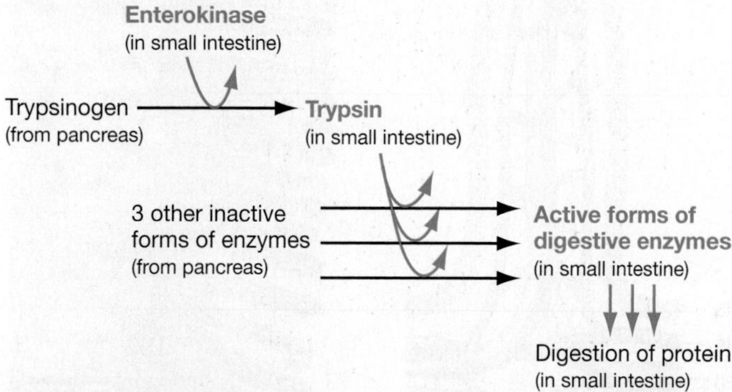

FIGURE 43.13 Enterokinase Triggers an Enzyme-Activation Cascade in the Small Intestine.

A classic experiment by William Bayliss and Ernest Starling, published in 1902, established how pancreatic enzymes are controlled. Bayliss and Starling began by cutting the nerves that connect to the pancreas and small intestine of a dog. Electrical signaling between the two organs was now impossible. But when the researchers introduced a weak HCl solution into the upper reaches of the animal's small intestine, to simulate the arrival of material from the stomach, its pancreas secreted in response.

This observation was startling: The small intestine had successfully signaled the pancreas that food had arrived, even though the nerves connecting the two organs had been cut.

Starling hypothesized that a chemical messenger must be involved, and that the chemical messenger must originate in the small intestine and travel to the pancreas via the blood. He tested this idea by cutting off a small piece of the small intestine, grinding it up, and injecting the resulting solution into a vein in the animal's neck. Minutes later, the pancreas sharply increased secretion.

Bayliss and Starling had discovered the first **hormone**—a chemical messenger that influences physiological processes at very low concentrations. The molecule they detected, which they called **secretin**, is produced by the small intestine in response to the arrival of food from the stomach.

Follow-up work showed that secretin's primary function is to induce a flow of bicarbonate ions (HCO_3^-) from the pancreas to the small intestine. The bicarbonate is important because it neutralizes the acid arriving from the stomach.

Researchers also discovered a second hormone produced in the small intestine, called cholecystokinin (pronounced *ko-la-sis-toe-KIN-in*), that induces secretion from the liver as well as the pancreas. **Cholecystokinin** (literally, "bile-bag-mover") stimulates the secretion of digestive enzymes from the pancreas and the secretion of molecules from the gallbladder that are involved in processing lipids.

Hormones are involved in stomach function as well. For example, after being stimulated by nerves or the arrival of food, certain stomach cells produce the hormone **gastrin**. In response, parietal cells begin secreting HCl.

HOW ARE CARBOHYDRATES DIGESTED AND TRANSPORTED? In addition to manufacturing protein-digesting enzymes, the pancreas produces nucleases and an amylase that is similar to the salivary enzyme introduced earlier. **Nucleases** digest the RNA and DNA in food; pancreatic amylase continues the digestion of carbohydrates that began in the mouth. Carbohydrate digestion ends with the production of monosaccharides such as glucose.

When digestion of proteins and carbohydrates is complete in the small intestine, the resulting slurry is a treasure trove of nutrients and water mixed with indigestible plant fibers from food and bacterial cells that live symbiotically in the gut. What molecular mechanisms make it possible for epithelial cells to transport monosaccharides like glucose from the lumen of the small intestine into the bloodstream?

It is logical to predict two general principles about nutrient absorption: (1) It is highly selective, meaning proteins in the plasma membranes of microvilli are responsible for bringing specific nutrients into the cell; and (2) it is active, meaning it requires an expenditure of ATP to bring nutrients into the epithelium across a concentration gradient.

Work over the past several decades has shown that both predictions are correct. One of the key results grew out of a series of experiments during the 1980s, which established that glucose absorption depends on the presence of an electrochemical gradient favoring an influx of sodium ions into the epithelium. Based on this observation, biologists hypothesized that the apical membranes of these cells must contain a series of cotransporters—proteins that would bring a nutrient molecule into the cell along with sodium ions. Work focused on glucose, because it is such a fundamentally important nutrient. To confirm that a sodium-glucose cotransporter exists, investigators set out to find the gene that codes for the hypothesized membrane protein.

The researchers began by purifying mRNAs from rabbit intestinal cells (**Figure 43.14**), which presumably were transcribing the cotransporter genes. Then the team separated the mRNAs by size via gel electrophoresis (see **BioSkills 9** in Appendix A) and injected one of each type of mRNA into a series of frog eggs—cells that do not normally transport glucose. The frog cells translated the rabbit mRNAs into proteins.

When tested, one of the experimental eggs was able to import Na^+ and glucose in tandem. Based on these data, the biologists inferred that this egg had received the mRNA for the rabbit Na^+-glucose cotransporter. Using techniques introduced in Chapter 19, the researchers made a DNA copy of the mRNA, determined the sequence of the gene, and inferred the amino acid sequence of the membrane protein.

The discovery of the Na^+-glucose cotransporter inspired a three-step model for glucose absorption:

1. Na^+/K^+-ATPase (sodium-potassium pumps) in the basolateral membrane of the epithelial cells creates an electrochemical gradient that favors the entry of Na^+.
2. Glucose from digested food enters the cell along with sodium via the cotransporter.
3. Glucose diffuses into nearby blood vessels through a glucose carrier in the basolateral membrane.

If this configuration of pumps, cotransporters, and carriers sounds familiar, there is a good reason: The same combination of membrane proteins occurs in the proximal tubule of the kidney, where the proteins are responsible for the reabsorption of sodium and glucose from urine (see Chapter 42).

Follow-up work showed that in the small intestine—just as in the proximal tubule—other cotransporters are responsible for the absorption of other nutrients, with specific channels and carriers in the basolateral membrane responsible for their transport to the blood.

EXPERIMENT

QUESTION: How is glucose transported into epithelial cells of the small intestine?

HYPOTHESIS: Glucose enters epithelial cells along with sodium ions via a Na^+-glucose cotransporter protein.

NULL HYPOTHESIS: Glucose transport does not depend on Na^+ transport.

EXPERIMENTAL SETUP:

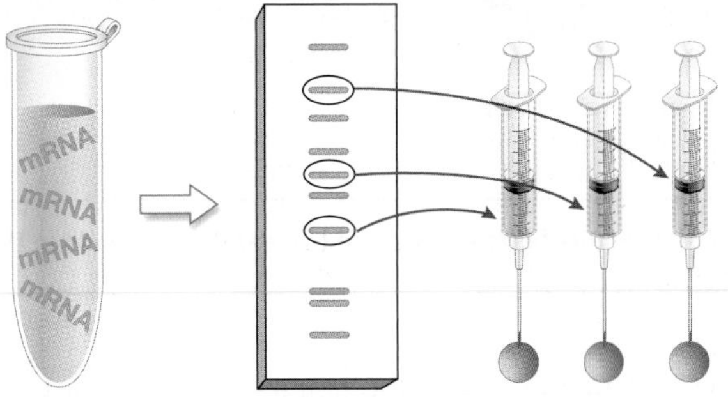

1. Purify mRNA from intestinal cells.

2. Separate mRNAs by size via gel electrophoresis.

3. Inject individual mRNAs into frog eggs. Test each egg—can it absorb Na^+ and glucose?

PREDICTION: An egg will be able to absorb Na^+ and glucose, because it received the mRNA that codes for the Na^+-glucose cotransporter.

PREDICTION OF NULL HYPOTHESIS: None of the eggs will be able to absorb Na^+ and glucose.

RESULTS:

Na^+, glucose in medium	Na^+, glucose in medium	Na^+, glucose in medium
Egg 1	Egg 2	Egg 3
ABSORBED	Not absorbed	Not absorbed

CONCLUSION: The egg that absorbs Na^+ and glucose received the mRNA from the Na^+-glucose cotransporter gene.

FIGURE 43.14 The Experimental Protocol for Locating the Na^+-Glucose Cotransporter Gene.

SOURCE: Wright, E. M. 1993. The intestinal Na^+/glucose cotransporter. *Annual Review of Physiology* 55: 575–589.

✔**QUESTION** Why did the researchers inject the RNAs into frog eggs, instead of into rabbit epithelial cells?

DIGESTING LIPIDS: BILE AND TRANSPORT The pancreatic secretions include digestive enzymes that act on fats as well as proteins and carbohydrates. Like the lingual lipase added to saliva in the mouth, the enzyme **pancreatic lipase** breaks certain bonds present in complex fats and results in the release of fatty acids and other small lipids.

Recall from Chapter 6 that fats are insoluble in water. As a result, they tend to form large globules as they are churned in the stomach. Before pancreatic lipase can act, large fat globules that emerge from the stomach must be broken up—a process known as **emulsification**.

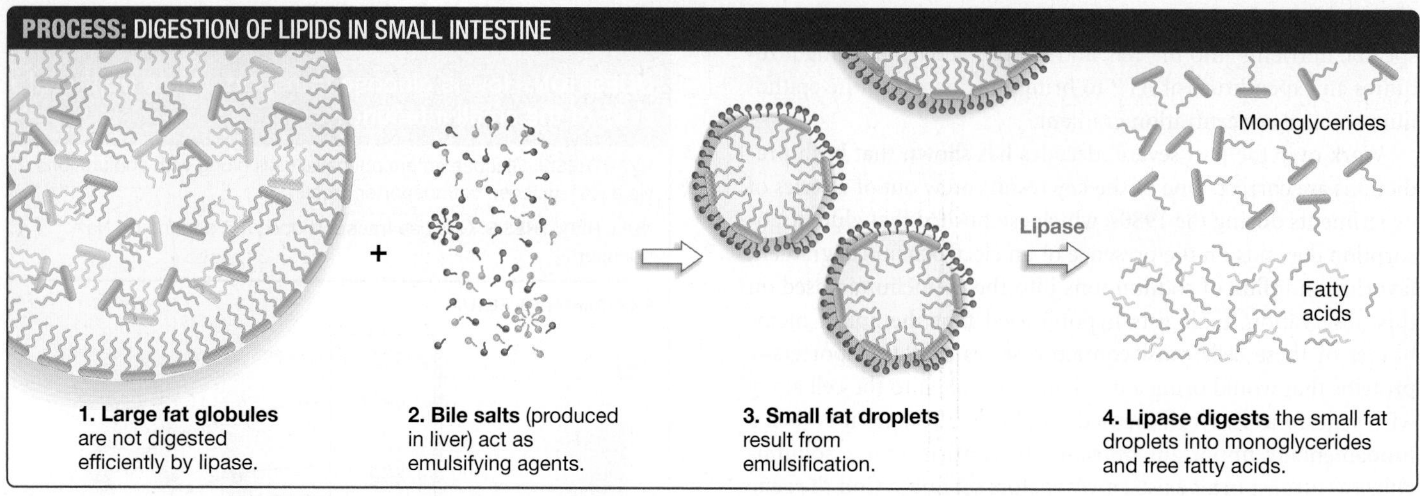

FIGURE 43.15 Emulsifying Agents and Lipases Digest Lipids in the Small Intestine. Once bile salts break up large fat globules, lipase can digest fats efficiently.

In the small intestine, emulsification results from the action of small lipids called bile salts. As **Figure 43.15** shows, bile salts function like the detergents that researchers use to break up the lipids in plasma membranes (see Chapter 6).

Bile salts are synthesized in the **liver**, an organ that performs an array of functions related to digestion, and secreted in a complex solution called **bile**, which is stored in the **gallbladder**. When bile enters the small intestine, it raises the pH and emulsifies fats. Once fats are broken into small globules with high surface area, they can be attacked by enzymes and digested. **Table 43.3** summarizes the major digestive enzymes.

What happens to the monoglycerides and fatty acids released by lipase activity? An answer emerged when researchers injected radioactively labeled fatty acids into the small intestines of laboratory rats. Most of the radioactive label entered epithelial cells and attached to a protein named **fatty-acid binding protein**. Later, other researchers established that a fatty-acid binding protein also occurs in the membranes of these cells.

Once fatty-acid binding proteins bring lipids into the cell, they are processed into protein-coated globules called **chylomicrons**. Chylomicrons diffuse into lacteals—such as the green vessel in Figure 43.12—near the epithelial cells. The lacteals merge with larger lymph vessels, which merge with larger blood vessels. In this way, fats enter the bloodstream without clogging small blood vessels. Eventually, the products of fat digestion end up in adipose tissue and other tissues.

HOW IS WATER ABSORBED? When solutes from digested material are brought into the epithelium of the small intestine via active transport, water follows passively by osmosis. This is an important mechanism for (1) absorbing water that has been ingested, and (2) reclaiming liquid that was secreted into the digestive tract in the form of saliva, mucus, and pancreatic fluid.

This mechanism of water absorption inspired an important medical strategy called oral rehydration therapy. If a patient has diarrhea, clinicians frequently prescribe dilute solutions of glucose and electrolytes to be taken orally. When the glucose in the drink is absorbed in the small intestine through sodium-glucose cotransporters, enough water and sodium follow to prevent the life-threatening effects of dehydration. This simple medication saves thousands of lives every year. ✔**If you understand this concept, you should be able to predict two effects of a molecule that selectively blocks the sodium-glucose cotransporter.**

To review how carbohydrates, proteins, and fats are digested and absorbed, go to the study area at *www.masteringbiology.com.*

 **Web Activity** The Digestion and Absorption of Food

The Cecum and Appendix

Some birds house cellulose-digesting bacteria and protists in their crop, and ruminants house cellulose-digesting symbionts in a modified stomach. But in rabbits, many rodents, some marsupials, and certain leaf-eating primates, fermentation by symbionts occurs in an organ called the cecum.

The **cecum** is an outpocketing of the digestive tract located at the start of the large intestine. A dead-ended or "blind" sac, the cecum is greatly enlarged in species that use it as a fermentation chamber for processing cellulose. Rabbits and some other species that ferment food in the cecum are also able to excrete the structure's contents in pellets, which are then re-ingested and passed through the digestive tract a second time. This practice is called **coprophagy** ("excrement-eating").

In humans, the cecum is dramatically reduced in size compared to most other primates, and functions in defense against invading bacteria and viruses instead of as a fermentation vat. Because its size and function differ from those of a cecum, it is called the **appendix.**

The Large Intestine

By the time digested material reaches the large intestine of a human, a large amount of water (~5 liters per day in total) and virtually all of the available nutrients have been absorbed. The primary function of the **large intestine** is to compact the wastes that remain and absorb enough water to form feces.

SUMMARY TABLE 43.3 **Digestive Enzymes in Mammals**

Digestion is accomplished by the enzymes listed here, by HCl produced in the stomach in response to the hormone gastrin, and by bile salts from the liver. Bile salts are stored in the gallbladder. They are released in response to the hormone cholecystokinin and emulsify fats in the small intestine.

	Where Synthesized	Regulation	Function
Carboxypeptidase	Pancreas	Released in response to cholecystokinin from small intestine; activated by trypsin	In small intestine, breaks peptide bonds in polypeptides—releasing amino acids
Chymotrypsin	Pancreas	Released in response to cholecystokinin from small intestine; activated by trypsin	In small intestine, breaks peptide bonds in polypeptides—releasing amino acids
Elastase	Pancreas	Released in response to cholecystokinin from small intestine; activated by trypsin	In small intestine, breaks peptide bonds in polypeptides—releasing amino acids
Lingual lipase	Salivary glands	Released in response to taste and smell stimuli	In mouth, breaks bonds in fats—releasing fatty acids and monoglycerides
Nucleases	Pancreas	Released in response to cholecystokinin from small intestine	In small intestine, break apart nucleic acids—releasing nucleotides
Pancreatic amylase	Pancreas	Released in response to cholecystokinin from small intestine	In small intestine, breaks apart carbohydrates—releasing sugars
Pancreatic lipase	Pancreas	Released in response to cholecystokinin from small intestine	In small intestine, breaks bonds in fats—releasing fatty acids and monoglycerides
Pepsin	Stomach	Released in inactive form (pepsinogen); activated by low pH in stomach lumen	In stomach, breaks peptide bonds between certain amino acids in proteins—releasing polypeptides
Salivary amylase	Salivary glands	Released in response to taste and smell stimuli	In mouth, breaks apart carbohydrates—releasing sugars
Trypsin	Pancreas	Released in inactive form (trypsinogen) in response to cholecystokinin from small intestine; activated by hormone enterokinase from small intestine	In small intestine, breaks specific peptide bonds in polypeptides—releasing amino acids

These processes occur in the **colon**—the main section of the structure. Feces are held in the **rectum**, which is the final part of the large intestine, until they can be excreted. Although the kidneys are responsible for maintaining water balance in the body, water absorption in the large intestine is important for individuals to remain well hydrated.

AQUAPORINS PLAY A KEY ROLE IN WATER REABSORPTION To identify the mechanism of water absorption in the large intestine, researchers have focused on aquaporins, which were introduced in Chapter 6. Recall that **aquaporins** are water channels in plasma membranes that provide a mechanism for increasing the rate of water movement via osmosis. The best studied of these proteins, called AQP1 for aquaporin 1, is common in the nephrons of the kidney. AQP1 is one of the proteins responsible for water reabsorption from urine along the osmotic gradient described in Chapter 42.

To date, four distinct aquaporins have been found in the large intestines of rats, mice, and humans. The aquaporins called AQP3 and AQP4, for example, are located in the basolateral membrane of cells in the epithelium of the large intestine in mice. Researchers are working to unravel exactly how the activity of these aquaporins is regulated. To date, the molecular mechanisms responsible for water absorption in the large intestine are not as well understood as are those in the collecting duct of nephrons.

VARIATION IN STRUCTURE AND FUNCTION It is not unusual for animal species to lack a colon and rectum entirely. For example, Chapter 42 pointed out that in insects, the posteriormost portion of the digestive tract, called the hindgut, functions to reabsorb water and ions and form feces. Among vertebrates, the various lineages of fish have no large intestine at all.

In contrast, horses, elephants, tapirs, and certain other plant-eating species have extremely large colons. In these species, the colon is packed with symbiotic bacteria and protists that digest cellulose and ferment the glucose that is released. Plant-eating animals can house cellulose-digesting symbionts in the esophagus (crop), stomach, cecum, or large intestine.

Another striking structural variation occurs in the rectum. In amphibians, reptiles, and birds, the urine that is produced by the kidneys empties into an enlarged portion of the large intestine called the **cloaca**. Instead of having two orifices for excretion, these species have one.

CHECK YOUR UNDERSTANDING

If you understand that . . .

- In humans and many other animals, digestion begins with the mechanical breakdown of food into small pieces that are then acted on by acids, enzymes, emulsifying agents, and other chemical treatments.
- Distinct compartments within a digestive tract have distinct structures and functions.
- Nutrients are actively and selectively absorbed through specific membrane proteins in epithelial cells of the digestive tract.

✓ You should be able to . . .

1. Explain how each compartment in the human digestive tract aids the ingestion and digestion of food, absorption of nutrients, and excretion of wastes.
2. Predict the consequences of treating a person with drugs that inhibit the release of bile salts, that inactivate trypsin, or that block the action of the Na^+-glucose cotransporter in the epithelial cells of the small intestine.

Answers are available in Appendix B.

43.4 Nutritional Homeostasis—Glucose as a Case Study

When digestion is complete, amino acids, fatty acids, ions, and sugars enter the bloodstream and are delivered to the cells that need them. Too much of a nutrient, or too little, can be problematic or even fatal, however.

The illness **diabetes mellitus** is a classic example of nutrient imbalance. People with diabetes experience abnormally high levels of glucose in their blood. Over the course of a lifetime, chronically elevated blood glucose can lead to an array of complications including blindness, impaired circulation, and heart failure. What causes the imbalance?

The Discovery of Insulin

In 1879, researchers removed the pancreas from a dog and observed that diabetes developed. This experiment suggested that the pancreas secretes a compound involved in removing glucose from the blood.

When other investigators cut up pancreatic tissues and injected extracts into diabetic dogs, however, they observed no response. Frustrated, they realized that digestive enzymes in the pancreas were probably destroying the active agent during the extraction process.

Then in 1921, Frederick Banting and Charles Best conducted a breakthrough experiment:

1. They began by tying off a dog's pancreatic duct—the tube where digestive enzymes collect and flow to the small intestine. The logic was that blocking the secretion of digestive enzymes might kill the cells that synthesize them.
2. The investigators waited several weeks for the cells near the duct to die and then removed the gland.
3. They froze the pancreatic tissue, ground it up, and injected an extract into a diabetic dog.
4. To their delight, the dog's blood-sugar levels stabilized, and the dog became more active and healthy looking.

After Banting and Best repeated the experiment and observed the same result, they grew increasingly confident that they had located the source of the molecule responsible for controlling diabetes. The molecule came to be called insulin.

Insulin's Role in Homeostasis

Insulin is a hormone that is produced in the pancreas when blood-glucose levels are high. It travels through the bloodstream and binds to receptors on cells throughout the body (see Chapter 47 for more detail on the structure and function of hormones and other chemical signals).

In response, cells that have insulin receptors increase their rate of glucose uptake and processing. Specifically:

- Cells in the liver and skeletal muscle synthesize more glycogen from glucose monomers.
- Cells that store lipids synthesize more storage forms of fat, using glucose as a precursor.

As a result, glucose levels in the blood decline.

If blood-glucose levels fall, as they do after hard exercise or when food is lacking, cells in the pancreas secrete a hormone called **glucagon**. In response to glucagon, cells in the liver catabolize glycogen and produce glucose via **gluconeogenesis** (the synthesis of glucose from non-carbohydrate compounds). As a result, glucose levels in the blood rise (**Figure 43.16**).

Insulin and glucagon interact to form a negative feedback system capable of achieving homeostasis with respect to glucose concentrations in the blood. To review this homeostatic system, go to the study area at *www.masteringbiology.com*.

(MB) **BioFlix™** Homeostasis: Regulating Blood Sugar, **Web Activity** Understanding Diabetes Mellitus

Diabetes Can Take Several Forms

Diabetes mellitus develops in people who (1) do not synthesize insulin or (2) have defective versions of the receptor for insulin. The first condition is called type 1 diabetes mellitus, or insulin-dependent diabetes; the second condition is type 2 diabetes mellitus, or non-insulin-dependent diabetes. In both cases, effector cells do not receive the signal that would result in a drop in blood-glucose levels.

Glucose imbalance has a direct effect on urine formation. Normally, signals from insulin keep blood-glucose levels low

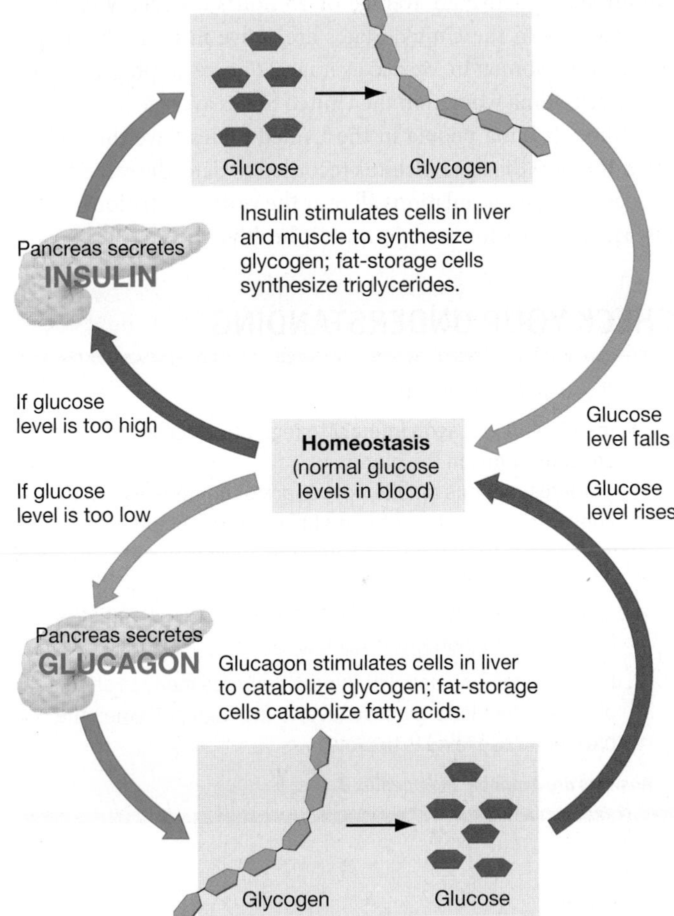

FIGURE 43.16 Insulin and Glucagon Provide Negative Feedback in a Homeostatic System. Both insulin and glucagon are secreted by cells in the pancreas but have opposite effects on blood-glucose concentrations.

✔**QUESTION** Which arrow is disrupted in individuals with type 2 diabetes mellitus? Which arrow is disrupted in individuals with type 1 diabetes mellitus?

enough that all of the glucose can be reabsorbed from the filtrate formed in the kidney. But when blood glucose levels are high, so much glucose enters the nephron that it cannot all be reabsorbed. High glucose concentrations in the filtrate increase its osmolarity and prevent water from being reabsorbed by osmosis. More water leaves the body, leading to high urine volume in both types of diabetes.

Recall from Chapter 42 that a disease called diabetes insipidus develops in people whose collecting ducts fail to reabsorb water. As a result, these individuals also produce large amounts of urine.

Production of copious urine occurs in all three types of diabetes, and inspired the illness's names. The word diabetes means "to run through"; water "runs through" people with diabetes. In addition, the word insipidus means "tasteless"; mellitus means "honeyed (sweet)." Before chemical methods of analyzing urine were available, physicians would distinguish between diabetes insipidus and diabetes mellitus by tasting the patient's urine. The test was definitive because glucose levels in the urine differ markedly in the two types of illness.

Currently, type 1 diabetes mellitus is treated with insulin injections and careful attention to diet; type 2 diabetes is managed primarily through prescribed diets and monitoring blood-glucose levels, as well as taking drugs that increase cellular responsiveness to insulin. The challenge is to achieve homeostasis with respect to blood glucose in the absence of the body's normal regulatory mechanisms.

✔If you understand the difference between type 1 and type 2 diabetes mellitus, you should be able to explain how, in individuals with each disease and without disease, insulin receptors in the plasma membrane of liver cells interact with insulin and how the liver cells respond to high glucose concentrations in the blood.

The Type 2 Diabetes Mellitus Epidemic

An epidemic of diabetes mellitus is currently under way in certain human populations. In the United States, for example, the incidence of diabetes in adults jumped 90 percent between 1997 and 2007.

In the U.S., about 10.7 percent of individuals aged 20 and over are diabetic. When only African Americans in this age group are considered, the prevalence of diabetes jumps to 14.7 percent. In both instances, about 90–95% of cases are type 2 diabetes.

Because there is a strong association between the prevalence of type 2 diabetes mellitus in parents and children, researchers have long suspected that some individuals have a genetic predisposition for developing the disease. Using techniques introduced in Chapters 19 and 20, researchers have now identified alleles at four different genes that predispose individuals to type 2 diabetes.

However, there is also strong evidence that environmental conditions have an important impact on the incidence of type 2 diabetes. As an example, consider the Pima Indians of North America.

The Pima consist of two main populations—one in southwest Arizona in the United States and one in a remote area of the Sierra Madre mountains of Mexico. The bars graphed in **Figure 43.17a** on page 858 indicate the average percent of individuals with type 2 diabetes, by gender, in Mexican people who are not members of the Pima and in the two populations of Pima. Although researchers have been unable to find any significant genetic differences between the two groups of Pima, the data indicate there are dramatic differences in the incidence of type 2 diabetes mellitus.

The incidence of type 2 diabetes is correlated with obesity. As the data graphed in **Figure 43.17b** show, Pima from the U.S. are more likely to be obese than individuals in the other two populations in this study. A person who has a body mass index greater than or equal to 30 is considered obese. The **body mass index** is calculated as weight (kg) divided by height (m) squared.

The differences in body mass index among the Pima are explained by differences in physical activity: Pima men in Mexico

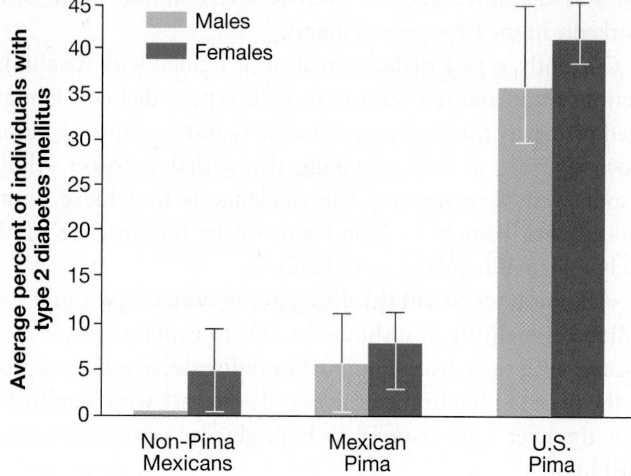

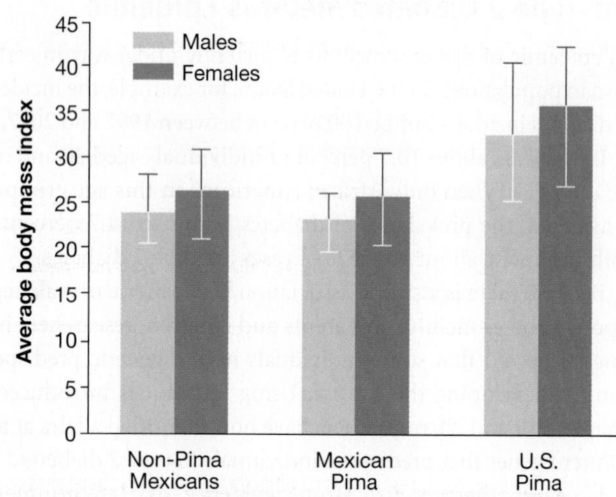

are physically active an average of 33 hours a week, while their counterparts in the United States are active just 12.1 hours per week. Pima women in Mexico average 22 hours of physical activity a week; Pima women in the United States average just 3.1.

Although Pima people in the United States have the highest rates of type 2 diabetes mellitus recorded to date, the same trends are occurring in populations all over the world. Nutrition-related diseases have become a major public health concern.

CHECK YOUR UNDERSTANDING

If you understand that . . .

- Insulin and glucagon interact to regulate glucose concentrations in the blood.
- Diabetes mellitus results from a lack of homeostasis in the concentration of the nutrient glucose in the blood.

✔ You should be able to . . .

1. Explain the similarities and differences between diabetes insipidus and diabetes mellitus.
2. Predict what a person with type 1 diabetes mellitus should do when the blood-glucose level is too high and when the blood-glucose level is too low.

Answers are available in Appendix B.

FIGURE 43.17 The Incidence of Type 2 Diabetes Mellitus Is Correlated with Obesity.

✔**QUESTION** In many human populations, incidence of type 2 diabetes mellitus has been increasing rapidly. Why?

CHAPTER 43 REVIEW

For media, go to the study area at www.masteringbiology.com

Summary of Key Concepts

Animals require an array of nutrients to stay healthy, including specific amino acids, vitamins, and elements, as well as organic compounds that act as building blocks in chemical synthesis or have high potential energy.

- The diets of animals include fats and carbohydrates that provide energy, proteins that furnish amino acids, vitamins that serve as coenzymes and perform other functions, ions required for water balance and for nerve and muscle function, and elements that are incorporated into molecules synthesized by cells.
- To determine the levels of nutrients that are needed for a particular animal species to sustain normal activities, researchers monitor the relationship between nutrient intake, the levels of nutrients maintained in the body, and health.

✔ You should be able to design an experiment to document the effects of magnesium deprivation in laboratory mice.

There is usually a close correspondence between the structure of an animal's mouthparts and the function of those mouthparts in capturing and processing food.

- All animals are heterotrophs.
- Most animals are mass feeders that obtain food by seizing and manipulating it with mouthparts such as teeth, jaws, or beaks or with special toxin-injecting organs.
- Natural selection has modified mouthparts in different species to act as efficient tools for obtaining particular types of food.

✔ You should be able to predict the types of mouthparts and the nature of the digestive tract in a mammal that eats only nectar from flowers.

Digestion occurs in the digestive tract, which is compartmentalized into organs that have specialized functions in the ingestion

and digestion of food, absorption of nutrients and water, or excretion of wastes.

- In most animals, the digestive tract begins at the mouth and ends at the anus.
- In many species, chemical digestion of food also begins in the mouth. In mammals, an enzyme in saliva called amylase begins to hydrolyze the bonds linking glucose monomers in starch, glycogen, and other carbohydrates.
- Once food is swallowed, it passes down the esophagus via peristalsis.
- Digestion continues in the stomach. In humans, the stomach is a highly acidic environment that denatures proteins and in which the enzyme pepsin begins the cleavage of peptide bonds that link amino acids.
- Food passes from the stomach into the small intestine, where it is mixed with secretions from the pancreas and liver.
- Secretions from the liver and pancreas are triggered by the hormones cholecystokinin and secretin, which are produced in the small intestine.
- In the small intestine, carbohydrate digestion is continued by pancreatic amylase; fats are emulsified by bile salts and digested by lipase; protein digestion is completed by a suite of pancreatic enzymes that are activated by enterokinase.
- Cells that line the small intestine absorb the nutrients released by digestion. In many cases, uptake is driven by an electrochemical gradient established by Na^+/K^+-ATPase that favors a flow of Na^+ into the cell.
- As solutes leave the lumen of the small intestine and enter cells, water follows by osmosis.
- Water reabsorption is completed in the large intestine, where feces form.
- The structure of organs in the digestive tract varies widely among species, in ways that support efficient processing of the food a particular species ingests.

✔ You should be able to explain why gastric bypass surgery, which makes the stomach smaller and allows food to bypass part of the small intestine, often leads to weight loss.

MB **Web Activity** The Digestion and Absorption of Food

Lack of homeostasis with respect to nutrients such as glucose can cause disease.

- Maintaining homeostasis with respect to nutrients is critical to health.
- Diabetes mellitus develops when concentrations of glucose in the blood are chronically too high.
- Type 1 diabetes mellitus is caused by a defect in the production of insulin—a hormone secreted by the pancreas that promotes the uptake of glucose from the blood.
- Type 2 diabetes mellitus is caused by a defect in the insulin receptor on the surface of cells.
- The development of type 2 diabetes is correlated with obesity and is reaching epidemic proportions in some populations.

✔ You should be able to explain why blood glucose concentration eventually falls when a person with diabetes mellitus eats a candy bar.

MB **BioFlix™** Homeostasis: Regulating Blood Sugar, **Web Activity** Understanding Diabetes Mellitus

Questions

✔ TEST YOUR KNOWLEDGE

Answers are available in Appendix B

1. What does secretin stimulate?
   a. secretion of HCO_3^- from the pancreas
   b. secretion of HCl from the stomach epithelium
   c. secretion of digestive enzymes from the pancreas
   d. uptake of glucose from the bloodstream by cells throughout the body
2. What is the structure and function of the cecum in rabbits?
   a. a large, blind sac that functions as a fermentation vat
   b. an enlarged section of the esophagus that functions as a fermentation vat
   c. one of four chambers in the stomach; functions as a fermentation vat
   d. enlarged portion of the large intestine that functions as a fermentation vat
3. What is an incomplete digestive tract?
   a. a digestive tract that is missing a major compartment—for example, fish lack a large intestine
   b. a ruminant-type tract, where food is regurgitated and re-chewed before digestion is completed
   c. a reduced or vestigial tract like that found in animals that absorb nutrition directly across the body wall
   d. a digestive system with a single opening for ingestion and excretion
4. In mammals, how and where are carbohydrates digested?
   a. by lipases in the small intestine
   b. by pepsin and HCl in the stomach
   c. by aquaporins in the large intestine
   d. by amylases in the mouth and small intestine
5. What role do bile salts play in the digestion of complex fats?
   a. They catalyze the cleavage of bonds leading to the release of fatty acids and other small lipids.
   b. They emulsify lipids, meaning that large masses of fat molecules are broken into smaller masses.
   c. They include fatty-acid binding proteins, which are involved in fat absorption.
   d. They activate the enzymes that are responsible for digesting fats.
6. How is water absorbed in the small intestine of mammals?
   a. through aquaporins
   b. The exact mechanism is not known.
   c. by sodium cotransporters
   d. by osmosis (following solutes)

✔TEST YOUR UNDERSTANDING

Answers are available in Appendix B

1. Explain how the structure of each of the following digestive tract compartments correlates with its function: the bird crop, cow rumen, and elephant large intestine.
2. Why is it logical that digestive enzymes are produced in an inactive form and then activated in the lumen of the digestive tract?
3. Do you accept the conclusion that cichlid pharyngeal jaws are adaptations that increase feeding efficiency? Why or why not?
4. Fish lack a large intestine. What is the adaptive significance of this trait?
5. Explain why oral rehydration therapy works.
6. Explain how insulin and glucagon provide negative feedback in a homeostatic system.

✔APPLYING CONCEPTS TO NEW SITUATIONS

Answers are available in Appendix B

1. Predict how the nutritional requirements of female mammals change during pregnancy and breastfeeding.
2. Predict the problems that would result from defects in each of the following molecules: pancreatic amylase, pepsin, fatty-acid binding protein, and aquaporins.
3. Scientists who backed the hypothesis that secretion from the pancreas is under nervous control strenuously objected to the experiment that led to the discovery of hormonal control. They claimed that the experiment was inconclusive because it was very likely that not all of the nerves that contact the small intestine had been cut. The biologists who did the experiment replied that even if not all the nerves had been cut, the result was still valid. In your opinion, who is correct? Why?
4. Among vertebrates, the large intestine occurs only in lineages that are primarily terrestrial (amphibians, reptiles, and mammals). Propose a hypothesis to explain this observation.

During intense exercise, animal circulatory systems deliver large amounts of oxygen to tissues and remove large amounts of carbon dioxide. This chapter explores how gas exchange occurs in animals that live in aquatic or terrestrial environments.

Gas Exchange and Circulation 44

Animal cells are like factories that run 24 hours a day. Inside the plasma membrane, the chemical reactions that sustain life produce a steady stream of wastes. Those reactions also require a steady input of raw materials.

Chapter 42 analyzed how waste materials are removed from the body and excreted; Chapter 43 examined how nutrients enter the body. This chapter focuses on two major questions:

1. How are two of the most important molecules in the economy of the cell—the oxygen (O_2) required for cellular respiration and the carbon dioxide (CO_2) produced by cellular respiration—exchanged with the environment?
2. How are these gases—along with wastes, nutrients, and other types of molecules—transported throughout the body?

Understanding gas exchange and circulation is fundamental to understanding how animals work. If either process fails, the consequences are dire. Consider the maladies that can develop when gas exchange or circulation are disrupted in humans: anemia, pneumonia, tuberculosis, malaria, heart disease, and stroke are just a few examples.

Let's begin with an overview of animal respiratory and circulatory systems, then plunge into the details of how gases are exchanged and transported.

KEY CONCEPTS

- Animals have to take in oxygen and expel carbon dioxide to sustain cellular respiration and stay alive. Terrestrial animals and aquatic animals face different challenges in performing gas exchange.
- Gas-exchange organs maximize the rate of O_2 and CO_2 diffusion by (1) presenting a large, thin surface area to the environment, and (2) maintaining a steep partial-pressure gradient that favors entry of O_2 and elimination of CO_2.
- Blood is a specialized tissue that transports gases, along with nutrients and wastes, in some animals. Hemoglobin is an oxygen-carrying protein that is extremely efficient at taking up oxygen in the lungs and delivering it to tissues.
- Circulatory systems use the pressure generated by one or more hearts to transport blood and other substances throughout the body.

44.1 The Respiratory and Circulatory Systems

When the mitochondria inside animal cells are producing ATP via cellular respiration, they consume oxygen and produce carbon dioxide. To support continued ATP production, cells have to obtain oxygen and expel excess carbon dioxide continuously (see **The Big Picture: Energy**, located after Chapter 10).

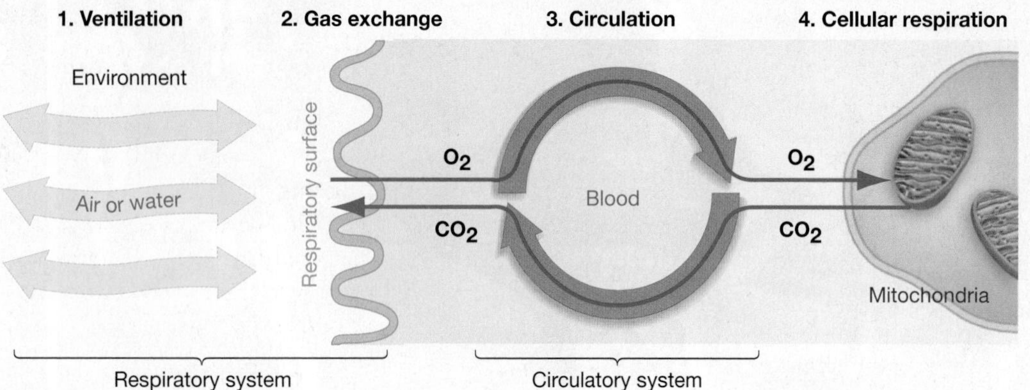

FIGURE 44.1 Gas Exchange Involves Ventilation, Circulation, and Respiration. In animals, oxygen and carbon dioxide are exchanged across the surface of a lung, a gill, the skin, or some other gas-exchange organ. In many species these gases are transported to and from cells—where gas exchange again takes place—in a fluid tissue such as blood.

How does gas exchange occur between an animal's environment and its mitochondria? In most cases, gas exchange involves the four steps illustrated in **Figure 44.1**:

1. Ventilation occurs when air or water moves through a specialized gas-exchange organ, such as lungs or gills.
2. Gas exchange takes place as O_2 and CO_2 diffuse between air or water and the blood at the respiratory surface.
3. Through circulation, the dissolved O_2 and CO_2 are transported throughout the body—along with nutrients, wastes, and other types of molecules—via the circulatory system.
4. Gas exchange between blood and cells occurs in tissues, where cellular respiration has led to low O_2 levels and high CO_2 levels.

Steps 1 and 2 are accomplished by the **respiratory system**: the collection of cells, tissues, and organs responsible for gas exchange between the individual and its environment. In essence, a respiratory system consists of structures for conducting air or water to a surface where gas exchange takes place.

In some animals the gas-exchange surface is the skin, but in most species it is located in a specialized organ like the lungs of tetrapods, the tracheae of insects, or the gills found in mollusks, arthropods, and fish. Section 44.3 analyzes the structure and function of gills, tracheae, and lungs in detail.

Step 3 in Figure 44.1 is usually accomplished by a **circulatory system**, which moves O_2, CO_2, and other materials around the body. In most species, a specialized, liquid transport tissue is propelled throughout the body by a muscular heart, through a system of vessels.

Given this broad overview of gas exchange and circulatory systems, let's plunge into the details of how they work—starting with the question of how oxygen and carbon dioxide move between an animal's body and its environment.

44.2 Air and Water as Respiratory Media

Gas exchange between the environment and cells is based on diffusion. Under normal conditions, oxygen concentrations are relatively high in the environment and low in tissues, while carbon dioxide levels are relatively high in tissues and low in the environment. Thus, oxygen tends to move from the environment into tissues, and carbon dioxide tends to move from tissues into the environment.

How much oxygen and carbon dioxide are present in the atmosphere versus the ocean? What factors influence how quickly these gases move by diffusion?

How Do Oxygen and Carbon Dioxide Behave in Air?

As **Figure 44.2a** shows, the atmosphere is composed primarily of nitrogen (N_2) and oxygen, with trace amounts of argon and CO_2. Nitrogen and argon are not important to animals living at sea level and are usually ignored in analyses of gas exchange.

The data in Figure 44.2a are actually slightly misleading, however. To understand why, consider that the percentage of O_2 in the atmosphere does not vary with elevation. The atmosphere at the top of Mt. Everest is composed of 21 percent oxygen just as it is at sea level. The key difference is that far fewer molecules of oxygen and other atmospheric gases are present at high elevations than at sea level. Air at the top of Mt. Everest is less dense than air at sea level, so less oxygen is present.

To understand how gases move by diffusion, it is important to express their presence in terms of partial pressures instead of percentages. Pressure is a type of force. A **partial pressure** is the pressure of a particular gas in a mixture of gases.

To calculate the partial pressure of a particular gas, multiply the fractional composition of that gas by the total pressure exerted by the entire mixture.[1]

For example, **Figure 44.2b** shows that the total atmospheric pressure at sea level is 760 mm Hg (millimeters of mercury). If you multiply this value by 0.21, which is the fraction of air that is O_2, you obtain a partial pressure of oxygen, abbreviated P_{O_2}, at sea level of 160 mm Hg. Because the atmospheric pressure is only about 250 mm Hg at the top of Mt. Everest, the P_{O_2} there is only $0.21 \times 250 = 53$ mm Hg.

[1]This calculation is valid because the total pressure in a mixture of gases is the sum of the partial pressures of all the individual gases. This relationship is called Dalton's law.

(a) Which gases make up the atmosphere?

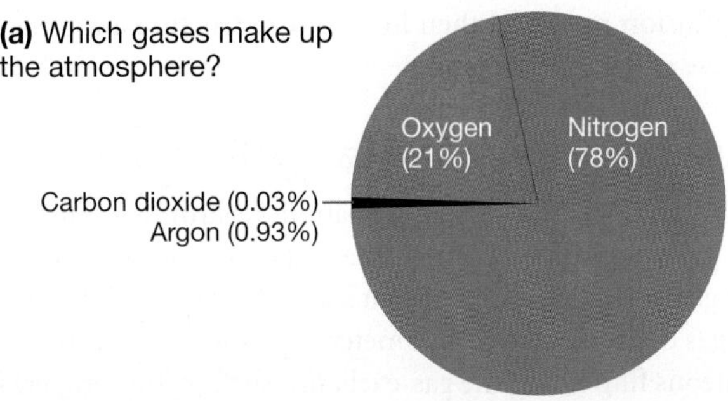

(b) The partial pressure of oxygen falls with increasing elevation.

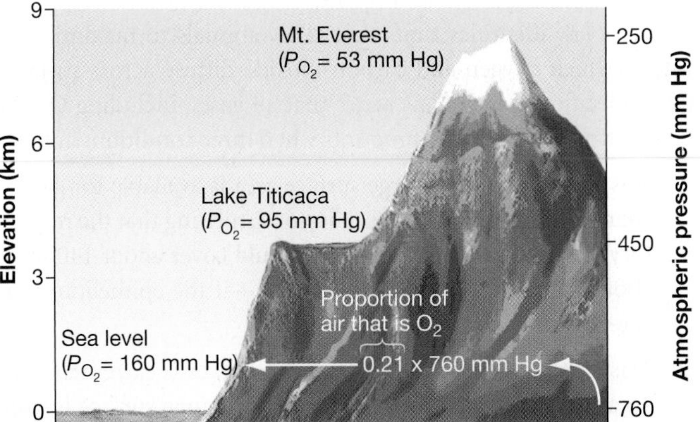

FIGURE 44.2 Oxygen Makes Up 21 Percent of the Atmosphere, but Its Partial Pressure Varies Widely. (a) Earth's atmosphere is dominated by nitrogen and oxygen. Oxygen makes up 21 percent of the atmosphere at all elevations. **(b)** Atmospheric pressure, and thus oxygen partial pressure (P_{O_2}), fall with increasing elevation.

Oxygen and carbon dioxide diffuse between the environment and cells along their respective partial-pressure gradients, just as solutes diffuse along their concentration gradients. In both air and water, O_2 and CO_2 move from regions of high partial pressure to regions of low partial pressure. It is hard to breathe at the top of Mt. Everest because the partial pressure of oxygen is low—meaning that the diffusion gradient between the atmosphere and your lung tissues is small.

How Do Oxygen and Carbon Dioxide Behave in Water?

To obtain oxygen, water breathers face a much more challenging environment than air breathers do. Aquatic animals live in an environment that contains much less oxygen than the environments inhabited by terrestrial animals. At 15°C, a liter of air can contain up to 209 mL of O_2, while a liter of water may contain a maximum of only 7 mL of O_2. To extract a given amount of oxygen, an aquatic animal has to process 30 times more water than the amount of air a terrestrial animal breathes.

In addition, water is about a thousand times denser than air and flows much less easily. As a result, water breathers have to expend much more energy to ventilate their respiratory surfaces than do air breathers.

WHAT AFFECTS THE AMOUNT OF GAS IN A SOLUTION? Oxygen and carbon dioxide diffuse into water from the atmosphere, but the amount of gas that dissolves depends on several factors:

- *Solubility of the gas in water* Oxygen has very low solubility in water. Only 0.003 mL of oxygen dissolves in 100 mL of water for each 1-mm-Hg increase in oxygen partial pressure. Because of this low solubility, blood contains a molecule that binds to oxygen and delivers it to tissues. Without this carrier molecule, the rate of blood flow to tissues would have to increase dramatically to meet oxygen demand.
- *Temperature of the water* As the temperature of water increases, the amount of gas that dissolves in it decreases. Other things being equal, warm-water habitats have much less oxygen available than cold-water habitats do. For a fish, breathing in warm water is comparable to a land-dwelling animal breathing at high elevation.
- *Presence of other solutes* Because seawater has a much higher concentration of solutes than does freshwater, seawater can hold less dissolved gas. At 10°C, up to 8.02 mL of O_2 can be present per liter of freshwater versus only 6.35 mL of O_2 per liter of seawater. As a result, freshwater habitats tend to be more oxygen rich than marine environments.
- *Partial pressure of the gas in contact with the water* Gases move from regions of high partial pressure to regions of low partial pressure. So if the partial pressure in a liquid exceeds that in the adjacent gas, the gas will "boil" out of the liquid. This is what happens when the cap is removed from a bottle of carbonated beverage. The partial pressure of carbon dioxide in the newly opened drink is much higher than it is in the atmosphere.

WHAT AFFECTS THE AMOUNT OF OXYGEN AVAILABLE IN AN AQUATIC HABITAT? The partial pressure of oxygen varies in different types of aquatic habitats, just as it varies with altitude on land. In addition to the four factors listed above, there are other important considerations that impact oxygen's availability in water.

For example, habitats with large numbers of photosynthetic organisms tend to be relatively oxygen rich. In contrast, habitats where most organisms live off existing organic material tend to be oxygen poor, because cellular respiration depletes the available oxygen.

The most important issue in oxygen availability, however, is often surface area. Surface area has a large impact on oxygen's ability to diffuse into water. For example:

- Shallow ponds and streams tend to be much better oxygenated than deep bodies of water, because shallower bodies have a higher ratio of surface area to volume.
- Unless currents mix water almost continuously, water near the surface has much higher oxygen content than water near the bottom of the same habitat.
- Rapids, waterfalls, and other types of whitewater are the most highly oxygenated of all aquatic environments, because a large surface area is exposed to the atmosphere as water splashes

over rocks and logs and because air bubbles are incorporated into the water.

- Oxygen content is extremely low in bogs and other stagnant-water habitats, because the small amounts of oxygen that diffuse into stagnant water are quickly used up by decomposers that use oxygen as an electron acceptor in cellular respiration.

Now let's consider the structure and function of ventilatory organs. How do the gills of fish, the tracheae of insects, and the lungs of mammals cope with the differences between air and water?

CHECK YOUR UNDERSTANDING

If you understand that . . .

- O_2 and CO_2 move from regions of high partial pressure to regions of low partial pressure.
- The partial pressure of oxygen in a body of water depends largely on its surface area and temperature.
- Water breathing is much more difficult than air breathing, in part because the partial pressure of oxygen in water is much lower than its partial pressure in air.

✔ You should be able to . . .

1. Explain why oxygen partial pressures are relatively high in mountain streams and relatively low at the ocean bottom.
2. Explain whether a large, medium, or small amount of air should be bubbled into aquaria containing warm-water species, vigorous algal growth, or sedentary animals.

Answers are available in Appendix B.

44.3 Organs of Gas Exchange

Many small animals lack specialized gas-exchange organs, such as gills or lungs. Instead, they obtain O_2 and eliminate CO_2 by diffusion across the body surface. This is possible because their size and shape give them an extraordinarily high surface area to volume ratio (see Chapter 41). For sponges, jellyfish, flatworms, and other species, diffusion across the body surface is rapid enough to fulfill their requirements for taking in O_2 and expelling CO_2.

Most of these animals are restricted to living in wet environments, however. For gas exchange to take place efficiently, the surface where it occurs has to be thin, and thin skin is prone to water loss. Living in wet or humid environments allows animals to exchange gases across their outer surface while avoiding dehydration.

In contrast, animals that are large or that live in dry habitats need some sort of specialized respiratory organ. Respiratory organs provide a greater surface area for gas exchange—enough to meet the demands of a large body filled with cells. In terrestrial animals, respiratory organs are located inside the body, which helps to minimize water loss.

Biologists have long marveled at the efficiency of gills and lungs. To appreciate why, let's examine the physical factors that control diffusion rates and then look at the structure and function of these respiratory organs.

Physical Parameters: The Law of Diffusion

In 1855 Adolf Fick derived an equation regarding diffusion, based on the results of experiments he had performed on the behavior of gases. **Fick's law of diffusion** states that the rate of diffusion of a gas depends on five parameters: the solubility of the gas in the aqueous film lining the gas-exchange surface, the temperature, the surface area available for diffusion, the difference in partial pressures of the gas across the gas-exchange surface, and the thickness of the barrier to diffusion (**Figure 44.3**).

Fick's law identifies traits that allow animals to maximize the rate at which oxygen and carbon dioxide diffuse across surfaces. Specifically, Fick's law states that all gases, including O_2 and CO_2, diffuse in the largest amounts when three conditions are met:

1. A is large, meaning a large surface area is available for gas exchange. Based on Fick's law, it is not surprising that the respiratory surface in the human lungs would cover about 140 m^2—about a quarter of a basketball court—if the epithelium were spread flat.
2. D is small, meaning the respiratory surface is extremely thin. To appreciate just how thin the gas-exchange surface is, consider that researchers who asked bicyclists to ride up a steep hill for seven minutes observed a marked increase in blood on the respiratory surface of the athletes' lungs. To explain this observation, the researchers proposed that the cyclists' high heart rates (up to 177 beats/minute during their sprint up the hill) increased blood pressure to the point at which thin-walled vessels in the lungs ruptured and leaked blood into the structure.
3. $P_2 - P_1$ is large, meaning the partial-pressure gradient of the gas across the surface is large. High partial-pressure gradients are maintained in part by having an efficient circulatory system in close contact with the gas-exchange surface. When blood flows close to the respiratory surface, oxygen is rapidly taken away from the area where inward diffusion is occurring, and carbon dioxide is rapidly brought into the area where outward diffusion is occurring. As a result, $P_2 - P_1$ stays high.

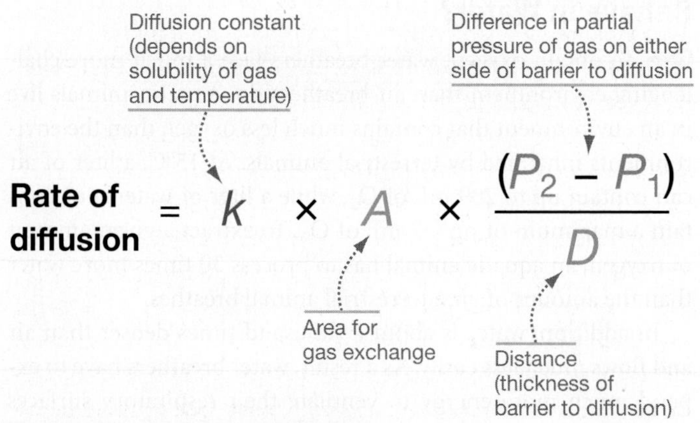

FIGURE 44.3 Fick's Law of Diffusion.

(a) External gills are in direct contact with water.

(b) Internal gills must have water brought to them.

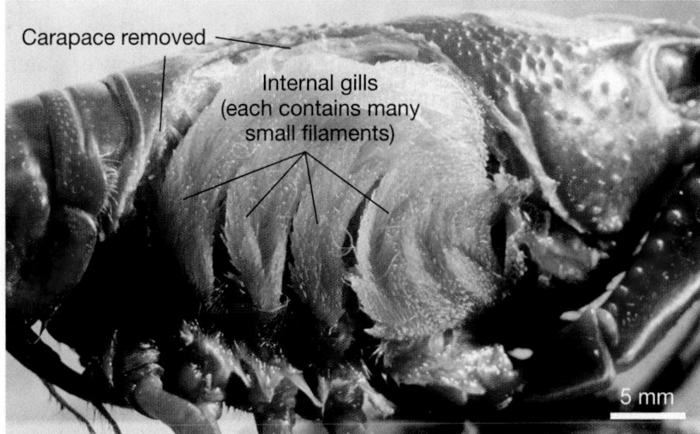

FIGURE 44.4 Gills Can Be External or Internal. (a) Nudibranchs are marine snails with gills that are outside the main body wall. Their gills are protected by toxins. **(b)** Crayfish gills are located inside the main body wall. A portion of the crayfish exoskeleton has been removed to expose the gills.

✓**QUESTION** What are the advantages and disadvantages of external versus internal gills in terms of fitness?

What other aspects of gill and lung structure affect diffusion rate? To answer this question, let's delve into the anatomy of these respiratory organs.

How Do Fish Gills Work?

Gills are outgrowths of the body surface or throat that are used for gas exchange in aquatic animals. Gills are efficient solutions to the problems posed by water breathing, primarily because they present a large surface area for oxygen to diffuse across a thin epithelium.

In some species of invertebrates, such as the nudibranch mollusk (**Figure 44.4a**), gills project from the body surface and contact the surrounding water directly. In other invertebrate species, such as the crayfish (**Figure 44.4b**), gills are located inside the exoskeleton or body wall. If gills are internal, water must be driven over them by cilia, the limbs, or other specialized structures.

In contrast to the diversity of gills found in invertebrates, the gills of bony fishes are all similar in structure. Fish gills are located on both sides of the head and in teleosts (see Chapter 34) consist of four arches, as **Figure 44.5** shows.

HOW DO FISH VENTILATE THEIR GILLS? To move water through their gills so gas exchange can take place, most fish open and close their mouths and a stiff flap of tissue, the **operculum**, that covers the gills. The pumping action of the mouth and operculum creates a pressure gradient that moves water over the gills.

In contrast, tuna and other fish that are particularly fast swimmers force water through their gills by swimming with their mouths open. This process is called ram ventilation.

Regardless of how fish gills are ventilated, water flows in one direction over gills. More specifically, water passes through long, thin structures called **gill filaments** that extend from each gill arch. Each gill filament is composed of hundreds or thousands of **gill lamellae**. Gill lamellae are sheetlike structures, shown in detail at the bottom of Figure 44.5. Note that a bed of small blood vessels called capillaries runs through each lamella.

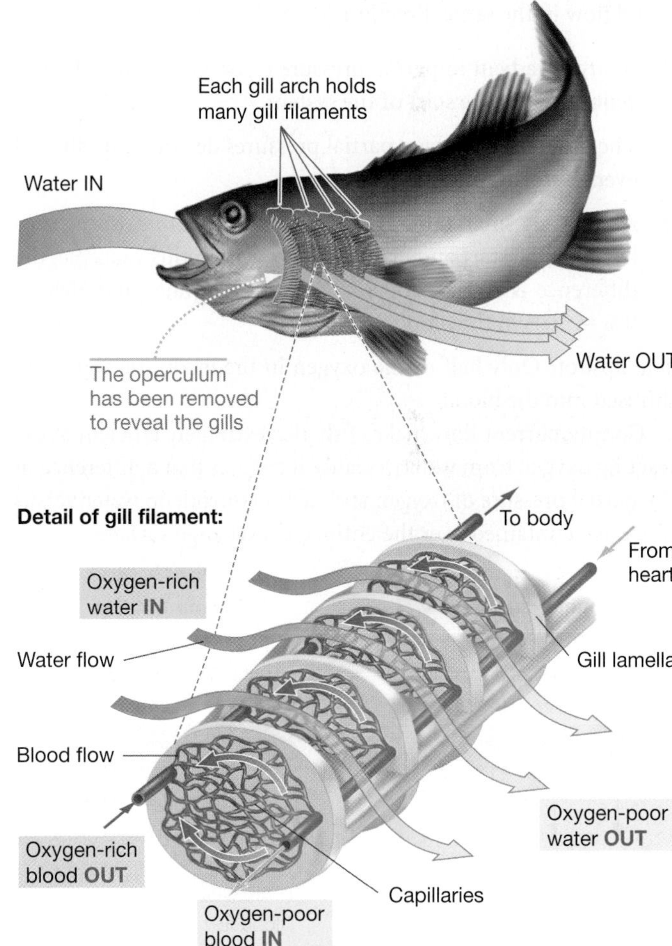

FIGURE 44.5 Fish Gills Are a Countercurrent Exchange System. In the lamellae of fish gills, water and blood flow in opposite directions.

THE FISH GILL IS A COUNTERCURRENT SYSTEM The one-way flow of water through gill lamellae has a profound impact on gill function, for a simple reason: The flow of blood through the capillary bed in each lamella is in the opposite direction to the flow of water. As a result, each lamella functions as a countercurrent exchanger.

Recall from Chapter 41 that countercurrent exchangers are based on two adjacent fluids flowing in opposite directions. **Figure 44.6** illustrates why the countercurrent flow is so critical.

Note two key points about the left side of the figure, where water and blood flow in opposite directions:

1. A slight gradient in partial pressure of oxygen (here, 10 percent) exists along the entire length of the lamella.
2. A large difference in oxygen partial pressure exists between the start and end of the system. In this example, the difference is 100% − 15% = 85% in the water and 90% − 5% = 85% in the blood.

The upshot? Most of the oxygen in the incoming water has diffused into the blood.

Now look at the right side of the figure, where water and blood flow in the same direction.

1. A large gradient in partial pressure of oxygen (here, 100 percent) exists at the start of the system.
2. The gradient in oxygen partial pressures declines rapidly and eventually disappears.
3. A relatively small difference in oxygen partial pressure exists between the start and end of the system. In this example, the difference is 100% − 50% = 50% in the water and 50% − 0% = 50% in the blood.

The upshot? Only half of the oxygen in the incoming water has diffused into the blood.

Countercurrent flow makes fish gills extremely efficient at extracting oxygen from water, because it ensures that a difference in the partial pressure of oxygen and carbon dioxide in water versus blood is maintained over the entire gas-exchange surface.

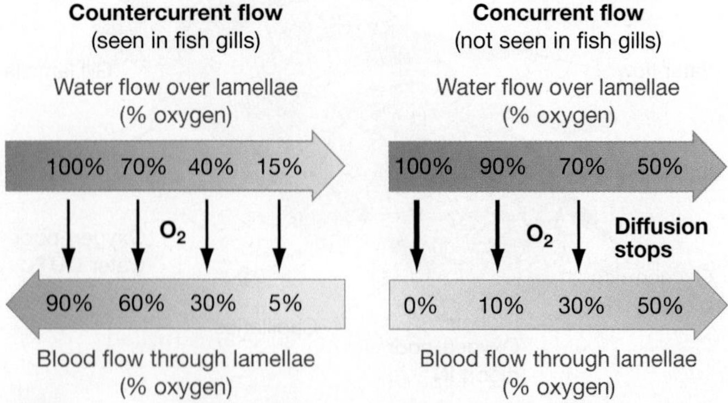

FIGURE 44.6 Countercurrent Exchange Is Much More Efficient than Concurrent Exchange. In the countercurrent system of fish gills, oxygen is transferred along the entire length of the capillaries. If "concurrent flow" occurred, oxygen transfer would stop, because the partial-pressure gradient driving diffusion would fall to zero partway along the length of the capillary.

The effect of countercurrent exchange is to maximize the $P_2 - P_1$ term in Fick's law of diffusion, averaged over the entire gill surface. Based on this observation, biologists cite countercurrent exchange as another example of how gills are optimized for efficient gas exchange.

How Do Insect Tracheae Work?

Air and water are dramatically different ventilatory media, because they have different densities, viscosities, and abilities to

EXPERIMENT

QUESTION: Does physical activity affect air movement through the tracheal system?

HYPOTHESIS: Air moves through the tracheal system faster during physical activity.

NULL HYPOTHESIS: Physical activity does not affect the rate of air movement.

EXPERIMENTAL SETUP:

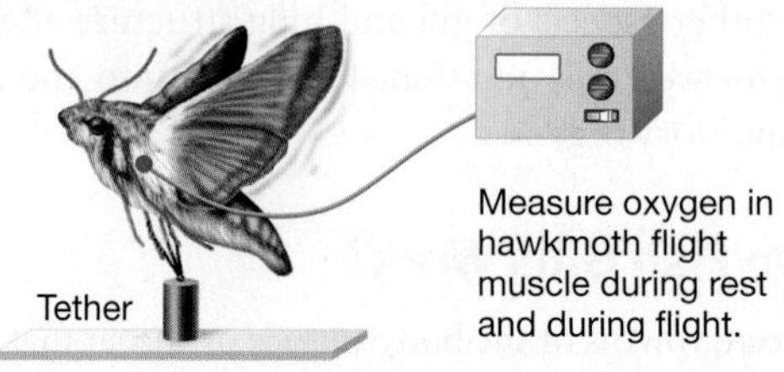

PREDICTION: Flying will increase ventilation of tracheal system, causing increase in P_{O_2} in wing muscle.

PREDICTION OF NULL HYPOTHESIS: Flying will not increase ventilation of tracheal system; P_{O_2} in flight muscle will decline steadily during flight.

RESULTS:

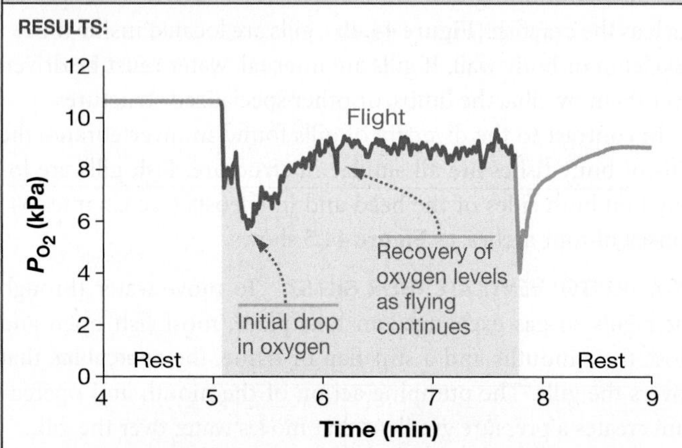

CONCLUSION: Muscular contractions may help ventilate the tracheal system in at least some insects.

FIGURE 44.7 Evidence That the Tracheal System Is Ventilated during Movement. On the graph's *y*-axis, kPa stands for a unit of pressure—or force per unit area—called the kilopascal.

SOURCE: Komai, Y. 1998. Augmented respiration in a flying insect. *Journal of Experimental Biology* 201: 2359–2366.

✔**QUESTION** Why was it important for the researcher to repeat this experiment with several different individuals?

hold oxygen and carbon dioxide. In addition, the consequences of exposing the gas-exchange surface to air versus water differ.

In aquatic habitats, ventilation tends to disrupt water and electrolyte balance, and homeostasis must be maintained by an active osmoregulatory system. As Chapter 42 explained, osmosis causes marine animals to lose water across their gas-exchange surface and freshwater animals to gain water. Diffusion tends to cause marine animals to gain sodium, chloride, and other ions, and freshwater animals to lose them.

In contrast, breathing leads to a loss of water by evaporation in terrestrial environments. How do terrestrial animals minimize water loss while maximizing the efficiency of gas exchange?

To answer this question, consider the tracheal system of insects. Recall from Chapter 42 that **tracheae** are an extensive system of tubes located well within the body. They connect to the exterior through openings called **spiracles**, which can be closed to minimize the loss of water by evaporation. The ends of tracheae are tiny, fluid filled, and highly branched. In this way, the tracheal system transports air close enough to cells for gas exchange to take place directly across their plasma membranes.

Recent research on the tracheal system has focused on how air moves into and out of the tubes through spiracles. Is simple diffusion efficient enough to ventilate the system, or is some type of breathing mechanism involved?

The short answer to this question is that breathing movements may play a role in at least some insects. Consider a study on the sweet potato hawkmoth. A researcher set out to investigate how the amount of O_2 delivered to muscles changes during flight. To document the partial pressure of oxygen in flight muscles, he inserted a needlelike electrode into the wing muscles of a hawkmoth. The electrode was attached to an instrument that measured the partial pressure of oxygen, and the hawkmoth was tethered to a stand (**Figure 44.7**).

The biologist recorded P_{O_2} as the insect rested, then stimulated the moth to fly by exposing it to wind. The "Results" section of Figure 44.7 shows how P_{O_2} changed during one such experiment. (The procedure was repeated on several individuals, and the same pattern was observed.)

To read this graph, put your finger on the orange line and move it to the right, to where the insect started flying and the line changes to red. Note that P_{O_2} levels in the flight muscles dropped initially. This is not surprising, because flight is an energetically demanding activity. As flying continued, however, P_{O_2} levels recovered until they were nearly as high as when they were at rest.

To explain this observation, biologists propose that tracheae alternately open and close as the muscles around them contract and relax. As a result, the volume of the tracheal system changes. This is a key point: If the volume occupied by a fixed amount of gas expands, its pressure decreases. But if the volume that a fixed amount of gas occupies declines, its pressure increases.

If the volume of the tracheal system increases when muscles relax, pressure inside the system goes down and air from the atmosphere will rush in. But if volume decreases when muscles contract, then pressure inside the system increases and gas will move out (**Figure 44.8**).

In effect, then, insects may have a breathing mechanism that ventilates the respiratory surface when demand for oxygen is high.

How Do Vertebrate Lungs Work?

In most terrestrial vertebrates, air enters the body through both the nose and mouth. A tube known as the **trachea** (not to be confused with the tracheae of insects) carries the inhaled air to narrower tubes called **bronchi** (singular: **bronchus**). The bronchi branch off into yet narrower tubes, the **bronchioles.** The organs of ventilation, the lungs, enclose the bronchioles and part of the bronchi (**Figure 44.9a** on page 868).

Lungs are infoldings of the throat that are used for gas exchange. In addition to being found in terrestrial vertebrates—amphibians, reptiles (including birds), and mammals—they occur in many fish and certain invertebrates.

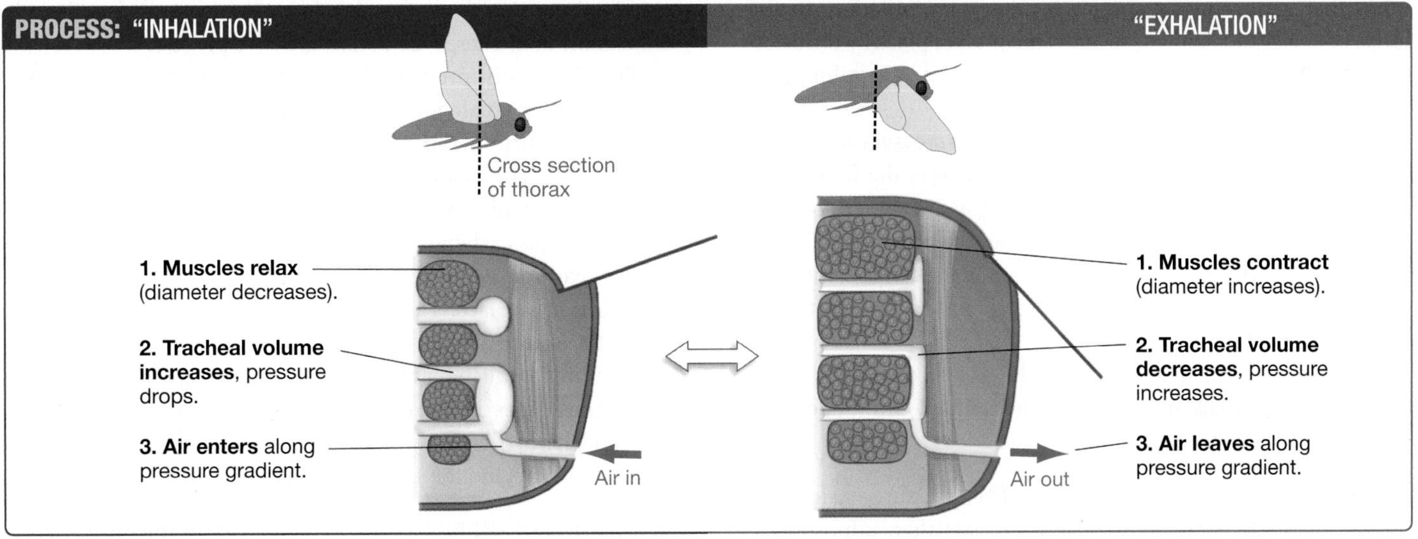

FIGURE 44.8 Muscle Activity May Ventilate Tracheae. To explain the result in Figure 44.7, biologists propose that tracheae change volume as wings beat (drawings are schematic).

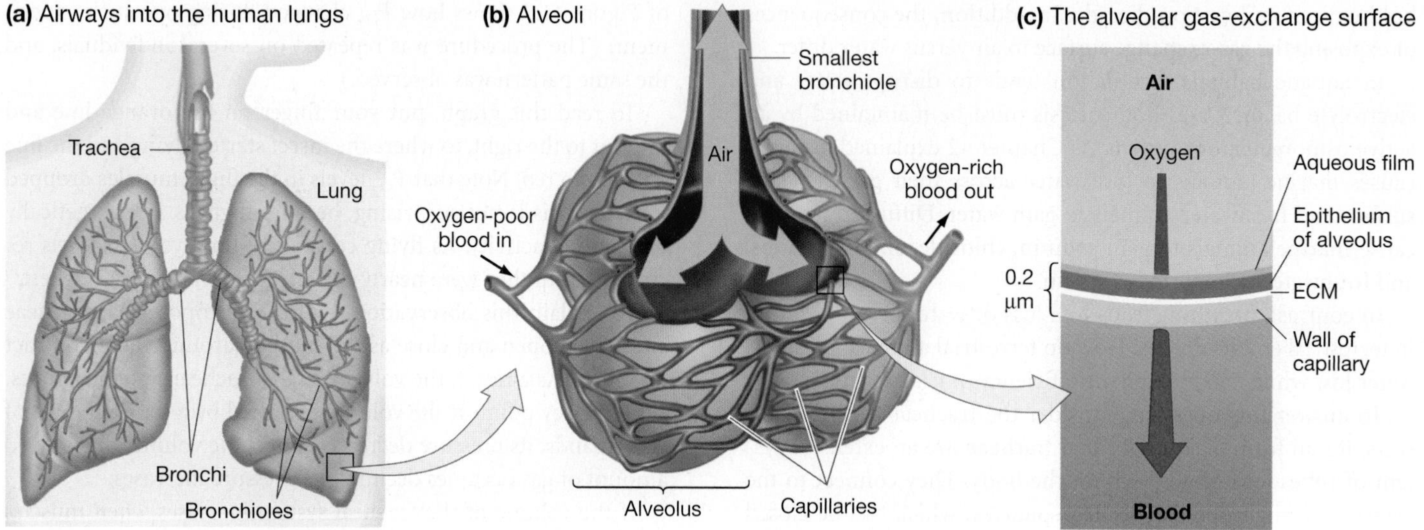

FIGURE 44.9 Lungs Offer a Large, Thin Surface Area for Gas Exchange between Air and Blood. **(a)** The human respiratory tract branches repeatedly from the largest airway, the trachea, to the smallest, called bronchioles. Gas exchange does not occur across these airways. The system of airways ends in clusters of tiny sacs called alveoli. **(b)** Alveoli are covered with capillary networks and **(c)** are the site of gas exchange.

LUNG STRUCTURE AND VENTILATION VARY AMONG SPECIES The amount of lung surface area available for gas exchange varies a great deal among species. In frogs and other amphibians, the lung is a simple sac lined with blood vessels. The lungs of mammals, in contrast, are finely divided into tiny sacs called **alveoli** (singular: **alveolus; Figure 44.9b**). Each human lung contains approximately 150 million alveoli, which give mammalian lungs about 40 times more surface area for gas exchange than an equivalent volume of frog lung tissue.

As **Figure 44.9c** shows, an alveolus provides an interface between air and blood that consists of a thin aqueous film, a layer of epithelial cells, some extracellular matrix (ECM) material, and the wall of a capillary. In the human lung, this barrier to diffusion is only 0.2 μm thick—about 1/200th of the thickness of this page.

In addition to total surface area, the other major feature of lungs that varies among species is mode of ventilation. In the lungs of snails and spiders, air movement takes place by diffusion only. Vertebrates, in contrast, actively ventilate their lungs by pumping air via muscular contractions.

One mechanism for pumping air is **positive pressure ventilation**, used by frogs and related animals. A frog lowers the floor of its throat, increasing its volume and drawing in air from the atmosphere, through the nasal passages, and into the oral cavity. The animal then closes the nasal passages and contracts its throat muscles. These actions increase the pressure on air in the oral cavity and force it into the lungs.

In effect, frogs push air into their lungs. In contrast, humans and other mammals pull air into their lungs. How does this **negative pressure ventilation** work?

VENTILATION OF THE HUMAN LUNG The pressure inside the human chest cavity is about 5 mm Hg less than atmospheric pressure. The negative pressure surrounding the lung is just enough to keep it expanded. If a wound penetrates the chest wall and the pressure differential between the chest cavity and the atmosphere disappears, the lung on the side of the injury will collapse like a deflated balloon.

Humans ventilate their lungs by changing the pressure within the chest cavity between about −5 mm Hg and −8 mm Hg relative to the atmosphere. As **Figure 44.10a** shows, inhalation is based on increasing the volume of the chest cavity and thus lowering the pressure. The change is caused by a downward motion of the thin muscular sheet called the **diaphragm** and an outward motion of the ribs. As the pressure surrounding the lungs drops, air flows into the airways along a pressure gradient.

Exhalation, in contrast, is a passive process when an individual is at rest. This is possible because (1) the lung is **elastic**—it returns to its original, collapsed shape if it is not stretched or compressed—and (2) the volume of the chest cavity decreases as the diaphragm and rib muscles relax. During exercise, though, exhalation is an energy-demanding, active process.

The changes in pressure that occur during negative pressure ventilation are analogous to changing the volume of a jar, as shown in **Figure 44.10b**.

About 450 mL of air moves into and out of the lungs in an average breath. Only about two-thirds of this volume actually participates in gas exchange, however, because 150 mL of the air occupies **dead space**—portions of the air passages that do not have a respiratory surface. The trachea and bronchi shown in Figure 44.9a, for example, represent dead space.

Breathing is much more efficient during exercise, when the chest cavity undergoes larger changes in volume. When a person is breathing hard, over 2500 mL of air can move with each inhalation-exhalation cycle, but the 150 mL in dead space stays the same.

VENTILATION OF THE BIRD LUNG Flight is one of the most energy-demanding activities performed by animals. Even so, some birds fly tens of thousands of kilometers during annual migrations.

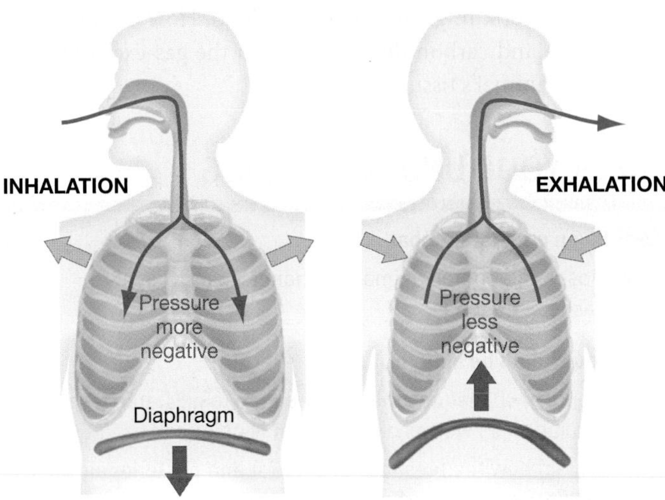

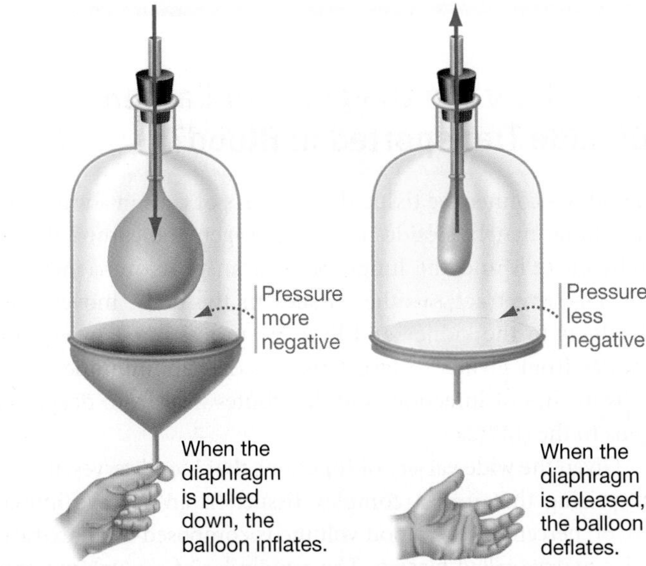

FIGURE 44.10 Changes in the Volume of the Chest Cavity Drive Negative Pressure Ventilation. (a) Inhalation: When the diaphragm and rib muscles contract, the volume of the lung cavity increases, lowering pressure. In response, the lungs expand and air flows in. Exhalation: When the diaphragm and rib muscles relax, the volume of the lung cavity decreases, causing internal pressure to increase. In response, lung volume decreases—due to elasticity of the lungs—and air flows out. **(b)** A model of negative pressure ventilation.

✔**QUESTION** What happens when a person sighs or takes a deep breath?

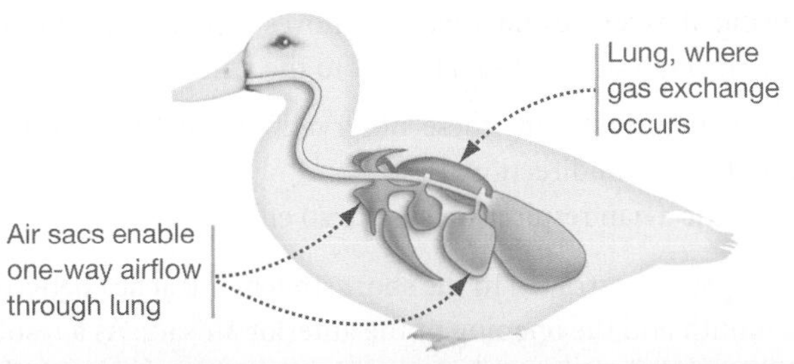

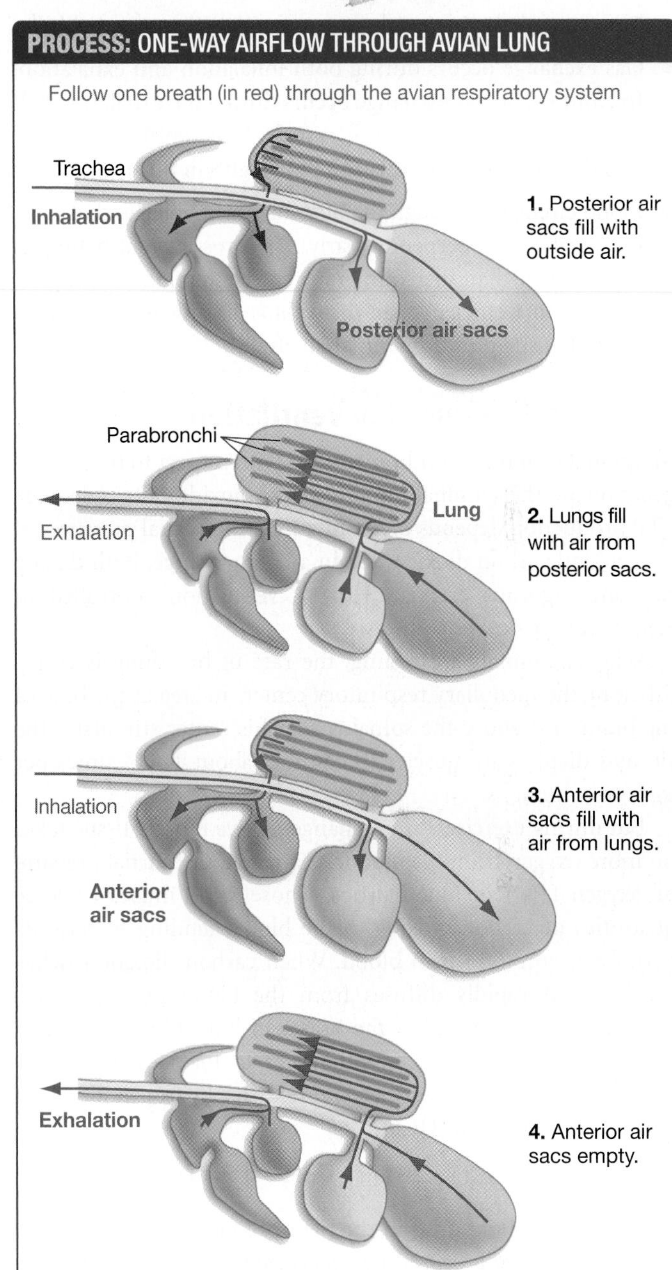

FIGURE 44.11 Air Flows in One Direction through the Bird Lung.

Even more impressive, geese regularly fly over the top of Mount Everest. At this elevation, the partial pressure of oxygen is so low that most humans would immediately black out if they were flown there and dropped off. How are birds able to extract enough oxygen from the atmosphere to support long flights and flights at high elevations?

Figure 44.11 shows why respiration is so efficient in birds.

1. During inhalation, air flows through the trachea and enters two large air sacs posterior to the lungs.
2. During exhalation, air leaves the posterior air sacs and enters tiny, branching airways, called parabronchi, in the posterior portion of the lung.
3. During the next inhalation, air moves into parabronchi in the anterior part of the lung and on to a system of air sacs anterior to the lung.

4. During the next exhalation, air moves out of the anterior sacs, through the trachea, and out to the atmosphere.

The key conclusion from these observations? Airflow through the avian lung is unidirectional.

Why is the avian respiratory system so efficient?

- Dead space is restricted to the short stretch of trachea between the mouth and the opening of the anterior air sacs. As a result, birds use inhaled air much more efficiently than mammals do.
- Gas exchange occurs during both inhalation and exhalation. In contrast, no gas exchange occurs during the exhalation half of the respiratory cycle in mammals. Bird ventilation resembles the continuous ventilation of fish gills in this respect.
- Blood circulates through the bird lung in capillaries that cross the parabronchi perpendicularly. This crosscurrent pattern is less efficient than the countercurrent circulation of fish gills, but far more efficient than the weblike arrangement of capillaries that surrounds mammalian alveoli.

Homeostatic Control of Ventilation

An animal is in trouble if homeostasis with respect to blood oxygenation or the elimination of carbon dioxide fails. Adequate ATP production depends on maintaining the partial pressures of oxygen and carbon dioxide within a narrow range, both during rest and vigorous exercise. How is ventilation controlled to achieve this critical homeostasis?

When mammals are resting, the rate of breathing is established by the medullary respiratory center, an area at the base of the brain, just above the spinal cord. This center stimulates the rib and diaphragm muscles to contract about 12–14 times per minute in humans.

But during exercise, things change. Active muscle tissue takes up more oxygen from the blood. As a result, the partial pressure of oxygen (P_{O_2}) in blood drops. Those same muscles release quantities of carbon dioxide to the blood, tending to raise its partial pressure (P_{CO_2}) in blood. When carbon dioxide reaches the brain, it rapidly diffuses from the blood into the cerebrospinal fluid that bathes the brain. In both blood and cerebrospinal fluid, CO_2 reacts with water to form carbonic acid (H_2CO_3), which then dissociates to release a hydrogen ion (H^+) and a bicarbonate ion (HCO_3^-):

$$CO_2 + H_2O \rightleftharpoons H_2CO_3 \rightleftharpoons H^+ + HCO_3^-$$

The result is a slight drop in the pH of blood and cerebrospinal fluid. The change is sensed by specialized neurons located near the large arteries that travel from the heart into the neck and to the base of the brain itself.

Signals from these nerves or from pH detectors in the medullary respiratory center cause the breathing rate to increase. The resulting rise in ventilation rate increases the rate of oxygen delivery to the tissues and the rate at which carbon dioxide is eliminated from the body, restoring P_{O_2} and P_{CO_2} to their resting levels. This control system is so effective that it can maintain stable blood levels of oxygen and carbon dioxide even during intense exercise.

Next, let's look in greater depth at blood. How does it transport oxygen and carbon dioxide between the gas-exchange surface and an animal's tissues?

CHECK YOUR UNDERSTANDING

If you understand that . . .

- Most large-bodied animals exchange gases via gills, tracheae, or lungs.

✔ You should be able to . . .

1. Identify two features that are common to gills, tracheae, and lungs, as well as one trait that is unique to each.
2. Explain how the presence of anterior and posterior air sacs facilitate one-way airflow through the bird lung.

Answers are available in Appendix B.

44.4 How Are Oxygen and Carbon Dioxide Transported in Blood?

Blood is a connective tissue that consists of cells in a watery extracellular matrix. Besides carrying oxygen and carbon dioxide between cells and the lungs, blood transports nutrients from the digestive tract to other tissues in the body, moves waste products to the kidney and liver for processing, conveys hormones from glands to target tissues, delivers immune system cells to sites of infection, and distributes heat from deeper organs to the surface.

Given the wide variety of functions that blood serves, it is not surprising that it is a complex tissue. In an average human, 50–60 percent of the blood volume is composed of an extracellular matrix called **plasma**. The remainder of the volume comprises cells and cell fragments that are collectively called formed elements.

The formed elements in blood include platelets, several types of white blood cells, and red blood cells.

- **Platelets** are cell fragments that minimize blood loss from ruptured blood vessels. Platelets release material that assists in forming the blockages known as clots.
- **White blood cells** are part of the immune system. As Chapter 49 will explain in detail, white blood cells fight infections.
- **Red blood cells** transport oxygen from the lungs to tissues throughout the body. They also play a critical role in transporting carbon dioxide from tissues to the lungs. In humans, red blood cells make up 99.9 percent of the formed elements.

The human body synthesizes new red blood cells at the rate of 2.5 million per second to replace old red blood cells, which die at the same rate. These red blood cells live about 120 days. They develop along with white blood cells and platelets from stem cells located in the tissue inside bone (bone marrow).

Vertebrates other than mammals transport oxygen in red blood cells that retain their nuclei. But in mammals, red blood cells lose their nuclei as they mature, along with their mitochondria and most other organelles. Mammalian red blood cells are essentially bags filled with approximately 280 million copies of the oxygen-carrying molecule **hemoglobin**.

Structure and Function of Hemoglobin

Even though oxygen is not highly soluble in water, it is often found in high concentrations in blood. Blood has a high oxygen-carrying capacity because O_2 readily binds to the hemoglobin molecules in red blood cells.

The evolution of hemoglobin was a key event in the diversification of animals. By increasing the oxygen-carrying capacity of blood, hemoglobin made it possible for cellular respiration rates to increase. High rates of ATP production, in turn, support high rates of growth, movement, digestion, and other activities.

Hemoglobin is a tetramer, meaning that it consists of four polypeptide chains (**Figure 44.12**). Each of the four polypeptide chains binds to a nonprotein group called **heme**. Each heme molecule, in turn, contains an iron ion (Fe^{2+}) that can bind to an oxygen molecule. As a result, each hemoglobin molecule can bind up to four oxygen molecules. In blood, 98.5 percent of the oxygen is bound to hemoglobin; only 1.5 percent is dissolved in plasma.

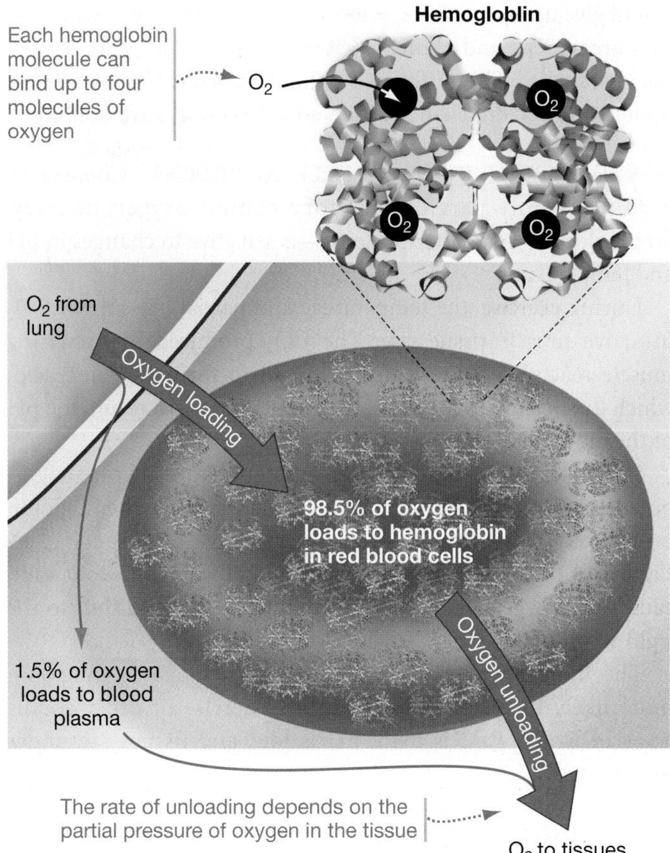

FIGURE 44.12 Hemoglobin Transports Oxygen to Tissues.

WHAT IS COOPERATIVE BINDING? Blood leaving the human lungs has a P_{O_2} of about 100 mm Hg, while muscles and other tissues have P_{O_2} levels of about 40 mm Hg at rest. This partial-pressure difference creates a diffusion gradient that unloads O_2 from hemoglobin to the tissues.

Researchers who studied the dynamics of O_2 unloading in tissues found the pattern shown in **Figure 44.13**. This graph plots the percent saturation of hemoglobin in red blood cells versus the P_{O_2} levels in the blood within tissues. If saturation is 100 percent, it means that every possible binding site in hemoglobin contains an oxygen molecule.

The graph in Figure 44.13 is called an **oxygen-hemoglobin equilibrium curve** or an oxygen dissociation curve. Note that the *x*-axis plots the partial pressure of oxygen in tissues. In effect, this represents "demand." Oxygen-depleted tissues, where demand for oxygen is high, are on the left-hand part of the horizontal axis; oxygen-rich tissues are toward the right. The *y*-axis, in contrast, plots the percentage of hemoglobin molecules in blood that are saturated with hemoglobin—a measure of "supply," or how readily hemoglobin gives up oxygen. Each 25-percent change in saturation corresponds to an average of one additional oxygen molecule bound per hemoglobin molecule or one oxygen molecule delivered to tissues.

The most remarkable feature of the oxygen-hemoglobin equilibrium curve is that it is sigmoidal, or S-shaped. The pattern occurs because the binding of each successive oxygen molecule to a subunit of the hemoglobin molecule causes a conformational change in the protein that makes the remaining subunits much more likely to bind oxygen.

This phenomenon is called **cooperative binding**. Conversely, the loss of a bound oxygen molecule changes hemoglobin's conformation in a way that makes the loss of additional oxygen molecules more likely.

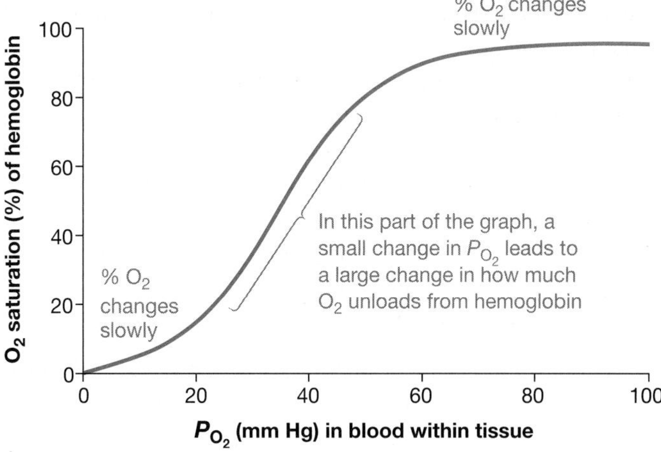

FIGURE 44.13 The Oxygen-Hemoglobin Equilibrium Curve.
A sigmoidal curve like this has three distinct regions.

(a) With cooperative binding, large amounts of O_2 are delivered to resting and exercising tissues.

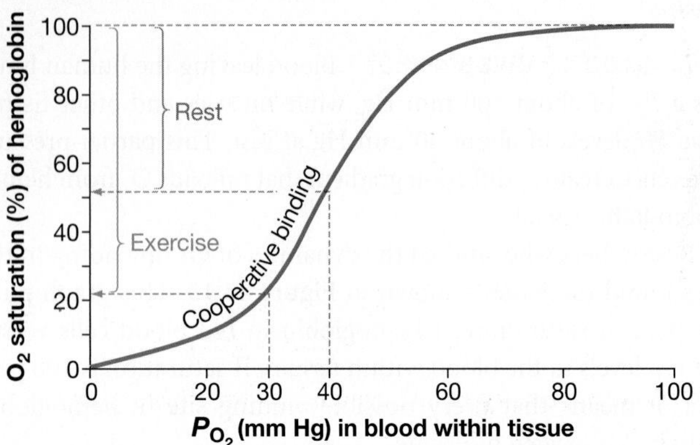

(b) Without cooperative binding, smaller amounts of O_2 would be delivered to resting and exercising tissues.

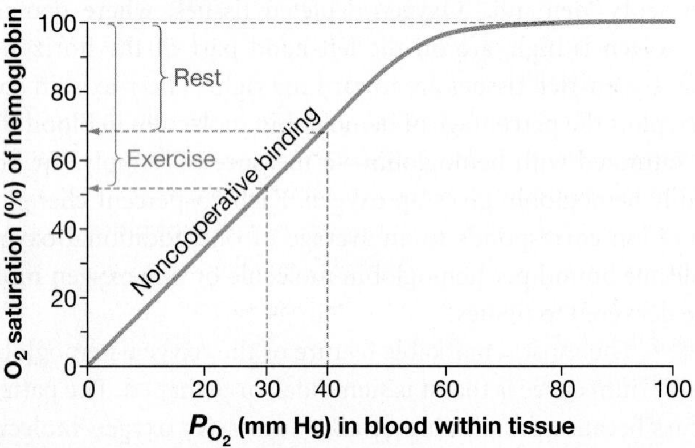

FIGURE 44.14 Cooperative Binding of O_2 by Hemoglobin Results in Greater O_2 Delivery than Noncooperative Binding. Effects of **(a)** cooperative versus **(b)** noncooperative oxygen binding. Note that hemoglobin is almost 100 percent saturated with oxygen when it arrives at tissues.

WHY IS COOPERATIVE BINDING IMPORTANT? To understand why cooperative binding is important, use **Figure 44.14a** to figure out what happens to hemoglobin saturation when oxygen demand in tissues changes.

Let's begin with some basic observations. When blood arrives at tissues, hemoglobin saturation is close to 100 percent. At rest, tissue P_{O_2} is typically about 40 mm Hg. But during exercise, cells are using so much oxygen in cellular respiration that tissue P_{O_2} drops to about 30 mm Hg. Now:

- Put your finger on the *x*-axis at 40 mm Hg, trace the dotted line up until it hits the saturation curve, and check where this point is on the *y*-axis. The answer is about 52 percent.
- The following statement should make sense to you: When tissues are at rest, hemoglobin gives up about half of its oxygen (about 48 percent) to the tissues.
- Now put your finger on the *x*-axis at 30 mm Hg, trace the dotted line up, and check hemoglobin saturation at this point on the curve. The answer is about 22 percent.
- You should be able to make the following conclusion: When tissues are exercising, hemoglobin gives up about 78 percent of its oxygen to tissues.

Here's the punchline: In response to a relatively small change in tissue P_{O_2}, there is a relatively large change in the percentage saturation of hemoglobin. The large change occurs because the saturation curve is extremely steep in the range of P_{O_2} values commonly observed in tissues. The curve is steep because of cooperative binding. Cooperative binding is important because it makes hemoglobin exquisitely sensitive to changes in the P_{O_2} of tissues.

✔If you understand this concept, you should be able to explain the consequences of an oxygen-hemoglobin equilibrium curve with a middle section that is even steeper than the one shown in Figure 44.14a.

If cooperative binding did not occur, all four subunits of hemoglobin would load or unload oxygen independently of each other. They would lose or gain oxygen in direct proportion to the partial pressure of oxygen in the blood, until the molecule was either completely saturated or unsaturated with oxygen. There would be a linear relationship between hemoglobin saturation and tissue P_{O_2}—not a sigmoidal one.

Figure 44.14b shows noncooperative binding. As the dotted lines on this graph indicate, a relatively small change in oxygen delivery would occur when tissue P_{O_2} changes from its resting level of about 40 mm Hg to 30 mm Hg. Specifically, hemoglobin would give up about 100% − 68% = 32% of its oxygen when tissues are at rest, and about 100% − 52% = 48% of its oxygen during exercise. This difference—about 16 percent—is much less than the 30 percent change observed with cooperative binding.

HOW DO TEMPERATURE AND PH AFFECT HEMOGLOBIN? Cooperative binding is only part of the story behind oxygen delivery. Hemoglobin—like other proteins—is sensitive to changes in pH and temperature.

During exercise, the temperature and partial pressure of CO_2 in active muscle tissue rise. The CO_2 produced by exercising muscle reacts with the water in blood to form carbonic acid, which dissociates and releases a hydrogen ion. As a result, the pH of the blood in exercising muscle drops.

Decreases in pH and increases in temperature alter hemoglobin's conformation. These shape changes make hemoglobin more likely to release O_2 at any given value of tissue P_{O_2}. As **Figure 44.15** shows, this phenomenon, known as the **Bohr shift**, causes the oxygen-hemoglobin equilibrium curve to shift to the right when pH declines.

The Bohr shift is important because it makes hemoglobin more likely to release oxygen during exercise or other conditions in which P_{CO_2} is high, pH is low, and tissues are under oxygen stress.

OXYGEN DELIVERY IS EXTREMELY EFFICIENT To appreciate cooperative binding and the Bohr shift in action, consider an experiment on how the oxygen transport system in rainbow trout responds to sustained exercise.

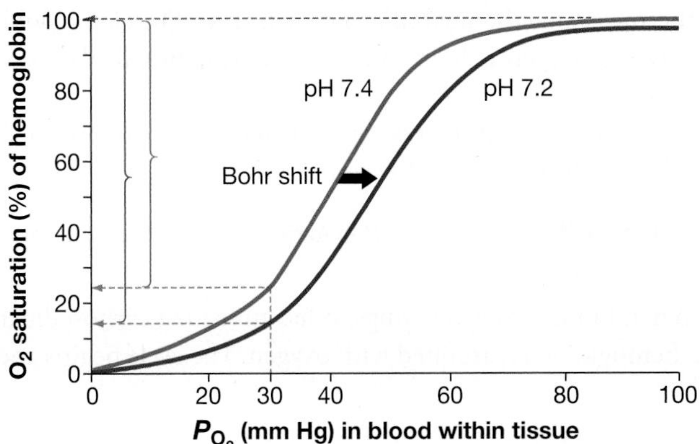

FIGURE 44.15 The Bohr Shift Makes Hemoglobin More Likely to Release Oxygen to Tissues with Low pH. As pH drops, oxygen becomes less likely to stay bound to hemoglobin at all values of tissue P_{O_2}. Exercising tissues have lower pH than resting tissues and thus receive more oxygen from hemoglobin.

✔**EXERCISE** Estimate how much more oxygen is released from hemoglobin at pH 7.2 than at pH 7.4 when the tissue has a P_{O_2} of 30 mm Hg.

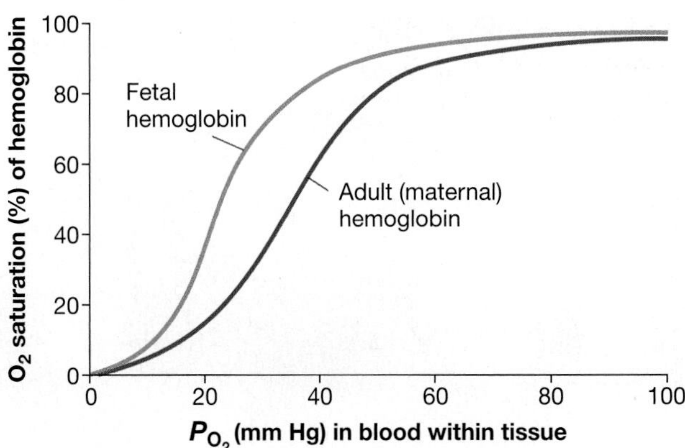

FIGURE 44.16 Some Types of Hemoglobin Bind Oxygen More Tightly than Others.

To begin, biologists had fish swim continuously against a current in a water tunnel. As the researchers increased the speed of the current and thus the swimming speed of the fish, they periodically sampled the O_2 content of arterial and venous blood. Arterial blood is freshly oxygenated and moving away from the gills; venous blood is returning to the gills from the rest of the body.

Not surprisingly, the biologists found that arterial O_2 levels remained fairly constant as swimming speed increased—meaning the gills continued to work efficiently enough to saturate hemoglobin with oxygen.

In contrast, the O_2 content of venous blood, which had undergone gas exchange with the tissues, dropped steadily as swimming speed increased. When the fish had reached their maximum sustainable speed, virtually all the oxygen available in hemoglobin in the venous blood had been extracted.

The data show that in hard-working tissues, the combination of increased temperature, lower pH, and lower P_{O_2} caused hemoglobin to become almost completely deoxygenated. Oxygen-delivery systems based on hemoglobin are extremely efficient.

COMPARING HEMOGLOBINS Hemoglobin molecules from different individuals or species may vary in ways that affect fitness—the ability to survive and produce offspring. As an example, consider the oxygen-hemoglobin equilibrium curves in **Figure 44.16.** The curve in dark red is from a pregnant woman; the curve in light red is from a fetus she is carrying.

The hemoglobin found in fetuses is encoded by different genes than adult hemoglobin and has a distinctive structure and function. Specifically, the oxygen-hemoglobin equilibrium curve for fetal hemoglobin is shifted to the left with respect to the curve for adult hemoglobin.

Recall that the rightward shift called the Bohr effect means that hemoglobin is more likely to release oxygen. The leftward shift in Figure 44.16 means that fetal hemoglobin is less likely to give up oxygen—it binds oxygen more tightly. Stated another way, fetal hemoglobin has a higher affinity for oxygen than adult hemoglobin at every P_{O_2}.

This is critically important. In the placenta, hemoglobin from the mother's blood gets close to hemoglobin from the fetus's blood. Because fetal hemoglobin has such a high affinity for oxygen, there is a transfer of oxygen from the mother's blood to the fetus's blood. The difference in hemoglobin structure and function ensures an adequate supply of oxygen as the fetus develops.

CO_2 Transport and the Buffering of Blood pH

The carbon dioxide that is produced by cellular respiration in the tissues enters the blood, where it reacts with water to form carbonic acid. Recall that the resulting drop in blood pH stimulates an increase in breathing rate. Rapid exhalation of CO_2 then counteracts the drop in blood pH.

Homeostasis with respect to blood pH is reinforced by an elegant series of events that takes place inside red blood cells. Biologists were able to work out what was happening when they discovered large amounts of the enzyme **carbonic anhydrase** in red blood cells.

THE ROLE OF CARBONIC ANHYDRASE AND HEMOGLOBIN Recall from Chapter 43 that carbonic anhydrase catalyzes the formation of carbonic acid from carbon dioxide in water. Consequently, CO_2 that diffuses into red blood cells is quickly converted to bicarbonate ions and protons. Thus, most CO_2 is transported in blood (specifically in plasma) in the form of HCO_3^-—the bicarbonate ion. The same reaction occurs in the plasma surrounding red blood cells, but much more slowly in the absence of the enzyme.

Why is the carbonic anhydrase activity in red blood cells so important? The answer has two parts.

1. The protons produced by the enzyme-catalyzed reaction induce the Bohr shift, which makes hemoglobin more likely to release oxygen.

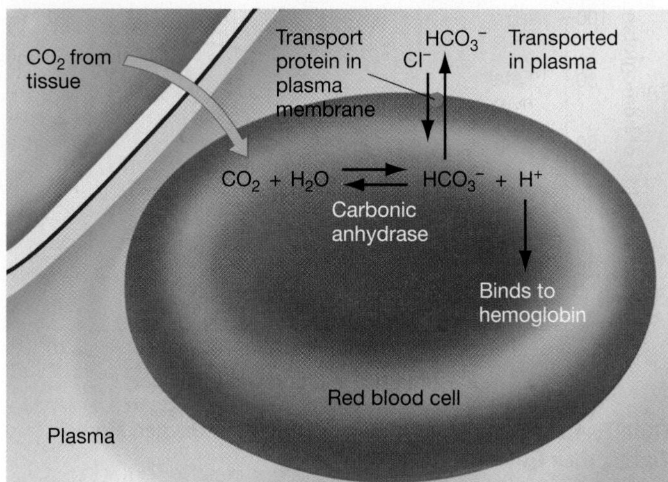

FIGURE 44.17 Carbonic Anhydrase Is Key to CO_2 Transport in Blood. When CO_2 diffuses into red blood cells, carbonic anhydrase quickly converts it to carbonic acid, which dissociates into bicarbonate ion (HCO_3^-) and a proton (H^+). This reaction maintains the partial-pressure gradient favoring the entry of CO_2 into red blood cells. The protons produced by the reaction bind to deoxygenated hemoglobin. Most CO_2 in blood is transported to the lungs in the form of HCO_3^-.

✔**QUESTION** This diagram shows the sequence of events in tissues. Explain what happens when the red blood cell in the diagram reaches the lungs.

2. The partial pressure of CO_2 in blood drops when carbon dioxide is converted to soluble bicarbonate ions, maintaining a strong partial-pressure gradient favoring the entry of CO_2 into red blood cells.

Thus, carbonic anhydrase activity makes CO_2 uptake from tissues more efficient (**Figure 44.17**).

Once bicarbonate ions form in the red blood cell, they diffuse into the blood plasma along a concentration gradient. In contrast, the protons produced by the reaction stay inside red blood cells. Relatively small amounts of H^+ build up in the plasma, so pH remains stable.

What happens to these protons? When hemoglobin is not carrying oxygen molecules, it has a high affinity for protons. As a result, it takes up much of the H^+ that is produced by the dissociation of carbonic acid. Hemoglobin acts as a **buffer**—a compound that minimizes changes in pH.

WHAT HAPPENS IN THE LUNGS? When deoxygenated blood reaches the alveoli, the environment changes dramatically. In the lungs, a partial-pressure gradient favors the diffusion of CO_2 from plasma and red blood cells to the atmosphere. As CO_2 diffuses from the blood into the alveoli, P_{CO_2} in blood declines.

The drop in blood P_{CO_2} reverses the chemical reactions that occurred in tissues:

1. Hydrogen ions leave their binding sites on hemoglobin.
2. Protons react with bicarbonate to form CO_2.
3. CO_2 diffuses into the alveoli and is exhaled from the lungs.

In the meantime, hemoglobin has picked up O_2. Hemoglobin's affinity for oxygen is high in alveoli because blood pH rises as P_{CO_2} declines.

To review the events that occur as O_2 and CO_2 are transported in blood, go to the study area at *www.masteringbiology.com.*

(MB) **BioFlix™** Gas Exchange, **Web Activity** Gas Exchange in the Lungs and Tissues

When blood leaves the lungs, it has unloaded carbon dioxide and hemoglobin is saturated with oxygen. The cycle begins anew.

CHECK YOUR UNDERSTANDING

If you understand that . . .

- In blood, oxygen is bound to hemoglobin and transported inside red blood cells. Carbon dioxide is converted to bicarbonate ions and transported in plasma.
- Hemoglobin has several properties that make it an effective transport protein, including cooperative binding of oxygen, the Bohr shift response to low pH, and the ability to bind the protons that are generated when carbon dioxide is converted to bicarbonate ions.

✔ **You should be able to . . .**

Predict how the oxygen-hemoglobin saturation curves measured in Tibetan people, whose ancestors have lived at high elevations for thousands of years, compare to curves from people whose ancestors have lived at sea level for many generations.

Answers are available in Appendix B.

44.5 The Circulatory System

According to Fick's law, differences in the partial pressure of gases are only part of the story when it comes to understanding diffusion rates. Surface area—*A* in the equation featured in Figure 44.3—also plays a key role.

How do animals maximize the surface area available for diffusion of gases and other key solutes?

- Animals that are only a few millimeters in size, like rotifers and tardigrades, have a small enough volume that diffusion over their body surface is adequate to keep them alive.
- Jellyfish and corals have a large, highly folded gastrovascular cavity that offers a large surface area for exchange of molecules with the environment.
- The flattened bodies of flatworms and tapeworms give these animals a high surface area/volume ratio (see Chapter 41). In these species, molecules are exchanged with the environment directly across the outer body surface.
- Diffusion across the body wall also occurs in roundworms, where gas exchange is facilitated by muscular contractions in their body wall. Diffusion is enhanced as roundworms circulate fluids by sloshing them back and forth.

In larger animals, however, the problem of providing a large surface area for diffusion is solved by a circulatory system. A circulatory system carries transport tissues called blood or hemolymph into close contact with every cell in the body. In this case, "close contact" is a distance of about 0.1 mm between the smallest blood vessels and cells within tissues. Diffusion is efficient at this scale.

To explore how circulatory systems work, let's start by distinguishing the two most basic types of systems—open and closed.

What Is an Open Circulatory System?

In an **open circulatory system**, a fluid connective tissue called hemolymph is actively pumped throughout the body in blood vessels. The hemolymph is not confined exclusively to blood vessels, however. Instead, hemolymph comes into direct contact with tissues. As a result, the molecules being exchanged between hemolymph and tissues do not have to diffuse across the wall of a blood vessel. Hemolymph transports wastes and nutrients and may also contain oxygen-carrying pigments, some cells, and clotting agents.

Figure 44.18 illustrates the open circulatory system in a clam. Note that a muscular organ called the **heart** pumps hemolymph into blood vessels that empty into an open, fluid-filled space. Hemolymph is returned to the heart when the heart relaxes and its internal pressure drops below that in the body cavity. General body movements also move hemolymph to and from the heart(s).

Because it moves throughout the volume of the body, hemolymph is under relatively low pressure in an open circulatory system. As a result, hemolymph flow rates may also be low. These features make open circulatory systems most suitable for relatively sedentary organisms, which do not have high oxygen demands.

Insects, with their rapid movements and more active lifestyles, are an obvious exception to this rule. Insects overcome the limitations imposed by low hemolymph pressure via their tracheal respiratory system, which delivers oxygen directly to the tissues.

Another characteristic of open circulatory systems is that without discrete, continuous vessels, the flow of hemolymph cannot be directed toward tissues that have a high oxygen demand and CO_2 buildup. An open circulatory system moves hemolymph throughout an animal's body in much the same way that a ceiling fan moves air throughout a room in a house.

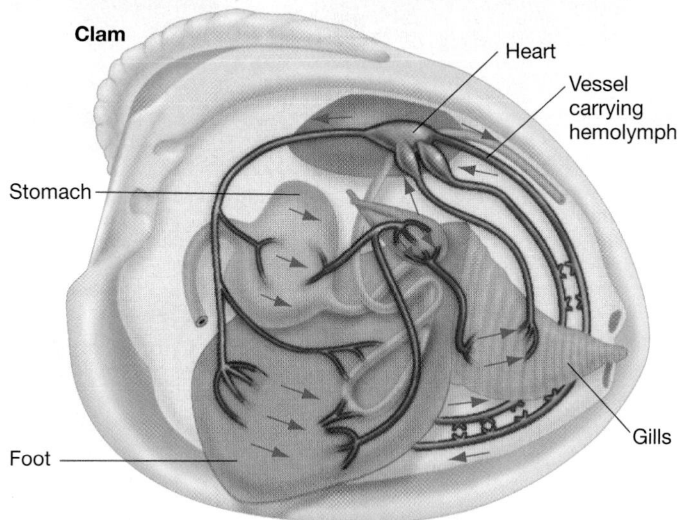

FIGURE 44.18 Open Circulatory Systems in Clams.

Crustaceans are an important exception to this rule, however. Even though their circulatory system is classified as open, these species have a network of small vessels that can preferentially send hemolymph to tissues with the highest oxygen demands.

What Is a Closed Circulatory System?

In a **closed circulatory system**, blood flows in a continuous circuit through the body under pressure generated by a heart. Because blood is confined to vessels, a closed system can generate enough pressure to maintain a high flow rate.

Blood flow can also be directed in a precise way in a closed circulatory system. For example, blood can be shunted to leg muscles during exercise, the intestines after a meal, or regions of the brain engaged in particular mental tasks.

WHICH LINEAGES HAVE CLOSED CIRCULATORY SYSTEMS? Closed circulatory systems are found in vertebrates and a few other lineages where individuals tend to be active. Earthworms and other annelids, for example, have a closed circulatory system and exchange gases with the environment across their thin, moist skin, which has a dense supply of capillaries. As a result, annelids are able to obtain and circulate enough oxygen to support intense muscular activity. Most live as active burrowers and hunters.

A similar situation occurs in squid, octopuses, and other cephalopods that hunt down prey. The closed circulatory system of these mollusks generates high rates of blood flow, which oxygenates their muscles well enough to support rapid movements and a predatory lifestyle.

Closed circulatory systems contain an array of blood vessels, each of which has a distinct structure and function. Let's review the major types of blood vessels, and then consider how the vessels of a closed circulatory system interact with the lymphatic system.

TYPES OF BLOOD VESSELS An enormous amount of tubing is required to distribute blood within "diffusion distance" of every cell in the body. If all the blood vessels in a human body were laid end to end, they would stretch about 100,000 km (over 60,000 miles).

Blood vessels are classified as follows:

- **Arteries** are tough, thick-walled vessels that take blood away from the heart under high pressure. Small arteries are called **arterioles**.
- **Capillaries** are vessels whose walls are just one cell thick, allowing gases and other molecules to exchange with tissues in networks called **capillary beds**.
- **Veins** are vessels that return blood to the heart(s) under low pressure. Small veins are called **venules**.

The structure of arteries, capillaries, and veins correlates closely with their function in a closed circulatory system. For example, the heart ejects blood into a large artery, usually called the **aorta**. All arteries have both muscle fibers and elastic fibers in their walls, but elastic fibers dominate the walls of the aorta. As a result, the aorta can expand when blood enters it under high pressure from the heart.

When contraction of the heart ends, the diameter of the aorta returns to its resting state. This elastic response propels blood away from the heart and augments the force generated by heart contraction. Similar types of secondary pumping action occur to some extent in other arteries as well. This feature helps maintain forward blood flow in the period between heart contractions.

In the walls of arterioles, muscle fibers called **sphincters** wrap around the circumference of the vessel. When these muscle fibers relax, the arteriole diameter increases, resistance to flow in the arteriole is reduced, and blood flow increases in the tissues served by the vessel. But when these arteriolar sphincters constrict, they increase resistance to flow, slow blood passage, and thus divert blood flow for use by other tissues. In this way, blood flow to tissues can be carefully regulated by signals from the nervous system.

Capillaries are the smallest blood vessels. Their walls are only one cell layer thick, and they are just wide enough to let red blood cells through one at a time (**Figure 44.19a**). Because capillaries are extremely thin and because they form an extremely dense network throughout the body, they are the structure where gases, nutrients, and wastes are exchanged between the blood and other tissues.

In some organs, such as the liver, the walls of capillaries contain numerous small openings, which further diminish the barrier to diffusion between blood and the tissues. Despite their thinness, it is rare for capillaries to rupture, because blood pressure drops dramatically as blood passes through arterioles on its way to capillary beds.

Veins carry blood back to the heart after the blood passes through capillaries. Because blood is under relatively low pressure by the time it exits the tissues, veins have thinner walls and larger interior diameters than arteries do (**Figure 44.19b**).

Blood flow in veins is speeded by muscle activity in the extremities, including the hands and feet, which compresses large veins. Larger veins also contain one-way **valves**, which are thin flaps of tissue that prevent any backflow of blood.

All veins contain some muscle fibers, which contract in response to signals from the nervous system, decreasing the diameter and overall volume of the vessels. Blood pressure in a closed circulatory system is regulated, in part, by actively adjusting the volume of blood contained within the veins.

WHAT IS INTERSTITIAL FLUID? The relatively high operating pressure of closed circulatory systems, combined with the thinness of capillaries, produces a small but steady leakage of fluid from these blood vessels into the surrounding space. The area between cells is called interstitial space; the fluid that fills it is **interstitial fluid**. Blood cells are retained within capillaries, so interstitial fluid resembles plasma in its electrolyte composition.

Why does interstitial fluid build up? In 1896 Ernest Starling proposed that two forces were at work (**Figure 44.20**):

1. There is an outward-directed hydrostatic force in capillaries, created by the pressure on blood generated by the heart. This force is analogous to the pressure that drives water through the wall of a leaky garden hose.

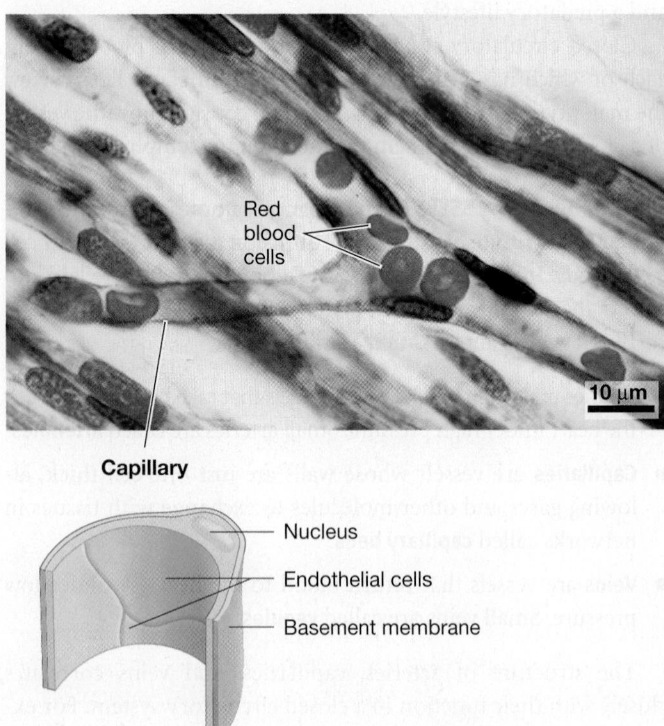

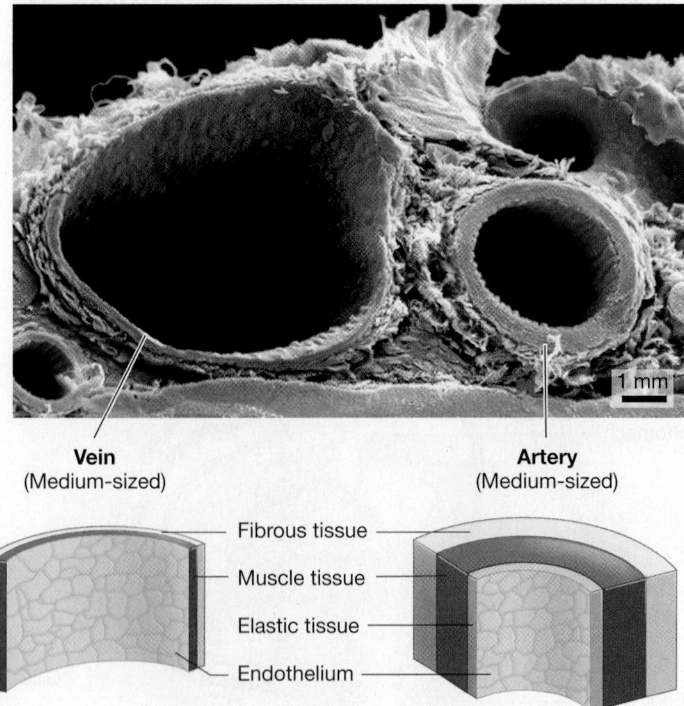

FIGURE 44.19 Capillaries, Veins, and Arteries. Notice the differences in relative wall thickness and overall size. The vein and artery in part (b) have been colorized blue and red respectively.

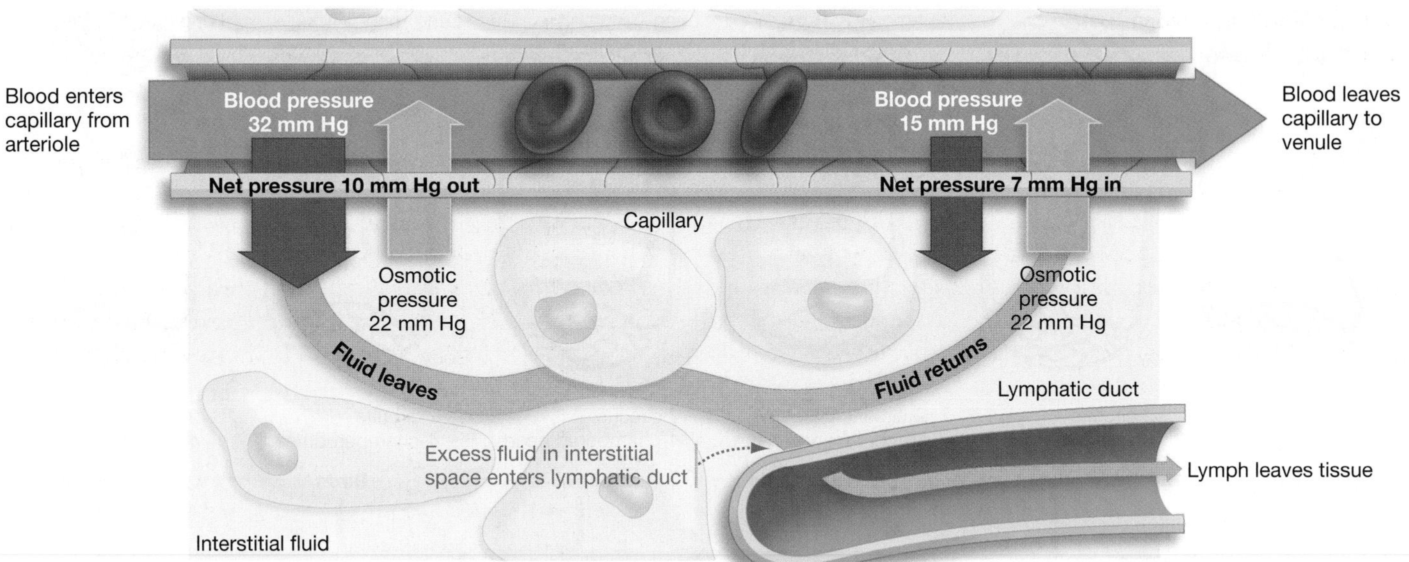

FIGURE 44.20 Pressure Differences in Capillaries Create Interstitial Fluid and Lymph. The balance of blood pressure and osmotic forces favors fluid loss from the beginning (inflow end) of capillaries and fluid recovery at the other (outflow) end. Fluid that is not recovered by the capillaries is transported out of the tissue as lymph, which eventually rejoins the blood circulation.

2. There is also an inward-directed osmotic force in capillaries, created by the higher concentration of solutes in the blood plasma than in the interstitial space.

Starling reasoned that at the end of the capillary nearest to an arteriole, the hydrostatic force would exceed the osmotic force. If so, then fluid would move out of the capillary into the interstitial space. But because blood pressure drops as fluid passes through a long, thin tube, Starling proposed that at the venous end of the capillary, the inward-directed osmotic force would exceed the outward-directed hydrostatic force. Thus, the fluid that was lost at the arteriolar end of the capillary would be largely reclaimed at the venous end.

Note the adverb "largely," however—not all interstitial fluid is reabsorbed by capillaries.

THE ROLE OF THE LYMPHATIC SYSTEM Starling proposed that any interstitial fluid that was not reclaimed in the capillary would be collected in the **lymphatic system**: a collection of thin-walled, branching tubules called lymphatic ducts or vessels. Interstitial fluid that enters the lymphatic ducts is called **lymph**.

Starling's model was based on three key observations about lymphatic ducts:

1. They permeate all tissues.
2. They eventually join with one another, like the tributaries of a river, to form larger vessels.
3. The largest lymphatic vessels return excess interstitial fluid, in the form of lymph, to the major veins entering the heart.

The model also rested on a key assumption, however: For Starling's ideas about events in the capillaries, interstitial fluid, and lymphatic ducts to work, the total solute concentrations in plasma and in interstitial fluid must be different enough to bring fluid into capillaries via osmosis.

Specifically, the fluid that leaks out of capillaries must have low osmolarity. For this to be true, capillaries have to act as filters that retain large protein molecules and other solutes as fluid leaks out into the surrounding tissues.

To test this assumption, a group of biologists placed small tubes in the lymphatic ducts of animals to collect lymph. Then the team introduced radioactively labeled proteins of various sizes into the blood inside veins. They found that even though small proteins crossed readily from the blood to the lymph, larger proteins did not. Enough large proteins remained in the capillaries to set up the osmotic gradient required by Starling's model.

Follow-up work has shown that the most important of these large proteins is **albumin**. Albumin's large size and net negative charge effectively exclude it from interstitial fluid. The presence of albumin in blood keeps solute concentrations high, maintaining a strong osmotic gradient that brings fluid back into capillaries.

As a result, the total daily formation of lymph is limited to just 2–5 percent of plasma volume. This amount is easily taken up by lymph vessels and returned to the blood.

How Does the Heart Work?

In animals with closed circulatory systems, the heart contains at least two chambers. There is at least one thin-walled **atrium** (plural: **atria**), which receives blood, and at least one thick-walled **ventricle**, which generates the force required to propel blood out of the heart and through the circulatory system. Atria are separated from ventricles by atrioventricular valves.

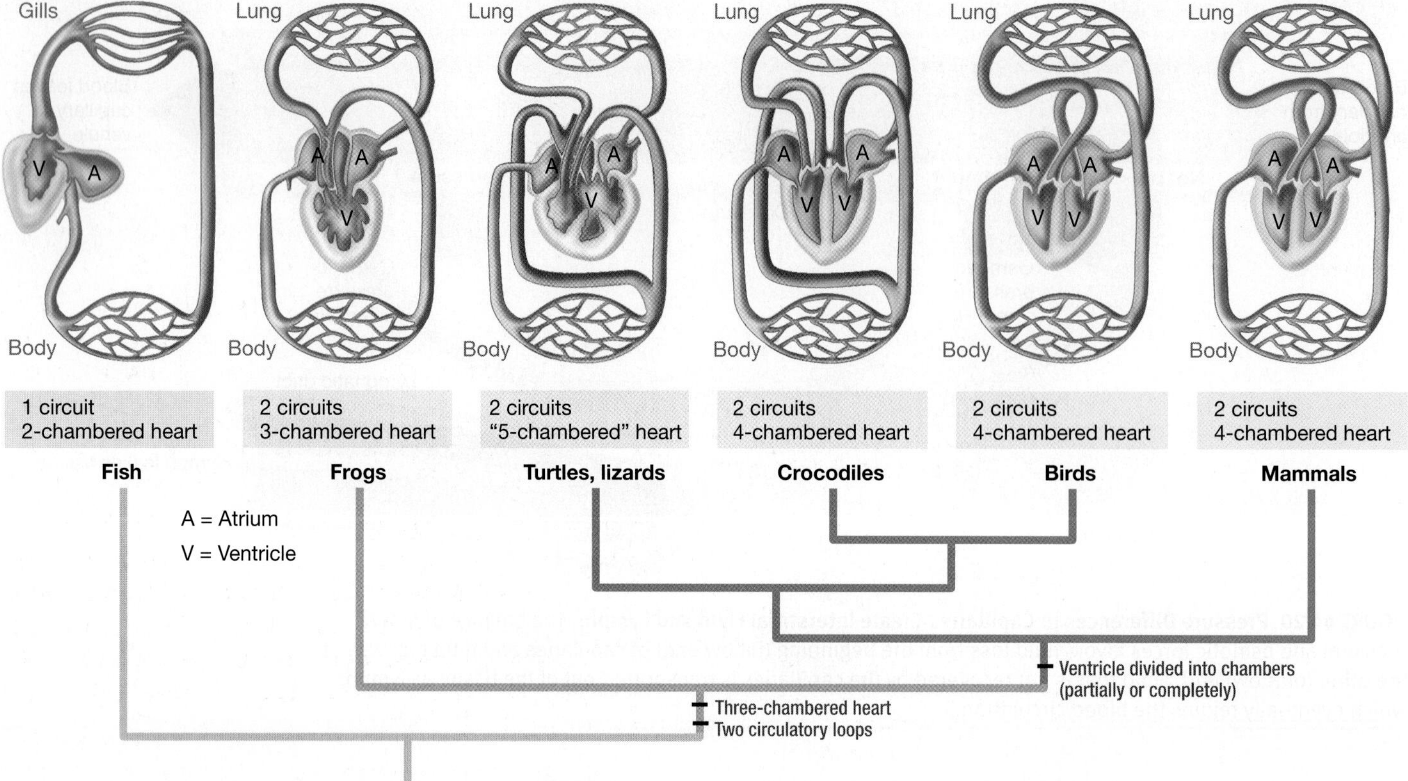

FIGURE 44.21 The Evolution of the Vertebrate Circulatory System. "A" denotes the atria—chambers that receive blood coming into the heart from the body and the lungs. "V" denotes the ventricles—chambers that pump blood out to the lungs and the body.

The phylogenetic tree in **Figure 44.21** shows the evolutionary relationships among some major vertebrate groups and a simplified sketch of the heart and circulatory system for each lineage. Two points are particularly important to note:

1. The number of distinct chambers in the heart increased as vertebrates diversified. Fish hearts have two chambers; amphibians have three; turtles and lizards have five (if the partially divided ventricle is counted as three chambers); crocodiles, birds, and mammals have four.
2. In fish, the circulatory system forms a single circuit—one loop services the gills and the body. In other lineages, there are separate circuits to the lungs and to the body.

WHY DID MULTICHAMBERED HEARTS AND MULTIPLE CIRCULATIONS EVOLVE? To understand these evolutionary patterns, let's first consider the relatively simple heart and circulatory system found in fish. A two-chambered heart and single circuit are adequate in fish. Even though blood pressure drops as blood passes through the gills—due to the mechanical resistance to flow that occurs in the gills' capillary beds—blood pressure stays high enough to move blood throughout the body. This is largely because fish live in the neutrally buoyant environment of water, where gravity does not have a large impact on blood flow.

In contrast, gravity has a much larger effect on circulation in land-dwelling vertebrates. Gravity is particularly important in opposing blood flow to elevated portions of the body. To overcome gravity in terrestrial environments, blood must be pumped at high pressure. However, the capillaries and alveoli of the lungs are too thin to withstand high pressures.

The successful solution in land-dwelling vertebrates was the evolution of two separate pumping circuits:

- The **pulmonary circulation** is a lower-pressure circuit to and from the lung.
- The **systemic circulation** is a higher-pressure circuit to and from the rest of the body.

The pulmonary and systemic circulations are completely separated in the four-chambered hearts of birds and mammals, which are essentially two fish hearts side by side.

The pulmonary and systemic circulations are only partially separated in the three-chambered hearts of amphibians and the "five-chambered" hearts of turtles and lizards, however. In addition, some of these animals have a bypass vessel running from the right side of the ventricle directly into the systemic circulation. This bypass vessel is also observed in the unusual four-chambered hearts of crocodiles.

The bypass vessels observed in certain lineages have an important function: They shunt blood from the pulmonary to the systemic circulation when an animal is underwater. As a result, the individual greatly reduces blood flow to the lungs when it is not breathing.

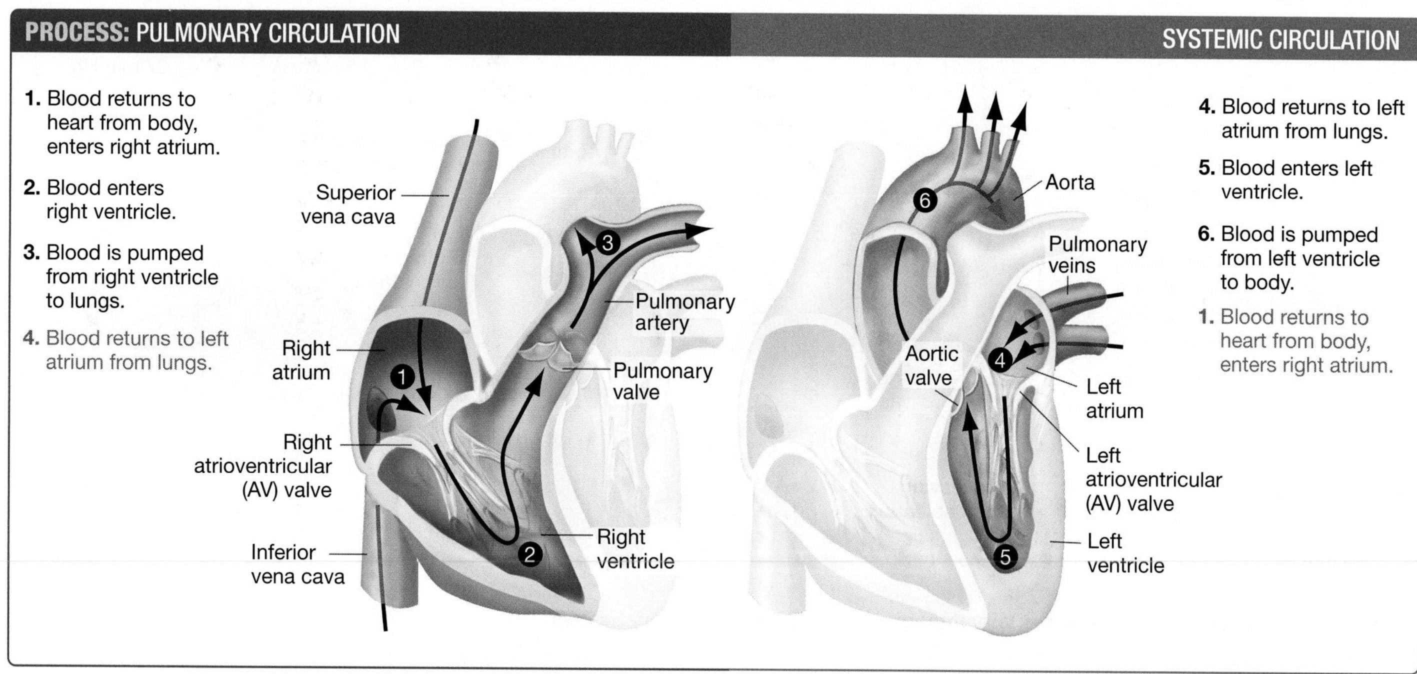

FIGURE 44.22 The Human Heart. Blood flows through a four-chambered heart in the sequence shown.

THE HUMAN HEART Your heart is located in the chest cavity between your lungs, and is roughly the size of your fist. As **Figure 44.22** shows, the circulatory system returns blood from the body to the right atrium of the human heart. This blood is low in oxygen, and arrives via two large veins called the inferior (lower) and superior (upper) **venae cavae** (singular: **vena cava**).

When the muscles that line the right atrium contract, they send deoxygenated blood to the right ventricle. The right ventricle, in turn, contracts and sends blood out to the lungs, via the **pulmonary artery**. In this way, the right ventricle powers the movement of blood through the pulmonary circulation. Blood that returns to the heart from the body is sent to the lungs.

Blood flows from atrium to ventricle to artery only one way, because one-way valves separate the heart's chambers from each other and from the adjacent blood vessels. As Figure 44.22 indicates, the valves are flaps, oriented in a way that ensures a one-way flow of blood with little or no backflow. If heart valves are damaged or defective, the resulting backflow can be heard through a stethoscope. The backflow reduces the organ's efficiency and is called a **heart murmur**.

After blood circulates through the capillary beds in the lung's alveoli and becomes oxygenated, it returns to the heart through the **pulmonary veins**. The oxygenated blood enters the left atrium.

When the left atrium contracts, it pushes blood into the left ventricle. The walls of the left ventricle are so thick with muscle cells that their contraction sends oxygenated blood at high pressure through the aorta and into the arteries and capillaries that form the systemic circulation.

Figure 44.23 on page 880 summarizes the flow pattern through the human circulatory system and the blood gas concentrations at various points in the pulmonary and systemic circulations. Notice that blood vessels are called arteries or veins according to the direction of blood flow relative to the heart, not because of the oxygen content of the blood in them. Thus, the pulmonary artery is called an artery because it takes blood away from the heart, even though this blood is low in oxygen.

In healthy individuals, gas exchange in the lungs and tissues is rapid relative to the rate at which blood flows through capillaries. This situation is beneficial: It ensures that the maximum amount of oxygen is taken up and the maximum amount of carbon dioxide is released. As Figure 44.23 indicates, the partial pressures of oxygen and carbon dioxide in the alveoli and the pulmonary veins are equal.

Similarly, the slow passage of blood through capillaries means that a maximum amount of oxygen is released from blood to tissues and a maximum amount of carbon dioxide is taken up. As a result, partial pressures for oxygen and carbon dioxide in the tissues and the systemic veins are equal.

THE CARDIAC CYCLE Although the preceding discussion suggests that each chamber in the heart acts independently, this is not the case. The contraction phase of the atria and the ventricles, called **systole**, is closely coordinated with their relaxation phases, or **diastole**.

For example, both atria contract simultaneously about 0.1 second ahead of ventricular contraction. Because both ventricles are still in diastole and relaxed when the atria contract, the ventricles fill with blood from the atria. Then the ventricles enter systole and contract.

This sequence of contraction and relaxation is called the **cardiac cycle**. It consists of one diastole and one systole for both atria and ventricles.

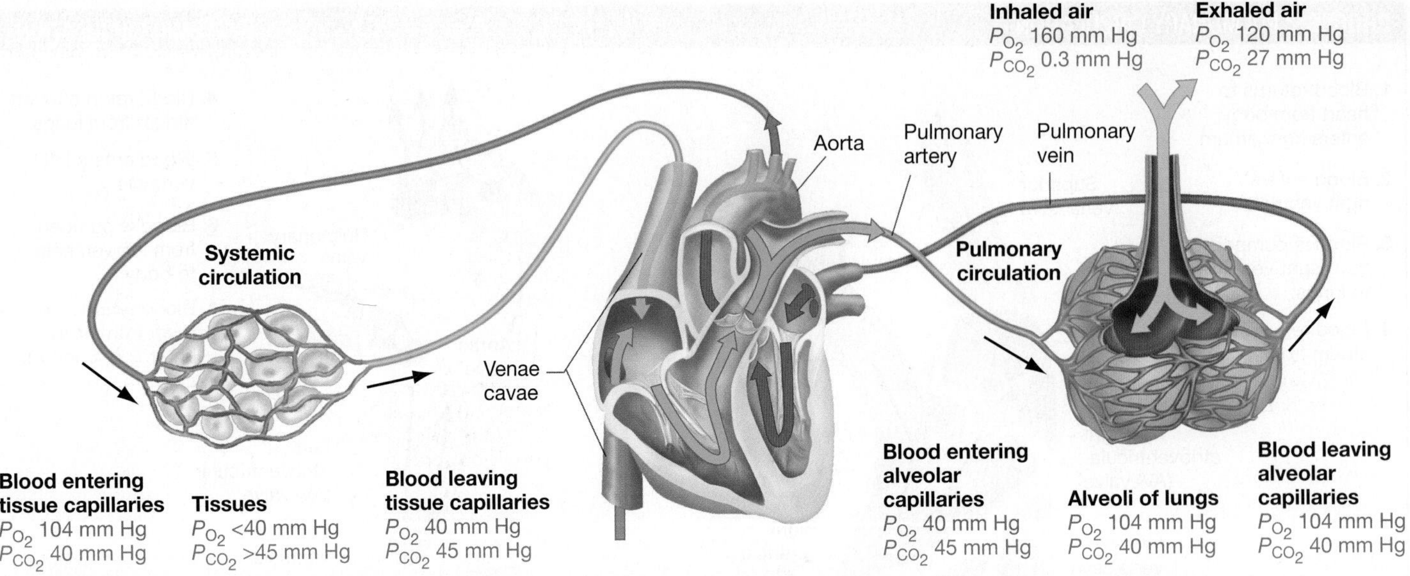

FIGURE 44.23 Partial Pressures of Gases Vary throughout the Human Circulatory System.

✔**QUESTION** Why are the partial pressures of oxygen and carbon dioxide in exhaled air intermediate in magnitude between the partial pressures in inhaled and alveolar air?

Ventricular contraction (ventricular systole) leads to a rapid increase in pressure within both ventricles, as recorded in the purple line in **Figure 44.24**. Blood is ejected into the pulmonary artery and the aorta when ventricular pressure equals the pressure within each respective artery. Blood pressure measured in the systemic arterial circulation at the peak of ventricular ejection into the aorta is called the **systolic blood pressure**. Blood pressure measured just prior to ventricular ejection is called the **diastolic blood pressure.**

Clinicians report blood pressure measurements in fractional notation, with systolic pressure as the numerator and diastolic pressure as the denominator. People with blood pressures consistently higher than 140/90 mm Hg have high blood pressure, or **hypertension.**

Hypertension is a serious concern to physicians because it can lead to a variety of defects in the heart and circulatory systems. Abnormally high blood pressure puts mechanical stress on arteries. If the walls of an artery fail, the individual may experience heart attack, stroke, kidney failure, and burst or dilated blood vessels.

To review the anatomy of the mammalian heart and how the cardiac cycle works, go to the study area at *www.masteringbiology.com.*

(MB) **Web Activity** The Human Heart

ELECTRICAL ACTIVATION OF THE HEART Like other muscle cells, cardiac muscle cells contract in response to electrical signals. In invertebrates, the electrical impulses that trigger contraction come directly from the nervous system. But in vertebrates, a group of cells in the heart itself is responsible for generating the initial signal.

The cells that initiate contraction in the vertebrate heart are known as **pacemaker cells.** They are located in a region of the right atrium called the **sinoatrial (SA) node.**

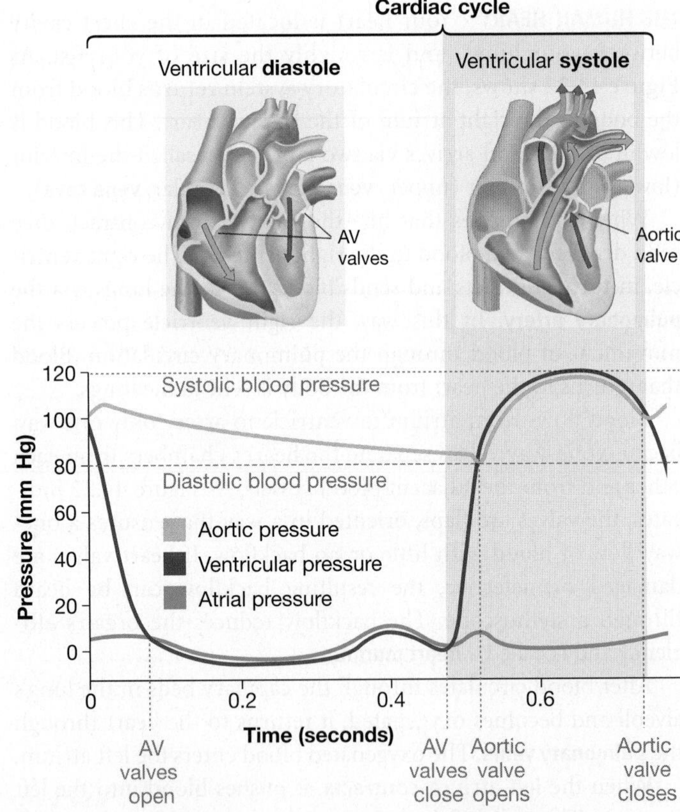

FIGURE 44.24 Blood Pressure Changes during the Cardiac Cycle. These data show the pressures created by the left ventricle and left atrium in the cardiac cycle. In this case, the blood pressure measured in the upper arm would be 120/80 mm Hg. The right ventricle and pulmonary artery pressures follow a similar pattern, but the blood pressure measured in the pulmonary artery would be much lower—closer to 25/8 mm Hg.

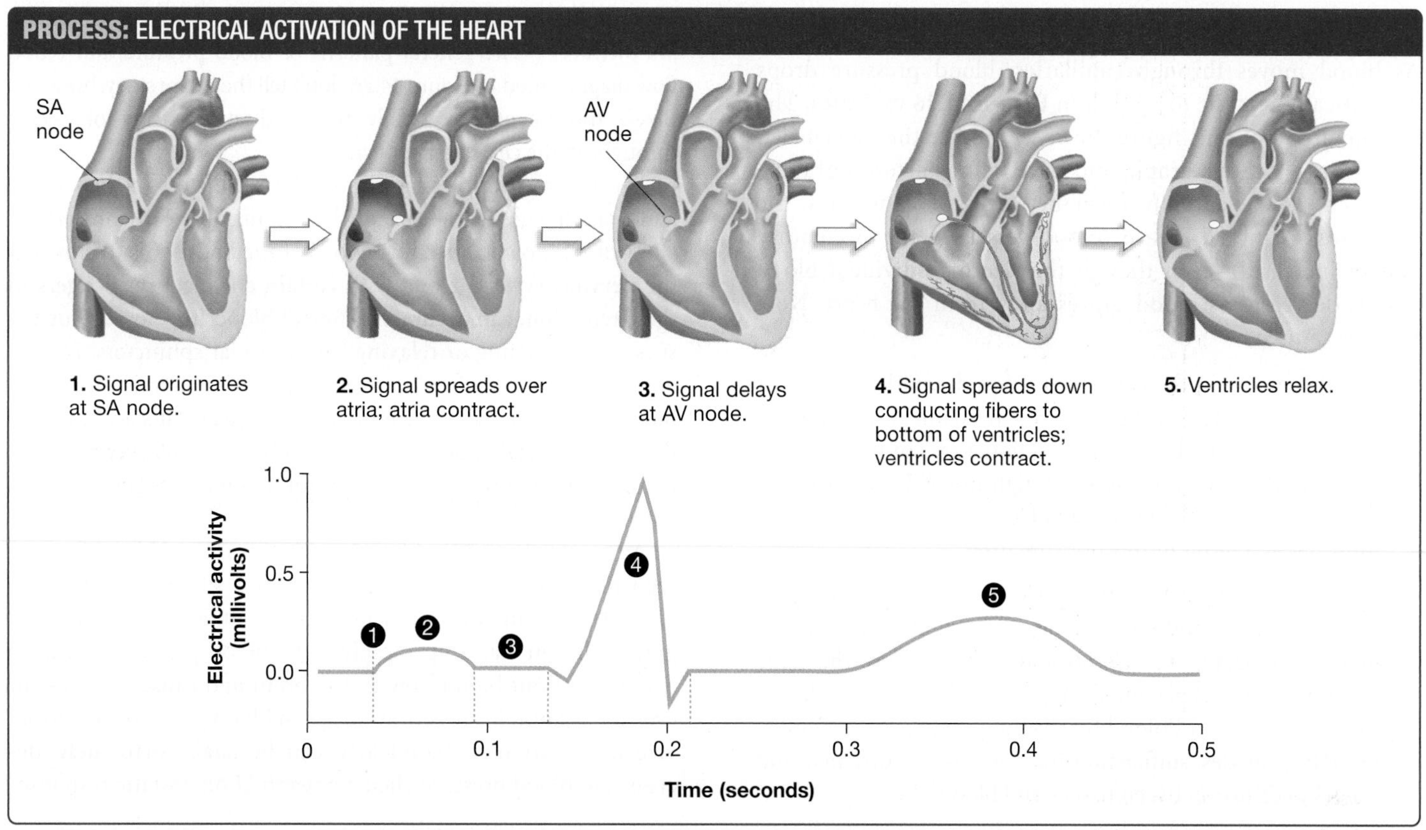

FIGURE 44.25 Sequential Electrical Activation Leads to Coordinated Contraction of the Heart.

The SA node and the muscle cells of the heart receive input from the nervous system and from chemical messengers carried in the blood. These inputs are important for regulating both the heart rate and the strength of ventricular contraction. In this way, the amount of blood moving through the circulatory system varies in response to electrical signals and hormones. During the "fight or flight" response introduced in Chapter 47, for example, a chemical signal called epinephrine causes both heart rate and contraction strength to increase—sending more blood through the body in preparation for rapid movement.

Even though heart rate and contraction strength are carefully regulated, a vertebrate heart will beat even if all nerves supplying it are severed. Why? An electrical impulse that stimulates contraction is generated in the SA node and rapidly conducted throughout the right and left atria. The signal spreads quickly from cell to cell because of a striking property of cardiac muscle cells: They form physical and electrical connections with each other.

All cardiac muscle cells branch to contact several other cardiac muscle cells, join end to end with these neighboring cells, and connect to them by specialized structures called **intercalated discs**. Because these discs contain numerous gap junctions—cell-to-cell connections described in Chapter 8—electrical signals pass directly from one cardiac muscle cell to the next.

To understand electrical activation of the heart, follow the orange line on the graph in **Figure 44.25**. This line is called an **electrocardiogram**, or **EKG**—a recording of the electrical events that occur over the course of a cardiac cycle. The drawings above the graph show where key events are happening.

1. The SA node originates a signal.
2. The signal from the SA node is propagated over the atria. As a result, the atria contract simultaneously and fill the ventricles.
3. The signal from the atria is conducted to an area of the heart called the **atrioventricular (AV) node**. The AV node delays the signal slightly before passing it to the ventricles. The delay allows the ventricles to fill completely with blood before they contract.
4. After the delay, the electrical impulse is rapidly transmitted through specialized fibers in the muscular wall that separates the ventricles. The impulse spreads through both ventricles, causing them to contract as the atria relax. The ventricles empty efficiently because the signal and the resulting muscular contraction move from the bottom up to the top of each ventricle—toward the arteries that allow blood to exit.
5. The final electrical event occurs as the ventricles relax and their cells recover—restoring their electrical state prior to contraction.

An EKG recording is generated by amplifying the overall electrical signal conducted from the heart to the chest wall through the tissues of the body. By inspecting an EKG, physicians can diagnose disturbances of heart rhythm and detect damage to the heart muscle.

Patterns in Blood Pressure and Blood Flow

As blood moves through capillaries, blood pressure drops dramatically—as the top graph in **Figure 44.26** indicates. The bottom graph on this figure shows why. Trace the line labeled "Total Area" on this graph, and note that as arteries branch, rebranch, and eventually form networks of capillaries, the total cross-sectional area of blood vessels in the circulatory system increases—even though the size of individual blood vessels decreases as blood moves away from the heart. Note two key points:

1. Blood pressure in the capillaries drops dramatically (on the top graph) because the amount of mechanical resistance to flow is a function of total cross-sectional area of the vessels—meaning that friction losses are high in capillary beds. The pulsing nature of the pressure noted in the larger arteries diminishes as a result of this pressure drop.
2. As the line labeled "Velocity" in the bottom graph indicates, the velocity of blood flow also decreases significantly in capillary beds relative to arteries and veins, because the same amount of fluid is passing through a much larger area. Recall that the slow rate of blood flow through capillaries is important: It provides sufficient time for gases, nutrients, and wastes to diffuse between tissues and blood.

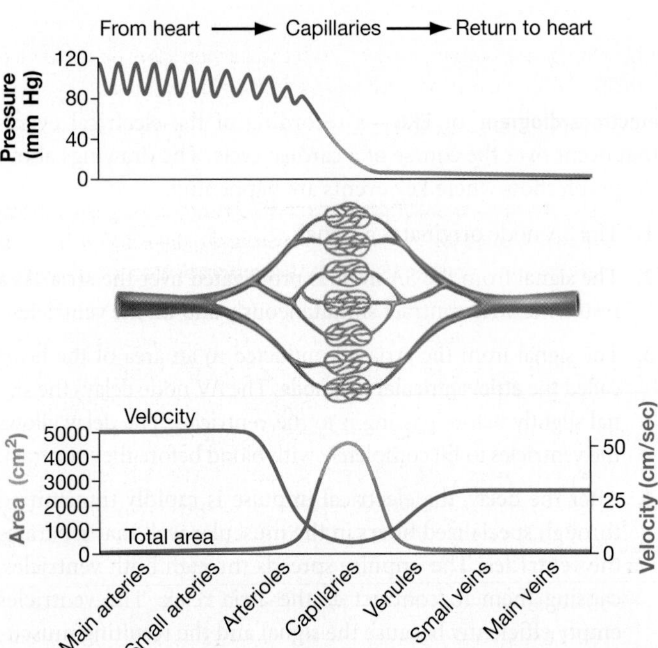

FIGURE 44.26 Blood Pressure Drops Dramatically in the Circulatory System. The top graph shows how blood pressure changes as blood leaves the heart and travels through arteries, capillaries, and veins, as in the branching pattern of the middle diagram. In arteries near the heart, each heartbeat causes fluctuations in blood pressure. These pressure pulses disappear in the capillaries, so blood flows there at a steady speed. The bottom graph plots the total area of blood vessels shown in the diagram, as well as the velocity of blood flow through the vessels.

WHY IS REGULATION OF BLOOD PRESSURE AND BLOOD FLOW IMPORTANT? The general patterns of blood pressure and blood flow diagrammed in Figure 44.26 don't tell the entire story, however. Blood movement is carefully regulated at an array of points throughout the circulatory system.

Recall that arterioles have muscular sphincters in their walls, and that changes in the size of these sphincters can constrict or allow blood flow to specific regions of the body. This means that the nervous system, along with certain chemical messengers in the circulation, can accurately control blood flow to various tissues by contracting or relaxing the arteriolar sphincters. For example, arterioles in the skin may dilate during exercise, diverting blood flow to the skin to eliminate excess heat. This accounts for the flushed facial appearance induced by vigorous exercise. Regulating blood flow is important for maintaining homeostasis with respect to body temperature.

As another example of how blood pressure and blood flow are regulated, consider what happens if you sit long enough for blood to pool in your lower extremities, under the influence of gravity. If you stand up rapidly, your blood pressure can drop enough to limit blood flow to the brain and cause dizziness or even a blackout. More serious drops in blood pressure, due to severe dehydration or blood loss, can be fatal. Fortunately, decreases in blood pressure elicit a powerful homeostatic response.

HOMEOSTATIC CONTROL OF BLOOD PRESSURE Recall from Chapter 41 that all homeostatic responses involve (1) sensors that detect the change in condition, (2) an integrator that processes information about the change, and (3) effectors that diminish the impact of the change.

Specialized pressure-sensing nerve cells called **baroreceptors** detect decreases in blood pressure. Baroreceptors are found in the walls of the heart and the major arteries, both as they leave the heart and as they enter the neck. This distribution is logical, because the head tends to be the most elevated part of land-dwelling animals and the preservation of blood flow to the brain is the highest priority.

When baroreceptors transmit signals to the brain indicating a serious fall in blood pressure, a rapid, three-component effector response ensues:

1. Cardiac output—the volume of blood leaving the left ventricle—increases. This is due to an increase in heart rate and an increase in stroke volume—meaning, the amount of blood ejected from the ventricles during each cardiac cycle. (Cardiac output = heart rate × stroke volume.)
2. Arterioles that serve the capillaries of tissues such as the skin and intestines, which can endure short-term restrictions in their blood supply without damage, constrict. Arteriolar constriction diverts blood to more critical organs.
3. Veins constrict, decreasing their overall volume. Because more than half of the blood in the circulatory system is contained within the veins, constriction of these vessels shifts blood volume toward the heart and arteries to maintain blood pressure and flow to vital organs.

This coordinated response is mediated both by a portion of the nervous system called the sympathetic nervous system (Chapter 45) and by hormones (Chapter 47) produced by the adrenal glands. Sympathetic nerves and the hormones involved in blood pressure regulation deliver their messages directly to the SA node to increase heart rate, the ventricles of the heart to increase stroke volume, the sphincters of the arterioles to modify their resistance, and the muscular walls of the veins to modify their total volume.

The homeostatic response to falling blood pressure resembles the **fight-or-flight response**, which is triggered by a threatening stimulus (see Chapter 47). During the fight-or-flight response, an organism prepares to defend itself or rapidly escape from danger. Part of this response involves directing blood to the brain and muscles in preparation for quick action. The side effects include the "cold sweat" and nausea often induced by a fearful situation. These feelings occur because blood flow is directed away from the skin and intestines.

CHECK YOUR UNDERSTANDING

If you understand that . . .

- Animal circulatory systems may be open or closed, but both types of systems circulate blood or hemolymph via pressure generated by one or more hearts.
- In closed systems, regulated changes in the diameter of blood vessels can direct blood to specific regions, and overall blood pressure is carefully regulated through changes in cardiac output.

✓ **You should be able to . . .**

1. Explain how the mammalian lymphatic system and circulatory system interact.
2. Make a labeled diagram showing how blood circulates through the mammalian heart.

Answers are available in Appendix B.

CHAPTER 44 REVIEW

For media, go to the study area at www.masteringbiology.com 

Summary of Key Concepts

Animals have to take in oxygen and expel carbon dioxide to sustain cellular respiration and stay alive. Terrestrial animals and aquatic animals face different challenges in performing gas exchange.

- As media for exchanging oxygen and carbon dioxide, air and water are dramatically different.
- Compared with water, air contains much more oxygen and is much less dense and viscous. As a result, terrestrial animals have to process a much smaller volume of air to extract the same amount of O_2, and don't have to work as hard to do so.
- Both terrestrial and aquatic animals pay a price for exchanging gases: Land-dwellers lose water to evaporation; freshwater animals lose ions and gain excess water; marine animals gain sodium and chloride and lose water.

✓ You should be able to explain why no water breathers have the extremely high rates of metabolism required for endothermy.

Gas-exchange organs maximize the rate of O_2 and CO_2 diffusion by (1) presenting a large, thin surface area to the environment, and (2) maintaining a steep partial-pressure gradient that favors entry of O_2 and elimination of CO_2.

- The structure of gills, tracheae, lungs, and other gas-exchange organs minimizes the cost of ventilation while maximizing the rate at which O_2 and CO_2 diffuse.
- Consistent with predictions made by Fick's law of diffusion, respiratory epithelia tend to be extremely thin and folded to increase surface area.
- In fish gills, countercurrent exchange ensures that the partial-pressure differences between O_2 and CO_2 in water and blood are high over the entire length of the ventilatory surface.
- In bird lungs, structural adaptations lead to a high ratio of useful ventilatory space to dead space.
- Breathing rate is regulated, to keep the carbon dioxide content of the blood stable during both rest and exercise.

✓ You should be able to explain why no large animals have gas exchange occurring only across the skin.

Blood is a specialized tissue that transports gases, along with nutrients and wastes, in some animals. Hemoglobin is an oxygen-carrying protein that is extremely efficient at taking up oxygen in the lungs and delivering it to tissues.

- The tendency of hemoglobin to give up oxygen varies as a function of the P_{O_2} in surrounding tissue in a sigmoidal fashion. As a result, a relatively small change in tissue causes a large change in the amount of oxygen released from hemoglobin.
- Oxygen binds less tightly to hemoglobin when pH is low. Because CO_2 tends to react with water to form carbonic acid, the existence of high CO_2 partial pressures in exercising muscle tissues lowers their pH and makes oxygen less likely to stay bound to hemoglobin and more likely to be released into tissues.
- The CO_2 that diffuses into red blood cells from tissues is rapidly converted to carbonic acid by the enzyme carbonic anhydrase. The protons that are released as carbonic acid dissociates bind to deoxygenated hemoglobin. In this way, hemoglobin acts as a buffer that takes protons out of solution and prevents large fluctuations in blood pH.

✓ You should be able to explain how oxygen and carbon dioxide are transported in blood in the icefish native to the Antarctic, since these fish lack hemoglobin.

MB **BioFlix™** Gas Exchange, **Web Activity** Gas Exchange in the Lungs and Tissues

Circulatory systems use the pressure generated by one or more hearts to transport blood and other substances throughout the body.

- In many animals, blood or hemolymph moves through the body via a circulatory system consisting of a pump (heart) and vessels.
- In open circulatory systems, overall pressure is low and tissues are bathed directly in hemolymph.
- In closed circulatory systems, blood is contained in vessels that form a continuous circuit. Containment of blood allows higher pressures and flow rates, as well as the ability to direct blood flow accurately to tissues that need it the most.
- In organisms with a closed circulatory system, a lymphatic system returns excess fluid that leaks from the capillaries back to the circulation.
- In mammals and birds, a four-chambered heart pumps blood into two circuits, which separately serve the lungs and the rest of the body.
- The cardiac cycle is controlled by electrical signals that originate in the heart itself.
- Heart rate, cardiac output, and contraction of both arterioles and veins are regulated by chemical signals and by electrical signals from the brain.

✔ You should be able to describe the circulatory systems expected to be found in terrestrial animals with high activity rates.

MB **Web Activity** The Human Heart

Questions

✔ TEST YOUR KNOWLEDGE

Answers are available in Appendix B

1. O_2 will diffuse from blood to tissue faster in response to which of the following conditions?
   a. an increase in the P_{O_2} of the tissue
   b. a decrease in the P_{O_2} of the tissue
   c. an increase in the thickness of the capillary wall
   d. a decrease in the surface area of the capillary
2. Which of the following does blood *not* do?
   a. transport O_2 and CO_2
   b. distribute body heat
   c. produce red blood cells and other formed elements
   d. buffer against pH changes
3. In insects, what is the adaptive significance of spiracles?
   a. They open and close during flight or other types of movement, so function as a "breathing" mechanism.
   b. They open into the body cavity, allowing direct contact between hemolymph and tissues.
   c. They are thin and highly branched, so offer a large surface area for gas exchange.
   d. They close off tracheae to minimize water loss.
4. Which of the following is *not* an advantage of breathing air over breathing water?
   a. Air is less dense than water, so it takes less energy to move during ventilation.
   b. Oxygen diffuses faster through air than it does through water.
   c. The oxygen content of air is greater than that of an equal volume of water.
   d. Air breathing leads to high evaporation rates from the respiratory surface.
5. Which of the following promotes oxygen release from hemoglobin?
   a. a decrease in temperature
   b. a decrease in CO_2 levels
   c. a decrease in pH
   d. a decrease in carbonic anhydrase
6. An open circulatory system is less efficient than a closed circulatory system in what respect?
   a. It is harder to deliver O_2 to specific tissues based on need.
   b. Hemolymph does not contain oxygen-binding molecules.
   c. There is no heart to pump the blood.
   d. In closed systems, body movements cannot help circulate the blood.

✔ TEST YOUR UNDERSTANDING

Answers are available in Appendix B

1. Compare and contrast the respiratory and circulatory systems of an insect and a human.
2. Describe the changes in oxygen delivery that occur as a person changes from a resting state to intense exercise.
3. Explain how most carbon dioxide is transported in the blood. In humans, why don't intense exercise and rapid production of CO_2 lead to a rapid reduction in blood pH?
4. Why is ventilation in birds considered much more efficient than the respiratory system of humans and other mammals?
5. Explain how each parameter in Fick's law of diffusion is reflected in the structure of the mammalian lung.
6. Why did separate systemic and pulmonary circulations evolve in species that have the high-pressure circulatory system required for rapid movement of blood?

✔ APPLYING CONCEPTS TO NEW SITUATIONS

Answers are available in Appendix B

1. Researchers who compared the total amount of ventilatory surface area in the gills of various species of fish found that species that do a lot of high-speed swimming have a larger total surface area than do slow-moving species of the same size. Interpret this pattern in light of the theory of evolution by natural selection.
2. At high elevations, certain cells in humans increase the production of 2,3-diphosphoglycerate (DPG). In blood, DPG shifts the oxygen-hemoglobin equilibrium curve to the right. How does this shift affect O_2 unloading and loading?
3. Carp are fish that thrive in stagnant-water habitats with low-oxygen partial pressures. Compared with the hemoglobin of many other fish species, carp hemoglobin has an extremely high affinity for O_2. Is this trait adaptive? Explain your answer.
4. The carbon monoxide (CO) found in furnace and engine exhaust and in cigarette smoke binds to the heme groups in hemoglobin 210 times more tightly than does O_2. Explain why exposure to large doses of CO can lead to suffocation.

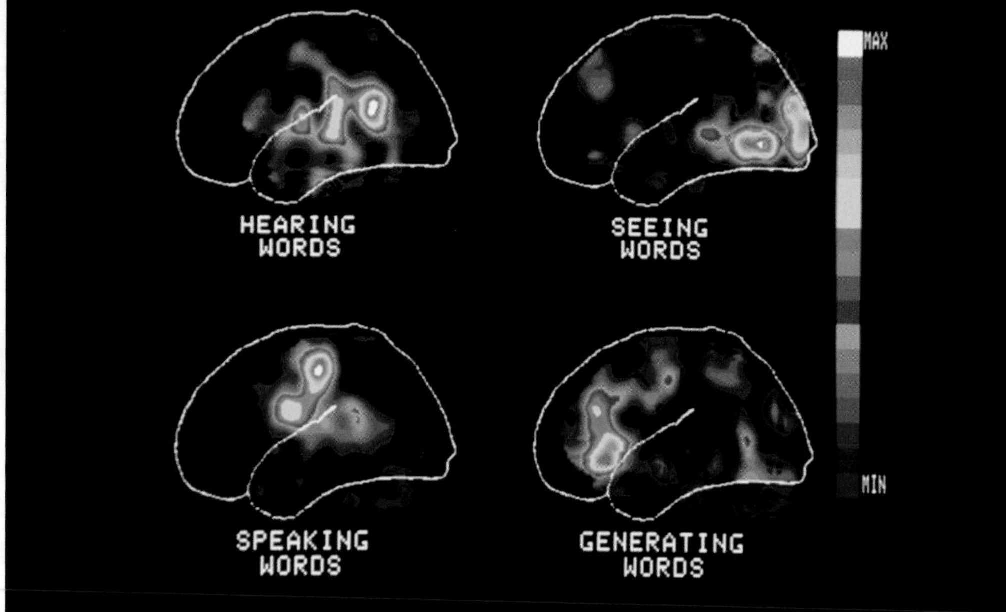

Positron-emission tomography (PET) scans showing changes in the activity of brain cells during different tasks, such as remembering words and speaking them aloud. Different brain areas are specialized for performing different tasks.

Electrical Signals in Animals 45

Most students and professional biologists are attracted to the study of electrical signaling, or neurobiology, because they want to understand the human brain and higher-order processes like consciousness, intelligence, emotion, learning, and memory. But early in the history of neurobiology, researchers realized that the human brain—with its billions of cells—was much too complex to study productively.

Instead, biologists did the same thing they did in the early days of studying cells, genetics, and evolution: They started simple. Until recently, much of the most productive work in neurobiology has focused on how individual nerve cells, or **neurons**, work.

Neurons conduct information in the form of electrical signals. Electrical signals are fast and precise. Neurons conduct them from point to point in the body at speeds of up to 200 m/sec (450 mph). Electrical signaling is a crucial aspect of information processing—one of the five attributes of life introduced in Chapter 1.

Most neurons work in the same general way. Early research on the electrical properties of a single neuron laid a broad foundation for more recent studies of how the brain works.

This chapter works the same way. After a brief overview of nervous systems, we'll focus on the neuron itself and analyze the membrane proteins that dictate its electrical properties. By the end of the chapter, you'll be considering how the brain is organized and how phenomena like memory work.

45.1 Principles of Electrical Signaling

As Chapter 32 emphasized, the evolution of neurons—along with muscle cells—was a key event in the diversification of animals. All animals except sponges have neurons and muscle cells. Neurons transmit electrical signals; muscles can respond to electrical signals by contracting.

KEY CONCEPTS

- Neurons are cells that transmit electrical signals used in communication. Their plasma membranes carry a voltage, called a membrane potential, due to differences in the concentrations of ions on the inner and outer surfaces.
- Action potentials are all-or-none changes in membrane potential that serve as electrical signals. During an action potential, an inflow of sodium ions is followed by an outflow of potassium ions.
- At synapses, the electrical signal from a neuron triggers the release of a chemical signal—a neurotransmitter. When the neurotransmitter arrives at an adjacent neuron, there is a change in that cell's membrane potential.
- Most animals have a central nervous system (CNS) and a peripheral nervous system (PNS). PNS neurons receive sensory information and transmit it to the CNS for processing. The CNS then sends signals to muscles, glands, or other tissues via PNS neurons.

✓ When you see this checkmark, stop and test yourself. Answers are available in Appendix B.

The evolution of neurons and muscle cells made rapid movement possible. They were morphological innovations that enabled animals to evolve into today's flatworms, squid, beetles, sharks, and birds.

Chapter 32 also pointed out that neurons are organized into two basic types of nervous systems:

- The diffuse arrangement of cells called a **nerve net**, found in cnidarians (jellyfish, hydra, anemones) and ctenophores (comb jellies).
- A **central nervous system (CNS)** that includes large numbers of neurons aggregated into clusters called ganglia.

In most cases, animals with a CNS have a large cerebral ganglion, or brain, located in their anterior end. You might recall from Chapter 32 that this phenomenon is known as cephalization—the evolution of a bilaterally symmetric body with information gathering and processing structures located at the head end.

Cephalization made animals into efficient eating and moving machines: they faced the environment in one direction, with sensory appendages taking in information and sending it to a nearby brain for processing. After integrating information from an array of sensory cells, the brain sends electrical signals to muscles or other tissues that respond to the stimulus.

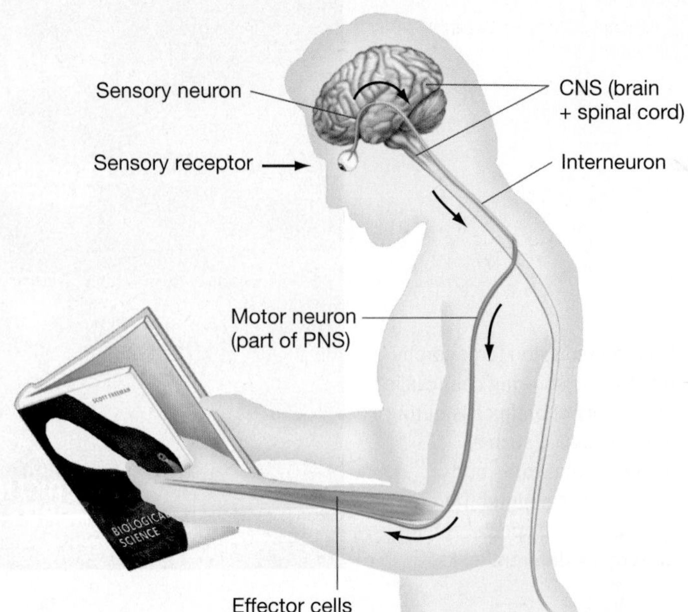

FIGURE 45.1 The Brain Integrates Sensory Information and Sends Signals to Effector Cells. In most cases, sensory neurons send information to the brain. There, the electrical signals are integrated with information from other sources. Once integration is complete, a response is sent to effector cells through motor neurons.

Types of Neurons in the Nervous System

The sensory cells that are responsible for information-gathering respond to light, sound, touch, or other stimuli. Sensory receptors in the skin, eyes, ears, and nose transmit streams of data about the external environment. Sensory cells inside the body monitor conditions that are important in homeostasis, such as blood pH and oxygen levels. In this way, sensory cells monitor conditions both outside and inside the body.

As **Figure 45.1** shows, a sensory receptor cell transmits the information it receives by means of a nerve cell known as a **sensory neuron**. (In many cases, the receptor cell itself also acts as a sensory neuron.)

In vertebrates, the sensory neuron sends the information to neurons in the brain or spinal cord via nerves; in other animals nerves send the information to the brain. **Nerves** are long, tough strands of nervous tissue. They contain thousands of projections from neurons and carry information to and from the brain and spinal cord.

Together, the brain and spinal cord form the CNS of vertebrates. One function of the CNS is to integrate information from many sensory neurons. Cells in the CNS called **interneurons** (literally, "between-neurons") perform this integration.

Interneurons also make connections between sensory neurons and **motor neurons**, which are nerve cells that send signals to effector cells in glands or muscles. Recall from Chapter 41 that effectors are structures that respond to stimuli.

All neurons and other components of the nervous system that are outside the CNS are considered part of the **peripheral nervous system**, or **PNS**. Section 45.4 describes the structure and function of the PNS in more detail.

Typically, sensory information from receptors in the PNS is sent to the CNS, where it is processed. Then a response is transmitted back to appropriate parts of the body. When you stub your toe, pain receptors in your toe relay sensory information to the brain, which then modifies your movements to avoid further injury. You might hop a bit, for example, and then limp to avoid further contact with that toe.

The Anatomy of a Neuron

Neurons are difficult to study, because they are small, transparent, and morphologically complex. So when Camillo Golgi discovered that some neurons become visible when samples of preserved tissue are treated with a solution containing silver nitrate, his finding was a major advance. The year was 1898.

Through the early decades of the twentieth century, the work of Golgi and Santiago Ramón y Cajal revealed several important points about the anatomy of neurons. Most neurons have the same three parts, shown in **Figure 45.2a**:

1. a cell body, which contains the nucleus;
2. a highly branched group of relatively short projections called dendrites; and
3. a long projection called an axon, which may or may not branch.

Dendrites are rarely more than 2 mm long, but axons can be over a meter in length. The number of dendrites and their arrangement vary greatly from cell to cell. Further, many brain cells have only dendrites and lack axons.

A **dendrite** receives a signal from the axons of adjacent cells; a neuron's **axon** sends a signal to the dendrites and cell bodies of other neurons (**Figure 45.2b**). Dendrites collect signals; axons

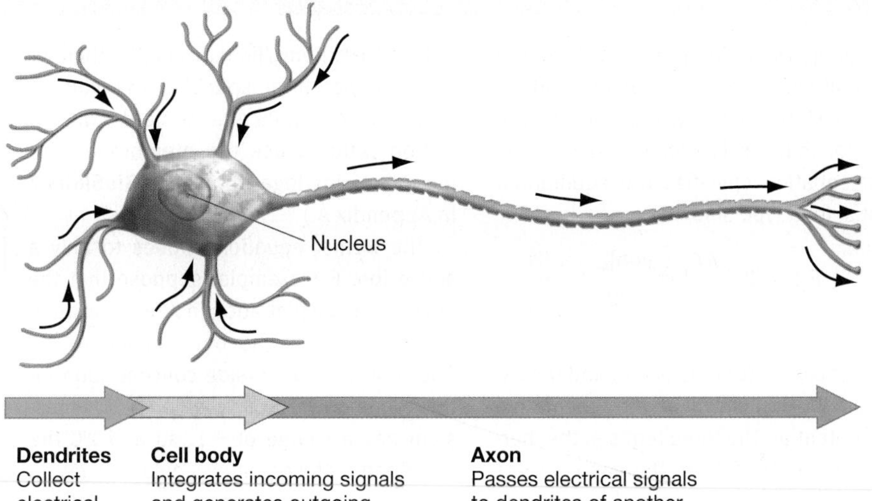

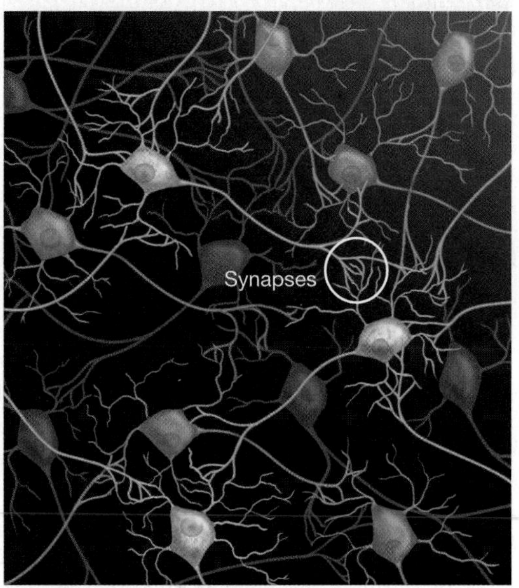

FIGURE 45.2 How Does Information Flow in a Neuron? (a) The structure of a generalized neuron. **(b)** Most neurons receive inputs from several different neurons and send projections to several different neurons.

pass them on. The **cell body**, or **soma**, also receives signals from axons. The cell body is where incoming signals are integrated and an outgoing signal is sent to the axon.

Ramón y Cajal maintained that the plasma membrane of each neuron is distinct, and that the membranes of axons and dendrites meet at junctions called synapses. This hypothesis was confirmed in the 1950s, when images from electron microscopes showed that only a small subset of neurons make direct, cytoplasm-to-cytoplasm connections. In a nervous system, most interactions occur where the plasma membranes of two neurons meet at synapses.

Given that neurons transmit electrical signals, how do they produce them?

An Introduction to Membrane Potentials

Ions carry an electric charge. In virtually all cells, the cytoplasm and extracellular fluids adjacent to the plasma membrane contain unequal distributions of ions. As a result, cells are inherently electrical in nature.

A difference of electrical charge between any two points creates a difference in **electrical potential**, or a **voltage**. If the positive and negative charges on ions that exist on the two sides of a plasma membrane do not balance each other, the membrane will have an electrical potential.

When an electrical potential exists on either side of a plasma membrane, the separation of charges is called a **membrane potential**. If there is a large separation of charges across the membrane, the membrane potential is large.

It is important to remember that membrane potentials refer only to a separation of charge immediately adjacent to the plasma membrane, on either side of the membrane. Even if there is a large membrane potential, there may be no charge separation slightly farther from the membrane.

UNITS AND SIGNS Membrane potentials are measured in units called millivolts. A **millivolt (mV)** is 1/1000 of a volt; a **volt** is the standard unit of electrical potential. As a comparison, an AA battery that you buy in the store has an electrical potential of 1500 mV between its + and − terminals. In neurons, membrane potentials are typically about 70–80 mV.

By convention, membrane potentials are always expressed as inside-relative-to-outside. Because there are usually more negatively charged ions and fewer positively charged ions on the inside surface of a membrane relative to its outside surface, membrane potentials are usually negative in sign.

ELECTRICAL POTENTIAL, ELECTRIC CURRENTS, AND ELECTRICAL GRADIENTS Membrane potentials are a form of potential energy. When a membrane potential exists, the ions on both sides of the membrane have potential energy. Recall that potential energy is energy based on the position of matter.

To convince yourself that ions have potential energy when a membrane potential exists, consider what would happen if the membrane were removed. Ions would spontaneously move from the area of like charge to the area of unlike charge—causing a flow of charge. This flow of charge, called an **electric current**, would occur because like charges repel and unlike charges attract.

As Chapter 6 emphasized, however, ions move across membranes in response to concentration gradients as well as charge gradients. Therefore, a membrane potential also includes energy stored as the concentration gradient of charged ions on the two sides of the membrane. Recall that the combination of an electric gradient and a concentration gradient is an **electrochemical gradient**. What do all of these facts have to do with neurons?

BOX 45.1 QUANTITATIVE METHODS: Using the Nernst Equation to Calculate Equilibrium Potentials

An equilibrium potential exists whenever there is a concentration gradient for an ion and the membrane is permeable to that ion. At the equilibrium potential, the rate at which an ion moves across the membrane down its concentration gradient is equal to the rate at which it moves, in the opposite direction, down its electrical gradient. It's the voltage at which the concentration and electrical gradients acting on an ion balance out.

The ion concentration gradients that contribute to equilibrium potentials are due to the action of Na^+/K^+-ATPase. In this way, chemical energy in the form of ATP is converted to electrical energy in the form of a membrane potential.

To calculate the amount of energy involved, biologists use the Nernst equation—a formula that converts the energy stored in a concentration gradient to the energy stored as an electrical potential. The concentration gradient for an ion is symbolized $[\text{ion}]_o/[\text{ion}]_i$, where $[\text{ion}]_o$ and $[\text{ion}]_i$ are the concentrations of the ion outside and inside the cell, respectively. The electrical potential for that ion is symbolized E_{ion}. The Nernst equation specifies the equilibrium potential for a type of ion as

$$E_{\text{ion}} = 2.3\frac{RT}{zF}\log\frac{[\text{ion}]_o}{[\text{ion}]_i}$$

In this expression, z is the valence of the ion (for instance, $+1$ for potassium) and the expression RT/F is known as the thermodynamic potential. The three terms in the thermodynamic potential are the gas constant (R), which acts as a constant of proportionality; the absolute temperature (T), measured in kelvins; and the Faraday constant (F), which specifies the amount of charge carried by one mole of an ion with a valence of $+1$ or -1.

Note that the Nernst equation is based on the base-10 logarithm of the ion concentration ratio, $[\text{ion}]_o/[\text{ion}]_i$. Thus, the thermodynamic potential specifies the voltage required to balance a tenfold concentration ratio across the membrane. (For more on using logarithms, see **BioSkills 7** in Appendix A.)

The Nernst equation applies to only a single ion. For example, suppose that the potassium concentration inside the axon of a squid neuron has been measured as 400 mM, yet the outside concentration for this ion is only 20 mM (Table 45.1). Potassium has a charge of $+1$, so at 20°C the RT/zF part of the equation yields $+25$ mV. In this case, the equilibrium potential for potassium becomes

$$E_K = 2.3 \times 25 \text{ mV} \times \log\frac{20 \text{ mM}}{400 \text{ mM}}$$

$$E_K = 58 \text{ mV} \times \log 0.05 = -75 \text{ mV}$$

The minus sign indicates that the interior of the axon is negatively charged with respect to the exterior.

Repeating this process for Na^+ ions and Cl^- ions yields the following results for the squid axon:

$$E_{Na} = 58 \text{ mV} \times \log\frac{440 \text{ mM}}{50 \text{ mM}} = +54.8 \text{ mV}$$

$$E_{Cl} = -58 \text{ mV} \times \log\frac{560 \text{ mM}}{51 \text{ mM}} = -60 \text{ mV}$$

Again, notice that the equilibrium potential given by the Nernst equation is calculated independently for each ion.

TABLE 45.1 **Concentration of Important Ions across a Squid Neuron's Plasma Membrane at Rest**

Ions	Cytoplasm Concentration	Extracellular Concentration	Equilibrium Potential
Na^+	50 mM	440 mM	+ 54.8 mV
K^+	400 mM	20 mM	−75 mV
Cl^-	51 mM	560 mM	−60 mV
Organic anions	385 mM	—	—

How Is the Resting Potential Maintained?

When a neuron is not transmitting an electrical signal but is merely sitting in extracellular fluid at rest, the membrane potential across its membrane is called the **resting potential**. To understand why the resting potential exists, consider the distribution of the various ions and other charged molecules on the two sides of the neuron's plasma membrane, shown in **Figure 45.3**:

- The interior side of the membrane has relatively low concentrations of sodium (Na^+) and chloride (Cl^-) ions, a relatively large concentration of potassium ions (K^+), and some organic anions—amino acids and other organic molecules that drop a proton and carry a negative charge.
- In the extracellular fluid, sodium and chloride ions predominate.

If each type of ion diffused across the membrane in accordance with its concentration gradient, organic anions and K^+ would leave the cell while Na^+ and Cl^- would enter.

Ions cannot cross phospholipid bilayers readily, however. They cross plasma membranes efficiently in only three ways (see Chapter 6):

1. flowing along their electrochemical gradient through an **ion channel**—a protein that forms a pore in the membrane and allows a specific ion to pass through; or
2. carried, via a membrane cotransporter protein or antiporter protein, with an ion that experiences a strong electrochemical gradient; or
3. pumped against an electrochemical gradient by a membrane protein that hydrolyzes ATP.

How does the distribution of ions illustrated in Figure 45.3 come to be? The answer hinges on two types of membrane proteins: an ion channel that allows potassium ions to move, and a pump that actively transports sodium and potassium ions. Let's consider each in turn.

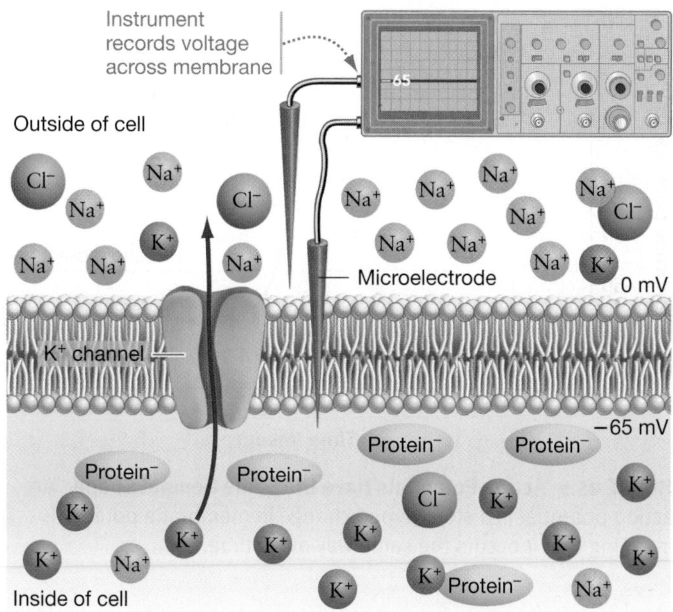

FIGURE 45.3 Neurons Have a Resting Potential. In resting neurons, the membrane is selectively permeable to K^+. As K^+ leaves the cell along its concentration gradient, the inside of the membrane becomes negatively charged relative to the outside. To measure a neuron's membrane potential, researchers insert a microelectrode into the cell and compare that reading with the reading outside the cell.

✔**QUESTION** Will K^+ continue to leave the cell indefinitely? Explain why or why not.

THE K^+ LEAK CHANNEL When a neuron is not actively involved in transmitting an electrical signal, the most common type of channel that is open is one that admits potassium ions. Therefore, resting neurons are most permeable to K^+ ions, which cross the membrane easily along their concentration gradient. The potassium channels involved are sometimes called **leak channels**, because they allow K^+ to leak out of the cell.

As K^+ moves from the interior of the cell to the exterior through potassium channels, the inside of the cell becomes more and more negatively charged relative to the outside. This buildup of negative charge inside the cell begins to attract K^+ and counteract the concentration gradient that had favored the movement of K^+ out.

As a result of these counteracting influences, the membrane reaches a voltage at which there is an equilibrium between the concentration gradient that moves K^+ out and the electrical gradient that moves K^+ in. At this voltage, there is no longer a net movement of K^+. This voltage is called the **equilibrium potential** for K^+. **Box 45.1** discusses how equilibrium potentials for individual ions are calculated.

Although Cl^- and Na^+ cross the plasma membrane much less readily than does K^+, some movement of these ions also occurs through a small number of open ion channels that are selective for each ion. As a result, each type of ion has an equilibrium potential. The membrane as a whole has a membrane potential that combines the effects of the individual ions.

THE ROLE OF THE NA^+/K^+-ATPASE In addition to the passive movement of ions that takes place when neurons are at rest, Na^+/K^+-ATPase actively pumps Na^+ out of the cell and K^+ into the cell. More specifically, the sodium-potassium pump uses the energy gained when it receives a phosphate group from ATP to move three Na^+ ions out of the cell and two K^+ ions into the cell (**Figure 45.4**).

Active transport via Na^+/K^+-ATPase ensures that eventually the concentration of K^+ is much higher inside the cell than outside, while the concentration of Na^+ is lower inside than outside. In addition to setting up concentration gradients that affect K^+ and Na^+, the pump establishes an electrical gradient: The

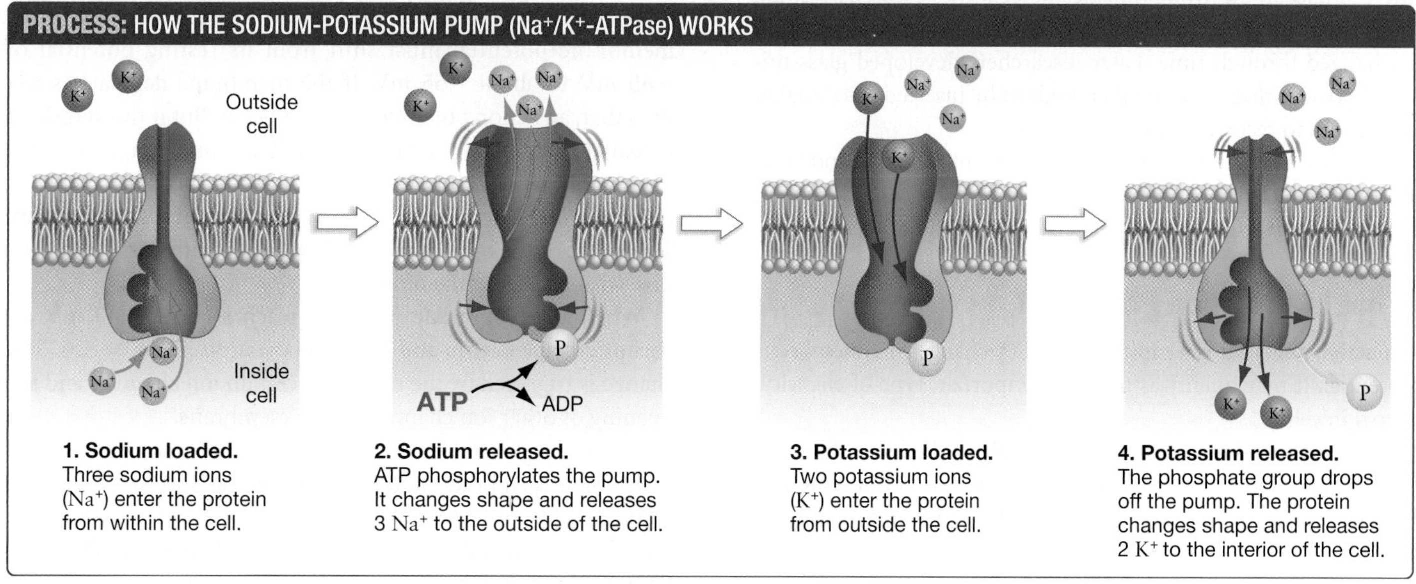

FIGURE 45.4 In Neurons, Na^+/K^+-ATPase Imports Potassium Ions and Exports Sodium Ions. By following radioactive Na^+ and K^+, biologists found that the pump transports 3 Na^+ out of the cell for every 2 K^+ brought in. Notice that this pump operates via a conformational change.

outward movement of three positive charges and inward movement of two positive charges makes the interior of the membrane less positive (more negative) than the outside.

Thus, the neuron has a negative resting membrane potential. The resting potential represents energy stored as concentration and electrical gradients in a series of ions. ✔If you understand this concept, you should be able to explain why both the Na^+/K^+-ATPase and the K^+ "leak" channels are vital to establishing the resting potential.

To review the ions, gradients, channels, and pumps involved in establishing and maintaining the resting potential, go to the study area at *www.masteringbiology.com.*

Web Activity Membrane Potentials

Using Microelectrodes to Measure Membrane Potentials

During the 1930s and 1940s, A. L. Hodgkin and Andrew Huxley focused on what has become a classic model system in the study of electrical signaling: the axons of squid.

Squid live in the ocean and are preyed on by fish and whales. When a squid is threatened, electrical signals travel down the axon to muscle cells. When these muscles contract, water is expelled from a cavity in the squid's body. As a result, the squid lurches away from danger by jet propulsion (see Chapter 33). The extremely rapid electrical signal is an adaptation that helps squid avoid predation.

Hodgkin and Huxley decided to study the squid's axon simply because it is so large. Many of the axons found in humans are a mere 2 μm in diameter, but the squid axon is about 500 μm in diameter—large enough that the researchers could record membrane potentials by inserting a wire down its length.

By measuring the voltage difference between the wire inside the cell and an electrode outside the cell, the researchers could record the voltage across the plasma membrane and observe how it changed through time. Later researchers developed glass microelectrodes that were small enough to be inserted into smaller nerve cells to record membrane voltage.

With their relatively simple early equipment, Hodgkin and Huxley were able to record the axon's resting potential, and document that it can be disrupted by an event called the action potential.

What Is an Action Potential?

An **action potential** is a rapid, temporary change in a membrane potential. It may qualify as the most important type of electrical signal in cells.

Although Hodgkin and Huxley initially studied it in the squid giant axon, subsequent work has shown that the action potential has the same general characteristics in all species and in all types of neurons.

A THREE-PHASE SIGNAL **Figure 45.5** shows the form of the action potential that Hodgkin and Huxley recorded from the squid's giant axon—the signal that allows these animals to jet away from predators. Put your finger on the figure's "Resting potential" label and then trace the graph to the right. The action potential has three distinct phases:

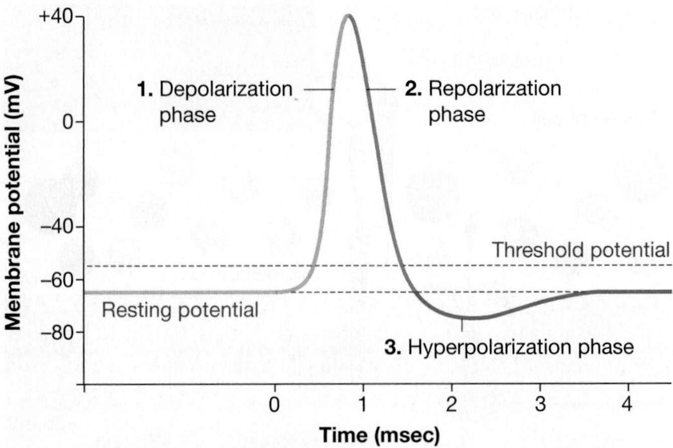

FIGURE 45.5 Action Potentials Have the Same General Shape. An action potential is a stereotyped change in membrane potential—meaning that it occurs the same way every time.

1. **Depolarization** of the membrane. A membrane is said to be polarized if the charges on the two sides are different. Depolarization means that the membrane becomes less polarized than before. During the depolarization phase, the membrane potential changes from highly negative to positive.
2. A rapid **repolarization**, which changes the membrane potential from positive back to negative.
3. A **hyperpolarization** phase, when the membrane is slightly more negative than the resting potential.

All three phases of an action potential occur in a few milliseconds.

For an action potential to begin in a squid giant axon, the membrane potential must shift from its resting potential of −65 mV to about −55 mV. If the membrane depolarizes less than that, an action potential does not occur. But if this **threshold potential** is reached, certain channels in the axon membrane open and ions rush into the axon, following their electrochemical gradients. The current flow causes further depolarization. The inside of the membrane becomes less negative and then positive with respect to the outside.

When the membrane potential reaches about +40 mV, an abrupt change occurs and the repolarization phase begins. The change is triggered by the closing of certain ion channels and the opening of other ion channels in the membrane.

An action potential occurs because specific ion channels in the plasma membrane open or close in response to changes in voltage. An action potential always has the same three-phase form, even though the size of the resting potential, threshold potential, and peak depolarization may vary among species.

AN "ALL-OR-NONE" SIGNAL THAT PROPAGATES Hodgkin and Huxley made other important observations about the action potential. In addition to being fast and having three distinct phases, it is an all-or-none event.

- There is no such thing as a partial action potential.
- All action potentials for a given neuron are identical in magnitude and duration.
- Action potentials are propagated down the length of the axon.

For example, when an impulse was recorded at a particular point on a squid axon, an action potential that was identical in shape and size would be observed farther down the same axon soon afterward. Neurons are said to have **excitable membranes**, because neurons are capable of generating action potentials that propagate rapidly along the length of their axons.

Taken together, these observations suggested a mechanism for electrical signaling. In the nervous system, information is coded in the form of action potentials that travel along axons. The frequency of action potentials—not their size—is the meaningful signal. In the squid's giant axon, action potentials signal muscles to contract. As a result, the animal escapes from danger.

CHECK YOUR UNDERSTANDING

If you understand that . . .

- The plasma membranes of neurons carry a resting potential because Na^+/K^+-ATPase pumps Na^+ out of the cell and K^+ into the cell and because the membrane is selectively permeable to K^+ ions, which leak out.
- The action potential is a three-phase, all-or-none signal that moves down the length of a neuron.

✔ You should be able to . . .

1. Predict what would happen to the resting potential of a squid axon if potassium leak channels were blocked.
2. Explain why only the frequency of action potentials—not their size—contains information.

Answers are available in Appendix B.

45.2 Dissecting the Action Potential

Which ions are involved in the currents that form the action potential? Is Na^+, Cl^-, or K^+ responsible for the depolarization and repolarization phases of the event?

Hodgkin made a crucial start in answering this question when he realized that +40 mV was close to the equilibrium potential for Na^+ in the squid giant axon. If sodium channels opened early in the action potential, then Na^+ should flow into the neuron until the membrane potential was about +40 mV. How could this hypothesis be tested?

Distinct Ion Currents Are Responsible for Depolarization and Repolarization

To understand the currents responsible for the action potential, Hodgkin and Huxley recorded electrical activity in squid axons that were bathed in solutions containing different concentrations of ions.

- Washing Na^+ out of the solution surrounding the axon abolished action potentials.
- When they used solutions with various concentrations of Na^+, the peak of the action potential tracked the equilibrium potential of Na^+. If Na^+ concentration outside the cell was high, the peak was high. If Na^+ concentration outside the cell was low, the peak was low.

These experiments furnished strong support for the hypothesis that the action potential begins when Na^+ flows into the neuron. Sodium ions are responsible for the depolarization phase.

What happens during the repolarization phase? Using radioactive K^+, Hodgkin and Huxley showed that there was a strong flow of potassium ions out of the cell during the repolarization phase.

The action potential consists of a strong inward flow of sodium ions followed by a strong outward flow of potassium ions. ✔If you understand this concept, you should be able to add labels that read, "Sodium channels open—Na^+ enters" and "Potassium channels open—K^+ leaves" to Figure 45.5.

How Do Voltage-Gated Channels Work?

The action potential depends on **voltage-gated channels**—membrane proteins that open and close in response to changes in membrane voltage. The shape of a voltage-gated channel, and thus its ability to admit ions, changes in response to the charges present at the membrane surface. **Figure 45.6** shows a simple model of how voltage-gated sodium channels change as a function of membrane potential.

Hodgkin and Huxley confirmed that voltage-gated channels exist using a technique called voltage clamping. **Voltage clamping** allows researchers to hold an axon at any voltage and record the electrical currents that occur. When the researchers held the squid axon at various voltages, different currents resulted. These

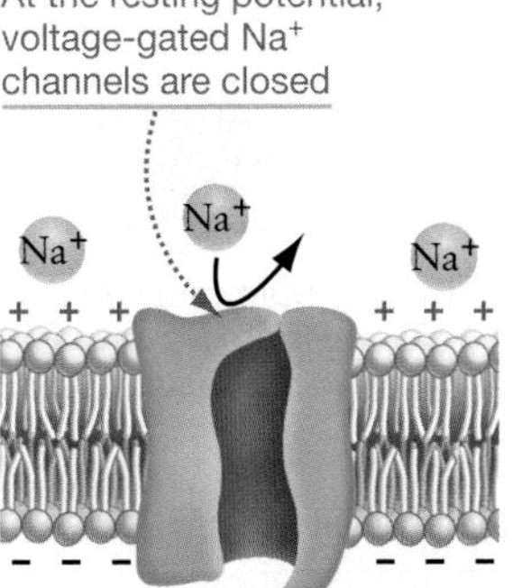

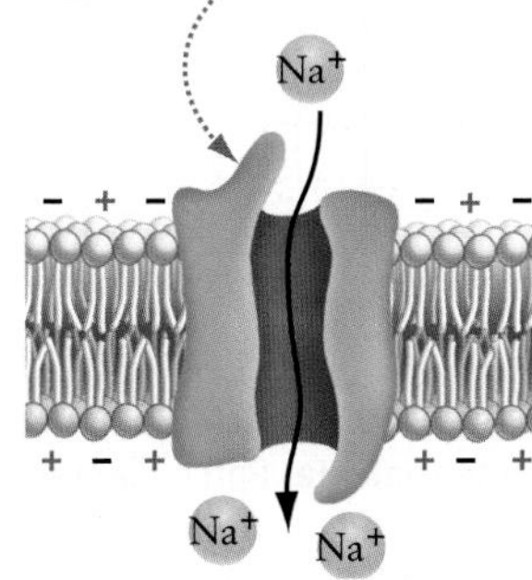

FIGURE 45.6 The Shape of a Voltage-Gated Channel Depends on the Membrane Potential. Changes in the conformation of voltage-gated channels are responsible for changes in a neuron's permeability to Na^+ and K^+.

experiments supported the hypothesis that the behavior of the ion channels depends on voltage.

PATCH CLAMPING AND STUDIES OF SINGLE CHANNELS Studying individual ion channels became possible when Erwin Neher and Bert Sakmann perfected a technique known as **patch clamping**. As **Figure 45.7a** shows, the researchers touched a membrane with a fine-tipped microelectrode and applied suction to capture a single ion channel within the tip of the microelectrode. The suction seals the membrane against the glass so that no current leaks out.

Using this technique, the researchers documented the currents that flowed through individual channels. The recordings shown in **Figure 45.7b**, from three sodium channels and three potassium channels, show how current flows through individual channels over the course of an action potential. The data in the figure make several important points:

- Voltage-gated channels are either open or closed. There is no gradation in channel behavior. This conclusion is based on the shape of the current records—current flow starts and stops instantly, and the size of the current is always the same.
- Sodium channels open quickly after depolarization. They stay open for about a millisecond, close, and remain inactive for 1 to 2 milliseconds. That explains why the cell can repolarize: Once the sodium channels close, they have a lag before they can open again.
- Potassium channels open with a delay after depolarization. They continue to flip open and closed until the membrane repolarizes. Once the membrane returns to the resting potential, potassium channels remain closed.

(a) Patch clamping isolates a single ion channel.

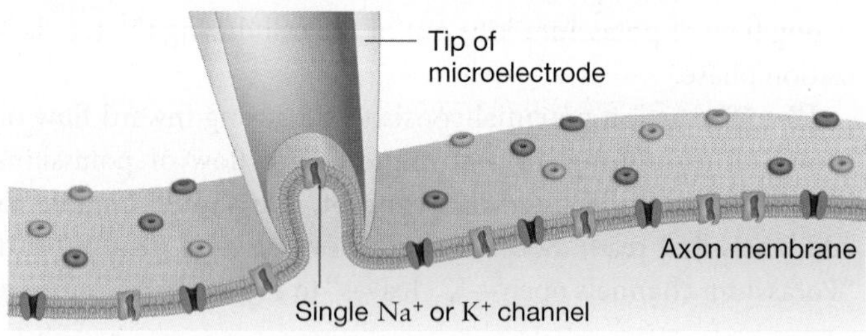

(b) Currents through isolated channels can be measured during an action potential.

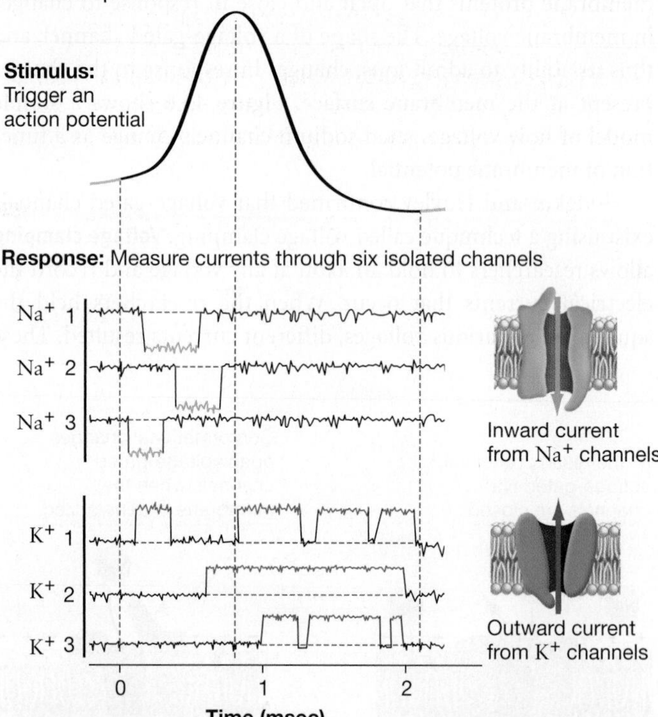

FIGURE 45.7 Patch Clamping Provided Insights into the Action Potential. **(a)** Patch clamping depends on the use of extremely fine-tipped microelectrodes. The goal is to isolate one channel and record from it. **(b)** Current records show that voltage-gated channels are either open or closed, sodium channels open quickly after depolarization, and potassium channels open with a delay after polarization. No current flows through either type of channel at the resting potential.

POSITIVE FEEDBACK OCCURS DURING DEPOLARIZATION More detailed experiments on Na$^+$ channels also explained why the action potential is an all-or-none event. The key observation was that Na$^+$ channels are more likely to open as a membrane depolarizes. As a result, an initial depolarization leads to the opening of more Na$^+$ channels, which depolarizes the membrane further, which leads to the opening of additional Na$^+$ channels.

The opening of Na$^+$ channels exhibits **positive feedback**—meaning that the occurrence of an event makes the same event more likely to recur. When a fuse is lit, for example, the heat generated by the oxidation reaction accelerates the reaction itself, which generates still more heat and leads to additional oxidation reactions and keeps the fuse burning. Positive feedback is rare in organisms because it often leads to uncontrolled events. The opening of Na$^+$ channels during an action potential is one of the few examples known.

✓If you understand how sodium and potassium channels work, you should be able to explain **(1)** why positive feedback occurs in the opening of Na$^+$ channels, **(2)** why Na$^+$ stops flowing across a membrane during an action potential, and **(3)** why K$^+$ channels start opening.

USING NEUROTOXINS TO IDENTIFY CHANNELS AND DISSECT CURRENTS In addition to using voltage clamping and patch clamping, researchers have used poisons that target neurons, from sources as diverse as venomous snakes and foxglove plants, to explore the dynamics of voltage-gated channels. **Neurotoxins** are poisons that affect neuron function—often resulting in convulsions, paralysis, or unconsciousness.

For example, when biologists treated giant axons from lobsters with the tetrodotoxin found in puffer fish, they found that the resting potential in treated neurons was normal, but action potentials were abolished. More specifically, the outward-directed K$^+$ current was normal but the inward-directed Na$^+$ flow was wiped out. Researchers concluded that puffer-fish toxin blocks the voltage-gated Na$^+$ channel, probably by binding to a specific site on the channel protein.

How Is the Action Potential Propagated?

To explain how action potentials propagate down an axon, Hodgkin and Huxley suggested the model illustrated in **Figure 45.8a**.

Step 1 The influx of Na^+ at the start of an action potential causes charge to spread away from sodium channels. Positive charges inside the cell are repulsed by the influx of Na^+, and negative charges are attracted to the Na^+.

Step 2 As positive charges are pushed farther from the initial sodium channels, they depolarize adjacent portions of the membrane.

Step 3 Nearby voltage-gated Na^+ channels pop open in response to depolarization. Positive feedback occurs, and a full-fledged action potential results.

In this way, an action potential is continuously regenerated as it moves down the axon (**Figure 45.8b**). The signal does not diminish as it moves, because the response is all or none.

Why don't action potentials propagate back up the axon? To answer this question, recall that Na^+ channels are **refractory**—that is, once they have opened and closed, they are less likely to open again for a short period. Action potentials are propagated in one direction only, because sodium channels "downstream" of the site are not in the refractory state.

The hyperpolarization phase, in which the membrane is more negative than the resting potential—also keeps the charge that spreads "upstream" from triggering an action potential in that direction.

AXON DIAMETER AFFECTS SPEED Understanding how the action potential propagates helped researchers explain why the squid's axons are so large. When sodium ions enter the interior of an axon at the start of an action potential, they can spread down the membrane or leak back out through open sodium channels. Compared to small axons, large axons have relatively few sodium channels per unit of membrane surface. As a result, less current leaks back out of a large axon than a small axon, and the charge spreads farther down the membrane.

The upshot is that the squid's giant axon and other large-diameter neurons transmit action potentials much faster than small axons can. The squid axon's large size is an adaptation that makes particularly rapid signaling possible.

(a) PROCESS: PROPAGATION OF ACTION POTENTIAL

Neuron
Axon

1. Na^+ enters axon.

2. Charge spreads; membrane "downstream" depolarizes.

Depolarization at next ion channel

3. Downstream voltage-gated channel opens in response to depolarization.

FIGURE 45.8 Action Potentials Propagate because Charge Spreads down the Membrane. **(a)** An action potential starts with an inflow of Na^+. The influx of positive charge attracts negative charges inside the cell and repels positive charges. As a result, positive charge spreads away from the channel where the Na^+ enters, and depolarizes nearby regions of the neuron. Voltage-gated Na^+ channels open in response. **(b)** The action potential spreads down the axon as a wave of depolarization, but there is no loss of signal because the all-or-none action potential regenerates itself as it travels.

(b) Action potential spreads as a wave of depolarization.

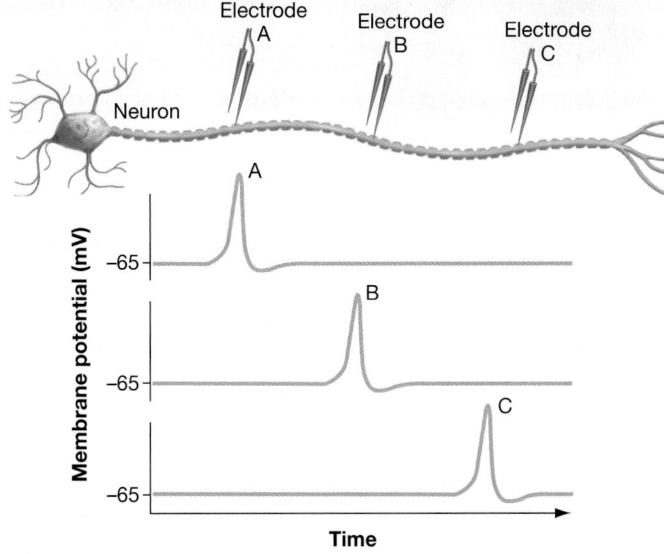

MYELINATION AFFECTS SPEED Analyzing charge spread also helped biologists explain the phenomenon called myelination. Relatively few vertebrates have giant axons. Instead, in vertebrates—and some invertebrates—the membranes of specialized accessory cells wrap around the axons of neurons and increase the efficiency of action potential propagation.

In the central nervous system, the specialized accessory cells are called **oligodendrocytes**. In the peripheral nervous system, described in Section 45.4, the cells are **Schwann cells** (**Figure 45.9a**). Oligodendrocytes and Schwann cells are two of several types of nervous system cells that support neurons. Collectively, these accessory cells are called **glia**.

When oligodendrocytes or Schwann cells wrap around an axon, they form a **myelin sheath**, which acts as a type of electrical insulation. As charge spreads down an axon, the myelin sheath prevents charge in the form of ions from leaking back out across the plasma membrane of the neuron.

Consequently, the influx of charge that results from an action potential is able to spread unimpeded until it hits an unmyelinated section of the axon, called a **node of Ranvier** (**Figure 45.9b**). The node has a dense concentration of voltage-gated Na^+ channels, so action potentials can occur.

Electrical signals jump down a myelinated axon much faster than they can move down an unmyelinated axon. In an unmyelinated axon, sodium and potassium channels are found in all locations and action potentials occur continuously down its length. Myelination is interpreted as an adaptation that makes rapid transmission of electrical signals possible in axons that have a small diameter.

To drive the importance of myelination home, consider what happens when it is decreased. If myelin degenerates, the transmission of electrical signals slows considerably. The disease **multiple sclerosis (MS)** develops as damage to myelin increases and electrical signaling is impaired, causing the muscles to weaken and coor-

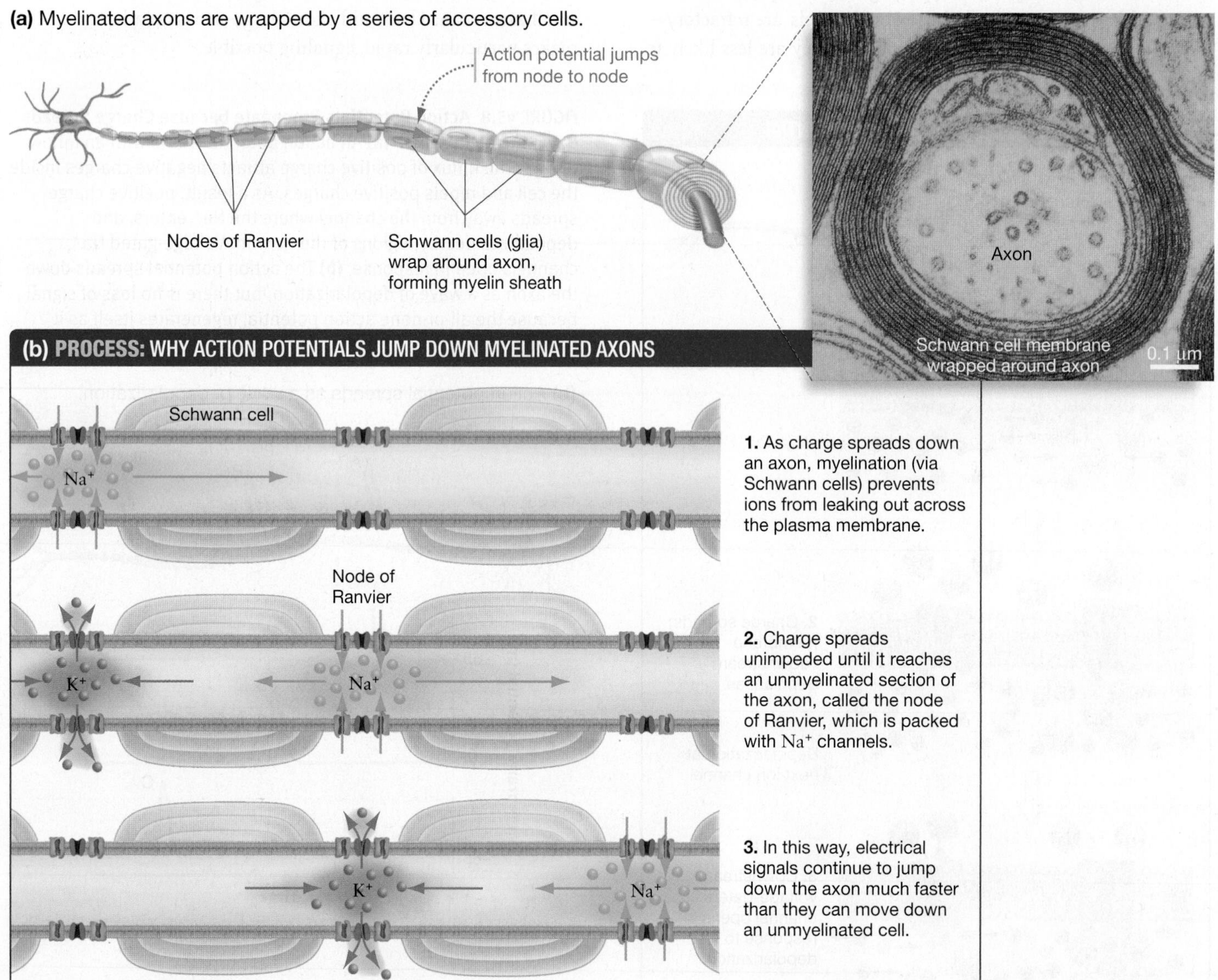

FIGURE 45.9 Action Potentials Propagate Quickly in Myelinated Axons.

dination to lessen. The symptoms of MS are highly variable; in severe cases the disease progresses and can be crippling.

What happens once an action potential has traveled the length of the axon? In most neurons the membrane at the end of the axon approaches the membrane of another neuron's dendrite, and the two surfaces are separated by a tiny gap. What happens when an action potential arrives at this interface between cells?

CHECK YOUR UNDERSTANDING

If you understand that . . .

- During an action potential, membrane voltage undergoes rapid changes due to an influx of sodium ions, followed by an outflow of potassium ions.
- Action potentials propagate down an axon because inrushing sodium ions depolarize adjacent portions of the membrane.

✔ You should be able to . . .

1. Explain why the action potential is an all-or-none phenomenon.
2. Predict what would happen if sodium channels continued to open once membrane depolarization was complete.

Answers are available in Appendix B.

45.3 The Synapse

The cytoplasm of most neurons is not directly connected to the cytoplasm of other neurons. Based on this observation, there must be some indirect mechanism that transmits electrical signals from cell to cell, across their plasma membranes.

In the 1920s, Otto Loewi showed that this indirect mechanism involves **neurotransmitters**. Neurotransmitters are chemical messengers that transmit information from one neuron to another neuron, or from a neuron to a target cell in a muscle or gland.

Loewi knew that signals from the vagus nerve slow the heart. To test the hypothesis that the signal from nerve to muscle is delivered by a chemical, he performed the experiment diagrammed in **Figure 45.10**.

First, Loewi isolated the vagus nerve and heart of a frog. As predicted, the heart rate slowed when he stimulated the nerve electrically. Next, he took the solution that bathed the first heart and applied it to another, isolated heart, and showed that the second heart rate slowed as well.

This result provided strong evidence for the chemical transmission of electrical signals. The vagus nerve had released a neurotransmitter into the bath.

Synapse Structure and Neurotransmitter Release

When transmission electron microscopy became available in the 1950s, biologists finally understood the physical nature of the interface, or **synapse**, between neurons—the place where

EXPERIMENT

QUESTION: How is information transferred from one neuron to another?

HYPOTHESIS: Molecules called neurotransmitters carry information from one neuron to the next.

NULL HYPOTHESIS: Information is not transferred between neurons in the form of molecules.

EXPERIMENTAL SETUP:

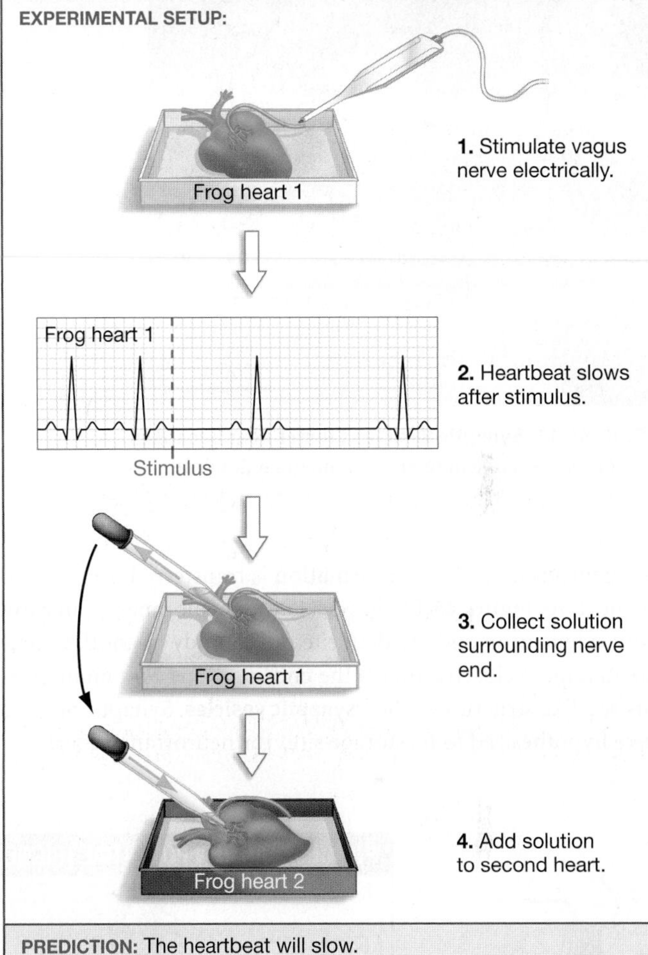

PREDICTION: The heartbeat will slow.

PREDICTION OF NULL HYPOTHESIS: There will be no change in heartbeat.

RESULTS:

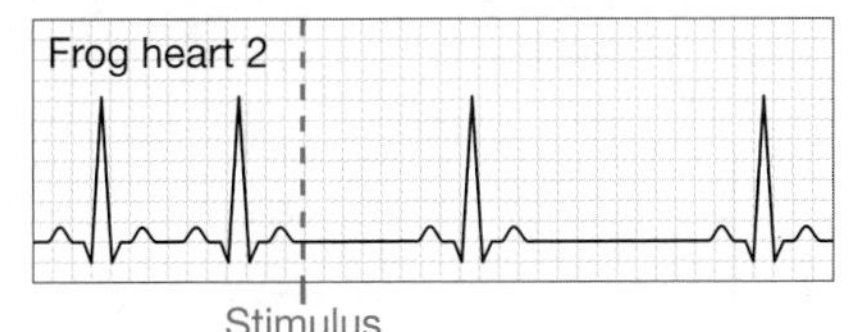

Heartbeat slows after solution is added.

CONCLUSION: The vagus nerve releases molecules that slow heartbeat. Neurotransmitters carry information.

FIGURE 45.10 Experimental Evidence for the Existence of Neurotransmitters.

SOURCE: Loewi, O. 1921. Über humorale Übertragbarkeit der Herznervenwirkung. *Pflügers Archiv European Journal of Physiology* 189: 239–242.

✔QUESTION What would be an appropriate control for this experiment?

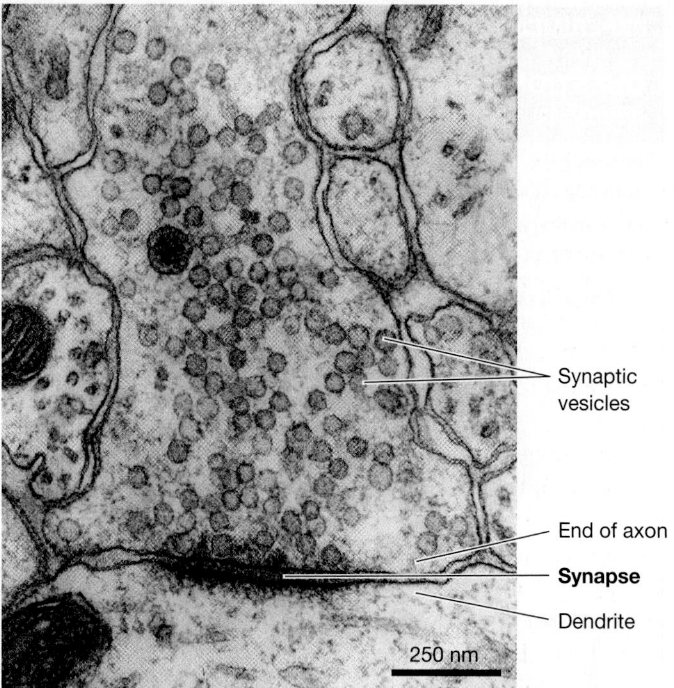

FIGURE 45.11 Synaptic Vesicles Cluster near Synapses. A cross section of the site where an axon meets a dendrite.

two neurons meet—and information is transferred from one to the next. As **Figure 45.11** shows, (1) the membranes of an axon from one neuron and the dendrite or cell body of another neuron juxtapose closely, and (2) the ends of axons contain numerous sac-like structures called **synaptic vesicles**. Synaptic vesicles were hypothesized to be storage sites for neurotransmitters.

Anatomical observations such as these, combined with chemical studies of the synapse, led to the model of synaptic transmission illustrated in **Figure 45.12**. Notice that the "sending" cell is the **presynaptic neuron** and the "receiving" cell is the **postsynaptic neuron.**

Step 1 The action potential arrives at the end of the axon.

Step 2 The depolarization created by an action potential opens voltage-gated calcium channels located near the synapse, in the presynaptic membrane. The electrochemical gradient for Ca^{2+} results in the inflow of calcium ions through the open channels.

Step 3 In response to the increased calcium concentration inside the axon, synaptic vesicles fuse with the membrane and release a neurotransmitter into the gap between the cells. This gap is called the **synaptic cleft**. The delivery of neurotransmitters into the cleft is an example of exocytosis, a process introduced in Chapter 7.

Step 4 Neurotransmitters bind to receptors on the postsynaptic cell, leading to changes in the membrane potential of the postsynaptic cell and possibly triggering the start of an action potential there.

Step 5 The response ends as the neurotransmitter is broken down and taken back up by the presynaptic cell.

Is the model correct? Let's begin by analyzing the role of neurotransmitters.

What Do Neurotransmitters Do?

To establish that a molecule functions as a neurotransmitter, researchers have to provide evidence for the following three criteria:

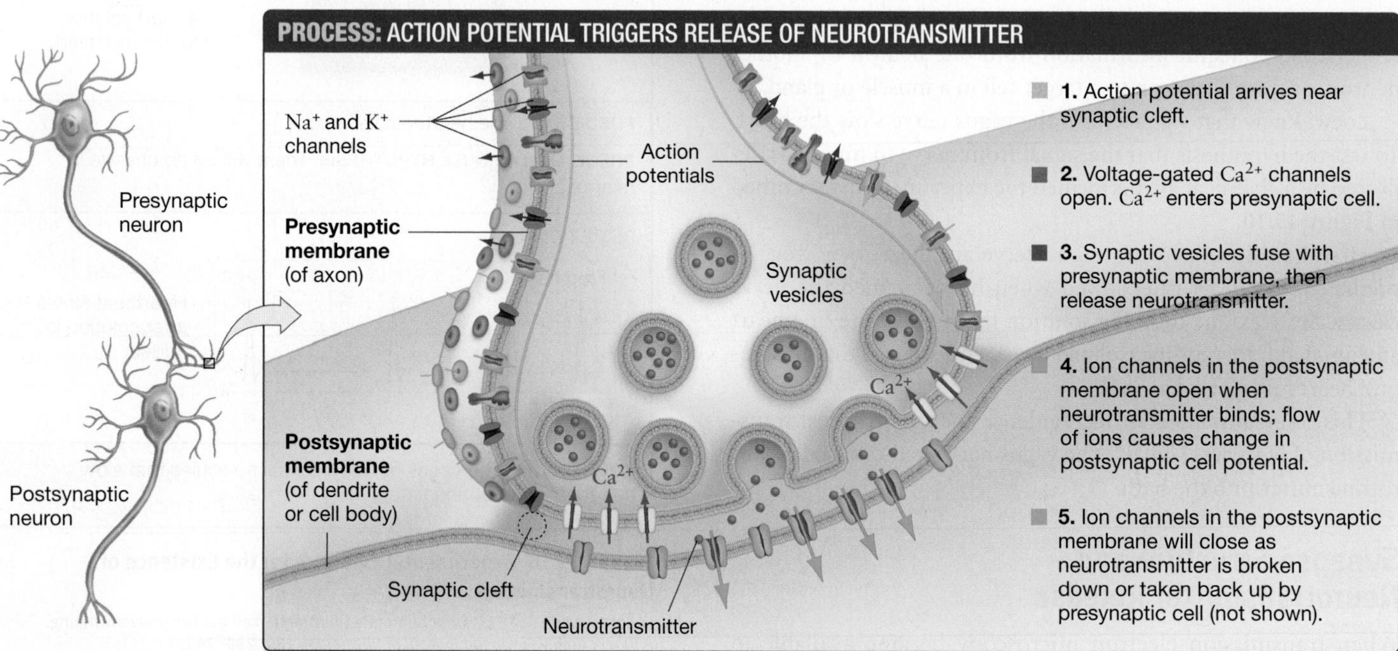

FIGURE 45.12 Neurons Meet and Transfer Information at Synapses. The sequence of events that occurs when an action potential arrives at a synapse.

1. The neurotransmitter is present at the synapse and released in response to an action potential.
2. It binds to a receptor on a postsynaptic cell.
3. It is taken up or degraded.

For example, researchers can look for neurotransmitters by stimulating a neuron, collecting the molecules that are released, and analyzing them chemically. To find the receptor for a particular neurotransmitter, researchers can attach a radioactive atom or other type of label to the chemical messenger, and add the labeled molecule to neurons. Once the labeled transmitter has bound to its receptor, the receptor protein can be isolated and analyzed.

Using techniques such as these, biologists have discovered and characterized a wide array of neurotransmitters and receptors, some of which are listed in **Table 45.2** on page 898.

By patch-clamping receptors, biologists confirmed that many neurotransmitters function as ligands. A **ligand** is a molecule that binds to a specific site on a receptor molecule. Many neurotransmitters are ligands that bind to receptors called **ligand-gated channels**. These are channel proteins that open in response to binding by a specific ligand—just as voltage-gated channels open in response to a change in voltage.

When a neurotransmitter binds to a ligand-gated ion channel in the postsynaptic membrane, the channel opens and admits a flow of ions along an electrochemical gradient. In this way, the neurotransmitter's chemical signal is transduced to an electrical signal—a change in the membrane potential of the postsynaptic cell. ✔If you understand this concept, you should be able to imagine a membrane with a resting potential of −65 mV and explain what happens when a ligand-gated ion channel opens and allows chloride ions to leave the cell.

Not all neurotransmitters bind to ion channels, however. Some receptors activate enzymes that lead to the production of a second messenger in the postsynaptic cell. Recall from Chapter 8 that **second messengers** are chemical signals produced inside a cell in response to a chemical signal that arrives at the cell surface.

The second messengers induced by neurotransmitters may trigger changes in enzyme activity, gene transcription, or membrane potential. Chapter 47 explores the cellular role of second messengers in detail.

Postsynaptic Potentials

What happens when a neurotransmitter binds to a receptor and opens an ion channel in the postsynaptic cell?

If the receptors at the synapse admit sodium ions in response to the arrival of the neurotransmitter, the postsynaptic membrane depolarizes (**Figure 45.13a**). In most cases, depolarization makes an action potential in the postsynaptic cell more likely. Changes in the postsynaptic cell that make action potentials more likely are called **excitatory postsynaptic potentials (EPSPs)**.

If the receptors at the synapse lead to an outflow of potassium ions or an inflow of chloride ions in the postsynaptic cell, the postsynaptic membrane hyperpolarizes—making action potentials less likely to occur in the postsynaptic cell (**Figure 45.13b**). Changes in the postsynaptic cell that make action potentials less likely are called **inhibitory postsynaptic potentials (IPSPs)**. IPSPs make the membrane potential more negative.

If an EPSP and an IPSP occur at the same time in the same place, they cancel each other out (**Figure 45.13c**). Synapses can also be modulatory—meaning that they modify a neuron's response to other EPSPs or IPSPs.

POSTSYNAPTIC POTENTIALS ARE GRADED It is critical to realize that, unlike action potentials, EPSPs and IPSPs are not all-or-none events. Instead, they are graded in size.

The size of an EPSP or IPSP depends on the amount of neurotransmitter that is released at the synapse. Greater transmitter release leads to a larger EPSP or IPSP. Both types of signal are short lived because neurotransmitters do not bind irreversibly to channels in the postsynaptic cell. Instead, they are quickly inactivated or taken up by the presynaptic cell and recycled.

If either the amount or life span of neurotransmitters is altered, the normal functioning of neurons is altered. The street drugs cocaine and amphetamine, for example, exert their effects by inhibiting the uptake and recycling of particular neurotransmitters (see Table 45.2).

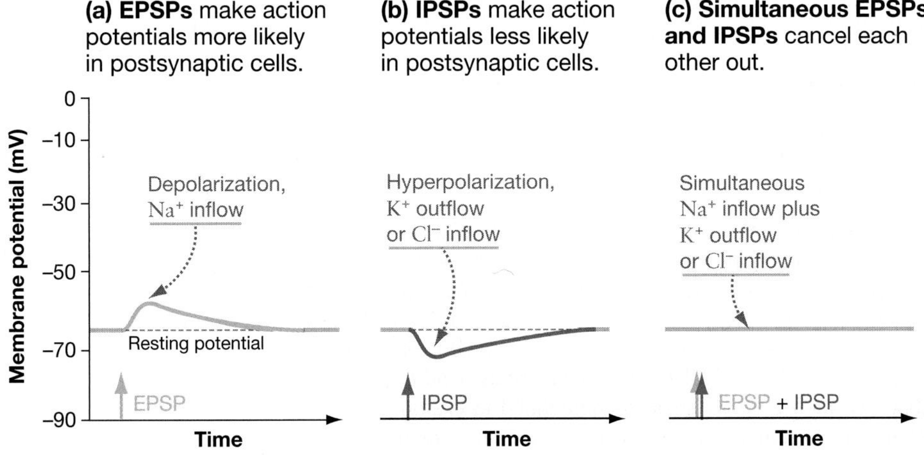

FIGURE 45.13 Events at the Synapse May Lead to Depolarization or Hyperpolarization of the Postsynaptic Membrane. These recordings show changes in the membrane potential of a postsynaptic neuron with the arrival of signals that cause **(a)** depolarization, **(b)** hyperpolarization, or **(c)** no change, as a result of simultaneous depolarizing and hyperpolarizing signals cancelling each other out.

TABLE 45.2 **Categories of Neurotransmitters**

Excitatory neurotransmitters make action potentials more likely in postsynaptic cells.
Inhibitory neurotransmitters make action potentials less likely; modulatory neurotransmitters modify the response at other synapses.
Drugs that prevent reuptake of neurotransmitters increase their activity.

Neurotransmitter	Site of Action	Action	Drugs That Interfere
Acetylcholine	Neuromuscular junction, some CNS pathways	Excitatory (inhibitory in some parasympathetic neurons)	• Botulism toxin blocks release • Black widow spider venom increases, then eliminates, release • α-bungarotoxin (in some snake venoms) binds to receptor
Monoamines			
Norepinephrine	Sympathetic neurons, some CNS pathways	Excitatory or inhibitory	• Ritalin (used for attention deficit hyperactivity disorder) increases release • Some antidepressants prevent reuptake
Dopamine	Many CNS pathways	Excitatory or modulatory	• Cocaine prevents reuptake • Amphetamines prevent reuptake
Serotonin	Many CNS pathways	Inhibitory or modulatory	• MDMA (ecstasy) causes increased release
Amino Acids			
Glutamate	Many CNS pathways	Excitatory	• PCP (angel dust) blocks receptor
Gamma-aminobutyric acid (GABA)	Some CNS pathways	Inhibitory	• Ethanol mimics response to GABA
Peptides			
Endorphins, enkephalins, substance P	Used in sensory pathways (pain)	Excitatory, modulatory, or inhibitory	

SUMMATION AND THRESHOLD How do EPSPs and IPSPs affect the postsynaptic cell? As **Figure 45.14a** shows, the dendrites and the cell body of a neuron typically make hundreds or thousands of synapses with other cells. At any instant, the EPSPs and IPSPs that occur at each of these synapses lead to short-lived surges of charge in the dendrites and cell body of the postsynaptic cell.

If an IPSP and EPSP occur close together in space or time, the changes in membrane potential tend to cancel each other out. But if several EPSPs occur close together in space or time, they sum and make the neuron more likely to fire an action potential (**Figure 45.14b**). The additive nature of postsynaptic potentials is termed **summation.**

(a) Most neurons receive information from many other neurons.

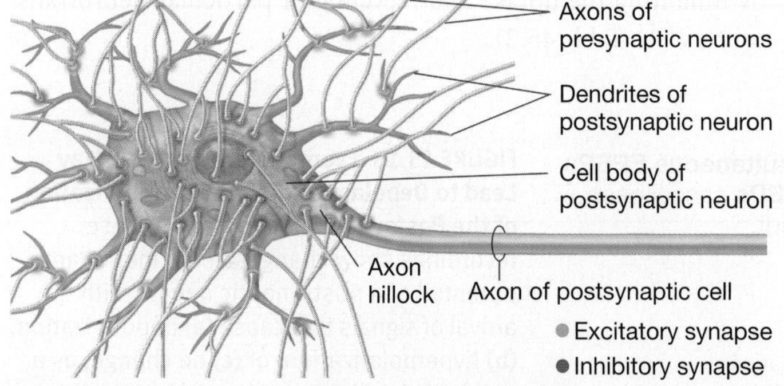

(b) Postsynaptic potentials sum.

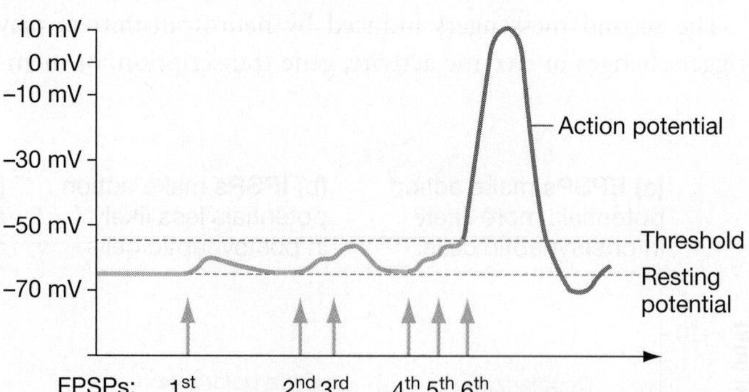

FIGURE 45.14 Neurons Integrate Information from Many Synapses. (a) The dendrites and cell body of a neuron typically receive signals from hundreds or thousands of other neurons. **(b)** When action potentials arrive close together in time from the same axon or from different axons with synapses close to one another, the postsynaptic potentials sum. In this example, the first excitatory signal is insufficient to generate an action potential. Two excitatory signals arriving close together cause summation but do not reach the threshold for generating an action potential. Three excitatory signals arriving closely spaced sum to exceed the threshold. If excitatory postsynaptic potentials depolarize the axon hillock past threshold, enough channels open to trigger an action potential. This example is simplified—in reality, hundreds or thousands of IPSPs and EPSPs sum to determine action potential frequency.

The sodium channels that trigger action potentials in the postsynaptic cell are located near the start of the axon at a site called the **axon hillock** (see Figure 45.14a). As IPSPs and EPSPs are received and interact throughout the dendrites and cell body, charge spreads to the axon hillock. If the membrane at the axon hillock depolarizes past the threshold potential, enough sodium channels open to trigger positive feedback and an action potential. Once an action potential starts at the axon hillock, it propagates down the axon to the next synapse.

Summation is critically important. Because neurons receive input from many synapses, and because IPSPs and EPSPs sum, information in the form of electrical signals is modified at the synapse before being passed along. An action potential is not necessarily transmitted from one neuron to the next—the response by the postsynaptic cell depends on the information it receives from a wide array of neurons.

To review (1) the molecules and channels responsible for the action potential and (2) the structure and function of the synapse, go to the study area at *www.masteringbiology.com*.

MB **BioFlix™** How Neurons Work, **BioFlix™** How Synapses Work, **Web Activity** Action Potentials

CHECK YOUR UNDERSTANDING

If you understand that . . .

- At a synapse, electrical information in the form of changes in membrane voltage is transduced to chemical information in the form of released neurotransmitters.
- Binding of a neurotransmitter to its receptor causes a change in the membrane potential of the postsynaptic cell.

✔ **You should be able to . . .**

Explain why the presence of synapses (instead of direct electrical connections) is important to a neuron's ability to integrate and process information from many different neurons.

Answers are available in Appendix B.

45.4 The Vertebrate Nervous System

The first three sections of this chapter examined electrical signaling at the level of molecules, membranes, and individual cells. This is how the study of the nervous system started, and is called a reductionist approach. Reductionism is common—biologists often start to figure out how something works by studying the component parts.

The goal of this section, however, is to discuss electrical signaling at the level of tissues, organs, and systems. Once researchers understand how individual components work, they want to explore how they interact to create a functioning system.

To begin, let's consider the overall anatomy of the vertebrate nervous system. Then we can ask how researchers explore the function of the most complex organ known: the human brain. The chapter concludes by returning to the molecular level and introducing recent work on learning and memory.

What Does the Peripheral Nervous System Do?

Recall from Section 45.1 that the central nervous system (CNS) is made up of the brain and spinal cord and is concerned primarily with integrating information. The peripheral nervous system (PNS) is made up of neurons outside the CNS.

What functions do the cells of the PNS control? Anatomical and functional studies indicate that the PNS consists of two systems with distinct functions:

1. an **afferent division**, which transmits sensory information to the CNS; and
2. an **efferent division**, which carries commands from the CNS to the body.

Neurons in the afferent division monitor conditions inside and outside the body. Once information from afferent neurons has been processed in the CNS, neurons in the efferent division carry signals that allow the body to respond to changed conditions in an appropriate way. The afferent and efferent divisions carry out sensory and motor functions, respectively.

As **Figure 45.15** shows, the afferent and efferent divisions are part of a hierarchy of PNS functions. The efferent division is further divided into a **somatic nervous system**, which controls movement, and an **autonomic nervous system**, which controls internal processes such as digestion and heart rate.

- The somatic nervous system carries out voluntary responses, which are under conscious control, with skeletal muscles serving as the effectors.
- The autonomic nervous system carries out involuntary responses, which are not under conscious control, with smooth muscle, cardiac muscle, and several glands serving as the effectors.

In effect, the somatic system responds to external stimuli and governs behavior, while the autonomic system responds to internal stimuli and controls the activity of internal organs and glands.

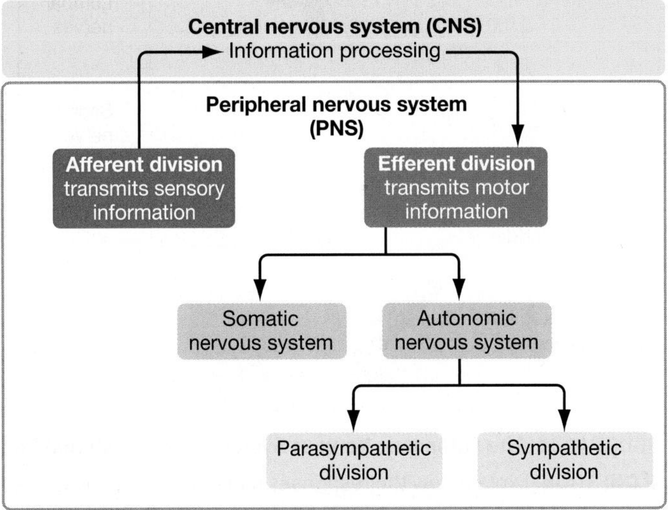

FIGURE 45.15 The PNS Comprises Distinct Components.

Many organs and glands are served by two functionally distinct types of autonomic nerves, summarized in **Figure 45.16**: One inhibits activity; the other promotes it.

- Nerves in the **parasympathetic nervous system** promote "rest and digest" functions that conserve or restore energy. For example, the parasympathetic nerves that connect to the heart slow it down, while those that serve the digestive tract stimulate its activity.
- Nerves in the **sympathetic nervous system** typically prepare organs for stressful "fight or flight" situations. Sympathetic nerves speed up the heart rate, stimulate the release of glucose from the liver, and inhibit action by digestive organs.

Functional Anatomy of the CNS

Parasympathetic nerves originate at the base of the brain or the base of the spinal cord. Most sympathetic nerves also originate in

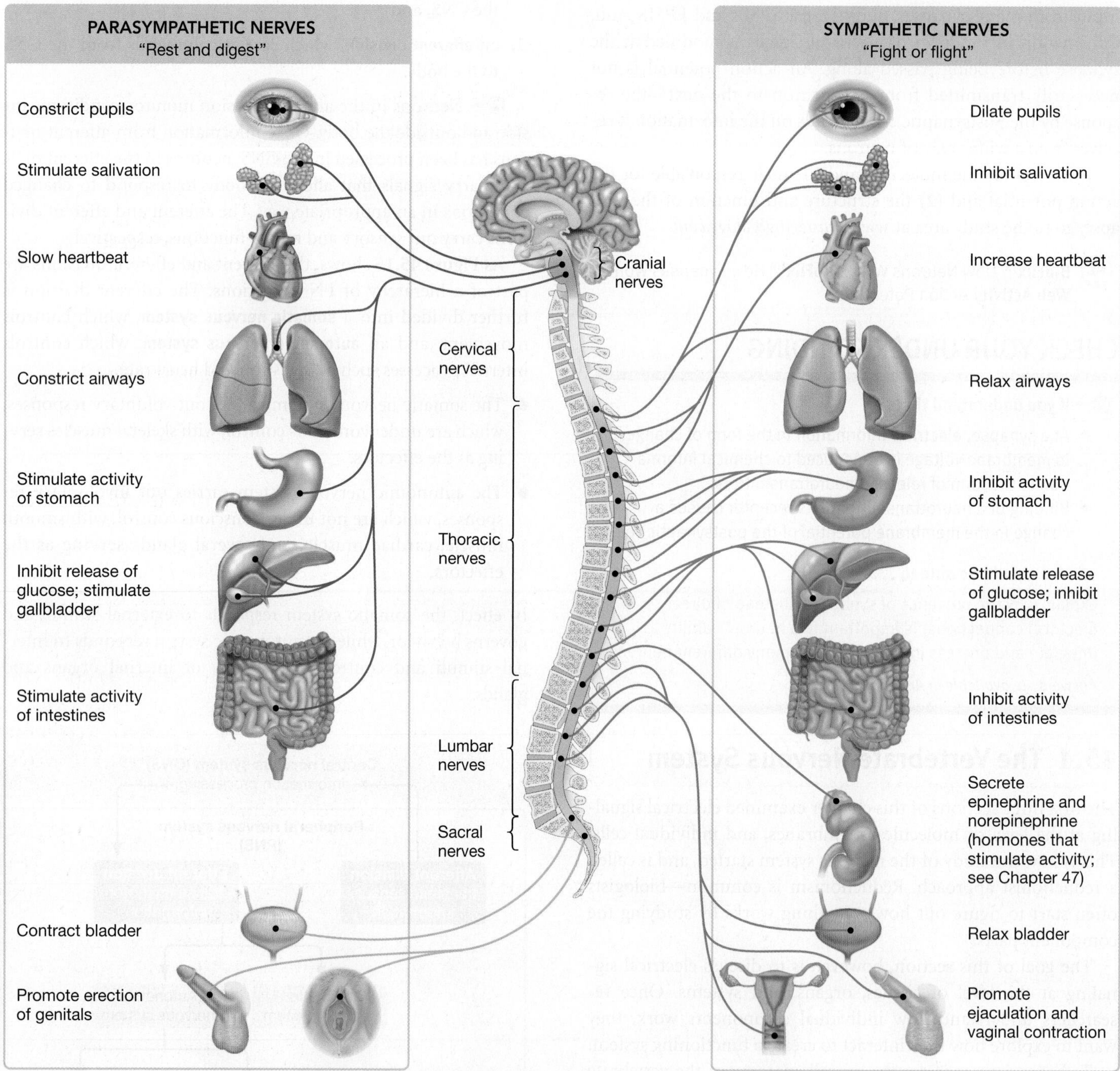

FIGURE 45.16 The Autonomic Nervous System Controls Internal Processes.

✔**QUESTION** Explain how the responses listed here for pupils, heartbeat, and liver support the "rest and digest" versus "fight or flight" functions.

the spinal cord, but they emerge along the middle of its length. Similarly, most sensory and motor neurons in the somatic nervous system project to or from the spinal cord.

In effect, then, the spinal cord serves as an information conduit that collects and transmits information from throughout the body. With a few exceptions, such as the reflex that occurs when you touch a hot burner, virtually all the information that travels to or from the spinal cord is sent to the brain for processing.

The brain is far and away the most complex organ found in animals. Researchers estimate that the human brain has 100 billion neurons, each making thousands of synaptic connections with other neurons. How do biologists even begin to study such a fantastically complex structure? They begin with general anatomy.

GENERAL ANATOMY OF THE BRAIN Nineteenth-century anatomists established that the brain is made up of the four structures labeled in **Figure 45.17**: the cerebrum, cerebellum, diencephalon, and brain stem. Each has a distinct function.

- The **cerebrum** makes up the bulk of the brain, is divided into left and right hemispheres, and is involved in conscious thought and memory.
- The **cerebellum** coordinates complex motor patterns.
- The **diencephalon** relays sensory information to the cerebellum and controls homeostasis.
- The **brain stem** connects the brain to the spinal cord, and is the autonomic center for regulating the heart, lungs, and digestive system.

Each cerebral hemisphere has four major areas, or lobes: the **frontal lobe**, the **parietal lobe**, the **occipital lobe**, and the **temporal lobe** (**Figure 45.18**). The two hemispheres are connected by a thick band of axons called the **corpus callosum**.

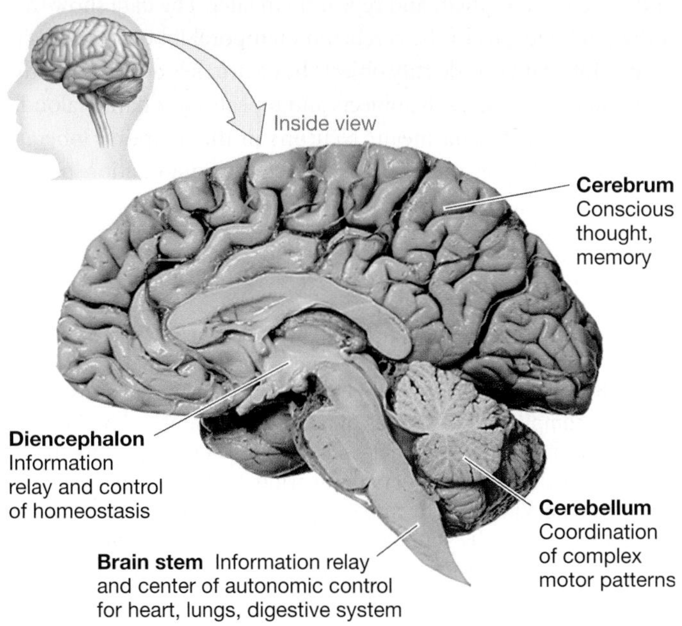

FIGURE 45.17 The Human Brain Contains Four Distinct Structures.

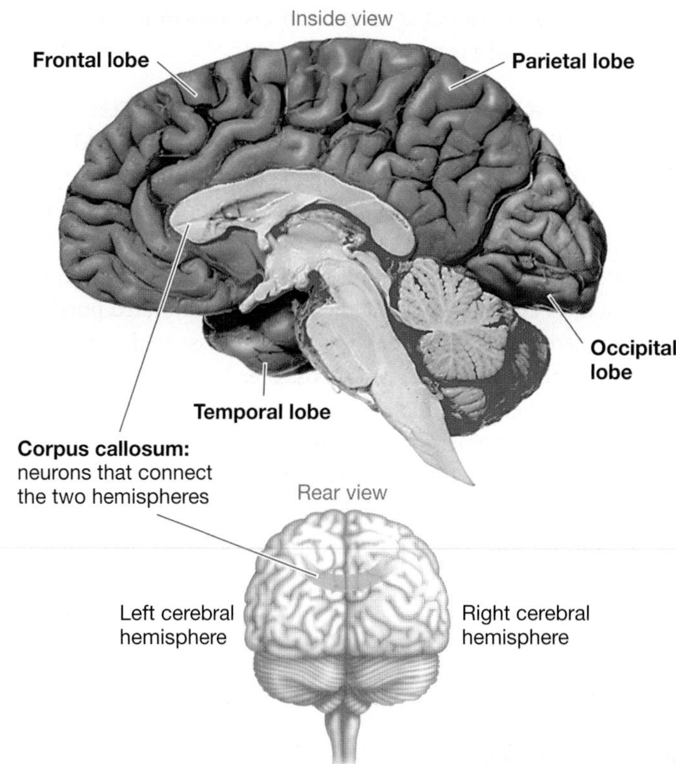

FIGURE 45.18 The Human Cerebrum Has Four Lobes and Two Hemispheres.

✔**EXERCISE** On your own head, point to each of the labeled areas.

What tools do researchers use to explore the function of each area within the cerebrum?

MAPPING FUNCTIONAL AREAS IN THE CEREBRUM I: LESION STUDIES Early work on brain function studied people with specific mental deficits caused by areas of brain damage, or lesions. Paul Broca, for example, studied an individual who could understand language but could not speak. After the person's death in 1861, Broca examined the patient's brain and discovered a damaged area in the left frontal lobe of the cerebrum. Broca hypothesized that this region is responsible for speech. More generally, he formulated the hypothesis that specific regions of the brain are specialized for coordinating particular functions.

Broca's claim that functions are localized to specific brain areas has been verified through extensive efforts to map the cerebrum. In some cases, advances were made by studying people who had to have portions of their brains removed.

In 1953, for example, surgeons treated a 27-year-old man named Henry Gustav Moliason for life-threatening seizures by removing a small portion of his temporal lobe. The man recovered (he died in December 2008, at age 83), had normal intelligence, and vividly remembered his childhood, but he had no short-term memory. Brenda Milner, who studied this individual for over 40 years, had to introduce herself to him every time they met; he could not even recognize a recent picture of himself. Based on case histories and studies of memory in laboratory animals, a consensus has emerged that several aspects of memory localize to interior sections of the temporal lobe.

MAPPING FUNCTIONAL AREAS IN THE CEREBRUM II: ELECTRICAL STIMULATION OF CONSCIOUS PATIENTS Wilder Penfield pioneered a different approach to studying brain function by working with severe epileptics—people suffering from seizures. These individuals were scheduled to have seizure-prone areas of their brains surgically removed. While the patients were awake and under a local anesthetic, Penfield electrically stimulated portions of their cerebrums. His goal was to map essential areas that should be spared from removal if possible.

When Penfield stimulated specific areas, patients reported sensations or movement in particular regions of the body. Penfield was able to map the sensory regions of the cerebrum shown in **Figure 45.19**, as well as the adjacent motor regions. Brain surgeons still use this technique to map critical areas near tumors and seizure-prone areas.

Perhaps the most striking of Penfield's findings was that, on occasion, patients would respond to stimulation of their temporal lobe by having what appeared to be flashbacks. After one region was stimulated, a woman said, "I hear voices. It is late at night around the carnival somewhere—some sort of traveling circus. . . . I just saw lots of big wagons that they used to haul animals in."

Was this a memory, stored in a small set of neurons that Penfield happened to stimulate? The hypothesis that memories are stored in specialized cells is intensely controversial. As critics have pointed out, Penfield's results are difficult to interpret because he was working with people who suffered from severe brain dysfunction. In addition, Penfield's patients sometimes described the same memory when other cells were stimulated after the original area had been surgically removed.

Have other approaches to studying memory been productive?

How Does Memory Work?

Learning is an enduring, usually adaptive change in behavior that results from a specific experience in an individual's life. **Memory** is the retention of learned information. Learning and memory are thus closely related and are often studied in tandem. As an introduction to how researchers explore these phenomena, let's first examine work that focuses on neurons and then review research at the molecular level.

RECORDING FROM SINGLE NEURONS DURING MEMORY TASKS How do the action potentials generated by a cell change as learning and memory take place? Researchers have attempted to answer this question by recording from individual neurons in the temporal lobes of humans.

Before operating on patients who were still awake and about to undergo surgery to remove seizure-prone areas of their brains, physicians have projected words or names of objects on a screen and asked the individuals to read them silently, read them aloud, and/or remember them and repeat them later. The data show that individual neurons in the cerebrum's temporal lobe are relatively quiet while patients identify objects but extremely active when the individual remembers the objects and repeats their names aloud.

What do such data mean? Neurons in the temporal lobe are most active during memory tasks. So how could action potentials from particular cells make memory possible?

(a) Top view of cerebrum

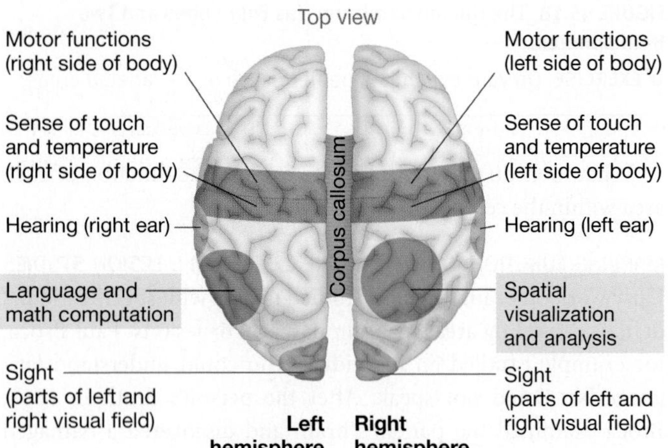

(b) Cross section through area responsible for sense of touch and of temperature

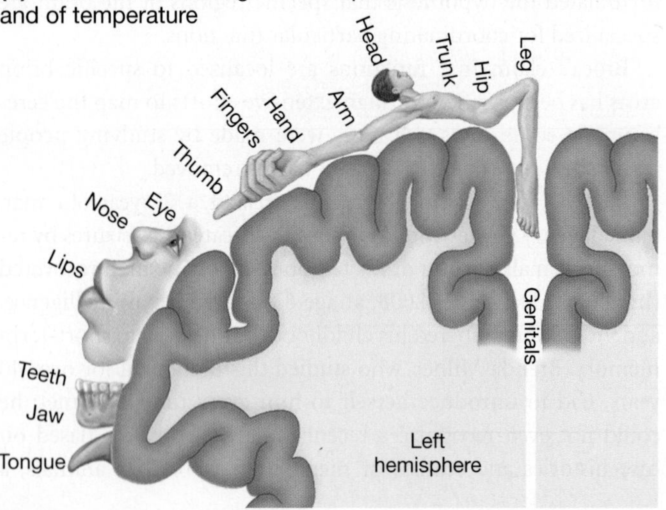

FIGURE 45.19 Specific Brain Areas Have Specific Functions.
(a) Map of the brain, in top view (as if the person is looking at the top of the page), showing the functions of some major regions. The map was compiled from studies of people with damaged brain areas or with brain regions that were removed surgically. Note that the corpus callosum is not actually as wide as shown here.
(b) Researchers mapped the area responsible for the sense of touch and of temperature by stimulating neurons in the brains of patients who were awake. The size of the icons corresponds to the amount of brain area devoted to sensing those parts.

✔**QUESTION** Is there a correlation between the size of the brain area devoted to sensing a particular body part and the size of that body part? Explain.

(a) Sea slug *Aplysia californica*

Siphon

1 cm

(b) Gill-withdrawal reflex protects the gills during an attack.

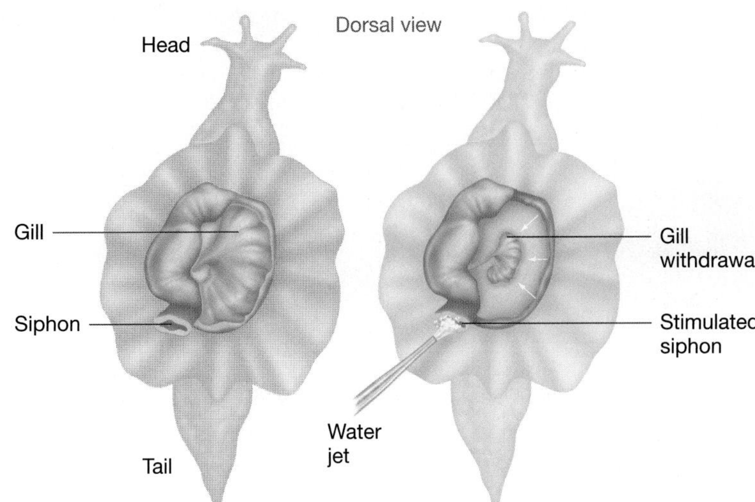

FIGURE 45.20 The Gill-Withdrawal Reflex in *Aplysia* Is a Model System in Learning and Memory.

DOCUMENTING CHANGES IN SYNAPSES Research on the molecular basis of memory is based on two fundamental ideas. First, learning and memory must involve some type of short-term or long-term change in the neurons responsible for these processes. This change could be structural or chemical in nature. Structural changes might include modifications in the number of synapses that a particular neuron makes. Chemical changes might involve alterations in the amount of neurotransmitter released at certain synapses or changes in the number of receptors present in postsynaptic cells. Second, it will be much easier to understand these changes if an extremely simple system of neurons can be studied.

To explore the molecular basis of learning and memory, Eric Kandel's group has focused on the sea slug *Aplysia californica* (**Figure 45.20a**). Much of their work has explored the reflex diagrammed in **Figure 45.20b**. When a structure in *Aplysia* called the siphon is touched—for example, by a stream of water—the individual responds by withdrawing its gill. The reflex involves a sensory neuron that is activated by touch and a motor neuron that projects to a gill muscle. Retracting the gill protects it from predators.

Early work established that this simple reflex is modified by learning. For example, *Aplysia* also withdraw their gills when their tails receive an electrical shock. If shocks to the tail are combined with a very light touch to the siphon—too light to normally get a response on its own—an *Aplysia* will learn to withdraw its gills in response to a light siphon touch alone.

Follow-up studies on this reflex showed that the neurons involved in learning release the neurotransmitter **serotonin**, which causes an EPSP in the motor neuron to the gill. Repeated application of serotonin mimics what happens at the synapse during learning, when the neurons fire repeatedly. As **Figure 45.21** shows, experimental application of serotonin leads to higher EPSPs, meaning that the motor neuron is more likely to generate action potentials. These results suggest that in *Aplysia*, changes in the nature of the synapse form the molecular basis of learning and memory. This process is termed **synaptic plasticity**.

Recently Kandel's team replicated these results with sensory and motor neurons growing in culture. **Figure 45.22a** on page 904 shows two *Aplysia* motor neurons on a culture plate. Each motor neuron is receiving synapses from a sensory neuron. To mimic the learning process, the investigators applied serotonin to one synapse five times over a short period. When they stimulated the sensory neuron a day later, they found a huge increase in EPSPs (**Figure 45.22b**). The experimental cells had also established additional synapses. The structure and behavior of the experimental neuron had changed, based on its experience. Neurons that had not received the repeated stimulation with serotonin had normal postsynaptic responses and numbers of synapses.

Results like these reinforce a growing consensus that learning and memory involve both molecular and structural changes in synapses. Further, most researchers now agree that at least some aspects of long-term memory involve changes in gene expression. Chapter 47 explores how the chemical messengers called hormones cause changes in gene expression in target cells. But before investigating how hormones work, let's focus on the electrical signals involved in vision, hearing, taste, and movement—the subject of Chapter 46.

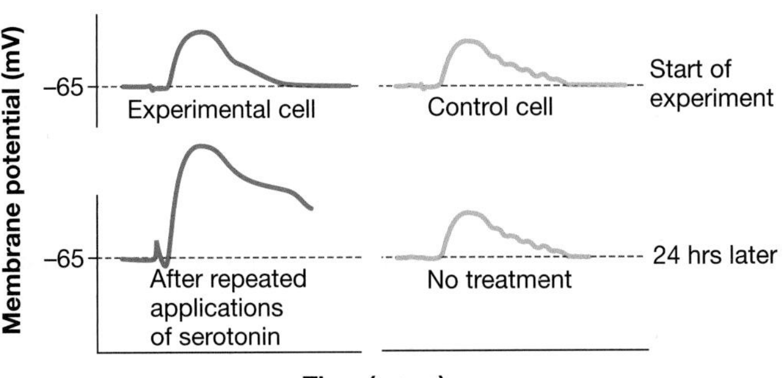

FIGURE 45.21 Repeated Application of Serotonin Changes the Behavior of Postsynaptic Neurons.

(a) Apply neurotransmitters to neurons in culture.

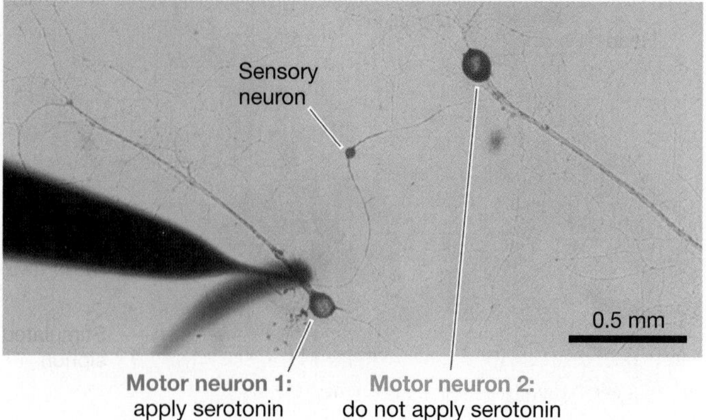

(b) Repeating *Aplysia* experiment in culture

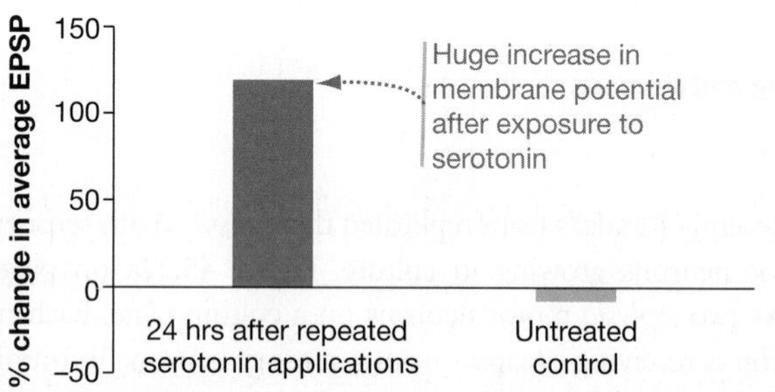

FIGURE 45.22 Learning and Memory Involve Changes in Synapses. **(a)** An in vitro study of interactions between *Aplysia* motor neurons and sensory neurons. **(b)** Histograms documenting the percentage increase in postsynaptic potentials that occurs after repeated application of serotonin, as in Figure 45.21.

CHECK YOUR UNDERSTANDING

If you understand that . . .

- The CNS and PNS work together to gather information about the external and internal environments, process that information, and signal muscles, glands, and other tissues to make appropriate responses to the information.
- Both the CNS and PNS are organized into structurally distinct regions that have specific functions.

✔ **You should be able to . . .**

1. Diagram the divisions of the PNS.
2. Describe the research strategies that allowed biologists to localize particular functions to specific regions in the brain.

Answers are available in Appendix B.

CHAPTER 45 REVIEW

For media, go to the study area at www.masteringbiology.com

Summary of Key Concepts

Neurons are cells that transmit electrical signals used in communication. Their plasma membranes carry a voltage, called a membrane potential, due to differences in the concentrations of ions on the inner and outer surfaces.

- All neurons have a cell body and multiple short dendrites that receive electrical signals from other cells. Most neurons also have axons that transmit electrical signals to other neurons or to effector cells in glands or muscles.
- Studies of the squid giant axon neurons established that neurons have a resting potential created by the sodium-potassium pump and potassium leak channels. When Na^+/K^+-ATPase hydrolyzes ATP, it transports 3 Na^+ out of the cell and 2 K^+ in.

✔ You should be able to diagram the plasma membrane of a neuron. Add symbols to show the relative concentrations of Na^+, K^+, and Cl^-. Add labels indicating the role of K^+ leak channels and the Na^+/K^+-ATPase.

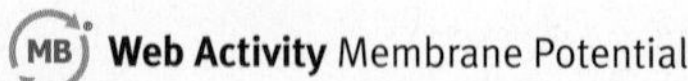

Action potentials are all-or-none changes in membrane potential that serve as electrical signals. During an action potential, an inflow of sodium ions is followed by an outflow of potassium ions.

- Studies of the squid axon established that the action potential is a rapid, all-or-none change in membrane potential.
- An action potential begins with an inflow of Na^+ that depolarizes the membrane. An outflow of K^+ follows and repolarizes the membrane.
- Both Na^+ and K^+ flow through voltage-gated channels.
- As charge spreads from the site of an action potential, the nearby membrane is depolarized enough to trigger additional Na^+ inflows and propagate the signal.
- Propagation takes place most rapidly in large-diameter axons or myelinated axons.

✔ You should be able to explain why every action potential in a neuron is identical.

MB **BioFlix™** How Neurons Work, **BioFlix™** How Synapses Work, **Web Activity** Action Potentials

At synapses, the electrical signal from a neuron triggers the release of a chemical signal—a neurotransmitter. When the neurotransmitter arrives at an adjacent neuron, there is a change in that cell's membrane potential.

- When action potentials arrive at a synapse, synaptic vesicles fuse with the axon's membrane and deliver neurotransmitters that bind to receptors on the membrane of a postsynaptic cell.

- One class of receptors functions as ligand-gated channels. In response to binding by a neurotransmitter, the channels open and admit ions that depolarize or hyperpolarize the postsynaptic cell's membrane.
- Postsynaptic potentials from nearby synapses sum.
- If the membrane at the axon hillock depolarizes to a threshold value, an action potential is triggered.

✔ You should be able to explain how summation in the postsynaptic cell relates to the claim that the cell body integrates information from many different synapses.

Most animals have a central nervous system (CNS) and a peripheral nervous system (PNS). PNS neurons receive sensory information and transmit it to the CNS for processing. The CNS then sends signals to muscles, glands, or other tissues via PNS neurons.

- The CNS consists of the brain and spinal cord (in vertebrates); the PNS consists of all nervous system components outside the CNS.
- In vertebrates, the PNS contains somatic and autonomic components. The somatic PNS is responsible for sensing external stimuli and effecting movement; the autonomic system monitors internal conditions and effects changes in the activity of organs.
- Early efforts to map functional regions of the brain depended on analyzing deficits in individuals with brain lesions or on stimulating certain regions of the cerebrum.
- Efforts to understand higher brain functions such as learning and memory form the current focus of research on the CNS.
- To date, research has established that learning and memory are based on modifications in synapses. After learning takes place, certain neurons release more or less neurotransmitter, or make additional synapses, in response to stimulation.

✔ You should be able to describe what an animal would be like if synapses were "fixed" early in development and unchangeable.

Questions

✔ TEST YOUR KNOWLEDGE

Answers are available in Appendix B

1. Which ion leaks across a neuron's membrane to help create the resting potential?
   a. Ca^{2+} **b.** K^+ **c.** Na^+ **d.** Cl^-
2. Why did the squid axon become a model system for studying electrical signaling in animals?
   a. Its action potentials are particularly large and frequent.
   b. It is the tissue from which researchers initially isolated Na^+/K^+-ATPase.
   c. Squids are abundant and easy to obtain.
   d. It was large enough to support intracellular recording by the first microelectrodes.
3. How does myelination affect the propagation of an action potential?
   a. It speeds propagation by increasing the density of voltage-gated channels.
   b. It speeds propagation by increasing electrochemical gradients favoring Na^+ entry.
   c. It speeds propagation because charge does not leak out of the membrane as it spreads down the axon.
   d. It slows down propagation because Na^+ channels exist only at unmyelinated nodes (nodes of Ranvier).
4. In a neuron, what creates the electrochemical gradient favoring the outflow of K^+ when the cell is at rest?
   a. Na^+/K^+-ATPase
   b. voltage-gated K^+ channels
   c. voltage-gated Na^+ channels
   d. ligand-gated Na^+/K^+ channels
5. Why do biologists say that positive feedback occurs during an action potential?
   a. The action potential is an all-or-none event, meaning that once it starts, it goes to completion.
   b. The opening of potassium channels repolarizes the membrane, making it less likely that sodium channels will open and depolarize the membrane.
   c. Once sodium channels open and begin to depolarize the membrane, they become more likely to open and cause further depolarization.
   d. Sodium channels are refractory—once they have opened, they are less likely to open again for a few milliseconds.
6. Why is memory thought to involve changes in particular synapses?
   a. In some systems, an increased release of neurotransmitters occurs after learning takes place.
   b. In some systems, the type of neurotransmitter released at the synapse changes after learning takes place.
   c. When researchers stimulated certain neurons electrically, individuals replayed memories.
   d. People who lack short-term memory have specific deficits in synapses within the brain regions responsible for memory.

✔ TEST YOUR UNDERSTANDING

Answers are available in Appendix B

1. Explain why the Na^+/K^+-ATPase is "electrogenic"—meaning that it creates a voltage across a membrane.
2. Draw a graph of an action potential and label the axes. Label the parts of the graph and explain which ion flow or flows are responsible for each part.
3. Explain the difference between a ligand-gated K^+ channel and a voltage-gated K^+ channel.
4. Why does summation occur in postsynaptic cells?
5. Compare and contrast the somatic and autonomic components of the PNS.
6. Compare and contrast the sympathetic and parasympathetic components of the autonomic nervous system.

APPLYING CONCEPTS TO NEW SITUATIONS

Answers are available in Appendix B

1. Explain why drugs that prevent neurotransmitters from being taken back up by the presynaptic cell have dramatic effects on the activity of postsynaptic neurons.
2. Discuss the pros and cons of lesion studies and electrical stimulation of conscious patients in determining the functions of particular brain structures.
3. The data points on the graph to the right record the movement of potassium ions (μSiemens is a unit of conductance) observed when neuron membranes were voltage-clamped at the array of voltages indicated on the *x*-axis. The "Neuron with venom" data were generated by adding black mamba poison to the solution bathing the neuron. (Black mambas are snakes.) What can you conclude from the data?
4. In some species, researchers are able to identify individual neurons in the brain and record the action potentials they produce. Using these data, how can they infer the function of particular neurons?

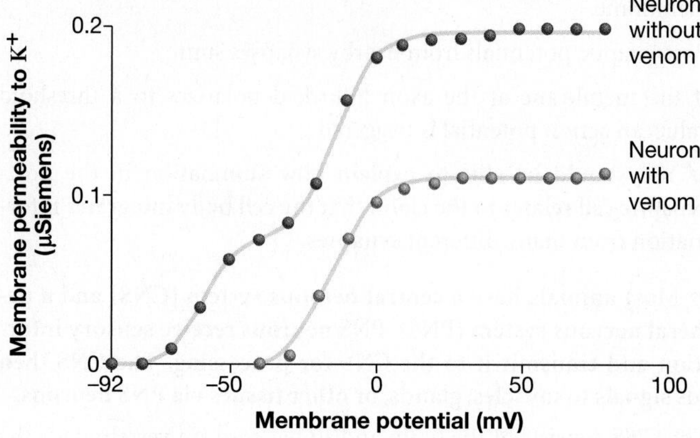

In many species of moth, males have much larger antennae than females do. Receptor cells on the males' feathery antennae detect airborne chemical signals that are produced by sexually mature females. As a result, males can locate females in total darkness.

Animal Sensory Systems and Movement 46

Many adult moths are active at night when it is difficult or impossible to see. Instead of looking for a mate, sexually mature female moths release a chemical attractant called a pheromone into the air. A **pheromone** is a small molecule that acts as a signal between individuals. Male moths of the same species can detect even a single molecule of the pheromone, due to the receptor cells located on their large, feathery antennae. In response to an airborne gradient of pheromone molecules, males fly toward a female.

As they patrol in search of these airborne pheromones, however, male moths are hunted by bats. Like moths, bats are active almost exclusively at night. Instead of hunting by sight, like a falcon or a cheetah, bats hunt with the aid of sonar. Bats emit a train of high-pitched sounds as they fly and then listen for echoes that indicate the direction and shape of objects in their path. If the object is a moth, the bat flies toward it, catches the individual in its mouth, and eats it.

Some moth species can hear bat calls, however. When moths detect sounds from an onrushing bat, they tumble out of the sky in chaotic escape flights.

If you were out at night as these dramas unfolded, at best you might be dimly aware that bats and moths were flying about. Humans cannot smell moth pheromones or hear the sounds that bats emit when flying. It took decades of careful experimentation for biologists to understand how moths and bats sense the world around them and how they move in response to the information they receive.

Sensing changes in the environment and moving in response to this information is fundamental to how animals work. Let's begin with a basic question: How are sounds, smells, and other stimuli transformed into a signal that the brain can understand?

KEY CONCEPTS

- Sensory receptor cells transduce stimuli to changes in membrane potential. Action potentials are sent to the brain, where the signals are processed and integrated.
- Hearing is based on sensory receptor cells that move in response to sound waves of a particular frequency.
- Vision is based on sensory receptor cells that contain a light-absorbing pigment bound to a protein. The pigment changes conformation when it absorbs light.
- Taste and smell sensations are registered by membrane proteins that act as ion channels or receptors for particular molecules.
- In many cases, animals respond to sensory stimuli by moving. Movement is based on antagonistic muscle groups that act on a skeleton. Muscle contraction occurs when myosin proteins move down the length of actin fibers.

46.1 How Do Sensory Organs Convey Information to the Brain?

As a moth flies through the night, its brain receives streams of signals from an array of sensory organs. Antennae provide information about the concentration of pheromones; ears located on various parts of the body send data on the presence of high-pitched sounds; detectors for balance and gravity transmit signals about the body's orientation in space.

Each type of sensory information is detected by a sensory neuron or by a specialized receptor cell that makes a synapse with a sensory neuron. As **Figure 46.1** shows, the moth's nervous system integrates the sensory input—information from sensory neurons—and responds with motor output, via electrical signals, to specific muscle groups (effectors).

The ability to sense a change in the environment depends on three processes:

1. **transduction**, or the conversion of an external stimulus to an internal signal in the form of an action potential;
2. amplification of the signal; and
3. transmission to the central nervous system (CNS).

The first step in the sequence requires a sensory receptor cell to convert light, sound, touch, or some other signal into an electrical signal. Sensory receptors are located throughout the body and are categorized by the type of stimulus:

- **Nociceptors** sense harmful stimuli such as tissue injury.
- **Thermoreceptors** detect changes in temperature.
- **Mechanoreceptors** respond to distortion caused by pressure.
- **Chemoreceptors** perceive specific molecules.
- **Photoreceptors** respond to particular wavelengths of light.
- **Electroreceptors** detect electric fields.

With such a broad range of sensory detectors available, it is no wonder that animals can monitor and respond to a wide array of changes in their environments. And this list is not exhaustive. Many birds, sea turtles, and other animals can sense changes in magnetic fields and use Earth's magnetic field as an aid in navigation; some species of birds can sense changes in barometric pressure.

Now, how do sensory cells receive information from the environment and report it to the brain, so an appropriate response can occur?

Sensory Transduction

When most sensory cells are in the resting state, the inside of the plasma membrane is more negative than the exterior. If ion flows cause the interior to become more positive (less negative), the membrane is **depolarized**. If changes in ion channels cause the cell interior to become more negative than the resting potential, the membrane is **hyperpolarized**.

Figure 46.2a shows the membrane potential from a sound-receptor cell. If you put your finger on the red line and trace to the right, you'll notice that when the experimenter played a sound, the sound-receptor cell depolarized for a short time in response. Other sensory cells work in a similar way.

Although sensory receptors can detect a remarkable variety of stimuli, they all transduce sensory input—including light, sounds, touch, and odors—to a change in membrane potential. In this way, different types of information are transduced to a common type of signal—one that can be interpreted by the brain.

If a sensory stimulus induces a large change in a sensory receptor's membrane potential, there is a change in the firing rate of action potentials sent to the brain. The amount of depolarization that occurs in a sound-receptor cell, for example, is proportional to the loudness of the sound. If the depolarization passes threshold, enough voltage-gated sodium channels open to trigger action potentials that are sent to the brain.

Recall from Chapter 45 that all action potentials from a given neuron are identical in size and shape. **Figure 46.2b** graphs the action potential "firing rate" recorded from a sound-receptor cell, when sounds at various frequencies were played at two distinct intensities. Notice that loud sounds induce a higher frequency of action potentials than do soft sounds. In this way, receptor cells provide information about the intensity of a stimulus.

But if all types of external stimuli are converted to electrical signals in the form of action potentials, and if all action poten-

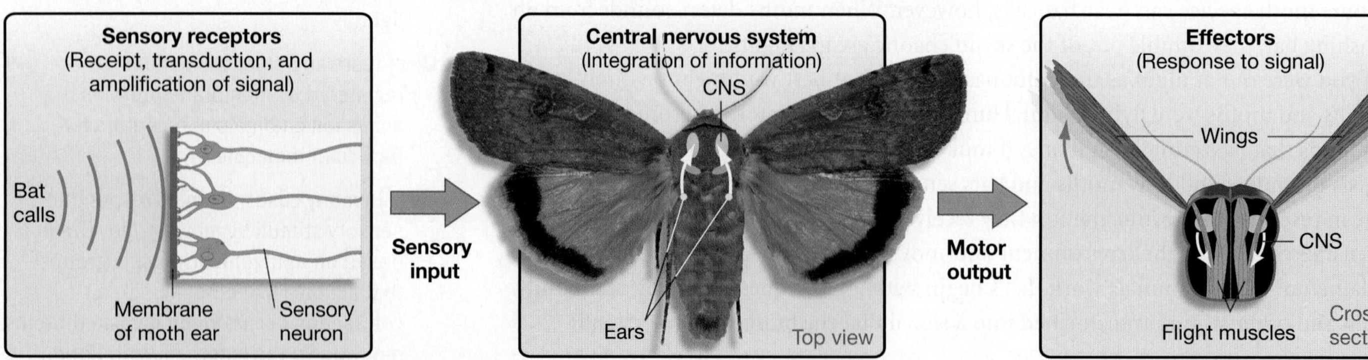

FIGURE 46.1 Sensory Systems, the CNS, and Effectors Such as Muscles Are Linked. Sensory neurons relay information about conditions inside and outside an animal to the central nervous system. After integrating information from many sensory neurons, the CNS sends signals to muscles.

(a) Sound-receptor cells depolarize in response to sound.

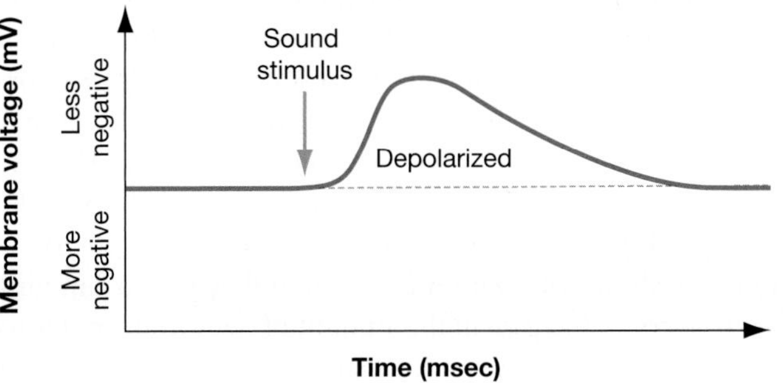

(b) Sound-receptor cells respond more strongly to louder sounds.

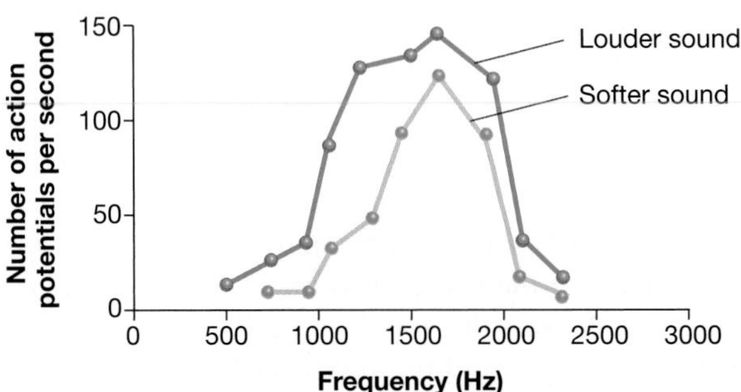

FIGURE 46.2 Sensory Inputs Change the Membrane Potential of Receptor Cells. (a) In response to sensory stimuli, ions flow across the membranes of receptor cells and either depolarize or hyperpolarize the membrane. **(b)** The frequency of action potentials from a receptor transmits information about the nature and intensity of the sensory stimulus.

tials are alike in size and duration, how does the brain interpret the information properly?

Transmitting Information to the Brain

There are two keys to understanding how the brain interprets sensory information. First, receptor cells tend to be highly specific. For example, each receptor cell in a human ear responds best to certain frequencies of sound. Some receptors are more sensitive to low-pitched sounds at a frequency of 1000 Hz (hertz, or cycles per second—a unit of frequency); others respond best to high-pitched sounds such as 8000 Hz.

The receptor-cell response depicted in Figure 46.2b, for example, is strongest for sounds at about 1650 Hz. At this frequency, the maximum number of action potentials per second emerge from the receptor. In this way, the pattern of action potentials from a cell contains information about the frequency of sound that is being received, its intensity, and how long the stimulus lasts.

The second key point: Each type of sensory neuron sends its signal to a specific portion of the brain. Axons from sensory neurons in the human ear project to a particular area at the side of the brain, but axons from sensory receptors in the eye deliver action potentials to another area at the back of the brain. Different regions of the brain are specialized for interpreting different types of stimuli.

Now that the basic principles of sensory reception and transduction have been introduced, let's delve into the details of four sensory systems that are particularly well understood: hearing, vision, taste, and smell.

46.2 Hearing

Animals have a wide variety of mechanisms for sensing changes in pressure. Crabs, for example, have a fluid-filled organ that helps them sense the pressure created by gravity. The organ, known as a **statocyst**, is lined with pressure-receptor cells and contains a small calcium-rich structure. The calcium-rich particle normally rests on the bottom of the organ. But if the crab is tipped or flipped over, this structure presses against receptors that are *not* on the bottom of the organ. When the brain receives action potentials from these cells, it responds by activating muscles that restore the animal to its normal posture.

It's also common for animals to have cells that are responsible for detecting direct physical pressure on skin, as well as pressure-receptor cells that monitor how far muscles or blood vessels are stretched. But the best-studied type of pressure-sensing is called hearing.

Hearing is the ability to sense the wavelike changes in air pressure called sound. A sound consists of waves of pressure in air or in water. The number of pressure waves that occur in one second is called the **frequency** of the sound. We perceive different sound frequencies as different **pitches**.

Hearing and other pressure-sensing systems found in animals are based on the same mechanism. Let's briefly examine the general nature of a mechanoreceptor cell that responds to pressure, then investigate the specific structures involved in vertebrate hearing.

How Do Sensory Cells Respond to Sound Waves and Other Forms of Pressure?

The mechanoreceptors responsible for sensing sound and vibrations in the environment are relatively simple in design. In every case, direct physical pressure on a plasma membrane or distortion by bending changes the conformation of ion channels in the membrane and causes the channels to open or close.

In response to a change in ion flow through channel proteins, the membrane depolarizes or hyperpolarizes. The result is a new pattern of action potentials from a sensory neuron.

THE STRUCTURE OF HAIR CELLS In vertebrates, ion channels that respond to pressure are found in hair cells. **Hair cells** are pressure-receptor cells, illustrated in **Figure 46.3a** on page 910, named for their stiff outgrowths called **stereocilia** (singular: **stereocilium**). The "hairy-looking" stereocilia are microvilli that are reinforced by actin filaments.

Many hair cells also have a single **kinocilium**, a true cilium that contains a 9 + 2 arrangement of microtubules introduced in Chapter 7. Hair cells are found in the ears of land-dwelling vertebrates and the lateral line system in many species of fish.

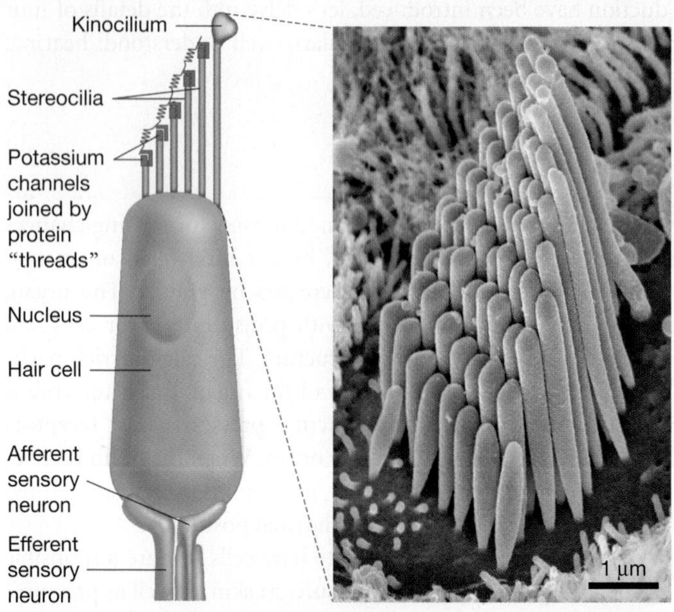

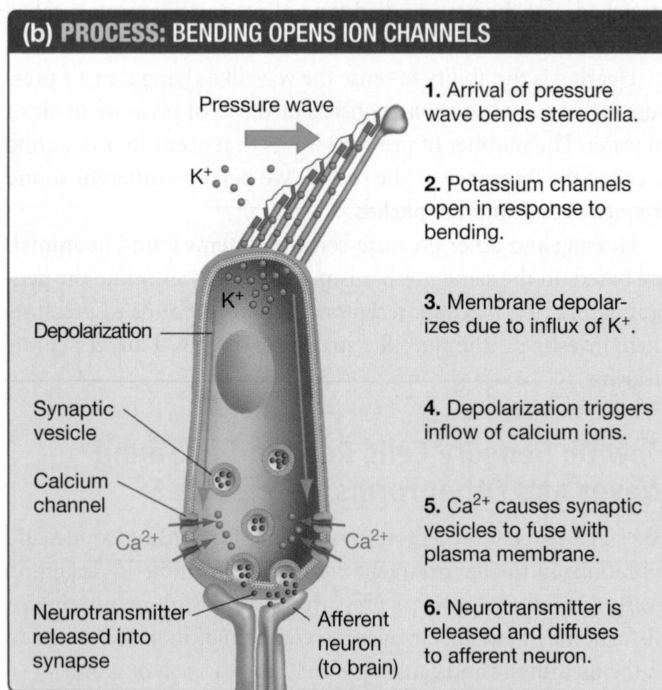

FIGURE 46.3 Hair Cells Transduce Sound Waves to Electrical Signals.

As Figure 46.3a shows, the stereocilia in hair cells are arranged in order of increasing height; if a kinocilium is present, it is the tallest of all the projections. All these structures extend into a fluid-filled chamber.

SIGNAL TRANSDUCTION IN HAIR CELLS If stereocilia are bent in the direction of the kinocilium in response to pressure (**Figure 46.3b**), the membranes of the structure are distorted and potassium ion (K^+) channels in the stereocilia open. This is the common theme in pressure-sensing cells: Bending opens ion channels.

Recall from Chapter 45 that the opening of K^+ channels usually causes an outflow of K^+ that hyperpolarizes neurons. Hair-cell plasma membranes respond differently, because they are bathed by extracellular fluid with an extraordinarily high K^+ concentration. As a result, the equilibrium potential for K^+ in hair cells is 0 mV instead of the −85 mV in a typical neuron. The resting potential of the hair-cell plasma membrane is −70 mV, so K^+ rushes in and causes a depolarization of approximately 20 mV.

In hair cells, depolarization causes an inflow of calcium ions, which triggers an increase in the amount of neurotransmitter released at the synapse between the hair cell and a sensory neuron. The end result is excitation of the postsynaptic cell, meaning that it becomes more likely to fire an action potential to the brain, via an afferent sensory neuron. You might recall, from Chapter 45, that afferent neurons are part of the peripheral nervous system and conduct information to the CNS.

If sound pressure waves bend stereocilia the other way, however, the K^+ channels close and the cell hyperpolarizes by 5 mV. Hyperpolarization of the hair cell decreases the amount of neurotransmitter released at the synapse and inhibits the sensory neuron, meaning that the neuron becomes less likely to trigger action potentials. Pressure that is perpendicular to the stereocilia does not change the activity of the ion channels.

How can bending affect ion channels? Electron micrographs show that tiny threads connect the tips of stereocilia to each other. One hypothesis contends that when the stereocilia are bent, the threads somehow pull open ion channels in the wall of the next tallest stereocilium—like tiny trapdoors (see Figure 46.3b, step 2). This hypothesis remains to be confirmed, however. Researchers still do not fully understand how the ion channels involved in pressure reception work.

The Mammalian Ear

To understand how changes in the membrane potential of hair cells result in hearing, let's focus on the human ear as a case study. The human ear has three sections: the **outer ear**, **middle ear**, and **inner ear** (**Figure 46.4**). A membrane separates each section from the others.

To trace the path of sound through the ear, study the lower part of Figure 46.4. The outer ear, which projects from the head, collects incoming pressure waves and funnels them into a tube known as the ear canal. At the end of the ear canal, the waves strike the **tympanic membrane**, or eardrum, which separates the outer ear from the middle ear.

The repeated cycles of air compression cause the tympanic membrane to vibrate back and forth with the same frequency as the sound wave. The vibrations are passed to three tiny bones called the **ear ossicles**, which vibrate against one another in response. The last ossicle, the **stapes** (pronounced *STAY-peez*), vibrates against a membrane. That membrane, called the **oval window**, separates the middle ear from the inner ear. The oval window oscillates in response and generates waves in the fluid inside a chamber known as the **cochlea** (pronounced *KOK-lee-ah*). These pressure waves are sensed by hair cells in the cochlea.

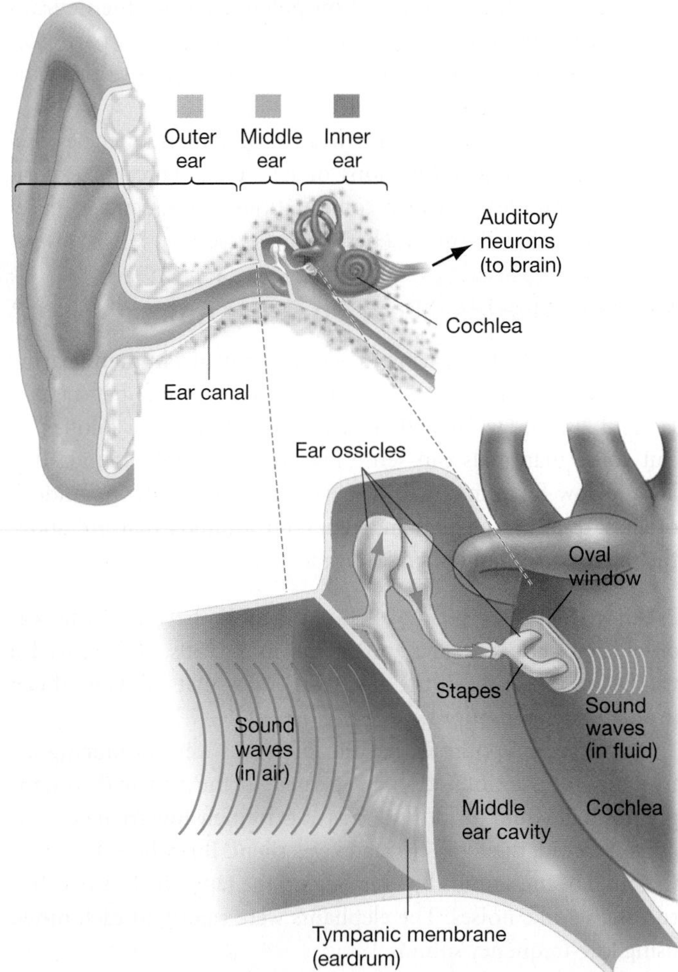

FIGURE 46.4 Mammals Have an Outer Ear, Middle Ear, and Inner Ear. The middle ear starts with the tympanic membrane and ends in the oval window of the cochlea.

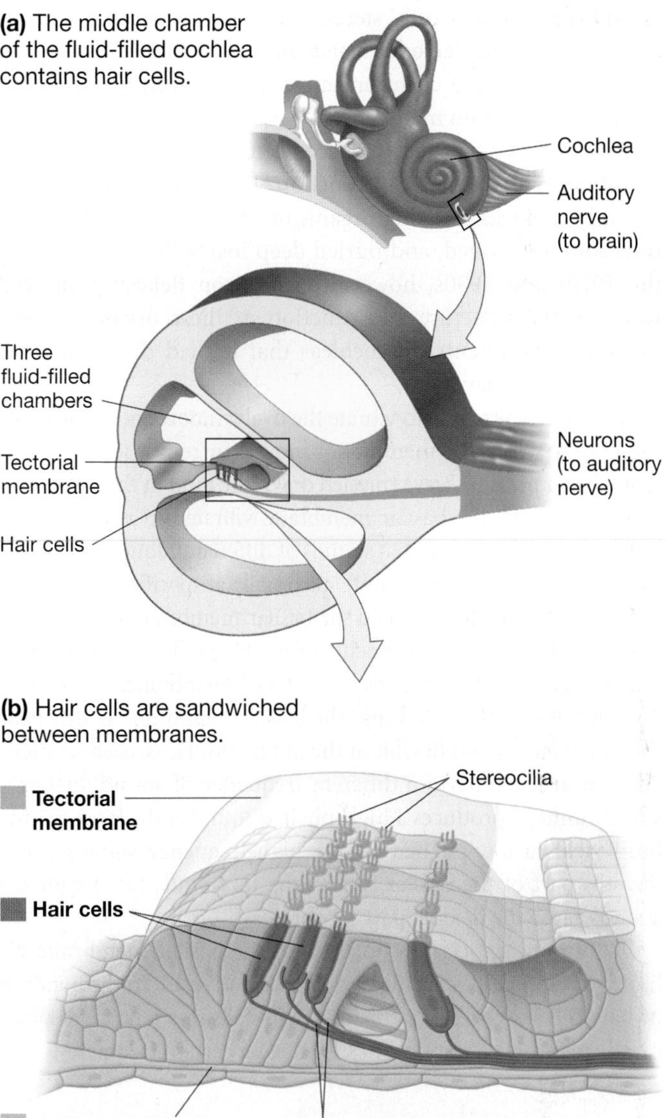

FIGURE 46.5 In the Human Cochlea, Hair Cells Are Sandwiched between Membranes. (a) The cochlea contains fluid-filled chambers separated by membranes. Hair cells are located in the middle chamber and **(b)** are sandwiched between the basilar membrane and the tectorial membrane.

In effect, the ear translates airborne waves into waterborne waves. The system seems extraordinarily complex, though, for such a simple result. Why doesn't the outer ear canal lead directly to the oval window? Why have a middle ear at all?

THE MIDDLE EAR AMPLIFIES SOUNDS Biologists began to understand the function of the middle ear when they recognized two key aspects of its structure. First, the size difference between the tympanic membrane and the oval window is important. Because the tympanic membrane is about 15 times larger than the oval window, the amount of vibration induced by sound waves increases by a factor of 15 by the time it reaches the oval window. This phenomenon is similar to using the same amount of force to bang on a very large door versus a very small door.

In addition, the three ossicles act as levers that further amplify vibrations at the tympanic membrane. Reptiles have just one ear ossicle instead of three so this levering action can occur. The overall effect in mammals is to amplify sound by a factor of 22—meaning that soft sounds are amplified enough to stimulate hair cells in the cochlea. Biologists interpret the mammalian middle ear as an adaptation for increasing sensitivity to sound.

To summarize, the mammalian outer ear transmits sound waves from the environment to the middle ear; the middle ear amplifies these waves enough to stimulate the hair cells within the cochlea of the inner ear.

If all hair cells responded equally to all frequencies of sound, we would be able to perceive only one pitch. Everyone's voice—indeed, every noise—would sound the same. How can identical hair cells distinguish different frequencies?

THE COCHLEA DETECTS THE FREQUENCY OF SOUNDS As the cross section in **Figure 46.5a** shows, the cochlea has a set of internal membranes that divide it into three chambers. Hair cells form rows in the middle chamber. The bottom of each hair cell connects to a structure called the **basilar membrane** (**Figure 46.5b**).

In addition, the hair cells' stereocilia touch yet another, smaller surface called the **tectorial membrane**. (The kinocilium is not present in a mature cochlear hair cell.) In effect, hair cells are sandwiched between membranes.

Researchers struggled for decades to understand how these membranes affect hair-cell function. It is virtually impossible to study cochleas in living organisms, because the cochleas are tiny, complex, coiled, and buried deep inside the skull. During the 1920s and 1930s, however, Georg von Békésy pioneered work on the structure and function of these organs by performing experiments on cochleas that he had dissected from fresh human cadavers.

Von Békésy was able to vibrate the oval window and record how the cochlea's internal membranes moved in response. He found that when a pressure wave traveled down the fluid in the upper and lower chambers, the basilar membrane vibrated in response. His key finding, though, was that sounds of different frequencies caused the basilar membrane to vibrate maximally at specific points along its length (**Figure 46.6**). When the basilar membrane vibrated in a particular location, the stereocilia of the hair cells there were bent one way and then the other by the tectorial membrane.

Von Békésy also noted that the basilar membrane is stiff near the oval window and flexible at the other end. Thus, each segment vibrates in response to a different frequency of sound. Just as a stiff drumhead produces a high-pitched sound and a loose drumhead yields a low-pitched sound, high-frequency sounds cause the stiff part of the basilar membrane to vibrate; low-frequency sounds cause the flexible part to vibrate.

To summarize, certain portions of the basilar membrane vibrate in response to specific frequencies and result in the bending of hair-cell stereocilia. In this way, hair cells in a particular place on the membrane respond to sounds of a certain frequency.

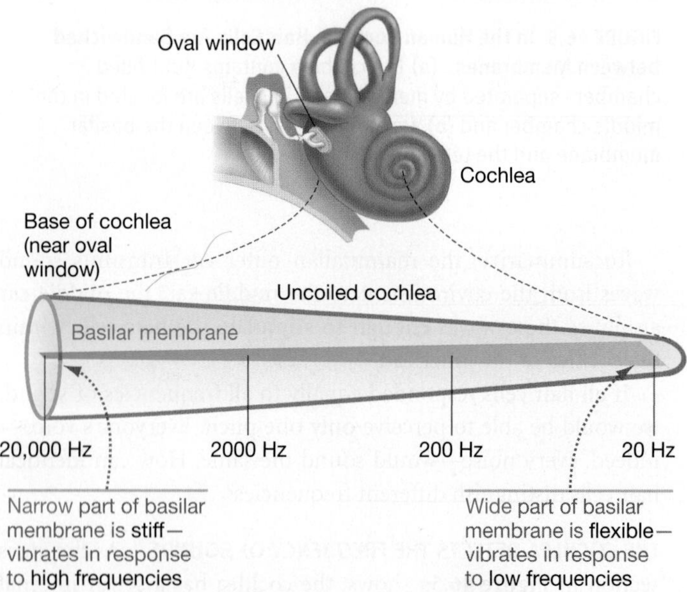

FIGURE 46.6 The Basilar Membrane Varies in Stiffness.
The numbers below the basilar membrane are in units of frequency called Hertz (Hz).

When the brain receives action potentials from the neurons associated with each hair cell, it interprets them as a particular pitch—meaning a specific frequency of sound. The result is the sense we call hearing.

Complex sounds contain a wide variety of frequencies and trigger particular combinations of hair cells. Through experience, the brain learns which combinations of frequencies represent music, a fire alarm, or a best friend's voice.

Sensory Worlds: What Do Other Animals Hear?

Compared with the hearing of many mammals, human hearing is not particularly acute. Humans can hear sounds between 20 Hz and 20,000 Hz (20,000 Hz is equal to 20 kHz, or kilohertz). But some mammals can sense low-frequency infrasounds that are too low for humans to hear (*infra* means below or under); others are aware of high-frequency ultrasounds that are above the range of human hearing (*ultra* means beyond).

ELEPHANTS DETECT INFRASOUND When Katherine Payne was observing elephants at a zoo in the mid-1980s, she noticed a throbbing sensation in the air. Payne knew that infrasound can produce such sensations.

To test the hypothesis that the elephants were producing infrasonic vocalizations, Payne returned to the zoo with microphones that could pick up sounds at extremely low frequency. At normal speed, the tape she made was silent. But when she raised the pitch of the sounds by speeding up the tape, she heard a chorus of cow-like noises. The elephants were calling to each other, using low-frequency sounds.

According to follow-up research, elephants have the best infrasonic hearing of any land mammal. Because infrasound can travel exceptionally long distances, biologists hypothesize that infrasonic calls allow wild elephants to coordinate their movements when they are miles apart.

Recent research suggests that in addition to detecting infrasound via their large ears, elephants use their feet to detect a by-product of infrasound: seismic vibrations traveling in the ground.

BATS DETECT ULTRASOUND Ultrasonic hearing in bats was discovered in the late 1930s, when Donald Griffin borrowed the only ultrasonic apparatus then in existence from Robert Galambos, a fellow graduate student. Griffin used the machine to demonstrate that flying bats constantly emit ultrasounds. In follow-up experiments, he documented that a bat with cotton in its ears, or with its mouth taped shut, crashed into walls when released in a room. Blindfolded bats, in contrast, never crashed.

Griffin and Galambos concluded that bats use sound echoes (sonar) to navigate. This concept, termed **echolocation**, was an outlandish idea at the time. When Galambos described it at a meeting in 1940, another scientist shook him by the shoulders and said, "You can't really mean that!"

More recent research has shown that dolphins, shrews, and certain other animals besides bats use sonar. In fact, it is likely that at least some of these species perceive shapes with their ears better than they do with their eyes.

THE LATERAL LINE SYSTEM IN FISH AND AMPHIBIANS Hair cells found in the mammalian ear allow these animals to sense the changes in air pressure that we call sound. But in fishes and amphibians, hair cells allow individuals to sense pressure changes in water.

In most fishes and larval amphibians, groups of hair cells are located in a line that runs the length of the body. This sensory organ is called the **lateral line system**. If changes in water pressure—from waves, an animal swimming by, or some other force—cause kinocilia to bend, the distortion leads to a change in action potentials sent to the brain. In this way, most aquatic animals get information about pressure changes at specific points along the length of the body.

CHECK YOUR UNDERSTANDING

If you understand that . . .

- Hearing is a type of pressure detection that begins when the stereocilia on hair cells bend in response to changes in pressure. The bending movement opens ion channels and results in a change in membrane potential.
- The mammalian ear comprises specialized structures that function in transmitting and amplifying sound, and in responding to specific frequencies, as well as recording changes in intensity.

✔ You should be able to . . .

1. Describe the functions of the outer ear, middle ear, and inner ear of a mammal.
2. Predict the type of hearing loss that results from each of the following: a punctured eardrum, a mutation that results in dramatically shortened stereocilia, and an age-related loss of flexibility in the basilar membrane.

Answers are available in Appendix B.

46.3 Vision

Most animals have a way to sense light. The organs involved range from simple light-sensitive eyespots in flatworms to the sophisticated, image-forming eyes of vertebrates, cephalopod mollusks, and arthropods.

Variation in the structure of light-sensing organs illustrates an important general principle about the sensory abilities of animals: In most cases, a species' sensory abilities correlate with the environment it lives in and its mode of life—how it finds food and mates. Eyes and other sensory structures are adaptations that allow individuals to thrive in a particular environment. Salamanders that live in meadows and forests have sophisticated eyes; salamanders that live in lightless caves have no functional eyes at all.

Keep this point in mind as we delve into the details of how insects and vertebrates see.

The Insect Eye

Insects have a **compound eye**, composed of hundreds or thousands of light-sensing columns called **ommatidia**. As **Figure 46.7** shows, each ommatidium has a lens that focuses light onto a small number of receptor cells—usually four. The receptor cells, in turn, send axons to the brain.

Each ommatidium contributes information about one small piece of the visual field, not unlike a single pixel on a computer monitor. Thus, the more ommatidia in a compound eye, the better the resolution—meaning the resolving power, or ability to distinguish objects.

In addition, the presence of many light-sensing columns makes species with compound eyes particularly good at detecting movement. Insects that hunt by sight, such as damselflies and dragonflies, have particularly large numbers of ommatidia.

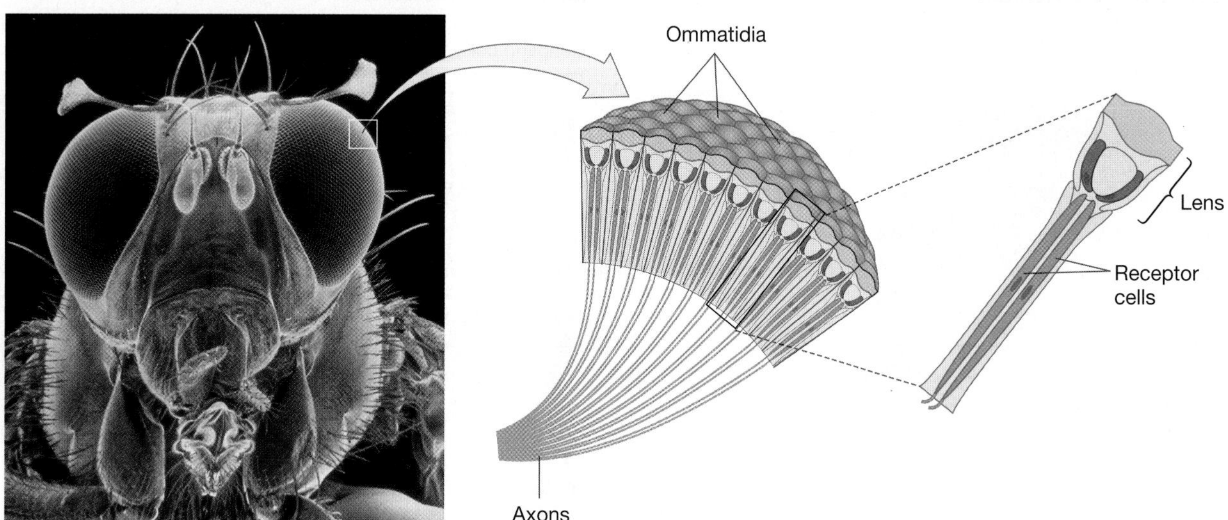

FIGURE 46.7 In the Compound Eyes of Insects, Each Ommatidium Sees Part of the World. The micrograph in part (a) is colorized to match the drawings in part (b).

The Vertebrate Eye

Compound eyes are found in insects, crustaceans, and certain other arthropods. Because they appear only in species that are part of the same monophyletic group (see Chapter 27), researchers conclude that this type of eye structure evolved just once—in an ancestor of today's arthropods.

In contrast, the **camera eye**—a structure with a single lens that focuses incoming light onto a layer of many receptor cells—evolved independently in two widely divergent groups: the cephalopod mollusks (squid and octopuses) and the vertebrates. Let's examine the vertebrate version of the camera eye more closely.

THE STRUCTURE OF THE CAMERA EYE **Figure 46.8a** shows the major structures in a typical vertebrate eye:

- The outermost layer of the structure is a tough rind of white tissue called the sclera. This is the "white of the eye."
- The front of the sclera forms the **cornea**, a transparent sheet of connective tissue.
- The **iris** is a colored, round muscle just inside the cornea. The iris can contract or expand to control the amount of light entering the eye.
- The **pupil** is the hole in the center of the iris.
- Light enters the eye through the cornea and passes through the pupil and a curved, clear **lens**.
- Together, the cornea and lens focus incoming light onto the retina in the back of the eye. The **retina** contains a thin layer of light-sensitive cells and several layers of neurons.

Figure 46.8b provides a closer look at the retina, which is attached to the rest of the eye by a single layer of pigmented epithelial cells. From back to front, the retina comprises three distinct cell layers:

1. The sensory cells that respond to light, the photoreceptors, are held in place by the pigmented epithelium.
2. Photoreceptors synapse with an intermediate layer of connecting neurons called bipolar cells.
3. Cells in the intermediate neuron layer connect with one another and with the neurons called **ganglion cells**, which form the innermost layer of the retina. The axons of the ganglion cells project to the brain via the **optic nerve**.

A FLAWED STRUCTURE? Although the vertebrate eye is sometimes claimed to be a "perfect" adaptation, optometrists have cataloged a large array of common defects. In addition, even vertebrate eyes that are in ideal condition have a curious flaw: the blind spot.

Vertebrate eyes have a blind spot because there are no photoreceptor cells where the optic nerve leaves the retina. If light falls in this area, there are no sensory cells available to respond. No signal is sent to the brain, so the stimulus isn't seen.

In cephalopod mollusks (squid and octopus), however, a better camera-type eye evolved independently. In cephalopods, the photoreceptor cells form a continuous layer on the inside of the retina. The ganglion cells that transmit messages to the brain are located behind the sensory cells—not in front of them, as in the vertebrate eye (see Figure 46.8b). As a result, the optic nerve does not interrupt the layer of photoreceptor cells. In this respect, cephalopods receive more complete visual information about their environment than vertebrates do.

Why do vertebrates have a flawed design? The leading hypothesis is historical constraint. As Chapter 24 pointed out, the traits observed in today's organisms evolved from traits that occurred in their ancestors. Apparently, the vertebrate camera eye happened to evolve from an ancestor with photoreceptor cells located in an interior layer.

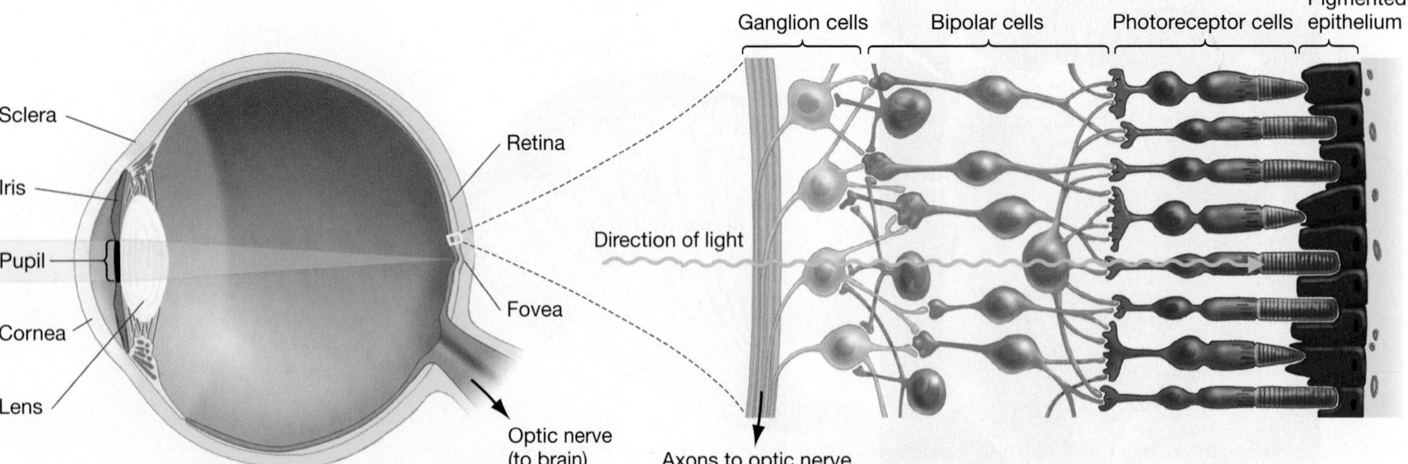

FIGURE 46.8 Camera Eyes Have a Single Lens That Focuses Incoming Light on Receptor Cells. (a) Light passes through the pupil of the eye and is focused onto the retina by the cornea and lens. **(b)** The photoreceptor cells that respond to light are in the "outermost" layer of the retina, furthest from the light source.

WHAT DO RODS AND CONES DO? Early anatomists established that the photoreceptors in vertebrate eyes come in two distinct types: small rod-shaped or cone-shaped cells called **rods** and **cones** (**Figure 46.9a**). When technical advances allowed changes in the membrane potentials of these cells to be recorded, it became clear that rods and cones differ in function as well as structure.

Rods are sensitive to dim light but not to color. Cones, in contrast, are much less sensitive to faint light but are stimulated by different wavelengths (that is, colors). These discoveries explained why night vision is largely black and white—at night, the rods do most of the work.

Rods dominate most of the retina, but one small spot in the center of the retina has only cones. This is the **fovea** (see Figure 46.8a). When people focus on an object, their eyes move so that the image falls on the fovea of each eye. Based on these observations, biologists concluded that the high density of cones in the fovea maximizes the resolution of the image.

HOW DO RODS AND CONES DETECT LIGHT? As Figure 46.9a shows, rods and cones have segments that are packed with membrane-rich disks. The membranes contain large quantities of a transmembrane protein called **opsin**. Each opsin molecule is associated with a molecule of the pigment **retinal**. In rod cells, the two-molecule complex is called **rhodopsin** (**Figure 46.9b**).

Experiments with isolated retinal, opsin, and rhodopsin molecules confirmed that retinal changes shape when it absorbs light. Specifically, the number-11 carbon in the retinal molecule changes from the *cis* conformation to the *trans* conformation (**Figure 46.9c**). Retinal is a light switch.

The shape change that occurs in retinal triggers a series of events that culminate in a different stream of action potentials being sent to the brain. The sequence of events is unusual, though, because the receipt of a light stimulus does not open ion channels or trigger the release of a neurotransmitter to a sensory neuron.

In vertebrates, the molecular basis of vision is a shape change in retinal that shuts down an existing ion channel and decreases the amount of neurotransmitter being released to the sensory neuron. In rod cells, electrical activity across the membrane, as well as neurotransmitter release, are maximized in the dark. Exposure to light transmits information by *shutting down* both processes.

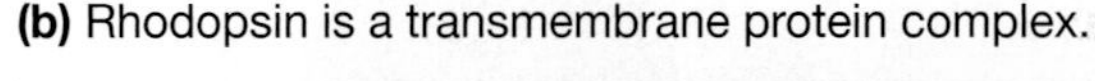

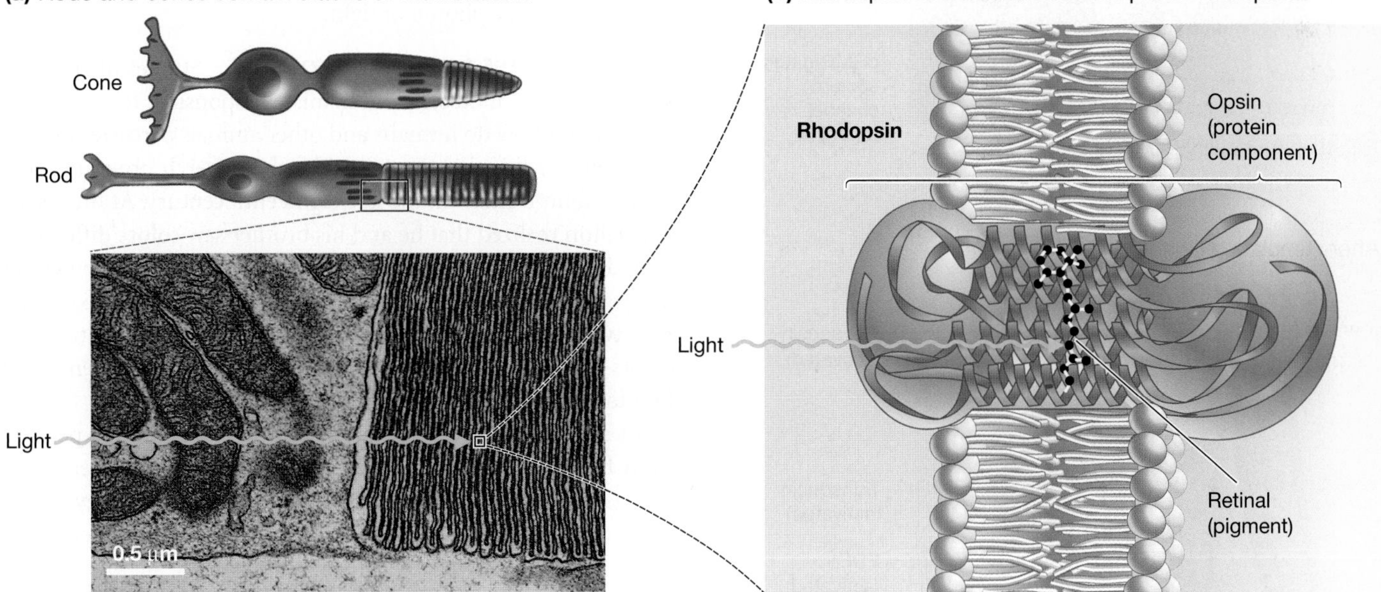

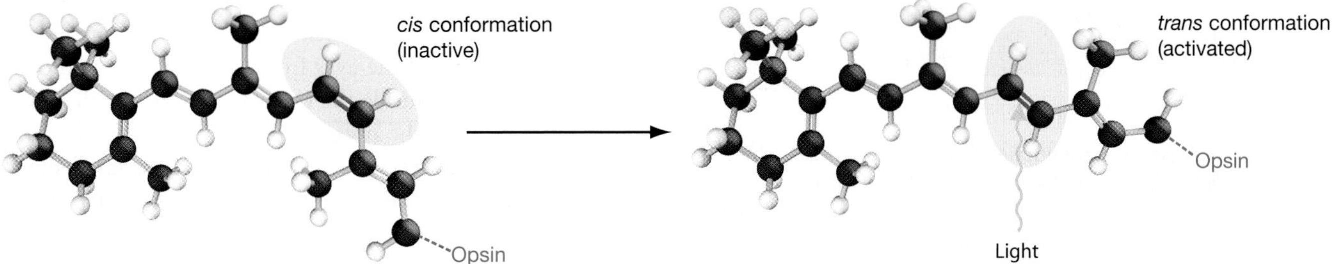

FIGURE 46.9 Rods and Cones Are Packed with Transmembrane Proteins That Contain the Pigment Retinal.
(a) Rods and cones have membranous disks containing thousands of opsin molecules. **(b)** Each opsin holds one retinal molecule. **(c)** Retinal changes conformation when it absorbs light. In response, opsin also changes shape.

✔**QUESTION** Explain why the change from *cis* to *trans* shown in part (c) would affect opsin's function.

Figure 46.10 shows how shutting down happens:

1. Rhodopsin is activated when light causes retinal to change shape from the *cis* to *trans* conformation.
2. Rhodopsin activation causes a membrane-bound molecule called transducin to activate the enzyme phosphodiesterase (PDE).

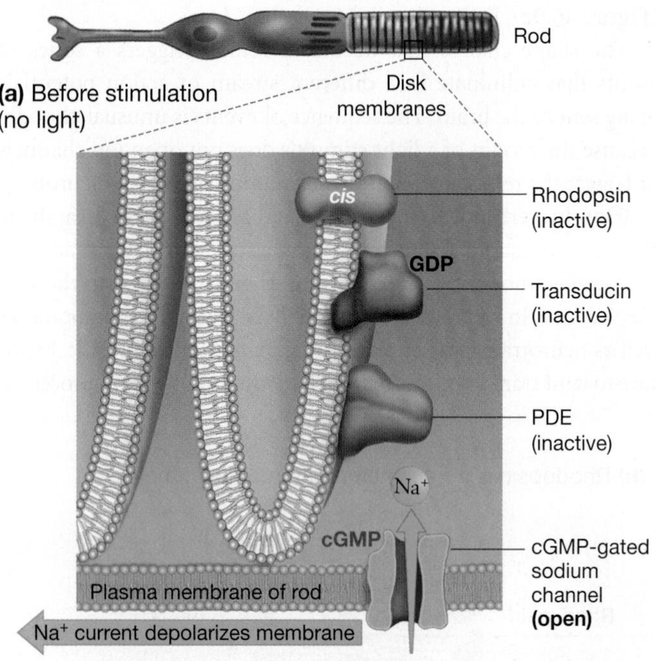

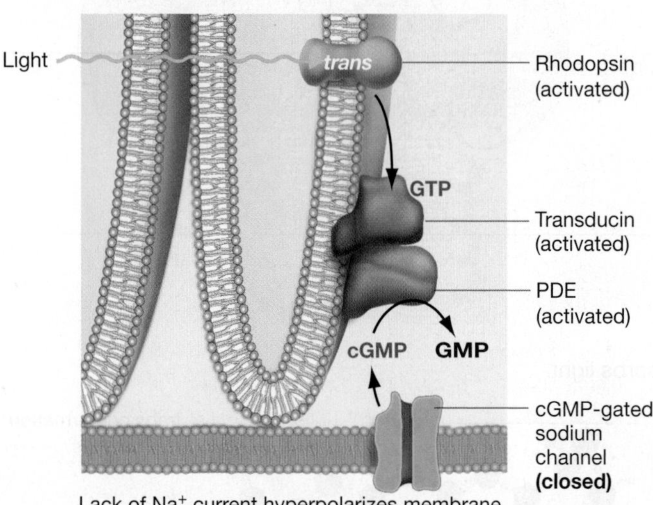

FIGURE 46.10 A Signal Transduction Pathway Connects Light Absorption with Changes in Membrane Potential. (a) An unstimulated photoreceptor. Notice that sodium ions flow into the cell when light is *not* being received. **(b)** A stimulated photoreceptor. When rhodopsin is activated, it leads to a reduction in cGMP concentration. With less cGMP available, cGMP-gated sodium channels close and the membrane hyperpolarizes.

✔**QUESTION** The inflow of sodium ions into a photoreceptor cell is called "the dark current." Why?

3. PDE breaks down a nucleotide called cyclic guanosine monophosphate (cGMP) to guanosine monophosphate (GMP);
4. As cGMP levels decline, cGMP-gated sodium channels in the plasma membrane of the rod cell close.
5. When sodium channels close, Na^+ entry decreases and the membrane hyperpolarizes.

✔If you understand this sequence of events, you should be able to explain why cGMP acts as both a second messenger and a ligand in this system, and how transducin compares to the G proteins introduced in Chapter 8. You should also be able to explain why shutting Na^+ channels results in hyperpolarization.

In response to the ensuing change in membrane potential, smaller quantities of the neurotransmitter glutamate are discharged at the synapse. The decrease in neurotransmitter indicates to the postsynaptic cell, called a **bipolar cell**, that the rod absorbed light. As a result, a new pattern of action potentials is sent to the brain, via neurons called ganglion cells. Axons from ganglion cells are bundled into a structure called the optic nerve.

This system is exquisitely sensitive: Biologists have recorded a measurable change in the membrane potentials of rod cells in response to a single photon of light.

COLOR VISION: THE PUZZLE OF DALTON'S EYE Studies of rhodopsin revealed the molecular mechanism responsible for light perception. But how do humans and other animals perceive color?

To answer this question, consider the research program initiated by John Dalton[1] in the late eighteenth century. At the age of 26, Dalton realized that he and his brother saw colors differently than did other people. To them, red sealing wax and green laurel leaves appeared to be the same color, and a rainbow exhibited only two hues. Dalton and his brother could not differentiate the colors red and green. This condition is called red-green color blindness (**Figure 46.11**).

In a lecture delivered in 1794, Dalton explained his perceptions by hypothesizing that red wavelengths failed to reach his retinas. Further, he hypothesized that because a normal eyeball is filled with clear fluid, and because blue fluids absorb red light, his defective vision resulted from the presence of bluish fluid rather than clear fluid in his eyes.

To test this hypothesis, Dalton left instructions that his eyes should be removed after his death and examined to see if the fluid inside was blue. When he died 50 years later, an assistant dutifully removed the eyes from Dalton's corpse and examined them. The fluid inside the eye was not blue at all, however, but slightly yellow—the normal color for an older person. Further, when the back was cut off one eye and colored objects were viewed through the lens, the objects looked perfectly normal. Dalton's hypothesis was incorrect.

[1]Dalton was an accomplished physicist. He was the first proponent of the atomic theory and formulated Dalton's law on the partial pressures of gases (introduced in Chapter 44). Red-green color blindness is sometimes called Daltonism in his honor.

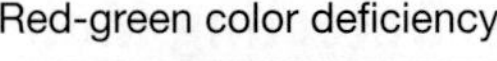

FIGURE 46.11 People with Color Deficiencies See Colors Differently than Do People without Deficiencies. These images show what a person with red-green color blindness would see, compared to a person with intact color receptors.

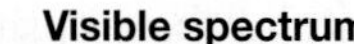

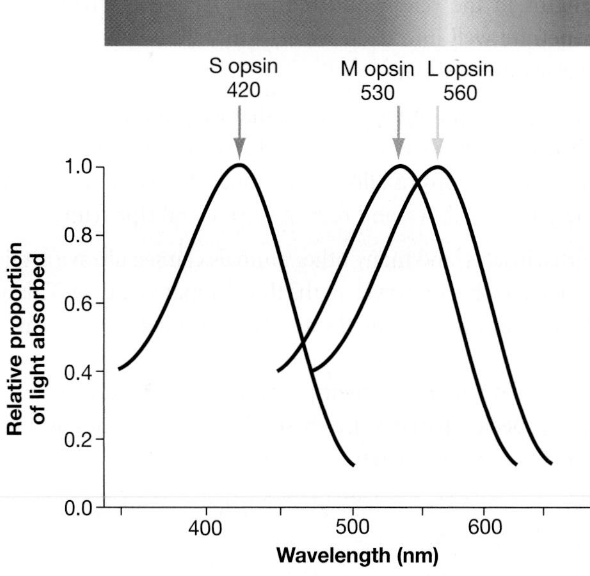

FIGURE 46.12 Color Vision Is Possible because Different Opsins Absorb Different Wavelengths of Light. Each human cone cell contains one of three different types of opsin. Each opsin has a different range of absorbed wavelengths.

✔**QUESTION** The retinal molecules in S, M, and L opsins are identical. How is it possible that they respond to different wavelengths of light?

COLOR VISION: MULTIPLE OPSINS What caused Dalton's color blindness? The key to answering this question was the discovery that the human retina contains three types of color-sensitive photoreceptors: blue, green, and red cones, named for the colors that they best perceive.

To follow up on this result, biologists analyzed opsin molecules from the three cell types and found that each had a distinct amino acid sequence. The three proteins are called the blue, green, and red opsins (or S, M, and L, for short, medium, and long wavelengths, respectively). Although retinal is the light-absorbing molecule in all photoreceptor cells, various types of opsins respond only to distinct wavelengths of light.

Based on these results, biologists hypothesized that the brain distinguishes colors by combining signals initiated by the three classes of opsins. **Figure 46.12**, for example, graphs how much light is absorbed, across a range of wavelengths, by the L, M, and S opsins of humans. Notice that a wavelength of 560 nm stimulates L cones strongly, M cones to an intermediate degree, and S cones hardly at all. In response to the corresponding signals from these cells, the brain perceives the color yellow.

Does this hypothesis explain Dalton's color blindness? According to the data in Figure 46.12, wavelengths from green to red do not stimulate S opsin at all. It is thus unlikely that S opsin is involved in red-green color blindness. Did Dalton fail to distinguish red and green because his M or L cones were defective? Research has shown that red-green color-blind people lack either functional M or L cones, or both. Was the same true of Dalton?

This question was answered in the 1990s, when the genes for the M and L opsins were sequenced. Remarkably, Dalton's eyes had been preserved. Researchers managed to extract DNA from the 150-year-old tissue and analyze his M opsin genes and L opsin genes. They found that Dalton had a normal *L* allele but lacked a functional *M* allele. As a result, he did not have green-sensitive cones. The puzzle of Dalton's color vision was solved.

To review the structure and function of the vertebrate eye, go to the study area at *www.masteringbiology.com.*

 **Web Activity** The Vertebrate Eye

Sensory Worlds: Do Other Animals See Color?

What about other animals—do they see color the way humans do? The answer is probably not. Animals that are active at night have relatively few cone cells and many rods, giving them high sensitivity to light but poor color vision. Numerous vertebrate and invertebrate species have four or more types of opsins and probably perceive a world of colors that is much richer than ours.

In general, the types of opsins found in a species correlates with the environment it inhabits and its mode of life. For example:

- A marine fish called the coelacanth (pronounced *SEE-luh-kanth*), which lives in water 200 meters deep, has two opsins that respond to the blue region of the spectrum (wavelengths of 478 nm and 485 nm). As a result, coelacanths perceive several distinct hues of blue that we would perceive as a single color. The existence of these opsins is logical, because wavelengths in the yellow and red parts of the spectrum do not penetrate well into deep water—only blue light exists in the coelacanth's habitat.

- In humans and other primates that eat fruit, two of the three opsins are sensitive to wavelengths around 550 nm. The presence of these opsins allows individuals to distinguish between the greens, yellows, and reds of unripe and ripe fruits.
- Birds, insects, and many other animals can see ultraviolet light, which has shorter wavelengths than humans can see. This ability important. Certain flowers have ultraviolet patterns that serve as signals for insect pollinators; many birds have strong ultraviolet patterns in their plumage that are invisible to us. In some species, females use these ultraviolet patches as a criterion for selecting mates.

Opsins are adaptations—their structure and function vary among species.

CHECK YOUR UNDERSTANDING

If you understand that . . .

- In vertebrates, light detection begins when retinal changes shape after absorbing light.
- If opsin changes conformation in response to movement by retinal, a series of events results in the closing of sodium channels in the membrane of a photoreceptor cell. The resulting change in membrane voltage triggers a change in neurotransmitter release from the receptor cell.

✔ You should be able to . . .

1. Explain why retinal can be thought of as a "light switch."
2. Predict the type of vision loss that results from each of the following: a tear in the fovea, a mutation that knocks out the gene for S opsin, and an age-related clouding of the lens.

Answers are available in Appendix B.

46.4 Taste and Smell

The senses of taste (**gustation**) and smell (**olfaction**) originate in chemoreceptors. Chemoreceptors can detect the presence of particular molecules because they undergo a change in membrane potential when a specific compound is present. In this way, information about the presence of a particular chemical is transduced to an electrical signal in the body.

Until recently, taste and smell were poorly understood in comparison with vision and hearing. It is easy to see why. Eyes and ears respond to relatively simple stimuli—light and sound waves. The tongue and nose, in contrast, respond to thousands of different chemicals. Consequently, taste and smell were difficult to study until techniques became available for identifying how particular molecules bind to certain receptors. But in the past 15 years, research on the chemosenses has exploded.

Taste: Detecting Molecules in the Mouth

The chemoreceptor cells that sense taste are clustered in structures known as **taste buds**. Although humans have taste buds scattered around the mouth and throat, most taste buds are located on the tongue (**Figure 46.13**). A taste bud contains about 100 spindle-shaped taste cells, which are taste receptors that synapse onto sensory neurons.

How do these receptors work on a molecular level, and how do they produce the sensation of taste? Early taste research focused on the hypothesis that four "basic tastes" existed: salty, sour, bitter, and sweet.

SALT AND SOUR Researchers who analyzed the membrane proteins in taste cells found strong evidence that salt and sour sensations result from the activity of ion channels.

- The sensation of saltiness is due primarily to sodium ions (Na^+) dissolved in food. These ions flow into certain taste cells through open Na^+ channels and depolarize the cells' membranes.
- Sourness is due to the presence of protons (H^+), which flow directly into certain taste cells through H^+ channels, and depolarize the membrane.

The sour taste of grapefruit and other citrus fruits, for example, results from the release of protons by citric acid. In general, the lower the pH of a food, the more it depolarizes a taste cell's plasma membrane and the more sour the food tastes.

Compared to salt and sour, the molecular mechanisms responsible for the sensations of bitterness and sweetness have been much more difficult to identify. Researchers have only recently been able to document that certain food molecules actually bind to specific receptors on taste cells to cause bitter and sweet tastes.

WHY DO MANY DIFFERENT FOODS TASTE BITTER? Bitterness has been difficult to understand because molecules with very different structures are all perceived as bitter. How is this possible?

An answer began to emerge after researchers confirmed that some humans genetically lack the ability to taste certain bitter

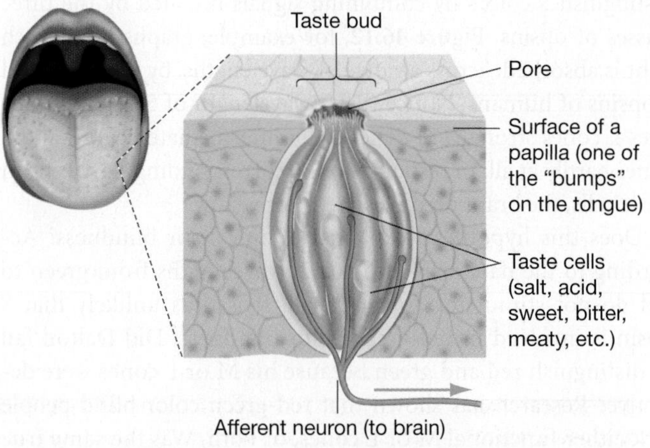

FIGURE 46.13 Taste Buds Contain Many Types of Chemoreceptors.

substances. In 1931, Arthur Fox was synthesizing phenylthiocarbamide (PTC) and accidentally blew some of it into the air. A nearby colleague complained of a bitter taste in his mouth, but Fox could not taste anything. Follow-up research confirmed that the ability to taste PTC is inherited and polymorphic. About 25 percent of United States citizens cannot sense the molecule.

To find the gene responsible for this trait, biologists compared the distribution of genetic markers observed in "tasters" and "nontasters." The mapping effort recently narrowed down this gene's location to several candidate chromosomal segments. (Chapter 19 introduced this type of gene hunt.) In one of the regions, researchers found a family of 40 to 80 genes that encode transmembrane receptor proteins.

Follow-up work has documented that each protein in the family binds to a different type of bitter molecule. A taste cell, however, can have many different receptor proteins from this family. As a result, many different molecules can depolarize the same cell and cause the sensation of bitterness.

Why are so many genes devoted to detecting bitterness? Many of the molecules that bind to these receptors are found in toxic plants; most animals react to bitter foods by spitting them out and avoiding them in the future. In essence, bitterness indicates "this food is dangerous; don't swallow it."

The proliferation of genes responsible for detecting bitterness hints that the trait is extremely important—meaning that individuals with the capacity to detect a wide array of bitter compounds produce more offspring than do individuals that lack this ability.

WHAT IS THE MOLECULAR BASIS OF SWEETNESS AND OTHER TASTES? Inspired by progress on bitter receptors, research teams used a similar approach—analyzing mutant mice that could *not* sense sweetness—to search for sweetness receptors.

In humans and mice, three closely related membrane receptors are responsible for detecting sweetness as well as glutamate and other amino acids. The glutamate receptor is responsible for the sensation called **umami**, which is the meaty taste of the molecule monosodium glutamate (MSG). The three receptors combine to form an array of multiunit proteins, some of which respond to sweetness; others to amino acids.

Recent work has solved a long-standing question about sweetness—why so many different types of sugars trigger the same sensation. As it turns out, a single receptor has binding sites for an array of sugars, meaning that a variety of molecules can stimulate action potentials from the same receptor.

Currently, several teams are studying how the different taste sensations are conveyed to the brain and interpreted. Although taste is beginning to reveal its secrets, the complete story will probably not be known for many years.

Olfaction: Detecting Molecules in the Air

Taste allows animals to assess the quality of their food before swallowing it. Smell, in contrast, allows animals to monitor airborne molecules that convey information. Wolves and domestic dogs, for example, can distinguish millions of different odors at vanishingly small concentrations. The molecules that cause odor contain information about the movements and activities of prey and other members of their own species.

How does the sense of smell work? When odor molecules, or odorants, reach the nose, they diffuse into a mucus layer in the roof of the nose (**Figure 46.14**). There, they activate olfactory receptor neurons via membrane-bound receptor proteins. Axons from these neurons project up to the **olfactory bulb**, the part of the brain where olfactory signals are processed and interpreted.

Understanding the anatomy of the odor recognition system was a relatively simple task. Understanding how receptor neurons distinguish one molecule from another was much more difficult. Initially, investigators hypothesized that receptors respond to a small set of "basic odors," such as musky, floral, minty, and so on. The idea was that each basic odor would be detected by its own type of receptor.

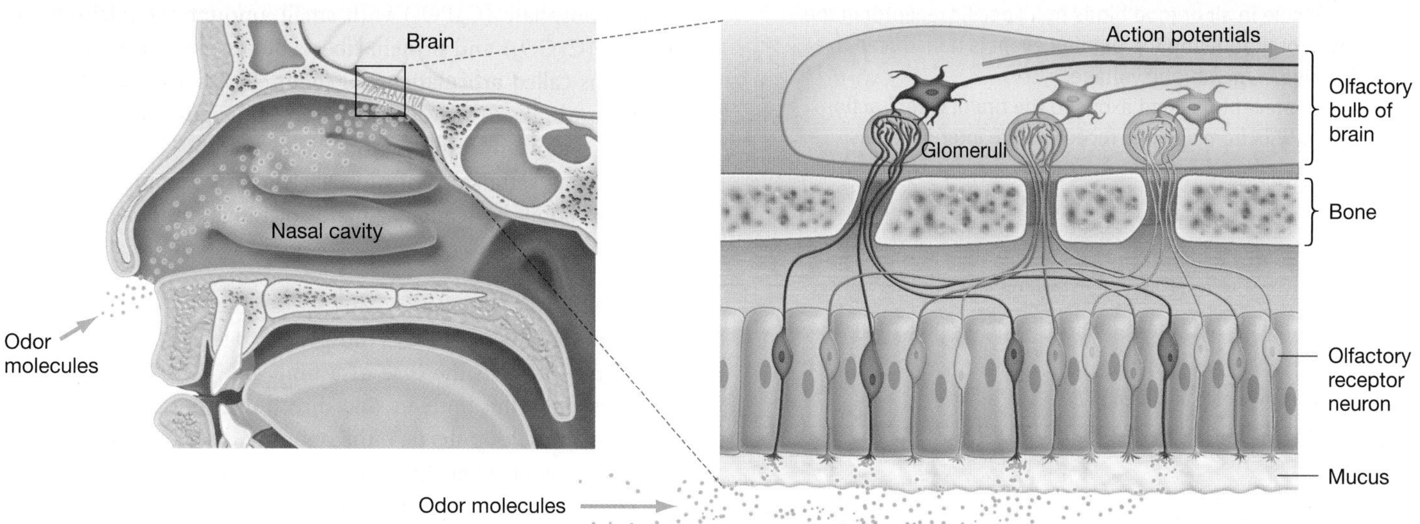

FIGURE 46.14 In Mammals, Chemoreceptor Cells in the Nose Respond to Specific Odorants. Each of the chemosensory neurons in the nose carries one type of odor-receptor protein on its dendrites. Sensory neurons with the same receptor project to the same glomerulus, or section within the olfactory bulb of the brain.

In 1991, however, Linda Buck and Richard Axel demolished the basic-odor hypothesis with an astonishing result: They discovered a gene family in mice—containing hundreds of distinct coding regions—that encodes receptor proteins on the surface of odor-receptor neurons. Follow-up experiments confirmed that each receptor protein binds to a small set of molecules.

Further work established that most, if not all, vertebrates have this family of genes. Mammals typically have 500 to 1000 of these genes, meaning mammals can produce 500 to 1000 different odor-receptor proteins. In humans, however, about half of these genes have mutations that probably render them nonfunctional. This observation may explain why the sense of smell is so poor in humans compared with that of other mammals.

Buck and Axel's announcement inspired a series of questions. How many different receptors occur in the membrane of each neuron involved in odor reception? How does the brain make sense of the input from so many different receptors?

Recently, Buck determined that each olfactory neuron has only one type of receptor and neurons with the same type of receptor are linked to distinct regions in the olfactory bulb of the brain. These regions are called **glomeruli** (meaning "little balls").

Follow-up work from Axel's lab indicated that particular smells in mice are associated with the activation of a certain subset of the 2000 glomeruli in the brain. For example, the activation of clumps 130, 256, and 1502 might be perceived as the smell "cinnamon."

In essence, then, the sense of smell is similar to the visual system's use of three cones to perceive many colors, but on a much larger scale. Research on this complex and impressive sense continues at a furious pace.

CHECK YOUR UNDERSTANDING

If you understand that . . .

- In most cases, chemoreception occurs when a specific molecule in air or food binds to a specific receptor in the nose or mouth and the binding event is transduced to a change in membrane voltage.
- Chemoreceptors send axons to the brain, where action potentials from specific receptors are interpreted as particular smells or tastes.

✓ You should be able to . . .

1. Discuss why a loss in chemosensory ability occurs when you burn your tongue with extremely hot food.
2. Explain why dogs have a better sense of smell than people do.

Answers are available in Appendix B.

46.5 Movement

The first four sections of this chapter focused on how animals sense different aspects of their environment. Acquiring information is only a first step, however. Sensing a change in the environment is useless unless an animal can respond to it in an appropriate way—usually by moving.

When bats and moths hear ultrasonic frequencies, for example, they alter their flight paths and activities in a way that allows them to catch food or avoid being eaten. Flight and other types of motion are based on muscle contractions in conjunction with a skeleton.

It's important to recognize that **locomotion**, or the movement of an entire animal, is only one type of movement, however. Sessile organisms, such as barnacles and sea anemones, may not undergo locomotion, but they rely on movement of the heart muscle or gills to perform gas exchange and circulate body fluids.

To explore locomotion in detail, let's briefly examine the skeleton's role as a scaffold for muscle attachment, then analyze the structure and function of vertebrate muscle cells and tissues.

Skeletons

Skeletons provide attachment sites for muscles and a support system for the body's soft tissues. As Chapters 32–34 indicated, three types of skeletons are found in animals:

- **Exoskeletons** are hard, hollow structures that envelop the body.
- **Hydrostatic skeletons** use the pressure of enclosed body fluids to support the body.
- **Endoskeletons** are hard structures inside the body.

Chapters 32–34 introduced exoskeletons and hydrostatic skeletons in detail; the focus here is on the structure and function of endoskeletons.

ENDOSKELETON STRUCTURE Endoskeletons are composed of connective tissues called cartilage and bone. Recall from Figure 41.3c that **cartilage** is made up of cells scattered in a gelatinous matrix of polysaccharides and protein fibers. In this skeletal system, cartilage provides padding between bones.

Bone is made up of cells in a hard extracellular matrix of calcium phosphate ($CaPO_4$) with small amounts of calcium carbonate ($CaCO_3$) and protein fibers. Bones meet and interact at locations called **articulations**, or **joints**. Bones articulate in ways that allow limbs to swivel, hinge, or pivot (**Figure 46.15**).

Bones move in response to force exerted by skeletal muscles. In vertebrates, the ends of most skeletal muscles are attached to different bones by **tendons**—bands of tough, fibrous connective tissue.

ENDOSKELETON FUNCTION Muscles can exert force only by contracting, so pairs of muscles must work together to move a bone back and forth. In the case of a limb, one muscle or group of muscles pulls the limb in one direction; the other muscle or group of muscles pulls it in the opposite direction. This pairing of muscles is called an antagonistic muscle group.

The muscle that swings two long bones in an arc toward each other is a **flexor**; the muscle that straightens them out is an **extensor** (**Figure 46.16a**). For example, the hamstring muscles in the back of your thigh flex your lower leg; the quadriceps muscles in the front of your thigh extend it. When the hamstrings

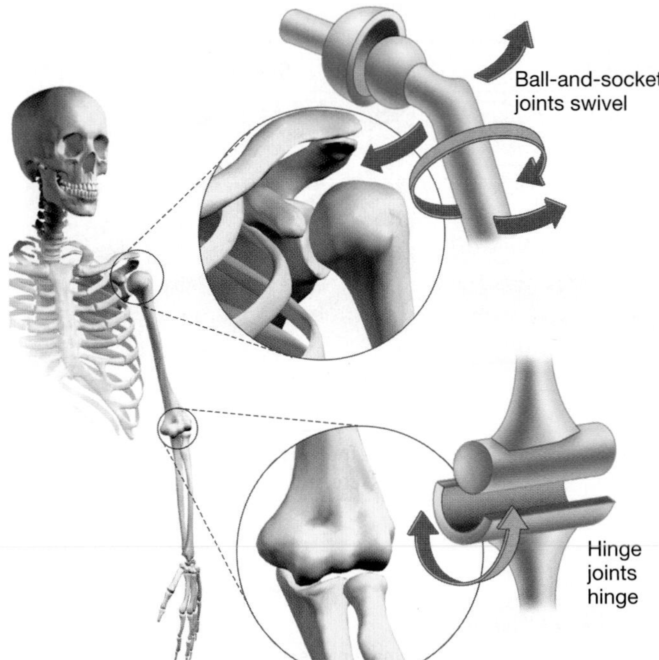

FIGURE 46.15 Bones Articulate at Joints. Bones articulate in ways that make specific types of movement possible.

contract, the quadriceps relax; when the quadriceps contract, the hamstrings relax. The hamstrings and quadriceps together make up an antagonistic muscle group.

Locomotion in animals with exoskeletons and hydrostatic skeletons is also based on antagonistic muscle groups. Animals with exoskeletons have paired flexor-extensor muscles inside their leg joints (**Figure 46.16b**). Many animals with hydrostatic skeletons have both circumferential muscles and longitudinal muscles, and these muscles work in concert to shorten and lengthen parts of the body (**Figure 46.16c**).

In all vertebrate animals, the movements of paired muscles are coordinated by motor neurons that originate in the brain or spinal cord. These motor neurons project from processing centers that receive input from sensory systems.

Motor neuron activity changes in response to information about balance, smells, sights, and sounds. In this way, an animal's movements are directly tied to the functioning of its sensory systems. The interplay of sensory input and motor output results in the coordinated behaviors we call running, eating, swimming, and flying.

Muscle Types

Early anatomical studies confirmed that vertebrates have three distinct types of muscle tissue: (**1**) skeletal muscle, (**2**) cardiac muscle, and (**3**) smooth muscle. These three muscle types were introduced briefly in Chapter 41; **Table 46.1** on page 922 provides a more detailed comparison.

Although all three muscle cell types contract in response to electrical stimulation and actin-myosin interactions, they differ in overall morphology as follows:

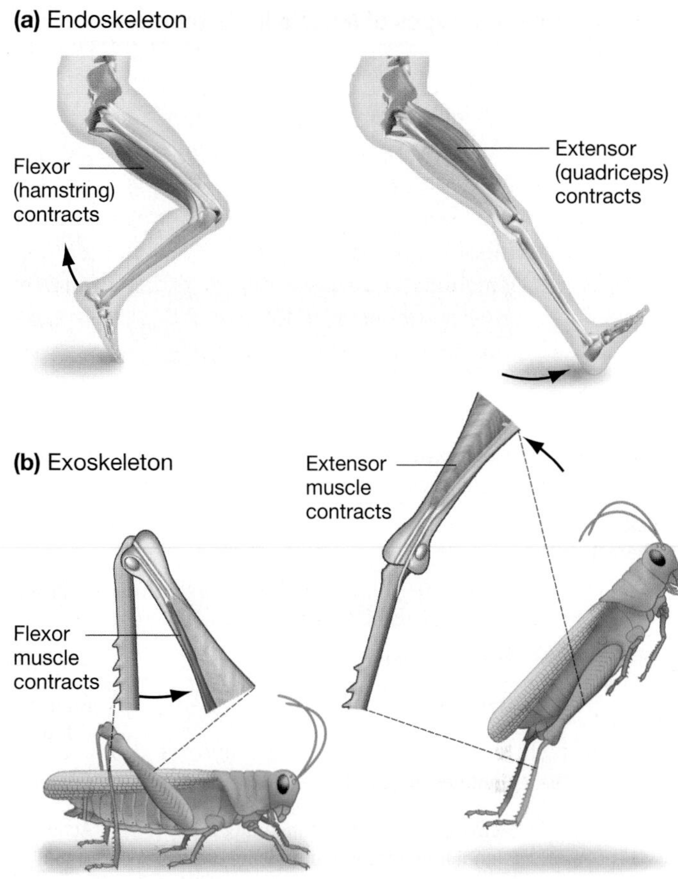

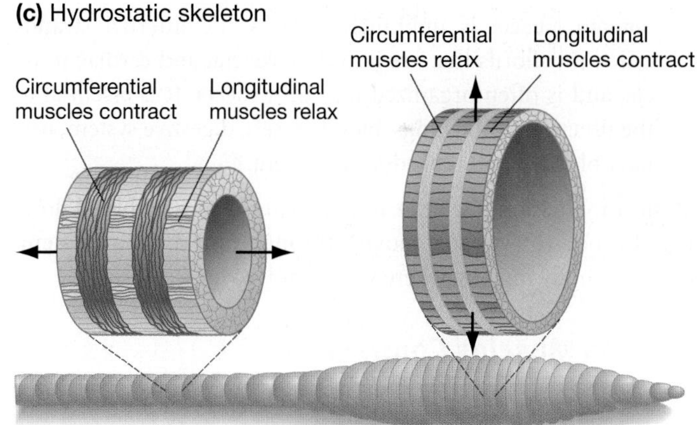

FIGURE 46.16 Antagonistic Muscle Groups Flex or Extend Skeletal Elements.

1. **Skeletal muscle** consists of unbranched multinucleate cells—they contain multiple nuclei. Because this tissue appears to be striated, or striped, it is also known as **striated muscle**.
2. **Cardiac muscle** contains branched cells whose ends are connected via specialized regions called intercalated discs. In an intercalated disc, protein-lined openings called gap junctions provide a direct cytoplasmic connection between adjacent muscle cells. As Chapter 44 noted, intercalated discs are critical to the flow of electrical signals from cell to cell and to the coordination of the heartbeat.

SUMMARY TABLE 46.1 **Types of Muscle in Vertebrates**

	Skeletal Muscle	Cardiac Muscle	Smooth Muscle
	25 μm	25 μm	25 μm
Location	Attached to bones	Heart	Intestines, arteries, other
Function	Move skeleton	Pump blood	Move food, help regulate blood pressure, etc.
Cell characteristics	Nuclei	Intercalated discs	
	Multinucleate	1 or 2 nuclei	Single nucleus
	Unbranched	Branched; intercalated discs form direct cytoplasmic connection end to end	Unbranched
	Contains myofibrils	Contains myofibrils	No myofibrils
	Activity is "voluntary," meaning that signal from motor neuron is required	Activity is "involuntary," meaning that signal from motor neuron is not required	Activity is "involuntary," meaning that signal from motor neuron is not required

3. **Smooth muscle** is unbranched, lacks the internal strands called myofibrils that are found in skeletal and cardiac muscle, and is often organized into thin sheets. It is essential to the function of the lungs, blood vessels, digestive system, urinary bladder, and reproductive system.

In combination, the three muscle types keep your blood flowing, the food you've eaten moving through your digestive system, and your body moving where you want it to go.

How Do Muscles Contract?

Early microscopists established that the muscle tissue found in vertebrate limbs and hearts is composed of slender fibers. A **muscle fiber** is a long, thin muscle cell. Within each of the muscle cells from limbs are many small strands called **myofibrils**. How do these strands work?

MYOFIBRIL STRUCTURE When viewed with the electron microscope, the myofibrils inside a skeletal muscle cell look striped (**Figure 46.17**). The pattern is caused by alternating light-dark units called **sarcomeres**, which repeat down the length of a myofibril.

The sarcomeres in a myofibril shorten as the cell contracts. They lengthen when the cell relaxes and an external force—exerted by the antagonistic muscle of the pair—stretches the muscle. Based on these observations, it became clear that the question of how muscles contract simplifies to the question of how sarcomeres contract.

THE SLIDING-FILAMENT MODEL In 1954, biologists Hugh Huxley and Jean Hanson noticed that the relationships between different parts of the sarcomere—the points labeled A–D in Figure 46.17—change during contraction.

- Points A and B move closer to each other during contraction, as do points C and D.
- The distance from point A to point C does not change, and the distance from point B to point D does not change.

To explain these observations, Huxley and Hanson proposed the **sliding-filament model** illustrated in **Figure 46.18**. The hypothesis was that the banding patterns in the sarcomere are actually caused by two types of long filaments—**thick filaments** and **thin filaments**—and that these filaments slide past one another during a contraction. Note that thin filaments extend from A to C and thick filaments extend from B to D. Follow-up research has shown that the Huxley-Hanson model is correct in almost every detail.

Structurally, thin filaments are similar to the microfilaments introduced in Chapter 7. You might recall that microfilaments are prominent parts of the cytoskeleton. Thin filaments are composed of two coiled chains of the globular protein **actin**. One end of each thin filament is bound to a structure called the **Z disk**, which forms the wall of the sarcomere. Z disks contain a protein that binds tightly to actin, anchoring the thin filament. The other end of a thin filament is free to interact with thick filaments.

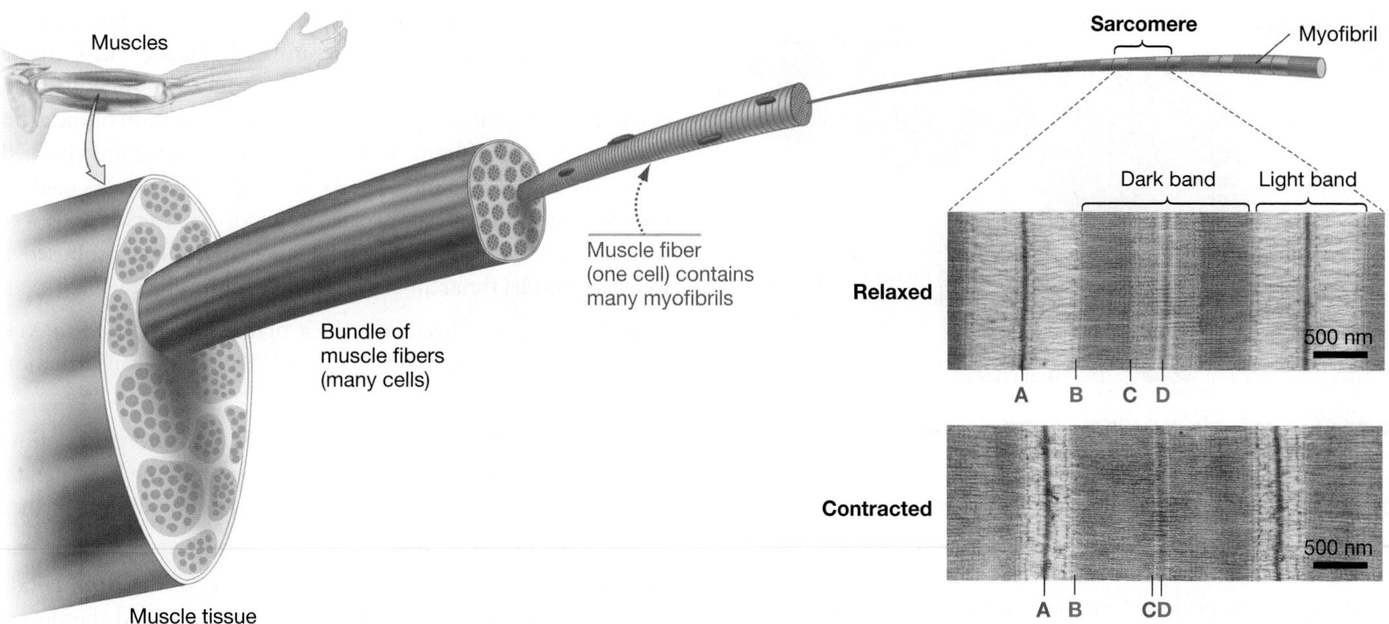

FIGURE 46.17 Muscle Cells Contain Many Myofibrils, Which Contain Many Sarcomeres. Skeletal muscle cells (fibers) have a striped appearance due to repeating sarcomeres, which are units of alternating light-dark bands. When a sarcomere contracts, the distances between some of the points labeled A, B, C, and D change.

Thick filaments are composed of multiple strands of a long protein called **myosin** and are anchored to the middle of the sarcomere. They are free at both ends to interact with thin filaments.

As predicted, thick and thin filaments interact in a way that allows them to slide past one another. For example, when researchers isolated actin filaments and myosin molecules and mixed them together on a glass slide in the presence of ATP, they observed actin sliding past myosin.

To appreciate how the sliding-filament model works, consider the following analogy: Two large trucks are parked 50 m apart, facing each other. Each has a long rope attached to the front bumper. Teams of burly weightlifters grab onto each rope and pull, hand over hand, so that the two trucks roll toward one another. ✔If you understand the model, you should be able to explain which elements in the analogy represent the Z disks, the thick filaments, and the thin filaments.

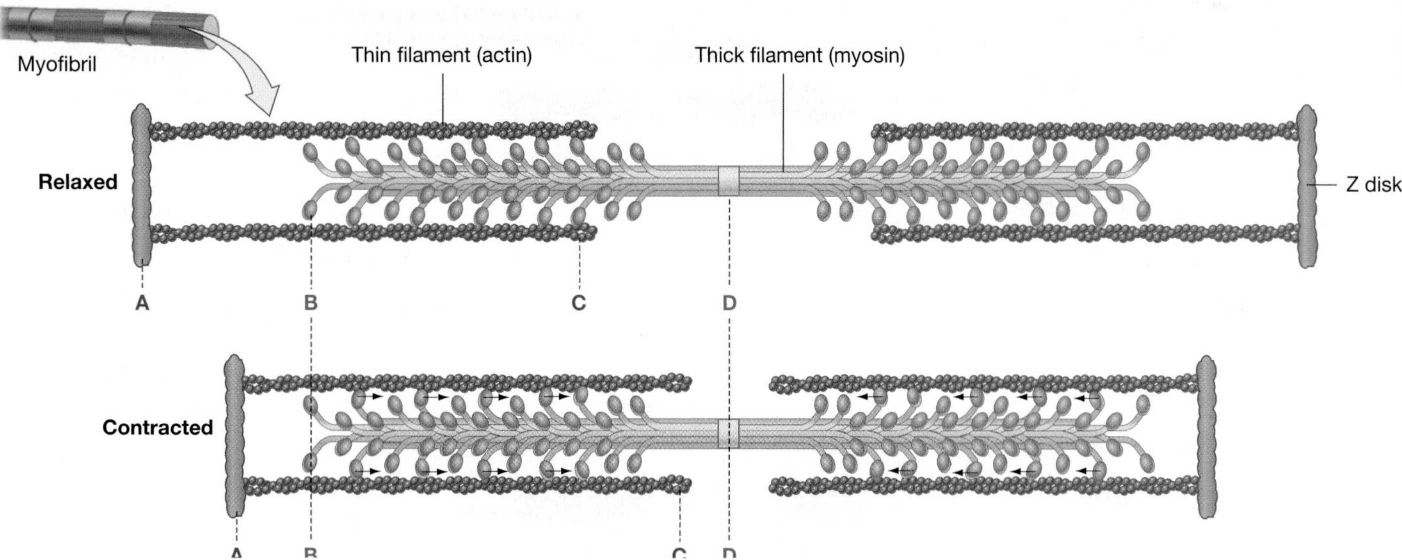

FIGURE 46.18 The Sliding-Filament Model of Sarcomere Contraction. Points A, B, C, and D are the same as those labeled on the sarcomere photographs in Figure 46.17.

✔**QUESTION** According to the model shown here, why is the dark band in a sarcomere (see Figure 46.17) dark and the light band light?

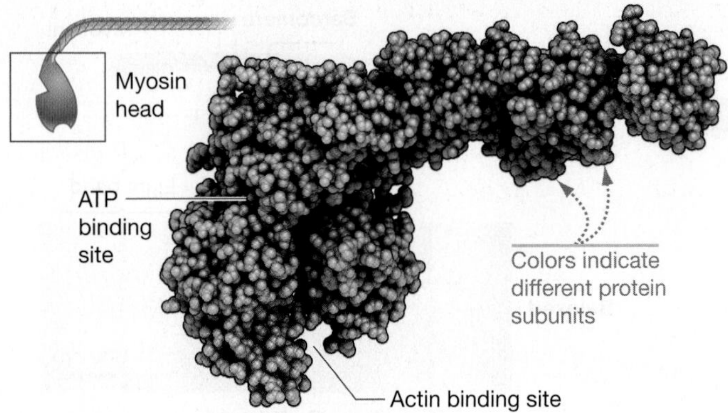

FIGURE 46.19 Myosin's "Head" Binds ATP and Actin.

HOW DO ACTIN AND MYOSIN INTERACT? How does this sliding action occur at the molecular level? Early work on the three-dimensional structure of myosin revealed that each molecule has a head region that projects from the main body of the thick filament. The myosin head can bind to actin, and the head region can catalyze the hydrolysis of ATP into ADP and a phosphate ion. These results suggested that myosin—not actin—was the site of active movement.

In addition, electron microscopy revealed that myosin and actin are locked together shortly after an animal dies and its muscles enter the stiff state known as rigor mortis. Because ATP is unavailable in dead tissue, the data suggested that ATP is involved in getting myosin to release from actin once the two molecules have bound to each other.

Later, Ivan Rayment and colleagues solved the detailed three-dimensional structure of the myosin head (**Figure 46.19**), using the X-ray crystallographic techniques introduced in **BioSkills 10** in Appendix A. Rayment's group determined the location of the actin binding site and examined how the protein's structure changed when ATP or ADP bound to it. As predicted, the protein's conformation changed significantly when bound to ATP versus ADP.

Based on these data, Rayment and co-workers proposed a four-step model for actin-myosin interaction (**Figure 46.20**):

Step 1 The myosin head of a thick filament is attached to ATP but not to actin in a thin filament. (This is the conformation that occurs when a muscle is relaxed.)

Step 2 When ATP is hydrolyzed to ADP and inorganic phosphate, the neck of the myosin straightens and the head pivots. The myosin head then binds to a new actin subunit farther down the thin filament.

Step 3 When inorganic phosphate is released, the neck bends back to its original position. This bending, called the power stroke, moves the entire thin filament.

Step 4 After ADP is released, a new ATP molecule binds to the myosin and causes myosin to release from actin.

As ATP binding, hydrolysis, and release continue, the two ends of the sarcomere are pulled closer together. (The transition from step 4 to step 1 cannot occur after death. Rigor mortis sets in as ATP supplies run out.)

The same basic mechanism is responsible for contraction in cardiac and smooth muscle cells, the amoeboid movement observed in

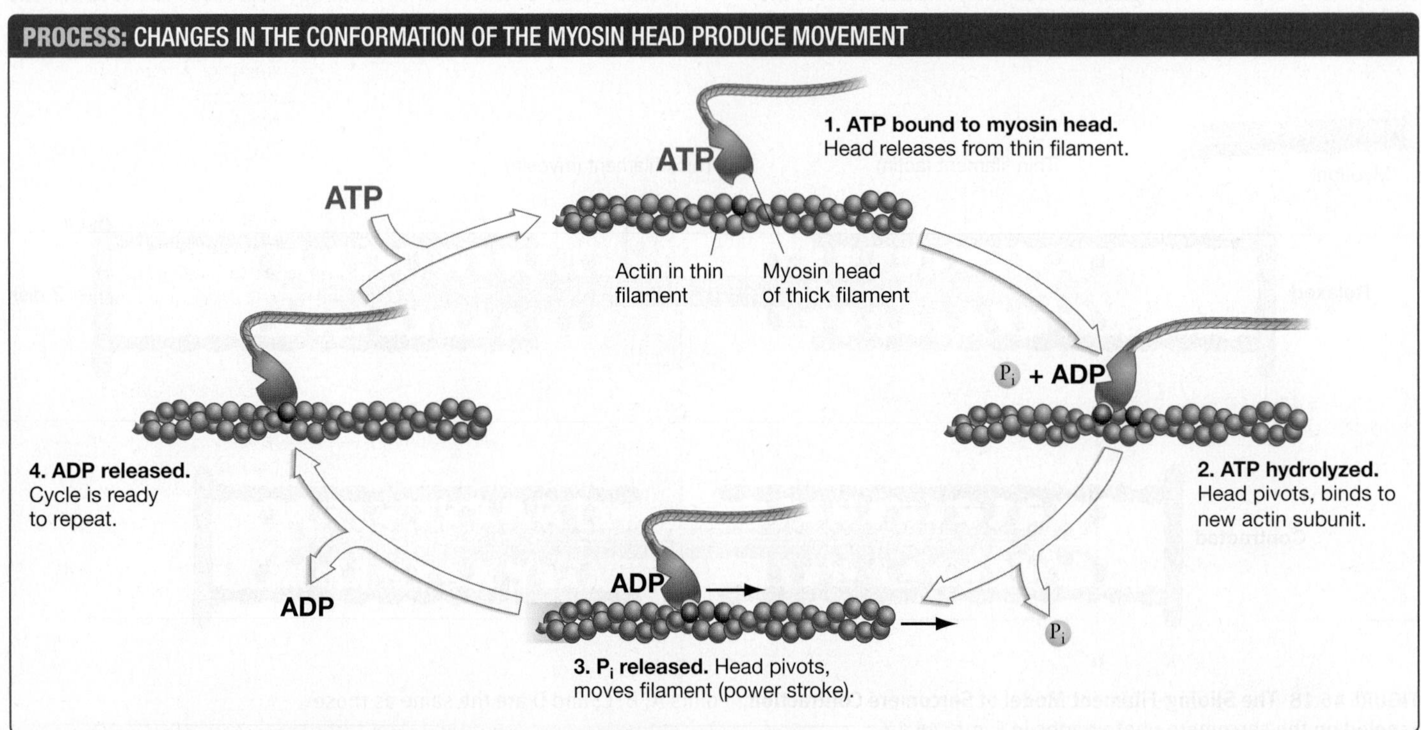

FIGURE 46.20 Myosin and Actin Interact during Muscle Contraction. Summary of the current model of how myosin and actin interact as a sarcomere contracts. The four steps repeat rapidly.

amoebae and slime molds (see Chapter 29) and the streaming of cytoplasm observed in algae and land plants. Actin and myosin have played a critical role in the diversification of eukaryotes, because they make movement possible in the absence of cilia and flagella.

HOW DOES RELAXATION OCCUR? ATP is almost always available in living muscles. How do our muscles ever stop contracting and relax?

In addition to containing actin, thin filaments contain two key proteins called **tropomyosin** and **troponin**. Tropomyosin and troponin work together to block the myosin binding sites on actin. The myosin-actin interaction cannot occur, and thick and thin filaments cannot slide past each other. Relaxation can occur (**Figure 46.21a**).

But when calcium ions bind to troponin, the troponin-tropomyosin complex moves in a way that exposes the myosin binding sites on actin. As **Figure 46.21b** shows, myosin then binds and contraction can begin.

How are calcium ions released so that contraction can begin? The answer turned out to have several steps, beginning with the arrival of an electrical signal from a motor neuron.

AN OVERVIEW OF EVENTS AT THE NEUROMUSCULAR JUNCTION **Figure 46.22** summarizes what happens when an action potential from a motor neuron arrives at a muscle cell and initiates contraction:

1. Action potentials trigger the release of the neurotransmitter **acetylcholine (ACh)** from the motor neuron into the synaptic cleft between the motor neuron and the muscle cell.
2. Acetylcholine diffuses across the synaptic cleft and binds to ACh receptors on the plasma membrane of the muscle cell. By recording voltage changes in muscle cells, biologists showed that a membrane depolarization occurs in response to ACh release. If enough ACh is applied to a muscle cell, depolarization triggers an action potential in the fiber itself.

(a) Tropomyosin and troponin work together to block the myosin binding sites on actin.

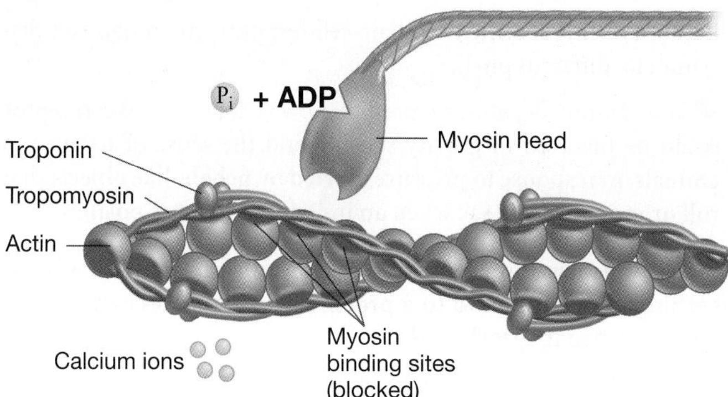

(b) When a calcium ion binds to troponin, the troponin-tropomyosin complex moves, exposing myosin binding sites.

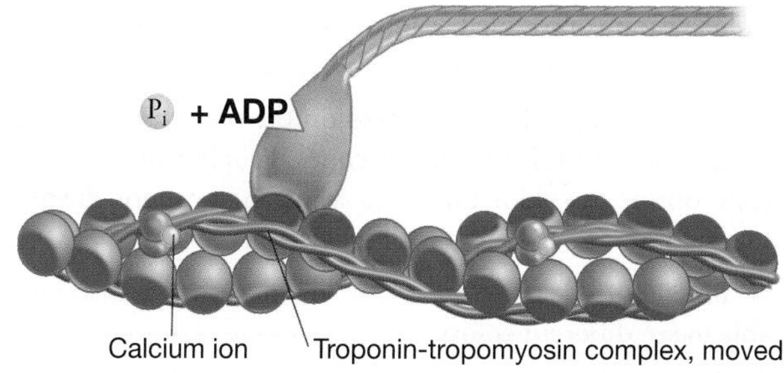

FIGURE 46.21 Troponin and Tropomyosin Regulate Muscle Activity.

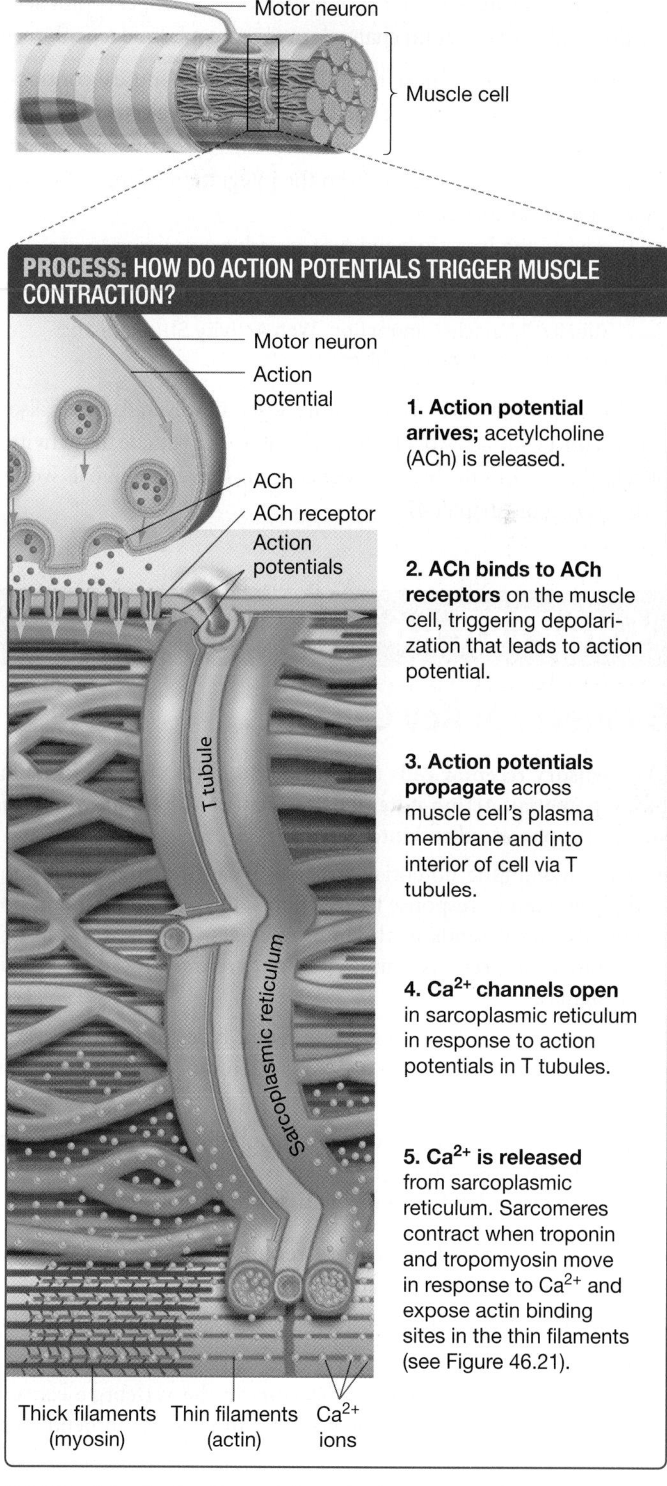

FIGURE 46.22 Action Potentials at the Neuromuscular Junction Trigger the Release of Ca^{2+}, Which Binds to Troponin-Tropomyosin and Allows Myosin to Form a Cross-Bridge with Actin.

3. The action potentials propagate along the length of the muscle fiber and spread into the interior of the fiber via invaginations of the muscle cell membrane called **T tubules**. (The *T* stands for *transverse*, meaning "extending across.")
4. T tubules intersect with extensive sheets of smooth endoplasmic reticulum called the **sarcoplasmic reticulum**. When an action potential passes down a T tubule and reaches one of these intersections, a protein in the T tubule membrane changes conformation and opens calcium channels in the sarcoplasmic reticulum.

Events at the neuromuscular junction produce muscle contraction and movement—one of the key responses to electrical signals from peripheral neurons, which in turn are triggered by signals from the CNS, which in turn result from the integration of input from a wide array of sensory cells.

To review the basic principles of muscle contraction, go to the study area at *www.masteringbiology.com.*

BioFlix™ Muscle Contraction, **Web Activity** Structure and Contraction of Muscle Fibers

Electrical signals are not the entire story, however. Animals also depend on a wide array of chemical signals to coordinate the activity of cells throughout the body. Understanding how these signals work is the subject of Chapter 47.

CHECK YOUR UNDERSTANDING

If you understand that . . .

- Antagonistic muscle groups work in conjunction with a skeleton to produce movement.
- Many types of movement are based on interactions between actin and myosin.
- Muscle cells contract in response to action potentials, which trigger chemical changes that allow actin and myosin to interact.

✓ You should be able to . . .

1. Describe the sliding-filament model.
2. Predict the effect on muscle function of drugs that have the following actions: increase ACh release at the neuromuscular junction, prevent conformational changes in troponin, and block uptake of calcium ions into the sarcoplasmic reticulum.

Answers are available in Appendix B.

CHAPTER 46 REVIEW

For media, go to the study area at www.masteringbiology.com

Summary of Key Concepts

Sensory receptor cells transduce stimuli to changes in membrane potential. Action potentials are sent to the brain, where the signals are processed and integrated.

- If the membrane potential of a sensory cell is altered substantially enough in response to a stimulus, the pattern of action potentials that it sends to the brain changes. In this way, sensory stimuli as different as sound and light are transduced to electrical signals.
- The brain is able to distinguish different types of stimuli because axons from different types of sensory neurons project to different regions of the brain.

✓ You should be able to suggest a hypothesis to explain why people who have had limbs amputated experience "phantom pain"—the perception that their missing tissue hurts.

Hearing is based on sensory receptor cells that move in response to sound waves of a particular frequency.

- Pressure receptors detect direct physical stimulation, including stimulation from sound.
- Hair cells, the major sensory detectors in the vertebrate ear, undergo a change in membrane potential in response to bending of their stereocilia.
- Sound waves of a certain frequency cause a certain part of the cochlea's basilar membrane to vibrate. Hair cells at this location fire action potentials in response to the vibration.
- Because of this specificity in hair-cell response, mammals can discriminate different pitches.

✓ You should be able to predict how a hair-cell–like receptor could be involved in gravity sensing and the sense of balance in animals, in response to pressure exerted by pebble-like objects that roll or move inside a sac when an individual changes position.

Vision is based on sensory receptor cells that contain a light-absorbing pigment bound to a protein. The pigment changes conformation when it absorbs light.

- In the vertebrate eye, photoreceptors are located in rods and cones. Although these two cell types differ in structure and function, both contain rhodopsin molecules that consist of retinal paired with opsin.
- The rhodopsin found in rods is stimulated by even the faintest light.
- Color vision is possible because cones contain opsins that respond to specific wavelengths of light absorbed by retinal.
- Humans distinguish colors based on the pattern of stimulation of three types of opsins found in cones. People who lack one of the cone opsins are color blind, meaning they cannot distinguish as many colors as people with all three opsins can.

✓ You should be able to explain why different animal species are able to see different colors.

Taste and smell sensations are registered by membrane proteins that act as ion channels or receptors for particular molecules.

- Chemoreceptors detect the presence of certain foodborne or airborne molecules.
- Taste buds contain taste cells with membrane proteins that respond to toxins, salt, acid, and other types of molecules in food. Sodium ions and protons enter taste cells via channels and depolarize the membrane directly, producing the sensations of saltiness and acidity. Sugars and toxic compounds bind to membrane receptors and trigger action potentials that are interpreted by the brain as sweetness and bitterness, respectively.
- Smell, or olfaction, is used to scan molecules from the outside environment. Airborne chemicals are detected by hundreds of different odor-receptor proteins located in the membranes of receptor cells in the nose.

✓ You should be able to explain why people with nasal congestion complain that food tastes bland, and why the brain can perceive so many different flavors based on inputs from just four or five basic types of taste receptors.

In many cases, animals respond to sensory stimuli by moving. Movement is based on antagonistic muscle groups that act on a skeleton. Muscle contraction occurs when myosin proteins move down the length of actin fibers.

- All animal muscles use the same basic mechanism for contraction.
- Muscles shorten when thick filaments comprised of myosin slide past thin filaments comprised of actin, in a series of binding events mediated by the hydrolysis of ATP.
- Calcium ions play an essential role in muscle contraction, by making the actin in thin filaments available for binding by myosin.
- In animals with exoskeletons or endoskeletons, muscles are usually arranged in opposing pairs of flexors and extensors. Many animals with hydrostatic skeletons have muscles arranged in opposing pairs of circular and longitudinal bands.

✓ You should be able to predict the primary symptom of botulism, which occurs when a toxin prevents release of ACh from the neuromuscular junction, and explain your answer.

(MB) **BioFlix™** Muscle Contraction, **Web Activity** Structure and Contraction of Muscle Fibers

Questions

✓ TEST YOUR KNOWLEDGE

Answers are available in Appendix B

1. What is echolocation?
   - **a.** use of echoes from high-frequency vocalizations to "see" objects
   - **b.** use of extremely low-frequency vocalizations to communicate over long distances
   - **c.** vision based on input from many independent lenses, functioning like pixels on a computer screen
   - **d.** variation in the structure of opsin proteins, which allows animals to see different colors
2. In the human ear, why do different hair cells respond to different frequencies of sound?
   - **a.** Waves of pressure move through the fluid in the cochlea.
   - **b.** Hair cells are "sandwiched" between membranes.
   - **c.** Receptors in the stereocilia of each hair cell are different; each receptor protein responds to a certain range of frequencies.
   - **d.** Because the basilar membrane varies in stiffness, it vibrates in certain places in response to certain frequencies.
3. Which of the following comparisons of rods and cones is *false*?
   - **a.** Most human eyes have one type of rod and three types of cones.
   - **b.** Rods are more sensitive to dim light than cones are.
   - **c.** There are more rods than cones in the fovea.
   - **d.** Both rods and cones use retinal and opsins to detect light.
4. Which of the following statements about taste is *true*?
   - **a.** Sweetness is a measure of the concentration of hydrogen ions in food.
   - **b.** Sodium ions from foods can directly depolarize certain taste cells.
   - **c.** All bitter-tasting compounds have a similar chemical structure.
   - **d.** Membrane receptors are involved in detecting acids.
5. In muscle cells, myosin molecules continue moving along actin molecules as long as
   - **a.** ATP is present and troponin is not bound to Ca^{2+}.
   - **b.** ADP is present and tropomyosin is released from intracellular stores.
   - **c.** ADP is present and intracellular ACh is high.
   - **d.** ATP is present and intracellular Ca^{2+} is high.
6. Which of the following is critical to the function of exoskeletons, endoskeletons, and hydrostatic skeletons?
   - **a.** Muscles interact with the skeleton in antagonistic groups.
   - **b.** Muscles attach to each of these types of skeleton via tendons.
   - **c.** Muscles extend joints by pushing them.
   - **d.** Segments of the body or limbs are extended when paired muscles relax in unison.

✓ TEST YOUR UNDERSTANDING

Answers are available in Appendix B

1. When a sound and an odor triggers a change in the pattern of action potentials from a sensory cell, how does the brain know which sense is which when the action potentials reach the brain?
2. Give three examples of how the sensory abilities of an animal correlate with its habitat or method of finding food and mates.
3. How did the discovery of odor-receptor genes affect our understanding of how the sense of smell works?
4. Explain how the bending of stereocilia in a hair cell and binding by a bitter-tasting molecule can both result in an ion channel opening.
5. Scientists generally think that a "good hypothesis" is one that is reasonable, testable, and inspires further research into the field. Using these criteria, was Dalton's hypothesis about color vision a good hypothesis? Was it correct? Explain your answer.
6. How did data on sarcomere structure inspire the sliding-filament model of muscle contraction? Explain why the observation that muscle cells contain many mitochondria and extensive smooth endoplasmic reticulum turned out to be logical once the molecular mechanism of muscular contraction was understood.

✔ APPLYING CONCEPTS TO NEW SITUATIONS

Answers are available in Appendix B

1. Myasthenia gravis is a disease that develops in humans when the immune system produces proteins that bind to the acetylcholine (ACh) receptors in muscles. The primary symptom of myasthenia gravis is muscle weakness. Why?

2. Houseflies have about 800 ommatidia in each of their compound eyes. Dragonflies, in contrast, have up to 10,000 ommatidia per eye. Houseflies feed by lapping up watery material from piles of excrement or rotting carcasses, which they locate by scent. Dragonflies are aerial predators and hunt by sight. How would you test the hypothesis that the large numbers of ommatidia in dragonflies makes them more efficient hunters?

3. Skeletal muscles contain two general types of cells. "Slow-twitch fibers" produce ATP via cellular respiration. They are reddish because they have a high concentration of myoglobin—a protein that delivers oxygen to electron transport chains during cellular respiration. Slow-twitch fibers support endurance exercise. "Fast-twitch fibers," in contrast, produce ATP primarily via fermentation. Fast-twitch fibers do not have large quantities of myoglobin and are light in color. They support extremely fast contraction but tire easily. Chickens, turkeys, and other ground-dwelling birds can run long distances but escape from predators by flying in short, extremely fast bursts. Based on these observations, explain the distribution of "dark meat" and "white meat" in these birds. Most other birds do not have white meat. Why?

4. When looking at faint stars through a telescope, astronomers focus their eyes just to the side of the object. Instead of landing on the fovea, then, the star's image falls next to the fovea. With this technique, faint objects pop into view. They vanish if looked at directly, however. Explain what is going on.

The spectacular transformation that occurs during insect metamorphosis is triggered by the chemical signals called hormones.

Chemical Signals in Animals 47

In response to sights, sounds, and other sensory stimuli, an animal's nervous system sends rapid messages, in the form of action potentials, to precise locations in the body. In many cases, these messages result in muscle contractions and movement.

In response to changes in external or internal conditions, cells in the central nervous system (CNS) or the endocrine system release certain molecules. These molecules produce longer-term responses in a broad range of tissues and organs.

The **endocrine system** is a collection of organs and cells that secrete chemical signals into the bloodstream. A chemical signal that circulates through body fluids and affects distant target cells is called a **hormone**.

As you read this, a large suite of hormones is coursing through your circulatory system. These molecules are regulating the maturation of sperm or eggs by your reproductive system, changing the composition of the urine forming in your kidneys, and controlling the release of digestive enzymes in your gastrointestinal tract. Earlier in your life, changes in hormone concentrations led to the dramatic physical changes associated with puberty.

The goal of this chapter is to explore how hormones and other types of chemical signals work in animals. Together, animal nervous systems and endocrine systems process information about the environment—one of the five key attributes of life introduced in Chapter 1.

Let's begin with an overview of chemical signaling systems, and then plunge into analyzing how hormones regulate the activity of target cells.

KEY CONCEPTS

- Animals use at least six major types of chemical signals. Hormones are chemical signals that are present in tiny concentrations and travel throughout the body to affect target cells.
- The information carried by hormones helps animals develop as embryos, undergo sexual maturation, respond to environmental change, and achieve homeostasis.
- The production of a hormone is tightly regulated by input from the nervous system and by other hormones.
- Some hormones bind to receptors inside target cells and change gene expression. Other hormones bind to receptors at the cell surface and lead to changes in protein activation.

47.1 Cell-to-Cell Signaling: An Overview

Animal chemical signals are present in extremely low concentrations but can have enormous effects on their target cells. Unlike action potentials, which are electrical impulses that have a short-term effect on a single cell or on a small population of adjacent cells, the messages that chemical signals carry have a relatively long-lasting effect.

✔ When you see this checkmark, stop and test yourself. Answers are available in Appendix B.

In combination, electrical and chemical signals allow animals to coordinate the activities of cells throughout the body. They are the mechanism responsible for maintaining trillions of cells as the integrated unit we call an individual.

Major Categories of Chemical Signals

The chemical signals found in animals have diverse structures and functions. **Table 47.1** summarizes how biologists go about organizing the diversity of chemical signals, based on where the molecules originate and where they act.

Notice that the names for most of the categories use the Greek word root *crin*, meaning "separated." Its use captures something essential about how chemical signals act: They are released from cells and thus are separated from them.

SUMMARY TABLE 47.1 **Six Categories of Chemical Signals in Animals**

Type of chemical signal	Source and target
Autocrine signals act on the same cell that secretes them.	
Paracrine signals diffuse locally and act on neighboring cells.	
Endocrine signals are hormones carried between cells by blood or other body fluids.	
Neural signals diffuse a short distance between neurons.	
Neuroendocrine signals are released from neurons but are carried by blood or other body fluids and act on distant cells.	
Pheromones are released into the environment and act on a different individual.	

It's important to note that these six classes of chemical messenger do not coincide with six structurally distinct classes of molecules. For example, the endocrine signals found in a particular organism routinely belong to several families of chemical compounds, ranging from amino acid derivatives to lipids. And a particular family of molecules—say, peptides or the lipids called steroids—may function as endocrine, autocrine, and paracrine signals in the same individual.

AUTOCRINE SIGNALS ACT ON THE SAME CELL THAT SECRETES THEM Translated literally, autocrine means "same-separated." The name is appropriate because **autocrine** signals affect the same cell that releases them.

Perhaps the best-studied autocrine signals are **cytokines** ("cell-movers"). Most cytokines amplify the response of a cell to a stimulus. An example is interleukin 2, which is synthesized and released by a type of white blood cell called a T cell in the course of fighting an infection. Interleukin 2 activates T cells to help eliminate the infection. It also causes the cells to divide repeatedly, producing more activated T cells for host defense.

PARACRINE SIGNALS ACT ON NEIGHBORING CELLS Translated literally, **paracrine** means "beside-separated." Paracrine signals diffuse locally and act on target cells beside the source cell. Cytokines, for example, may act as paracrine signals as well as autocrine signals, because they can trigger responses by other cells of the immune system.

It is common, in fact, to observe that a single chemical messenger can be assigned to more than one category of signal, based on its mode of action. Like some cytokines, the cell-cell signals named **insulin**, **glucagon**, and **somatostatin** cross categories. These molecules are produced by three distinct populations of cells within a region of the pancreas called the islets of Langerhans. The molecules act on nearby pancreatic cells as paracrine signals and ensure a smooth, steady response to changing blood-glucose levels. But they also act as hormones—meaning they target distant cells—in controlling the concentration of glucose in the blood.

ENDOCRINE SIGNALS ARE HORMONES **Endocrine** ("inside-separated") signals are carried between distant cells by blood or other body fluids. The cells that produce endocrine signals may be organized into discrete organs called **glands** or may be interspersed among the cells of other organs—such as the islets of Langerhans in the pancreas.

Because hormones are well studied and particularly important to understanding how animals work, they serve as the focus of this chapter. Many or most of the principles discussed are relevant to the other five categories of animal cell-cell signals as well, however.

NEURAL SIGNALS ARE NEUROTRANSMITTERS **Neural** signaling was introduced in Chapter 45. You might recall that when an action potential arrives at a synapse, it triggers the release of neurotransmitters that bind to receptors on the postsynaptic cell and induce a change in membrane potential—altering the tendency for the postsynaptic cell to fire action potentials. Changes in neural signaling are important in learning and memory.

(a) Endocrine pathway

Stimulus
Endocrine cell
Endocrine signal
Effector cell
Response
Feedback inhibition

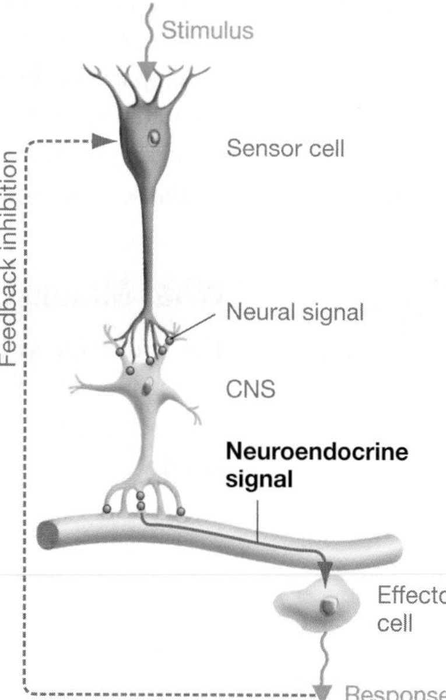

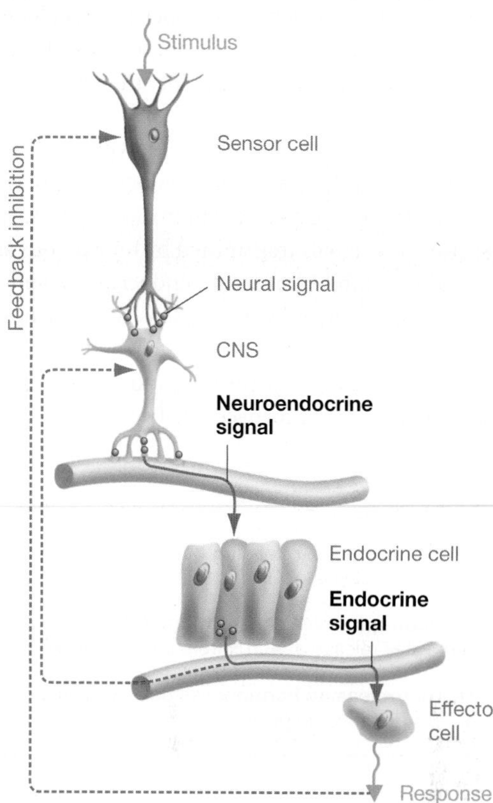

FIGURE 47.1 Hormones Act via Three Pathways and Are Regulated by Negative Feedback.

Neural signaling can be very fast, because neurotransmitters have to diffuse only a short distance—the tiny gap between two neurons, called the synaptic cleft. Neural signals are also short-lived, because the signaling molecules are broken down or taken back up by the presynaptic cell.

NEUROENDOCRINE SIGNALS ACT AT A DISTANCE Even though they are released from neurons, **neuroendocrine** ("nerve-inside-separated") signals are considered hormones. They share a key attribute with endocrine signals: They act on distant cells. Unlike neural signals, they do not act at the adjacent synapse.

Antidiuretic hormone (ADH; also called vasopressin) is a particularly well-studied neuroendocrine signal. ADH is produced by neurons that originate in the hypothalamus of the brain. But instead of acting as a neural signal, ADH acts on cells in the collecting duct of the kidney to help regulate water excretion (see Chapter 42). The source cells and target cells are inside the same body, but separated.

PHEROMONES ACT ON DIFFERENT INDIVIDUALS Pheromones are not "endo" signals. Instead, a **pheromone** is released into the environment and acts on a different individual.

In sea urchins, insects, and many other species, pheromones help coordinate reproduction in males and females or function in attracting mates. Chapter 46, for example, began with a description of how female moths release molecules into the environment that serve to attract males interested in courtship.

Other interindividual signaling molecules, such as the trail-marking compounds released by food-gathering ants, are also considered pheromones. Mice and other rodents even have a specialized region in the nose that is packed with pheromone-responsive chemosensory cells. Humans have a reduced version of the same region, and there is compelling evidence that at least some human pheromones exist. To date, however, none have been isolated and identified.

Hormone Signaling Pathways

In plants, sensory cells perceive changes in the environment and broadcast a hormonal signal that triggers an appropriate response from effector cells. Some animal hormones are also sent directly from endocrine cells to effector cells, in response to a stimulus (**Figure 47.1a**). But more frequently, hormonal signaling in animals involves additional steps.

In many or most cases, information about external or internal conditions is gathered by receptors and then integrated by neurons in the central nervous system (CNS) prior to the production of a hormonal signal. Neurons in the CNS respond by releasing neuroendocrine signals that:

1. act on effector cells directly (**Figure 47.1b**); or
2. stimulate cells in the endocrine system, which respond by producing a hormone (**Figure 47.1c**).

The second response is more common.

All three types of signaling pathway—direct from an endocrine cell, direct from CNS, or CNS-to-endocrine-system—are regulated by **negative feedback**, or **feedback inhibition**. When feedback inhibition occurs, the product of a process inhibits its production. As Chapter 41 pointed out, negative feedback is a key to homeostasis.

In the endocrine pathway, the product of the effector cells feeds back on endocrine cells, lowering production of the hormone and down-regulating the response (Figure 47.1a). An effector's response also feeds back to cells that initiate the neuroendocrine and neuroendocrine-to-endocrine pathways. A change in input from these cells then lowers production of the signal and reduces the response (see Figures 47.1b and 47.1c). Neuroendocrine-to-endocrine signaling pathways have an additional layer of regulation, because the endocrine signal usually inhibits production of the neuroendocrine signal (Figure 47.1c).

The take-home message? The nervous system and endocrine system are tightly integrated. Endocrine signals are released in response to electrical signals; in turn, endocrine signals modulate the electrical signals transmitted by the nervous system.

Feedback inhibition in the endocrine system is analogous to temperature control by a heat-sensitive thermostat. If the temperature is too high, the thermostat sends a signal that turns the furnace off; if the temperature is too low, the thermostat sends a signal that turns the furnace on. The result is a constant air temperature. In animal cell-cell signaling, feedback inhibition reduces production and/or secretion of the hormone.

✔If you understand this concept, you should be able to explain which parts of the thermostat analogy correspond to the sensory input, CNS, cell-to-cell signal, and effector in a hormone signaling pathway. You should also be able to predict what happens when feedback inhibition fails in a hormone signaling pathway.

What Makes Up the Endocrine System?

The endocrine system is the collection of cells, tissues, and organs responsible for hormone production and secretion. Organs that secrete a hormone into the bloodstream are called **endocrine glands**.

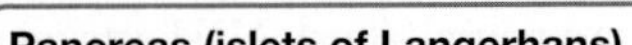

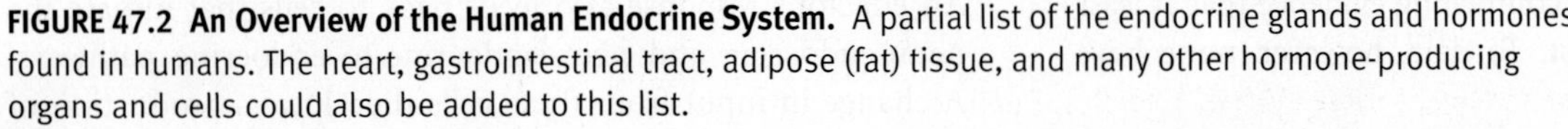

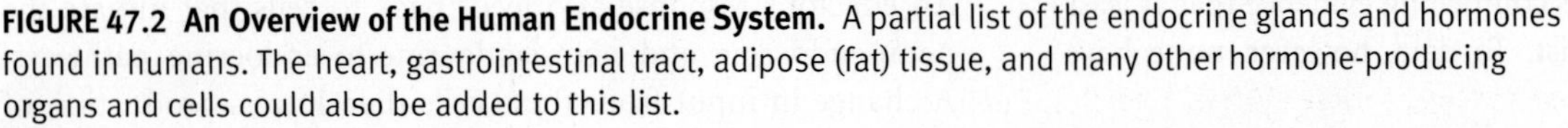

FIGURE 47.2 An Overview of the Human Endocrine System. A partial list of the endocrine glands and hormones found in humans. The heart, gastrointestinal tract, adipose (fat) tissue, and many other hormone-producing organs and cells could also be added to this list.

The tissues and organs that make up the endocrine system vary widely among animals. For example, neurons that manufacture and secrete hormones are particularly important in insects, where they regulate molting, metamorphosis, and other processes. Salmon have an unusual gland that secretes a hormone responsible for regulating calcium ion concentration.

Even within one species, the diversity of endocrine system components can be impressive. For example, **Figure 47.2** shows major human glands with endocrine functions.

- The **hypothalamus** is a region deep within the brain.
- The **pituitary gland** sits just below the hypothalamus and has distinct anterior and posterior regions.
- The **thyroid gland** is situated in the neck.
- The four **parathyroid glands** are embedded in the thyroid gland.
- The two **kidneys** lie in the posterior part of the abdominal cavity.
- The two **adrenal glands** sit atop the kidneys and have an outer cortex and a central medulla.
- The endocrine component of the **pancreas** is located in the anterior part of the abdominal cavity.
- The paired **ovaries** (in females) or **testes** (in males) are in or suspended below the pelvic cavity, respectively.

In many cases, however, hormone-secreting cells are not organized into discrete glands. Instead, they are located in other organs. For example, the intestine produces secretin, the heart produces atrial natriuretic hormone, and the cells of fat tissue produce leptin. Secretin stimulates the exocrine portion of the pancreas, **atrial natriuretic hormone** causes the kidney to excrete salt, and **leptin** helps regulate the amount of fat stored in the body.

Finally, it's important to note that not all glands in the body are part of the endocrine system. **Exocrine glands**, in contrast to endocrine glands, deliver their secretions through outlets called ducts into a space other than the circulatory system. Most of the digestive glands introduced in Chapter 43 are either exocrine glands, such as the salivary glands, or mixed endocrine and exocrine glands, such as the pancreas. The exocrine portions of the pancreas secrete digestive enzymes through ducts into the intestine. The endocrine portion of the pancreas consists of cells that secrete insulin and glucagon directly into the bloodstream.

At first glance, the diversity of hormones, glands, and effects can seem overwhelming, especially considering that Figure 47.2 represents just a partial catalog for a single species. The picture is simplified somewhat, however, by recognizing that most animal hormones belong to one of just three major structural families: polypeptides, amino acid derivatives, or steroids.

Chemical Characteristics of Hormones

Figure 47.3 illustrates the three major classes of chemicals that can act as hormones in animals:

1. polypeptides, which are chains of amino acids linked by peptide bonds (see Chapter 3);
2. amino acid derivatives; and
3. steroids, which are a family of lipids distinguished by a four-ring structure (see Chapter 6).

The secretin produced in the small intestine is a polypeptide; epinephrine is synthesized in the adrenal medulla from the amino acid tyrosine; and cortisol is synthesized in the adrenal cortex from the steroid cholesterol.

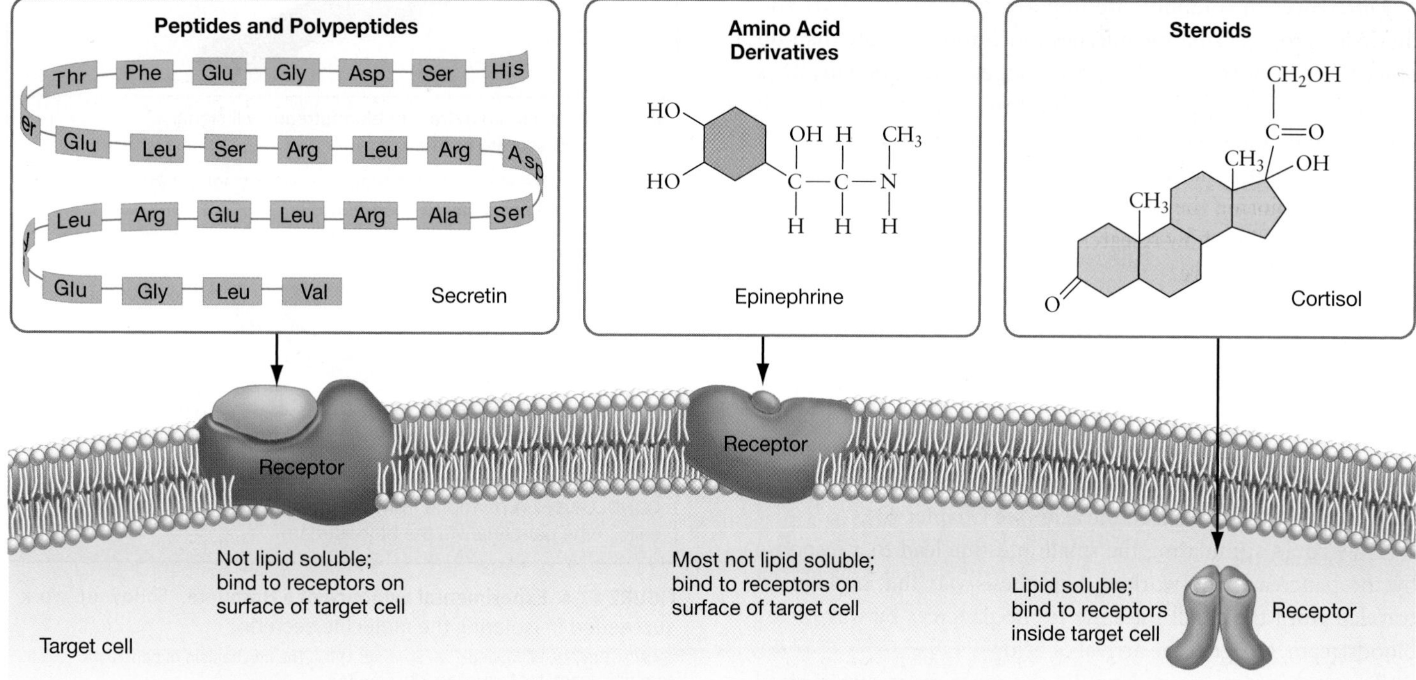

FIGURE 47.3 Most Animal Hormones Belong to One of Three Chemical Families.

HORMONE CONCENTRATIONS ARE LOW, BUT THEIR EFFECTS ARE LARGE Hormones vary widely in structure but share a common characteristic: they have profound effects on individuals, even though they are present at vanishingly small concentrations. As an example, consider work on **growth hormone (GH)**, also known as somatotropin.

Several researchers noted that rats and other laboratory animals stopped growing when their pituitary glands were removed. Based on this observation, it was widely suspected that the pituitary produces a chemical signal that promotes cell division and other aspects of growth.

To test this hypothesis, a research group purified a polypeptide from cow pituitary glands, injected the polypeptide into lab rats, and documented rapidly accelerated growth. When the researchers injected 0.01 mg of the molecule each day for nine days into rats that lacked pituitary glands, the width of the growth plates in the rats' leg bones increased by 50 percent. The individuals also gained an average of 10 g compared with rats that lacked a pituitary and did not receive the hormone treatment. Stated another way, an additional 0.09 mg of hormone led to a weight gain of 10,000 mg.

Further, 1 kg of cow pituitary tissue yielded a mere 0.04 g of growth hormone. By mass, the hormone makes up just four one-thousandths of 1 percent of the cow pituitary.

ONLY SOME HORMONES CAN CROSS CELL MEMBRANES Given that small amounts of polypeptide, amino-acid–derived, and steroid hormones have large effects on the activity of cells, organs, and systems, how do the three types of hormones differ? The major difference is that steroids are lipid soluble, while polypeptides and most amino acid derivatives are not (see Figure 47.3).

An important exception to this rule is the hormone **thyroxine**, which is produced by the thyroid gland. Thyroxine is derived from the amino acid tyrosine but is lipid soluble.

Differences in solubility are important because steroids and thyroxine cross plasma membranes much more readily than do other types of hormones. To affect a target cell, all polypeptides and most amino acid derivatives bind to a receptor on the cell surface. Lipid-soluble hormones, in contrast, can diffuse through the plasma membrane and bind to receptors inside the cell.

Before exploring the consequences of this difference in more detail, let's consider how researchers discovered the array of hormones introduced here.

How Do Researchers Identify a Hormone?

Research on animal hormones began in earnest in the early 1900s, when researchers found that adding dilute hydrochloric acid (HCl) to the small intestine, to mimic the arrival of acidic material from the stomach, stimulated the pancreas to secrete compounds that neutralized the acid (see Chapter 43).

How could stimulating the small intestine lead to a response by the pancreas? The working hypothesis was that a chemical traveled from the small intestine to the pancreas by way of the bloodstream, to signal the arrival of acid.

To test this idea, researchers did the experiment summarized in **Figure 47.4**. An extract from the small intestine was injected into blood vessels in a dog's neck. A short time later, the pancreas secreted an alkaline solution. This was strong evidence that the extract from the small intestine contained a cell-cell signaling molecule. The hormone was later purified and named **secretin**.

EXPERIMENT

QUESTION: How could stimulating the small intestine with acid cause the pancreas to secrete buffers that neutralize the acid?

HYPOTHESIS: A chemical messenger from the small intestine reaches the pancreas via the bloodstream.

NULL HYPOTHESIS: There is no blood-borne chemical messenger from the small intestine that acts on the pancreas.

EXPERIMENTAL SETUP:

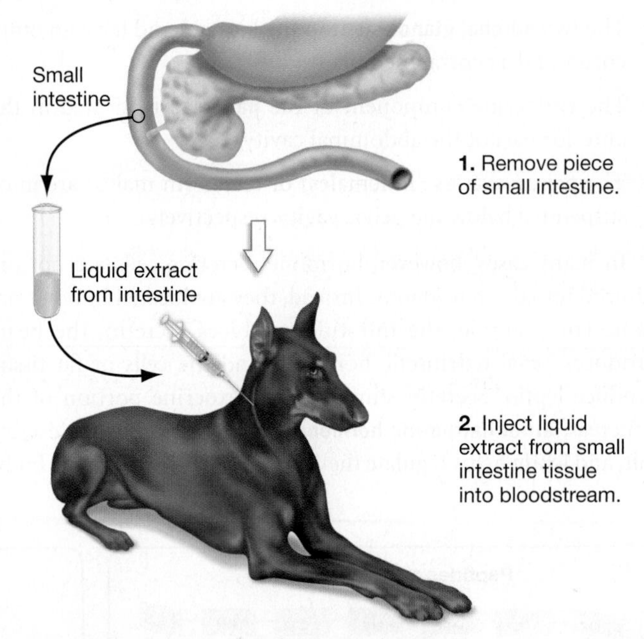

PREDICTION: Liquid extract in bloodstream will signal pancreas to secrete buffer.

PREDICTION OF NULL HYPOTHESIS: Liquid extract in bloodstream will not signal pancreas to secrete buffer.

RESULTS:

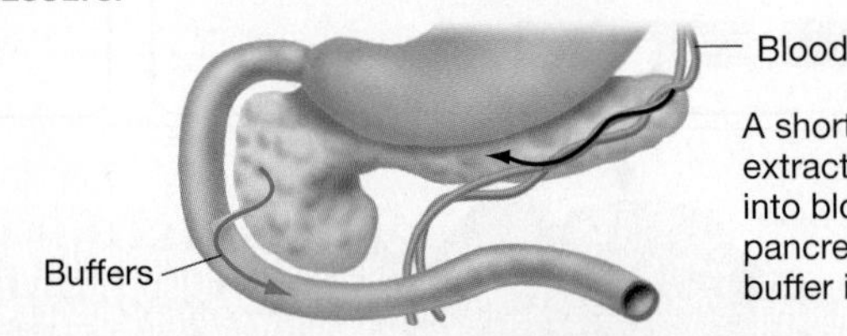

CONCLUSION: A hormone from the small intestine (secretin) signals cells in the pancreas via the bloodstream.

FIGURE 47.4 Experimental Evidence of a Hormone. Follow-up work succeeded in isolating the molecule secretin.

SOURCE: Bayliss, W. M. and E. H. Starling. 1902. The mechanism of pancreatic secretion. *Journal of Physiology* 28: 325–353.

✔**QUESTION** What would be an appropriate control in this experiment?

In essence, using liquid extracts from suspected endocrine glands is a way to add a suspected hormone and evaluate the effect. The opposite strategy has also been productive: observing the consequences of losing specific organs or tissues. For example, removing the adrenal glands of an animal rapidly leads to death due to low blood sodium, low blood sugar, and low blood pressure. But injecting adrenal extracts into the blood of animals lacking adrenal glands corrects these abnormalities. Results like these provided strong evidence that hormones secreted by the adrenal glands help regulate blood sugar concentrations and blood pressure.

Documenting an association between a particular gland or hormone and an effect in the body is just a first step, however. To understand hormone action, researchers have to figure out how these signals help animals stay alive and produce offspring.

47.2 What Do Hormones Do?

At the beginning of this chapter, you read that hormones are chemical messengers. If so, what do hormones "say"?

A first step in answering this question is to recognize that a single hormone can exert a variety of effects. For example, thyroxine stimulates metabolism in humans and thus oxygen consumption throughout the body. But it also promotes growth, increases heart rate, and stimulates the synthesis of many important macromolecules.

A second step in grasping what hormones do is to recognize that several different hormones may affect the same aspect of physiology. Insulin, glucagon, epinephrine, and cortisol all influence glucose levels in the blood.

Some hormones have extremely diverse effects; the functions of other hormones appear to overlap. These observations begin to make sense when hormone action is viewed in the context of the whole organism. Hormones coordinate the activities of cells in response to three situations: (1) development, reproduction, and growth; (2) environmental challenges; and (3) homeostasis.

Let's analyze each situation in turn.

How Do Hormones Direct Developmental Processes?

In animals, as in plants, hormones play a key role in regulating growth and development. Growth hormones and sex hormones play crucial roles in promoting cell division, increasing overall body size, and promoting sexual differentiation as an individual matures; certain hormones direct the development of particular cells and tissues at critical junctures in an individual's life.

Let's explore two of the most dramatic examples of hormonal control—metamorphosis in amphibians and in insects—and then survey other developmental processes that are affected by hormone action.

T_3's ROLE IN AMPHIBIAN METAMORPHOSIS Frogs, toads, and salamanders are called amphibians ("double-lives") because in most species juveniles live in water while adults live on land. The process of changing from an immature, aquatic tadpole to a sexually mature, terrestrial frog, toad, or salamander is an example of **metamorphosis** ("change-form"; **Figure 47.5**).

Two sets of complementary experiments, published in 1912 and 1916, established that frog metamorphosis depends on thyroid hormones. Researchers could induce frog tadpoles to undergo metamorphosis by feeding them ground-up thyroid glands from horses; they could prevent metamorphosis by surgically removing the tadpoles' thyroid glands.

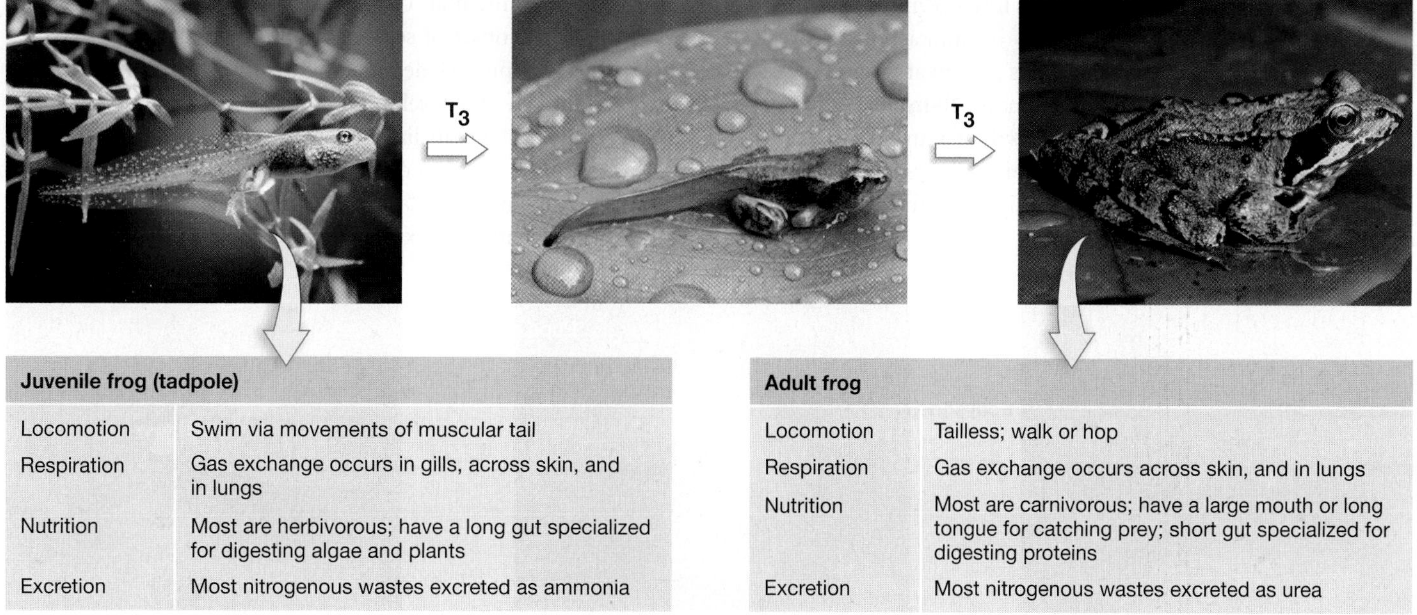

Juvenile frog (tadpole)	
Locomotion	Swim via movements of muscular tail
Respiration	Gas exchange occurs in gills, across skin, and in lungs
Nutrition	Most are herbivorous; have a long gut specialized for digesting algae and plants
Excretion	Most nitrogenous wastes excreted as ammonia

Adult frog	
Locomotion	Tailless; walk or hop
Respiration	Gas exchange occurs across skin, and in lungs
Nutrition	Most are carnivorous; have a large mouth or long tongue for catching prey; short gut specialized for digesting proteins
Excretion	Most nitrogenous wastes excreted as urea

FIGURE 47.5 Amphibian Metamorphosis Is a Continuous Process. When metamorphosis begins in a frog, toad, or salamander, the individual stays active and feeding. The continuous and gradual transition from juvenile to adult is mediated by T_3.

Follow-up work showed that the thyroid hormone **triiodothyronine**, or T_3, is responsible for most of the changes observed in metamorphosis. T_3 is produced in response to a signal from the brain: thyroid-stimulating hormone produced in the pituitary gland.

In juvenile amphibians, cells respond to increased levels of T_3 in one of three ways:

1. By growing and forming new structures, such as legs.
2. By dying, as structures—such as a tadpole's tail—disintegrate.
3. Or, by changing structure and function. For example, changes in existing cells are responsible for the switch from a tadpole's long intestine, specialized for digesting plant material, to an adult's short intestine, specialized for digesting insects and other prey. In the liver, cells respond to T_3 by manufacturing the enzymes required to excrete urea—the nitrogenous waste product released by adults—instead of the ammonia produced by tadpoles.

✔If you understand the basic principles of hormone action, you should be able to suggest a hypothesis explaining why different frog cells can respond to T_3 in such different ways.

HORMONE INTERACTIONS REGULATE INSECT METAMORPHOSIS Chapter 32 introduced a remarkable type of juvenile-to-adult transition in insects called holometabolous metamorphosis. In species that undergo this process, juveniles are called larvae. Insect larvae look completely different from adults, live in different habitats, and eat different food.

As larvae grow, they undergo a series of molts in which they shed their old exoskeleton, expand their bodies, and produce a new exoskeleton. After a specific number of these juvenile molts, however, they secrete a tough case called a pupa. Inside the pupal case, specific populations of larval cells give rise to a completely new adult body. The rest of the larval body is torn down (**Figure 47.6**).

In insects, metamorphosis depends on interactions between two hormones. If **juvenile hormone (JH)** is present at a high concentration in the larva, surges of the hormone **20-hydroxyecdysone**, usually called **ecdysone**, induce the growth of a juvenile insect via molting. But if JH levels are low, ecdysone triggers a complete remodeling of the body—metamorphosis—and the transition to adulthood and sexual maturity.

SEXUAL DEVELOPMENT AND ACTIVITY IN VERTEBRATES In mammals and other vertebrates, long distance cell-to-cell signals play key roles as embryos develop. Hormones also direct anatomical and physiological changes that occur later in life. Some of the most important of these changes involve the reproductive organs.

Events early in development dictate whether the sex organs, or **gonads**, of a vertebrate embryo become male (testes) or female (ovaries). This process is called primary sex determination. In mammals, primary sex determination does not depend on hormone action.

Once testes or ovaries develop, though, they begin producing different hormones. In human males, the early testes produce two hormones:

- A steroid hormone called **testosterone** induces early development of the male reproductive tract.
- A polypeptide hormone called **Müllerian inhibitory substance** inhibits development of the female reproductive tract.

In females, reproductive organs called ovaries produce the steroid hormone **estradiol**, which is in the family of molecules called **estrogens**. Estradiol is required for further development of the female reproductive tract.

Sex hormones also play a key role in the juvenile-to-adult transition. When humans reach early adolescence, for example, surges of sex hormones lead to the physical and emotional changes associated with **puberty**. These developmental changes create the adult phenotype and the ability to produce offspring.

In boys, surges of sex hormones lead to changes that include enlargement of the penis and testes and growth of facial and body hair. In girls, increased concentrations of estradiol lead to the enlargement of breasts, the onset of menstruation, and other changes.

The sex hormones continue to play a key role in adults. Most long-lived animals, for example, reproduce seasonally. In many species, environmental cues such as increasing day length, warmth, or the onset of seasonal rains trigger the release of sex hormones. Chapter 51 details how this flush of testosterone or estrogen induces the development of seasonal traits, such as sexual receptivity in female lizards.

Even though humans do not breed seasonally, sex hormones are instrumental in regulating sperm production and the menstrual cycle. Chapter 48 explores these processes in detail.

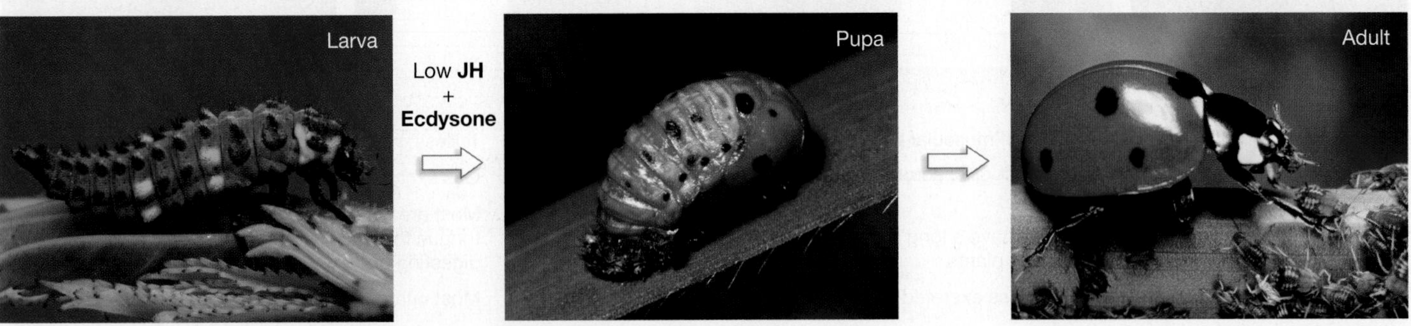

FIGURE 47.6 Insect Metamorphosis Occurs during a Resting Stage. When metamorphosis begins in a holometabolous insect, the individual enters a resting stage called the pupa.

GROWTH IN MAMMALS In humans and other mammals, the long bones in the limbs and the vertebrae of the spinal column must grow for full adult height to be achieved. This growth is stimulated by growth hormone (GH) produced in the pituitary gland. Chapter 19 introduced GH and its role in pituitary dwarfism. GH regulates growth factors, which are signals that control the cell cycle (see Chapter 11).

Puberty is associated with a growth spurt because the effect of growth hormone on the human skeleton is enhanced by the action of sex hormones, which are produced in increased amounts during adolescence. Even though growth hormone and sex hormones are produced in low levels throughout life, bone growth stops when sex hormone concentrations fall at the end of puberty.

How Do Hormones Coordinate Responses to Environmental Change?

The stimuli that hormones respond to can be simple or complex. Digestive hormones are a good example of how hormones function in simple stimulus-and-response circuits.

When acidic food material passes from the stomach to the upper reaches of the small intestine, the stimulus induces intestinal cells to release secretin and cholecystokinin into the bloodstream.

- Secretin induces the pancreas to secrete an alkaline solution that neutralizes acid.
- **Cholecystokinin** signals the pancreas to secrete digestive enzymes into the small intestine. Cholecystokinin also causes the gallbladder to eject bile salts into the intestine to emulsify fats.

In this way, digestive hormones signal the arrival of food and regulate the release of molecules that aid digestion. But what about more complex environmental stimuli?

SHORT-TERM RESPONSES TO STRESS When a person is thrust into a dangerous or unpredictable situation, hormones are involved in both the short-term and long-term responses.

The short-term reaction, called the **fight-or-flight response**, is triggered by the sympathetic nervous system (see Chapter 45). If you were being chased by a grizzly bear, action potentials from your sympathetic nerves would stimulate your adrenal medulla and lead to the release of **epinephrine**, also known as **adrenaline.** (The Greek word roots *epi* and *nephron* mean "top-kidney"; the Latin word roots *ad* and *renal* also mean "top-kidney.")

To determine how epinephrine affects the body, researchers injected human volunteers with a saline solution—as a control—or epinephrine. Each point graphed in the "Results" section of **Figure 47.7** represents data from one of the volunteers; the data in the table are average values from the seven study participants.

The data indicate dramatic increases in an array of physiological processes: concentrations of free fatty acids and glucose in the blood, pulse rate, blood pressure, and oxygen consumption by the brain. In addition, the volunteers reported strong subjective feelings of anxiety and excitement.

EXPERIMENT

QUESTION: How does epinephrine affect the body?

HYPOTHESIS: Epinephrine causes changes involved in the fight-or-flight response.

NULL HYPOTHESIS: Epinephrine does not affect the fight-or-flight response.

EXPERIMENTAL SETUP:

Control — Epinephrine

Saline solution

1. Inject human volunteers with saline solution or epinephrine.

2. Document changes in fatty-acid and glucose concentrations in blood, pulse rate, blood pressure, and oxygen consumption in brain.

PREDICTION: Epinephrine increases fatty-acid and glucose concentrations in blood, pulse rate, blood pressure, and brain oxygen consumption relative to controls.

PREDICTION OF NULL HYPOTHESIS: No differences in physiological state of individuals based on molecule injected.

RESULTS:

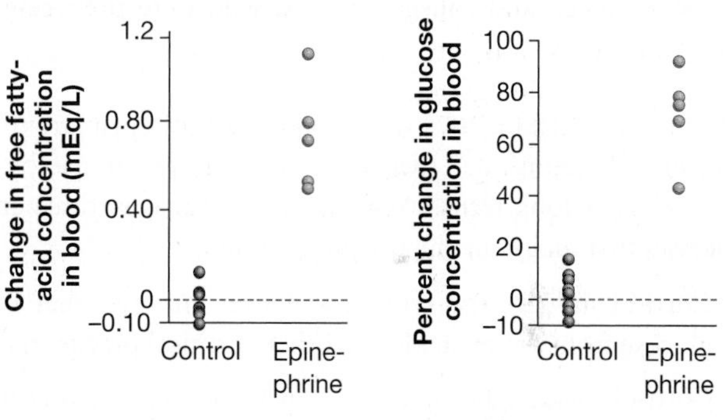

	Control	Epinephrine
Pulse rate (beats/min)	78.3	89.6
Blood pressure (average, mm Hg)	90.9	108.7
O_2 consumption in brain (cc O_2/100 g/min)	3.41	4.16

CONCLUSION: Epinephrine causes an array of changes associated with the fight-or-flight response.

FIGURE 47.7 Epinephrine Prepares the Body for Action.

SOURCES: King, B. D., L. Sokoloff, and R. L. Wechsler. 1952. The effects of *l*-epinephrine and *l*-nor-epinephrine upon cerebral circulation and metabolism in man. *Journal of Clinical Investigation* 31: 273–279. Mueller, P. S. and D. Horwitz. 1962. Plasma free fatty acid and blood glucose responses to analogues of norepinephrine in man. *Journal of Lipid Research* 3: 251–255.

✔**QUESTION** Why did researchers bother to inject volunteers in the control group with saline? Why inject them with anything?

Other experiments showed that epinephrine redirects blood away from the skin and digestive system and toward the heart, brain, and muscles. Epinephrine also relaxes smooth muscle and thereby opens blood vessels—increasing blood delivery to target tissues.

Taken together, the responses to epinephrine lead to a state of heightened alertness and increased energy use that prepares the body for rapid, intense action such as fighting or fleeing. By coordinating the activities of cells in many organs and systems throughout the body, epinephrine prepares an individual to cope with a life-threatening situation.

LONG-TERM RESPONSES TO STRESS If you have ever experienced the fight-or-flight response, you may recall that the state is short lived. Once an epinephrine "rush" wears off, most people feel exhausted and want to rest and eat.

What happens if the stress continues and turns into a long-term condition? In the course of a lifetime, it is not unusual for a person to experience periods of starvation or fasting, prolonged emotional distress, or chronic illness. How do hormones help humans and other animals cope with extended stress?

Early studies of long-term stress in human subjects suggested a role for the hormone **cortisol**, which is produced in the adrenal cortex. Increased levels of cortisol were found in airplane pilots and crew members during long flights, athletes who were training for intense contests, parents of children undergoing treatment for cancer, and college students who were preparing for final exams. Why?

WHAT DOES CORTISOL DO? In humans, cortisol's primary role is to ensure the continued availability of glucose for use by the brain, during long-term stress. Cortisol manages three main processes that maintain glucose production:

1. Cortisol induces the synthesis of liver enzymes that make glucose from amino acids and other chemical precursors.
2. Cortisol makes adipose tissue—fat tissue—and resting muscles resistant to insulin. Insulin normally stimulates **adipocytes** and resting muscle cells to remove glucose from the bloodstream. But when cortisol makes these cells resistant to insulin, glucose is reserved for use by the brain and exercising muscles.
3. Cortisol promotes the release of fatty acids—the body's major fuel molecules—from adipose tissue, for use by the heart and muscles.

Because of its importance in regulating blood glucose, cortisol is referred to as a **glucocorticoid**.

The long-term stress response comes at a high price, however, as any victim of a serious injury or illness knows.

- Glucocorticoids make amino acids available for glucose synthesis by promoting the degradation of contractile proteins in muscle. The resulting loss of muscle mass may cause severe weakness.
- Glucocorticoids impair wound healing and suppress immune and inflammatory responses. These processes are costly in terms of energy use, but reducing them makes the body more susceptible to infection.

The overall concept here is that the long-term stress response is a compromise—a fitness trade-off (see Chapter 41). The fuel requirements of the brain are met at the expense of other tissues and organs.

LONG-TERM STRESS IN NONHUMAN ORGANISMS Glucocorticoids may mediate the response to stressful environmental challenges in species other than humans. For example, you might recall from Chapter 42 that when salmon move from freshwater to salt water, individuals gain sodium ions via diffusion and lose water across the gills by osmosis. Salmon counteract these stresses by actively pumping sodium and chloride ions out of the gills.

The chloride cells that perform this pumping are induced to proliferate by a burst of cortisol. Cortisol release occurs as individuals move downstream toward the ocean. If this release fails to occur, salmon die shortly after reaching salt water.

How Are Hormones Involved in Homeostasis?

Chapter 41 introduced the concept of homeostasis, the maintenance of relatively constant physical and chemical conditions inside the body. Glance back at Figure 41.13 and recall that homeostatic systems depend on:

1. a sensory receptor that monitors conditions relative to a normal value, or set point;
2. an integrator that processes information from the sensor; and
3. effector cells that return conditions to the set point.

In homeostatic systems, messages often travel from integrators to effectors in the form of hormones.

LEPTIN AND ENERGY RESERVES Healthy animals keep energy in reserve for use during periods of decreased food availability. This energy reserve typically takes the form of the lipid triglyceride (see Chapter 6). **Triglyceride** is an effective energy-storage molecule because large amounts of ATP can be generated when its three fatty-acid subunits are oxidized.

Although some triglyceride is present in muscle cells, most is stored in adipocytes. These cells make up the fat bodies found in insects and other species and the adipose tissue of mammals. Even a lean 70-kg human male stores enough energy in adipose tissue to survive for over 30 days without eating.

Is there a homeostatic system that regulates triglyceride stores? An answer began to emerge in the 1970s, when biologists began studying mutations in mouse genes called *obese* and *diabetic*. Homozygous *obese* (*ob/ob*) and *diabetic* (*db/db*) mice eat large quantities of food and move much less than heterozygous or wild-type siblings do. The homozygotes also become extremely obese (**Figure 47.8**).

To test the hypothesis that the *obese* and *diabetic* gene products are involved in cell-cell signaling, researchers turned to an experimental technique called **parabiosis**, which works as follows:

- Two closely related animals are surgically united by suturing the pelvis, shoulders, and abdominal walls together. The skin of the two animals is joined over the surgical connection.

FIGURE 47.8 In Mice, Mutations in the *obese* and *diabetic* Genes Can Cause Obesity. These mice are siblings from the same litter. At the *ob* gene, the lean mouse is heterozygous and the obese mouse is homozygous. Note that the phenotype of this *ob/ob* mouse is the same as a *db/db* mouse.

- The newly created twin animal is allowed to recover from the procedure.
- Within a short time, capillaries form between the two parabiotic partners. As a result, the two individuals develop a shared circulatory system.
- The united circulatory system permits the passage of certain hormones and other long-lived molecules between the two partners, but not molecules that are rapidly metabolized, such as glucose and fatty acids.
- No new nerves grow between the two animals. As a result, they can influence one another through endocrine signals but not through electrical signals.

When researchers performed parabiosis between *db/db*, *ob/ob*, and lean mice, the results were striking (**Figure 47.9**):

1. When a *db/db* animal was joined to either a lean animal or an *ob/ob* animal, the *db/db* mouse continued to eat and grow normally, but its partner stopped eating, lost weight, and eventually died of apparent starvation.
2. When *ob/ob* animals were joined to lean animals, the *ob/ob* mice ate less food and gained weight less rapidly than did *ob/ob* mice joined to other *ob/ob* mice.

In addition, when two *db/db* animals were joined, both partners ate and grew as expected—both became obese. There was no difference between the two in survival.

To interpret these results, biologists hypothesized that mice produce a satiation, or "stop-eating," hormone—a negative feedback signal in homeostasis with respect to fat stores. The interpretation was that *db/db* mice lack the receptor required for the hormone to affect target cells.

As *db/db* mice got fatter and fatter, they would produce more and more hormone—to no avail. This model explained why the

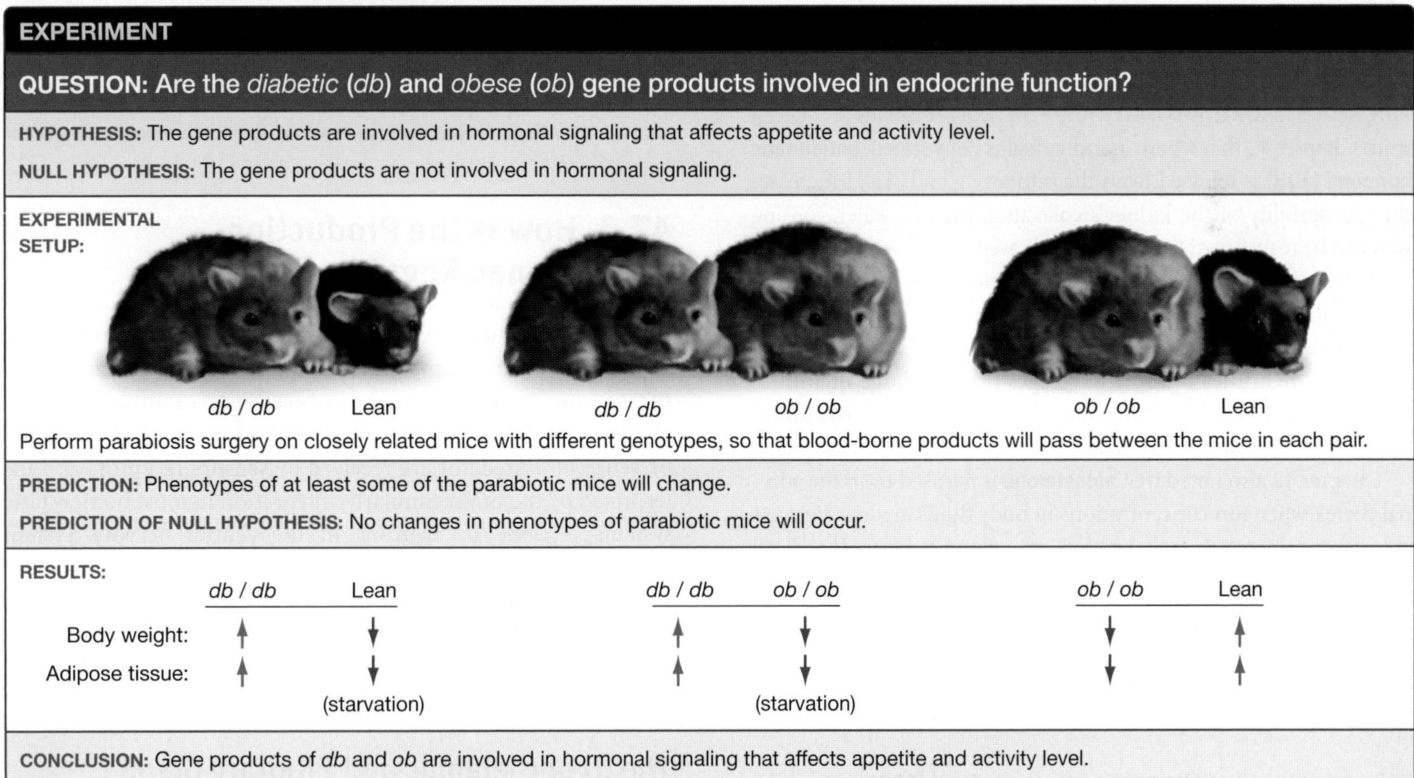

FIGURE 47.9 Parabiosis of Genetically Obese and Lean Mice Provides Evidence of a "Satiation Hormone."

SOURCE: Coleman, D. L. 1973. Effects of parabiosis of obese with diabetes and normal mice. *Diabetologia* 9: 294–298.

✔**QUESTION** Suppose a mouse were doubly homozygous *(ob/ob db/db)*. Would its phenotype be the same as or different from singly homozygous individuals? Explain your reasoning.

signal from *db/db* mice greatly reduced food intake in their parabiotic partners but had no effect on the *db/db* mouse.

The satiation hormone itself was postulated to be encoded by the *obese* gene. As a result, *ob/ob* mice do not produce the hormone. They respond to the signal if it is available from a normal partner, however.

A key idea here is that the two genotypes produce the same phenotype because they disable different parts of the same hormone-signaling system.

This model was confirmed in 1994, when the *obese* gene product, called **leptin**, was shown to be a polypeptide hormone. Leptin is secreted into the blood by adipocytes and interacts with a specific receptor located in many tissues, including areas of the brain known to control feeding behavior.

In mice, leptin injections correct the obesity of *ob/ob* (leptin-deficient) mice but not of *db/db* (leptin receptor–deficient) individuals. Unfortunately, leptin injections are not helpful for the vast majority of obese humans.

Follow-up work has shown that leptin levels in the blood vary in proportion to total adipose tissue mass. When adipose mass falls below a set point, the leptin level in the blood also falls. The brain senses the decrease in leptin level and generates both an increase in appetite and a decrease in energy expenditure. These responses promote eating and restore energy balance.

When sufficient food intake has occurred to restore triglyceride stores, leptin levels rise. The result is diminished appetite, increased energy expenditure, and the stabilization of adipose tissue mass. This is a striking example of homeostasis achieved by feedback inhibition.

ADH, ALDOSTERONE, AND WATER AND ELECTROLYTE BALANCE Recall from Chapter 42 that when an individual is dehydrated, **antidiuretic hormone (ADH)** is released from the pituitary gland. ADH increases the permeability of the kidney's collecting ducts to water, causing water to be reabsorbed from urine and saved.

ADH is instrumental in achieving homeostasis with respect to water balance. For example, the ethanol in alcoholic beverages inhibits the release of ADH from the pituitary. People who imbibe large quantities of these beverages produce large quantities of dilute urine. The resulting water loss can lead to dehydration and nausea—symptoms associated with an alcoholic hangover.

Chapter 42 also noted that **aldosterone** is released from the adrenal cortex when ion concentrations in body fluids are low. Because aldosterone increases reabsorption of sodium ions in the distal tubules of the kidney, it plays a key role in homeostasis with respect to electrolyte concentrations and overall volume of body fluids. Adrenal hormones with this effect are called **mineralocorticoids**.

ADH saves water; aldosterone saves sodium. Together, they are key players in maintaining water and electrolyte balance.

EPO AND OXYGEN AVAILABILITY **Erythropoietin (EPO)** is a crucial element in the homeostatic system for blood oxygen levels. When blood oxygen levels fall, the kidneys and other tissues release EPO, which stimulates the production of red blood cells. The more red blood cells, the higher the oxygen-carrying capacity of blood is. If you moved to a location at high elevation and experienced chronic oxygen deficit, your body would respond by releasing EPO and increasing red blood cell concentration.

Unfortunately, some endurance athletes have turned to EPO injections as a way to increase the oxygen-carrying capacity of their blood and give themselves a competitive edge. The practice is dangerous, as well as illegal. The increased viscosity of blood in EPO-abusers is accentuated during exercise, when blood plasma volume drops due to dehydration. The combination of high blood viscosity and low blood volume can impair blood flow through capillaries, increasing the risk of tissue damage and blood clotting. If clots form in blood vessels that lead to the heart or brain, heart attack or stroke may occur.

EPO abuse is thought to be responsible for the collapse and death of several cyclists during races in the mid-1990s. If so, EPO can give the sports phrase "sudden death" new meaning.

CHECK YOUR UNDERSTANDING

If you understand that . . .

- Hormones usually function in directing development and sexual maturation, preparing an individual for environmental challenges, or achieving homeostasis.

✓ You should be able to . . .

1. Explain how the various changes induced by elevated cortisol result in a coordinated response to long-term stress.
2. Explain how the overall effect of epinephrine increases the individual's chances of surviving and reproducing.

Answers are available in Appendix B.

47.3 How Is the Production of Hormones Regulated?

Most hormones are released in response to an environmental cue or a message from an integrator in a homeostatic system. Often, the nervous system is closely involved. For example, environmental cues that signal the onset of the breeding season or the presence of a predator are received by sensory receptors and interpreted by the brain. Similarly, integration in most homeostatic systems is done by neurons in the central nervous system (CNS)—the brain and spinal cord (see Chapter 45).

Based on these observations, the short answer to the question posed in the title of this section is simple: In many cases, hormone production is directly or indirectly controlled by the nervous system.

The Hypothalamus and Pituitary Gland

The pituitary, located at the base of the brain, is directly connected to the brain region called the hypothalamus. This physical link between the hypothalamus and pituitary is the basis of the connection between the CNS and the endocrine system.

The pituitary has two distinct segments: the **anterior pituitary** and the **posterior pituitary**. In 1930 a biologist documented the consequences of removing the entire pituitary in laboratory rats: The animals stopped growing, could not maintain a normal body temperature, and suffered atrophy (shrinkage) of their genitals, thyroid glands, and adrenal cortexes. Not surprisingly, their life span shortened dramatically.

These experiments suggested that, in addition to secreting growth hormone, the pituitary secretes hormones that regulate the production of a wide variety of other hormones. Based on this observation, the pituitary is nicknamed the "master gland." As a case study, let's look at the pituitary hormone that acts on the adrenal glands.

CONTROLLING THE RELEASE OF GLUCOCORTICOIDS Early work on rats suggested that a molecule from the pituitary gland affects the adrenal gland. This molecule soon came to be called **adrenocorticotropic hormone**, or **ACTH**. (*Adreno* refers to the adrenal glands; *cortico* refers to the outer portion, or cortex, of the gland; and *tropic* means "affecting the activity of.")

ACTH was purified and characterized in 1943. When human volunteers are injected with ACTH, cortisol levels in their blood rise (**Figure 47.10a**). This result supports the hypothesis that ACTH is a regulatory hormone. The adrenal cortex secretes glucocorticoids in response to ACTH released from the pituitary.

What regulates ACTH? Biologists from two laboratories independently showed that ACTH is released in response to a molecule produced by the hypothalamus. After years of effort, a different team of researchers succeeded in purifying a peptide—just 41 amino acids long—called **corticotropin-releasing hormone (CRH)**. When the hypothalamus releases CRH, it stimulates cells in the anterior pituitary to secrete ACTH into the bloodstream.

FEEDBACK INHIBITION BY GLUCOCORTICOIDS What *stops* glucocorticoid secretion? The key is to recognize that glucocorticoids themselves suppress ACTH production by the pituitary gland. Glucocorticoids accomplish feedback inhibition—they suppress their own production.

When human volunteers are injected with cortisol, ACTH levels in their bloodstream drop dramatically (**Figure 47.10b**). Cortisol also inhibits release of CRH from the hypothalamus. Thus, if glucocorticoid levels become too high, ACTH levels fall. But if glucocorticoid levels become too low, ACTH levels rise and drive a compensatory increase in glucocorticoid production. **Figure 47.10c** summarizes the relationships among CRH, ACTH, and the glucocorticoid cortisol.

What happens when feedback inhibition fails? Certain pituitary tumors diminish the ability of cortisol to suppress ACTH production, leading to persistently high blood ACTH and cortisol levels. The result is **Cushing's disease**: an unrelenting stress response that depletes the body's protein reserves. It is fatal if not treated.

PATTERNS IN GLUCOCORTICOID RELEASE Under ordinary circumstances, the production of CRH by the hypothalamus displays a daily rhythm, with the highest level in the early morning hours. This drives a corresponding daily rhythm in ACTH production and blood cortisol level.

(a) Results of injecting ACTH into human volunteers

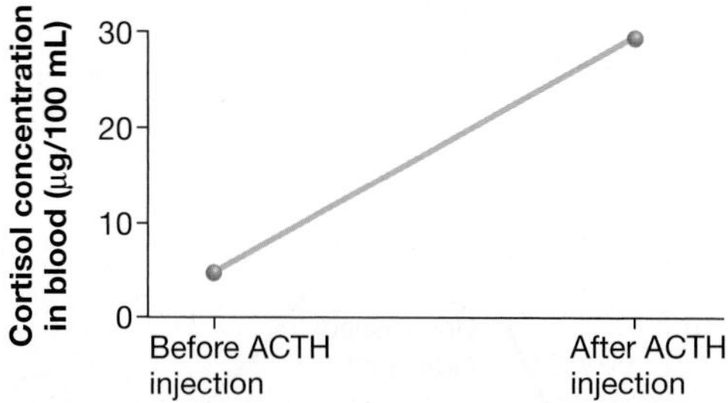

(b) Results of injecting cortisol into human volunteers

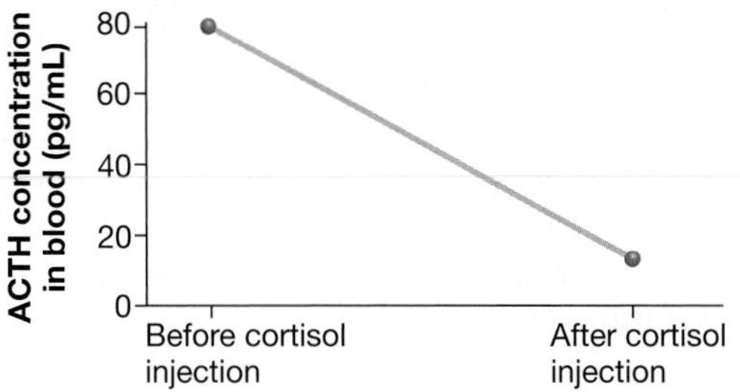

(c) Feedback inhibition by cortisol on ACTH

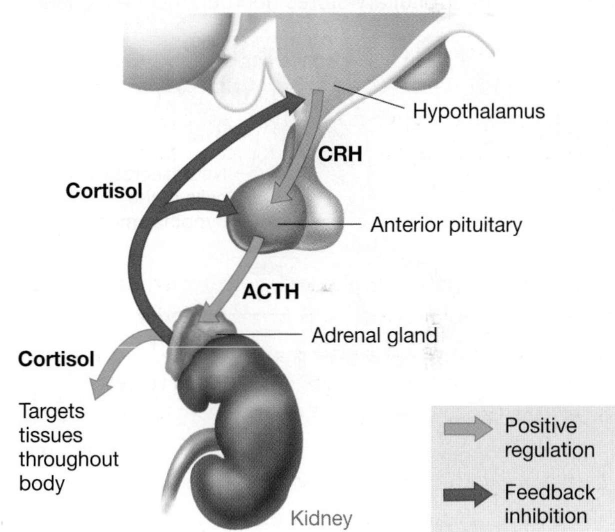

FIGURE 47.10 Feedback Inhibition by Cortisol on ACTH Release. **(a)** ACTH stimulates release of cortisol. **(b)** Cortisol reduces ACTH. **(c)** The interaction between cortisol, ACTH, and CRH is an example of feedback inhibition.

✔**EXERCISE** How would you use the data in part (a) to devise a test for adrenal failure in humans?

Ordinarily, the morning peak in blood cortisol levels coincides with arousal and initiation of the day's activities—with the effect of saving glucose for use by the brain. As an aside, the unpleasant symptoms of jet lag are due to the daily cortisol rhythm being out of synchrony with local time for several days after you arrive in a new time zone.

When the brain processes stimuli that produce pain or anxiety, however, it initiates a long-term stress response and a sustained

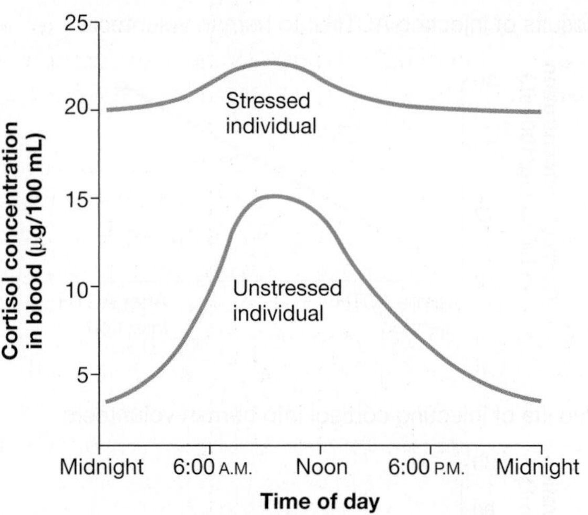

FIGURE 47.11 Blood Cortisol Levels Have a Daily Rhythm. Blood cortisol levels are usually highest in the early morning and drop in the late afternoon and evening. Under stressful circumstances, the morning rise still occurs but blood cortisol levels are much higher overall than in unstressed individuals.

increase in CRH production. Increased CRH production causes blood ACTH and cortisol levels to remain much higher throughout the day than they are in the unstressed state, as the graphs in **Figure 47.11** show. In analyzing these data, note that stressed individuals still show increased cortisol levels in the morning, though not the dramatic swing observed in unstressed individuals.

THE HYPOTHALAMIC-PITUITARY AXIS—AN OVERVIEW The CRH-ACTH-glucocorticoid relationship is just one of many hormone systems based on interactions among the hypothalamus, pituitary, and target glands or cells. The **hypothalamic-pituitary axis** actually forms two anatomically distinct systems (**Figure 47.12**). The anterior pituitary develops from cells in an embryo's mouth and throat lining; the posterior pituitary is an extension of the brain.

Two distinct populations of neurons in the hypothalamus influence the posterior versus anterior sections of the pituitary gland. Both types of hypothalamic neurons synthesize and release hormones and are called **neurosecretory cells.** The neurosecretory cells release hormones under the control of brain regions responsible for integrating information about the external or internal environment. For example, information about an upcom-

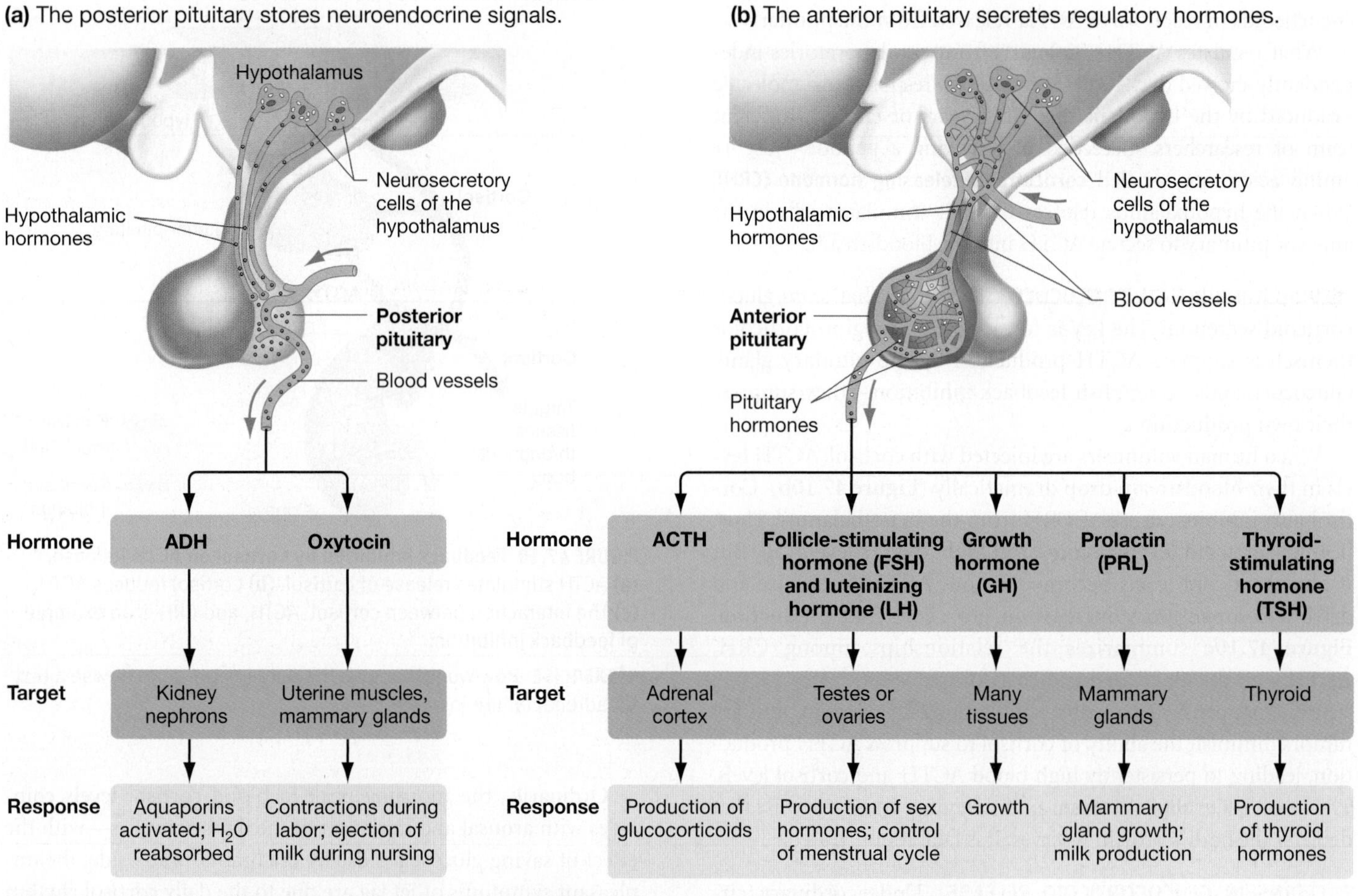

(a) The posterior pituitary stores neuroendocrine signals.

Hypothalamus

Neurosecretory cells of the hypothalamus

Hypothalamic hormones

Posterior pituitary

Blood vessels

Hormone	**ADH**	**Oxytocin**
Target	Kidney nephrons	Uterine muscles, mammary glands
Response	Aquaporins activated; H_2O reabsorbed	Contractions during labor; ejection of milk during nursing

(b) The anterior pituitary secretes regulatory hormones.

Neurosecretory cells of the hypothalamus

Hypothalamic hormones

Blood vessels

Anterior pituitary

Pituitary hormones

Hormone	**ACTH**	**Follicle-stimulating hormone (FSH) and luteinizing hormone (LH)**	**Growth hormone (GH)**	**Prolactin (PRL)**	**Thyroid-stimulating hormone (TSH)**
Target	Adrenal cortex	Testes or ovaries	Many tissues	Mammary glands	Thyroid
Response	Production of glucocorticoids	Production of sex hormones; control of menstrual cycle	Growth	Mammary gland growth; milk production	Production of thyroid hormones

FIGURE 47.12 The Hypothalamus and Pituitary Interact Closely. (a) Developmentally and anatomically, the posterior pituitary is an extension of the hypothalamus. Neurosecretory cells in the hypothalamus extend directly into the posterior pituitary and secrete ADH (vasopressin) and oxytocin. **(b)** The hypothalamus and the anterior pituitary communicate indirectly, via blood vessels. Hormones produced by neurosecretory cells in the hypothalamus travel in the blood to the anterior pituitary, where they control the release of pituitary hormones.

ing exam or athletic contest might trigger action potentials that lead to the release of CRH.

THE POSTERIOR PITUITARY Even though both parts of the gland contain neurosecretory cells, the anterior pituitary and posterior pituitary function in different ways. As Figure 47.12a indicates, the posterior portion of the pituitary is an extension of the hypothalamus itself.

Neurosecretory cells that project from the hypothalamus produce the hormones ADH and oxytocin, which are then stored in the posterior pituitary. From there, ADH and oxytocin are released into the bloodstream. This is an example of the neuroendocrine pathway of hormone action. Recall that ADH aids in the reabsorption of water by the kidneys. **Oxytocin** helps induce labor and milk release in females.

THE ANTERIOR PITUITARY In contrast to the situation in the posterior pituitary, the hypothalamus and anterior pituitary are connected indirectly. Neurosecretory cells from the hypothalamus secrete stimulatory or inhibitory neuroendocrine signals into blood vessels, which then carry the signals to the anterior pituitary. In response, the anterior pituitary alters the secretion of hormones that enter the bloodstream and act on target tissues or glands. This is an example of the neuroendocrine-to-endocrine pathway of hormone action.

Many of the hormones produced by the anterior pituitary stimulate the production of other hormones—justifying the structure's designation as the master gland. The anterior pituitary hormones include ACTH; **follicle-stimulating hormone (FSH)** and **luteinizing hormone (LH)**, which are involved in producing sex hormones and regulating the menstrual cycle; GH; **prolactin**, which stimulates mammary gland growth and milk production; and **thyroid-stimulating hormone (TSH)**, which triggers the production of thyroid hormones.

To review the major structures in the endocrine system and get a closer look at the structure and function of the posterior and anterior pituitary, go to the study area at *www.masteringbiology.com.*

MB **Web Activity** Endocrine System Anatomy

Control of Epinephrine by Sympathetic Nerves

When biologists analyze how the nervous system and endocrine system interact to control the release of epinephrine, the distinction between the two systems begins to blur. Section 47.2 introduced how epinephrine acts as an endocrine signal. During the fight-or-flight response, sympathetic nerves trigger the release of epinephrine from the adrenal medulla into the bloodstream. But in addition, some sympathetic nerves release the related molecule **norepinephrine** directly onto target cells.

In effect, the endocrine system broadcasts a similar messenger by secreting it into the bloodstream while the nervous system delivers a chemical messenger directly to particular cells. Epinephrine and norepinephrine, which differ from one another only by the presence of an additional methyl group on epinephrine, are members of the family of molecules called **catecholamines**.

Catecholamines function as neurotransmitters as well as hormones. Similarities between hormones and neurotransmitters do not end there, however. In some cases, their mode of action is similar.

Chapter 45 noted that some neurotransmitters initiate changes in gene expression in neurons. You might recall that in the sea slug *Aplysia*, repeated application of serotonin led to gene activation and changes in the behavior of the synapse. By altering synapses in this way, neurotransmitters play a central role in learning and memory.

Similarly, many hormones exert their effects by activating particular genes in target cells. Understanding how gene activation occurs in response to neurotransmitters and hormones is the subject of intense research at laboratories around the world. It is also the subject of Section 47.4.

CHECK YOUR UNDERSTANDING

 **If you understand that . . .**

- Hormone concentrations are tightly regulated—in some cases by negative feedback, in other cases by stimulatory or inhibitory signals from the hypothalamus.

✓ **You should be able to . . .**

1. Explain how feedback inhibition occurs in production of ACTH.
2. Discuss the relationship between processing centers in the brain, neurosecretory cells in the hypothalamus, and hormone-secreting cells in the anterior pituitary.

Answers are available in Appendix B.

47.4 How Do Hormones Act on Target Cells?

The key to understanding how hormones act on target cells is to recognize that only some animal hormones are lipid soluble and cross plasma membranes readily (see Figure 47.3). More specifically, steroid hormones are small lipids that enter cells without difficulty. But the peptide and polypeptide hormones—as well as most amino acid derivatives—do not cross plasma membranes easily because of their large size and electrical charge.

Differences in the lipid solubility of hormones are important because they influence where a target cell receives the chemical message. Steroids often act inside the cell, while most amino acid derivatives and all polypeptides act at the cell surface.

To compare these two distinct paths of hormone action, let's consider how estrogen and epinephrine affect target cells. As a steroid and a nonsteroid, they serve as model systems for target cell responses to hormonal signals.

Steroid Hormones Bind to Intracellular Receptors

Estrogens are steroids that direct the development of female secondary sex characteristics in many animal species. In humans and other mammals, the most important estrogen is the molecule estradiol (formally, 17β-estradiol).

Because of its importance in reproduction by humans and domesticated animals, estradiol's mode of action has been the topic of intense investigation for over 50 years. How do target cells receive the signal carried by estradiol?

IDENTIFYING THE ESTRADIOL RECEPTOR In 1964, biologists succeeded in isolating the estradiol receptor in laboratory rats. **Figure 47.13** indicates the experimental approach. In essence, the research team introduced labeled hormone molecules into females, then used sucrose density centrifugation—a technique introduced in **BioSkills 11** in Appendix A—to separate molecules in target cells by size.

When centrifugation was complete, the biologists found that the labeled estradiol was concentrated in a narrow band comprised of the labeled hormone bound to its receptor. After purifying the receptor molecule, they found that proteinase enzymes could destroy it. Based on this result, they inferred that the estradiol receptor was a protein.

Follow-up experiments established that the estradiol receptor is located in the nucleus but is not associated with the nuclear envelope. Further, the receptor is found only in estradiol target tissues, including the uterus, hypothalamus, and mammary glands. The latter finding was particularly exciting, because it clarified how hormones act in a tissue-specific way.

This is a crucial point: Hormones are broadcast throughout the body via the bloodstream, but they act only on cells that express the appropriate receptor. Target cells respond to a particular hormone because they contain a receptor for that hormone.

Later, biologists found that the gene for the estradiol receptor is similar to the genes that encode receptors for the glucocorticoids, testosterone, and other steroid hormones. This result suggested that all steroid receptors are descended from an ancestral receptor molecule, and that the binding of any steroid hormone to its receptor affects the target cell in a similar way.

What happens once the hormone-receptor complex forms inside the nucleus of a target cell?

DOCUMENTING CHANGES IN GENE EXPRESSION During the 1970s and 1980s, work in several laboratories suggested that estradiol and other steroid hormones affect gene transcription after they bind to their receptors. For example, researchers injected laboratory animals with estradiol or other steroid hormones and documented changes in the mRNAs and proteins produced in target cells. These data showed that steroid hormones can cause dramatic changes in the amount or timing of mRNA production by a large number of genes.

How? The estradiol receptor, like other members of the **steroid-hormone receptor** family, has a distinctive DNA-binding region called a zinc finger. DNA-binding domains are sections of a protein that make physical contact with DNA. The presence of a zinc finger in the estradiol receptor suggested that once estradiol had bound to it, the hormone-receptor complex might affect gene expression by binding directly to DNA.

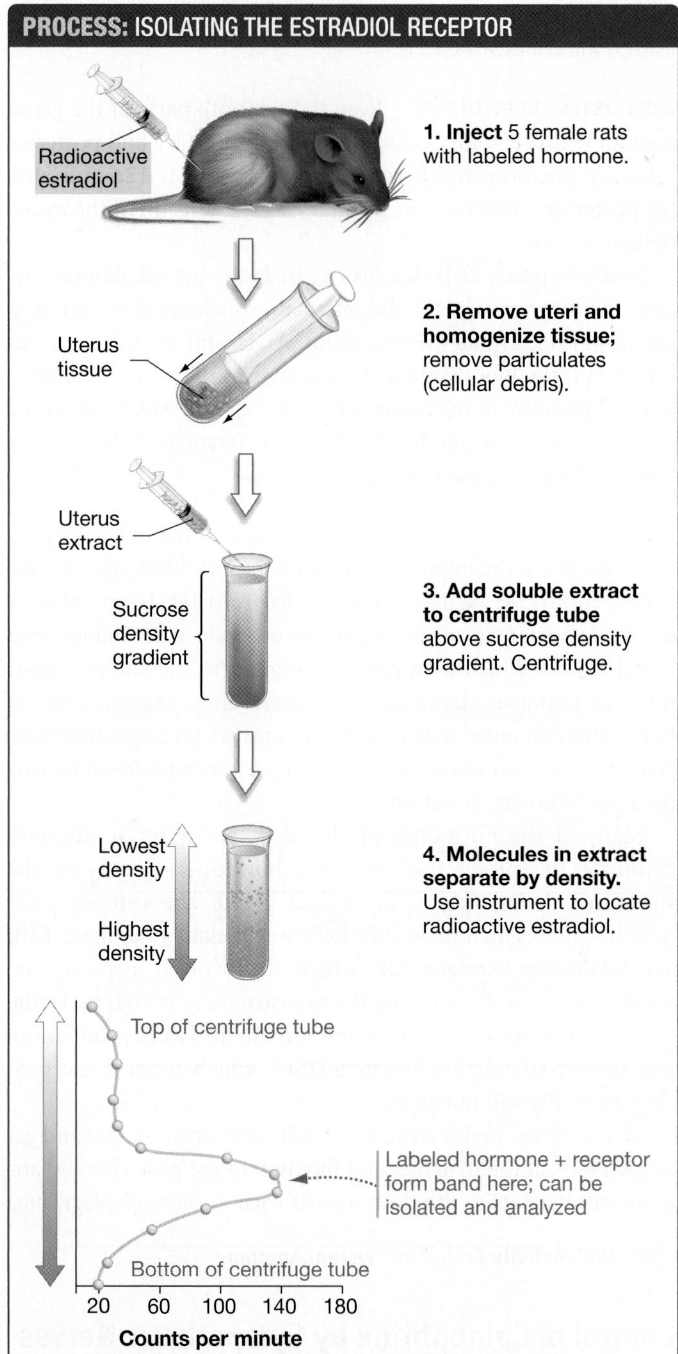

FIGURE 47.13 Labeled Hormones Can be Used to Find Hormone Receptors. If radioactive estradiol binds to its receptor in the uterus, the hormone-receptor complex should form a distinct band of radioactivity when molecules from uterine cells are separated by centrifugation.

Follow-up work confirmed that steroid hormone-receptor complexes bind to specific sites in DNA called **hormone-response elements.** Hormone-response elements are located just "upstream" (in the 5′ direction) from the start of target genes. Gene expression changes when a regulatory protein such as a steroid hormone-receptor complex binds to the hormone-response element for that gene.

Figure 47.14 summarizes the current model of how steroid hormones affect target cells.

Step 1 Estradiol or another steroid hormone enters a cell.

Step 2 In target cells, the hormone binds to its receptor. The binding event causes a conformational change in the receptor.

Step 3 The hormone-receptor complex binds to DNA and stimulates transcription.

Step 4 Many mRNAs are produced.

Step 5 Each mRNA is translated many times.

Because each hormone-receptor complex leads to the production of many copies of the gene product, the signal from the hormone is amplified. In this way, a small number of hormone molecules produces a large change in the activity of target cells and tissues.

Currently, researchers are following up on the discovery that at least some mammals, including humans, have two distinct receptors for estradiol. As experiments continue, it will be interesting to learn whether these two receptors trigger different responses to the same hormone.

Hormones That Bind to Cell-Surface Receptors

Unlike steroids, most polypeptide hormones and amino-acid–derived hormones are not lipid soluble. For these molecules to affect a cell, they must bind to receptors on the cell surface. Because the messenger never enters the target cell, its message must be transduced—changed into a form that is active inside the cell. Recall from Chapter 8 that this phenomenon is known as **signal transduction**.

To explore how signal transduction occurs, let's first examine hormone receptors that reside in the plasma membrane, then explore the molecules that process the message inside the cell. In both cases, we'll focus on epinephrine as a model system.

IDENTIFYING THE EPINEPHRINE RECEPTOR In 1948 a biologist published an exhaustive set of studies on how epinephrine affects dogs, cats, rats, and rabbits. The responses fell into two distinct categories, depending on the tissue being considered. To explain this observation, the researcher suggested that epinephrine binds two distinct types of receptor. He called these hypothetical proteins the alpha receptor and the beta receptor.

Follow-up work with molecules that block epinephrine receptors documented that there are actually two types of alpha receptors and two types of beta receptors. Thus, there are four distinct epinephrine receptors. Each is found in a distinct tissue type, and each induces a different response from the cell.

The discovery of four epinephrine receptors reinforces the concept of tissue specificity observed in experiments on the estradiol receptor. Hormones are transmitted throughout the body, not unlike a cell phone signal that is broadcast through the atmosphere. But their message is received only by cells with the appropriate receptor—just as a cell phone signal is received only by equipment with the appropriate antenna. Since there are four distinct epinephrine receptors, the same hormone can trigger different effects in different cells. What happens once epinephrine binds to one of these receptors?

WHAT ACTS AS THE SECOND MESSENGER? Signal transduction occurs when a chemical message at the cell surface triggers a response inside the cell. Cell-surface receptors "read" hormonal messages and initiate an appropriate response.

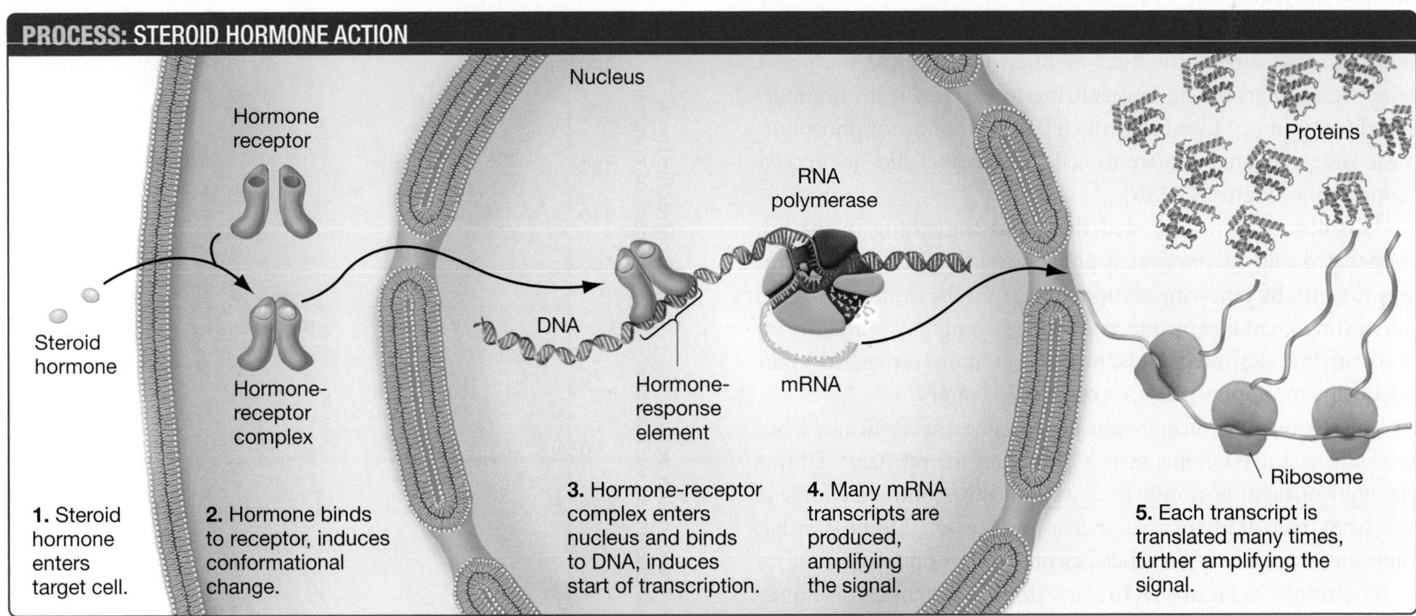

FIGURE 47.14 Steroid Hormones Bind to Receptors inside Target Cells and Change Gene Expression.

✔**QUESTION** Tamoxifen is a drug that blocks estrogen receptors in breast tissue cells. (Estrogen stimulates growth of breast cells, so tamoxifen is often prescribed as a treatment for breast cancer.) On this diagram, which arrow does tamoxifen block?

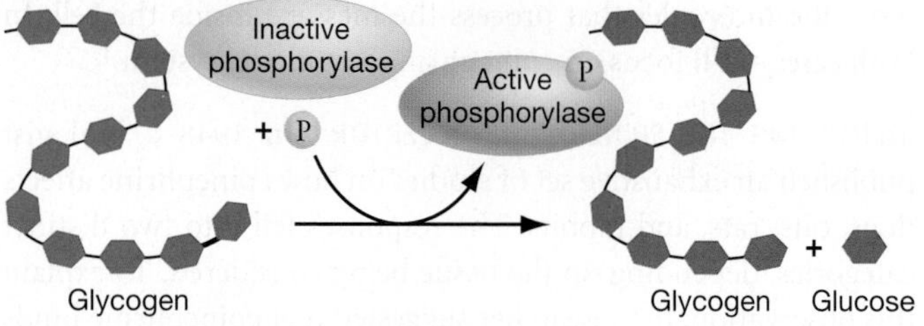

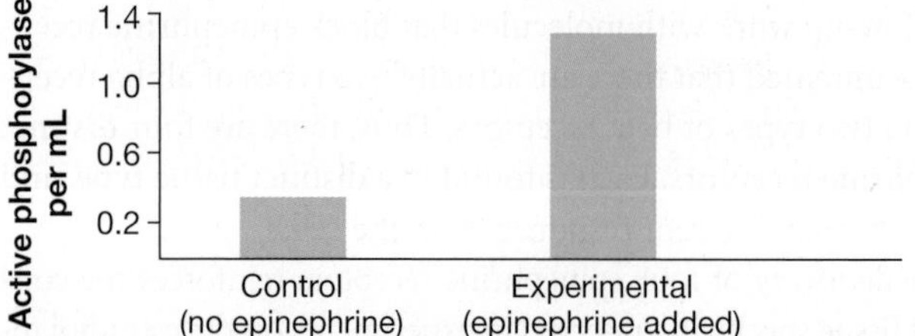

FIGURE 47.15 Epinephrine Activates the Enzyme That Catalyzes the Formation of Glucose from Glycogen. **(a)** Phosphorylase is activated when an enzyme adds a phosphate group to it. **(b)** When epinephrine is added to cell-free extracts from liver tissue, the amount of activated phosphorylase increases dramatically.

How does a signal from epinephrine increase glucose levels in the blood? To answer this question, biologists focused on the enzyme **phosphorylase**, which catalyzes a reaction that cleaves glucose molecules off glycogen (**Figure 47.15a**). Phosphorylase exists in active and inactive forms; the enzyme switches between these states when it is phosphorylated or dephosphorylated by another enzyme.

Phosphorylase is present in liver cells—the primary source of blood glucose during the fight-or-flight response. As predicted, when researchers added epinephrine to extracts from homogenized (ground up) liver cells, much larger amounts of phosphorylase were activated relative to cell extracts that did not receive epinephrine (**Figure 47.15b**).

This observation suggested there was something in the homogenized cells that activated phosphorylase when epinephrine was present. By purifying components from the liver cell extracts and testing them one by one, researchers eventually found the ingredient that activated phosphorylase: a molecule called cyclic adenosine monophosphate, or **cyclic AMP (cAMP)**.

The role of cAMP in epinephrine signaling was confirmed when researchers studied epinephrine's effects on the rat heart. During the fight-or-flight response, heart rate and the contractile force of the heart rise dramatically—increasing cardiac output. Cardiac output is a measure of the efficiency of the heart's pumping.

The graphs in **Figure 47.16** show three ways that cardiac muscle cells respond to epinephrine. In each graph, the *x*-axis plots time, in seconds, after epinephrine is applied.

- *Top* Almost immediately, there is a striking increase in cAMP levels inside the cells.
- *Middle* A few seconds later, the contractile force of the cells increases, peaking about 18 seconds after the hormone's arrival.
- *Bottom* Phosphorylase activity also rises, but peaks later—about 40 seconds after the signal.

To capture the importance of cAMP in triggering these effects, biologists refer to it as a second messenger. Recall from Chapter 8 that a **second messenger** is a nonprotein signaling molecule that increases in concentration inside a cell in response to a received signal—a molecule that binds at the surface.

A PHOSPHORYLATION CASCADE How does cAMP transfer the cell-surface signal to phosphorylase? Follow-up work showed that cAMP binds to an enzyme called cAMP-dependent protein kinase A. This enzyme responds by phosphorylating the enzyme phosphorylase kinase, which then phosphorylates phosphorylase.

This chain of events, called a **signal transduction cascade**, is initially triggered by the synthesis of cAMP. cAMP is produced from ATP in a reaction that is catalyzed by the enzyme adenylyl cyclase. Adenylyl cyclase is activated by a G protein (see Chapter 8), which is activated when epinephrine binds to its receptor.

To pull all these reactions together, researchers proposed the model for epinephrine action diagrammed in **Figure 47.17**. In studying this model, it is crucial to recognize two points:

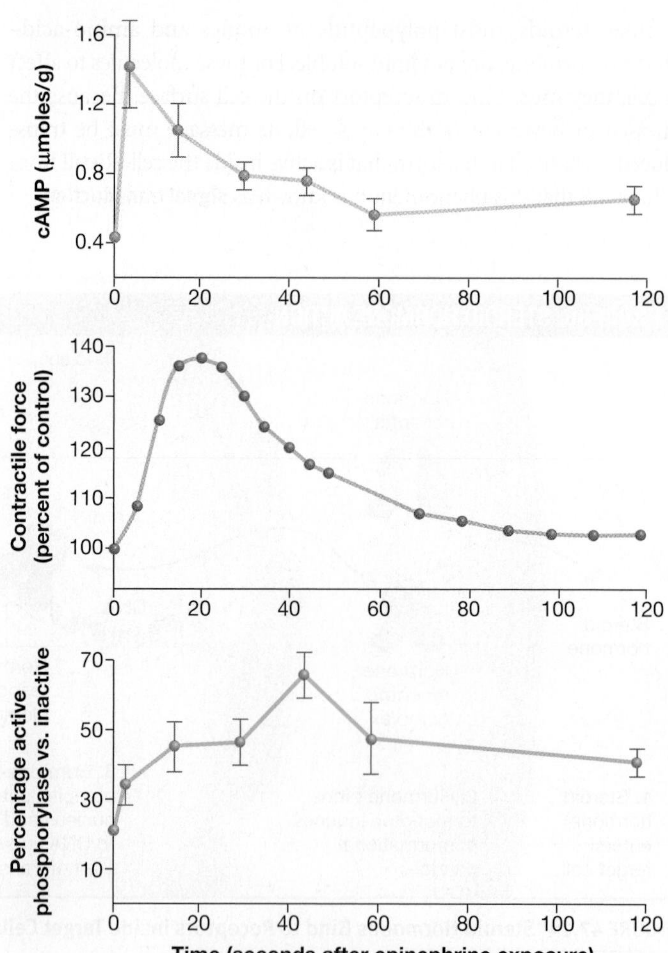

FIGURE 47.16 The Chemistry and Activity of Heart-Muscle Cells Change in Response to Epinephrine.

1. cAMP transmits the signal from the cell surface to the signaling cascade.
2. Together, cAMP production and the subsequent phosphorylation events amplify the original signal from epinephrine.

To drive the latter point home, consider that, in response to stimulation by the hormone-receptor complex, adenylyl cyclase is thought to catalyze the formation of at least 100 molecules of cAMP. In turn, each of these cAMP molecules activates many molecules of cAMP-dependent protein kinase A. Subsequently, each protein kinase molecule activates many molecules of phosphorylase kinase, and so on.

In this way, the binding of just a single molecule of epinephrine may trigger the release of millions or even billions of glucose molecules. Amplification through a signal transduction cascade explains why tiny amounts of hormones can have such huge effects on an individual.

The model in Figure 47.17 was inspired by experiments on the epinephrine receptor called the beta-1 receptor. But other researchers showed that when epinephrine binds to an alpha-1 receptor, a completely different signal transduction event occurs. In this and many other receptor systems, calcium ions (Ca^{2+}) serve as the second messenger in conjunction with a second-messenger molecule called IP_3. Diacylglycerol (DAG) and 3′, 5′-cyclic GMP (cGMP) are also common second messengers in hormone response systems.

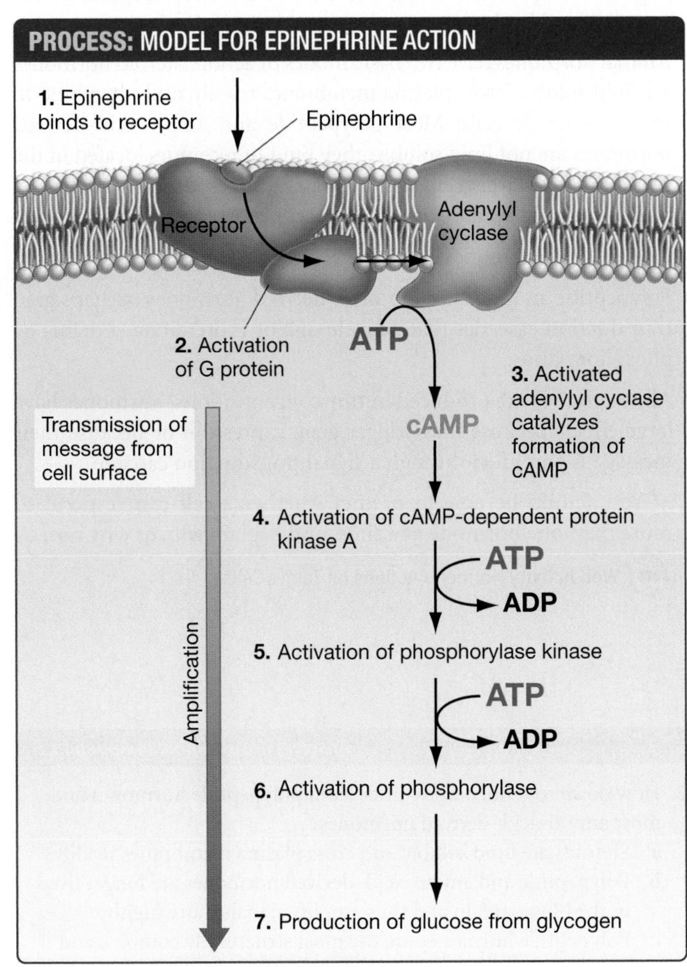

FIGURE 47.17 Epinephrine Triggers a Signal Transduction Cascade. Epinephrine's signal is amplified at each of steps 3–6.

To review second messengers, amplification, and other fundamental concepts about hormone action, go to the study area at *www.masteringbiology.com.*

 **Web Activity** Hormone Actions on Target Cells

Why Do Different Target Cells Respond in Different Ways?

Researchers are increasingly impressed with the diversity and complexity of signal transduction cascades. For example, target cells that have the same receptor protein may have different second messengers, different genes available for transcription, or different enzyme systems that are available for activation. As a result, the same hormone and receptor can give rise to different responses in different target cells.

This finding helps explain one of the most fundamental observations about hormones: The same chemical messenger can trigger different responses in cells from different organs or in cells at different developmental stages. The reason is that the cells contain different receptors, second messengers, amplification steps, protein kinases, enzymes, or transcriptionally active genes.

To summarize this section, steroid hormones tend to exert their effects through changes in gene expression. In contrast, polypeptide and amino-acid–derived hormones activate a specific protein or set of proteins, usually by phosphorylation. Steroid hormones activate transcription factors that lead to the production of new proteins; nonsteroid hormones trigger signal transduction cascades that activate existing proteins.

CHECK YOUR UNDERSTANDING

If you understand that . . .

- Hormones act on target cells by binding to receptors.
- The response to a hormone is tissue specific because only certain cells contain receptors for particular hormones.
- In response to binding by lipid-soluble hormones, hormone-receptor complexes bind to DNA and induce changes in gene expression.
- In response to binding by lipid-insoluble hormones, activated cell-surface receptors induce the production of second messengers or the activation of signal transduction cascades that result in the phosphorylation of existing proteins.

✔ **You should be able to . . .**

1. Predict what would happen to the estradiol response if an individual had mutations that changed the DNA sequence of its hormone-response element.
2. Suppose a steroid hormone bound to a cell-surface receptor. Explain how its mode of action would compare to a polypeptide hormone.

Answers are available in Appendix B.

CHAPTER 47 REVIEW

For media, go to the study area at www.masteringbiology.com 

Summary of Key Concepts

Animals use at least six major types of chemical signals. Hormones are chemical signals that are present in tiny concentrations and travel throughout the body to affect target cells.

- Hormones are chemical messengers that are released from neurons or cells of the endocrine system, circulate in the blood or other body fluids, and trigger a response in distant target cells containing an appropriate receptor.
- Hormones have a variety of chemical structures. Most animal hormones are polypeptides, amino acid derivatives, or steroids.

✔ You should be able to explain the relationship between electrical signals from the nervous system and chemical signals from the nervous system and endocrine system, as they work in combination to coordinate the body's response to environmental change.

The information carried by hormones helps animals develop as embryos, undergo sexual maturation, respond to environmental change, and achieve homeostasis.

- In conjunction with the nervous system, hormones coordinate the activities of diverse cells and tissues. A single hormone may affect a wide array of cells and tissues and induce a variety of responses.
- Estradiol is an example of a hormone that regulates development and sexual maturation. Estradiol stimulates the formation of female sex characteristics in human embryos and the maturation of these tissues in adolescence.
- Epinephrine and cortisol are examples of hormones that help individuals cope with environmental changes. These hormones activate the short-term and long-term responses to stress, triggering the fight-or-flight response or inducing changes that conserve glucose for use by the brain.
- Hormones are involved in a wide array of homeostatic interactions. For example, hormones are involved in directing cells that modify the concentrations of water, sodium ions, and other components of the blood and interstitial fluid. Homeostasis with respect to triglyceride stores is also under endocrine control.

✔ You should be able to explain why hormones—instead of electrical signals from the nervous system—are primarily responsible for regulating responses to environmental change, embryonic and sexual development, and homeostasis.

The production of a hormone is tightly regulated by input from the nervous system and by other hormones.

- In many cases, the release of a hormone is regulated by chemical messengers from the anterior pituitary.
- Hormone-secreting cells in the anterior pituitary are regulated by hormones released by the hypothalamus region of the brain.
- The long-term stress response is a well-studied example of hormone regulation. The brain responds to long-term stress by triggering the release of the hypothalamic hormone CRH. CRH activates the release of ACTH by the pituitary gland, which stimulates the production of cortisol by cells in the adrenal cortex. Because cortisol inhibits the production of ACTH and CRH, the chain of events is regulated by feedback inhibition.

✔ You should be able to predict the consequences if negative feedback fails to occur after the release of CRH.

MB **Web Activity** Endocrine System Anatomy

Some hormones bind to receptors inside target cells and change gene expression. Other hormones bind to receptors at the cell surface and lead to changes in protein activation.

- Animal hormones have two basic modes of action. Steroid hormones are lipid soluble, cross plasma membranes readily, and often bind to receptors inside cells. Most polypeptide and amino-acid–derived hormones are not lipid soluble; they bind to receptors located in the membranes of target cells. In both cases, the response to a hormone is tissue specific because only certain cells express certain receptors.
- Most steroid hormones act by inducing a change in gene expression.
- Polypeptide and most amino-acid–derived hormones trigger signal transduction cascades that activate one or more target proteins by phosphorylation.
- Although they are produced in tiny concentrations, hormones have large effects because they trigger gene expression or because their message is amplified through a signal transduction cascade.

✔ You should be able to predict whether a cell can respond to more than one hormone at a time, and explain why or why not.

MB **Web Activity** Hormone Actions on Target Cells

Questions

✔ TEST YOUR KNOWLEDGE

Answers are available in Appendix B

1. Both epinephrine and cortisol are involved in the body's response to stress. How do the two molecules differ?
   a. Cortisol is an amino acid derivative; epinephrine is a steroid.
   b. Cortisol binds to receptors on the plasma membranes of target cells; epinephrine binds to receptors in the interior of target cells.
   c. Epinephrine mediates the short-term response; cortisol mediates the short- and long-term responses.
   d. Cortisol controls the release of epinephrine from the adrenal glands.

2. How do steroid hormones differ from polypeptide hormones and most amino-acid–derived hormones?
   a. Steroids are lipid soluble and cross plasma membranes readily.
   b. Polypeptide and amino-acid–derived hormones are longer lived in the bloodstream and thus amplify signals more highly.
   c. Polypeptide hormones are the most structurally complex and induce permanent changes in target cells.
   d. Only polypeptide and steroid hormones bind to receptors in the plasma membrane.

3. What is signal transduction?
   a. the binding of a steroid hormone-receptor complex to DNA
   b. the release of a hormone from the anterior pituitary in response to a hypothalamic hormone
   c. the release of a hormone from the posterior pituitary in response to action potentials from the hypothalamus
   d. the production of a second chemical messenger inside a cell in response to the binding of a hormone at the cell surface

4. Which of the following developmental processes is *not* controlled by hormones?
   a. the initial development of male and female gonads, soon after fertilization
   b. overall growth
   c. molting in insects and other invertebrate animals
   d. metamorphosis in insects and other invertebrate animals

5. What is a hormone-response element?
   a. a receptor for a steroid hormone
   b. a receptor for a polypeptide hormone
   c. a segment of DNA where a hormone-receptor complex binds
   d. an enzyme that is activated in response to hormone binding and produces a second messenger

6. In hormone systems, when does feedback inhibition occur?
   a. when the presence of a hormone inhibits its release
   b. when the presence of a hormone stimulates its release
   c. when a second messenger triggers the phosphorylation of an inhibitory protein
   d. when a hormone from the hypothalamus inhibits the release of a hormone from the anterior pituitary

✓ TEST YOUR UNDERSTANDING

Answers are available in Appendix B

1. Compare the structure and function of the anterior and posterior pituitary glands.
2. Compare and contrast the modes of action of steroid and nonsteroid hormones.
3. The pituitary is often referred to as the "master gland." Why?
4. Why is the observation that one hormone may bind to more than one type of receptor important?
5. Hormones are present in tiny concentrations yet have large effects on target cells and on the individual as a whole. How is this possible?
6. Why can leptin be considered a "satiation signal"? What role does leptin play in homeostasis?

✓ APPLYING CONCEPTS TO NEW SITUATIONS

Answers are available in Appendix B

1. Suppose that during a detailed anatomical study of a marine invertebrate, you found a small, previously undescribed structure. How would you test the hypothesis that the structure is a gland that releases one or more hormones?
2. Cortisone is a glucocorticoid that suppresses inflammation and other aspects of wound healing. Cortisone was once widely used to treat athletes with joint injuries and people with arthritis. In the short term, cortisone is extremely effective in reducing swelling and pain. Over time, however, physicians found that repeated large doses of cortisone had damaging side effects. Predict what these side effects are. Explain your logic.
3. You are a physician supervising a patient's recovery from the surgical removal of the posterior pituitary. Name a hormone that you will have to administer to this patient artificially. Which symptom(s) will you monitor to assess whether the dosage and timing of your injections are having the desired effect?
4. Suppose that a researcher announces the discovery of a hormone that affects the metabolism of fats in lab rats. Preliminary data indicate that the hormone is a polypeptide, about 50 amino acids long. How would you go about isolating the receptor for this hormone? Using fat cells growing in vitro, how could you test the hypothesis that hormone binding causes a change in a second messenger?

UNIT 8

HOW ANIMALS WORK

The swollen, red rump of this female Hamadryas baboon indicates that she is about to produce an egg. She will probably mate with several males before the egg is fertilized. This chapter discusses how hormones regulate the female reproductive cycle and the consequences of multiple mating.

48 Animal Reproduction

KEY CONCEPTS

- The reproductive systems of animals are highly variable. Some species switch between asexual and sexual reproduction. When sexual reproduction occurs, fertilization may be external or internal and egg development may take place inside or outside the mother's body, depending on the species.
- In humans, the male reproductive system includes structures specialized for producing and storing sperm, synthesizing other components of semen, or transporting and delivering semen. The female reproductive system includes structures specialized for producing eggs, receiving sperm, and nourishing offspring during early development.
- In humans, hormones from the pituitary gland and female reproductive organs regulate the menstrual cycle. These hormones interact via positive or negative feedback. Pregnancy is maintained by hormonal signals from the embryo and the mother's reproductive organs.

All the cells, tissues, organs, and systems introduced in Unit 8 exist for one reason: They allow animals to survive long enough and gather enough resources to reproduce. Stated another way, producing offspring is the reason that adaptations exist. Reproduction is the unconscious goal of virtually everything that an animal does. As Chapter 1 noted, replication is a fundamental attribute of life.

Evolution by natural selection explains why animals reproduce; the goal of this chapter is to explore *how* reproduction occurs. Section 48.1 introduces research on animals that cycle between asexual and sexual modes of reproduction and explains how both modes occur at the cellular level. Section 48.2 surveys the diverse ways in which animals accomplish fertilization once gametes have formed. The final three sections focus on mammalian reproduction, using humans as the primary model organism.

The chapter includes topics of urgent practical interest, such as human birth control methods and the recent and dramatic declines in the death rate of human mothers during childbirth. Understanding and manipulating animal reproductive systems is an important issue for physicians, veterinarians, farmers, zookeepers, and many others in biology-related professions.

48.1 Asexual and Sexual Reproduction

Several earlier chapters have explored how asexual and sexual reproduction differ. When reproduction is asexual, it occurs without fusion of gametes. **Asexual reproduction** is usually based on mitosis and results in offspring that are genetically identical to their parent. **Sexual reproduction**, in contrast, is based on meiosis and fusion of gametes. Due to genetic recombination during meiosis and the fusion of haploid gametes—usually from different parents—during fertilization, sexual reproduction results in offspring that are genetically different from each other and from their parents.

✔ When you see this checkmark, stop and test yourself. Answers are available in Appendix B.

Let's first consider how animals make genetically identical copies of themselves, then review the mechanisms responsible for sexual reproduction.

How Does Asexual Reproduction Occur?

In thousands of animal species, individuals can **clone** themselves—that is, produce large numbers of identical copies of themselves asexually.

(a) **Budding** in hydra

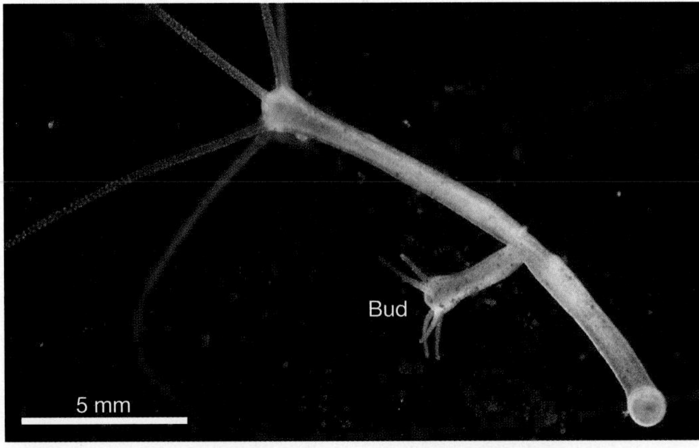

(b) **Fission** in anemones

(c) **Parthenogenesis** in lizards

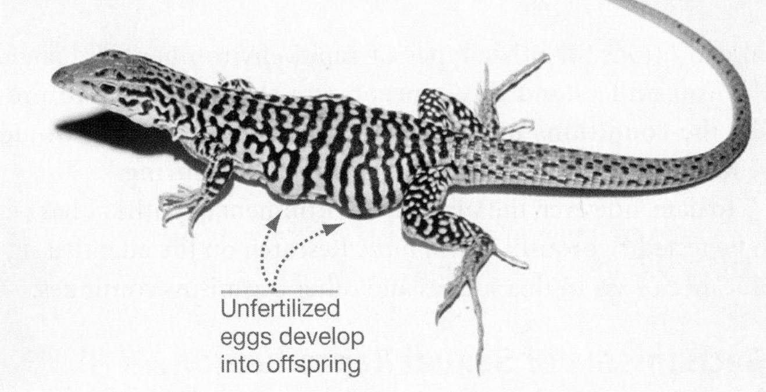

FIGURE 48.1 Mechanisms of Asexual Reproduction in Animals Are Diverse.

There are three main mechanisms of asexual reproduction:

- When **budding** occurs, an offspring begins to form within or on a parent (**Figure 48.1a**). The process is complete when the offspring—a miniature version of the parent—breaks free and begins to grow on its own.
- During **fission**, an individual simply splits into two or more descendants (**Figure 48.1b**).
- **Parthenogenesis** (literally, "virgin-origin") occurs when female offspring develop from unfertilized eggs. In most cases, parthenogenetic eggs are produced by mitosis, or by meiosis that occurs after chromosome number has doubled, but without crossing over and recombination. In both cases, offspring are genetically identical to the mother. If a species reproduces exclusively via parthenogenesis, no males exist. Parthenogenesis occurs in a wide diversity of lineages, including certain guppies, crustaceans, rotifers, and lizards (**Figure 48.1c**).

Many animal species regularly switch between reproducing asexually and reproducing sexually. Why?

Switching Reproductive Modes: A Case History

Daphnia are crustaceans that live in freshwater habitats throughout the world. In a typical year, *Daphnia* produce only diploid female offspring throughout the spring and summer, via parthenogenesis. The eggs produced by parthenogenesis develop in a structure called a brood pouch (**Figure 48.2**), and are released when the female molts her exoskeleton.

In late summer or early fall, however, many *Daphnia* females begin producing unfertilized, asexually produced eggs that develop into males as well as females. Once the males have matured, sexual reproduction ensues: Haploid **sperm** (male gametes) produced by meiosis in the males fertilize haploid **eggs** (female gametes) that females produce via meiosis.

Fertilization is the fusion of sperm and egg. In *Daphnia*, fertilized eggs are released into a durable case that falls to the bottom of the pond or lake for the winter. In spring, the sexually produced offspring hatch and begin reproducing asexually.

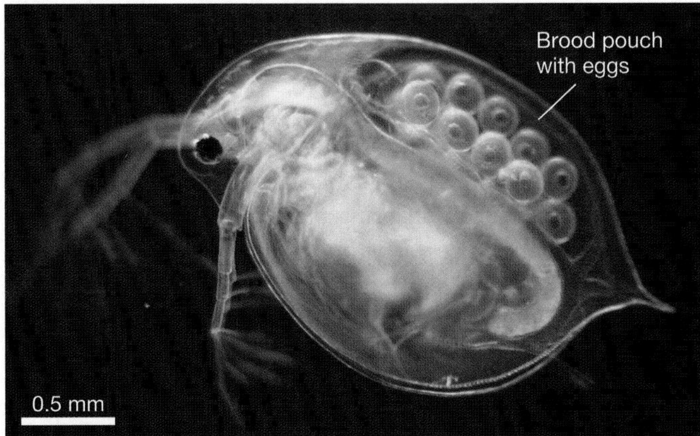

FIGURE 48.2 Female *Daphnia* Can Lay Parthenogenetic Eggs. *Daphnia* produce diploid eggs asexually.

Biologists try to explain observations like this at two levels (see Chapter 51):

- **Ultimate causation** addresses *why* a trait occurs, in terms of its effect on fitness. Researchers who work at the ultimate level try to understand the evolutionary history of traits.
- **Proximate causation** addresses *how* a trait is produced. When researchers identify the genetic, developmental, hormonal, or neural mechanisms responsible for a phenotype, they are working at the proximate level.

Let's consider work on proximate aspects of the asexual-sexual switch in *Daphnia*, then consider ultimate causation.

WHAT ENVIRONMENTAL CUES TRIGGER THE SWITCH? For decades, most researchers contended that day length triggered the asexual-sexual switch. The idea was that the shortening days of late summer or fall affected sensors in the brain, and these receptors produced electrical or hormonal signals that induced the production of males and haploid eggs.

In 1965, however, biologists showed that high population densities are also a factor. Researchers who brought *Daphnia pulex* populations into the lab and kept day length constant found a strong, positive correlation between population density and the percentage of females reproducing sexually. When you read the graph in **Figure 48.3a**, note that the *y*-axis is logarithmic (see **BioSkills 7** in Appendix A). The highest proportion of sexually reproducing offspring is found at high density.

Another group of investigators extended this result by pinpointing the specific aspects of crowding that affected the animals. These biologists brought a closely related species—*D. magna*—into the laboratory and altered day length, the amount of food available to individuals, and the quality of the water they occupied. (To vary water quality, the investigators used either clean water or water taken from tanks where *D. magna* were being maintained at high density.)

As **Figure 48.3b** shows, individuals in the study population switched to sexual reproduction only if they were exposed to poor-quality water, low food availability, *and* short day lengths. In short, *D. magna* needs three different cues from the environment to switch to sexual reproduction. Two of these cues were associated with high population density; the third is associated with the onset of winter.

WHY DO *DAPHNIA* SWITCH BETWEEN ASEXUAL AND SEXUAL REPRODUCTION? *Daphnia* appear to start sexual reproduction when conditions worsen. Why?

The leading hypothesis to answer this question acknowledges that sexually produced offspring are genetically diverse. When the environment changes, genetically variable offspring are likely to survive better and reproduce more than offspring that are identical to their parents.

This hypothesis is consistent with research introduced in Chapter 12. You might recall that sexual reproduction in a snail species is associated with high rates of parasite infection. Parasites evolve rapidly—changing the environment for their hosts.

Genetically variable offspring have higher fitness in environments with rapidly evolving parasites, deteriorating physical conditions, or other types of rapid environmental change.

✓If you understand this concept, you should be able to predict the conditions under which asexual reproduction would be found as the only method of producing offspring.

To date, however, the variable-environment hypothesis has yet to be tested rigorously in *Daphnia*. Research on the adaptive significance of sex in this species and other organisms continues.

(a) Sexual reproduction is more common in crowded populations of *Daphnia pulex* than in sparse populations.

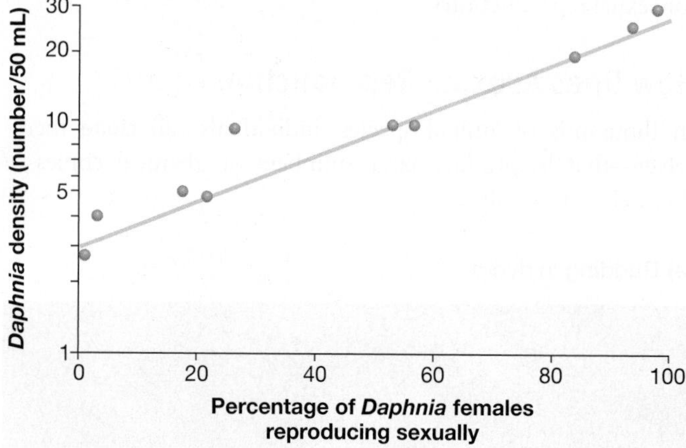

(b) In *Daphnia magna*, which environmental cues trigger the switch to sexual reproduction?

Water quality	Food concentration	Day length	% Sexual broods
Clean	Low	Short	0
Crowded	Low	Short	44
Clean	Low	Long	0
Crowded	Low	Long	0
Clean	High	Short	0
Crowded	High	Short	0
Clean	High	Long	0
Crowded	High	Long	0

FIGURE 48.3 In *Daphnia*, Environmental Cues Signal the Switch from Asexual to Sexual Reproduction. **(a)** There is a strong, positive correlation between the percentage of females that reproduce sexually versus density (plotted on a log scale here). **(b)** Environmental conditions were varied experimentally for *Daphnia*. "Crowded" water was taken from tanks containing dense populations.

✓**QUESTION** How would you go about determining which molecule or molecules in crowded water serve as a signal that triggers sexual reproduction?

Mechanisms of Sexual Reproduction: Gametogenesis

The mitotic cell divisions, meiotic cell divisions, and developmental events that produce male and female gametes, or sperm

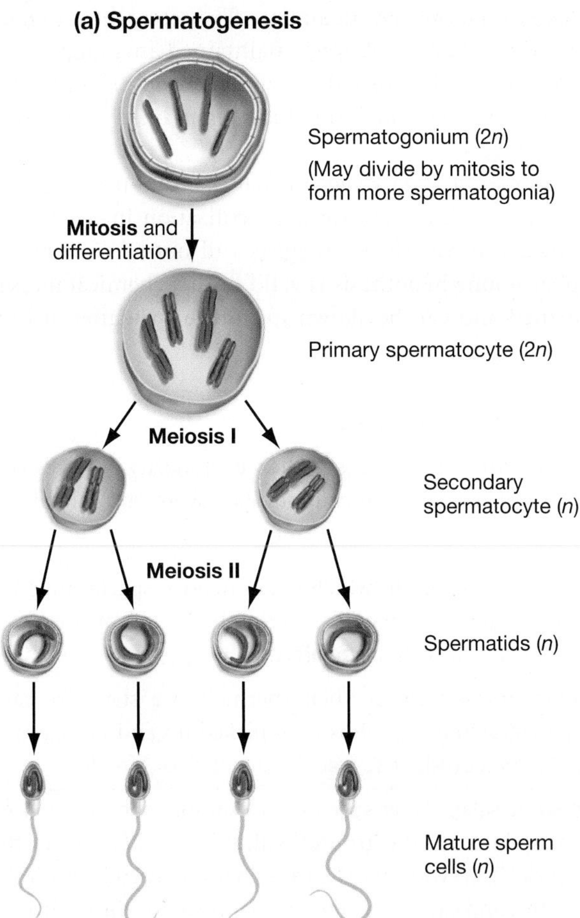

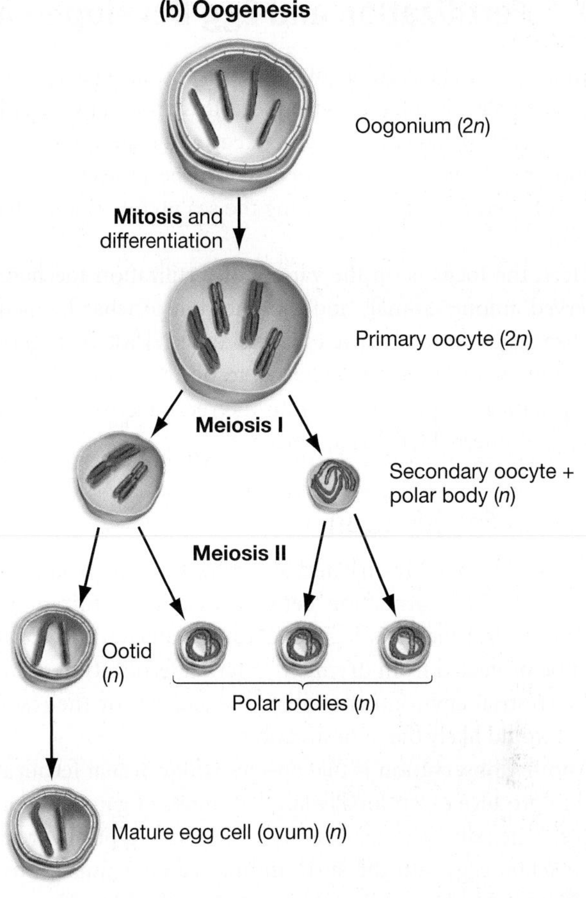

FIGURE 48.4 Gametogenesis in Humans.

and eggs, are collectively called **gametogenesis** (see Chapter 22). **Spermatogenesis** is the formation of sperm; **oogenesis** is the formation of eggs.

In the vast majority of animals, gametogenesis occurs in a sex organ, or **gonad**. Male gonads are called **testes**; female gonads are called **ovaries**. Early in development, reproductive cells known as germ cells enter the testes and ovaries and give rise to diploid cells that will undergo gametogenesis.

SPERMATOGENESIS IN MAMMALS **Figure 48.4a** summarizes the events that take place during spermatogenesis in humans. Note that in the male gonad, diploid cells called **spermatogonia** (singular: **spermatogonium**) divide by mitosis. Some of the resulting cells continue to function as spermatogonia; others change to form specialized cells that are committed to producing sperm.

In males, the specialized cells produced by spermatogonia are called **primary spermatocytes**. They undergo meiosis I and produce two **secondary spermatocytes**, which then undergo meiosis II. The result is four haploid cells called **spermatids**.

Each haploid spermatid matures into a sperm—a cell that is specialized for carrying a haploid genome from the male through the female reproductive tract, and fertilizing an egg. The production of spermatogonia, primary spermatocytes, and sperm occurs continuously throughout a man's adult life.

OOGENESIS IN MAMMALS The top of **Figure 48.4b** highlights an important similarity between spermatogenesis and oogenesis: In the female gonad, diploid cells called **oogonia** (singular: **oogonium**) divide by mitosis. Some of the resulting cells continue to function as oogonia; others change to form specialized cells that are committed to producing an egg. In this sense, oogonia and spermatogonia are similar.

However, subsequent steps in gametogenesis are markedly different in human females. The specialized cells produced by an oogonium are called **primary oocytes**. When these cells undergo meiosis, only one of the four haploid products, known as an **ovum**, matures into an egg. The other cells produced by meiosis in females have a tiny amount of cytoplasm and do not mature into eggs. Because the distribution of cytoplasm is so unequal during each meiotic division in females, the smaller cells are called **polar bodies**. (Recall that polar refers to inequality or opposites.)

In addition, the production of primary oocytes stops early in development in many mammals; in humans, it stops before birth. And in humans and many other mammals, the primary oocytes enter prophase of meiosis I during embryogenesis, but then stop developing for months, years, or decades.

To review the steps in spermatogenesis and oogenesis, go to the study area at *www.masteringbiology.com.*

Web Activity Human Gametogenesis

48.2 Fertilization and Egg Development

Fertilization is the joining of a sperm and an egg to form a diploid **zygote**. Chapter 22 introduced this process by describing how a sperm makes contact with an egg and penetrates the egg's membrane. That discussion focused on the molecular mechanisms of sperm-egg binding, using the sea urchin as a model organism.

Here, the focus is on the variety of fertilization mechanisms observed among animals and the question of what happens to fertilized eggs. In many species, individuals release their gametes into their environment and external fertilization occurs. In other animals, males deposit sperm into the reproductive tracts of females and internal fertilization occurs.

External Fertilization

Most animals that rely on external fertilization live in aquatic environments. The correlation between external fertilization and aquatic environments is logical, because gametes and embryos must be protected from drying. If external fertilization occurred in a terrestrial environment, either the gametes or the resulting zygote would likely die of desiccation.

Another observation is that species with external fertilization tend to produce exceptionally large numbers of gametes. For example, a female sea star *Asterias amurensis* typically releases 100,000,000 eggs into the surrounding seawater during spawning. Males release many times that number of sperm. The leading hypothesis to explain this pattern is that the probability of a sperm and egg meeting in an ocean or lake is small unless large numbers of gametes are present.

If sperm and eggs from different individuals must be released into the environment synchronously (at the same time) for external fertilization to work, how is gamete release coordinated? The answer has two parts.

1. Gametogenesis usually occurs in response to environmental cues, such as lengthening days and warmer water temperatures, that indicate a favorable season for breeding.
2. Gametes are released in response to specific cues from individuals of the same species.

In fishes and other aquatic animals with well-developed eyes and external fertilization, spawning is often the culmination of an elaborate courtship ritual between a male and female. In contrast, courtship behavior appears to be much less important—or even absent—in species such as clams, sea urchins, and sea cucumbers. How do animals that cannot see their mates time the release of their gametes?

Recent research indicates that the chemical messengers called **pheromones** (see Chapter 47) might be involved in synchronizing gamete release. For example, biologists maintained two groups of sea cucumbers under natural conditions of light and temperature for 15 months. In one treatment, individuals were kept in isolated tanks; in another, they were kept in tanks with other sea cucumbers. The researchers found that all the individuals that were maintained in groups released gametes during the normal spawning period. In contrast, only about 10 percent of the individuals maintained in isolation released gametes.

Although these data suggest that pheromones might be involved in coordinating external fertilization in sea stars, they are not definitive. Most biologists will not be convinced that the pheromone hypothesis is valid until a chemical messenger is purified and can be shown to induce spawning in isolated individuals.

Internal Fertilization

Internal fertilization occurs in the vast majority of terrestrial animals as well as in a significant number of aquatic animals. Internal fertilization occurs in one of two ways:

1. after copulation, in which males deposit sperm directly into the female reproductive tract with the aid of a copulatory organ, usually called a **penis**; or
2. when males package their sperm into a structure called a **spermatophore**, which is then picked up and placed into the female's reproductive tract by the male or female.

In some salamander species, for example, the male places the spermatophore on the ground within its territory. Later the female picks it up with her **cloaca**, a chamber used by both the reproductive and excretory systems that opens to the environment. In this case, the result is internal fertilization without any direct physical contact between the sexes.

WHAT IS SPERM COMPETITION? In terms of understanding animal behavior, perhaps the most important insight about internal fertilization originated with Geoff Parker. In 1970 Parker published experiments on dung flies that confirmed the existence of **sperm competition**—competition between sperm from different males to fertilize the eggs of the same female.

Parker's experiments consisted of a series of matings between one female and two males. In each experiment, the two males were selected in such a way that Parker could distinguish between their offspring. The proportion of offspring fathered by each male was not 50:50. Instead, whichever male was last to copulate fathered an average of 85 percent of the offspring produced.

These data not only suggested that sperm were competing to fertilize eggs but indicated that, in this experiment, the second male won. Follow-up research has confirmed that **second-male advantage** is widespread, although not universal, in insects and some other animal groups. How does it occur?

SPERM COMPETITION AND SECOND-MALE ADVANTAGE To explore why second-male advantage occurs, biologists turned to the fruit fly *Drosophila melanogaster*. A research group introduced a gene into male fruit flies that produced sperm with green tails (**Figure 48.5**). When a mating by a green-spermed

EXPERIMENT

QUESTION: How does the "second-male advantage" occur in sperm competition?

HYPOTHESIS: In sperm storage areas, sperm from the second male displaces sperm from the first male.

NULL HYPOTHESIS: The mechanism does not involve sperm displacement from sperm storage areas.

EXPERIMENTAL SETUP:

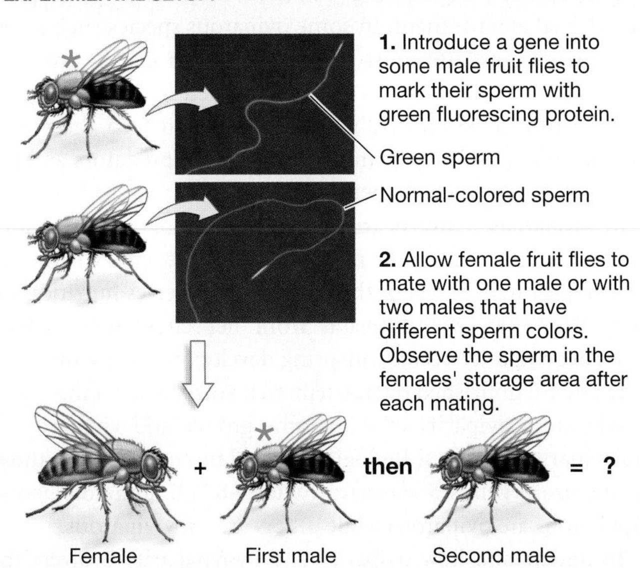

PREDICTION: When females mate twice, few sperm from the first male remain in storage.

PREDICTION OF NULL HYPOTHESIS: When females mate twice, most or all of the sperm deposited by the first male is still present.

RESULTS:

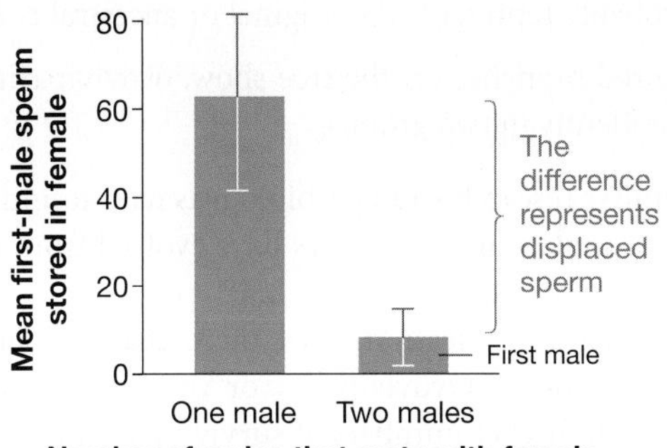

CONCLUSION: If a female mates a second time, most sperm from the first male she mated with disappears.

FIGURE 48.5 Experimental Evidence for Second-Male Advantage during Sperm Competition in *Drosophila*. The graph shows the average number of green sperm stored in females when males with green sperm were the only male to mate or the first of two males to mate.

SOURCE: Price, C. S. C., K. A. Dyer, and J. A. Coyne. 1999. Sperm competition between *Drosophila* males involves both displacement and incapacitation. *Nature* 400: 449–452.

✔**QUESTION** Based on these data, is the claim that sperm from the second male physically displaces sperm from the first male valid?

male was followed with a second mating—by a male having normal-colored sperm—many fewer green-tailed sperm were found in the female's sperm-storage area compared with the number observed when no second mating took place.

To interpret this finding, the biologists suggested that the sperm of the second male physically dislodged the first male's gametes from the female sperm storage area and inserted themselves in their place. The researchers also showed that the fluid that accompanies sperm during fertilization is able to displace stored sperm from competing males. These two mechanisms resulted in the second male's sperm fertilizing most of the eggs laid.

WHY IS TESTES SIZE VARIABLE AMONG SPECIES? Research on sperm competition has recently contributed another major result. In species in which females routinely mate with multiple males before laying eggs or giving birth, males have extraordinarily large testes for their size and produce proportionately larger numbers of sperm.

The leading hypothesis to explain this observation is that fertilization is similar to a lottery in which each sperm represents a ticket. The more tickets a male enters in the competition, the higher his chance of "winning" fertilizations and passing his alleles on to the next generation. Males with exceptionally large testes produce exceptionally large numbers of sperm, and are more likely to win the lottery.

The lottery model has been challenged, however, by evidence that females often store sperm and exert control over which sperm are successful in fertilization. In other words, females do not always accept the results of sperm competition passively.

Females of some species actively choose which male performs the last copulation before fertilization takes place. In other species, females physically eject sperm from undesirable males. This phenomenon has been dubbed cryptic female choice. The name is appropriate because the selection of sperm by females is hidden from males.

Unusual Aspects of Mating

Studies on animal sexual reproduction have documented some of the most remarkable behaviors and structures observed in nature. The list that follows is just a sample, offered simply to illustrate the diversity of mating arrangements in animals.

- *Femmes fatales* When Australian redback spiders mate, the male does a somersault after inserting his penis-like organ. The somersault places his dorsal surface in front of the female's mouthparts. In many cases the female responds by eating the male. (Biologists refer to females who cannibalize males as femmes fatales—a phrase used to describe murderous human females in movies.) Cannibalized males copulate longer and fertilize more eggs than non-cannibalized males, probably because sperm transfer continues until the meal is over.
- *Giant sperm* In the fruit fly *Drosophila bifurca*, males average 1.5 mm in total body length. Their sperm, however, are each 6 cm long. The coiled, bulky sperm fill the female's sperm storage area and make it impossible for another male's sperm to enter.

- *False penises* Although very few bird species have a penis, the red-billed buffalo weaver has a false penis that becomes erect during copulation (**Figure 48.6**). It does not function in sperm transfer, but appears to stimulate the female's reproductive tract and trigger an orgasm-like state in males. Female buffalo weavers have a similar, but smaller, organ of unknown function.
- *Infidelity* When researchers assess paternity in bird species that appear to be monogamous, they find that up to 60 percent of nests contain at least one offspring fathered by a male that is not mated to the resident female. In most cases, females actively solicit copulations from males holding nearby territories.
- *Love darts* Many snails and slugs are **hermaphroditic**, meaning an individual has both male and female gonads. During mating, two individuals simultaneously receive and deposit sperm to fertilize each other's eggs. In some species, individuals fire mucus-covered "love darts" from their genitalia into the mating partner. A substance in the mucus increases the chance that sperm will successfully fertilize eggs. In other species, the penis frequently becomes stuck in the female reproductive tract and is bitten off by one partner.
- *Hypodermic insemination* Some bedbugs have hypodermic penises, meaning that they function like a hypodermic needle. Males force the organ through the female's abdominal wall and deposit sperm directly into her body cavity.

Did the process of sexual selection, introduced in Chapter 25, lead to the evolution of these structures and behaviors? If so, how and why? In many or even most cases, the answers to these questions are not known.

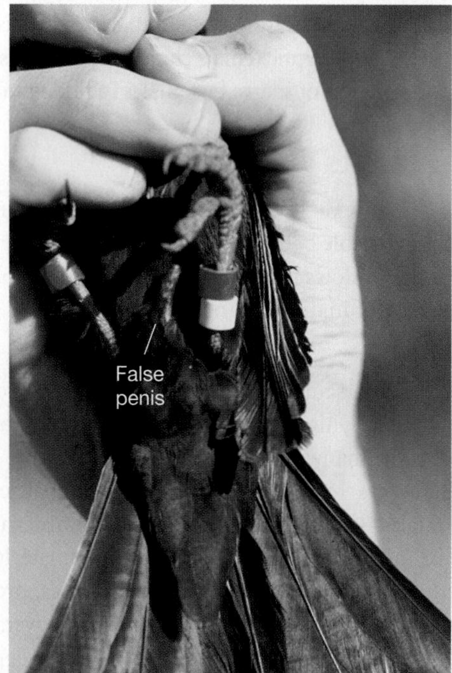

FIGURE 48.6 A False Penis in a Weaverbird. The organ shown here is in the erect state. It is "false" because it is not inserted into the female's reproductive tract and does not transmit semen.

Why Do Some Females Lay Eggs while Others Give Birth?

For females, many aspects of fertilization and egg-laying vary among species: a female's size or age at first breeding; the number and size of eggs she produces; the number of times she reproduces over the course of a lifetime. Here let's consider a different aspect of variation in animal reproduction: whether eggs are laid outside the body or retained inside the body.

In **oviparous** ("egg-bearing") animals, the embryo develops in the external environment. In some oviparous species such as sea stars, sea urchins, and most insects, no further care is provided by the parents; the eggs and embryos are left to fend for themselves. Birds, however, incubate their eggs and feed the young after hatching; fish may guard their eggs from predators and fan the clutches to oxygenate them.

In **viviparous** ("live-bearing") species, embryonic development takes place within the mother's body. The embryo attaches to the reproductive tract of the mother and receives nutrition directly from her—via diffusion from her circulatory system. When **ovoviviparity** occurs, offspring develop inside the mother's body but are nourished by nutrient-rich yolk stored in the egg.

Why does oviparity exist in some groups and viviparity or ovoviviparity in others? Biologists tackled this question by studying the lizard genus *Sceloporus* (**Figure 48.7a**). Some *Sceloporus* populations are oviparous while others are ovoviviparous.

To understand how oviparity and ovoviviparity evolved, the biologists analyzed a phylogenetic tree—based on molecular and morphological data—of many *Sceloporus* species (**Figure 48.7b**). Two conclusions should make sense to you (see **BioSkills 3** in Appendix A for help with interpreting phylogenetic trees):

1. Because most *Sceloporus* populations are oviparous, egg laying probably represents the original or ancestral condition.
2. As the red branches on the tree show, ovoviviparity evolved independently in two groups.

Using a similar research strategy, biologists have found that ovoviviparous populations of sea stars have evolved from oviparous populations on several occasions.

Why did natural selection favor these changes between egg laying and live birth? Ovoviviparity or viviparity should evolve when it leads to higher numbers of surviving young.

Researchers have long hypothesized that natural selection favoring live birth should be especially strong in cold habitats. Low temperatures slow the development of embryos, so in cold habitats, it might be advantageous for females to retain eggs inside their bodies so that offspring can develop at a more favorable temperature.

Several lines of evidence support this hypothesis:

- The ovoviviparous *Sceloporus* species live in the highlands of the southwestern United States and central Mexico, where temperatures can be cool.
- In at least one oviparous species of lizard, lowering nest temperatures leads to a higher percentage of deformed offspring.

(a) *Sceloporus* lizard (female)

(b) Phylogeny of *Sceloporus* from central Mexico

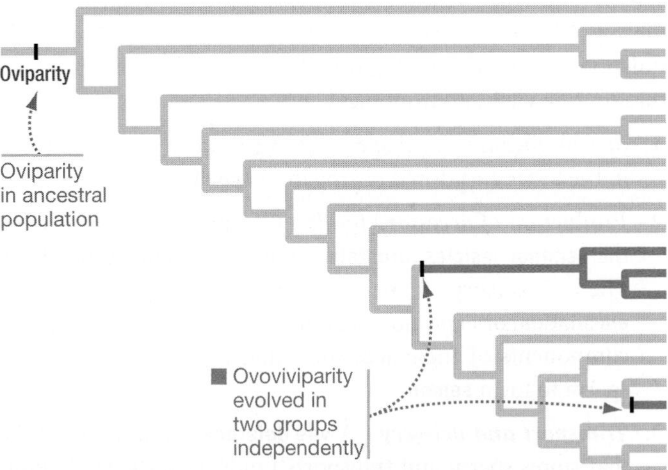

FIGURE 48.7 Ovoviviparity Has Evolved More than Once in *Sceloporus* Lizards. (a) *Sceloporus* is a genus of lizards from desert areas in North America. **(b)** Each twig on this phylogenetic tree represents a *Sceloporus* species or population from central Mexico.

- In the same lizard species, offspring that developed in colder nests run more slowly than do individuals of the same age that developed at higher temperatures.

Research into why live birth has evolved in certain populations continues.

Sceleporus lizards have been a productive group to study on this question because the trait has evolved multiple times in extremely closely related species—making it easier to find patterns between live birth and other characteristics, such as living in cold habitats. The issue is much more difficult to study in mammals, where viviparity evolved just once. As Chapter 34 pointed out, the monotremes (duck-billed platypus and echidna) are oviparous, while marsupials and placental mammals are viviparous. It is still not clear why, and how, viviparity evolved in mammals prior to the marsupial-placentals split. Is the cold-habitats hypothesis relevant? To date, nobody knows.

CHECK YOUR UNDERSTANDING

If you understand that . . .

- Some of the most basic aspects of reproduction are variable among animal populations or species: whether reproduction is sexual or asexual, whether fertilization is external or internal, and whether fertilized eggs develop in the environment or in the mother's body.
- When females mate with more than one, sperm from different males compete to fertilize eggs. Females may also select sperm from different males.

You should be able to . . .

1. Describe the advantages and disadvantages of oviparity and viviparity.
2. Suppose that female fruit flies can release a molecule called spermkillerene into their sperm-storage areas, and design an experiment to test the hypothesis that it can destroy sperm.

Answers are available in Appendix B.

48.3 Reproductive Structures and Their Functions

The first two sections of this chapter considered (1) the broad contrast between asexual and sexual reproduction and (2) general patterns in fertilization and egg care. Now let's explore the mechanics of sexual reproduction in more detail. The first task is to understand the anatomy of the male and female reproductive systems—focusing primarily on mammals.

The Male Reproductive System

In humans, the external anatomy of the male reproductive system consists of the scrotum and the penis. The saclike **scrotum** holds the testes; the penis functions as the organ of copulation prior to internal fertilization.

Biologists who have compared these structures among animal species have been struck by their variability and complexity. For example, a scrotum occurs only in certain mammal species. Whales, elephants, hedgehogs, moles, and many other mammal groups lack the structure entirely and instead have testes that are located well within the abdominal cavity. Among species that do have a scrotum, the structure's size and shape vary from tiny to prominent. In numerous species of primates, the scrotum is brightly colored and appears to function in courtship or other sexual display behaviors.

HOW DOES EXTERNAL ANATOMY AFFECT SPERM COMPETITION? Diversity in scrotal morphology pales in comparison with variation in the structure of the penis or other types of male **genitalia**. (The term genitalia refers to copulatory organs that have distinct parts.) In many groups of insects and spiders, for example, closely related species are morphologically identical except for their distinctive genitalia. Why are these organs so diverse?

The leading hypothesis on this question is that certain genitalia shapes give males an advantage when sperm competition occurs. To test this idea, biologists measured the size of the spines found on the tips of male genitalia in a species of seed beetle (**Figure 48.8a**), then documented which males were most successful, in terms of fertilizing eggs, when females mated with two males.

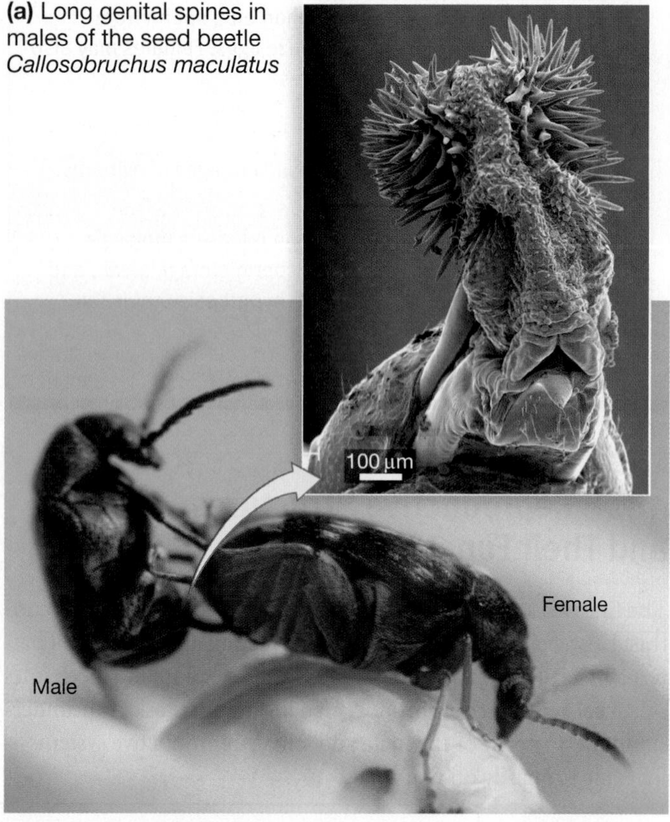

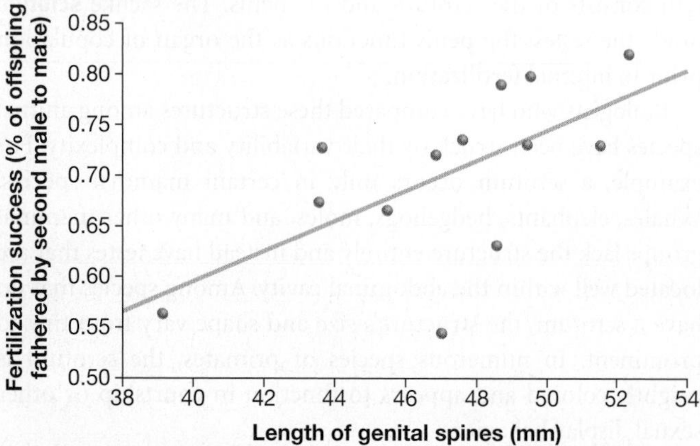

FIGURE 48.8 In Insects and Spiders, Male Genitalia May Vary among Species and Individuals and May Affect Reproductive Success. **(a)** An elaborate male reproductive structure in a seed beetle, used to transfer sperm to the female. The spikey structures at the top of the structure are called genital spines. **(b)** Data from experiments on reproductive success during sperm competition.

The *x*-axis in **Figure 48.8b** plots variation in the length of genital spines; the *y*-axis plots the proportion of eggs fertilized by the second male to copulate with a female, during experimental matings. The data points on the graph indicate genital spine length in each second-to-mate male tested. Note that second-to-mate males with the longest genital spines fathered a higher percentage of offspring than did others. The mechanism appears to be the ability of long spines to stick in the female's reproductive tract and prolong copulation time, even if the female is ready for copulation to end.

The punchline? Size matters, in seed beetles. Natural selection that occurs during sperm competition may explain why genitalia are so diverse among insect and spider species.

EXTERNAL AND INTERNAL ANATOMY OF HUMAN MALE REPRODUCTIVE ORGANS Although numerous structures are involved, the reproductive system in human males has just three basic functional components. **Figure 48.9** shows the relevant structures in side view and front view.

1. ***Spermatogenesis and sperm storage*** Sperm are produced in the testes and stored nearby in the **epididymis**.
2. ***Production of accessory fluids*** Complex solutions form in the **seminal vesicles**, **prostate gland**, and **bulbourethral gland**. These accessory fluids are added to sperm prior to **ejaculation**, or expulsion from the body. **Table 48.1** lists some components of these accessory fluids, which combine with sperm to form **semen**.
3. ***Transport and delivery*** A **vas deferens** is a muscular tube that stores sperm and transports fluids from the epididymis to the short **ejaculatory duct**. The semen then enters the **urethra**, a longer tube that passes through the penis and services both the reproductive and urinary systems in males. The semen is expelled during ejaculation.

The composition of the accessory fluids varies widely among animals. In many insects, spiders, and vertebrates, molecules in the accessory fluids congeal after they arrive in the female reproductive tract and plug it. Experiments have shown that these copulatory plugs can serve as an effective deterrent to future matings. In some species, though, females or second males actively remove them.

Another diverse aspect of male internal anatomy is a bone inside the penis called the **baculum**. Some mammal species, including humans, lack this feature. But among rodents, the shape of the baculum is so variable that it can be used to distinguish among species. And in seals, baculum size correlates with mating system—the mating practices observed in a species. For example, in seal species in which females routinely mate with several males before becoming pregnant, males have not only large testes for their size, but a large baculum as well. The testes and baculum are much smaller in species in which females mate with a single male. In all species in which it appears, the baculum helps stiffen the penis during copulation.

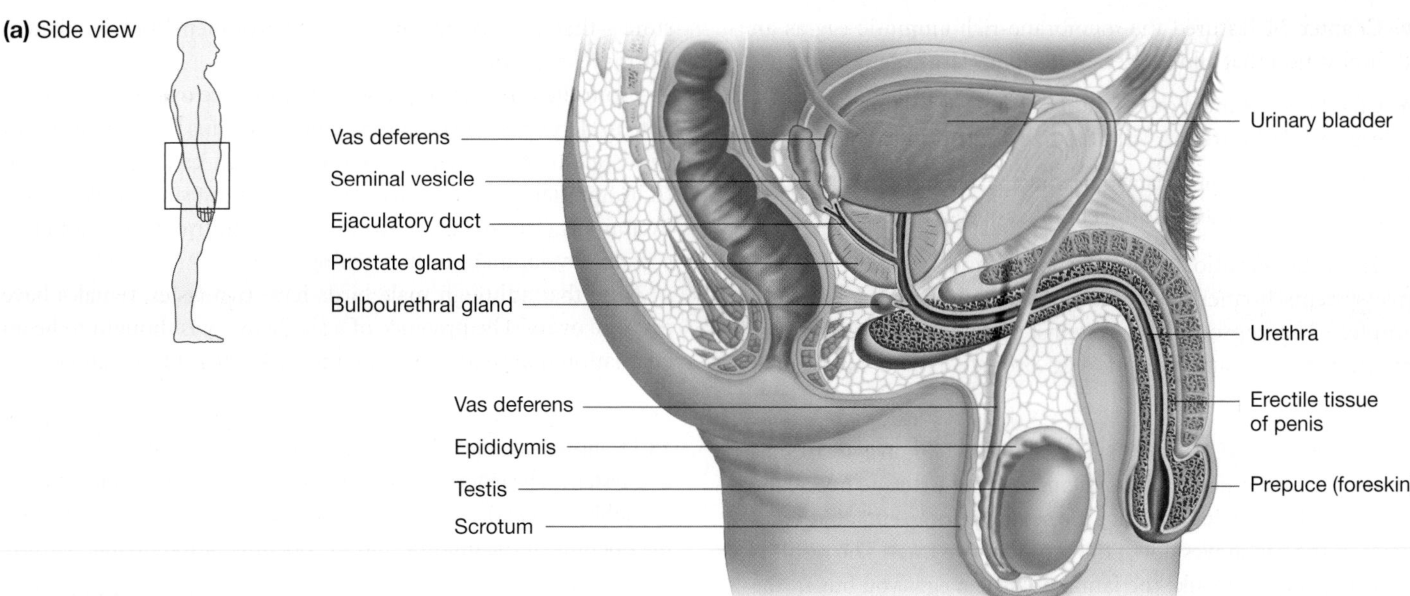

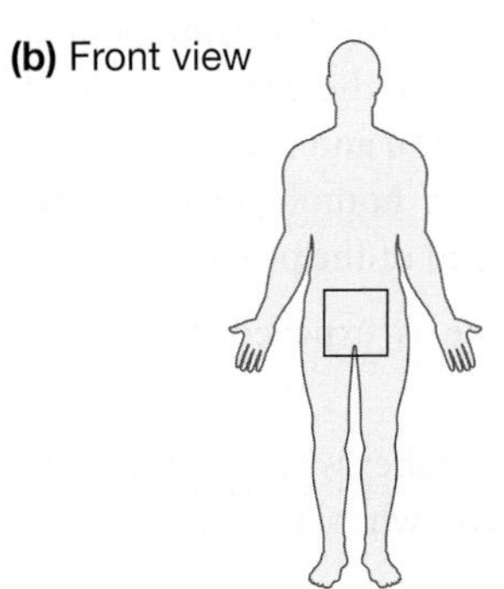

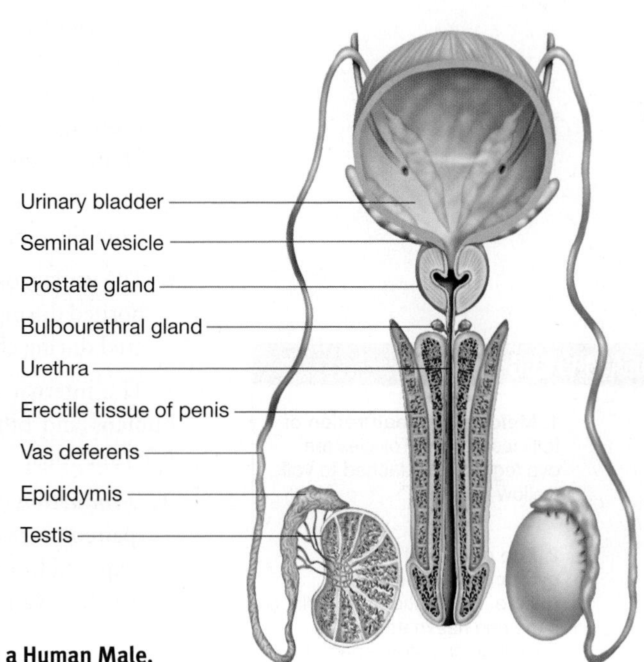

FIGURE 48.9 Anatomy of the Reproductive Tract in a Human Male.

SUMMARY TABLE 48.1 **Accessory Fluids in Human Semen**

Source	Content	Function
Seminal vesicles	Fructose (a sugar)	Source of chemical energy for sperm movement
	Prostaglandins	Stimulate smooth-muscle contractions in uterus
Prostate gland	Antibiotic compound	Prevent urinary tract infections in males?
	Citric acid	Nutrient used by sperm
Bulbourethral gland	Alkaline mucus	Lubricates tip of penis; neutralizes acids in urethra

The Female Reproductive System

In animals, the most important part of the female reproductive system is the ovary—where meiosis occurs and mature egg cells, or ova, are produced. In the vast majority of species, the mature egg cell is a membrane-bound structure consisting of a haploid nucleus, a full complement of other organelles, and a large supply of nutrients in the form of yolk.

Earlier chapters have analyzed aspects of egg structure:

- Chapter 22 introduced the extensive outer layers and the plasma membranes of sea urchin eggs, which play a key role in binding sperm from the same species and initiating fertilization.

- Chapter 34 featured the membrane-rich amniotic egg as an innovation that allowed tetrapods to lay large eggs.
- Chapter 41 pointed out that the amount of yolk present in an egg correlates strongly with the size of offspring at hatching.
- Chapter 41 presented data indicating that females face a trade-off between egg size and the total number of eggs they can produce.

To probe variation in egg structure and female reproductive systems further, let's consider two highly specialized examples: The reproductive systems of female birds and humans. Birds lay an amniotic egg protected by a hard shell; humans are viviparous.

THE REPRODUCTIVE TRACT OF FEMALE BIRDS **Figure 48.10** diagrams the bird reproductive system. Starting with the ovary, the labels on the drawing indicate the sequence of events that takes place as the egg moves down the reproductive tract. The result is a hard-shelled egg—like the familiar chicken eggs you buy in the store—that can be laid into the environment and incubated. Birds are oviparous.

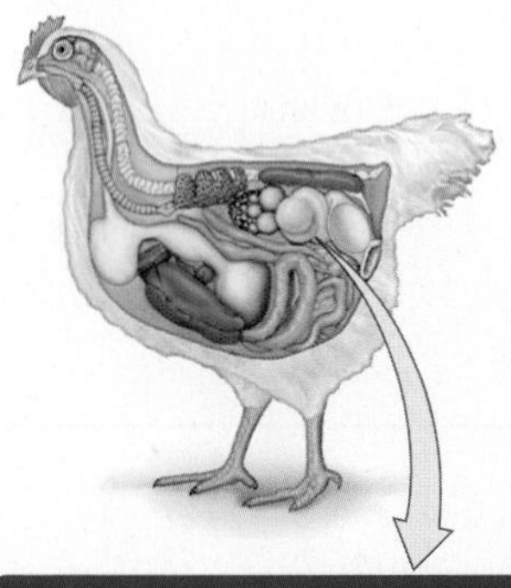

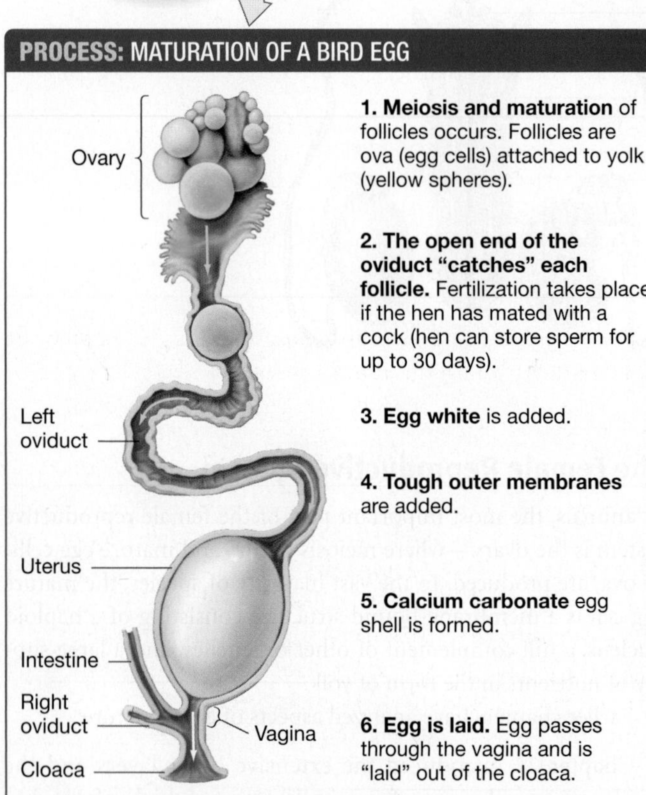

FIGURE 48.10 Structure and Function of the Reproductive Tract in a Female Bird.

From the time the egg is released from the ovary until the zygote undergoes mitosis and the embryo begins to develop, a bird egg is a single cell. The ostrich egg, which can be over 15 cm (6 in.) in diameter, contains one of the largest single cells known in animals. The structure stores enough nutrients and water to sustain development until hatching.

Note that although male birds have two testes, females have just one ovary. The presence of a single ovary is thought to be an adaptation that reduces weight and makes flight more efficient.

EXTERNAL AND INTERNAL ANATOMY OF HUMAN FEMALES **Figure 48.11** shows side and front views of the human female reproductive system. The external anatomy features the **labia minora** (singular: **labium minus**) and the **labia majora** (singular: **labium majus**), the opening of the urethra, and the opening of the vagina.

- The labia are folds of skin that cover the urethral and vaginal openings.
- The **clitoris** is an organ that develops from the same population of embryonic cells that gives rise to the penis in males. It becomes erect during sexual stimulation and is covered with a protective sheath called the *prepuce*, homologous with the prepuce (foreskin) that covers the end of the penis.
- The urethral opening, where urine is expelled, is separate from the reproductive structures.
- The **vagina**, or birth canal, is the chamber where semen is deposited during sexual intercourse and where the baby is delivered during childbirth.

The internal anatomy of the female reproductive system in humans and other mammals is dominated by structures with two functions:

1. ***Production and transport of eggs*** Eggs are produced in the paired ovaries. During **ovulation**, a developing oocyte (egg) is expelled from the ovary and enters the **oviduct**, also known as the **fallopian tube**, where fertilization may take place. Fertilized eggs are then transported from the oviduct to the muscular sac called the **uterus**.
2. ***Development of offspring*** The uterus is where embryonic development takes place. During childbirth, the developed embryo (now called a fetus) passes through the opening of the uterus, known as the **cervix**, and into the vagina.

To understand ovulation and embryonic development, it's essential to investigate the hormones that regulate them. Next let's look at research into sex hormones in mammals.

48.4 The Role of Sex Hormones in Mammalian Reproduction

Chapter 47 introduced the sex hormones **testosterone** and **estradiol**; the latter belongs to a class of hormones known as **estrogens**. Recall that testosterone and estradiol are steroids that

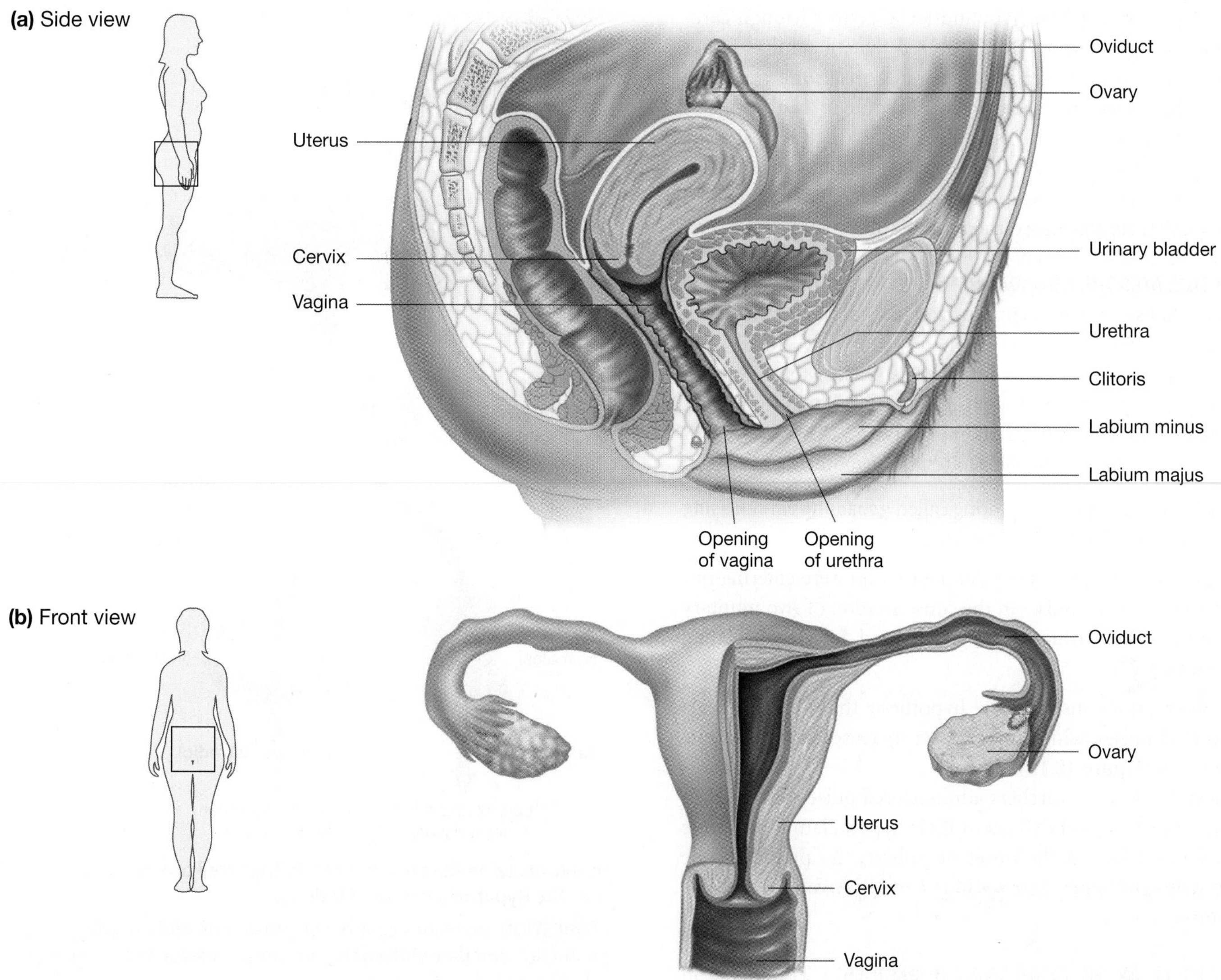

FIGURE 48.11 Anatomy of the Reproductive Tract in a Human Female.

✔**QUESTION** Generate a hypothesis to explain why male and female gonads are paired in mammals. How would you test your hypothesis?

bind to receptors inside the cytoplasm or nucleus of target cells. The resulting hormone-receptor complexes bind to DNA and trigger changes in gene expression.

Testosterone and estradiol are classified as gonadal hormones because they are produced in the gonads. Most testosterone is synthesized in specialized cells inside the testes; most estradiol and other estrogens are synthesized in the ovaries. More specifically, the female sex hormones are produced by cells that surround each developing egg. These surrounding cells form a structure called a **follicle** around the egg.

As Chapter 47 pointed out, the human sex hormones play a key role in three events:

1. development of the reproductive tract in embryos;
2. maturation of the reproductive tract during the transition from childhood to adulthood; and
3. regulation of spermatogenesis and oogenesis in adults.

To explore the action of sex hormones, let's take a closer look at their role in the transition from juveniles to adults.

Which Hormones Control Puberty in Mammals?

Puberty is the process that leads to sexual maturation in humans; **Table 48.2** on page 963 outlines the events that physicians use to track its progress. Chapter 47 pointed out that in amphibians, the juvenile-to-adult transition is triggered by the hormone T_3 (triiodothyronine); in insects the transition to adulthood occurs in response to ecdysone. But in humans, the transition is directed by increased levels of gonadal hormones—testosterone in boys and estradiol in girls.

A group of physicians recently offered dramatic evidence for testosterone's role: a 2-year-old boy who showed signs of entering puberty. His symptoms included an enlarged penis, the development of pubic hair, and facial acne. Through careful interviewing, the doctors discovered that the child's father was a bodybuilder who was smearing a testosterone cream on his own shoulders and arms in an attempt to build muscle mass. The child was apparently absorbing enough testosterone from being carried by his father to trigger puberty-like symptoms. All his symptoms except for penile enlargement subsided once his father discontinued the testosterone cream.

WHAT REGULATES THE GONADAL HORMONES? Some researchers suggested that sex hormone production is regulated by the hypothalamic-pituitary axis introduced in Chapter 47. Recall that chemical signals from the hypothalamus lead to the release of regulatory hormones from the pituitary gland, which then cause the release of hormones from other glands.

Two advances made it possible to test this hypothesis rigorously:

1. Researchers isolated a hormone called **gonadotropin-releasing hormone (GnRH)** from the hypothalamus.
2. Investigators noted that boys and girls who were entering puberty experienced pulses in the concentration of two pituitary hormones, **luteinizing hormone (LH)** and **follicle-stimulating hormone (FSH)**.

These observations inspired the hypothesis that GnRH directs LH and FSH pulses, which then trigger increases in testosterone and estradiol (**Figure 48.12**).

To test this idea, researchers administered pulses of GnRH to boys and girls who had deficits in their hypothalamus and were experiencing delays in the onset of puberty. As predicted, the GnRH treatment induced surges in LH and FSH, followed by puberty onset.

WHAT REGULATES THE REGULATORY HORMONES? The model in Figure 48.12 raises a question: What triggers GnRH increases at the appropriate age? Although this question remains unanswered, there is some evidence that nutritional state is involved. For example:

- The current average age for the onset of menstruation in females in the United States is slightly over 12 years. This is much earlier than the average age of 17 years in the United States during the eighteenth and nineteenth centuries, when the general nutritional state of the population was poorer.
- Among girls living today, individuals with large fat stores tend to enter puberty earlier than do girls who are thin.

Research continues on how age, nutritional condition, and perhaps other factors interact to promote the release of GnRH.

If you recall the discussion in Chapter 47 of how the adrenal hormone cortisol is controlled, however, you might suspect that Figure 48.12's model of sex-hormone regulation is simplified. Chapter 47 emphasized that many hormones participate in negative feedback—also called feedback inhibition—meaning that the presence of a hormone inhibits the factor that triggers its release.

Do sex hormones participate in negative feedback? The short answer to this question is yes. To appreciate the details, let's investigate hormonal control of the human menstrual cycle.

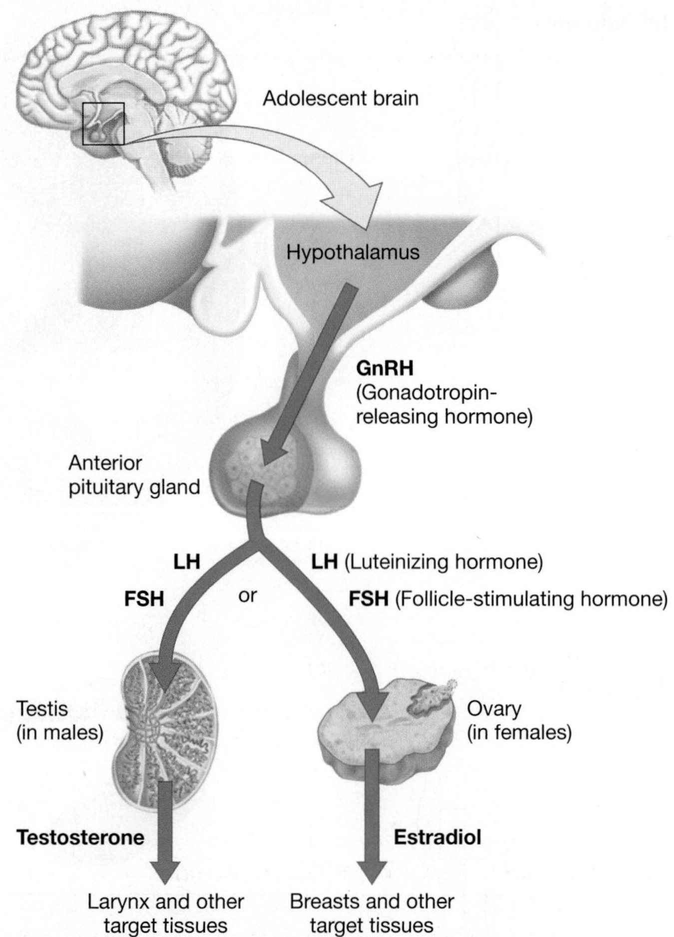

FIGURE 48.12 In Humans, Puberty Is Triggered by Hormones from the Hypothalamus and Pituitary.

✔**QUESTION** How does control of testosterone and estradiol production compare with control of cortisol release by the adrenal gland? (See Figure 47.10c.)

Which Hormones Control the Menstrual Cycle in Mammals?

Figure 48.13 illustrates the sequence of events in the human ovary during the **menstrual cycle**, a monthly reproductive cycle. Although the cycle's length varies among women, 28 days is about average.

In conjunction with changes in the ovary illustrated in the figure, the lining of the uterus undergoes a dramatic thickening and regression. Ultimately, part of the uterine lining sloughs off and is expelled through the vagina.

Day 0 in the menstrual cycle is marked by the beginning of **menstruation**—the expulsion of the uterine lining. The remainder of the cycle has two distinct phases:

1. ***Follicular phase*** A follicle matures during the **follicular phase**, which lasts an average of 14 days. Primary oocytes complete meiosis I during this phase. Ovulation occurs when the follicle is mature and releases its secondary oocyte into the oviduct. Researchers are still unsure what controls which ovary—right or left—will release an egg during ovulation.

TABLE 48.2 **Changes That Take Place during Puberty**

Pediatricians use the five stages of puberty listed here to diagnose the developmental stage of their patients.

Type	Boys	Girls
Stage 1 (Prepubertal)	No sexual development	No sexual development
Stage 2	Testes enlarge Body odor	Breast budding, first pubic hair Body odor Height spurt
Stage 3	Penis enlarges First pubic hair Ejaculation (wet dreams)	Breasts enlarge Pubic hair darkens, becomes curlier Vaginal discharge
Stage 4	Continued enlargement of testes and penis Penis and scrotum deepen in color Pubic hair curlier and coarser Height spurt Male breast development (temporary)	Onset of menstruation Nipple is distinct from surrounding areola Pelvis begins to widen
Stage 5 (Fully mature adult)	Pubic hair extends to inner thighs Increases in height slow, then stop Increased muscle mass	Pubic hair extends to inner thighs Increases in height slow, then stop Deposition of fat in hips, buttocks, thighs, and breasts

2. ***Luteal phase*** The **luteal phase** begins with ovulation and averages 14 days in length. Its name was inspired by the formation and subsequent degeneration of a structure called the **corpus luteum** ("yellowish body") from the ruptured follicle. Meiosis II occurs only if the ovulated cell is actually fertilized.

The regular occurrence of a menstrual cycle throughout the year makes human females extremely unusual among mammals. Although some mammals ovulate multiple times during the year, most ovulate only during a prescribed breeding season—often in response to environmental cues such as the onset of spring or fall; less often in response to cues from males.

In addition, only humans and other great apes menstruate. In the vast majority of mammals, the lining of the uterus is reabsorbed if pregnancy does not occur. Females that do not menstruate have an **estrous cycle** and are sexually receptive only during estrus—also known as being "in heat."

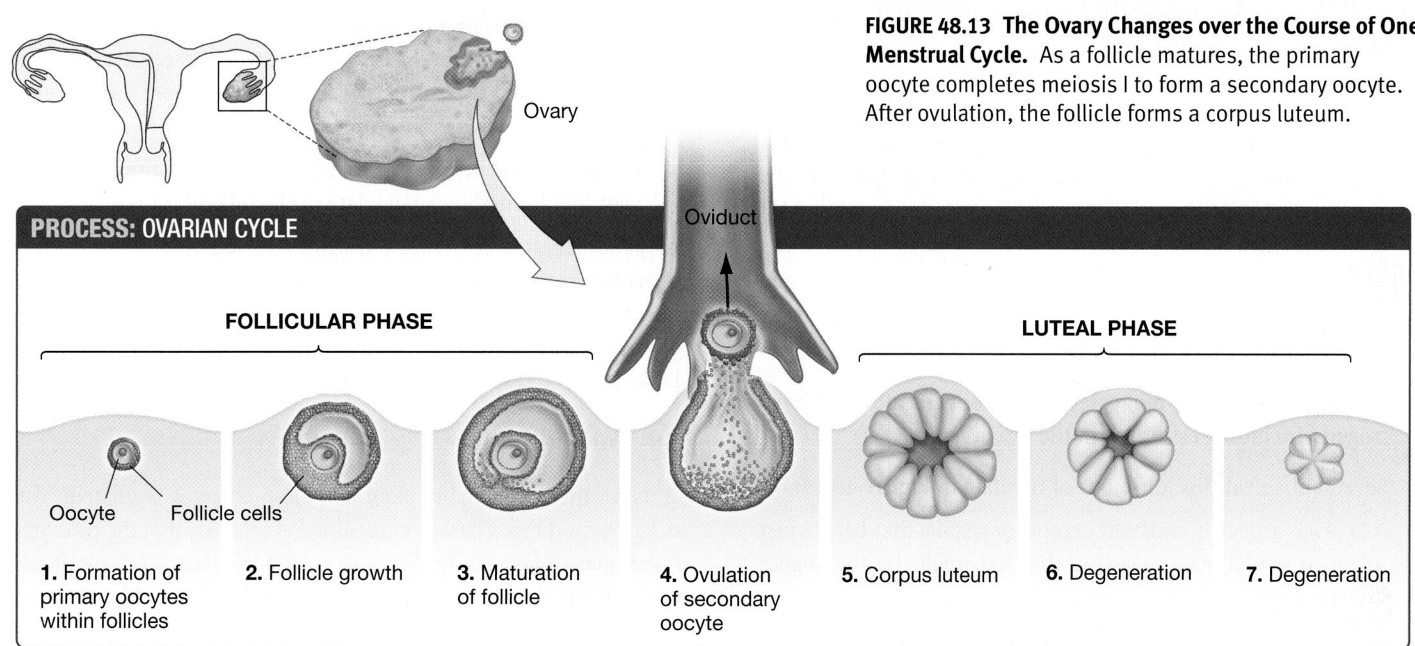

FIGURE 48.13 The Ovary Changes over the Course of One Menstrual Cycle. As a follicle matures, the primary oocyte completes meiosis I to form a secondary oocyte. After ovulation, the follicle forms a corpus luteum.

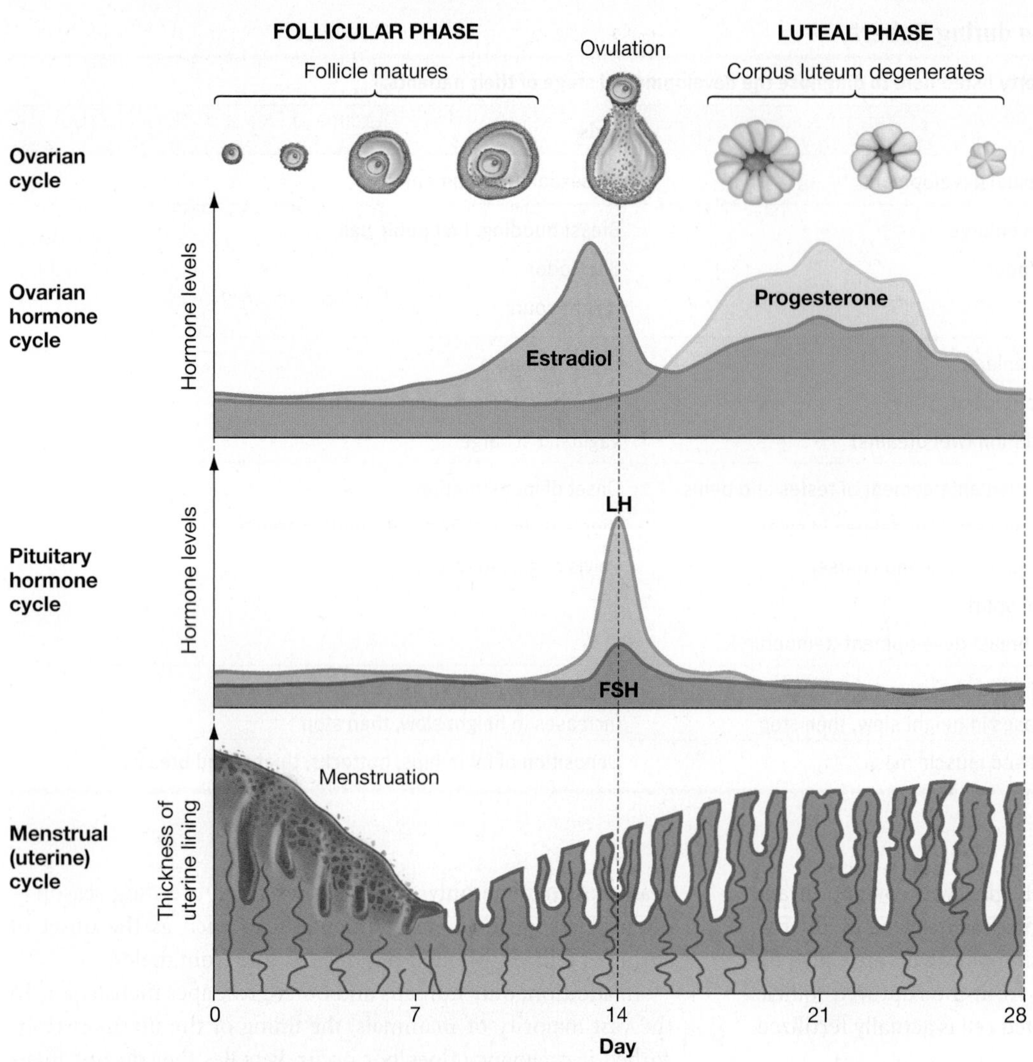

FIGURE 48.14 Changes That Occur during a Menstrual Cycle in a Human Female.

But whether an estrous or menstrual cycle occurs, the basic sequence of events, with a follicular phase preceding ovulation and a luteal phase following ovulation, is shared among mammals. As it turns out, hormonal control of the estrous and menstrual cycles is also similar.

HOW DO PITUITARY AND OVARIAN HORMONES CHANGE DURING A MENSTRUAL CYCLE? By monitoring hormone concentrations in the blood or urine of a large number of women over the course of the menstrual cycle, researchers were able to document dramatic changes in the concentrations of estradiol and several other hormones. During each cycle:

- LH and FSH are produced in the anterior pituitary gland in response to GnRH.
- The steroid hormone **progesterone** is produced along with estrogens, including estradiol, in the ovaries.

Several observations jump out of the data in **Figure 48.14**:

1. LH levels are fairly constant except for a spike that begins just prior to ovulation, suggesting that LH might be the trigger for this event.
2. FSH concentrations are relatively high during the follicular phase and low during the luteal phase, though they also make a small spike prior to ovulation.
3. Progesterone is present at very low levels during the follicular phase but at high levels during the luteal phase. This observation suggested that progesterone might support the maturation of the thickened uterine lining.
4. Estradiol concentrations change in a much more complex way, highlighted by a peak late in the follicular phase.

In general, estradiol surges during the follicular phase while progesterone surges during the luteal phase. Similar patterns are observed in other mammals, ranging from marsupials to mice. To review the hormonal changes that occur before and after ovulation, go to the study area at *www.masteringbiology.com.*

MB **Web Activity** Human Reproduction

Now the question is: How are these changes regulated? You might hypothesize, based on reading Chapter 47, that the pituitary hormones released in response to GnRH trigger the release of gonadal hormones, and that feedback occurs. If so, you'd be correct.

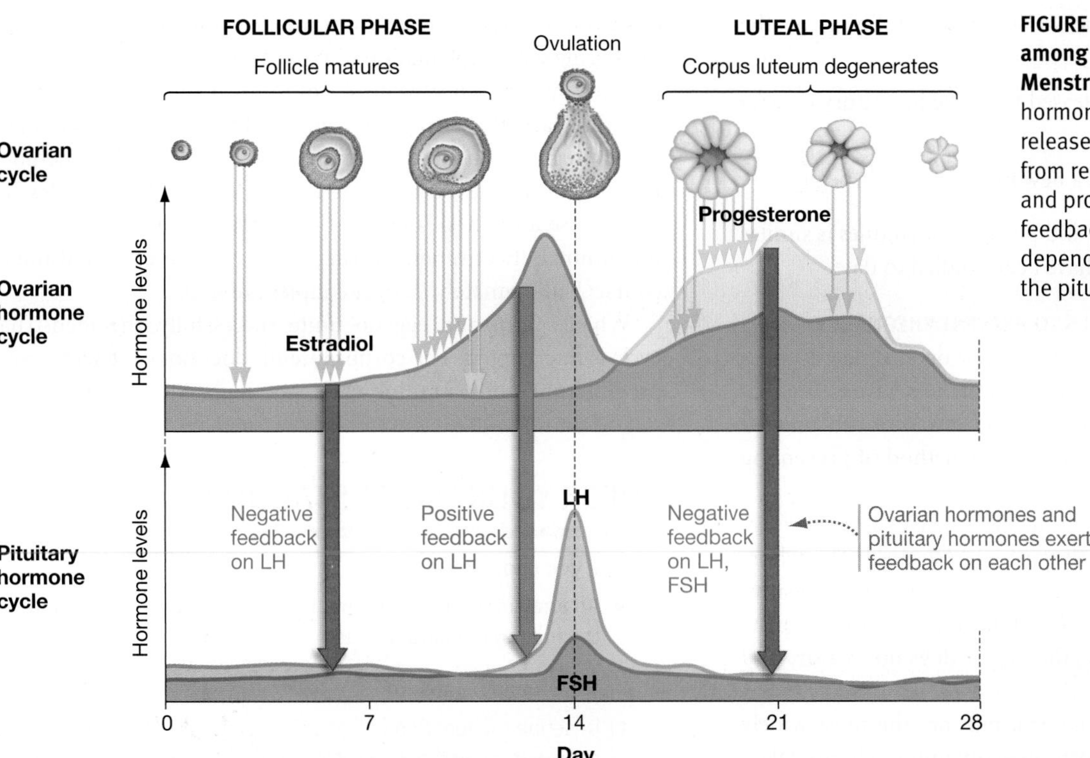

FIGURE 48.15 Complex Interactions among Hormones Regulate the Menstrual Cycle. The pituitary hormones FSH and LH control the release of estradiol and progesterone from reproductive tissues. Estradiol and progesterone may exert negative feedback, positive feedback, or both—depending on their concentration—on the pituitary hormones.

HOW DO PITUITARY AND OVARIAN HORMONES INTERACT? Experiments helped establish that changes in the concentration of estradiol and progesterone affect the release of the pituitary hormones LH and FSH. Researchers worked with three volunteers whose ovaries had been removed because of cancerous growths or other problems. The women were receiving low, maintenance-level doses of estradiol, which appeared to exert negative feedback on LH and FSH.

But when the investigators injected the women with larger doses of estradiol or with progesterone, the situation changed dramatically. For example, a large increase in estradiol stimulated a dramatic spike in LH levels. This suggested that positive feedback was occurring. High levels of estradiol *increase* the release of LH, even though low doses of estradiol suppress it. This is a key observation: When feedback occurs, estradiol's effect on the pituitary hormones is dosage-dependent. In contrast, progesterone injections appeared to inhibit both FSH and LH. Progesterone exerts only negative feedback on the pituitary hormones.

To summarize the interplay between LH, FSH, estradiol, and progesterone, let's start at day 0 and follow key events as the cycle progresses (**Figure 48.15**).

Day 0–7

- As the uterus is shedding much of its lining, a follicle is beginning to develop in one ovary under the influence of FSH.
- The follicle produces estradiol and a small amount of progesterone.
- While its levels are still relatively low, estradiol suppresses LH secretion through negative feedback inhibition.

Day 8–14

- As the follicle grows, its production of estradiol gradually increases. The increase in estradiol stimulates mitosis and an increase in cell number in the uterine lining.
- The enlarged follicle produces large quantities of estradiol, which begin to exert positive feedback on LH secretion.
- Positive feedback results in a spike in LH levels, just after estradiol concentrations peak.
- The LH surge triggers ovulation and ends the follicular phase.

✔If you understand how estradiol's effect on LH changes from negative regulation to positive regulation, you should be able to predict the effect of (1) injecting a female volunteer with high estradiol concentrations early in a cycle, and (2) using a drug to keep estradiol production low throughout the cycle.

Day 15–21

- As the corpus luteum develops from the remains of the ruptured follicle, it secretes large amounts of progesterone and small quantities of estradiol, in response to LH.
- The rise in progesterone lowers production of LH and FSH and activates the thickened uterine lining, creating a spongy tissue with a well-developed blood supply. In this way, progesterone fosters an environment that supports embryonic development if fertilization occurs.

Day 22–28

- If fertilization does not occur, the corpus luteum degenerates.
- Progesterone levels fall as the corpus luteum shrinks.

- When progesterone declines, the thickened lining of the uterus degenerates.
- GnRH, LH, and FSH are released from the inhibitory control that progesterone exerts.
- LH and FSH levels rise, and a new menstrual cycle begins.

The interplay between ovarian and pituitary hormones is similar in other mammal species that have been studied to date.

MANIPULATING HORMONE LEVELS TO PREVENT PREGNANCY Data on hormonal control of the menstrual cycle opened new avenues in birth control research. Specifically, researchers have found that manipulating levels of progesterone and estradiol can prevent ovulation and serve as a safe and effective method of preventing unwanted pregnancies.

Hormone-based birth-control methods deliver synthetic versions of progesterone, or progesterone and estradiol. These hormones suppress the release of GnRH and FSH through negative feedback inhibition. Because an LH spike does not occur during the follicular phase of the cycle, the follicle does not mature and ovulation does not occur.

In the United States, birth control pills are the most widely used contraceptive method. Hormone-containing pills are taken for three weeks and then stopped for one week to allow menstruation to occur.

But as **Table 48.3** indicates, hormone-based methods are just one of several approaches to preventing pregnancy. Other methods work by preventing sperm from contacting the oocyte, or by interfering with implantation of the embryo.

The "Percent Effectiveness" column on the far right of the table indicates the average percentage of women who do not become pregnant during one year of typical use of that method. Percent effectiveness usually increases dramatically if couples are able to use a method "perfectly"—meaning, exactly as specified for optimal effectiveness, during every episode of sexual intercourse. Unfortunately, perfect couples are rare.

When sperm and egg do unite successfully, the menstrual cycle is interrupted. The corpus luteum does not degenerate, and progesterone and estradiol levels stay high. Menstruation does not occur. Instead, the mother is now pregnant.

CHECK YOUR UNDERSTANDING

If you understand that . . .

- An array of hormones interacts in complex ways to regulate the human menstrual cycle.

✓ You should be able to . . .

1. Describe the function of FSH and explain why FSH levels increase at the end of one cycle and the start of the next cycle.
2. Predict the consequences of a drug that inhibits the release of FSH.

Answers are available in Appendix B.

SUMMARY TABLE 48.3 Comparing Methods of Contraception

"Percent Effectiveness" indicates the average percentage of women who do not become pregnant during one year of typical use.

Type	Name	Mode of Action	Percent Effectiveness
Hormone-based methods	The Pill, the Patch, the Ring, the Shot, the Implant	Provides continuous or cyclical delivery of progesterone, or progesterone plus estradiol	92 to 99.9*
Barrier methods	Condom	Covers penis and prevents sperm from entering vagina	85
	Female condom	Covers labia, vagina, and cervix and prevents sperm from entering uterus	79
	Diaphragm	Covers cervix and prevents sperm from entering uterus	84
	Sponge	Covers cervix and prevents sperm from entering uterus; also contains a molecule that immobilizes sperm	84
	Spermicide	Foam or jelly covers cervix and prevents sperm from entering uterus; contains a molecule that immobilizes sperm	71
Behavioral methods	Rhythm method	Couple refrains from vaginal intercourse around time of ovulation	80
	Withdrawal	Man withdraws penis prior to ejaculation	73
Prevent implantation	Intrauterine device (IUD)	Small T-shaped structure inserted into uterus; mode of action not known	99
	Emergency contraception	Delivers progesterone or progesterone and estradiol after unprotected vaginal intercourse	92
	Mifepristone (RU-486)	Blocks progesterone receptors so menstruation occurs even after fertilization and implantation	92

*Depends on delivery system used.

48.5 Pregnancy and Birth in Mammals

Viviparity allows the mother to provide a warm, protected environment for offspring during early development. Oviparous species that guard or incubate their eggs also provide warm, safe surroundings for their young.

Any investment that parents make in an offspring comes at a cost, however: The more a mother invests in each offspring, the lower the number of offspring she can produce. Chapter 41 explored this fitness trade-off between the quantity and quality of offspring.

Pregnancy and **lactation**—providing milk that nourishes offspring after birth—represent some of the most extreme forms of parental care known in animals. And in some mammal species, parental care continues long after lactation ends. Humans, for example, are largely or completely dependent on their parents for protection and nutrition until puberty or young adulthood.

Let's examine how mammals make this investment, starting with marsupials and then turning to the eutherian mammals.

Gestation and Early Development in Marsupials

You might recall from Chapter 34 that mammals comprise three major monophyletic groups: monotremes, eutherians, and marsupials. Although all mammals are endothermic, have fur, and lactate, the three lineages are distinguished by their mode of reproduction:

- **Monotremes** lay eggs and incubate them until hatching.
- In **eutherians**, mothers carry offspring for relatively long periods of development during gestation and nourish them via a placenta.
- In **marsupials**, the corpus luteum is not maintained as it is in eutherians, and no placenta forms.

In marsupials, young are ejected from the mother's body at the end of the estrous cycle. As a result, they are far less developed than eutherian mammal offspring which complete **gestation**—the developmental period that takes place inside the mother.

The jaws, gut, lungs, and forelimbs of a newly born kangaroo are relatively well developed at birth. As a result, the offspring is able to climb from its mother's vagina to a nipple, which is usually enclosed in a pouch created by a flap of skin (**Figure 48.16a**). The offspring clamps onto the nipple and continues to develop, fed by the mother's milk (**Figures 48.16b** and **48.16c**).

Even after growing large enough to leave the pouch and begin moving and feeding on its own, offspring will return to the pouch for protection. Marsupial mothers invest a great deal in their offspring, even though a relatively short period of development takes place inside their bodies.

Major Events during Human Pregnancy

Marsupials and eutherians differ sharply in terms of how long the developing embryo is retained inside the mother's body. As a model organism in eutherian reproduction, let's consider humans.

When a secondary oocyte is released from the human ovary, the cell is viable for less than 24 hours. Human sperm, in contrast, remain capable of fertilizing an egg for up to five days. Therefore, sexual intercourse in humans has to occur less than five days prior to ovulation, or immediately after ovulation, for pregnancy to result.

Although an ejaculate may contain hundreds of millions of sperm, most die as they travel through the uterus. Only 100 to 300 actually succeed in reaching the oviduct, where fertilization takes place.

Chapter 22 detailed how the sperm and oocyte (egg) interact when they meet. You might recall that when sperm and oocyte meet, enzymes released from the head of the sperm create a path through the material surrounding the oocyte membrane. Once the membranes of the oocyte and sperm have fused, the oocyte nucleus completes meiosis II. The two nuclei then unite to form a diploid zygote.

IMPLANTATION Smooth-muscle contractions in the oviduct gradually move the zygote toward the uterus. As it travels, the cell begins to divide by mitosis. By the time it reaches the lining of the uterus, the embryo consists of a hollow ball of cells. It then undergoes **implantation**—meaning that it becomes embedded in the thickened, vascularized wall of the uterus. It will stay in the uterus for approximately 270 days (9 months).

Once the embryo is implanted in the uterine lining, its cells begin synthesizing and secreting the hormone **human chorionic gonadotropin (hCG)**. hCG is a chemical messenger that prevents

(a) Red kangaroo at birth

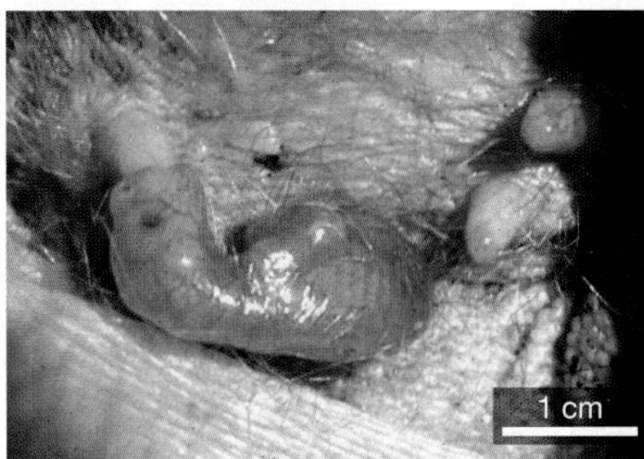

(b) 4 weeks after birth

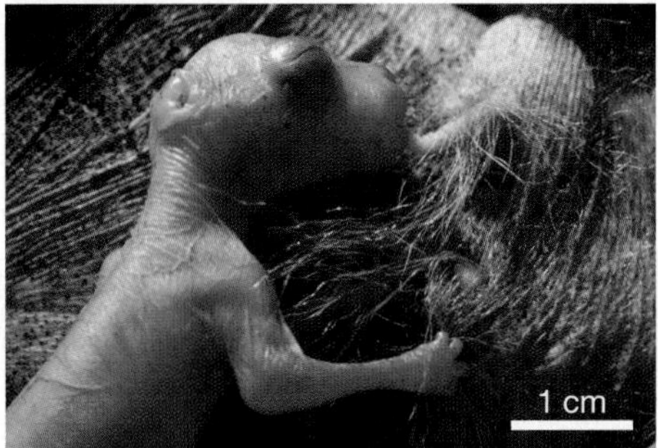

(c) 4 months after birth

FIGURE 48.16 Marsupials Trade a Long Gestation Period for a Long Lactation Period. Relative to eutherians, marsupial offspring spend a short time inside the mother and a relatively long time being fed milk after birth and being protected in the mother's pouch.

the corpus luteum from degenerating. When hCG is present, the ovary continues secreting progesterone and the menstrual cycle is arrested. Enough hCG is produced by the embryo and excreted in the mother's urine to be used as an indicator in pregnancy tests.

THE FIRST TRIMESTER Human gestation is divided into 3-month stages called trimesters. Not long after implantation is complete, mass movements of cells result in the formation of the three major embryonic tissues called ectoderm, endoderm, and mesoderm (see Chapter 22). By 8 weeks of age, these tissues have differentiated into the various organs and systems of the body. Also by this time, the heart has begun pumping blood through a circulatory system. The embryo at this stage is called a **fetus** (**Figure 48.17a**).

Early in development, the embryonic ectoderm contributes to several important membranes. One of these membranes, the **amnion**, completely surrounds the embryo. The amnion eventually fills with amniotic fluid, which provides the embryo with a protective cushion.

The other key event in the first trimester is the formation of the **placenta**. This organ, which starts to form on the uterine wall a few weeks after implantation, is composed of tissues from both mother and embryo (see Chapter 34). Because the placenta contains a dense supply of blood vessels from the mother, it is the primary source of nutrition for the growing fetus. Arteries transport blood from the circulatory system of the fetus, through the **umbilical cord**, to an extensive capillary bed in the placenta. In this way, the placenta provides a large surface area for the exchange of gases, nutrients, and wastes between maternal and fetal blood, even though maternal and fetal blood do not commingle.

The placenta secretes a variety of hormones, including large amounts of progesterone and estrogens. Because these hormones suppress the release of GnRH, LH, and FSH through negative feedback, they prevent the maturation and ovulation of additional follicles. By the end of the first trimester, the placenta is producing more than enough progesterone to replace the amount that had been produced by the corpus luteum, which has degenerated by this time. In essence, the placenta takes over from the corpus luteum in terms of secreting the hormones required to maintain the pregnancy.

✔If you understand how hormones influence pregnancy, you should be able to explain (1) why women who produce low levels of progesterone from the corpus luteum are prone to miscarriage, and (2) why progesterone therapy to help these women can be safely discontinued after the first trimester.

THE SECOND AND THIRD TRIMESTERS After the fetal organs and placenta form during the first trimester, the remainder of development focuses on growth (**Figures 48.17b** and **48.17c**). During the last weeks of pregnancy, the brain and lungs undergo particularly dramatic growth and development. If a baby is born prematurely, intervention may be required to keep the baby alive until the lungs can complete their development.

The machinery and level of hospital care required by premature infants emphasize just how superbly adapted mothers are to nourishing a growing fetus in the uterus. It costs hundreds of thousands of dollars for health care providers to do what mothers do naturally in the last trimester. Let's take a closer look at this critical aspect of pregnancy.

How Does the Mother Nourish the Fetus?

In oviparous and ovoviviparous species, mothers produce relatively large eggs that contain all of the nutrients and fluids that the embryo needs to develop until hatching. But in some viviparous species, eggs are relatively small and contain almost no nutrients.

(a) 1st trimester

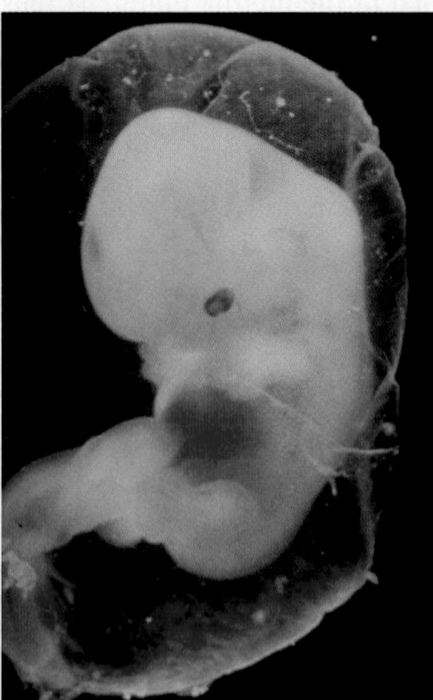

(b) 2nd trimester

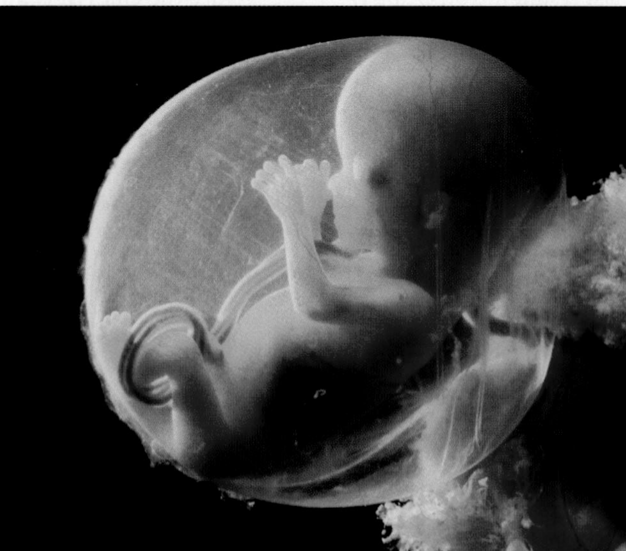

(c) 3rd trimester

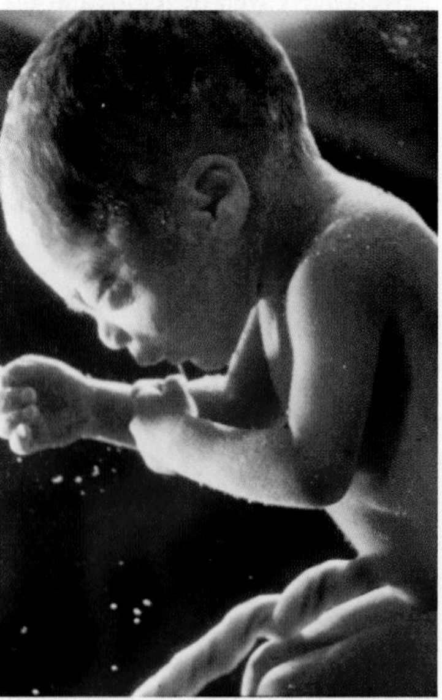

FIGURE 48.17 Development of the Human Fetus.

In species such as humans, the developing embryo depends on the mother's body for oxygen, chemical energy in the form of sugars, amino acids and other raw materials for growth, and waste removal. What physiological changes occur in human mothers to accommodate these demands?

OXYGEN EXCHANGE BETWEEN MOTHER AND FETUS During pregnancy, a mother's respiratory and circulatory systems change in ways that increase the efficiency of nutrient transfer and gas exchange with the fetus. For example:

- A woman's total blood volume expands by as much as 50 percent during pregnancy.
- To accommodate the increase in blood volume, maternal blood vessels dilate (widen) and blood pressure drops.
- The mother's heart enlarges and beats faster, increasing her total cardiac output by almost 50 percent.
- The mother's breathing rate and breathing volume increase to accommodate the fetus's demand for oxygen and its production of carbon dioxide.

In addition, important adaptations heighten the efficiency of gas exchange between the mother and the embryo, in the placenta. In many species, maternal and fetal blood flows in the placenta occur in a countercurrent fashion (**Figure 48.18a**). As Chapter 41 explained, countercurrent flows maintain a concentration gradient that increases the efficiency of diffusion or other types of exchange. The data for partial pressures of oxygen in Figure 48.18a are from experiments on sheep.

Countercurrent flow does not occur in the human placenta, so fetal blood does not become as highly oxygenated as sheep blood. Still, oxygen exchange between mother and fetus is efficient. For example, maternal arteries in humans empty into a space at the junction of the maternal and fetal portions of the placenta. This space is packed with small projections called villi, which contain the fetal blood vessels. Thus, a large surface area from the fetus is bathed with highly oxygenated maternal blood. The fetal villi are analogous to the alveoli of the lungs (Chapter 44), which provide a large surface area for gas exchange.

Figure 48.18b illustrates another adaptation that increases the rate of oxygen delivery to a human fetus. Notice that the axes on this graph give the partial pressure of oxygen in blood or tissue versus the percentage of hemoglobin that holds oxygen at that pressure. (Recall from Chapter 44 that this type of graph is an **oxygen-hemoglobin equilibrium curve**.)

The key point on this graph is that the data for fetal hemoglobin are shifted to the left of the data for adult hemoglobin. As a result, the fetus's blood always has a higher affinity for oxygen than does the mother's blood. This means that even if the partial pressure of oxygen is the same in the mother and fetus, oxygen will move from the mother's blood to the fetus's blood, because it is held more tightly by the fetal hemoglobin.

Biologists interpret this pattern as an adaptation. The high oxygen affinity of fetal hemoglobin ensures that the fetus is always able to acquire oxygen from the mother.

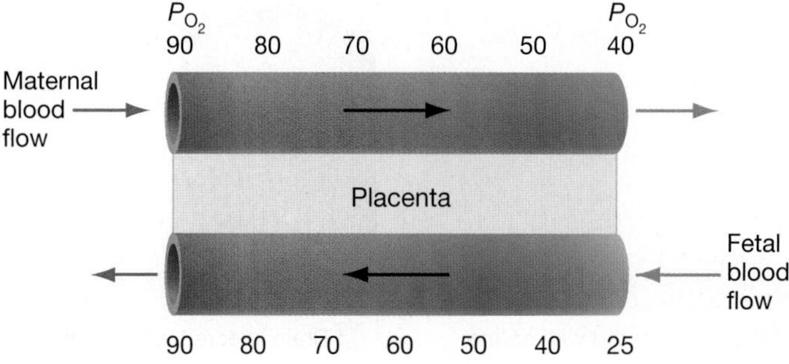

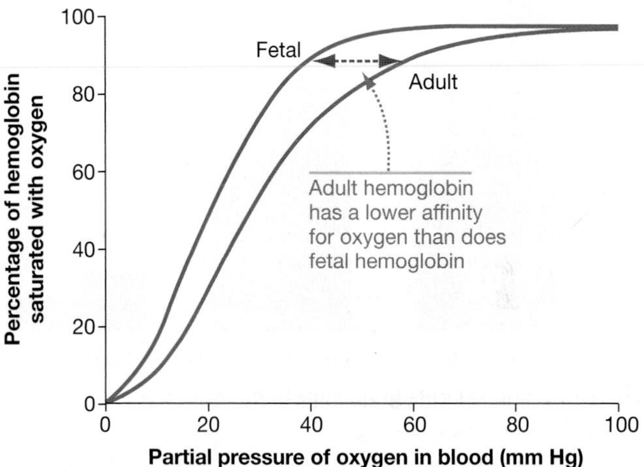

FIGURE 48.18 Adaptations That Increase Delivery of Oxygen to a Mammalian Fetus.

TOXIC CHEMICALS CAN BE TRANSFERRED FROM MOTHER TO FETUS Mothers and embryos exchange more than nutrients and wastes—they can also exchange dangerous molecules. As an example, consider the thalidomide tragedy. During the 1950s, hundreds of children were affected by their mothers' consumption of the tranquilizer thalidomide. The molecule diffused into the fetal bloodstream and caused birth defects—often a dramatic shortening of the arms.

Although thalidomide is now banned for use by pregnant women, alcohol use continues to affect newborns. Compared to children of mothers who do not drink alcohol, children of mothers who imbibe ethanol are at high risk for hyperactivity, severe learning disabilities, and depression. Collectively, these symptoms are termed **fetal alcohol syndrome (FAS)**.

To investigate why FAS occurs, researchers simulated what happens when a pregnant mother ingests alcoholic beverages—exposing offspring to ethanol. They gave one group of newborn (7-day old) rats two injections of a harmless salt solution at a dosage of 2.5 g per kg body weight, and a second group identical injections except that the salt solution contained 20 percent ethanol. For a female human of average weight, this would be equivalent to drinking 2.5 cans of beer or glasses of wine.

(a) Brain sections from newborn rats

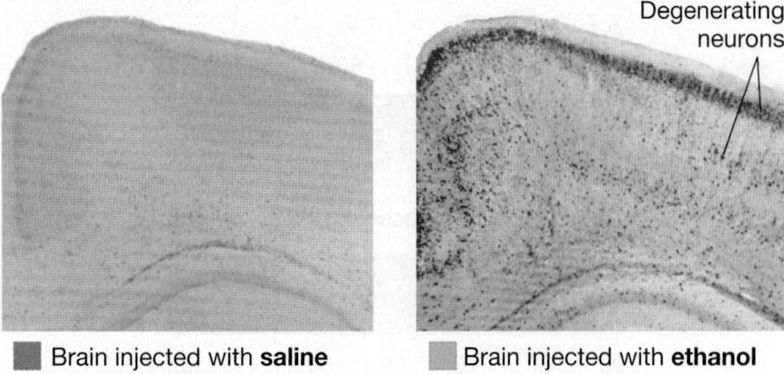

(b) Weights of brain tissue from newborn rats

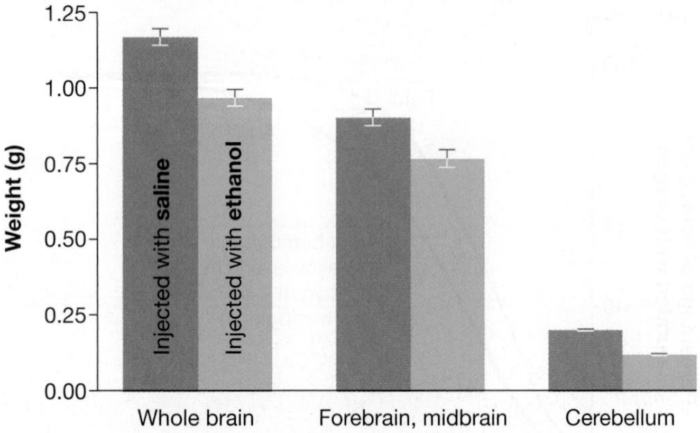

FIGURE 48.19 Ethanol Kills Brain Cells in Newborn Rats.

The photos of brain sections in **Figure 48.19a** were taken a day after the injections. The black specks in the brains of the ethanol-treated rats indicate degenerating neurons. As the bar graphs in **Figure 48.19b** show, the ethanol-treated rats had smaller brains than the saline-treated individuals did.

The messages of these data are that (1) ethanol destroys growing neurons, and (2) even a single, "moderate" dose of alcohol can have a devastating effect. Based on results like these, public health officials strongly advise pregnant women against drinking *any* alcohol.

Birth

Although the mechanisms responsible for triggering the birthing process are not completely understood, the posterior pituitary hormone **oxytocin** is important in stimulating smooth-muscle cells in the uterine wall to begin contractions. The contractions that expel the fetus from the uterus constitute **labor.**

When the birthing process starts, the uterus contracts at relatively low frequency. Then, as **Figure 48.20** shows:

1. The cervix at the base of the uterus begins to open, or dilate. Once the cervix is fully dilated, uterine contractions become more forceful, longer, and more frequent.
2. The fetus is expelled through the cervix and into the vagina.
3. After the baby is delivered, the placenta remains attached to the uterine wall. At this point, caregivers clamp and cut the umbilical cord that connects the child and the placenta. When the mother delivers the placenta and accompanying membranes, birth is complete.

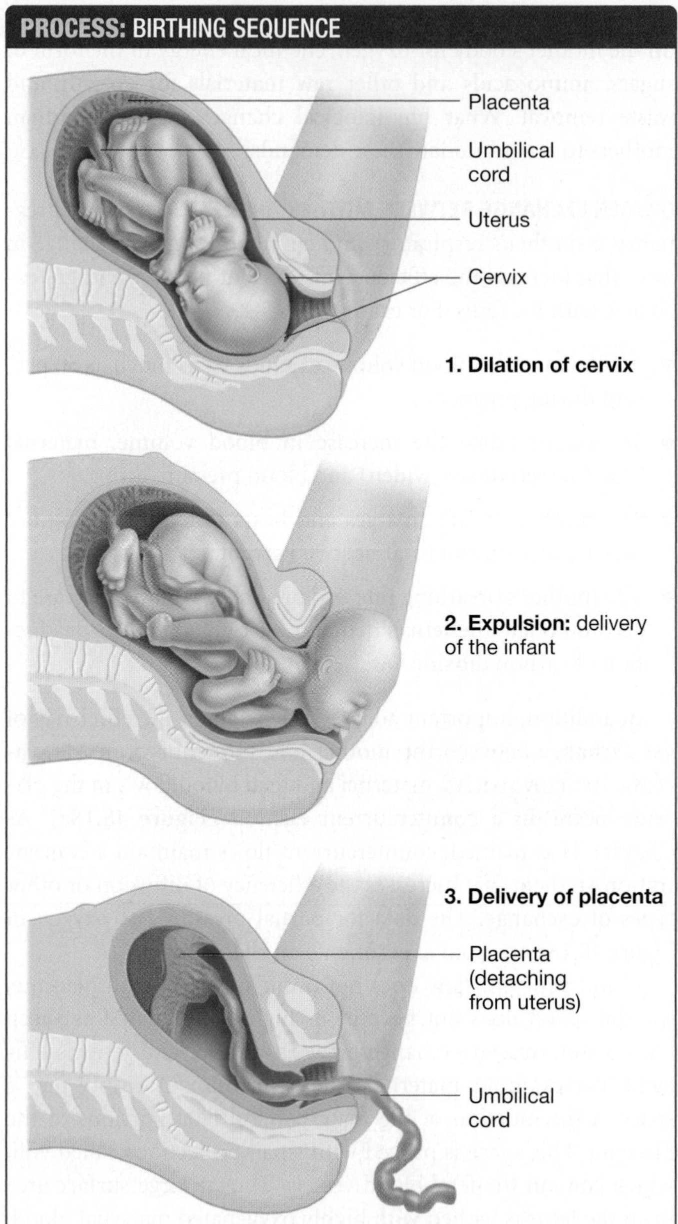

FIGURE 48.20 Major Events during Human Birth.

✔QUESTION Human babies are normally born headfirst and face down, as shown. Predict the consequences of a baby positioned face up, or feetfirst, in the birth canal.

Although this description sounds straightforward, in reality a large number of complications are possible. For example, **Figure 48.21** shows the number of Swedish mothers who died, per 100,000 live births, in five-year intervals between 1760 and 1980. In 1760, approximately 1.4 mothers died for every 100 infants successfully delivered. In most cases, the cause of death was blood loss or infection following delivery. Note that the steepest drop in mortality rate occurred with the introduction of hand-washing in the late 1870s.

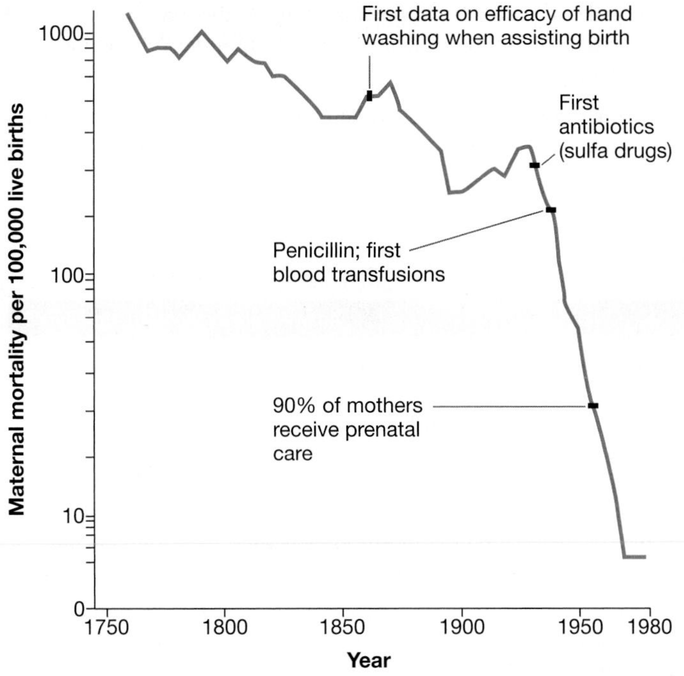

As a result of sterile techniques, antibiotics, and blood transfusion technology, Sweden's mortality rate has now declined to less than 0.007 percent. Improved nutrition, sanitation, and medical care have also reduced infant mortality rates in many countries.

The huge decline in the rate of death associated with childbirth qualifies as one of the great triumphs of modern medicine. Unfortunately, because many developing nations lack sterile facilities and antibiotics, the mortality of mothers and infants remains high in those countries.

FIGURE 48.21 Maternal Mortality in Childbirth Has Decreased Dramatically. These data are from Sweden; the scale on the vertical axis is logarithmic. *Note:* Data showing that hand-washing dramatically reduced maternal mortality during childbirth were published in 1861, but doctors resisted the idea until well after Pasteur had published the germ theory of disease in 1878.

✔**QUESTION** If a woman living in Sweden in 1760 had 10 children over the course of her lifetime, what was her overall chance of dying in childbirth?

CHAPTER 48 REVIEW

For media, go to the study area at www.masteringbiology.com

Summary of Key Concepts

The reproductive systems of animals are highly variable. Some species switch between asexual and sexual reproduction. When sexual reproduction occurs, fertilization may be external or internal and egg development may take place inside or outside the mother's body, depending on the species.

- Asexual reproduction produces offspring that are genetically identical to the parent. In contrast, sexually produced offspring are genetically unique because of recombination during meiosis and the fusion of haploid gametes from different parents during fertilization.
- In human males, spermatogenesis is continuous throughout adult life, but in human females all primary oocytes are formed early in development. Meiosis in females is arrested for long periods of time, and cell division after meiosis is so unequal in females that just one egg—not four—is produced from each primary oocyte.
- Fertilization is external in many aquatic animals but internal in almost all terrestrial species.
- When sperm competition occurs, males have large testes relative to their body size, and the last male to mate usually fathers a disproportionately large number of offspring.
- Once eggs are fertilized, females may lay the eggs or retain them and give birth to live offspring. In some groups at least, viviparity may be an adaptation that increases the survival of young in cold habitats.

✔ You should be able to explain why fish that care for their young are expected to produce fewer eggs per year than fish that do not care for their young, and also whether sperm competition is common in fish.

 **Web Activity** Human Gametogenesis

In humans, the male reproductive system includes structures specialized for producing and storing sperm, synthesizing other components of semen, or transporting and delivering semen. The female reproductive system includes structures specialized for producing eggs, receiving sperm, and nourishing offspring during early development.

- In humans, the external anatomy of the male reproductive tract consists of the penis and scrotum.
- The external anatomy of the human female reproductive tract consists of the vaginal opening and clitoris.

✔ You should be able to predict how surgeons alter the male and female reproductive tracts of human volunteers to sterilize them.

In humans, hormones from the pituitary gland and female reproductive organs regulate the menstrual cycle. These hormones interact via positive or negative feedback. Pregnancy is maintained by hormonal signals from the embryo and the mother's reproductive organs.

- In mammals, GnRH from the hypothalamus triggers the release of FSH and LH from the pituitary gland. The pituitary hormones regulate the production of the gonadal hormones, testosterone and estradiol, in the testes and ovaries, respectively.
- During the human menstrual cycle, estradiol and progesterone exert negative and positive feedback on the production of FSH and LH. Interactions between the pituitary and ovarian hormones are responsible for regulating cyclical changes in the ovaries and uterus.

- If fertilization occurs, the developing embryo secretes the hormone hCG, which arrests the menstrual cycle and allows pregnancy to continue.
- During the first trimester, the embryo becomes implanted in the thickened uterine wall, organs develop, and the nutritive organ called the placenta forms.
- To make rapid growth possible during the second and third trimesters, the mother's heart rate and pumping volume increase. Nutrients and gases are exchanged efficiently in the placenta.

✔ **You should be able to explain why administration of progesterone is an effective method of birth control in humans.**

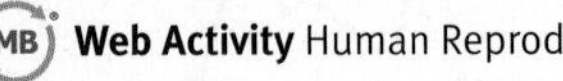

Web Activity Human Reproduction

Questions

✔ TEST YOUR KNOWLEDGE

Answers are available in Appendix B

1. What term describes the mode of asexual reproduction in which offspring develop from unfertilized eggs?
   a. parthenogenesis
   b. budding
   c. regeneration
   d. fission
2. In sperm competition, what is "second-male advantage"?
   a. the observation that when females mate with two males, each male fertilizes the same number of eggs
   b. the observation that when females mate with two males, the second male fertilizes most of the eggs
   c. the observation that females routinely mate with at least two males before laying eggs or becoming pregnant
   d. the observation that accessory fluids prevent matings by second males—for example, by forming copulatory plugs
3. How are the human penis and clitoris similar?
   a. They develop from the same population of embryonic cells.
   b. Both develop during the earliest stages of puberty.
   c. Both contain the urethra.
   d. Both produce accessory fluids required during sexual intercourse.
4. In hormone production, how do the follicle and corpus luteum compare?
   a. Both produce primarily estradiol.
   b. Both produce primarily progesterone.
   c. The follicle produces more estradiol than progesterone; the corpus luteum produces more progesterone than estradiol.
   d. The follicle produces mostly progesterone; the corpus luteum produces estradiol and progesterone.
5. What pituitary hormones are involved in regulating the human menstrual cycle?
   a. GnRH and LH
   b. estradiol and progesterone
   c. oxytocin and LH
   d. FSH and LH
6. The corpus luteum is retained upon fertilization due to the presence of which hormone?
   a. LH
   b. estradiol
   c. progesterone
   d. human chorionic gonadotropin (hCG)

✔ TEST YOUR UNDERSTANDING

Answers are available in Appendix B

1. What is the fundamental difference between sexual and asexual reproduction, in terms of the characteristics of offspring?
2. Summarize the experimental evidence that *Daphnia* require three cues to trigger sexual reproduction. Discuss what these cues indicate about the environment. Generate a hypothesis for why sexual reproduction is adaptive for these animals.
3. Compare and contrast spermatogenesis with oogenesis in humans. How do these processes differ with respect to numbers of cells produced, gamete size, and timing of the second meiotic division?
4. Give examples of negative feedback and positive feedback in hormonal control of the human menstrual cycle. Why can high estradiol levels be considered a "readiness" signal from a follicle?
5. Why are pregnant mothers advised to refrain from drinking alcoholic beverages?
6. Suppose that females of a certain insect species routinely mate with two males and choose their mates based on courtship displays or other traits. Predict whether females would be choosier about the first male or the second male that they mate with, or equally choosy about both males. Explain the logic behind your prediction.

✔ APPLYING CONCEPTS TO NEW SITUATIONS

Answers are available in Appendix B

1. Researchers have recently developed methods for cloning mammals. In effect, biologists can now induce asexual reproduction in species that do not normally reproduce asexually. Suppose this practice becomes so widespread in the future that most of the sheep in the world become genetically identical. Discuss some possible consequences of this development.
2. The text claims that species with external fertilization produce extraordinarily large numbers of gametes. How would you test this hypothesis rigorously? In answering, assume that you have data on the average number of gametes produced by different species, along with data on their average body size.
3. Suppose that you've been given a chance to study several populations of *Sceloporus* lizards—some of which are viviparous, and others of which are oviparous. Which populations would you expect to produce especially large eggs, relative to their overall body size? How would you go about testing this prediction?
4. When marsupial eggs are developing, a shell membrane appears for a short time and then disintegrates. State a hypothesis to explain this observation.

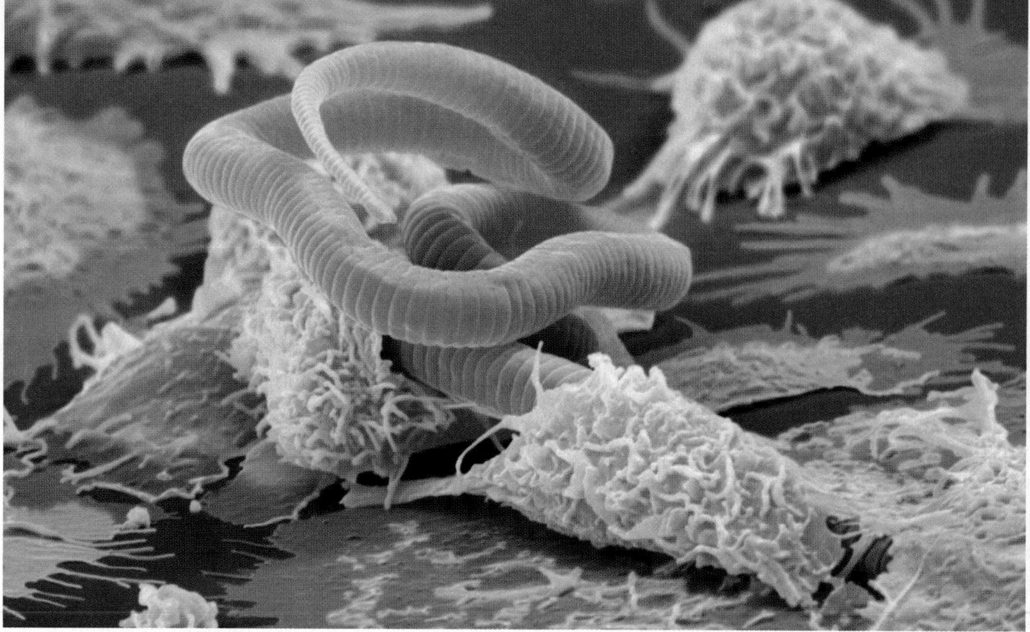

Colorized scanning electron micrograph of human immune system cells attacking *Wuchereria bancrofti*—a parasitic nematode that clogs lymph vessels and causes the dramatic swelling called elephantiasis. This chapter explores how immune system cells are able to recognize parasitic worms, bacteria, and viruses as foreign and eliminate them.

The Immune System in Animals 49

Disease threatens every animal. Humans, for example, are victimized by thousands of different disease-causing bacteria and viruses and hundreds of parasitic worms, fungi, and protists.

Given the ability of these pathogens to cause illness and death, it is remarkable that so many animals stay healthy for most of their lives. To understand how, biologists have investigated three observations:

1. Wounds usually heal, even if they become infected.
2. Most people who contract a bacterial or viral illness eventually recover, even without help from medications.
3. People who acquire bacterial or viral infections and recover are frequently immune (literally, "exempt") to that disease—that is, they do not contract the same disease in the future. **Immunity** is a resistance to or protection against disease-causing pathogens.

This last observation is particularly interesting. When plague struck Athens in 430 B.C., Thucydides noted that only people who had recovered from the illness could nurse the sick, because they did not become ill a second time. In the middle ages, Chinese and Turkish practitioners protected people from smallpox by intentionally exposing them to the dried crusts of smallpox pustules taken from infected individuals. These are examples of **immunization**—the conferring of immunity to a particular disease.

In the late 1700s Edward Jenner refined this immunization technique. In Jenner's day, milkmaids were considered pretty because their faces were not pockmarked with scars from smallpox infections. Jenner knew that cows suffered from a smallpox-like disease called cowpox. He reasoned that milkmaids became immune to smallpox because they had been exposed to cowpox while milking cows.

To test this hypothesis, Jenner inoculated a boy with fluid from a cowpox pustule. Later he inoculated the same child with fluid from a smallpox pustule. That is, he

KEY CONCEPTS

- The innate immune response to infection is mounted by leukocytes that respond in a nonspecific way to pathogens. In contrast, the adaptive immune response is mounted by lymphocytes that respond to specific antigens—often a protein from a pathogen.
- The adaptive immune response begins when certain lymphocytes are activated in a regulated, step-by-step process triggered by interaction with a specific antigen.
- Activated lymphocytes either destroy infected cells or produce antibodies that tag pathogens for destruction.

✔ When you see this checkmark, stop and test yourself. Answers are available in Appendix B.

inoculated the boy with cowpox pathogens, and then with smallpox pathogens.

As predicted, the boy did not contract smallpox. Jenner's technique was named **vaccination**. (The Latin root *vacca* means "cow.") Vaccination is the introduction of a weakened or altered pathogen to prime the body's immune system, so it fights later infections effectively.

Why does vaccination work? What molecules and cells allow us to fight off a bacterial or viral infection and achieve immunity?

To answer these questions, we need to explore how the human immune system recognizes and eliminates pathogens. Suppose, for example, that you were on a crowded bus several days ago next to someone with a persistent cough and runny nose, and 30 minutes later you stumbled while sprinting across campus and skinned your elbow. Now you realize that the wound on your arm has gotten sore and red. Worse yet, you are developing a sore throat, runny nose, and fever. Your elbow has become infected with bacteria, and your upper respiratory tract is supporting a growing population of influenza, or "flu," virus. Let's start by analyzing what is happening in the wound at your elbow.

49.1 Innate Immunity

When biologists began analyzing the immune system, they realized that certain immune system cells are ready to respond to invaders at all times; other components must be activated first.

- Cells that are always ready confer **innate immunity**.
- Cells that are selectively activated to eliminate a specific pathogen confer **adaptive immunity**.

The key to understanding the two types of immunity is to recognize that the cells involved provide different responses to antigens.

An **antigen** is any foreign molecule that can initiate an immune system response. Most antigens are proteins or glycoproteins from bacteria, viruses, or other invaders, but foreign carbohydrates, nucleic acids, and lipids can also function as antigens.

The cells involved in innate immunity are nonspecific in their response to antigens: They respond in the same way to all antigens. In contrast, cells involved in adaptive immunity respond in an extremely specific way to each particular strain of bacterium, virus, or fungus that enters the body.

In combination, the innate and adaptive immune responses form a powerful system for protecting individuals against a formidable and ever-changing array of parasites.

To launch this investigation into immune system function, let's first focus on how the body prevents entry by foreign invaders. Then we can consider what happens if some do get in.

Barriers to Entry

The most effective way to avoid getting sick is to avoid contact with pathogens. In humans and many other animals, the most important barrier to pathogen entry is the skin.

In addition to providing a tough physical barrier, human skin offers a chemical deterrent. The oil secreted by skin cells is converted to fatty acids by bacterial cells that live on the surface. These bacteria are non-pathogenic. The fatty acids lower the pH of the skin surface to about 5, creating a dry, acidic environment that prevents the growth of most bacterial species that might be a threat to the body.

The bodies of insects and other animals with exoskeletons are also difficult for pathogens to enter because they are covered with a tough layer called the cuticle, along with a layer of wax (see Chapter 42).

HOW ARE OPENINGS IN THE BODY PROTECTED? An animal's body has gaps in the surface where the digestive tract, reproductive tract, gas-exchange surfaces, and sensory organs make contact with the environment. As **Figure 49.1** shows, these types of gaps are protected by mucus or other features that discourage pathogen entry.

Airways (lining of trachea)
Most pathogens are trapped in mucus before they can reach the lungs. Beating cilia sweep pathogens up and out of the airway.

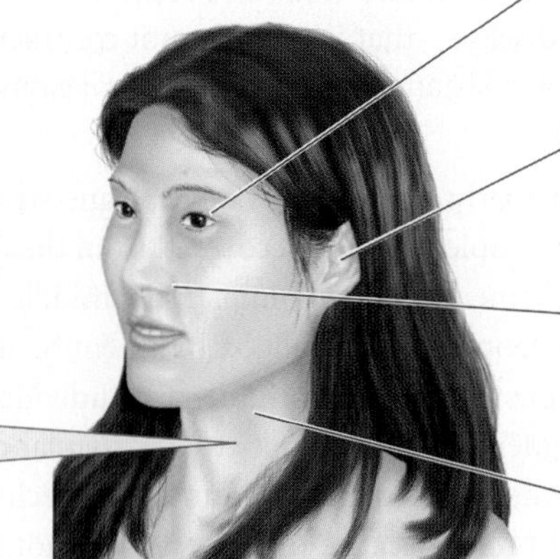

FIGURE 49.1 How Does the Body Keep Pathogens Out? The scanning electron micrograph of mucus-secreting and ciliated cells has been colorized.

✔**QUESTION** Why is a combination of cilia and sticky secretions effective at trapping pathogens?

Mucus is a solution secreted by cells within the epithelium. It is rich in proteoglycans—molecules that consist of large polysaccharides bonded to proteins. The bodies of slugs, snails, earthworms, and other soft-bodied animals are covered with a layer of mucus that shields their epithelial cells from pathogens in the environment.

Mucus is equally important in vertebrates. For example, many of the pathogens that you breathe in or ingest while eating or drinking stick to the mucus that lines your airways and digestive tract. Pathogens that are stuck in mucus cannot come in contact with the plasma membranes of epithelial tissues. In mammals, many of these pathogens are either coughed out or swallowed and killed in the acidic environment of the stomach. In the respiratory tract, they are swept out of harm's way by the beating of cilia (see Figure 49.1).

Gaps in the body that are not covered with mucous layers, such as the eyes, are often protected by other types of secretions. Your ears are protected by waxy secretions, and your eyes by tears that contain the enzyme **lysozyme**. Lysozyme acts as an antibiotic by digesting bacterial cell walls.

HOW DO PATHOGENS GAIN ENTRY? With regularity, preventive measures fail and pathogens gain entry to tissues beneath the skin. Flu viruses, for example, have an enzyme on their surface that disrupts the mucous lining of the respiratory tract. When the outer surface of the virus makes contact with a host cell beneath the mucous layer, the virus is able to enter the cell and begin an infection.

Falls, wounds, and other types of physical trauma provide another important mode of entry. When the skin is broken, bacteria and other pathogens gain direct access to the tissues inside.

To viruses, bacteria, and fungi, your body is a tropical paradise, brimming with resources. Within hours, they can begin growing. What happens then?

The Innate Immune Response

When bacteria enter your body at the site of a wound, the cells pictured in **Figure 49.2** respond by implementing the **innate immune response**—the body's nonspecific response to pathogens. As a group, these cells are called white blood cells to distinguish them from red blood cells. More formally, they are known as **leukocytes** ("white-cells").

Leukocytes include macrophages, neutrophils, and other cells involved in the innate response as well as the lymphocytes responsible for adaptive immunity. Like the red blood cells that carry oxygen in the bloodstream, leukocytes are produced in the bone marrow.

The leukocytes involved in the innate response are alerted to the presence of foreign invaders by molecules that are found on the surfaces of pathogens, but not on host cells. Some of the cells in the innate system have proteins on their plasma membranes that bind to these pathogen-specific compounds. When these proteins, known as **pattern-recognition receptors**, are activated by molecules from pathogens, the cells respond. Pattern-recognition receptors are an "on" switch.

(a) Mast cells secrete signals that increase blood flow.

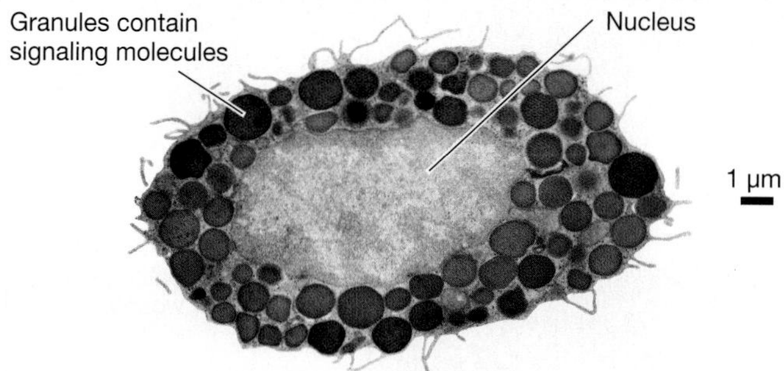

(b) Neutrophils ingest and kill pathogens.

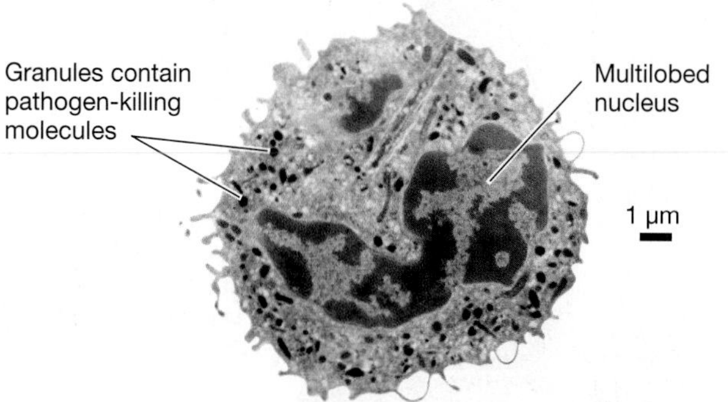

(c) Macrophages recruit other cells and ingest and kill pathogens.

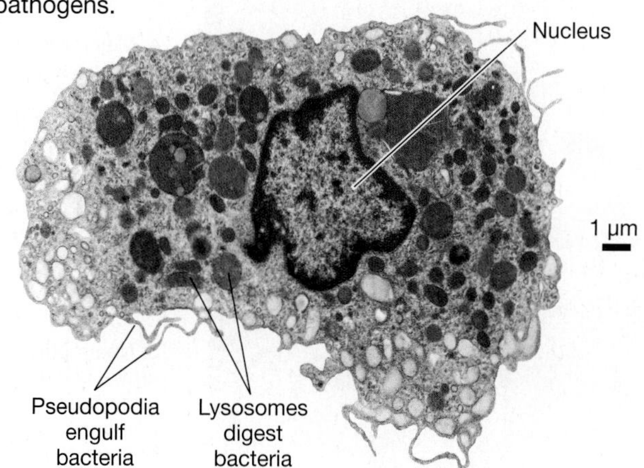

FIGURE 49.2 Cells of the Innate Immune Response. Transmission electron micrographs illustrating a few of the leukocytes responsible for innate immunity. Colors refer to Figure 49.3.

THE INFLAMMATORY RESPONSE IN HUMANS **Figure 49.3** on page 976 summarizes the major steps in the **inflammatory** ("inflames") **response**, a multistep innate immune response observed in an array of animals.

Step 1 A break in the skin allows bacteria to enter the body. If capillaries and other small blood vessels are broken by the wound, blood leaves.

Step 2 Blood components called **platelets** release proteins that form clots and lessen bleeding. Other clotting proteins in the blood form cross-linked structures that help wall off the wound and reduce blood loss.

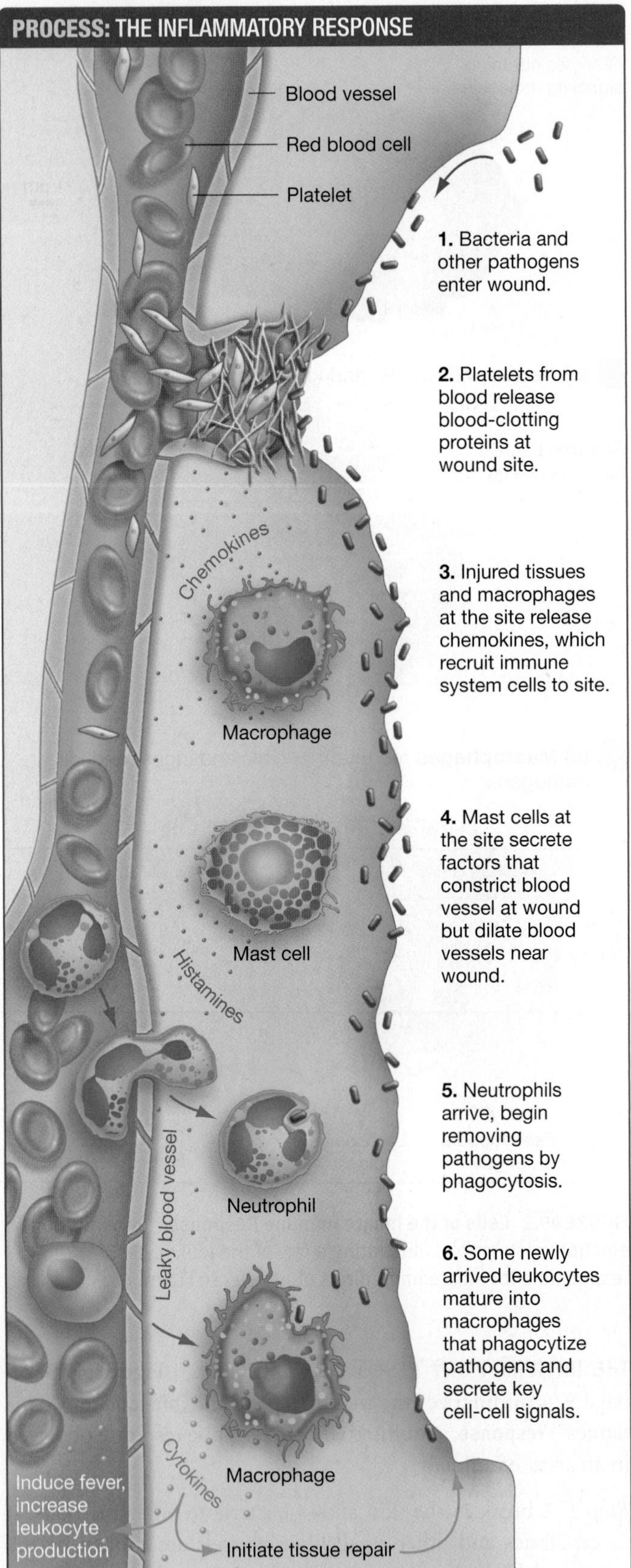

FIGURE 49.3 The Inflammatory Response of Innate Immunity Has Many Elements.

✔**QUESTION** Only one blood vessel is shown, for simplicity. If others were added, where would they be constricted and where would they be dilated? Explain why this is important.

Step 3 Wounded tissues and leukocytes called **macrophages** secrete signaling molecules called **chemokines** ("chemical-movers"). Chemokines are important because they form a gradient that marks a path to the wound site.

Step 4 **Mast cells** release chemical messengers that constrict blood vessels near the wound—reducing blood flow and thus blood loss. Mast cells also release **histamine** and other signaling molecules that induce blood vessels slightly farther from the wound to dilate and become more permeable.

Step 5 The combination of dilated blood vessels and a chemokine gradient is like a 911 call that provides specific directions to the scene of a fire. Leukocytes called neutrophils move out of dilated blood vessels and migrate to the site of the infection. **Neutrophils** are major players in the innate response. They destroy invading cells by phagocytosis ("eating-cell")—meaning that they engulf them. Once invading cells are inside neutrophils, they are killed with a complex array of toxic compounds. These molecules include lysozyme—which degrades bacterial cell walls—free radicals, nitric oxide (NO), and molecules called reactive oxygen intermediates (ROIs), including hydrogen peroxide (H_2O_2).

Step 6 Cells that will mature into macrophages arrive. In addition to phagocytizing bacteria at the wound, the growing population of macrophages secretes chemical messengers called **cytokines** ("cell-movers") that attract other immune system cells to the site, stimulate bone marrow to make and release additional neutrophils and macrophages, induce fever—an elevated body temperature that aids in healing—and activate cells involved in tissue repair and wound healing.

This overview actually simplifies the situation—there are many other cell types and cell-cell signals involved in responding to bacteria and viruses at a site of infection.

The site of inflammation often becomes swollen due to increased numbers of cells and fluids in the area, red and warm due to increased blood flow, and painful due to signals from pain receptors.

The inflammatory response continues until all foreign material is eliminated and the wound is repaired. **Table 49.1a** summarizes key cells in the response, while **Table 49.1b** summarizes some of the key molecules involved. To complete your review of the inflammatory response, you should also go to the study area at *www.masteringbiology.com.*

(MB) **Web Activity** The Inflammatory Response

INNATE IMMUNITY IN INVERTEBRATES Humans and other vertebrates are not the only animals with sophisticated innate immune responses. Innate responses make up the entire immune system observed in millions of species of invertebrates.

The spectacular abundance and diversity of invertebrate animals support the hypothesis that their innate immune systems provide efficient protection against invading bacteria, viruses, and fungi. For example:

SUMMARY TABLE 49.1 The Innate Immune System

(a) Key Cells

Name	Primary Function
Mast cells	Release signals that increase blood flow to wound site
Neutrophils	Kill invading cells via phagocytosis
Macrophages	Release cytokines that recruit other cells to wound site; kill invading cells via phagocytosis

(b) Key Signaling Molecules

Name	Produced By	Received By	Message/Function
Histamine	Mast cells	Blood vessels	High concentration next to wound constricts blood vessels; low concentration farther from wound dilates blood vessels
Chemokines*	Injured tissues and macrophages in tissues	Neutrophils and macrophages	Mark path to wound; promote dilation and increased permeability of blood vessels
Cytokines other than chemokines	Macrophages	Leukocytes	Mark path to wound
		Bone marrow	Increase production of macrophages and neutrophils
		CNS	Induce fever by increasing set point for control of body temperature
		Local tissues	Stimulate cells involved in wound repair

*Note that chemokines are a subset of cytokines.

- If a pathogen succeeds in entering the main body cavity of an insect through a wound or other break in the barriers to entry, cells respond by synthesizing and secreting peptides with potent antibacterial or antifungal properties.
- Sea stars have specialized cells that are similar in function to neutrophils and macrophages—they secrete cytokines or engulf and destroy pathogens by phagocytosis.

But in general, biologists still have a great deal to learn about defense systems in invertebrates.

What happens when the innate immune system of a vertebrate fails to contain and eliminate invading pathogens?

CHECK YOUR UNDERSTANDING

If you understand that . . .

- The innate immune response occurs when macrophages and mast cells that reside in tissues, and neutrophils that circulate in the blood, react in a nonspecific way to signals from invading pathogens.

✔ You should be able to . . .

Describe which events in the innate response mimic steps that first-aid workers use when initially treating a wound:
(1) applying direct pressure to close blood vessels, and
(2) applying bandages containing compounds that halt blood flow from the wound.

Answers are available in Appendix B.

49.2 The Adaptive Immune Response: Recognition

In vertebrates, the adaptive immune system responds to invading viruses and other pathogens. The **adaptive immune response,** also known as the acquired immune response, is based on interactions between specific immune system cells and a specific antigen.

Given the array of pathogens that exist, an animal is almost certain to be exposed to an enormous variety of antigens in the course of its lifetime. Is there a limit to how many different antigens its adaptive immune system responds to?

Research conducted in the early 1920s answered this question. Researchers synthesized organic compounds that do not exist in nature, injected the novel molecules into rabbits, and observed whether they activated the animals' adaptive immune system. To the amazement of the scientists, the rabbits were able to mount an immune response to antigens that did not exist in nature.

More specifically, the animals produced a different antibody against each antigen. **Antibodies** are proteins that are produced and secreted by certain lymphocytes and that bind to a specific part of a specific antigen. The take-home message was that the immune system can produce an almost limitless array of antibodies.

In combination with insights from Jenner's work on vaccines—reviewed in the preceding section—early work on antibodies focused on four key characteristics of the adaptive immune response:

1. *Specificity* Antibodies and other components of the adaptive immune system bind only to specific sites on specific antigens.

2. ***Diversity*** The adaptive response recognizes an almost limitless array of antigens.
3. ***Memory*** The adaptive response can be reactivated quickly if it recognizes antigens from a previous infection.
4. ***Self-nonself recognition*** Molecules that are produced by the individual do not act as antigens, meaning that the adaptive immune system can distinguish between self and nonself. Nonself molecules are antigens; self molecules are not.

What molecular mechanisms give the adaptive immune system these characteristics? Let's begin by reviewing the cells and organs that are responsible for the adaptive immune response.

An Introduction to Lymphocytes

The cells that carry out the major features of the adaptive immune response are called **lymphocytes**. Lymphocytes originate in the immune system and are subsequently activated and transported inside it. Where does each phase of a lymphocyte's life occur?

STRUCTURES OF THE IMMUNE SYSTEM The colored structures in **Figure 49.4** are the major components of the immune system in humans. As the labels on the figure indicate, each of these structures plays a key role in the life of a lymphocyte.

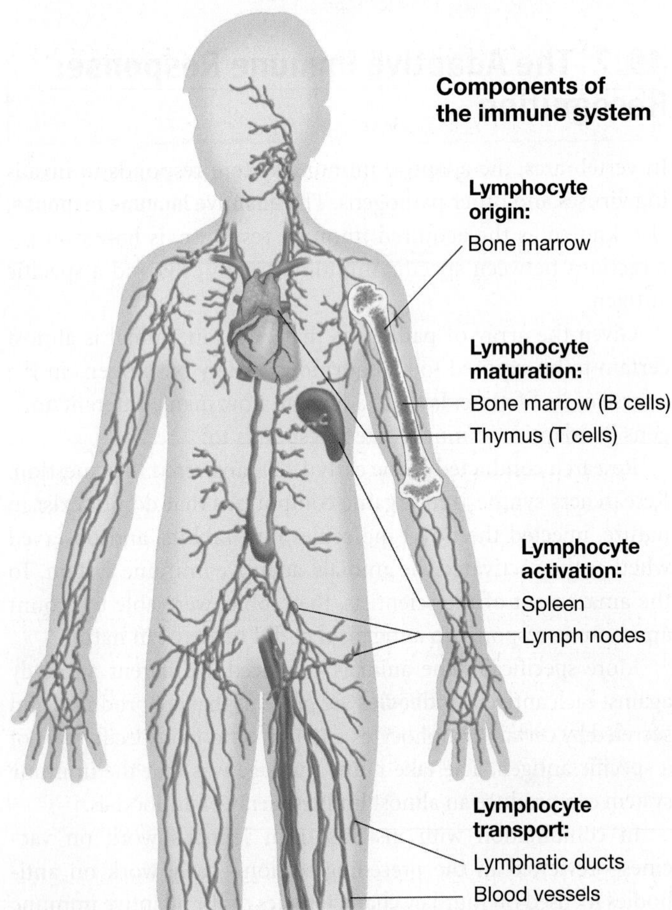

FIGURE 49.4 Lymphocytes Are Formed, Activated, and Transported in the Immune System.

- ***Lymphocyte origin*** All lymphocytes and blood cells are produced in **bone marrow**, tissue that fills the internal cavities in bones. In some mammal species, lymphocytes can also originate in the **spleen**—a lymphatic organ located in the abdominal cavity.
- ***Lymphocyte maturation*** Lymphocytes called B cells mature in the bone marrow where they originate. But lymphocytes called T cells mature in the **thymus**—an organ located in the upper part of the chest of vertebrates (just behind the breastbone in humans).
- ***Lymphocyte activation*** Lymphocytes recognize antigens and become activated in the spleen and lymph nodes. In addition to its function in the immune response, the spleen destroys old blood cells and stores iron from red blood cells. **Lymph nodes** are small, oval organs that are located all around the body. Lymph nodes filter the lymph passing through them. Recall from Chapter 44 that **lymph** is a mixture of fluid and lymphocytes. The liquid portion of lymph originates in fluid that is forced out of capillaries by blood pressure.
- ***Lymphocyte transport*** Lymphocytes circulate through the blood and the secondary organs of the immune system—lymph nodes, spleen, and lymphatic ducts. Lymphatic ducts are thin-walled, branching tubules that transport lymph throughout the body in the **lymphatic system**.

In addition to being found in bone marrow, the thymus, spleen, lymph nodes, lymphatic ducts, and blood vessels, large numbers of lymphocytes, along with other leukocytes, are associated with skin cells and with epithelial tissues that secrete mucus—primarily in the digestive tract and respiratory tract. Collectively, the immune system cells found in the gut and respiratory organs are called **mucosal-associated lymphoid tissue (MALT)**. Leukocytes in the skin and MALT are important because they guard points of entry against pathogens.

LYMPHOCYTES MUST BE ACTIVATED Lymphocytes are normally in a resting, or inactive, state as they circulate through the bloodstream and lymphatic system or reside in skin or MALT. As **Figure 49.5a** shows, inactive lymphocytes have a large nucleus, little cytoplasm, few mitochondria, and a ruffled membrane. Over the course of a day, an inactive lymphocyte might migrate through the spleen, enter the blood, cross over into the lymphatic vessels, migrate through a lymph node, return to the blood, and so on.

If an inactive lymphocyte does not encounter the antigen to which it is programmed to respond, the cell eventually dies. But if a lymphocyte does encounter the appropriate antigen, it becomes activated. When activated, the type of lymphocyte called a B cell has a massive amount of rough ER and a large number of mitochondria (**Figure 49.5b**). Recall from Chapter 7 that many of the proteins synthesized in a cell's rough ER are inserted into the plasma membrane or secreted from the cell. The increased rough ER in this activated lymphocyte suggests that it is manufacturing and possibly secreting proteins. In general, activation produces dramatic changes in lymphocytes.

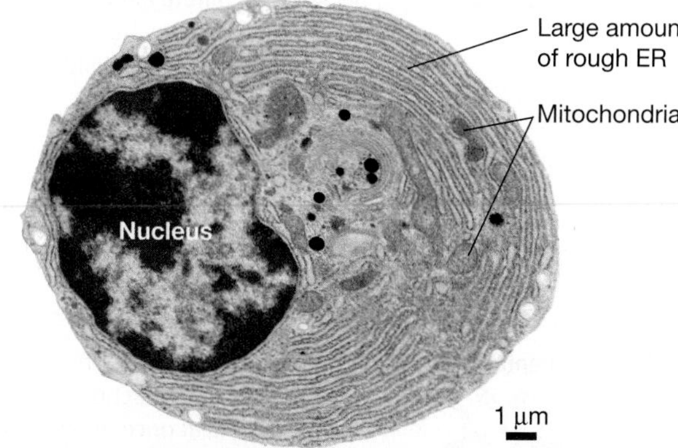

FIGURE 49.5 Lymphocytes Exist in Two States: Inactive and Activated. (a) This inactive lymphocyte—in this case, a type called a B cell—has a small amount of cytoplasm and few organelles. **(b)** This activated lymphocyte (also a B cell) has extensive rough endoplasmic reticulum and many mitochondria—suggesting that a great deal of protein synthesis is taking place.

The Discovery of B Cells and T Cells

In 1956 a group of researchers provided an important insight into the adaptive immune system, by accident. The biologists were investigating the immune system's response to the bacterium *Salmonella typhimurium*, a common cause of food poisoning in humans. To do this work, they needed to produce and isolate antibodies to a particular antigen from *S. typhimurium*—an antigen that is toxic. Their plan was to inject a large number of chickens with the antigen and collect the antibodies that the treated animals produced in response.

In addition to injecting many normal chickens, though, the biologists happened to include some chickens that had undergone experimental removal of an organ called the bursa. Six of the nine chickens that lacked a bursa and that were injected with the antigen died. The other three birds without a bursa survived but failed to produce antibodies to the antigen. In contrast, chickens with the bursa intact produced large quantities of antibodies and survived. To make sense of these results, the researchers proposed that the bursa is critical for antibody production and that antibodies are important in neutralizing antigens.

Not long after this observation was published, three groups of scientists independently conducted a related experiment. To explore the function of the thymus in mammals, these groups removed the organ from newborn mice. Mice lacking a thymus developed pronounced defects in their immune systems. For example, when pieces of skin from other mice were grafted onto the experimental individuals, their immune systems did not recognize the tissue as foreign. In contrast, individuals with an intact thymus quickly mounted an immune response that killed the foreign skin cells.

The results of these and follow-up experiments showed that lymphocytes from the bursa and thymus have different functions. The two types of lymphocytes became known as bursa-dependent and thymus-dependent lymphocytes, or B cells and T cells, respectively.

- **B cells** produce antibodies. (The lymphocytes in Figure 49.5 are B cells.) Later work showed that in humans and other species that lack a bursa, B cells mature in bone marrow.
- **T cells** are involved in an array of functions, including recognizing and killing host cells that are infected with a virus.

Now let's turn to one of the most fundamental questions in immunology: How are B cells and T cells able to recognize so many different antigens?

The Clonal-Selection Theory

By the 1960s biologists understood that:

- B cells can produce antibodies to a seemingly limitless number of antigens.
- Each antibody is specific to an antigen.
- The immune response intensifies through time after an infection begins.
- The immune response is "remembered"—meaning individuals do not get sick at all, or recover extremely quickly, if they are exposed to the same pathogen again.

To explain these patterns, researchers developed the **clonal-selection theory**, which made several key claims about how the adaptive immune system works:

1. ***Antigens are recognized by receptors on B cells and T cells.*** Each lymphocyte formed in the bone marrow or thymus has thousands of copies of a unique receptor on its surface. The receptor is a membrane protein and recognizes one antigen.
2. ***Lymphocytes are activated by antigen-receptor binding.*** When the receptor on a lymphocyte binds to an antigen, the lymphocyte switches from a quiescent to an activated state.
3. ***Activated lymphocytes are cloned.*** An activated lymphocyte divides and makes many identical copies of itself. In this way, specific cells are selected and cloned in response to an infection.
4. ***Activated lymphocytes endure.*** Some of the cloned cells descended from an activated lymphocyte persist long after the pathogen is eliminated. As a result, the cloned cells are able to respond quickly and effectively if the infection recurs.

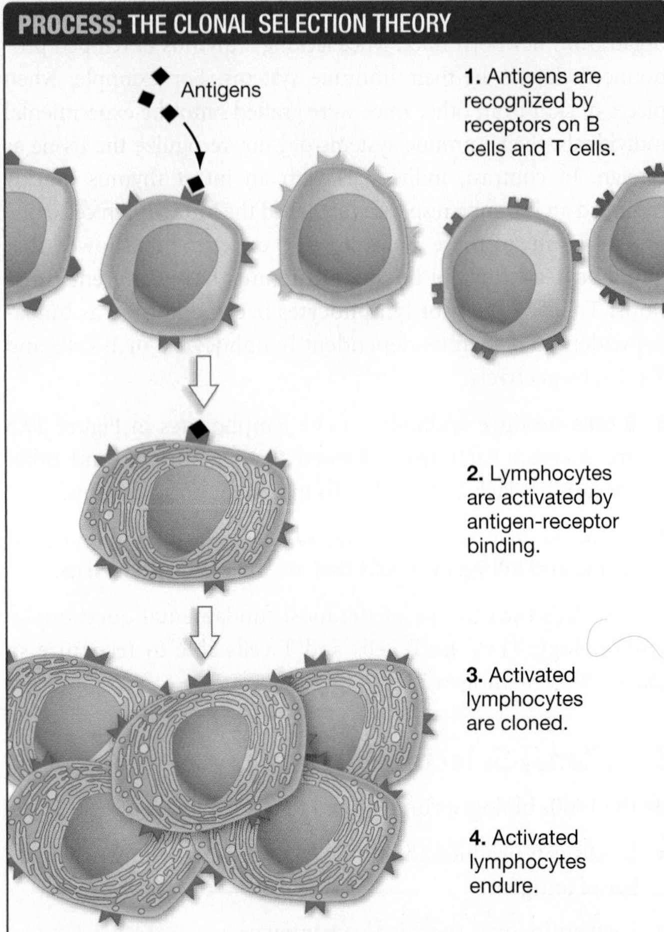

FIGURE 49.6 The Clonal-Selection Theory. The clonal-selection theory maintains that certain lymphocytes are "selected" by binding to an antigen, and that the selected cells then multiply.

Figure 49.6 summarizes these points. Let's explore the theory in detail, beginning with a look at the B-cell and T-cell receptors—the molecules that start the adaptive immune response.

THE DISCOVERY OF B-CELL RECEPTORS To test the prediction that lymphocytes have unique receptors on their surfaces, researchers injected experimental animals with radioactively labeled antigens. This strategy was similar to the experiments with labeled estradiol that allowed biologists to isolate that hormone's receptor (see Chapter 47).

As predicted, the labeled antigens bound to a protein on the surface of B cells. Chemical analysis of this **B-cell receptor (BCR)** showed that the protein has the same structure as the antibodies in the blood produced by B cells.

The BCR has three distinct components (**Figure 49.7a**). The first is a protein called the **light chain**. The second component is roughly twice the size of the light chain and is called the **heavy chain**. Each BCR has two copies of the light chain and two copies of the heavy chain. The third component is within the heavy chain: transmembrane domains that anchor the protein in the plasma membrane of the B cell.

TABLE 49.2 **Five Classes of Immunoglobulins**

Name	Structure (secreted form)	Function
IgG	Monomer	The most abundant type of secreted antibody. Circulates in blood and interstitial fluid. Protects against bacteria, viruses, and toxins
IgD	Monomer	Present on membranes of immature B cells; rarely secreted. Serves as BCR.
IgE	Monomer	Secreted in minute amounts. Involved in response to parasitic worms. Also responsible for hypersensitive reaction that produces allergies.
IgA	Dimer	Most common antibody in breast milk, tears, saliva, and the mucus lining the respiratory and digestive tracts. Prevents bacteria and viruses from attaching to mucous membranes; helps immunize breastfed newborns
IgM	Pentamer	First type of secreted antibody to appear during an infection. Binds many antigens at once; effective at clumping viruses and bacteria so that they can be killed.

B-cell antibodies are identical in structure to the BCR, except that they lack the transmembrane domains. Instead of being inserted into the plasma membrane, antibodies are secreted from the cell and circulate throughout the body.

Both the BCR and the antibodies produced by B cells belong to a family of proteins called the **immunoglobulins (Igs)**. Immunoglobulins are critically important to the adaptive immune response.

Table 49.2 shows the five classes of immunoglobulin proteins that act as B-cell receptors or antibodies. The five types are symbolized IgG, IgD, IgE, IgA, and IgM. Each class is distinguished by unique amino acid sequences in the heavy-chain region, and each has a distinct function in the immune response.

THE DISCOVERY OF T-CELL RECEPTORS It took much longer for researchers to isolate and characterize the **T-cell receptor (TCR)**. It turns out that the TCR only binds to antigens that have been (1) modified by other cells and (2) presented, or displayed, on the plasma membranes of these cells. For a TCR to bind to an antigen, the foreign protein has to undergo a complex process called **antigen presentation**.

This is a fundamentally important distinction. B cells bind to antigens directly; T cells bind only to antigens that are displayed by other cells.

Other data showed that the TCR is composed of two protein chains: an alpha (α) chain and a beta (β) chain (**Figure 49.7b**). The overall shape of the TCR is similar to the "arm" of an antibody or BCR molecule.

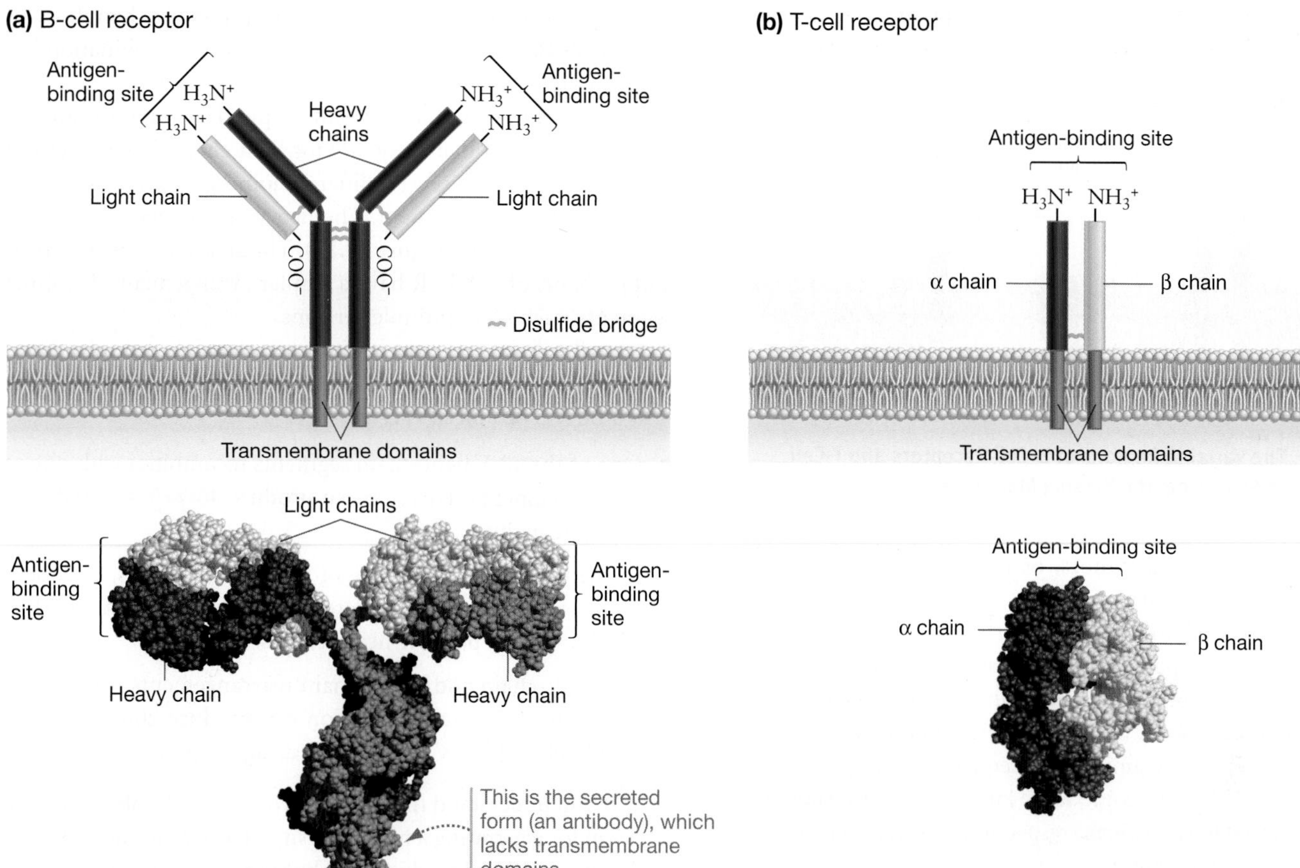

FIGURE 49.7 B-Cell Receptors and T-Cell Receptors Have Transmembrane Domains and Antigen-Binding Sites. **(a)** Schematic model of the B-cell receptor, which is shaped like a Y. The antibodies and receptors produced by each B cell are identical, except that antibodies lack the transmembrane domain and are secreted. **(b)** The shape of the T-cell receptor resembles one "arm" of the Y-shaped B-cell receptor.

ANTIBODIES AND RECEPTORS BIND TO EPITOPES Antibodies, BCRs, and TCRs do not bind to the entire antigen. Instead, they bind to a selected region of the antigen called an **epitope**.

To understand the relationship between an antigen and an epitope, consider that every bacterium, virus, fungus, and protist is made up of a large number of different molecules. Each of these molecules is an antigen, because each is foreign to your cells. In turn, each antigen may have many different epitopes, where binding by antibodies and lymphocyte receptors actually takes place.

Figure 49.8 illustrates a protein called hemagglutinin, which is found on the surface of the influenza virus. This antigen has six epitopes, identified by the blue and red lines in the figure. Each epitope is recognized by a particular antibody, BCR, or TCR. It is not unusual for an antigen to have between 10 and 100 different epitopes.

How do the immunoglobulins recognize specific epitopes? The answer to this question came through detailed studies of the BCR's heavy and light chains.

WHAT IS THE MOLECULAR BASIS OF ANTIBODY SPECIFICITY AND DIVERSITY? In the 1950s, biologists developed an important model system for studying BCR and antibody production. The

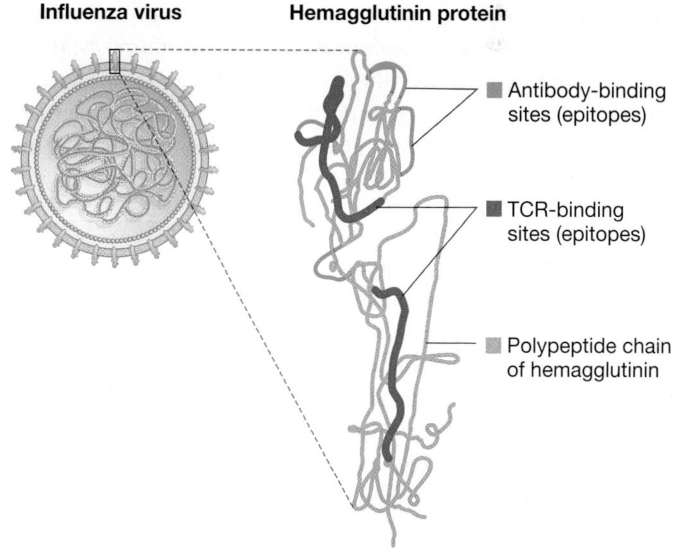

FIGURE 49.8 Most Antigens Have Multiple Binding Sites for B-Cell Receptors, Antibodies, and T-Cell Receptors. The envelope of the influenza virus includes the protein hemagglutinin. This version of hemagglutinin has four different sites where antibodies bind and two distinct places where T-cell receptors bind.

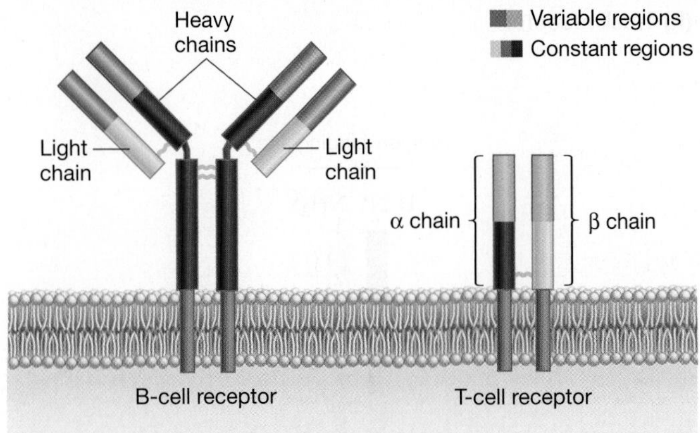

FIGURE 49.9 The Variable Regions of B-Cell Receptors and T-Cell Receptors Face Away from the Plasma Membrane.

cells involved were B-cell tumors, or myelomas, that could be grown in laboratory culture.

Each type of myeloma produces a single type of antibody. When researchers compared the light chains produced by different myelomas, they found that light chains have a segment where the amino acid sequence is virtually identical among light chains and a segment where the amino acid sequence is unique. These light-chain segments have come to be known as the **constant (C) regions** and **variable (V) regions**, respectively. Heavy chains also have a C region and a V region.

Figure 49.9 makes two important points:

1. The V regions of a BCR are adjacent and face away from the plasma membrane.
2. TCRs also have V and C regions, arranged in the same manner.

The presence of unique amino acid sequences in the V regions of every BCR and TCR explains why each of these proteins binds to a unique epitope: Receptors with different amino acid sequences bind to different epitopes. Your body can respond to an almost limitless number of antigens because there is a virtually limitless number of different BCRs, antibodies, and TCRs. How does all this variation come to be?

THE DISCOVERY OF GENE RECOMBINATION In 1965, W. J. Dryer and J. Claude Bennett proposed a fantastic-sounding explanation for how immunoglobulin genes code for so many different variable regions and thus so many different proteins. They hypothesized that as a lymphocyte is maturing, a segment from a variable-region gene is cut and combined with a segment from the constant-region gene. Further, they proposed that this cutting and pasting is done in a different way in each lymphocyte. The result is that each lymphocyte has a novel "*V* + *C* gene" for the light-chain protein.

This type of DNA cutting and pasting had never been observed, however. As a result, most researchers considered the hypothesis wildly implausible.

In 1976, however, Nobumichi Hozumi and Susumu Tonegawa showed that the amount of DNA in the *V* + *C* region of mature lymphocytes is shorter than it is in immature lymphocytes (**Figure 49.10**). This is exactly what the gene-recombination hypothesis predicts.

Hozumi and Tonegawa's result inspired a flurry of studies. This work showed that the genes for light chains have dozens of different *V* segments, several different joining (*J*) segments, and a single constant (*C*) segment. The heavy-chain gene also has diversity (*D*) segments (**Figure 49.11**). The genes that encode the α and β chains of the TCR have a similar arrangement of distinct segments, each with multiple versions.

As a B cell matures, the various gene regions are mixed and matched to produce unique receptors. As an example, consider how a BCR is produced (**Figure 49.12**):

- One of the 40 *V* light-chain segments recombines with one of the 5 *J* segments. This step can produce 40 × 5 = 200 different light chains.
- In the heavy chain, any one of the 51 *V* segments, 27 *D* segments, and 6 *J* segments can recombine, giving a total of 51 × 27 × 6 = 8262 possible heavy chains.
- The light-chain and heavy-chain rearrangements occur independently; DNA recombination can produce 200 × 8262 = 1,652,400 = 1.65×10^6 different antigen-specific BCRs.

✔If you understand these events, you should be able to devise a system for generating a high diversity of lottery numbers that is analogous to gene recombination, but uses only the digits 0–9.

In addition, gene segments do not always join precisely during DNA recombination. Some variation occurs where the *V* and *D* segments join and where the *D* and *J* segments join. As a result, an estimated 10^{10} to 10^{14} different BCRs can form in a single individual. TCR production is just as diverse.

Due to gene recombination, each BCR, antibody, and TCR has a unique amino acid sequence—enabling it to bind a unique epitope on an antigen. The charges and geometry at each surface make the receptor-epitope fit extremely specific. Gene recombination makes the adaptive immune system both specific and diverse.

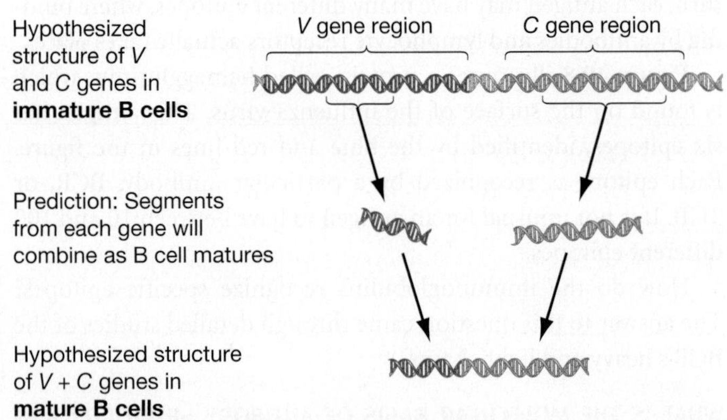

FIGURE 49.10 The Gene Recombination Hypothesis for Antibody Formation in B Cells.

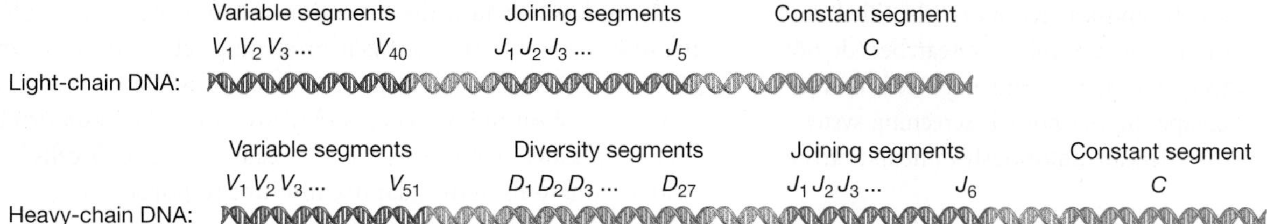

FIGURE 49.11 In Immature Lymphocytes, Immunoglobulin Genes Consist of Many Segments.

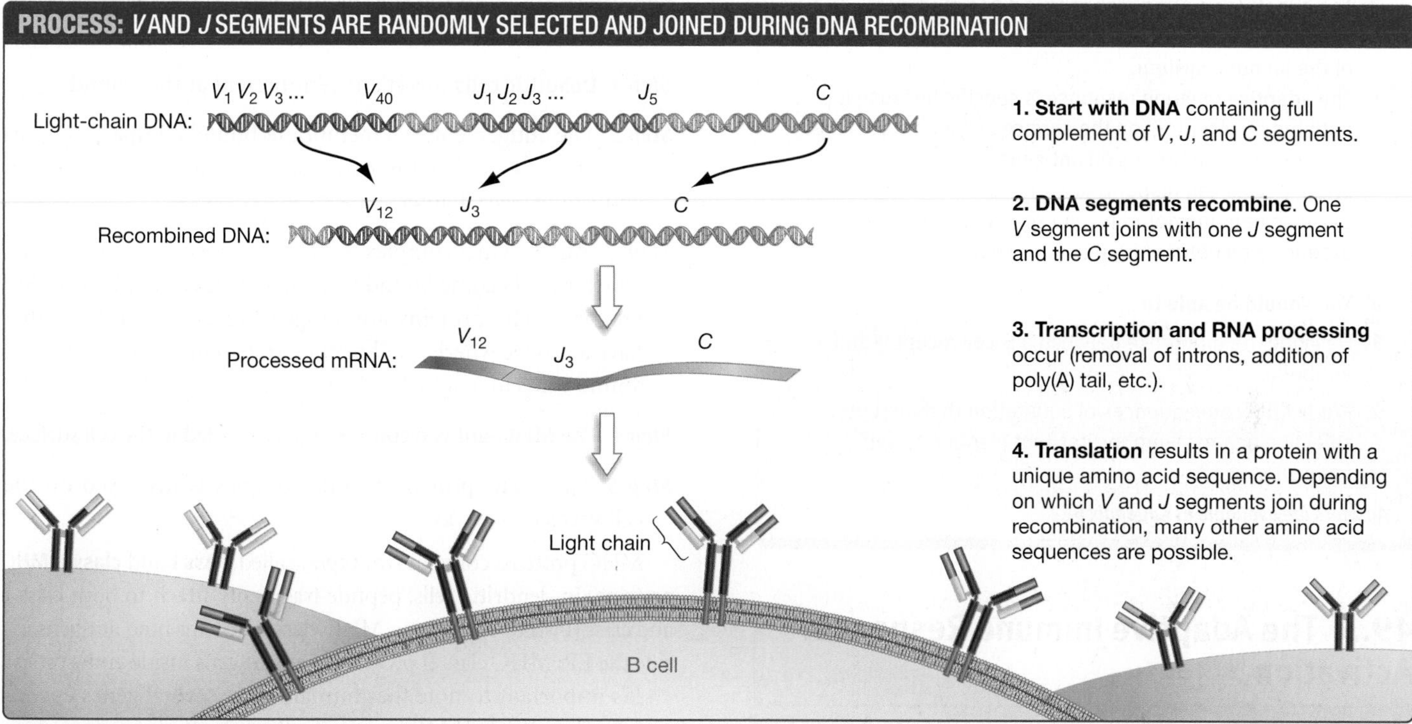

FIGURE 49.12 As Lymphocytes Mature, Immunoglobulin Genes Recombine to Form a Single Gene. In a mature lymphocyte, the final light-chain gene might consist of the V_{12}, J_2, and C segments spliced together; the final heavy-chain gene might consist of the V_{48}, D_{22}, J_1, and C segments spliced together.

How Does the Immune System Distinguish Self from Nonself?

How does the immune system ensure that B-cell and T-cell receptors don't respond to molecules that are part of normal host cells? If a receptor responded to a **self molecule**—that is, a molecule belonging to the host—the receptor would trigger an immune response. An anti-self reaction such as this is known as **autoimmunity**. Autoimmune reactions can lead to immune system cells destroying parts of the host's own body. In most cases, disease results.

- Multiple sclerosis (MS) results from the production of anti-self T cells that attack the myelin sheath of nerve fibers (see Chapter 45). Because damage to myelin reduces the efficiency of nerve signaling, muscular and coordination problems result.
- Rheumatoid arthritis develops when T cells and antibodies alter the lining of joints, causing painful inflammation.
- Type 1 diabetes mellitus occurs when T cells attack and kill insulin-secreting cells in the pancreas, resulting in a lack of insulin and inability to regulate blood glucose levels (see Chapter 43).

Because autoimmune diseases are relatively rare, biologists predicted that there must be some mechanism for eliminating young B cells and T cells that have receptors for self molecules. This prediction was confirmed by injecting B cells and T cells with anti-self receptors into mice, and finding that the injected lymphocytes were eliminated. Follow-up work showed that if B cells and T cells maturing in the bone marrow and thymus have anti-self receptors, most are destroyed or permanently inactivated before they leave these organs.

How are lymphocytes with anti-self receptors identified and eliminated? The answer is not known. Similarly, researchers do not have a good handle on why certain self-recognizing T cells or antibody-secreting cells can escape the self-nonself screening system. As a result, autoimmune disorders are notoriously difficult to treat.

CHECK YOUR UNDERSTANDING

If you understand that . . .

- The adaptive immune response is performed by lymphocytes that circulate in the lymph, blood, and organs of the immune system.
- The adaptive immune response is specific because it is initiated by receptors on the surface of lymphocytes that bind to unique epitopes on antigens.
- Lymphocytes can respond to a wide array of antigens because immunoglobulin gene segments recombine to produce a receptor unique to each cell.

✓ You should be able to . . .

1. Describe the difference between a B-cell receptor and an antibody.
2. Predict the consequences of a mutation that disrupts selection against lymphocytes that respond to self molecules.

Answers are available in Appendix B.

49.3 The Adaptive Immune Response: Activation

When the cell-surface receptor of a B cell or T cell binds to an antigen, the event is like pushing the button that launches a "smart bomb"—a missile that is programmed to destroy a specific target. Receptor-antigen binding unleashes a series of events that ultimately destroys the antigen. The lethal power of activated B cells and T cells is nothing short of awesome.

The activation of the relevant B cells and T cells is a carefully controlled, stepwise process. The mechanism is reminiscent of the precautions that nations with powerful missiles take to avoid accidental deployment. For the most dangerous weapons, the signal to launch is checked and cross-checked, using a series of codes and signals. In the immune system, the checking and cross-checking occurs through protein-protein interactions on the surfaces of cells, and the release and receipt of cytokines and other signaling molecules.

T-Cell Activation

T lymphocyte activation begins when antigens are taken up by a specific type of leukocyte or an infected cell, cut into pieces, packaged with specific cell proteins, and then transferred to the cell surface. Once antigens are presented in this way, T cells can bind to them via the TCR and begin their transformation from an inactive to an activated state.

To understand how the activation system works, let's explore the interaction between antigen-presenting cells and the T lymphocytes called $CD8^+$ T cells. T cells are classified as $CD4^+$ or $CD8^+$, based on the presence of key proteins called **CD4** or **CD8** on their plasma membranes. $CD4^+$ T cells and $CD8^+$ T cells have distinct functions in the adaptive immune response.

ANTIGEN PRESENTATION BY MHC PROTEINS If the innate immune response is overwhelmed and bacteria begin to multiply rapidly at a wound, leukocytes known as dendritic ("tree-like") cells are recruited to the site. **Figure 49.13** shows what happens when these cells arrive at the scene.

Step 1 **Dendritic cells** ingest antigens present at the wound.

Step 2 The antigen enters a membrane-bound compartment inside the cell—either the endoplasmic reticulum (ER) or an endosome (see Chapter 7).

Step 3 An enzyme complex breaks the proteins into pieces, which then become bound to a **major histocompatibility (MHC) protein.** MHC proteins are antigen-presenting proteins that have a groove where small peptide fragments, typically 8 to 20 amino acids in length, bind.

Step 4 The MHC-antigen complex is transported to the cell surface.

Step 5 The MHC protein-peptide complex is displayed on the cell surface.

MHC proteins come in two types, called **class I** and **class II MHC proteins.** In dendritic cells, peptide fragments attach to both class I and class II MHC complexes. MHC class I proteins bind antigens inside the ER; MHC class II proteins bind antigens inside endosomes.

It's important to note that humans have several genes encoding class I and class II MHC proteins. As a result, you can produce several distinct proteins of each type. In addition, the MHC genes are among the most polymorphic of any genes known—meaning that many different alleles exist in the population (see Chapter 13). Because so many distinct alleles exist, most individuals are heterozygous for the class I and class II MHC genes. Heterozygous individuals produce an even greater diversity of class I and class II MHC proteins. A wide array of MHC proteins means that a wide array of foreign peptides can be bound and presented—so cells can trigger an efficient response to many different pathogens.

In addition, virtually all cells in the body display pieces of "self"-proteins, bound to MHC class I proteins, on their surfaces. These are signs that indicate, "I'm a self-cell." Foreign cells that do not display these signs are destroyed by components of the innate immune system.

HOW DO T CELLS RESPOND TO ANTIGEN-PRESENTING CELLS? **Figure 49.14** illustrates what happens once a dendritic cell displays an MHC protein-foreign peptide complex. In the lymph organs, antigen-displaying dendritic cells interact with T cells. $CD8^+$ T cells interact with MHC–class-I–bound antigens on dendritic cells; $CD4^+$ T cells interact with MHC–class-II–bound antigens.

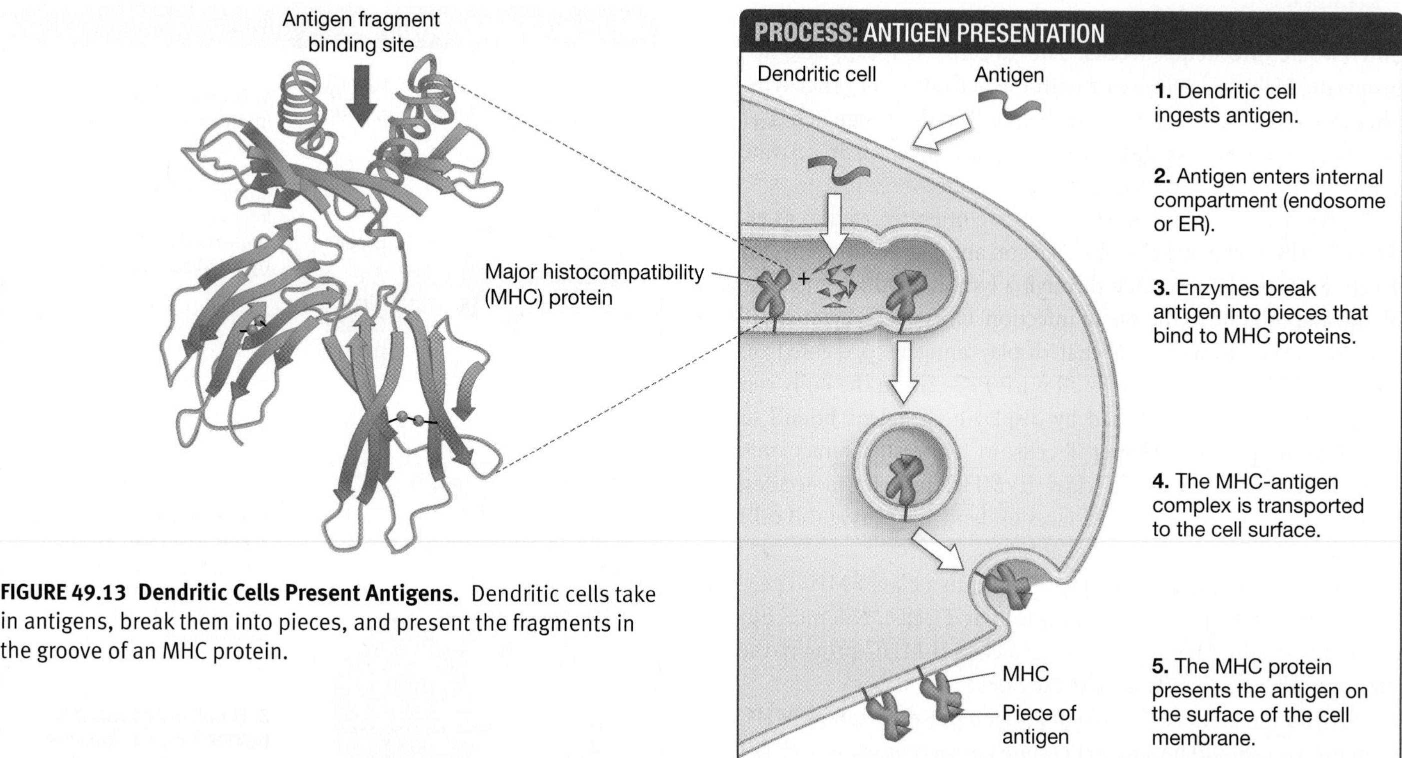

FIGURE 49.13 Dendritic Cells Present Antigens. Dendritic cells take in antigens, break them into pieces, and present the fragments in the groove of an MHC protein.

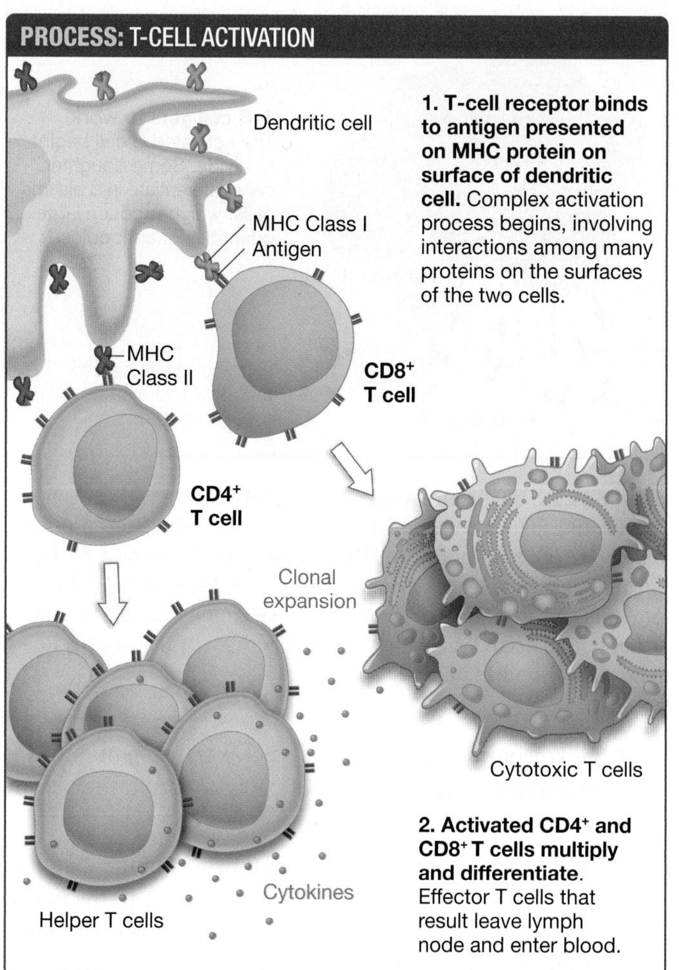

FIGURE 49.14 T Cells Are Activated after Interacting with Antigens Presented by Dendritic Cells. If a T cell has a receptor that is complementary to the MHC-antigen complex presented on the surface of a dendritic cell, the T cell will be activated.

To the T cell, the antigen-presenting cell carries the message "I've found antigen—response required." If the TCR binds to an epitope on the antigen, the activation process starts in response to interactions with proteins on the dendritic cell. In some cases, activation of $CD8^+$ T cells also requires interactions with cytokines produced by activated $CD4^+$ T cells.

An activated T cell divides to produce a series of genetically identical daughter cells. This event, clonal expansion, is a crucial step in the adaptive immune response. It leads to a large population of lymphocytes capable of responding specifically to the antigen that has entered the body.

✔If you understand why clonal expansion is important, you should be able to explain why cancer chemotherapy—which destroys rapidly dividing cells—suppresses the ability of the patient's immune system to respond to infections.

CYTOTOXIC T CELLS AND HELPER T CELLS When activated $CD8^+$ T cells undergo clonal expansion, the daughter cells develop into **cytotoxic** ("cell-poison") **T cells**, also known as cytotoxic T lymphocytes (CTLs) or killer T cells. The adjectives cytotoxic and killer are appropriate: $CD8^+$ T cells kill cells that are infected with a virus or other pathogen in their cytoplasm.

In contrast, the daughter cells of activated CD4$^+$ lymphocytes differentiate into **helper T cells**. The adjective helper is also appropriate: Helper T cells assist with the activation of other lymphocytes. There are two types of helper T cells, designated T_H1 and T_H2, and they have distinct functions: T_H1 cells help activate cytotoxic T cells; T_H2 cells help activate B cells.

Helper T cells and cytotoxic T cells are often referred to as effector T cells. Following clonal expansion and maturation, effector T cells leave the lymph node through a lymphatic duct, enter the blood, and migrate to the site of infection. Once there, cytotoxic T cells interact only with cells that display antigens presented on class I MHC proteins. Virtually all nucleated cells in the body can indicate that they are infected by displaying antigens bound to class I MHC proteins. Helper T cells, in contrast, interact only with antigens presented on class II MHC protein molecules, which are found only on the surfaces of dendritic cells and B cells and other leukocytes that present antigens.

When a cell displays an antigen bound to a class I MHC protein, it sends a simple message to cytotoxic T cells: "Kill me." But if a leukocyte displays an antigen on a class II MHC protein, the message to helper T cells is "Activate now."

✔If you understand the role of antigen presentation by MHC proteins, you should be able to (**1**) state which cells express class I versus class II MHC proteins, (**2**) state which types of T cells respond to peptides bound to class I versus class II MHC proteins, and (**3**) summarize the message provided by MHC-peptide complexes on a dendritic cell, an infected cell, and a B cell.

Effector T cells are ready for action. A key part of the adaptive immune response is now under way.

B-Cell Activation and Antibody Secretion

CD8$^+$ and CD4$^+$ lymphocytes are activated by interactions with dendritic cells and other leukocytes that present antigens. How are B cells activated? The answer involves several steps (**Figure 49.15**):

Step 1 The BCRs on B cells interact directly with bacterial or viral antigens that are floating free in lymph or blood. Once a free antigen is bound, B cells internalize the molecule, digest it into fragments, and load them into class II proteins. As a result, a B cell that encounters its antigen displays the antigen's epitopes on its own surface, with the peptides cradled in the groove of a class II MHC protein.

Step 2 When an activated CD4$^+$ T_H2 cell with a complementary receptor arrives, it binds to the antigen-MHC complex on the B cell. The interaction between a B cell and a helper T cell supplies activation signals that stimulate the helper T cell.

Step 3 The helper T cell responds by releasing stimulatory cytokines that activate the B cell.

Step 4 The B cell divides and forms daughter cells. Some of the daughters differentiate into activated B lymphocytes called **plasma cells**. Plasma cells produce and secrete large quantities of antibodies. Recall that antibodies are identical to B-cell receptors, except that they lack a transmembrane domain and are secreted instead of being found in the plasma membrane.

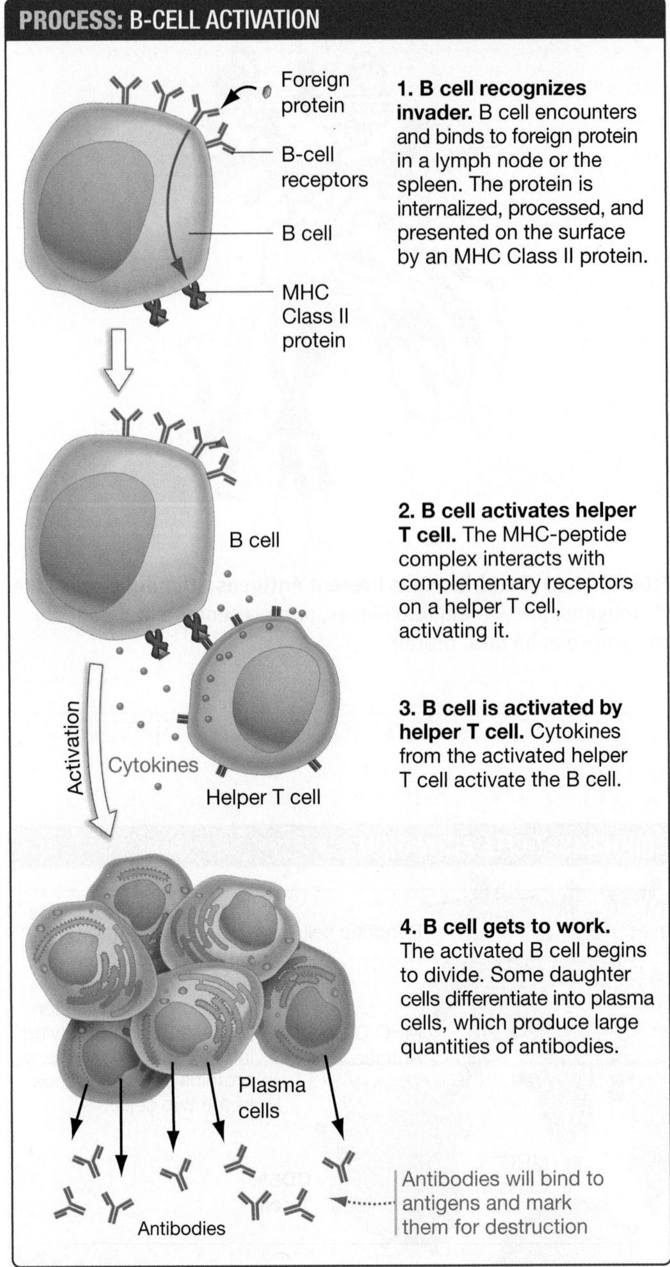

FIGURE 49.15 B Cells Are Activated after Interacting with Helper T Cells. B cells are activated in a series of steps. Once receptors on a helper T cell have bound to MHC-antigen complexes on a B cell, the helper T cell releases cytokines that activate the B cell. (Other proteins on the surface of a B cell and T cell must also interact for activation to occur.) Activation eventually causes plasma cells to produce antibodies.

Once B cells are activated, the adaptive immune response gains an important dimension. Antibodies specific to the invading bacterium or virus begin to circulate in the blood. The adaptive immune system has recognized an antigen and initiated its response. For pathogens, the results are usually devastating.

CHECK YOUR UNDERSTANDING

If you understand that . . .

- T cells are activated when their receptor binds to an antigen presented by the MHC proteins on a dendritic cell or an infected cell.
- B cells are activated when their receptor binds to an antigen, the antigen is presented by an MHC protein on their surface, and they are stimulated by a helper T cell activated by the same antigen.
- Activated lymphocytes undergo dramatic changes in morphology and divide rapidly.

✔ **You should be able to . . .**

Generate a hypothesis to explain the observation that humans who are heterozygous for the genes encoding MHC proteins tend to be healthier than individuals who are homozygous for these genes.

Answers are available in Appendix B.

49.4 The Adaptive Immune Response: Culmination

In combination with the leukocytes involved in the innate immune system, the cells of the adaptive immune system are almost always successful in eliminating threats from bacteria, parasites, fungi, and viruses.

To understand how the adaptive response actually kills pathogens, let's return to our earlier examples of a bacterial infection in a wounded elbow and an upper respiratory tract infection caused by the flu virus. How do activated B cells and T cells eliminate these invaders?

How Are Bacteria and Other Foreign Cells Killed?

During the innate immune response to pathogenic cells that enter a wound, macrophages and dendritic cells at the site phagocytize some of the invaders. In addition to killing the foreign cells, these leukocytes process and present antigens via the MHC class I and class II proteins. Macrophages at the site of infection display epitopes on their surfaces, and dendritic cells move to the lymph nodes to interact with T cells. If the epitopes are recognized by helper T cells, the adaptive immune response is activated.

If an activated T_H1 cell binds to these antigen-laden macrophages, two things happen: First, the phagocytic activity of the macrophages is enhanced. Second, the T_H1 cells secrete cytokines that recruit additional phagocytic cells to the site—increasing the inflammatory response.

In addition, antibodies from plasma cells begin coating bacteria, fungi, and other foreign cells. In many cases, antibody action causes **agglutination**, or clumping of cells. Each antibody has at least two binding sites, so a single antibody can bind epitopes on foreign cells and link them, forming a clump (**Figure 49.16**).

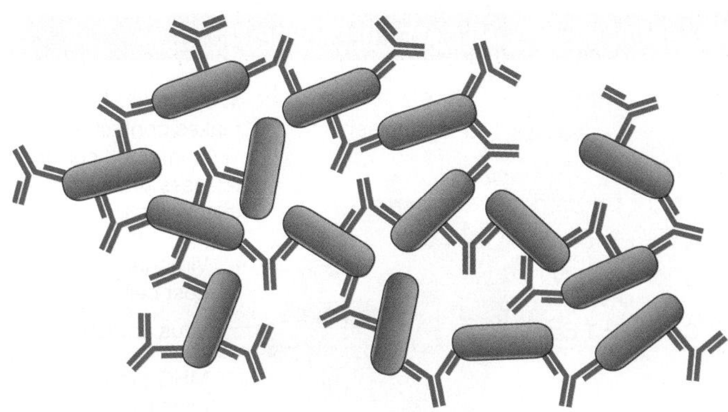

FIGURE 49.16 Antibodies Cause Agglutination. Clumped pathogens are unlikely to infect host tissues, and are readily destroyed.

Clumped and single cells that are tagged with antibodies are readily destroyed by macrophages via phagocytosis. Antibodies that are bound to antigens also stimulate a lethal group of proteins called the **complement system**. Complement proteins circulate in the bloodstream and assemble at antigen-antibody complexes. When complement proteins activate, they punch deadly holes in the plasma membranes of invading cells.

Within a few days, this combination of killing mechanisms—armies of phagocytic cells and complement proteins that hone in on antibody-tagged bacteria—usually eliminates all foreign cells.

How Are Viruses Destroyed?

The adaptive immune system has an array of mechanisms to dispose of bacteria. It also has two major ways to eliminate viruses.

1. The **cell-mediated response** involves cytotoxic T cells (activated $CD8^+$ cells) and takes place at the surface of infected cells.
2. The **humoral response** involves antibodies and takes place in blood and lymph. (The Latin root *humor* means "fluid.")

Let's take a closer look at both.

THE CELL-MEDIATED RESPONSE LEADS TO CELL SUICIDE Infected cells respond to the arrival of a virus by processing antigens from the invader. MHC class I proteins attach to the viral antigens and present the antigens on the surface of the infected cell. Almost all nucleated cells in the body express MHC class I proteins and have the ability to signal that they are infected.

Cells that display viral antigens bound to class I MHC molecules are effectively waving a flag that says, "I'm infected. If you destroy me, you'll destroy them."

As activated $CD8^+$ cells migrate into the area, they recognize and bind to the antigen's epitopes and the MHC protein displayed on infected cells. Once binding occurs, the $CD8^+$ cell secretes molecules that assemble on the surface of the infected cell's plasma membrane and produce pores. Chemicals released by the cytotoxic T cell then enter the cell and activate a self-destruct response (**Figure 49.17a** on page 988).

Once the infected cell dies, the cytotoxic T cell releases it and seeks out another infected cell to kill. Over time, all virus-infected cells are eliminated.

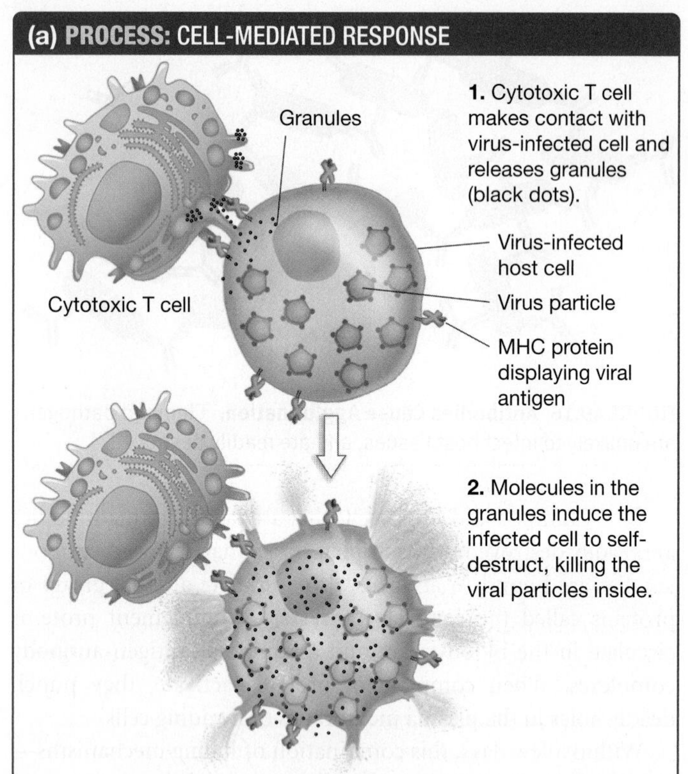

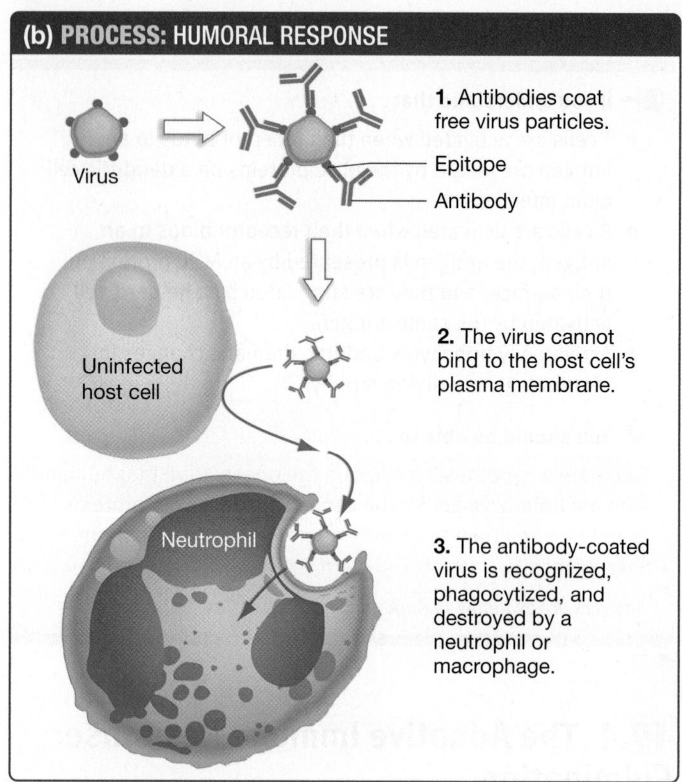

FIGURE 49.17 Lymphocytes Eliminate Viruses via the (a) Cell-Mediated Response and the (b) Humoral Response.

Because viruses can reproduce only inside host cells, the cell-mediated response limits the spread of the infection by preventing new generations of virus particles from maturing.

THE HUMORAL RESPONSE COATS PATHOGENS WITH ANTIBODIES Even as cytotoxic T lymphocytes swing into action, plasma cells produce antibodies to viral proteins. **Figure 49.17b** shows the consequences.

- In most cases, the most effective antibodies are those that bind to epitopes on the surface of the virus.

Two things happen once antibodies bind to the outside of a virus:

- The virus is blocked from infecting host cells.
- Macrophages and other phagocytic cells recognize the antibody-coated particles and phagocytize them.

Eventually, the virus population is reduced to zero.

Why Does the Immune System Reject Foreign Tissues and Organs?

Antibodies and cytotoxic T cells have devastating effects on invading pathogens. Unfortunately, they are equally deadly in response to tissues or organs that are introduced into a patient to heal a wound or cure a disease.

Consider the problems that can arise with blood transfusions. You might recall from Chapter 13 that certain individuals have red blood cells with membrane glycoproteins called A and B. These molecules act as antigens if they are introduced into a person whose own blood cells lack those glycoproteins.

For example, if you have type A blood, it means that your blood cells have the A antigen. If your blood is transfused into a person who lacks the A antigen—meaning someone who has type B or type O blood—the recipient's immune system will recognize the A antigen as foreign and mount a devastating response against it. For a blood transfusion to be successful, this recipient has to receive blood that lacks the A and B glycoproteins entirely or that contains the same antigens found in his or her own blood.

Similar problems arise in organ transplants, except that the antigenic molecules in foreign organs are the MHC proteins found on the surfaces of their cells. To prevent strong immune reactions to a transplanted kidney, heart, or liver, physicians do two things:

1. obtain the organ to be transplanted from a sibling or other donor whose MHC proteins are extremely similar in structure to those of the recipient; and
2. treat the recipient with drugs that suppress the immune response.

Thanks to steady improvements in drug development and in systems for matching MHC types between donors and recipients, the success rate for organ transplants has improved dramatically in recent years.

As the blood transfusion and organ transplant examples show, the immune system rejects foreign tissues because they contain nonself proteins—antigens. To your T cells and B cells, a blood transfusion or an organ transplant is indistinguishable from a massive influx of bacteria, viruses, or other foreign invaders.

Responding to Future Infections: Immunological Memory

In addition to producing the cells that implement the humoral and cell-mediated responses, activated B cells and T cells produce specialized daughter cells called memory cells. **Memory cells** do not participate in the initial adaptive response, or **primary immune response**. Instead, they provide a surveillance service after the original infection has been cleared. Memory cells remain in the spleen and lymph nodes for years or decades, ready to provide a rapid response should an infection with the same antigen recur.

The production of memory cells is a hallmark of the vertebrate immune response. It occurs only to a limited degree, if at all, in invertebrates.

THE SECONDARY RESPONSE IS STRONG AND FAST If the same antigen enters the body a second time, memory cells recognize certain epitopes of the antigen and trigger a second adaptive response, or **secondary immune response**. The launching of a secondary immune response by means of memory cells is known as **immunological memory.**

The secondary immune response is faster and more efficient than the primary response. It is faster because the presence of memory T and B cells increases the likelihood that lymphocytes with the correct antigen-specific receptors will find the antigen and activate quickly. It is more efficient because some of the memory B cells that respond to the returning antigen migrate to a specialized area in the lymph node called the germinal center. There the DNA sequences that code for the variable region of the immunoglobulin gene undergo rapid mutations that modify the receptors produced by the memory cell. Memory B cells with receptors that bind best to the antigen's epitope live and produce daughter cells; those that bind to the antigen less effectively die.

This process of **somatic hypermutation** fine-tunes the immune response. The antibodies that result from somatic hypermutation bind to the antigen more tightly than the antibodies produced by plasma cells during the primary immune response. As the secondary immune response proceeds and somatic hypermutation continues, better-fitting antibodies are produced.

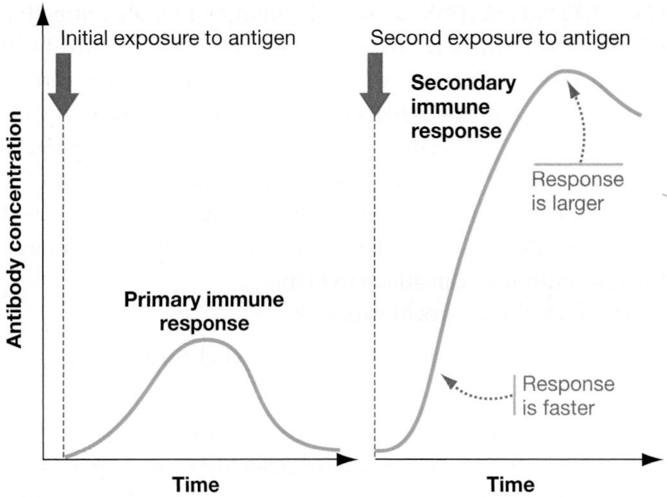

FIGURE 49.18 The Secondary Immune Response Is Faster and Stronger than the Primary Response. The data shown here summarize what happens when biologists inject a mouse with an antigen, document changes in antibody concentration over time, inject the same individual with the same antigen later, and measure the response in antibody concentration.

✔**EXERCISE** The primary response peaks in about 2 weeks; the secondary response peaks in 3–5 days. Add appropriate labels to scale the *x*-axes.

Figure 49.18 compares the rate of antibody production during the first and second exposures to a virus. The graphs are based on data that researchers collect by inoculating a laboratory mouse with influenza virus, collecting blood from the mouse every other day, and then measuring the amount of antiviral antibodies present in the fluid component of the blood.

To review immunological memory, starting with how B cells and T cells are activated, how clonal expansion works, and how daughter cells differentiate into plasma cells, helper T cells, and cytotoxic T lymphocytes, study **Table 49.3.** Then go to the study area at *www.masteringbiology.com.*

 **Web Activity** The Adaptive Immune Response

SUMMARY TABLE 49.3 **Activation and Function of Adaptive Immune System Cells**

Type of Lymphocyte	Method of Activation	Cells That Result from Activation and Clonal Expansion	Function of Resulting Cells
B cell	Receptor binds to free antigen, then interacts with T_H2 cell	Plasma cells	Secrete antibodies
		Memory B cells	Participate in secondary response (secrete antibodies)
$CD4^+$ T cell	Receptor binds to antigen-MHC class II protein complex on dendritic cell or other antigen-presenting cell	T_H1 helper T cells	Activate cytotoxic T cells, regulate inflammatory response
		T_H2 helper T cells	Activate B cells
		Memory T cells	Participate in secondary response
$CD8^+$ T cell	Receptor binds to antigen-MHC class I protein complex on infected cell; may also interact with T_H1 cells	Cytotoxic T cells	Kill infected host cells
		Memory T cells	Participate in secondary response

WHY DOES VACCINATION WORK? In addition to highlighting the effectiveness of the secondary response, the data in Figure 49.18 explain why vaccination is an effective defense against certain pathogens. A **vaccine** contains antigens from a pathogen or a killed or weakened version of the pathogen itself. The antigens are usually components of a virus's exterior—the capsid of a nonenveloped virus or the envelope proteins from an enveloped virus (see Chapter 35). Exterior proteins are effective antigens because antibodies can attach to them.

There are three types of vaccines:

1. *Subunit vaccines* consist of isolated viral proteins. Familiar examples include the hepatitis B and influenza vaccines.
2. *Inactivated viruses* have been damaged by chemical treatments—often exposure to formaldehyde—or exposure to ultraviolet light. They do not cause infections, but are antigenic. If you have been vaccinated for hepatitis A or polio, you may have received an inactivated virus.
3. *Attenuated viruses* are also called "live" virus vaccines because they consist of complete virus particles. Researchers make these viruses harmless by culturing them on cells from species other than the normal host. In adapting to growth on the atypical cells, the viruses usually lose the ability to grow rapidly in their normal host cells. The smallpox, polio, and measles vaccines consist of attenuated viruses.

After vaccination (inoculation with a vaccine), the body mounts a primary immune response that produces memory cells. If a second infection occurs later, these memory cells respond quickly and eliminate the threat before illness appears. Vaccinations function like fire drills or earthquake preparedness exercises—they prepare the immune system for a specific threat.

Edward Jenner's vaccination strategy worked because the antigens presented by cowpox are extremely similar to the antigens presented by smallpox. As a result, exposing people to cowpox virus elicited the production of memory cells that effectively thwarted future infections by smallpox virus.

CHECK YOUR UNDERSTANDING

If you understand that . . .

- The antibodies produced by activated B cells mark invading pathogens for destruction.
- Activated cytotoxic T cells recognize epitopes displayed by infected cells and kill them before the pathogens inside can replicate.
- Memory cells produced during B-cell and T-cell activation remain in the body and provide secondary immunity against future infections.

✔ You should be able to . . .

Explain the relationship between vaccination, an infection, and the primary and secondary immune responses.

Answers are available in Appendix B.

Unfortunately, viruses such as influenza and the human immunodeficiency virus (HIV) mutate so rapidly that they present the immune system with a constantly changing array of epitopes. Memory cells that were effective during a previous infection by these viruses are unlikely to bind to the changed epitopes and trigger a response to a later infection. As a result, it is extremely difficult to design an effective vaccine against these agents.

Currently, the only "cure" for HIV infection is prevention; to keep up with rapidly mutating flu populations, flu vaccines must be redesigned and administered every year. The high mutation rates observed in these viruses are thought to be an adaptation that helps them escape detection by the immune system.

49.5 What Happens When the Immune System *Doesn't* Work Correctly?

The vertebrate immune system is a marvel of adaptation. When people are well rested and well nourished, and when basic sanitation limits exposure to pathogens that are transmitted in contaminated drinking water, the immune system is able to defeat the vast majority of infections without medical intervention (see data in Figure 28.2).

To drive this point home, consider what happens when the immune system fails. You have already reviewed some important autoimmune disorders, which develop when T cells or antibodies mistakenly attack self cells. What happens when the immune system stops working entirely?

Immunodeficiency Diseases

As Chapter 19 noted, children who are born with severe combined immunodeficiency (SCID) have a genetic defect in one of the enzymes responsible for DNA recombination in maturing lymphocytes. As a result, they are unable to generate normal T-cell and B-cell receptors.

Because TCRs and BCRs are fundamental to adaptive immune system function, the response is badly impaired in these children. Afflicted individuals suffer debilitating illness from infections that other children fight off easily, and die before they are two years old.

Similarly, people who are infected with the **human immunodeficiency virus (HIV)** suffer from a progressive failure of the immune system. As Chapter 35 detailed, HIV infects and kills $CD4^+$ T cells and macrophages. Recall that these cells are required for both the humoral and cell-mediated immune responses. As the infection continues, populations of $CD4^+$ T cells gradually decline to a point where the immune system can no longer mount an effective response to infection.

Eventually, HIV-infected people develop **AIDS—acquired immune deficiency syndrome**. They succumb to illnesses that physicians almost never see in people with healthy immune systems.

Allergies

What happens when the immune system *overreacts* to an antigen? For some people, walking through a field of ragweed or petting a cat is a prescription for developing a runny nose, itchy eyes, and labored breathing. For other people, being stung by a bee or eating a peanut is a life-threatening experience. Why are some people allergic to certain molecules?

An **allergy**, or allergic reaction, is an abnormal response to an antigen. For reasons that are still not clear, certain people produce the IgE class of antibodies in response to specific molecules found in cat dander, nuts, plant pollen, or other products. Molecules that trigger this response are called **allergens** instead of antigens.

In people who are not allergic, IgE antibodies are produced only in response to infections by worms. Recent research has also shown that in areas where infections with intestinal worms are common, allergies are almost unknown. These data suggest that allergies don't occur if IgEs are expressed normally—that is, in response to worm infections.

The presence of IgE antibodies in an allergic response is important because it triggers a series of events known as the **hypersensitive reaction**. When a person is first exposed to an allergen, receptors on leukocytes called mast cells and basophils bind to the IgE antibodies that are produced. Once this binding event occurs, the cells (and the person) are said to be sensitized.

If the person is later exposed to the same allergen, the sensitized cells rapidly produce large quantities of histamine, cytokines, chemokines, and other compounds. In response to these molecules, blood vessels dilate and become more permeable, smooth-muscle cells contract, and mucus-producing cells secrete.

- *Hay fever* is a common allergic response to the allergens in grass or ragweed or birch-tree pollen, and results in symptoms like runny nose, watery eyes, and mild wheezing.
- *Hives* occurs when tissues respond to an allergen by reddening, swelling, and itching.
- *Asthma* develops if the hypersensitive response is localized to the respiratory passages; the swelling that results constricts the airway.

If the immune response is more severe, blood vessels can dilate to the point where blood pressure plummets, oxygen delivery to the brain is reduced dramatically, and the person loses consciousness. In addition, severe smooth-muscle contractions in the digestive system and respiratory system can induce vomiting, diarrhea, and complete constriction of the airway passages. This combination of events, known as anaphylactic shock, is lethal.

The most effective treatment for anaphylactic shock is to inject the affected individual with epinephrine. Epinephrine is a hormone that relaxes smooth muscle (see Chapter 47). As a result, it opens up blood vessels and counteracts the effects of histamine and other agents of the hypersensitive response. People who are hypersensitive to bee stings or other allergens routinely carry small syringes filled with epinephrine. If they face a situation where anaphylactic shock is possible, they inject themselves.

What can be done about allergies? The number one approach is to avoid the allergen. If exposure does occur, drugs called antihistamines can sometimes reduce symptoms by blocking histamine receptors. In some cases corticosteroids applied to skin can reduce swelling. Longer-term solutions are still in the experimental stage, however. Developing effective treatments for allergies is an important research frontier in biomedicine.

CHAPTER 49 REVIEW

For media, go to the study area at www.masteringbiology.com 

Summary of Key Concepts

The innate immune response to infection is mounted by leukocytes that respond in a nonspecific way to pathogens. In contrast, the adaptive immune response is mounted by lymphocytes that respond to specific antigens—often a protein from a pathogen.

- The innate response occurs in the same way no matter which pathogens have entered the body, but the adaptive response is specific to the pathogen present.
- During the innate immune system's inflammatory response, mast cells release chemical messengers that increase blood flow to wounds or areas of tissue damage.
- Leukocytes are recruited to the site of an inflammatory response. Neutrophils respond to bacteria that stimulate their pattern-recognition receptors by phagocytizing the invading cells and destroying them. Macrophages also phagocytize pathogens and release chemical messengers that activate other leukocytes and raise body temperature.

✔ You should be able to explain what would happen if blood vessels near a wound did not dilate and become more permeable in response to chemical signals from leukocytes and damaged tissues.

MB **Web Activity** The Inflammatory Response

The adaptive immune response begins when certain lymphocytes are activated in a regulated, step-by-step process triggered by interaction with a specific antigen.

- Lymphocytes are activated when T-cell receptors and B-cell receptors recognize an epitope on an antigen.
- The receptor protein on the surface of a T cell or B cell is generated through DNA recombination—a rearrangement of gene segments.
- Because every receptor that results from DNA recombination is slightly different, the immune system is able to recognize and respond to an almost limitless array of antigens.

✔ You should be able to explain how a drug that binds to a particular epitope of an antigen and blocks it would affect the activation of T cells with receptors specific to that antigen.

Activated lymphocytes either destroy infected cells or produce antibodies that tag pathogens for destruction.

- During the cell-mediated response, host cells that are infected with a virus display antigens on their surface and are induced to self-destruct by cytotoxic T cells. As a result, the virus particles inside are destroyed and cannot contribute further to the infection.

- During the humoral response, activated B cells produce antibodies to specific epitopes on the pathogen that has invaded the body. Viruses that become coated with antibodies are unable to enter host cells and are destroyed by macrophages. Bacteria that are tagged with antibodies are destroyed by complement proteins or macrophages.
- Because memory cells are created as activated B cells and T cells undergo clonal expansion, the immune system is able to respond rapidly and effectively to future infections by the same antigen.
- A vaccine triggers the production of memory cells because it contains epitopes from antigens.

✔ **You should be able to explain why cowpox can be considered an attenuated virus vaccine.**

MB **Web Activity** The Adaptive Immune Response

Questions

✔ TEST YOUR KNOWLEDGE

Answers are available in Appendix B

1. What is the primary difference between the innate and adaptive responses?
   a. The innate response is modified over time; the adaptive response occurs in the same way throughout life.
   b. Only the adaptive response is triggered by antigens.
   c. The adaptive response is specific and is "remembered"; the innate response is nonspecific.
   d. There is no nonself recognition component in the innate response.
2. All of the following events are involved in the inflammatory response *except*:
   a. Cytotoxic T cells kill infected host cells.
   b. Neutrophils phagocytize pathogens that stimulate their pattern-specific receptors.
   c. Mast cells secrete chemical messengers that lead to increased blood flow.
   d. Macrophages release chemical messengers that lead to increased body temperature.
3. What is the difference between an epitope and an antigen?
   a. An epitope is any foreign substance; an antigen is a foreign protein.
   b. An epitope is the part of an antigen where an antibody or lymphocyte receptor binds.
   c. An antigen is the part of an epitope where an antibody or lymphocyte receptor binds.
   d. Antigens are recognized by B cells and antibodies; epitopes are recognized by T cells.
4. How do B-cell receptors and antibodies differ?
   a. B-cell receptors are made up of heavy chains; antibodies are made up of light chains.
   b. B-cell receptors are made up of light chains; antibodies are made up of heavy chains.
   c. Only antibodies include a variable region.
   d. Antibodies lack a transmembrane domain and are secreted.
5. How do memory cells become activated?
   a. They undergo somatic hypermutation.
   b. The same way that B cells and T cells are activated in the primary response.
   c. They undergo clonal expansion.
   d. They are stimulated by histamines.
6. In terms of morphology, T cells come in two major types. How are they distinguished?
   a. They have a CD4 or CD8 protein on their surfaces.
   b. They have extensive ER and many mitochondria.
   c. Only one type has pattern-recognition receptors.
   d. One type functions in the primary response; the other functions in the secondary response.

✔ TEST YOUR UNDERSTANDING

Answers are available in Appendix B

1. To a physician, the classical signs of the inflammatory response are rubor (reddening), calor (heat), dolor (pain), and tumor (swelling). Explain why each symptom occurs.
2. Compare and contrast the general structure of a B-cell receptor and a T-cell receptor (make a sketch of each). Explain how each interacts with an antigen.
3. What do vaccines need to contain in order to be effective? Why don't we have vaccines for HIV?
4. In the clonal-selection theory of adaptive immune system function, what do "clonal" and "selection" refer to?
5. Compare and contrast the interaction between (a) pathogens and the pattern-recognition receptors on leukocytes versus (b) antigens and BCRs or TCRs.
6. Explain how DNA recombination leads to the production of almost limitless numbers of different B-cell receptors, T-cell receptors, and antibodies.

✔ APPLYING CONCEPTS TO NEW SITUATIONS

Answers are available in Appendix B

1. If you were being treated for a badly skinned knee from a bicycle accident, health care workers would irrigate the wound with sterile water, scrub it thoroughly with warm, soapy water, and then apply antibiotics. Explain why these measures are effective.
2. Suppose you discover an antibody to the sodium-potassium pump. Describe how you could use this antibody to study the location of sodium-potassium pumps in various cell types.
3. It seems astonishing that the immune system can produce antibodies to compounds that were recently synthesized for the first time in the lab. Given that viruses and other pathogens evolve rapidly, explain why natural selection favored individuals with genes that made extensive antibody diversity possible.
4. During World War II, physicians discovered that if they removed undamaged skin from a patient and grafted it onto the site of a burn on the same patient, the tissue healed well. But if the grafted tissue came from a different individual, the skin graft was rejected. Explain why.

APPENDIX A BioSkills

BIOSKILLS 1

THE METRIC SYSTEM

The metric system is the system of units of measure used in every country of the world but three (Liberia, Myanmar, and the United States). It is also the basis of the SI system used in scientific publications.

The popularity of the metric system is based on its consistency and ease of use. These attributes, in turn, arise from the system's use of the base 10. For example, each unit of length in the system is related to all other measures of length in the system by a multiple of 10. There are 10 millimeters in a centimeter; 100 centimeters in a meter; 1000 meters in a kilometer.

Measures of length in the English system, in contrast, do not relate to each other in a regular way. Inches are routinely divided into 16ths; there are 12 inches in a foot; 3 feet in a yard; 5280 feet (or 1760 yards) in a mile.

TABLE B1.2 **Prefixes Used in the Metric System**

Prefix	Abbreviation	Definition
nano-	n	0.000 000 001 = 10^{-9}
micro-	μ	0.000 001 = 10^{-6}
milli-	m	0.001 = 10^{-3}
centi-	c	0.01 = 10^{-2}
deci-	d	0.1 = 10^{-1}
—	—	1 = 10^{0}
kilo-	k	1000 = 10^{3}
mega-	M	1 000 000 = 10^{6}
giga-	G	1 000 000 000 = 10^{9}

TABLE B1.1 **Metric System Units and Conversions**

Measurement	Unit of Measurement and Abbreviation	Metric System Equivalent	Converting Metric Units to English Units
Length	kilometer (km)	1 km = 1000 m = 10^3 m	1 km = 0.62 mile
	meter (m)	1 m = 100 cm	1 m = 1.09 yards = 3.28 feet = 39.37 inches
	centimeter (cm)	1 cm = 0.01 m = 10^{-2} m	1 cm = 0.3937 inch
	millimeter (mm)	1 mm = 0.001 m = 10^{-3} m	1 mm = 0.039 inch
	micrometer (μm)	1 μm = 10^{-6} m = 10^{-3} mm	
	nanometer (nm)	1 nm = 10^{-9} m = 10^{-3} μm	
Area	hectare (ha)	1 ha = 10,000 m^2	1 ha = 2.47 acres
	square meter (m^2)	1 m^2 = 10,000 cm^2	1 m^2 = 1.196 square yards
	square centimeter (cm^2)	1 cm^2 = 100 mm^2 = 10^{-4} m^2	1 cm^2 = 0.155 square inch
Volume	liter (L)	1 L = 1000 mL	1 L = 1.06 quarts
	milliliter (mL)	1 mL = 1000 μL = 10^{-3} L	1 mL = 0.034 fluid ounce
	microliter (μL)	1 μL = 10^{-6} L	
Mass	kilogram (kg)	1 kg = 1000 g	1 kg = 2.20 pounds
	gram (g)	1 g = 1000 mg	1 g = 0.035 ounce
	milligram (mg)	1 mg = 1000 μg = 10^{-3} g	
	microgram (μg)	1 μg = 10^{-6} g	
Temperature	Kelvin (K)*		K = °C + 273.15
	degrees Celsius (°C)		°C = $\frac{5}{9}$(°F − 32)
	degrees Fahrenheit (°F)		°F = $\frac{9}{5}$°C + 32

*Absolute zero is −273.15°C = 0 K

If you have grown up in the United States and are accustomed to using the English system, it is extremely important to begin developing a working familiarity with metric units and values. The tables and questions below should help you get started with this process.

✔Questions

1. Some friends of yours just competed in a 5-kilometer run. How many miles did they run?
2. An American football field is 120 yards long, while rugby fields are 144 meters long. In yards, how much longer is a rugby field than an American football field?
3. What is your normal body temperature in degrees Celsius? (Normal body temperature is 98.6°F.)
4. What is your current weight in kilograms?
5. A friend asks you to buy a gallon of milk. How many liters would you buy to get approximately the same volume?

BIOSKILLS 2

READING GRAPHS

Graphs are the most common way to report data, for a simple reason. Compared to reading raw numerical values in a table or list, a graph makes it much easier to understand what the data mean.

Learning how to read and interpret graphs is one of the most basic skills you'll need to acquire as a biology student. As when learning piano or soccer or anything else, you need to understand a few key ideas to get started and then have a chance to practice—a *lot*—with some guidance and feedback.

Getting Started

To start reading a graph, you need to do three things: read the axes, figure out what the data points represent—that is, where they came from—and think about the overall message of the data. Let's consider each in turn.

What Do the Axes Represent?

Graphs have two axes: one horizontal and one vertical. The horizontal axis of a graph is also called the *x*-axis or the abscissa. The vertical axis of a graph is also called the *y*-axis or the ordinate. Each axis represents a variable that takes on a range of values. These values are indicated by the ticks and labels on the axis. Note that each axis should *always* be clearly labeled with the unit or treatment it represents.

Figure B2.1 shows a scatterplot—a type of graph where continuous data are graphed on each axis. Continuous data can take an array of values over a range. In contrast, discrete data can take only a restricted set of values. If you were graphing the average height of men and women in your class, height is a continuous variable but gender is a discrete variable.

In the example in the figure, the *x*-axis represents time in units of generations of maize; the *y*-axis represents the average percentage of the dry weight of a maize kernel that is protein.

To create a graph, researchers plot the independent variable on the *x*-axis and the dependent variable on the *y*-axis (Figure B2.1a). The terms independent and dependent are used because the values on the *y*-axis depend on the *x*-axis values. In our example, the researchers wanted to show how the protein content of maize kernels in a study population changed over time. Thus, the protein concentration plotted on the *y*-axis depended on the year (generation) plotted on the *x*-axis. The value on the *y*-axis always depends on the value on the *x*-axis, but not vice versa.

In many graphs in biology, the independent variable is either time or the various treatments used in an experiment. In these cases, the *y*-axis records how some quantity changes as a function of time or as the outcome of the treatments applied to the experimental cells or organisms.

What Do the Data Points Represent?

Once you've read the axes, you need to figure out what each data point is. In our maize kernel example, the data point in Figure B2.1b represents the average percentage of protein found in a sample of kernels from a study population in a particular generation.

If it's difficult to figure out what the data points are, ask yourself where they came from—meaning, how the researchers got them. You can do this by understanding how the study was done and by understanding what is being plotted on each axis. The *y*-axis will tell you what they measured; the *x*-axis will usually tell you when they measured it or what group was measured. In some cases—for example, in a plot of average body size versus average brain size in primates—the *x*-axis will report a second variable that was measured.

What Is the Overall Trend or Message?

Look at the data as a whole, and figure out what they mean. Figure B2.1c suggests an interpretation of the maize kernel example. If the graph shows how some quantity changes over time, ask yourself if that quantity is increasing, decreasing, fluctuating up and down, or staying the same. Then ask whether the pattern is the same over time or whether it changes over time.

When you're interpreting a graph, it's extremely important to limit your conclusions to the data presented. Don't extrapolate beyond the data, unless you are explicitly making a prediction based on the assumption that present trends will continue. For example, you can't say that average % protein content was increasing in the population before the experiment started, or that it will continue to increase in the future. You can only say what the data tell you.

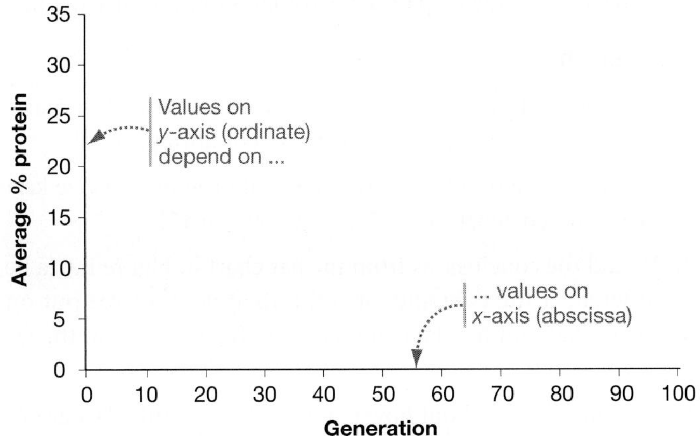

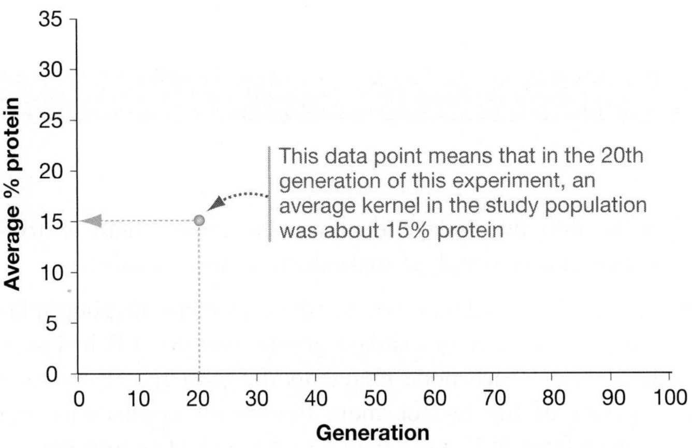

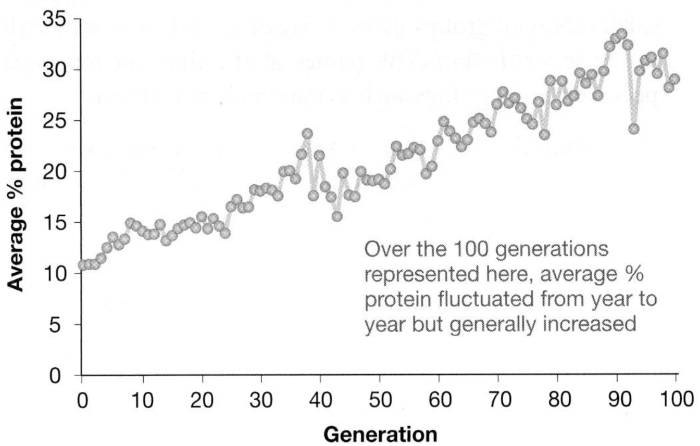

FIGURE B2.1 Scatterplots Are Used to Graph Continuous Data.

Types of Graphs

Most of the graphs in this text are scatterplots like the one shown here, where individual data points are plotted.

Sometimes the data points in a scatterplot will be by themselves, sometimes they will be connected by dot-to-dot lines to help make the overall trend clearer, as in this figure, and sometimes they will have a smooth line through them. A smooth line through data points—sometimes straight, sometimes curved—is a mathematical "line of best fit." A line of best fit represents a mathematical function that summarizes the relationship between the *x* and *y* variables. It is "best" in the sense of fitting the data points most precisely.

Scatterplots are the most appropriate type of graph when the data have a continuous range of values and you want to show individual data points. But you will also come across two other major types of graphs in this text:

- *Bar charts* plot data that have discrete or categorical values instead of a continuous range of values. In many cases the bars might represent different treatment groups in an experiment, as in **Figure B2.2a**. In this graph, the height of the bar indicates the average value.
- *Histograms* illustrate frequency data and can be plotted as numbers or percentages. **Figure B2.2b** shows an example where height is plotted on the *x*-axis and the number of students in a population is plotted on the *y*-axis. Each rectangle indicates the number of individuals in each interval of height, which reflects the relative frequency, in this population, of people whose heights are in that interval. The measurements could also be recalculated so that the *y*-axis would report the proportion of people in each interval. Then the sum of all the bars would equal 100 percent.

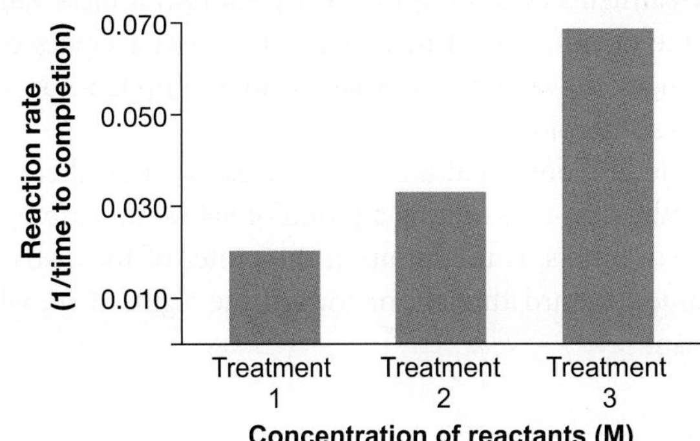

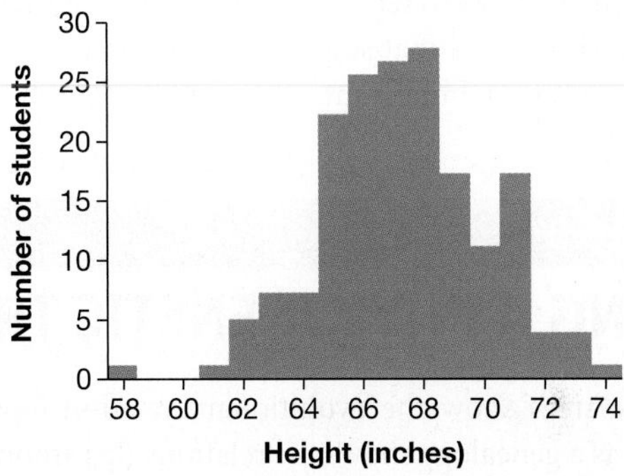

FIGURE B2.2 Bar Charts and Histograms. (a) Bar charts are used to graph data that are discontinuous or categorical. **(b)** Histograms show the distribution of frequencies or values in a population.

When you are looking at a bar chart that plots values from different treatments in an experiment, ask yourself if these values are the same or different. If the bar chart reports averages over discrete ranges of values, ask what trend is implied—as you would for a scatterplot.

When you are looking at a histogram, ask whether there is a "hump" in the data—indicating a group of values that are more frequent than others. Is the hump in the center of the distribution of values, toward the left, or toward the right? If so, what does it mean?

Getting Practice

Working with this text will give you lots of practice with reading graphs—they appear in almost every chapter. In many cases we've inserted an arrow to represent your instructor's hand at the whiteboard, with a label that suggests an interpretation or draws your attention to an important point on the graph. In other cases, you should be able to figure out what the data mean on your own or with the help of other students or your instructor.

✔Questions

1. What is the total change in average percent protein in maize kernels, from the start of the experiment until the end?
2. What was the trend in average percent protein in maize kernels between generation 37 and generation 42?
3. Would the conclusions from the bar chart in Figure B2.2a be different if the data and label for Treatment 3 were put on the far left and the data and label for Treatment 1 on the far right?
4. In Figure B2.2b, about how many students in this class are 70 inches tall?
5. What is the most common height in the class graphed in Figure B2.2b?

BIOSKILLS 3

READING A PHYLOGENETIC TREE

Phylogenetic trees show the evolutionary relationships among species, just as a genealogy shows the relationships among people in your family. They are unusual diagrams, however, and it can take practice to interpret them correctly.

To understand how evolutionary trees work, consider **Figure B3.1**. Notice that a phylogenetic tree consists of branches, nodes, and tips.

- Branches represent populations through time. In this text, branches are drawn as horizontal lines. In most cases the length of the branch is arbitrary and has no meaning, but in some cases branch lengths are proportional to time (if so, there will be a scale at the bottom of the tree). The vertical lines on the tree represent splitting events, where one group broke into two independent groups. Their length is arbitrary—chosen simply to make the tree more readable.
- Nodes (also called forks) occur where an ancestral group splits into two or more descendant groups (see point B in Figure B3.1). Thus, each node represents the most recent common ancestor of the two or more descendant populations that emerge from it. If more than two descendant groups emerge from a node, the node is called a polytomy (see node C).
- Tips (also called terminal nodes) are the tree's endpoints, which represent groups living today or a dead end—a branch ending in extinction. The names at the tips can represent species or larger groups such as mammals or conifers.

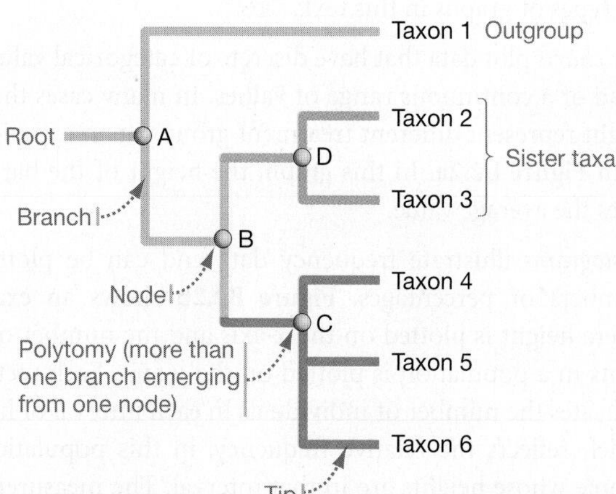

FIGURE B3.1 Phylogenetic Trees Have Roots, Branches, Nodes, and Tips.

✔**EXERCISE** Circle all four monophyletic groups present.

Recall from Chapter 1 that a taxon (plural: taxa) is any named group of organisms. A taxon could be a single species, such as *Homo sapiens*, or a large group of species, such as Primates. Tips connected by a single node on a tree are called sister taxa.

The phylogenetic trees used in this text are all rooted. This means that the first, or most basal, node on the tree—the one on the far left in this book—is the most ancient. To determine where the root on a tree occurs, biologists include one or more outgroup species when they are collecting data to estimate a particular phylogeny. An outgroup is a taxonomic group that is known to have diverged prior to the rest of the taxa in the study.

In Figure B3.1, "Taxon 1" is an outgroup to the monophyletic group consisting of taxa 2–6. A monophyletic group consists of an ancestral species and all of its descendants. The root of a tree is placed between the outgroup and the monophyletic group being studied. This position in Figure B3.1 is node A.

Understanding monophyletic groups is fundamental to reading and estimating phylogenetic trees. Monophyletic groups may also be called lineages or clades and can be identified using the "one-snip test": If you cut any branch on a phylogenetic tree, all of the branches and tips that fall off represent a monophyletic group. Using the one-snip test, you should be able to convince yourself that the monophyletic groups on a tree are nested. In Figure B3.1, for example, the monophyletic group comprising node A and taxa 1–6 contains a monophyletic group consisting of node B and taxa 2–6, which includes the monophyletic group represented by node C and taxa 4–6.

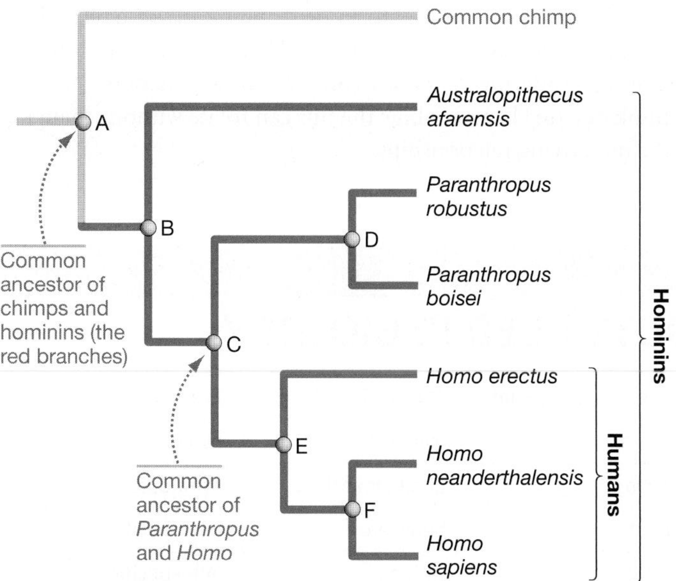

FIGURE B3.2 An Example of a Phylogenetic Tree. A phylogenetic tree showing the relationships of species in the monophyletic group called hominins.

✔**EXERCISE** All of the hominins walked on two legs—unlike chimps and all of the other primates. Add a mark on the phylogeny to show where upright posture evolved, and label it "origin of walking on two legs." Circle and label a pair of sister species. Label an outgroup to the monophyletic group called humans (species in the genus *Homo*).

To put all these new terms and concepts to work, consider the phylogenetic tree in **Figure B3.2**, which shows the relationships between common chimpanzees and six human and humanlike species that lived over the past 5–6 million years. Chimps functioned as an outgroup in the analysis that led to this tree, so the root was placed at node A. The branches marked in red identify a monophyletic group called the hominins.

To practice how to read a tree, put your finger at the tree's root, at the far left, and work your way to the right. At node A, the ancestral population split into two descendant populations. One of these populations eventually evolved into today's chimps; the other gave rise to the six species of hominins pictured. Now continue moving your finger toward the tips of the tree until you hit node C. It should make sense to you that at this splitting event, one descendant population eventually gave rise to two *Paranthropus* species, while the other became the ancestor of humans—species in the genus *Homo*. If multiple branches emerge from a node, creating a polytomy, it means that the populations split from one another so quickly that it is not possible to tell which split off earlier or later.

As you study Figure B3.2, you should consider a couple of important points.

1. There are many equivalent ways of drawing this tree. For example, this version shows *Homo sapiens* on the bottom. But the tree would be identical if the two branches emerging from node E were rotated 180°, so that the species appeared in the

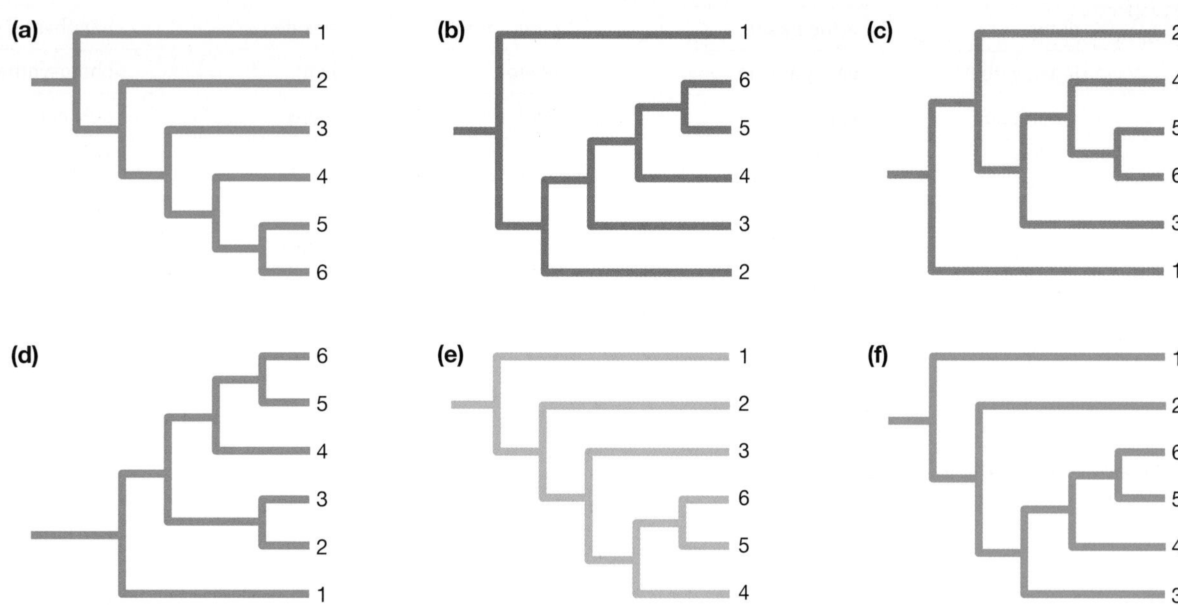

FIGURE B3.3 Alternative Ways of Drawing the Same Tree.

✔**QUESTION** Five of these six trees describe exactly the same relationships among taxa 1 through 6. Identify the tree that is different from the other five.

order *Homo sapiens, Homo neanderthalensis, Homo erectus.* Trees are read from root to tips, not from top to bottom or bottom to top.

2. No species on any tree is any higher or lower than any other. Chimps and *Homo sapiens* have been evolving exactly the same amount of time since their divergence from a common ancestor—neither species is higher or lower than the other. It is legitimate to say that more-ancient groups like *Australopithecus afarensis* have traits that are ancestral or more basal—meaning, that appeared earlier in evolution—compared to traits that appear in *Homo sapiens*, which are referred to as more derived.

Figure B3.3 on page B:5 presents a chance to test your tree-reading ability. Five of the six trees shown in this diagram are identical in terms of the evolutionary relationships they represent. One differs. The key to understanding the difference is to recognize that the ordering of tips does not matter in a tree—only the ordering of nodes (branch points) matters. You can think of a tree like a mobile: the tips can rotate without changing the underlying relationships.

BIOSKILLS 4

SOME COMMON LATIN AND GREEK ROOTS USED IN BIOLOGY

Greek or Latin Root	English Translation	Example Term
a, an	not	anaerobic
aero	air	aerobic
allo	other	allopatric
amphi	on both sides	amphipathic
anti	against	antibody
auto	self	autotroph
bi	two	bilateral symmetry
bio	life, living	bioinformatics
blast	bud, sprout	blastula
co	with	cofactor
cyto	cell	cytoplasm
di	two	diploid
ecto	outer	ectoparasite
endo	inner, within	endoparasite
epi	outer, upon	epidermis
exo	outside	exothermic
glyco	sugary	glycolysis
hetero	different	heterozygous
homo	alike	homozygous
hydro	water	hydrolysis
hyper	over, more than	hypertonic
hypo	under, less than	hypotonic
inter	between	interspecific
intra	within	intraspecific
iso	same	isotonic
logo, logy	study of	morphology
lyse, lysis	loosen, burst	glycolysis
macro	large	macromolecule
meta	change, turning point	metamorphosis
micro	small	microfilament
morph	form	morphology
oligo	few	oligopeptide
para	beside	parathyroid gland
photo	light	photosynthesis
poly	many	polymer
soma	body	somatic cells
sym, syn	together	symbioticm, synapsis
trans	across	translation
tri	three	trisomy
zygo	yoked together	zygote

✔Questions

Provide literal translations of the following terms:

1. heterozygote
2. glycolysis
3. morphology
4. trisomy

USING STATISTICAL TESTS AND INTERPRETING STANDARD ERROR BARS

When biologists do an experiment, they collect data on individuals in a treatment group and a control group, or several such comparison groups. Then they want to know whether the individuals in the two (or more) groups are different. For example, Chapter 2 introduces an experiment in which student researchers measured how fast a product formed when they set up a reaction with three different concentrations of reactants. Each treatment—meaning, each combination of reactant concentrations—was replicated many times.

Figure B5.1 graphs the average reaction rate for each of the three treatments in the experiment. Note that Treatments 1, 2, and 3 represent increasing concentrations of reactants. The thin "I-beams" on each bar indicate the standard error of each average. The standard error is a quantity that indicates the uncertainty in the calculation of an average.

For example, if two trials with the same concentration of reactants had a reaction rate of 0.075 and two trials had a reaction rate of 0.025, then the average reaction rate would be 0.50. In this case, the standard error would be large. But if two trials had a reaction rate of 0.051 and two had a reaction rate of 0.049, the average would still be 0.050, but the standard error would be small.

In effect, the standard error quantifies how confident you are that the average you've calculated is the average you'd observe if you did the experiment under the same conditions an extremely large number of times. It is a measure of precision.

Once they had calculated these averages and standard errors, the students wanted to answer a question: Does reaction rate increase when reactant concentration increases?

After looking at the data, you might conclude that the answer is yes. But how could you come to a conclusion like this objectively, instead of subjectively?

The answer is to use a statistical test. This can be thought of as a three-step process.

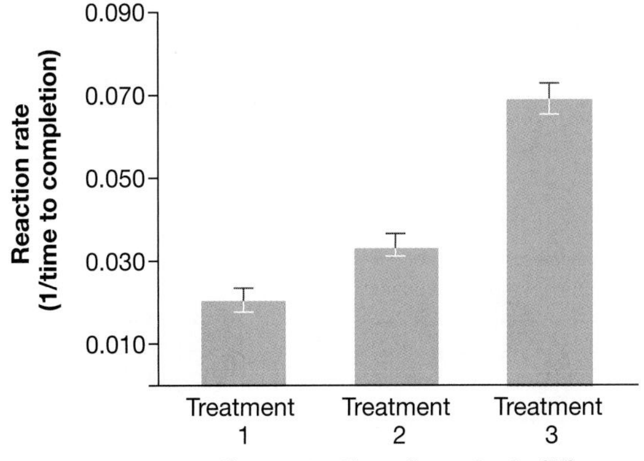

FIGURE B5.1 Standard Error Bars Indicate the Uncertainty in an Average.

1. Specify the null hypothesis, which is that reactant concentration has no effect on reaction rate.
2. Calculate a test statistic, which is a number that characterizes the size of the difference among the treatments. In this case, the test statistic compares the actual differences in reaction rates among treatments to the difference predicted by the null hypothesis. The null hypothesis predicts that there should be no difference.
3. The third step is to determine the probability of getting a test statistic as large as the one calculated just by chance. The answer comes from a reference distribution—a mathematical function that specifies the probability of getting various values of the test statistic if the null hypothesis is correct. (If you take a statistics course, you'll learn which test statistics and reference distributions are relevant to different types of data.)

You are very likely to see small differences among treatment groups just by chance—even if no differences actually exist. If you flipped a coin 10 times, for example, you are unlikely to get exactly five heads and five tails, even if the coin is fair. A reference distribution tells you how likely you are to get each of the possible outcomes of the 10 flips if the coin is fair, just by chance.

In this case, the reference distribution indicated that if the null hypothesis of no actual difference in reaction rates is correct, you would see differences as large as those observed only 0.01 percent of the time just by chance. By convention, biologists consider a difference among treatment groups to be statistically significant if you have less than a 5 percent probability of observing it just by chance. Based on this convention, the student researchers were able to claim that the null hypothesis is not correct for reactant concentration. According to their data, the reaction they studied really does happen faster when reactant concentration increases.

It is likely that you'll be doing actual statistical tests early in your undergraduate career. To use this text, though, you only need to be aware of what statistical testing does. And you should take care to inspect the standard error bars on graphs in this book. As a *very* rough rule of thumb, averages often turn out to be significantly different, according to an appropriate statistical test, if there is no overlap between two times the standard errors.

✔Question

Suppose you estimated the average height of students in your class by sampling two individuals at random. Then suppose you estimated the same quantity by sampling every individual that showed up for class on a particular day. Which estimate of the average is likely to have the smallest standard error, and why?

READING CHEMICAL STRUCTURES

If you haven't had much chemistry yet, learning basic biological chemistry can be a challenge. One of the stumbling blocks is simply being able to read chemical structures efficiently and understand what they mean. This skill will come much easier once you have a little notation under your belt and you understand some basic symbols.

Atoms are the basic building blocks of everything in the universe, just as cells are the basic building blocks of your body. Every atom has a 1- or 2-letter symbol. **Table B6.1** shows the symbols for most of the atoms you'll encounter in this book. You should memorize these. The table also offers details on how the atoms form bonds as well as how they are represented in visual models.

When atoms attach to each other by covalent bonding, a molecule forms. Biologists have a couple of different ways of representing molecules—you'll see each of these in the book and in class.

- A molecular formula like those in **Figure B6.1a** simply lists the atoms present in a molecule, with subscripts indicating how many of each atom are present. If the formula has no subscript, only one of that type of atom is present. A methane (natural gas) molecule, for example, can be written as CH_4. It consists of one carbon atom and four hydrogen atoms.
- Structural formulas like those in **Figure B6.1b** show which atoms in the molecule are bonded to each other, with each bond indicated by a dash. The structural formula for methane indicates that each of the four hydrogen atoms forms one covalent bond with carbon, and that carbon makes a total of four covalent bonds. Single covalent bonds are symbolized by a single dash; double bonds are indicated by two dashes.

Even simple molecules have distinctive shapes, because different atoms make covalent bonds at different angles. Ball-and-stick and space-filling models show the geometry of the bonds accurately.

TABLE B6.1 **Some Attributes of Atoms Found in Organisms**

Atom	Symbol	Number of Bonds It Can Form	Standard Color Code*
Hydrogen	H	1	white
Carbon	C	4	black
Nitrogen	N	3	blue
Oxygen	O	2	red
Sodium	Na	1	—
Magnesium	Mg	2	—
Phosphorus	P	5	orange or purple
Sulfur	S	2	yellow
Chlorine	Cl	1	—
Potassium	K	1	—
Calcium	Ca	2	—

*In ball-and-stick or space-filling models.

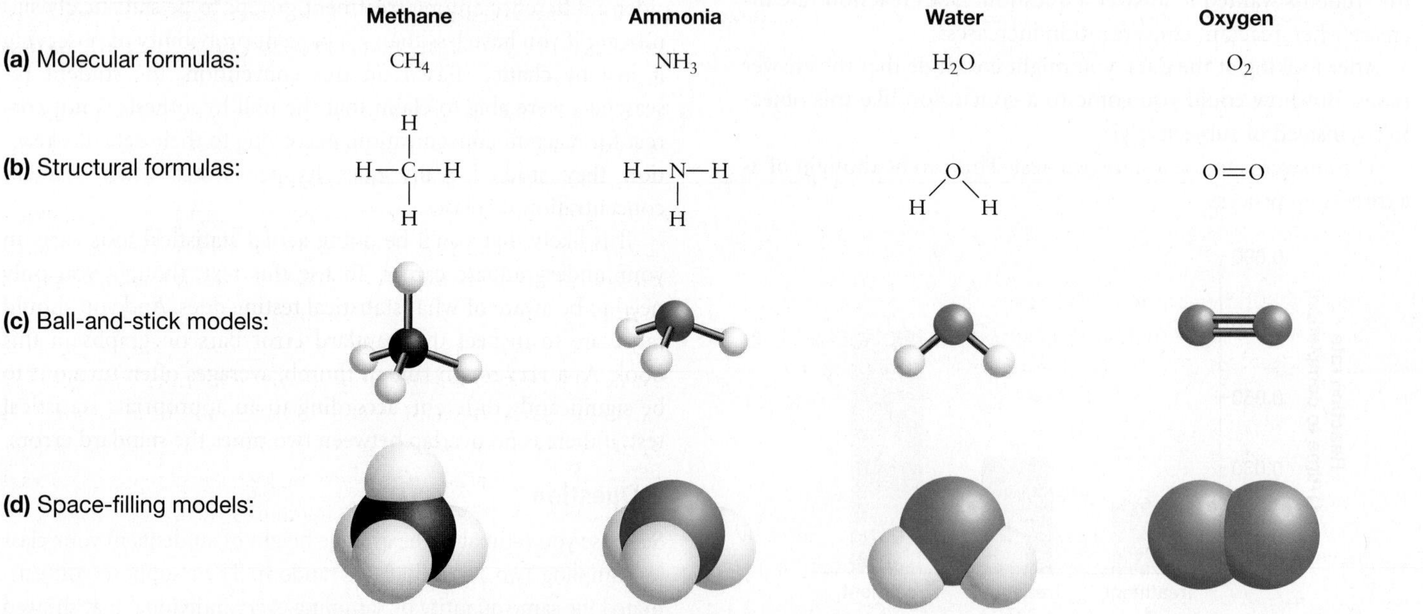

	Methane	Ammonia	Water	Oxygen
(a) Molecular formulas:	CH_4	NH_3	H_2O	O_2
(b) Structural formulas:	H—C—H (with H above and below C)	H—N—H (with H below N)	H—O—H (bent)	O=O
(c) Ball-and-stick models:				
(d) Space-filling models:				

FIGURE B6.1 Molecules Can Be Represented in Several Different Ways.

✔**EXERCISE** Carbon dioxide consists of a carbon atom that forms a double bond with each of two oxygen atoms, for a total of four bonds. It is a linear molecule. Write carbon dioxide's molecular formula, then draw its structural formula, a ball-and-stick model, and a space-filling model.

- In a ball-and-stick model, a stick is used to represent each covalent bond (see **Figure B6.1c**).
- In space-filling models, the atoms are simply stuck onto each other in their proper places (see **Figure B6.1d**).

To learn more about a molecule when you look at a chemical structure, ask yourself three questions:

1. ***Is the molecule polar—meaning that some parts are more negatively or positively charged than others?*** Molecules that contain nitrogen or oxygen atoms are often polar, because these atoms have such high electronegativity (see Chapter 2). This trait is important because polar molecules dissolve in water.
2. ***Does the structural formula show atoms that might participate in chemical reactions?*** For example, are there charged atoms or amino or carboxyl (—COOH) groups that might act as a base or an acid?
3. ***In ball-and-stick and especially space-filling models of large molecules, are there interesting aspects of overall shape?*** For example, is there a groove where a protein might bind to DNA, or a cleft where a substrate might undergo a reaction in an enzyme?

BIOSKILLS 7

USING LOGARITHMS

You have probably been introduced to logarithms and logarithmic notation in algebra courses, and you will encounter logarithms at several points in this course. Logarithms are a way of working with powers—meaning, numbers that are multiplied by themselves one or more times.

Scientists use exponential notation to represent powers. For example,

$$a^x = y$$

means that if you multiply a by itself x times, you get y. In exponential notation, a is called the base and x is called the exponent. The entire expression is called an exponential function.

What if you know y and a, and you want to know x? This is where logarithms come in. You can solve for exponents using logarithms. For example,

$$x = \log_a y$$

This equation reads, x is equal to the logarithm of y to the base a. Logarithms are a way of working with exponential functions. They are important because so many processes in biology (and chemistry and physics, for that matter) are exponential in nature. To understand what's going on, you have to describe the process with an exponential function and then use logarithms to work with that function.

Although a base can be any number, most scientists use just two bases when they employ logarithmic notation: 10 and e. Logarithms to the base 10 are so common that they are usually symbolized in the form $\log y$ instead of $\log_{10} y$. A logarithm to the base e is called a natural logarithm and is symbolized ln (pronounced *EL-EN*) instead of log. You write "the natural logarithm of y" as $\ln y$. The base e is an irrational number (like π) that is approximately equal to 2.718. Like 10, e is just a number. But both 10 and e have qualities that make them convenient to use in biology (and chemistry, and physics).

Most scientific calculators have keys that allow you to solve problems involving base 10 and base e. For example, if you know y, they'll tell you what $\log y$ or $\ln y$ are—meaning that they'll solve for x in our example above. They'll also allow you to find a number when you know its logarithm to base 10 or base e. Stated another way, they'll tell you what y is if you know x, and y is equal to e^x or 10^x. This is called taking an antilog. In most cases, you'll use the inverse or second function button on your calculator to find an antilog (above the log or ln key).

To get some practice with your calculator, consider the equation

$$10^2 = 100$$

If you enter 100 in your calculator and then press the log key, the screen should say 2. The logarithm tells you what the exponent is. Now press the antilog key while 2 is on the screen. The calculator screen should return to 100. The antilog solves the exponential function, given the base and the exponent.

If your background in algebra isn't strong, you'll want to get more practice working with logarithms—you'll see them frequently during your undergraduate career. Remember that once you understand the basic notation, there's nothing mysterious about logarithms. They are simply a way of working with exponential functions, which describe what happens when something is multiplied by itself a number of times—like cells that divide and then divide again and then again.

Using logarithms will also come up when you are studying something that can have a large range of values, like the concentration of hydrogen ions in a solution or the intensity of sound that the human ear can detect. In cases like this, it's convenient to express the numbers involved as exponents. Using exponents makes a large range of numbers smaller and more tractable. For example, instead of saying that hydrogen ion concentration in a solution can range from 1 to 10^{-14}, the pH scale allows you to simply say that it ranges from 1 to 14. Instead of giving the actual value, you're expressing it as an exponent. It just simplifies things.

✓Questions

In Chapter 52, you'll use the equation $N_t = N_0 e^{rt}$.

1. What type of function does this equation describe?
2. After taking the natural logarithm of both sides, how would you write the equation?

BIOSKILLS 8

MAKING CONCEPT MAPS

A concept map is a graphical device for organizing and expressing what you know about a topic. It has two main elements: (1) concepts that are identified by words or short phrases and placed in a box or circle, and (2) labeled arrows that physically link two concepts and explain the relationship between them. The concepts are arranged hierarchically on a page, with the most general concepts at the top and the most specific ideas at the bottom.

The combination of a concept, a linking word, and a second concept is called a proposition. Good concept maps also have cross-links—meaning, labeled arrows that connect different elements in the hierarchy, as you read down the page.

Concept maps were initially developed by Joseph Novak in the early 1970s and have proven to be an effective studying and learning tool. They can be particularly valuable if constructed by a group, or when different individuals exchange and critique concept maps they have created independently. Although concept maps vary widely in quality and can be graded using objective criteria, there are many equally valid ways of making a high-quality concept map on a particular topic.

When you are asked to make a concept map in this text, you will usually be given at least a partial list of concepts to use. As an example, suppose you were asked to create a concept map on experimental design and were given the following concepts: results, predictions, control treatment, experimental treatment, controlled (identical) conditions, conclusions, experiment, hypothesis to be tested, null hypothesis. One possible concept map is shown in **Figure B8.1**.

Good concept maps have four qualities:

- They exhibit an organized hierarchy, indicating how each concept on the map relates to larger and smaller concepts.
- The concept words are specific—not vague.
- The propositions are accurate.
- There is cross-linking between different elements in the hierarchy of concepts.

As you practice making concept maps, go through these criteria and use them to evaluate your own work, as well as the work of fellow students.

✔Exercises

1. In many cases, investigators contrast a hypothesis being tested with an alternative hypothesis that does not qualify as a null hypothesis. Add an "Alternative hypothesis" concept to the map in Figure B8.1, along with other concepts and labeled linking arrows needed to indicate its relationship to other information on the map.
2. Add a box for the concept "Statistical testing" (see **BioSkills 5**) along with appropriately labeled linking arrows.

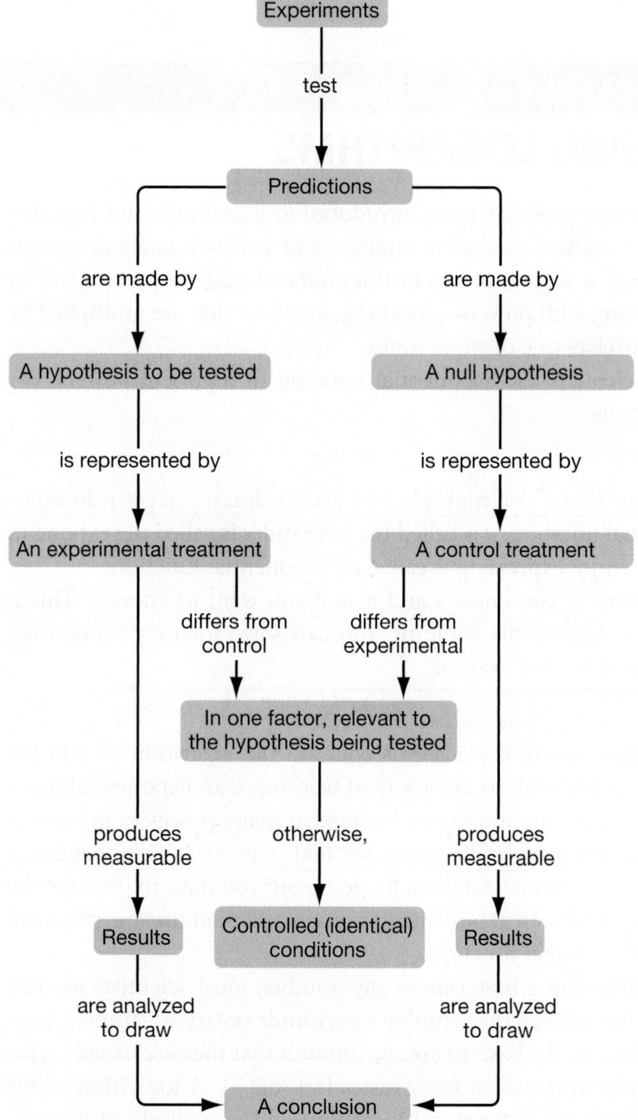

FIGURE B8.1 A Concept Map on Principles of Experimental Design.

SEPARATING AND VISUALIZING MOLECULES

To study a molecule, you have to be able to isolate it. This is a two-step process: the molecule has to be separated from other molecules in a mixture, and then physically picked out or located in a purified form. This BioSkill focuses on the techniques that biologists use to separate nucleic acids and proteins and then find the particular one they are interested in.

Using Electrophoresis to Separate Molecules

In molecular biology, the standard technique for separating proteins and nucleic acids is called gel electrophoresis or, simply, electrophoresis (literally "electricity-moving"). You may be using electrophoresis in a lab for this course, and you will certainly be analyzing data derived from electrophoresis in this text.

The principle behind electrophoresis is simple. Both proteins and nucleic acids carry a charge. As a result, these molecules move when placed in an electric field. Negatively charged molecules move toward the positive electrode; positively charged molecules move toward the negative electrode.

To separate a mixture of macromolecules so that each can be isolated and analyzed, researchers place the sample in a gelatinous substance. More specifically, the sample is placed in a "well"—a slot in a sheet or slab of the gelatinous substance. The "gel" itself consists of long molecules that form a matrix of fibers. The presence of the fibers keeps molecules in the sample from moving around randomly, but the gelatinous matrix also has pores through which the molecules can pass.

When an electrical field is applied across the gel, the molecules in the well move through the gel toward an electrode. Molecules that are smaller or more highly charged for their size move faster than do larger or less highly charged molecules. As they move, then, the molecules separate by size and by charge. Small and highly charged molecules end up at the bottom of the gel; large, less-charged molecules remain near the top.

An Example "Run"

Figure B9.1 shows the electrophoresis setup used in an experiment investigating how RNA molecules polymerize In this case, the investigators wanted to document how long RNA molecules became over time, when ribonucleotide triphosphates were present in a particular type of solution.

Step 1 shows how they loaded samples of macromolecules, taken on different days during the experiment, into wells at the top of the gel slab. This is a general observation: Each well holds a different sample. In this and many other cases, the researchers also filled a well with a sample containing fragments of known size, called a size standard or "ladder."

In step 2 the researchers immersed the gel in a solution that conducts electricity and applied a voltage across the gel. The molecules in each well started to run down the gel, forming a lane. After several hours of allowing the molecules to move, they removed the electric field (step 3). By then, molecules of different size and charge had separated from one another. In this case,

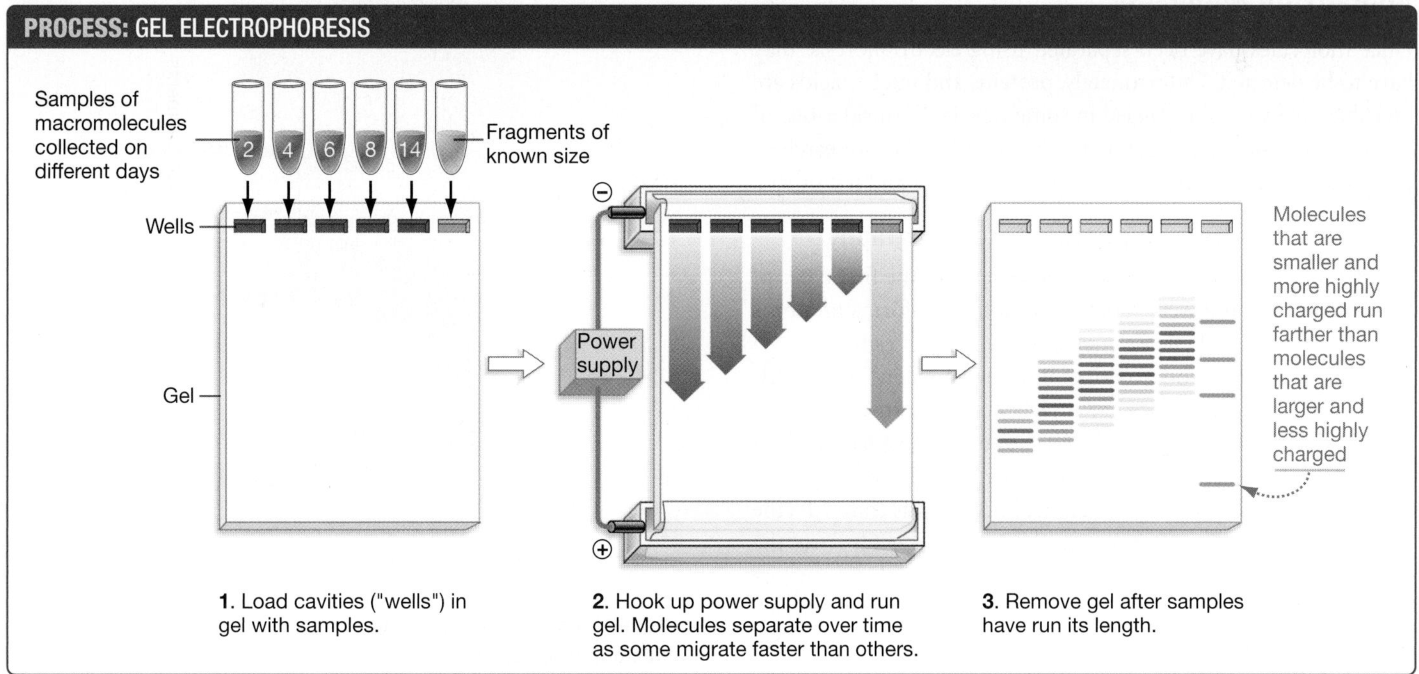

FIGURE B9.1 Macromolecules Can Be Separated via Gel Electrophoresis.

✔**QUESTION** DNA and RNA run toward the positive electrode. Why are these molecules negatively charged?

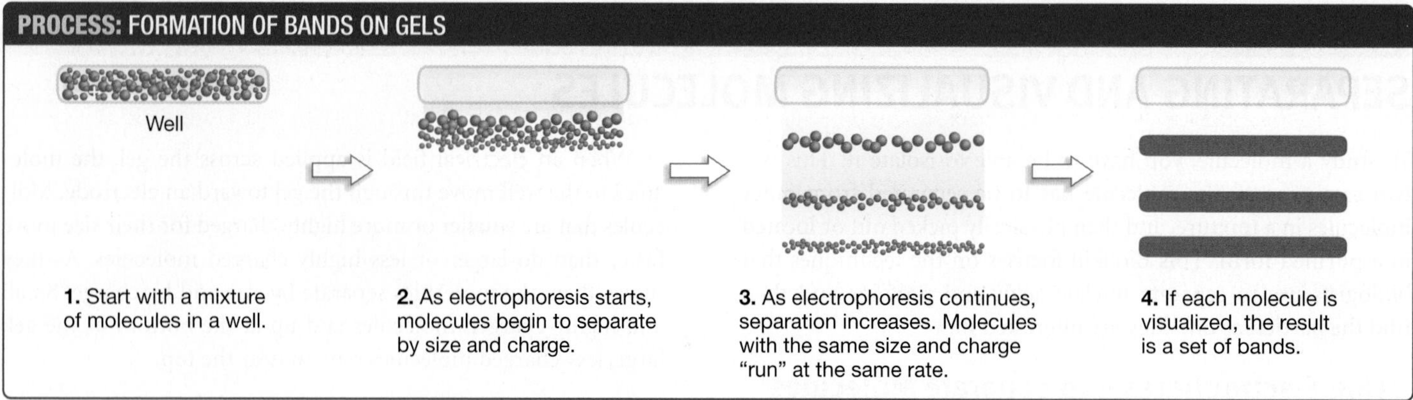

FIGURE B9.2 On a Gel, Alike Molecules Form Bands.

small RNA molecules had reached the bottom of the gel. Above them were larger RNA molecules, which had run more slowly.

Why Do Separated Molecules Form Bands?

When researchers visualize a particular molecule on a gel, using techniques described below, the image that results consists of bands: shallow lines that are as wide as a lane in the gel. Why?

To understand the answer, study **Figure B9.2**. The top panel shows the original mixture of molecules. In this cartoon, the size of each dot represents the size of each molecule. The key is to realize that the original sample contains many copies of each specific molecule, and that these copies run down the length of the gel together—meaning, at the same rate—because they have the same size and charge.

It's that simple: Alike molecules form a band because they stay together.

Visualizing Molecules

Once molecules have been separated using electrophoresis, they have to be detected. Unfortunately, proteins and nucleic acids are invisible unless they are tagged in some way. Let's consider two of the most common tagging systems, then consider how researchers can tag and visualize specific molecules of interest and not others.

Using Radioactive Isotopes and Autoradiography

When molecular biology was getting under way, the first types of tags in common use were radioactive isotopes—forms of atoms that are unstable and release energy in the form of radiation.

In the polymerization experiment diagrammed in Figure B9.1, for example, the researchers had attached a radioactive phosphorus atom to the monomers—ribonucleotide triphosphates—used in the original reaction mix. Once polymers formed, they contained radioactive atoms. When electrophoresis was complete, the investigators visualized the polymers by laying X-ray film over the gel. Because radioactive emissions expose film, a black dot appears wherever a radioactive atom is located in the gel. So many black dots occur so close together that the collection forms a dark band.

This technique for visualizing macromolecules is called autoradiography. The autoradiograph that resulted from the polymerization experiment is shown in **Figure B9.3**. The samples, taken on days 2, 4, 6, 8, and 14 of the experiment, are labeled along the bottom. The far right lane contains macromolecules of known size; this lane is used to estimate the size of the molecules in the experimental samples. The bands that appear in each sample lane represent the different polymers that had formed.

Reading a Gel

One of the keys to interpreting or "reading" a gel is to realize that darker bands contain more radioactive markers, indicating the presence of many radioactive molecules. Lighter bands contain fewer molecules.

To read a gel, then, you look for **(1)** the presence or absence of bands in some lanes—meaning, some experimental samples—versus others, and **(2)** contrasts in the darkness of the bands—meaning, differences in the number of molecules present.

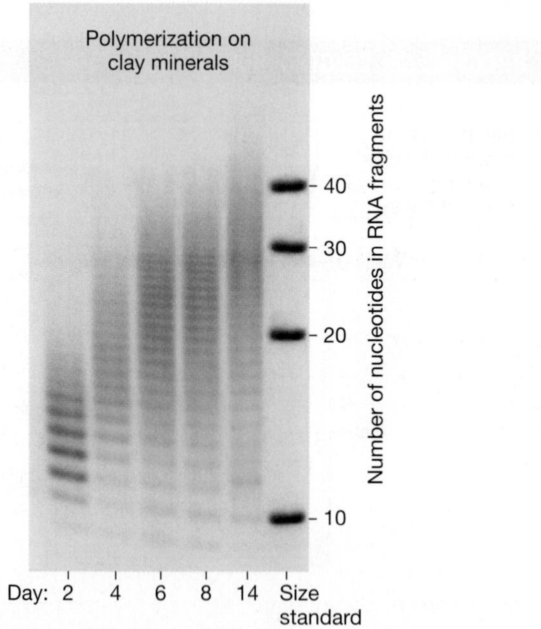

FIGURE B9.3 Autoradiography Is a Technique for Visualizing Macromolecules. The molecules in a gel can be visualized in a number of ways. In this case, the RNA molecules in the gel exposed an X-ray film because they had radioactive atoms attached. When developed, the film is called an autoradiograph.

For example, several conclusions can be drawn from the data in Figure B9.3. First, a variety of polymers formed at each stage. After the second day, for example, polymers from 12 to 18 monomers long had formed on the clay particles used in this experiment. Second, the overall length of polymers produced increased with time. At the end of the fourteenth day, most of the RNA molecules were between 20 and 40 monomers long.

Starting in the late 1990s and early 2000s, it became much more common to tag nucleic acids with fluorescent tags. Once electrophoresis is complete, the presence of fluorescence can be detected by exposing the gel to an appropriate wavelength of light; the fluorescent tag fluoresces or glows in response (fluorescence is explained in Chapter 10).

Fluorescent tags have important advantages over radioactive isotopes: (1) They are safer to handle. (2) They are faster—you don't have to wait hours or days for the radioactive isotope to expose a film. (3) They come in multiple colors, so you can tag several different molecules in the same experiment and detect them independently.

Using Nucleic Acid Probes

In many cases, researchers want to find one specific molecule—a certain DNA sequence, for example—in the collection of molecules on a gel. How is this possible? The answer hinges on using a particular molecule as a probe.

Chapter 19 explains how probes work in detail. Here it's enough to get the general idea: A probe is a marked molecule that binds specifically to your molecule of interest. The "mark" is often a radioactive atom or a fluorescent tag.

If you are looking for a particular DNA or RNA sequence on a gel, for example, you can expose the gel to a single-stranded probe that binds to the target sequence by complementary base pairing. Once it has bound, you can detect the band through autoradiography or fluorescence.

- Southern blotting is a technique for making DNA fragments that have been run out on a gel single-stranded, transferring them from the gel to a nylon filter, and then probing them to identify segments of interest. The technique was named after its inventor, Edwin Southern.
- Northern blotting is a technique for transferring RNA fragments from a gel to a nylon filter, and then probing them to detect target segments. The name is a lighthearted play on Southern blotting—the protocol from which it was derived.

Using Antibody Probes

How can researchers find a particular protein out of a large collection of different proteins? The answer is to use an antibody. An antibody is a protein that binds specifically to a section of a different protein. The structure of antibodies and their function in the immune system are explored in detail in Chapter 49.

To use an antibody as a probe, investigators attach a tag molecule—often an enzyme that catalyzes a color-forming reaction—to the antibody and allow it to react with proteins in a mixture. The antibody will stick to the specific protein that it binds to, and then can be visualized thanks to the tag it carries.

If the proteins in question have been separated by gel electrophoresis, the result is called a Western blot. The name Western is an extension of the Southern and Northern pattern.

✓Question

Suppose you've been given a gel that has been stained for "RNA X." One lane contains no bands. Two lanes have a band in the same location, even though one of the bands is barely visible and the other is extremely dark. The fourth lane has a light band located above the bands in the other lanes. Interpret the gel.

BIOSKILLS 10

BIOLOGICAL IMAGING: MICROSCOPY AND X-RAY CRYSTALLOGRAPHY

A lot of biology happens at levels that can't be detected with the naked eye. Biologists use an array of microscopes to study small multicellular organisms, individual cells, and the contents of cells. And to understand what individual macromolecules or multimolecular machines like ribosomes look like, researchers use data from a technique called X-ray crystallography.

You'll probably use dissecting microscopes and compound light microscopes to view specimens during your labs for this course, and throughout this text you'll be seeing images generated from other types of microscopy and from X-ray crystallographic data. One of the fundamental skills you'll be acquiring as an introductory student, then, is a basic understanding of how these techniques work. The key is to recognize that each approach for visualizing microscopic structures has strengths and weaknesses. As a result, each technique is appropriate for studying certain types or aspects of cells or molecules.

Light Microscopy

If you use a dissecting microscope during labs, you'll recognize that it works by magnifying light that bounces off a whole specimen—often a live organism. You'll be able to view the specimen in three dimensions, which is why these instruments are sometimes called stereomicroscopes, but the maximum magnification possible is only about 20 to 40 times normal size (20× to 40×).

To view smaller objects, you'll probably use a compound microscope. Compound microscopes magnify light that is passed *through* a specimen. The instruments used in introductory labs are usually capable of 400× magnifications; the most sophisticated compound microscopes available can achieve magnifications of about 2000×. This is enough to view individual bacterial or eukaryotic cells and see large structures inside cells, like condensed chromosomes (see Chapter 11). To prepare a specimen for viewing under a compound light microscope, the tissues or cells are usually sliced to create a section thin enough for light to pass through efficiently. The section is then dyed to increase contrast and make structures visible. In many cases, different types of dyes are used to highlight different types of structures.

Electron Microscopy

Until the 1950s, the compound microscope was the biologist's only tool for viewing cells directly. But the invention of the electron microscope provided a new way to view specimens. Two basic types of electron microscopy are now available: one that allows researchers to examine cross sections of cells at extremely high magnification, and one that offers a view of surfaces at somewhat lower magnification.

Transmission Electron Microscopy (TEM)

The transmission electron microscope is an extraordinarily effective tool for viewing cell structure at high magnification. TEM forms an image from electrons that pass through a specimen, just as a light microscope forms an image from light rays that pass through a specimen.

Biologists who want to view a cell under a transmission electron microscope begin by "fixing" the cell, meaning that they treat it with a chemical agent that stabilizes the cell's structure and contents while disturbing them as little as possible. Then the researcher permeates the cell with an epoxy plastic that stiffens the structure. Once this epoxy hardens, the cell can be cut into extremely thin sections with a glass or diamond knife. Finally, the sectioned specimens are impregnated with a metal—often lead. (The reason for this last step is explained shortly.)

Figure B10.1a outlines how the transmission electron microscope works. A beam of electrons is produced by a tungsten filament at the top of a column and directed downward. (All of the air is pumped out of the column, so that the electron beam isn't scattered by collisions with air molecules.) The electron beam passes through a series of lenses and through the specimen. The lenses are actually electromagnets, which alter the path of the beam much like a glass lens in a dissecting or compound microscope bends light. The lenses magnify and focus the image on a screen at the bottom of the column. There the electrons strike a coating of fluorescent crystals, which emit visible light in response—just like a television screen. When the microscopist moves the screen out of the way and allows the electrons to expose a sheet of black-and-white film, the result is a micrograph—a photograph of an image produced by microscopy.

The image itself is created by electrons that pass through the specimen. If no specimen were in place, all the electrons would pass through and the screen (and micrograph) would be uniformly bright. Unfortunately, cell materials by themselves would also appear fairly uniform and bright. This is because an atom's ability to deflect an electron depends on its mass. In turn, an atom's mass is a function of its atomic number. The hydrogen, carbon, oxygen, and nitrogen atoms that dominate biological molecules have low atomic numbers. This is why cell biologists must saturate cell sections with lead solutions. Lead has a high atomic number and scatters electrons effectively. Different macromolecules take up lead atoms in different amounts, so the

(a) Transmission electron microscopy: High magnification of cross sections

Tungsten filament (source of electrons)

Condenser lens

Specimen

Objective lens

Projector lens

Image on fluorescent screen

0.2 μm

Cross section of *E. coli* bacterium

(b) Scanning electron microscopy: Lower magnification of surfaces

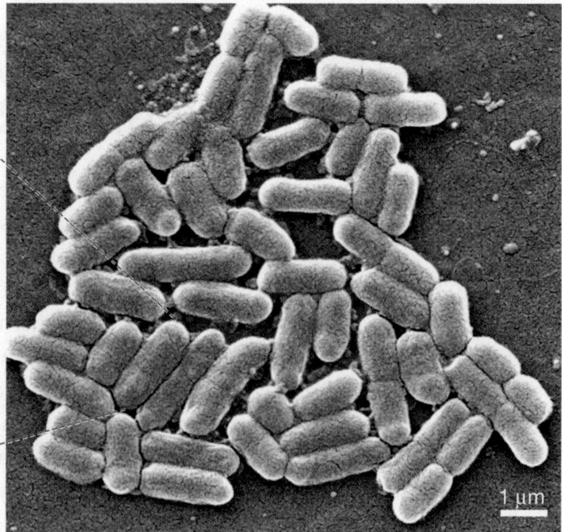

Surface view of *E. coli* bacteria

FIGURE B10.1 There Are Two Basic Types of Electron Microscopy.

metal acts as a "stain" that produces contrast. With TEM, areas of dense metal scatter the electron beam most, producing dark areas in micrographs.

The advantage of TEM is that it can magnify objects up to 250,000×—meaning that intracellular structures are clearly visible. The downsides are that researchers are restricted to observing dead, sectioned material, and they must take care that the preparation process does not distort the specimen.

Scanning Electron Microscopy (SEM)

The scanning electron microscope is the most useful tool biologists have for looking at the surfaces of structures. Materials are prepared for scanning electron microscopy by coating their surfaces with a layer of metal atoms. To create an image of this surface, the microscope scans the surface with a narrow beam of electrons. Electrons that are reflected back from the surface or that are emitted by the metal atoms in response to the beam then strike a detector. The signal from the detector controls a second electron beam, which scans a TV-like screen and forms an image magnified up to 50,000 times the object's size.

Because SEM records shadows and highlights, it provides images with a three-dimensional appearance (**Figure B10.1b**). It cannot magnify objects nearly as much as TEM can, however.

Studying Live Cells and Real-Time Processes

Until the 1960s, it was not possible for biologists to get clear, high-magnification images of living cells. But a series of innovations over the past 50 years has made it possible to observe organelles and subcellular structures in action.

The development of video microscopy, where the image from a light microscope is captured by a video camera instead of by an eye or a film camera, proved revolutionary. It allowed specimens to be viewed at higher magnification, because video cameras are more sensitive to small differences in contrast than are the human eye or still cameras. It also made it easier to keep live specimens functioning normally, because the increased light sensitivity of video cameras allows them to be used with low illumination, so specimens don't overheat. And when it became possible to digitize video images, researchers began using computers to remove out-of-focus background material and increase image clarity.

A more recent innovation was the use of a fluorescent molecule called green fluorescent protein, or GFP, which allows researchers to tag specific molecules or structures and follow their movement over time. GFP is naturally synthesized in jellyfish that fluoresce, or emit light. By affixing GFP molecules to another protein and then inserting it into a cell, investigators can follow the protein's fate over time and even videotape its movement. For example, researchers have videotaped GFP-tagged proteins being transported from the rough ER through the Golgi apparatus and out to the plasma membrane. This is cell biology: the movie.

GFP's influence has been so profound that the researchers who developed its use in microscopy were awarded the 2008 Nobel prize in Chemistry.

Visualizing Structures in 3-D

The world is three-dimensional. To understand how microscopic structures and macromolecules work, it is essential to understand their shape and spatial relationships. Consider three techniques currently being used to reconstruct the 3-D structure of cells, organelles, and macromolecules.

- *Confocal microscopy* is carried out by mounting cells that have been treated with one or more fluorescing tags on a microscope slide and then focusing a beam of ultraviolet light at a specific depth within the specimen. The fluorescing tag emits visible light in response. A detector for this light is then set up at exactly the position where the emitted light comes into focus. The result is a sharp image of a precise plane in the cell being studied (**Figure B10.2**). By altering the focal plane, a researcher can record images from an array of depths in the specimen; a computer can then be used to generate a 3-D image of the cell.
- *Electron tomography* uses a transmission electron microscope to generate a 3-D image of an organelle or other subcellular structure. The specimen is rotated around a single axis, with the researcher taking many "snapshots." The individual images are then pieced together with a computer. This technique has provided a much more accurate view of mitochondrial structure than was possible using traditional TEM (see Chapter 7).
- *X-ray crystallography, or X-ray diffraction analysis,* is the most widely used technique for reconstructing the 3-D structure of molecules. As its name implies, the procedure is based on bombarding crystals of a molecule with X-rays. X-rays are scattered in precise ways when they interact with the electrons surrounding the atoms in a crystal, producing a diffraction pattern that can be recorded on X-ray film or other types of

(a) Conventional fluorescence image of single cell

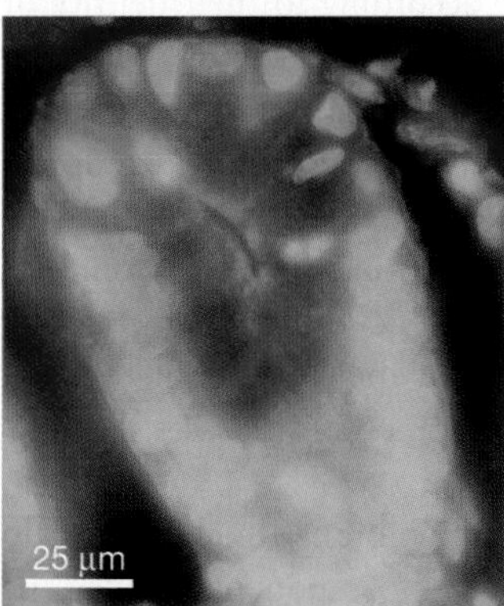

(b) Confocal fluorescence image of same cell

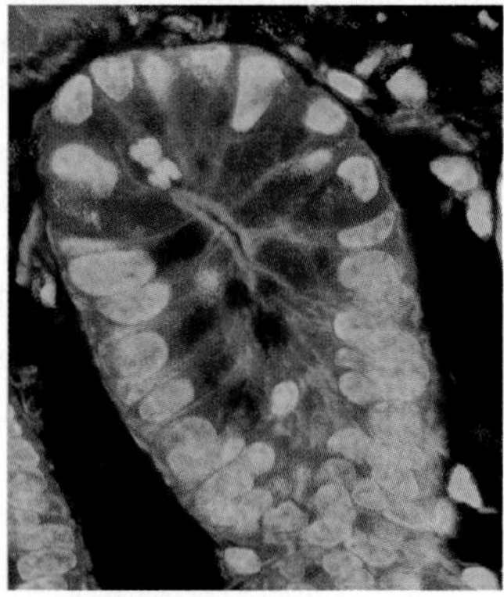

FIGURE B10.2 Confocal Microscopy Provides Sharp Images of Living Cells. **(a)** The conventional image of this mouse intestinal cell is blurred, because it results from light emitted by the entire cell. **(b)** The confocal image is sharp, because it results from light emitted at a single plane inside the cell.

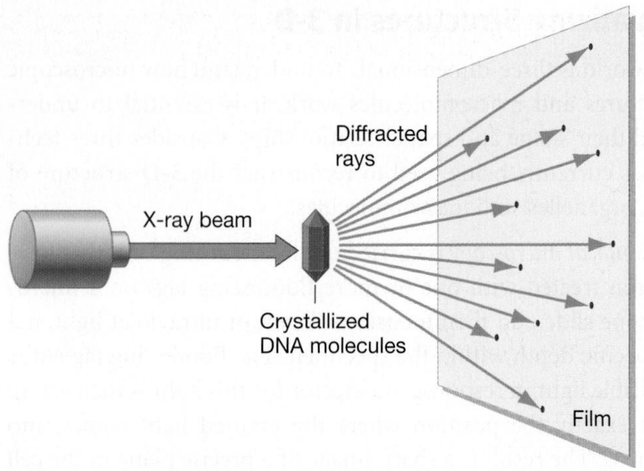

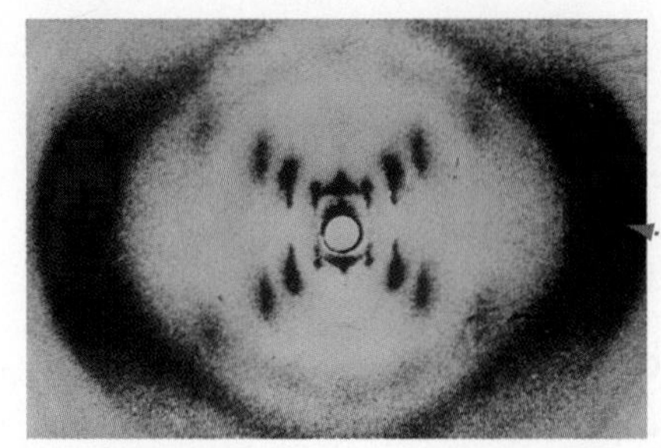

FIGURE B10.3 X-Ray Crystallography. When crystallized molecules are bombarded with X-rays, the radiation is scattered in distinctive patterns. The photograph at the right shows an X-ray film that recorded the pattern of scattered radiation from DNA molecules.

detectors (**Figure B10.3**). By varying the orientation of the X-ray beam as it strikes a crystal and documenting the diffraction patterns that result, researchers can construct a map representing the density of electrons in the crystal. By relating these electron-density maps to information about the primary structure of the nucleic acid or protein, a 3-D model of the molecule can be built. Virtually all of the molecular models used in this book were built from X-ray crystallographic data.

✔Questions

1. Suppose you are looking at a transmission electron micrograph of a cancerous cell from the human liver. No mitochondria are present. Does this mean that the cell lacks mitochondria?
2. X-ray crystallography is time consuming and technically difficult. Why is the effort to understand the structure of biological molecules worthwhile? What's the payoff?

BIOSKILLS 11

SEPARATING CELL COMPONENTS BY CENTRIFUGATION

Biologists use a technique called differential centrifugation to isolate specific cell components. Differential centrifugation is based on breaking cells apart to create a complex mixture and then separating components in a centrifuge. A centrifuge accomplishes this by spinning cells in a solution that allows molecules and other cell components to separate according to their density or size and shape. The individual parts of the cell can then be purified and studied in detail, in isolation from other parts of the cell.

The first step in preparing a cell sample for centrifugation is to release the organelles and cell components by breaking the cells apart. This can be done by putting them in a hypotonic solution, by exposing them to high-frequency vibration, by treating cells with a detergent, or by grinding them up. Each of these methods breaks apart plasma membranes and releases the contents of the cells.

The resulting pieces of plasma membrane quickly reseal to form small vesicles, often trapping cell components inside. The solution that results from the homogenization step is a mixture of these vesicles, free-floating macromolecules released from the cells, and organelles. A solution such as this is called a cell extract or cell homogenate.

When a cell homogenate is placed in a centrifuge tube and spun at high speed, the components that are in solution tend to move outward, along the red arrow in **Figure B11.1a**. The effect is similar to a merry-go-round, which seems to push you outward in a straight line away from the spinning platform. In response to this outward-directed force, the solution containing the cell homogenate exerts a centripetal (literally, "center-seeking") force that pushes the homogenate away from the bottom of the tube. Larger, denser molecules or particles resist this inward force more readily than do smaller, less dense ones and so reach the bottom of the centrifuge tube faster.

To separate the components of a cell extract, researchers often perform a series of centrifuge runs. Steps 1 and 2 of **Figure B11.1b** illustrate how an initial treatment at low speed causes larger, heavier parts of the homogenate to move below smaller, lighter parts. The material that collects at the bottom of the tube is called the pellet, and the solution and solutes left behind form the supernatant ("above swimming"). The supernatant is placed in a fresh tube and centrifuged at increasingly higher speeds and longer durations. Each centrifuge run continues to separate cell components based on their size and density.

To accomplish even finer separation of macromolecules or organelles, researchers frequently follow up with centrifugation at extremely high speeds. One strategy is based on filling the centrifuge tube with a series of sucrose solutions of increasing

(a) How a centrifuge works

When the centrifuge spins, the macromolecules tend to move toward the bottom of the centrifuge tube (red arrow)

The solution in the tube exerts a centripetal force, which resists movement of the molecules to the bottom of the tube (blue arrow)

Motor

Very large or dense molecules overcome the centripetal force more readily than smaller, less dense ones. As a result, larger, denser molecules move toward the bottom of the tube faster.

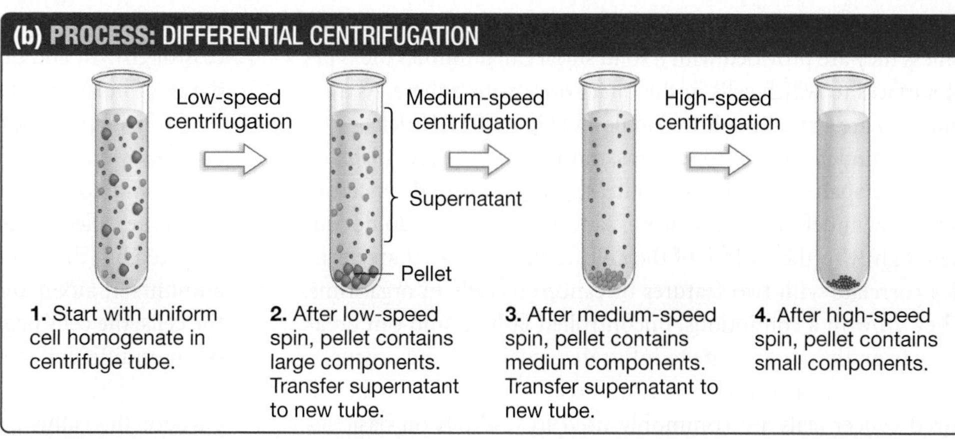

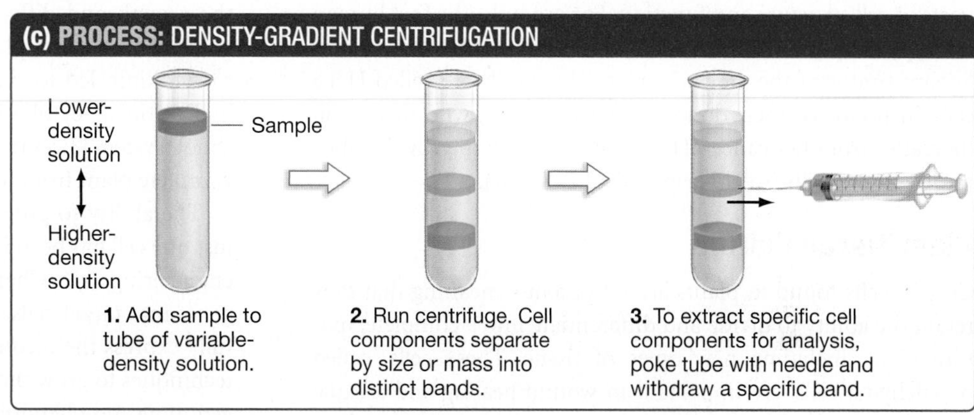

FIGURE B11.1 Cell Components Can Be Separated by Centrifugation. (a) The forces inside a centrifuge tube allow cell components to be separated. **(b)** Through a series of centrifuge runs made at increasingly higher speeds, an investigator can separate fractions of a cell homogenate by size via differential centrifugation. **(c)** A high-speed centrifuge run can achieve extremely fine separation among cell components by density-gradient centrifugation.

density (**Figure B11.1c**). The density gradient allows cell components to separate on the basis of small differences in size, shape, and density. When the centrifuge run is complete, each cell component occupies a distinct band of material in the tube, based on how quickly each moves through the increasingly dense gradient of sucrose solution during the centrifuge run. A researcher can then collect the material in each band for further study.

✓Questions

1. Electrophoresis separates molecules by charge and size. What is the physical basis for separating molecules or cell components via centrifugation?
2. Compared to other types of centrifugation, why would extremely high-speed centrifugation allow similar molecules to separate?

BIOSKILLS 12

CELL AND TISSUE CULTURE METHODS

For researchers, there are important advantages to growing plant and animal cells and tissues outside the organism itself. Cell and tissue cultures provide large populations of a single type of cell or tissue and the opportunity to control experimental conditions precisely.

Animal Cell Culture

The first successful attempt to culture animal cells occurred in 1907, when a researcher cultivated amphibian nerve cells in a drop of fluid from the spinal cord. But it was not until the 1950s and 1960s that biologists could routinely culture plant and animal cells in the laboratory. The long lag time was due to the difficulty of re-creating conditions that exist in the intact organism precisely enough for cells to grow normally.

To grow in culture, animal cells must be provided with a liquid mixture containing the nutrients, vitamins, and hormones that stimulate growth. Initially, this mixture was serum, the liquid portion of blood; now serum-free media are available for certain cell types. Serum-free media are preferred because they are much more precisely defined chemically than serum.

In addition, many types of animal cells will not grow in culture unless they are provided with a solid surface that mimics the types of surfaces to which cells in the intact organisms adhere. As a result, cells are typically cultured in flasks (**Figure B12.1a**, left).

Even under optimal conditions, though, normal cells display a finite life span in culture. In contrast, many cultured cancerous cells grow indefinitely. In culture, cancerous cells also do not adhere tightly to the surface of the culture flask. These characteristics correlate with two features of cancerous cells in organisms: They grow in a continuous, uncontrolled fashion and can break away from the original site to infiltrate new tissues and organs.

Because of their immortality and relative ease of growth, cultured cancer cells are commonly used in research on basic aspects of cell structure and function. For example, the first human cell type to be grown in culture was isolated in 1951 from a malignant tumor of the uterine cervix. These cells are called HeLa cells in honor of their donor, Henrietta Lacks, who died soon thereafter from her cancer. HeLa cells continue to grow in laboratories around the world (Figure B12.1a, right).

Plant Tissue Culture

Certain cells found in plants are totipotent—meaning that they retain the ability to divide and differentiate into a complete, mature plant, including new types of tissue. These cells, called parenchyma cells, are important in wound healing and asexual reproduction. But they also allow researchers to grow complete adult plants in the laboratory, starting with a small number of parenchyma cells.

Biologists who grow plants in tissue culture begin by placing parenchyma cells in a liquid or solid medium containing all the nutrients required for cell maintenance and growth. In the early days of plant tissue culture, investigators found not only that specific growth signals called hormones were required for successful growth and differentiation but also that the relative abundance of hormones present was critical to success.

The earliest experiments on hormone interactions in tissue cultures were done with tobacco cells in the 1950s by Folke Skoog and co-workers. These researchers found that when the hormone called auxin was added to the culture by itself, the cells enlarged but did not divide. But if the team added roughly equal amounts of auxin and another growth signal called cytokinin to the cells, the cells began to divide and eventually formed a callus, or an undifferentiated mass of parenchyma cells.

By varying the proportion of auxin to cytokinins in different parts of the callus and through time, the team could stimulate the growth and differentiation of root and shoot systems and produce whole new plants (**Figure B12.1b**). A high ratio of auxin to cytokinin led to the differentiation of a root system, while a high ratio of cytokinin to auxin led to the development of a shoot system. Eventually Skoog's team was able to produce a complete plant from just one parenchyma cell.

The ability to grow whole new plants in tissue culture from just one cell has been instrumental in the development of genetic engineering (see Chapter 19). Researchers insert recombinant genes into target cells, test the cells to identify those that successfully express the recombinant genes, and then use tissue culture techniques to grow those cells into adult individuals with a novel genotype and phenotype.

✔Questions

1. What is a limitation of how experiments on HeLa cells are interpreted?
2. State a disadvantage of doing experiments on plants that have been propagated from single cells growing in tissue culture.

(a) Animal cell culture: immortal HeLa cancer cells

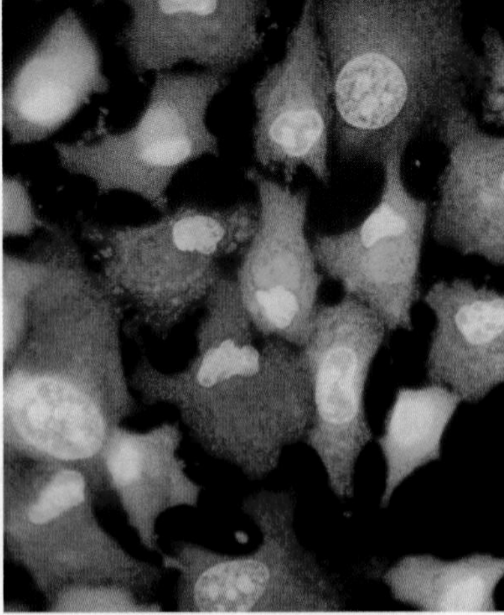

(b) Plant tissue culture: tobacco callus

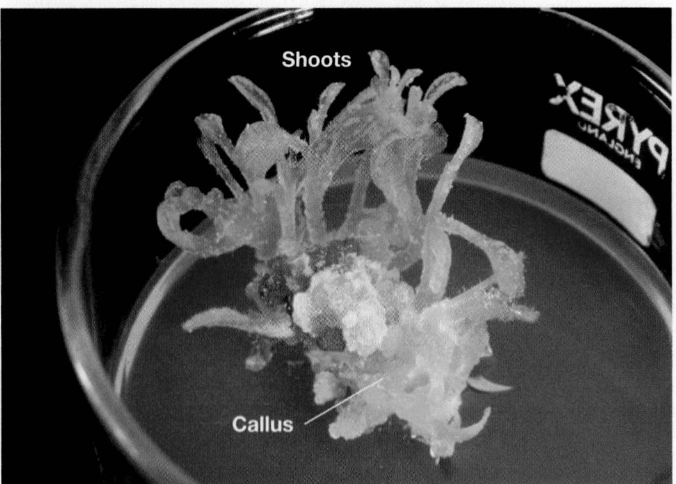

FIGURE B12.1 Animal and Plant Cells Can Be Grown in the Lab.

BIOSKILLS 13

COMBINING PROBABILITIES

In several cases in this text, you'll need to combine probabilities from different events in order to solve a problem. One of the most common applications is in genetics problems. For example, Punnett squares work because they are based on two fundamental rules of probability. Each rule pertains to a distinct situation.

The Both-And Rule

The both-and rule—also known as the product rule or multiplication rule—applies when you want to know the probability that two or more independent events occur together. Let's use the rolling of two dice as an example. What is the probability of rolling two sixes? These two events are independent, because the probability of rolling a six on one die has no effect on the probability of rolling a six on the other die. (In the same way, the probability of getting a gamete with allele *R* from one parent has no effect on the probability of getting a gamete with allele *R* from the other parent. Gametes fuse randomly.)

The probability of rolling a six on the first die is 1/6. The probability of rolling a six on the second die is also 1/6. The probability of rolling a six on *both* dice, then, is $1/6 \times 1/6 = 1/36$. In other words, if you rolled two dice 36 times, on average you would expect to roll two sixes once.

In the case of a cross between two parents heterozygous at the *R* gene, the probability of getting allele *R* from the father is 1/2 and the probability of getting *R* from the mother is 1/2. Thus, the probability of getting both alleles and creating an offspring with genotype *RR* is $1/2 \times 1/2 = 1/4$.

The Either-Or Rule

The either-or rule—also known as the sum rule or addition rule—applies when you want to know the probability of an event happening when there are several different ways for the same event or outcome to occur. In this case, the probability that the event will occur is the sum of the probabilities of each way that it can occur.

For example, suppose you wanted to know the probability of rolling either a one or a six when you toss a die. The probability of drawing each is 1/6, so the probability of getting one or the other is $1/6 + 1/6 = 1/3$. If you rolled a die three times, on average you'd expect to get a one or a six once.

In the case of a cross between two parents heterozygous at the *R* gene, the probability of getting an *R* allele from the father and an *r* allele from the mother is $1/2 \times 1/2 = 1/4$. Similarly, the probability of getting an *r* allele from the father and an *R* allele from the mother is $1/2 \times 1/2 = 1/4$. Thus, the combined probability of getting the *Rr* genotype in either of the two ways is $1/4 + 1/4 = 1/2$.

✔Questions

1. Suppose that four students each toss a coin. What is the probability of four "tails"?
2. After a single roll of a die, what is the probability of getting either a two, a three, or a six?

BIOSKILLS 14

MODEL ORGANISMS

Research in biological science starts with a question. In most cases, the question is inspired by an observation about a cell or an organism. To answer it, biologists have to study a particular species. Study organisms are often called model organisms, because investigators hope that they serve as a model for what is going on in a wide array of species.

Model organisms are chosen because they are convenient to study and because they have attributes that make them appropriate for the particular research proposed. They tend to have some common characteristics:

- ***Short generation time and rapid reproduction*** This is important because it makes it possible to produce offspring quickly and perform many experiments in a short amount of time—you don't have to wait long for individuals to grow.
- ***Large numbers of offspring*** This is particularly important in genetics, where many offspring phenotypes and genotypes need to be assessed to get a large sample size.
- ***Small size, simple feeding and habitat requirements*** These attributes make it relatively cheap and easy to maintain individuals in the lab.

The following notes highlight just a few model organisms supporting current work in biological science.

Escherichia coli

Of all model organisms in biology, perhaps none has been more important than the bacterium *Escherichia coli*—a common inhabitant of the human gut. The strain that is most commonly worked on today, called K-12 (**Figure B14.1a** on page B:20), was originally isolated from a hospital patient in 1922.

During the last half of the twentieth century, key results in molecular biology originated in studies of *E. coli*. These results include the discovery of enzymes such as DNA polymerase, RNA polymerase, DNA repair enzymes, and restriction endonucleases; the elucidation of ribosome structure and function; and the initial characterization of promoters, regulatory transcription factors, regulatory sites in DNA, and operons. In many cases, initial discoveries made in *E. coli* allowed researchers to confirm that homologous enzymes and processes existed in an array of organisms, often ranging from other bacteria to yeast, mice, and humans.

The success of *E. coli* as a model for other species inspired Jacques Monod's claim that "Once we understand the biology of *Escherichia coli*, we will understand the biology of an elephant." The genome of *E. coli* K-12 was sequenced in 1997, and the strain continues to be a workhorse in studies of gene function, biochemistry, and particularly biotechnology. Much remains to be learned, however. Despite over 60 years of intensive study, the function of about a third of the *E. coli* genome is still unknown.

In the lab, *E. coli* is usually grown in suspension culture, where cells are introduced to a liquid nutrient medium, or on plates containing agar—a gelatinous mix of polysaccharides. Under optimal growing conditions—meaning before cells begin to get crowded and compete for space and nutrients—a cell takes just 30 minutes on average to grow and divide. At this rate, a single cell can produce a population of over a million descendants in just 10 hours. Except for new mutations, all of the descendant cells are genetically identical.

Dictyostelium discoideum

The cellular slime mold *Dictyostelium discoideum* is not always slimy, and it is not a mold—meaning a type of fungus. Instead, it is an amoeba. Amoeba is a general term that biologists use to characterize a unicellular eukaryote that lacks a cell wall and is extremely flexible in shape. *Dictyostelium* has long fascinated biologists because it is a social organism. Independent cells sometimes aggregate to form a multicellular structure.

Under most conditions, *Dictyostelium* cells are haploid (n) and move about in decaying vegetation on forest floors or other habitats. They feed on bacteria by engulfing them whole. When these cells reproduce, they can do so sexually by fusing with another cell then undergoing meiosis, or asexually by mitosis, which is more common. If food begins to run out, the cells begin to aggregate. In many cases, tens of thousands of cells cohere to form a 2-mm-long mass called a slug (**Figure B14.1b**). (This is not the slug that is related to snails.)

After migrating to a sunlit location, the slug stops and individual cells differentiate according to their position in the slug. Some form a stalk; others form a mass of spores at the tip of the stalk. (A spore is a single cell that develops into an adult organism, but it is not formed from gamete fusion like a zygote.) The entire structure, stalk plus mass of spores, is called a fruiting body. Cells that form spores secrete a tough coat and represent a durable resting stage. The fruiting body eventually dries out, and the wind disperses the spores to new locations, where more food might be available.

Dictyostelium has been an important model organism for investigating questions about eukaryotes:

- Cells in a slug are initially identical in morphology but then differentiate into distinctive stalk cells and spores. Studying this process helped biologists better understand how cells in plant and animal embryos differentiate into distinct cell types.
- The process of slug formation has helped biologists study how animal cells move and how they aggregate as they form specific types of tissues.
- When *Dictyostelium* cells aggregate to form a slug, they stick to each other. The discovery of membrane proteins responsible for cell-cell adhesion helped biologists understand some of the general principles of multicellular life highlighted in Chapter 8.

Arabidopsis thaliana

In the early days of biology, the best-studied plants were agricultural varieties such as maize (corn), rice, and garden peas. When biologists began to unravel the mechanisms responsible for oxygenic photosynthesis in the early to mid-1900s, they relied on green algae that were relatively easy to grow and manipulate in the lab—often the unicellular species *Chlamydomonas reinhardii*—as an experimental subject.

Although crop plants and green algae continue to be the subject of considerable research, a new model organism emerged in the 1980s and now serves as the preeminent experimental subject in plant biology. That organism is *Arabidopsis thaliana*, commonly known as thale cress or wall cress (**Figure B14.1c**).

Arabidopsis is a member of the mustard family, or Brassicaceae, so it is closely related to radishes and broccoli. In nature it is a weed—meaning a species that is adapted to thrive in habitats where soils have been disturbed.

One of the most attractive aspects of working with *Arabidopsis* is that individuals can grow from a seed into a mature, seed-producing plant in just four to six weeks. Several other attributes make it an effective subject for study: It has just five chromosomes, has a relatively small genome with limited numbers of repetitive sequences, can self-fertilize as well as undergo cross-fertilization, can be grown in a relatively small amount of space and with a minimum of care in the greenhouse, and produces up to 10,000 seeds per individual per generation.

Arabidopsis has been instrumental in a variety of studies in plant molecular genetics and development, and it is increasingly popular in ecological and evolutionary studies. In addition, the entire genome of the species has now been sequenced, and studies have benefited from the development of an international "*Arabidopsis* community"—a combination of informal and formal associations of investigators who work on *Arabidopsis* and use regular meetings, e-mail, and the Internet to share data, techniques, and seed stocks.

Saccharomyces cerevisiae

When biologists want to answer basic questions about how eukaryotic cells work, they often turn to the yeast *Saccharomyces cerevisiae.*

S. cerevisiae is unicellular and relatively easy to culture and manipulate in the lab (**Figure B14.1d**). In good conditions, yeast cells grow and divide almost as rapidly as bacteria. As a result, the

(a) Bacterium *Escherichia coli* (strain K-12)

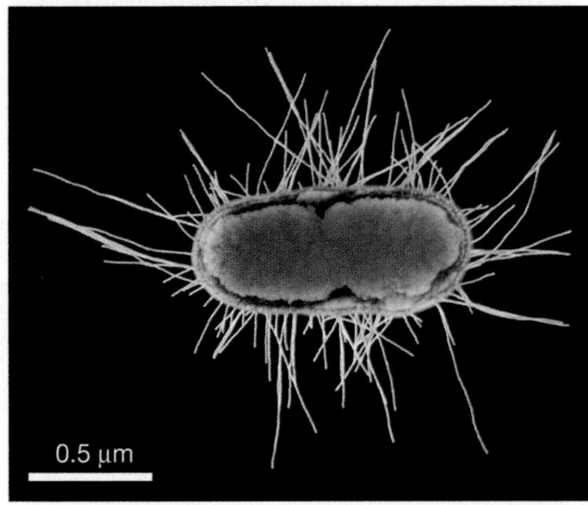

(b) Slime mold *Dictyostelium discoideum*

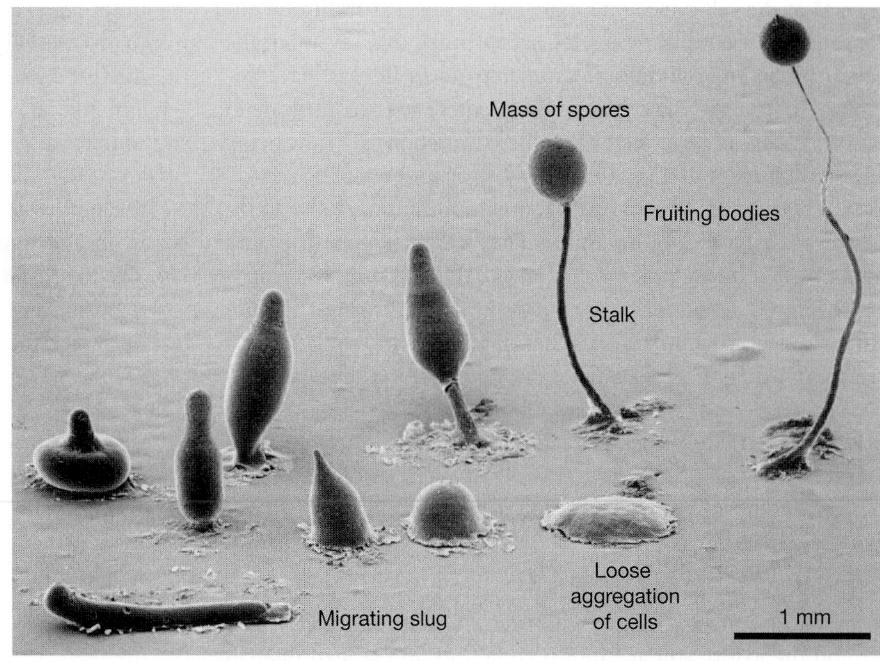

(c) Thale cress *Arabidopsis thaliana*

(e) Fruit fly *Drosophila melanogaster*

(f) Roundworm *Caenorhabditis elegans*

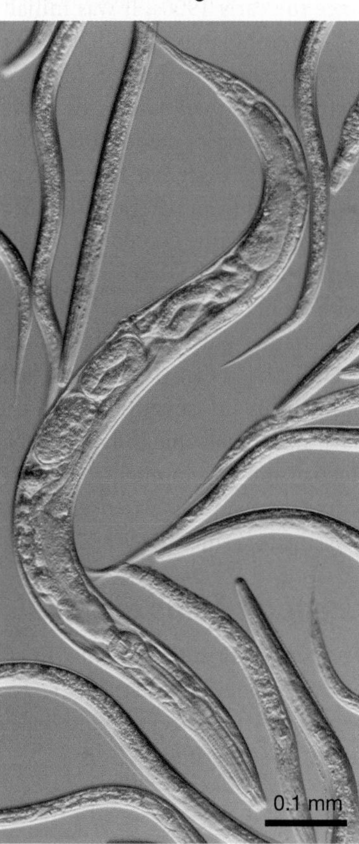

(d) Yeast *Saccharomyces cerevisiae*

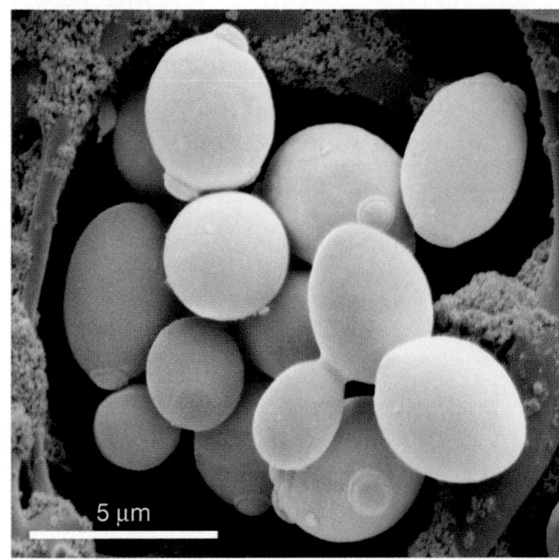

(g) Mouse *Mus musculus*

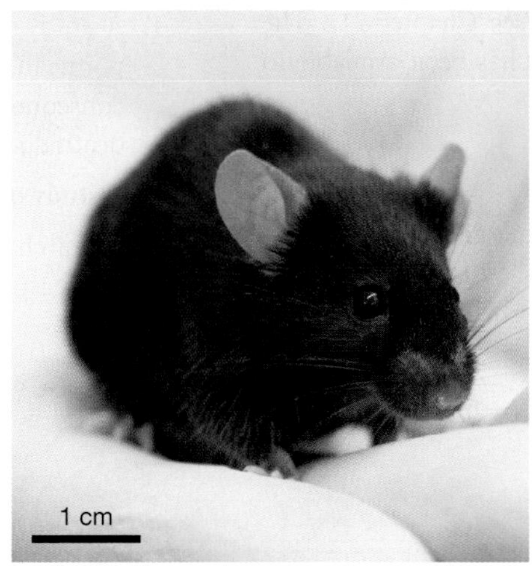

FIGURE B14.1 Model Organisms.

✔**QUESTION** *E. coli* is grown at a temperature of 37°C. Why?

species has become the organism of choice for experiments on control of the cell cycle and regulation of gene expression in eukaryotes. For example, research has confirmed that several of the genes controlling cell division and DNA repair in yeast have homologs in humans; and when mutated, these genes contribute to cancer. Strains of yeast that carry these mutations are now being used to test drugs that might be effective against cancer.

S. cerevisiae has become even more important in efforts to interpret the genomes of organisms like rice, mice, zebrafish, and humans. It is much easier to investigate the function of particular genes in *S. cerevisiae* by creating mutants or transferring specific alleles among individuals than it is to do the same experiments in mice or zebrafish. Once the function of a gene has been established in yeast, biologists can look for the homologous gene in other eukaryotes. If such a gene exists, they can usually infer that it has a function similar to its role in *S. cerevisiae.* It was also the first eukaryote with a completely sequenced genome.

Drosophila melanogaster

If you walk into a biology building on any university campus around the world, you are almost certain to find at least one lab where the fruit fly *Drosophila melanogaster* is being studied (**Figure B14.1e**).

Drosophila has been a key experimental subject in genetics since the early 1900s. It was initially chosen as a focus for study by T. H. Morgan because it can be reared in the laboratory easily and inexpensively, matings can be arranged, the life cycle is completed in less than two weeks, and females lay a large number of eggs. These traits made fruit flies valuable subjects for breeding experiments designed to test hypotheses about how traits are transmitted from parents to offspring (see Chapter 13).

More recently, *Drosophila* has also become a key model organism in the field of developmental biology. The use of flies in developmental studies was inspired in large part by the work of Christianne Nüsslein-Volhard and Eric Wieschaus, who in the 1980s isolated flies with genetic defects in early embryonic development. By investigating the nature of these defects, researchers have gained valuable insights into how various gene products influence the development of eukaryotes (see Chapter 21). The complete genome sequence of *Drosophila* has been available to investigators since the year 2000.

Caenorhabditis elegans

The roundworm *Caenorhabditis elegans* emerged as a model organism in developmental biology in the 1970s, due largely to work by Sydney Brenner and colleagues. (*Caenorhabditis* is pronounced *see-no-rab-DIE-tiss.*)

C. elegans was chosen for three reasons: **(1)** Its cuticle (soft outer layer) is transparent, making individual cells relatively easy to observe (**Figure B14.1f**); **(2)** adults have exactly 959 nonreproductive cells; and, most important, **(3)** the fate of each cell in an embryo can be predicted because cell fates are invariant among individuals. For example, when researchers examine a 33-cell *C. elegans* embryo, they know exactly which of the 959 cells in the adult will be derived from each of those 33 embryonic cells.

In addition, *C. elegans* are small (less than 1 mm long), are able to self-fertilize or cross-fertilize, and undergo early development in just 16 hours. The entire genome of *C. elegans* has now been sequenced.

Mus musculus

Because the house mouse *Mus musculus* is the most important model organism among mammals, it is especially prominent in biomedical research—where researchers need to work on individuals with strong genetic and developmental similarities to humans.

The house mouse was an intelligent choice of model organism in mammals because it is small and thus relatively inexpensive to maintain in captivity, and because it breeds rapidly. A litter can contain 10 offspring and generation time is only 12 weeks—meaning that several generations can be produced in a year. Descendants of wild house mice have been selected for docility and other traits that make them easy to handle and rear; these populations are referred to as laboratory mice (**Figure B14.1g**).

Some of the most valuable laboratory mice are strains with distinctive, well-characterized genotypes. Inbred strains are virtually homogenous genetically (see Chapter 25) and are useful in experiments where gene-by-gene or gene-by-environment interactions have to be controlled. Other populations carry mutations that knock out genes and cause diseases similar to those observed in humans. These individuals are useful for identifying the cause of genetic diseases and testing drugs or other types of therapies.

✓Questions

Which model organisms described here would be the best choice for the following studies? In each case, explain your reasoning.

1. A study of why specific cells in an embryo die at certain points in normal development. One goal is to understand the consequences for the individual when programmed cell death does not occur when it is supposed to.
2. A study of proteins that are required for cell-cell adhesion.
3. Research on a gene suspected to be involved in the formation of breast cancer in humans.

APPENDIX B Answers

CHAPTER 1

Check Your Understanding (CYU)

CYU p. 5 The data points would all be about 11 percent, indicating no change in average kernel protein content over time. **CYU p. 8** From the sequence data provided, Species A and B differ only in one nucleotide of the rRNA sequence (position 10 from left). Species C differs from Species A and B in four nucleotides (positions 1, 2, 9, and 10). A correctly drawn phylogenetic tree would indicate that Species A and B appear to be closely related, with Species C being more distantly related [see Figure A1.1]. **CYU p. 12** The key here is to test predation rates during the hottest part of the day (when desert ants actually feed) versus other parts of the day. The experiment would best be done in the field, where natural predators are present. One approach would be to capture a large number of ants, divide the group in two, and measure predation rates (number of ants killed per hour) when they are placed in normal habitat during the hottest part of the day versus an hour before (or after). (1) The control group here is the normal condition—ants out during the hottest part of the day. If you didn't include a control, a critic could argue that predation did or did not occur because of your experimental setup or manipulation, not because of differences in temperature. (2) You would need to make sure that there is no difference in body size, walking speed, how they were captured and maintained, or other traits that might make the ants in the two groups more or less susceptible to predators. They should also be put out in the same habitat, so the presence of predators is the same in the two treatments.

You Should Be Able To (YSBAT)

YSBAT p. 5 Average kernel protein content declined, from 11 percent to a much lower value. (The experiment is still continuing; to date the low-selection line is about 5 percent.) **YSBAT p. 12** (1) A critic could argue that the ants weren't navigating normally, because they had been caught and released and transferred to a new channel. (2) A critic could argue that the ants can't navigate normally on their manipulated legs.

Caption Questions and Exercises

Figure 1.2 If Pasteur had done any of the things listed, he would have had more than one variable in his experiment. This would allow critics to claim that he got different results because of the differences in broth types, heating, or flask types—not the difference in exposure to preexisting cells. The results would not be definitive. **Figure 1.4** Molds and other fungi are more closely related to green algae because they differ from plants at two positions (5 and 8 from left), but differ from green algae at only one position (8). **Figure 1.6** The eukaryotic cell is roughly 10 times (more exactly, 8.5 times) the size of the prokaryotic cell. **Figure 1.8** You should be skeptical, because the results could be due to the single individual being sick or injured or unusual in some other way.

Summary of Key Concepts

KC 1.1 Dead cells cannot maintain a difference in conditions inside the cell versus outside, replicate, use energy, or process information. **KC 1.2** Observations on thousands of diverse species supported the claim that all organisms consist of cells. The hypothesis that all cells come from preexisting cells was supported when Pasteur showed that new cells do not arise and grow in a boiled liquid unless they are introduced from the air. **KC 1.3** If seeds with higher protein content leave the most offspring, then individuals with low protein in their seeds will become rare over time. **KC 1.4** A newly discovered species can be classified as a member of the Bacteria if the sequence of its rRNA contains some features found only in Bacteria. The same logic applies to classifying a new species in the Archaea or Eukarya. **KC 1.5** (1) A hypothesis is an explanation of how the world works; a prediction is an outcome you should observe if the hypothesis is correct. (2) Experiments are convincing because they measure predictions from two opposing hypotheses. Both predicted actions cannot occur, so one hypothesis will be supported while the other will not.

Test Your Knowledge

1. d; **2.** d; **3.** c; **4.** b; **5.** d; **6.** c

Test Your Understanding

1. That the entity they discovered replicates, processes information, acquires and uses energy, is cellular, and that its populations evolve. **2.** [See Figure A1.2] **3.** The rule ensured that two different organisms would never end up with the same name. **4.** Over time, traits that increased the fitness of individuals in this habitat became increasingly more frequent in the population. **5.** Individuals with certain traits selected, in the sense that they produce the most offspring. **6.** Yes. If evolution is defined as "change in the characteristics of a population over time," then those organisms that are most closely related should have experienced less change over time. On a phylogenetic tree, species with substantially similar rRNA sequences would be diagrammed with a closer common ancestor—one that had the sequences they inherited—than the ancestors shared between species with dissimilar rRNA sequences.

Applying Concepts to New Situations

1. A scientific theory is not a guess—it is an idea whose validity can be tested with data. Both the cell theory and the theory of evolution have been validated by large bodies of observational and experimental data. **2.** If all eukaryotes living today have a nucleus, then it is logical to conclude that the nucleus arose in a common ancestor of all eukaryotes, indicated by the arrow you should have added to the figure on page 14 [see Figure A1.3]. If it had arisen in a common ancestor of Bacteria or Archaea, then species in those groups would have had to lose the trait—an unlikely event. **3.** The data set was so large and diverse that it was no longer reasonable to argue that noncellular life-forms would be discovered. **4.** Yes. The heritable traits that confer resistance to HIV should increase over time.

CHAPTER 2

Check Your Understanding (CYU)

CYU p. 21 [See Figure A2.1] **CYU p. 33** (1) Gibbs equation: $\Delta G = \Delta H - T\Delta S$. ΔG symbolizes the change in the Gibbs free energy. ΔH represents the difference in potential energy between the products and the reactants. T represents the temperature (in degrees Kelvin) at which the reaction is taking place. ΔS symbolizes the change in entropy (disorder). (2) When ΔH is negative—meaning that the reactants have lower potential energy than the products—and when ΔS is positive, meaning that the products have higher entropy (are more disordered) than the reactants.

You Should Be Able To (YSBAT)

YSBAT p. 22 (1) $^{\delta+}H{-}O^{\delta-}{-}H^{\delta+}$. (2) If water were linear, the partial negative charge on oxygen would have partial positive charges on either side. Compared to the actual, bent molecule, the partial negative charge would be much less exposed and less able to participate in hydrogen bonding. **YSBAT p. 26** The pH 8 solution has 100 times fewer protons than the pH 6 solution. **YSBAT p. 30** (1) As T increases, ΔG is more likely to be negative, making the reaction spontaneous. (2) Exothermic reactions may be nonspontaneous if they result in a decrease in entropy—meaning that the products are more ordered than the reactants (ΔS is negative).

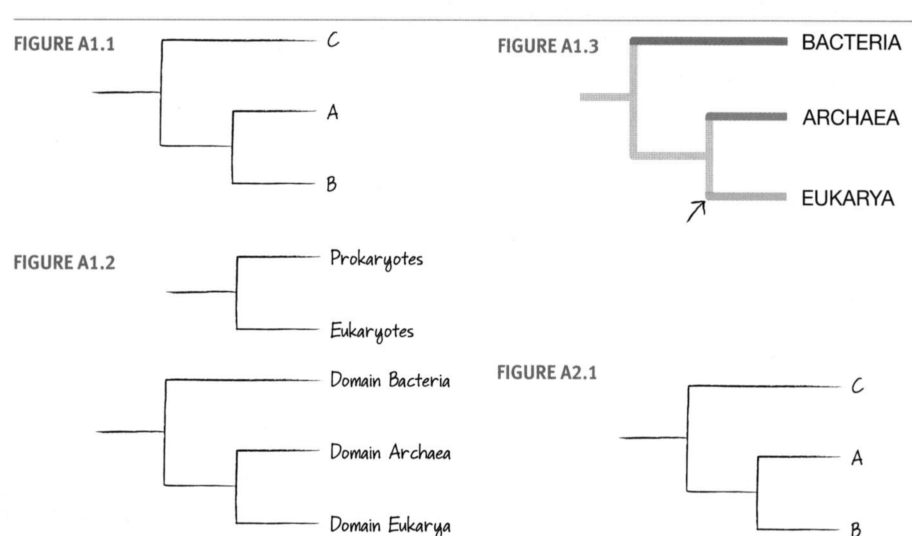

Caption Questions and Exercises

Figure 2.7 Oxygen and nitrogen have high electronegativity. They hold shared electrons more tightly than C, H, and many other atoms, resulting in polar bonds. **Figure 2.13** Oils are nonpolar. They have long chains of carbon atoms bonded to hydrogen atoms, which share electrons evenly because their electronegativities are similar. When an oil and water are mixed, the polar water molecules interact with each other via hydrogen bonding much more strongly than they interact with the nonpolar oil molecules, which interact with themselves instead. **Figure 2.16** The coffee becomes less acidic because milk is more alkaline (pH 6.5) than black coffee (pH 5). **Table 2.2** (Row 1) "Cause": Electrostatic attraction between partial charges on water molecules and opposite charges on ions; hydrogen bonds between water and other polar molecules. (Row 2) "Biological Consequences": Ice floats on denser liquid water. If it were denser and sank, oceans and lakes would fill with ice. (Row 4) "Cause": Liquid water must absorb lots of heat energy to break hydrogen bonds and change to a gas. One possible concept map relating the structure of water to its properties is shown below [see Figure A2.2]. **Figure 2.19** [See Figure A2.3] **Figure 2.22** The reaction rate at a specific set of reactant concentrations, averaged over many replicates. **Table 2.3** All the functional groups in Table 2.3, except the sulfhydryl group (–SH), are highly polar. The sulfhydryl group is only very slightly polar.

Summary of Key Concepts

KC 2.1 [See Figure A2.4] **KC 2.2** Because these functional groups are polar, water forms hydrogen bonds with them. **KC 2.3** Chemical energy—meaning, potential energy stored in chemical bonds—is transformed to heat. **KC 2.4** Spontaneous chemical reactions lead toward disorder and a release of energy. Cells are full of highly ordered molecules that contain high-energy bonds. Energy must be added to make these molecules and maintain life. **KC 2.5** Electrons in carbon-carbon bonds are held loosely and equally between the carbon atoms, whereas electrons in a carbon-oxygen bond are held tightly by the oxygen. Because of this difference, molecules with carbon-carbon bonds have more potential energy than carbon dioxide with its two carbon-oxygen bonds. Carbon-carbon bonds are often present in large, highly ordered molecules. The entropy of a group of such molecules, a measure of their disorder, is much less than the entropy of a group of simple, less-ordered molecules such as carbon dioxide.

Test Your Knowledge

1. b; **2.** a; **3.** a; **4.** d; **5.** d; **6.** d

Test Your Understanding

1. The reaction lowers the pH of the solution by releasing extra H^+ into the solution. If additional CO_2 is added, the sequence of reactions would be driven to the right, which would make the ocean more acidic. **2.** No. Shells that are farther from the protons (positive charges) in the nucleus house electrons that have greater potential energy than shells closer to the nucleus. **3.** [See Figure A2.5] The electron orbitals surrounding oxygen are oriented in the shape of a tetrahedron. Two of these orbitals are occupied by unshared electrons and the other two are shared with hydrogen atoms, resulting in the bent shape of the water molecule. Due to its greater electronegativity, oxygen attracts the shared electrons more strongly than does hydrogen, so it has a partial negative charge while the hydrogen atoms have a partial positive charge. **4.** In a covalent bond, the shared electrons are extremely close to two nuclei. In a hydrogen bond, the (+) and (–) charges are only partial, and the attraction is at a much greater distance. **5.** Water forms large numbers of hydrogen bonds, which must be broken before the molecules can start moving faster—meaning that their temperature has increased. **6.** The overall shape of an organic molecule is determined by its carbon framework. The functional groups attached to the carbons determine the molecule's chemical behavior, because these groups are likely to interact with other molecules.

Applying Concepts to New Situations

1. In CO_2, the double bonds between the carbon and each of the oxygen atoms lock the molecule into a linear shape that is nonpolar. In water, the partial charges on

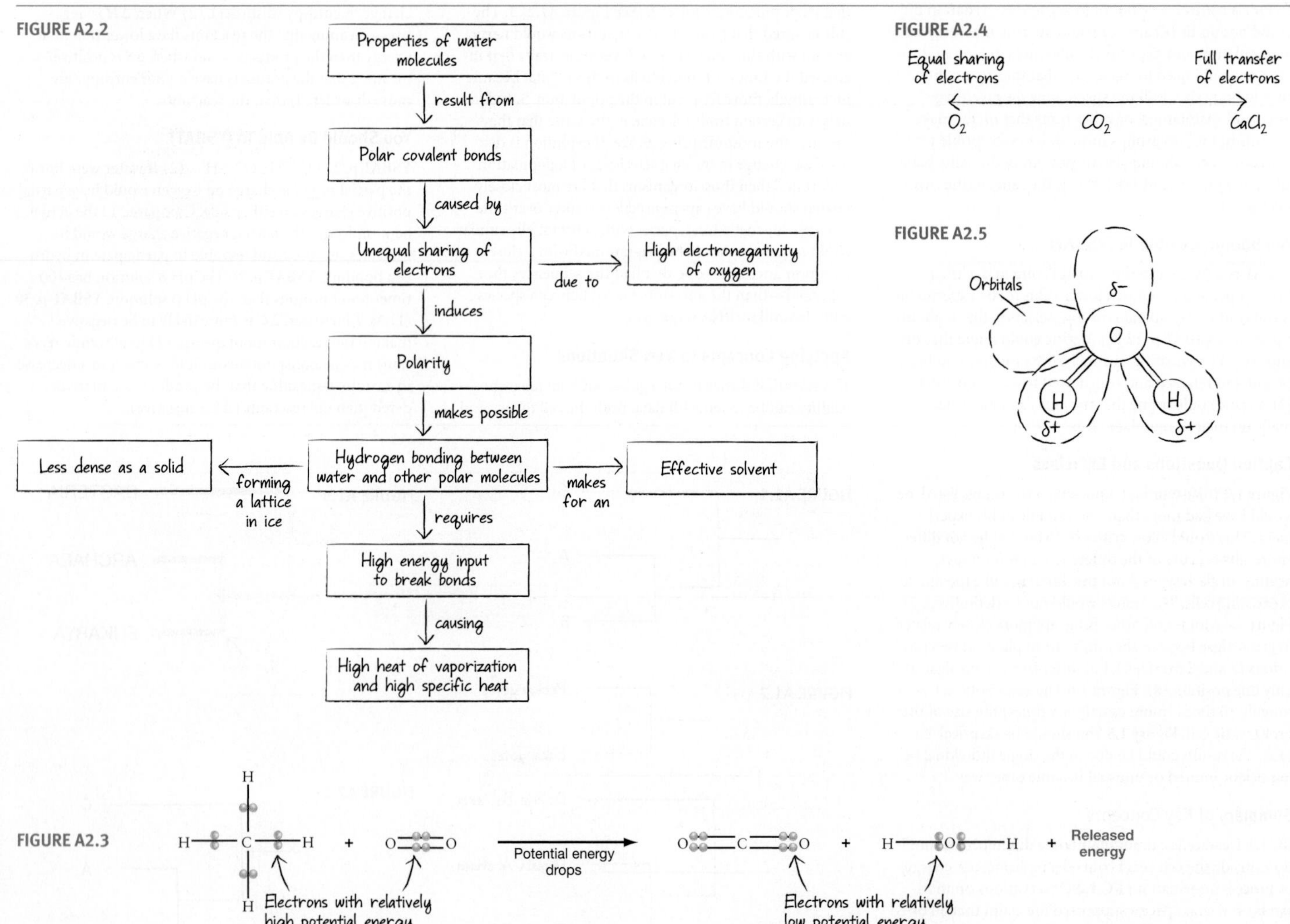

oxygen and hydrogen allow water to form hydrogen bonds with other polar or charged substances. In addition, the single covalent bonds in water are much easier to break than the double covalent bonds in carbon dioxide, making it more reactive. **2.** When oxygen is covalently bonded to carbon or hydrogen, the bond electrons spend more time near the oxygen atom. In contrast, carbon and hydrogen have roughly similar electronegativities, and they tend to share the electrons of a covalent bond more equally. **3.** The Sun will burn out when the mass of all of its component atoms is finally converted to energy. **4.** In hot weather, water absorbs large amounts of heat due to its high specific heat and high heat of vaporization. In cold weather, water releases the large amount of heat that it has absorbed.

CHAPTER 3

Check Your Understanding (CYU)

CYU p. 44 [See Figure A3.1] **CYU p. 51** Secondary, tertiary, and quaternary structure all depend on bonds and other interactions between amino acids that are linked in a chain (primary structure). **CYU p. 56** (1) Substrates can react only if they meet in a precise orientation. (2) The shape change alters the shape of the active site so that substrates no longer fit or can't be oriented correctly for a reaction to occur.

You Should Be Able To (YSBAT)

YSBAT p. 50 The sequence of amino acids in the two polypeptide chains constitutes the primary structure. Each chain contains α-helices and a β-pleated sheet (secondary structures), and is folded into a specific three-dimensional shape (tertiary structure). The overall three-dimensional shape of the dimeric Cro protein, formed by association of two identical polypeptides, constitutes its quaternary structure. **YSBAT p. 53** More—because catalysts lower the activation energy required for the reaction to proceed, more molecules at any given temperature have enough kinetic energy to supply the activation energy. **YSBAT p. 54** (1) orienting substrates, (2) transition state, (3) R-groups, (4) structure.

Caption Questions and Exercises

Figure 3.1 The water-filled flask is the ocean; the gas-filled flask is the atmosphere; the condensed water droplets are rain; the electrical sparks are lightning. **Figure 3.3** The green R-groups contain mostly C and H, which have roughly equal electronegativities. Electrons are evenly shared in C—H bonds and C—S bonds, so the groups are nonpolar. Most of the pink R-groups have a highly electronegative oxygen atom with a partial negative charge, making them polar. Cysteine has a sulfur that is slightly more electronegative than hydrogen, so it will be less polar than the other pink groups. **Figure 3.6** A polar covalent bond. The electrons are shared in the C—N bond, but because N is more electronegative than C, the bond is polar. **Figure 3.16** In order for a chemical reaction to proceed, the kinetic energy of the reactant molecules and atoms must be greater than the activation energy. Only the molecules to the right of the line you drew will be able to react. **Figure 3.18** No—a catalyst only changes the activation energy, not the overall free energy change. **Figure 3.22** [See Figure A3.2]

Summary of Key Concepts

KC 3.1 Many possible correct answers, including: the presence of an active site in an enzyme that is precisely shaped to fit a substrate or substrates in the correct orientation for a reaction to occur; the "donut" shape of porin allowing certain substances to pass through it; the butterfly shape of TATA-box binding protein being precisely the right size for a DNA molecule to fit. **KC 3.2** R-groups containing partial charges or full charges can form hydrogen bonds with water, make the amino acid soluble. Nonpolar R-groups do not interact with water and make the amino acid insoluble. **KC 3.3** [See Figure A3.3] **KC 3.4** When a catalyst is present, the reactants and products don't change—only the transition state changes. Thus, the overall free energy change in the reaction is the same—only the activation energy changes.

Test Your Knowledge

1. d; **2.** c; **3.** a; **4.** b; **5.** a; **6.** b

Test Your Understanding

1. The shape of reactant molecules (the key) fits into the active site of an enzyme (the lock). The model assumed that the enzyme is rigid; in fact it is flexible and dynamic. **2.** It will fold in on itself, away from water, when placed in an aqueous solution. **3.** Both are mechanisms that regulate enzymes; the difference is whether the regulatory molecule binds at the active site (competitive inhibition) or away from the active site (allosteric regulation). **4.** Energy is required. Polymerization is a nonspontaneous reaction because the product molecules have lower entropy and greater potential energy than the reactants. **5.** Proteins are highly variable in overall shape and chemical properties due to variation in the composition of R-groups and the array of secondary through quaternary structures that are possible. This variation allows them to fulfill many different roles in the cell. Diversity in the shape and reactivity of active sites makes them effective catalysts. **6.** The function of an enzyme depends on the shape and chemical reactivity of its active site. Temperature and pH affect enzyme function because they can change the shape and chemical reactivity of the active site—either by disrupting bonds or interfering with acid-base reactions.

Applying Concepts to New Situations

1. Yes—within limits. As Figure 3.23 shows, some bacteria thrive in very hot or highly acidic environments on Earth. But extremely low pHs or high temperatures are likely to denature proteins. **2.** The data suggest that the enzyme and substrate form a transition state that requires a change in the shape of the active site, with each movement corresponding to one reaction. **3.** Without the coenzyme, the free radical–containing transition state would not be stabilized and the reaction rate would drop dramatically. **4.** Residues in the active site changed in a way that made the drugs less likely to bind and prevent the normal substrate from binding.

FIGURE A3.1

FIGURE A3.2

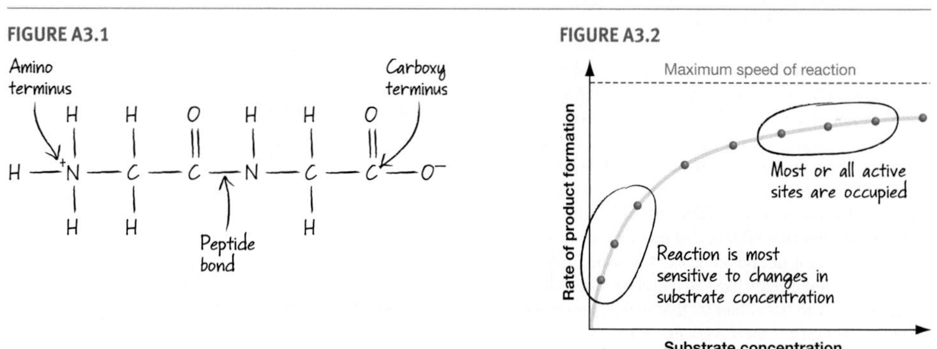

FIGURE A3.3

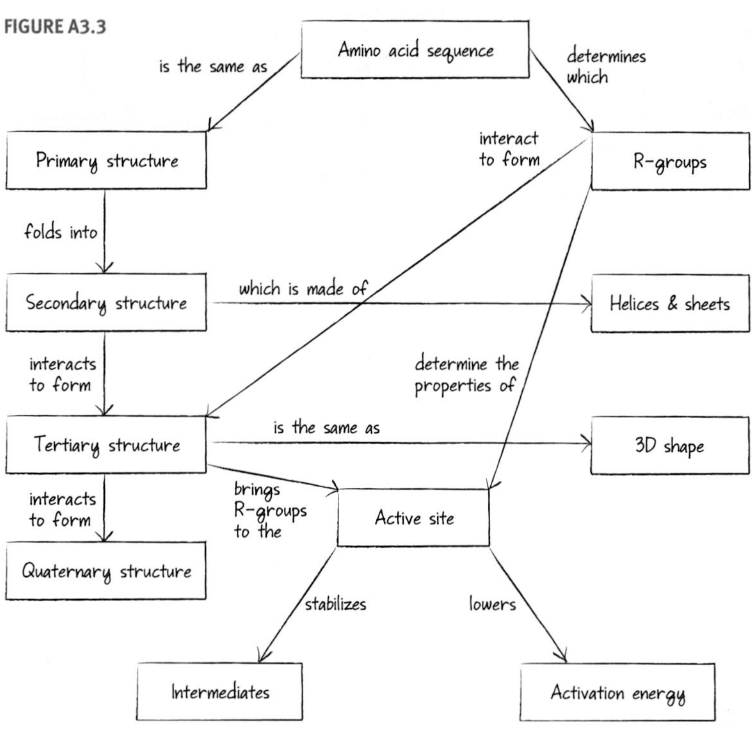

CHAPTER 4

Check Your Understanding (CYU)

CYU p. 62 [See Figure A4.1] **CYU p. 66** [See Figure A4.2]

You Should Be Able To (YSBAT)

YSBAT p. 60 [See Figure A4.3] **YSBAT p. 64** If the two strands are parallel, the nitrogenous bases are not aligned in a way that allows hydrogen bond formation. No helix would form.

Caption Questions and Exercises

Figure 4.3 5′ UAGC 3′. **Figure 4.9** Endergonic—energy must be added (as heat) for the reaction to occur.

Summary of Key Concepts

KC 4.1 (1) Both polymerize via energy-demanding condensation reactions that add a monomer to a growing chain. But proteins polymerize via formation of peptide bonds while nucleic acids polymerize via formation of phosphodiester linkages. (2) Proteins have an amino terminus at one end and a carboxyl terminus at the other end, while nucleic acids have a 5′ end (with an unlinked phosphate group) and a 3′ end (with an unlinked hydroxyl group). (3) Both have a "backbone"; in proteins it consists of amino and carbonyl groups, while in nucleic acids it consists of phosphate groups and sugars. **KC 4.2** C-G pairs involve three hydrogen bonds, so they are more stable than A-T pairs with just two bonds. **KC 4.3** A single-stranded RNA molecule has unpaired bases that can pair up with other bases on the same RNA strand, thereby folding the molecule into stem-and-loop configurations with a particular three-dimensional shape. The unpaired bases also can bind to another RNA molecule, linking the two together in a quaternary structure. Because DNA molecules are double stranded, with no unpaired bases, internal folding and interaction with other DNA molecules is not possible. **KC 4.4** If the copied ribozymes are not exactly like the template molecule, then some are likely to have changes that make them more efficient as catalysts. If there were no variation—meaning, if copying were exact—there could be no improvement in efficiency or other change over time.

Test Your Knowledge

1. c; **2.** d; **3.** a; **4.** b; **5.** a; **6.** c

Test Your Understanding

1. [See Figure A4.4] **2.** The addition of the phosphate groups raises the potential energy of the resulting monomers (nucleoside triphosphates) enough that the polymerization reaction is exergonic. **3.** Nucleic acids are directional because the two ends of the polymer are different. One end has a free phosphate group on a 5′ carbon; the other end has a free hydroxyl group bonded to a 3′ carbon. **4.** DNA is a more stable molecule than RNA because it lacks a hydroxyl group on the 2′ carbon and is therefore resistant to degradation, and because the two sugar-phosphate backbones are linked by many hydrogen bonds between nitrogenous bases. DNA is more stable than proteins because it is symmetrical and has few exposed chemical groups that could participate in chemical reactions. **5.** Hairpin structures form when folding of the sugar-phosphate backbone allows the bases in one part of an RNA strand to align with bases in another segment of the same RNA strand in an antiparallel fashion, so they can hydrogen-bond and form a double helix.
6. DNA has limited catalytic ability because it (1) lacks functional groups that can stabilize transition states and (2) has a regular structure that is not conducive to forming active sites. RNA molecules can catalyze some reactions because they (1) have exposed hydroxyl functional groups that can stabilize transition states and (2) can fold into shapes that create active sites. Proteins are the most effective catalysts because (1) amino acids have a wide range of chemical reactivity, and (2) they fold into an enormous diversity of shapes that can create active sites.

Applying Concepts to New Situations

1. Yes—if the complementary bases lined up over the entire length of the two strands, they would twist into a double helix analogous to a DNA molecule. The same types of hydrogen bonds and hydrophobic interactions would occur as observed in the "stem" portions of hairpins in single-stranded RNA. **2.** An RNA replicase would undergo replication and be able to evolve. It would process information in the sense of copying itself, and would use energy to drive endergonic polymerization reactions. It would not be bound by a membrane, however, and would not be able to acquire energy. It would best be considered as an intermediate step between nonlife and "true life," as outlined in Chapter 1. **3.** It is unlikely that bases would align properly for hydrogen bonding to occur, so hydrophobic interactions would probably be more important. **4.** No single right answer (these are opinion questions). Note that when biologists design studies about the origin of life, they base their experimental conditions on the best available current knowledge about early Earth conditions. New information about these conditions might require a revision of the experiment that leads to different results and conclusions.

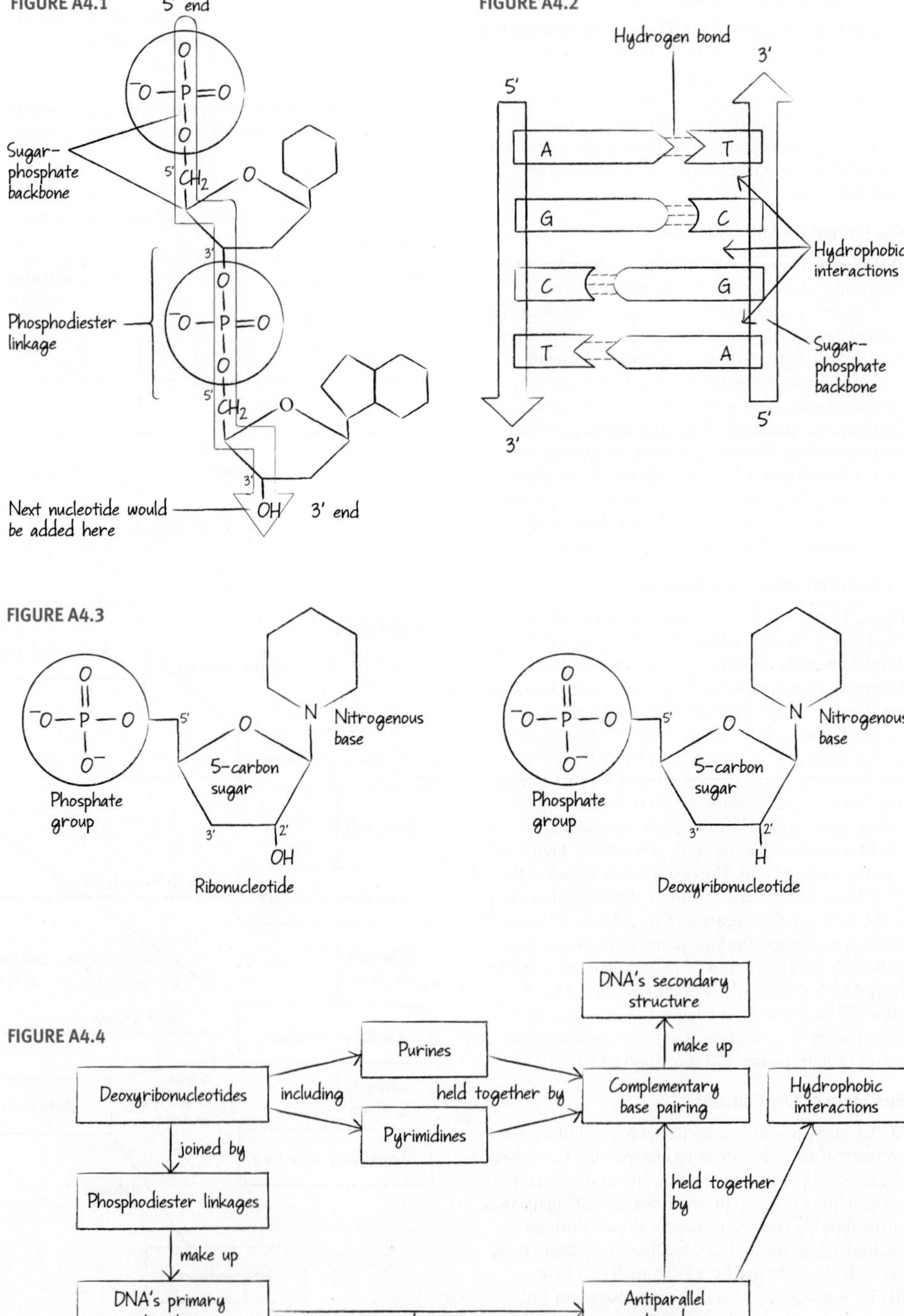

CHAPTER 5

Check Your Understanding (CYU)

CYU p. 73 [See Figure A5.1] **CYU p. 76** They could differ in (1) location of linkages (e.g., 1,4 or 1,6); (2) types of linkages (e.g., α or β); (3) the sequence of the monomers (e.g., two galactose then two glucose, versus alternating galactose and glucose); and/or (4) whether the four monomers are linked in a line or whether they branch. **CYU p. 79** (1) Aspect 1: The β-1,4-glycosidic linkages in these molecules are difficult to degrade. Aspect 2: When individual molecules of these carbohydrates align, bonds form between them and produce tough fibers or sheets. (2) Most are probably being broken down into glucose, which in turn is being broken down in reactions that lead to the synthesis of ATP. In short, the carbohydrates are providing you with chemical energy.

Caption Questions and Exercises

Figure 5.2 [See Figure A5.2] **Figure 5.7** All of the C—C and C—H bonds should be circled.

Summary of Key Concepts

KC 5.1 Molecules have to interact in an extremely specific orientation in order for a reaction to occur. Changing the location of a functional group by even one carbon can mean that the molecule will undergo completely different types of reactions. **KC 5.2** The orientations of the α- versus β-glycosidic linkages are different, so the molecules in a chain end up in a spiral versus linear arrangement. **KC 5.3** (1) Polysaccharides used for energy storage are formed entirely from glucose monomers joined by α-glycosidic linkages; structural polysaccharides are comprised of glucose or other sugars joined by β-glycosidic linkages. (2) The monomers in energy-storage polysaccharides are linked in a spiral arrangement; the monomers in structural polysaccharides are linked in a linear arrangement. (3) Energy-storage polysaccharides may branch; structural polysaccharides do not. (4) Individual chains of energy-storage polysaccharides do not associate with each other; adjacent chains of structural polysaccharides are linked by hydrogen bonds or covalent bonds.

Test Your Knowledge

1. d; **2.** a; **3.** c; **4.** a; **5.** d; **6.** b

Test Your Understanding

1. Carbohydrates are ideal for signaling the identity of the cell because they are so diverse structurally. This diversity enables them to serve as very specific identity tags for cells. **2.** When you compare the glucose monomers in an α-1,4-glycosidic linkage versus a β-1,4-glycosidic linkage, the linkages are located on opposite sides of the plane of the glucose rings (e.g., "above" versus "below" the plane), and the glucose monomers are linked in the same orientation versus having every other glucose flipped in orientation. β-1,4-glycosidic linkages are much more difficult for enzymes to break, so they resist degradation. **3.** Starch and glucose both consist of glucose monomers joined by α-1,4-glycosidic linkages, and both function as storage carbohydrates. Starch is composed of an unbranched amylose and branched amylopectin, while glycogen is even more highly branched. **4.** The electrons in the C=O bonds of carbon dioxide molecules are held tightly by the highly electronegative oxygen atoms, so have low potential energy. The electrons in the C—C and C—H bonds of carbohydrates are shared equally and have much higher potential energy. **5.** They have β-1,4-glycosidic linkages, which are resistant to degradation and put glucose monomers in positions where hydrogen bonds can form between adjacent strands, forming strong fibers. **6.** Glycogen has α-1,4-glycosidic linkages with α-1,6-glycosidic linkages at branch points; cellulose has β-1,4-glycosidic linkages and is not branched—instead, it forms hydrogen bonds with adjacent cellulose molecules. Glycogen is an energy-storage molecule; cellulose is a structural polysaccharide.

Applying Concepts to New Situations

1. Carbohydrates are energy-storage molecules, so minimizing their consumption may reduce total energy intake. Lack of available carbohydrate also forces the body to use fats for energy, reducing the amount of fat that is stored. **2.** Lactose is made up of glucose and galactose. **3.** Amylase breaks down the starch in the cracker into glucose monomers, which stimulate the sweet receptors in your tongue. **4.** When bacteria contact lysozyme, their cell walls, which contain peptidoglycan, begin to degrade, leading to the death of the bacteria.

CHAPTER 6

Check Your Understanding (CYU)

CYU p. 85 (1) Fats consist of three fatty acids linked to glycerol; steroids have a distinctive four-ring structure with a side group attached; phospholipids have a hydrophilic, phosphate-containing "head" region and a hydrocarbon tail. (2) In cholesterol, the hydrocarbon steroid rings are hydrophobic; the hydroxyl group is hydrophilic. In phospholipids, the phosphate-containing head group is hydrophilic; the fatty acid chains are hydrophobic. **CYU p. 89** [See Table A6.1] **CYU p. 92** [See Figure A6.1] **CYU p. 99** Passive transport does not require an

FIGURE A5.1

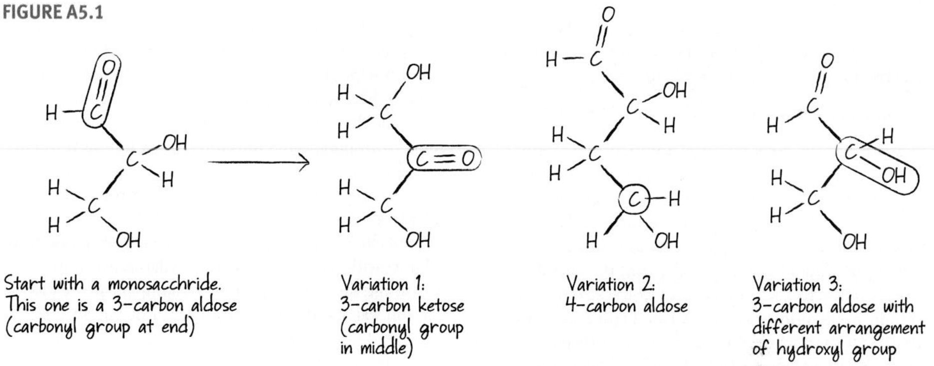

Start with a monosacchride. This one is a 3-carbon aldose (carbonyl group at end)

Variation 1: 3-carbon ketose (carbonyl group in middle)

Variation 2: 4-carbon aldose

Variation 3: 3-carbon aldose with different arrangement of hydroxyl group

FIGURE A5.2

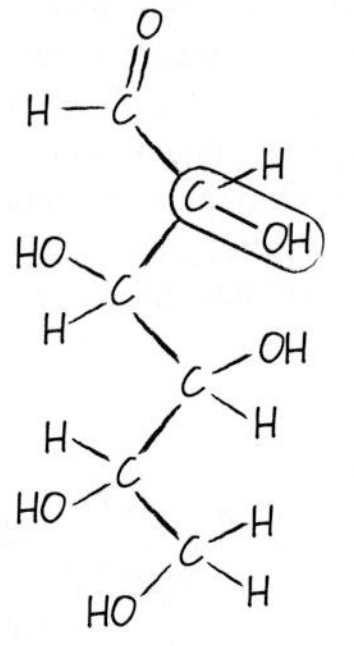

FIGURE A6.1

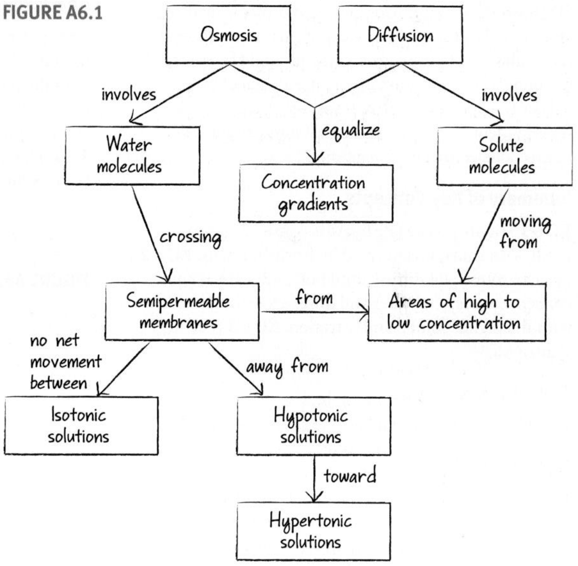

TABLE A6.1

Factor	Effect on permeability	Reason
Temperature	Decreases as temperature decreases.	Hydrophobic tails move more slowly, increasing the strength of hydrophobic interactions (membrane is more dense).
Cholesterol	Decreases as cholesterol content increases.	Cholesterol molecules fill in the spaces between the fatty-acid tails, making the membrane more tightly packed.
Length of hydrocarbon tails	Decreases as length of hydrocarbon tails increases.	Longer hydrocarbon tails have stronger hydrophobic interactions, making the tails pack together more tightly.
Saturation of hydrocarbon tails	Decreases as degree of saturation increases.	Saturated fatty acids have straight hydrocarbon tails that pack together and form many hydrophobic interactions, leaving few gaps.

expenditure of energy—it happens as a result of existing concentration or electrical gradients. Active transport is active in the sense of requiring energy. In cotransport, a second ion or molecule is transported against its electrochemical gradient with ("co") an ion that is transported along its electrochemical gradient.

You Should Be Able To (YSBAT)

YSBAT p. 84 Fats have many C—C and C—H bonds, which have high free energy. They are hydrophobic because they contain long hydrocarbon chains and lack any highly polar groups. **YSBAT p. 86** No, because both are relatively large molecules that are polar or carry a charge. **YSBAT p. 91** (1) Add a large number of green dots to the inside of the liposome. (2) Add a large number of green dots to the solution outside the liposome. **YSBAT p. 94** Your arrow should point out of the cell. There is no concentration gradient, but the outside has a net positive charge, which favors outward movement of negative ions. **YSBAT p. 96** [See Figure A6.2]

Caption Questions and Exercises

Figure 6.3 At the polar hydroxyl group in cholesterol and the polar head group in phospholipids. **Figure 6.14** Higher, because less water would have to move to the right side to achieve equilibrium. **Figure 6.21** The electrical gradient would reverse, because the inside of the membrane would be more positive than the outside. The concentration gradient for sodium would also reverse; sodium would diffuse to the exterior. **Figure 6.22** No—the 10 replicates where no current was recorded probably represent instances where the CFTR protein was damaged and not functioning properly. (In general, no experimental method works "perfectly.") **Figure 6.27** "Diffusion": description as given; no proteins involved. "Facilitated diffusion": Passive movement of ions or molecules that cannot cross a phospholipid bilayer readily, along a concentration gradient. Facilitated by channels or transporters. "Active transport": Active (energy-demanding) movement of ions or molecules against an electrochemical gradient.

Summary of Key Concepts

KC 6.1 Highly permeable bilayers consist of phospholipids with short, unsaturated hydrocarbon tails. **KC 6.2** (1) The solute will diffuse until both sides are at equal concentrations. (2) Water will diffuse toward the side with the higher solute concentration. **KC 6.3** [See Figure A6.3]

Test Your Knowledge

1. b; **2.** a; **3.** b; **4.** d; **5.** c; **6.** d

Test Your Understanding

1. No, because they have no polar end to interact with water. Instead, these lipids would collect and float on the surface of water, or collect in droplets suspended in water, reducing their interaction with water to a minimum. **2.** Hydrophilic, phosphate-containing head groups interact with water; hydrophobic, fatty-acid tails associate with each other. A bilayer has a lower potential energy and thus is more stable than are independent phospholipids in solution. **3.** Ethanol's polarity reduces the speed at which it can cross a membrane, but its small size and lack of charge increase the speed at which it crosses membranes. Ethanol probably crosses a membrane more slowly than water (which is also polar, but smaller) but faster than larger polar molecules, and much faster than charged molecules. **4.** If no membrane separates the two solutions, then the constant, random motion of solute and water molecules causes even mixing of the two solutions. No net diffusion of water molecules will occur. **5.** Only hydrophobic amino acids interact with the nonpolar lipid tails in the interior of a phospholipid bilayer. Figure 3.3 lists the following nonpolar, hydrophobic amino acids: glycine, alanine, valine, leucine, isoleucine, methionine, phenylalanine, tryptophan, and proline. According to Table 3.1, tyrosine and cysteine are also relatively hydrophobic, suggesting that they might also be found in the interior of transmembrane proteins. **6.** CO_2 will diffuse into the cell; water will move out via osmosis. Na^+ and K^+ will enter the cell through gramicidin by facilitated diffusion. Cl^- will not move.

Applying Concepts to New Situations

1. Flip-flops should be rare, because they require polar head group to pass through the hydrophobic portion of the lipid bilayer. To test this prediction, you could monitor the number of dyed phospholipids that transfer from one side of the membrane to the other in a given period of time. **2.** The kinks in unsaturated hydrocarbon tails keep membranes fluid, even at low temperature, because they prevent extensive hydrophobic interactions. In hot environments, it would be advantageous for phospholipids to have saturated tails, to prevent membranes from being too fluid. **3.** Adding a methyl group makes a drug more hydrophobic and thus more likely to pass through a lipid bilayer. Adding a charged group make it hydrophilic and reduces its ability to pass through the lipid bilayer. **4.** Detergents are amphipathic. Their hydrophobic ends interact with grease while their hydrophilic ends interact with water. This forms a bridge between the grease and the water, effectively making the grease dissolve in water and allowing it to be washed away.

CHAPTER 7

Check Your Understanding (CYU)

CYU p. 105 (1) Photosynthetic membranes increase food production by providing a large surface area to hold the pigments and enzymes required for photosynthesis. (2) The presence of a magnetic mineral, which changes position as the cell moves through a magnetic field, allows the cell to move in a directed way. (3) The layer of thick, strong material stiffens the cell and provides protection from mechanical damage. **CYU p. 115** (1) Both organelles contain specific sets of enzymes, and the interior of lysosomes is acidic. Peroxisomes contain catalase and other enzymes that process fatty acids and toxins via oxidation reactions. Lysosomal enzymes digest macromolecules, releasing monomers that can be recycled into new macromolecules. (2) From top to bottom, the cells in the last column should read as follows: administrative/information hub, protein factory, large molecule manufacturing and shipping (protein synthesis and folding center, protein finishing and shipping line, fat factory, waste processing and recycling center), fatty-acid processing and detox center, warehouse, power station, food-manufacturing facility, support beams, perimeter fencing with secured gates, and leave blank. **CYU p. 122** (1) Proteins that lack a signal sequence will not be delivered to the ER. They are released into the cytosol. (2) The proteins will not be secreted—they will be found in lysosomes. **CYU p. 128** The products will not be able to move through the system normally, because their microtubule "roads" are absent.

You Should Be Able To (YSBAT)

YSBAT p. 128 (1) The microtubules at the bottom of the axoneme would slide to the right, but the axoneme would not bend. (2) Nothing would happen (dynein wouldn't move).

Caption Questions and Exercises

Figure 7.1 [See Figure A7.1] **Figure 7.15** The vesicles deliver digestive enzymes that are required for the mature lysosome to function. **Figure 7.16** Storing the toxins in vacuoles prevents the toxins from damaging the plant's own organelles and cells. **Figure 7.19** The cell wall is outside the plasma membrane. **Figure 7.21** [See Figure A7.2] **Figure 7.23** "Prediction": The labeled tail region fragments or the labeled core region fragments of the nucleoplasmin protein will be found in the cell nucleus. "Prediction of null hypothesis": No labeled fragments of the nucleoplasmin protein will be found in the nucleus of the cell. "Conclusion": The "Send to nucleus" zip code is in the tail region of the nucleoplasmin protein.

FIGURE A6.2

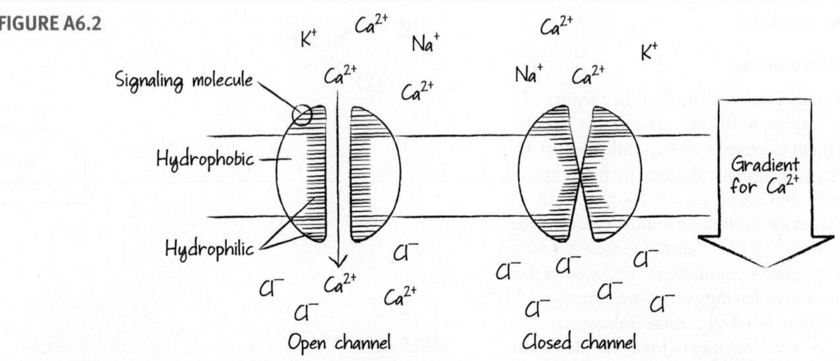

FIGURE A6.3

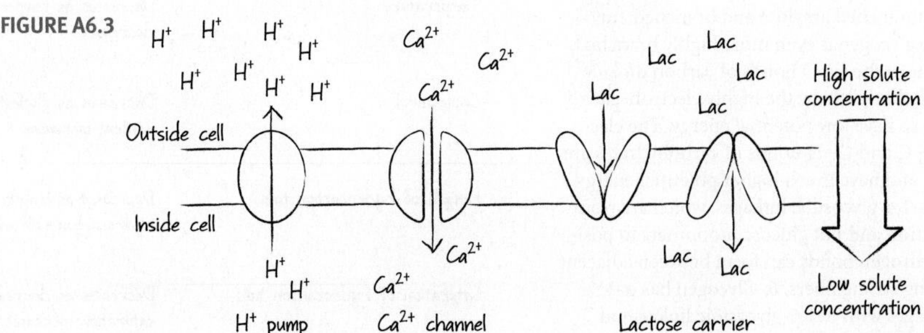

Summary of Key Concepts

KC 7.1 (1) They will be unable to make ATP and will die. (2) They will be limp and less able to resist attack by predators. In many environments they will be unable to resist the osmotic pressure of water entering the cytoplasm and will burst. (3) They will be unable to synthesize new proteins and will die. **KC 7.2** As a group, proteins have complex and highly diverse shapes and chemical properties. Individual proteins, then, can recognize and bind to molecular zip codes in a very specific way (like a key fitting into a lock). **KC 7.3** Actin filaments are made up of two strands of actin monomers, microtubules are made up of tubulin protein dimers that form a tube, intermediate filaments are made up of a variety of different protein monomers. Actin filaments and microtubules exhibit polarity (or directionality), with new subunits constantly being added or subtracted at either end (but added faster to the plus end). All three elements provide structural support, but only actin filaments and microtubules are involved in movement and cell division. **KC 7.4** Micrographs of cells at increasing time intervals after the pulse treatment represent time-lapse photography—a set of still images that show change through time.

Test Your Knowledge

1. a; **2.** c; **3.** c; **4.** b; **5.** a; **6.** d

Test Your Understanding

1. All cells are bound by a plasma membrane, are filled with cytoplasm, carry their genetic information (DNA) in chromosomes, and contain ribosomes (the sites of protein synthesis). Some prokaryotes have organelles not found in plants or animals, such as a magnetite-containing structure. Plant cells have chloroplasts, vacuoles, and a cell wall. Animal cells contain lysosomes and lack a cell wall. **2.** Ribosome in cytoplasm (protein is synthesized) → Rough ER (protein is folded and initially modified) → Transport vesicle → Golgi apparatus (protein is glycosylated; has molecular tag indicating destination) → Transport vesicle → Plasma membrane → Extracellular space. **3.** When a globular "head" section of kinesin binds and releases ATP, it undergoes a conformational change that swings it forward and binds it to the microtubule. The two globular head regions alternate between swinging and binding movements, causing the protein to "walk" down the microtubule. **4.** Changes in the length of microfilaments and microtubules contribute to changes in the size and shape of the cytoskeleton. **5.** Plasma membrane proteins are produced by the endomembrane system. Cells that function in membrane transport processes should have an extensive endomembrane system, including RER. **6.** All microtubules have the same basic structure, consisting of tubulin dimers that form a tube with plus and minus ends. The microtubules involved in structural support have no associated motor proteins; those involved in vesicle movement and flagella do. The microtubules that make up flagella are grouped in a 9 doublets + 2 single-microtubules arrangement. Structural and vesicle transport microtubules are single structures.

Applying Concepts to New Situations

1. Since mannose-6-phosphate serves as a lysosome tag in the endomembrane system, material that is ingested by phagocytic cells of the immune system might be targeted to lysosomes in the same way. To test this hypothesis, you could expose immune system cells to a population of bacteria that have labeled proteins, isolate the proteins after ingestion, and test to see whether they now have a mannose-6-phosphate tag attached. **2.** The proteins must receive a molecular zip code that binds to a receptor on the surface of peroxisomes. They could diffuse randomly to peroxisomes, or be transported in a directed way by vesicles or specialized motor proteins. **3.** (a) "Housekeeping"—destruction of macromolecules

FIGURE A7.1

Nucleoid

Nucleoid

FIGURE A7.2

(a) Animal pancreatic cell: Exports digestive enzymes.

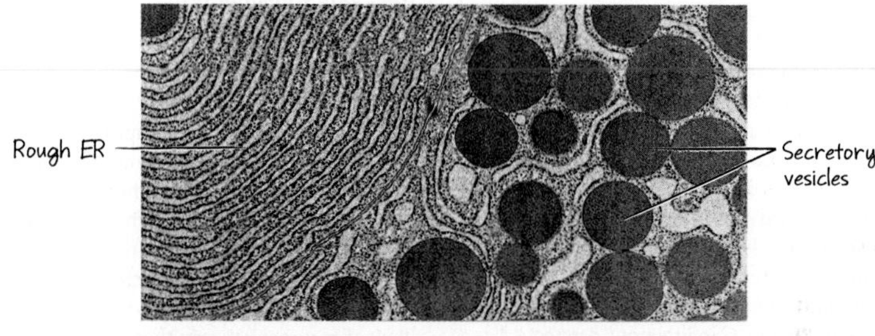

(b) Animal testis cell: Exports lipid-soluble signals.

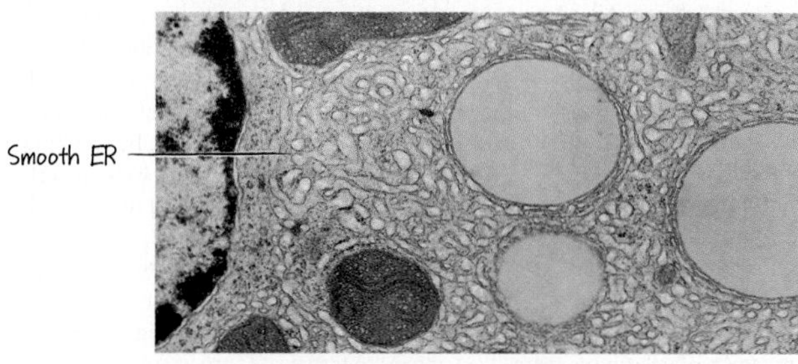

(c) Plant leaf cell: Manufactures ATP and sugar.

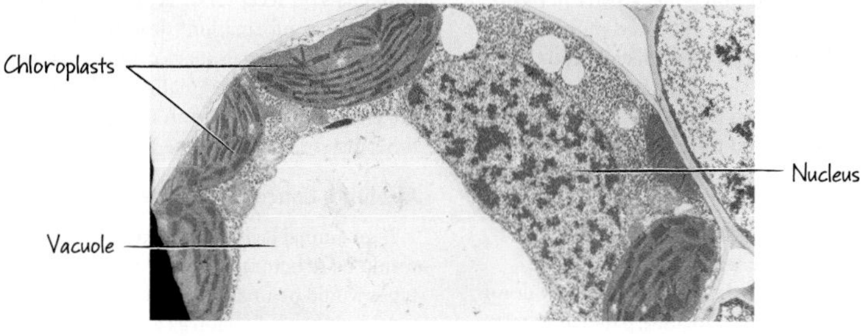

(d) Plant root cell: Stores starch.

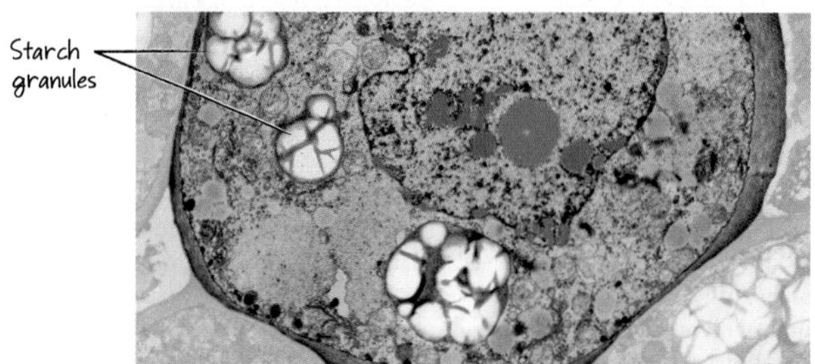

that are dangerous or no longer needed. (b) Structural support (e.g., wood is made up of cells with extensive cell walls). (c) Detoxification of dangerous compounds via oxidation reactions, or processing of stored fatty acids. **4.** The cell wall prevents the cells from bursting in hypotonic (aqueous) environments, or protects the cell from damage. Expose mutant and normal individuals to (1) a hypotonic environment, and (2) a eukaryote or virus that routinely attacks and kills this archaeal species. If the first hypothesis is correct, the mutant cells will burst but the normal cells will live. If the second hypothesis is correct, more mutant cells will die than normal cells.

CHAPTER 8

Check Your Understanding (CYU)

CYU p. 134 Plant cell walls and animal ECMs are both fiber composites. In plant cell walls the fiber component consists of cross-linked cellulose fibers and the ground substance is pectin. In animal ECMs the fiber component consists of collagen fibrils and the ground substance is gel-forming polysaccharides. **CYU p. 139** (1) The three structures differ in composition, but their function is similar. The middle lamella in plants is composed of pectins that glue adjacent cells together. Tight junctions are made up of membrane proteins that line up and "stitch" adjacent cells together. Desmosomes are "rivet-like" structures composed of proteins that link the cytoskeletons of adjacent cells. (2) The plasma membranes of adjacent plant cells are continuous at plasmodesmata, so cytoplasm, smooth endoplasmic reticulum, and other cell components are shared. Gap junctions connect adjacent animal cells by forming protein-lined pores. The openings allow ions and small molecules to be shared between cells. **CYU p. 146** (1) Each cell-cell signal binds to a specific receptor protein. A cell can respond to a signal only if it has the appropriate receptor. Different types of cells have different types of receptors. (2) Signals are amplified if one or more steps in a signal transduction pathway, involving either second messengers or a phosphorylation cascade, results in the activation of multiple downstream molecules.

You Should Be Able To (YSBAT)

YSBAT p. 138 (1) Yes—molecules can pass through a middle lamella because it is made up of gelatinous material that is not watertight. (2) Developing muscle cells could not adhere normally and muscle tissue would not form properly. The embryo would die. **YSBAT p. 143** The spy is the signal that arrives at the receptor (the castle gate). The guard is the G protein–linked receptor in the plasma membrane, and the queen is the G protein. The commander of the guard is the enzyme that catalyzes production of a second messenger (the soldiers). **YSBAT p. 144** (1) The red dominos (RTK components) would be the first dominos in the chain and the first to be tipped to initiate the domino cascade. The black domino (Ras) would be the second domino in line. Then there would be green, blue, pink, and yellow dominos (other kinases) set up in branching patterns to represent how Ras phosphorylates and activates a number of different kinases. Each of those colored dominos would be followed by a line of dominos of the same color representing the target proteins of those kinases. (2) Tipping the red domino would knock down the black domino (Ras) and would initiate a branching domino cascade that is dependent on the black domino's falling. The red domino represents how RTK activates Ras and how Ras initiates a cascade of phosphorylation.

Caption Questions and Exercises

Figure 8.9 "Prediction of null hypothesis": Cells will not adhere, or will adhere randomly with respect to cell type and species. **Figure 8.10** "Prediction": Cells treated with an antibody that blocks membrane proteins involved in adhesion will not adhere. "Prediction of null hypothesis": All cells will adhere normally. **Figure 8.12** Top: Tight junctions; Middle: Desmosome; Bottom: Gap junctions.

Summary of Key Concepts

KC 8.1 (1) Adjacent cells fall apart from each other. (2) Nearby cells fall apart from each other; both cells and tissues are weaker and more susceptible to damage. **KC 8.2** Epithelium separates compartments, creating an inside and outside. If tight junctions degrade, materials from the outside will diffuse in and materials from the inside will diffuse out. **KC 8.3** Adrenalin binds to both heart and liver cells, but the activated receptors trigger different signal transduction pathways and lead to different cell responses. **KC 8.4** If only one step existed in a signal transduction pathway, it would be difficult for the process to be regulated by many different pathways. Multiple steps provide points for different pathways to regulate the response.

Test Your Knowledge

1. b; **2.** b; **3.** a; **4.** b; **5.** d; **6.** c

Test Your Understanding

1. The cross-linked fiber components withstand tension; the ground substance withstands compression. **2.** If each enzyme in the cascade phosphorylates many enzymes in the next step of the cascade, the initial signal will be amplified many times over. **3.** Although both are made up of membrane-spanning proteins, tight junctions seal adjacent animal cells together while gap junctions allow a flow of material between them. Middle lamellae are pectin-rich layers that glue adjacent plant cells together; plasmodesmata are gaps in the cell walls of adjacent cells where the plasma membrane is continuous, allowing exchange of materials between them. **4.** When dissociated cells from two sponge species were mixed, the cells sorted themselves into distinct aggregates that contained only cells of the same species. By blocking membrane proteins with antibodies and isolating cells that would *not* adhere, researchers found that specialized groups of proteins, including cadherins, are responsible for selective adhesion. **5.** Signal crosses plasma membrane and binds to intracellular receptor (reception) → receptor changes conformation, and signal-receptor complex moves to target site (processing) → signal-receptor complex binds to a target molecule (e.g., a gene or membrane pump), which changes its activity (response) → signal falls off receptor or is destroyed; receptor changes to inactive conformation (deactivation). **6.** Information from different signals may conflict or be reinforcing. "Cross-talk" between signaling pathways allows cells to integrate information from many signals at the same time, instead of responding to each signal in isolation.

Applying Concepts to New Situations

1. If an animal lacked an extracellular matrix, its cells would be structurally weak. Aggregations of similar cell types would be unable to form tissues, so individuals would probably develop as a mass of poorly connected or unconnected cells. A plant species with a similar disability would experience similar problems, but would also burst if placed in a hypotonic environment. **2.** (a) The response would have to be extremely local—the activated signal-receptor complex would have to affect nearby proteins. (b) Little or no amplification could occur, because there could never be more than one molecule that triggers the response. (c) The only way to regulate the response would be to block the receptor or make it more responsive to the signal. **3.** Chitin forms chains that can cross-link with one another. Therefore, the fungal cell wall likely has a structure similar to that of the plant cell wall (see Figure 8.2, left) except that chitin, rather than cellulose, is the major fiber component. **4.** Antibody binding would prevent activation of the associated G protein. Even if the appropriate signal were present, no second messenger could be produced and no cell response could occur. Permanent binding by a drug would either activate the G protein constantly or block activation completely.

CHAPTER 9

Check Your Understanding (CYU)

CYU p. 153 (1) In part, because its three phosphate groups have four negative charges in close proximity. The electrons repulse each other, raising their potential energy. (2) Electrons in C—O bonds are held more tightly than electrons in C—H bonds, so they have lower potential energy. **CYU p. 161** (1) and (2) are combined with the answer to CYU p. 166 below. (3) Start with 12 dimes on glucose. (These dimes represent the 12 electrons that will be moved to electron carriers during redox reactions throughout glycolysis and the citric acid cycle.) Move two dimes to the NADH box generated by glycolysis and the other 10 dimes to the pyruvate box. Then move these 10 dimes through the pyruvate dehydrogenase box, placing two of them in the NADH box next to pyruvate dehydrogenase and the remaining eight dimes in the acetyl CoA box. Next move the eight dimes in the acetyl CoA box through the citric acid cycle, placing six of them in the NADH box and two in the $FADH_2$ box generated during the citric acid cycle. (4) These boxes are marked with stars in the diagram. **CYU p. 166** [See Figure A9.1] To illustrate the chemiosmotic mechanism, take the dimes (electrons) piled on the NADH and $FADH_2$ boxes and move them through the ETC. While moving these dimes, also move pennies from the mitochondrial matrix to the intermembrane space. As the dimes exit the ETC, add them to oxygen to generate water. Once all of the pennies have been pumped by the ETC into the intermembrane space, move them through ATP synthase back into the mitochondrial matrix to fuel the formation of ATP. **CYU p. 168** Electron acceptors such as oxygen have a much lower electronegativity than pyruvate. Donating an electron to O_2 causes a greater drop in potential energy, making it possible to generate much more ATP per molecule of glucose.

You Should Be Able To (YSBAT)

YSBAT p. 150 (1) Ribulose and carbon dioxide will not react because the reaction is endergonic and cannot proceed without the input of energy. (2) It is "activated" because it now has high potential energy. (3) The endergonic reaction (ribulose + CO_2) is coupled to the exergonic reaction (ATP hydrolysis) via the phosphorylation of ribulose. **YSBAT p. 152** (1) They are oxidized. (2) They are reduced. (3) Oxygen. (4) The reactants have higher energy. "Energy" should be added to the product side of the equation with "Release of energy" written below it. **YSBAT p. 163** "Indirect" is accurate because the energy released during glucose oxidation is stored in reduced electron carriers (instead of being used to produce ATP directly). The energy released as electrons from these carriers flow through the ETC is used to generate a proton gradient across a membrane. These protons then diffuse down their concentration gradient through ATP synthase located in the membrane, which drives ATP synthesis.

Caption Questions and Exercises

Figure 9.6 In glucose, put two dots in the middle of each C—C or C—H bond, and two dots next to the O in the C—O bond. In O_2, put four electrons in the middle of the double bond. In CO_2, put four electrons next to each O. In H_2O, put two electrons near O in each O—H bond. The electrons are held more tightly in the products than in the reactants, so potential energy has decreased. The difference in potential energy is released, so the reaction is exergonic. **Figure 9.8** *Glycolysis:* "What goes in" = glucose, NAD^+, ADP, inorganic phosphate; "What comes out" = pyruvate, NADH, ATP. *Pyruvate processing:* "What goes in" = pyruvate, NAD^+; "What comes out" = NADH, CO_2, acetyl CoA. *Citric acid cycle:* "What goes in" = acetyl CoA, NAD^+, FADH, GDP or ADP, inorganic phosphate; "What comes out" = NADH, $FADH_2$, ATP or GTP, CO_2. *Electron transport and chemiosmosis:* "What goes in" = NADH, $FADH_2$, O_2; "What comes out" = ATP, H_2O, NAD^+, FAD. **Figure 9.12** If the regulatory site had a higher affinity for ATP than the active site, then ATP would always be bound at the regulatory site, and glycolysis would always proceed at a very slow rate. **Figure 9.14** "Positive control": AMP, NAD^+, CoA (reaction substrates). "Negative control by feedback inhibition": acetyl CoA, NADH, ATP (reaction products). **Figure 9.16** An allosteric regulator binds somewhere other than the active site and causes shape changes that slow the reaction or speed it up. A competitive inhibitor binds at the active site and keeps the substrate from binding. **Figure 9.18** Production of NADH and $FADH_2$. (In the line representing free energy, there are greater drops at each occurrence of NADH and $FADH_2$ versus ATP formation.) **Figure 9.20** Yes, because the artificial vesicles include only an ATP-synthesizing enzyme and a proton pump that creates a proton-motive force. The only energy source in this experiment is the proton gradient, so the result strongly supports the chemiosmotic hypothesis. (Note: This experiment does not test whether all ATP production in the ETC is through chemiosmosis. It does not preclude the possibility that the ETC also performs some substrate-level phosphorylation.) **Figure 9.21** The proton gradient arrow should start above in the inner membrane space and point down across the membrane into the mitochondrial matrix. *Complex I:* "What goes in" = NADH and H^+; "What comes out" = NAD^+, e^-, H^+. *Complex II:* "What goes in" = $FADH_2$; "What comes out" = FAD, e^-. *Complex III:* "What goes in" = e^-; "What comes out" = e^-. *Complex IV:* "What goes in" = e^-, H^+, O_2; "What comes out" = H_2O, H^+. **Figure 9.26** [See Figure A9.2]

Summary of Key Concepts

KC 9.1 Phosphorylation adds two negative charges in a small area. The electrical repulsion that results raises the protein's potential energy and its tertiary structure. **KC 9.2** The free-energy drop from glucose to oxygen (or another electron acceptor, during cellular respiration) is much greater than the free energy drop from glucose to pyruvate (during fermentation). Thus, there is more free energy available to use in synthesizing ATP. **KC 9.3** NADH would decrease if a drug poisoned the acetyl CoA-to-citrate enzyme, but increase if a drug poisoned ATP synthase. **KC 9.4** [See Figure A9.3] **KC 9.5** Organisms that produce ATP by fermentation grow more slowly than those that produce ATP via cellular respiration

FIGURE A9.1

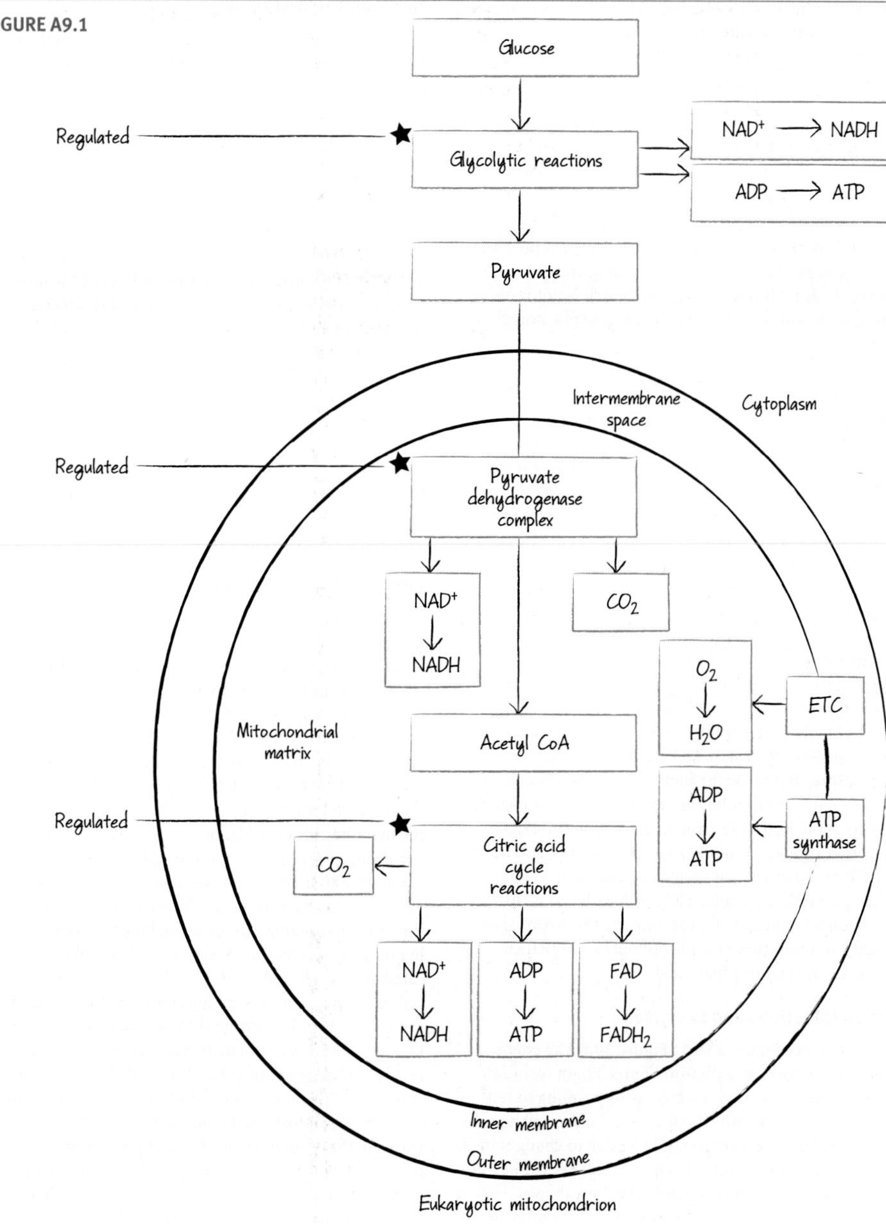

FIGURE A9.2

10 reactions of glycolysis
Citric acid cycle
Glucose

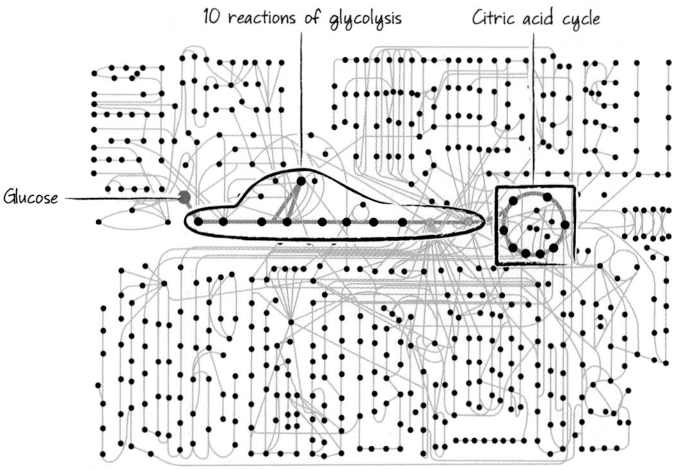

FIGURE A9.3

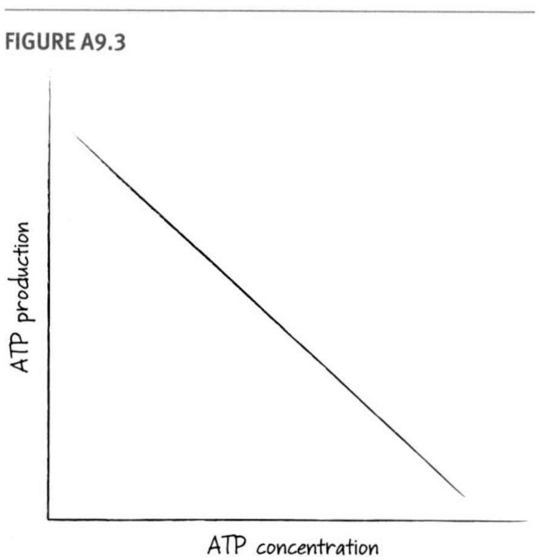

simply because fermentation produces fewer ATP molecules per glucose molecule than cellular respiration does.

Test Your Knowledge

1. b; **2.** b; **3.** d; **4.** c; **5.** a; **6.** b

Test Your Understanding

1. NADH and $FADH_2$ get their electrons from the intermediates of glycolysis, pyruvate processing, and the TCA cycle and deliver them to the ETC, where they reduce O_2. **2.** Both processes produce ATP from ADP and P_i, but substrate-level phosphorylation occurs when enzymes remove a "high-energy" phosphate from a substrate and directly transfer it to ADP, while oxidative phosphorylation is based on electrons moving through an ETC and production of a proton-motive force that drives ATP synthase. **3.** Aerobic respiration is much more productive because oxygen has extremely high electronegativity compared to other electron acceptors, resulting in a greater release of energy during electron transport and more proton pumping. **4.** Glycolysis → Pyruvate processing → TCA cycle → ETC and chemiosmosis. The first three steps are responsible for glucose oxidation; the final step produces the most ATP. **5.** Electron transport makes oxidative phosphorylation possible (oxidative phosphorylation occurs when electrons have been transported through the ETC and a proton gradient has been established). ATP synthase consists of an F_o unit and F_1 unit joined by a stalk. When protons flow through the F_o unit, the stalk and F_1 unit spin. The motion drives the synthesis of ATP from ADP and P_i. **6.** Stored carbohydrates can be broken down into glucose that enters the glycolytic pathway. If carbohydrates are absent, products from fat and protein catabolism can be used to fuel cellular respiration or fermentation. If ATP is plentiful, anabolic reactions use intermediates of the glycolytic pathway and the TCA cycle to synthesize carbohydrates, fats, and proteins.

Applying Concepts to New Situations

1. When complex IV is blocked, electrons can no longer be transferred to oxygen, the final acceptor, and cellular respiration stops. Fermentation could keep glycolysis going, but it is inefficient and unlikely to fuel a cell's energy needs over the long term. Cells that lack the enzymes required for fermentation would die first. **2.** Because mitochondria with few cristae would have fewer electron transport chains and ATP synthase molecules, they would produce much less ATP than mitochondria with numerous cristae. **3.** When oxygen is unavailable for cellular respiration, yeast cells switch to fermentation, which occurs in the cytosol. They are unlikely to expend large amounts of energy and materials to maintain mitochondria. **4.** The potential energy drop between glucose and pyruvate is relatively small. When pyruvate acts as the electron acceptor during fermentation, a great deal of potential energy remains in the alcohol or other fermentation products.

CHAPTER 10

Check Your Understanding (CYU)

CYU p. 179 When a pigment absorbs energy and an electron is excited in response, the event is analogous to striking the tuning fork. When an excited electron transmits resonance energy to another pigment molecule, the event is analogous to touching a vibrating tuning fork to another and making it vibrate. **CYU p. 184** (1) Your model should conform to the relationships shown in Figure 10.13. (2) Put four dimes on the water-splitting reaction ($2\ H_2O \rightarrow 4\ H^+ + O_2$). When four photons hit the PSII antenna complex, move the dimes to pheophytin, then to the PSII ETC (plastoquinone and the cytochrome complex). The dimes go to plastocyanin, then to the antenna complex of PSI. When four photons strike, move the dimes to the PSI ETC and ferrodoxin, then combine them with $NADP^+ + 2\ H^+$ to form NADPH. **CYU p.190** (1) (a) C_4 plants use PEP carboxylase to fix CO_2 into organic acids in mesophyll cells. These organic acids are then pumped into bundle-sheath cells, where they release carbon dioxide to rubisco. (b) CAM plants take in CO_2 at night, and have enzymes that fix it into organic acids stored in the central vacuoles of photosynthesizing cells. During the day, the organic acids are processed to release CO_2 to rubisco. (c) By diffusion through a plant's stomata when they are open. (2) Sucrose can be transported to other parts of the plant because it is water soluble. Starch is not water soluble, so it cannot be transported. It acts as a storage product, instead.

You Should Be Able To (YSBAT)

YSBAT p. 181 Photosystem II is very similar to the ETC in mitochondria. Both use the energy released from electron transfer to pump protons, generating a proton gradient that is used to generate ATP through ATP synthase. The high-energy electrons for PSII come from chlorophyll via water, however, while they come from NADH and $FADH_2$ for the ETC. The eventual products of PSII are ATP and oxygen; the eventual products of the mitochondrial ETC are ATP and water. **YSBAT p. 182** Light → Antenna complex → Reaction center → Pheophytin → ETC → Proton gradient → ATP synthase. Electrons from water are donated to the reaction center. When they absorb resonance energy and reduce pheophytin, they convert electromagnetic energy to chemical energy. **YSBAT p. 183** (1) Plastocyanin transfers electrons that move through PSII to the reaction center of PSI. (2) After they are excited by a photon and donated to the initial electron acceptor. (3) The electrons that originate from PSII's reaction center end up in NADPH—they are not cycled back to form water. **YSBAT p. 186** Rubisco joins CO_2 with RuBP to form a six-carbon intermediate that is immediately hydrolyzed to yield two molecules of 3-phosphoglycerate, which participate in reactions that yield G3P.

Caption Questions and Exercises

Figure 10.5 [See Figure A10.1] **Figure 10.7** The energy state corresponding to a photon of green light would be located between the energy states corresponding to red and blue photons. **Figure 10.10** Yes—otherwise, changes in the rate of photosynthesis could be due to changes in the density of photosynthetic cells, not differences in wavelengths received. **Figure 10.17** The researchers didn't have any basis on which to predict these intermediates. They needed to perform the experiment to identify them. **Figure 10.24** One arrow should point from sucrose to starch, another from starch to sucrose. Note that the interconversion requires transport between the cytosol and chloroplast.

Summary of Key Concepts

KC 10.1 The Calvin cycle depends on the ATP and NADPH produced by PSI and PSII, so it is not independent of light. **KC 10.2** Oxygen is produced by a critical step in photosynthesis: splitting water to provide electrons to PSII. If oxygen production increases, it means that more electrons are moving through the photosystems. **KC 10.3** The "final" product of the reaction sequence reacts to form the molecule that is the substrate for the initial reaction in the sequence. **KC 10.4** Both forms of rubisco would increase carbohydrate production: Oxygen would not compete with CO_2 for binding to the active site of rubisco and the competing reactions of photorespiration would not occur, or rubisco would fix 10 times as much CO_2 in a given period of time.

Test Your Knowledge

1. d; **2.** c; **3.** a; **4.** c; **5.** b; **6.** d

Test Your Understanding

1. In PSII, it occurs when excited electrons from the reaction center are accepted by pheophytin, reducing it. In PSI, it occurs when excited electrons from the reaction center are accepted by ferredoxin. **2.** CO_2 is fixed when rubisco catalyzes the reaction between carbon dioxide and ribulose bisphosphate to form 3-phosphoglycerate. Subsequent reactions that produce sugar require phosphorylation by ATP and high-energy electrons from NADPH. **3.** PSII and PSI both have reaction centers, an energy transformation step, and an ETC. PSII produces ATP while PSI produces NADPH, and only PSII splits water to obtain electrons. Plastocyanin transfers electrons between the two photosystems, connecting them. **4.** Photorespiration occurs when levels of CO_2 are low and O_2 are high. Less sugar is produced because (1) CO_2 doesn't participate in the initial reaction catalyzed by rubisco, and (2) when rubisco catalyzes the reaction with O_2 instead, one of the products is eventually broken down to CO_2 in a process that uses ATP. **5.** See Figure 10.22b and Figure 10.23. **6.** Photosynthesis in chloroplasts produces glucose, which is processed in mitochondria to produce ATP.

Applying Concepts to New Situations

1. Both organelles have a double membrane, extensive internal membranes, ETCs that include cytochromes and quinones, and ATP synthase. Only chloroplasts have chlorophyll and other pigment molecules that absorb light. Chloroplasts use NADPH as an electron carrier; mitochondria use NADH and $FADH_2$. The Calvin cycle occurs only in chloroplasts; the TCA cycle occurs only in mitochondria. **2.** Because CO_2 and O_2 levels minimized the impact of photorespiration when rubisco evolved, the hypothesis is credible. But once O_2 levels increased, any change in rubisco that minimized photorespiration would give individuals a huge advantage over organisms with "old" forms of rubisco. There has been plenty of time for such changes to occur, making the "holdover" hypothesis less credible. **3.** Carotenoids should be located close to the chlorophyll molecules in the reaction center, so they can pass energy along and neutralize free radicals that could damage chlorophyll. One way to test this hypothesis is to isolate thylakoid membranes and test for the presence of carotenoid and chlorophyll molecules. **4.** No—they are unlikely to have the same complement of photosynthetic pigments. Different wavelengths of light are available in various layers of a forest and water depths. It is logical to predict that plants and algae have pigments that absorb the available wavelengths efficiently. One way to test this hypothesis would be to isolate pigments from species in different locations, and test the absorbance spectra of each.

THE BIG PICTURE: ENERGY

Check Your Understanding (CYU), p. 192

1. Photosynthesis uses H_2O as a substrate and releases O_2 as a by-product; cellular respiration uses O_2 as a sub-

FIGURE A10.1

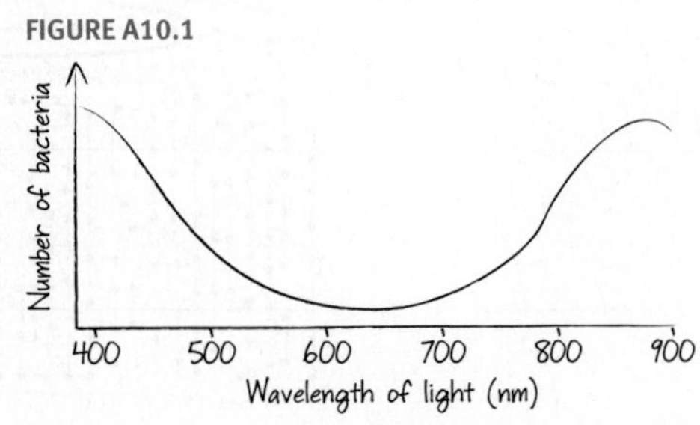

strate and releases H_2O as a by-product. **2.** Photosynthesis uses CO_2 as a substrate; cellular respiration releases CO_2 as a by-product. **3.** CO_2 fixation would essentially stop; CO_2 would continue to be released by cellular respiration. CO_2 levels in the atmosphere would increase rapidly, and production of new plant tissue would cease—meaning that most animals would quickly starve to death. **4.** ATP "is used by" the Calvin cycle; photosystem I "yields" NADPH.

CHAPTER 11

Check Your Understanding (CYU)

CYU p. 202 (1) [See Figure A11.1] (2) Interphase, prophase, prometaphase, metaphase, anaphase, telophase. Sister chromatids condense in prophase, move to the middle of the cell in metaphase, break apart in anaphase. **CYU p. 206** (1) $G_1 \rightarrow S \rightarrow G_2 \rightarrow M \rightarrow G_1$, etc. Checkpoints occur at the end of G_1 and G_2 and during M phase. (2) MPF Cdk levels are fairly constant throughout the cycle, but MPF cyclin increases. MPF activity increases at the end of G_2, initiating M phase, and declines at the end of M phase.

You Should Be Able To (YSBAT)

YSBAT p. 197 (1) Genes are segments of chromosomes that code for proteins. (2) Chromosomes are made of chromatin. (3) Sister chromatids are copies of the same chromosome, joined together. **YSBAT p. 200** [See Table A11.1] **YSBAT p. 204** MPF activates proteins that get mitosis under way. MPF consists of cyclin and a Cdk, and is inactivated by phosphorylation (at two sites) and activated by dephosphorylation (at one site). Enzymes that degrade cyclin reduce MPF levels. **YSBAT p. 208** [See Figure A11.2]

Caption Questions and Exercises

Figure 11.6 Unreplicated. **Figure 11.10** Fuse two interphase cells. If the regulatory molecule hypothesis is correct, neither cell should start M phase. But if the fusion event itself is the trigger, then at least one cell should start M phase. **Figure 11.13** In effect, MPF turns itself off after it is activated. If this didn't happen, the cell would undergo mitosis again right away. **Figure 11.15** After 1990, the death rate from lung cancer decreased in men but not in women. One hypothesis to explain the data is that rates of cigarette smoking declined in men but increased in women.

Summary of Key Concepts

KC 11.1 The gap phases give the cell time to replicate organelles and grow prior to division, as well as perform the normal functions required to stay alive. **KC 11.2** Kinetochore microtubules originate in the cytoplasm but chromosomes are in the nucleus. If the nuclear envelope did not disintegrate, the microtubules could not reach the chromosomes. **KC 11.3** The cell cycle will move forward, from G_1 to S phase, triggering cell division. **KC 11.4** A wide variety of defects, and multiple defects, cause a cell to become cancerous. No one drug or therapy can cure all of the defects involved.

Test Your Knowledge

1. b; **2.** d; **3.** d; **4.** d; **5.** c; **6.** c

Test Your Understanding

1. [See Figure A11.3] For daughter cells to have identical complements of chromosomes, all the chromosomes must be replicated during the S phase, the spindle apparatus must connect with the kinetochores of each replicated chromosome in prometaphase, and the sister chromatids of each replicated chromosome must separate in anaphase. **2.** One possible concept map is shown

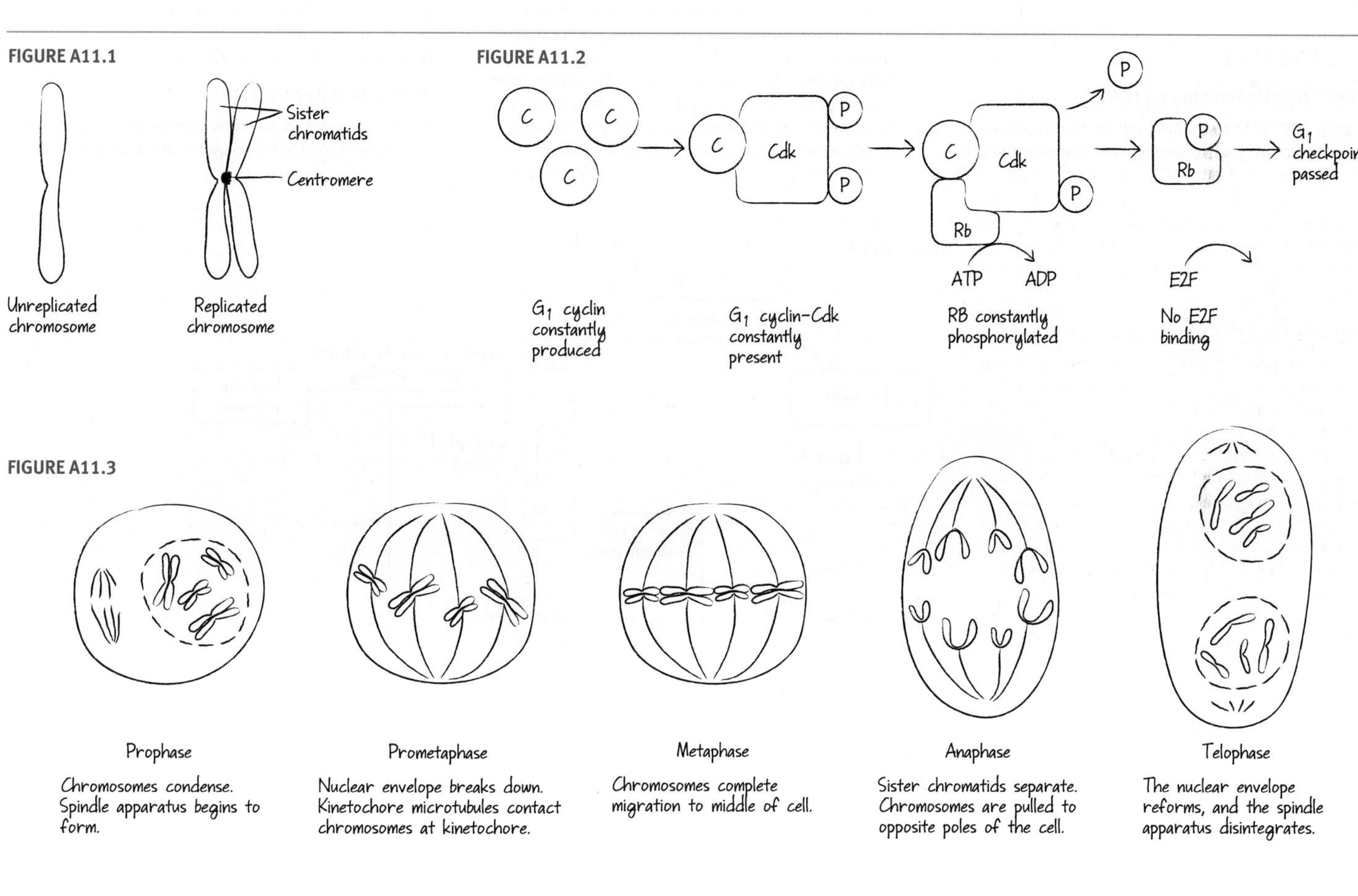

TABLE A11.1

	Prophase	Prometaphase	Metaphase	Anaphase	Telophase
Mitotic spindles	Grow	Contact chromosomes	Move chromosomes	Shorten	Break down
Nuclear envelope	Disintegrates	Nonexistent	Nonexistent	Nonexistent	Re-forms
Chromosomes	Condense	Attach to kinetochore microtubules	Move to metaphase plate	Sister chromatids separate	Collect at opposite poles

below [see Figure A11.4]. **3.** Cell fusion experiments suggested that something in mitotic cells initiated mitosis in other cells. Microinjection experiments suggested that this "something" is in the cytoplasm of M-phase cells. **4.** Protein kinases phosphorylate other proteins. Phosphorylation changes a protein's shape, altering its function (activating or inactivating it). As a result, protein kinases regulate the function of other proteins. **5.** Cyclin concentrations cycle during the cell cycle. At high concentration, cyclins bind to a specific cyclin-dependent kinase (or Cdk), forming an active protein kinase, such as MPF. **6.** If a cancer has not yet metastasized, the tumor can be completely removed by surgery.

Applying Concepts to New Situations

1. It is efficient: If the cell is not going to divide, there is no reason to invest energy in replication. **2.** A single cell with many identical nuclei in it. **3.** Labeled cells are in S phase, so 4 hours passed between the end of S phase and the start of M phase. **4.** Cancer requires many defects. Older cells have had more time to accumulate defects. Individuals with a genetic predisposition to cancer start out with some cancer-related defects, but this does not mean that the additional defects required for cancer to occur will develop.

CHAPTER 12

Check Your Understanding (CYU)

CYU p. 220 (1) Use four long and four short pipe cleaners (or pieces of cooked spaghetti) to represent the chromatids of two replicated homologous chromosomes (four total chromosomes). Mark two long and two short ones with a colored marker pen to distinguish maternal and paternal copies of these chromosomes. Twist identical pipe cleaners (e.g., the two long colored ones) together to simulate replicated chromosomes. Arrange the pipe cleaners to depict the different phases of meiosis I as follows: *Early prophase I:* Align sister chromatids of each homologous pair to form two tetrads. *Late prophase I:* Form one or more crossovers between non-sister chromatids in each tetrad. (This is hard to simulate with pipe cleaners—you'll have to imagine that each chromatid now contains both maternal and paternal segments.) *Metaphase I:* Line up homologous pairs (the two pairs of short pipe cleaners and the two pairs of long pipe cleaners) at the metaphase plate. *Anaphase I:* Separate homologs. Each homolog still consists of sister chromatids joined at the centromere. *Telophase I and cytokinesis:* Move homologs apart to depict formation of two haploid cells, each containing a single replicated copy of two different chromosomes. (2) During anaphase I, homologs (not sister chromatids, as in mitosis) are separated, making the cell products of meiosis I haploid. **CYU p. 223** (1) [See Figure A12.1] Maternal chromosomes are white and paternal chromosomes are black. Daughter cells with many other possible combinations of chromosomes than shown could result from meiosis of this parent cell. (2) Asexual reproduction generates no appreciable genetic diversity. Self-fertilization is preceded by meiosis so it generates gametes, through crossing over and independent assortment, that have combinations of alleles not present in the parent. Outcrossing generates the most genetic diversity among offspring because it produces new combinations of alleles from two different individuals.

You Should Be Able To (YSBAT)

YSBAT p. 213 $n = 12$; the organism is diploid with $2n = 24$ (in females; 23 in males). **YSBAT p. 214** [See Figure A12.2] Because the two sister chromatids are identical and attached, it is sensible to consider them as parts of a single chromosome. **YSBAT p. 216** [See Figure A12.3] **YSBAT p. 218** Crossing over would not occur and the daughter cells produced by meiosis would be diploid, not haploid. There would be no reduction division. **YSBAT p. 221** Each gamete would inherit either all maternal or all paternal chromosomes. This would limit genetic variation in the offspring by precluding the many possible gametes containing various combinations of maternal and paternal chromosomes.

Caption Questions and Exercises

Figure 12.11 [See Figure A12.4] **Figure 12.12** Asexually: 64 (16 individuals from generation three produce 4 offspring per individual). Sexually: 16 (8 individuals from generation three form 4 couples; each couple produces 4 offspring). **Figure 12.13** Sampling widely reduces the possibility that the results are due to an environmental factor—other than the presence of parasites—that differs between sexually and asexually reproducing populations.

Summary of Key Concepts

KC 12.1 Haploid cells cannot undergo meiosis, because there is no homolog to synapse with—haploid cells can-

FIGURE A11.4

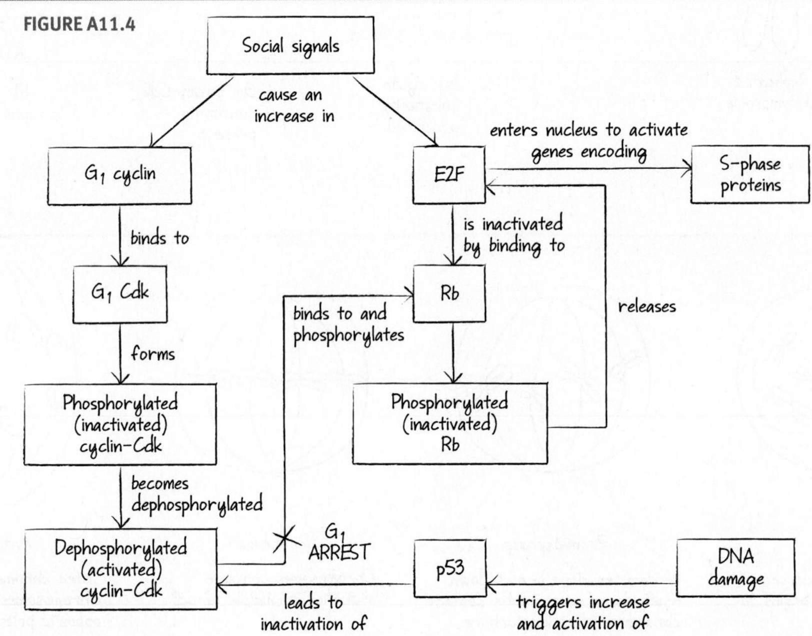

FIGURE A12.1

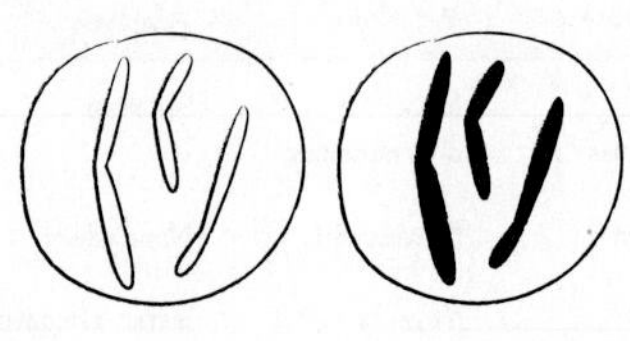

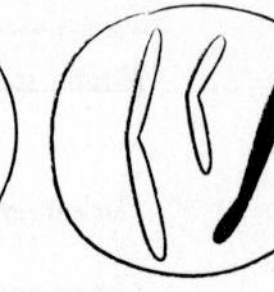
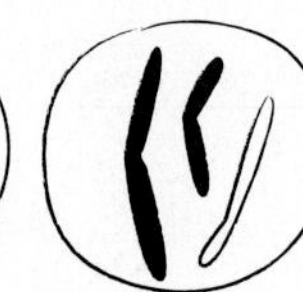

not undergo a reduction in ploidy. **KC 12.2** Monozygotic twins are genetically identical—the cells that originally gave rise to the two individuals were the product of mitosis. Dizygotic twins each arise from gametes that are the product of meiosis. Genetically, they are no more alike than any pair of siblings. **KC 12.3** Sexual reproduction will likely occur during seasons when conditions are changing rapidly, because genetically diverse offspring may have an advantage in the new conditions. **KC 12.4** The resulting gametes will be diploid instead of haploid.

Test Your Knowledge

1. b; **2.** a; **3.** b; **4.** b; **5.** d; **6.** a

Test Your Understanding

1. Homologous chromosomes are similar in size, shape, and gene content, and originate from different parents. Sister chromatids are exact copies of a chromosome that are generated when chromosomes are replicated (S phase of the cell cycle). **2.** Refer to Figure 12.6 as a guide for this exercise. The four pens represent the chromatids in one replicated homologous pair; the four pencils, the chromatids in a different homologous pair. To simulate meiosis II, make two "haploid cells"—each with a pair of pens and a pair of pencils representing two replicated chromosomes (one of each type in this species). Line them up in the middle of the cell, then separate the two pens and the two pencils in each cell such that one pen and one pencil goes to each of four daughter cells. **3.** Meiosis I is a reduction division because homologs separate—daughter cells have just one of each type of chromosome instead of two. Meiosis II is not a reduction division because sister chromatids separate—daughter cells have unreplicated chromosomes instead of replicated chromosomes, but still just one of each type. **4.** Tetraploids produce diploid gametes, which combine with a haploid gamete from a diploid individual to form a triploid offspring. Mitosis proceeds normally in triploid cells because replicated chromosomes align independently before sister chromatids separate. But during meiosis in a triploid, homologous chromosomes can't pair up correctly. The third set of chromosomes does not have a homologous partner to pair with. **5.** Asexually produced individuals are genetically identical, so if one is susceptible to a new disease, all are. Sexually produced individuals are genetically unique, so if a new disease strain evolves, at least some plants are likely to be resistant to it. **6.** If homologs or sister chromatids do not separate normally, the daughter cells will receive one too many chromosomes or one too few. The resulting chromosome sets are considered unbalanced because they do not have the normal number of copies of each gene.

Applying Concepts to New Situations

1. The gibbon would have 22 chromosomes in each gamete, and the siamang would have 25. Each somatic cell of the offspring would have 47 chromosomes. The offspring should be sterile because it has an unbalanced set of chromosomes, so not all of the chromosomes would have a homolog to pair with in meiosis prophase I. **2.** One in eight. **3.** Aneuploidy is the major cause of spontaneous abortion. If spontaneous abortion is rare in older women, it would result in a higher incidence of aneuploid conditions such as Down syndrome in older women, as recorded in the figure. **4.** (a) Such a study might be done in the laboratory, controlling conditions in identical tanks. A population of snails can be established in each tank. A parasite could be added to one tank, and then changes in the frequency of sexual reproduction in the snail populations can be monitored and observed over time. (b) In nature, all of the forces that can act to produce change in a population are present. A laboratory test is only an approximation of what occurs in the wild. For example, sexually reproduced offspring might only have an advantage when the parasite-ridden population is also stressed by food shortages or high or low temperatures. If so, an experiment in the lab with ample food and preferred temperatures would give an incorrect result.

CHAPTER 13

Check Your Understanding (CYU)

CYU p. 239 See answers to Genetics Problems on p. A:14. **CYU p. 243** (1) [See **Figure A13.1**] Segregation of alleles occurs when homologs that carry those alleles are separated during anaphase I. One allele ends up in each

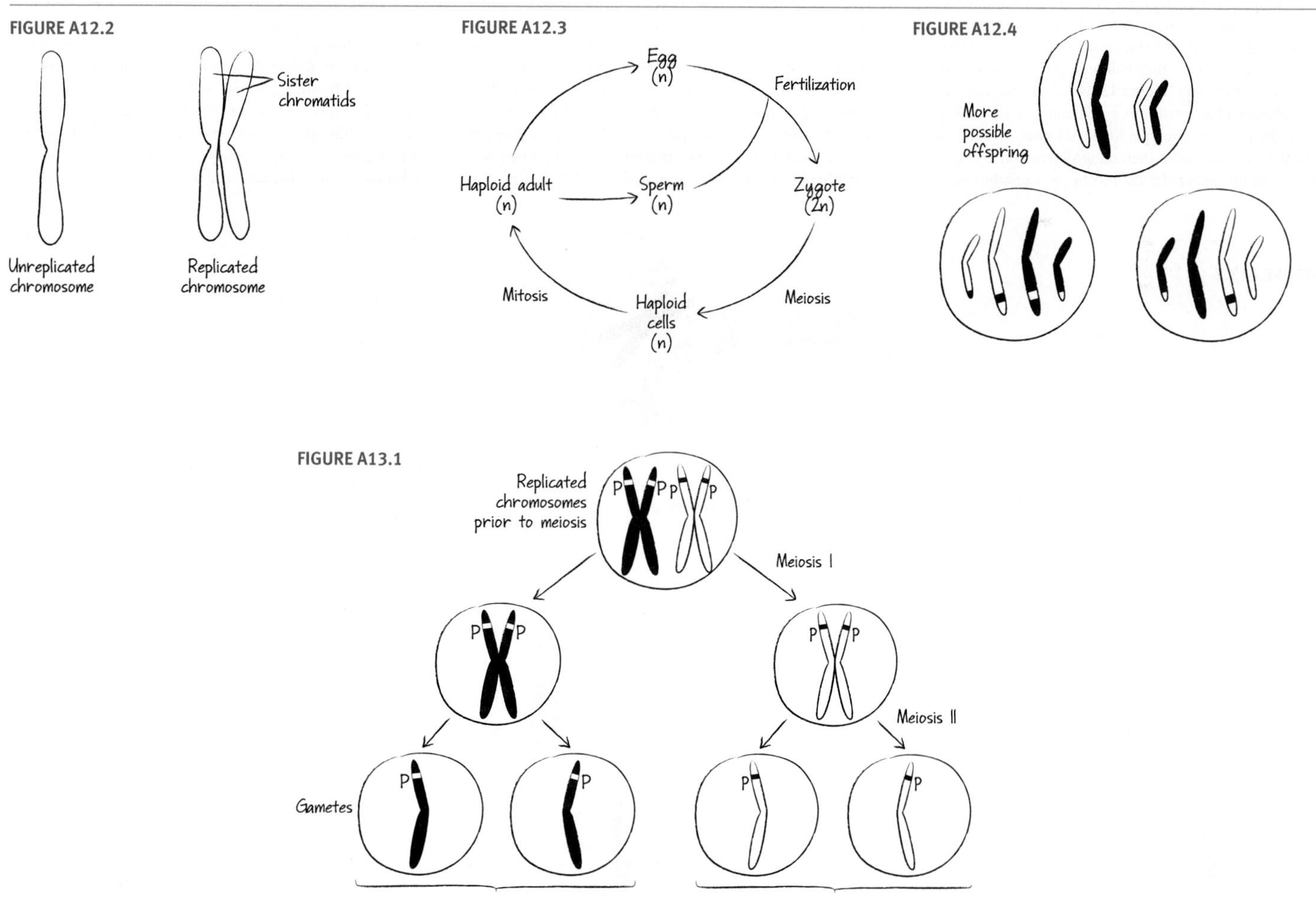

daughter cell. (2) [See Figure A13.2] Independent assortment occurs because homologous pairs line up randomly at the metaphase plate during metaphase I. The figure shows two alternative arrangements of homologs in metaphase I. As a result, it is equally possible for a gamete to receive the following four combinations of alleles: *YR, Yr, yR, yr*. **CYU p. 250** (1) The comb phenotype results from interactions between alleles at two different genes, not a single gene. Matings between rose and pea comb chickens produce F_2 offspring that may have a new combination of alleles, and thus new phenotypes.
(2) Kernel color in wheat is influenced by alleles at many different genes, not a single gene. F_2 offspring have a normal distribution of phenotypes, not a 3:1 ratio.

You Should Be Able To (YSBAT)

YSBAT p. 236 Punnett squares are an efficient way to predict offspring genotypes from parents of known genotype. The phenotype ratios are 1:1 round:wrinkled; the genotype ratios are 1:1 *Yy:yy*. **YSBAT p. 238** *AABb* → *AB* and *Ab*. *PpRr* → *PR, Pr, pR,* and *pr*. *AaPpRr* → *APR, APr, ApR, Apr, aPR, apR, apr,* and *aPr*.

Caption Questions and Exercises

Figure 13.3 An experiment is a failure if you didn't learn anything from it. That is not the case here. **Figure 13.4** No—the outcome (the expected offspring genotypes that the Punnett square generates) will be the same.
Figure 13.12 X^{WY}, X^{Wy}, X^{wY}, X^{wy}. **Figure 13.13** Random chance (or perhaps red-eyed, gray-bodied males don't survive well). **Figure 13.17** The gene colored orange is *ruby*; the gene colored blue is *miniature wings*.
Figure 13.22 If both parents are heterozygous (carriers), on average 1/4 of their children will be affected. If the allele is rare, then it is unlikely for the spouse of an affected (homozygous) parent to be a carrier—hence, children will not be affected—they will only get one allele from their affected parent. **Figure 13.24** All affected females would have all affected sons and daughters, no matter what the father's genotype. Affected fathers would transmit the allele to half of their daughters, so on average half would be affected if the mother was unaffected.

Summary of Key Concepts

KC 13.1 *B* and *b*. **KC 13.2** *BR, Br, bR,* and *br*, in equal proportions. **KC 13.3** The *B* and *b* alleles are located on different but homologous chromosomes, which separate into different daughter cells during meiosis I. The *BbRr* notation indicates that the *B* and *R* genes are on different chromosomes. As a result, the chromosomes line up independently of each other in metaphase of meiosis I. The *B* allele is equally likely to go to a daughter cell with *R* as with *r*; likewise the *b* allele is equally likely to go to a daughter cell with *R* as with *r*. **KC 13.4** Instead of two alleles on the same chromosome being transmitted together, crossing over separates them so they are transmitted independently of each other. They are no longer physically linked.

Genetics Questions

1. d; **2.** b; **3.** a; **4.** d; **5.** a; **6.** a; **7.** d; **8.** b; **9.** b; **10.** d

Genetics Problems

1. 3/4; 1/256 (see **BioSkills 13** in Appendix A); 1/2 (the probabilities of transmitting the alleles or having sons does not change over time). **2.** Your answer to the first three parts should conform to the F_1 and F_2 crosses diagrammed in Figure 13.5b, except that different alleles and traits are being analyzed. The recombinant gametes would be *Yi* and *yI*. Yes—there would be some individuals with round, green seeds and with wrinkled, yellow seeds. **3.** Cross 1: non-crested (*Cc*) × non-crested (*Cc*) = 14 non-crested (*C_*); 7 crested (*cc*). Cross 2: crested (*cc*) × crested (*cc*) = 22 crested (*cc*). Cross 3: non-crested (*Cc*) × crested (*cc*) = 7 non-crested (*Cc*); 6 crested (*cc*). Non-crested (*C*) is the dominant allele. **4.** This is a dihybrid cross that yields progeny phenotypes in a 9:3:3:1 ratio. Let *O* stand for the allele for orange petals and *o* the allele for yellow petals; let *S* stand for the allele for spotted petals and *s* the allele for unspotted petals. Start with the hypothesis that *O* is dominant to *o*, that *S* is dominant to *s*, that the two genes are found on different chromosomes so they assort independently, and that the parent individual's genotype is *OoSs*. If you do a Punnett square for the *OoSs* × *OoSs* mating, you'll find that progeny phenotypes should be in the observed 9:3:3:1 proportions. **5.** Let *D* stand for the normal allele and *d* for the allele responsible for Duchenne-type muscular dystrophy. The woman's family has no history of the disease, so her genotype is almost certainly *DD*. The man is not afflicted, so he must be *DY*. (The trait is X-linked, so he has only one allele; the "*Y*" stands for the Y chromosome.) Their children are not at risk. The man's sister could be a carrier, however—meaning she has the genotype *Dd*. If so, then half of the second couple's male children are likely to be affected. **6.** Your stages of meiosis should look like Figure 12.6, except with $2n = 4$ instead of $2n = 6$. The *A* and *a* alleles could be on the red and blue versions of the longest chromosome, and the *B* and *b* alleles could be on the red and blue versions of the smallest chromosomes. The places you draw them are the locations of the *A* and *B* genes, but each chromosome has only one allele. Each pair of red and blue chromosomes is a homologous pair. Sister chromatids bear the same allele (e.g., both sister chromatids of the long blue chromosomes might bear the *a* allele). Chromatids from the longest and shortest chromosomes are not homologous. To identify the events that result in the principles of segregation and independent assortment, see Figures 13.7 and 13.8 and substitute *A, a,* and *B, b* for *R, r* and *Y, y*. **7.** Half of their offspring should have the genotype iI^A and the type A blood phenotype. The other half of their offspring should have the genotype iI^B and the type B blood phenotype. Second case: the genotype and phenotype ratios would be 1:1:1:1 I^AI^B (type AB) : I^Ai (type A) : I^Bi (type B) : *ii* (type O). **8.** Because the children of Tukan and Valco had no eyes and smooth skin, you can conclude that the allele for eyelessness is dominant to eyes and the allele for smooth skin is dominant to hooked skin. *E* = eyeless, *e* = two eye sets, *S* = smooth skin, *s* = hooked skin. Tukan is *eeSS*; Valco is *EEss*. The children are all *EeSs*. Grandchildren with eyes and smooth skin are *eeS_*. Assuming that the genes are on different chromosomes, one-fourth of the children's gametes are *ee* and three-fourths are *S_*. So ¼ *ee* × ¾ *S_* × 32 = 6 children would be expected to have two sets of eyes and smooth skin. **9.** Although the mothers were treated as children by a reduction of dietary phenylalanine, they would have accumulated phenylalanine and its derivatives once they

FIGURE A13.2

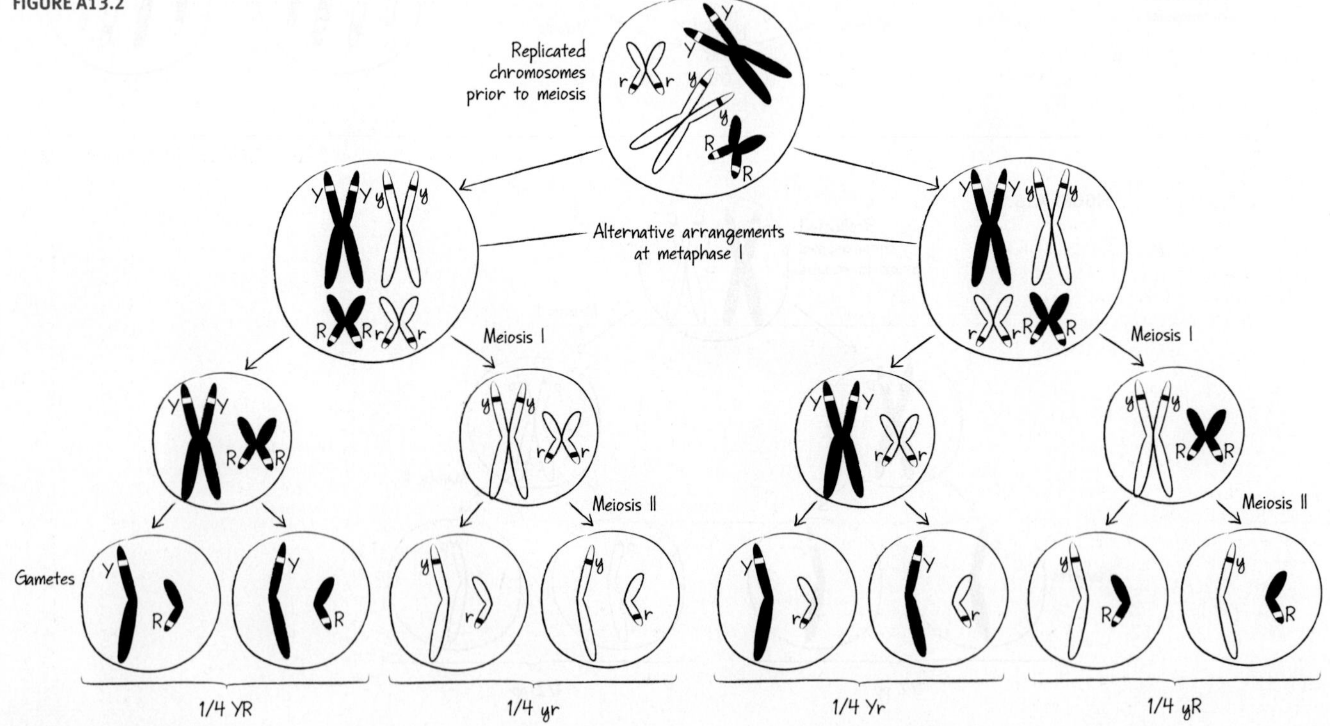

went off the low-phenylalanine diet as young adults. Children born of such mothers were therefore exposed to high levels of phenylalanine during pregnancy. For this reason, a low-phenylalanine diet is recommended for such mothers throughout the pregnancy. **10.** According to Mendel's model, palomino individuals should be heterozygous at the locus for coat color. If you mated palomino individuals, you would expect to see a combination of chestnut, palomino, and cremello offspring. If blending inheritance occurred, however, all of the offspring should be palomino. **11.** Because this is an X-linked trait, the father who has hemophilia could not have passed the trait on to his son. Thus, the mother in couple 1 must be a carrier and have passed the recessive allele on to her son, who is XY and affected. To educate a jury about the situation, you should draw what happens to the X and Y during meiosis, then make a drawing showing the chromosomes in couple 1 and couple 2, with a Punnett square showing how these chromosomes get passed to the affected and unaffected children. **12.** The curved-wing allele is autosomal recessive; the lozenge-eye allele is sex-linked (specifically, X-linked) recessive. Let *L* be the allele for long wings and *l* be the allele for curved wings; let X^R be the allele for red eyes and X^r the allele for lozenge eyes. The female parent is LlX^RX^r; the male parent is LlX^RY. **13.** Albinism indicates the absence of pigment, so let *b* stand for an allele that gives the absence of blue and *y* for an allele that gives the absence of yellow pigment. If blue and yellow pigment blend to give green, then both green parents are *BbYy*. The green phenotype is found in *BBYY*, *BBYy*, *BbYY*, and *BbYy* offspring. The blue phenotype is found in *BByy* or *Bbyy* offspring. The yellow phenotype is observed in *bbYY* or *bbYy* offspring. Albino offspring are *bbyy*. The phenotypes of the offspring should be in the ratio 9:3:3:1 as green:blue:yellow:albino. Two types of crosses yield *BbYy* F_1 offspring: *BByy* × *bbYY* (blue × yellow) and *BBYY* × *bbyy* (green × albino). **14.** Autosomal dominant.

CHAPTER 14

Check Your Understanding (CYU)

CYU p. 268 (1) DNA polymerase adds only nucleotides to the free 3′-OH on a strand. Primase synthesizes a short RNA sequence that provides the free 3′ end necessary for DNA polymerase to start working. (2) The helix is already opened and the downstream portions relaxed. **CYU p. 271** (1) Telomerase is not needed—bacterial chromosomes don't shorten after replication because they are circular. DNA polymerase can synthesize a complementary strand all the way around. (2) Otherwise there would be no template to synthesize the extension of a single DNA strand.

You Should Be Able To (YSBAT)

YSBAT p. 264 [See Figure A14.1] The new strands grow in opposite directions, each in $5' \rightarrow 3'$ direction. **YSBAT p. 266** Helicase, topoisomerase, single-strand DNA-binding proteins, primase, and DNA polymerase are all required for leading-strand synthesis. If any one of these proteins is nonfunctional, then DNA replication will not occur. **YSBAT p. 267** [See Figure A14.2] If DNA ligase was defective, then the leading strand would be continuous, and the lagging strand would have gaps in it where the Okazaki fragments had not been joined.

Caption Questions and Exercises

Figure 14.2 The lack of radioactive protein in the pellet (after centrifugation) is strong evidence; they could also make micrographs of infected bacterial cells before and after agitation. **Figure 14.3** 5′ TAG 3′. **Figure 14.5** The same two bands should appear, but the upper band (DNA containing only ^{14}N) should get bigger and darker and the lower band (hybrid DNA) should get smaller and lighter in color, as there is less and less heavy DNA in each succeeding generation. **Figure 14.13** As long as the RNA template could bind to the "overhanging" section of single-stranded DNA, any sequence could produce a longer strand. For example, 5′ CCCAUUCCC 3′ would work just as well. **Figure 14.15** They are much lower in energy. **Figure 14.17** Exposure to UV radiation can cause formation of thymine dimers. If thymine dimers are not repaired, they represent mutations. If such mutations occur in genes controlling the cell cycle, cells can grow abnormally, resulting in cancers.

Summary of Key Concepts

KC 14.1 Complementary bases hydrogen bond, and hydrophobic interactions occur between bases stacked inside the double helix. **KC 14.2** The bases added during DNA replication are shown in color type.

Original DNA: CAATTACGGA
GTTAATGCCT

Replicated DNA: CAATTACGGA
GTTAATGCCT
CAATTACGGA
GTTAATGCCT

KC 14.3 DNA polymerase synthesizes DNA by adding deoxyribonucleotides to the free 3′-OH group of an existing nucleotide sequence (primer), using single-stranded DNA as a template. Primase synthesizes short RNA sequences, using single-stranded DNA as a template. In contrast to DNA polymerase, primase can begin synthesis de novo, that is, without a primer. Telomerase adds deoxyribonucleotides to the unreplicated 3′ end of the lagging strand, using as the template a short RNA molecule that is associated with the enzyme. **KC 14.4** If errors in DNA aren't corrected, they represent mutations. When DNA repair systems fail, the mutation rate increases. As the mutation rate increases, the chance that one or more cell cycle genes will be mutated increases. Mutations in these genes often result in uncontrolled cell division, ultimately leading to cancer.

Test Your Knowledge

1. d; **2.** c; **3.** d; **4.** b; **5.** c; **6.** d

Test Your Understanding

1. Whether DNA or proteins were labeled. **2.** There are two replication forks at each point of origin—replication proceeds in both directions at the same time. **3.** On the lagging strand, DNA polymerase moves away from the replication fork. When helicase unwinds a new section of DNA, primase must build a new primer on the lagging strand (closer to the fork) and another polymerase molecule must begin synthesis at this point. On the leading strand, DNA polymerase moves in the same direction as helicase, so synthesis can continue without interruption from one primer (at the origin of replication). **4.** Telomerase binds to the 3′ overhang at the end of a chromosome. Once bound, it begins catalyzing the addition of deoxyribonucleotides to the overhang in the $5' \rightarrow 3'$ direction, lengthening the overhang. This allows primase, DNA polymerase, and ligase to catalyze the addition of deoxyribonucleotides to the lagging strand in the $5' \rightarrow 3'$ direction, restoring the lagging strand to its original length. **5.** First, DNA polymerases are highly selective in matching complementary bases. Second, DNA polymerase III proofreads by recognizing mismatched bases, cutting out the incorrect base, and replacing it with the correct base. Third, proteins of the mismatch repair system excise incorrect bases and insert the proper ones. **6.** DNA is a symmetrical double helix. A mismatch (e.g., A-G or C-T) or damage to a base pair (e.g., thymine dimer) will cause a distortion or kink in the molecule that is recognizable by repair enzymes. Because the bases on the two strands are complementary, repair enzymes are able to replace the damaged bases easily and accurately.

Applying Concepts to New Situations

1. [See Figure A14.3] Primase would not have to build primers on the leading and lagging strands, because DNA polymerase III would be able to use a "raw" single-stranded template. It is likely that telomerase would still

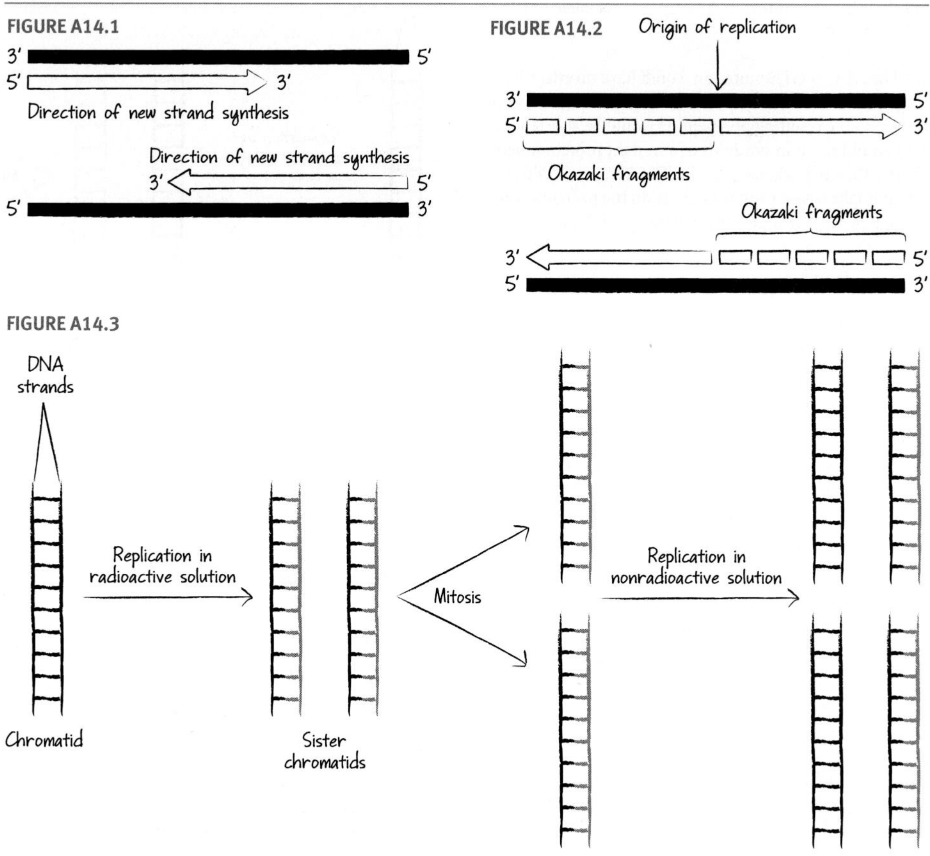

be required because, like primase, DNA polymerase would probably not have room to attach to the last end of the lagging strand, leaving a single-stranded overhang. **2.** (a) In **Figure A.14** below, the gray lines represent DNA strands containing radioactivity. Following replication in a radioactive solution, one DNA strand of each sister chromatid is radioactive. (b) [See **Figure A14.4**] **3.** (a) The double mutant of both *uvrA* and *recA* is most sensitive to UV light; the single mutants are in between; and the wild type is least sensitive. (b) The *recA* gene contributes more to UV repair through most of the UV dose levels. But at very high UV doses, the *uvrA* gene is somewhat more important than the *recA* gene. **4.** Many types of cancer arise because of mutations, especially those in genes involved in controlling cell growth. If chemicals that increase mutation rates in bacteria also increase mutation rates in humans, then the probability of mutations in cell control genes, and hence of cancer formation, also increases.

CHAPTER 15

Check Your Understanding (CYU)

CYU p. 282 If an error in transcription changes the sequence of bases in an mRNA, it may change the protein that is subsequently produced—and thus change the cell's phenotype. Changes in the sequence of other types of RNAs might also affect the phenotype. **CYU p. 285** (1) Remember that DNA and mRNA sequences are written in the $5' \rightarrow 3'$ direction unless indicated otherwise. Thus, to convert DNA triplets (codons) to mRNA codons, the DNA sequence is read "backwards" ($3' \rightarrow 5'$) and U (rather than T) is the base transcribed from A.

DNA codon	mRNA codon	Amino acid
ATA	UAU	Tyrosine
GTA	UAC	Tyrosine
TTA	UAA	Stop
GCA	CGU	Arginine

(2) The ATA → GTA mutation would have no effect on the protein. The ATA → TTA mutation introduces a stop codon, so the resulting polypeptide would be shortened. This would result in synthesis of a mutant protein much shorter than the original protein. The ATA → GCA mutation might have a profound effect on the protein's conformation because arginine's structure is different from tyrosine's.

You Should Be Able To (YSBAT)

YSBAT p. 285 (1) The codons in Figure 15.5 are translated correctly. (2) [See **Figure A15.1**] (3) There are many possibilities (just pick alternate codons for one or more of the amino acids); one is an mRNA sequence of (running $5' \rightarrow 3'$) GCG-AAC-GAU-UUC-CAG. To get the corresponding DNA sequence, write this sequence but substitute Ts for Us: GCG-AAC-GAT-TTC-CAG. This strand also runs in the $5' \rightarrow 3'$ direction. Now write the complementary bases, which will be in the $3' \rightarrow 5'$ direction: CGC-TTC-CTA-AAG-GTC. When this second strand is transcribed by RNA polymerase, it will produce the mRNA given with the proper $5' \rightarrow 3'$ orientation.

Caption Questions and Exercises

Figure 15.1 No, it could not make citrulline from ornithine without enzyme 2. Yes, it would no longer need enzyme 2 to make citrulline. **Figure 15.2** Many possibilities: strain of fungi used, exact method for creating mutants and harvesting spores to plant, exact growing conditions (temperature, light, recipe for growth medium—including concentration of supplemented amino acids), objective criteria for determining growth or no-growth. **Figure 15.5** $4 \times 4 \times 4 \times 4 = 256$ amino acids. A 4-base code is unlikely to evolve because only 20 amino acids are commonly used in the synthesis of proteins—there is no selective advantage for a code to be more complicated than necessary. **Figure 15.7** [See **Figure A15.1**] **Figure 15.8** The DNA sequence changes only at a single location, or point.

Summary of Key Concepts

KC 15.1 Not all genes code for polypeptides. A better statement might be something like, "one-gene, one RNA or polypeptide product." **KC 15.2** Transcription means to copy, and mRNA is a short-lived copy of the information in DNA. Translation means to change languages, and proteins are a different "chemical language" than nucleic acids (DNA and RNA). **KC 15.3** If a change in a DNA sequence doesn't change the amino acid specified by the corresponding mRNA codon, then the change in genotype doesn't change the phenotype. Many changes in the third positions of codons do not change the phenotype, because the genetic code is redundant. **KC 15.4** The new mutation produced light coat colors, which allowed beach-dwelling mice to be camouflaged and survive better. If the same mutation occurred in a woodland-dwelling population, the light-colored individuals would probably be spotted by predators and eaten.

Test Your Knowledge

1. d; **2.** a; **3.** d; **4.** a; **5.** d; **6.** b

Test Your Understanding

1. In Morse code, different combinations of dots and dashes code for the complexity of the English language. In the genetic code, different combinations of bases code for the complexity of proteins in the cell. **2.**

$$\text{Substrate 1} \xrightarrow{A} \text{Substrate 2} \xrightarrow{B} \text{Substrate 3} \xrightarrow{C}$$

$$\text{Substrate 4} \xrightarrow{D} \text{Substrate 5} \xrightarrow{E} \text{BiologicalSciazine}$$

Substrate 3 would accumulate. Hypothesis: The individuals have a mutation in the gene for enzyme D. **3.** They supported an important prediction of the hypothesis: Losing a gene (via mutation) resulted in loss of an enzyme. **4.** It puts the three-nucleotide codons out of register. If the first base is deleted from the sequence ATG-CGA-GAC-TTA, then the reading frame becomes TGC-GAG-ACT-TA. **5.** In a triplet code, addition or deletion of 1–2 bases disrupts the reading frame "downstream" of the mutation site(s), resulting in a dysfunctional protein. But addition or deletion of 3 bases restores the reading frame—the normal sequence is disrupted only between the first and third mutation. The resulting protein is altered, but may still be able to function normally. Only a triplet code would show these patterns. **6.** A point mutation changes the nucleotide sequence of an existing allele, creating a new one, so it always changes the genotype. But because the genetic code is redundant, some point mutations do not change the resulting mRNA codons and thus do not change the protein product.

Applying Concepts to New Situations

1. [See **Figure A15.2**] **2.** Every copying error would result in a mutation that would change the amino acid sequence of the protein and would likely affect its function. **3.** Before the central dogma was understood, DNA was known to be the hereditary material, but no one knew how particular sequences of bases resulted in the production of RNA and protein products. The central dogma clarified

FIGURE A14.4

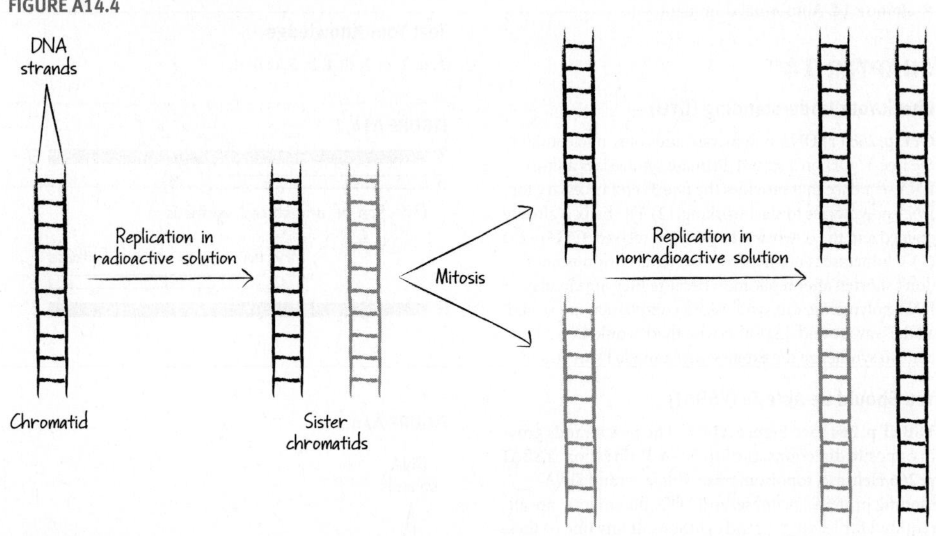

FIGURE A15.1

mRNA sequence:
5′ AUG-CCC-CUG-GAG-GGG-GUU-AGA-CAU 3′

Amino acid sequence:
Met-Pro-Leu-Glu-Gly-Val-Arg-His

FIGURE A15.2

Bottom DNA strand:
5′ AACTT-TAC(start)-GGG-CAA-ACC-ACT-AGC-CCA-ATG-TCG-ATC(stop)-AGTTTC 3′

mRNA sequence:
5′ AUG-CCC-GUU-UGG-AGA-UCG-GGU-UAC-AGC-UAG 3′

Amino acid sequence:
Met-Pro-Val-Trp-Arg-Ser-Gly-Tyr-Ser

how genotypes produce phenotypes. **4.** A triplet code that overlapped by one base might work as follows:

```
5′ AUGUUACGGAAUUGA 3′
     GUU
       AUC
         CGG
           GAA
             AUU
               UGA
```

CHAPTER 16

Check Your Understanding (CYU)

CYU p. 293 (1) Different types of sigma proteins have different amino acid sequences and overall structures, so they are able to bind to stretches of DNA that differ in their base sequence. (2) NTPs are required because the three phosphate groups raise the monomer's potential energy enough to make the polymerization reaction exergonic. **CYU p. 295** (1) The subunits contain both RNA (the "ribonucleo" in the name) and proteins. (2) The cap and tail protect mRNAs from degradation and facilitate translation. **CYU p. 304** E is for Exit—the site where empty tRNAs are ejected; P is for Peptidyl (or peptide bond)—the site where peptide bond formation takes place; A is for Aminoacyl—the site where aminoacyl tRNAs enter.

You Should Be Able To (YSBAT)

YSBAT p. 299 (1) The amino acid attaches on the top left of the L-shaped structure. (2) The anticodon is antiparallel in orientation to the mRNA codon, and contains the complementary bases.

Caption Questions and Exercises

Figure 16.1 RNA is synthesized in the 5′ → 3′ direction; the DNA template is "read" 3′ → 5′. **Figure 16.5** There would be no loops—the molecules would match up exactly. **Figure 16.12** If the amino acids stayed attached to the tRNAs, the gray line in the graph would stay high and the green line low. If the amino acids were transferred to some other cell component, the gray line would decline but the green line would be low.

Summary of Key Concepts

KC 16.1 The new sequence would probably not bind sigma or basal transcription factors as effectively as the original sequence. If so, transcription rate at that gene would probably decrease. **KC 16.2** Bacterial genes lack introns, so their RNA products do not require splicing. And because bacterial mRNAs are translated immediately, they do not need a cap and tail. **KC 16.3** These RNAs transfer amino acids to growing polypeptides; they transfer information in nucleic acids (an mRNA codon) to a protein product. **KC 16.4** An aminoacyl tRNA synthetase catalyzes the addition of an amino acid to a tRNA, forming an aminoacyl tRNA. **KC 16.5** One possible concept map is shown below [see **Figure A16.1**].

Test Your Knowledge

1. c; **2.** c; **3.** d; **4.** c; **5.** c; **6.** d

Test Your Understanding

1. Basal transcription factors bind to promoter sequences in eukaryotic DNA and facilitate the positioning of RNA polymerase. As part of the RNA polymerase holoenzyme, sigma binds to a promoter sequence in bacterial DNA and initiates transcription by the core enzyme. **2.** If the wobble rules did not exist, it would take one tRNA for each amino-acid-specifying codon in the genetic code. This is inefficient. The wobble rules allow a single tRNA to match several mRNA codons, reducing the inefficiency caused by redundancy in the genetic code. **3.** Eukaryotic mRNAs contain introns which must be removed by splicing. Splicing takes place in the nucleus and is accomplished by a large complex called a spliceosome, composed of many snRNPs. **4.** After a peptide bond forms between the polypeptide and the amino acid held by the tRNA in the A site, the ribosome moves down the mRNA. As it does, an empty tRNA leaves the E site. The now-empty tRNA that was in the P site enters the E site; the tRNA holding the polypeptide chain moves from the A site to the P site, and a new aminoacyl tRNA enters the A site. **5.** The ribosome's active site is made up of RNA, not protein. **6.** The separation allows the aminoacyl tRNA to fit into the ribosome correctly, such that the anticodon binds to the mRNA and the amino acid fills the active site.

Applying Concepts to New Situations

1. Ribonucleases degrade mRNAs that are no longer needed by the cell. If an mRNA for a hormone that increased heart rate were never degraded, the hormone would be produced continuously and heart rate would stay elevated—a dangerous situation. **2.** Cordycepin triphosphate has a triphosphate group on the 5′ carbon, but no 3′-OH group. If RNA were built 3′ → 5′, then the incoming nucleoside triphosphates would have to have an OH group on the 3′ carbon in order for a phosphodiester bond to be formed. Since cordycepin lacks a 3′-OH, it could not be added to the 5′ end of an RNA chain growing in the 3′ → 5′ direction. **3.** The regions most crucial to the ribosome's function should be the most highly conserved: the active site, the E-, P-, and A-sites, and the site where mRNAs initially bind. Other regions should be more variable among species. **4.** The most likely locations are one of the grooves or channels where RNA, DNA, and ribonucleotides move through the enzyme—plugging one of them would prevent transcription.

CHAPTER 17

Check Your Understanding (CYU)

CYU p. 314 (1) The genes for metabolizing lactose should be expressed only when lactose is available. (2) [See **Figure A17.1**] **CYU p. 316** (1)

FIGURE A16.1

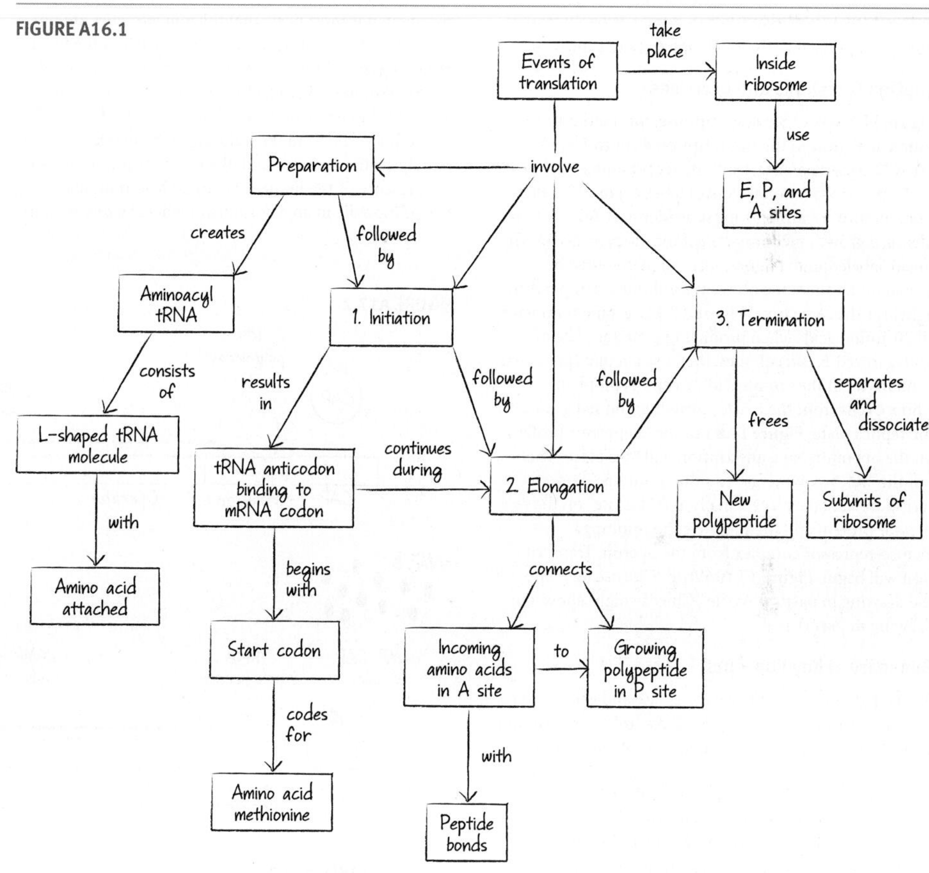

FIGURE A17.1

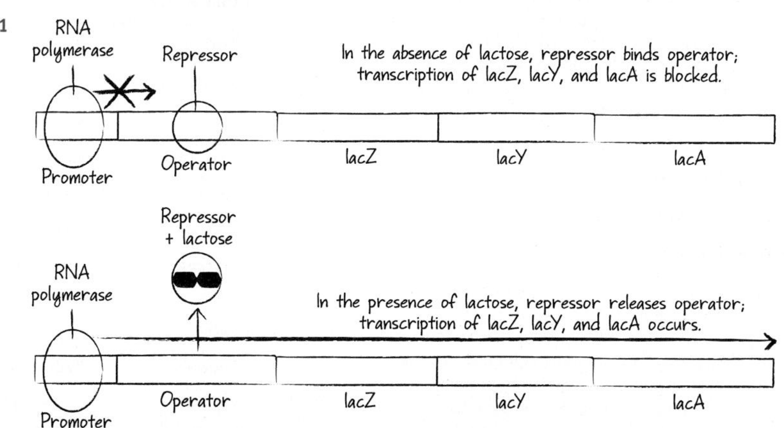

[See Figure A17.2] (2) Two conditions must be met for the *lac* operon to be expressed at a high rate. First, glucose levels must be low. Absence of glucose exerts positive control, because the cAMP-CAP complex binds to the CAP site and increases transcription rate. Second, lactose must be present. Lack of lactose exerts negative control, because the repressor remains attached to the operator and reduces transcription rate.

You Should Be Able To (YSBAT)

YSBAT p. 310 [See Table A17.1] **YSBAT p. 312** *lacZ* codes for the β-galactosidase enzyme, which breaks the disaccharide lactose into glucose and galactose. *lacY* codes for the lactose permease enzyme, which transports lactose into the bacterial cell. *lacI* codes for a protein that shuts down production of the other *lac* products. When lactose is absent, the *lacI* product prevents transcription. This is logical because there is no reason for the cell to make β-galactosidase and lactose permease if there is no lactose to metabolize. But when lactose is present, it interacts with *lacI* in some way so that *lacZ* and *lacY* are induced (their transcription can occur). When lactose is present, the enzymes that metabolize it are expressed.

Caption Questions and Exercises

Figure 17.1 Write "Slowest response, most efficient resource use" next to the transcriptional control label. Write "Fastest response, least efficient resource use" next to the post-translational control label. **Figure 17.2** Plates from all three treatments must be identical and contain identical growth medium, except for the presence of the sugars labeled in the figure. Also, all plates must be grown under the same physical conditions (temperature, light) for the same time. **Figure 17.3** Use a medium with all 20 amino acids when producing a master plate of mutagenized *E. coli* colonies, then use a replica plate that contains all of the amino acids except tryptophan. Choose cells from the master plate that did *not* grow on the replica plate. **Figure 17.8** Put the "Repressor protein" on the operator. No transcription will take place. Then put the "RNA polymerase" on the promoter. No transcription will take place. Finally, put "lactose" on the repressor protein and then remove the resulting lactose-repressor complex from the operon. Transcription will begin. **Figure 17.10** Write "Glucose low" above the drawing in part (a). Write "Glucose high" above the drawing in part (b).

Summary of Key Concepts

KC 17.1 Production of β-galactosidase and galactosidase permease are under transcriptional control—transcription depends on the action of regulatory proteins. The activity of the repressor and CAP—the regulatory proteins—are under post-translational control. **KC 17.2** After a dessert, glucose concentrations in your gut should be high, causing the *lac* operon in *E. coli* to shut down. After drinking milk, glucose concentrations should be low and lactose concentrations high, causing the *lac* operon in *E. coli* cells to be highly expressed. **KC 17.3** (1) The operator is the parking brake; the repressor locks it in place, and the inducer releases it. (2) The CAP-binding site is the gas pedal; the CAP-cAMP complex is a heavy foot on it.

Test Your Knowledge

1. b; **2.** d; **3.** b; **4.** c; **5.** a; **6.** c

Test Your Understanding

1. The glycolytic enzymes are always needed in the cell, because they are required to produce ATP, and ATP is always needed. **2.** Positive control means that a regulatory protein, when present, causes transcription to increase. Negative control means that a regulatory protein, when present, prevents transcription. **3.** The combination of positive and negative control allows *E. coli* to shut down the *lac* operon when glucose is present or lactose absent, but activate it when glucose is absent and lactose present. If only negative control occurred, the operon would be activated when glucose is present. If only positive control occurred, the operon would be activated when lactose is absent. **4.** cAMP levels rise when glucose is low. If glucose is low, the cell is in danger of starving unless it can switch to another food source. **5.** CAP acts like a receptor for cAMP—when cAMP binds to it, its activity changes. **6.** Unlike lactose, glucose can enter the glycolytic pathway directly—without being converted to another molecule. Less ATP and fewer enzymes are needed to acquire and use glucose as an energy source.

Applying Concepts to New Situations

1. Set up cultures with individuals that all come from the same colony of toluene-tolerating bacteria. Half of the cultures should have toluene as the only source of carbon; half should have glucose or another common source of carbon as well as toluene. Cells will grow in both cultures if they are able to use toluene as a source of carbon. **2.** Because toluene is not common in the environment, it is most likely that the toluene genes are under negative control. That is, the presence of toluene—by removing some brake on expression of the enzyme's genes—would trigger production of whatever enzymes are needed to break down toluene. **3.** The *LacI*^{S} mutants would not respond to the inducer, so the repressor would remain on the operator and block transcription even in the presence of the inducer, lactose. These mutants would soon die in an environment where lactose was the only sugar present. **4.** Cells with functioning β-galactosidase are blue; cells that are not producing normal β-galactosidase are white. The sequence of the β-galactosidase gene in structural mutants would differ from the sequence of the wild-type gene, whereas the sequence of the β-galactosidase gene in regulatory mutants would be the same as the sequence of the wild-type gene.

CHAPTER 18

Check Your Understanding (CYU)

CYU p. 323 (1) The basic double helical structure of DNA is the same, but the DNA in eukaryotic cells is complexed with histones to form chromatin, which is tightly packed into 30-nm fibers attached to scaffolding proteins. In bacteria, DNA is associated with histone-like proteins but is not organized into complex structures. (2) Addition of acetyl or methyl groups to histones can cause chromatin to open or close. Different patterns of acetylation or methylation will determine which genes in muscle cells versus brain cells can be transcribed and which are not available for transcription. **CYU p. 326** (1) Bacterial regulatory sequences are found close to the promoter; eukaryotic regulatory sequences can be close to the promoter or far from it. Bacterial regulatory proteins interact directly with RNA polymerase to initiate or prevent transcription; eukaryotic regulatory proteins influence transcription by altering chromatin structure or binding to the basal transcription complex through mediator proteins. (2) Certain regulatory proteins open

FIGURE A17.2

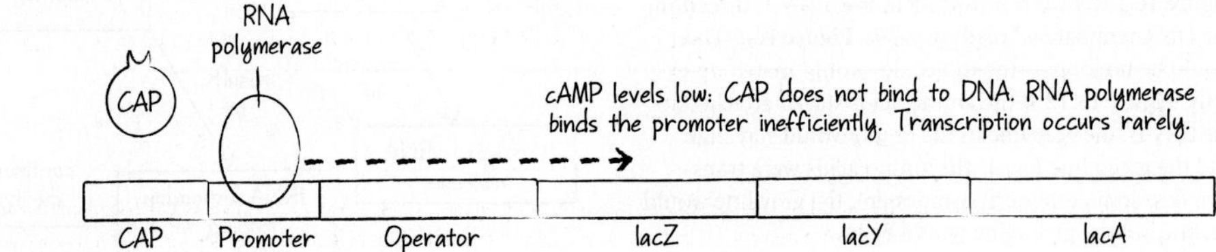

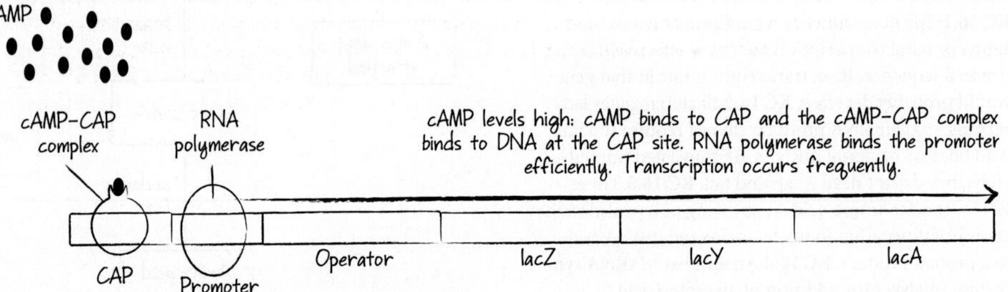

TABLE A17.1

Name	Function
lacZ	Gene for β-galactosidase
lacY	Gene for galactosidase permease
Operator	Binding site for repressor
Promoter	Binding site for RNA polymerase
CAP site	Binding site for CAP–cAMP
Repressor	Shuts down transcription
CAP	Increases transcription rate when bound to cAMP
cAMP	Produced when glucose is low; binds to CAP and activates it
Lactose	Binds to repressor and stimulates transcription (removes negative control)
Glucose	At low concentration, increases transcription (via increased production of cAMP)

chromatin at muscle- or brain-specific genes, and then activate or repress the transcription of cell-type-specific genes. Muscle-specific genes are expressed only if muscle-specific regulatory proteins are produced and activated. **CYU p. 330** (1) It became clear that a single gene can code for multiple products instead of a single one. (2) miRNAs interfere with mRNAs by targeting them for destruction or preventing them from being translated. **CYU p. 333** (1) Many different types of mutations can disrupt control of the cell cycle and initiate cancer. These mutations can affect any of the six levels of control over gene regulation outlined in Figure 18.1. (2) The p53 protein is responsible for shutting down the cell cycle in cells with damaged DNA. If the protein does not function, then cells with damaged DNA—and thus many mutations—continue to divide. If these cells have mutations in genes that regulate the cell cycle, then they may continue to divide in an uncontrolled fashion.

You Should Be Able To (YSBAT)

YSBAT p. 326a The basal transcription factors found in muscle and nerve cells are similar or identical; the regulatory transcription factors found in each cell type are different. **YSBAT p. 326b** DNA forms loops when distant regulatory regions, such as silencers and enhancers, are brought close to the promoter through binding of regulatory transcription factors to the mediator complex. **YSBAT p. 329** Alternative splicing does not occur in bacteria because bacterial genes do not contain introns—in bacteria, each gene codes for a single product. Alternative splicing is part of step 3 in Figure 18.1. **YSBAT p. 330** Step 4 and step 5. RNA interference either (1) decreases the life span of mRNAs or (2) inhibits translation.

Caption Questions and Exercises

Figure 18.4 Acetylation of histones decondenses chromatin and allows transcription to begin, so HATs are elements in positive control. Deacetylation condenses the chromatin and inactivates transcription, so HDACs are elements in negative control. **Figure 18.6** This treatment allowed them to measure the normal amount of antibody mRNA produced, for comparison with the other treatments. **Figure 18.7** A typical eukaryotic gene usually contains introns and is regulated by multiple enhancers. Bacterial operons lack introns and enhancers. The promoter-proximal element found in some eukaryotic genes is comparable to the CAP binding site in the *lac* operon of bacteria. Bacterial operons have a single promoter but code for more than one protein; eukaryotic genes code for a single product.

Summary of Key Concepts

KC 18.1 Because eukaryotic RNAs are not translated as soon as transcription occurs, it is possible for RNA processing to occur, which creates variation in the mRNAs produced from a primary RNA transcript and their life span. **KC 18.2** The default state of eukaryotic genes is "off," because the highly condensed state of the chromatin makes DNA unavailable to RNA polymerase. **KC 18.3** [See Figure A18.1] **KC 18.4** Alternative splicing makes it possible for a single gene to code for multiple products. **KC 18.5** Cancer is a disease that results from accumulated mutations. Radiation damages DNA and increases the rate of mutation. Mutations accumulate as cells undergo repeated divisions, so are more common in older people.

Test Your Knowledge

1. c; **2.** a; **3.** d; **4.** d; **5.** b; **6.** d

Test Your Understanding

1. (a) Enhancers and the CAP site are similar because both are sites in DNA where regulatory proteins bind. They are different because enhancers generally are located at great distances from the promoter, whereas the CAP site is located near the promoter. (b) Promoter-proximal elements and the *lac* operon operator are both regulatory sites in DNA located close to the promoter. (c) Basal transcription factors and sigma are proteins that must bind to the promoter before RNA polymerase can initiate transcription. They differ because sigma is part of the RNA polymerase holoenzyme, while the basal transcription complex recruits RNA polymerase to the promoter. **2.** If changes in the environment cause changes in how spliceosomes function, then the RNAs and proteins produced from a particular gene could change in a way that helps the cell cope with the new environmental conditions. **3.** (a) Enhancers and silencers are both regulatory sequences located at a distance from the promoter. Enhancers bind regulatory transcription factors that activate transcription; silencers bind regulatory proteins that shut down transcription. (b) Promoter-proximal elements and enhancers are both regulatory sequences that bind positive regulatory transcription factors. Promoter-proximal elements are located close to the promoter; enhancers are far from the promoter. (c) Transcription factors bind to regulatory sites in DNA; the mediator complex does not bind to DNA but instead forms a bridge between regulatory transcription factors and basal transcription factors. **4.** RNA interference happens when complementary base pairing occurs between a small single-stranded miRNA and a single-stranded mRNA. **5.** Certain chemical modifications of histones mark sections of chromatin for transcription activation or repression. These modifications are passed to daughter cells when cells divide. A certain array of chromatin modifications or "histone codes" is associated with the production of muscle- or nerve-specific proteins. **6.** Lack of functional p53 protein prevents damaged cells from being destroyed or shut down (which would prevent their continuing to divide). As a result, cells with damaged DNA and thus many mutations continue to divide, possibly initiating cancerous growth.

Applying Concepts to New Situations

1. In eukaryotic cells, histones are involved in the basic folding of DNA molecules in chromatin. Mutations in histones are likely to interfere with this folding, and thus the expression of large numbers of genes. Individuals with mutant forms of histones are likely to die as a result—meaning that histone structure should be highly conserved over time. **2.** Tissues that divide more often must replicate their DNA more often; thus they are more susceptible to mutation. Because the cell cycle is accelerated in these cell types, it is logical to expect that control over the cell cycle could easily be lost. **3.** You could treat a culture of DNA-damaged cells with a drug that stops transcription and then compare them with untreated DNA-damaged cells. If transcriptional control regulates p53 levels, then the p53 level would be lower in the treated cells versus control cells. If control of p53 levels is post-translational, then in both cultures the p53 level would be the same. *Other approaches:* add labeled NTPs to damaged cells and see if they are incorporated into mRNAs for p53; or add labeled amino acids to damaged cells and see if they are incorporated into completed p53 proteins; or add labeled phosphate groups and see if they are added to p53 proteins. **4.** The miRNA would bind to viral RNA inside infected cells and prevent the viral RNA from working, thus stopping the infection.

THE BIG PICTURE: GENETIC INFORMATION

Check Your Understanding (CYU), p. 336

1. Star = DNA, mRNA, proteins. **2.** RNA "is reverse transcribed by" reverse transcriptase "to form" DNA. **3.** E = splicing, etc.; E = meiosis and sexual reproduction (along with their links). **4.** Chromatin "makes up" chromosomes; independent assortment and recombination "contribute to" high genetic diversity.

CHAPTER 19

Check Your Understanding (CYU)

CYU p. 344 (1) When the endonuclease makes a staggered cut in a palindromic sequence and the strands separate, the single-stranded bases that are left will bind ("stick") to the single-stranded bases left where the endonuclease cut the same palindrome at a different location. (2) The word probe means to examine thoroughly. A DNA probe "examines" a large set of sequences thoroughly, and binds to one—the one that has a complementary base sequence. **CYU p. 346** (1) Denaturation makes DNA single stranded so the primer can bind to the sequence during the annealing step. Once the primer is in place, *Taq* polymerase can synthesize the rest of the strand during the extension step. It is a "chain reaction" because the products of each reaction cycle are used in the next reaction cycle—this is why the number of copies doubles in each cycle. (2) One of many possible

FIGURE A18.1

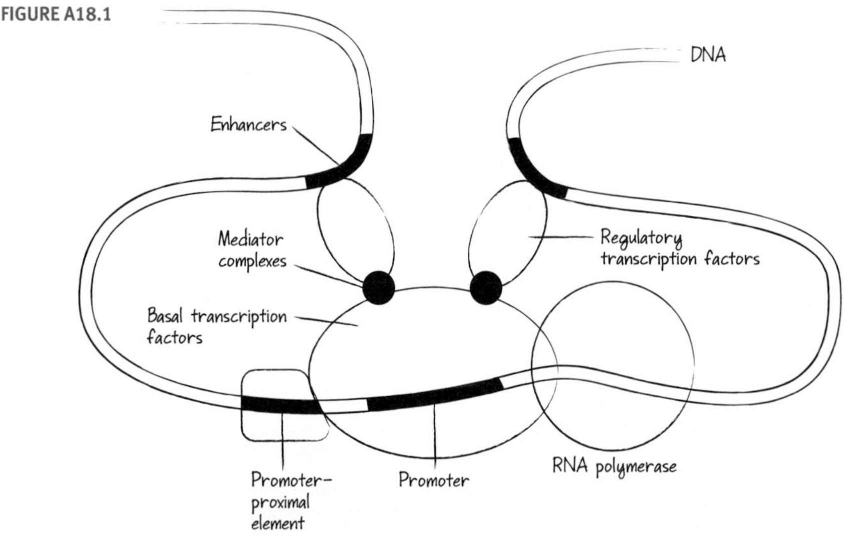

answers is shown below [see Figure A19.1]. **CYU p. 348** If ddNTPs were present at high concentration, they would almost always be incorporated—meaning that only fragments from the first complementary base in the sequence would be produced. **CYU p. 351** Start with a genetic map with as many polymorphic markers as possible. Get the genotype at these markers for a large number of individuals who have the same type of alcoholism—one that is thought to have a genetic component—as well as a large number of unaffected individuals. Look for particular versions of a marker that is almost always found in affected individuals. Genes that contribute to a predisposition to alcoholism will be near that marker.

You Should Be Able To (YSBAT)

YSBAT p. 342a Isolate the DNA and cut it into small fragments with EcoR1, which leaves sticky ends. Cut copies of a plasmid or other vector with EcoR1. Mix the fragments and plasmids under conditions that promote complementary base pairing by sticky ends of fragments and plasmids. Use DNA ligase to catalyze formation of phosphodiester bonds and seal the sequences. **YSBAT p. 342b** The probe must be single stranded so that it will bind by complementary base pairing to the target DNA, and it must be labeled so that it can be detected. The probe will only base pair with fragments that include a sequence complementary to the probe's sequence. A probe with the sequence 5′ AATCG 3′ will bind to the region of the target DNA that has the sequence 5′ AATCG 3′ as shown below:

```
        5′ AATCG 3′
3′ TCCGGTTAGCATTACCATTTT 5′
```

YSBAT p. 348 Using a map with many markers makes it more likely that there will be one very tightly linked to the gene of interest—meaning that a version of the marker will almost always be associated with the phenotype you are tracking.

Caption Questions and Exercises

Figure 19.4 No—many times, because there were many copies of each type of mRNA present in the pituitary cells and many pituitary cells were used to prepare the library. **Figure 19.7** The polymerase will begin at the 3′ end of each primer. On the top strand in part (b), it will move to the left; on the bottom strand it will move to the right. As always, synthesis is in the 5′ to 3′ direction. **Figure 19.10** The insertion will probably disrupt the gene and have serious consequences for the cell and potentially the individual.

Summary of Key Concepts

KC 19.1 When "sticky ends" anneal, the recombinant product would not be stable—it would tend to break apart because no phosphodiester bond would form between the DNA fragments. **KC 19.2** PCR advantages: It is relatively fast and easy, and can amplify a DNA sequence that is rare in the sample. PCR disadvantage: It requires knowledge of sequences on either side of the target gene, so primers can be designed. Plasmid advantages: No knowledge of the sequence is required. Plasmid disadvantages: It is slower and technically more difficult than PCR. **KC 19.3** The length of each fragment is dictated by where a ddNTP was incorporated into the growing strand, and each ddNTP corresponds to a base on the template strand. Thus, the sequence of fragment sizes corresponds to the sequence on the template DNA. **KC 19.4** Both affected and unaffected individuals have the same marker, so you have no way of knowing that it is close to the gene you are interested in. **KC 19.5** Genes can be inserted into the Ti plasmids, and the recombinant plasmids transferred to plant cells.

Test Your Knowledge

1. c; **2.** b; **3.** d; **4.** a; **5.** b; **6.** d

Test Your Understanding

1. When a restriction endonuclease cuts a "foreign gene" sequence and a plasmid, the same sticky ends are created on the excised foreign gene and the cut plasmid. After the sticky ends on the foreign gene and the plasmid anneal, DNA ligase catalyzes closure of the DNA backbone, sealing the foreign gene into the plasmid DNA. [See **Figure A19.2**] **2.** Vectors hold a piece of foreign DNA and carry it into an intact host cell. A "perfect" vector (1) has well-characterized restriction endonuclease sites, (2) can be inserted into host cells easily, (3) expresses recombinant genes in the host cell in a controlled way, and (4) doesn't damage the host cell. **3.** A cDNA library is a collection of complementary DNAs made from all the mRNAs present in a certain group of cells. A cDNA library from a human nerve cell would be different from one made from a human muscle cell, because nerve cells and muscle cells express many different, cell type–specific genes. **4.** Genetic markers are genes or other loci that have known locations in the genome. When these locations are diagrammed, they represent the physical relationships among landmarks—in other words, they form a map. **5.** The genes for making β-carotene need to be expressed in the endosperm. Adding an endosperm-specific promoter adjacent to the genes ensures that they are expressed in the endosperm. In general, including specific promoter and enhancer sequences with recombinant genes helps researchers control where and when the recombinant genes are expressed. **6.** PCR and cellular DNA synthesis are similar in the sense of producing copies of a template DNA. Both rely on primers and DNA polymerase. The major difference between the two is that PCR copies only a specific target sequence, while the entire genome is copied during cellular DNA synthesis.

Applying Concepts to New Situations

1. You could use a computer program to identify likely promoter sequences in the sequence data, and then look for sequences just downstream that have an AUG start codon and a stop codon about 1500 base pairs apart. **2.** Yes—you would transform cells that are the precursors to sperm or eggs, so that the recombinant DNA is found in mature sperm or eggs and passed on to offspring. **3.** In both techniques, researchers use an indicator to identify either a gene of interest or a colony of bacteria with a particular trait. The problem is the same—picking one particular thing (a certain gene or a cell with a particular mutation) out of a large collection. **4.** (a) Primer 1b binds to the top right strand and would allow DNA polymerase to synthesize the top strand across the target gene. Primer 1a, however, binds to the upper left strand and would allow DNA polymerase to synthesize the upper strand *away* from the target gene. Primer 2a binds to the bottom left strand and would allow DNA polymerase to synthesize the bottom strand across the target gene. Primer 2b, however, binds to the bottom right strand and would allow DNA polymerase to synthesize away from the target gene. (b) She could use primer 1b with primer 2a.

CHAPTER 20

Check Your Understanding (CYU)

CYU p. 364 (1) Parasites don't need genes that code for enzymes required to synthesize molecules they acquire from their hosts. (2) If two closely related species inherited the same gene from a common ancestor, the genes should be similar. But if one species acquired the same gene from a distantly related species via lateral gene transfer, then the genes should be much less similar. **CYU p. 370** (1) Unequal crossover and mistakes in DNA synthesis are so common in simple tandem repeats that new alleles—consisting of the same chromosome region with different repeat numbers—are created almost every generation, producing a molecular fingerprint. (2) Unequal crossover occurs when homologous chromosomes

FIGURE A19.1

```
5′ CATGACTATTACGTATCGGGTACTATGCTATCGATCTAGCTACGCTAGCT 3′
3′ GTACTGATAATGCATAGCCCATGATACGATAGCTAGATCGATGCGATCGA 5′
```

Probe #1, which will anneal to the 3′ end of the top strand:

```
5′ AGCTAGCGTAGCTAGATCGAT 3′
```

Probe #2, which will anneal to the 3′ end of the bottom strand:

```
5′ CATGACTATTACGTATCGGGG 3′
```

The primers will bind to the separated strands of the parent DNA sequence as follows:

```
5′ CATGACTATTACGTATCGGGTACTATGCTATCGATCTAGCTACGCTAGCT 3′
                                 3′ TAGCTAGATCGATGCGATCGA 5′

5′ CATGACTATTACGTATCGGG 3′
3′ GTACTGATAATGCATAGCCCATGATACGATAGCTAGATCGATGCGATCGA 5′
```

FIGURE A19.2

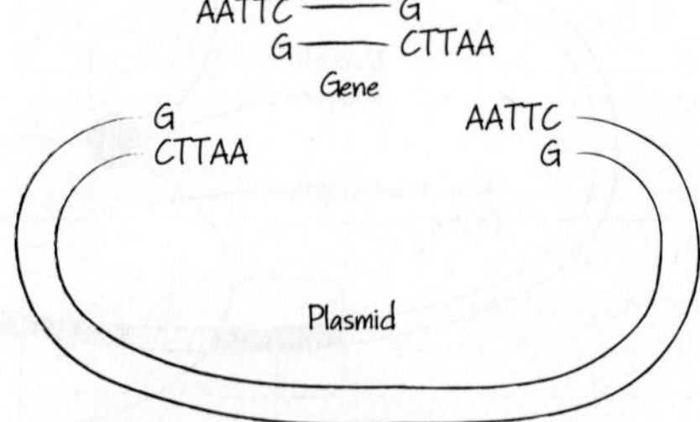

Sticky ends on cut plasmid and gene to be inserted are complementary and can base pair

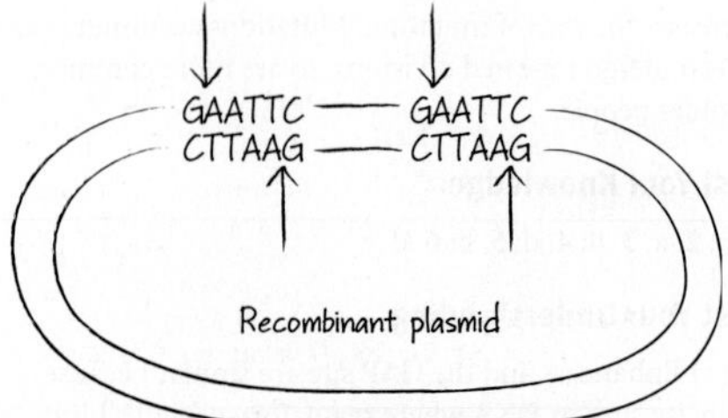

DNA ligase forms four phosphodiester bonds between gene and plasmid DNA at the points indicated by arrows

align incorrectly during synapsis. When crossing over occurs within a misaligned section of chromosome, one homolog ends up with less DNA and the other ends up with more DNA. The extra DNA is made up of sequences from the other chromosome that are duplicates of the DNA on the original strand.

You Should Be Able To (YSBAT)

YSBAT p. 360 If no overlap occurred, there would be no way of ordering the fragments correctly. You would only be able to put fragments in random order—not the correct order. **YSBAT p. 371** Start with a microarray containing exons from a large number of human genes. Isolate mRNAs from brain tissue and liver tissue, and make labeled cDNAs from each. Probe the microarray with each type of probe, and record where binding occurs. Binding events identify exons that are transcribed in each type of tissue. Compare the results to identify genes that are expressed in brain but not liver, or liver but not brain.

Caption Questions and Exercises

Figure 20.2 Shotgun sequencing is based on fragmenting the genome into many small pieces. **Figure 20.3** Because the mRNA codons that match up with each strand are oriented in the 5′ to 3′ direction. **Figure 20.8** Chromosome 1: $\psi\beta2$-ϵ-Gγ-Aγ-$\psi\beta1$-δ-$\psi\beta2$-ϵ-Gγ-Aγ-$\psi\beta1$-δ-β Chromosome 2: β

Summary of Key Concepts

KC 20.1 If a search of human gene sequence databases revealed a gene that was similar in base sequence, and if follow-up work confirmed that the mouse and human genes were similar in their pattern of exons and introns and regulatory sequences, then the researchers could claim that they are homologous. **KC 20.2** DNA can move from one prokaryotic species to another packaged in plasmids or viruses, or by the direct uptake of DNA fragments from the environment (transformation). **KC 20.3** Junk DNA refers to noncoding regions of DNA—sequences that are not transcribed into RNA. The function of repeated sequences is not yet clear, but many noncoding sequences are interesting because they are relicts from transposable elements. **KC 20.4** Expression of the gene may be involved in the development of cancer.

Test Your Knowledge

1. c; **2.** a; **3.** d; **4.** b; **5.** d; **6.** d

Test Your Understanding

1. Computer programs are used to scan sequences in both directions to find ATG start codons, a gene-sized logical sequence with recognizable codons, and then a stop codon. One can also look for characteristic promoter, operator, and other regulatory sites. It is more difficult to identify open reading frames in eukaryotes because their genomes are so much larger and because of the presence of introns and repeated sequences. **2.** To have the potential to thrive in different environments, a cell requires a large array of enzymes and proteins—to use different sources of food, cope with different temperature or pH conditions and so on—and thus a large array of genes. **3.** They use the transcription and translation machinery of the host cell to copy themselves and insert themselves into the host's genome, and this may have a negative effect on the host cell. **4.** Each individual has a DNA fingerprint—a unique collection of microsatellite and minisatellite alleles (length variants). Because alleles are inherited, closely related individuals share more alleles than more distantly related individuals. **5.** A DNA microarray experiment identifies which genes are being expressed in a particular cell at a particular time. If a series of experiments shows that different genes are expressed in cells at different times or under different conditions, it implies that expression was turned on or off in response to changes in age or changes in conditions. **6.** Homology is a similarity among different species that is due to their inheritance from a common ancestor. If a newly sequenced gene is found to be homologous with a known gene of a different species, it is assumed that the gene products have similar function. Duplicated genes are homologous because they share a common ancestor gene.

Applying Concepts to New Situations

1. If "gene A" is not necessary for existence, it can be lost by an event like unequal crossing over (on the chromosome with deleted segments) with no ill effects on the organism. In fact, individuals who have lost unnecessary genes are probably at a competitive advantage, because they no longer have to spend time and energy copying and repairing unused genes. **2.** If the grave was authentic, it might include two very different parental patterns along with three children whose patterns each represented a mix between the two parents. The other unrelated individuals would have patterns not shared by anyone else in the grave. **3.** The genes may have similar DNA sequences because the viral gene was incorporated into the genome of a human host in the past, and subsequently passed on to future generations. **4.** You would expect that the livers and blood of chimps and humans would function similarly, but that strong differences occur in brain function. The microarray data support this prediction and suggest that even though the brain proteins might have similar sequences, chimp and human brains are different because certain genes are turned on or off at different times and expressed in different amounts.

CHAPTER 21

Check Your Understanding (CYU)

CYU p. 379 Biologists have been able to grow entire plants from a single differentiated cell taken from an adult. They have also succeeded at producing animals by transferring the nucleus of a fully differentiated cell to an egg whose nucleus has been removed. **CYU p. 384** (1) Bicoid is a transcription factor, so cells that experience a high versus medium versus low concentration transcribe different amounts of Bicoid-regulated genes. (2) Bicoid proteins control the expression of gap genes, which—in effect—tell cells which third of the embryo they are in along the anterior-posterior axis. Gap genes control the expression of pair-rule genes, which organize the embryo into individual segments. Pair-rule genes control expression of segment polarity genes, which establish an anterior-posterior polarity to each individual segment. Segment polarity genes turn on homeotic genes, which then trigger genes for producing segment-specific structures like wings or legs.

Caption Questions and Exercises

Figure 21.2 It would look like the embryo of the left side of part (b)—a normal phenotype. **Figure 21.5** In cells that are expressing a particular gene, the gene is transcribed into RNA. In situ hybridization locates these copies of RNA, identifying which cells are expressing the gene. **Figure 21.10** Genes are usually considered homologous if they have similar structure—meaning similar DNA sequence and exon-intron structure.

Summary of Key Concepts

KC 21.1 The seedling would grow very slowly and be small, but development should proceed normally otherwise. **KC 21.2** Parenchyma cells in the stem still contain a complete set of genes, so if they can be de-differentiated and received appropriate signals from other cells, they could start producing vascular-tissue-specific proteins. **KC 21.3** Different concentrations of transcription factors or cell-cell signals will cause differences in the amount of expression from certain genes—so a high concentration means something different than a low concentration. **KC 21.4** The idea is that a single molecule can set up the anterior-posterior or apical-basal axis of the embryo. Once the axis is established, a sequence of events follows in which cells get increasingly more precise information about where they are located along that axis—based on the original concentration of the master regulator. **KC 21.5** When the patterns of expression of two key genes involved in limb versus rib formation are compared in chick and snake embryos, they differ. In snakes, the "form ribs here" pattern occurs instead of the "form legs here" pattern.

Test Your Knowledge

1. b; **2.** d; **3.** a; **4.** c; **5.** a; **6.** a

Test Your Understanding

1. They occur in both plants and animals, and are responsible for the changes that occur as an embryo develops. **2.** If the transplanted nucleus has undergone some permanent change or loss of genetic information during development, it should not be able to direct the development of a viable adult. But if the transplanted nucleus is genetically equivalent to the nucleus of a fertilized egg, then it should be capable of directing the development of a new individual. **3.** The researchers exposed adult flies to treatments that induce mutations, then looked for embryos with defects in the anterior-posterior body axis or body segmentation. The embryos had mutations in genes required for body axis formation and segmentation. **4.** Development—specifically differentiation—depends on changes in gene expression. Changes in gene expression depend on differences in regulatory transcription factors. **5.** Differentiation is triggered by the presence or absence of external signals. These signals trigger the production of transcription factors, which induce other transcription factors, and so on—a sequence that constitutes a regulatory cascade—as development progresses. At each step in the cascade, a new subset of genes is activated—resulting in a step-by-step progression from undifferentiated to fully differentiated cells. **6.** *Hox* genes were first discovered in *Drosophila* and have been identified in nearly every animal. The number of *Hox* genes varies across species, but their arrangement on the chromosome and pattern of expression in the embryo are strikingly similar across animal species.

Applying Concepts to New Situations

1. There would be more Bicoid protein farther toward the posterior and less in the anterior—meaning that the anterior segments would be "less anterior" in their characteristics and the posterior segments would be "more anterior." **2.** The result means that the human gene can be transcribed in worms and that its protein product has performed the same function in worms as in humans. This is strong evidence that the genes are homologous, and that their function has been evolutionarily conserved. **3.** The stem cells would have to be stimulated with nerve-specific signals to trigger their differentiation into nerve cells. **4.** Homeotic genes—such as *Hox* genes—are responsible for triggering the production of structures like wings or legs in particular segments. One hypothesis to explain the variation is that changes in gene expression led to different numbers of segments that express leg-forming *Hox* genes.

CHAPTER 22

Check Your Understanding (CYU)

CYU p. 392 (1) If a mutation in either bindin or the egg-cell receptor for sperm prevented the proteins from binding to each other, then fusion of sperm and egg-cell membranes could not take place and fertilization would not occur. (2) The entry of a sperm triggers a wave of calcium ion release around the egg-cell membrane. Thus, calcium ions act as a signal that indicates "A sperm has entered the egg." In response to the increase in calcium ion concentration, cortical granules fuse with the egg-cell membrane and release their contents to the exterior, causing the fertilization envelope to develop. **CYU p. 394** If cytoplasmic determinants were distributed in equal concentration throughout the egg, all blastomeres would end up with the same types and concentrations of cytoplasmic determinants. **CYU p. 395** (1) Endoderm is in the interior, ectoderm is on the outside, mesoderm is in between. (2) These cells become endoderm that forms the gut lining. The direction of the gut helps define the anterior-posterior axis of the body—it connects mouth and anus.

You Should Be Able To (YSBAT)

YSBAT p. 391 If the bindin-like protein were blocked, it would not be able to bind to the egg-cell receptor for sperm. Fertilization would not take place.

Caption Questions and Exercises

Figure 22.4 The researchers didn't start out with a hypothesis about which egg-cell membrane protein bound to bindin. Instead, they tested every type of protein projecting from the egg-cell surface to see which one, if any, bound to bindin. **Figure 22.8** [See Figure A22.1] **Figure 22.9** Ectoderm forms most of the structures on the outside of the adult. Endoderm forms mostly interior structures—for example, the inner lining of the respiratory and digestive tracts. Mesoderm forms structures located between the ectoderm-derived and endoderm-derived structures, including bones and muscles and the bulk of most of the organ systems. **Figure 22.15** It made gene expression possible in non-muscle cells. A "general purpose" promoter can lead to transcription of an inserted cDNA in any type of cell, including fibroblasts.

Summary of Key Concepts

KC 22.1 Bindin and the egg-cell receptor for sperm bind to each other. For this to happen, their tertiary structures must fit together somewhat like the structures of a lock and key. **KC 22.2** If the master regulator were localized to a certain part of the egg cytoplasm or formed a Bicoid-like concentration gradient throughout the eggs, blastomeres would contain or not contain the regulator or contain a distinct concentration of it. **KC 22.3** Both cells would contain cell-cell signals and transcription factors specific to mesodermal cells, but the anterior cell would contain anterior-specific signals and regulatory transcription factors while the posterior cell would contain posterior-specific signals and regulators. **KC 22.4** Cell-cell signals from the notochord direct the formation of the neural tube; subsequently, cell-cell signals from the notochord, neural tube, and ectoderm direct the differentiation of somite cells. Notochord cells die later in development. Cells expand during neural tube formation. Somite cells move to new positions, proliferate, and differentiate.

Test Your Knowledge

1. d; **2.** c; **3.** c; **4.** d; **5.** c; **6.** a

Test Your Understanding

1. Eggs contain stores of nutrients. **2.** If more than one sperm nucleus entered the egg, the resulting cell would contain more than two copies of each chromosome. Mitosis could not occur correctly and the embryo would be deformed or die. **3.** They contain different types and/or concentrations of cytoplasmic determinants. **4.** They give rise to all of the tissue and organs of the adult (from the Latin *germen*, meaning shoot or sprout). **5.** When transplanted early in development, somite cells become the cell type associated with their new location. But when transplanted later in development, somite cells become the cell type associated with the original location. These observations indicate that the same cell is not committed to a particular fate until later in development. **6.** If the gene for MyoD is expressed in non-muscle cells, the cells begin producing muscle-specific proteins. It triggers differential gene expression typical of muscle cells.

Applying Concepts to New Situations

1. The interactions that allow proteins to bind to each other are extremely specific—certain amino acids have to line up in certain positions with specific charges and/or chemical groups exposed. Sperm-egg binding is species specific because each form of bindin binds to a different egg cell receptor for sperm. **2.** Translation is not needed for cleavage to occur normally—meaning that all the proteins needed for early cleavage are present in the egg. **3.** The yellow pigment is a cytoplasmic determinant that triggers a gene regulatory cascade leading to differentiation of muscle cells. **4.** The embryo cannot yet feed, so all of its activity is fueled by the nutrient stores in the egg. It is not taking in a lot of new molecules that would allow it to get bigger.

CHAPTER 23

Check Your Understanding (CYU)

CYU p. 404 (1) In the male reproductive organs of a mature flower, cells undergo meiosis to produce haploid cells. Each of these haploid cells divides by mitosis to produce a pollen grain containing haploid cells. One of the cells in the pollen grain divides again by mitosis to form two sperm cells. (2) They ensure that the plant is fertilized by a pollen grain from the same species. If any pollen could germinate on any stigma, pollen from a pine tree could grow on the stigma of an orchid. **CYU p. 406** (1) Cells located on the outside of the embryo become epidermal tissue; cells in the interior become vascular tissue; cells in between become ground tissue. (2) The *MONOPTEROS* gene product would be overactivated. The embryonic cells would get information indicating that they are in the shoot apex, where MONOPTEROS levels are high. As a result, the root would not develop normally. **CYU p. 409** [See Figure A23.1] If a new supply of fertilizer (cow dung) appeared near the plant's left side, the roots on that side would extend toward the nutrients. The shoot system should grow in response to the new nutrients but, all other things being equal (e.g., no shading of plant from sunlight), shoot growth should be symmetrical. **CYU p. 411** (1) The A protein might function as a transcriptional repressor by binding to a regulatory site near the *C* gene, thereby preventing transcription of the *C* gene. (2) Like *Hox* genes, *MADS-box* genes code for DNA-binding proteins that function as transcriptional regulators. Both Hox proteins and MADS-box proteins are involved in regulatory transcription cascades that lead to development of specific structures in specific locations. Although they perform similar functions, *Hox* genes and *MADS-box* genes evolved independently and differ in their base sequences. Thus, these genes and their protein products are not homologous.

You Should Be Able To (YSBAT)

YSBAT p. 410 If four genes coded for flower structure and each gene specified one of the four organs, then each mutant should lack an element in only one of the whorls. If each of three genes is expressed in two adjacent whorls, then four different gene expression patterns are possible—one for each of the four different whorls.

FIGURE A22.1

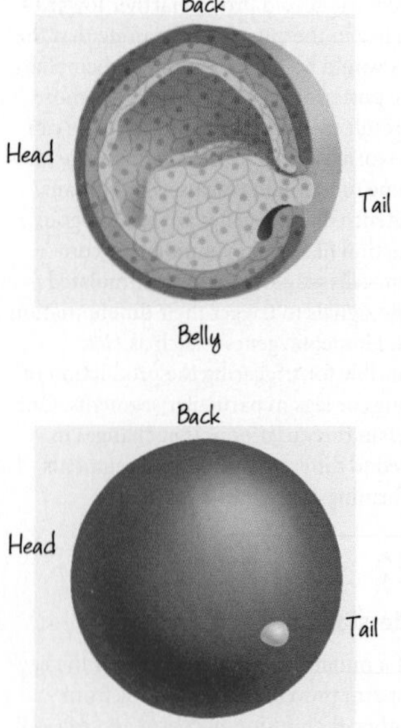

FIGURE A23.1

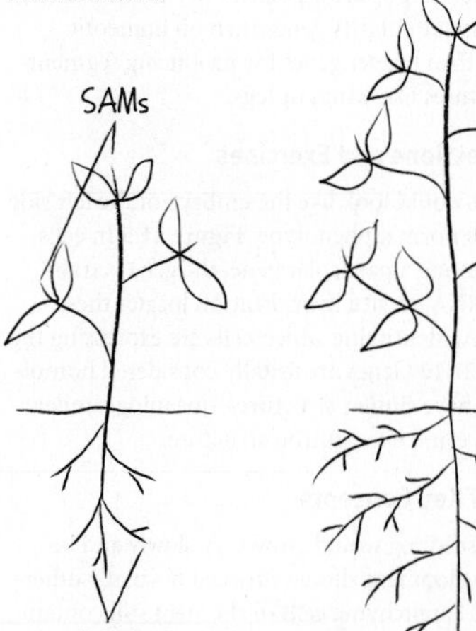

FIGURE A23.2

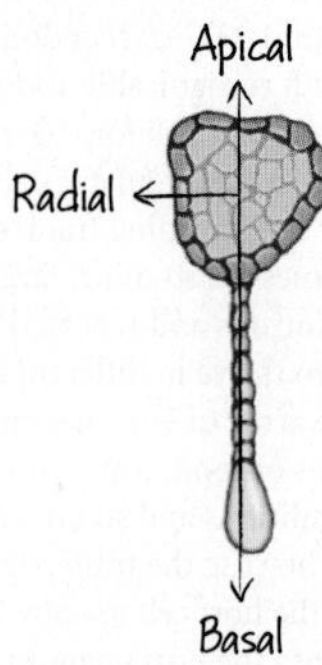

Caption Questions and Exercises

Figure 23.3 In both cases, protein-protein interactions between male and female structures in two different individuals prevent fertilization between members of different species. **Figure 23.5** [See Figure A23.2] **Figure 23.9** Many animals have three body axes analogous to the three axes of leaves: anterior-posterior (like the leaf's proximal-distal), dorsal-ventral (like the leaf's adaxial-abaxial), and a left-right axis (like the leaf's lateral axis). The main plant body, in contrast, has only two axes: apical-basal and radial.

Summary of Key Concepts

KC 23.1 Bicoid defines the anterior-posterior axis of the fly embryo; auxin defines the apical-basal axis of the plant embryo. Both work via concentration gradients. Bicoid is a regulatory transcription factor while auxin is a cell-cell signal. **KC 23.2** There are two fusion events between sperm and haploid cells in the female reproductive structure: one sperm fuses with the egg and leads to the development of an embryo; the other sperm fuses with two (or more) nuclei and leads to the development of endosperm. **KC 23.3** [See Figure A23.3] **KC 23.4** The presence of multiple SAMs and RAMs provides plants with flexibility in adapting to environmental changes and acquiring sunlight and nutrients, thereby increasing their ability to grow and reproduce. For example, multiple SAMs allow a plant to extend shoots (branches) in multiple directions to receive optimal sunlight, and multiple RAMs allow it to extend roots in multiple directions to acquire optimal water and nutrients. Also, a plant would be in trouble if its only SAM or RAM got eaten. **KC 23.5** Homeotic mutants are defined by a phenotype in which one body structure is replaced by another. *Arabidopsis* plants with a defective *B* gene have flowers in which petals are replaced by sepals and stamens are replaced by carpels.

Test Your Knowledge

1. a; **2.** b; **3.** d; **4.** d; **5.** c; **6.** a

Test Your Understanding

1. Proteins on the surface of a pollen grain interact with proteins on the surface of the stigma. Signals from the egg direct the growth of the pollen tube. **2.** Animal cells that lead to sperm or egg formation are "sequestered" early in development, so that they are involved only in reproduction and undergo few rounds of mitosis. Because mutations occur in every round of mitosis, plant eggs and sperm should contain many more mutations than animal eggs and sperm. **3.** Both meristems and stem cells can produce differentiated cells in an adult. Meristems are much more flexible than stem cells, as they can form all parts of the plant, and they increase the size of the body. Animal stem cells can usually develop into only a few differentiated cell types, and only replace lost or damaged cells. **4.** Just like the three tissue layers in plant embryos, the tissues produced in the SAMs and RAMs of a 300-year-old oak tree can differentiate into all of the specialized cell types found in a mature plant. **5.** The ABC model predicted that the *A*, *B*, and *C* genes would each be expressed in specific whorls in the developing flower. The experimental data showed that the prediction was correct. **6.** Cell proliferation is responsible for expansion of meristems and growth in size. Cell death occurs when leaves drop, and when three or four meiotic products die during gametogenesis in female tissues. Asymmetrical cell proliferation along with cell expansion is responsible for giving the adult plant a particular shape. Differentiation is responsible for generating functional tissues and organs in the adult plant. Cell-cell signals are responsible for setting up the apical-basal axis of the embryo and for guiding the pollen tube to the egg.

Applying Concepts to New Situations

1. If an individual is likely to die, it should throw all of its remaining resources into reproduction. **2.** Because vegetative growth is continuous and occurs in all directions (unrestricted), it could be considered indeterminate. However, reproductive growth produces mature reproductive organs and stops when those organs are complete. Because its duration is limited, it could be considered determinant. **3.** High light conditions could trigger changes in gene expression that reduced the rate of cell proliferation during leaf development. In contrast, low light conditions might trigger changes in gene expression that increase the rate of cell proliferation. **4.** [See Figure A23.4] In a young oak tree, SAMs would be evenly distributed in the upper parts of the plant and the RAMs would be evenly distributed in the tips of the roots. But after 50 years in the situation described above, the RAMs would be concentrated on the right side of the plant's roots, enabling the roots to grow into the area where water is available from the leaky pipe next to the building. The SAMs would be concentrated on the left side of the plant, allowing the upper parts of the tree to grow away from billboard, where light is available.

CHAPTER 24

Check Your Understanding (CYU)

CYU p. 429 (1) *Postulate 1:* Traits vary within a population. *Postulate 2:* Some of the trait variation is heritable. *Postulate 3:* There is variation in reproductive success (some individuals produce more offspring than others). *Postulate 4:* Individuals with certain heritable traits produce the most offspring. The first two postulates describe heritable variation; the second two describe differential reproductive success. (2) Beak size and shape and body size vary among individual finches, in part because of differences in their genotypes (some alleles lead to larger or narrower beaks, for example). When a drought hit, individuals with deep beaks survived better and produced more offspring than individuals with shallow beaks. **CYU p. 432** (1) When certain individuals are selected, their traits do not change—they simply produce more offspring than other individuals.
(2) Adaptations are not optimal solutions to challenges posed by a particular environment because they are compromised by (1) the necessity of meeting many challenges at the same time (an adaptive "solution" to one problem—such as flying faster—may make another problem worse—such as being maneuverable in flight); (2) lack of "optimal" alleles or presence of alleles that affect more than one trait; and (3) the necessity of selecting only preexisting variation in traits.

You Should Be Able To (YSBAT)

YSBAT p. 425 (1) Relapse occurred because the few bacteria remaining after drug therapy were not eliminated by the patient's weakened immune system and began to reproduce quickly. (2) No—almost all of the cells present at the start of the infection would have been resistant to the drug. **YSBAT p. 430** In biology, an adaptation is any heritable trait that increases an individual's ability to produce offspring in a particular environment. In everyday English, adaptation is often used to refer to an individual's nonheritable adjustment to meet an environmental challenge, a phenomenon that biologists call acclimation. The phenotypic changes resulting from acclimation are not passed on to offspring.

Caption Questions and Exercises

Figure 24.4 The theory of special creation would claim that the fossil and extant sloths were both created 6000 years ago, but that the fossil species became extinct during the flood in Noah's time, described in the Bible. It is not clear how the theory of special creation would explain transitional features observed in the fossil record. **Figure 24.5** If vestigial traits result from inheritance of acquired characteristics, some individuals must have lost the traits during their own lifetimes and passed the reduced traits on to their offspring. For example, a certain monkey's long tail might have been bitten off by a predator, or an ape's hair might have been pulled out of its skin by a rival during a fight. The new traits would then somehow have passed to the individuals' eggs and sperm, resulting in shorter-tailed and less-hairy offspring, until humans with a coccyx and goose bumps resulted. **Figure 24.16** "Prediction": Beak measurements were different before and after the drought. "Prediction of null hypothesis": No difference in beak measurements before and after the drought.

Summary of Key Concepts

KC 24.1 Under the theory of special creation, changes in *Mycobacterium* populations would be explained as individual creative events governed by an intelligent creator. Under the theory of evolution by inheritance of acquired characters, changes in *Mycobacterium* populations would be explained by the cells trying to transcribe genes in the presence of the drug, and their *rpoB* gene becoming altered as a result. **KC 24.2** In biology, fitness is the ability of an individual to produce offspring, relative to that ability in other individuals in the population.

FIGURE A23.3

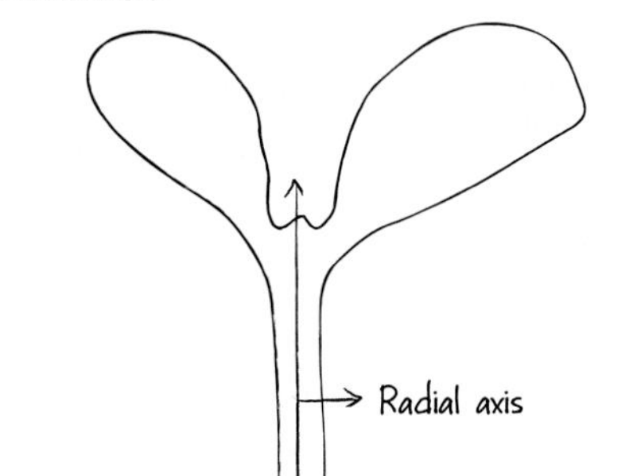

FIGURE A23.4

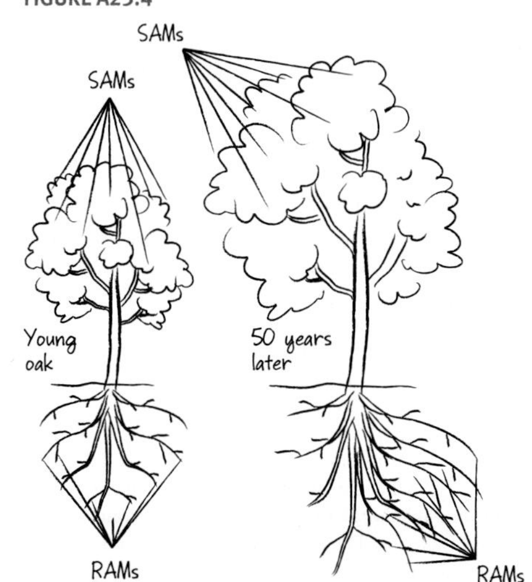

In everyday English, fitness is a physical attribute that is acquired as a result of practice or exercise. **KC 24.3** Brain size in *H. sapiens* might be constrained by the need for babies to pass through the mother's birth canal, by the energy required to maintain a large brain as an adult, or by lack of genetic variation for even larger brain size. Flying speed might be constrained by loss of maneuverability (and thus less success in hunting), the energy demands of extremely rapid flight, or lack of genetic variation for even faster flight.

Test Your Knowledge

1. b; **2.** d; **3.** a; **4.** d; **5.** a; **6.** c

Test Your Understanding

1. The theory of evolution by natural selection predicts that when a population's environment changes, individuals with certain traits will produce the most offspring, and the frequency of those traits will increase in the population. These changes should accumulate over long periods of time, eventually resulting in the formation of a new species that is geographically close to its ancestral species, and a fossil record should reflect changes in traits over time. The theory of special creation predicts that species do not change over time (or do so only due to divine intervention). Finally, the theory of acquired characters predicts that changes that occur in individuals in response to the environment are passed on to the next generation. **2.** Mutation produces new genetic variations, at random, without any forethought as to which variations might prove adaptive in the future. Individuals with mutations that are disadvantageous won't produce many offspring, but individuals with beneficial mutations will produce many offspring. **3.** The evidence for within-patient evolution is that DNA sequences at the start and end of treatment were identical except for a single nucleotide change in the *rpo* gene. If rifampin were banned, it is likely that *rpoB* mutant strains would have had lower fitness in the drug-free environment and would not continue to increase in frequency in *M. tuberculosis* populations. **4.** Typological thinking is based on the idea that species are unchanging and that any differences between individuals within a species are unimportant. Population thinking treats variation among individuals as critically important. That variation is what natural selection acts on, or selects. Population thinking was a radical break with typological thinking. **5.** Yes—the predictions made by a theory can be tested with indirect evidence in convincing ways. For example, if evolutionary theory predicts that certain transitional fossils should be found in rocks of a certain age, then researchers can search in those rocks and see if those fossils are indeed found. New groups of species found on island chains should be each others' closest relatives. And so on. **6.** Organisms do not necessarily become more complex over time; there are many examples of traits that have been lost. The biggest and strongest organisms do not necessarily produce the most offspring because they may not be the best adapted to the environment at that time.

Applying Concepts to New Situations

1. The theory of evolution fits the six criteria as follows: (1) and (2): It provides a common underlying mechanism responsible for puzzling observations such as homology, geographic proximity of similar species, the law of succession in the fossil record, vestigial traits, and extinctions. (3) and (4): It suggests new lines of research to test predictions about the outcome of changing environmental conditions in populations, about the presence of transitional forms in the fossil record, and so on. (5) It is a simple idea that explains the tremendous diversity of living and fossil organisms and why species continue to change today. (6) The realization that all organisms are related by common descent and that none are higher or lower than others was a surprise. **2.** It is well documented that people who are healthy and well fed grow taller than people who are sick and malnourished. In the absence of data indicating that alleles associated with increased height have increased in frequency recently, it is more logical to hypothesize that the observed change is due to changes in the environment—not evolutionary (genetic) changes. **3.** Compare the sequences of the 20 genes from many samples of preserved human tissue with the sequences of the same genes from a large sample of currently living humans. If evolution has occurred, the frequency of alleles correlated with "tallness" should be significantly greater in living humans than in those who lived a hundred years ago. **4.** The ability to tan is an adaptation because it is passed on from one generation to the next. The tan itself is not passed on, but the ability to tan is passed on; therefore it is heritable and can be classified as an adaptation.

CHAPTER 25

Check Your Understanding (CYU)

CYU p. 440 Given the observed genotype frequencies, the observed allele frequencies are freq(A_1) = 0.574 + ½(0.339) = 0.744; freq(A_2) = ½(0.339) + 0.087 = 0.256. Given these allele frequencies, the genotype frequencies expected under the Hardy-Weinberg principle are A_1A_1: $0.744^2 = 0.554$; A_1A_2: $2(0.744 \times 0.256) = 0.381$; A_2A_2: $0.256^2 = 0.066$. There are 4 percent too few heterozygotes observed, relative to the expected proportion. One of the assumptions of the Hardy-Weinberg principle is not met at this gene in this population, at this time. **CYU p. 446** (1) When allele frequencies fluctuate randomly up and down, sooner or later the frequency of an allele will hit 0. That allele thus is lost from the population, and the other allele at that locus is fixed. (2) In small populations, sampling error is large. For example, the accidental death of a few individuals would have a large impact on allele frequencies. **CYU p. 455** (1) Sperm are small and hence relatively cheap to produce, whereas eggs are large and require a large investment of resources to produce. (2) Sperm are inexpensive to produce, so reproductive success for males depends on their ability to find mates—not on their ability to find resources to produce sperm. The opposite pattern holds for females. Sexual selection is based on variation in ability to find mates, so is more intense in males—leading to more exaggerated traits.

You Should Be Able To (YSBAT)

YSBAT p. 444 If allele frequencies are changing due to drift, the populations in Table 25.1 would behave like the simulated populations in Figure 25.6—frequencies would drift up and down over time, and diverge. **YSBAT p. 451** The proportions of homozygotes should increase, and the proportions of heterozygotes should decrease. **YSBAT p. 452** (1) More recessive deleterious alleles are found in homozygotes and eliminated by natural selection. (2) If there are few or no deleterious recessives in a population, there is less inbreeding depression.

Caption Questions and Exercises

Table 25.1 The observed allele frequencies, calculated from the observed genotype frequencies, are 0.43 for *M* and 0.57 for *N*. The expected genotypes, calculated from the observed allele frequencies under the Hardy-Weinberg principle, are 0.185 for *MM*; 0.49 for *MN*; 0.325 for *NN*. **Figure 25.6** [See **Figure A25.1**] **Figure 25.8** Original population: freq(A_1) = (9 + 9 + 11) / 54 = 0.54; New population: freq(A_1) = (2 + 2 + 1) / 6 = 0.83. The frequency of A_1 has increased dramatically. **Figure 25.11** Yes—the frozen cells are traces of organisms that lived in the past. **Figure 25.13** The difference between number of flowers in individuals from the two types of matings represents inbreeding depression. This difference increases with age, which means that inbreeding depression increases with age. **Figure 25.16** An allele in males, because a successful male can have as many as 100 offspring—10 times more than a successful female.

Summary of Key Concepts

KC 25.1 There will be an excess of observed genotypes containing the favored allele compared to the proportion expected under Hardy-Weinberg proportions. **KC 25.2** Genetic drift will rapidly reduce genetic variation in small populations. Because captive individuals are usually housed under optimal conditions, selection will not be as intense as it would be in the wild. Alleles that allow individuals to thrive in captivity will increase; alleles that might lower fitness may not be eliminated. **KC 25.3** It increased homozygosity of recessive alleles associated with diseases such as hemophilia. As a result, the royal families were plagued by genetic diseases. **KC 25.4** In this case, reproductive success in females will depend on the number of male mates they can attract. It is reasonable to predict that female red-necked phalaropes are under intense sexual selection for traits that can help them attract mates and are more brightly colored than males.

Test Your Knowledge

1. b; **2.** a; **3.** a; **4.** b; **5.** d; **6.** c

Test Your Understanding

1. Selection may decrease genetic variation or maintain it (e.g., if heterozygotes are favored). Drift reduces it. Gene flow may increase or decrease it, depending on whether immigrants bring new alleles or emigrants remove alleles. Mutation increases it. **2.** Selective pressures often change over time or in different areas occupied by the same species. Even if selective pressures do not vary, mutation continually introduces new alleles. **3.** Sexual selection is most intense on the sex that makes the least investment in the offspring, so that sex tends to have exaggerated traits that make individuals successful in competition for mates. In this way, sexual dimorphism evolves. **4.** The prediction of a null hypothesis states what you should observe if the hypothesis you are testing is not correct. If you are testing the hypothesis that allele frequencies are changing or that mating is nonrandom with respect to a certain gene, the Hardy-Weinberg principle furnishes the appropriate null. **5.** Even if individuals in a small population mate at random or attempt to avoid inbreeding, over time all individuals in a small population are closely related. There are simply not enough nonrelatives to mate with. **6.** Alleles associated with extreme phenotypes are eliminated; alleles associated with intermediate phenotypes increase in frequency.

FIGURE A25.1

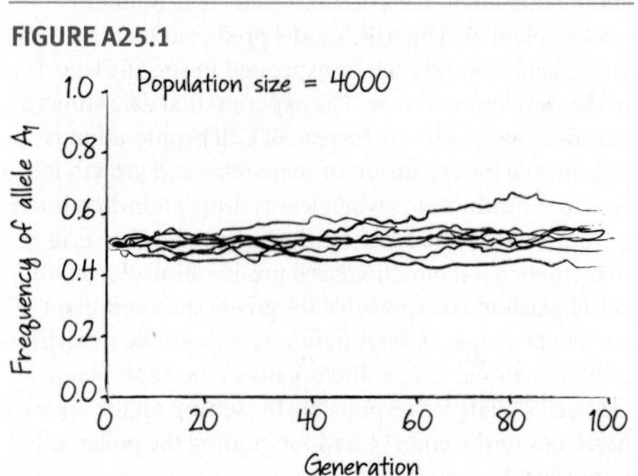

Applying Concepts to New Situations

1. The frequency of heterozygotes is $2p_1p_2$, or $2 \times 0.01 \times 0.99$, which is 0.0198. **2.** Inbreeding causes a decrease in heterozygotes, and a loss of fitness due to the increased numbers of homozygotes for deleterious alleles. Introducing new genetic variants through gene flow will increase allelic diversity in the population and result in more heterozygotes. **3.** If males never lift a finger to help females raise children, the fundamental asymmetry of sex is pronounced and sexual dimorphism should be high. If males invest a great deal in raising offspring, then the fundamental asymmetry of sex is small and sexual dimorphism should be low. **4.** Captive-bred young, transferred adults, and habitat corridors could be introduced to counteract the damaging effects of drift and inbreeding in the small, isolated populations. It is important, though, to introduce individuals only from similar habitats and connect patches of similar habitat with corridors—to avoid introducing alleles that lower fitness.

CHAPTER 26

Check Your Understanding (CYU)

CYU p. 462 (1) Biological species concept: Reproductive isolation means that no gene flow is occurring. Morphological species concept: If populations are evolving independently, they may have evolved morphological differences. Phylogenetic species concept: If populations are evolving independently, they should have synapomorphies that identify them as independent twigs on the tree of life. (2) Biological species concept: It cannot be used to evaluate fossils, species that reproduce asexually, or species that do not occur in the same area and therefore never have the opportunity to mate. Morphological species concept: It cannot identify species that differ in traits other than morphology and is subjective by nature—it can lead to differences of opinion that cannot be resolved by data. Phylogenetic species concept: Reliable phylogenetic information exists only for a small number of organisms. **CYU p. 468** Use one of the cases given in the chapter to illustrate colonization (Galápagos finches), vicariance (shrimp), or habitat specialization (apple maggot flies). In each case, drift will cause allele frequencies to change randomly. Differences in the habitats occupied by the isolated populations will cause allele frequencies to change under natural selection.

You Should Be Able To (YSBAT)

YSBAT p. 465 Selection will favor individuals that (1) breed on apples and have enzymes that are good at digesting apple fruit and that function well as larvae develop in warm temperatures and spend a relatively long time overwintering, and (2) breed on hawthorns and have enzymes that are good at digesting hawthorn fruit and that function well as larvae develop in cool temperatures and spend a relatively short time overwintering. **YSBAT p. 467a** A diploid individual experiences a defect in meiosis resulting in the formation of diploid gametes. The individual self-fertilizes, producing tetraploid offspring. The tetraploid individuals self-fertilize or mate with other tetraploid individuals, producing a tetraploid population. **YSBAT p. 467b** A tetraploid ($4n$) species gives rise to diploid ($2n$) gametes, and a diploid species gives rise to haploid (n) gametes. When a haploid gamete (n) fertilizes a diploid gamete ($2n$), a triploid offspring results. If this individual self-fertilizes, then a population of hexaploid wheat is formed.

Caption Questions and Exercises

Figure 26.3 Save one from the Atlantic Coast and one from the Gulf Coast. Their DNA sequences differ more than do subspecies within a geographic region, so you would be preserving more genetic diversity. **Figure 26.7** It creates genetic isolation—the precondition for divergence. **Figure 26.11** If the same processes were at work in the past as are at work in an experiment with living organisms, then the outcome of the experiment should be a valid replication of the outcome that occurred in the past.

Summary of Key Concepts

KC 26.1 (Many possibilities) Catch a group of fruit flies that are living in a lab, separate the individuals into different cages, and expose the two new populations to dramatically different environmental conditions—heat, lighting, food sources. In essence: create genetic isolation and then conditions for divergence due to drift and selection. **KC 26.2** Human populations would not be considered separate species under the biological concept because human populations can successfully interbreed. They would not be considered separate species under the morphological species concept because all human populations have the same basic morphology. (Although human races differ in minor superficial attributes such as skin color and hair texture, they have virtually identical anatomy and physiology in all other regards.) Nor would human populations be separate species under the phylogenetic species concept because they all arose from a very recent common ancestor. (DNA comparisons have revealed that human races are remarkably similar genetically and do not differ enough genetically to qualify even for subspecies status.) **KC 26.3** The sample experiment described above is an example of vicariance, because the experimenter fragmented the habitat into isolated cages. (If your own experiment differed from the one described above, then a different mechanism of speciation may have been involved.) **KC 26.4** Hybrid individuals will increase in frequency, and natural selection will favor parents that hybridize. The two parent populations may go extinct if hybrids live in the same environment. If the hybrids occupy a different environment, a new species may form.

Test Your Knowledge

1. a; **2.** b; **3.** d; **4.** a; **5.** a; **6.** a

Test Your Understanding

1. Direct observation: Galápagos finch colonization, apple maggot flies, *Tragopogon* allopolyploids, maidenhair ferns. Indirect evidence: snapping shrimp. **2.** On the basis of biological and morphospecies criteria (breeding range and morphological traits), six different species/subspecies of seaside sparrows were recognized. However, based on phylogenetic criteria (comparison of gene sequences), biologists concluded that seaside sparrows represent only two monophyletic groups. **3.** The colonizing population is typically small, and genetic drift has a large effect on smaller populations. If the newly colonized habitat differs from the original one, natural selection will favor individuals with alleles that increase fitness in the new environment. **4.** Some flies breed on apple fruits, others breed on hawthorn fruits. This reduces gene flow. Because apple fruits and hawthorn fruits have different scents and other characteristics, selection is causing the populations to diverge. **5.** Cells that go through many rounds of mitosis often accumulate errors, including mistakes that produce a tetraploid cell. If such a cell becomes part of a developing flower, and goes through meiosis, it will produce diploid gametes that can fuse to form polyploidy offspring. This is less likely to happen in animals because the future reproductive cells undergo fewer rounds of mitosis, so they have fewer chances to become polyploid. **6.** If the species have lived in the same area for a long time, there has been opportunity for selection to favor traits that prevent hybridization. This is reinforcement. If the species do not live in the same area, then there has been no opportunity for selection to favor traits that prevent hybridization.

Applying Concepts to New Situations

1. Decreasing. Gene flow tends to equalize allele frequencies among populations. **2.** These data cast doubt on the hypothesis. If founder events trigger speciation, then at least some speciation events due to introductions should be in progress. **3.** Two things: (1) Some flies happen to have alleles that allow them to respond to apple scents; others happen to have alleles that allow them to respond to hawthorn scents. There is no "need" involved. (2) The alleles for scent response exist; they are not acquired by spending time on the fruit. **4.** If the populations and habitat fragments are small enough, the species is likely to dwindle to extinction due to inbreeding and loss of genetic variation or catastrophes like a severe storm or a disease outbreak. If the populations survive, they are likely to diverge into new species because they are genetically isolated and because the habitats may differ.

CHAPTER 27

Check Your Understanding (CYU)

CYU p. 479 Hair and limb structures in humans and whales are examples of homology because they are traits that can be traced to a common ancestor. All mammals have hair and similar limb bone structure. However, extensive hair loss and advanced social behavior in whales and humans are examples of homoplasy. These traits are not common to all mammalian species and likely arose independently during the evolution of specific mammalian lineages. **CYU p. 488** (1) Doushantuo fossils are microscopic and include sponges and embryos from unknown species. Ediacaran fossils include sponges, jellyfish, and comb jellies along with traces of many other unidentified animals. Most were filter feeders. Burgess Shale fossils include every major animal lineage. The species present are larger, have much more complex morphology, and lived in a wide array of niches. (2) There were many resources available because no other animals (or other types of organisms) existed to exploit them. Also, the evolution of new species in new niches made new niches available for predators. Morphological innovations like limbs and complex mouthparts were important because they made it possible for animals to live in habitats other than the benthic area. **CYU p. 492** (1) Evidence includes the high content of iridium in sedimentary rock formed at the K-T boundary, shocked quartz and microtektite in rock layers that date 65 mya, and a large crater that was found off the coast of Mexico's Yucatán peninsula with microtektite in the crater's walls. Any of these three could be considered convincing, as they are known to form only at impact sites. Taken together, they are particularly convincing. (2) Mass extinctions are caused by such rapid and unusual environmental changes that any species is "lucky" to survive—it is not possible to adapt to such fast changes and such unusual environments.

You Should Be Able To (YSBAT)

YSBAT p. 479 Because whales and hippos share SINEs 4–7 and no other species has these four, they are synapomorphies that define them as a monophyletic group. The similarity is unlikely to be due to homoplasy because the chance of four SINEs inserting in exactly the same place in two different species, independently, is almost astronomically small. **YSBAT p. 485** In habitats on the mainland, there are more competitors and so species are already using the niches (resources) that are available to silverswords in Hawaii and *Anolis* lizards on Caribbean islands.

Caption Questions and Exercises

Figure 27.5 [See Figure A27.1] **Figure 27.16** The dinosaurs went extinct during the end-Cretaceous.

Summary of Key Concepts

KC 27.1 Humans are the only mammal that walks upright on two legs. **KC 27.2** They are the hardest structures in vertebrates and plants, so fossilize most readily. **KC 27.3** They could eat food that was available off the substrate. And once animals lived off the bottom, it created an opportunity for other animals who could eat them to evolve. **KC 27.4** Human-induced extinctions today are not due to poor adaptation to the normal environment. Humans are changing habitats rapidly and in unusual ways, so it is not possible for most species to adapt to these changes rapidly enough to avoid going extinct.

Test Your Knowledge

1. c; **2.** a; **3.** b; **4.** d; **5.** d; **6.** d

Test Your Understanding

1. The fossil record is biased because recent, abundant organisms with hard parts that live underground or in environments where sediments are being deposited are most likely to fossilize. Even so, it is the only data available on what organisms that lived in the past looked like, and where they lived. **2.** Diverse animal forms are found in the Cambrian that are not present in earlier strata. An enormous amount of diversity appeared in a time frame that was short compared to the sweep of earlier Earth history. **3.** Homoplasy is rare relative to homology, so phylogenetic trees that minimize the total change required are usually more accurate. In the case of the artiodactyl astralagus, parsimony was misleading because the astralagus was lost when whales evolved limblessness—creating two changes (a loss following a gain) instead of one (a gain). **4.** (Many answers are possible.) Adaptive radiation of *Anolis* lizards after they colonized new islands in the Caribbean. *Hypothesis:* After lizards arrived on each new island, where there were no predators or competitors, they rapidly diversified to occupy four distinct types of habitats on each island. (Many answers are possible.) Adaptive radiation during the Cambrian period following a morphological innovation. *Hypothesis:* Additional *Hox* genes made it possible to organize a large, complex body; the evolution of complex mouthparts and limbs made it possible for animals to move and find food in new ways. **5.** Environments all over the world deteriorated rapidly—large parts of the ocean lacked oxygen, many coastal habitats disappeared due to change in sea level, and little oxygen was available in the atmosphere. The underlying cause of these changes is not known. **6.** If a trait evolved in a common ancestor, then all of its descendants should share that trait, unless it is lost.

Applying Concepts to New Situations

1. Place the corpse in an environment in which decomposition is slow. One possibility is a swamp or bog; another is a beach or other coastal environment where mud or sand are being deposited. **2.** The fossil record and phylogenetic trees of extinct species indicate that whales evolved from a semiaquatic ancestor. Phylogenetic trees of living species indicate that hippos and whales share a recent common ancestor—meaning that one of the semiaquatic organisms that gave rise to today's whales also gave rise to the semiaquatic species that are today's hippos. **3.** It would be helpful to have (1) a large crater or other physical evidence from a major impact dated at 251 Mya, (2) shocked quartz and microtektites in rock layers dating to the end-Permian era, and (3) high levels of iridium or other elements common in space rocks and rare on Earth, dated to the time of the extinctions. **4.** Oxygen is an effective final electron acceptor in cellular respiration because of its high electronegativity. Organisms that use it as a final electron acceptor can produce more usable energy than organisms that do not use oxygen, but only if it is available. With more available energy, aerobic organisms can grow larger and move faster.

THE BIG PICTURE: EVOLUTION

Check Your Understanding (CYU), p. 494

1. Circle = inbreeding, sexual selection, natural selection, genetic drift, mutation, and gene flow. **2.** Adaptation "increases" fitness; synapomorphies "identify branches on" the tree of life. **3.** Several answers possible: e.g., the flower in angiosperms, the pharyngeal jaw in cichlid fishes, feathers and flight in birds. **4.** Genetic drift, mutation, and gene flow "are random with respect to" fitness.

CHAPTER 28

Check Your Understanding (CYU)

CYU p. 504 (1) Conditions should mimic a spill—sand or stones with a layer of crude oil or seawater with oil floating on top. Add samples, from sites contaminated with oil, that might contain cells capable of using molecules in oil as electron donors or electron acceptors. Other conditions (temperature, pH, etc.) should be realistic. (2) Use direct sequencing. Choose a well-studied gene, such as 16S RNA, with reliable PCR primers. After isolating DNA from a soil sample, do a PCR reaction and clone the resulting genes into plasmids grown in *E. coli* cells, so that each culture contains a different gene. Once many copies are available, sequence the genes and use the data to place the original organisms on the tree of life. **CYU p. 512** Eukaryotes can only (1) fix carbon via the Calvin-Benson pathway, (2) use aerobic respiration with organic compounds as electron donors, and (3) perform oxygenic photosynthesis. Among bacteria and archaea, there is much more diversity in pathways for carbon fixation, respiration, and photosynthesis, along with many more fermentation pathways.

You Should Be Able To (YSBAT)

YSBAT p. 506 cyanobacteria—photoautotrophs; *Clostridium aceticum*—chemoorganoautotroph; *Nitrosomonas* sp.—chemolithotrophs; heliobacteria—photoheterotrophs; *Escherichia coli*—chemoorganoheterotroph; *Beggitoa*—chemolithotrophic heterotrophs **YSBAT p. 513a, YSBAT p. 513b, YSBAT p. 514a, YSBAT p. 514b, YSBAT p. 515** [See Figure A28.1]

Caption Questions and Exercises

Figure 28.1 Prokaryotic—as it would require just one evolutionary change, the origin of the nuclear envelope in Eukarya. If it were eukaryotic, it would require that the nuclear envelope was lost in both Bacteria and Archaea—two changes and less parsimonious (see Chapter 27). **Table 28.1** [See Figure A28.2] **Figure 28.2** Improved nutrition made people better able to fight off disease, and improved sanitation lowered transmission of disease-causing bacteria. **Figure 28.4** Weak—different culture conditions may have revealed different species. **Figure 28.9** Table 28.5 contains the answers. For example, for organisms called sulfate reducers you would have H_2 as the electron donor, SO_4^{2-} as the electron acceptor, and H_2S as the reduced by-product. For humans the electron donor is glucose ($C_6H_{12}O_6$), the electron acceptor is O_2, and the reduced by-product is water (H_2O). **Table 28.5** Organotrophs use sugars, which are organic compounds, as their electron donor—getting energy by "feeding" on them. Sulfate reducers use sulfate ions as their electron acceptors, thus reducing the sulfate ion. Methanogens generate methane as a by-product. **Figure 28.11** Aerobic respiration. More free energy is released when oxygen is the final electron acceptor than when any other molecule is used, so more ATP can be produced and used for growth.

FIGURE A27.1

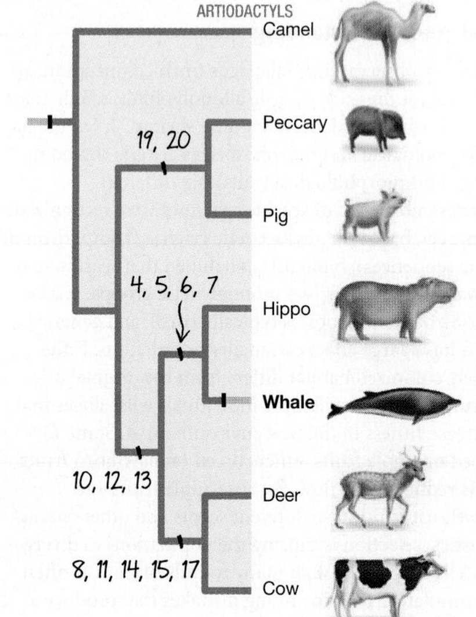

FIGURE A28.1

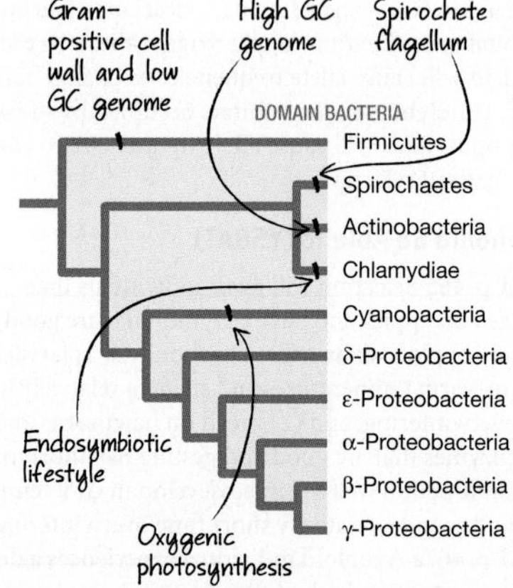

FIGURE A28.2

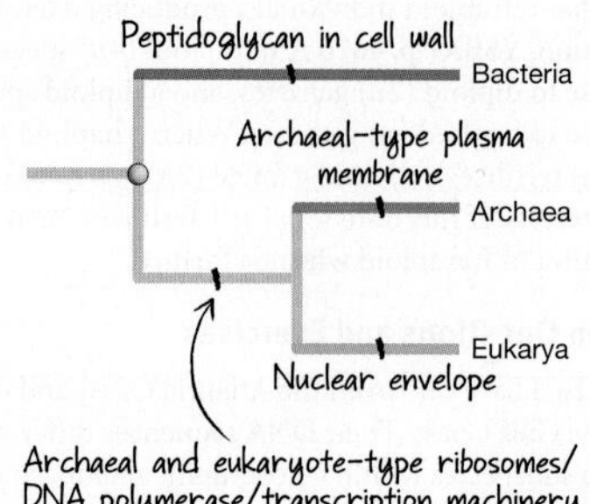

Summary of Key Concepts

KC 28.1 Eukaryotes have a nuclear envelope that encloses their chromosomes; bacteria and archaea do not. Bacteria have cell walls that contain peptidoglycan, and archaea have phospholipids containing isoprene subunits in their plasma membranes. Thus, the exteriors of a bacterium and archaeon are radically different. Archaea and eukaryotes also have similar machinery for processing genetic information. **KC 28.2** They are compatible, and thrive in each others' presence—one species' waste product is the other species' food. **KC 28.3** If bacteria and archaea did not exist, then (1) the atmosphere would have little or no oxygen, and (2) almost all nitrogen would exist in molecular form (the gas N_2).

Test Your Knowledge

1. c; **2.** b; **3.** d; **4.** d; **5.** b; **6.** c

Test Your Understanding

1. An electron donor provides the potential energy required to produce ATP. **2.** Yes. The array of substances that bacteria and archaea can use as electron donors, electron acceptors, and fermentation substrates, along with the diversity of ways that they can fix carbon and perform photosynthesis, allows them to live just about anywhere. **3.** You can sample and amplify genes—getting DNA sequence data that allow you to place unseen species on the tree of life. **4.** Large amounts of potential energy are released and ATP produced when oxygen is the electron acceptor, because oxygen is so electronegative. Large body size and high growth rates are not possible without large amounts of ATP. **5.** [See Figure A28.3] The following can serve as both electron donors and by-products: CH_4, NO_2^-, H_2S. The following can serve as both electron acceptors and by-products: SO_4^{2-}, CO_2, and NO_3^-. **6.** They are paraphyletic because the prokaryotes include some (bacteria and archaea) but not all (eukaryotes) groups derived from the common ancestor of all organisms living today.

Applying Concepts to New Situations

1. This result supports their hypothesis, because the drug poisons the enzymes of the electron transport chain and prevents electron transfer to Fe^{3+}, which is required to drive magnetite synthesis. If magnetite had still formed, another explanation would have been needed. **2.** Hypothesis: A high rate of tooth cavities in Western children is due to an excess of sucrose in the diet, which is absent from the diets of East African children. To test this hypothesis, have East African children switched to a diet that contains sucrose and monitor the presence of *S. mutans*. Have Western children switch to a diet lacking sucrose and monitor the presence of *S. mutans*. **3.** Look in waters or soil polluted with benzene-containing compounds. Put samples from these environments in culture tubes where benzene is the only source of carbon. Monitor the cultures and study the cells that grow efficiently. **4.** They should get energy from reduced organic compounds "stolen" from their hosts.

CHAPTER 29

Check Your Understanding (CYU)

CYU p. 523 (1) *Plasmodium* species are transmitted to humans by mosquitoes. If mosquitoes can be prevented from biting people, they cannot spread the disease. (2) Iron added → primary producers (photosynthetic protists and bacteria) bloom → more carbon dioxide taken up from atmosphere during photosynthesis → consumers bloom, eat primary producers → bodies of primary producers and consumers fall to bottom of ocean → large deposits of carbon-containing compounds form on ocean floor. **CYU p. 526** (1) Opisthokonts have a flagellum at the base or back of the cell; alveolate cells contain unique support structures called alveoli; stramenopiles have straw-like hairs on their flagella. (2) In direct sequencing, DNA is isolated directly from the environment and analyzed to place species on the tree of life. It is not necessary to actually see the species being studied.

You Should Be Able To (YSBAT)

YSBAT p. 526a [See Figure A29.1 on p. A:28] **YSBAT p. 526b** Membrane infoldings observed in bacterial species today support the hypothesis's plausibility—they confirm that the initial steps could have actually happened. The continuity of the nuclear envelope and ER are consistent with the hypothesis, which predicts that the two structures are derived from the same source (infolded membranes). **YSBAT p. 528** A photosynthetic bacterium (e.g., a cyanobacterium) could have been engulfed by a larger eukaryotic cell. If it was not digested, it could continue to photosynthesize and supply sugars to the host cell. **YSBAT p. 531** The chloroplast genes label should come off of the branch that leads to cyanobacteria. (If you had a phylogeny just of the cyanobacteria, the chloroplast branch would be located somewhere inside.) **YSBAT p. 532** The acquisition of the mitochondrion and the chloroplast represent the transfer of entire genomes, and not just single genes, to a new organism. **YSBAT p. 533** Yes—when food is scarce or population density is high, the environment is changing rapidly (deteriorating). Offspring that are genetically unlike their parents may be better able to cope with the new and challenging environment. **YSBAT p. 536a** (1) Alternation of generations refers to a life cycle in which there are multicellular haploid phases and multicellular diploid phases. A gametophyte is the multicellular haploid phase; the sporophyte is the multicellular diploid phase. A spore is a cell that grows into a multicellular individual, but is not produced by fusion of two cells. A zygote is a cell that grows into a multicellular individual, but *is* produced by fusion of two cells (gametes). Gametes are haploid cells that fuse to form a zygote. (2) **[See Figure A29.2 on p. A:28] YSBAT p. 536b, YSBAT p. 536c, YSBAT p. 536d, YSBAT p. 537a, YSBAT p. 537b, YSBAT p. 537c [See Figure A29.1 on p. A:28] YSBAT p. 537d** It is most likely that the two types of amoebae evolved independently. The alternative hypothesis is that the common ancestor of alveolates, stramenopiles, rhizarians, plants, opisthokonts, and amoebozoa were amoeboid, and that this growth form was lost many times. **[See Figure A29.1 on p. A:28] YSBAT p. 538a, YSBAT p. 538b [See Figure A29.1 on p. A:28] YSBAT p. 539a** If euglenids could take in food via phagocytosis (ingestive feeding), then it would have provided a mechanism by which a smaller photosynthetic protist could have been engulfed and incorporated into the cell via secondary endosymbiosis. **YSBAT p. 539b, YSBAT p. 540a, YSBAT p. 540b, YSBAT p. 541a, YSBAT p. 541b, YSBAT p. 542a, YSBAT p. 542b, YSBAT p. 543 [See Figure A29.1 on p. A:28]**

Caption Questions and Exercises

Figure 29.9 Two—one derived from the original bacterium and one derived from the eukaryotic cell that engulfed the bacterium. **Table 29.3** Yes—green algae, euglenids, and chlorarachniophytes all have chlorophyll *a* and *b*, as predicted by the hypothesis that a green algal chloroplast was transferred to the ancestor of euglenids and of chlorarachniophytes. Red algae have chlorophyll *a*, and chromalveolates (which include the brown algae, diatoms, and dinoflagellates) have chlorophyll *a* and *c*. If the hypothesis is correct, chlorophyll *c* must have evolved in an ancestor of the chromalveolates independently of the acquisition of chlorophyll *a* from red algae.

Summary of Key Concepts

KC 29.1 Photosynthetic protists use CO_2 and light to produce sugars and other organic compounds, so they furnish the first or primary source of organic material in an ecosystem. **KC 29.2** (1) Outside membrane was from host eukaryote; inside from engulfed cyanobacterium. (2) From the outside in, the four membranes are derived from the eukaryote that engulfed a chloroplast-containing eukaryote, the plasma membrane of the eukaryote that was engulfed, the outer membrane of the engulfed cell's chloroplast, and the inner membrane of its chloroplast. **KC 29.3** Each set of pigments absorbs most strongly in a certain part of the electromagnetic spectrum. If different species have different pigments, they can live in close proximity and perform photosynthesis without competing for the same wavelengths of light. **KC 29.4** A gametophyte is haploid, and a sporophyte is diploid.

Test Your Knowledge

1. b; **2.** b; **3.** a; **4.** b; **5.** d; **6.** b

FIGURE A28.3

TABLE 28.5 **Some Electron Donors and Acceptors Used by Bacteria and Archaea**

Electron Donor	Electron Acceptor	By-Products: From Electron Donor	By-Products: From Electron Acceptor	Category*
Sugars	O_2	CO_2	H_2O	Organotrophs
H_2 or organic compounds	SO_4^{2-}	H_2O or CO	H_2S or S^{2-}	Sulfate reducers
H_2	CO_2	H_2O	CH_4	Methanogens
CH_4	O_2	CO_2	H_2O	Methanotrophs
S^{2-} or H_2S	O_2	SO_4^{2-}	H_2O	Sulfur bacteria
Organic compounds	Fe^{3+}	CO_2	Fe^{2+}	Iron reducers
NH_3	O_2	NO_2^-	H_2O	Nitrifiers
Organic compounds	NO_3^-	CO_2	N_2O, NO, or N_2	Denitrifiers (or nitrate reducers)
NO_2^-	O_2	NO_3^-	H_2O	Nitrosifiers

*The name biologists use to identify species that use a particular metabolic strategy.

Test Your Understanding

1. An extensive cytoskeleton is required to shape the cell membrane so that the pseudopod can form and surround the food item. A cell wall is not flexible enough to wrap around the food item as the plasma membrane does. **2.** Because all eukaryotes living today have cells with a nuclear envelope, it is valid to infer that their common ancestor also had a nuclear envelope. Because bacteria and archaea do not have a nuclear envelope, it is valid to infer that the trait arose in the common ancestor of eukaryotes. **3.** Meiosis is required for alternation of generations because it converts a diploid phase of the life cycle to a haploid phase. Without it, the generations wouldn't "alternate." **4.** The host cell provided a protected environment and carbon compounds for the endosymbiont; the endosymbiont provided increased ATP from the carbon compounds. **5.** It confirmed a fundamental prediction made by the hypothesis, and could not be explained by any alternative hypothesis. **6.** All alveolates have alveoli, which are unique structures that function in supporting the cell. Among alveolates there are species that are (1) ingestive feeders, photosynthetic, or parasitic, and (2) move using cilia, flagella, or a type of amoeboid movement.

Applying Concepts to New Situations

1. The observation suggests that eukaryotes did not acquire mitochondria until there was enough oxygen present to make aerobic respiration efficient. (Oxygen is also poisonous to cells at high concentration, so mitochondria gave the early eukaryotes a way to "detoxify" oxygen.) **2.** If the apicoplast that is found in *Plasmodium* (the organism that causes malaria) is genetically similar to chloroplasts, and if glyphosate poisons chloroplasts, it is reasonable to hypothesize that glyphosate will poison the apicoplast and potentially kill the *Plasmodium.* This would be a good treatment strategy for malaria because humans have no chloroplasts, provided that the glyphosate produces no other effects that would be detrimental to humans. **3.** Primary producers usually grow faster when CO_2 concentration increases, but to date, they have not grown fast enough to make CO_2 levels drop—CO_2 levels have been increasing steadily over decades. **4.** Given that lateral gene transfer can occur at different points in a phylogenetic history, specific genes can become part of a lineage by a different route from that taken by other genes of the organism. In the case of chlorophyll *a*, its history traces back to a bacterium being engulfed by a protist and forming a chloroplast.

CHAPTER 30

Check Your Understanding (CYU)

CYU p. 552 (1) Green algae and land plants share an array of morphological traits that are synapomorphies, including the chlorophylls they contain, (2) green algae appear before land plants in the fossil record, and (3) on phylogenetic trees estimated from DNA sequence data, green algae and land plants share a most recent common ancestor, with green algae being the initial groups to diverge and land plants diverging subsequently. **CYU p. 566** (1) Cuticle prevents water loss from the plant; vascular tissue moves water up from the soil and moves photosynthetic products down to the roots. (2) [See Figure A30.1]

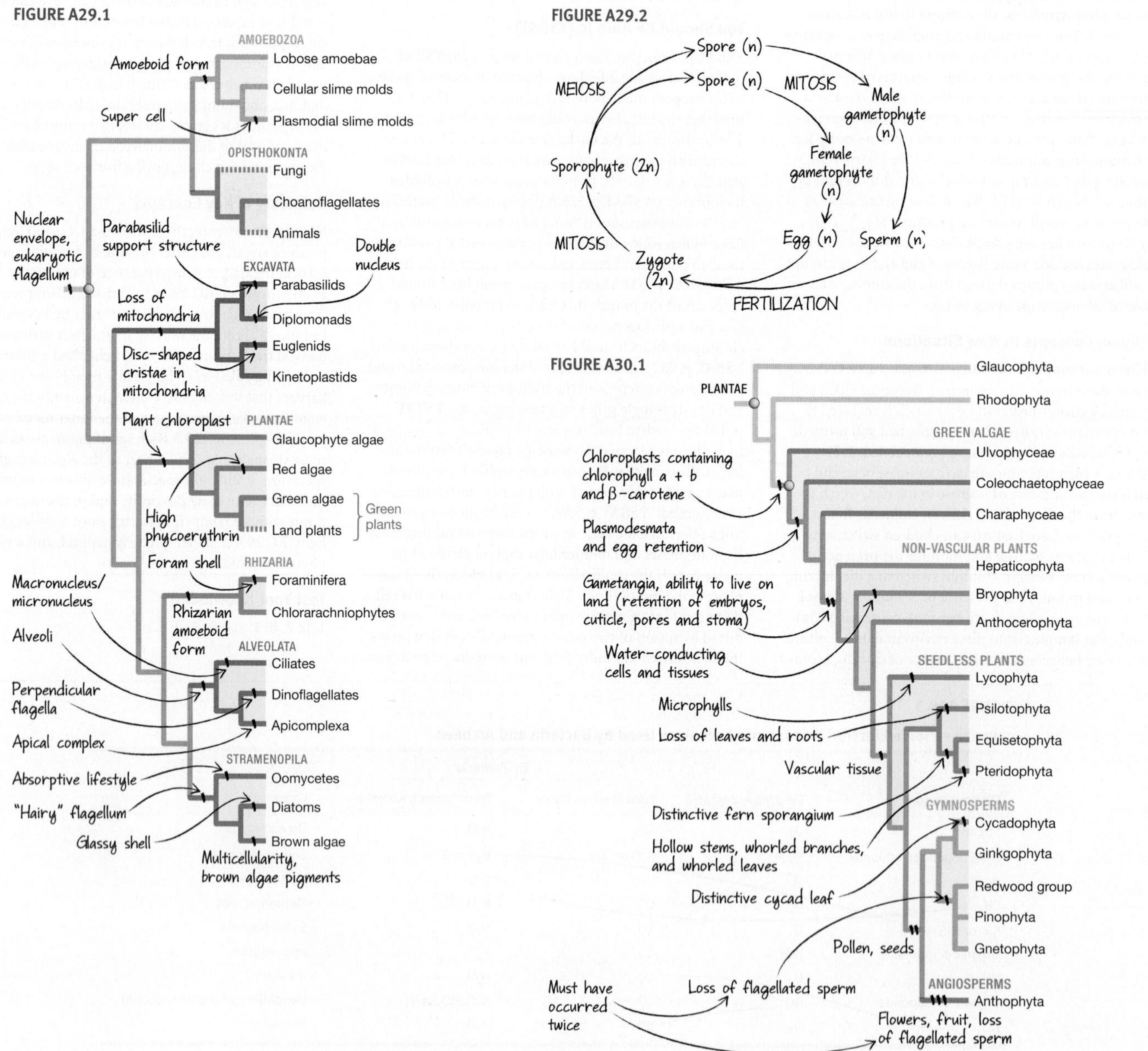

You Should Be Able To (YSBAT)

YSBAT p. 558 Alternation of generations occurs when there are multicellular haploid individuals and multicellular diploid individuals in a life cycle. A sporophyte is a multicellular diploid individual that produces spores by meiosis. A spore is a single cell that is not formed by fusion of two cells, and that grows by mitosis into a multicellular adult. **YSBAT p. 559** In the hornwort photo, the sporophyte is the spike-like green and brown structure; the gametophyte is the leafy-looking structure underneath. The horsetail gametophyte is the microscopic individual on the left; the sporophyte is the much larger individual on the right. **YSBAT p. 569a, YSBAT p. 569b, YSBAT p. 570, YSBAT p. 571, YSBAT p. 572a, YSBAT p. 572b, YSBAT, p. 573, YSBAT p. 574** [See Figure A30.1]

Caption Questions and Exercises

Figure 30.7 If you find the common ancestor of all the green algae (at the base of the Ulvophyceae), the lineages that are collectively called green algae don't include that common ancestor and all of its descendants—only some of its descendants. The same is true for non-vascular plants (the common ancestor here is at the base of the Hepaticophyta) and seedless vascular plants (the common ancestor here is at the base of the Lycophyta). **Figure 30.9** Water flows more easily through a short, wide pipe than through a long, skinny one because there is less resistance from the walls of the pipe. **Figure 30.10** To map an innovation on a phylogenetic tree, biologists determine the location(s) on the tree that is consistent with all descendants from that point having the innovation (unless, in rare cases, the innovation was lost in a few descendants). **Figure 30.15** There are multicellular haploid stages and multicellular diploid stages in these plants. **Figure 30.19** Gymnosperm gametophytes are microscopic, so they are even smaller than fern gametophytes. The gymnosperm gametophyte is completely dependent on the sporophyte for nutrition, while fern gametophytes are not. Fern gametophytes are photosynthetic and even supply nutrition to the young sporophytes. **Figure 30.20** Consistent—the fossil data suggest that gymnosperms evolved earlier, and gymnosperms have larger gametophytes than angiosperms. **Figure 30.22** It tests the hypothesis that the presence of yarn on the spur changes pollinator behavior and thus reproductive success. **Figure 30.24** [See Figure A30.1]

Summary of Key Concepts

KC 30.1 Rates of soil formation will drop; rates of soil loss will increase. **KC 30.2** Because so many adaptations are required—starting with cuticle and pores or stomata—it is unlikely that the transition from water to land would happen more than once. Stated another way, the aquatic and terrestrial environments are so different for plants that the transition was difficult and thus unlikely. **KC 30.3** Both spores and seeds have a tough, protective coat, so can survive while being dispersed to a new location. Seeds have the advantage of carrying a store of nutrients with them—when a spore germinates, it has to make its own food via photosynthesis right away.

Test Your Knowledge

1. c; **2.** b; **3.** d; **4.** c; **5.** c; **6.** a

Test Your Understanding

1. Plants build and hold soils required for human agriculture and forestry, and increase water supplies that humans can use for drinking, irrigation, or industrial use. Plants release oxygen that we breathe. **2.** Cuticle prevents water loss from leaves but also prevents entry of CO_2 required for photosynthesis. Stomata allow CO_2 to diffuse but can close to minimize water loss. Liverwort pores allow gas exchange but cannot be closed if conditions become dry. Liverworts that lack pores have a cuticle that is thin enough to allow some gas exchange. **3.** They provided the support needed for plants to grow upright and not fall over in response to wind or gravity. Erect growth allowed plants to compete for light. **4.** (1) Gametangia are found in all land plant groups except angiosperms; (2) transfer cells are found in all land plants; (3) pollen is found in gymnosperms and angiosperms; (4) seeds are found in gymnosperms and angiosperms, (5) fruit is found in angiosperms. **5.** In a gametophyte-dominant life cycle, the gametophyte is larger and longer lived than the sporophyte and produces most of the nutrition. In a sporophyte-dominant life cycle, the sporophyte generation is the larger, longer-lived, and photosynthetic phase of the life cycle. **6.** Homosporous plants produce a single type of spore that develops into a gametophyte that produces both egg and sperm. Heterosporous plants produce two different types of spores that develop into two different gametophytes that produce either egg or sperm. In a tulip, the microsporangium is found within the stamen, and the megasporangium is found within the ovule. Microspores divide by mitosis to form male gametophytes (pollen grains); megaspores divide by mitosis to form the female gametophyte.

Applying Concepts to New Situations

1. Homosporous plants produce a single type of spore that develops into a gametophyte that produces both egg and sperm. Heterosporous plants produce two different types of spores that develop into two different gametophytes that produce either egg or sperm. In a tulip, the microsporangium is found within the stamen, and the megasporangium is found within the ovule. Microspores divide by mitosis to form male gametophytes (pollen grains); megaspores divide by mitosis to form the female gametophyte. **2.** The combination represents a compromise between efficiency and safety. Tracheids can still transport water if vessels become blocked by air bubbles. **3.** A "reversion" to wind pollination might be favored by natural selection because it is costly to produce a flower that can attract animal pollinators. Because wind-pollinated species grow in dense clusters, they can maximize the chance that the wind will carry pollen from one individual to another (less likely if the individuals are far apart). Wind-pollinated deciduous trees flower in early spring before their developing leaves begin to block the wind. **4.** Alter one characteristic of a flower and present the flower to the normal pollinator. As a control, present the normal (unaltered) flower to the normal pollinator. Record the amount of time the pollinator spends in the flower, the amount of pollen removed, or some other measure of pollination success. Repeat for other altered characteristics. Analyze the data to determine which altered characteristic affects pollination success the most.

CHAPTER 31

Check Your Understanding (CYU)

CYU p. 586 (1) Because they are made up of a network of thin, branching hyphae, mycelia have a large surface area, which makes absorption efficient. (2) Swimming spores and gametes, zygosporangia, basidia, asci. **CYU p. 594** (1) When birch tree seedlings are grown in the presence and absence of EMF, individuals denied their normal EMF are not able to acquire sufficient nitrogen and phosphorus. Isotope-tracing experiments also show that sugars are transferred from host plant to EMF and that nitrogen and phosphorus are transferred from EMF to host plant. (2) Meiosis and production of haploid spores.

You Should Be Able To (YSBAT)

YSBAT p. 591 Human sperm and egg undergo plasmogamy followed by karyogamy during fertilization, but heterokaryosis does not occur in humans or other eukaryotes besides fungi. **YSBAT p. 594, YSBAT p. 596a, YSBAT p. 596b, YSBAT p. 596c, YSBAT p. 597, YSBAT p. 598a, YSBAT p. 598b** [See Figure A31.1]

Caption Questions and Exercises

Figure 31.10 Labeled-nutrient experiments identify which nutrients are exchanged and in which direction. They explain *why* plants do better in the presence of mycorrhizae, and why the fungus also benefits.

FIGURE A31.1

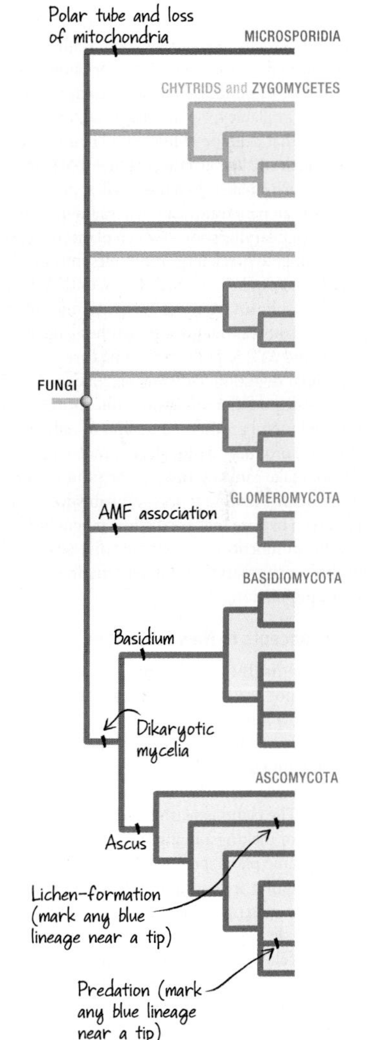

Figure 31.11 [See Figure A31.2] **Figure 31.13** The haploid mycelium. **Figure 31.14** Mycelia produced by asexual reproduction are genetically identical to their parent; mycelia produced by sexual reproduction are genetically different from both parents (each spore has a unique genotype).

Summary of Key Concepts

KC 31.1 Loss of mycorrhizal fungi would decrease nutrient delivery to plants and reduce their growth inside the experimental plots, compared to control plots with intact fungi. Also, lack of fungi would slow decay of dead plant material, causing a dramatic buildup of dead organic material. **KC 31.2** Absorption has to occur across a surface (usually via transport proteins in the plasma membrane). Other things being equal, more absorption will occur across a large surface area than a small surface area. **KC 31.3** Haploid hyphae fuse, forming a heterokaryotic mycelium. When karyogamy occurs, a diploid nucleus forms—just as when gametes fuse.

Test Your Knowledge

1. a; **2.** d; **3.** c; **4.** b; **5.** b; **6.** d

Test Your Understanding

1. Along with a few bacteria, fungi are the only organisms that can digest wood completely. If the wood is not digested, carbon remains trapped in wood. Without fungi, CO_2 would be tied up and unavailable for photosynthesis, and the presence of undecayed organic matter would reduce the space available for plants to grow. **2.** Fungi produce enzymes that degrade cellulose and lignin. **3.** Plant roots have much smaller surface area than EMF or AMF. Hyphae are much smaller than the smallest portions of plant roots so can penetrate dead material more efficiently. Extracellular digestion—which plant roots cannot do—allows fungi to break large molecules into small compounds that can be absorbed. **4.** DNA, RNA, and ATP all contain phosphorus. Without adequate amounts of phosphorus, plants cannot grow by synthesizing more DNA, RNA, and ATP. **5.** Both compounds are processed via extracellular digestion. Different enzymes are involved, however. The degradation of lignin is uncontrolled and does not yield useful products; digestion of cellulose is controlled and produces useful glucose molecules. **6.** Most fungi do not make gametes. Instead, the products of specific alleles identify mating types—probably to encourage fusion between hyphae that are dissimilar genetically, resulting in the production of genetically diverse offspring. It is possible for thousands of different mating-type alleles to exist in a population.

Applying Concepts to New Situations

1. (1) Confirm that the chytrid fungus is found only in sick frogs and not healthy frogs. (2) Isolate the chytrid fungus and grow it in a pure culture. (3) Expose healthy frogs to the cultured fungus and see if they become sick. (4) Isolate the fungus from the experimental frogs, grow it in culture, and test whether it is the same as the original fungus. **2.** The claim is reasonable because fungi have so many ways of making a living from plant tissues. For example, the same plant species could have several species of fungi that are endophytic, mycorrhizal, or parasitic, as well as an array of species that break down its tissues when it dies. **3.** Each of the different cellulase enzymes attacks cellulose in a different way, so producing all the enzymes together increases the efficiency of the fungus in breaking down cellulose completely. It is likely that lignin peroxidase is produced along with cellulases, so they can act in concert to degrade wood. To test this idea, you could harvest enzymes secreted from a mycelium before and after it contacts wood in a culture dish, and see if the cellulases and lignin peroxidase appear together once the mycelium begins growing on the wood. **4.** Collect a large array of colorful mushrooms that are poisonous and capture mushroom-eating animals, such as squirrels. Present a hungry squirrel with a choice of mushrooms that have been dyed or painted a drab color versus treated with a solution that is identical to the dye or paint used but uncolored. Record which mushrooms the squirrel eats. Repeat the test with many squirrels and many mushrooms.

CHAPTER 32

Check Your Understanding (CYU)

CYU p. 609 (1) Bilateral symmetry led to the evolution of long, slender body plans. Triploblasty gave rise to an inner tube (gut), an outer tube (body lining), and muscles and organs in between. The coelom functions as a hydrostatic skeleton to facilitate movement in soft-bodied animals that lack limbs. (2) When an animal moves through an environment in one direction, its ability to acquire food and perceive and respond to threats is greater if its feeding, sensing, and information-processing structures are at the leading end. **CYU p. 617** (1) The mouthparts of deposit feeders are relatively simple, as they simply gulp relatively soft material. The mouthparts of mass feeders are more complex, because they have to tear off and process chunks of relatively hard material. (2) Gametes that are shed into aquatic environments can float or swim. This cannot happen on land, so internal fertilization is more common.

You Should Be Able To (YSBAT)

YSBAT p. 606 If you do the exercise correctly, the balloon should wriggle. The wriggling movements would produce movement if the balloon (or animal) were surrounded by water or soil. **YSBAT p. 618** The origin of epithelial tissue should be marked on the same branch as multicellularity. **YSBAT p. 619** The origin of cnidocytes should be marked on the Cnidaria branch. **YSBAT p. 620** The origin of cilia-powered swimming should be marked on the Ctenophora branch.

Caption Questions and Exercises

Figure 32.6 The animal would get shorter and fatter. **Figure 32.9** The label and bar should go on the branch to the left of the clam shell and "Mollusk" label. **Figure 32.19** Stained *Dll* gene products would be located in the legs of the insect but would not be concentrated anywhere in the onychophoran and segmented worm. **Figure 32.22** No—there is no multicellular haploid form.

Summary of Key Concepts

KC 32.1 It should increase dramatically—animals would no longer be consuming plant material. **KC 32.2** Your drawing should show the tube-within-a-tube design of a worm, with one end labeled head and containing the mouth and brain. From the brain, one or two major nerve tracts should run the length of the body. The gut should go from the mouth to the other end, ending in the anus. In between the gut and outer body wall, there should be blocks or layers of muscle. **KC 32.3** During gastrulation, the initial pore becomes the mouth in protostomes, whereas in deuterostomes the pore becomes the anus. The protostome coelom is formed from blocks of mesodermal tissue that form cavities. The deuterostome coelom forms when mesodermal tissue around the gut pinches off. **KC 32.4** Natural selection has produced eye structures that function well in a particular species' habitat, and mollusks live in a wide array of habitats. Some mollusks live buried in sand and have no eyes—eyes would have no function in this habitat. Mollusks like squid and octopuses live in open water and hunt prey. They have complex eyes that form images and help them find food. **KC32.5** Juveniles look like miniature adults and feed on the same foods as adults. Larvae look much different from adults, live in different habitats, and eat different foods. The difference is important because larvae do not compete with adults for food.

FIGURE A31.2

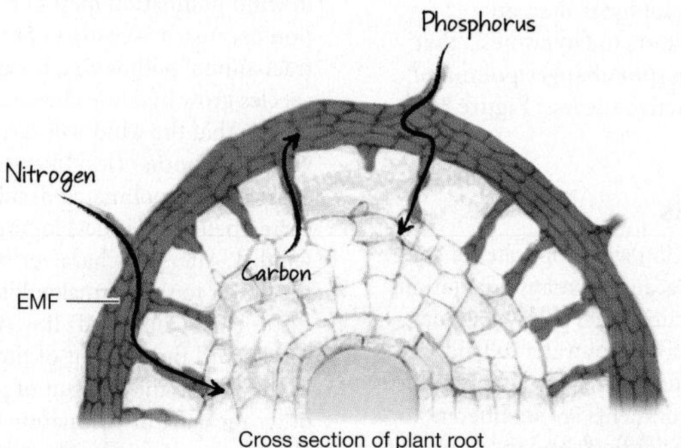

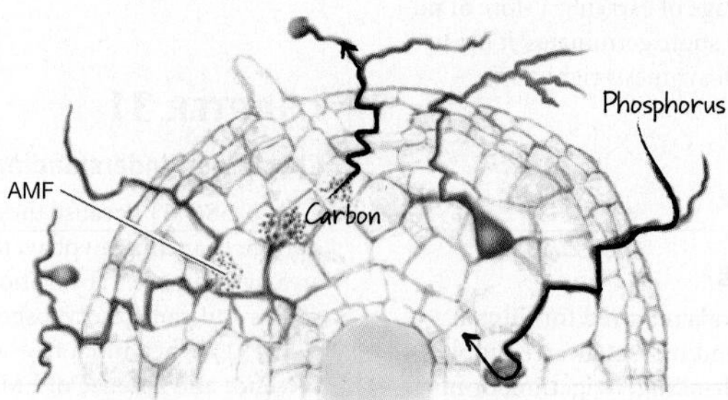

Test Your Knowledge

1. a; **2.** d; **3.** b; **4.** a; **5.** d; **6.** c

Test Your Understanding

1. Diploblasts have two types of embryonic tissue; triploblasts have three. Mesoderm made it possible for an enclosed, muscle-lined cavity to develop, creating a coelom. **2.** A hydrostatic skeleton is a pressurized, fluid-filled chamber (the coelom) that is surrounded by muscles. Contractions on either side of the body change the pressure of the fluid in the hydrostatic skeleton and thus the shape of the coelom. When muscle contractions are coordinated throughout the length of the animal, the shape of the hydrostatic skeleton changes and results in movement. **3.** There are many unicellular organisms that are heterotrophic, but they can consume only small packets of food. Animals are multicellular, so they are larger and can consume larger packets of food. **4.** A long, hollow structure that is sharp enough to pierce the wall of the stem; a simple, muscular mouth opening that can gulp soft, decaying material; sharp, stiff structures that can move in a scissoring action; a long, hollow structure that can be inserted into the flower. **5.** In oviparous species, the mother adds nutrient-rich yolk to the egg that nourishes the developing embryo. In viviparous species, the mother transfers nutrients directly from her body to the growing embryo. **6.** Radially symmetric organisms encounter the environment in many directions, so it is advantageous to have an equal number of neurons located throughout the body. Bilaterally symmetric organisms encounter the environment in one direction, so it is advantageous to have most neurons clustered in the head region.

Applying Concepts to New Situations

1. Yes—if the same gene is found in nematodes and humans, it was likely found in the common ancestor of protostomes and deuterostomes. If so, then fruit flies should also have this gene. **2.** Mosquitoes, because larval and adult mosquitoes feed on different food sources in different habitats and do not compete with one another. You could test this prediction by collecting data on the total number of mosquito versus tick species, their abundance, and geographic distribution. **3.** It should resemble a nerve net, because echinoderms need to take in and process information from multiple directions—not just one. **4.** Web-building spiders are some of the only terrestrial animals that suspension feed.

CHAPTER 33

Check Your Understanding (CYU)

CYU p. 630 (1) Arthropods have a tube-within-a-tube body plan with a drastically reduced coelom. They have a hemocoel body cavity that holds internal organs and body fluids. The body is segmented, with segments grouped into tagmata. They have jointed limbs and an exoskeleton made of chitin. Compared to unjointed limbs, jointed limbs allow animals to move faster and with more precision. (2) Supporting the body (and moving) without support from water. Preventing the body from drying out. Facilitating gas exchange.

You Should Be Able To (YSBAT)

YSBAT p. 631a, YSBAT p. 631b, YSBAT p. 632, YSBAT p. 633, YSBAT p. 637, YSBAT p. 638 [See Figure A33.1]

Caption Questions and Exercises

Figure 33.3 The tuft is the cluster of ciliated tentacles in part (a); the wheel is the ring of cilia in part (b). **Figure 33.6** Different phyla of worms have different feeding strategies, different mouthparts, and eat different foods, so they are not in direct competition for food. **Figure 33.22** [See Figure A33.2]

Summary of Key Concepts

KC 33.1 All of the phyla that are considered protostomes share a pattern of early development found in no other animals. Based on this observation, it is logical to claim that this pattern of development arose in the common ancestor of these phyla—meaning that it is a synapomorphy. **KC 33.2** Flatworms likely lost the adaptation of a coelom because as they evolved a thin, flattened body plan, the coelom was no longer needed for internal gas circulation or movement. Arthropods have a drastically reduced coelom because they have evolved limbs and complex musculature for movement and no longer require the coelom to provide a hydrostatic skeleton. **KC 33.3** (1) Because larvae and adults live in a different habitat, young and adults do not compete for food. (2) Because larvae can swim, they can disperse to new locations.

Test Your Knowledge

1. d; **2.** a; **3.** c; **4.** c; **5.** a; **6.** b

Test Your Understanding

1. Ecdysozoans grow by molting; lophotrochozoans grow by incremental additions to their bodies. In addition, some lophotrochozoan phyla have lophophores and/or trochophore larvae. Both groups are bilaterally symmetric triploblasts with the protostome pattern of development. **2.** Multiple times, because phylogenies show that segmented groups evolved from unsegmented ancestors in both lophotrochozoans (annelids) and ecdysozoans (arthropods). **3.** The ability to live on land offered the opportunity of exploiting new habitats and food sources, with minimal competition from other animals. **4.** An exoskeleton provided a stiff surface for muscle attachment—facilitating rapid, precise movement—as well as protection from enemies. **5.** The ability to fly allowed insects to disperse to new habitats and find new food sources efficiently. **6.** The leading hypothesis is variation in mouthpart structure and feeding strategies.

Applying Concepts to New Situations

1. [See Figure A33.3] **2.** If the ancestors of brachiopods and mollusks lived in similar habitats and experienced natural selection that favored similar traits, then they would have evolved to have similar forms and habitats. This is called convergent evolution (see Chapter 27). **3.** Aplacophora are probably basal—meaning that they would branch off between the ancestral form at the base of the tree and the rest of the groups. Because they have reduced forms of key molluscan synapomorphies, it is likely that they are the least-derived of the Mollusca. Because they lack a shell for muscle attachment and a well-developed foot, they probably move slowly via undulating movements and live in benthic habitats. **4.** Agree—it is likely that the common ancestor of arthropods was marine, and that terrestrial forms evolved from aquatic ancestors independently in crabs, isopods, and insects.

CHAPTER 34

Check Your Understanding (CYU)

CYU p. 650 (1) If it's an echinoderm, it should have five-part radial symmetry, a calcium carbonate endoskeleton just underneath the skin, and a water vascular system (e.g., visible podia). **CYU p. 659** (1) Jaws allow animals to capture food efficiently and process it by crushing or tearing. (2) The increased cushioning from enclosed fluids provided mechanical support, and the increased surface area made transport of gases and other materials more efficient.

FIGURE A33.1

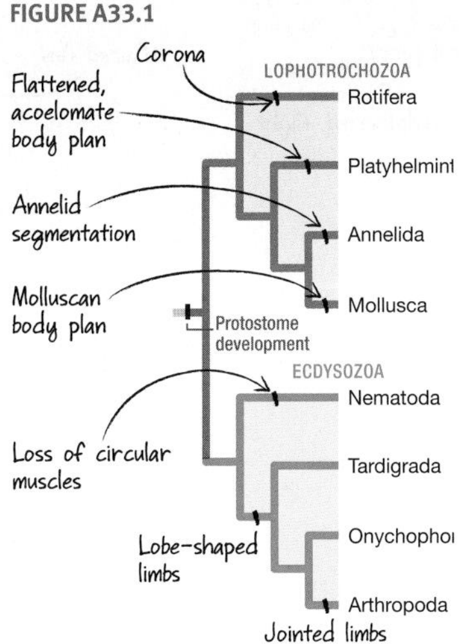

FIGURE A33.2

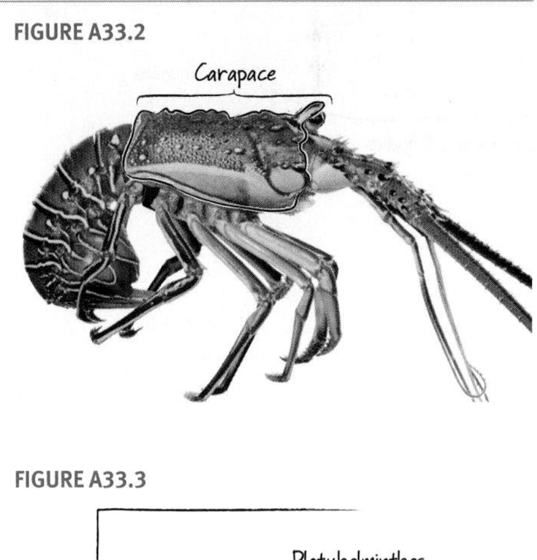

FIGURE A33.3

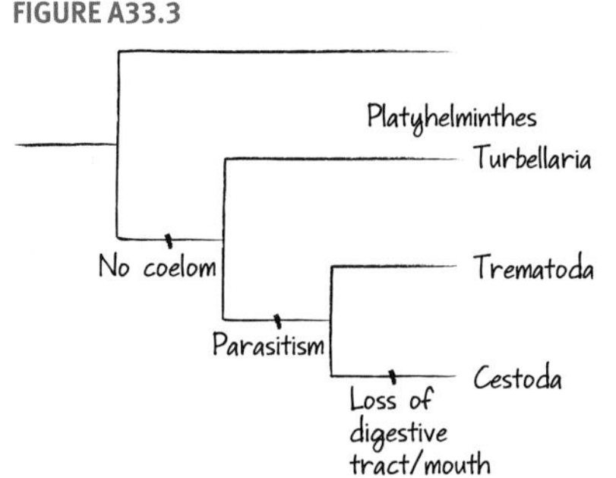

You Should Be Able To (YSBAT)

YSBAT p. 649a, YSBAT p. 649b, YSBAT p. 650 [See Figure A34.1] **YSBAT p. 655** [See Figure A34.2] **YSBAT p. 660** [See Figure A34.3] **YSBAT p. 662a, YSBAT p. 662b, YSBAT p. 663, YSBAT p. 664** [See Figure A34.4] **YSBAT p. 665, YSBAT p. 666a, YSBAT p. 666b, YSBAT p. 667a, YSBAT p. 667b, YSBAT p. 668** [See Figure A34.3]

Caption Questions and Exercises

Figure 34.1 Invertebrates are a paraphyletic group. The group does not include all the descendants of the common ancestor because vertebrates are excluded. **Figure 34.12** Mammals and birds are equally related to amphibians, because birds and mammals share a common ancestor, and this ancestor shares a common ancestor with amphibians. **Figure 34.14** Yes—even a rudimentary jaw could have been useful to better grasp prey or to change the size of the mouth. **Figure 34.16** If this phylogeny were estimated on the basis of limb traits, it would be based on the *assumption* that the limb evolved from fish fins in a series of steps. But because the phylogeny was estimated from other types of data, it is legitimate to use it to analyze the evolution of the limb without any assumptions about how limbs evolved. **Figure 34.20** They are smaller—their function in an amniotic egg has been taken over by the placenta. **Figure 34.37** Four hominin species existed 2.2 mya, five existed 1.8 mya, and four existed 100,000 years ago. **Figure 34.38** The forehead became much larger with the face becoming "flatter"; the brow ridges are less prominent in later skulls than in earlier skulls.

Summary of Key Concepts

KC 34.1 (1) Mussels and clams will increase dramatically; (2) kelp density will increase. **KC 34.2** Like today's lungfish, the earliest tetrapods could have used their limbs for pulling themselves along the substrate in shallow-water habitats. **KC 34.3** The earliest fossils of *H. sapiens* are found in Africa, and phylogenetic analyses of living human populations indicate that the most basal groups are all African.

Test Your Knowledge

1. a; **2.** a; **3.** d; **4.** a; **5.** d; **6.** b

Test Your Understanding

1. It is an enclosed, fluid-filled structure surrounded by muscle. Muscular contraction forces water into the tube feet, resulting in extension of the podia. **2.** Pharyngeal gill slits function in suspension feeding. The notochord furnishes a simple endoskeleton that stiffens the body; electrical signals that coordinate movement are carried by the dorsal hollow nerve cord to the muscles in the tail, which beats back and forth to make swimming possible. Cephalochordates and ascidians are chordates but they are not vertebrates. **3.** If jaws are derived forms of gill arches, then the same genes and the same cells should be involved in the development of the jaw and the gill arches. **4.** Homologous genes are involved in the formation of the fins of ray-finned fish and the limbs of tetrapods. This observation supports a prediction of the fins-to-limbs hypothesis. **5.** The hominins fulfill the criteria for an adaptive radiation: Over a short time interval, many species that occupy an array of foods and habitats evolved. Changes in tooth and jaw structure, tool use, and body size suggest that different hominin species exploited different types of food. **6.** Increased parental care allows offspring to be better developed, and thus have increased chances of survival, before they have to live on their own.

Applying Concepts to New Situations

1. If confirmed, it means that pharyngeal gill slits were present in the earliest echinoderms and lost later. **2.** No—most of the feathered dinosaurs known from fossils did not fly. (The first feathers may have functioned in display or as insulation.) **3.** Xenoturbellidans either retain traits that were present in the common ancestor or all deuterostomes, or they have lost many complex morphological characteristics. **4.** Independently. Birds are a highly derived lineage of reptiles, and most reptiles are not endothermic, so it is logical to infer that the common ancestor of birds and mammals was also not endothermic.

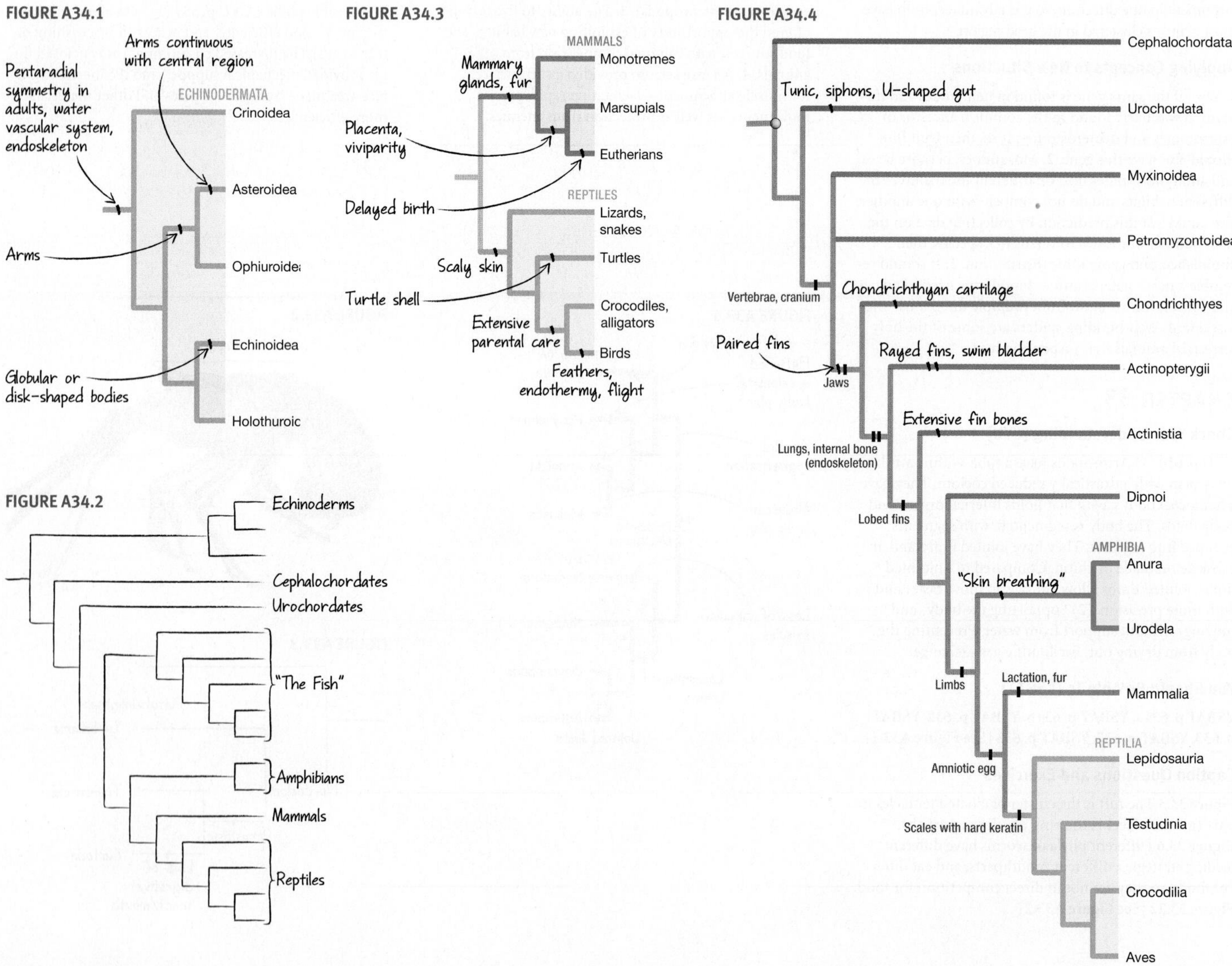

CHAPTER 35

Check Your Understanding (CYU)

CYU p. 686 HIV binds to CD4 and a co-receptor on the surface of T cells. This binding allows the viral envelope and the T-cell plasma membrane to fuse. The viral capsid then enters the cell. **CYU p. 689** (1) After the genome is copied to form positive-sense RNA, the transcript has to be copied to form a negative-sense, single-stranded genome. (2) Much more dangerous—with airborne transmission, a noninfected individual can become infected simply by inhaling virus particles released from an infected individual. This mode of transmission is much more efficient than transmission via body fluids.

You Should Be Able To (YSBAT)

YSBAT p. 680 In the lytic cycle, viral genes are transmitted to a new generation of virions. In lysogenic growth, viral genes are transmitted to daughter cells of the host cell. **YSBAT p. 686** [See Figure A35.1]

Caption Questions and Exercises

Figure 35.1 Many answers are possible: Herpes simplex 2: genitalia; sexually transmitted; HIV: immune system; sexually transmitted; Influenza virus: respiratory tract; coughing/sneezing (airborne); Chicken pox (varicella zoster virus): skin; direct contact. **Figure 35.8** Lysogenic bacteriophages are similar to transposable elements because they insert their DNA into the host cell chromosome and are passed on to daughter cells. They are unlike transposable elements because they can switch to a lytic cycle. **Figure 35.9** No—it only shows the CD4 is required. (Subsequent work showed that other proteins are involved as well.) **Figure 35.14** Budding viruses may disrupt the integrity of the host-cell plasma membrane enough to kill the cell. **Figure 35.15** You should have the following bars and labels: on branch to HIV-2 (sooty mangabey to human); on branch to HIV-1 strain O (chimp to human); on branch to HIV-1 strain N (chimp to human); on branch to HIV-1 strain M (chimp to human).

Summary of Key Concepts

KC 35.1 There is heritable variation among virions, due to random changes that occur as their genomes are copied. There is also differential reproductive success among virions in their ability to successfully infect host cells. This differential success is due to the presence of certain heritable traits. **KC 35.2** As the virus's envelope proteins are produced, they are inserted into the host cell plasma membrane. During budding, the capsid becomes surrounded by host cell membrane that includes the envelope proteins. **KC 35.3** (1) Fusion inhibitors block viral envelope proteins or host cell receptors; (2) protease inhibitors prevent processing/assembly of viral proteins; (3) reverse transcriptase inhibitors block reverse transcriptase, preventing replication of genome; (4) use of condoms or practice of monogamy prevents transmission. **KC 35.4** To replicate an ssDNA genome, a viral DNA polymerase copies the genome into a complementary strand, which then is used as a template to generate an ssDNA that is identical to the original viral genome.

Test Your Knowledge

1. b; **2.** c; **3.** d; **4.** b; **5.** b; **6.** d

Test Your Understanding

1. HIV has an envelope on its outer surface; adenovirus has only a capsid on its outer surface. A virus with an envelope exits host cells by budding. A virus that lacks an envelope exits host cells by lysis (or other mechanisms that don't involve budding). **2.** T4 bacteriophage should have a larger genome than HIV does, because T4 is much more complex morphologically—it should have genes that code for the proteins required for its head and tail regions. **3.** Growth rate is much higher during lytic growth, but viral DNA can increase during latent/lysogenic growth if the host cell is actively dividing and transmitting viral genomes to daughter cells. **4.** Viruses rely on host-cell enzymes to replicate, whereas bacteria do not. Therefore, many drugs designed to disrupt the virus life cycle cannot be used because they would kill host cells as well. Only viral-specific proteins are good targets for drug design. **5.** Each major hypothesis to explain the origin of viruses—from transposable-element-like sequences, from symbiotic bacteria, and from RNA genomes present early in Earth history—is associated with a different type of genome: single-stranded DNA or possibly single-stranded RNA, double-stranded DNA, and single- or double-stranded RNA, respectively. **6.** The phylogenies of SIVs and HIVs show that the two shared common ancestors, but that SIVs are ancestral to the HIVs. Also, there are plausible mechanisms for SIVs to be transmitted to humans through butchering or contact with pets, but fewer or no plausible mechanisms for HIVs to be transmitted to monkeys or chimps.

Applying Concepts to New Situations

1. Culture the *Staphylococcus* strain outside the human host and then add the virus to determine whether the virus kills the bacterium efficiently. Then test the virus on cultured human cells to determine whether the virus harms human cells. Then test the virus on monkeys or other animals to determine if it is safe. Finally, the virus could be tested on human volunteers. **2.** Prevention is currently the most cost-effective program but does not help people who are already infected. Treatment with effective drugs not only prolongs lives but also reduces virus loads in infected people, so that they have less chance of infecting others. **3.** Eukarya evolved from ancestors that did not contain genes encoding restriction endonucleases. Or, methylation of DNA performs other important functions in eukaryotic gene expression (see Chapter 18), so is not appropriate as a guard against endonuclease action. **4.** Viruses cannot be considered to be alive by the definition given in (a), because viruses are not capable of replicating by themselves. By the definition given in (b), it can be argued that viruses are alive because they store, maintain, replicate, and use genetic information—although they cannot perform all these tasks on their own.

FIGURE A35.1

(+)ssRNA —translation→ Viral proteins

(−)ssRNA —RNA replicase→ mRNA —translation→ Viral proteins

dsRNA —transcription→ mRNA —translation→ Viral proteins

dsDNA or ssDNA —transcription→ mRNA —translation→ Viral proteins

(+)ssRNA —reverse transcription→ dsDNA —transcription→ mRNA —translation→ Viral proteins

CHAPTER 36

Check Your Understanding (CYU)

CYU p. 704 (1) [See Figure A36.1] (2) The generalized body of a plant has a taproot with many lateral roots and broad leaves. Examples of deviations from this generalized body structure include: Modified roots: Unlike taproots, *fibrous roots* do not have one central root, and *adventitious roots* arise from stems. Modified stems: *Stolons* grow along the soil and grow roots and leaves at each node. *Rhizomes* grow horizontally underground. Both of these modified stems function in asexual reproduction. Modified leaves: The *needle-shaped leaves* of cacti lose less water to transpiration than do typical

FIGURE A36.1

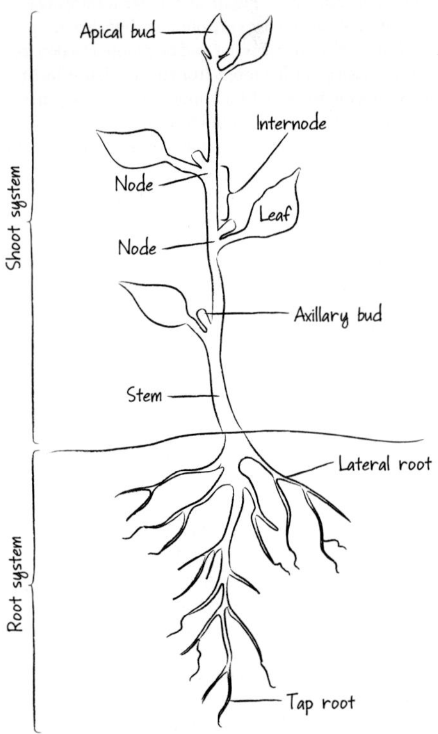

broad leaves, and they protect the plant from predation. *Tendrils* on climbing plants are modified leaves that do not photosynthesize but wrap around trees or other substrates to facilitate climbing. **CYU p. 712** (1) The apical meristem gives rise to the three primary meristems. The apical meristem is a single mass of cells localized at the tip of a root or shoot; the primary meristems are localized in distinctive locations behind the apical meristem. (2) Epidermal cells are flattened and lack chloroplasts; they secrete the cuticle (in shoots) or extend water and nutrient-absorbing root hairs (in roots) and protect the plant. Parenchyma cells are "workhorse cells" found throughout the plant body; they perform photosynthesis and synthesize and/or store materials. Tracheids are long, thin cells with pits in their secondary cell walls; they are found in xylem and are dead when mature; they conduct water and solutes up the plant. Sieve-tube elements are long, thin cells found in phloem; they are alive when mature and conduct sugars and other solutes up and down the plant. **CYU p. 715** (1) Primary growth increases the length of roots and shoots, and secondary growth increases their width. The function of primary growth is to extend the reach of the root and shoot system and thus increase a plant's ability to absorb light and acquire carbon dioxide, water, and nutrients. The function of secondary growth is to increase the amount of lateral conducting tissue and provide the structural support required for extending primary growth at the top of the plant. (2) The rings are small on the shaded side and large on the sunny side.

You Should Be Able To (YSBAT)

YSBAT p. 713a, YSBAT p. 713b, YSBAT p. 713c, YSBAT p. 714 [See Figure A36.2]

Caption Questions and Exercises

Figure 36.2 New branches and leaves should develop to the right (the plant will also lean that way); new lateral roots will develop to the left. **Figure 36.4** Lawn grasses are shallow rooted, so when the top part of soil dries out, the plants cannot grow. **Figure 36.5** For aerobic respiration, which generates the ATP needed to keep the cells alive. **Figure 36.7** The individuals are genetically identical—thus, any differences in size or shape in the different habitats are due to phenotypic plasticity and not genetic differences. **Figure 36.9** Large surface area provides more area to capture photons, but also more area from which to lose water and be exposed to potentially damaging tearing forces from wind. **Figure 36.20** Many possible answers; for example, in leaf cells, genes for forming chloroplasts would be expressed. These genes wouldn't be expressed in the root cells. **Figure 36.25** The leading hypothesis is that they lack many organelles so that the space inside the cell is available to transport nutrients. **Figure 36.27** You should have labeled a light band as early wood, and the dark band to its right as late wood; both bands together comprise one growth ring.

Summary of Key Concepts

KC 36.1 Phenotypic plasticity is more important in (1) environments where conditions vary because it gives individuals the ability to change the growth pattern of their roots, shoots, and stems to access sunlight, water, and other nutrients as the environment changes; and (2) in long-lived species because it gives individuals a mechanism to change their growth pattern as the environment changes throughout their lifetime. **KC 36.2** (1) Plants that live in warm, arid climates would have large root systems to increase their chances of accessing water, and small shoot systems because water is lost from surfaces exposed to air. (2) Plants that live in very wet climates would have small root systems because water is plentiful, and large shoot systems because the abundant water would support more aboveground growth especially in light-limiting environments such as forests. **KC 36.3** (1) In both cases, the structure would stop growing; (2) the shoot would lack epidermal cells and would die. **KC 36.4** (1) The plant would not produce secondary xylem and phloem, so its girth and transport ability would be reduced, though it would still produce bark. (2) The trunk would have much wider tree rings on one side than the other side.

Test Your Knowledge

1. b; **2.** b; **3.** a; **4.** d; **5.** c; **6.** a

Test Your Understanding

1. The general function of both systems is to acquire resources: The shoot system captures light and carbon dioxide; the root system absorbs water and nutrients. Vascular tissue is continuous throughout both the shoot and root systems. Diversity in roots and shoots enables plants of different species to live together in the same environment without directly competing for resources. **2.** Continuous growth enhances phenotypic plasticity because it allows plants to grow and respond to changes or challenges in their environment (such as changes in light and water availability). **3.** Cactus spines are modified leaves; thorns are modified stems. **4.** Cuticle reduces water loss; stomata facilitate gas exchange. Plants from wet habitats should have a relatively large number of stomata and thin cuticle. Plants living in dry habitats should have relatively few stomata and thick cuticle. **5.** Parenchyma cells in the ground tissue perform photosynthesis and/or synthesize and store materials. In vascular tissue, parenchyma cells in "rays" conduct water and solutes across the stem; other parenchyma cells differentiate into the sieve-tube members and companion cells in phloem. **6.** Cells produced to the inside of the vascular cambium differentiate into secondary xylem; cells produced to the outside of the vascular cambium differentiate into secondary phloem.

Applying Concepts to New Situations

1. Fewer sclerenchyma cells, because sclerenchyma cells have extremely tough walls and these populations have been selected to be tender. Humans select individuals with fewer sclerenchyma cells to be the parents of the next generation, so the frequency of sclerenchyma cells will decline over time. **2.** Asparagus—stem; Brussels sprouts—lateral bud; celery—petiole; spinach—leaf (petiole and blade); carrots—taproot; potato—modified stem. **3.** They grow continuously, so do not have alternating groups of large and small cells that form rings. **4.** Girdling disrupts transport of solutes in secondary phloem, and damages (but probably doesn't eliminate) the transport of water and solutes in secondary xylem. The tree starves.

CHAPTER 37

Check Your Understanding (CYU)

CYU p. 721 (1) Wet soils have almost no pressure potential, while dry soils have a highly negative pressure potential because the few water molecules present cling to soil particles. (2) Salty soils have extremely low solute potentials compared to typical soils, because the concentration of solutes is high. **CYU p. 727** (1) At the air-water interface under a stoma, hydrogen bonding is asymmetrical—the water molecules can bond only with water molecules below them. This pulls the water molecules down—creating tension at the surface. When transpiration occurs, there are few water molecules at the surface, which makes the asymmetry in bonding even more pronounced—increasing tension. (2) When a nearby stoma closes, the humidity inside the air space increases, transpiration decreases, and surface tension on menisci decreases. The same changes occur when a rain shower starts. When air outside the leaf dries, though, the opposite occurs: The humidity inside the air space decreases, transpiration increases, and surface tension on menisci increases. **CYU p. 734** (1) In early spring, the water potential of phloem sap near leaves is low because developing leaves are using sucrose much faster than they make it; root cells are releasing sucrose so the water potential of phloem cells in roots is high. (2) Midday in summer, the water potential of phloem sap near leaves is high because leaves are making much more sucrose than they consume. Root cells are storing sugar, however, so the water potential of phloem in roots is low.

You Should Be Able To (YSBAT)

YSBAT p. 719 (1) Add a solute (e.g., salt or sugar) to the left side of the U-tube, so that solute concentration is higher than on the right side. (2) Water would move from the right side of the U-tube into the left side. **YSBAT p. 720** Pull the plunger up. **YSBAT p. 723** The individual would not be able to exclude certain solutes from the xylem. If they were transported throughout the plant in high enough concentrations, they could damage tissues.

Caption Questions and Exercises

Figure 37.2 It increases (becomes less negative), which reduces the solute potential gradient between the two

FIGURE A36.2

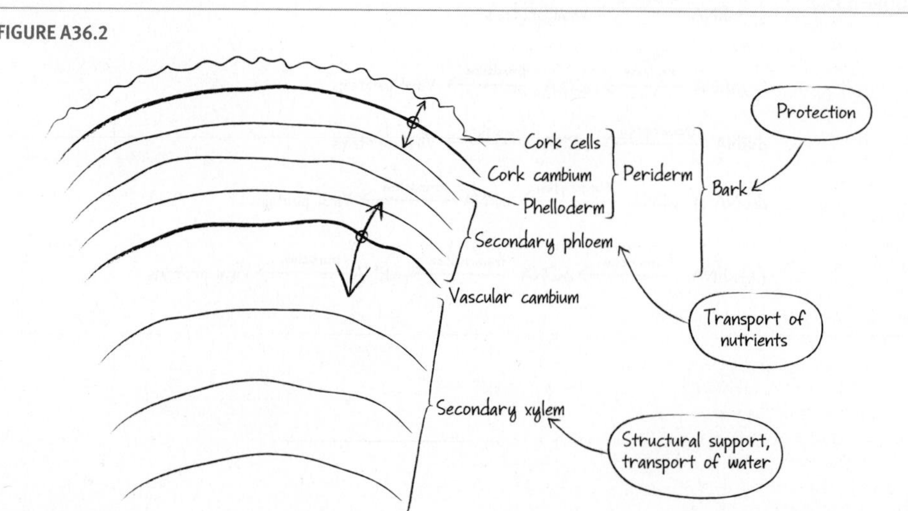

sides. **Figure 37.7** [See Figure A37.1] **Figure 37.10** Your labels should read as follows: Atmosphere: $\psi = -95.2$ MPa; Leaves: $\psi = -0.8$ MPa; Roots: $\psi = -0.6$ MPa; Soil: $\psi = -0.3$ MPa. **Figure 37.12** The line on the graph would go up, with large jumps occurring each time the light level was turned up, because increased transpiration rates would increase negative pressure (tension) at the leaf surface, and thus increase the water potential gradient. **Figure 37.17** [See Figure A37.2]

Summary of Key Concepts

KC 37.1 Because xylem cells are dead at maturity, there are no plasma membranes for water to cross and thus no solute potential. The only significant force acting on water in xylem is pressure. **KC 37.2** It drops, because transpiration rates from stomata decrease, reducing the water potential gradient between roots and leaves.
KC 37.3 To create high pressure in phloem near sources and low pressure near sinks, water has to move from xylem to phloem near sources and from phloem back to xylem near sinks. If xylem were not close to phloem, this water movement could not occur and the pressure gradient in phloem wouldn't exist.

Test Your Knowledge

1. d; **2.** c; **3.** b; **4.** a; **5.** b; **6.** c

Test Your Understanding

1. [See Figure A37.3] **2.** The force responsible for root pressure is the high solute potential of root cells at night, when they continue to accumulate ions from soil, with water following by osmosis. The mechanism is active, in the sense that ions are imported against their concentration gradient. Capillarity is passive; it is driven by adhesion of water molecules to the sides of xylem cells that creates a pull upwards, and cohesion with water molecules below. Cohesion-tension is also passive, because it is driven by the loss of water molecules from menisci in leaves, which creates a large negative pressure (tension). **3.** Water moves up xylem because of transpiration-induced tension created at the air-water interface under stomata, which is communicated down to the roots by the cohesion of water molecules. Sap flows in phloem because of a pressure gradient that exists between source and sink cells, driven by differences in sucrose concentration and flows of water into or out of nearby xylem. **4.** Aphid A is attacking a source, and Aphid B is attacking a sink. If sucrose concentrations are higher in phloem sap near sources, then aphids should be found primarily near mature leaves. (It's actually more common to find them near apical meristems, though—probably because the tissues are not as stiff and strong and difficult to pierce.) **5.** By pumping protons out, companion cells create a strong electrochemical gradient favoring entry of protons. A cotransport protein (a symporter) uses this proton gradient to import sucrose molecules *against* their concentration gradient. **6.** When it germinates—sucrose stored inside the seed is released to the growing embryo, which cannot yet make enough sucrose to feed itself.

Applying Concepts to New Situations

1. Plants have to gain CO_2 for photosynthesis to occur, which means that stomata have to be open. But transpiration occurs whenever stomata are open and the surrounding air is drier than the inside of the leaf. A similar "side effect" occurs when terrestrial animals breathe—they have to take dry air into their lungs, where water evaporates from the body and is lost. **2.** Plants would not grow as quickly or as tall, because the taller they grew, the more energy they would need to use to transport water and thus the less they would have available for growth. **3.** Because they cannot readily replace water that is lost to transpiration, they have to close stomata. During a heat wave, transpiration cannot cool the plant's tissues. They bake to death. **4.** Closing aquaporins will slow or stop movement of water from cells into the xylem and out of the plant via transpiration. Because water does not leave the cells, they are able to maintain turgor (normal solute potentials).

CHAPTER 38

Check Your Understanding (CYU)

CYU p. 748 (1) Proton pumps establish an electrical gradient across root hair membranes, with the inside of the membrane being much more negative than the outside.

FIGURE A37.1

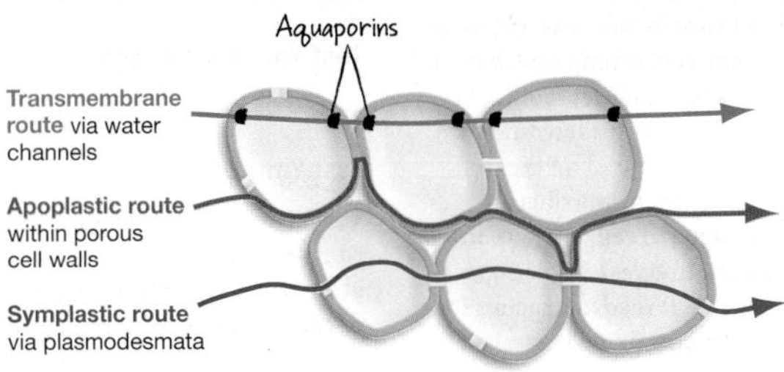

FIGURE A37.2

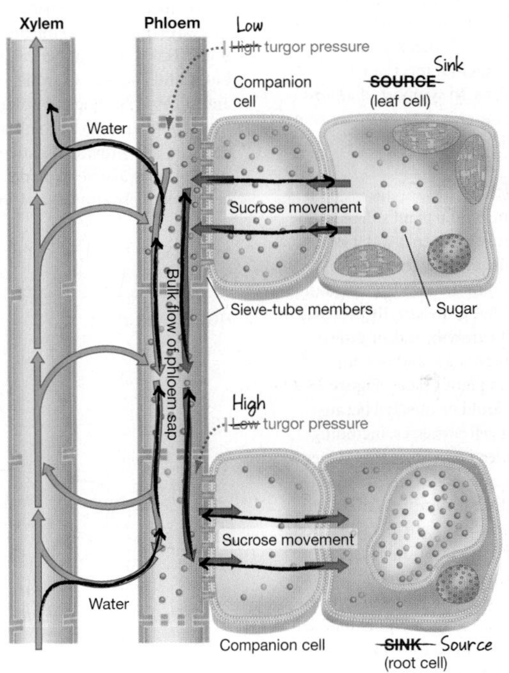

FIGURE A37.3

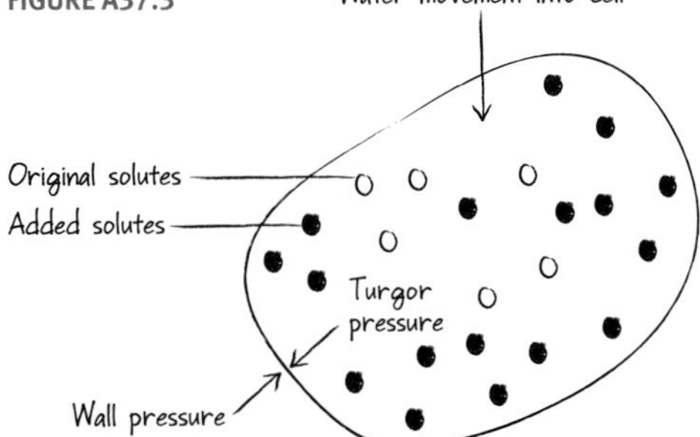

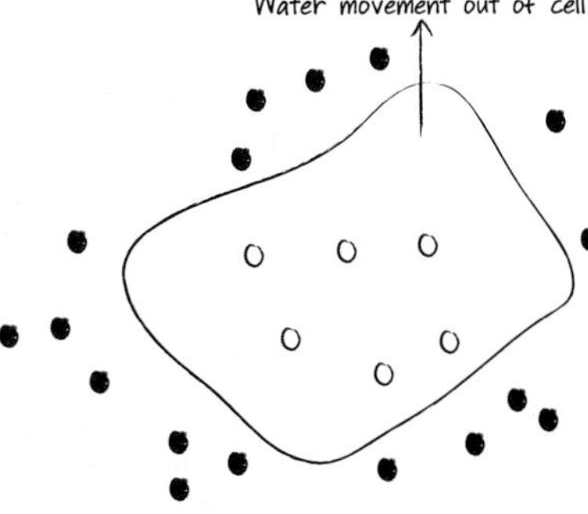

Cations follow this electrical gradient into the cell. (2) [See Figure A38.1] **CYU p. 750** (1) Many possible answers, including the following: If N-containing ions are abundant in the soil, the energetic cost of maintaining nitrogen-fixing bacteria may outweigh the benefits in terms of increased nitrogen availability. If mycorrhizal fungi provide a plant with nitrogen-containing ions, the energetic cost of maintaining nitrogen-fixing bacteria may outweigh the benefits in terms of increased nitrogen availability. Plants that grow near species with N-fixing bacteria might gain nitrogen by absorbing nitrogen-containing ions after root nodules die, or by "stealing it" (roots of different species will often grow together). There could be a genetic constraint (see Chapter 24): Plant species may simply lack the alleles required to manage the relationship. Note: To be considered correct, your hypothesis should be plausible (based on the underlying biology) and testable. (2) Many possible answers, including: Because the bacterial cell enters root cells, it is important for the plant to have reliable signals that it is not a parasitic bacterium. It is advantageous for the plant to be able to reject nitrogen-fixing bacteria if usable nitrogen is already abundant in the soil. The signals have no function.

You Should Be Able To (YSBAT)

YSBAT p. 745 (1) If there is no voltage across the root hair membrane, there is no electrical gradient favoring entry of cations, and absorption stops. (2) There is no route for cations to cross the root hair membrane, so absorption stops. **YSBAT p. 746** (1) If there is no proton gradient across the root hair membrane, there is no gradient favoring entry of protons, so symport of anions stops. (2) There is no route for anions to cross the root hair membrane, so absorption stops.

Caption Questions and Exercises

Figure 38.1 Point out that most of a plant's mass is due to carbon-containing molecules, but that water does not contain carbon. To test the hypothesis that most of a plant's mass comes from CO_2 in the atmosphere, design an experiment where plants are grown in the presence or absence of CO_2 and compare growth, and/or grow plants in the presence of labeled CO_2 and document the presence of labeled carbon in plant tissues. **Figure 38.2** Proteins and nucleic acids would be affected because they contain nitrogen. Most cell processes, including photosynthesis, would be affected because they depend on proteins. **Figure 38.3** Because soil is so complex, it would be difficult to defend the assumption that the soils with and without added copper are identical except for the difference in copper concentrations. Also, it would be better to compare treatments that had no copper versus normal amounts of copper, instead of comparing with and without added copper. **Figure 38.8** One cation (a proton) is being exchanged for another (calcium or magnesium). **Figure 38.10** The proton gradient arrow in (b) should begin above the membrane and cross the membrane pointing down. The electrical gradient arrow in (c) should begin above the membrane (marked positive) and cross the membrane pointing down (toward negative). **Figure 38.17** Mistletoe is parasitic (it extracts water and certain nutrients from host trees); bromeliads are not (they don't affect the fitness of their host plants).

Summary of Key Concepts

KC 38.1 Add nitrogen to an experimental plot and compare growth within the plot to growth of a similar plot nearby. Do many replicates of the comparison, to convince yourself and others that the results are not due to unusual circumstances in one or a few plots. **KC 38.2** A higher membrane voltage would allow cations in soil to cross root-hair membranes more readily through membrane channels and would allow anions to enter more readily via symporters. Nutrient absorption should increase. **KC 38.3** Grow pea plants in the presence of rhizobia, exposed to air with (1) N_2 containing the heavy isotope of nitrogen, and (2) radioactive carbon dioxide. Allow the plant to grow and then analyze the rhizobia and plant tissues. If the rhizobia-plant interaction is mutualistic, then ^{15}N-containing compounds and radioactive-carbon-containing compounds should be observed in both rhizobia and plant tissues. As a control, grow pea plants in the presence of the labeled compounds but without rhizobia. If the mutualism hypothesis is correct, the plant should contain labeled carbon but no heavy nitrogen.

Test Your Knowledge

1. b; **2.** d; **3.** d; **4.** c; **5.** b; **6.**d

Test Your Understanding

1. (1) Grow a large sample of corn plants with a solution containing all essential nutrients needed for normal growth and reproduction. Grow another set of genetically identical (or similar) corn plants in a solution containing all essential nutrients except iron. Compare growth, color, and seed production (and/or other attributes) in the two treatments. (2) Grow corn plants in solutions containing all essential nutrients in normal amounts, but with varying concentrations of iron. Determine the lowest iron concentration at which plants exhibit normal growth rates. **2.** If nutrients in soil are scarce, there is intense natural selection favoring alternative ways of obtaining nutrients—for example, by digesting insects or stealing nutrients. **3.** Higher in soils with both sand and clay. Clay fills spaces between sand grains and slows (1) passage of water through soil and away from roots, and (2) leaching of anions in soil water. Clay particles also provide negative charges that retain cations so they are available to plants. The presence of sand keeps soil loose enough to allow roots to penetrate. **4.** H^+-ATPases in the plasma membrane pump protons out of the cell—making the inside of the membrane less positive (more negative) and the outside of the membrane more positive. By convention, the voltage on a membrane is expressed as inside-relative-to-outside. **5.** Some metal ions are poisonous to plants, and high levels of other ions are toxic. Passive mechanisms of ion exclusion—such as a lack of ion channels that allow passage into root cells—do not require an expenditure of ATP. Active mechanisms of exclusion—such as production of metallothioneins—require an expenditure of ATP. **6.** (1) The amounts required and the mechanisms of absorption are different. Much larger amounts of C, H, and O are needed than mineral nutrients. CO_2 enters plants via stomata and water is absorbed passively along a water potential gradient; mineral nutrients enter via active uptake in root hairs or via mycorrhizae. (2) Macronutrients are required in relatively large quantities and usually function as components of macromolecules; micronutrients are required in relatively small quantities and usually function as enzyme cofactors.

Applying Concepts to New Situations

1. The water that he used may have contained many of the macro- and micronutrients that were incorporated into plant mass. To test this hypothesis, conduct an experiment of the same design but have one treatment with pure water and one treatment with water containing solutes ("hard water"). Compare the percentage of soil mass incorporated into the willow tree in both treatments. Willow may be unusual. To test this hypothesis, repeat the experiment with willow and several other species under identical conditions, and compare the percentage of soil mass incorporated into the plants. Van Helmont's measurements were inaccurate. Repeat the experiment. **2.** Acid rain inundates the soil with protons, which bind with the negatively charged clay parti-

FIGURE A38.1

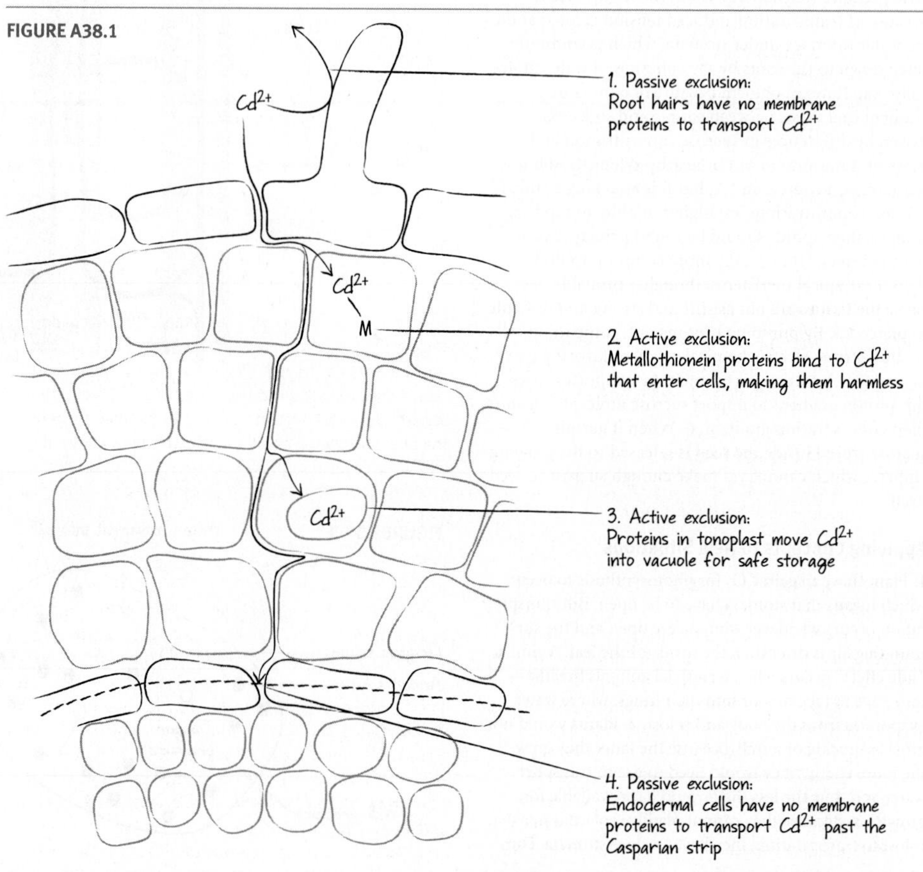

cles—displacing the cations that are normally bound there. Cations may then be leached away. **3.** If the hypothesis is correct, vanadate should inhibit phosphorus uptake. To test this hypothesis, grow plants in solution with radioactively labeled phosphorus without vanadate and in solution with vanadate. Measure the amount of labeled phosphorus in root cells with and without vanadate. **4.** Alder decomposes much faster, because it provides more nitrogen to the fungi and bacteria that are responsible for decomposition—meaning that they can grow faster.

CHAPTER 39

Check Your Understanding (CYU)

CYU p. 757 (1) The only cells that can respond to a specific environmental signal or hormone are cells that have an appropriate receptor for that signal or hormone. (2) Once a receptor is activated, it triggers production of many second messengers or the activation of many proteins in a phosphorylation cascade. **CYU p. 762** (1) Using a syringe or other device, apply auxin to the west side of each stem, near the base. **CYU p. 764** (1) Lettuce grows best in full sunlight. If a seed receives red light, it indicates that the seed is in full sunlight—meaning that conditions for germination are good. But if a seed receives far-red light, it indicates that the seed is in shade—meaning that conditions for germination are poor. (2) Far-red light indicates shade, and stem-elongation gives plants a chance to grow above nearby plants into full sunlight. **CYU p. 766** (1) In both roots and shoots, auxin is redistributed asymmetrically in response to a signal. In phototropism, auxin is redistributed to the shaded side of the plant in response to blue light. In gravitropism, auxin is redistributed to the lower part of the root or shoot in response to gravity. (2) In root cells, auxin concentrations indicate the direction of gravity. The gravitropic response is triggered by changes in the distribution of auxin in root-tip cells. **CYU p. 776** (1) Short term, stomata should close and photosynthesis should stop. Long term, the plants will not grow well, compared to individuals that do not receive extra ABA, or die. (2) Their stems should elongate rapidly, compared to plants that do not receive extra GAs. **CYU p. 781** (1) Because specific *R* gene products recognize and match specific proteins produced by pathogens, having a wide array of *R* alleles allows individuals to recognize and respond to a wide array of pathogens. (2) If herbivores are not present, synthesizing large quantities of proteinase inhibitors wastes resources (ATP and substrates) that could be used for growth and reproduction.

You Should Be Able To (YSBAT)

YSBAT p. 777 Virtually all bananas have the same *R* alleles. If a pathogen evolves that is not identified by the existing *R* alleles, all bananas would be susceptible to the new disease.

Caption Questions and Exercises

Figure 39.4 These cells served as controls. If the radioactive band appeared in cells that did not have the *PHOT1* gene, it clearly could not be a *PHOT1* band. If the band appeared in cells that had not been exposed to blue light, then *PHOT1* would not be acting as a blue-light receptor. **Figure 39.5** For "Tip removed": cut the tip and replace it, to control for the hypothesis that the wound itself influences bending (not the loss of the tip). For "Tip covered": cover the tip with a transparent cover instead of opaque one, to control for the hypothesis that the cover itself affects bending (not excluding light). **Figure 39.7** Two possibilities: (1) none of the decapitated shoots would bend, or (2) the blocks from the completely divided tip would bend much less than the partially divided tip, because more cells were disrupted. **Table 39.1** The average germination rate of lettuce seeds last exposed to red light is about 99 percent. The average for seeds last exposed to far-red light is about 50 percent. Both of these values are much higher than the germination rate of buried seeds that receive no light at all; their germination rate is only 9 percent. **Figure 39.10** A signal in the form of a light wavelength is changed into a signal in the form of a protein's shape. In this way, a signal from outside the cell is changed to a signal inside the cell. **Figure 39.12** (One possibility) Direct pressure from a statolith changes the shape of the receptor protein. The shape change triggers a signal transduction cascade leading to a gravitropic cell response. **Figure 39.14** In nature, this response is most likely to occur in windy environments. A reduction in height might keep plants beneath the path of the wind, and stockier plants would be better able to resist bending when blown. **Figure 39.19** The dwarfed individuals can put more of their available energy into reproduction because they are using less energy for growth than taller individuals. **Figure 39.21** Without these data, a critic could argue that the stomata closed in response to water potentials in the leaf, not a signal from the roots. **Figure 39.24** The transported nutrients are conserved in the plant instead of being lost when leaves fall. Cessation of chlorophyll synthesis saves plants the energy and nutrients that would have gone into producing chlorophyll in leaves that soon will fall off. **Figure 39.28** The molecule travels throughout the shoot system (and root system).

Summary of Key Concepts

KC 39.1 They do not perform photosynthesis and therefore do not need to switch behavior based on being located in full sunlight versus shade. **KC 39.2** An external signal arrives at a receptor cell/protein (a person sees a barn on fire); the person calls (the receptor transduces the signal); the dispatcher rings a siren (a second messenger is produced or phosphorylation cascade occurs); the fire department members get to the pumper truck (gene expression or some other cell activity changes). **KC 39.3** Sun-adapted (high-light) plants are likely to have a significant positive phototropic response (i.e., growth toward light), whereas shade-adapted (low-light) plants are less likely to respond phototropically. **KC 39.4** They need to be transported throughout the plant (because cells throughout the body need to respond to the signal)—not in a single direction.

Test Your Knowledge

1. d; **2.** a; **3.** b; **4.** b; **5.** c; **6.** b

Test Your Understanding

1. The P_{fr}-P_r switch in phytochromes lets plants know whether they are in shade or sunlight, and triggers responses that allow sun-adapted plants to avoid shade. Phototropins are activated by blue light, which indicates full sunlight, and trigger responses that allow plants to photosynthesize at full capacity. **2.** If *PHOT1* were inserted into a plant, a critic could contend that it was being phosphorylated by a plant protein. But because the *PHOT1* was in an insect cell, this alternative hypothesis was not credible. **3.** *Transduce* means "to convert energy from one form to another." There are many examples: a touch may be converted to an electrochemical potential, light energy or pressure into a shape change in a protein, and so on. **4.** (Many possibilities) Cells or tissue involved: Auxin directs gravitropism in roots but phototropism in shoots. Developmental stage or age: GA breaks dormancy in seeds but extends stems in older individuals. Concentration: Large concentrations of auxin in stems promote cell elongation; small concentrations do not. Other hormones: The concentration of GA relative to ABA influences whether germination proceeds. **5.** Receptor cells sense changes in the availability of blue light received. They respond by changing the distribution of auxin. A change in auxin concentration causes target cells in the stem to grow, and results in the stem bending toward the source of blue light. In this way, plants can respond to changes in shading by growing toward areas where light is available. **6.** Ethylene promotes senescence. Fruits exposed to ethylene ripen faster than fruits that are not exposed to ethylene; increased ethylene sensitivity in leaves (combined with reduced auxin concentrations) causes abscission.

Applying Concepts to New Situations

1. Small-seeded plants need to perform photosynthesis early in seedling development or they will starve. Therefore, they need to germinate in direct sunlight. The food reserves in large-seeded plants can support seedling growth for a relatively long time without photosynthesis. Therefore, they do not need to germinate in direct sunlight. **2.** The experiment provides evidence that cytokinins stimulate cell division in lateral buds—suggesting that large amounts of cytokinins can overcome apical dominance. **3.** In these seeds, (a) little or no ABA is present, or (b) ABA is easily leached from the seeds, so its inhibitory effects are eliminated. Test hypothesis (a) by determining whether ABA is present in the seeds. Test hypothesis (b) by comparing the amount of ABA present before and after running water—enough to mimic the amount of rain that falls in a tropical rain forest over a few days or weeks—over the seeds. **4.** Yes—the experiments are consistent with the hypothesis that ABA must be on the surface of cells to trigger a response, meaning that the receptor must be located there.

CHAPTER 40

Check Your Understanding (CYU)

CYU p. 792 (1) Both are diploid cells that divide by meiosis to produce spores. The megasporocyte produces a megaspore (female spore); a microsporocyte produces a microspore (male spore). (2) Both are multicellular individuals that produce gametes by mitosis. The female gametophyte is larger than the male gametophyte and produces an egg; the male gametophyte produces sperm. **CYU p. 795** (1) Insects feed on nectar and/or pollen in flowers. Flowers provide food rewards to insects to encourage visitation. The individuals that attract the most pollinators produce the most offspring. (2) One product of double fertilization is the zygote, which will eventually grow by mitosis into a mature sporophyte. The other product is the endosperm nucleus, which will grow by mitosis to form a source of nutrients for the embryo.

You Should Be Able To (YSBAT)

YSBAT p. 786 (1) The microsporocyte is the male spore; the megasporocyte is the female spore. (2) In angiosperms the female gametophyte stays within the ovary at the base of the flower even after it matures and produces an egg cell. Fertilization and seed development take place in the same location. **YSBAT p. 795** The endosperm nucleus in the central cell is triploid; the zygote is diploid; the synergid and other cells remaining from the female gametophyte are haploid. **YSBAT p. 796** The protoderm gives rise to the epidermal tissue; the ground meristem gives rise to the ground tissue; the procambium gives rise to the vascular tissue.

Caption Questions and Exercises

Figure 40.5 An appropriate control treatment would be to graft a leaf from a plant that had never been exposed to a short-night photoperiod onto a new plant that also had never been exposed to the correct conditions for flowering. If the hypothesis is correct, this grafted leaf should not induce flowering. **Figure 40.9** A gametophyte is the multicellular individual that produces gametes by mitosis. The embryo sac conforms to this definition because it is a multicellular form that produces an egg by mitosis. **Figure 40.10** The pollen grain is a gametophyte because it is multicellular and produces sperm by mitosis.

Figure 40.17 Hypothesis: Fruit changes color when it ripens as a signal to fruit eaters. The color change is advantageous because seeds are not mature in unripe fruit, and fruit eaters disperse mature seeds in ripe fruit. **Figure 40.21** Beans—their cotyledons are aboveground.

Summary of Key Concepts

KC 40.1 The sporophyte is diploid, the gametophyte is haploid, spores and gametes are haploid, and zygotes are diploid. **KC 40.2** There would be four embryo sacs in each ovule, and thus four eggs. **KC 40.3** The carpel is the female portion of the flower and contains the ovary. The ovary contains ovules, which house the female gametophytes. Once fertilization has occurred, the ovary develops into the fruit and the ovules develop into seeds within the fruit.

Test Your Knowledge

1. b; **2.** c; **3.** a; **4.** d; **5.** d; **6.** c

Test Your Understanding

1. Asexual reproduction is a quick, efficient way for a plant to reproduce large numbers of offspring. The disadvantage is that offspring are genetically identical to the parent, and thus vulnerable to the same diseases and only able to thrive in habitats similar to those inhabited by the parent. **2.** Megasporocytes (female) and microsporocytes (male) undergo meiosis to generate megaspores and microspores, respectively. These divide mitotically to give rise to female and male gametophytes—the embryo sac and pollen grain, respectively. **3.** Sepals are the outermost structure in a flower and usually protect the structures within; petals are located just inside the sepals and usually function to advertise the flower to pollinators. In wind- and bee-pollinated flowers, sepals probably do not vary much in overall function or general structure. Petals are often lacking in wind-pollinated species; they tend to be broad and flat (to provide a landing site) and colored purple or blue or yellow in bee-pollinated species (often with ultraviolet markings). **4.** Outcrossing increases genetic diversity among offspring, making them more likely to thrive if environmental conditions change from the parental generation. However, outcrossing requires that cross-pollination is successful. Self-fertilization results in relatively low genetic diversity among offspring but ensures that pollination is successful. **5.** The female gametophyte develops inside the ovule; the ovary develops into the pericarp of the fruit after fertilization. A mature ovule contains both gametophyte and sporophyte tissue; the ovary and carpel are all sporophyte tissue. **6.** The endosperm of a corn seed and the cotyledons of a bean seed both contain nutrients needed for the seed to germinate. The bean cotyledons have absorbed the nutrients of the endosperm.

Applying Concepts to New Situations

1. Because the island is near the equator, photoperiod will not provide much information about when conditions are optimal for germination. Presumably, daily temperature fluctuations will also remain relatively constant. Based on the information provided, the most logical hypotheses are that flowering occurs at the onset of the rainy season and that dry, dormant seeds germinate when exposed to moisture—cues that indicate that conditions for growth are good. **2.** In response to "cheater" plants, insects should be under intense selection pressure to detect and avoid species that do not offer a food reward. Individuals would have to be able to distinguish scents, colors, or other traits that identify cheaters. In response to "cheater" insects, plants possessing unusually thick petals or sepals that prevented insects from bypassing the anthers on their way to the nectaries would be more successful at producing seed in the next generation. **3.** One possibility: Under identical growth conditions, pollinate many individuals of the same species with two pollen grains, and many individuals of this species with one pollen grain. Measure and compare the rates of pollen tube growth. **4.** Acorns have a large, edible mass and are probably animal dispersed (e.g., by squirrels that store them and forget some). Cherries have an edible fruit and are probably animal dispersed. Burrs stick to animals and are dispersed as the animals move around. Milkweed seeds float in wind. To estimate the distance that each type of seed is dispersed from the parent, (1) set up "seed traps" to capture seeds at various distances from the parent plant, (2) sample locations at various distances from the parent and analyze young individuals—using techniques introduced in Chapter 20 to determine if they are offspring from the parent being studied, (3) mark seeds, if possible, and re-find them after dispersal.

CHAPTER 41

Check Your Understanding (CYU)

CYU p. 810 *Connective tissues* consist of cells embedded in a matrix. The density of the matrix determines the rigidity of the connective tissue and its function in padding/protection, structural support, or transport. *Nervous tissue* is composed of neurons and support cells. Neurons have long projections that function as "biological wires" for transmitting electrical signals. *Muscle tissue* has cells that can contract and functions in movement. *Epithelial tissue* has tightly packed layers of cells with distinct apical and basal surfaces. They protect underlying tissues and regulate the entry and exit of materials into these tissues. **CYU p. 814** (1) As three-dimensional size increases, volume increases more quickly than does surface area, so the surface area/volume ratio decreases. (2) They are inversely proportional (negatively correlated). On a per gram basis, small animals have higher basal metabolic rates than large animals. **CYU p. 819** (1) Endotherms can remain active during the winter and at night and sustain high levels of aerobic activities such as running or flying, but require large amounts of food energy. Ectotherms need much less food and can devote a larger proportion of their food intake to reproduction, but have a hard time maintaining high activity levels at night or in cold weather. (2) **[See Figure A41.1]**

You Should Be Able To (YSBAT)

YSBAT p. 812 A newborn.

Caption Questions and Exercises

Figure 41.1 There would be no relationship between the variables. In all three cases, the graph would be a flat line. **Figure 41.3** Loose connective tissues have a soft matrix that is effective for cushioning and protecting organs. Dense connective tissues have a fiber-rich matrix that allows them to link adjacent structures. Structural connective tissues have a stiff or hard matrix that allows them to withstand bending or compressing forces. Fluid connective tissues are effective in transport because liquids move easily and carry solutes. **Figure 41.4** The projections allow electrical signals to be transmitted over long distances. **Figure 41.10** The dog has to eat more because it has a higher mass-specific metabolic rate than a human. **Figure 41.11** There should be no oxygen uptake on either side of the chamber.

Summary of Key Concepts

KC 41.1 Loosening tight junctions would compromise the barrier between the external and internal environments. It would allow water and other substances (including toxins) to pass more readily across the epithelial boundary. **KC 41.2** King Kong is endothermic and his huge mass would generate a great deal of heat. His relatively small surface area would not be able to dissipate the heat—especially if it were covered with fur. **KC 41.3** Individuals could acclimatize by normal homeostatic mechanisms (increased sweating, seeking shade, etc.). Individuals with alleles that allowed them to function better at higher temperatures would produce more offspring than individuals without those alleles. Over time, this would lead to adaptation via natural selection. **KC 41.4** They multiply the amount of heat or material exchanged between the two countercurrent flows, compared to a "con-current" system.

Test Your Knowledge

1. b; **2.** d; **3.** b; **4.** d; **5.** a; **6.** a

Test Your Understanding

1. It is where environmental changes are sensed first. As an effector, it has a large surface area for losing or gaining heat. **2.** A warm frog can move, breathe, and digest much faster than a cold frog, because enzymes work faster (rates of chemical reactions increase) at 35°C than at 5°C. A warm frog needs much more food, though, to support this high metabolic rate. **3.** *Absorptive regions* have numerous folds and projections, which increase their surface area for absorption via diffusion. *Capillaries* have a high surface area because they are thin and highly branched, making exchange of fluids and gases efficient. *Beaks of Galápagos finches* have sizes and shapes that correlate with the type of food. Large beaks are used to crack large seeds; long, thin beaks are used to pick insects off surfaces; etc. *Fish gills* are thin, flattened structures with a large surface area, which facilitates the efficient exchange of gases and wastes. **4.** A larger sphere has a relatively smaller surface area for its interior volume than does a smaller sphere, because surface area increases with size at a lower rate than volume does. Diffusion will be more efficient in the case of the smaller sphere. **5.** In the morning, an ant should move to a sunlit spot to gain heat by conduction from the ground and radiation striking its body directly. In midday, the ant should avoid overheating by losing heat from wet body surfaces, standing in breezy areas to lose heat by convection, or retreating to the cool burrow to lose heat by conduction and avoid gaining heat by radiation. **6.** The feedback causes a response that counters the current state of the system. It is negative in the sense of reducing the difference between the current state and the set point.

Applying Concepts to New Situations

1. Large individuals require more food, so are more susceptible to starvation if food sources can't be defended (e.g., when seeds are scattered around). They may also be slower and/or more visible to predators. **2.** You would have to show changes in the frequencies of alleles whose products are affected by temperature—for example, increases in the frequencies of alleles for enzymes that operate best at higher temperature.

FIGURE A41.1

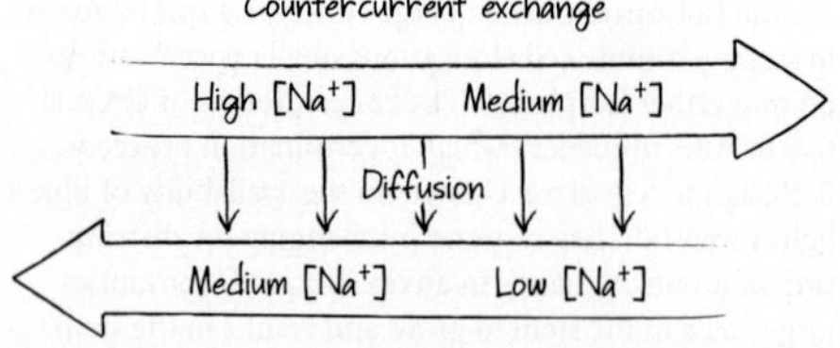

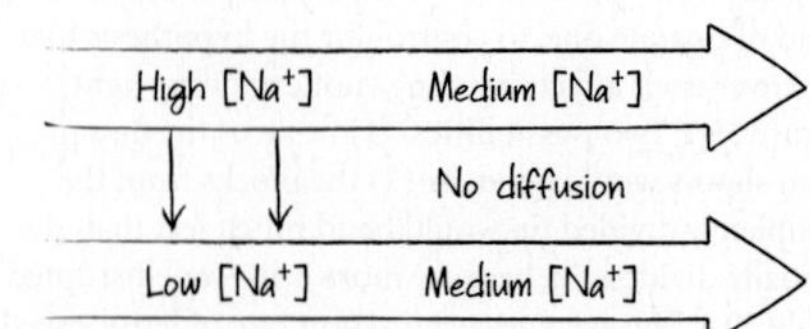

3. The heat-dissipating system should have a very high surface area/volume ratio. Based on biological structures, it should be flattened (thin) and highly folded, branched, and/or contain tubelike projections. Another approach inspired by biological systems would be to set up a countercurrent heat exchanger with fluid heated by the engine. **4.** Buy the turtle. Turtles are ectothermic, meaning they do not expend energy to regulate their body temperature. Mice are endothermic, meaning they do expend energy to regulate body temperature. Mice will therefore have a higher basal metabolic rate and consume more food.

CHAPTER 42

Check Your Understanding (CYU)

CYU p. 828 Without the "master gradient" established by the pump, sodium ions and chloride ions cannot move out of the epithelial cells into the surrounding seawater. Salt will build up in the shark's tissues. **CYU p. 831** (1) Uric acid is insoluble in water and can be excreted without much water loss. (2) When electrolytes are reabsorbed in the hindgut epithelia, water follows along an osmotic gradient. **CYU p. 838** (1) Aldosterone stimulates Na^+ reabsorption in pre-urine in the distal tubule. This increases the Na^+ level in the blood and leads to a more dilute (hypotonic) urine. (2) Water intake increases blood pressure and filtration rate in the renal corpuscle, and leads to lowered electrolyte concentration in the blood and filtrate and production of dilute urine. Eating large amounts of salt results in concentrated, hypertonic urine. Water deprivation triggers ADH release and the production of concentrated, hypertonic urine.

You Should Be Able To (YSBAT)

YSBAT p. 833 Blood contains cells and large molecules as well as electrolytes and wastes; the filtrate contains only electrolytes and wastes. **YSBAT p. 835** If sodium reabsorption is inhibited, then less water will be reabsorbed along an osmotic gradient and more urine will be produced. **YSBAT p. 837** (1) Less, because the osmotic gradient is not as steep. (2) Lower, because more water has been retained. (3) Less, because the concentration gradient is not as steep. **YSBAT p. 838** Ethanol consumption inhibits water reabsorption, leading to a larger volume of less concentrated urine. Nicotine consumption increases water reabsorption, leading to a smaller volume of more concentrated urine.

Caption Questions and Exercises

Figure 42.1 The purple and white molecules are moving down their concentration gradients, and the concentration of red molecules does not affect the concentration of purple or white molecules. **Figure 42.6** The sodium-potassium ATPase does active transport; the sodium-chloride-potassium cotransporter does secondary active transport; the chloride and potassium channels do passive transport. **Figure 42.8** The underside of the abdomen is shaded by the grasshopper's body. This location is relatively cool and should reduce the loss of water during respiration.

Summary of Key Concepts

KC 42.1 In the ocean, river otters must excrete the additional salt that is taken up from saltier marine foods and seawater. In freshwater, their bodies must conserve electrolytes and excrete the water taken in during drinking. **KC 42.2** A sodium gradient establishes an osmotic gradient that moves water. It also sets up an electrochemical gradient that can move ions against their electrochemical gradient by secondary transport through a cotransporter, or with their electrochemical gradient by passive diffusion through a channel.
KC 42.3 The desert locust—to save water, it has to actively reabsorb ions in the hindgut so that almost no water is lost during excretion. **KC 42.4** The longer the loop of Henle, the steeper the osmotic gradient in the medulla. Steep osmotic gradients allow a great deal of water to be reabsorbed as urine passes through the collecting duct.

Test Your Knowledge

1. c; **2.** a; **3.** d; **4.** d; **5.** a; **6.** c

Test Your Understanding

1. In salt water, a salmon gains NaCl by diffusion and loses water by osmosis. It replaces water by drinking and secretes ions through chloride cells in its gill epithelium. In freshwater, it gains water by osmosis and loses NaCl by diffusion. Epithelial cells in the gills take up ions, it stops drinking, and produces copious amounts of urine to rid itself of excess water. **2.** Mitochondria are needed to produce ATP. A key component of salt regulation in fish is the Na^+/K^+-ATPase, which requires ATP to function. Because ATP fuels establishment of the ion gradients necessary for the cotransport mechanisms utilized by chloride cells as well as other epithelial cells, an abundance of mitochondria would be expected. **3.** Microvilli increase the surface area available for pumps, cotransporters, and channels.
4. The wax layer should be thinner in tropical insects, because the animals are under less osmotic stress (less evaporation due to high humidity). You could test this prediction by measuring the thickness of the wax layer in two closely related species from each habitat, or in a wide array of species from each habitat. Insects that live in extremely humid habitats may lack the ability to close the openings to their respiratory passages; animals that live in humid habitats may excrete primarily ammonia or urea and have relatively short loops of Henle. In each case, you could test these predictions by comparing individuals of closely related species that live in dry versus humid habitats. **5.** In both cases, the initial filtrate is isotonic with blood and reabsorption of certain solutes depends on Na^+/K^+-ATPase activity and is energetically costly. One difference is that the initial filtrate forms by active pumping of ions followed by osmosis in insects versus by filtration in the renal corpuscle of mammals. Another difference is that mammals make use of a countercurrent exchanger in the loop of Henle versus direct pumping of ions in the insect hindgut. **6.** Ammonia is toxic and must be diluted with large amounts of water to be excreted safely. Urea and uric acid are safer and do not have to be excreted with large amounts of water, but are more expensive to produce in terms of energy expenditure. Fish excrete ammonia; mammals excrete urea; insects excrete uric acid. You would expect the embryos inside terrestrial eggs to excrete uric acid, as it is the least toxic and is insoluble in water.

Applying Concepts to New Situations

1. The countercurrent flow removes salt from the tissues around the ascending limb, raising the osmolarity of blood, and then water from the tissues around the descending limb, along an osmotic gradient. The countercurrent arrangement maintains a concentration and osmotic gradient all along the length of the nephron.
2. The observation that the Na^+ concentration decreased by 30 percent, even as volume also decreases, indicates that Na^+ ions were reabsorbed to a greater degree than water. The observation that the urea concentration increased by 50 percent indicates that urea was not reabsorbed. **3.** Increased salt concentrations in freshwater environments put organisms under a new kind of osmotic stress. They may not have chloride cells or other adaptations that allow them to rid themselves of additional salt. **4.** Their urine volume is much higher (actually 10 times).

CHAPTER 43

Check Your Understanding (CYU)

CYU p. 856 (1) *Mouth:* Food is taken in; teeth physically break down food into smaller particles; salivary amylase begins to break down carbohydrates; lingual lipase initiates the digestion of fats. *Esophagus:* Food is moved to the stomach via peristaltic contractions. *Stomach:* HCl denatures proteins; pepsin begins to digest them. *Small intestine:* Pancreatic enzymes complete the digestion of carbohydrates, proteins, lipids, and nucleic acids. Most of the water and all of the nutrients are absorbed here. *Large intestine:* Water is reabsorbed and feces are formed. *Anus:* Feces accumulate in the rectum and are expelled out the anus. (2) If the release of bile salts is inhibited, fats would not be digested and absorbed efficiently, and they would pass into the large intestine. The individual would likely produce fatty feces and lose weight over time. Trypsin inactivation would inhibit protein digestion and reduce absorption of amino acids, likely leading to weight loss and muscle atrophy due to protein deprivation. If the Na^+-glucose cotransporter were blocked, then glucose would not be absorbed into the bloodstream. The individual would likely become sluggish from lack of energy. **CYU p. 858** (1) Both lead to high urine volume due to reduced water reabsorption in the kidneys, but for different reasons. In diabetes mellitus, less water is reabsorbed because glucose concentrations in the filtrate from blood are high. In diabetes insipidus, less water is reabsorbed because of defects in the collecting ducts. (2) When the blood sugar of individuals with type I diabetes mellitus gets too high, they should inject themselves with insulin, which will trigger the absorption and storage of glucose by their cells. When their blood glucose gets too low, they should eat something with a high concentration of sugar (such as orange juice or a candy bar) to increase their blood glucose levels quickly.

You Should Be Able To (YSBAT)

YSBAT p. 851 Nutrient absorption occurs in the small intestine, but not in the esophagus or stomach, and the rate of absorption increases with available surface area. **YSBAT p. 854** (1) Decreased energy yield from foods due to reduced glucose absorption, (2) increased feces production due to the passing of unabsorbed glucose into the colon, (3) watery feces due to decreased water reabsorption in the large intestine (lower osmotic gradient). As an aside, also increased flatulence due to the metabolism of unabsorbed glucose by bacteria that produce methane as a waste product. **YSBAT p. 857** Individuals with type I diabetes have normal insulin receptors on their liver cells, but do not produce and release insulin from the pancreas. Individuals with type II diabetes produce and release insulin, but their insulin receptors are defective so liver cells cannot respond to insulin. In both types of diseased individuals, high glucose concentrations in the blood remain high. In individuals without disease, both insulin production and insulin receptors are normal, so liver cells respond to insulin by taking up glucose from the blood and storing it.

Caption Questions and Exercises

Figure 43.3 Your label should point to the rightmost two bones in the illustration. **Figure 43.14** Frog eggs don't normally make the sodium-glucose cotransporter protein, so the researchers could be confident that if the protein appeared, it was from the injected RNA. They would not be confident of this if the RNAs were injected into rabbit epithelial cells, where the protein was probably already present. **Figure 43.16** In type 2 diabetes mellitus, the top, black arrow from glucose (in the blood) to glycogen (inside liver and muscle cells) is disrupted. In type 1 diabetes mellitus, the green arrow on the top left, indicating insulin produced by the pancreas, is disrupted. **Figure 43.17** The leading hypothesis is an increased incidence of obesity, which is linked to development of type 2 diabetes mellitus.

Summary of Key Concepts

KC 43.1 Form two groups of mice selected from a common population, so that individuals have similar genetic characteristics and history. One group (the control group) would receive a regular diet of rat chow, and the other group (the experimental group) would receive a diet of rat chow that was deficient in magnesium. By comparing the two groups on a number of behavioral and physical measures, you could determine the effects of magnesium deficiency. **KC 43.2** Their mouth or tongue would be modified to make probing of flowers and sucking nectar efficient. They would not require teeth, and their digestive tract would be relatively simple. For example, the stomach would not have to pulverize food or digest large amounts of protein, and the large intestine would not have to process and store large amounts of bulky waste. **KC 43.3** By making the stomach smaller, gastric bypass surgery decreases the amount of food that can be comfortably contained in the stomach and digested. By bypassing a portion of the small intestine, the surgery decreases the absorption of food. (As an aside, people who have this surgery are at risk for serious nutrient deficiencies, chronic intestinal upset, ulcers, and flatulence.) **KC 43.4** The excess glucose is eventually eliminated in urine. (This takes much longer than sequestering the excess glucose in liver, muscle, or adipose cells.)

Test Your Knowledge

1. a; **2.** a; **3.** d; **4.** d; **5.** b; **6.** d

Test Your Understanding

1. The bird crop is an enlarged sac that can hold quickly ingested food; in leaf-eating species it is filled with symbiotic organisms and functions as a fermentation vessel. The cow rumen is an enlarged portion of the stomach; the elephant large intestine is enlarged relative to other species. Both structures are filled with symbiotic organisms and function as fermentation vessels. **2.** Digestive enzymes break down macromolecules (proteins, carbohydrates, nucleic acids, lipids). If they weren't produced in an inactive form, they would destroy the cells that produce and secrete them. **3.** Most biologists accept the hypothesis, for two reasons: (1) the structure is unique among fish in being effective at biting, and (2) its surface correlates with the type of food each species eats. **4.** In most terrestrial vertebrates, the primary function of the large intestine is water reabsorption; fish do not need to reabsorb water. **5.** When an individual ingests a solution of glucose and electrolytes, the solutes are absorbed. Water follows by osmosis, preventing dehydration. **6.** Insulin triggers negative feedback to high blood glucose concentrations by stimulating glucose uptake by several types of cells. Glucagon triggers negative feedback to low blood glucose concentrations by stimulating glucose release by several types of cells.

Applying Concepts to New Situations

1. Because they are feeding growing embryos and newborns, it is likely that female mammals during pregnancy and breastfeeding would require higher levels of almost every nutrient—particularly calcium, used by offspring for bone growth. **2.** Individuals with defects in pancreatic amylase would not be able to complete carbohydrate digestion and probably would be lethargic (low energy) and experience weight loss. Pepsin defects would reduce or eliminate protein digestion in the stomach, leading to severe amino acid deficiency. Defects in the fatty-acid binding protein would reduce or eliminate fatty-acid absorption in the small intestine, likely producing fatty feces, weight loss, and diarrhea. Aquaporin defects would prevent water reabsorption in the large intestine and lead to diarrhea and dehydration. **3.** The result is still valid because the injection was correlated with secretion—if lack of injection was correlated with lack of secretion. The criticism is also somewhat valid, however, because the researchers couldn't rule out the hypothesis that signaling from nerves plays some sort of role, too. **4.** Terrestrial animals are exposed to increased risk of water loss, and the large intestine is where water reabsorption occurs. In most cases fish do not need to reabsorb large amounts of water from their feces.

CHAPTER 44

Check Your Understanding (CYU)

CYU p. 864 (1) Oxygen partial pressure is high in mountain streams because the water is cold, mixes constantly, and has a high surface area (due to white water). Oxygen partial pressure is low at the ocean bottom because the area is far from the surface where gas exchange takes place and there is relatively little mixing. (2) *Warm-water species:* Large amount of air because the oxygen-carrying capacity of warm water is low. *Vigorous algal growth:* Small amount of air because algae contribute oxygen to the water through photosynthesis. *Sedentary animals:* Small amount of air because sedentary animals require relatively little oxygen. **CYU p. 870** (1) Common features include large surface area, short diffusion distance (a thin gas exchange membrane), and a mechanism that keeps fresh air or water moving over the gas-exchange surface. Only fish gills use a countercurrent exchange mechanism; only tracheae deliver oxygen directly to cells without using a circulatory system; only lungs contain "dead space"—areas that are not involved in gas exchange. (2) The anterior and posterior air sacs provide additional compartments for inhaled and exhaled air; they are arranged so that air moves through them and the lungs in one direction. **CYU p. 874** The saturation curve for Tibetans should be shifted to the left relative to people from sea level—meaning that their hemoglobin has a higher affinity for oxygen at all partial pressures. **CYU p. 883** (1) If the blood plasma that leaks out of the capillaries is not reabsorbed in capillaries, it enters lymphatic vessels that eventually merge with blood vessels. (2) [See Figure A44.1]

You Should Be Able To (YSBAT)

YSBAT p. 872 There would be an even larger change in the oxygen saturation of hemoglobin in response to an even smaller change in the partial pressure of oxygen.

Caption Questions and Exercises

Figure 44.4 External gills are ventilated passively and are efficient because they are in direct contact with water. They are exposed to predators and mechanical damage, however. Internal gills are protected but have to be ventilated by some type of active mechanism for water flow. **Figure 44.7** Using several to many different animals increases confidence that the results are true for most or all individuals in the population, and not due to one or a few unusual individuals or circumstances. **Figure 44.10** Sighing increases pressure in the lung cavity, forcing air out of the lungs more rapidly and completely than occurs during normal exhalation. When taking a deep breath, volume in the lung cavity is increased, resulting in lower pressure and causing more air to enter the lungs than during normal inhalation. **Figure 44.15** According to data in the figure, the oxygen saturation of hemoglobin is about 15 percent for blood at pH 7.2 and about 25 percent at pH 7.4. Therefore, about 85 percent of the oxygen is released from hemoglobin at pH 7.2, but only about 75 percent of the oxygen is released at pH 7.4. **Figure 44.17** In the lungs, a strong partial pressure gradient favors diffusion of dissolved CO_2 from blood into the alveoli. As the partial pressure of CO_2 in the blood declines, hydrogen ions leave hemoglobin and react with bicarbonate to form more CO_2, which then diffuses into the alveoli and is exhaled from the lungs. **Figure 44.23** Air from the alveoli mixes with air in the "dead space" in the bronchi and trachea on its way out of the body. This dead-space air is from the previous inhalation (P_{O_2} = 160 mm Hg; P_{CO_2} = 0.3), so when the alveolar air mixes with the dead-space air, the partial pressures in the exhaled air achieve levels intermediate between that of inhaled and alveolar air.

Summary of Key Concepts

KC 44.1 It is harder to extract oxygen from water than it is to extract it from air. This limits the metabolic rate of water breathers. **KC 44.2** Large animals have a relatively small body surface area relative to their volume. If they had to rely solely on gas exchange across their skin, they would not have enough skin surface area to exchange the volume of oxygen needed to meet their metabolic needs. **KC 44.3** Cold water carries more oxygen than warm water, so icefish blood can carry enough oxygen to supply the tissues with oxygen even in the absence of hemoglobin. The oxygen and carbon dioxide are simply dissolved in the blood. **KC 44.4** Their circulatory systems should have relatively high pressures—achieved by independent systemic and pulmonary circulations powered by a four-chambered heart—to maximize the delivery of oxygenated blood to metabolically active tissues.

Test Your Knowledge

1. b; **2.** c; **3.** d; **4.** d; **5.** c; **6.** a

Test Your Understanding

1. The insect tracheal system delivers O_2 directly to respiring cells, but in humans, O_2 is first taken up by the blood of the circulatory system, which then delivers it to the cells. The human respiratory system also has specialized muscles devoted to ventilating the lungs. The open circulatory system of the insect is a low-pressure system, which makes use of pumps (hearts) and body movements to circulate hemolymph. The closed circulatory system of humans is a high-pressure system, which can respond to rapid changes in O_2 demand by tissues. **2.** During exercise, P_{O_2} decreases in the tissues and P_{CO_2} increases. The increase in P_{CO_2} lowers tissue pH. The drop in P_{O_2} and pH causes more oxygen to be released from hemoglo-

FIGURE A44.1

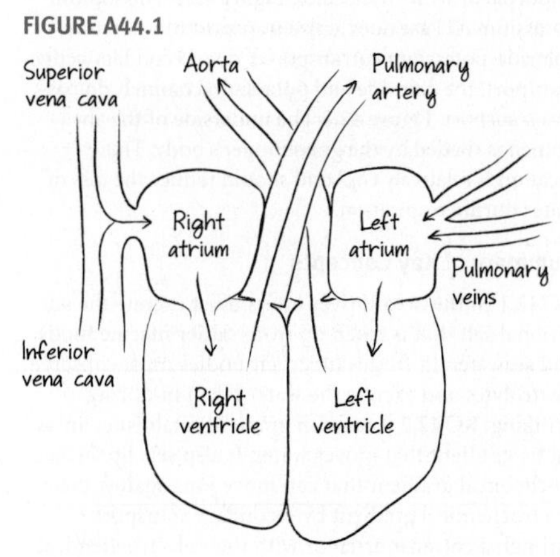

bin. **3.** In blood, CO_2 is converted to carbonic acid, which dissociates into a proton and a bicarbonate ion. Because the protons bind to deoxygenated hemoglobin, they do not cause a dramatic drop in blood pH (hemoglobin acts as a buffer). Also, small decreases in blood pH trigger a homeostatic response that increases breathing rate and expulsion of CO_2. **4.** Airflow through bird lungs is unidirectional, so it allows for continuous ventilation of gas-exchange surfaces with fresh, oxygenated air. The trachea and bronchi in a mammalian lung do not have a gas-exchange surface, and bidirectional airflow in these species means that "stale" air has to be expelled before "fresh" air can be inhaled. As a result, the alveoli are not ventilated continuously. **5.** Lungs increase the temperature of the air and are moist to allow greater solubility of gases (increasing k); alveoli present a large surface area (large A), the epithelium of alveoli is thin (small D), and constant delivery of deoxygenated blood to alveoli maintains a steep partial pressure gradient favoring diffusion of oxygen into the body ($P_2 - P_1$ is high). **6.** If the pulmonary circulation was under pressure as high as that found in the systemic circulation, large amounts of fluid would be forced out of capillaries in the lungs. There is a conflict between the thin surface required for efficient gas exchange and thick blood vessels required to withstand high pressure.

Applying Concepts to New Situations

1. In species that depend on rapid movement to chase down prey or perform other key actions, individuals with larger gill surfaces would absorb more O_2 from water and likely produce more offspring than individuals with smaller gill-surface areas. In slow-moving species, there would be no advantage to increasing the surface area in the gills, so such an adaptation would not be preferentially selected for. **2.** A shift to the right of the oxygen-hemoglobin dissociation curve represents a decrease in the affinity of hemoglobin for O_2. Thus DPG promotes the release of O_2 from hemoglobin into the tissues. **3.** Yes—the trait compensates for the small P_{O_2} gradient between stagnant water and the blood of the carp. **4.** Since O_2 cannot compete as well as CO for binding sites in hemoglobin, O_2 transport decreases. As oxygen levels in the blood drop, tissues (particularly the brain) become deprived of oxygen, and suffocation occurs.

CHAPTER 45

Check Your Understanding (CYU)

CYU p. 891 (1) The resting potential would fall (be much less negative) because K^+ could no longer leak out of the cell. (2) The size of an action potential from a particular neuron does not vary, so it cannot contain information. Only the frequency of action potentials from the same neuron varies. **CYU p. 895** (1) Once the threshold level of depolarization is attained, the probability that the voltage-gated sodium channels will open approaches 100 percent. But below threshold, the massive opening of Na^+ channels does not occur. This is why the action potential is "all or none." (2) Na^+ would continue to move into the cell as voltage-gated potassium channels opened and potassium began to diffuse out of the cell. As a result, repolarization would take much longer. **CYU p. 899** If neurons made direct electrical connections, action potentials would simply travel from one neuron to the next. The presence of synapses allows information from many different neurons to affect the activity of a postsynaptic neuron, through the process of summation. **CYU p. 904** (1) [See Figure A45.1] (2) 1. Study individuals with brain damage and correlate the location of the defect with a deficit in mental or physical function. 2. Directly stimulate brain areas in conscious patients during brain surgery and record the response.

You Should Be Able To (YSBAT)

YSBAT p. 890 The Na^+/K^+-ATPase makes the inside of the membrane less positive (more negative) than the outside, and generates a concentration gradient favoring movement of K^+ out of the cell, through the K^+ leak channel. As K^+ leaves, the resting potential becomes even more negative. **YSBAT p. 891** [See Figure A45.2] **YSBAT p. 892** (1) Na^+ channels open when the cell membrane depolarizes, and as more channels open, more Na^+ flows into the cell, further depolarizing the cell, causing more voltage-gated Na^+ channels to open. (2) After opening, the voltage-gated Na^+ channels close and temporarily cannot reopen. Also, sodium reaches its equilibrium potential, so no net force is available to drive Na^+ movement. (3) K^+ channels open in response to membrane depolarization. **YSBAT p. 897** The inside of the cell becomes more positive (depolarized). This shifts the membrane potential closer to the threshold, making it easier for the postsynaptic cell to fire an action potential.

Caption Questions and Exercises

Figure 45.3 No—as K^+ leaves the cell along its concentration gradient, the interior of the cell becomes more negative. As a result, an electrical gradient favoring movement of K^+ into the cell begins to counteract the concentration gradient favoring movement of K^+ out of the cell. Eventually, the two opposing forces balance out, and there is no net movement of K^+. **Figure 45.10** (1) Get a solution taken from the synapse between the heart muscle and the vagus nerve *without* the nerve being stimulated. Expose a second heart to this solution. There should be no change in heart rate. **Figure 45.16** In the "rest and digest" mode, pupils take in less light stimulation, the heartbeat slows to conserve energy, and liver conserves glucose and promotes digestion by stimulating release of gallbladder products. In the "fight or flight" mode, the pupils open to take in more light, the heartbeat increases to support muscle activity, and the liver releases glucose and inhibits digestion. **Figure 45.18** Point to your forehead and top of your head for the frontal lobe, the top and top rear of your head for the parietal lobe, the back of your head for the occipital lobe, and the sides of your head (just above your ear openings) for the temporal lobes. **Figure 45.19** No—for example, part (b) indicates that the size of the brain area devoted to the trunk is no bigger than the size of the brain area devoted to the thumb.

FIGURE A45.1

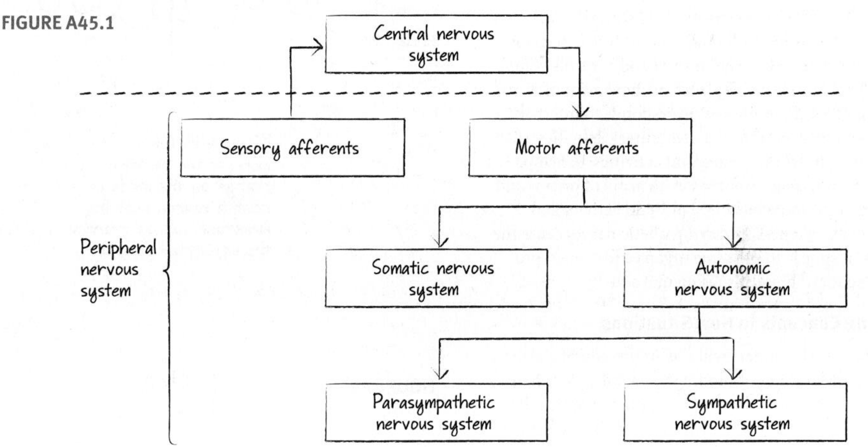

FIGURE A45.2

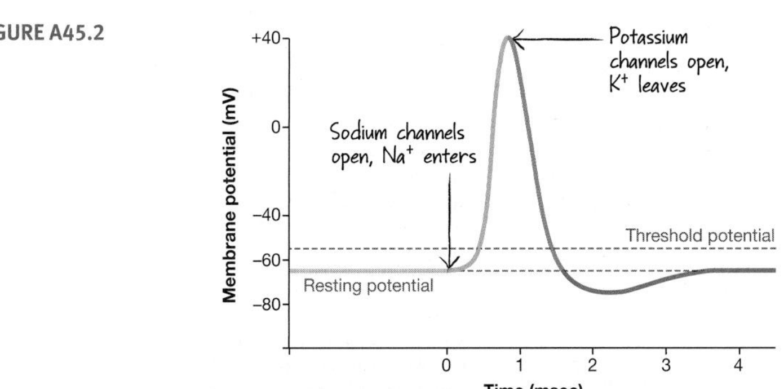

Summary of Key Concepts

KC 45.1 [See Figure A45.3] **KC 45.2** Because the behavior of voltage-gated Na^+ and K^+ channels does not change. Every time a membrane depolarizes past threshold and all of the voltage-gated Na^+ channels are open, the same amount of Na^+ will enter the cell and the cell will depolarize to the same extent. In addition, K^+ channels open on the same time course and allow the same amount of K^+ to leave the cell, meaning that it repolarizes to the same extent. **KC 45.3** Summation means that input from many neurons is involved in triggering a response from a postsynaptic cell. Thus, an action potential from a postsynaptic cell requires integration of input from many synapses. **KC 45.4** The animal's ability to learn—and perhaps remember—would be impaired or nonexistent.

Test Your Knowledge

1. b; **2.** d; **3.** c; **4.** a; **5.** c; **6.** a

Test Your Understanding

1. The Na^+/K^+-ATPase pumps 3 Na^+ out of the cell for every 2 K^+ it brings in. Since more positive charges leave the cell than enter it, there is a difference in charge on the two sides of the membrane and thus a voltage. **2.** [See Figure A45.4] **3.** The ligand-gated channel opens or closes in response to binding by a small molecule (for example, a neurotransmitter); the voltage-gated channel opens or closes in response to changes in membrane potential. **4.** EPSPs and IPSPs are based on flows of ions, which change the membrane potential in the postsynaptic cell. If a flow of ions at one point depolarizes the cell but a nearby flow of ions hyperpolarizes the cell, then in combination the flows of ions cancel each other out. But if two adjacent flows of ions depolarize the cell, then the total amount of ion movement—and thus the total change in membrane potential—sums. **5.** The somatic system responds to external stimuli and controls voluntary skeletal muscle activity, such as movement of arms and legs. The autonomic system responds to internal stimuli and controls internal involuntary activities, such as digestion, heart rate, and gland activities. **6.** Sympathetic nerves trigger responses in an array of organs and tissues that promote increased physical activity and heightened awareness. Parasympathetic nerves cause the opposite response in the same organs and tissues, promoting reduced physical and mental activity.

Applying Concepts to New Situations

1. The neurotransmitters will stay in the synaptic cleft longer so the amount of binding to ligand-gated channels will increase. Ion flows into the postsynaptic cell will increase dramatically, affecting its membrane potential and likelihood of firing action potentials. **2.** Diseased or damaged brains may not respond like healthy and undamaged brains. Also, the extent of lesions and the exact location of electrical stimulation may be difficult to determine, making correlations between the regions affected and the response imprecise. **3.** Because the current is reduced from normal levels, but not eliminated completely, there must be more than one type of potassium channel present. (The poison probably knocks out just one specific type of channel.) **4.** Record from the neuron while the individual is performing various tasks—feeding, moving, courting, etc.

CHAPTER 46

Check Your Understanding (CYU)

CYU p. 913 (1) The outer ear collects sound waves from the environment and directs them into the ear canal. The middle ear amplifies sound. The inner ear contains hair cells that transduce the information in sound waves to action potentials that are sent to the brain. (2) A punctured eardrum wouldn't vibrate correctly and would result in hearing loss at all frequencies in the affected ear. If the stereocilia are too short to come into contact with the tectorial membrane, vibration of the basilar membrane will not cause them to bend, and sound will not be detected. A loss in basilar membrane flexibility would result in the inability to hear lower-pitched sounds, such as human speech. **CYU p. 918** (1) Retinal acts like an on-off switch that indicates whether light has fallen on a rod cell. When retinal absorbs light, it changes shape. The shape change triggers events that result in a change in action potentials that signals that light has been absorbed. (2) A tear in the fovea would likely result in blurred vision in the affected eye. The mutation would produce blue-purple color blindness because this opsin responds to wavelengths in that region of the spectrum. A clouded lens would reduce the amount of light that reaches the retina, reducing visual sensitivity. **CYU p. 920** (1) Extremely hot food damages taste receptors, so proteins would no longer be able to respond to their chemical triggers. (2) Dogs have more than twice the number of odor receptors as humans. **CYU p. 926** (1) In a sarcomere, thick myosin filaments are sandwiched between thin actin filaments. When the heads on myosin contact actin and change conformation, they pull the actin filaments toward one another, shortening the whole sarcomere. (2) Increased ACh release would result in an increased rate of muscle cell contraction. Preventing conformational changes in troponin would prevent muscle contraction. Blocking the uptake of calcium ions into the sarcoplasmic reticulum would lead to sustained muscle contraction.

You Should Be Able To (YSBAT)

YSBAT p. 916 cGMP is a ligand that opens sodium channels and a second messenger in the sense that lack of it carries a signal from activated transducin. Transducin is like a G protein because it switches from "off" to "on" in response to a receptor protein, and activates a key pro-

FIGURE A45.3

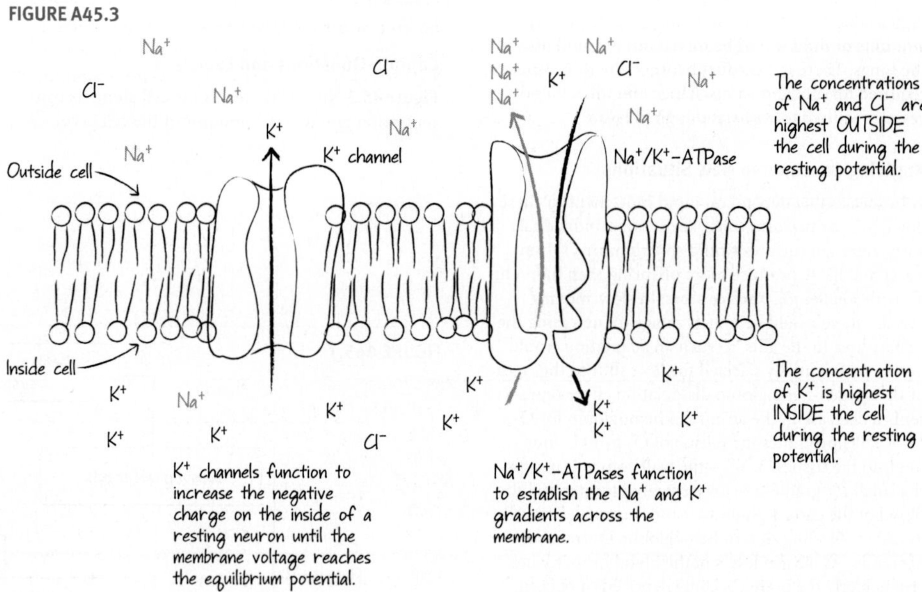

FIGURE A45.4

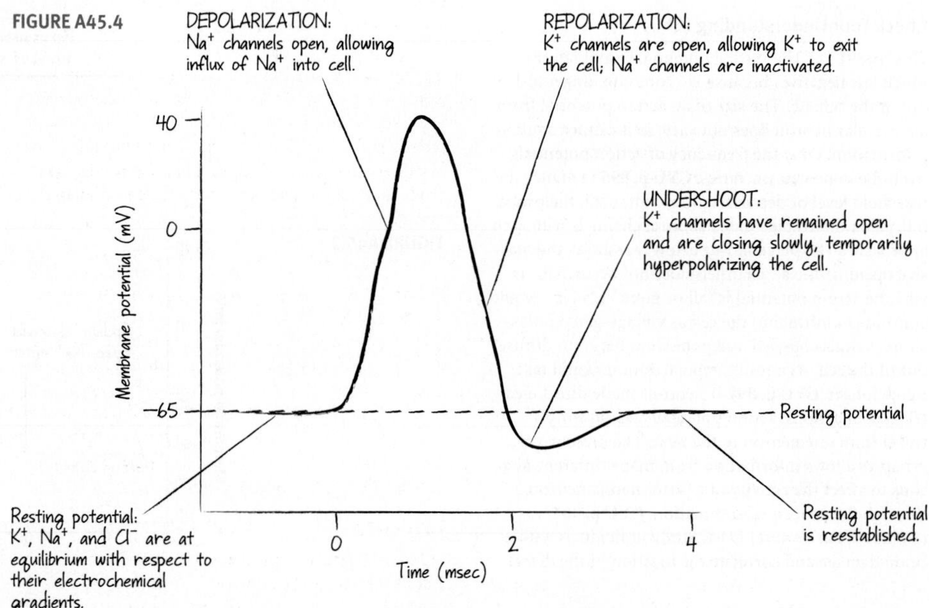

tein in response. Closing Na^+ channels stops the entry of positive charges into the cell, making the inside of the cell more negative relative to the outside. **YSBAT p. 923** The trucks are the Z disks, the ropes are the thin filaments, and the burly weightlifters are the thick filaments.

Caption Questions and Exercises

Figure 46.9 The change in retinal's shape would induce a change in opsin's shape. A protein's function correlates with its shape, so opsin's function is likely to change as well. **Figure 46.10** This ion flow occurs when the photoreceptor cell is not receiving light—when the cell is in the dark. **Figure 46.12** The S, M, and L opsin proteins are each different in structure. These structural differences affect the ability of retinal to absorb specific frequencies of light and change shape. **Figure 46.18** The dark band includes thin filaments and a dense concentration of bulbous structures extending from the thick filament; the light band consists of just thin filaments.

Summary of Key Concepts

KC 46.1 The sensory neurons that used to serve that limb are cut, but may still fire action potentials that stimulate brain areas associated with awareness of pain in the missing limb. **KC 46.2** If dense, starch-filled amyloplasts exerted pressure on hair-cell-like receptors, the stereocilia would bend, causing a change in the membrane potential of the receptors and thus transducing the signal from gravity. **KC 46.3** Because different animal species have opsin molecules in their cone cells that respond to different wavelengths of light absorbed by retinal, they see different colors. **KC 46.4** Smells are part of the sensation of flavor, and people with nasal congestion cannot smell. In addition to smell being important, the small number of taste receptors can send many different frequencies of action potentials to the brain, and the taste receptors can be stimulated in many different combinations. **KC 46.5** Paralysis, because ACh has to bind to its receptors on the membrane of postsynaptic muscle fibers for action potentials to propagate in the muscle and cause contraction.

Test Your Knowledge

1. a; **2.** d; **3.** c; **4.** b; **5.** d; **6.** a

Test Your Understanding

1. The brain can distinguish sensory stimuli because the axons from different sensory neurons go to different and specific areas of the brain. For example, the axons of all the olfactory neurons that have the same receptor proteins go to the same location in the olfactory bulb, resulting in the sensation of smelling a particular odor.
2. Many possibilities, including several blue opsins allowing coelacanths to distinguish the blue wavelengths present in their deep-sea habitat, infrasound hearing allowing elephants to hear over long distances, ultrasonic hearing allowing bats to hunt by echolocation and moths to avoid bat predators, red/yellow opsins allowing fruit-eating mammals to distinguish ripe from unripe fruit. **3.** It answered the question of why so many different smells can be detected and identified separately. Each type of odorant has its own receptor. **4.** Ion channels in hair cells are thought to open in response to physical distortion caused by the bending of stereocilia. Ion channels in taste receptors, in contrast, change shape and open when certain bitter-tasting molecules bind to them. **5.** Dalton's hypothesis was reasonable because blue fluid absorbs red light and would prevent it from reaching the retina, and could be tested by dissecting his eyes. Although the hypothesis was rejected, it inspired a rigorous test and required researchers to think of alternative explanations. **6.** The key observation was that the banding pattern of sarcomeres changed during contraction. Even though the entire unit became shorter, only some portions moved relative to each other. This observation suggested that some portions of the structure slid past other portions. Muscle fibers have to have many mitochondria because large amounts of ATP are needed to power myosin heads to move along actin filaments; large amounts of calcium stored in smooth ER are needed to initiate contraction by binding to troponin.

Applying Concepts to New Situations

1. If ACh receptors are covered, they cannot bind to ACh and become activated. Muscles will have trouble contracting. **2.** Block all but a few hundred ommatidia in the center of each compound eye of many dragonflies with an opaque material, and observe any differences that might occur in their ability to detect and pursue prey compared to individuals whose ommatidia had been covered by a transparent material. **3.** Ground-dwelling birds have large amounts of "white meat" in their breasts, used for powering short, rapid flights, but "dark meat" in their legs, used for endurance running. Birds that rely on long, sustained flights need "dark meat" in the breast muscles that power flight. **4.** When a person focuses directly on an object, the image lands on the fovea, which contains relatively few rods. But if the image lands beside the fovea, there are many more rod cells. Rod cells are specialized for detecting dim light.

CHAPTER 47

Check Your Understanding (CYU)

CYU p. 940 (1) By reducing immune system function, promoting release of fatty acids from storage cells and use of amino acids from muscles for energy production, and preventing release of glucose in response to signals from insulin, cortisol conserves glucose supplies for use by the brain. (2) Increased heart rate, glucose and fatty-acid release, and blood pressure all support rapid action in response to danger. If an animal were not able to respond quickly and at the maximum level, it might not escape predators or successfully challenge rivals. **CYU p. 943** (1) ACTH triggers the release of cortisol, but cortisol inhibits ACTH release by blocking the release of CRH from the hypothalamus and suppressing ACTH production. High levels of cortisol tend to lower cortisol levels in the future. (2) Processing centers in the brain are responsible for synthesizing a wide array of sensory input. To start a response to this sensory input, they stimulate neurosecretory cells in the hypothalamus. Secretions from these cells travel to the anterior pituitary, where they trigger the production and release of hormones that control hormones that stimulate the appropriate activity. **CYU p. 947** (1) The steroid-hormone receptor complex would probably fail to bind to the hormone-response element. If so, then gene expression would not change—the arrival of the hormone would have little or no effect on the target cell. (2) The mode of action would be similar—hormone binding would have to trigger a signal transduction event that resulted in production of a second messenger or a phosphorylation cascade.

You Should Be Able To (YSBAT)

YSBAT p. 932 Room temperature is analogous to the sensory input; the thermostat, to the CNS; the thermostat signal, to the cell-to-cell signal; and the furnace, to the effector. If feedback inhibition fails, hormone production will not stop and the response will continue indefinitely. **YSBAT p. 936** The cells could have (1) different receptors, (2) different signal transduction systems, or (3) different response systems (genes or proteins that can be activated).

Caption Questions and Exercises

Figure 47.4 Injecting the dog with a liquid extract from a different part of the body, or with a solution that is similar in chemical composition to the extract from the pancreas (e.g., similar pH or ions present) but lacking molecules produced by the pancreas. **Figure 47.7** The saline injection controlled for any stress induced by the injection procedure and for introducing additional fluids. **Figure 47.9** Same—a defect in both signal and receptor has the same effect—no response—as a defective signal alone or a defective receptor alone. **Figure 47.10** Inject ACTH and monitor cortisol levels. If cortisol does not increase, adrenal failure is likely. **Figure 47.14** The arrow that joins the signal and the receptor. The subsequent events also fail to occur.

Summary of Key Concepts

KC 47.1 The production and/or release of hormones are controlled, directly or indirectly, by electrical or chemical signals from the nervous system. A good example is the hypothalamic-pituitary axis, where neuroendocrine signals regulate endocrine signals. **KC 47.2** Electrical signals are short lived and localized. All of the responses listed in the question are long-term changes in the organism that require responses by tissues and organs throughout the body. **KC 47.3** The hypothalamus would continue to release CRH, stimulating the pituitary to release ACTH, which in turn would stimulate continued release of cortisol from the adrenal cortex. Sustained, high circulating levels of cortisol have negative effects. **47.4** A cell can respond to more than one hormone at a time if the receptors for the hormones are different and the cell produces receptors for each hormone.

Test Your Knowledge

1. c; **2.** a; **3.** d; **4.** a; **5.** c; **6.** a

Test Your Understanding

1. The posterior pituitary is an extension of the hypothalamus; the anterior pituitary is independent—it communicates with the hypothalamus via chemical signals in blood vessels. The posterior pituitary is a storage area for hypothalamic hormones; the anterior pituitary synthesizes and releases an array of hormones in response to releasing hormones from the hypothalamus.
2. Steroid hormones act directly, and alter gene expression. They bind to receptors inside the cell, forming a complex that binds to DNA and activates transcription. Nonsteroid hormones act indirectly, and activate proteins. They bind to receptors on the cell surface and trigger production of a second messenger or a phosphorylation cascade, ending in activation of proteins already present in the cell. **3.** The pituitary produces hormones that regulate hormone production in other glands. **4.** This is one of the reasons that the same hormone can trigger different effects in different tissues. For example, epinephrine binds to four different types of receptors in different tissues—eliciting a different response from each. **5.** The hormonal signal is amplified through production of many copies of a second messenger, the activation of many proteins through a phosphorylation cascade, or the transcription and translation of many mRNAs. **6.** When fat stores are high, leptin release acts on the brain to reduce feeding. In this way, it helps maintain weight and energy balance—preventing obesity in mice.

Applying Concepts to New Situations

1. There are two basic strategies: (1) remove the structure from some individuals, and compare their behavior and condition to untreated individuals in the same environment, or (2) make a liquid extract from the structure, inject it into some individuals, and compare their behavior and condition to individuals in the same environment who were injected with water or a saline solution.
2. If sustained high cortisone levels suppress wound healing, individuals receiving large doses might become more susceptible to disease or heal slowly. **3.** The patient will be unable to produce oxytocin and ADH. Although the patient could manage without oxytocin, artificial

administration of ADH would be required to maintain water balance. You would need to monitor the individual's total water intake and the water content of the urine, along with symptoms of dehydration or water retention (such as swelling or high blood pressure) and increase or decrease ADH dosage accordingly. **4.** Label the hormone and introduce it into an experimental animal. Analyze fat cells by isolating cell components, testing for the presence of the label, and eventually isolating just the labeled hormone-receptor complex. To test the second messenger hypothesis, treat fat cells growing in culture with the hormone and monitor levels of cAMP, Ca^{2+}, and other known second messengers.

CHAPTER 48

Check Your Understanding (CYU)

CYU p. 957 (1) Oviparity usually requires less energy input from the mother after egg laying, and mothers do not have to carry eggs around as long—meaning that they can lay more eggs and be more mobile. Mothers have to produce all the nutrition required by the embryo prior to egg laying, however, and eggs may not be well protected after laying. Viviparity usually increases the likelihood that the developing offspring will survive until birth, but limits the number of young that can be produced to the space available in the mother's reproductive tract. If viviparous young can be nourished longer than oviparous young, then they may be larger and more capable of fending for themselves. (2) Divide a population of sperm into two groups. Subject one group of sperm to spermkillerene at concentrations observed in the female reproductive tract (the experimental group) but not the other group (the control). Document the number of sperm that are still alive over time in both groups. **CYU p. 966** (1) FSH triggers maturation of an ovarian follicle. It rises at the end of a cycle because it is no longer inhibited by progesterone, which stops being produced at high levels when the corpus luteum degenerates. (2) The drug would keep FSH levels low, meaning that follicles would not mature and would not begin producing estradiol and progesterone. The uterine lining would not thicken.

You Should Be Able To (YSBAT)

YSBAT p. 952 Asexual reproduction would be expected in environments where conditions change little over time. **YSBAT p. 965** (1) There should be an LH spike early in the cycle, followed by early ovulation. (2) LH levels should remain low—no mid-cycle spike and no ovulation. **YSBAT p. 968** (1) The uterine lining may not be maintained adequately—if it degenerates, a miscarriage is likely. (2) After the first trimester, the placenta produces enough progesterone to maintain the pregnancy, so supplementation is no longer needed.

Caption Questions and Exercises

Figure 48.3 Isolate and identify the molecules found in crowded water. Test each molecule by adding it, at the same concentration found in "crowded water," to clean water occupied by a single *Daphnia* and recording whether the female produces a male-containing brood. Repeat with many test females, and for each molecule identified. As a control, record the number of male-containing broods produced in clean water. **Figure 48.5** No—the data are consistent with the displacement hypothesis, but there is no direct evidence for it. There are other plausible explanations for the data. **Figure 48.11** (Note: There is more than one possible answer—this is an example.) Having two gonads is an "insurance policy" against loss or damage. To test this idea, surgically remove one gonad from a large number of male and female rats. Do a similar operation on a large number of similar male and female rats but do not remove either gonad. Once the animals have recovered, place them in a barn or other "natural" setting and let them breed. Compare reproductive success of individuals with paired vs. unpaired gonads. **Figure 48.12** It is similar. In both cases, the hypothalamus produces a releasing factor (GnRH or CRH) that acts on the pituitary. The releasing factor stimulates release of regulatory hormones from the anterior pituitary (LH and FSH or ACTH). These hormones travel via the bloodstream and act on the gonads or adrenals to induce the release of hormones from these glands. In both cases, the hormones are involved in negative feedback control of the regulatory hormones from the pituitary. **Figure 48.20** The birth will be more difficult, as limbs or other body parts can get caught. **Figure 48.21** A mother's chance of dying in childbirth in 1760 was a little over 1000 in 100,000 live births, or close to 1.0 percent. If she gave birth 10 times, she would have a 10 percent chance of dying in childbirth.

Summary of Key Concepts

KC 48.1 Parental care demands resources (time, nutrients) that cannot be used to produce more eggs. Sperm competition is not likely when external fertilization occurs, and most fishes use external fertilization. **KC 48.2** A surgeon can ligate (tie off) the fallopian tubes in a woman to stop the delivery of the egg to the uterus and can cut the vas deferens in a man to stop the delivery of sperm into the semen. **KC 48.3** Progesterone suppresses the release of GnRH and FSH through negative feedback. When progesterone levels are high, an LH spike does not occur, ovulation is prevented, and no egg is available for fertilization.

Test Your Knowledge

1. a; **2.** b; **3.** a; **4.** c; **5.** d; **6.** d

Test Your Understanding

1. Every offspring that is produced sexually is genetically unique; every offspring that is produced asexually is genetically identical to its parent. **2.** *Daphnia* females only produced males when they were exposed to short day lengths, *and* water from crowded populations, *and* low food levels. These conditions are likely to occur in the fall. Sexual reproduction could be adaptive in fall if genetically variable offspring are better able to thrive in conditions that occur the following spring. **3.** Spermatogenesis generates four haploid sperm cells from each primary spermatocyte; oogenesis produces only one haploid egg cell from each primary oocyte. Egg cells are much larger than sperm cells because they contain more cytoplasm. In males, the second meiotic division occurs right after the first meiotic division, but in females it is delayed until fertilization. **4.** LH triggers release of estradiol, but at low levels estradiol inhibits further release of LH. LH and FSH trigger release of progesterone, but progesterone inhibits further release of LH and FSH. High levels of estrogen trigger release of more LH. The follicle can produce high levels of estrogen only if it has grown and matured—meaning that it is ready for ovulation to occur. **5.** Ethanol can cross the placenta, enter the fetus, and adversely affect development. Ethanol abuse by pregnant women is correlated with fetal alcohol syndrome (FAS), a condition caused by loss of neurons in developing embryos. **6.** Females should be choosier about their second mate because the sperm of the second male has an advantage in sperm competition.

Applying Concepts to New Situations

1. If all sheep were genetically identical, it is less likely that any individuals could survive a major adverse event (e.g., a disease outbreak or other environmental challenge) than a population composed of genetically diverse individuals, which are likely to vary in their ability to cope with the new conditions. **2.** Compare the number of gametes produced by species that have similar body sizes, but undergo external versus internal fertilization. By comparing similar-sized animals, you would be able to rule out the possibility that differences in gamete production are due to differences in body size. **3.** Oviparous populations should produce larger eggs than viviparous populations. Because oviparous species deposit eggs in the environment, the eggs must contain all the nutrients and water required for the entire period of embryonic development. But because embryos in viviparous species receive nourishment directly from the mother, it is likely that their eggs are smaller. **4.** The shell membrane is a vestigial trait (see Chapter 24)—specifically, an "evolutionary holdover" from an ancestor of today's marsupials that laid eggs.

CHAPTER 49

Check Your Understanding (CYU)

CYU p. 977 Applying direct pressure constricts blood vessels, mimicking the effect of histamine released from mast cells. Applying bandages impregnated with platelet-recruiting compounds mimics the effect of chemokines released from injured tissues and macrophages, which attract circulating leukocytes and platelets to facilitate blood clotting and wound repair. **CYU p. 984** (1) They are identical, except that a B-cell receptor has a transmembrane domain that allows it to be located in the B-cell plasma membrane. (2) Lymphocytes that respond to self molecules would circulate in the blood, and trigger immune system responses to self cells—leading to their destruction. The mutation would probably be fatal. **CYU p. 987** Individuals who are heterozygous for MHC genes have greater variability in their MHC proteins than do individuals who are homozygous for these genes. The increased variability in MHC proteins allows a greater variety of antigens to be presented to T cells, which thus would be able to recognize, attack, and eliminate a wider range of pathogens compared with T cells in homozygotes. As a result, heterozygous individuals would likely suffer fewer infections than homozygous individuals. **CYU p. 990** A vaccination is a "false alarm" that triggers a primary response from the immune system. The memory cells that result produce the secondary response when and if an authentic infection occurs.

You Should Be Able To (YSBAT)

YSBAT p. 982 Have the lottery numbers contain a large number of digits—say 5, or even 10. To generate a number, pick a number from 0–9 for each of the 5 or 10 places. If there are 5 places, there would be $10 \times 10 \times 10 \times 10 \times 10 = 10{,}000$ different numbers. If there were 10 places, there would be 10^{10} (10 billion) different numbers. **YSBAT p. 985** Rapidly dividing B and T cells may be destroyed by chemotherapy, along with cancer cells. If clonal expansion cannot occur, antibody production and other aspects of the immune response will be dampened and possibly ineffective. **YSBAT p. 986** (1) All nucleated cells in the body express Class I MHC proteins, whereas only B cells and some other leukocytes express Class II MHC proteins. (2) Cytotoxic T cells interact only with cells that display antigens presented on Class I MHC proteins. Helper T cells interact only with antigens presented on Class II MHC proteins. (3) An MHC-peptide displayed on an infected cell means "kill me"; displayed on a dendritic cell it means "activate killer T cells"; displayed on a B cell it means "helper T cells should activate me."

Caption Questions and Exercises

Figure 49.1 Pathogens stick to mucus; contaminated mucus is then swept to a disposal point by cilia. **Figure 49.3** Blood vessels near the wound would be constricted, which restricts blood loss. Blood vessels slightly farther from the wound would be dilated, which increases blood flow and the delivery rate of platelets, neutrophils, and macrophages to the surrounding area. **Figure 49.18** [See Figure A49.1]

Summary of Key Concepts

KC 49.1 Leukocytes could not move to the damaged tissue efficiently to clean up debris, remove pathogens, and begin the repair process. **KC 49.2** The drug would inhibit the activation of T and B cells by that antigen, because it would prevent an interaction between the epitope and the appropriate TCRs and BCRs. The cell-mediated response would also fail if the drug blocked the interaction between cytotoxic T cells and infected cells displaying the antigen. **KC 49.3** Cowpox is similar enough to smallpox to be antigenic, but cannot grow effectively enough in humans to cause disease.

Test Your Knowledge

1. c; **2.** a; **3.** b; **4.** d; **5.** b; **6.** a

Test Your Understanding

1. Rubor (reddening) is due to increased blood flow to the infected or injured area. Mast cells trigger dilation of vessels by releasing histamine and other signaling molecules. Calor (heat) is the result of fever, which is activated by the release of cytokines from macrophages that are in the area of infection. Dolor (pain) occurs when tissue damage stimulates pain receptors. Tumor (swelling) occurs due to dilation of blood vessels triggered by histamines and other compounds released by macrophages. **2.** Both the BCR and TCR interact with antigens via binding sites located in the variable regions of their polypeptide chains. A TCR is like one "arm" of a BCR. [See Figure A49.2] **3.** Vaccines have to contain an antigen that can stimulate an appropriate primary immune response. Vaccines have not worked for HIV because the antigens on this virus are constantly changed through mutation, rendering it unrecognizable to memory cells generated following vaccination. **4.** "Clonal" refers to the cloning—producing many exact copies—of cells that are "selected" by the binding of their receptor to an antigen. **5.** Pattern-recognition receptors on leukocytes bind to surface molecules that are present on many pathogens. BCRs and TCRs bind to particular epitopes of antigens. Pattern-recognition-receptor binding is nonspecific; BCR and TCR binding is specific. **6.** By mixing and matching different combinations of gene segments from the variable and joining regions of the light-chain immunoglobulin gene, along with diversity regions of the heavy-chain gene, lymphocytes end up with a unique sequence for both chains in the BCR and TCR.

Applying Concepts to New Situations

1. Irrigating the wound removes dirt and debris that may contain pathogens. Scrubbing with soapy water removes more pathogens and may also kill them (soap may disrupt plasma membranes enough to be fatal). Antibiotics are toxic to pathogenic bacteria. **2.** If the antibody had a fluorescent molecule or other type of label attached, you could treat various types of cells with it, examine them under the microscope, and determine the location of the antibody-pump complexes. **3.** Natural selection favors individuals that can create a large array of antibodies, because the high mutation rates and rapid evolution observed in pathogens means that they will constantly present the immune system with new antigens. If these antigens were not recognized by the immune system, the pathogens would multiply freely and kill the individual. **4.** Tissue from the patient's own body is marked with the major histocompatibility (MHC) proteins that the patient's immune system recognizes as self. This results in the preservation, rather than the destruction, of the transplanted tissue. Tissue from a different person is marked with MHC proteins that are unique to that individual. The body of the transplant recipient will recognize the MHC proteins (and other molecules) of the donor as foreign, resulting in a full immune response and rejection of the grafted tissue.

CHAPTER 50

Check Your Understanding (CYU)

CYU p. 1001 (1) They are shallow, so a relatively large proportion of the water receives enough sunlight to support photosynthesis. In addition, nutrients are available from rivers that contribute nutrients from inland and upwellings that bring nutrients up from the ocean bottom. (2) It sinks, and the lower-density water below it rises, bringing nutrients to the surface. **CYU p. 1008** Tropical dry forests are probably less productive than tropical wet forests because they have less water available to support photosynthesis during some periods of the year. **CYU p. 1012** Your answer will depend on where you live. For example, if global warming continues at the current projected rates, several effects can be expected. If you live in a coastal area, you can expect water levels to rise. In general, plant communities will change because average temperature and moisture and variation in temperature and moisture will change.

You Should Be Able To (YSBAT)

YSBAT p. 997 The littoral zone, because light is abundant and nutrients are available from the substrate. **YSBAT p. 998** Bogs are nitrogen-poor, so plants that are able to capture and digest insects have a large advantage. This advantage does not exist in marshes and swamps, where nutrients are more readily available. **YSBAT p. 999** Cold water contains more oxygen than warm water, so it can support much more cellular respiration in fish.

YSBAT p. 1000 Species that live in estuaries must be able to tolerate variable salinity; marsh species do not. The abiotic environment (salt concentration) is so different that few species grow well in both habitats. **YSBAT p. 1001** No—the aphotic zone is lightless, so natural selection favors individuals that do not invest energy in developing and maintaining eyes. **YSBAT p. 1003** Vines and epiphytes increase productivity because they are photosynthetic organisms that fill space between small trees and large trees—they capture light and use nutrients that might not be used in a forest that lacked vines and epiphytes. **YSBAT p. 1004** Most leaves have a large surface area to capture light. Light is abundant in deserts, however, and a leaf with a large surface area would be susceptible to high water loss and/or overheating. **YSBAT p. 1005** Vegetation is continuous in a grassland, so a fire carries better. In a desert, the fire is likely to run out of fuel. **YSBAT p. 1006** There is a continuous grass cover and scattered trees. (A biome like this is called a savannah.) **YSBAT p. 1007** Boreal forests should move north. **YSBAT p. 1008** High elevations present an abiotic environment (precipitation and temperature) that is similar to the conditions in arctic tundras. As a result, the plant species present will have similar adaptations to cope with these physical conditions. **YSBAT p. 1010** [See Figure A50.1]

Caption Questions and Exercises

Figure 50.4 Freshwater, because light penetrates much deeper. **Figure 50.23** Dry, because dry air is descending here. This is consistent with the very low rainfall amounts in Barrow, Alaska, shown in Figure 50.21. **Figure 50.27** Visible wavelengths enter the chamber

FIGURE A49.1

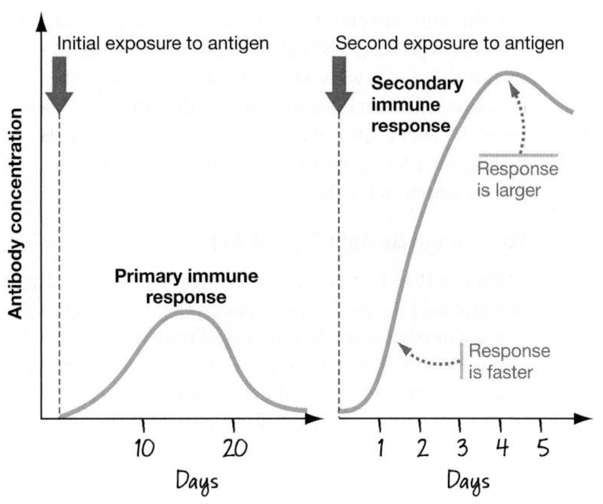

FIGURE A49.2

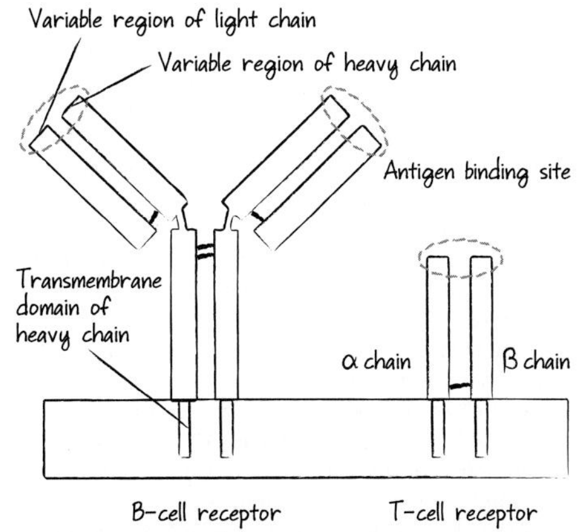

FIGURE A50.1

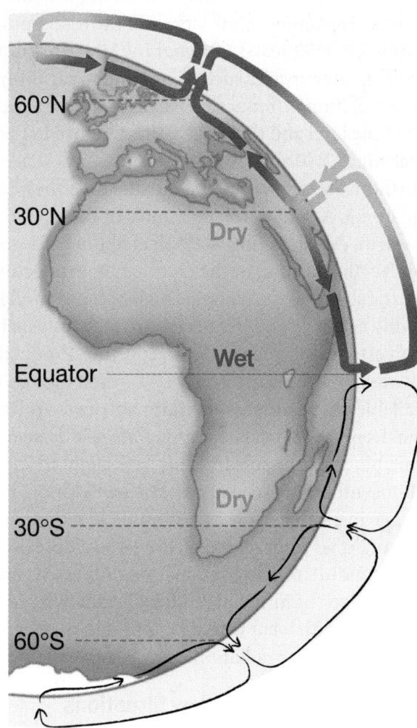

from the top and through the glass, warming the plants and ground inside the chamber. Some of this heat energy is retained within the mini-greenhouse. **Figure 50.28** They address the question posed because they capture important aspects of a plant community. NPP indicates how much biomass is available for other organisms to use; species diversity indicates how many different species are present. **Figure 50.34** (Several possible answers; one possibility is provided.) If water is short, there is not enough available to support a continuous cover of photosynthetic tissue aboveground. Roots of bunchgrasses can spread out and harvest water below spaces that are "bare" aboveground.

Summary of Key Concepts

KC 50.1 Populations are made up of individual organisms; communities are made up of populations of different species; ecosystems are made up of groups of communities (along with the abiotic environment). **KC 50.2** Increased evaporation from the oceans should increase precipitation on land—at least in some places. If precipitation does not increase, then increased transpiration rates will put plants under water stress and reduce productivity. **KC 50.3** All latitudes would get equal amounts of sunlight all year round. There would be no seasons and no changes in climate with latitude. **KC 50.4** It would increase, because cattle would no longer succumb to the disease carried by tsetse flies. (In Africa, cattle are limited by a biotic condition, not abiotic conditions.)

Test Your Knowledge

1. b; **2.** a; **3.** d; **4.** b; **5.** a; **6.** c

Test Your Understanding

1. Organismal, population, community, and ecosystem. Examples of possible questions: *Organismal:* How do humans cope with extremely hot (or cold) weather conditions? *Population:* How large will the total human population be in 50 years? *Community:* How are humans affecting prey species such as cod and tuna? *Ecosystem:* How will human-induced changes in global temperature affect sea level and thus organisms that live in the intertidal zone? **2.** More—in June the Sun's rays strike the Northern Hemisphere more directly than they do the equator. **3.** A Hadley cell just south of the equator results in warm, dry air falling to Earth at about 30 degrees south—the location of the Outback. **4.** Productivity in intertidal and neritic zones is high because sunlight is readily available and because nutrients are available from estuaries and deep-ocean currents. Productivity in the oceanic zone is extremely low—even though light is available at the surface—because nutrients are scarce. The deepest part of the oceanic zone may have nutrients available from the substrate but lacks light to support photosynthesis. **5.** As you increase in elevation, biomes change in a similar way as increasing in latitude (e.g., you might go from temperate forest to a boreal-type forest to tundra). **6.** The combination of low oxygen, low pH, and lack of nitrogen reduces productivity in bogs compared with what would be found in marshes or lakes with similar solar radiation.

Applying Concepts to New Situations

1. Yes—as Mars orbits the Sun, the part of the planet that is tilted toward the Sun is warmest because it receives the most direct sunlight. But as the planet orbits, that portion of the planet begins to tilt away from the Sun. Its temperature drops as a result, producing something like Earth's cycle of winter and summer. **2.** Mountaintops are cold because as air rises, it expands; this expansion cools the air, making the mountaintop cold. Also, there is less land surface nearby to absorb solar radiation and heat up. **3.** The hypothesis is that winds are usually from the northeast, and that higher elevations in the middle of the islands cause a rain shadow in the southwest corner. **4.** Edinburgh is surrounded by water. Because water has a high specific heat, it keeps temperatures moderate. Moscow, in contrast, is surrounded by a large landmass that does not have the same moderating effects on temperature as the ocean does.

CHAPTER 51

Check Your Understanding (CYU)

CYU p. 1022 If a sitter-like allele in bee-eaters was favored when population density is low, then it should be at high frequency in small colonies. If a rover-like allele was favored when population density is high, then it should be at high frequency in large colonies. **CYU p. 1025** (1) In the experiment, females came into breeding condition slowly if they were exposed only to springlike light conditions, but much more quickly if they were exposed to springlike light conditions *and* displaying males. Both are required for maximum effect. (2) Many possible answers, but based on the experiments with *Anolis*, it would be legitimate to hypothesize that they have to be exposed to springlike light and temperature conditions as well as courtship displays from males. **CYU p. 1031** (1) Auditory communication allows individuals to communicate over long distances but can be heard by predators. Olfactory communication is effective in the dark and scents can continue to carry information long after the signaler has left, but scents do not carry long distances. Visual communication is effective during the day but can be seen by predators. (2) The ability to detect deceit protects the individual from fitness costs (e.g., being eaten, having a mate's eggs fertilized by someone else); avoiding or punishing "liars" should also lower the frequency of deceit (because it becomes less successful). Alleles associated with detecting and avoiding or punishing "liars" should be favored by natural selection and increase in frequency. **CYU p. 1034** (1) Altruism is likely when B is high, as when a predator is about to pounce; r is high, as when a close relative is threatened; C is low, as when the caller is far from the predator. Altruism is unlikely under the opposite conditions. (2) For reciprocal altruism to work, individuals have to interact repeatedly and remember who has helped whom in the past.

You Should Be Able To (YSBAT)

YSBAT p. 1033 Humans often live near kin, are good at recognizing kin, and in many cases can give kin resources or protection that increase fitness. It is not known whether kin selection occurs in plants. In many cases, limited seed dispersal means that close relatives live together—making kin selection more likely. But it is not known whether plants can recognize kin and confer benefits preferentially to kin.

Caption Questions and Exercises

Figure 51.1 Several legitimate hypotheses are possible. Examples at the proximate level: The increased size of the antennae provide space for more sensory neurons; or large antennae allow individuals to sense a larger area. Examples at the ultimate level: large antennae help individuals navigate better and thus escape predation; or large antennae allow individuals to forage more successfully in the dark. **Figure 51.5** Bring females into the lab and give them the same food and housing conditions as the treatment groups, but expose them to artificial lighting that simulates the short daylight conditions of winter. **Figure 51.7** Because individuals were chosen for the different treatments at random, the investigator could claim that on average, the only thing that differed among individuals in the different groups was average tail length. **Figure 51.10** [See Figure A51.1] **Figure 51.14** Between first cousins, $r = \frac{1}{2} \times \frac{1}{2} \times \frac{1}{2} = \frac{1}{8}$. **Figure 51.15** The control would be to drag an object of similar size through the colony, at similar times of day. This would test the hypothesis that prairie dogs are reacting to the presence of the experimenter and the disturbance—not a predator.

Summary of Key Concepts

KC 51.1 At the proximate level, different alleles of the *for* gene affect the probability that an individual will move or stay after feeding. At the ultimate level, roving is favored when population density is high; sitting is favored when population density is low. **KC 51.2** Individuals have the capacity to act altruistically or non-altruistically toward kin. Hamilton's rule specifies the conditions—in terms of the costs and benefits of the act and the degree of relationship between the participants—under which altruism is favored. **KC 51.3** Roving is expensive energetically and could make individuals more susceptible to predation, but could allow individuals to find unexploited food sources. Sitting is inexpensive energetically, but may restrict individuals to areas where food supplies have been depleted. **KC 51.4** If only males that are well fed and free of parasites and disease are able to produce a brightly colored dewlap and do the bobbing display vigorously, then these traits would be a reliable indication of their condition. **KC 51.5** If migration does not occur when extra food is supplied, it suggests that the "decision" to migrate depends on conditions. **KC 51.6** Deceitful communication should decrease. If large males are gone, there are few nests available and more female-mimic males would compete at them. Most female-mimic males would have no (or fewer) eggs to fertilize and no male to take care of the eggs they did fertilize. There would be less natural selection pressure favoring deceitful communication. **KC 51.7** (1) Long-lived, so that extensive interactions with kin and nonrelatives are possible; (2) good memory, to record reciprocal interactions; (3) kin nearby, to make inclusive fitness gains possible; (4) ability to share fitness benefits (e.g., food) between individuals.

Test Your Knowledge

1. c; **2.** d; **3.** a; **4.** c; **5.** d; **6.** c

Test Your Understanding

1. At the ultimate level, homing to a safe den increases survival. The proximate cause for this behavior is still not fully known, but appears to involve the ability to sense changes in Earth's magnetic field. **2.** When optimal foraging occurs, an individual maximizes the benefit of foraging in terms of energy gain and minimizes the cost of foraging in terms of time, energy expended, and risk of predation. **3.** Only males that are in good shape (well-nourished and free of disease) have the resources required to produce a trait like a long tail. **4.** Longer day lengths indicate the arrival of spring, when renewed plant growth and insect activity make more food available. Courtship displays from males indicate the presence of males that can fertilize eggs. **5.** A map provides information about the spatial relationships of places on

FIGURE A51.1

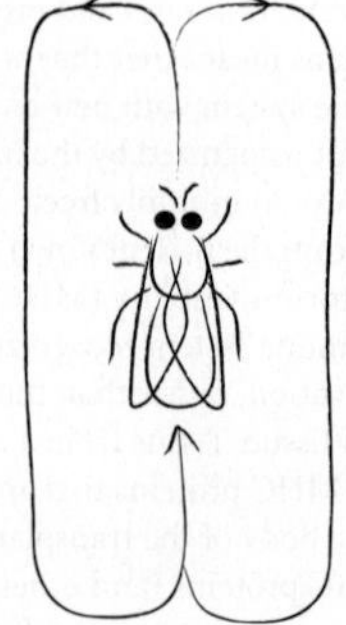

a landscape; a compass allows you to orient the map—so that it aligns with the actual landscape. Animals have been shown to use the Sun's position, the position of the North star, and information from Earth's magnetic field as a compass (a way to find North). **6.** Kin selection can only act on kin; reciprocal altruism can occur between nonrelatives. Kin selection should be more common, because it potentially can occur any time that kin interact. Reciprocal altruism should be relatively rare because it is only possible when unrelated individuals interact repeatedly, have resources to share, and can remember the outcomes of past interactions.

Applying Concepts to New Situations

1. One possibility would be to set up a testing arena with food sources at various distances from a food source where adult flies are originally released. Test adults with a rover versus sitter genotype individually, and record how far each flies over the course of a set time interval. If the rover and sitter alleles affect foraging movements in adults, then rovers should be more likely to move and move farther than sitters. **2.** Simulate the onset of rains after a dry period, by using a hose to sprinkle water on the habitat where lizards are being maintained. **3.** Individuals have an r of 1/2 with full siblings and an r of 1/8 with first cousins. If the biologist will lose his life, he needs to save two siblings to keep the "lost copy" of his altruism alleles in the population but eight first cousins to maintain a copy. **4.** People would be expected to donate blood under two conditions: (1) They either received blood before or expect to have a transfusion in the future, and/or (2) they receive some other benefit in return, such as a good reputation among people who might be able to help them in other ways.

CHAPTER 52

Check Your Understanding (CYU)

CYU p. 1046 One possibility is competition for food. To test this idea, compare carrying capacity in identical vials that differ only in the amount of food added. (You could also test a hypothesis of oxygen limitation by adding more oxygen to one set of vials, or test the hypothesis of space limitation by doubling the volume but keeping the amount of food and oxygen the same.) **CYU p. 1053** (1) In developed nations, there are roughly equal numbers of individuals in each age class because the fertility rate has been constant and survivorship high for many years. In developing countries there are many more children and young people than older people, because the fertility rate and survivorship have been increasing. (2) Because survivorship is high in most human populations, changes in overall growth rate depend almost entirely on fertility rates.

You Should Be Able To (YSBAT)

YSBAT p. 1041 (1) Compared to northern populations, fecundity should be high and survivorship low. (2) Fecundity can be much higher if females lay eggs instead of retaining them in their bodies and giving birth to live young. **YSBAT p. 1055** The population appears to be increasing over time: from the original 1000 females, there are 2308 females in the third year.

Caption Questions and Exercises

Figure 52.2 [See Figure A52.1] **Figure 52.4** [See Table A52.1] **Figure 52.8** At high population density, competition for food limits the amount of energy available to female song sparrows for egg production. **Figure 52.12** Large-scale field experiments like this are extremely expensive and difficult to implement—it was not practical to replicate all treatments. **Table 52.2** From 1999 to 2009, $6800 = 6000\ \lambda^{10}$ (see Eq. 52.5 in Box 52.2). Solving, $\lambda = 1.013$. If $e^r = 1.013$, then $r = 0.0125$. Now use the expression $N_t = N_0e^{rt}$ (Eq. 52.7 in Box 52.2) to solve the problems.

Number of years required to add 1 billion people to 2009 level:

$$7800 = 6800e^{0.0125t}$$
$$ln(1.15) = 0.0125t$$
$$t = \sim 11 \text{ years}$$

Number of years required to double population from 2009 level:

$$13{,}600 = 6800e^{0.0125t}$$
$$ln(2) = 0.0125t$$
$$t = \sim 5.5 \text{ years}$$

Figure 52.17 [See Figure A52.2 on p. A:48]

Summary of Key Concepts

KC 52.1 A life table represents a snapshot of how a particular population is growing. As conditions change, survivorship and fecundity may change. In ancient Rome, for example, survivorship was probably low and fecundity high—women started reproducing at a young age and did not live long, on average. In Rome today, the population has high survivorship and low fecundity. **KC 52.2** [See Figure A52.3 on p. A:48] **KC 52.3** Fewer children are being born per female, but there are many more females of reproductive age due to high fecundity rates in the previous generation. **KC 52.4** Small, isolated populations are likely to be wiped out by bad weather, a disease outbreak, or changes in the habitat. For example, primrose populations may die out when a gap is shaded. But in a metapopulation, migration between the individual small populations helps reestablish subpopulations and maintain the overall size of the metapopulation.

Test Your Knowledge

1. b; **2.** d; **3.** a; **4.** a; **5.** d; **6.** c

Test Your Understanding

1. Equation 52.4: After one breeding interval, the population size is equal to the original population size times the discrete growth rate. Equation 52.5: After t breeding intervals, the population size is equal to the original population size times the discrete growth rate multiplied by itself t times (i.e., raised to the tth power). **2.** Species with type I survivorship curves have high survivorship until old age, so population growth is based on the number of offspring produced—not on how many of the

FIGURE A52.1

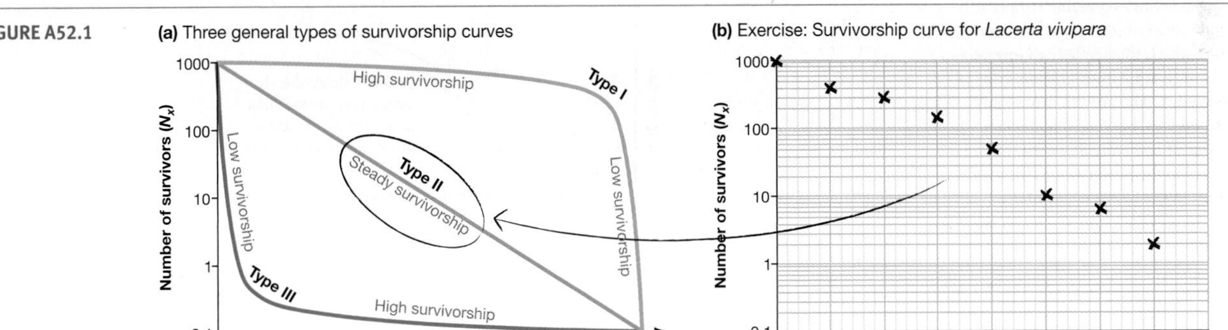

TABLE A52.1

Trait	Life-History Continuum		
	Left	Middle	Right
Growth habit	Herbaceous	Shrub	Tree
Disease- and predator-fighting ability	Low	Medium	High
Seed size	Small	Medium	Large
Seed number	Many	Moderate	Few

offspring survive. Species with type III survivorship curves exhibit very high juvenile mortality, but high survivorship of those few individuals able to reach adulthood. Because only the very few adults can reproduce, any changes to survivorship at the adult level will have a large effect on the population as a whole. **3.** The population has undergone near-exponential growth recently because advances in nutrition, sanitation, and medicine have allowed humans to live at high density without suffering from decreased survivorship and fecundity. Eventually, however, growth rates must slow as density-dependent effects increase death rates and lower birth rates. Experiments on snowshoe hares have shown that food availability and predation are density-dependent factors that influence population growth. **4.** (a) Corridors allow individuals to move between subpopulations, increasing gene flow and making it possible to recolonize habitats where populations have been lost. (b) Maintaining unoccupied habitat makes it possible for the habitat to be recolonized. **5.** As a sexually transmitted disease, AIDS will reduce the number of sexually active adults. If the epidemic continues unabated, the numbers of both reproductive-age adults and children will decline, causing a top-heavy age distribution dominated by older adults and the elderly. [See Figure A52.4] **6.** Lambda is simple to calculate but only describes growth over a discrete time interval. Determining R_0 requires knowledge of all age-specific birth and death rates. r indicates the growth rate at any instant, so is independent of generation time and can be applied to any population. r_{max} gives the population growth rate in the absence of density-dependent limitation; r is the actual growth rate, which is usually affected by density-dependent factors.

Applying Concepts to New Situations

1. Fewer older individuals will be left in the population; there will be relatively more young individuals. If too many older individuals are taken, growth rate may decline sharply as reproduction stops or slows. (But if relatively few older individuals are taken, more resources are available to younger individuals and their survivorship and fecundity, and the population's overall growth rate, may increase.) **2.** The sunflowers and beetles are both metapopulations. To preserve them, you must preserve as many of the sunflower patches as possible (or plant more) and maintain corridors (which may be smaller sunflower patches) along which the beetles can migrate between the patches. **3.** Large-scale immigration into this population is projected. **4.** The growth rate of any population is female dependent, because females can produce a limited number of offspring at a time, regardless of how many males are in the population. This means that there are "extra" men in the Chinese population—they will not participate in reproduction. Growth rate should decrease, as a result.

CHAPTER 53

Check Your Understanding (CYU)

CYU p. 1070 (1) The individuals do not choose or try to have traits that reduce competition—they simply have those traits (or not). Resource partitioning just happens, because individuals with traits that allow them to exploit different resources produce more offspring, which also have those traits. (Recall from Chapter 24 that natural selection occurs on individuals, but adaptive responses such as resource partitioning are properties of populations.) (2) When species interact via consumption, a trait that gives one species an advantage will exert natural selection on individuals of the other species who have traits that reduce that advantage. This reciprocal adaptation will continue indefinitely. An example is the interaction of *Plasmodium* with the human immune system: The human immune system has evolved the

FIGURE A52.2

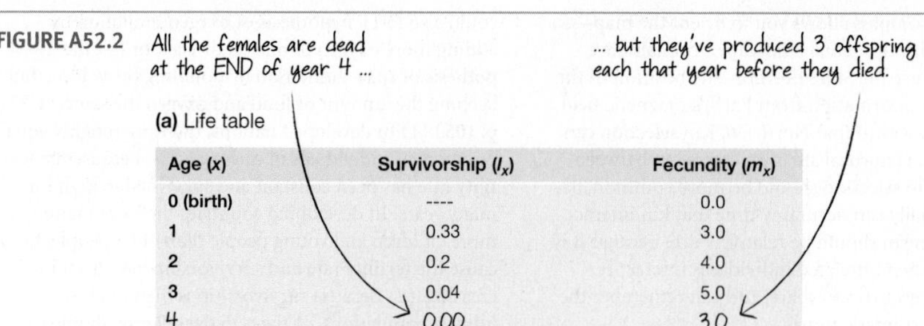

(a) Life table

Age (x)	Survivorship (l_x)	Fecundity (m_x)
0 (birth)	----	0.0
1	0.33	3.0
2	0.2	4.0
3	0.04	5.0
4	0.00	3.0

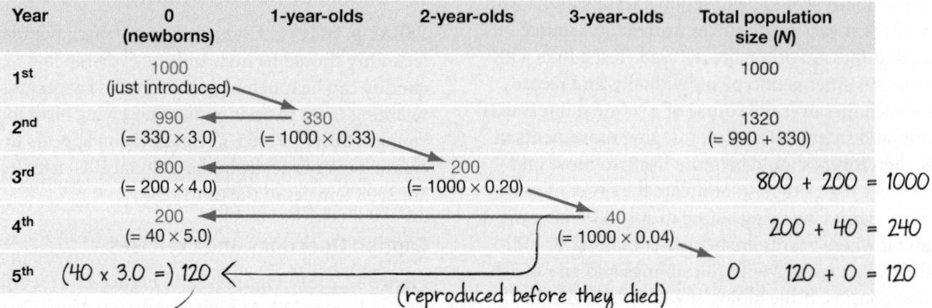

(b) Fate of first-generation females

Year	0 (newborns)	1-year-olds	2-year-olds	3-year-olds	Total population size (N)
1st	1000 (just introduced)				1000
2nd	990 (= 330 × 3.0)	330 (= 1000 × 0.33)			1320 (= 990 + 330)
3rd	800 (= 200 × 4.0)		200 (= 1000 × 0.20)		800 + 200 = 1000
4th	200 (= 40 × 5.0)			40 (= 1000 × 0.04)	200 + 40 = 240
5th	(40 x 3.0 =) 120			0	120 + 0 = 120

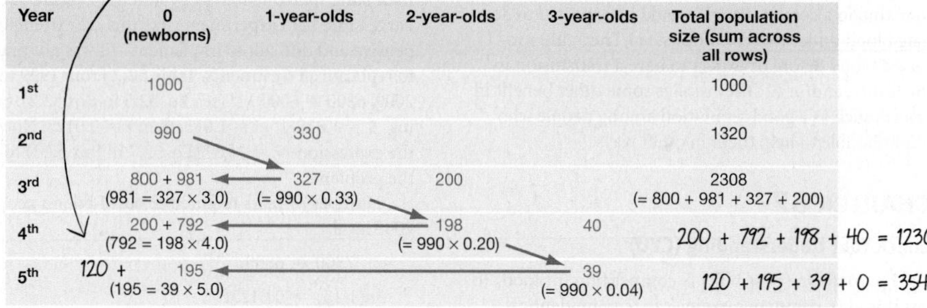

(c) Fate of first- and second-generation females

Year	0 (newborns)	1-year-olds	2-year-olds	3-year-olds	Total population size (sum across all rows)
1st	1000				1000
2nd	990	330			1320
3rd	800 + 981 (981 = 327 × 3.0)	327 (= 990 × 0.33)	200		2308 (= 800 + 981 + 327 + 200)
4th	200 + 792 (792 = 198 × 4.0)		198 (= 990 × 0.20)	40	200 + 792 + 198 + 40 = 1230
5th	120 + 195 (195 = 39 × 5.0)			39 (= 990 × 0.04)	120 + 195 + 39 + 0 = 354

FIGURE A52.3

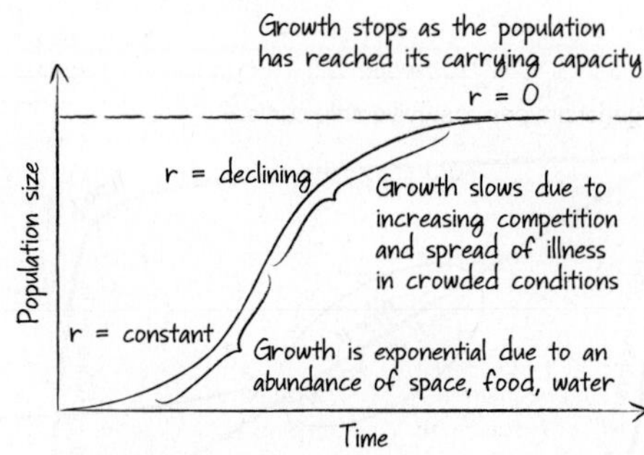

FIGURE A52.4

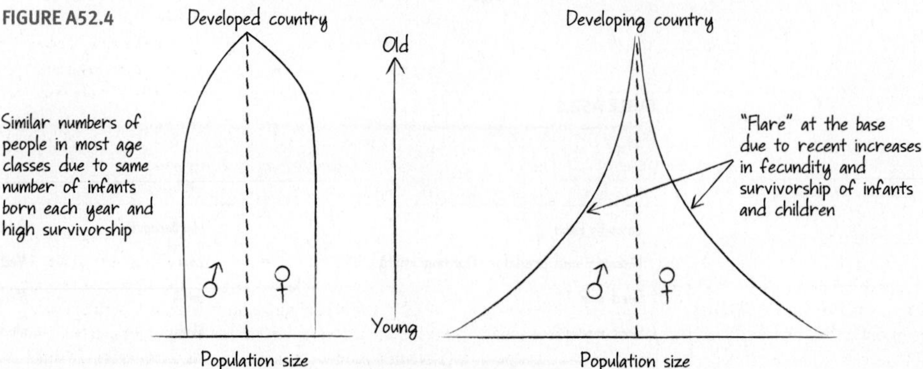

ability to detect proteins from the *Plasmodium* and kill infected cells; in response, *Plasmodium* has evolved different proteins that the immune system does not detect. **CYU p. 1077** (1) The shade provided by early successional species increases humidity, and decomposition of their tissues adds nutrients and organic material to the soil. These conditions favor growth by later successional species, which can outcompete the early successional species. (2) The presence or absence of a plant species where nitrogen-fixation occurs would dramatically alter nutrient conditions, and thus the speed of succession and the types of species that could become established. For example, species that require high nitrogen would be favored on sites where alder grew, while species that can tolerate low nitrogen would thrive on sites where alder is absent.

You Should Be Able To (YSBAT)

YSBAT p. 1079a The calculations for Community 1 are as follows:

Species A	**Species B**
$p_A = 10/18 = 0.55$	$p_B = 1/18 = 0.055$
$\ln 0.55 = -0.598$	$\ln 0.055 = -2.9$
$-0.598 \times 0.55 = -0.329$	$-2.9 \times 0.055 = -0.16$

Species C	**Species D**
$p_C = 1/18 = 0.055$	$p_D = 3/18 = 0.167$
$\ln 0.055 = -2.9$	$\ln 0.167 = -1.78$
$-2.9 \times 0.055 = -0.16$	$-1.78 \times 0.167 = -0.299$

Species E	**Species F**
$p_E = 2/18 = 0.11$	$p_F = 1/18 = 0.055$
$\ln 0.11 = -2.21$	$\ln 0.055 = -2.9$
$-2.2 \times 0.11 = -0.243$	$-2.9 \times 0.055 = -0.16$

Summing the values of $p \times \ln p$ for each species and multiplying by -1 gives the Shannon index of species diversity for Community 1:

$$(-1)[(-0.329) + (-0.16) + (-0.16) + (-0.299) + (-0.243) + (-0.16)] = 1.351$$

Similar calculations would give species diversity values of 1.79 for Community 2 and 1.61 for Community 3.
YSBAT p. 1079b
Community 1: Species richness = 12; Diversity = 2.04
Community 2: Species richness = 12; Diversity = 2.49
Community 3: Species richness = 10; Diversity = 2.3
YSBAT p. 1079c Species richness doubled in each community because the number of species doubled. Although the species diversity values differ from those in Figure 53.25, the communities with more equal numbers of individuals in each species still have the higher diversity values.

Caption Questions and Exercises

Figure 53.6 If he had done the treatments on different rocks, a critic could argue that differences in survival were due to differences in the nature of the rocks—not differences in competition. **Figure 53.7** The species partition, or divide up, resources. In this way, evolution leads to a reduction or elimination of competition. **Figure 53.12** This experimental design tested the hypothesis that mussels can sense the presence of crabs. If the crabs had been fed mussels, a critic could argue that the mussels detect the presence of damaged mussels—not crabs. (As it turns out, the mussels can do both. The experimenters did the same experiment with broken mussel shells in the chamber instead of a crab-fed fish.) **Figure 53.13** Beavers must represent the greater overall threat to cottonwoods, because the trees have evolved a compound that deters further beaver attack but does not deter beetle larvae. Getting the trunk cut down probably reduces fitness more than getting leaves eaten. **Figure 53.15** Put equal numbers of infected and uninfected ants in a pen that includes a bird predator. Count how many of each type are eaten over time. **Figure 53.16** Many possible answers are reasonable hypotheses. For example, acacia trees may expend energy in producing the large bulbs in which the ants live and the food they eat. The *Crematogaster* ants may expend energy and risk injury or death in repelling herbivores. The cleaner shrimp may occasionally get injured or eaten by the fish they are cleaning, or their diets may be somewhat restricted by the association. The host fish may miss meals or be at greater risk of predation when they are undergoing cleaning by the shrimp. In both examples of mutualism, however, the overall associations are positive for both parties. **Figure 53.17** These steps controlled for the alternative hypothesis that differences in treehopper survival were due to differences in the plants they occupied—not the presence or absence of ants. **Figure 53.18** Predictable, at least to a degree. **Figure 53.24** The effect would be to make the island more remote, which would lower the rate of immigration and move the whole immigration curve downward. The rate of extinction would increase, shifting the curve upward. The overall effect would be to decrease the number of species because the island would have effectively become more remote.

Summary of Key Concepts

KC 53.1 A mutualistic relationship becomes a parasitic one if one of the species stops receiving a benefit. The treehopper-ant mutualism becomes parasitic in years when spiders are rare, because the treehoppers no longer derive a benefit but pay a fitness cost (producing honeydew that the ants eat). **KC 53.2** All species at Glacier Bay must be able to survive in a cold climate with the local amount of precipitation. The species earliest in the succession must also be able to grow on rock exposed as the glacier melts. But chance, historical differences in seed sources, and presence or absence of alder (and nitrogen-fixation) created differences in the species present. **KC 53.3** Lakes are abundant in northern latitudes but rare in the tropics. The presence of much greater amounts of habitat supports higher species richness for breeding shorebirds at high latitudes.

Test Your Knowledge

1. d; **2.** a; **3.** b; **4.** d; **5.** c; **6.** d

Test Your Understanding

1. Yes—the treehopper-ant mutualism is parasitic or mutualistic depending on conditions; competition can evolve into no competition over time if niche differentiation occurs; arms races mean that the outcome of host-parasite interactions can change over time, and so forth (many other examples). **2.** *Top-down control:* Herbivore populations are kept at relatively low levels by predation and disease. *Bottom-up control:* Plant matter is low in nitrogen or defended by chemicals that are toxic to herbivores. These hypotheses are certainly not mutually exclusive. For example, resprouted cottonwoods produce compounds that make them less edible for beavers, which is a plant defense. In addition, these cottonwoods might not provide enough nitrogen for beavers to grow quickly, and species that prey on beavers or make them sick could be keeping their population under control. **3.** (a) If community composition is predictable, then the species present should not change over time. But if composition is not predictable, then there should be significant changes in the species present over time. (b) If community composition is predictable, then two sites with identical abiotic factors should develop identical communities. If community composition is not predictable, then sites with identical abiotic factors should develop variable communities. In most tests, the data best match the predictions of the "not predictable" hypothesis, though communities show elements of both. **4.** Disturbance is an event that removes biomass. Compared to low-frequency fires, high-frequency fires would tend to be less severe (less fuel builds up) and would tend to exert more intense natural selection for adaptations to resist the effects of fire. Compared to low-severity fires, high-severity fires would open up more space for pioneering species and would tend to exert more intense natural selection for adaptations to resist the effects of fire. **5.** Early successional species are adapted to disperse to new environments (small seeds) and grow and reproduce quickly (reproduce at an early age, grow quickly). They can tolerate severe abiotic conditions (high temperature, low humidity, low nutrient availability) but have little competitive ability. These species are able to enter a new environment (with no competitors) and thrive. **6.** The idea is that high productivity will lead to high population density of consumers, leading to competition and intense natural selection favoring niche differentiation that leads to speciation.

Applying Concepts to New Situations

1. Natural selection will favor orchid individuals that have traits that resist bee attack: thicker flower walls, nectar storage in a different position, a toxin in the flower walls, or other. Individuals could also be favored if their anthers were in a position that accomplished pollination, even if bees eat through the walls of the nectar-storage structure. **2.** Set fires or let natural fires burn, to match the time interval between fires recorded prior to the arrival of European settlers. Note that because fires have been suppressed for so long, the initial fires will have to be carefully controlled or fuel will have to be removed manually—for example by selective logging. The logic is that species in this habitat are adapted to a disturbance regime of frequent, low-intensity fires. For them to thrive, these conditions have to be re-created. **3.** The exact answer will depend on the location of the campus. The first species to appear must possess good dispersal ability, rapid growth, quick reproductive periods, and tolerance for very harsh and severe conditions. The two-acre plot is likely to be colonized first by pioneer species that have very "weedy" characteristics. But once colonization is under way, the course of succession will depend more on how the various species interact with each other. The presence of one species can inhibit or facilitate the arrival and establishment of another. For example, an early-arriving species might provide the shade and nutrients required by a late-arriving species. The site's history and nearby ecosystems may influence which species appear at each stage; for instance, an undisturbed ecosystem nearby could be a source for native species. The pattern and rate of this succession is also influenced by the overall environmental conditions affecting it. Only species with traits appropriate to the local climate are likely to colonize the site. **4.** One reasonable experiment would involve constructing artificial ponds and introducing different numbers of plankton species to different ponds, but the same total number of individuals. (Any natural immigration to the ponds would have to be prevented.) After a period of time, remove all of the plankton and measure the biomass present. Make a graph with number of species on the *x*-axis and total biomass on the *y*-axis. If the hypothesis is correct, the line of best fit through the data should have a positive slope.

CHAPTER 54

Check Your Understanding (CYU)

CYU p. 1092 (1) At each trophic level, most of the energy that is consumed is lost to heat, metabolism, or other maintenance activities, which leaves only a small percentage of energy for biomass production (growth and reproduction). (2) More food will be available to primary consumers, resulting in increases in biomass at each trophic level. An increase in biomass entering decomposer food chains will also increase decomposer populations. **CYU p. 1098** Perhaps the most direct impacts concern rates of groundwater replenishment. Converting biomes into farms or suburbs decreases groundwater recharge and increases runoff. Irrigation pumps groundwater to the surface, where much of it runs off into streams.

You Should Be Able To (YSBAT)

YSBAT p. 1086 To grow a kilogram of beef, you have to first grow 10 kilograms of grain or grass and feed it to the cow. Only 10 percent of this 10 kg will be used for growth and reproduction—the other 9 kg is used for maintenance or lost as heat. **YSBAT p. 1087a** Crustaceans and fish are ectothermic, so they are much more efficient at converting primary production into the biomass in their bodies than endothermic birds and mammals. **YSBAT p. 1087b** Bigger: elk to wolf, mule deer to wolf, aspen/cottonwood/willows to beaver, mice to rough-legged hawk. Smaller: mice to coyote, aspen/cottonwood/willows to elk and mule deer. **YSBAT p. 1094** The uptake arrow was removed—there were no plants to take up nutrients from the soil nutrient pool.

Caption Questions and Exercises

Figure 54.3 The percentage of energy used for maintenance would be much lower because they are ectothermic and sedentary. The percentage excreted would also be much lower because their diet contains little indigestible material. Thus the percentage of energy available for growth and reproduction would be much higher. **Figure 54.10** The diatoms' bodies sink, removing nutrients from the surface ocean and feeding organisms in the deeper ocean. **Figure 54.13** One logical hypothesis is that the total amount exported increases as tree roots and other belowground organic material decay and begin to wash into the stream; the amount exported should begin to decline as nitrate reserves become exhausted—eventually, there is no more nitrate to wash away. **Figure 54.14** Much more water should evaporate. One possibility is that more water vapor will be blown over land and increase precipitation over land. **Figure 54.21** Photosynthesis increases in summer, resulting in removal of CO_2 from the atmosphere.

Summary of Key Concepts

KC 54.1 [See Figure A54.1] **KC 54.2** A food chain consisting of only primary producers and primary decomposers would minimize energy loss. If organisms at higher trophic levels were present, the efficiency of energy transfer would be highest if they are ectothermic, move little, and have extremely efficient digestion. **KC 54.3** Decomposition is slowest in cold, wet temperatures on land and in anoxic conditions in the ocean (or in freshwater). Stagnant water often becomes anoxic as decomposers use up available oxygen and it is not replenished by diffusion from the atmosphere. **KC 54.4** Many possible answers, including: (1) reducing carbon dioxide emissions by finding alternatives to fossil fuels should decrease the amount of carbon dioxide released into the atmosphere, (2) recycling programs decrease carbon dioxide emissions from manufacturing plants and power plants, (3) reforestation can tie up carbon dioxide in wood; (4) iron fertilization could increase NPP in the open ocean—meaning that more CO_2 would be used.

Test Your Knowledge

1. a; **2.** d; **3.** a; **4.** c; **5.** c; **6.** b

Test Your Understanding

1. A food chain shows just a single connection between organisms at different trophic levels; a food web shows many or most of the connections that exist. The arrows represent the movement of chemical energy, obtained through feeding. **2.** If predation by cougars succeeds in reducing the deer population, then trees and low-growing plant species should increase. More (uneaten) biomass will remain in primary producers, meaning that more will be available for other primary consumers besides deer. **3.** Warmer, wetter climates speed decomposition; cool temperatures slow it. Lower amounts of nitrogen and oxygen also slow decomposition by limiting decomposer growth. Wood is slow to decompose because it contains lignin. Decomposition regulates nutrient availability because it releases nutrients from detritus and allows them to reenter the food web. **4.** Bleaching occurs when corals release symbiotic algae that perform photosynthesis; acidification is occurring as increased CO_2 in the atmosphere leads to increased carbonic acid in the ocean and a lower pH. Corals can starve if bleaching is extensive; acidification slows the growth of corals because it makes formation of calcium carbonate skeletons more difficult. **5.** Positive feedbacks include increased forest fires (due to drier, hotter summers) and increased rates of decomposition in tundras (due to warming); negative feedbacks include increased NPP due to warmer temperatures, increased precipitation, and/or increased access to CO_2 for photosynthesis. **6.** The open ocean has almost no nutrient input from the land and has little upwelling to supply nutrients from the deep ocean. In contrast, intertidal and coastal areas receive large inputs of nutrients from rivers as well as from upwellings from ocean depths.

Applying Concepts to New Situations

1. One possibility is to add radioactive phosphorus to the water, then follow this isotope through the primary producers, primary consumers, and secondary consumers in the system by measuring the amount of radioisotope in the tissues of organisms at each trophic level. **2.** If large amounts of organic material were produced in the open ocean and then moved to nearshore areas by currents, then decomposition of the material might use up all available oxygen and lead to creation of anoxic dead zones. **3.** Without herbivores, there is no link in the nitrogen cycle between primary producers and secondary consumers. All of the plant nitrogen would go to the primary decomposers and back into the soil. If decomposition is rapid enough, nitrogen would cycle quickly between primary producers, decomposers, the soil, and back to primary producers. **4.** Atmospheric oxygen would increase due to extensive photosynthesis, but carbon dioxide levels would decrease because little decomposition was occurring. The temperature would drop because fewer greenhouse gases would be trapping heat reflected from the Earth's surface.

CHAPTER 55

Check Your Understanding (CYU)

CYU p. 1109 Do an all-taxon survey: organize experts and volunteers to collect, examine, and identify all species present. This could include direct sequencing studies (see Chapter 29) to document the bacteria and archaea present in different habitats on campus. **CYU p. 1117** (1) Fragmentation reduces habitat quality by creating edges that are susceptible to invasion and loss of species, due to changed abiotic conditions. Genetic problems occur inside fragments as species become inbred and/or lose genetic diversity via genetic drift. (2) Using the equation $S = cA^z$, then $S = 19(100{,}000^{0.20}) = 190$ prior to habitat destruction and $S = 19(10{,}000^{0.20}) = 120$ afterwards, for a total loss of 70 species. Using the same equation for a z of 0.25 suggests that 338 species existed prior to extinction and 190 exist afterwards, a difference of 148. Twice as much diversity was lost due to the increased z. **CYU p. 1120** Establish a large number of study plots on the same hillside. Randomly assign the plots to contain 0, 1, 2, 4, 16, 32, or 64 species (or some other combination, to test a range of species richness), with several replicates for each level of species richness. Prior to planting, document soil depth and soil nutrient levels. Over time, maintain each plot by weeding to keep the original species richness intact. After several years, document soil depth and soil nutrient levels. Compare values as a function of species richness.

Caption Questions and Exercises

Figure 55.6 Nearly all endangered species are threatened by more than one factor and therefore appear on the graph more than once, leading to species totals greater than 100 percent. **Figure 55.10** Because the treatments and study areas were assigned at random, there is no bias in terms of picking certain areas that are unusual in terms of their biomass or species diversity. They should represent a random sample of biomass and species diversity in fragments of various sizes versus intact forest. **Figure 55.11** Draw a horizontal line at lambda of 1.0. **Figure 55.14** Null hypothesis: NPP is not affected by species richness or functional diversity of species. Prediction: NPP will be greater in plots with more species and more functional groups. Prediction of null: There will be no difference in NPP based on species richness or number of functional groups. **Figure 55.15** This line represents no change in biomass, which would mean the

FIGURE A54.1

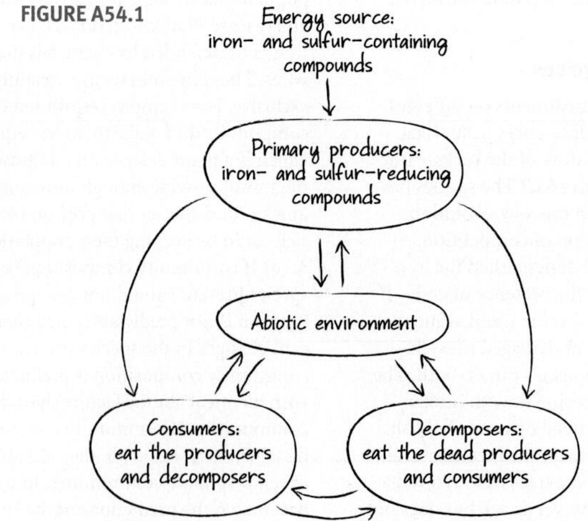

community was completely resistant to drought—that is, unaffected. **Figure 55.16** [See Figure A55.1]

Summary of Key Concepts

KC 55.1 By sequencing the genes present in a sample from a habitat, you get a broad view of the allelic diversity present—even if you don't know which species are there. **KC 55.2** Small, geographically isolated populations have low genetic diversity due to lack of gene flow and genetic drift and can suffer from inbreeding depression. Low genetic diversity makes them vulnerable to changes in their habitat, and small population size makes them susceptible to catastrophic events such as storms, disease outbreaks, or fires. **KC 55.3** If many species are present, and each uses the available resources in a unique way, then a larger proportion of all available resources should be used—leading to higher biomass production. **KC 55.4** 11percent is already protected, but a mass extinction event is still going on. This is because 11 percent is not sufficient to save most species and because many or most protected areas are not located in biodiversity or conservation hotspots.

Test Your Knowledge

1. d; **2.** a; **3.** b; **4.** d; **5.** c; **6.** a

Test Your Understanding

1. The threats have increased, and instead of most endangered species occurring on islands and being threatened by introduced species and overexploitation, now most threatened species are on continents (where habitat destruction is reducing habitats to small islands) and are in trouble due to habitat destruction and climate change. **2.** Level of resource use—because resources are only used at the rate at which they are replenished. **3.** By comparing the number of species estimated to reside in the park before and after the survey, biologists will have an idea of how much actual species diversity is being underestimated in other parts of the world—where research was at the level of Great Smoky Mountains National Park before the survey. The limitation is that it may be difficult to extrapolate from the data—no one knows if the situation at this park is typical of other habitats. **4.** Species–area curves specify how many species are expected to occur in habitats of various size. After projecting amounts of habitat loss, biologists can read a species–area curve to estimate how many species will be left in the amount of area that remains. **5.** In experiments with species native to North American grasslands, study plots that have more species produce more biomass, change less during a disturbance (drought), and recover from a disturbance faster than study plots with fewer species. **6.** Wildlife corridors facilitate the movement of individuals. Corridors allow areas to be recolonized if a species is lost and the introduction of new alleles that can counteract genetic drift and inbreeding in small isolated populations.

Applying Concepts to New Situations

1. Ecosystem services are beneficial effects that ecosystems have on the abiotic environment (soil, air, water quality, climate) for humans. No one owns or pays for these services, so no one has a vested interest in maintaining them. This is one of the primary reasons that ecosystems are destroyed, even though the services they offer are valuable. **2.** Catalog existing biodiversity at the genetic, species, and ecosystem levels. Using these data, find areas that have the highest concentration of biodiversity at each level. Protect as many of these areas as possible, and connect them with corridors of habitat. (In essence, protect and connect the biodiversity hotspots.) **3.** No "correct" answer. One argument is that conservationists could lobby officials in Brazil and Indonesia to learn from the mistakes made in developed nations, and preserve enough forests to maintain biodiversity and ecosystem services intact—avoiding the expenditures that developed countries had to make to clean up pollution and restore ecosystems and endangered species. **4.** Species with specialized food or habitat requirements, large size (and thus large requirements for land area), small population size, and low reproductive rate are vulnerable to extinction. A good example is the koala bear. Species that are particularly resistant to pressure from humans possess the opposite of many of these traits. A good example of a resistant species is the Norway rat.

THE BIG PICTURE: ECOLOGY

Check Your Understanding (CYU), p. 1126

1. Habitat loss "may eliminate" species; elimination of species (1) may disrupt community structure, (2) may reduces species richness, and (3) may reduce primary productivity. **2.** CO_2 "increases/contributes to" global warming; global warming (1) may disrupt community structure, (2) may reduce species richness, and (3) may reduce primary productivity. **3.** Populations "make up" species; ecosystems "are based on" primary productivity; water temperature, etc. "dictate species that can be found in certain" aquatic ecosystems; CO_2 "concentrations affect" climate.

APPENDIX A: BIOSKILLS

BIOSKILLS 1 (p. B:2) 1. 3.1 miles **2.** 36.96 yards **3.** 37°C **4.** Multiply your weight in pounds by 1/2.2 (0.45). **5.** 4 **BIOSKILLS 2 (p. B:4) 1.** about 18% **2.** a dramatic drop (almost 10%) **3.** No—the order of presentation in a bar chart does not matter (though it's convenient to arrange the bars in a way that reinforces the overall message). **4.** about 11 **5.** 68 inches **BIOSKILLS 3 Figure B3.1** [See Figure AB.1] **Figure B3.2** [See Figure AB.2]

FIGURE A55.1

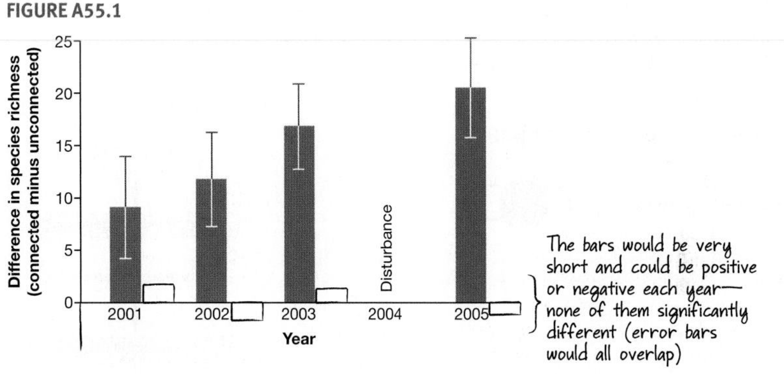

FIGURE AB.1

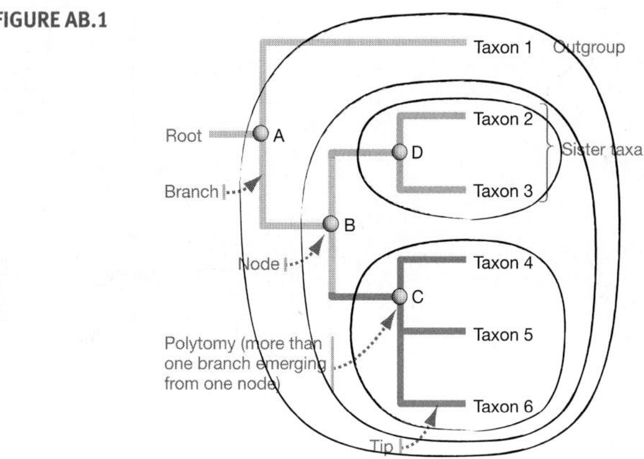

FIGURE AB.2

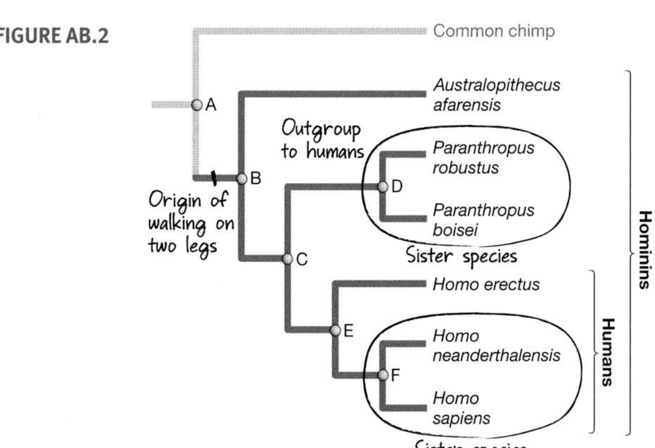

Figure B3.3 d **BIOSKILLS 4 (p. B:6) 1.** "different-yoked-together" **2.** "sugary-loosened" **3.** "study-of-form" **4.** "three-bodies" **BIOSKILLS 5 (p. B:7)** The estimate based on the larger sample—the more replicates or observations you have, the more precise your estimate of the average should be. **BIOSKILLS 6 Figure B6.1** [See **Figure AB.3**] **BIOSKILLS 7 (p. B:9) 1.** exponential **2.** $\ln N_t = \ln N_0 + rt$ **BIOSKILLS 8 (p. B:10) 1.** [See **Figure AB.4**] **2.** [See **Figure AB.4**] **BIOSKILLS 9 Figure B9.1** DNA and RNA are acids that tend to drop a proton in solution, giving them a negative charge. **(p. B:13)** The lane with no band comes from a sample where RNA X is not present. The same size RNA X is present in the next two lanes, but the light band has very few copies while the dark band has many. In the fourth lane, the band is formed by a smaller version of RNA X, with relatively few copies present. **BIOSKILLS 10 (p. B:16) 1.** No—it's just that no mitochondria happened to be present in this section sliced through the cell. **2.** Understanding a molecule's structure is often critical to understanding how it functions in cells. **BIOSKILLS 11 (p. B:17) 1.** size and/or density **2.** The use of high speeds generates the much higher forces needed to separate molecules that are similar in size and mass. **BIOSKILLS 12 (p. B:18) 1.** It may not be clear that the results are relevant to noncancerous cells that are not growing in cell culture—that is, that the artificial conditions mimic natural conditions. **2.** It may not be clear that the results are relevant to individuals that developed normally, from an embryo—that is, that the artificial conditions mimic natural conditions. **BIOSKILLS 13 (p. B:19) 1.** $\frac{1}{2} \times \frac{1}{2} \times \frac{1}{2} \times \frac{1}{2} = \frac{1}{16}$ **2.** $\frac{1}{6} + \frac{1}{6} + \frac{1}{6} = \frac{1}{2}$ **BIOSKILLS 14 Figure B14.1** This is human body temperature—the natural habitat of *E. coli.* **(p. B:22) 1.** *Caenorhabditis elegans* would be a good possibility, because the cells that normally die have already been identified. You could find mutant individuals that lacked normal cell death; you could compare the resulting embryos to normal embryos and be able to identify exactly which cells change as a result. **2.** Any of the multicellular organisms in the list would be a candidate, but *Dictyostelium discoideum* might be particularly interesting because cells only stick to each other during certain points in the life cycle. **3.** *Mus musculus*—as the only mammal in the list, it is the organism most likely to have a gene similar to the one you want to study.

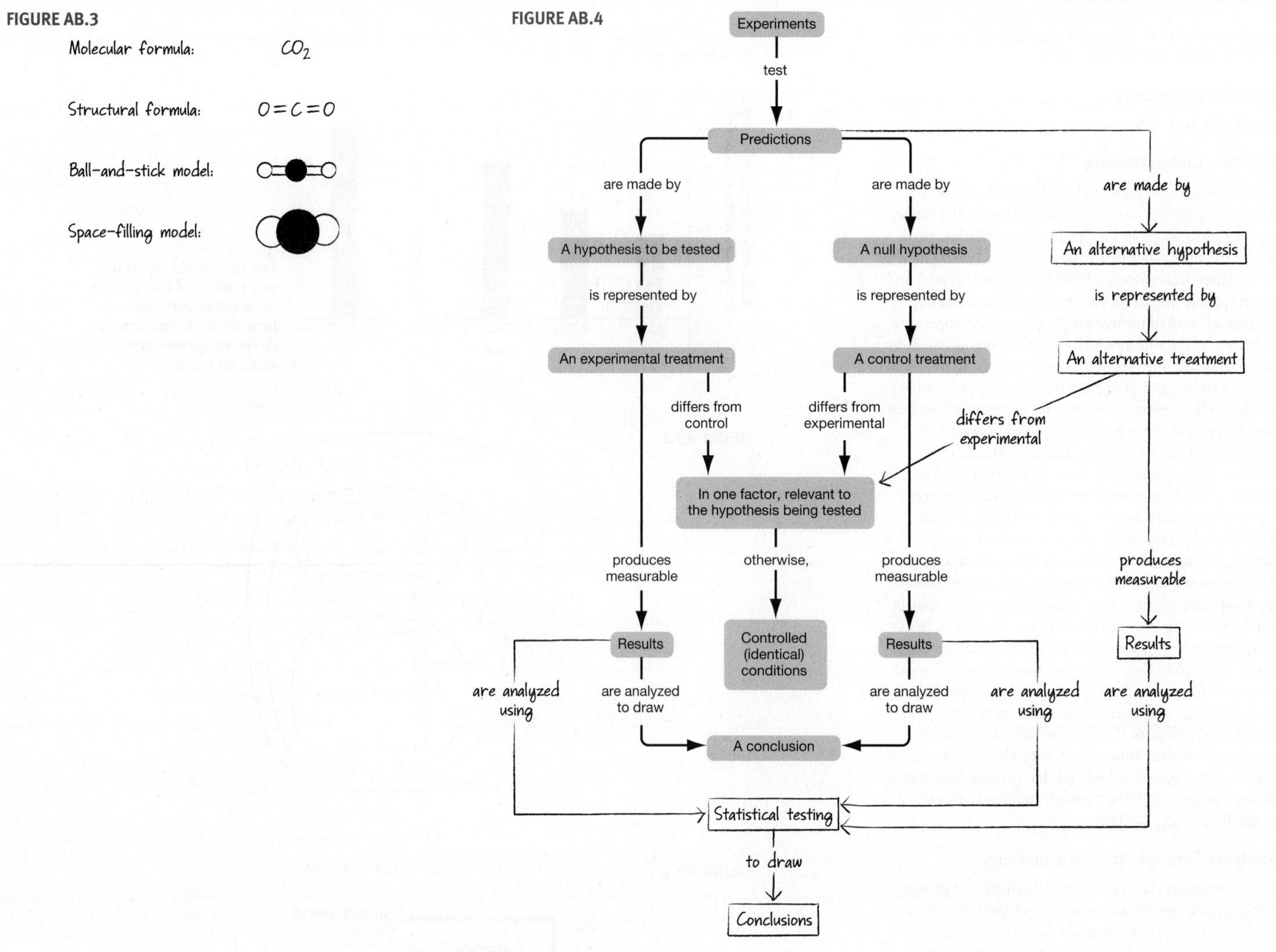

Glossary

5′ cap A chemical grouping, consisting of 7-methylguanylate and three phosphate groups, that is added to the 5′ end of newly transcribed messenger RNA molecules.

20-hydroxyecdysone See **ecdysone**.

abdomen A region of the body; in insects, one of the three prominent body regions called tagmata.

abiotic Not alive (e.g., air, water, and soil). Compare with **biotic**.

aboveground biomass The total mass of living plants in an area, excluding roots.

abscisic acid (ABA) A plant hormone that inhibits cell elongation and stimulates leaf shedding and dormancy.

abscission In plants, the normal (often seasonal) shedding of leaves, fruits, or flowers.

abscission zone The region at the base of a petiole that thins and breaks during dropping of leaves.

absorption In animals, the uptake of ions and small molecules derived from food across the lining of the intestine and into the bloodstream.

absorption spectrum The amount of light of different wavelengths absorbed by a pigment. Usually depicted as a graph of light absorbed versus wavelength. Compare with **action spectrum**.

acclimation, acclimatization Gradual physiological adjustment of an organism to new environmental conditions that occur naturally or as part of a laboratory experiment.

acetyl CoA A molecule produced by oxidation of pyruvate (the final product of glycolysis) in a reaction catalyzed by pyruvate dehydrogenase. Can enter the citric acid cycle and also is used as a carbon source in the synthesis of fatty acids, steroids, and other compounds.

acetylation Addition of an acetyl group (CH_3COO^-) to a molecule.

acetylcholine (ACh) A neurotransmitter, released by nerve cells at neuromuscular junctions, that triggers contraction of muscle cells. Also used as a neurotransmitter between neurons.

acid Any compound that gives up protons or accepts electrons during a chemical reaction or that releases hydrogen ions when dissolved in water.

acid-growth hypothesis The hypothesis that auxin triggers elongation of plant cells by inducing the synthesis of proton pumps whose activity makes the cell wall more acidic, leading to expansion of the cell wall and an influx of water.

acoelomate An animal that lacks an internal body cavity (coelom). Compare with **coelomate** and **pseudocoelomate**.

acquired immune deficiency syndrome (AIDS) A human disease characterized by death of immune system cells (in particular helper T cells and macrophages) and subsequent vulnerability to other infections. Caused by the human immunodeficiency virus (HIV).

acquired immunity Immunity to a particular pathogen or other antigen conferred by antibodies and activated B and T cells following exposure to the antigen. Is characterized by specificity, diversity, memory, and self-nonself recognition. Also called *acquired immune response*. Compare with **innate immunity**.

acrosome A caplike structure, located on the head of a sperm cell, that contains enzymes capable of dissolving the outer coverings of an egg.

ACTH See **adrenocorticotropic hormone**.

actin A globular protein that can be polymerized to form filaments. Actin filaments are part of the cytoskeleton and constitute the thin filaments in skeletal muscle cells.

actin filament A long fiber, about 7 nm in diameter, composed of two intertwined strands of polymerized actin protein; one of the three types of cytoskeletal fibers. Involved in cell movement. Also called a *microfilament*. Compare with **intermediate filament** and **microtubule**.

action potential A rapid, temporary change in electrical potential across a membrane, from negative to positive and back to negative. Occurs in cells, such as neurons and muscle cells, that have an excitable membrane.

action spectrum The relative effectiveness of different wavelengths of light in driving a light-dependent process such as photosynthesis. Usually depicted as a graph of some measure of the process versus wavelength. Compare with **absorption spectrum**.

activation energy The amount of energy required to initiate a chemical reaction; specifically, the energy required to reach the transition state.

active site The portion of an enzyme molecule where substrates (reactant molecules) bind and react.

active transport The movement of ions or molecules across a plasma membrane or organelle membrane against an electrochemical gradient. Requires energy (e.g., from hydrolysis of ATP) and assistance of a transport protein (e.g., pump).

adaptation Any heritable trait that increases the fitness of an individual with that trait, compared with individuals without that trait, in a particular environment.

adaptive immunity Immunity to a particular pathogen or other antigen conferred by antibodies and activated B and T cells following exposure to the antigen. Is characterized by specificity, diversity, memory, and self-nonself recognition. Also called *adaptive immune response*. Compare with **innate immunity**.

adaptive radiation Rapid evolutionary diversification within one lineage, producing numerous descendant species with a wide range of adaptive forms.

adenosine diphosphate (ADP) A molecule consisting of adenine, a sugar, and two phosphate groups. Addition of a third phosphate group produces adenosine triphosphate (ATP).

adenosine triphosphate (ATP) A molecule consisting of adenine, a sugar, and three phosphate groups that can be hydrolyzed to release energy. Universally used by cells to store and transfer energy.

adenylyl cyclase An enzyme that can catalyze the formation of cyclic AMP (cAMP) from ATP. Involved in controlling transcription of various operons in prokaryotes and in some eukaryotic signal-transduction pathways.

adhesion The tendency of certain dissimilar molecules to cling together due to attractive forces. Compare with **cohesion**.

adipocyte A fat cell.

adipose tissue A type of connective tissue whose cells store fats.

ADP See **adenosine diphosphate**.

adrenal glands Two small endocrine glands that sit above each kidney. The outer portion (cortex) secretes several steroid hormones; the inner portion (medulla) secretes epinephrine and norepinephrine.

adrenaline See **epinephrine**.

adrenocorticotropic hormone (ACTH) A peptide hormone, produced and secreted by the anterior pituitary, that stimulates release of steroid hormones (e.g., cortisol, aldosterone) from the adrenal cortex.

adult A sexually mature individual.

adventitious root A root that develops from a plant's shoot system instead of from the plant's root system.

aerobic Referring to any metabolic process, cell, or organism that uses oxygen as an electron acceptor. Compare with **anaerobic**.

afferent division The part of the nervous system, consisting mainly of sensory neurons, that transmits information about the internal and external environment to the central nervous system. Compare with **efferent division**.

age class All the individuals of a specific age in a population.

age structure The proportion of individuals in a population that are of each possible age.

age-specific fecundity The average number of female offspring produced by a female in a certain age class.

agglutination Clumping together of cells, typically caused by antibodies.

aggregate fruit A fruit (e.g., raspberry) that develops from a single flower that has many separate carpels. Compare with **multiple** and **simple fruit**.

AIDS See **acquired immune deficiency syndrome**.

albumen A solution of water and protein (particularly albumins), found in amniotic eggs, that nourishes the growing embryo. Also called *egg white*.

albumin A class of large proteins found in plants and animals, particularly in the albumen of eggs and in blood plasma.

alcohol fermentation Catabolic pathway in which pyruvate produced by glycolysis is converted to ethanol in the absence of oxygen.

aldosterone A hormone produced in the adrenal cortex that stimulates the kidney to conserve salt and water and promotes retention of sodium.

alimentary canal See **digestive tract**.

allele A particular version of a gene.

allergen A molecule (antigen) that triggers an allergic response (an allergy).

allergy An abnormal response to an antigen, usually characterized by dilation of blood vessels, contraction of smooth muscle cells, and increased activity by mucous-secreting cells.

allopatric speciation The divergence of populations into different species by physical isolation of populations in different geographic areas. Compare with **sympatric speciation**.

allopatry Condition in which two or more populations live in different geographic areas. Compare with **sympatry**.

allopolyploidy (adjective: allopolyploid) The state of having more than two full sets of chromosomes (polyploidy) due to hybridization between different species. Compare with **autopolyploidy**.

allosteric regulation Regulation of a protein's function by binding of a regulatory molecule, usually to a specific site distinct from the active site, causing a change in the protein's shape.

α-amylase See **amylase**.

α-helix (alpha-helix) A protein secondary structure in which the polypeptide backbone coils into a spiral shape stabilized by hydrogen bonds between atoms.

alternation of generations A life cycle involving alternation of a multicellular haploid stage (gametophyte) with a multicellular diploid stage (sporophyte). Occurs in most plants and some protists.

alternative splicing In eukaryotes, the splicing of primary RNA transcripts from a single gene in different ways to produce different mature mRNAs and thus different polypeptides.

altruism Any behavior that has a cost to the individual (such as lowered survival or reproduction) and a benefit to the recipient. See **reciprocal altruism**.

alveolus (plural: alveoli) One of the tiny air-filled sacs of a mammalian lung.

ambisense virus A virus whose genome contains both positive-sense and negative-sense sequences.

amino acid A small organic molecule with a central carbon atom bonded to an amino group ($-NH_3$), a carboxyl group ($-COOH$), a hydrogen atom, and a side group. Proteins are polymers of 20 common amino acids.

aminoacyl tRNA A transfer RNA molecule that is covalently bound to an amino acid.

aminoacyl tRNA synthetase An enzyme that catalyzes the addition of a particular amino acid to its corresponding tRNA molecule.

ammonia (NH_3) A small molecule, produced by the breakdown of proteins and nucleic acids, that is very toxic to cells. Is a strong base that gains a proton to form the ammonium ion (NH_4^+).

amnion The membrane within an amniotic egg that surrounds the embryo and encloses it in a protective pool of fluid (amniotic fluid).

amniotes A major lineage of vertebrates (Amniota) that reproduce with amniotic eggs. Includes all reptiles (including birds) and mammals.

amniotic egg An egg that has a watertight shell or case enclosing a membrane-bound water supply (the amnion), food supply (yolk sac), and waste sac (allantois).

amoeboid motion See **cell crawling**.

amphibians A lineage of vertebrates many of whom breathe through their skin and feed on land but lay their eggs in water. Represent the earliest tetrapods; include frogs, salamanders, and caecilians.

amphipathic Containing hydrophilic and hydrophobic elements.

amylase Any enzyme that can break down starch by catalyzing hydrolysis of the glycosidic linkages between the glucose residues.

amyloplasts Dense, starch-storing organelles that settle to the bottom of plant cells and that may be used as gravity detectors.

anabolic pathway Any set of chemical reactions that synthesizes larger molecules from smaller ones. Generally requires an input of energy. Compare with **catabolic pathway**.

anadromous Having a life cycle in which adults live in the ocean (or large lakes) but migrate up freshwater streams to breed and lay eggs.

anaerobic Referring to any metabolic process, cell, or organism that uses an electron acceptor other than oxygen, such as nitrate or sulfate. Compare with **aerobic**.

anaphase A stage in mitosis or meiosis during which chromosomes are moved to opposite ends of the cell.

anatomy The study of the physical structure of organisms.

ancestral trait A trait found in ancestors.

aneuploidy (adjective: aneuploid) The state of having an abnormal number of copies of a certain chromosome.

angiosperm A flowering vascular plant that produces seeds within mature ovaries (fruits). The angiosperms form a single lineage. Compare with **gymnosperm**.

animal A member of a major lineage of eukaryotes (Animalia) whose members typically have a complex, large, multicellular body, eat other organisms, and are mobile.

animal model Any disease that occurs in a nonhuman animal and has many parallels to a similar disease of humans. Studied by medical researchers in hope that findings may apply to human disease.

anion A negatively charged ion.

annelids Members of the phylum Annelida (segmented worms). Distinguished by a segmented body and a coelom that functions as a hydrostatic skeleton. Annelids belong to the lophotrochozoan branch of the protostomes.

annual Referring to a plant whose life cycle normally lasts only one growing season—less than one year. Compare with **perennial**.

anoxygenic Referring to any process or reaction that does not produce oxygen. Photosynthesis in purple sulfur and purple nonsulfur bacteria, which does not involve photosystem II, is anoxygenic. Compare with **oxygenic**.

antenna (plural: antennae) A long appendage that is used to touch or smell.

antenna complex Part of a photosystem, containing an array of chlorophyll molecules and accessory pigments, that receives energy from light and directs the energy to a central reaction center during photosynthesis.

anterior Toward an animal's head and away from its tail. The opposite of posterior.

anterior pituitary The part of the pituitary gland containing endocrine cells that produce and release a variety of peptide hormones in response to other hormones from the hypothalamus. Compare with **posterior pituitary**.

anther The pollen-producing structure at the end of a stamen in flowering plants (angiosperms).

antheridium (plural: antheridia) The sperm-producing structure in most land plants except angiosperms.

anthropoids One of the two major lineages of primates, including apes, humans, and all monkeys. Compare with **prosimians**.

antibiotic Any substance, such as penicillin, that can kill or inhibit the growth of bacteria.

antibody An immunoglobulin protein, produced by B cells, that can bind to a specific part of an antigen, tagging it for attack by the immune system. All antibody molecules have a similar Y-shaped structure and, in their monomer form, consist of two identical light chains and two identical heavy chains.

anticodon The sequence of three bases (triplet) in a transfer RNA molecule that can bind to a mRNA codon with a complementary sequence.

antidiuretic hormone (ADH) A peptide hormone, secreted from the posterior pituitary gland, that stimulates water retention by the kidney. Also called *vasopressin*.

antigen Any foreign molecule, often a protein, that can stimulate a specific response by the immune system.

antigen presentation Process by which small peptides, derived from ingested particulate antigens (e.g., bacteria) or intracellular antigens (e.g., viruses in infected cell) are complexed with MHC proteins and transported to the cell surface where they are displayed and can be recognized by T cells.

antiparallel Describing the opposite orientation of the strands in a DNA double helix with one strand running in the $5' \rightarrow 3'$ direction and the other in the $3' \rightarrow 5'$ direction.

antiporter A carrier protein that allows an ion to diffuse down an electrochemical gradient, using the energy of that process to transport a different substance in the opposite direction *against* its concentration gradient. Compare with **symporter**.

antiviral Any drug or other agent that can kill or inhibit the transmission or replication of viruses.

anus In a multicellular animal, the end of the digestive tract where wastes are expelled.

aorta In terrestrial vertebrates, the major artery carrying oxygenated blood away from the heart.

aphotic zone Deep water receiving no sunlight. Compare with **photic zone**.

apical Toward the top. In plants, at the tip of a branch. In animals, on the side of an epithelial layer that faces the environment and not other body tissues. Compare with **basal**.

apical bud A bud at the tip of a stem, where growth occurs to lengthen the stem.

apical dominance Inhibition of lateral bud growth by the apical meristem at the tip of a plant branch.

apical meristem A group of undifferentiated plant cells, at the tip of a stem or root, that is responsible for primary growth. Compare with **lateral meristem**.

apical–basal axis The long or vertical, shoot-to-root axis of a plant.

apomixis The formation of mature seeds without fertilization occurring; a type of asexual reproduction.

apoplast In plant roots, a continuous pathway through which water can flow, consisting of the porous cell walls of adjacent cells and the intervening extracellular space. Compare with **symplast**.

apoptosis Series of genetically controlled changes that lead to death of a cell. Occurs frequently during embryological development and later may occur in response to infections or cell damage. Also called *programmed cell death*.

appendix A blind sac (having only one opening) that extends from the colon in some mammals.

aquaporin A type of channel protein through which water can move by osmosis across a plasma membrane.

arbuscular mycorrhizal fungi (AMF) Fungi whose hyphae enter the root cells of their host plants. Also called *endomycorrhizal fungi*.

Archaea One of the three taxonomic domains of life consisting of unicellular prokaryotes distinguished by cell walls made of certain polysaccharides not found in bacterial or eukaryotic cell walls, plasma membranes composed of unique isoprene-containing phospholipids, and ribosomes and RNA polymerase similar to those of eukaryotes. Compare with **Bacteria** and **Eukarya**.

archegonium (plural: archegonia) The egg- producing structure in most land plants except angiosperms.

arteriole One of the many tiny vessels that carry blood from arteries to capillaries.

artery Any thick-walled blood vessel that carries blood (oxygenated or not) under relatively high pressure away from the heart to organs throughout the body. Compare with **vein**.

arthropods Members of the phylum Arthropoda. Distinguished by a segmented body; a hard, jointed exoskeleton; paired appendages; and an extensive body cavity called a hemocoel. Arthropods belong to the ecdysozoan branch of the protostomes.

articulation A movable point of contact between two bones of a skeleton. See **joint**.

artificial selection Deliberate manipulation by humans, as in animal and plant breeding, of the genetic composition of a population by allowing only individuals with desirable traits to reproduce.

ascocarp A large, cup-shaped reproductive structure produced by some ascomycete fungi. Contains many microscopic asci, which produce spores.

ascus (plural: asci) Specialized spore-producing cell found at the ends of hyphae in sac fungi (ascomycetes).

asexual reproduction Any form of reproduction resulting in offspring that are genetically identical to the parent. Includes binary fission, budding, and parthenogenesis. Compare with **sexual reproduction**.

asymmetric competition Ecological competition between two species in which one species suffers a much greater fitness decline than the other. Compare with **symmetric competition**.

atomic number The number of protons in the nucleus of an atom, giving the atom its identity as a particular chemical element.

ATP See **adenosine triphosphate**.

ATP synthase A large membrane-bound protein complex in chloroplasts, mitochondria, and some bacteria that uses the energy of protons flowing through it to synthesize ATP. Also called *F_oF_1 complex.*

atrial natriuretic hormone An animal hormone that stimulates excretion of salt from the kidneys.

atrioventricular (AV) node A region of the heart between the right atrium and right ventricle where electrical signals from the atrium are slowed briefly before spreading to the ventricle. This delay allows the ventricle to fill with blood before contracting. Compare with **sinoatrial (SA) node**.

atrium (plural: atria) A thin-walled chamber of the heart that receives blood from veins and pumps it to a neighboring chamber (the ventricle).

autocrine Relating to a chemical signal that affects the same cell that produced and released it.

autoimmunity A pathological condition in which the immune system attacks self cells or tissues of an individual's own body.

autonomic nervous system The part of the peripheral nervous system that controls internal organs and involuntary processes, such as stomach contraction, hormone release, and heart rate. Includes parasympathetic and sympathetic nerves. Compare with **somatic nervous system**.

autophagy The process by which damaged organelles are surrounded by a membrane and delivered to a lysosome to be destroyed.

autopolyploidy (adjective: autopolyploid) The state of having more than two full sets of chromosomes (polyploidy) due to a mutation that doubled the chromosome number.

autoradiography A technique for detecting radioactively labeled molecules separated by gel electrophoresis by placing an unexposed film over the gel. A black dot appears on the film wherever a radioactive atom is present in the gel. Also can be used to locate labeled molecules in fixed tissue samples.

autosomal inheritance The inheritance patterns that occur when genes are located on autosomes rather than on sex chromosomes.

autosome Any chromosome that does not carry genes involved in determining the sex of an individual.

autotroph Any organism that can synthesize reduced organic compounds from simple inorganic sources such as CO_2 or CH_4. Most plants and some bacteria and archaea are autotrophs. Also called *primary producer.* Compare with **heterotroph**.

auxin Indoleacetic acid, a plant hormone that stimulates phototropism and some other responses.

avirulence (*avr*) genes Genes in pathogens encoding proteins that trigger a defense response in plants. Compare with **resistance (*R*) genes**.

axillary bud A bud that forms in the angle between a leaf and a stem and may develop into a lateral (side) branch. Also called *lateral bud.*

axon A long projection of a neuron that can propagate an action potential and transmit it to another neuron.

axon hillock The site in a neuron where an axon joins the cell body and where action potentials are first triggered.

axoneme A structure found in eukaryotic cilia and flagella and responsible for their motion; composed of two central microtubules surrounded by nine doublet microtubules (9 + 2 arrangement).

B cell A type of leukocyte that matures in the bone marrow and, with T cells, is responsible for adaptive immunity. Produces antibodies and also functions in antigen presentation. Also called *B lymphocyte.*

BAC library A collection of all the sequences found in the genome of a species, inserted into bacterial artificial chromosomes (BACs).

background extinction The average rate of low-level extinction that has occurred continuously throughout much of evolutionary history. Compare with **mass extinction**.

Bacteria One of the three taxonomic domains of life consisting of unicellular prokaryotes distinguished by cell walls composed largely of peptidoglycan, plasma membranes similar to those of eukaryotic cells, and ribosomes and RNA polymerase that differ from those in archaeans or eukaryotes. Compare with **Archaea** and **Eukarya**.

bacterial artificial chromosome (BAC) An artificial version of a bacterial chromosome that can be used as a cloning vector to produce many copies of large DNA fragments.

bacteriophage Any virus that infects bacteria.

baculum A bone inside the penis usually present in mammals with a penis that lacks erectile tissue.

balancing selection A pattern of natural selection in which no single allele is favored in all populations of a species at all times. Instead, there is a balance among alleles in terms of fitness and frequency.

ball-and-stick model A representation of a molecule where atoms are shown as balls—colored and scaled to indicate the atom's identity—and covalent bonds are shown as rods or sticks connecting the balls in the correct geometry.

bar coding The use of well-characterized gene sequences to identify species.

bark The protective outer layer of woody plants, composed of cork cells, cork cambium, and secondary phloem.

baroreceptors Specialized nerve cells in the walls of the heart and certain major arteries that detect changes in blood pressure and trigger appropriate responses by the brain.

basal Toward the base. In plants, at the base of a branch where it joins the stem. In animals, on the side of an epithelial layer that abuts underlying body tissues. Compare with **apical**.

basal body A structure of nine pairs of microtubules arranged in a circle at the base of eukaryotic cilia and flagella where they attach to the cell. Structurally similar to a centriole.

basal lamina A thick, collagen-rich extracellular matrix that underlies most epithelial tissues (e.g., skin) in animals.

basal metabolic rate (BMR) The total energy consumption by an organism at rest in a comfortable environment. For aerobes, often measured as the amount of oxygen consumed per hour.

basal transcription complex A large multi-protein structure that assembles near the promoter of eukaryotic genes and initiates transcription. Composed of basal transcription factors, TATA-binding protein, coactivators, and RNA polymerase.

basal transcription factor General term for proteins, present in all cell types, that bind to eukaryotic promoters and help initiate transcription. Compare with **regulatory transcription factor**.

base Any compound that acquires protons or gives up electrons during a chemical reaction or accepts hydrogen ions when dissolved in water.

basidium (plural: basidia) Specialized spore-producing cell at the ends of hyphae in club fungi.

basilar membrane The membrane in the vertebrate cochlea on which the bottom portion of hair cells sit.

basolateral Toward the bottom and sides. In animals, the side of an epithelial layer that faces other body tissues and not the environment.

Batesian mimicry A type of mimicry in which a harmless or palatable species resembles a dangerous or poisonous species. Compare with **Müllerian mimicry**.

B-cell receptor (BCR) An immunoglobulin protein (antibody) embedded in the plasma membrane of mature B cells and to which antigens bind.

beak A structure that exerts biting forces and is associated with the mouth; found in birds, cephalopods, and some insects.

behavior Any action by an organism.

beneficial In genetics, referring to any mutation, allele, or trait that increases an individual's fitness.

benign tumor A mass of abnormal tissue that grows slowly or not at all, does not disrupt surrounding tissues, and does not spread to other organs. Benign tumors are not cancers. Compare with **malignant tumor**.

benthic Living at the bottom of an aquatic environment.

benthic zone The area along the bottom of an aquatic environment.

β-pleated sheet (beta-pleated sheet) A protein secondary structure in which the polypeptide backbone folds into a sheetlike shape stabilized by hydrogen bonding.

bilateral symmetry An animal body pattern in which there is one plane of symmetry dividing the body into a left side and a right side. Typically, the body is long and narrow, with a distinct head end and tail end. Compare with **radial symmetry**.

bilaterian A member of a major lineage of animals (Bilateria) that are bilaterally symmetrical at some point in their life cycle, have three embryonic germ layers, and have a coelom. All protostomes and deuterostomes are bilaterians.

bile A complex solution produced by the liver, stored in the gall bladder, and secreted into the intestine. Contains steroid derivatives called bile salts that are responsible for emulsification of fats during digestion.

biodiversity The diversity of life considered at three levels: genetic diversity (variety of alleles in a population, species, or group of species); species diversity (variety and relative abundance of species present in a certain area); and ecosystem diversity (variety of communities and abiotic components in a region).

biodiversity hot spot A region that is extraordinarily rich in species.

biofilm A hard, polysaccharide-rich layer secreted by bacterial cells, which allows them to attach to a surface.

biogeochemical cycle The pattern of circulation of an element or molecule among living organisms and the environment.

biogeography The study of how species and populations are distributed geographically.

bioinformatics The field of study concerned with managing, analyzing, and interpreting biological information, particularly DNA sequences.

biological fitness See **fitness**.

biological species concept The definition of a species as a population or group of populations that are reproductively isolated from other groups. Members of a species have the potential to interbreed in nature to produce viable, fertile offspring but cannot produce viable, fertile hybrid offspring with members of other species. Compare with **morphospecies** and **phylogenetic species concept**.

bioluminescence The emission of light by a living organism.

biomagnification In animal tissues, an increase in the concentration of particular molecules that may occur as those molecules are passed up a food chain.

biomass The total mass of all organisms in a given population or geographical area; usually expressed as total dry weight.

biome A large terrestrial ecosystem characterized by a distinct type of vegetation and climate.

bioprospecting The effort to find commercially useful compounds by studying organisms—especially species that are poorly studied to date.

bioremediation The use of living organisms, usually bacteria or archaea, to degrade environmental pollutants.

biotechnology The application of biological techniques and discoveries to medicine, industry, and agriculture.

biotic Living, or produced by a living organism. Compare with **abiotic**.

bipedal Walking primarily on two legs.

bipolar cell A cell in the vertebrate retina that receives information from one or more photoreceptors and passes it to other bipolar cells or ganglion cells.

bivalves A lineage of mollusks that have two shells, such as clams and mussels.

bladder A mammalian organ that holds urine until it can be excreted.

blade The wide, flat part of a plant leaf.

blastocoel Fluid-filled cavity in the blastula of many animal species.

blastocyst Specialized type of blastula in mammals. A spherical structure composed of trophoblast cells on the exterior and a cluster of cells (the inner cell mass), which fills part of the interior space.

blastomere A small cell created by cleavage divisions in early animal embryos.

blastopore A small opening (pore) in the surface of an early vertebrate embryo, through which cells move during gastrulation.

blastula In vertebrate development, a hollow ball of cells (blastomere cells) that is formed by cleavage of a zygote and immediately undergoes gastrulation. See **blastocyst**.

blood A type of connective tissue consisting of red blood cells and leukocytes suspended in a fluid portion called plasma.

blood pressure See **diastolic blood pressure** and **systolic blood pressure**.

body mass index (BMI) A mathematical relationship of weight and height used to assess obesity in humans. Calculated as weight (kg) divided by the square of height (m^2).

body plan The basic architecture of an animal's body, including the number and arrangement of limbs, body segments, and major tissue layers.

bog A wetland that has no or almost no water flow, resulting in very low oxygen levels and acidic conditions.

Bohr shift The rightward shift of the oxygen-hemoglobin dissociation curve that occurs with decreasing pH. Results in hemoglobin being more likely to release oxygen in the acidic environment of exercising muscle.

bone A type of vertebrate connective tissue consisting of living cells and blood vessels within a hard extracellular matrix composed of calcium phosphate ($CaPO_4$) and small amounts of calcium carbonate ($CaCO_3$) and protein fibers.

bone marrow The soft tissue filling the inside of long bones containing stem cells that develop into red blood cells and leukocytes throughout life.

Bowman's capsule The hollow, double-walled cup-shaped portion of a nephron that surrounds a glomerulus in the vertebrate kidney.

brain A large mass of neurons located in the head region of an animal, that is involved in information processing; may also be called the cerebral ganglion.

brain stem The most posterior portion of the vertebrate brain, connecting to the spinal cord and responsible for autonomic body functions such as heart rate, respiration, and digestion.

braincase See **cranium**.

branch (1) A part of a phylogenetic tree that represents populations through time. (2) Any extension of a plant's shoot system.

brassinosteroids A family of steroid hormones found in plants.

bronchiole One of the small tubes in mammalian lungs that carry air from the bronchi to the alveoli.

bronchus (plural: bronchi) In mammals, one of a pair of large tubes that lead from the trachea to each lung.

bryophytes Members of several phyla of green plants that lack vascular tissue including liverworts, hornworts, and mosses. Also called *non-vascular plants.*

budding Asexual reproduction via outgrowth from the parent that eventually breaks free as an independent individual; occurs in yeasts and some invertebrates.

buffer A substance that, in solution, acts to minimize changes in the pH of that solution when acid or base is added.

bulbourethral glands In male mammals, small paired glands at the base of the urethra that secrete an alkaline mucus (part of semen), which lubricates the tip of the penis and neutralizes acids in the urethra during copulation. In humans, also called *Cowper's glands.*

bulk flow The directional movement of a substantial volume of fluid due to pressure differences, such as movement of water through plant phloem and movement of blood in animals.

bundle-sheath cell A type of cell found around the vascular tissue (veins) of plant leaves.

C_3 photosynthesis The most common form of photosynthesis in which atmospheric CO_2 is used to form 3-phosphoglycerate, a three-carbon sugar.

C_4 photosynthesis A variant type of photosynthesis in which atmospheric CO_2 is first fixed into four-carbon sugars, rather than the three-carbon sugars of classic C_3 photosynthesis. Enhances photosynthetic efficiency in hot, dry environments, by reducing loss of oxygen due to photorespiration.

cadherin Any of a class of cell-surface proteins involved in cell adhesion and important for coordinating movements of cells during embryological development.

callus In plants, a mass of undifferentiated cells that can generate roots and other tissues necessary to create a mature plant.

Calvin cycle In photosynthesis, the set of light-independent reactions that use NADPH and ATP formed in the light-dependent reactions to drive the fixation of atmospheric CO_2 and reduction of the fixed carbon, ultimately producing sugars. Also called *carbon fixation* and *light-independent reactions.*

calyx All of the sepals of a flower.

CAM See **crassulacean acid metabolism**.

cambium (plural: cambia) See **lateral meristem**.

Cambrian explosion The rapid diversification of animal body types and lineages that occured between the species present in the Ediacaran faunas (565–542 mya) and the Cambrian faunas (525–515 mya).

camera eye The type of eye in vertebrates and cephalopods, consisting of a hollow chamber with a hole at one end (through which light enters) and a sheet of light-sensitive cells against the opposite wall.

cAMP See **cyclic AMP**.

cancer General term for any tumor whose cells grow in an uncontrolled fashion, invade nearby tissues, and spread to other sites in the body.

canopy The uppermost layers of branches in a forest (i.e., those fully exposed to the Sun).

CAP binding site A DNA sequence upstream of certain prokaryotic operons to which catabolite activator protein can bind, increasing gene transcription.

capillarity The tendency of water to move up a narrow tube due to surface tension, adhesion, and cohesion.

capillary One of the numerous small, thin-walled blood vessels that permeate all tissues and organs, and allow exchange of gases and other molecules between blood and body cells.

capillary bed A thick network of capillaries.

capsid A shell of protein enclosing the genome of a virus particle.

carapace In crustaceans, a large platelike section of the exoskeleton that covers and protects the cephalothorax (e.g., a crab's "shell").

carbohydrate Any of a class of molecules that contain a carbonyl group, several hydroxyl groups, and several to many carbon-hydrogen bonds. See **monosaccharide** and **polysaccharide**.

carbon cycle, global The worldwide movement of carbon among terrestrial ecosystems, the oceans, and the atmosphere.

carbon fixation See **Calvin cycle**.

carbonic anhydrase An enzyme that catalyzes the formation of carbonic acid (H_2CO_3) from carbon dioxide and water.

carboxylic acids Organic acids with the form R-COOH (a carboxyl group).

cardiac cycle One complete heartbeat cycle, including systole and diastole.

cardiac muscle The muscle tissue of the vertebrate heart. Consists of long branched fibers that are electrically connected and that initiate their own contractions; not under voluntary control. Compare with **skeletal** and **smooth muscle**.

carnivore (adjective: carnivorous) An animal whose diet consists predominantly of meat. Most members of the mammalian taxon Carnivora are carnivores. Some plants are carnivorous, trapping and killing small animals, then absorbing nutrients from the prey's body. Compare with **herbivore** and **omnivore**.

carotenoid Any of a class of accessory pigments, found in chloroplasts, that absorb wavelengths of light not absorbed by chlorophyll; typically appear yellow, orange, or red. Includes carotenes and xanthophylls.

carpel The female reproductive organ in a flower. Consists of the stigma, to which pollen grains adhere; the style, through which pollen grains move; and the ovary, which houses the ovule. Compare with **stamen**.

carrier A heterozygous individual carrying a normal allele and a recessive allele for an inherited trait; does not display the phenotype of the trait but can pass the recessive gene to offspring.

carrier protein A membrane protein that facilitates diffusion of a small molecule (e.g., glucose) across the plasma membrane by a process involving a reversible change in the shape of the protein. Also called *carrier* or *transporter*.

carrying capacity (*K*) The maximum population size of a certain species that a given habitat can support.

cartilage A type of vertebrate connective tissue that consists of relatively few cells scattered in a stiff matrix of polysaccharides and protein fibers.

Casparian strip In plant roots, a waxy layer containing suberin, a water-repellent substance, that prevents movement of water through the walls of endodermal cells, thus blocking the apoplastic pathway.

cast A type of fossil, formed when the decay of a body part leaves a void that is then filled with minerals that later harden.

catabolic pathway Any set of chemical reactions that breaks down larger, complex molecules into smaller ones, releasing energy in the process. Compare with **anabolic pathway**.

catabolite activator protein (CAP) A protein that can bind to the CAP binding site upstream of certain prokaryotic operons, facilitating binding of RNA polymerase and stimulating gene expression.

catabolite repression A type of positive transcriptional control in which the end product of a catabolic pathway inhibits further transcription of the gene encoding an enzyme early in the pathway.

catalysis (verb: catalyze) Acceleration of the rate of a chemical reaction due to a decrease in the free energy of the transition state, called the activation energy.

catalyst Any substance that increases the rate of a chemical reaction without itself undergoing any permanent chemical change.

catecholamines A class of small compounds, derived from the amino acid tyrosine, that are used as hormones or neurotransmitters. Include epinephrine, norepinephrine, and dopamine.

cation A positively charged ion.

cation exchange In botany, the release (displacement) of cations, such as magnesium and calcium from soil particles, by protons in acidic soil water. The released cations are available for uptake by plants.

CD4 A membrane protein on the surface of some T cells in humans. $CD4^+$ T cells can give rise to helper T cells.

CD8 A membrane protein on the surface of some T cells in humans. $CD8^+$ T cells can give rise to cytotoxic T cells.

Cdk See **cyclin-dependent kinase**.

cDNA See **complementary DNA**.

cDNA library A set of cDNAs from a particular cell type or stage of development. Each cDNA is carried by a plasmid or other cloning vector and can be separated from other cDNAs. Compare with **genomic library**.

cecum A blind sac between the small intestine and the colon. Is enlarged in some species (e.g., rabbits) that use it as a fermentation vat for digestion of cellulose.

cell A highly organized compartment bounded by a thin, flexible structure (plasma membrane) and containing concentrated chemicals in an aqueous (watery) solution. The basic structural and functional unit of all organisms.

cell body The part of a neuron that contains the nucleus and where incoming signals are integrated. Also called the *soma*.

cell crawling A form of cellular movement involving actin filaments in which the cell produces bulges (pseudopodia) that stick to the substrate and pull the cell forward. Also called *amoeboid motion*.

cell culture See **culture**.

cell cycle Ordered sequence of events in which a eukaryotic cell replicates its chromosomes, evenly partitions the chromosomes to two daughter cells, and then undergoes division of the cytoplasm.

cell-cycle checkpoint Any of several points in the cell cycle at which progression of a cell through the cycle can be regulated.

cell division Creation of new cells by division of pre-existing cells.

cell-mediated (immune) response The type of immune response that involves generation of cytotoxic T cells from $CD8^+$ T cells. Defends against pathogen-infected cells, cancer cells, and transplanted cells. Compare with **humoral (immune) response**.

cell membrane See **plasma membrane**.

cell plate A double layer of new plasma membrane that appears in the middle of a dividing plant cell; ultimately divides the cytoplasm into two separate cells.

cell sap An aqueous solution found in the vacuoles of plant cells.

cell theory The theory that all organisms are made of cells and that all cells come from preexisting cells.

cell wall A fibrous layer found outside the plasma membrane of most bacteria and archaea and many eukaryotes.

cellular respiration A common pathway for production of ATP, involving transfer of electrons from compounds with high potential energy (often NADH and $FADH_2$) to an electron transport chain and ultimately to an electron acceptor (often oxygen).

cellulose A structural polysaccharide composed of β-glucose monomers joined by β-1,4-glycosidic linkages. Found in the cell wall of algae, plants, bacteria, fungi, and some other groups.

Cenozoic era The most recent interval of geologic time, beginning 65.5 million years ago, during which mammals became the dominant vertebrates and angiosperms became the dominant plants.

central dogma The long-accepted hypothesis that information in cells flows in one direction: DNA codes for RNA, which codes for proteins. Exceptions are now known (e.g., retroviruses).

central nervous system (CNS) The brain and spinal cord of vertebrate animals. Compare with **peripheral nervous system (PNS)**.

centriole One of two small cylindrical structures, structurally similar to a basal body, found together within the centrosome near the nucleus of a eukaryotic cell.

centromere Constricted region of a replicated chromosome where the two sister chromatids are joined and the kinetochore is located.

centrosome Structure in animal and fungal cells, containing two centrioles, that serves as a microtubule-organizing center for the cell's cytoskeleton and for the spindle apparatus during cell division.

cephalization The formation of a distinct anterior region (the head) where sense organs and a mouth are clustered.

cephalochordates One of the three major chordate lineages (Cephalochordata), comprising small, mobile organisms that live in marine sands; also called *lancelets* or *amphioxi*. Compare with **urochordates** and **vertebrates**.

cephalopods A lineage of mollusks including the squid, octopuses, and nautiluses. Distinguished by large brains, excellent vision, tentacles, and a reduced or absent shell.

cerebellum Posterior section of the vertebrate brain that is involved in coordination of complex muscle movements, such as those required for locomotion and maintaining balance.

cerebrum The most anterior section of the vertebrate brain. Divided into left and right hemispheres and four lobes: parietal lobe, involved in complex decision making (in humans); occipital lobe, receives and interprets visual information; parietal lobe, involved in integrating sensory and motor functions; and temporal lobe, functions in memory, speech (in humans), and interpreting auditory information.

cervix The narrow passageway between the vagina and the uterus of female mammals.

chaetae (singular: chaeta) Bristle-like extensions found in some annelids.

channel A protein that forms a pore in a cell membrane. The structure of most channels allows them to admit just one or a few types of ions or molecules.

character displacement The tendency for the traits of similar species that occupy overlapping ranges to change in a way that reduces interspecific competition.

chelicerae A pair of clawlike appendages found around the mouth of certain arthropods called chelicerates (spiders, mites, and allies).

chemical bond An attractive force binding two atoms together. Covalent bonds, ionic bonds, and hydrogen bonds are types of chemical bonds.

chemical carcinogen Any chemical that can cause cancer.

chemical energy The potential energy stored in covalent bonds between atoms.

chemical equilibrium A dynamic but stable state of a reversible chemical reaction in which the forward reaction and reverse reactions proceed at the same rate, so that the concentrations of reactants and products remain constant.

chemical evolution The theory that simple chemical compounds in the ancient atmosphere and ocean combined via spontaneous chemical reactions to form larger, more complex substances, eventually leading to the origin of life and the start of biological evolution.

chemical reaction Any process in which one compound or element is combined with others or is broken down; involves the making and/or breaking of chemical bonds.

chemiosmosis An energetic coupling mechanism whereby energy stored in an electrochemical proton gradient (proton-motive force) is used to drive an energy-requiring process such as production of ATP.

chemokine Any of several chemical signals that attract leukocytes to a site of tissue injury or infection.

chemolithotroph An organism that produces ATP by oxidizing inorganic molecules with high potential energy such as ammonia (NH_3) or methane (CH_4). Also called *lithotroph.* Compare with **chemoorganotroph**.

chemoorganotroph An organism that produces ATP by oxidizing organic molecules with high potential energy such as sugars. Also called *organotroph.* Compare with **chemolithotroph**.

chemoreceptor A sensory cell or organ specialized for detection of specific molecules or classes of molecules.

chiasma (plural: chiasmata) The X-shaped structure formed during meiosis by crossing over between non-sister chromatids in a pair of homologous chromosomes.

chitin A structural polysaccharide composed of *N*-acetylglucosamine monomers joined end to end by β-1,4-glycosidic linkages. Found in cell walls of fungi and many algae, and in external skeletons of insects and crustaceans.

chitons A lineage of marine mollusks that have a protective shell formed of eight calcium carbonate plates.

chlorophyll Any of several closely related green pigments, found in chloroplasts and photosynthetic protists, that absorb light during photosynthesis.

chloroplast A chlorophyll-containing organelle, bounded by a double membrane, in which photosynthesis occurs; found in plants and photosynthetic protists. Also the location of amino acid, fatty acid, purine, and pyrimidine synthesis.

choanocyte A specialized flagellated feeding cell found in choanoflagellates (protists that are the closest living relatives of animals) and sponges (the oldest animal phylum).

cholecystokinin A peptide hormone secreted by cells in the lining of the small intestine. Stimulates the secretion of digestive enzymes from the pancreas and of bile from the liver and gallbladder.

chordates Members of the phylum Chordata, deuterostomes distinguished by a dorsal hollow nerve cord, pharyngeal gill slits, a notochord, a dorsal hollow nerve cord, and a post-anal tail. Include vertebrates, cephalochordata, and urochordata.

chromatid One of the two identical strands composing a replicated chromosome that is connected at the centromere to the other strand.

chromatin The complex of DNA and proteins, mainly histones, that compose eukaryotic chromosomes. Can be highly compact (heterochromatin) or loosely coiled (euchromatin).

chromatin remodeling The process by which the DNA in chromatin is unwound from its associated proteins to allow transcription or replication. May involve chemical modification of histone proteins or reshaping of the chromatin by large multi-protein complexes in an ATP-requiring process.

chromosome Gene-carrying structure consisting of a single long molecule of DNA and associated proteins (e.g., histones). Most prokaryotic cells contain a single, circular chromosome; eukaryotic cells contain multiple noncircular (linear) chromosomes located in the nucleus.

chromosome theory of inheritance The principle that genes are located on chromosomes and that patterns of inheritance are determined by the behavior of chromosomes during meiosis.

chylomicron A ball of protein-coated lipids used to transport the lipids through the bloodstream.

cilium (plural: cilia) One of many short, filamentous projections of some eukaryotic cells containing a core of microtubules. Used to move the cell and/or to move fluid or particles along a stationary cell. See **axoneme**.

circadian clock An internal mechanism found in most organisms that regulates many body processes (sleep-wake cycles, hormonal patterns, etc.) in a roughly 24-hour cycle.

circulatory system The system in animals responsible for moving oxygen, carbon dioxide, and other materials (hormones, nutrients, wastes) around the body.

cisternae (singular: cisterna) Flattened, membrane-bound compartments that make up the Golgi apparatus.

citric acid cycle A series of eight chemical reactions that starts with citrate (citric acid, when protonated) and ends with oxaloacetate, which reacts with acetyl CoA to form citrate—forming a cycle that is part of the pathway that oxidizes glucose to CO_2. Also known as the *Krebs cycle, tricarboxylic acid cycle,* and *TCA cycle.*

clade See **monophyletic group**.

cladistic approach A method for constructing a phylogenetic tree that is based on identifying the unique traits of each monophyletic group. Compare with **phenetic approach**.

Class I MHC protein An MHC protein that is present on the plasma membrane of virtually all nucleated cells and functions in presenting antigen to $CD8^+$ T cells.

Class II MHC protein An MHC protein that is present only on the plasma membrane of dendritic cells, macrophages, and B cells and functions in presenting antigen to $CD4^+$ T cells.

cleavage In animal development, the series of rapid mitotic cell divisions, with little cell growth, that produces successively smaller cells and transforms a zygote into a multicellular blastula, or blastocyst in mammals.

cleavage furrow A pinching-in of the plasma membrane that occurs as cytokinesis begins in animal cells and deepens until the cytoplasm is divided.

climate The prevailing long-term weather conditions in a particular region.

climax community The stable, final community that develops from ecological succession.

clitoris A small rod of erectile tissue in the external genitalia of female mammals. Is formed from the same embryonic tissue as the male penis and has a similar function in sexual arousal.

cloaca An opening to the outside used by the excretory and reproductive systems in a few mammals and many nonmammalian vertebrates.

clonal-selection theory The dominant explanation of the development of adaptive immunity in vertebrates. According to the theory, the immune system retains a vast pool of inactive lymphocytes, each with a unique receptor for a unique antigen. Lymphocytes that encounter their antigens are stimulated to divide (selected and cloned), producing daughter cells that combat infection and confer immunity.

clone (1) An individual that is genetically identical to another individual. (2) A lineage of genetically identical individuals or cells. (3) As a verb, to make one or more genetic replicas of a cell or individual.

cloning vector A plasmid or other agent used to transfer recombinant genes into cultured host cells. Also called simply *vector.*

closed circulatory system A circulatory system in which the circulating fluid (blood) is confined to blood vessels and flows in a continuous circuit. Compare with **open circulatory system**.

club mosses Common name for species in the land plant lineage Lycophyta.

cnidocyte A specialized stinging cell found in cnidarians (e.g., jellyfish, corals, and anemones) that is used in capturing prey.

cochlea The organ of hearing in the inner ear of mammals, birds, and crocodilians. A coiled, fluid-filled tube containing specialized pressure-sensing neurons (hair cells) that detect sounds of different pitches.

coding strand See **non-template strand**.

codominance An inheritance pattern in which heterozygotes exhibit both of the traits seen in either kind of homozygous individual.

codon A sequence of three nucleotides in DNA or RNA that codes for a certain amino acid or that initiates or terminates protein synthesis.

coefficient of relatedness (*r*) A measure of how closely two individuals are related. Calculated as the probability that an allele in two individuals is inherited from the same ancestor.

coelom An internal, usually fluid-filled, body cavity that is lined with mesoderm.

coelomate An animal that has a true coelom. Compare with **acoelomate** and **pseudocoelomate**.

coenocytic Containing many nuclei and a continuous cytoplasm through a filamentous body, without the body being divided into distinct cells. Some fungi are coenocytic.

coenzyme A small organic molecule that is a required cofactor for an enzyme-catalyzed reaction. Often donates or receives electrons or functional groups during the reaction.

coenzyme A (CoA) A nonprotein molecule that is required for many cellular reactions involving transfer of acetyl groups ($-COCH_3$).

coenzyme Q A nonprotein molecule that shuttles electrons between membrane-bound complexes in the mitochondrial electron transport chain. Also called *ubiquinone* or *Q.*

coevolution A pattern of evolution in which two interacting species reciprocally influence each other's adaptations over time.

coevolutionary arms race A series of adaptations and counter-adaptations observed in species that interact closely over time and affect each other's fitness.

cofactor A metal ion or small organic compound that is required for an enzyme to function normally. May be bound tightly to an enzyme or associate with it transiently during catalysis.

cohesion The tendency of certain like molecules (e.g., water molecules) to cling together due to attractive forces. Compare with **adhesion**.

cohesion-tension theory The theory that water movement upward through plant vascular tissues is due to loss of water from leaves (transpiration), which pulls a cohesive column of water upward.

cohort A group of individuals that are the same age and can be followed through time.

coleoptile A modified leaf that covers and protects the stems and leaves of young grasses.

collagen A fibrous, pliable, cable-like glycoprotein that is a major component of the extracellular matrix of animal cells. Various subtypes differ in their tissue distribution.

collecting duct In the vertebrate kidney, a large straight tube that receives filtrate from the distal tubules of several nephrons. Involved in the regulated reabsorption of water.

collenchyma cell In plants, an elongated cell with cell walls thickened at the corners that provides support to growing plant parts; usually found in strands along leaf veins and stalks. Compare with **parenchyma cell** and **sclerenchyma cell**.

colon The portion of the large intestine where feces are formed by compaction of wastes and reabsorption of water.

colony An assemblage of individuals. May refer to an assemblage of semi-independent cells or to a breeding population of multicellular organisms.

commensalism (adjective: commensal) A symbiotic relationship in which one organism (the commensal) benefits and the other (the host) is not harmed. Compare with **mutualism** and **parasitism**.

communication In ecology, any process in which a signal from one individual modifies the behavior of another individual.

community All of the species that interact with each other in a certain area.

companion cell In plants, a cell in the phloem that is connected via numerous plasmodesmata to adjacent sieve-tube members. Companion cells provide materials to maintain sieve-tube members and function in the loading and unloading of sugars into sieve-tube members.

compass orientation A type of navigation in which movement occurs in a specific direction.

competition In ecology, the interaction of two species or two individuals trying to use the same limited resource (e.g., water, food, living space). May occur between individuals of the same species (intraspecific competition) or different species (interspecific competition).

competitive exclusion principle The principle that two species cannot coexist in the same ecological niche in the same area because one species will out-compete the other.

competitive inhibition Inhibition of an enzyme's ability to catalyze a chemical reaction via a nonreactant molecule that competes with the substrate(s) for access to the active site.

complement system A set of proteins that circulate in the bloodstream and can form holes in the plasma membrane of bacteria, leading to their destruction.

complementary base pairing The association between specific nitrogenous bases of nucleic acids stabilized by hydrogen bonding. Adenine pairs only with thymine (in DNA) or uracil (in RNA), and guanine pairs only with cytosine.

complementary DNA (cDNA) DNA produced in the laboratory using an RNA transcript as a template and reverse transcriptase; corresponds to a gene but lacks introns. Also produced naturally by retroviruses.

complementary strand A newly synthesized strand of RNA or DNA that has a base sequence complementary to that of the template strand.

complete digestive tract A digestive tract with two openings, usually called a mouth and an anus.

complete metamorphosis See **holometabolous metamorphosis**.

compound eye An eye formed of many independent light-sensing columns (ommatidia); occurs in arthropods. Compare with **simple eye**.

concentration gradient Difference across space (e.g., across a membrane) in the concentration of a dissolved substance.

condensation reaction A chemical reaction in which two molecules are joined covalently with the removal of an —OH from one and an —H from another to form water. Also called a *dehydration reaction*. Compare with **hydrolysis**.

conduction (1) Direct transfer of heat between two objects that are in physical contact. Compare with **convection**. (2) Transmission of an electrical impulse along the axon of a nerve cell.

cone cell A photoreceptor cell with a cone-shaped outer portion that is particularly sensitive to bright light of a certain color. Also called simply *cone*. Compare with **rod cell**.

connective tissue An animal tissue consisting of scattered cells in a liquid, jellylike, or solid extracellular matrix. Includes bone, cartilage, tendons, ligaments, and blood.

conservation biology The effort to study, preserve, and restore threatened populations, communities, and ecosystems.

constant (C) region The portion of an antibody's light chains or heavy chains that has the same amino acid sequence in the antibodies produced by every B cell of an individual. Compare with **variable (V) region**.

constitutive Always occurring; always present. Commonly used to describe enzymes and other proteins that are synthesized continuously or mutants in which one or more genetic loci are constantly expressed due to defects in gene control.

constitutive defense A defensive trait that is always manifested even in the absence of a predator or pathogen. Also called *standing defense*. Compare with **inducible defense**.

constitutive mutant An abnormal (mutated) gene that produces a product at all times, instead of under certain conditions only.

consumer See **heterotroph**.

consumption Predation or herbivory.

continental shelf The portion of a geologic plate that extends from a continent to under seawater.

continuous strand See **leading strand**.

control In a scientific experiment, a group of organisms or samples that do not receive the experimental treatment but are otherwise identical to the group that does.

convection Transfer of heat by movement of large volumes of a gas or liquid. Compare with **conduction**.

convergent evolution The independent evolution of analogous traits in distantly related organisms due to adaptation to similar environments and a similar way of life.

cooperative binding The tendency of the protein subunits of hemoglobin to affect each other's oxygen binding such that each bound oxygen molecule increases the likelihood of further oxygen binding.

coprophagy The eating of feces.

copulation The act of transferring sperm from a male directly into a female's reproductive tract.

coral reef A large assemblage of colonial marine corals that usually serves as shallow-water, sunlit habitat for many other species as well.

core enzyme The enzyme responsible for catalysis in a multi-part holoenzyme.

co-receptor Any membrane protein that acts with some other membrane protein in a cell interaction or cell response.

cork cambium One of two types of lateral meristem, consisting of a ring of undifferentiated plant cells found just under the cork layer of woody plants; produces new cork cells on its outer side. Compare with **vascular cambium**.

cork cell A waxy cell in the protective outermost layer of a woody plant.

corm A rounded, thick underground stem that can produce new plants via asexual reproduction.

cornea The transparent sheet of connective tissue at the very front of the eye in vertebrates and some other animals. Protects the eye and helps focus light.

corolla All of the petals of a flower.

corona The cluster of cilia at the anterior end of a rotifer.

corpus callosum A thick band of neurons that connects the two hemispheres of the cerebrum in the mammalian brain.

corpus luteum A yellowish structure in an ovary that secretes progesterone. Is formed from a follicle that has recently ovulated.

cortex (1) The outermost region of an organ, such as the kidney or adrenal gland. (2) In plants, a layer of ground tissue found outside the vascular bundles and pith of a plant stem.

cortical granules Small enzyme-filled vesicles in the cortex of an egg cell. Involved in formation of the fertilization envelope after fertilization.

corticotropin-releasing hormone (CRH) A peptide hormone, produced and secreted by the hypothalamus, that stimulates the anterior pituitary to release ACTH.

cortisol A steroid hormone, produced and secreted by the adrenal cortex, that increases blood glucose and prepares the body for stress. The major glucocorticoid hormone in some mammals. Also called *hydrocortisone*.

cost-benefit analysis Decisions or analyses that weigh the fitness costs and benefits of a particular action.

cotransport Transport of an ion or molecule against its electrochemical gradient, in company with an ion or molecule being transported with its electrochemical gradient. Also called *secondary active transport*.

cotransporter A transmembrane protein that facilitates diffusion of an ion down its previously established electrochemical gradient and uses the energy of that process to transport some other substance, in the same or opposite direction, *against* its concentration gradient. Also called *secondary active transporter*. See **antiporter** and **symporter**.

cotyledon The first leaf, or seed leaf, of a plant embryo. Used for storing and digesting nutrients and/or for early photosynthesis.

countercurrent exchanger In animals, any anatomical arrangement that allows the maximum transfer of heat or a soluble substance from one fluid to another. The two fluids must be flowing in opposite directions and have a heat or concentration gradient between them.

covalent bond A type of chemical bond in which two atoms share one or more pairs of electrons. Compare with **hydrogen bond** and **ionic bond**.

cranium A bony, cartilaginous, or fibrous case that encloses and protects the brain of vertebrates. Forms part of the skull. Also called *braincase*.

crassulacean acid metabolism (CAM) A variant type of photosynthesis in which CO_2 is stored in organic acids at night when stomata are open and then released to feed the Calvin cycle during the day when stomata are closed. Helps reduce water loss and oxygen loss by photorespiration in hot, dry environments.

cristae (singular: crista) Sac-like invaginations of the inner membrane of a mitochondrion. Location of the electron transport chain and ATP synthase.

Cro-Magnon A prehistoric European population of modern humans (*Homo sapiens*) known from fossils, paintings, sculptures, and other artifacts.

crop A storage organ in the digestive systems of certain vertebrates.

cross-fertilization A mating that combines gametes from different individuals, as opposed to combining gametes from the same individual (self-fertilization).

crossing over The exchange of segments of non-sister chromatids between a pair of homologous chromosomes that occurs during meiosis I.

cross-pollination Pollination of a flower by pollen from another individual, rather than by self-fertilization. Also called *crossing.*

cross-talk Interactions among signaling pathways that modify a cellular response.

crustaceans A lineage of arthropods that includes shrimp, lobster, and crabs. Many have a carapace (a platelike portion of the exoskeleton) and mandibles for biting or chewing.

cryptic species A species that cannot be distinguished from other species by easily identifiable morphological traits.

culture In cell biology, a collection of cells or a tissue growing under controlled conditions, usually in suspension or on the surface of a dish on solid growth medium.

cup fungus See **sac fungus**.

Cushing's disease A human endocrine disorder caused by loss of feedback inhibition of cortisol on ACTH secretion. Characterized by high ACTH and cortisol levels and wasting of body protein reserves.

cuticle A protective coating secreted by the outermost layer of cells of an animal or a plant.

cyanobacteria A lineage of photosynthetic bacteria formerly known as blue-green algae. Likely the first life-forms to carry out oxygenic photosynthesis.

cyclic AMP (cAMP) Cyclic adenosine monophosphate; a small molecule, derived from ATP, that is widely used by cells in signal transduction and transcriptional control.

cyclic photophosphorylation Path of electron flow during the light-dependent reactions of photosynthesis in which photosystem I transfers excited electrons back to the electron transport chain of photosystem II, rather than to $NADP^+$. Also called *cyclic electron flow.* Compare with **Z scheme**.

cyclin One of several regulatory proteins whose concentrations fluctuate cyclically throughout the cell cycle.

cyclin-dependent kinase (Cdk) Any of several related protein kinases that are active only when bound to a cyclin. Involved in control of the cell cycle.

cytochrome *c* (cyt *c*) A soluble iron-containing protein that shuttles electrons between membrane-bound complexes in the mitochondrial electron transport chain.

cytokines A diverse group of autocrine signaling proteins, secreted largely by cells of the immune system, whose effects include stimulating leukocyte production, tissue repair, and fever. Generally function to regulate the intensity and duration of an immune response.

cytokinesis Division of the cytoplasm to form two daughter cells. Typically occurs immediately after division of the nucleus by mitosis or meiosis.

cytokinins A class of plant hormones that stimulate cell division and retard aging.

cytoplasm All of the contents of a cell, excluding the nucleus, bounded by the plasma membrane.

cytoplasmic determinant A regulatory transcription factor or signaling molecule that is distributed unevenly in the cytoplasm of the egg cells of many animals and that directs early pattern formation in an embryo.

cytoplasmic streaming The directed flow of cytosol and organelles that facilitates distribution of materials within some large plant and fungal cells. Occurs along actin filaments and is powered by myosin.

cytoskeleton In eukaryotic cells, a network of protein fibers in the cytoplasm that are involved in cell shape, support, locomotion, and transport of materials within the cell. Prokaryotic cells have a similar but much less extensive network of fibers.

cytosol The fluid portion of the cytoplasm.

cytotoxic T cell An effector T cell that destroys infected cells and cancer cells. Is descended from an activated $CD8^+$ T cell that has interacted with antigen on an infected cell or cancer cell. Also called *cytotoxic T lymphocyte (CTL)* and *killer T cell.* Compare with **helper T cell**.

dalton (Da) A unit of mass equal to 1/12 the mass of one carbon-12 atom; about the mass of 1 proton or 1 neutron.

day-neutral plant A plant whose flowering time is not affected by the relative length of day and night (the photoperiod). Compare with **long-day** and **short-day plant**.

dead space Portions of the air passages that are not involved in gas exchange with the blood, such as the trachea and bronchi.

deciduous Describing a plant that sheds leaves or other structures at regular intervals (e.g., each fall)

decomposer See **detritivore**.

decomposer food chain An ecological network of detritus, decomposers that eat detritus, and predators and parasites of the decomposers.

definitive host The host species in which a parasite reproduces sexually. Compare with **intermediate host**.

dehydration reaction See **condensation reaction**.

deleterious In genetics, referring to any mutation, allele, or trait that reduces an individual's fitness.

demography The study of factors that determine the size and structure of populations through time.

denaturation (verb: denature) For a macromolecule, loss of its three-dimensional structure and biological activity due to breakage of hydrogen bonds and disulfide bonds, usually caused by treatment with excess heat or extreme pH conditions.

dendrite A short extension from a neuron's cell body that receives signals from other neurons.

dendritic cell A type of leukocyte that ingests and digests foreign antigens, moves to a lymph node, and presents the antigens displayed on its membrane to $CD4^+$ T cells.

dense connective tissue A type of connective tissue, distinguished by having an extracellular matrix dominated by collagen fibers.

density dependent In population ecology, referring to any characteristic that varies depending on population density.

density independent In population ecology, referring to any characteristic that does not vary with population density.

deoxyribonucleic acid (DNA) A nucleic acid composed of deoxyribonucleotides that carries the genetic information of a cell. Generally occurs as two intertwined strands, but these can be separated. See **double helix**.

deoxyribonucleoside triphosphate (dNTP) A monomer that can be polymerized to form DNA. Consists of deoxyribose, a base (A, T, G, or C), and three phosphate groups; similar to a nucleotide, but with two more phosphate groups.

deoxyribonucleotide See **nucleotide**.

depolarization Change in membrane potential from its resting negative state to a less negative or a positive state; a normal phase in an action potential. Compare with **hyperpolarization**.

deposit feeder An animal that eats its way through a food-containing substrate.

derived trait A trait that is clearly homologous with a trait found in an ancestor, but which has a new form.

dermal tissue system The tissue forming the outer layer of an organism. In plants, also called *epidermis;* in animals, forms two distinct layers: *dermis* and *epidermis.*

descent with modification The phrase used by Darwin to describe his hypothesis of evolution by natural selection.

desmosome A type of cell-cell attachment structure, consisting of cadherin proteins, that binds the cytoskeletons of adjacent animal cells together. Found where cells are strongly attached to each other. Compare with **gap junction** and **tight junction**.

detergent A type of small amphipathic molecule used to solubilize hydrophobic molecules in aqueous solution.

determination In embryogenesis, progressive changes in a cell that commit it to a particular cell fate. Once a cell is fully determined, it can differentiate only into a particular cell type (e.g., liver cell, brain cell).

detritivore An organism whose diet consists mainly of dead organic matter (detritus). Various bacteria, fungi, and protists are detritivores. Also called *decomposer.*

detritus A layer of dead organic matter that accumulates at ground level or on seafloors and lake bottoms.

deuterostomes A major lineage of animals that share a pattern of embryological development, including formation of the anus earlier than the mouth, and formation of the coelom by pinching off of layers of mesoderm from the gut. Includes echinoderms and chordates. Compare with **protostomes**.

developmental homology A similarity in embryonic form, or in the fate of embryonic tissues, that is due to inheritance from a common ancestor.

diabetes insipidus A human disease caused by defects in the kidney's system for conserving water. Characterized by production of large amounts of dilute urine.

diabetes mellitus A human disease caused by defects in insulin production (type I) or the response of cells to insulin (type II). Characterized by abnormally high blood glucose levels and large amounts of glucose-containing urine.

diaphragm An elastic, sheetlike structure. In mammals, the muscular sheet of tissue that separates the chest and abdominal cavities. Contracts and moves downward during inhalation, expanding the chest cavity.

diastole The portion of the heartbeat cycle during which the atria or ventricles of the heart are relaxed. Compare with **systole**.

diastolic blood pressure The force exerted by blood against artery walls during relaxation of the heart's left ventricle. Compare with **systolic blood pressure**.

dicot Any plant that has two cotyledons (embryonic leaves) upon germination. The dicots do not form a monophyletic group. Also called *dicotyledonous plant.* Compare with **eudicot** and **monocot**.

dideoxy sequencing A laboratory technique for determining the exact nucleotide sequence of DNA. Relies on the use of dideoxynucleotide triphosphates (ddNTPs), which terminate DNA replication.

diencephalon The part of the mammalian brain that relays sensory information to the cerebellum and functions in maintaining homeostasis.

differential centrifugation Procedure for separating cellular components according to their size and density by spinning a cell homogenate in a series of centrifuge runs. After each run, the supernatant is removed from the deposited material (pellet) and spun again at progressively higher speeds.

differential gene expression Expression of different sets of genes in cells with the same genome. Responsible for creating different cell types.

differentiation The process by which a relatively unspecialized cell becomes a distinct specialized cell type (e.g., liver cell, brain cell) usually by changes in gene expression. Also called *cell differentiation.*

diffusion Spontaneous movement of a substance from a region of high concentration to one of low concentration (i.e., down a concentration gradient).

digestion The physical and chemical breakdown of food into molecules that can be absorbed into the body of an animal.

digestive tract The long tube that begins at the mouth and ends at the anus. Also called *alimentary canal, gastrointestinal (GI) tract,* or the *gut.*

dihybrid cross A mating between two parents that are heterozygous for both of the two genes being studied.

dikaryotic Having two nuclei.

dimer An association of two molecules, which may be identical (homodimer) or different (heterodimer).

dioecious Describing an angiosperm species that has male and female reproductive structures on separate plants. Compare with **monoecious**.

diploblast (adjective: diploblastic) An animal whose body develops from two basic embryonic cell layers—ectoderm and endoderm. Compare with **triploblast**.

diploid (1) Having two sets of chromosomes ($2n$). (2) A cell or an individual organism with two sets of chromosomes, one set inherited from the maternal parent and one set from the paternal parent. Compare with **haploid**.

direct sequencing A technique for identifying and studying microorganisms that cannot be grown in culture. Involves detecting and amplifying copies of certain specific genes in their DNA, sequencing these genes, and then comparing the sequences with the known sequences from other organisms.

directional selection A pattern of natural selection that favors one extreme phenotype with the result that the average phenotype of a population changes in one direction. Generally reduces overall genetic variation in a population.

disaccharide A carbohydrate consisting of two monosaccharides (sugar residues) linked together.

discontinuous strand See **lagging strand**.

discrete trait An inherited trait that exhibits distinct phenotypic forms rather than the continuous variation characteristic of a quantitative trait such as body height.

dispersal The movement of individuals from their place of origin (birth, hatching) to a new location.

disruptive selection A pattern of natural selection that favors extreme phenotypes at both ends of the range of phenotypic variation. Maintains overall genetic variation in a population. Compare with **stabilizing selection**.

distal tubule In the vertebrate kidney, the convoluted portion of a nephron into which filtrate moves from the loop of Henle. Involved in the regulated reabsorption of sodium and water. Compare with **proximal tubule**.

disturbance In ecology, any event that disrupts a community, usually causing loss of some individuals or biomass from it.

disturbance regime The characteristic disturbances that affect a given ecological community.

disulfide bond A covalent bond between two sulfur atoms, typically in the side groups of some amino acids (e.g., cysteine). Often contributes to tertiary structure of proteins.

DNA See **deoxyribonucleic acid**.

DNA cloning Any of several techniques for producing many identical copies of a particular gene or other DNA sequence.

DNA fingerprinting Any of several methods for identifying individuals by unique features of their genomes. Commonly involves using PCR to produce many copies of certain simple sequence repeats (microsatellites) and then analyzing their lengths.

DNA library See **cDNA library** and **genomic library**.

DNA ligase An enzyme that joins pieces of DNA by catalyzing formation of a phosphodiester bond between the pieces.

DNA microarray A set of single-stranded DNA fragments, representing thousands of different genes, that are permanently fixed to a small glass slide. Can be used to determine which genes are expressed in different cell types, under different conditions, or at different developmental stages.

DNA polymerase Any enzyme that catalyzes synthesis of DNA from deoxyribonucleotides.

domain (1) A section of a protein that has a distinctive tertiary structure and function. (2) A taxonomic category, based on similarities in basic cellular biochemistry, above the kingdom level. The three recognized domains are Bacteria, Archaea, and Eukarya.

dominant Referring to an allele that determines the phenotype of a heterozygous individual. Compare with **recessive**.

dopamine A catecholamine neurotransmitter that functions mainly in a part of the mammalian brain involved with muscle control. Also functions as a hypothalamic inhibitory hormone that inhibits release of prolactin from the interior pituitary; also called *prolactin-inhibiting hormone (PIH).*

dormancy A temporary state of greatly reduced, or no, metabolic activity and growth in plants or plant parts (e.g., seeds, spores, bulbs, and buds).

dorsal Toward an animal's back and away from its belly. The opposite of ventral.

double fertilization An unusual form of reproduction seen in flowering plants, in which one sperm nucleus fuses with an egg to form a zygote and the other sperm nucleus fuses with two polar nuclei to form the triploid endosperm.

double helix The secondary structure of DNA, consisting of two antiparallel DNA strands wound around each other.

Down syndrome A human developmental disorder caused by trisomy of chromosome 21.

downstream In genetics, the direction in which RNA polymerase moves along a DNA strand. Compare with **upstream**.

dynein Any one of a class of motor proteins that use the chemical energy of ATP to "walk" along an adjacent microtubule. Dyneins are responsible for bending of cilia and flagella, play a role in chromosome movement during mitosis, and can transport certain organelles.

early endosome A small membrane-bound vesicle, formed by endocytosis, that is an early stage in the formation of a lysosome.

ecdysone An insect hormone that triggers either molting (to a larger larval form) or metamorphosis (to the adult form), depending on the level of juvenile hormone.

ecdysozoans A major lineage of protostomes (Ecdysozoa) that grow by shedding their external skeletons (molting) and expanding their bodies. Includes arthropods, insects, crustaceans, nematodes, and centipedes. Compare with **lophotrochozoans**.

echinoderms A major lineage of deuterostomes (Echinodermata) distinguished by adult bodies with five-sided radial symmetry, a water vascular system, and tube feet. Includes sea urchins, sand dollars, and sea stars.

echolocation The use of echoes from vocalizations to obtain information about locations of objects in the environment.

ecology The study of how organisms interact with each other and with their surrounding environment.

ecosystem All the organisms that live in a geographic area, together with the nonliving (abiotic) components that affect or exchange materials with the organisms; a community and its physical environment.

ecosystem diversity The variety of biotic components in a region along with abiotic components, such as soil, water, and nutrients.

ecosystem services Alterations of the physical components of an ecosystem by living organisms, especially beneficial changes in the quality of the atmosphere, soil, water, etc.

ecotourism Tourism that is based on observing wildlife or experiencing other aspects of natural areas.

ectoderm The outermost of the three basic cell layers in most animal embryos; gives rise to the outer covering and nervous system. Compare with **endoderm** and **mesoderm**.

ectomycorrhizal fungi (EMF) Fungi whose hyphae form a dense network that covers their host plant's roots but do not enter the root cells.

ectoparasite A parasite that lives on the outer surface of the host's body.

ectotherm An animal that does not use internally generated heat to regulate its body temperature. Compare with **endotherm**.

effector Any cell, organ, or structure with which an animal can respond to external or internal stimuli. Usually functions, along with a sensor and integrator, as part of a homeostatic system.

efferent division The part of the nervous system, consisting primarily of motor neurons, that carries commands from the central nervous system to the body.

egg A mature female gamete and any associated external layers (such as a shell). Larger and less mobile than the male gamete. In animals, also called *ovum.*

ejaculation The release of semen from the copulatory organ of a male animal.

ejaculatory duct A short duct connecting the vas deferens to the urethra, through which sperm move during ejaculation.

elastic Referring to a structure (e.g., lungs) with the ability to stretch and then spring back to its original shape.

electric current A flow of electrical charge past a point. Also called *current.*

electrical potential Potential energy created by a separation of electric charges between two points. Also called *voltage.*

electrocardiogram (EKG) A recording of the electrical activity of the heart, as measured through electrodes on the skin.

electrochemical gradient The combined effect of an ion's concentration gradient and electrical (charge) gradient across a membrane that affects the diffusion of ions across the membrane.

electrolyte Any compound that dissociates into ions when dissolved in water. In nutrition, refers to the major ions necessary for normal cell function.

electromagnetic spectrum The entire range of wavelengths of radiation extending from short wavelengths (high energy) to long wavelengths (low energy). Includes gamma rays, X-rays, ultraviolet, visible light, infrared, microwaves, and radio waves (from short to long wavelengths).

electron acceptor A reactant that gains an electron and is reduced in a reduction-oxidation reaction.

electron carrier Any molecule that readily accepts electrons from and donates electrons to other molecules.

electron donor A reactant that loses an electron and is oxidized in a reduction-oxidation reaction.

electron microscope See **scanning electron microscope** and **transmission electron microscope**.

electron shell A group of orbitals of electrons with similar energies. Electron shells are arranged in roughly concentric layers around the nucleus of an atom, with electrons in outer shells having more energy than those in inner shells. Electrons in the outermost shell, the valence shell, often are involved in chemical bonding.

electron transport chain (ETC) Any set of membrane-bound protein complexes and smaller soluble electron carriers involved in a coordinated series of redox reactions in which the potential energy of electrons transferred from reduced donors is successively decreased and used to pump protons from one side of a membrane to the other.

electronegativity A measure of the ability of an atom to attract electrons toward itself from an atom to which it is bonded.

electroreceptor A sensory cell or organ specialized to detect electric fields.

element A substance, consisting of atoms with a specific number of protons, that cannot be separated into or broken down to any other substance. Elements preserve their identity in chemical reactions.

elongation (1) The process by which messenger RNA lengthens during transcription. (2) The process by which a polypeptide chain lengthens during translation.

elongation factors Proteins involved in the elongation phase of translation, assisting ribosomes in the synthesis of the growing peptide chain.

embryo A young developing organism; the stage after fertilization and zygote formation.

embryo sac The female gametophyte in flowering plants that exhibit alternation of generations.

embryogenesis The process by which a single-celled zygote becomes a multicellular embryo.

Embryophyta An increasing popular name for the lineage called land plants.

embryophyte A plant that nourishes its embryos inside its own body. All land plants are embryophytes.

emergent vegetation Any plants in an aquatic habitat that extend above the surface of the water.

emerging disease Any infectious disease, often a viral disease, that suddenly afflicts significant numbers of humans for the first time; often due to changes in the host species for a pathogen or host population movements.

emigration The migration of individuals away from one population to other populations. Compare with **immigration**.

emulsification (verb: emulsify) The dispersion of fat into an aqueous solution. Usually requires the aid of an amphipathic substance such as a detergent or bile salts, which can break large fat globules into microscopic fat droplets.

endangered species A species whose numbers have decreased so much that it is in danger of extinction throughout all or part of its range.

endemic species A species that lives in one geographic area and nowhere else.

endergonic Referring to a chemical reaction that requires an input of energy to occur and for which the Gibbs free-energy change (ΔG) > 0. Compare with **exergonic**.

endocrine Relating to a chemical signal (hormone) that is released into the bloodstream by a producing cell and acts on a distant target cell.

endocrine system All of the glands and tissues that produce and secrete hormones into the bloodstream.

endocrine gland A gland that secretes hormones directly into the bloodstream or interstitial fluid instead of into ducts. Compare with **exocrine gland**.

endocytosis General term for any pinching off of the plasma membrane that results in the uptake of material from outside the cell. Includes phagocytosis, pinocytosis, and receptor-mediated endocytosis. Compare with **exocytosis**.

endoderm The innermost of the three basic cell layers in most animal embryos; gives rise to the digestive tract and organs that connect to it (liver, lungs, etc.). Compare with **ectoderm** and **mesoderm**.

endodermis In plant roots, a cylindrical layer of cells that separates the cortex from the vascular tissue.

endomembrane system A system of organelles in eukaryotic cells that performs most protein and lipid synthesis. Includes the endoplasmic reticulum (ER), Golgi apparatus, and lysosomes.

endomycorrhizal fungi See **arbuscular mycorrhizal fungi (AMF)**.

endoparasite A parasite that lives inside the host's body.

endophyte (adjective: endophytic) A fungus that lives inside the aboveground parts of a plant in a symbiotic relationship. Compare with **epiphyte**.

endoplasmic reticulum (ER) A network of interconnected membranous sacs and tubules found inside eukaryotic cells. See **rough** and **smooth endoplasmic reticulum**.

endoskeleton Bony and/or cartilaginous structures within the body that provide support. Examples are the spicules of sponges, the plates in echinoderms, and the bony skeleton of vertebrates. Compare with **exoskeleton**.

endosome See **early** and **late endosome**.

endosperm A triploid ($3n$) tissue in the seed of a flowering plant (angiosperm) that serves as food for the plant embryo. Functionally analogous to the yolk in some animal eggs.

endosymbiont An organism that lives in a symbiotic relationship inside the body of its host.

endosymbiosis An association between species in which one lives inside the cell or cells of the other.

endosymbiosis theory The theory that mitochondria and chloroplasts evolved from prokaryotes that were engulfed by host cells and took up a symbiotic existence within those cells, a process termed primary endosymbiosis. In some eukaryotes, chloroplasts originated by secondary endosymbiosis, that is, by engulfing a chloroplast-containing protist and retaining its chloroplasts.

endotherm An animal whose primary source of body heat is internally generated heat. Compare with **ectotherm**.

endothermic Referring to a chemical reaction that absorbs heat. Compare with **exothermic**.

energetic coupling In cellular metabolism, the mechanism by which energy released from an exergonic reaction (commonly, hydrolysis of ATP) is used to drive an endergonic reaction.

energy The capacity to do work or to supply heat. May be stored (potential energy) or available in the form of motion (kinetic energy).

enhancer A regulatory sequence in eukaryotic DNA that may be located far from the gene it controls or within introns of the gene. Binding of specific proteins to an enhancer enhances the transcription of certain genes.

enrichment culture A method of detecting and obtaining cells with specific characteristics by placing a sample, containing many types of cells, under a specific set of conditions (e.g., temperature, salt concentration, available nutrients) and isolating those cells that grow rapidly in response.

entropy (*S*) A quantitative measure of the amount of disorder of any system, such as a group of molecules.

envelope, viral A membrane-like covering that encloses some viruses and their capsid coats, shielding them from attack by the host's immune system.

environmental sequencing The inventory of all the genes in a community or ecosystem by sequencing, analyzing, and comparing the genomes of the component organisms.

enzyme A protein catalyst used by living organisms to speed up and control biological reactions.

epicotyl In some embryonic plants, a portion of the embryonic stem that extends above the cotyledons.

epidemic The spread of an infectious disease throughout a population in a short time period. Compare with **pandemic**.

epidermis The outermost layer of cells of any multicellular organism.

epididymis A coiled tube wrapped around the testis in reptiles, birds, and mammals. The site of the final stages of sperm maturation and storage.

epigenetic inheritance Pattern of inheritance involving differences in phenotype that are not due to changes in the nucleotide sequence of genes.

epinephrine A catecholamine hormone, produced and secreted by the adrenal medulla, that triggers rapid responses relating to the fight-or-flight response. Also called *adrenaline.*

epiphyte (adjective: epiphytic) A nonparasitic plant that grows on trees or other solid objects and is not rooted in soil.

epithelium (plural: epithelia) An animal tissue consisting of sheet-like layers of tightly packed cells that lines an organ, a duct, or a body surface. Also called *epithelial tissue.*

epitope A small region of a particular antigen to which an antibody, B-cell receptor, or T-cell receptor binds.

equilibrium potential The membrane potential at which there is no net movement of a particular ion into or out of a cell.

ER signal sequence A short amino acid sequence that marks a polypeptide for transport to the endoplasmic reticulum where synthesis of the polypeptide chain is completed and the signal sequence removed. See **signal recognition particle**.

erythropoietin (EPO) A peptide hormone, released by the kidney in response to low blood oxygen levels, that stimulates the bone marrow to produce more red blood cells.

esophagus The muscular tube that connects the mouth to the stomach.

essential amino acid An amino acid that an animal cannot synthesize and must obtain from the diet. May refer specifically to one of the eight essential amino acids of adult humans: isoleucine, leucine, lysine, methionine, phenylalanine, threonine, tryptophan, and valine.

essential nutrient Any chemical element, ion, or compound that is required for normal growth, reproduction, and maintenance of a living organism and that cannot be synthesized by the organism.

ester linkage The covalent bond formed by a condensation reaction between a carboxyl group (—COOH) and a hydroxyl group (—OH). Ester linkages join fatty acids to glycerol to form a fat or phospholipid.

estradiol The major estrogen produced by the ovaries of female mammals. Stimulates development of the female reproductive tract, growth of ovarian follicles, and growth of breast tissue.

estrogens A class of steroid hormones, including estradiol, estrone, and estriol, that generally promote female-like traits. Secreted by the gonads, fat tissue, and some other organs.

estrous cycle A female reproductive cycle, seen in all mammals except Old World monkeys and apes (including humans), in which the uterine lining is reabsorbed rather than shed in the absence of pregnancy, and the female is sexually receptive only briefly during mid-cycle (estrus). Compare with **menstrual cycle**.

estuary An environment of brackish (partly salty) water where a river meets the ocean.

ethylene A gaseous plant hormone that induces fruits to ripen, flowers to fade, and leaves to drop.

eudicot A member of a monophyletic group (lineage) of angiosperms that includes complex flowering plants and trees (e.g., roses, daisies, maples). All eudicots have two cotyledons, but not all dicots are members of this lineage. Compare with **dicot** and **monocot**.

Eukarya One of the three taxonomic domains of life consisting of unicellular organisms (most protists, yeast) and multicellular organisms (fungi, plants, animals) distinguished by a membrane-bound cell nucleus, numerous organelles, and an extensive cytoskeleton. Compare with **Archaea** and **Bacteria**.

eukaryote A member of the domain Eukarya; an organism whose cells contain a nucleus, numerous membrane-bound organelles, and an extensive cytoskeleton. May be unicellular or multicellular. Compare with **prokaryote**.

eutherians A lineage of mammals (Eutheria) whose young develop in the uterus and are not housed in an abdominal pouch. Also called *placental mammals*.

evaporation The energy-absorbing phase change from a liquid state to a gaseous state. Many organisms evaporate water as a means of heat loss.

evo-devo Research field focused on how changes in developmentally important genes have led to the evolution of new phenotypes.

evolution (1) The theory that all organisms on Earth are related by common ancestry and that they have changed over time, predominantly via natural selection. (2) Any change in the genetic characteristics of a population over time, especially, a change in allele frequencies.

ex situ conservation Preserving species outside of natural areas; e.g., in zoos, aquaria, or botanic gardens.

excitable membrane A plasma membrane that is capable of generating an action potential. Neurons, muscle cells, and some other cells have excitable membranes.

excitatory postsynaptic potential (EPSP) A change in membrane potential, usually depolarization, at a neuron dendrite that makes an action potential more likely.

exergonic Referring to a chemical reaction that can occur spontaneously, releasing heat and/or increasing entropy, and for which the Gibbs free-energy change $(\Delta G) < 0$. Compare with **endergonic**.

exocrine gland A gland that secretes some substance through a duct into a space other than the circulatory system, such as the digestive tract or the skin sufrace. Compare with **endocrine gland**.

exocytosis Secretion of intracellular molecules (e.g., hormones, collagen), contained within membrane-bounded vesicles, to the outside of the cell by fusion of vesicles to the plasma membrane. Compare with **endocytosis**.

exon A region of a eukaryotic gene that is translated into a peptide or protein. Compare with **intron**.

exoskeleton A hard covering secreted on the outside of the body, used for body support, protection, and muscle attachment. Examples are the shell of mollusks and the outer covering (cuticle) of arthropods. Compare with **endoskeleton**.

exothermic Referring to a chemical reaction that releases heat. Compare with **endothermic**.

exotic species A nonnative species that is introduced into a new area. Exotic species often are competitors, pathogens, or predators of native species.

expansins A class of plant proteins that actively increase the length of the cell wall when the pH of the wall falls below 4.5.

exponential population growth The accelerating increase in the size of a population that occurs when the growth rate is constant and density independent. Compare with **logistic population growth**.

expressed sequence tag (EST) A portion of a transcribed gene (synthesized from an mRNA in a cell), used to find the physical location of that gene in the genome.

extant species A species that is living today.

extensor A muscle that pulls two bones farther apart from each other, as in the extension of a limb or the spine. Compare with **flexor**.

extinct Said of a species that has died out.

extracellular digestion Digestion that takes place outside of an organism, as occurs in many fungi that make and secrete digestive enzymes.

extracellular matrix (ECM) A complex meshwork of proteins (e.g., collagen, fibronectin) and polysaccharides secreted by animal cells and in which they are embedded.

extremophile A bacterium or archaean that thrives in an "extreme" environment (e.g., high-salt, high-temperature, low-temperature, or low-pressure).

F_1 generation First filial generation. The first generation of offspring produced from a mating (i.e., the offspring of the parental generation).

facilitated diffusion Movement of a substance across a plasma membrane down its concentration gradient with the assistance of transmembrane carrier proteins or channel proteins.

facilitation In ecological succession, the phenomenon in which early-arriving species make conditions more favorable for later-arriving species. Compare with **inhibition** and **tolerance**.

facultative aerobe Any organism that can perform aerobic respiration when oxygen is available to serve as an electron acceptor but can switch to fermentation when it is not.

FAD/FADH$_2$ Oxidized and reduced forms, respectively, of flavin adenine dinucleotide. A nonprotein electron carrier that functions in the citric acid cycle and oxidative phosphorylation.

fallopian tube A narrow tube connecting the uterus to the ovary in humans, through which the egg travels after ovulation. Site of fertilization and cleavage. In nonhuman animals, called *oviduct*.

fat A lipid consisting of three fatty acid molecules joined by ester linkages to a glycerol molecule. Also called *triacylglycerol* or *triglyceride*.

fatty acid A lipid consisting of a hydrocarbon chain bonded to a carboxyl group (—COOH) at one end. Used by many organisms to store chemical energy; a major component of animal and plant fats.

fatty-acid binding protein Proteins that bind to fatty acids and enable them to be transported into cells.

fauna All the animals characteristic of a particular region, period, or environment.

feather A specialized skin outgrowth, composed of β-keratin, present in all birds and only in birds. Used for flight, insulation, display, and other purposes.

feces The waste products of digestion.

fecundity The average number of female offspring produced by a single female in the course of her lifetime.

feedback inhibition A type of metabolic control in which high concentrations of the product of a metabolic pathway inhibit one of the enzymes early in the pathway. A form of negative feedback.

fermentation Any of several metabolic pathways that make ATP by transferring electrons from a reduced compound such as glucose to a final electron acceptor other than oxygen. Allows glycolysis to proceed in the absence of oxygen.

ferredoxin In photosynthetic organisms, an iron- and sulfur-containing protein in the electron transport chain of photosystem I. Can transfer electrons to the enzyme $NADP^+$ reductase, which catalyzes formation of NADPH.

fertilization envelope A physical barrier that forms around a fertilized egg in amphibians and some other animals. Formed by an influx of water under the vitelline membrane.

fertilization Fusion of the nuclei of two haploid gametes to form a zygote with a diploid nucleus.

fetal alcohol syndrome A condition, marked by hyperactivity, severe learning disabilities, and depression, thought to be caused by exposure of an individual to high blood alcohol concentrations during embryonic development.

fetus In live-bearing animals, the unborn offspring after the embryonic stage, which usually are developed sufficiently to be recognizable as belonging to a certain species. In humans, from 9 weeks after fertilization until birth.

fiber In plants, a type of elongated sclerenchyma cell that provides support to vascular tissue. Compare with **sclereid**.

fibronectin An abundant protein in the extracellular matrix that binds to other ECM components and to integrins in plasma membranes; helps anchor cells in place. Numerous subtypes are found in different tissues.

Fick's law of diffusion A mathematical relationship that describes the rates of gas exchange in animal respiratory systems.

fight-or-flight response Rapid physiological changes that prepare the body for emergencies. Includes increased heart rate, increased blood pressure, and decreased digestion.

filament Any thin, threadlike structure, particularly (1) the threadlike extensions of a fish's gills or (2) the slender stalk that bears the anthers in a flower.

filter feeder See **suspension feeder**.

filtrate Any fluid produced by filtration, in particular the fluid ("pre-urine") in the nephrons of vertebrate kidneys.

filtration A process of removing large components from a fluid by forcing it through a filter. Occurs in a renal corpuscle of the vertebrate kidney, allowing water and small solutes to pass from the blood into the nephron.

finite rate of increase (λ) The rate of increase of a population over a given period of time. Calculated as the ending population size divided by the starting population size. Compare with **intrinsic rate of increase**.

first law of thermodynamics The principle of physics that energy is conserved in any process. Energy can be transferred and converted into different forms, but it cannot be created or destroyed.

fission (1) A form of asexual reproduction in which a prokaryotic cell divides to produce two genetically similar daughter cells by a process similar to mitosis of eukaryotic cells. Also called *binary fission*. (2) A form of asexual reproduction in which an animal splits into two or more individuals of approximately equal size; common among invertebrates.

fitness The ability of an individual to produce viable offspring relative to others of the same species.

fitness trade-off See **trade-off**.

fixed action pattern (FAP) Highly stereotyped behavior pattern that occurs in a certain invariant way in a certain species. A form of innate behavior.

flaccid Limp as a result of low internal pressure (e.g., a wilted plant leaf). Compare with **turgid**.

flagellum (plural: flagella) A long, cellular projection that undulates (in eukaryotes) or rotates (in prokaryotes) to move the cell through an aqueous environment. See **axoneme**.

flatworms Members of the phylum Platyhelminthes. Distinguished by a broad, flat, unsegmented body that lacks a coelom. Flatworms belong to the lophotrochozoan branch of the protostomes.

flavin adenine dinucleotide See **$FAD/FADH_2$**.

flexor A muscle that pulls two bones closer together, as in the flexing of a limb or the spine. Compare with **extensor**.

floral meristem A group of undifferentiated plant cells that can give rise to the four organs making up a flower.

florigen In plants, a protein hormone that is synthesized in leaves and transported to the shoot apical meristem where it stimulates flowering.

flower In angiosperms, the part of a plant that contains reproductive structures. Typically includes a calyx, a corolla, and one or more stamens and/or carpels. See **perfect** and **imperfect flower**.

fluid connective tissue A type of connective tissue, distinguished by having a liquid extracellular matrix.

fluid feeder An animal that feeds by sucking or mopping up liquids such as nectar, plant sap, or blood.

fluid-mosaic model The widely accepted hypothesis that the plasma membrane and organelle membranes consist of proteins embedded in a fluid phospholipid bilayer.

fluorescence The spontaneous emission of light from an excited electron falling back to its normal (ground) state.

follicle An egg cell and its surrounding ring of supportive cells in a mammalian ovary.

follicle-stimulating hormone (FSH) A peptide hormone, produced and secreted by the anterior pituitary, that stimulates (in females) growth of eggs and follicles in the ovaries or (in males) sperm production in the testes.

follicular phase The first major phase of a menstrual cycle during which follicles grow and estrogen levels increase; ends with ovulation.

food Any nutrient-containing material that can be consumed and digested by animals.

food chain A relatively simple pathway of energy flow through a few species, each at a different trophic level, in an ecosystem. Might include, for example, a primary producer, a primary consumer, a secondary consumer, and a decomposer. Compare with **food web**.

food web Any complex pathway along which energy moves among many species at different trophic levels of an ecosystem.

foot One of the three main parts of the mollusk body; a muscular appendage, used for movement and/or burrowing into sediment.

foraging Searching for food.

forebrain One of the three main regions of the vertebrate brain; includes the cerebrum, thalamus, and hypothalamus. Compare with **hindbrain** and **midbrain**.

fossil Any trace of an organism that existed in the past. Includes tracks, burrows, fossilized bones, casts, etc.

fossil record All of the fossils that have been found anywhere on Earth and that have been formally described in the scientific literature.

founder effect A change in allele frequencies that often occurs when a new population is established from a small group of individuals (founder event) due to sampling error (i.e., the small group is not a representative sample of the source population).

fovea In the vertebrate eye, a portion of the retina where incoming light is focused; contains a high proportion of cone cells.

free energy See **Gibbs free-energy change**.

free radical Any substance containing one or more atoms with an unpaired electron. Unstable and highly reactive.

frequency The number of wave crests per second traveling past a stationary point. Determines the pitch of sound and the color of light.

frequency-dependent selection A pattern of selection in which certain alleles are favored only when they are rare; a form of balancing selection.

fronds The large leaves of ferns.

frontal lobe In the vertebrate brain, one of the four major areas in the cerebrum.

fruit In flowering plants (angiosperms), a mature, ripened plant ovary (or group of ovaries), along with the seeds it contains and any adjacent fused parts. See **aggregate, multiple,** and **simple fruit**.

fruiting body A structure formed in some prokaryotes, fungi, and protists for spore dispersal; usually consists of a base, a stalk, and a mass of spores at the top.

functional genomics The study of how a genome works, that is, when and where specific genes are expressed and how their products interact to produce a functional organism.

functional group A small group of atoms bonded together in a precise configuration and exhibiting particular chemical properties that it imparts to any organic molecule in which it occurs.

fundamental niche The ecological space that a species occupies in its habitat in the absence of competitors. Compare with **realized niche**.

fungi A lineage of eukaryotes that typically have a filamentous body (mycelium) and obtain nutrients by absorption.

fungicide Any substance that can kill fungi or slow their growth.

G protein Any of various peripheral membrane proteins that bind GTP and function in signal transduction. Binding of a signal to its receptor triggers activation of the G protein, leading to production of a second messenger or initiation of a phosphorylation cascade.

G_1 phase The phase of the cell cycle that constitutes the first part of interphase before DNA synthesis (S phase).

G_2 phase The phase of the cell cycle between synthesis of DNA (S phase) and mitosis (M phase); the last part of interphase.

gallbladder A small pouch that stores bile from the liver and releases it as needed into the small intestine during digestion of fats.

gametangium (plural: gametangia) (1) The gamete-forming structure found in all land plants except angiosperms. Contains a sperm-producing antheridium and an egg-producing archegonium. (2) The gamete-forming structure of some chytrid fungi.

gamete A haploid reproductive cell that can fuse with another haploid cell to form a zygote. Most multicellular eukaryotes have two distinct forms of gametes: egg cells (ova) and sperm cells.

gametogenesis The production of gametes (eggs or sperm).

gametophyte In organisms undergoing alternation of generations, the multicellular haploid form that arises from a single haploid spore and produces gametes. A female gametophyte is commonly called an *embryo sac;* a male gametophyte, a *pollen grain*. Compare with **sporophyte**.

ganglia A mass of neurons in a centralized nervous system.

ganglion cell A neuron in the vertebrate retina that collects visual information from one or several bipolar cells and sends it to the brain via the optic nerve.

gap junction A type of cell-cell attachment structure that directly connects the cytoplasms of adjacent animal cells, allowing passage of water, ions, and small molecules between the cells. Compare with **desmosome** and **tight junction**.

gastrin A hormone produced by cells in the stomach lining in response to the arrival of food or to a neural signal from the brain. Stimulates other stomach cells to release hydrochloric acid.

gastrointestinal (GI) tract See **digestive tract**.

gastropods A lineage of mollusk distinguished by a large muscular foot and a unique feeding structure, the radula. Include slugs and snails.

gastrulation The process by which some cells on the outside of a young embryo move to the interior of the embryo, resulting in the three distinct germ layers (endoderm, mesoderm, and ectoderm).

gated channel A channel protein that opens and closes in response to a specific stimulus, such as the binding of a particular molecule or a change in the electrical charge on the outside of the membrane.

gel electrophoresis A technique for separating molecules on the basis of size and electric charge, which affect their differing rates of movement through a gelatinous substance in an electric field.

gemma (plural: gemmae) A small reproductive structure that is produced in some liverworts during the gametophyte phase and can grow into mature gametophyte.

gene A section of DNA (or RNA, for some viruses) that encodes information for building one or more related polypeptides or functional RNA molecules along with the regulatory sequences required for its transcription.

gene duplication The formation of an additional copy of a gene, typically by misalignment of chromosomes during crossing over. Thought to be an important evolutionary process in creating new genes.

gene expression Overall process by which the information encoded in genes is converted into an active product, most commonly a protein. Includes transcription and translation of a gene and in some cases protein activation.

gene family A set of genetic loci whose DNA sequences are extremely similar. Thought to have arisen by duplication of a single ancestral gene and subsequent mutations in the duplicated sequences.

gene flow The movement of alleles between populations; occurs when individuals leave one population, join another, and breed.

gene pool All of the alleles of all of the genes in a certain population.

gene therapy The treatment of an inherited disease by introducing normal alleles.

gene-for-gene hypothesis The hypothesis that there is a one-to-one correspondence between the resistance (*R*) loci of plants and the avirulence (*avr*) loci of pathogenic fungi; particularly, that *R* genes produce receptors and *avr* genes produce molecules that bind to those receptors.

generation The average time between a mother's first offspring and her daughter's first offspring.

genetic bottleneck A reduction in allelic diversity resulting from a sudden reduction in the size of a large population (population bottleneck) due to a random event.

genetic code The set of all 64 codons and the particular amino acids that each specifies.

genetic correlation A type of evolutionary constraint in which selection on one trait causes a change in another trait as well; may occur when the same gene(s) affect both traits.

genetic diversity The diversity of alleles in a population, species, or group of species.

genetic drift Any change in allele frequencies due to random events. Causes allele frequencies to drift up and down randomly over time, and eventually can lead to the fixation or loss of alleles.

genetic homology Similarities in DNA sequences or amino acid sequences that are due to inheritance from a common ancestor.

genetic map An ordered list of genes on a chromosome that indicates their relative distances from each other. Also called a *linkage map* or *meiotic map*. Compare with **physical map**.

genetic marker A genetic locus that can be identified and traced in populations by laboratory techniques or by a distinctive visible phenotype.

genetic model A set of hypotheses that explain how a certain trait is inherited.

genetic recombination A change in the combination of genes or alleles on a given chromosome or in a given individual. Also called *recombination*.

genetic screen Any of several techniques for identifying individuals with a particular type of mutation. Also called a *screen*.

genetic variation (1) The number and relative frequency of alleles present in a particular population. (2) The proportion of phenotypic variation in a trait that is due to genetic rather than environmental influences in a certain population in a certain environment.

genetics The field of study concerned with the inheritance of traits.

genitalia External copulatory organs.

genome All of the hereditary information in an organism, including not only genes but also other non-gene stretches of DNA.

genomic library A set of DNA segments representing the entire genome of a particular organism. Each segment is carried by a plasmid or other cloning vector and can be separated from other segments. Compare with **cDNA library**.

genomics The field of study concerned with sequencing, interpreting, and comparing whole genomes from different organisms.

genotype All of the alleles of every gene present in a given individual. May refer specifically to the alleles of a particular set of genes under study. Compare with **phenotype**.

genus (plural: genera) In Linnaeus' system, a taxonomic category of closely related species. Always italicized and capitalized to indicate that it is a recognized scientific genus.

geologic time scale The sequence of eons, eras, and periods used to describe the geologic history of Earth.

germ cell In animals, any cell that can potentially give rise to gametes. Also called *germ-line cells*.

germ layer In animals, one of the three basic types of tissue formed during gastrulation; gives rise to all other tissues. See **endoderm, mesoderm,** and **ectoderm**.

germ theory of disease The theory that infectious diseases are caused by bacteria, viruses, and other microorganisms.

germination The process by which a seed becomes a young plant.

gestation The duration of embryonic development from fertilization to birth in those species that have live birth.

gibberellins A class of hormones found in plants and fungi that stimulate growth. Gibberellic acid is one of the major gibberellins.

Gibbs free-energy change (ΔG) A measure of the change in potential energy and entropy that occurs in a given chemical reaction. $\Delta G < 0$ for spontaneous reactions and > 0 for nonspontaneous reactions.

gill Any organ in aquatic animals that exchanges gases and other dissolved substances between the blood and the surrounding water. Typically, a filamentous outgrowth of a body surface.

gill arch In aquatic vertebrates, curved region of tissue between the gills. Gills are suspended from the gill arches.

gill filament In fish, one of the many long, thin structures that extend from gill arches into the water and across which gas exchange occurs.

gill lamella (plural: gill lamellae) One of hundreds to thousands of sheetlike structures, each containing a capillary bed, that makes up a gill filament.

gland An organ whose primary function is to secrete some substance, either into the blood (endocrine gland) or into some other space such as the gut or skin (exocrine gland).

glia Collective term for several types of cells in nervous tissue that are not neurons and do not conduct electrical signals but provide support, nourishment, and electrical insulation and perform other functions. Also called *glial cells*.

global carbon cycle See **carbon cycle, global**.

global nitrogen cycle See **nitrogen cycle, global**.

global warming A sustained increase in Earth's average surface temperature.

global water cycle See **water cycle, global**.

glomalin A glycoprotein that is abundant in the hyphae of arbuscular mycorrhizal fungi; when hyphae decay it an important component of soil.

glomerulus (plural: glomeruli) (1) In the vertebrate kidney, a ball-like cluster of capillaries, surrounded by Bowman's capsule, at the beginning of a nephron. (2) In the brain, a ball-shaped cluster of neurons in the olfactory bulb.

glucagon A peptide hormone produced by the pancreas in response to low blood glucose. Raises blood glucose by triggering breakdown of glycogen and stimulating gluconeogenesis. Compare with **insulin**.

glucocorticoids A class of steroid hormones, produced and secreted by the adrenal cortex, that increase blood glucose and prepare the body for stress. Include cortisol and corticosterone. Compare with **mineralocorticoids**.

gluconeogenesis Synthesis of glucose from non-carbohydrate sources (e.g., proteins and fatty acids). Occurs in the liver in response to low insulin levels and high glucagon levels.

glucose Six-carbon monosaccharide whose oxidation in cellular respiration is the major source of ATP in animal cells.

glyceraldehyde-3-phosphate (G3P) The phosphorylated three-carbon compound formed as the result of carbon fixation in the first step of the Calvin cycle.

glycerol A three-carbon molecule that forms the "backbone" of phospholipids and most fats.

glycogen A highly branched storage polysaccharide composed of α-glucose monomers joined by 1,4- and 1,6-glycosidic linkages. The major form of stored carbohydrate in animals.

glycolipid Any lipid molecule that is covalently bonded to a carbohydrate group.

glycolysis A series of 10 chemical reactions that oxidize glucose to produce pyruvate and ATP. Used by all organisms as part of fermentation or cellular respiration.

glycoprotein Any protein with one or more covalently bonded carbohydrate groups.

glycosidic linkage The covalent bond formed by a condensation reaction between two sugar monomers; joins the residues of a polysaccharide.

glycosylation Addition of a carbohydrate group to a molecule.

glyoxisome Specialized type of peroxisome found in plant cells and packed with enzymes for processing the products of photosynthesis.

gnathostomes Animals with jaws.

Golgi apparatus A eukaryotic organelle, consisting of stacks of flattened membranous sacs (cisternae), that functions in processing and sorting proteins and lipids destined to be secreted or directed to other organelles. Also called *Golgi complex*.

gonad An organ that produces reproductive cells, such as a testis or an ovary.

gonadotropin-releasing hormone (GnRH) A peptide hormone, produced and secreted by the hypothalamus, that stimulates release of FSH and LH from the anterior pituitary.

grade In taxonomy, a group of species that share a position in an inferred evolutionary sequence of lineages but that are not a monophyletic group. Also called a *paraphyletic group*.

Gram stain A dye that distinguishes the two general types of cell walls found in bacteria. Used to routinely classify bacteria as Gram-negative or Gram-positive.

Gram-negative Describing bacteria that look pink when treated with a Gram stain. These bacteria have a cell wall composed of a thin layer of peptidoglycan and an outer phospholipid layer.

Gram-positive Describing bacteria that look purple when treated with a Gram stain. These bacteria have cell walls composed of a thick layer of peptidoglycan.

granum (plural: grana) In chloroplasts, a stack of flattened, membrane-bound vesicles (thylakoids) where the light reactions of photosynthesis occur.

gravitropism The growth or movement of a plant in a particular direction in response to gravity.

grazing food chain The ecological network of herbivores and the predators and parasites that consume them.

great apes See **hominids.**

green algae A paraphyletic group of photosynthetic organisms that contain chloroplasts similar to those in green plants. Often classified as protists, green algae are the closest living relatives of land plants and form a monophyletic group with them.

greenhouse gas An atmospheric gas that absorbs and reflects infrared radiation, so that heat radiated from Earth is retained in the atmosphere instead of being lost to space.

gross photosynthetic productivity The efficiency with which all the plants in a given area use the light energy available to them to produce sugars.

gross primary productivity In an ecosystem, the total amount of carbon fixed by photosynthesis, including that used for cellular respiration, over a given time period. Compare with **net primary productivity.**

ground meristem The middle layer of a young plant embryo. Gives rise to the ground tissue system.

ground tissue An embryonic tissue layer that gives rise to parenchyma, collenchyma, and sclerenchyma—tissues other than the epidermis and vascular tissue.

groundwater Any water below the land surface.

growth factor Any of a large number of signaling molecules that are secreted by certain cells and that stimulate other cells to divide or to differentiate.

growth hormone (GH) A peptide hormone, produced and secreted by the mammalian anterior pituitary, that promotes lengthening of the long bones in children and muscle growth, tissue repair, and lactation in adults. Also called *somatotropin.*

GTP See **guanosine triphosphate.**

guanosine triphosphate (GTP) A molecule consisting of guanine, a sugar, and three phosphate groups. Can be hydrolyzed to release free energy. Commonly used in many cellular reactions; also functions in signal transduction in association with G proteins.

guard cell One of two specialized, crescent-shaped cells forming the border of a plant stoma. Guard cells can change shape to open or close the stoma. See also **stoma.**

gustation The perception of taste.

guttation Excretion of water droplets from plant leaves in the early morning, caused by root pressure.

gymnosperm A vascular plant that makes seeds but does not produce flowers. The gymnosperms include four lineages of green plants (cycads, ginkgoes, conifers, and gnetophytes). Compare with **angiosperm.**

H^+-ATPase See **proton pump.**

habitat destruction Human-caused destruction of a natural habitat with replacement by an urban, suburban, or agricultural landscape.

habitat fragmentation The breakup of a large region of a habitat into many smaller regions, separated from others by a different type of habitat.

Hadley cell An atmospheric cycle of large-scale air movement in which warm equatorial air rises, moves north or south, and then descends at approximately 30° N or 30° S latitude.

hair cell A pressure-detecting sensory cell, found in the cochlea, that has tiny "hairs" (stereocilia) jutting from its surface.

hairpin A secondary structure in RNA consisting of a stable loop formed by hydrogen bonding between purine and pyrimidine bases on the same strand.

halophile A bacterium or archaean that thrives in high-salt environments.

Hamilton's rule The proposition that an allele for altruistic behavior will be favored by natural selection only if $Br > C$, where B = the fitness benefit to the recipient, C = the fitness cost to the actor, and r = the coefficient of relatedness between recipient and actor.

haploid (1) Having one set of chromosomes ($1n$). (2) A cell or an individual organism with one set of chromosomes. Compare with **diploid.**

haploid number The number of distinct chromosome sets in a cell. Symbolized as n.

Hardy-Weinberg principle A principle of population genetics stating that genotype frequencies in a large population do not change from generation to generation in the absence of evolutionary processes (e.g., mutation, migration, genetic drift, random mating, and selection).

heart A muscular pump that circulates blood throughout the body.

heart murmur A distinctive sound caused by backflow of blood through a defective heart valve.

heartwood The older xylem in the center of an older stem or root, containing protective compounds and no longer functioning in water transport.

heat Thermal energy that is transferred from an object at higher temperature to one at lower temperature.

heat of vaporization The energy required to vaporize 1 gram of a liquid into a gas.

heat-shock proteins Proteins that facilitate refolding of proteins that have been denatured by heat or other agents.

heavy chain The larger of the two types of polypeptide chains in an antibody molecule; composed of a variable (V) region, which contributes to the antigen-binding site, and a constant (C) region. Differences in heavy-chain constant regions determine the different classes of immunoglobulins (IgA, IgE, etc.). Compare with **light chain.**

helicase An enzyme that catalyzes the breaking of hydrogen bonds between nucleotides of DNA, "unzipping" a double-stranded DNA molecule.

helper T cell An effector T cell that secretes cytokines and in other ways promotes the activation of other lymphocytes. Is descended from an activated $CD4^+$ T cell that has interacted with antigen presented by dendritic cells, macrophages, or B cells.

heme A small molecule that binds to four polypeptides to form hemoglobin; contains an iron atom that can bind oxygen.

hemimetabolous metamorphosis A type of metamorphosis in which the animal increases in size from one stage to the next, but does not dramatically change its body form. Also called *incomplete metamorphosis.*

hemocoel A body cavity, present in arthropods and some mollusks, containing a pool of circulatory fluid (hemolymph) bathing the internal organs.

hemoglobin An oxygen-binding protein consisting of four polypeptide subunits, each containing an oxygen-binding heme group. The major oxygen carrier in mammalian blood.

hemolymph The circulatory fluid of animals with open circulatory systems (e.g., insects) in which the fluid is not confined to blood vessels.

hemophilia A human disease, caused by an X-linked recessive allele, that is characterized by defects in the blood-clotting system.

herbaceous Referring to a plant that is not woody.

herbivore (adjective: herbivorous) An animal that eats primarily plants and rarely or never eats meat. Compare with **carnivore** and **omnivore.**

herbivory The practice of eating plant tissues.

heredity The transmission of traits from parents to offspring via genetic information.

heritable Referring to traits that can be transmitted from one generation to the next.

hermaphrodite An organism that produces both male and female gametes.

heterokaryotic Describing a cell or fungal mycelium containing two or more nuclei that are genetically distinct.

heterospory (adjective: heterosporous) In seed plants, the production of two distinct types of spore-producing structures and thus two distinct types of spores: microspores, which become the male gametophyte, and megaspores, which become the female gametophyte. Compare with **homospory.**

heterotherm An animal whose body temperature varies markedly with environmental conditions. Compare with **homeotherm.**

heterotroph Any organism that cannot synthesize reduced organic compounds from inorganic sources and that must obtain them by eating other organisms. Some bacteria, some archaea, and virtually all fungi and animals are heterotrophs. Also called *consumer.* Compare with **autotroph.**

heterozygote advantage A pattern of natural selection that favors heterozygous individuals compared with homozygotes. Tends to maintain genetic variation in a population. Also called *heterozygote superiority.*

heterozygous Having two different alleles of a certain gene.

hexose A monosaccharide (simple sugar) containing six carbon atoms.

hibernation An energy-conserving physiological state, marked by a decrease in metabolic rate, body temperature, and activity, that lasts for a prolonged period (weeks to months). Occurs in some animals in response to winter cold and scarcity of food. Compare with **torpor.**

hindbrain One of the three main regions of the vertebrate brain; includes the cerebellum and medulla oblongata. Compare with **forebrain** and **midbrain.**

histamine A molecule released from mast cells during an inflammatory response that causes blood vessels to dilate and become more permeable.

histone One of several positively charged (basic) proteins associated with DNA in the chromatin of eukaryotic cells.

histone acetyl transferase (HAT) In eukaryotes, one of a class of enzymes that loosen chromatin structure by adding acetyl groups to histone proteins.

histone code The hypothesis that specific combinations of chemical modifications of histone proteins contain information that influences gene expression.

histone deacetylase (HDAC) In eukaryotes, one of a class of enzymes that recondense chromatin by removing acetyl groups from histone proteins.

HIV See human immunodeficiency virus (HIV).

holoenzyme A multipart enzyme consisting of a core enzyme (containing the active site for catalysis) along with other required proteins.

holometabolous metamorphosis A type of metamorphosis in which the animal completely changes its form. Also called *complete metamorphosis.*

homeosis Replacement of one body part by another normally found elsewhere in the body as the result of mutation in certain developmentally important genes (homeotic genes).

homeostasis (adjective: homeostatic) The array of relatively stable chemical and physical conditions in an animal's cells, tissues, and organs. May be achieved by the body's passively matching the conditions of a stable external environment (conformational homeostasis) or by active physiological processes (regulatory homeostasis) triggered by variations in the external or internal environment.

homeotherm An animal that has a constant or relatively constant body temperature. Compare with **heterotherm**.

homeotic gene Any gene that specifies a particular location within an embryo, leading to the development of structures appropriate for that location. Mutations in homeotic genes cause the development of extra body parts or body parts in the wrong places.

hominids Members of the family Hominidae, which includes humans and extinct related forms; chimpanzees, gorillas, and orangutans. Distinguished by large body size, no tail, and an exceptionally large brain. Also called *great apes*.

hominins Humans and extinct related forms; species in the lineage that branched off from chimpanzees and eventually led to humans.

homologous chromosomes In a diploid organism, chromosomes that are similar in size, shape, and gene content. Also called *homologs*.

homology (adjective: homologous) Similarity among organisms of different species due to their inheritance from a common ancestor. Features that exhibit such similarity (e.g., DNA sequences, proteins, body parts) are said to be homologous. Compare with **homoplasy**.

homoplasy Similarity among organisms of different species due to convergent evolution. Compare with **homology**.

homospory (adjective: homosporous) In seedless vascular plants, the production of just one type of spore. Compare with **heterospory**.

homozygous Having two identical alleles of a certain gene.

hormone Any of numerous different signaling molecules that circulate throughout the body in blood or other body fluids and can trigger characteristic responses in distant target cells at very low concentrations.

hormone-response element A specific sequence in DNA to which a steroid hormone-receptor complex can bind and affect gene transcription.

host cell A cell that has been invaded by an organism such as a parasite or a virus.

***Hox* genes** A class of homeotic genes found in several animal phyla, including vertebrates, that are expressed in a distinctive pattern along the anterior-posterior axis in early embryos and control formation of segment-specific structures.

human Any member of the genus *Homo*, which includes modern humans (*Homo sapiens*) and several extinct species.

human chorionic gonadotropin (hCG) A glycoprotein hormone produced by the human placenta from about week 3 to week 14 of pregnancy. Maintains the corpus luteum, which produces hormones that preserve the uterine lining.

Human Genome Project The multinational research project that sequenced the human genome.

human immunodeficiency virus (HIV) A retrovirus that causes AIDS (acquired immune deficiency syndrome) in humans.

humoral (immune) response The type of immune response that involves generation of antibody-secreting plasma cells from activated B cells. Defends against extracellular pathogens. Compare with **cell-mediated (immune) response**.

humus The completely decayed organic matter in soils.

Huntington's disease A degenerative brain disease of humans caused by an autosomal dominant allele.

hybrid The offspring of parents from two different strains, populations, or species.

hybrid zone A geographic area where interbreeding occurs between two species, sometimes producing fertile hybrid offspring.

hydrocarbon An organic molecule that contains only hydrogen and carbon atoms.

hydrogen bond A weak interaction between two molecules or different parts of the same molecule resulting from the attraction between a hydrogen atom with a partial positive charge and another atom (usually O or N) with a partial negative charge. Compare with **covalent bond** and **ionic bond**.

hydrogen ion (H^+) A single proton with a charge of 1+; typically, one that is dissolved in solution or that is being transferred from one atom to another in a chemical reaction.

hydrolysis A chemical reaction in which a molecule is split into smaller molecules by reacting with water. In biology, most hydrolysis reactions involve the splitting of polymers into monomers. Compare with **condensation reaction**.

hydrophilic Interacting readily with water. Hydrophilic compounds are typically polar compounds containing charged or electronegative atoms. Compare with **hydrophobic**.

hydrophobic Not interacting readily with water. Hydrophobic compounds are typically nonpolar compounds that lack charged or electronegative atoms and often contain many C—C and C—H bonds. Compare with **hydrophilic**.

hydroponic growth Growth of plants in liquid cultures instead of soil.

hydrostatic skeleton A system of body support involving fluid-filled compartments that can change in shape but cannot easily be compressed.

hydroxide ion (OH^-) An oxygen atom and a hydrogen atom joined by a single covalent bond and carrying a negative charge; formed by dissociation of water.

hyperpolarization Change in membrane potential from its resting negative state to an even more negative state; a normal phase in an action potential. Compare with **depolarization**.

hypersensitive reaction An intense allergic response by cells that have been sensitized by previous exposure to an allergen.

hypersensitive response In plants, the rapid death of a cell that has been infected by a pathogen, thereby reducing the potential for infection to spread throughout a plant. Compare with **systemic acquired resistance**.

hypertension Abnormally high blood pressure.

hypertonic Comparative term designating a solution that has a greater solute concentration, and therefore a lower water concentration, than another solution. Compare with **hypotonic** and **isotonic**.

hypha (plural: hyphae) One of the strands of a fungal mycelium (the meshlike body of a fungus). Also found in some protists.

hypocotyl The stem of a very young plant; the region between the cotyledon (embryonic leaf) and the radicle (embryonic root).

hypothalamic-pituitary axis The functional interaction of the hypothalamus and the pituitary gland, which are anatomically distinct but work together to regulate most of the other endocrine glands in the body.

hypothalamus A part of the brain that functions in maintaining the body's internal physiological state by regulating the autonomic nervous system, endocrine system, body temperature, water balance, and appetite.

hypothesis A proposed explanation for a phenomenon or for a set of observations.

hypotonic Comparative term designating a solution that has a lower solute concentration, and therefore a higher water concentration, than another solution. Compare with **hypertonic** and **isotonic**.

immigration The migration of individuals into a particular population from other populations. Compare with **emigration**.

immune system In vertebrates, the system whose primary function is to defend the body against pathogens. Includes several types of cells (e.g., lymphocytes and macrophages) and several organs where they develop or reside (e.g., lymph nodes and thymus).

immunity (adjective: immune) State of being protected against infection by disease-causing pathogens either by relatively nonspecific mechanisms (innate immunity) or by specific mechanisms triggered by exposure to a particular antigen (adaptive immunity).

immunization The conferring of immunity to a particular disease by artificial means.

immunoglobulin (Ig) Any of the class of proteins that function as antibodies.

immunological memory The ability of the immune system to "remember" an antigen and mount a rapid, effective response to a pathogen encountered years or decades earlier.

impact hypothesis The hypothesis that a collision between the Earth and an asteroid caused the mass extinction at the K-P boundary, 65 million years ago.

imperfect flower A flower that contains male parts (stamens) *or* female parts (carpels) but not both. Compare with **perfect flower**.

implantation The process by which an embryo buries itself in the uterine wall and forms a placenta. Occurs in mammals and a few other vertebrates.

in situ hybridization A technique for detecting specific DNAs and mRNAs in cells and tissues by use of labeled probes. Can be used to determine where and when particular genes are expressed in embryos.

inbreeding Mating between closely related individuals. Increases homozygosity of a population and often leads to a decline in the average fitness (inbreeding depression).

inbreeding depression In inbred offspring, fitness declines due to deleterious recessive alleles that are homozygous.

inclusive fitness The combination of (1) direct production of offspring (direct fitness) and (2) extra production of offspring by relatives in response to help provided by the individual in question (indirect fitness).

incomplete digestive tract A digestive tract that has just one opening.

incomplete dominance An inheritance pattern in which the heterozygote phenotype is a blend or combination of both homozygote phenotypes.

incomplete metamorphosis See **hemimetabolous metamorphosis**.

independent assortment, principle of The concept that each pair of hereditary elements (alleles of the same gene) behaves independently of other genes during meiosis. One of Mendel's two principles of genetics.

indeterminate growth A pattern of growth in which an individual continues to increase its overall body size throughout its life.

indicator plate A laboratory technique for detecting mutant cells by growing them on agar plates containing a compound that when metabolized by wild-type cells yields a colored product.

induced fit Change in the shape of the active site of an enzyme, as the result of the initial weak binding of a substrate, so that it binds substrate more tightly.

inducer A small molecule that triggers transcription of a specific gene, often by binding to and inactivating a repressor protein.

inducible defense A defensive trait that is manifested only in response to the presence of a consumer (predator or herbivore) or pathogen. Compare with **constitutive defense.**

infection thread An invagination of the membrane of a root hair through which beneficial nitrogen-fixing bacteria enter the roots of their host plants (legumes).

inflammatory response An aspect of the innate immune response, seen in most cases of infection or tissue injury, in which the affected tissue becomes swollen, red, warm, and painful.

inhibition In ecological succession, the phenomenon in which early-arriving species make conditions less favorable for the establishment of certain later-arriving species. Compare with **facilitation** and **tolerance.**

inhibitory postsynaptic potential (IPSP) A change in membrane potential, usually hyperpolarization, at a neuron dendrite that makes an action potential less likely.

initiation (1) In an enzyme-catalyzed reaction, the stage during which enzymes orient reactants precisely as they bind at specific locations within the enzyme's active site. (2) In DNA transcription, the stage during which RNA polymerase and other proteins assemble at the promoter sequence. (3) In RNA translation, the stage during which a complex consisting of a ribosome, a mRNA molecule, and an aminoacyl tRNA corresponding to the start codon is formed.

initiation factors A class of proteins that assist ribosomes in binding to a messenger RNA molecule to begin translation.

innate behavior Behavior that is inherited genetically, does not have to be learned, and is typical of a species.

innate immune response See **innate immunity.**

innate immunity A set of nonspecific defenses against pathogens that exist before exposure to an antigen and involves mast cells, neutrophils, and macrophages; typically results in an inflammatory response. Compare with **acquired immunity.**

inner cell mass (ICM) A cluster of cells in the interior of a mammalian blastocyst that undergo gastrulation and eventually develop into the embryo.

inner ear The innermost portion of the mammalian ear, consisting of a fluid-filled system of tubes that includes the cochlea (which receives sound vibrations from the middle ear) and the semicircular canals (which function in balance).

insulin A peptide hormone produced by the pancreas in response to high levels of glucose (or amino acids) in blood. Enables cells to absorb glucose and coordinates synthesis of fats, proteins, and glycogen. Compare with **glucagon.**

integral membrane protein Any membrane protein that spans the entire lipid bilayer. Also called *transmembrane protein.* Compare with **peripheral membrane protein.**

integrated pest management In agriculture or forestry, systems for managing insects or other pests that include carefully controlled applications of toxins, introduction of species that prey on pests, planting schemes that reduce the chance of a severe pest outbreak, and other techniques.

integrator A component of an animal's nervous system that functions as part of a homeostatic system by evaluating sensory information and triggering appropriate responses. See **effector** and **sensor.**

integrin Any of a class of cell-surface proteins that bind to fibronectins and other proteins in the extracellular matrix, thus holding cells in place.

intercalated disc A specialized junction between adjacent heart muscle cells that contains gap junctions, allowing electrical signals to pass between the cells.

intermediate disturbance hypothesis The hypothesis that moderate ecological disturbance is associated with higher species diversity than either low or high disturbance.

intermediate filament A long fiber, about 10 nm in diameter, composed of one of various proteins (e.g., keratins, lamins); one of the three types of cytoskeletal fibers. Form networks that help maintain cell shape and hold the nucleus in place. Compare with **actin filament** and **microtubule.**

intermediate host The host species in which a parasite reproduces asexually. Compare with **definitive host.**

interneuron A neuron that passes signals from one neuron to another. Compare with **motor neuron** and **sensory neuron.**

internode The section of a plant stem between two nodes (sites where leaves attach).

interphase The portion of the cell cycle between one mitotic (M) phase and the next. Includes the G_1 phase, S phase, and G_2 phase.

interspecific competition Competition between members of different species for the same limited resource. Compare with **intraspecific competition.**

interstitial fluid The plasma-like fluid found in the region (interstitial space) between cells.

intertidal zone The region between the low-tide and high-tide marks on a seashore.

intraspecific competition Competition between members of the same species for the same limited resource. Compare with **interspecific competition.**

intrinsic rate of increase (r_{max}) The rate at which a population will grow under optimal conditions (i.e., when birthrates are as high as possible and death rates are as low as possible). Compare with **finite rate of increase.**

intron A region of a eukaryotic gene that is transcribed into RNA but is later removed, so it is not translated into a peptide or protein. Compare with **exon.**

invasive species An exotic (nonnative) species that, upon introduction to a new area, spreads rapidly and competes successfully with native species.

inversion A mutation in which a segment of a chromosome breaks from the rest of the chromosome, flips, and rejoins with the opposite orientation as before.

invertebrates A paraphyletic group composed of animals without a backbone; includes about 95 percent of all animal species. Compare with **vertebrates.**

involuntary muscle Muscle that cannot respond to conscious thought.

ion An atom or a molecule that has lost or gained electrons and thus carries an electric charge, either positive (cation) or negative (anion), respectively.

ion channel A type of channel protein that allows certain ions to diffuse across a plasma membrane down an electochemical gradient.

ionic bond A chemical bond that is formed when an electron is completely transferred from one atom to another so that the atoms remain associated due to their opposite electric charges. Compare with **covalent bond** and **hydrogen bond.**

iris A ring of pigmented muscle just below the cornea in the vertebrate eye that contracts or expands to control the amount of light entering the eye through the pupil.

isotonic Comparative term designating a solution that has the same solute concentration and water concentration than another solution. Compare with **hypertonic** and **hypotonic.**

isotope Any of several forms of an element that have the same number of protons but differ in the number of neutrons.

joint A place where two components (bones, cartilages, etc.) of a skeleton meet. May be movable (an articulated joint) or immovable (e.g., skull sutures).

juvenile An individual that has adult-like morphology but is not sexually mature.

juvenile hormone An insect hormone that prevents larvae from metamorphosing into adults.

karyogamy Fusion of two haploid nuclei to form a diploid nucleus. Occurs in many fungi, and in animals and plants during fertilization of gametes.

karyotype The distinctive appearance of all of the chromosomes in an individual, including the number of chromosomes and their length and banding patterns (after staining with dyes).

keystone species A species that has an exceptionally great impact on the other species in its ecosystem relative to its abundance.

kidney In terrestrial vertebrates, one of a paired organ situated at the back of the abdominal cavity that filters the blood, produces urine, and secretes several hormones.

kilocalorie (kcal) A unit of energy often used to measure the energy content of food. A kcal of energy raises 1 g of water 1°C.

kin selection A form of natural selection that favors traits that increase survival or reproduction of an individual's kin at the expense of the individual.

kinesin Any one of a class of motor proteins that use the chemical energy of ATP to transport vesicles, particles, or chromosomes along microtubules.

kinetic energy The energy of motion. Compare with **potential energy.**

kinetochore A protein structure at the centromere where kinetochore microtubules attach to the sister chromatids of a replicated chromosome. Contains motor proteins that move a chromosome along a microtubule.

kinetochore microtubules Microtubules that form during mitosis and meiosis, and which extend from a spindle apparatus to an attachment point—the kinetochore—on a chromosome.

kinocilium (plural: kinocilia) A single cilium that juts from the surface of many hair cells and functions in detection of sound or pressure.

Klinefelter syndrome A syndrome seen in humans who have an XXY karyotype. People with this syndrome have male sex organs, may have some female traits, and are sterile.

knock-out allele A mutant allele that does not function at all, or an organism homozygous for such a mutation. Also called *null allele* or *loss-of-function allele.*

Koch's postulates Four criteria used to determine whether a suspected infectious agent causes a particular disease.

labia major (plural: labium majus) One of two outer folds of skin that protect the labia minora, clitoris, and vaginal opening of female mammals.

labia minora (plural: labium minus) One of two inner folds of skin that protect the opening of the urethra and vagina.

labor The strong muscular contractions of the uterus that expel the fetus during birth.

lactation (verb: lactate) Production of milk from mammary glands of mammals.

lacteal A small lymphatic vessel extending into the center of a villus in the small intestine. Receives chylomicrons containing fat absorbed from food.

lactic acid fermentation Catabolic pathway in which pyruvate produced by glycolysis is converted to lactic acid in the absence of oxygen.

lagging strand In DNA replication, the strand of new DNA that is synthesized discontinuously in a series of short pieces that are later joined together. Also called *discontinuous strand*. Compare with **leading strand**.

large intestine The distal portion of the digestive tract consisting of the cecum, colon, and rectum. Its primary function is to compact the wastes delivered from the small intestine and absorb enough water to form feces.

larva (plural: larvae) An immature stage of a species in which the immature and adult stages have different body forms.

late endosome A membrane-bound vesicle that arises from an early endosome and develops into a lysosome.

latency In viruses that infect animals, the ability to exist in a quiescence state without producing new virions.

lateral bud A bud that forms in the angle between a leaf and a stem and may develop into a lateral (side) branch. Also called *axillary bud*.

lateral gene transfer Transfer of DNA between two different species, especially distantly related species. Commonly occurs among bacteria and archaea via plasmid exchange; also can occur in eukaryotes via viruses and some other mechanisms.

lateral line system A pressure-sensitive sensory organ found in many aquatic vertebrates.

lateral meristem A layer of undifferentiated plant cells found in older stems and roots that is responsible for secondary growth. Also called *cambium* or *secondary meristem*. Compare with **apical meristem**.

lateral root A plant root extending from another, older root.

leaching Loss of nutrients from soil via percolating water.

leading strand In DNA replication, the strand of new DNA that is synthesized in one continuous piece, with nucleotides added to the 3′ end of the growing molecule. Also called *continuous strand*. Compare with **lagging strand**.

leaf The main photosynthetic organ of vascular plants.

leak channel Potassium channel that allows potassium ions to leak out of a neuron in its resting state.

learning An enduring change in an individual's behavior that results from specific experience(s).

leghemoglobin An iron-containing protein similar to hemoglobin. Found in root nodules of legume plants where it binds oxygen, preventing it from poisoning a bacterial enzyme needed for nitrogen fixation.

legumes Members of the pea plant family that form symbiotic associations with nitrogen-fixing bacteria in their roots.

lens A transparent, crystalline structure that focuses incoming light onto a retina or other light-sensing apparatus of an eye.

lenticels Spongy segments in bark that allow gas exchange between cells in a woody stem and the atmosphere.

leptin A hormone produced and secreted by fat cells (adipocytes) that acts to stabilize fat tissue mass in part by inhibiting appetite and increasing energy expenditure.

leukocytes Several types of blood cells, including neutrophils, macrophages, and lymphocytes, that circulate in blood and lymph and function in defense against pathogens. Also called *white blood cells*.

lichen A symbiotic association of a fungus and a photosynthetic alga.

life cycle The sequence of developmental events and phases that occurs during the life span of an organism, from fertilization to offspring production.

life history The sequence of events in an individual's life from birth to reproduction to death, including how an individual allocates resources to growth, reproduction, and activities or structures that are related to survival.

life table A data set that summarizes the probability that an individual in a certain population will survive and reproduce in any given year over the course of its lifetime.

ligand Any molecule that binds to a specific site on a receptor molecule.

ligand-gated channel An ion channel that opens or closes in response to binding by a certain molecule. Compare with **voltage-gated channel**.

light chain The smaller of the two types of polypeptide chains in an antibody molecule; composed of a variable (V) region, which contributes to the antigen-binding site, and a constant (C) region. Compare with **heavy chain**.

lignin A substance found in the secondary cell walls of some plants that is exceptionally stiff and strong. Most abundant in woody plant parts.

limiting nutrient Any essential nutrient whose scarcity in the environment significantly reduces growth and reproduction of organisms.

limnetic zone Open water (not near shore) that receives enough light to support photosynthesis.

lineage See **monophyletic group**.

LINEs (long interspersed nuclear elements) The most abundant class of transposable elements in human genomes; can create copies of itself and insert them elsewhere in the genome. Compare with **SINEs**.

linkage In genetics, a physical association between two genes because they are on the same chromosome; the inheritance patterns resulting from this association.

linkage map See **genetic map**.

lipase Any enzyme that can break down fat molecules into fatty acids and monoglycerides.

lipid Any organic subtance that does not dissolve in water, but dissolves well in nonpolar organic solvents. Lipids include fats, oils, phospholipids, and waxes.

lipid bilayer The basic structural element of all cellular membranes consisting of a two-layer sheet of phospholipid molecules whose hydrophobic tails are oriented toward the inside and hydrophilic heads, toward the outside. Also called *phospholipid bilayer*.

littoral zone Shallow water near shore that receives enough sunlight to support photosynthesis. May be marine or freshwater; often flowering plants are present.

liver A large, complex organ of vertebrates that performs many functions including storage of glycogen, processing and conversion of food and wastes, and production of bile.

lobe-finned fish Fish with fins supported by bony elements that extend down the length of the structure.

locomotion Movement of an organism under its own power.

locus (plural: loci) A gene's physical location on a chromosome.

logistic population growth The density-dependent decrease in growth rate as population size approaches the carrying capacity. Compare with **exponential population growth**.

long interspersed nuclear elements See **LINEs**.

long-day plant A plant that blooms in response to short nights (usually in late spring or early summer in the northern hemisphere). Compare with **day-neutral** and **short-day plant**.

loop of Henle In the vertebrate kidney, a long U-shaped loop in a nephron that extends into the medulla. Functions as a countercurrent exchanger to set up an osmotic gradient that allows reabsorption of water from a subsequent portion of the nephron.

loose connective tissue A type of connective tissue consisting of fibrous proteins in a soft matrix. Often functions as padding for organs.

lophophore A specialized feeding structure found in some lophotrochozoans and used in filter feeding.

lophotrochozoans A major lineage of protostomes (Lophotrochozoa) that grow by extending the size of their skeletons rather than by molting. Many phyla have a specialized feeding structure (lophophore) and/or ciliated larvae (trochophore). Includes rotifers, flatworms, segmented worms, and mollusks. Compare with **ecdysozoans**.

loss-of-function allele See **knock-out allele**.

lumen The interior space of any hollow structure (e.g., the rough ER) or organ (e.g., the stomach).

lung Any respiratory organ used for gas exchange between blood and air.

luteal phase The second major phase of a menstrual cycle, after ovulation, when the progesterone levels are high and the body is preparing for a possible pregnancy.

luteinizing hormone (LH) A peptide hormone, produced and secreted by the anterior pituitary, that stimulates estrogen production, ovulation, and formation of the corpus luteum in females and testosterone production in males.

lymph The mixture of fluid and white blood cells that circulates through the ducts and lymph nodes of the lymphatic system in vertebrates.

lymph node One of numerous small oval structures through which lymph moves in the lymphatic system. Filter the lymph and screen it for pathogens and other antigens. Major sites of lymphocyte activation.

lymphatic system In vertebrates, a body-wide network of thin-walled ducts (or vessels) and lymph nodes, separate from the circulatory system. Collects excess fluid from body tissues and returns it to the blood; also functions as part of the immune system.

lymphocytes Two types of leukocyte—B cells and T cells—that circulate through the bloodstream and lymphatic system and that are responsible for the development of acquired immunity.

lysogenic cycle A type of viral replication in which a viral genome enters a host cell, is inserted into the host's chromosome, and is replicated whenever the host cell divides. When activated, the viral DNA enters the lytic cycle, leading to production of new virus particles. Also called *lysogeny* or *latent growth*. Compare with **lytic cycle**.

lysosome A small organelle in an animal cell containing acids and enzymes that catalyze hydrolysis reactions and can digest large molecules. Compare with **vacuole**.

lysozyme An enzyme that functions in innate immunity by digesting bacterial cell walls. Occurs in saliva, tears, mucus, and egg white.

lytic cycle A type of viral replication in which a viral genome enters a host cell, new virus particles (virions) are made using host enzymes and eventually burst out of the cell, killing it. Also called *replicative growth*. Compare with **lysogenic cycle**.

macromolecule Any very large organic molecule, usually made up of smaller molecules (monomers) joined together into a polymer. The main biological macromolecules are proteins, nucleic acids, and polysaccharides.

macronutrient Any element (e.g., carbon, oxygen, nitrogen) that is required in large quantities for normal growth, reproduction, and maintenance of a living organism. Compare with **micronutrient**.

MADS box A DNA sequence that codes for a DNA-binding motif in proteins; present in floral organ identity genes in plants. Functionally similar sequences are found in some fungal and animal genes.

major histocompatibility protein See **MHC protein**.

maladaptive A trait that lowers fitness.

malaria A human disease caused by four species of the protist *Plasmodium* and passed to humans by mosquitoes.

malignant tumor A tumor that is actively growing and disrupting local tissues and/or is spreading to other organs. Cancer consists of one or more malignant tumors. Compare with **benign tumor**.

Malpighian tubules A major excretory organ of insects, consisting of blind-ended tubes that extend from the gut into the hemocoel. Filter hemolymph to form "pre-urine" and then send it to the hindgut for further processing.

mammals One of the two lineages of amniotes (vertebrates that produce amniotic eggs) distinguished by hair (or fur) and mammary glands. Includes the monotremes (platypuses), marsupials, and eutherians (placental mammals).

mammary glands Specialized exocrine glands that produce and secrete milk for nursing offspring. A diagnostic feature of mammals.

mandibles Any mouthpart used in chewing. In vertebrates, the lower jaw. In insects, crustaceans, and myriapods, the first pair of mouthparts.

mantle One of the three main parts of the mollusk body; the thick outer tissue that protects the visceral mass and may secrete a calcium carbonate shell.

Marfan syndrome A human syndrome involving increased height, long limbs and fingers, an abnormally shaped chest, and heart disorders. Probably caused by mutation in one pleiotropic gene.

marsh A wetland that lacks trees and usually has a slow but steady rate of water flow.

marsupials A lineage of mammals (Marsupiala) that nourish their young in an abdominal pouch after a very short period of development in the uterus.

mass extinction The extinction of a large number of diverse evolutionary groups during a relatively short period of geologic time (about 1 million years). May occur due to sudden and extraordinary environmental changes. Compare with **background extinction**.

mass feeder An animal that takes chunks of food into its mouth.

mass number The total number of protons and neutrons in an atom.

mast cell A type of leukocyte that is stationary (embedded in tissue) and helps trigger the inflammatory response to infection or injury, including secretion of histamine. Particularly important in allergic responses and defense against parasites.

maternal chromosome A chromosome inherited from the mother.

mechanoreceptor A sensory cell or organ specialized for detecting distortions caused by touch or pressure. One example is hair cells in the cochlea.

mediator complex regulatory proteins that form a physical link between regulatory transcription factors that are bound to DNA and the basal transcription complex

medium A liquid or solid in which cells can grow in vitro. Also called *growth medium.*

medulla The innermost part of an organ (e.g., kidney or adrenal gland).

medulla oblongata In vertebrates, a region of the brain stem that along with the cerebellum forms the hindbrain.

medusa (plural: medusae) The free-floating stage in the life cycle of some cnidarians (e.g., jellyfish). Compare with **polyp**.

megapascal (MPa) A unit of pressure (force per unit area), equivalent to 1 million pascals (Pa).

megasporangium (plural: megasporangia) In heterosporous species of plants, a spore-producing structure that produces megaspores, which go on to develop into female gametophytes.

megaspore In seed plants, a haploid (n) spore that is produced in a megasporangium by meiosis of a diploid ($2n$) megasporocyte; develops into a female gametophyte. Compare with **microspore**.

meiosis In sexually reproducing organisms, a special two-stage type of cell division in which one diploid ($2n$) parent cell produces four haploid (n) reproductive cells (gametes); results in halving of the chromosome number. Also called *reduction division.*

meiosis I The first cell division of meiosis, in which synapsis and crossing over occur, and homologous chromosomes are separated from each other, producing daughter cells with half as many chromosomes (each composed of two sister chromatids) as the parent cell.

meiosis II The second cell division of meiosis, in which sister chromatids are separated from each other. Similar to mitosis.

meiotic map See **genetic map**.

membrane potential A difference in electric charge across a cell membrane; a form of potential energy. Also called *membrane voltage.*

memory Retention of learned information.

memory cells A type of lymphocyte responsible for maintenance of immunity for years or decades after an infection. Descended from a B cell or T cell activated during a previous infection.

meniscus (plural: menisci) The concave boundary layer formed at most air-water interfaces due to surface tension.

menstrual cycle A female reproductive cycle seen in Old World monkeys and apes (including humans) in which the uterine lining is shed (menstruation) if no pregnancy occurs. Compare with **estrous cycle**.

menstruation The periodic shedding of the uterine lining through the vagina that occurs in females of Old World monkeys and apes, including humans.

meristem (adjective: meristematic) In plants, a group of undifferentiated cells that can develop into various adult tissues throughout the life of a plant.

mesoderm The middle of the three basic cell layers in most animal embryos; gives rise to muscles, bones, blood, and some internal organs (kidney, spleen, etc.). Compare with **ectoderm** and **endoderm**.

mesoglea A gelatinous material, containing scattered ectodermal cells, that is located between the ectoderm and endoderm of cnidarians (e.g., jellyfish, corals, and anemones).

mesophyll cell A type of cell, found near the surfaces of plant leaves, that is specialized for the light-dependent reactions of photosynthesis.

Mesozoic era The interval of geologic time, from 251 million to 65.5 million years ago, during which gymnosperms were the dominant plants and dinosaurs the dominant vertebrates. Ended with extinction of the dinosaurs.

messenger RNA (mRNA) An RNA molecule that carries encoded information, transcribed from DNA, that specifies the amino acid sequence of a polypeptide.

meta-analysis A comparative analysis of the results of many smaller, previously published studies.

metabolic pathway An ordered series of chemical reactions that build up or break down a particular molecule. Often, each reaction is catalyzed by a different enzyme.

metabolic rate The total energy use by all the cells of an individual. For aerobic organisms, often measured as the amount of oxygen consumed per hour.

metabolic water The water that is produced as a by-product of cellular respiration.

metabolism All the chemical reactions occurring in a living cell or organism.

metagenomics Sequencing of all or most of the genes present in an environment directly (also called environmental sequencing).

metallothioneins Small plant proteins that bind to and prevent excess metal ions from acting as toxins.

metamorphosis Transition from one developmental stage to another, such as from the larval to the adult form of an animal.

metaphase A stage in mitosis or meiosis during which chromosomes line up in the middle of the cell.

metaphase plate The plane along which chromosomes line up during metaphase of mitosis or meiosis; not an actual structure.

metapopulation A population made up of many small, physically isolated populations.

metastasis The spread of cancerous cells from their site of origin to distant sites in the body where they may establish additional tumors.

methanogen A prokaryote that produces methane (CH_4) as a by-product of cellular respiration.

methanotroph An organism that uses methane (CH_4) as its primary electron donor and source of carbon.

methyl salicylate (MeSA) A molecule that is hypothesized to function as a signal, transported among tissues, that triggers systematic acquired resistance in plants—a response to pathogen attack.

methylation The addition of a methyl ($-CH_3$) group to a molecule.

MHC protein One of a large set of mammalian cell-surface glycoproteins involved in marking cells as self and in antigen presentation to T cells. Also called *MHC molecule.* See **Class I** and **Class II MHC protein**.

microbe Any microscopic organism, including bacteria, archaea, and various tiny eukaryotes.

microbiology The field of study concerned with microscopic organisms.

microfilament See **actin filament**.

micrograph A photograph of an image produced by a microscope.

micronutrient Any element (e.g., iron, molybdenum, magnesium) that is required in very small quantities for normal growth, reproduction, and maintenance of a living organism. Compare with **macronutrient**.

micropyle The tiny pore in a plant ovule through which the pollen tube reaches the embryo sac.

microRNA (miRNA) A small, single-stranded RNA associated with proteins in an RNA-induced silencing complex. Can bind to complementary sequences in mRNA molecules, allowing the associated proteins to degrade the bound mRNA or inhibit its translation. See **RNA interference**.

microsatellite A noncoding stretch of eukaryotic DNA consisting of a repeating sequence 1- to 5-base pair long. Also called *simple sequence repeat.*

microsporangim (plural: microsporangia) In heterosporous species of plants, a spore-producing structure that produces microspores, which go on to develop into male gametophytes.

microspore In seed plants, a haploid (n) spore that is produced in a microsporangium by meiosis of a diploid ($2n$) microsporocyte; develops into a male gametophyte. Compare with **megaspore**.

microtubule A long, tubular fiber, about 25 nm in diameter, formed by polymerization of tubulin protein dimers; one of the three types of cytoskeletal fibers. Involved in cell movement and transport of materials within the cell. Compare with **actin filament** and **intermediate filament**.

microtubule-organizing center (MTOC) General term for any structure (e.g., centrosome and basal body) that organizes microtubules in cells.

microvilli (singular: microvillus) Tiny protrusions from the surface of an epithelial cell that increase the surface area for absorption of substances.

midbrain One of the three main regions of the vertebrate brain; includes sensory integrating and relay centers. Compare with **forebrain** and **hindbrain**.

middle ear The air-filled middle portion of the mammalian ear, which contains three small bones (ossicles) that transmit and amplify sound from the tympanic membrane to the inner ear. Is connected to the throat via the eustachian tube.

migration (1) In ecology, a cyclical movement of large numbers of organisms from one geographic location or habitat to another. (2) In population genetics, movement of individuals from one population to another.

millivolt (mV) A unit of voltage equal to 1/1000 of a volt.

mimicry A phenomenon in which one species has evolved (or learns) to look or sound like another species. See **Batesian mimicry** and **Müllerian mimicry**.

mineralocorticoids A class of steroid hormones, produced and secreted by the adrenal cortex, that regulate electrolyte levels and the overall volume of body fluids. Aldosterone is the principal one in humans. Compare with **glucocorticoids**.

minisatellite A noncoding stretch of eukaryotic DNA consisting of a repeating sequence that is 6 to 500 base pairs long. Also called *variable number tandem repeat (VNTR).*

mismatch repair The process by which mismatched base pairs in DNA are fixed.

missense mutation A point mutation (change in a single base pair) that causes a change in the amino acid sequence of a protein. Also called *replacement mutation.*

mitochondrial matrix Central compartment of a mitochondrion, which is lined by the inner membrane; contains the enzymes and substrates of the citric acid cycle and mitochondrial DNA.

mitochondrion (plural: mitochondria) A eukaryotic organelle that is bounded by a double membrane and is the site of aerobic respiration.

mitosis In eukaryotic cells, the process of nuclear division that results in two daughter nuclei genetically identical to the parent nucleus. Subsequent cytokinesis (division of the cytoplasm) yields two daughter cells.

mitosis-promoting factor (MPF) A complex of a cyclin and cyclin-dependent kinase that phosphorylates a number of specific proteins needed to initiate mitosis in eukaryotic cells.

mitotic (M) phase The phase of the cell cycle during which cell division occurs. Includes mitosis and cytokinesis.

model organism An organism selected for intensive scientific study based on features that make it easy to work with (e.g., body size, life span), in the hope that findings will apply to other species.

molarity A common unit of solute concentration equal to the number of moles of a dissolved solute in 1 liter of solution.

mole The amount of a substance that contains 6.022×10^{23} of its elemental entities (e.g., atoms, ions, or molecules). This number of molecules of a compound will have a mass equal to the molecular weight of that compound expressed in grams.

molecular chaperone A protein that facilitates the three-dimensional folding of newly synthesized proteins, usually by an ATP-dependent mechanism.

molecular formula A notation that indicates only the numbers and types of atoms in a molecule, such as H_2O for the water molecule. Compare with **structural formula**.

molecular weight The sum of the mass numbers of all of the atoms in a molecule; roughly, the total number of protons and neutrons in the molecule.

molecule A combination of two or more atoms held together by covalent bonds.

mollusks Members of the phylum Mollusca. Distinguished by a body plan with three main parts: a muscular foot, a visceral mass, and a mantle. Include bivalves (clams, oysters), gastropods (snails, slugs), chitons, and cephalopods (squid, octopuses). Mollusks belong to the lophotrochozoan branch of the protostomes.

molting A method of body growth, used by ecdysozoans, that involves the shedding of an external protective cuticle or skeleton, expansion of the soft body, and growth of a new external layer.

monocot Any plant that has a single cotyledon (embryonic leaf) upon germination. Monocots form a monophyletic group. Also called a monocotyledonous plant. Compare with **dicot**.

monoecious Describing an angiosperm species that has both male and female reproductive structures on each plant. Compare with **dioecious**.

monohybrid cross A mating between two parents that are both heterozygous for a given gene.

monomer A small molecule that can covalently bind to other similar molecules to form a larger macromolecule. Compare with **polymer**.

monophyletic group An evolutionary unit that includes an ancestral population and all of its descendants but no others. Also called a *clade* or *lineage.* Compare with **paraphyletic group**.

monosaccharide A small carbohydrate, such as glucose, that has the molecular formula $(CH_2O)_n$ and cannot be hydrolyzed to form any smaller carbohydrates. Also called *simple sugar.* Compare with **disaccharide** and **polysaccharide**.

monosomy Having only one copy of a particular type of chromosome.

monotremes A lineage of mammals (Monotremata) that lay eggs and then nourish the young with milk. Includes just three living species: the platypus and two species of echidna.

morphogen A molecule that exists in a concentration gradient and provides spatial information to embryonic cells.

morphospecies concept The definition of a species as a population or group of populations that have measurably different anatomical features from other groups. Also called *morphological species concept.* Compare with **biological** and **phylogenetic species concept**.

morphology The shape and appearance of an organism's body and its component parts.

motor neuron A nerve cell that carries signals from the central nervous system (brain and spinal cord) to an effector, such as a muscle or gland. Compare with **interneuron** and **sensory neuron**.

motor protein A class of proteins whose major function is to convert the chemical energy of ATP into motion. Includes dynein, kinesin, and myosin.

MPF See **mitosis-promoting factor**.

mRNA See **messenger RNA**.

mucigel A slimy substance secreted by plant root caps that eases passage of the growing root through the soil.

mucosal-associated lymphoid tissue (MALT) Collective term for lymphocytes and other leukocytes associated with skin cells and with mucus-secreting epithelial tissues in the gut and respiratory tract. Plays important role in preventing entry of pathogens into the body.

mucous cell A type of cell found in the epithelial layer of the stomach; responsible for secreting mucus into the stomach.

mucus (adjective: mucous) A slimy mixture of glycoproteins (called mucins) and water that is secreted in many animal organs for lubrication.

Müllerian inhibitory substance A peptide hormone secreted by the embryonic testis that causes regression (withering away) of the female reproductive ducts.

Müllerian mimicry A type of mimicry in which two (or more) harmful species resemble each other. Compare with **Batesian mimicry**.

multicellularity The state of being composed of many cells that adhere to each other and do not all express the same genes with the result that some cells have specialized functions.

multienzyme complex A group of enzymes that are physically attached to each other, even though each of the enzymes catalyzes a separate—but usually related—chemical reaction.

multiple allelism In a population, the existence of more than two alleles of the same gene.

multiple fruit A fruit (e.g., pineapple) that develops from many separate flowers and thus many carpels. Compare with **aggregate** and **simple fruit**.

multiple sclerosis (MS) A human autoimmune disease caused by the immune system attacking the myelin sheaths that insulate nerve axons.

muscle fiber A single muscle cell.

muscle tissue An animal tissue consisting of bundles of long, thin contractile cells (muscle fibers).

mutagen Any physical or chemical agent that increases the rate of mutation.

mutant An individual that carries a mutation, particularly a new or rare mutation.

mutation Any change in the hereditary material of an organism (DNA in most organisms, RNA in some viruses).

mutualism (adjective: mutualistic) A symbiotic relationship between two organisms (mutualists) that benefits both. Compare with **commensalism** and **parasitism**.

mycelium (plural: mycelia) A mass of underground filaments (hyphae) that form the body of a fungus. Also found in some protists and bacteria.

mycorrhiza (plural: mycorrhizae) A mutualistic association between certain fungi and most vascular plants, sometimes visible as nodules or nets in or around plant roots.

myelin sheath Multiple layers of myelin, derived from the cell membranes of certain glial cells, that is wrapped around the axon of a neuron, providing electrical insulation.

myoD A transcription factor that is critical for differentiation of muscle cells (short for "*myo*blast *d*etermination").

myofibril Long, slender structure composed of contractile proteins organized into repeating units (sarcomeres) in vertebrate heart muscle and skeletal muscle.

myosin Any one of a class of motor proteins that use the chemical energy of ATP to move along actin filaments in muscle contraction, cytokinesis, and vesicle transport.

myriapods A lineage of arthropods with long segmented trunks, each segment bearing one or two pairs of legs. Includes millipedes and centipedes.

NAD^+/NADH Oxidized and reduced forms, respectively, of nicotinamide adenine dinucleotide. A nonprotein electron carrier that functions in many of the redox reactions of metabolism.

$NADP^+$/NADPH Oxidized and reduced forms, respectively, of nicotinamide adenine dinucleotide phosphate. A nonprotein electron carrier that is reduced during the light-dependent reactions in photosynthesis and extensively used in biosynthetic reactions.

natural experiment A situation in which groups to be compared are created by an unplanned, natural change in conditions rather than by manipulation of conditions by researchers.

natural selection The process by which individuals with certain heritable traits tend to produce more surviving offspring than do individuals without those traits, often leading to a change in the genetic makeup of the population. A major mechanism of evolution.

nauplius A distinct planktonic larval stage seen in many crustaceans.

Neanderthal A recently extinct European species of hominid, *Homo neanderthalensis*, closely related to but distinct from modern humans.

nectar The sugary fluid produced by flowers to attract and reward pollinating animals.

nectary A nectar-producing structure in a flower.

negative control Of genes, when a regulatory protein shuts down expression by binding to DNA on or near the gene

negative feedback A self-limiting, corrective response in which a deviation in some variable (e.g., body temperature, blood pH, concentration of some compound) triggers responses aimed at returning the variable to normal. Compare with **positive feedback.**

negative pressure ventilation Ventilation of the lungs that is accomplished by "pulling" air into the lungs by expansion of the rib cage. Compare with **positive pressure ventilation.**

negative-sense virus A virus whose genome contains sequences complementary to those in the mRNA required to produce viral proteins. Compare with **ambisense virus** and **positive-sense virus.**

nematodes See **roundworms.**

nephron One of the tiny tubes within the vertebrate kidney that filter blood and concentrate salts to produce urine. Also called *renal tubule.*

neritic zone Shallow marine waters beyond the intertidal zone, extending down to about 200 meters, where the continental shelf ends.

nerve A long, tough strand of nervous tissue typically containing thousands of axons wrapped in connective tissue; carries impulses between the central nervous system and some other part of the body.

nerve cord A bundle of nerves extending from the brain along the dorsal (back) side of a chordate animal, with cerebrospinal fluid inside a hollow central channel. One of the defining features of chordates.

nerve net A nervous system in which neurons are diffuse instead of being clustered into large masses or tracts.

nervous tissue An animal tissue consisting of nerve cells (neurons) and various supporting cells.

net primary productivity (NPP) In an ecosystem, the total amount of carbon fixed by photosynthesis over a given time period minus the amount oxidized during cellular respiration. Compare with **gross primary productivity.**

net reproductive rate (R_0) The growth rate of a population per generation; equivalent to the average number of female offspring that each female produces over her lifetime.

neural Relating to nerve cells (neurons) and the nervous system.

neural tube A folded tube of ectoderm that forms along the dorsal side of a young vertebrate embryo and that will give rise to the brain and spinal cord.

neuroendocrine Referring to nerve cells (neurons) that release hormones into the blood or to such hormones themselves.

neuron A cell that is specialized for the transmission of nerve impulses. Typically has dendrites, a cell body, and a long axon that forms synapses with other neurons. Also called *nerve cell.*

neurosecretory cell A nerve cell (neuron) that produces and secretes hormones into the bloodstream. Principally found in the hypothalamus. Also called *neuroendocrine cell.*

neurotoxin Any substance that specifically destroys or blocks the normal functioning of neurons.

neurotransmitter A molecule that transmits electrical signals from one neuron to another or from a neuron to a muscle or gland. Examples are acetylcholine, dopamine, serotonin, and norepinephrine.

neutral In genetics, referring to any mutation or mutant allele that has no effect on an individual's fitness.

neutrophil A type of leukocyte, capable of moving through body tissues, that engulfs and digests pathogens and other foreign particles; also secretes various compounds that attack bacteria and fungi.

niche The particular set of habitat requirements of a certain species and the role that species plays in its ecosystem.

niche differentiation The change in resource use by competing species that occurs as the result of character displacement.

nicotinamide adenine dinucleotide See NAD^+/NADH.

nicotinamide adenine dinucleotide phosphate See $NADP^+$/NADPH.

nitrogen cycle, global The movement of nitrogen among terrestrial ecosystems, the oceans, and the atmosphere.

nitrogen fixation The incorporation of atmospheric nitrogen (N_2) into forms such as ammonia (NH_3) or nitrate (NO_3^-), which can be used to make many organic compounds. Occurs in only a few lineages of bacteria and archaea.

nociceptor A sensory cell or organ specialized to detect tissue damage, usually producing the sensation of pain.

Nod factors Molecules produced by nitrogen-fixing bacteria that help them recognize and bind to roots of legumes.

node (1) In animals, any small thickening (e.g., a lymph node). (2) In plants, the part of a stem where leaves or leaf buds are attached. (3) In a phylogenetic tree, the point where two branches diverge, representing the point in time when an ancestral group split into two or more descendant groups. Also called *fork.*

node of Ranvier One of the periodic unmyelinated sections of a neuron's axon at which an action potential can be regenerated.

nodule Lumplike structure on roots of legume plants that contain symbiotic nitrogen-fixing bacteria.

noncyclic electron flow See **Z scheme.**

nondisjunction An error that can occur during meiosis or mitosis in which one daughter cell receives two copies of a particular chromosome, and the other daughter cell receives none.

nonpolar covalent bond A covalent bond in which electrons are equally shared between two atoms of the same or similar electronegativity. Compare with **polar covalent bond.**

non-sister chromatids The chromatids of a particular type of chromosome (after replication) with respect to the chromatids of its homologous chromosome. Crossing over occurs between non-sister chromatids. Compare with **sister chromatids.**

non-template strand The strand of DNA that is not transcribed during synthesis of RNA. Its sequence corresponds to that of the mRNA produced from the other strand. Also called *coding strand.*

non-vascular plants See **bryophytes.**

norepinephrine A catecholamine used as a neurotransmitter in the sympathetic nervous system. Also is produced by the adrenal medulla and functions as a hormone that triggers rapid responses relating to the fight-or-flight response.

notochord A long, gelatinous, supportive rod down the back of a chordate embryo, below the developing spinal cord. Replaced by vertebrae in most adult vertebrates. A defining feature of chordates.

nuclear envelope The double-layered membrane enclosing the nucleus of a eukaryotic cell.

nuclear lamina A lattice-like sheet of fibrous nuclear lamins, which are one type of intermediate filaments. Lines the inner membrane of the nuclear envelope, stiffening the envelope and helping organize the chromosomes.

nuclear lamins Intermediate filaments that make up the nuclear lamina layer—a lattice-like layer inside the nuclear envelope that stiffens the structure.

nuclear localization signal (NLS) A short amino acid sequence that marks a protein for delivery to the nucleus.

nuclear pore An opening in the nuclear envelope that connects the inside of the nucleus with the cytoplasm and through which molecules such as mRNA and some proteins can pass.

nuclear pore complex A large complex of dozens of proteins lining a nuclear pore, defining its shape and transporting substances through the pore.

nuclease Any enzyme that can break down RNA or DNA molecules.

nucleic acid A macromolecule composed of nucleotide monomers. Generally used by cells to store or transmit hereditary information. Includes ribonucleic acid and deoxyribonucleic acid.

nucleoid In prokaryotic cells, a dense, centrally located region that contains DNA but is not surrounded by a membrane.

nucleolus In eukaryotic cells, specialized structure in the nucleus where ribosomal RNA processing occurs and ribosomal subunits are assembled.

nucleosome A repeating, bead-like unit of eukaryotic chromatin, consisting of about 200 nucleotides of DNA wrapped twice around eight histone proteins.

nucleotide A molecule consisting of a five-carbon sugar (ribose or deoxyribose), a phosphate group, and one of several nitrogen-containing bases. DNA and RNA are polymers of nucleotides containing deoxyribose (deoxyribonucleotides) and ribose (ribonucleotides), respectively. Equivalent to a nucleoside plus one phosphate group.

nucleotide excision repair The process of removing a damaged region in one strand of DNA and correctly replacing it using the undamaged strand as a template.

nucleus (1) The center of an atom, containing protons and neutrons. (2) In eukaryotic cells, the large organelle containing the chromosomes and surrounded by a double membrane. (3) A discrete clump of neuron cell bodies in the brain, usually sharing a distinct function.

null allele See **knock-out allele**.

null hypothesis A hypothesis that specifies what the results of an experiment will be if the main hypothesis being tested is wrong. Often states that there will be no difference between experimental groups.

nutrient A substance that an organism requires for normal growth, maintenance, or reproduction.

occipital lobe In the vertebrate brain, one of the four major areas in the cerebrum.

oceanic zone The waters of the open ocean beyond the continental shelf.

oil A fat that is liquid at room temperature.

Okazaki fragment Short segment of DNA produced during replication of the 5′ to 3′ template strand. Many Okazaki fragments make up the lagging strand in newly synthesized DNA.

olfaction The perception of odors.

olfactory bulb A bulb-shaped projection of the brain just above the nose. Receives and interprets odor information from the nose.

oligodendrocyte A type of glial cell that wraps around axons of some neurons in the central nervous system, forming a myelin sheath that provides electrical insulation. Compare with **Schwann cell**.

oligopeptide A chain composed of fewer than 50 amino acids linked together by peptide bonds. Often referred to simply as *peptide*.

ommatidium (plural: ommatidia) A light-sensing column in an arthropod's compound eye.

omnivore (adjective: omnivorous) An animal whose diet regularly includes both meat and plants. Compare with **carnivore** and **herbivore**.

oncogene Any gene whose protein product stimulates cell division at all times and thus promotes cancer development. Often is a mutated form of a gene involved in regulating the cell cycle. See **proto-oncogene**.

one-gene, one-enzyme hypothesis The hypothesis that each gene is responsible for making one (and only one) protein, in most cases an enzyme that catalyzes a specific reaction. Many exceptions to this hypothesis are now known.

oocyte A cell in the ovary that can undergoes meiosis to produce an ovum.

oogenesis The production of egg cells (ova).

oogonia (singular: oogonia) The diploid cells in an ovary that can divide by mitosis to create more oogonia and primary oocytes, which can undergo meiosis.

open circulatory system A circulatory system in which the circulating fluid (hemolymph) is not confined to blood vessels. Compare with **closed circulatory system**.

open reading frame (ORF) Any DNA sequence, ranging in length from several hundred to thousands of base pairs long, that is flanked by a start codon and a stop codon. ORFs identified by computer analysis of DNA may be functional genes, especially if they have other features characteristic of genes (e.g., promoter sequence).

operator In prokaryotic DNA, a binding site for a repressor protein; located near the start of an operon.

operculum The stiff flap of tissue that covers the gills of teleost fishes.

operon A region of prokaryotic DNA that codes for a series of functionally related genes and is transcribed from a single promoter into a polycistronic mRNA.

opsin A transmembrane protein that is covalently linked to retinal, the light-detecting pigment in rod and cone cells.

optic nerve A bundle of neurons that runs from the eye to the brain.

optimal foraging The concept that animals forage in a way that maximizes the amount of usable energy they take in, given the costs of finding and ingesting their food and the risk of being eaten while they're at it.

orbital The spherical region around an atomic nucleus in which an electron is present most of the time.

ORF See **open reading frame**.

organ A group of tissues organized into a functional and structural unit.

organ system Groups of tissues and organs that work together to perform a function.

organelle Any discrete, membrane-bound structure within a cell (e.g., mitochondrion) that has a characteristic structure and functions.

organic For a compound, containing carbon and hydrogen and usually containing carbon-carbon bonds. Organic compounds are widely used by living organisms.

organism Any living entity that contains one or more cells.

organogenesis A stage of embryonic development, just after gastrulation in vertebrate embryos, during which major organs develop from the three embryonic germ layers.

origin of replication The site on a chromosome at which DNA replication begins.

osmoconformer An animal that does not actively regulate the osmolarity of its tissues but conforms to the osmolarity of the surrounding environment.

osmolarity The concentration of dissolved substances in a solution, measured in moles per liter.

osmoregulation The process by which a living organism controls the concentration of water and salts in its body.

osmoregulator An animal that actively regulates the osmolarity of its tissues.

osmosis Diffusion of water across a selectively permeable membrane from a region of high water concentration (low solute concentration) to a region of low water concentration (high solute concentration).

osmotic potential See **solute potential**.

ossicles, ear In mammals, three bones found in the middle ear that function in transferring and amplifying sound from the outer ear to the inner ear.

ouabain A plant toxin that poisons the sodium-potassium pumps of animals.

outcrossing Reproduction by fusion of the gametes of different individuals, rather than self-fertilization. Typically refers to plants.

outer ear The outermost portion of the mammalian ear, consisting of the pinna (ear flap) and the ear canal. Funnels sound to the tympanic membrane.

outgroup A taxon that is closely related to a particular monophyletic group but is not part of it.

out-of-Africa hypothesis The hypothesis that modern humans (*Homo sapiens*) evolved in Africa and spread to other continents, replacing other *Homo* species without interbreeding with them.

oval window A membrane separating the fluid-filled cochlea from the air-filled middle ear through which sound vibrations pass from the middle ear to the inner ear in mammals.

ovary The egg-producing organ of a female animal, or the seed-producing structure in the female part of a flower.

oviduct See **fallopian tube**.

oviparous Producing eggs that are laid outside the body where they develop and hatch. Compare with **ovoviviparous** and **viviparous**.

ovoviviparous Producing eggs that are retained inside the body until they are ready to hatch. Compare with **oviparous** and **viviparous**.

ovulation The release of an ovum from an ovary of a female vertebrate. In humans, an ovarian follicle releases an egg at the end of the follicular phase of the menstrual cycle.

ovule In flowering plants, the structure inside an ovary that contains the female gametophyte and eventually (if fertilized) becomes a seed.

ovum (plural: ova) See **egg**.

oxidation The loss of electrons from an atom during a redox reaction, either by donation of an electron to another atom or by the shared electrons in covalent bonds moving farther from the atomic nucleus.

oxidative phosphorylation Production of ATP molecules from the redox reactions of an electron transport chain.

oxygen-hemoglobin equilibrium curve The graphical depiction of the percentage of hemoglobin in the blood that will bind to oxygen at various partial pressures of oxygen.

oxygenic Referring to any process or reaction that produces oxygen. Photosynthesis in plants, algae, and cyanobacteria, which involves photosystem II, is oxygenic. Compare with **anoxygenic**.

oxytocin A peptide hormone, secreted by the posterior pituitary, that triggers labor and milk production in females and that stimulates pair bonding, parental care, and affiliative behavior in both sexes.

p53 A tumor-suppressor protein (molecular weight of 53 kilodaltons) that responds to DNA damage by stopping the cell cycle and/or triggering apoptosis. Encoded by the *p53* gene.

pacemaker cell A specialized cardiac muscle cell in the sinoatrial (SA) node of the vertebrate heart that has an inherent rhythm and can generate an electrical impulse that spreads to other heart cells.

paleontologists Scientists who study the fossil record and the history of life.

Paleozoic era The interval of geologic time, from 542 million to 251 million years ago, during which fungi, land plants, and animals first appeared and diversified. Began with the Cambrian explosion and ended with the extinction of many invertebrates and vertebrates.

pancreas A large gland in vertebrates that has both exocrine and endocrine functions. Secretes digestive enzymes into a duct connected to the intestine and several hormones (notably, insulin and glucagon) into the bloodstream.

pancreatic lipase An enzyme that is produced in the pancreas and acts in the small intestine to break bonds in complex fats, releasing small lipids.

pandemic The spread of an infectious disease in a short time period over a wide geographic area and affecting a very high proportion of the population. Compare with **epidemic**.

parabiosis An experimental technique for determining whether a certain physiological phenomenon is regulated by a hormone, by surgically uniting two individuals so that hormones can pass between them.

paracrine Relating to a chemical signal that is released by one cell and affects neighboring cells.

paraphyletic group An evolutionary unit that includes an ancestral population and *some* but not all of its descendants. Paraphyletic groups are not meaningful units in evolution. Compare with **monophyletic group**.

parapodia (singular: parapodium) Appendages found in some annelids from which bristle-like structures (chaetae) extend.

parasite An organism that lives on or in a host species and that damages its host.

parasitism (adjective: parasitic) A symbiotic relationship between two organisms that is beneficial to one organism (the parasite) but detrimental to the other (the host). Compare with **commensalism** and **mutualism**.

parasitoid An organism that has a parasitic larval stage and a free-living adult stage. Most parasitoids are insects that lay eggs in the bodies of other insects.

parasympathetic nervous system The part of the autonomic nervous system that stimulates functions for conserving or restoring energy, such as reduced heart rate and increased digestion. Compare with **sympathetic nervous system**.

parathyroid glands Four small glands, located near or embedded in the thyroid gland of vertebrates, that secrete parathyroid hormone (PTH), a peptide hormone that increases blood calcium.

parenchyma cell In plants, a general type of cell with a relatively thin primary cell wall. These cells, found in leaves, the centers of stems and roots, and fruits, are involved in photosynthesis, starch storage, and new growth. Compare with **collenchyma cell** and **sclerenchyma cell**.

parental care Any action by which an animal expends energy or assumes risks to benefit its offspring (e.g., nest-building, feeding of young, defense).

parental generation The adult organisms used in the first experimental cross in a formal breeding experiment.

parietal cell A cell in the stomach lining that secretes hydrochloric acid.

parietal lobe In the vertebrate brain, one of the four major areas in the cerebrum.

parsimony The logical principle that the most likely explanation of a phenomenon is the most economical or simplest. When applied to comparison of alternative phylogenetic trees, it suggests that the one requiring the fewest evolutionary changes is most likely to be correct.

parthenogenesis Development of offspring from unfertilized eggs; a type of asexual reproduction.

partial pressure The pressure of one particular gas in a mixture; the contribution of that gas to the overall pressure.

particulate inheritance The observation that genes from two parents do not blend together to form a new physical entity in offspring, but instead remain separate or particle-like.

pascal (Pa) A unit of pressure (force per unit area).

passive transport Diffusion of a substance across a plasma membrane or organelle membrane. When this occurs with the assistance of membrane proteins, it is called facilitated diffusion.

patch clamping A technique for studying the electrical currents that flow through individual ion channels by sucking a tiny patch of membrane to the hollow tip of a microelectrode.

paternal chromosome A chromosome inherited from the father.

pathogen (adjective: pathogenic) Any entity capable of causing disease, such as a microbe, virus, or prion.

pattern formation The series of events that determines the spatial organization of an embryo, including alignment of the major body axes and orientation of the limbs.

pattern-recognition receptor One of a class of membrane proteins on leukocytes that bind to molecules on the surface of many bacteria. Part of the innate immune response.

PCR See **polymerase chain reaction**.

peat Semidecayed organic matter that accumulates in moist, low-oxygen environments such as bogs.

pectin A gelatinous polysaccharide found in the primary cell wall of plant cells. Attracts and holds water, forming a gel that helps keep the cell wall moist.

pedigree A family tree of parents and offspring, showing inheritance of particular traits of interest.

penis The copulatory organ of male mammals, used to insert sperm into a female.

pentose A monosaccharide (simple sugar) containing five carbon atoms.

PEP carboxylase An enzyme that catalyzes addition of CO_2 to phosphoenol pyruvate, a three-carbon compound, forming a four-carbon organic acid. Found in mesophyll cells of plants that perform C_4 photosynthesis.

pepsin A protein-digesting enzyme present in the stomach.

pepsinogen The precursor of the digestive enzyme pepsin. Is secreted from cells in the stomach lining and converted to pepsin by the acidic environment of the stomach lumen.

peptide See **oligopeptide**.

peptide bond The covalent bond (C—N) formed by a condensation reaction between two amino acids; links the residues in peptides and proteins.

peptidoglycan A complex structural polysaccharide found in bacterial cell walls.

perennial Describing a plant whose life cycle normally lasts for more than one year. Compare with **annual**.

perfect flower A flower that contains both male parts (stamens) and female parts (carpels). Compare with **imperfect flower**.

perforation In plants, a small hole in the primary and secondary cell walls of vessel elements that allow passage of water.

pericarp The part of a fruit, formed from the ovary wall, that surrounds the seeds and protects them. Corresponds to the flesh of most edible fruits and the hard shells of most nuts.

pericycle In plant roots, a layer of cells that give rise to lateral roots.

peripheral membrane protein Any membrane protein that does not span the entire lipid bilayer and associates with only one side of the bilayer. Compare with **integral membrane protein**.

peripheral nervous system (PNS) All the components of the nervous system that are outside the central nervous system (the brain and spinal cord). Includes the somatic nervous system and the autonomic nervous system.

peristalsis Rhythmic waves of muscular contraction that push food along the digestive tract.

permafrost A permanently frozen layer of icy soil found in most tundra and some taiga.

permeability The tendency of a structure, such as a membrane, to allow a given substance to diffuse across it.

peroxisome An organelle found in most eukaryotic cells that contains enzymes for oxidizing fatty acids and other compounds including many toxins, rendering them harmless. See **glyoxisome**.

petal One of the leaflike organs arranged around the reproductive organs of a flower. Often colored and scented to attract pollinators.

petiole The stalk of a leaf.

pH A measure of the concentration of protons in a solution and thus of acidity or alkalinity. Defined as the negative of the base-10 logarithm of the proton concentration: $pH = -\log[H^+]$.

phagocytosis Uptake by a cell of small particles or cells by pinching off the plasma membrane to form small membrane-bound vesicles; one type of endocytosis.

pharyngeal gill slits A set of parallel openings from the throat through the neck to the outside. A diagnostic trait of chordates.

pharyngeal jaw A secondary jaw in the back of the mouth, found in some fishes. Derived from modified gill arches.

phelloderm A component of bark; specifically, a tissue layer produced to the inside of the cork cambium in plants with secondary growth.

phelloderm In the stems of woody plants, a thin layer of cells located between the outer cork cells and inner cork cambium.

phenetic approach A method for constructing a phylogenetic tree by computing a statistic that summarizes the overall similarity among populations, based on the available data. Compare with **cladistic approach**.

phenology The timing of events during the year, in environments where seasonal changes occur.

phenotype The detectable physical and physiological traits of an individual, which are determined its genetic makeup. Also the specific trait associated with a particular allele. Compare with **genotype**.

phenotypic plasticity Within-species variation in phenotype that is due to differences in environmental conditions. Occurs more commonly in plants than animals.

phenylketonuria (PKU) A disease caused by the inability to process the amino acid phenylalanine.

pheophytin In photosystem II, a molecule that accepts excited electrons from a reaction center chlorophyll and passes them to an electron transport chain.

pheromone A chemical signal, released by one individual into the external environment, that can trigger responses in a different individual.

phloem A plant vascular tissue that conducts sugars; contains sieve-tube members and companion cells. Primary phloem develops from the procambium of apical meristems; secondary phloem, from the vascular cambium of lateral meristems. Compare with **xylem**.

phosphodiester linkage Chemical linkage between adjacent **nucleotide residues** in DNA and RNA. Forms when the phosphate group of one nucleotide condenses with the hydroxyl group on the sugar of another nucleotide. Also known as *phosphodiester bond*.

phosphofructokinase The enzyme that catalyzes synthesis of fructose-1,6-bisphosphate from fructose-6-phosphate, a key reaction (step 3) in glycolysis.

phospholipid A class of lipid having a hydrophilic head (a phosphate group) and a hydrophobic tail (one or more fatty acids). Major components of the plasma membrane and organelle membranes.

phosphorylase An enzyme that breaks down glycogen by catalyzing hydrolysis of the α-glycosidic linkages between the glucose residues.

phosphorylation (verb: phosphorylate) The addition of a phosphate group to a molecule.

phosphorylation cascade A series of enzyme-catalyzed phosphorylation reactions commonly used in signal transduction pathways to amplify and convey a signal inward from the plasma membrane.

photic zone In an aquatic habitat, water that is shallow enough to receive some sunlight (whether or not it is enough to support photosynthesis). Compare with **aphotic zone**.

photon A discrete packet of light energy; a particle of light.

photoperiodism Any response by an organism to the relative lengths of day and night (i.e., photoperiod).

photophosporylation Production of ATP molecules using the energy released as light-excited electrons flow through an electron transport chain during photosynthesis. Involves generation of a proton-motive force during electron transport and its use to drive ATP synthesis.

photoreceptor A molecule, a cell, or an organ that is specialized to detect light.

photorespiration A series of light-driven chemical reactions that consumes oxygen and releases carbon dioxide, basically reversing photosynthesis. Usually occurs when there are high O_2 and low CO_2 concentrations inside plant cells, often in bright, hot, dry environments when stomata must be kept closed.

photoreversibility A change in conformation that occurs in certain plant pigments when they are exposed to their preferred wavelengths of light and that triggers responses by the plant.

photosynthesis The complex biological process that converts the energy of light into chemical energy stored in glucose and other organic molecules. Occurs in plants, algae, and some bacteria.

photosystem One of two types of units, consisting of a central reaction center surrounded by antenna complexes, that is responsible for the light-dependent reactions of photosynthesis.

photosystem I A photosystem that contains a pair of P700 chlorophyll molecules and uses absorbed light energy to produce NADPH.

photosystem II A photosystem that contains a pair of P680 chlorophyll molecules and uses absorbed light energy to split water into protons and oxygen and to produce ATP.

phototroph An organism that produces ATP through photosynthesis.

phototropins A class of plant photoreceptors that detect blue light and initiate phototropic responses.

phototropism Growth or movement of an organism in a particular direction in response to light.

phylogenetic species concept The definition of a species as the smallest monophyletic group in a phylogenetic tree. Compare with **biological** and **morphospecies concept**.

phylogenetic tree A diagram that depicts the evolutionary history of a group of species and the relationships among them.

phylogeny The evolutionary history of a group of organisms.

phylum (plural: phyla) In Linnaeus' system, a taxonomic category above the class level and below the kingdom level. In plants, sometimes called a *division*.

physical map A map of a chromosome that shows the number of base pairs between various genetic markers. Compare with **genetic map**.

physiology The study of how an organism's body functions.

phytoalexin Any small compound produced by a plant to combat an infection (usually a fungal infection).

phytochrome A specialized plant photoreceptor that exists in two shapes depending on the ratio of red to far-red light and is involved in the timing certain physiological processes, such as flowering, stem elongation, and germination.

pigment Any molecule that absorbs certain wavelengths of visible light and reflects or transmits other wavelengths.

piloting A type of navigation in which animals use familiar landmarks to find their way.

pinocytosis Uptake by a cell of extracellular fluid by pinching off the plasma membrane to form small membrane-bound vesicles; one type of endocytosis.

pioneering species Those species that appear first in recently disturbed areas.

pit In plants, a small hole in the secondary cell walls of tracheids that allow passage of water.

pitch The sensation produced by a particular frequency of sound. Low frequencies are perceived as low pitches; high frequencies, as high pitches.

pith In the shoot systems of plants, ground tissue located to the inside of the vascular bundles.

pituitary gland A small gland directly under the brain that is physically and functionally connected to the hypothalamus. Produces and secretes an array of hormones that affect many other glands and organs.

placenta A structure that forms in the pregnant uterus from maternal and fetal tissues. Exchanges nutrients and wastes between mother and fetus, anchors the fetus to the uterine wall, and produces some hormones. Occurs in most mammals and in a few other vertebrates.

placental mammals See **eutherians**.

plankton Any small organism that drifts near the surface of oceans or lakes and swims little if at all.

Plantae The monophyletic group that includes red, green, and glaucophyte algae and land plants.

plantlet A small plant, particularly one that forms on a parent plant via asexual reproduction and drops, becoming an independent individual.

plasma The non-cellular portion of blood.

plasma cell An effector B cell, which produces large quantities of antibodies. Is descended from an activated B cell that has interacted with antigen.

plasma membrane A membrane that surrounds a cell, separating it from the external environment and selectively regulating passage of molecules and ions into and out of the cell. Also called *cell membrane*.

plasmid A small, usually circular, supercoiled DNA molecule independent of the cell's main chromosome(s) in prokaryotes and some eukaryotes.

plasmodesmata (singular: plasmodesma) Physical connection between two plant cells, consisting of gaps in the cell walls through which the two cells' plasma membranes, cytoplasm, and smooth ER can connect directly. Functionally similar to gap junctions in animal cells.

plasmogamy Fusion of the cytoplasm of two individuals. Occurs in many fungi.

plastocyanin A small protein that shuttles electrons from photosystem II to photosystem I during photosynthesis.

plastoquinone (PQ) A nonprotein electron carrier in the chloroplast electron transport chain. Receives excited electrons from pheophytin and passes them to more electronegative molecules in the chain. Also carries protons to the lumen side of the thylakoid membrane, generating a proton-motive force.

platelet A small membrane-bound cell fragment in vertebrate blood that functions in blood clotting. Derived from large cells in the bone marrow.

pleiotropy (adjective: pleiotropic) The ability of a single gene to affect more than one phenotypic trait.

ploidy The number of complete chromosome sets present. *Haploid* refers to a ploidy of 1; *diploid*, a ploidy of 2; *triploid*, a ploidy of 3; and *tetraploid*, a ploidy of 4.

podium (plural: podia) See **tube foot**.

point mutation A mutation that results in a change in a single nucleotide pair in a DNA molecule.

polar (1) Asymmetrical or unidirectional. (2) Carrying a partial positive charge on one side of a molecule and a partial negative charge on the other. Polar molecules are generally hydrophilic.

polar bodies The tiny, nonfunctional cells produced during meiosis of a primary oocyte, due to most of the cytoplasm going to the ovum.

polar covalent bond A covalent bond in which electrons are shared unequally between atoms differing in electronegativity, resulting in the more electronegative atom having a partial negative charge and the other atom, a partial positive charge. Compare with **nonpolar covalent bond**.

polar microtubules Microtubules that form during mitosis and meiosis, and which extend from a spindle apparatus and overlap with each other in the middle of the cell.

polar nuclei In flowering plants, the nuclei in the female gametophyte that fuse with one sperm nucleus to produce the endosperm. Most species have two.

pollen grain In seed plants, a male gametophyte enclosed within a protective coat.

pollen tube In flowering plants, a structure that grows out of a pollen grain after it reaches the stigma, extends down the style, and through which two sperm cells are delivered to the ovule.

pollination The process by which pollen reaches the carpel of a flower (in flowering plants) or reaches the ovule directly (in conifers and their relatives).

poly(A) tail In eukaryotes, a sequence of 100–250 adenine nucleotides added to the 3′ end of newly transcribed messenger RNA molecules.

polygenic inheritance The inheritance patterns that result when many genes influence one trait.

polymer Any long molecule composed of small repeating units (monomers) bonded together. The main biological polymers are proteins, nucleic acids, and polysaccharides.

polymerase chain reaction (PCR) A laboratory technique for rapidly generating millions of identical copies of a specific stretch of DNA by incubating the original DNA sequence of interest with primers, nucleotides, and DNA polymerase.

polymerization (verb: polymerize) The process by which many identical or similar small molecules (monomers) are covalently bonded to form a large molecule (polymer).

polymorphism (adjective: polymorphic) (1) The occurrence of more than one allele at a certain genetic locus in a population. (2) The occurrence of more than two distinct phenotypes of a trait in a population.

polyp The immotile (sessile) stage in the life cycle of some cnidarians (e.g., jellyfish). Compare with **medusa**.

polypeptide A chain of 50 or more amino acids linked together by peptide bonds. Compare with **oligopeptide** and **protein**.

polyploidy (adjective: polyploid) The state of having more than two full sets of chromosomes.

polyribosome A structure consisting of one messenger RNA molecule along with many attached ribosomes and their growing peptide strands.

polysaccharide A linear or branched polymer consisting of many monosaccharides joined by glycosidic linkages. Carbohydrate polymers with relatively few residues often are called *oligosaccharides*.

polyspermy Fertilization of an egg by multiple sperm.

population A group of individuals of the same species living in the same geographic area at the same time.

population density The number of individuals of a population per unit area.

population dynamics Changes in the size and other characteristics of populations through time.

population ecology The study of how and why the number of individuals in a population changes over time.

population thinking The ability to analyze trait frequencies, event probabilities, and other attributes of populations of molecules, cells, or organisms.

pore In land plants, an opening in the epithelium that allows gas exchange.

positive control Of genes, when a regulatory protein triggers expression by binding to DNA on or near the gene.

positive feedback A physiological mechanism in which a change in some variable stimulates a response that increases the change. Relatively rare in organisms but is important in generation of the action potential. Compare with **negative feedback**.

positive pressure ventilation Ventilation of the lungs that is accomplished by "pushing" air into the lungs by positive pressure in the mouth. Compare with **negative pressure ventilation.**

positive-sense virus A virus whose genome contains the same sequences as the mRNA required to produce viral proteins. Compare with **ambisense virus** and **negative-sense virus.**

posterior Toward an animal's tail and away from its head. The opposite of anterior.

posterior pituitary The part of the pituitary gland that contains the ends of hypothalamic neurosecretory cells and from which oxytocin and antidiuretic hormone are secreted. Compare with **anterior pituitary**.

postsynaptic neuron A neuron that receives signals, usually via neurotransmitters, from another neuron at a synapse. Muscle and gland cells also may receive signals from presynaptic neurons.

post-translational control Regulation of gene expression by modification of proteins (e.g., addition of a phosphate group or sugar residues) after translation.

postzygotic isolation Reproductive isolation resulting from mechanisms that operate after mating of individuals of two different species occurs. The most common mechanisms are the death of hybrid embryos or reduced fitness of hybrids.

potential energy Energy stored in matter as a result of its position or molecular arrangement. Compare with **kinetic energy**.

prebiotic soup A hypothetical solution of sugars, amino acids, nitrogenous bases, and other building blocks of larger molecules that may have formed in shallow waters or deep-ocean vents of ancient Earth and given rise to larger biological molecules.

Precambrian The interval between the formation of the Earth, about 4.6 billion years ago, and the appearance of most animal groups about 542 million years ago. Unicellular organisms were dominant for most of this era, and oxygen was virtually absent for the first 2 billion years.

predation The killing and eating of one organism (the prey) by another (the predator).

predator Any organism that kills other organisms for food.

prediction A measurable or observable result of an experiment based on a particular hypothesis. A correct prediction provides support for the hypothesis being tested.

pressure potential (ψ_P) A component of the potential energy of water caused by physical pressures on a solution. In plant cells, it equals the wall pressure plus turgor pressure. Compare with **solute potential (ψ_S)**.

pressure-flow hypothesis The hypothesis that sugar movement through phloem tissue is due to differences in the turgor pressure of phloem sap.

presynaptic neuron A neuron that transmits signals, usually by releasing neurotransmitters, to another neuron or to an effector cell at a synapse.

prezygotic isolation Reproductive isolation resulting from any one of several mechanisms that prevent individuals of two different species from mating.

primary cell wall The outermost layer of a plant cell wall, made of cellulose fibers and gelatinous polysaccharides, that defines the shape of the cell and withstands the turgor pressure of the plasma membrane.

primary consumer An herbivore; an organism that eats plants, algae, or other primary producers. Compare with **secondary consumer**.

primary decomposer A decomposer (detritivore) that consumes detritus from plants.

primary growth In plants, an increase in the length of stems and roots due to the activity of apical meristems. Compare with **secondary growth.**

primary immune response An acquired immune response to a pathogen that the immune system has not encountered before. Compare with **secondary immune response.**

primary oocyte The large diploid cell in an ovarian follicle that can initiate meiosis to produce a haploid ovum.

primary producer Any organism that creates its own food by photosynthesis or from reduced inorganic compounds and that is a food source for other species in its ecosystem. Also called *autotroph.*

primary RNA transcript In eukaryotes, a newly transcribed messenger RNA molecule that has not yet been processed (i.e., it has not received a 5′ cap or poly(A) tail, and still contains introns). Also called *pre-mRNA.*

primary spermatocyte A diploid cell in the testis that can initiate meiosis I to produce two secondary spermatocytes.

primary structure The sequence of amino acids in a peptide or protein; also the sequence of nucleotides in a nucleic acid. Compare with **secondary, tertiary,** and **quaternary structure.**

primary succession The gradual colonization of a habitat of bare rock or gravel, usually after an environmental disturbance that removes all soil and previous organisms. Compare with **secondary succession.**

primase An enzyme that synthesizes a short stretch of RNA to use as a primer during DNA replication.

primates The lineage of mammals that includes prosimians (lemurs, lorises, etc.), monkeys, and great apes (including humans).

primer A short, single-stranded RNA molecule that base pairs with the 5′ end of a DNA template strand and is elongated by DNA polymerase during DNA replication.

prion An infectious form of a protein that is thought to cause disease by inducing the normal form to assume an abnormal three-dimensional structure. Cause of spongiform encephalopathies, such as mad cow disease.

probe A radioactively or chemically labeled single-stranded fragment of a known DNA or RNA sequence that can bind to and thus detect its complementary sequence in a sample containing many different sequences.

proboscis A long, narrow feeding appendage through which food can be obtained.

procambium A group of cells in the center of a young plant embryo that gives rise to the vascular tissue.

product Any of the final materials formed in a chemical reaction.

productivity The total amount of carbon fixed by photosynthesis per unit area per year.

progesterone A steroid hormone produced in the ovaries and secreted by the corpus luteum after ovulation; causes the uterine lining to thicken.

programmed cell death See **apoptosis.**

prokaryote A member of the domain Bacteria or Archaea; a unicellular organism lacking a nucleus and containing relatively few organelles or cytoskeletal components. Compare with **eukaryote.**

prolactin A peptide hormone, produced and secreted by the anterior pituitary, that promotes milk production in female mammals and has a variety of effects on parental behavior and seasonal reproduction in other vertebrates.

prometaphase A stage in mitosis or meiosis during which the nuclear envelope breaks down and kinetochore microtubles attach to chromatids.

promoter A short nucleotide sequence in DNA that binds RNA polymerase, enabling transcription to begin. In prokaryotic DNA, a single promoter often is associated with several contiguous genes. In eukaryotic DNA, each gene generally has its own promoter.

promoter-proximal elements In eukaryotes, regulatory sequences in DNA that are close to a promoter and that can bind regulatory transcription factors.

proofreading The process by which a DNA polymerase recognizes and removes a wrong base added during DNA replication and then continues synthesis.

prophase The first stage in mitosis or meiosis during which chromosomes become visible and the spindle apparatus forms. Synapsis and crossing over occur during prophase of meiosis I.

prosimians One of the two major lineages of primates, including lemurs, tarsiers, pottos, and lorises. Compare with **anthropoids.**

prostate gland A gland in male mammals that surrounds the base of the urethra and secretes a fluid that is a component of semen.

protease An enzyme that can degrade proteins by cleaving the peptide bonds between amino acid residues.

proteasome A multi-molecular machine that destroys proteins that have been bound to ubiquitin.

protein A macromolecule consisting of one or more polypeptide chains composed of 50 or more amino acids linked together. Each protein has a unique sequence of amino acids and, in its native state, a characteristic three-dimensional shape.

protein kinase An enzyme that catalyzes the addition of a phosphate group to another protein, typically activating or inactivating the substrate protein.

proteinase inhibitors Defense compounds, produced by plants, that induce illness in herbivores by inhibiting digestive enzymes.

proteome The complete set of proteins produced by a particular cell type.

proteomics The systematic study of the interactions, localization, functions, regulation, and other features of the full protein set (proteome) in a particular cell type.

protist Any eukaryote that is not a green plant, animal, or fungus. Protists are a diverse paraphyletic group. Most are unicellular, but some are multicellular or form aggregations called colonies.

protoderm The exterior layer of a young plant embryo that gives rise to the epidermis.

proton pump A membrane protein that can hydrolyze ATP to power active transport of protons (H^+ ions) across a plasma membrane against an electrochemical gradient. Also called *H^+-ATPase*.

proton-motive force The combined effect of a proton gradient and an electric potential gradient across a membrane, which can drive protons across the membrane. Used by mitochondria and chloroplasts to power ATP synthesis via the mechanism of chemiosmosis.

proto-oncogene Any gene that normally encourages cell division in a regulated manner, typically by triggering specific phases in the cell cycle. Mutation may convert it into an oncogene.

protostomes A major lineage of animals that share a pattern of embryological development, including formation of the mouth earlier than the anus, and formation of the coelom by splitting of a block of mesoderm. Includes arthropods, mollusks, and annelids. Compare with **deuterostomes**.

proximal tubule In the vertebrate kidney, the convoluted section of a nephron into which filtrate moves from Bowman's capsule. Involved in the largely unregulated reabsorption of electrolytes, nutrients, and water. Compare with **distal tubule**.

proximate causation In biology, the immediate, mechanistic cause of a phenomenon (how it happens), as opposed to why it evolved. Also called *proximate explanation*. Compare with **ultimate causation**.

pseudogene A DNA sequence that closely resembles a functional gene but is not transcribed. Thought to have arisen by duplication of the functional gene followed by inactivation due to a mutation.

pseudopodium (plural: pseudopodia) A temporary bulge-like extension of certain cells used in cell crawling and ingestion of food.

puberty The various physical and emotional changes that an immature animal undergoes leading to reproductive maturity. Also the period when such changes occur.

pulmonary artery A short, thick-walled artery that carries oxygen-poor blood from the heart to the lungs.

pulmonary circulation The part of the circulatory system that sends oxygen-poor blood to the lungs. Is separate from the rest of the circulatory system (the systemic circulation) in mammals and birds.

pulmonary vein A short, thin-walled vein that carries oxygen-rich blood from the lungs to the heart.

pulse-chase experiment A type of experiment in which a population of cells or molecules at a particular moment in time is marked by means of a labeled molecule and then their fate is followed over time.

pump Any membrane protein that can hydrolyze ATP to power active transport of a specific ion or small molecule across a plasma membrane against its electrochemical gradient. See **proton pump**.

Punnett square A diagram that depicts the genotypes and phenotypes that should appear in offspring of a certain cross.

pupa (plural: pupae) A metamorphosing insect that is enclosed in a protective case.

pupil The hole in the center of the iris through which light enters a vertebrate or cephalopod eye.

pure line In animal or plant breeding, a strain of individuals that produce offspring identical to themselves when self-pollinated or crossed to another member of the same population. Pure lines are homozygous for most, if not all, genetic loci.

purifying selection Selection that lowers the frequency or even eliminates deleterious alleles.

purines A class of small, nitrogen-containing, double-ringed bases (guanine, adenine) found in nucleotides. Compare with **pyrimidines**.

pyrimidines A class of small, nitrogen-containing, single-ringed bases (cytosine, uracil, thymine) found in nucleotides. Compare with **purines**.

pyruvate dehydrogenase A large enzyme complex, located in the inner mitochondrial membrane, that is responsible for conversion of pyruvate to acetyl CoA during cellular respiration.

quantitative trait A heritable feature that exhibits phenotypic variation along a smooth, continuous scale of measurement (e.g., human height), rather than the distinct forms characteristic of discrete traits.

quaternary structure The overall three-dimensional shape of a protein containing two or more polypeptide chains (subunits); determined by the number, relative positions, and interactions of the subunits. Compare with **primary, secondary,** and **tertiary structure**.

quorum sensing Cell-cell signaling in bacteria, in which cells of the same species communicate via chemical signals. It is often observed that cell activity changes dramatically when the population reaches a threshold size, or quorum.

radial symmetry An animal body pattern in which there are least two planes of symmetry. Typically, the body is in the form of a cylinder or disk, with body parts radiating from a central hub. Compare with **bilateral symmetry**.

radiation Transfer of heat between two bodies that are not in direct physical contact. More generally, the emission of electromagnetic energy of any wavelength.

radicle The root of a plant embryo.

radula A rasping feeding appendage in gastropods (snails, slugs).

rain shadow The dry region on the side of a mountain range away from the prevailing wind.

range The geographic distribution of a species.

Ras protein A type of G protein that is activated by binding of signaling molecules to receptor tyrosine kinases and then initiates a phosphorylation cascade, culminating in a cell response.

rays In plant shoot systems with secondary growth, a lateral array of parenchyma cells produced by vascular cambium.

Rb protein A tumor-suppressor protein that helps regulate progression of a cell from the G_1 phase to the S phase of the cell cycle. Defects in Rb protein are found in many types of cancer.

reactant Any of the starting materials in a chemical reaction.

reaction center Centrally located component of a photosystem containing proteins and a pair of specialized chlorophyll molecules. Is surrounded by antenna complexes and receives excited electrons from them.

reactive oxygen intermediates (ROIs) Highly reactive oxygen-containing compounds that are used in plant and animal cells to kill infected cells and for other purposes.

reading frame The division of a sequence of DNA or RNA into a particular series of three-nucleotide codons. There are three possible reading frames for any sequence.

realized niche The ecological niche that a species occupies in the presence of competitors. Compare with **fundamental niche**.

receptor tyrosine kinase (RTK) Any of a class of cell-surface signal receptors that undergo phosphorylation after binding a signaling molecule. The activated, phosphorylated receptor then triggers a signal-transduction pathway inside the cell.

receptor-mediated endocytosis Uptake by a cell of certain extracellular macromolecules, bound to specific receptors in the plasma membrane, by pinching off the membrane to form small membrane-bound vesicles.

recessive Referring to an allele whose phenotypic effect is observed only in homozygous individuals. Compare with **dominant**.

reciprocal altruism Altruistic behavior that is exchanged between a pair of individuals at different points in time (i.e., sometimes individual A helps individual B, and sometimes B helps A).

reciprocal cross A breeding experiment in which the mother's and father's phenotypes are the reverse of that examined in a previous breeding experiment.

recombinant DNA technology A variety of techniques for isolating specific DNA fragments and introducing them into different regions of DNA and/or a different host organism.

recombinant Possessing a new combination of alleles. May refer to a single chromosome or DNA molecule, or to an entire organism.

recombination See **genetic recombination**.

rectal gland A salt-excreting gland in the digestive system of sharks, skates, and rays.

rectum The last portion of the digestive tract where feces are held until they are expelled.

red blood cell A hemoglobin-containing cell that circulates in the blood and delivers oxygen from the lungs to the tissues.

redox reaction Any chemical reaction that involves the transfer of one or more electrons from one reactant to another. Also called *reduction-oxidation reaction*.

reduction An atom's gain of electrons during a redox reaction, either by acceptance of an electron from another atom or by the electrons in covalent bonds moving closer to the atomic nucleus.

reduction-oxidation reaction See **redox reaction**.

reflex An involuntary response to environmental stimulation. May involve the brain (e.g., conditioned reflex) or not (e.g., spinal reflex).

refractory No longer responding to stimuli that previously elicited a response. For example, the tendency of voltage-gated sodium channels to remain closed immediately after an action potential.

regulatory genes DNA sequences that code for regulatory proteins—products that alter gene expression.

regulatory sequence, DNA Any segment of DNA that is involved in controlling transcription of a specific gene by binding certain proteins.

regulatory site A site on an enzyme to which a regulatory molecule can bind and affect the enzyme's activity; separate from the active site where catalysis occurs.

regulatory transcription factor General term for proteins that bind to DNA regulatory sequences (eukaryotic enhancers, silencers, and promoter-proximal elements), but not to the promoter itself, leading to an increase or decrease in transcription of specific genes. Compare with **basal transcription factor**.

reinforcement In evolutionary biology, the natural selection for traits that prevent interbreeding between recently diverged species.

release factors Proteins that can trigger termination of RNA translation when a ribosome reaches a stop codon.

renal corpuscle In the vertebrate kidney, the ball-like structure at the beginning of a nephron, consisting of a glomerulus and the surrounding Bowman's capsule. Acts as a filtration device.

replacement mutation See **missense mutation**.

replacement rate The number of offspring each female must produce over her entire life to "replace" herself and her mate, resulting in zero population growth. The actual number is slightly more than 2 because some offspring die before reproducing.

replica plating A method of identifying bacterial colonies that have certain mutations by transferring cells from each colony on a master plate to a second (replica) plate and observing their growth when exposed to different conditions.

replication fork The Y-shaped site at which a double-stranded molecule of DNA is separated into two single strands for replication.

replisome The multi-molecular machine that copies DNA; includes DNA polymerase, helicase, primase, and other enzymes.

repolarization Return to a normal membrane potential after it has changed; a normal phase in an action potential.

repressor Any regulatory protein that inhibits transcription.

reptiles One of the two lineages of amniotes (vertebrates that produce amniotic eggs) distinguished by adaptations for reproduction on land. Includes turtles, snakes and lizards, crocodiles and alligators, and birds. Except for birds, all are ectotherms.

resilience, community A measure of how quickly a community recovers following a disturbance.

resistance, community A measure of how much a community is affected by a disturbance.

resistance (*R*) genes Genes in plants encoding proteins involved in sensing the presence of pathogens and mounting a defensive response. Compare with **avirulence (*avr*) genes**.

respiratory system The collection of cells, tissues, and organs responsible for gas exchange between an animal and its environment.

resting potential The membrane potential of a cell in its resting, or normal, state.

restriction endonucleases Bacterial enzymes that cut DNA at a specific base-pair sequence (restriction site). Also called *restriction enzymes.*

retina A thin layer of light-sensitive cells (rods and cones) and neurons at the back of a camera-type eye, such as that of cephalopods and vertebrates.

retinal A light-absorbing pigment, derived from vitamin A, that is linked to the protein opsin in rods and cones of the vertebrate eye.

retrovirus A virus with an RNA genome that reproduces by transcribing its RNA into a DNA sequence and then inserting that DNA into the host's genome for replication.

reverse transcriptase A enzyme of retroviruses (RNA viruses) that can synthesize double-stranded DNA from a single-stranded RNA template.

rhizobia (singular: rhizobium) Members of the bacterial genus *Rhizobia;* nitrogen-fixing bacteria that live in root nodules of members of the pea family (legumes).

rhizoid The hairlike structure that anchors a bryophyte (non-vascular plant) to the substrate.

rhizome A modified stem that runs horizontally underground and produces new plants at the nodes (a form of asexual reproduction). Compare with **stolon**.

rhodopsin A transmembrane complex that is instrumental in detection of light by rods and cones of the vertebrate eye. Is composed of the transmembrane protein opsin covalently linked to retinal, a light-absorbing pigment.

ribonucleic acid (RNA) A nucleic acid composed of ribonucleotides that usually is single stranded and functions as structural components of ribosomes (rRNA), transporters of amino acids (tRNA), and translators of the message of the DNA code (mRNA).

ribonucleotide See **nucleotide**.

ribosomal RNA (rRNA) A RNA molecule that forms part of the structure of a ribosome.

ribosome A large complex structure that synthesizes proteins by using the genetic information encoded in messenger RNA strands. Consists of two subunits, each composed of ribosomal RNA and proteins.

ribosome binding site In a bacterial mRNA molecule, the sequence just upstream of the start codon to which a ribosome binds to initiate translation. Also called the *Shine-Dalgarno sequence.*

ribozyme Any RNA molecule that can act as a catalyst, that is, speed up a chemical reaction.

ribulose bisphosphate (RuBP) A five-carbon compound that combines with CO_2 in the first step of the Calvin cycle during photosynthesis.

RNA interference (RNAi) Degradation of an mRNA molecule or inhibition of its translation following its binding by a short RNA (microRNA) whose sequence is complementary to a portion of the mRNA.

RNA polymerase One of a class of enzymes that catalyze synthesis of RNA from ribonucleotides using a DNA template. Also called *RNA pol.*

RNA processing In eukaryotes, the changes that a primary RNA transcript undergoes in the nucleus to become a mature mRNA molecule, which is exported to the cytoplasm. Includes the addition of a 5′ cap and poly(A) tail and splicing to remove introns.

RNA replicase A viral enzyme that can synthesize RNA from an RNA template.

RNA See **ribonucleic acid**.

rod cell A photoreceptor cell with a rod-shaped outer portion that is particularly sensitive to dim light, but not used to distinguish colors. Also called simply *rod.* Compare with **cone cell**.

root (1) An underground part of a plant that anchors the plant and absorbs water and nutrients. (2) In a phylogenetic tree, the bottom, most ancient node.

root apical meristem (RAM) A group of undifferentiated plant cells at the tip of a plant root that can differentiate into mature root tissue.

root cap A small group of cells that covers and protects the tip of a plant root. Senses gravity and determines the direction of root growth.

root hair A long, thin outgrowth of the epidermal cells of plant roots, providing increased surface area for absorption of water and nutrients.

root pressure Positive (upward) pressure of xylem sap in the vascular tissue of roots. Is generated during the night as a result of the accumulation of ions from the soil and subsequent osmotic movement of water into the xylem.

root system The belowground part of a plant.

rough endoplasmic reticulum (rough ER) The portion of the endoplasmic reticulum that is dotted with ribosomes. Involved in synthesis of plasma membrane proteins, secreted proteins, and proteins localized to the ER, Golgi apparatus, and lysosomes. Compare with **smooth endoplasmic reticulum**.

roundworms Members of the phylum Nematoda. Distinguished by an unsegmented body with a pseudocoelom and no appendages. Roundworms belong to the ecdysozoan branch of the protostomes. Also called *nematodes.*

rRNA See **ribosomal RNA**.

rubisco The enzyme that catalyzes the first step of the Calvin cycle during photosynthesis: the addition of a molecule of CO_2 to ribulose bisphosphate.

ruminants A group of hoofed mammals (e.g., cattle, sheep, deer) that have a four-chambered stomach specialized for digestion of plant cellulose. Ruminants regurgitate the cud, a mixture of partially digested food and cellulose-digesting bacteria, from the largest chamber (the rumen) for further chewing.

salivary glands Vertebrate glands that secrete saliva (a mixture of water, mucus-forming glycoproteins, and digestive enzymes) into the mouth.

sampling error The accidental selection of a nonrepresentative sample from some larger population, due to chance.

saprophyte An organism that feeds primarily on dead plant material.

sapwood The younger xylem in the outer layer of wood of a stem or root, functioning primarily in water transport.

sarcomere The repeating contractile unit of a skeletal muscle cell; the portion of a myofibril located between adjacent Z disks.

sarcoplasmic reticulum Sheets of smooth endoplasmic reticulum in a muscle cell. Contains high concentrations of calcium, which can be released into the cytoplasm to trigger contraction.

saturated Referring to fats and fatty acids in which all the carbon-carbon bonds are single bonds. Such fats have relatively high melting points. Compare with **unsaturated**.

scanning electron microscope (SEM) A microscope that produces images of the surfaces of objects by reflecting electrons from a specimen coated with a layer of metal atoms. Compare with **transmission electron microscope**.

scarify To scrape, rasp, cut, or otherwise damage the coat of a seed. Necessary in some species to trigger germination.

Schwann cell A type of glial cell that wraps around axons of some neurons outside the brain and spinal cord, forming a myelin sheath that provides electrical insulation. Compare with **oligodendrocyte**.

scientific name The unique, two-part name given to each species, with a genus name followed by a species name—as in *Homo sapiens.* Scientific names are always italicized, and are also known as Latin names.

sclereid In plants, a type of sclerenchyma cell that usually functions in protection, such as in seed coats and nutshells. Compare with **fiber**.

sclerenchyma cell In plants, a cell that has a thick secondary cell wall and provides support; typically contains the tough structural polymer lignin and usually is dead at maturity. Includes fibers and sclereids. Compare with **collenchyma cell** and **parenchyma cell**.

screen See **genetic screen**.

scrotum A sac of skin, containing the testes, suspended just outside the abdominal body cavity of many male mammals.

second law of thermodynamics The principle of physics that the entropy of the universe or any closed system increases during any spontaneous process.

second messenger A nonprotein signaling molecule produced or activated inside a cell in response to stimulation at the cell surface. Commonly used to relay the message of a hormone or other extracellular signaling molecule.

secondary active transport Transport of an ion or molecule against its electrochemical gradient, in company with an ion or molecule being transported with its electrochemical gradient. Also called *cotransport.*

secondary active transporter A transmembrane protein that facilitates diffusion of an ion down its previously established electrochemical gradient and uses the energy of that process to transport some other substance, in the same or opposite direction, *against* its concentration gradient. Also called *cotransporter.* See also **antiporter** and **symporter**.

secondary cell wall The inner layer of a plant cell wall formed by certain cells as they mature. Provides support or protection.

secondary consumer A carnivore; an organism that eats herbivores. Compare with **primary consumer**.

secondary growth In plants, an increase in the width of stems and roots due to the activity of lateral meristems. Compare with **primary growth**.

secondary immune response The acquired immune response to a pathogen that the immune system has encountered before. Compare with **primary immune response**.

secondary metabolites Molecules that are closely related to compounds in key synthetic pathways, and that often function in defense.

secondary spermatocyte A cell produced by meiosis I of a primary spermatocyte in the testis. Can undergo meiosis II to produce spermatids.

secondary structure In proteins, localized folding of a polypeptide chain into regular structures (e.g., α-helix and β-pleated sheet) stabilized by hydrogen bonding between atoms of the backbone. In nucleic acids, elements of structure (e.g., helices and hairpins) stabilized by hydrogen bonding and other interactions between complementary bases. Compare with **primary, tertiary,** and **quaternary structure**.

secondary succession Gradual colonization of a habitat after an environmental disturbance (e.g., fire, windstorm, logging) that removes some or all previous organisms but leaves the soil intact. Compare with **primary succession**.

second-male advantage The reproductive advantage of a male who mates with a female last, after other males have mated with her.

secretin A peptide hormone produced by cells in the small intestine in response to the arrival of food from the stomach. Stimulates secretion of bicarbonate (HCO_3^-) from the pancreas.

sedimentary rock A type of rock formed by gradual accumulation of sediment, as in riverbeds and on the ocean floor. Most fossils are found in sedimentary rocks.

seed A plant reproductive structure consisting of an embryo, associated nutritive tissue (endosperm), and an outer protective layer (seed coat). In angiosperms, develops from the fertilized ovule of a flower.

seed bank A repository where seeds, representing many different varieties of domestic crops or other species, are preserved.

seed coat A protective layer around a seed that encases both the embryo and the endosperm.

segment A well-defined region of the body along the anterior-posterior body axis, containing similar structures as other, nearby segments.

segmentation Division of the body or a part of it into a series of similar structures; exemplified by the body segments of insects and worms and by the somites of vertebrates.

segmentation genes A group of genes that affect body segmentation in embryonic development. Includes gap genes, pair-rule genes, and segment polarity genes.

segregation, principle of The concept that each pair of hereditary elements (alleles of the same gene) separate from each other during the formation of offspring (i.e., during meiosis). One of Mendel's two principles of genetics.

selective adhesion The tendency of cells of one tissue type to adhere to other cells of the same type.

selective permeability The property of a membrane that allows some substances to diffuse across it much more readily than other substances.

selectively permeable membrane Any membrane across which some solutes can move more readily than others.

self Property of a molecule or cell such that immune system cells do not attack it, due to certain molecular similarities to other body cells.

self molecule A molecule that is synthesized by an organism and is a normal part of its cells and/or body; as opposed to non-self or foreign molecules.

self-fertilization In plants, the fusion of two gametes from the same individual to form a diploid offspring. Also called *selfing.*

semen The combination of sperm and accessory fluids that is released by male mammals and reptiles during ejaculation.

semiconservative replication The mechanism of replication used by cells to copy DNA. Results in each daughter DNA molecule containing one old strand and one new strand.

seminal vesicles In male mammals, paired reproductive glands that secrete a sugar-containing fluid into semen, which provides energy for sperm movement. In other vertebrates and invertebrates, often stores sperm.

senescence The process of aging.

sensor Any cell, organ, or structure with which an animal can sense some aspect of the external or internal environment. Usually functions, along with an integrator and effector, as part of a homeostatic system.

sensory neuron A nerve cell that carries signals from sensory receptors to the central nervous system. Compare with **interneuron** and **motor neuron**.

sepal One of the protective leaflike organs enclosing a flower bud and later supporting the blooming flower.

septum (plural: septa) Any wall-like structure. In fungi, septa divide the filaments (hyphae) of mycelia into cell-like compartments.

serotonin A neurotransmitter involved in many brain functions, including sleep, pleasure, and mood.

serum The liquid that remains when cells and clot material are removed from clotted blood. Contains water, dissolved gases, growth factors, nutrients, and other soluble substances. Compare with **plasma**.

sessile Permanently attached to a substrate; not capable of moving to another location.

set point A normal or target value for a regulated internal variable, such as body heat or blood pH.

severe combined immunodeficiency disease (SCID) A human disease characterized by an extremely high vulnerability to infectious disease. Caused by a genetic defect in the immune system.

sex chromosome Any chromosome carrying genes involved in determining the sex of an individual. Compare with **autosome**.

sex-linked inheritance Inheritance patterns observed in genes carried on sex chromosomes, so females and males have different numbers of alleles of a gene and may pass its trait only to one sex of offspring. Also called *sex-linkage.*

sexual dimorphism Any trait that differs between males and females.

sexual reproduction Any form of reproduction in which genes from two parents are combined via fusion of gametes, producing offspring that are genetically distinct from both parents. Compare with **asexual reproduction**.

sexual selection A pattern of natural selection that favors individuals with traits that increase their ability to obtain mates. Acts more strongly on males than females.

shell A hard protective outer structure.

Shine-Dalgarno sequence See **ribosome binding sequence**.

shoot apical meristem (SAM) A group of undifferentiated plant cells at the tip of a plant stem that can differentiate into mature shoot tissues.

shoot The combination of hypocotyl and cotyledons in a plant embryo, which will become the aboveground portions of the body.

shoot system The aboveground part of a plant comprising stems, leaves, and flowers (in angiosperms).

short interspersed nuclear elements See **SINEs**.

short tandem repeats (STRs) Relatively short DNA sequences that are repeated, one after another, down the length of a chromosome. The two major types are microsatellites and minisatellites.

short-day plant A plant that blooms in response to long nights (usually in late summer or fall in the northern hemisphere). Compare with **day-neutral** and **long-day plant**.

shotgun sequencing A method of sequencing genomes that is based on breaking the genome into small pieces, sequencing each piece separately, and then figuring out how the pieces are connected.

sieve plate In plants, a pore-containing structure at one end of a sieve-tube member in phloem.

sieve-tube member In plants, an elongated sugar-conducting cell in phloem that has sieve plates at both ends, allowing sap to flow to adjacent cells.

sign stimulus A simple stimulus that elicits an invariant, stereotyped behavioral response (fixed action pattern) from an animal. Also called a *releaser.*

signal In behavioral ecology, any information-containing behavior.

signal receptor Any cellular protein that binds to a particular signaling molecule (e.g., a hormone or neurotransmitter) and triggers a response by the cell. Receptors for water-soluble signals are transmembrane proteins in the plasma membrane; those for many lipid-soluble signals (e.g., steroid hormones) are located inside the cell.

signal recognition particle (SRP) A RNA-protein complex that binds to the ER signal sequence in a polypeptide as it emerges from a ribosome and transports the ribosome-polypeptide complex to the ER membrane where synthesis of the polypeptide is completed.

signal transduction cascade See **phosphorylation cascade**.

signal transduction The process by which a stimulus (e.g., a hormone, a neurotransmitter, or sensory information) outside a cell is amplified and converted into a response by the cell. Usually involves a specific sequence of molecular events, or signal transduction pathway.

silencer A regulatory sequence in eukaryotic DNA to which repressor proteins can bind, inhibiting transcription of certain genes.

silent mutation A mutation that does not detectably affect the phenotype of the organism.

simple eye An eye with only one light-collecting apparatus (e.g., one lens), as in vertebrates. Compare with **compound eye**.

simple fruit A fruit (e.g., apricot) that develops from a single flower that has a single carpel or several fused carpels. Compare with **aggregate** and **multiple fruit**.

simple sequence repeat See **microsatellite**.

SINEs (short interspersed nuclear elements) The second most abundant class of transposable elements in human genomes; can create copies of itself and insert them elsewhere in the genome. Compare with **LINEs**.

single nucleotide polymorphism (SNP) A site on a chromosome where individuals in a population have different nucleotides. Can be used as a genetic marker to help track the inheritance of nearby genes.

single-strand DNA-binding proteins (SSBPs) A class of proteins that attach to separated strands of DNA during replication or transcription, preventing them from re-forming a double helix.

sink Any tissue, site, or location where an element or a molecule is consumed or taken out of circulation (e.g., in plants, a tissue where sugar exits the phloem). Compare with **source**.

sinoatrial (SA) node A cluster of cardiac muscle cells, in the right atrium of the vertebrate heart, that initiates the heartbeat and determines the heart rate. Compare with **atrioventricular (AV) node**.

siphon A tubelike appendage of many mollusks, that is often used for feeding or propulsion.

sister chromatids The paired strands of a recently replicated chromosome, which are connected at the centromere and eventually separate during anaphase of mitosis and meiosis II. Compare with **non-sister chromatids**.

sister species Closely related species, which occupy adjacent branches in a phylogenetic tree.

skeletal muscle The muscle tissue attached to the bones of the vertebrate skeleton. Consists of long, unbranched muscle fibers with a characteristic striped (striated) appearance; controlled voluntarily. Also called *striated muscle*. Compare with **cardiac** and **smooth muscle**.

sliding-filament model The hypothesis that thin (actin) filaments and thick (myosin) filaments slide past each other, thereby shortening the sarcomere. Shortening of all the sarcomeres in a myofibril results in contraction of the entire myofibril.

small intestine The portion of the digestive tract between the stomach and the large intestine. The site of the final stages of digestion and of most nutrient absorption.

small nuclear ribonucleoproteins See **snRNPs**.

smooth endoplasmic reticulum (smooth ER) The portion of the endoplasmic reticulum that does not have ribosomes attached to it. Involved in synthesis and secretion of lipids. Compare with **rough endoplasmic reticulum**.

smooth muscle The unstriated muscle tissue that lines the intestine, blood vessels, and some other organs. Consists of tapered, unbranched cells that can sustain long contractions. Not voluntarily controlled. Compare with **cardiac** and **skeletal muscle**.

snRNPs (small nuclear ribonucleoproteins) Complexes of proteins and small RNA molecules that function in splicing (removal of introns from primary RNA transcripts) as components of spliceosomes.

sodium-potassium pump A transmembrane protein that uses the energy of ATP to move sodium ions out of the cell and potassium ions in. Also called *Na^+/K^+-ATPase*.

soil organic matter Organic (carbon-containing) compounds found in soil.

solute Any substance that is dissolved in a liquid.

solute potential (ψ_S) A component of the potential energy of water caused by a difference in solute concentrations at two locations. Also called *osmotic potential*. Compare with **pressure potential (ψ_P)**.

solution A liquid containing one or more dissolved solids or gases in a homogeneous mixture.

solvent Any liquid in which one or more solids or gases can dissolve.

soma See **cell body**.

somatic cell Any type of cell in a multicellular organism except eggs, sperm, and their precursor cells. Also called *body cells*.

somatic hypermutation Mutations that occur in the immunoglobulin genes of the immune system's memory cells, resulting in novel variation in the receptors that bind to antigens.

somatic nervous system The part of the peripheral nervous system (outside the brain and spinal cord) that controls skeletal muscles and is under voluntary control. Compare with **autonomic nervous system**.

somatostatin A hormone secreted by the pancreas and hypothalamus that inhibits the release of several other hormones.

somites Paired blocks of mesoderm on both sides of the developing spinal cord in a vertebrate embryo. Give rise to muscle tissue, vertebrae, ribs, limbs, etc.

sori In ferns, a cluster of spore-producing structures (sporangia).

source Any tissue, site, or location where a substance is produced or enters circulation (e.g., in plants, the tissue where sugar enters the phloem). Compare with **sink**.

space-filling model A representation of a molecule where atoms are shown as balls—color-coded and scaled to indicate the atom's identify—attached to each other in the correct geometry.

speciation The evolution of two or more distinct species from a single ancestral species.

species A distinct, identifiable group of populations that is thought to be evolutionarily independent of other populations and whose members can interbreed. Generally distinct from other species in appearance, behavior, habitat, ecology, genetic characteristics, etc.

species diversity The variety and relative abundance of the species present in a given ecological community.

species richness The number of species present in a given ecological community.

species–area relationship The mathematical relationship between the area of a certain habitat and the number of species that it can support.

specific heat The amount of energy required to raise the temperature of 1 gram of a substance by 1°C; a measure of the capacity of a substance to absorb energy.

sperm A mature male gamete; smaller and more mobile than the female gamete.

sperm competition Competition to fertilize eggs between the sperm of different males, inside the same female.

spermatid An immature sperm cell.

spermatogenesis The production of sperm. Occurs continuously in a testis.

spermatogonia (singular: spermatogonium) The diploid cells in a testis that can give rise to primary spermatocytes.

spermatophore A gelatinous package of sperm cells that is produced by males of species that have internal fertilization without copulation.

sphincter A muscular valve that can close off a tube, as in a blood vessel or a part of the digestive tract.

spicule Stiff spike of silica or calcium carbonate found in the body of many sponges.

spindle apparatus The array of microtubules responsible for contacting and moving chromosomes during mitosis and meiosis; includes kinetochore microtubules and polar microtubules.

spines In plants, modified leaves that are stiff and sharp and that function in defense.

spiracle In insects, a small opening that connects air-filled tracheae to the external environment, allowing for gas exchange.

spleen A dark red organ, found near the stomach of most vertebrates, that filters blood, stores extra red blood cells in case of emergency, and plays a role in immunity.

spliceosome In eukaryotes, a large, complex assembly of snRNPs (small nuclear ribonucleoproteins) that catalyzes removal of introns from primary RNA transcripts.

splicing The process by which introns are removed from primary RNA transcripts and the remaining exons are connected together.

sporangium (plural: sporangia) A spore-producing structure found in seed plants, some protists, and some fungi (e.g., chytrids).

spore (1) In bacteria, a dormant form that generally is resistant to extreme conditions. (2) In eukaryotes, a single cell produced by mitosis or meiosis (not by fusion of gametes) that is capable of developing into an adult organism.

sporophyte In organisms undergoing alternation of generations, the multicellular diploid form that arises from two fused gametes and produces haploid spores. Compare with **gametophyte**.

sporopollenin A watertight material that encases spores and pollen of modern land plants.

stabilizing selection A pattern of natural selection that favors phenotypes near the middle of the range of phenotypic variation. Reduces overall genetic variation in a population. Compare with **disruptive selection**.

stamen The male reproductive structure of a flower. Consists of an anther, in which pollen grains are produced, and a filament, which supports the anther. Compare with **carpel**.

standing defense See **constitutive defense**.

stapes The last of three small bones (ossicles) in the middle ear of vertebrates. Receives vibrations from the tympanic membrane and by vibrating against the oval window passes them to the cochlea.

starch A mixture of two storage polysaccharides, amylose and amylopectin, both formed from α-glucose monomers. Amylopectin is branched, and amylose is unbranched. The major form of stored carbohydrate in plants.

start codon The AUG triplet in mRNA at which protein synthesis begins; codes for the amino acid methionine.

statocyst A sensory organ of many arthropods that detects the animal's orientation in space (i.e., whether the animal is flipped upside down).

statolith A tiny stone or dense particle found in specialized gravity-sensing organs in some animals such as lobsters.

statolith hypothesis The hypothesis that amyloplasts (dense, starch-storing plant organelles) serve as statoliths in gravity detection by plants.

STATs See **signal transducers and activators of transcription**.

stem cell Any relatively undifferentiated cell that can divide to produce daughter cells identical to itself or more specialized daughter cells, which differentiate further into specific cell types.

stems Vertical, aboveground structures that make up the shoot system of plants.

stereocilium (plural: stereocilia) One of many stiff outgrowths from the surface of a hair cell that are involved in detection of sound by terrestrial vertebrates or of waterborne vibrations by fishes.

steroid A class of lipid with a characteristic four-ring structure.

steroid-hormone receptor One of a family of intracellular receptors that bind to various steroid hormones, forming a hormone-receptor complex that acts as a regulatory transcription factor and activates transcription of specific target genes.

sticky end The short, single-stranded ends of a DNA molecule cut by a restriction endonuclease. Tend to form hydrogen bonds with other sticky ends that have complementary sequences.

stigma The moist tip at the end of a flower carpel to which pollen grains adhere.

stolon A modified stem that runs horizontally over the soil surface and produces new plants at the nodes (a form of asexual reproduction). Compare with **rhizome**.

stoma (plural: stomata) Generally, a pore or opening. In plants, a microscopic pore on the surface of a leaf or stem through which gas exchange occurs.

stomach A tough, muscular pouch in the vertebrate digestive tract between the esophagus and small intestine. Physically breaks up food and begins digestion of proteins.

stop codon One of three mRNA triplets (UAG, UGA, or UAA) that cause termination of protein synthesis. Also called a *termination codon.*

strain A population of genetically similar or identical individuals.

stream A body of water that moves constantly in one direction.

striated muscle See **skeletal muscle**.

stroma The fluid matrix of a chloroplast in which the thylakoids are embedded. Site where the Calvin cycle reactions occur.

structural formula A two-dimensional notation in which the chemical symbols for the constituent atoms are joined by straight lines representing single (—), double (=), or triple (≡) covalent bonds. Compare with **molecular formula**.

structural gene A stretch of DNA that codes for a functional protein or functional RNA molecule, not including any regulatory sequences (e.g., a promoter, enhancer).

structural homology Similarities in organismal structures (e.g., limbs, shells, flowers) that are due to inheritance from a common ancestor.

style The slender stalk of a flower carpel connecting the stigma and the ovary.

subspecies A population that has distinctive traits and some genetic differences relative to other populations of the same species but that is not distinct enough to be classified as a separate species.

substrate (1) A reactant that interacts with an enzyme in a chemical reaction. (2) A surface on which a cell or organism sits.

substrate-level phosphorylation Production of ATP by transfer of a phosphate group from an intermediate substrate directly to ADP. Occurs in glycolysis and in the citric acid cycle.

succession In ecology, the gradual colonization of a habitat after an environmental disturbance (e.g., fire, flood), usually by a series of species. See **primary** and **secondary succession**.

sucrose A disaccharide formed from glucose and fructose. One of the two main products of photosynthesis.

sugar Synonymous with carbohydrate, though usually used in an informal sense to refer to small carbohydrates (monosaccharides and disaccharides).

sulfate reducer A prokaryote that produces hydrogen sulfide (H_2S) as a by-product of cellular respiration.

summation The additive effect of different postsynaptic potentials at a nerve or muscle cell, such that several subthreshold stimulations can cause an action potential.

supporting connective tissue A type of connective tissue, distinguished by having a firm extracellular matrix.

surface tension The cohesive force that causes molecules at the surface of a liquid to stick together, thereby resisting deformation of the liquid's surface and minimizing its surface area.

surfactant A mixture of phospholipids and proteins produced by lung cells that reduces surface tension, allowing the lungs to expand more.

survivorship On average, the proportion of offspring that survive to a particular age.

survivorship curve A graph depicting the percentage of a population that survives to different ages.

suspension feeder Any organism that obtains food by filtering small particles or small organisms out of water or air. Also called *filter feeder.*

sustainability The planned use of environmental resources at a rate no faster than the rate at which they are naturally replaced.

sustainable agriculture Agricultural techniques that are designed to maintain long-term soil quality and productivity.

swamp A wetland that has a steady rate of water flow and is dominated by trees and shrubs.

swim bladder A gas-filled organ of many ray-finned fishes that regulates buoyancy.

symbiosis (adjective: symbiotic) Any close and prolonged physical relationship between individuals of two different species. See **commensalism, mutualism,** and **parasitism**.

symmetric competition Ecological competition between two species in which both suffer similar declines in fitness. Compare with **asymmetric competition**.

sympathetic nervous system The part of the autonomic nervous system that stimulates fight-or-flight responses, such as increased heart rate, increased blood pressure, and decreased digestion. Compare with **parasympathetic nervous system**.

sympatric speciation The divergence of populations living within the same geographic area into different species as the result of their genetic (not physical) isolation. Compare with **allopatric speciation**.

sympatry Condition in which two or more populations live in the same geographic area, or close enough to permit interbreeding. Compare with **allopatry**.

symplast In plant roots, a continuous pathway through which water can flow through the cytoplasm of adjacent cells that are connected by plasmodesmata. Compare with **apoplast**.

symporter A carrier protein that allows an ion to diffuse down an electrochemical gradient, using the energy of that process to transport a different substance in the same direction *against* its concentration gradient. Compare with **antiporter**.

synapomorphy A shared, derived trait found in two or more taxa that is present in their most recent common ancestor but is missing in more distant ancestors. Useful for inferring evolutionary relationships.

synapse The interface between two neurons or between a neuron and an effector cell.

synapsis The physical pairing of two homologous chromosomes during prophase I of meiosis. Crossing over occurs during synapsis.

synaptic cleft The space between two communicating nerve cells (or between a neuron and effector cell) at a synapse, across which neurotransmitters diffuse.

synaptic plasticity Long-term changes in the responsiveness or physical structure of a synapse that can occur after particular stimulation patterns. Thought to be the basis of learning and memory.

synaptic vesicle A small neurotransmitter- containing vesicle at the end of an axon that releases neurotransmitter into the synaptic cleft by exocytosis.

synaptonemal complex A network of proteins that holds non-sister chromatids together during synapsis in meiosis I.

synthesis (S) phase The phase of the cell cycle during which DNA is synthesized and chromosomes are replicated.

systemic acquired resistance (SAR) A slow, widespread response of plants to a localized infection that protects healthy tissue from invasion by pathogens. Compare with **hypersensitive response**.

systemic circulation The part of the circulatory system that sends oxygen-rich blood from the lungs out to the rest of the body. Is separate from the pulmonary circulation in mammals and birds.

systemin A peptide hormone, produced by plant cells damaged by herbivores, that initiates a protective response in undamaged cells.

systole The portion of the heartbeat cycle during which the heart muscles are contracting. Compare with **diastole**.

systolic blood pressure The force exerted by blood against artery walls during contraction of the heart's left ventricle. Compare with **diastolic blood pressure**.

T cell A type of leukocyte that matures in the thymus and, with B cells, is responsible for acquired immunity. Involved in activation of B cells ($CD4^+$ helper T cells) and destruction of infected cells ($CD8^+$ cytotoxic T cells). Also called *T lymphocytes.*

T tubules Membranous tubes that extend into the interior of muscle cells. Propagate action potentials throughout a muscle cell and trigger release of calcium from the sarcoplasmic reticulum.

taiga A vast forest biome throughout subarctic regions, consisting primarily of short conifer trees. Characterized by intensely cold winters, short summers, and high annual variation in temperature.

taproot A large vertical main root of a plant.

taste bud Sensory structure, found chiefly in the mammalian tongue, containing spindle-shaped cells that respond to chemical stimuli.

TATA box A short DNA sequence in many eukaryotic promoters about 30 base pairs upstream from the transcription start site.

TATA-binding protein (TBP) A protein that binds to the TATA box in eukaryotic promoters and is a component of the basal transcription complex.

taxon (plural: taxa) Any named group of organisms at any level of a classification system.

taxonomy The branch of biology concerned with the classification and naming of organisms.

TBP See **TATA-binding protein.**

T-cell receptor (TCR) A transmembrane protein found on T cells that can bind to antigens displayed on the surfaces of other cells. Composed of two polypeptides called the alpha chain and beta chain. See **antigen presentation.**

tectorial membrane A membrane in the vertebrate cochlea that takes part in the transduction of sound by bending the stereocilia of hair cells in response to sonic vibrations.

telomerase An enzyme that replicates the ends of chromosome (telomeres) by catalyzing DNA synthesis from an RNA template that is part of the enzyme.

telomere The region at the end of a linear chromosome.

telophase The final stage in mitosis or meiosis during which sister chromatids (replicated chromosomes in meiosis I) separate and new nuclear envelopes begin to form around each set of daughter chromosomes.

temperate Having a climate with pronounced annual fluctuations in temperature (i.e., warm summers and cold winters) but typically neither as hot as the tropics nor as cold as the poles.

temperature A measurement of thermal energy present in an object or substance, reflecting how much the constituent molecules are moving.

template strand (1) The strand of DNA that is transcribed by RNA polymerase to create RNA. (2) An original strand of RNA used to make a complementary strand of RNA.

temporal lobe In the vertebrate brain, one of the four major areas in the cerebrum.

tendon A band of tough, fibrous connective tissue that connects a muscle to a bone.

tentacle A long, thin, muscular appendage of gastropod mollusks.

termination (1) In enzyme-catalyzed reactions, the final stage in which the enzyme returns to its original conformation and products are released. (2) In DNA transcription, the dissociation of RNA polymerase from DNA when it reaches a termination signal sequence. (3) In RNA translation, the dissociation of a ribosome from mRNA when it reaches a stop codon.

territory An area that is actively defended by an animal from others of its species.

tertiary consumers In a food chain or food web, organisms that feed on secondary consumers. Compare with **primary consumer** and **secondary consumer.**

tertiary structure The overall three-dimensional shape of a single polypeptide chain, resulting from multiple interactions among the amino acid side chains and the peptide backbone. Compare with **primary, secondary,** and **quaternary structure.**

testcross The breeding of an individual of unknown genotype with an individual having only recessive alleles for the traits of interest in order to infer the unknown genotype from the phenotypic ratios seen in offspring.

testis (plural: testes) The sperm-producing organ of a male animal.

testosterone A steroid hormone, produced and secreted by the testes, that stimulates sperm production and various male traits and reproductive behaviors.

tetrad The structure formed by synapsed homologous chromosomes during prophase of meiosis I.

tetrapod Any member of the taxon Tetrapoda, which includes all vertebrates with two pairs of limbs (amphibians, mammals, birds, and other reptiles).

texture A quality of soil, resulting from the relative abundance of different-sized particles.

theory A proposed explanation for a broad class of phenomena or observations.

thermal energy The kinetic energy of molecular motion.

thermocline A gradient (cline) in environmental temperature across a large geographic area.

thermophile A bacterium or archaean that thrives in very hot environments.

thermoreceptor A sensory cell or an organ specialized for detection of changes in temperature.

thermoregulation Regulation of body temperature.

thick filament A filament composed of bundles of the motor protein myosin; anchored to the center of the sarcomere. Compare with **thin filament.**

thigmotropism Growth or movement of an organism in response to contact with a solid object.

thin filament A filament composed of two coiled chains of actin and associated regulatory proteins; anchored at the Z disk of the sarcomere. Compare with **thick filament.**

thorax A region of the body; in insects, one of the three prominent body regions called tagmata.

thorn A modified plant stem shaped as a sharp protective structure. Helps protect a plant against feeding by herbivores.

threshold potential The membrane potential that will trigger an action potential in a neuron or other excitable cell. Also called simply *threshold.*

thylakoid A flattened, membrane-bound vesicle inside a plant chloroplast that functions in converting light energy to chemical energy. A stack of thylakoids is a granum.

thymus An organ, located in the anterior chest or neck of vertebrates, in which immature T cells generated in the bone marrow undergo maturation.

thyroid gland A gland in the neck that releases thyroid hormone (which increases metabolic rate) and calcitonin (which lowers blood calcium).

thyroid-stimulating hormone (TSH) A peptide hormone, produced and secreted by the anterior pituitary, that stimulates release of thyroid hormones from the thyroid gland.

thyroxine (T_4) A peptide hormone containing four iodine atoms that is produced and secreted by the thyroid gland. Acts primarily to increase cellular metabolism. In mammals, T_4 is converted to the more active hormone triiodothyronine (T_3) in the liver.

Ti plasmid A plasmid carried by *Agrobacterium* (a bacterium that infects plants) that can integrate into a plant cell's chromosomes and induce formation of a gall.

tight junction A type of cell-cell attachment structure that links the plasma membranes of adjacent animal cells, forming a barrier that restricts movement of substances in the space between the cells. Most abundant in epithelia (e.g., the intestinal lining). Compare with **desmosome** and **gap junction.**

tip The end of a branch on a phylogenetic tree. Represents a specific species or larger taxon that has not (yet) produced descendants—either a group living today or a group that ended in extinction. Also called *terminal node.*

tissue A group of similar cells that function as a unit, such as muscle tissue or epithelial tissue.

tolerance In ecological succession, the phenomenon in which early-arriving species do not affect the probability that subsequent species will become established. Compare with **facilitation** and **inhibition.**

tonoplast The membrane surrounding a plant vacuole.

top-down control The hypothesis that population size is limited by predators or herbivores (consumers).

topoisomerase An enzyme that cuts and rejoins DNA downstream of the replication fork, to ease the twisting that would otherwise occur as the DNA "unzips."

torpor An energy-conserving physiological state, marked by a decrease in metabolic rate, body temperature, and activity, that lasts for a short period (overnight to a few days or weeks). Occurs in some small mammals when the ambient temperature drops significantly. Compare with **hibernation.**

totipotent Capable of dividing and developing to form a complete, mature organism.

trachea (plural: tracheae) (1) In insects, one of the small air-filled tubes that extend throughout the body and function in gas exchange. (2) In terrestrial vertebrates, the airway connecting the larynx to the bronchi. Also called *windpipe.*

tracheid In vascular plants, a long, thin water-conducting cell that has gaps in its secondary cell wall, allowing water movement between adjacent cells. Compare with **vessel element.**

trade-off In evolutionary biology, an inescapable compromise between two traits that cannot be optimized simultaneously. Also called *fitness trade-off.*

trait Any heritable characteristic of an individual.

transcription The process by which RNA is made from a DNA template.

transcriptional control Regulation of gene expression by various mechanisms that change the rate at which genes are transcribed to form messenger RNA. In negative control, binding of a regulatory protein to DNA represses transcription; in positive control binding of a regulatory protein to DNA promotes transcription.

transcriptome The complete set of genes transcribed in a particular cell.

transduction Conversion of information from one mode to another. For example, the process by which a stimulus outside a cell is converted into a response by the cell.

transfer cell In land plants, a cell that transfers nutrients from a parent plant to a developing plant seed.

transfer RNA (tRNA) One of a class of RNA molecules that have an anticodon at one end and an amino acid binding site at the other. Each tRNA picks up a specific amino acid and binds to the corresponding codon in messenger RNA during translation.

transformation (1) Incorporation of external DNA into the genome. Occurs naturally in some bacteria; can be induced in the laboratory by certain processes. (2) Conversion of a normal cell to a cancerous one.

transgenic Referring to an individual plant or animal whose genome contains DNA introduced from another individual, either from the same or a different species.

transition state A high-energy intermediate state of the reactants during a chemical reaction that must be achieved for the reaction to proceed. Compare with **activation energy.**

transitional feature A trait that is intermediate between a condition observed in ancestral species and the condition observed in more derived species.

translation The process by which proteins and peptides are synthesized from messenger RNA.

translational control Regulation of gene expression by various mechanisms that alter the life span of messenger RNA or the efficiency of translation.

translocation (1) In plants, the movement of sugars and other organic nutrients through the phloem by bulk flow. (2) A type of mutation in which a piece of a chromosome moves to a nonhomologous chromosome. (3) The process by which a ribosome moves down a messenger RNA molecule during translation.

transmembrane protein Any membrane protein that spans the entire lipid bilayer. Also called *integral membrane protein.*

transmission The passage or transfer (1) of a disease from one individual to another or (2) of electrical impulses from one neuron to another.

transmission electron microscope (TEM) A microscope that forms an image from electrons that pass through a specimen. Compare with **scanning electron microscope**.

transpiration Water loss from aboveground plant parts. Occurs primarily through stomata.

transport protein Collective term for any membrane protein that enables a specific ion or small molecule to cross a plasma membrane. Includes carrier proteins and channel proteins, which carry out passive transport (facilitated diffusion), and pumps, which carry out active transport.

transporter See **carrier protein**.

transposable elements Any of several kinds of DNA sequences that are capable of moving themselves, or copies of themselves, to other locations in the genome. Include LINEs and SINEs.

tree of life A diagram depicting the genealogical relationships of all living organisms on Earth, with a single ancestral species at the base.

trichome A hairlike appendage that grows from epidermal cells of some plants. Trichomes exhibit a variety of shapes, sizes, and functions depending on species.

triglyceride See **fat**.

triiodothyronine (T_3) A peptide hormone containing three iodine atoms that is produced and secreted by the thyroid gland. Acts primarily to increase cellular metabolism. In mammals, T_3 has a stronger effect than does the related hormone thyroxine (T_4).

triose A monosaccharide (simple sugar) containing three carbon atoms.

triplet code A code in which a "word" of three letters encodes one piece of information. The genetic code is a triplet code because a codon is three nucleotides long and encodes one amino acid.

triploblast (adjective: triploblastic) An animal whose body develops from three basic embryonic cell layers: ectoderm, mesoderm, and endoderm. Compare with **diploblast**.

trisomy The state of having three copies of one particular type of chromosome.

tRNA See **transfer RNA**.

trochophore A larva with a ring of cilia around its middle that is found in some lophotrochozoans.

trophic cascade A series of changes in the abundance of species in a food web, usually caused by the addition or removal of a key predator.

trophic level A feeding level in an ecosystem.

trophoblast The exterior of a blastocyst (the structure that results from cleavage in embryonic development of mammals).

tropomyosin A regulatory protein present in thin (actin) filaments that blocks the myosin-binding sites on these filaments, thereby preventing muscle contraction.

troponin A regulatory protein, present in thin (actin) filaments, that can move tropomyosin off the myosin-binding sites on these filaments, thereby triggering muscle contraction. Activated by high intracellular calcium.

true navigation The type of navigation by which an animal can reach a specific point on Earth's surface.

trypsin A protein-digesting enzyme present in the small intestine that activates several other protein-digesting enzymes.

trypsinogen The precursor of protein-digesting enzyme trypsin. Secreted by the pancreas and activated by the intestinal enzyme enterokinase.

tube foot One of the many small, mobile, fluid-filled extensions of the water vascular system of echinoderms; the part extending outside the body is called a podium. Used in locomotion and feeding.

tuber A modified plant rhizome that functions in storage of carbohydrates.

tuberculosis A disease of the lungs caused by infection with the bacterium *Mycobacterium tuberculosis.*

tumor A mass of cells formed by uncontrolled cell division. Can be benign or malignant.

tumor suppressor A gene (e.g., *p53* and *Rb*) or the protein it encodes that prevents cell division, particularly when the cell has DNA damage. Mutated forms are associated with cancer.

tundra The treeless biome in polar and alpine regions, characterized by short, slow-growing vegetation, permafrost, and a climate of long, intensely cold winters and very short summers.

turgid Swollen and firm as a result of high internal pressure (e.g., a plant cell containing enough water for the cytoplasm to press against the cell wall). Compare with **flaccid**.

turgor pressure The outward pressure exerted by the fluid contents of a plant cell against its cell wall.

Turner syndrome A human genetic disorder caused by the presence of only one X chromosome and no Y chromosome ("XO"). Individuals with this condition are female but sterile.

turnover In lake ecology, the complete mixing of upper and lower layers of water that occurs each spring and fall in temperate-zone lakes.

tympanic membrane The membrane separating the middle ear from the outer ear in terrestrial vertebrates, or similar structures in insects. Also called the *eardrum.*

ubiquinone See **coenzyme Q**.

ulcer A hole in an epithelial layer that damages the underlying basement membrane and tissues.

ultimate causation In biology, the reason that a trait or phenomenon is thought to have evolved; the adaptive advantage of that trait. Also called *ultimate explanation.* Compare with **proximate causation**.

umami The taste of glutamate, responsible for the "meaty" taste of most proteins and of monosodium glutamate.

umbilical cord The cord that connects a developing mammalian embryo or fetus to the placenta and through which the embryo or fetus receives oxygen and nutrients.

unequal crossover An error in crossing over during meiosis I in which the two non-sister chromatids match up at different sites. Results in gene duplication in one chromatid and gene loss in the other.

universal tree The phylogenetic tree that includes all organisms.

unsaturated Referring to fats and fatty acids in which at least one carbon-carbon bond is a double bond. Double bonds produce kinks in the fatty acid chains and decrease the compound's melting point. Compare with **saturated**.

upstream In genetics, opposite to the direction in which RNA polymerase moves along a DNA strand. Compare with **downstream**.

ureter In vertebrates, a tube that transports urine from one kidney to the bladder.

urethra The tube that drains urine from the bladder to the outside environment. In male vertebrates, also used for passage of sperm during ejaculation.

uric acid A whitish excretory product of birds, reptiles, and terrestrial arthropods. Used to remove from the body excess nitrogen derived from the breakdown of amino acids. Compare with **urea**.

urochordates One of the three major chordate lineages (Urochordata), comprising sessile, filter-feeding animals that have a polysaccharide exoskeleton (tunic) and two siphons through which water enters and leaves; also called tunicates or sea squirts. Compare with **cephalochordates** and **vertebrates**.

uterus The organ in which developing embryos are housed in those vertebrates that give live birth. Common in most mammals and in some lizards, sharks, and other vertebrates.

vaccination The introduction into an individual of weakened, killed, or altered pathogens to stimulate development of acquired immunity against those pathogens.

vaccine A preparation designed to stimulate an immune response against a particular pathogen without causing illness. Vaccines consist of inactivated (killed) pathogens, live but weakened (attenuated) pathogens, or portions of a viral capsid (subunit vaccine).

vacuole A large organelle in plant and fungal cells that usually is used for bulk storage of water, pigments, oils, or other substances. Some vacuoles contain enzymes and have a digestive function similar to lysosomes in animal cells.

vagina The birth canal of female mammals; a muscular tube that extends from the uterus through the pelvis to the exterior.

valence electron An electron in the outermost electron shell, the valence shell, of an atom. Valence electrons tend to be involved in chemical bonding.

valence The number of unpaired electrons in the outermost electron shell of an atom; determines how many covalent bonds the atom can form.

valves In circulatory systems, flaps of tissue that prevent backward flow of blood, particularly in veins and between the chambers of the heart.

van der Waals interactions A weak electrical attraction between two hydrophobic side chains. Often contributes to tertiary structure in proteins.

variable (V) region The portion of an antibody's light chains or heavy chains that has a highly variable amino acid sequence and forms part of the antigen-binding site. Compare with **constant (C) region**.

variable number tandem repeat See **minisatellite**.

vas deferens (plural: vasa deferentia) A muscular tube that stores and transports semen from the epididymis to the ejaculatory duct. In nonhuman animals, called the *ductus deferens.*

vasa recta In the vertebrate kidney, a network of blood vessels that runs alongside the loop of Henle of a nephron. Functions in reabsorption of water and solutes from the filtrate.

vascular bundle A cluster of xylem and phloem strands in a plant stem.

vascular cambium One of two types of lateral meristem, consisting of a ring of undifferentiated plant cells inside the cork cambium of woody plants; produces secondary xylem (wood) and secondary phloem. Compare with **cork cambium**.

vascular tissue In plants, tissue that transports water, nutrients, and sugars. Made up of the complex tissues xylem and phloem, each of which contains several cell types.

vector A biting insect or other organism that transfers pathogens from one species to another. See also **cloning vector**.

vegetative organs The nonreproductive parts of a plant including roots, leaves, and stems.

vein Any blood vessel that carries blood (oxygenated or not) under relatively low pressure from the tissues toward the heart. Compare with **artery**.

veliger A distinctive type of larva, found in mollusks.

vena cava (plural: vena cavae) A large vein that returns oxygen-poor blood to the heart.

ventral Toward an animal's belly and away from its back. The opposite of dorsal.

ventricle (1) A thick-walled chamber of the heart that receives blood from an atrium and pumps it to the body or to the lungs. (2) One of several small fluid-filled chambers in the vertebrate brain.

venules Small veins (blood vessels that return blood to the heart).

vertebra (plural: vertebrae) One of the cartilaginous or bony elements that form the spine of vertebrate animals.

vertebrates One of the three major chordate lineages (Vertebrata), comprising animals with a dorsal column of cartilaginous or bony structures (vertebrae) and a skull enclosing the brain. Includes fishes, amphibians, mammals, reptiles, and birds. Compare with **cephalochordates** and **urochordates**.

vessel element In vascular plants, a short, wide water-conducting cell that has gaps through both the primary and secondary cell walls, allowing unimpeded passage of water between adjacent cells. Compare with **tracheid**.

vestigial trait Any rudimentary structure of unknown or minimal function that is homologous to functioning structures in other species. Vestigial traits are thought to reflect evolutionary history.

vicariance The physical splitting of a population into smaller, isolated populations by a geographic barrier.

villi (singular: villus) Small, fingerlike projections (1) of the lining of the small intestine or (2) of the fetal portion of the placenta adjacent to maternal arteries. Function to increase the surface area available for absorption of nutrients and gas exchange (in the placenta).

virion A single mature virus particle.

virulence The ability of a pathogen or parasite to cause disease and death.

virulent Referring to pathogens that can cause severe disease in susceptible hosts.

virus A tiny intracellular parasite that uses host cell enzymes to replicate; consists of a DNA or RNA genome enclosed within a protein shell (capsid). In enveloped viruses, the capsid is surrounded by a phospholipid bilayer derived from the host cell plasma membrane, whereas nonenveloped viruses lack this protective covering.

visceral mass One of the three main parts of the mollusk body; contains most of the internal organs and external gill.

visible light The range of wavelengths of electromagnetic radiation that humans can see, from about 400 to 700 nanometers.

vitamin An organic micronutrient that usually functions as a coenzyme.

vitelline envelope A fibrous sheet of glycoproteins that surrounds mature egg cells in many vertebrates. Surrounded by a thick gelatinous matrix (the jelly layer) in some species. In mammals, called the *zona pellucida*.

viviparous Producing live young (instead of eggs) that develop within the body of the mother before birth. Compare with **oviparous** and **ovoviviparous**.

volt (V) A unit of electrical potential (voltage).

voltage Potential energy created by a separation of electric charges between two points. Also called *electrical potential*.

voltage clamping A technique for imposing a constant membrane potential on a cell. Widely used to investigate ion channels.

voltage-gated channel An ion channel that opens or closes in response to changes in membrane voltage. Compare with **ligand-gated channel**.

voluntary muscle Muscle tissue that can respond to conscious thought.

wall pressure The inward pressure exerted by a cell wall against the fluid contents of a plant cell.

Wallace line A line that demarcates two areas in the Indonesian region, each of which is characterized by a distinct set of animal species.

water cycle, global The movement of water among terrestrial ecosystems, the oceans, and the atmosphere.

water potential (ψ) The potential energy of water in a certain environment compared with the potential energy of pure water at room temperature and atmospheric pressure. In living organisms, ψ equals the solute potential (ψ_S) plus the pressure potential (ψ_P).

water potential gradient A difference in water potential in one region compared with that in another region. Determines the direction that water moves, always from regions of higher water potential to regions of lower water potential.

water table The upper limit of the underground layer of soil that is saturated with water.

water vascular system In echinoderms, a system of fluid-filled tubes and chambers that functions as a hydrostatic skeleton.

watershed The area drained by a single stream or river.

Watson-Crick pairing See **complementary base-pairing**.

wavelength The distance between two successive crests in any regular wave, such as light waves, sound waves, or waves in water.

wax A class of lipid with extremely long hydrocarbon tails, usually combinations of long-chain alcohols with fatty acids. Harder and less greasy than fats.

weather The specific short-term atmospheric conditions of temperature, moisture, sunlight, and wind in a certain area.

weathering The gradual wearing down of large rocks by rain, running water, and wind; one of the processes that transform rocks into soil.

weed Any plant that is adapted for growth in disturbed soils.

wetland A shallow-water habitat where the soil is saturated with water for at least part of the year.

white blood cells See **leukocytes**.

wild type The most common phenotype seen in a population; especially the most common phenotype in wild populations compared with inbred strains of the same species.

wildlife corridor Strips of wildlife habitat connecting populations that otherwise would be isolated by man-made development.

wilt To lose turgor pressure in a plant tissue.

wobble hypothesis The hypothesis that some tRNA molecules can pair with more than one mRNA codon, tolerating some variation in the third base, as long as the first and second bases are correctly matched.

wood Xylem resulting from secondary growth. Also called *secondary xylem*.

xeroderma pigmentosum A human disease characterized by extreme sensitivity to ultraviolet light. Caused by an autosomal recessive allele that results in a defective DNA repair system.

X-linked inheritance Inheritance patterns for genes located on the mammalian X chromosome. Also called *X-linkage*.

X-ray crystallography A technique for determining the three-dimensional structure of large molecules, including proteins and nucleic acids, by analysis of the diffraction patterns produced by X-rays beamed at crystals of the molecule.

xylem A plant vascular tissue that conducts water and ions; contains tracheids and/or vessel elements. Primary xylem develops from the procambium of apical meristems; secondary xylem, or wood, from the vascular cambium of lateral meristems. Compare with **phloem**.

xylem sap The watery fluid found in the xylem of plants.

yeast Any fungus growing as a single-celled form. Also, a specific lineage of ascomycetes.

Y-linked inheritance Inheritance patterns for genes located on the mammalian Y chromosome. Also called *Y-linkage*.

yolk The nutrient-rich cytoplasm inside an egg cell; used as food for the growing embryo.

Z disk The structure that forms each end of a sarcomere. Contains a protein that binds tightly to actin, thereby anchoring thin filaments.

Z scheme Path of electron flow in which electrons pass from photosystem II to photosystem I and ultimately to $NADP^+$ during the light-dependent reactions of photosynthesis. Also called *noncyclic electron flow*.

zero population growth (ZPG) A state of stable population size due to fertility staying at the replacement rate for at least one generation.

zona pellucida The gelatinous layer around a mammalian egg cell. In other vertebrates, called the *vitelline envelope*.

zone of (cellular) division In plant roots, a group of apical meristematic cells just behind the root cap where cells are actively dividing.

zone of (cellular) elongation In plant roots, a group of young cells, located behind the apical meristem, that are increasing in length.

zone of (cellular) maturation In plant roots, a group of plant cells, located several centimeters behind the root cap, that are differentiating into mature tissues.

zygosporangium (plural: zygosporangia) The spore-producing structure in fungi that are members of the Zygomycota.

zygote The diploid cell formed by the union of two haploid gametes; a fertilized egg. Capable of undergoing embryological development to form an adult.

Credits

Photo Credits

Frontmatter p. ix Oscar Miller/SPL/Photo Researchers **p. xi** Photo by Jason Rick, supplied by Loren Rieseberg **p. xiv** Jeff Rotman/NPL/Minden Pictures **p. xvi** Frans Lanting/Minden Pictures **p. xxT** Natalie B. Fobes Photography **p. xxB** David Quillin

Chapter 1 Opener Marjorie Kibby **1.1** "Animalcules" observed by Anton van Leeuwenhoek, c1795, HIP/Art Resource, NY **1.6a** Samuel F. Conti and Thomas D. Brock **1.6b** Kwangshin Kim/Photo Researchers **1.7b** Michael Hughes/Laif/Aurora Photos **1.8R** From M. Wittlinger, R. Wehner, and H. Wolf, The ant odometer: Stepping on stilts and stumps, *Science* 312: 1965–1967, supporting online material (Jun 30 2006).

Chapter 2 Opener Frans Lanting/Corbis **2.1b** Dragan Trifunovic/Shutterstock **2.6c** Photos. com **2.11** Robert and Beth Plowes, ProteaPIX **2.14b** Dietmar Nill/Picture Press/Photolibrary **2.15c** Jan Vermeer/Minden Pictures **2.20** Shutterstock **2.20 inset** PhotoSpin/Alamy

Chapter 3 Opener Fritz Wilhelm (heisingart.com) **3.10a** Microworks/Phototake **3.10b** Walter Reinhart/ Phototake

Chapter 4 Opener SSPL/The Image Works **4.5** National Cancer Institute

Chapter 5 Opener Peter Arnold/Alamy

Chapter 6 6.5L James J. Cheetham, Carleton University **6.9a** iStockphoto **6.9b** iStockphoto **6.9c** Shutterstock **6.18** Harold Edwards/Visuals Unlimited **6.24T** David M. Phillips/Photo Researchers **6.24M** David M. Phillips/Photo Researchers **6.24B** Joseph F. Hoffman, Yale University School of Medicine

Chapter 7 Opener Torsten Wittmann/Photo Researchers **7.1** T. J. Beveridge/Visuals Unlimited **7.2** Gopal Murti/ Visuals Unlimited **7.3** Wanner/Eye of Science/Photo Researchers **7.4** Stanley C. Holt/Biological Photo Service **7.5** From David S. Goodsell, *TheMachinery of Life,* 2nd ed. (2009). © Springer-Verlag, New York. **7.7** Don W. Fawcett/ Photo Researchers **7.8** Don W. Fawcett/Photo Researchers **7.9** Don W. Fawcett/Photo Researchers **7.10** Biophoto Associates/Photo Researchers **7.11** Omikron/Photo Researchers **7.12** Don Fawcett, Daniel Friend, & Richard Wood/Photo Researchers **7.13** Gopal Murti/Visuals Unlimited **7.16** E. H. Newcomb & W. P. Wergin/Biological Photo Service **7.17** T.Kanaseki & Donald Fawcett/Visuals Unlimited **7.18** E. H. Newcomb & W. P. Wergin/Biological Photo Service **7.19** E. H. Newcomb & S. E. Frederick/ Biological Photo Service **7.20** By David S. Goodsell. Published in L. A. Moran et al., *Biochemistry* (1994). © Neil Patterson Publishers/Prentice-Hall. **7.21a** Don W. Fawcett/ Photo Researchers **7.21b** Don W. Fawcett/Photo Resarchers **7.21c** Biophoto Associates/Photo Researchers **7.21d** Dennis Kunkel/Visuals Unlimited **7.22** Don W. Fawcett/Photo Researchers **7.25a** James D. Jamieson, Yale University School of Medicine **7.25b/c** From J. D. Jamieson and G. E. Palade, *J. Cell Biol.* 34: 597–615 (1967). © The Rockefeller University Press. **7.30** Conly L. Rieder, Wadsworth Center, New York State Department of Health **7.31a/b** From *Mol. Biol. Cell* 9: cover (Dec 1998). © American Society for Cell Biology. Photo B. J. Schnapp. **7.32a** John E. Heuser, Washington University School of Medicine **7.33L** SPL/Photo Researchers **7.33R** Visuals Unlimited/Corbis **7.34a** Don W. Fawcett/Photo Researchers

Chapter 8 Opener Caroline Weight, G.I.T. Molecular Immunology, Institute of Food Research, Norwich, UK **8.1** William James Warren/Science Faction **8.2** Biophoto Associates/Photo Researchers **8.3b** Barry King, University of California, Davis/Biological Photo Service **8.5** SPL/Photo Researchers **8.6** Photo Researchers **8.7aL** Don W. Fawcett/Photo Researchers **8.7aR** Don W. Fawcett/Photo Researchers **8.8a** Don W. Fawcett/Photo Researchers **8.11a** E. H. Newcomb & W. P. Wergin/Biological Photo Service **8.11b** Don W. Fawcett/Photo Researchers **8.18** Janice Carr/CDC/Rodney M. Donlan

Chapter 9 Opener Richard Megna/Fundamental Photographs **9.7T** Shutterstock **9.7B** Shutterstock **9.13** Terry Frey, San Diego State University **9.22a** Science VU/B. Bhatnagar/Visuals Unlimited

Chapter 10 Opener Visuals Unlimited/Corbis **10.2T** John Durham/SPL/Photo Researchers **10.2M/B** Electron micrographs by W. P. Wergin, courtesy of E. H. Newcomb, University of Wisconsin, Madison **10.4b** Sinclair Stammers/ SPL/Photo Researchers **10.17L/R** James A. Bassham, Lawrence Berkeley Laboratory, UCB (retired) **10.20a** Jeremy Burgess/SPL/Photo Researchers

Chapter 11 Opener CNRI/Phototake **11.1a** From Walter Flemming, *Zellsubstanz, Kern, und Zelltheilung.* Leipzig: Verlag von F. C. W. Vogel, 1882. **11.1b** Conly Rieder, Wadsworth Center, New York State Department of Health. **11.1b** Photo Researchers **11.2T** Gopal Murti/Photo Researchers **11.2B** Biophoto Associates/Photo Researchers **11.5** Micrographs by Conly L. Rieder, Wadsworth Center, NYS Department of Health **11.6a** Ed Reschke/Peter Arnold **11.6b** Visuals Unlimited/Getty Images

Chapter 12 Opener David Phillips/The Population Council/Photo Researchers **12.6** Photos.com **12.7** Micrographs byEdward Novitski, courtesy of Charles Novitski

Chapter 13 Opener Brian Johnston **13.9a** Benjamin Prud'homme/Nicolas Gompel **13.9bL/bR** From J. Childress, R. Behringer, and G. Halder, "Learning to Fly: Phenotypic Markers in Drosophila," *Genesis* 43(1): cover illustration (2005). © Wiley-Liss. Photo Georg Halder, University of Texas. **13.15a** Robert Calentine/Visuals Unlimited

Chapter 14 Opener Gopal Murti/SPL/Photo Researchers **14.1b** Eye of Science/Photo Researchers **14.5** From M. Meselson and F. W. Stahl, *Proc. Natl. Acad. Sci. USA* 44: 671–682, fig 4 (Jul 15 1958). Photo Matthew S. Meselson, Harvard University. **14.7a** Gopal Murti/SPL/Photo Researchers

Chapter 15 Opener Alfred Pasieka/Photo Researchers **15.4a** Rod Williams/Nature Picture Library **15.4b** The Peromyscus Genetic Stock Center at the University of South Carolina. Photograph by Clint Cook. **15.9a/b** Peter Duesberg, University of California, Berkeley

Chapter 16 Opener Oscar Miller/SPL/Photo Researchers **16.5a** Bert W. O'Malley, Baylor College of Medicine **16.8a** From B. A. Hamkalo and O. L. Miller Jr., *Annu. Rev. Biochem.* 42: 379–96 (1973). © Annual Reviews. **16.9b** E. V. Kiseleva and Donald Fawcett/Visuals Unlimited

Chapter 17 Opener Stephanie Schuller/Photo Researchers **UN17.1** EDVOTEK, The Biotechnology Education Company (www.edvotek.com)

Chapter 18 18.2a Ada Olins & Don Fawcett/Photo Researchers **18.4T** Barbara Hamkalo, University of California, Irvine **18.4B** Victoria E. Foe, University of Washington

Chapter 19 Opener AJ/IRRI/Corbis **19.1b** Brady-Handy Photograph Collection (Library of Congress). Reproduction number: LC-DIG-cwpbh-02977. **19.12** Baylor College of Medicine/Peter Arnold **19.16bL/R** Golden Rice Humanitarian Board (www.goldenrice.org)

Chapter 20 Opener Sanger Institute/Wellcome Photo Library **20.7bL/R** Test results provided by GENDIA (www.gendia.net) **20.10** Agilent Technologies **20.11** Camilla M. Kao and Patrick O. Brown, Stanford University

Chapter 21 Opener Anthony Bannister/Gallo Images/Corbis **21.1a** From R. Merino et al., *Development* 126(23): 5515–5522, fig. 6 (1999). © The Company of Biologists. **21.1b** From K. Kuida et al., *Cell* 94: 325–337, figs. 2e and 2f (1998). © Elsevier Science Ltd. Photo Keisuke Kuida, Vertex Pharmaceuticals. **21.2** Roslin Institute/Phototake **21.3a** Richard Hutchings/Photo Researchers **21.3b** From S. Kulandavelu et al., Embryonic and neonatal phenotyping of genetically engineered mice. *ILAR Journal*, 47(2); 103–117, fig. 12a (2006). © National Academy of Sciences. **21.4a** F. Rudolf Turner, Indiana University **21.4b** Christiane Nusslein-Volhard, Max Planck Institute **21.5** Christiane Nusslein-Volhard, Max Planck Institute **21.6** Wolfgang Driever, University of Freiburg **21.7a/c** Jim Langeland, Stephen Paddock, and Sean Carroll, University of Wisconsin–Madison **21.7b** Stephen J. Small, New York University **21.8L** Visuals Unlimited/Getty Images **21.8M** David Scharf/Science Faction/Corbis **21.8R** Eye of Science/Photo Researchers **21.10a** F. Rudolf Turner, Indiana University **21.11** Anthony Bannister/NHPA **21.12a/b** From "The Origin of Form" by Sean B. Carroll, *Natural History* (Nov 2005); A. C. Burke, "*Hox* Genes and the Global Patterning of the Somitic Mesoderm," in C. Ordahl (ed.), *Somitogenesis, Part I* (2000). © Academic Press.

Chapter 22 Opener Yorgos Nikas/Photo Researchers **22.2** Holger Jastrow **22.5a** Michael Whitaker/SPL/Photo Researchers **22.5b** Victor D. Vacquier, Scripps Institution of Oceanography, University of California at San Diego **22.11a** Kathryn W. Tosney, University of Michigan **22.11b** Kathryn W. Tosney, The University of Miami

Chapter 23 Opener John Runions/Oxford Brookes University **23.3b** Cabisco/Visuals Unlimited **23.4b** Michael Clayton, Department of Botany, University of Wisconsin, Madison **23.8b** Ken Wagner/Phototake **23.10a/bT/bB** From M. Kim et al., The expression domain of *PHANTASTICA* determines leaflet placement in compound leaves, *Nature* 424: 438–443, fig. 1 (Jul 24 2003). © Macmillan Magazines Ltd. Photos Neelima Sinah. **23.12** Photos by John L. Bowman, University of California at Davis

Chapter 24 Opener Art Wolfe/Stone/Getty Images **24.2a** Sinclair Stammers/Photo Researchers **24.2b** From S. De Valais and R. N. Melchor, Ichnotaxonomy of bird-like footprints: an example from the Late Triassic–Early Jurassic of northwest Argentina. *J. Vertebr. Paleontol.* 28(1):145–159, fig 5c (2008). © Society of Vertebrate Paleontology. **24.2c** The Natural History Museum, London **24.3L** Gerd Weitbrecht for Landesbildungsserver Baden-Württemberg (www.schule-bw.de) **24.3R** iStockphoto **24.5aL** CMCD/ PhotoDisc/Getty Images **24.5aR** Vincent Zuber/Custom Medical Stock Photo **24.5bL** Mary Beth Angelo/Photo Researchers **24.5bR** Custom Medical Stock Photo **24.6a1** age fotostock/SuperStock **24.6a2** George D. Lepp/Photo Researchers **24.6a3** Tui De Roy/Minden Pictures **24.6a4** Stefan Huwiler/age fotostock **24.8L/M** From M. K. Richardson et al., *Anat. Embryol.* 196: 91–106, figs. 7 and 8 (1997). © Springer-Verlag. Photos Michael K. Richardson, Leiden University, and Ronan O'Rahilly, National Museum of Health and Medicine/Armed Forces Institute of Pathology. **24.8R** From M. K. Richardson et al., *Science* 280: 983 (in Letters) (May 15 1998). Photo Ronan O'Rahilly, National Museum of Health and Medicine/Armed Forces Institute of Pathology. **24.10** Walter J. Gehring, Biozentrum, University of Basel (retired) **24.15aL** G. Gerra & S. Sommazzi/www.justbirds.it **24.15aR** Tui De Roy/Minden Pictures **24.15b** Huw Rees Lewis **24.18T** From A. Abzhonov et al., *Bmp4* and Morphological Variation of Beaks in Darwin's Finches, *Science* 305: 1462–1465, fig. 1c (Sep 3

2004). **24.18B** From K. Petren et al., *Proc. R. Soc. B*, 266: 321–329, fig. 3 (1999). © The Royal Society. Photo Kenneth Petren, University of Cincinnati. **24.20** Erlend Haarberg/ NHPA/Photoshot

Chapter 25 Opener Phil Savoie/Minden Pictures **25.7a** André Karwath/Wikimedia Commons. Licensed under the Creative Commons Attribution Share Alike 2.5. **25.14a** Cyril Laubscher/Dorling Kindersley **25.15** Otorohanga Kiwi House, New Zealand **25.16a** Robert Rattner **25.17TL/TR** Robert & Linda Mitchell Photography **25.17BL/BR** Fotolia

Chapter 26 Opener Photo by Jason Rick, supplied by Loren Rieseberg **26.9bL** Michael Lustbader/Photo Researchers **26.9bR** From D. E. Soltis et al., Recent and recurrent polyploidy in *Tragopogon* (Asteraceae): Genetic, genomic, and cytogenetic comparisons. *Biol. J. Linn. Soc.* 82: 485–501 (2004). Photo Douglas and Pamela Soltis, University of Florida, Gainesville; Andrew Doust, Oklahoma State University. **26.10aT/M/B** H. Douglas Pratt/National Geographic Stock

Chapter 27 Opener Merv Feick (www.indiana9fossils .com) **27.7a** Stefan Piasecki, GEUS **27.7b** iStockphoto **27.7c/d** From The Virtual Petrified Wood Museum (petrifiedwoodmuseum.org). by Mike Viney **27.11LT** Gerald D. Carr **27.11LB/LR** Forest & Kim Starr **27.12a** Eladio Fernandez/Caribbean Nature Photography **27.12b** Jonathan B. Losos and Kevin de Quieroz, Washington University in St. Louis **27.13a** Colin Keates/ Dorling Kindersley, courtesy of the Natural History Museum, London **27.13b** John Davis **27.13c** Don P. Northup, www.africancichlidphotos .com **27.13d** James L. Amos/Corbis **27.14bTL** Chip Clark, National Museum of Natural History, Smithsonian Institution **27.14bTR** Ed Reschke/Peter Arnold **27.14bML** Simon Conway Morris, University of Cambridge **27.14bMR** J. Gehling, South Australian Museum **27.14bBL/BM/BR** Shuhai Xiao, Virginia Polytechnic Institute and State University **27.17bL/bR** Glen A. Izett, U.S. Geological Survey, Denver **27.17c** Peter H. Schultz, Brown University; Steven D'Hondt, University of Rhode Island Graduate School of Oceanography

Chapter 28 Opener John Hammond **28.3** Science VU/Visuals Unlimited **28.4** From S. V. Liu et al., *Science* 277:1106–1109, fig. 2 (1997). © AAAS. PhotoYul Roh, Oak Ridge National Laboratory. **28.7aT** Daniel Quilten/Visuals Unlimited **28.7aB** From H. N. Schulz, et al., *Science* 284:493–495, fig. 1b (Aug 4 1999). © AAAS. Photo Heide Schulz, Max-Planck-Institute for Marine Microbiology. **28.7bT/B** SciMAT/Photo Researchers **28.7cT** K. S. Kim/ Peter Arnold **28.7cB** Ralf Wagner **28.8a** R. Hubert, Iowa State University **28.10** Bruce J. Russell/BioMEDIA Associates **28.12** Wally Eberhart/Visuals Unlimited/Getty Images **28.15** Andrew Syred/Photo Researchers **28.16** Bill Schwartz/CDC **28.17** S. Amano, S. Miyadoh, and T. Shomura/The Society for Actinomycetes Japan **28.18** From J. L. Cocchiaro et al., Cytoplasmic lipid droplets are translocated into the lumen of the Chlamydia trachomatis parasitophorous vacuole, *Proc. Natl. Acad. Sci. USA* 105(27): 9379–9384 (Jul 8 2008). © National Academy of Sciences. **28.19** Peter Siver/Visuals Unlimited **28.20a** Yves V. Brun **28.20b** George Barron, University of Guelph, Canada **28.21** Kenneth M. Stedman/Portland State University **28.21 inset** Eye of Science/Photo Researchers **28.22** Doc Searls **28.22 inset** Priya DasSarma, Center of Marine Biotechnology, University of Maryland Biotechnology Institute

Chapter 29 Opener Wim van Egmond/Visuals Unlimited/Getty Images **29.2L** D. P. Wilson/FLPA/Minden Pictures **29.2M** Mark Conlin/V&W/imagequestmarine.com **29.2R** E. R. Degginger/Animals Animals–Earth Scenes **29.4** Pete Atkinson/Photographer's Choice/Getty Images **29.4 inset** D. Anderson, WHOI **29.6** Biophoto Associates/ Photo Researchers **29.11a** Steve Gschmeissner/Photo Researchers **29.11b** Biophoto Associates/Photo Researchers **29.11c** Andrew Syred/Photo Researchers **29.12a** Biophoto Associates/Photo Researchers **29.12b** Bruce Coleman/ Photoshot **29.15** Wim van Egmond/Visuals Unlimited/Getty Images **29.19** Scott Camazine/Photo Researchers **29.20** Dennis Kunkel/Visuals Unlimited **29.21** Tai-Soon Yong (www.atlas.or.kr) **29.22** Image by D. Patterson, provided by the microscope web project. **29.23** Francis Abbott/NPL/Minden Pictures **29.24** Peter Parks/ imagequestmarine.com **29.25** Image by D. Patterson, provided by the micro*scope web project. **29.26** Maurice LOIR (www.diatomloir.eu) **29.27** Dennis Kunkel/Visuals Unlimited/Corbis **29.28** Andrew Syred/SPL/Photo Researchers **29.29** Wim van Egmond/Visuals Unlimited **29.30** Joyce Photographics/Photo Researchers **T29.3(1)** Peter Siver/Visuals Unlimited/Getty Images **T29.3(2)** Visuals Unlimited/Corbis **T29.3(3)** Jeff Rotman/Stone/Getty Images **T29.3(4)** iStockphoto **T29.3(5)** Yuuji Tsukii, Protist Information Server (protist.i.hosei.ac.jp) **T29.3(6)** Dr. Kawachi of National Institute for Environmental Studies, Japan

Chapter 30 Opener Jean-Paul Ferrero/Auscape/Minden Pictures **30.1** Lynn Betts/USDA Natural Resources Conservation Service **30.2b** Hugh Iltis/The Doebley Lab, University of Wisconsin-Madison (teosinte.wisc.edu) **30.4L** Linda Graham, University of Wisconsin-Madison **30.4R** Lee W. Wilcox **30.5L** Lee W. Wilcox **30.5M** iStockphoto **30.5R** Rob Whitworth/GAP Photos/ Getty Images **30.11a/b** Lee W. Wilcox **30.12** Biodisc/Visuals Unlimited/Alamy **30.16a/b** Lee W. Wilcox **30.18** Carolina Biological Supply Company/Phototake **30.21a** Larry Deack/Wikimedia Commons **30.21b** Shutterstock **30.21c** Andreas Lander/dpa/Corbis **30.23a** Nigel Cattlin/ Holt Studios/Photo Researchers **30.23b** Shutterstock **30.25T1** Jerome Wexler/Visuals Unlimited/Getty Images **30.25T2** Keith Wheeler/SPL/Photo Researchers **30.25T3** iStockphoto**30.25T4** Noodle snacks (www.noodlesnacks.com/) via Wikipedia. This file is licensed under the Creative Commons Attribution Share Alike 3.0 and GNU Free Documentation License. **30.25B1** Shutterstock **30.25B2** ISM/Phototake **30.25B3** iStockphoto **30.25B4** Shutterstock **30.27a** Doug Allan/NPL/ Minden Pictures **30.27c** Lee W. Wilcox **30.27b** Wim van Egmond/Visuals Unlimited/Getty Images **30.28** Peter Verhoog/Foto Natura/Minden Pictures **30.29** Wim van Egmond **30.30** CAIP IMAGES © 2006 University of Florida (plants.ifas.ufl.edu/slidecol.html) **30.31** Adrian Davies/NPL/Minden Pictures **30.32a** Wildlife GmbH/Alamy **30.32b** Colin McPherson/Corbis **30.33** Steven P. Lynch, Lynch Images **30.34** Lee W. Wilcox **30.35** Frans Lanting/Corbis **30.36** Peter Arnold/Alamy **30.37L** David T. Webb, Botany, University of Hawaii **30.37R** Jami Tarris/ Workbook Stock/Getty Images **30.38** Lee W. Wilcox **30.39** iStockphoto **30.40a** Martin Land/SPL/Photo Researchers **30.40b** iStockphoto **30.41** Peter Chadwick/ Dorling Kindersley **30.42a/b** Lee W. Wilcox **30.43** Michael & Patricia Fogden/Minden Pictures **30.44a** Brian Johnston **30.44bL/R** Lee W. Wilcox

Chapter 31 Opener Richard Shiell/Animals Animals–Earth Scenes **31.1a** David Reed, Department of Animal and Plant Sciences, University of Sheffield **31.1b** Mycorrhizal Applications Inc. (www.mycorrhizae .com) **31.2** John Cang Photography **31.3a** Visuals Unlimited/Corbis **31.3b** iStockphoto **31.4a** Eye of Science/Photo Researchers **31.4b** Tony Brain/Photo Researchers **31.5bL** George Barron, University of Guelph, Canada **31.5bR** Biophoto Associates/ Photo Researchers **31.6a** Melvin Fuller/The Mycological Society of America **31.6b** James Richardson/Visuals Unlimited **31.6c** Biophoto Associates/Photo Researchers **31.6d** Nino Santamaria **31.10a/b** Mark Brundrett (http:// mycorrhizas.info) **31.13** J. I. Ronny Larsson, Professor of Zoology **31.14** Thomas J. Volk (TomVolkFungi.net) **31.15a** Gregory G. Dimijian/Photo Researchers **31.15b** David M. Phillips/Photo Researchers **31.16** Jim Deacon **31.17** Michel Poinsignon/NPL/Minden Pictures **31.18b** Daniel Mosquin **31.19** N. Allen and G. Barron, University of Guelph **31.20** Michael Fogden/OSF/Photolibrary

Chapter 32 Opener Bill Curtsinger/National Geographic Stock **32.1a** Satoshi Kuribayashi/Nature Production/Minden Pictures **32.1b** Mark Moffett/Minden Pictures **32.2** Walker England/Photo Researchers **32.3a** Roberto Rinaldi/NPL/Minden Pictures **32.3b** Ingo Arndt/Minden Pictures **32.8bT** Robert Steene/imagequestmarine.com **32.8bB** Russ Hopcroft/University of Alaska Fairbanks **32.12a** J. P. Ferrero/Jacana/Photo Researchers **32.12b** Todd Husman (www.soundaquatic.com) **32.13a** Pete Oxford/ Minden Pictures **32.13b** Heidi & Hans-Jurgen Koch/Minden Pictures **32.14a overlay** Tim Ridley/Dorling Kindersley, Courtesy of the Royal Veterinary College, University of London **32.14a background** Shutterstock **32.14b** Adrian Costea/Fotolia **32.14b inset** Josef Ramsauer and Prof. Dr. Robert Patzner, Salzburg University **32.15a** Satoshi Kuribayashi/OSF/Photolibrary **32.15b** Radius Images/ Corbis **32.16a** Andrew Syred/Photo Researchers **32.16b** Eye of Science/Photo Researchers **32.17a** Roland Seitre/Peter Arnold **32.17b** Jeff Foott/NPL/Minden Pictures **32.18a** D. P. Wilson/FLPA/Minden Pictures **32.18b** Sue Scott **32.19L/M/R** From G. Panganiban et al., *Proc. Natl. Acad. Sci. USA* 94(10): 5162–5166, figs. 1c and 1f (May 13 1997). © National Academy of Sciences. **32.20a** Shutterstock **32.20b** D. Parer & E. Parer-Cook/Auscape/Minden Pictures **32.21a** Nigel Cattlin/FLPA/Minden Pictures **32.21b inset** Frank Greenaway/Dorling Kindersley **32.21b** Kim Taylor/ NPL/Minden Pictures **32.23** Roberto Rinaldi/NPL/Minden Pictures **32.24** Gary Bell/Taxi/Getty Images **32.25** Gregory G. Dimijian/Photo Researchers **32.26** Matthew D. Hooge, University of Maine

Chapter 33 Opener Barbara Strnadova/Photo Researchers **33.3a** Peter Parks/imagequestmarine.com **33.3b** Carmel McDougall, University of Oxford **33.4a** Dave King/Dorling Kindersley **33.4b** Breck P. Kent/Animals Animals–Earth Scenes **33.8a** Amazon-Images/Alamy **33.8b** Wim van Egmond/Visuals Unlimited **33.8c** Wim van Egmond/Visuals Unlimited/Corbis **33.9a** Pankaj Bajpai **33.9b** Bates Littlehales/Animals Animals–Earth Scenes **33.9c** Jane Burton/NPL/Minden Pictures **33.11** Antonio Guillén Oterino **33.12a** Doug Anderson (doug.deep via Flickr) **33.12b** Manfred Kage/Peter Arnold **33.12c** Oliver Meckes/Photo Researchers **33.13a** J. W.Alker/Imagebroker/ Alamy **33.13b** Jasper Nance (www.flickr.com/photos/ nebarnix) **33.13c** Robert and Linda Mitchell **33.14a** Marevision/age fotostock **33.15a** Roger Steen/ imagequestmarine.com **33.15b** Daniela Wolf (www.taucher.li) **33.16** Kjell B. Sandved/Photo Researchers **33.17** Mark Conlin/VWPICS/age fotostock **33.18a** Shutterstock **33.18b** Andrew Syred/Photo Researchers **33.19a** Eye of Science/Photo Researchers **33.19b** C. Hough/Custom Medical Stock Photo **33.20** Mark Smith/Photo Researchers **33.21a** Holger Mette/Fotolia **33.21a inset** Holger Mette/Fotolia **33.21b** Manfred Kage/Peter Arnold **33.22a** iStockphoto **33.22b** National Geographic/Getty Images **33.23** Colin Milkins/OSF/ Photolibrary **T33.1(1)** Meul/ARCO/NPL/Minden Pictures **T33.1(2)** iStockphoto **T33.1(3)** iStockphoto **T33.1(4)** Trouncs/Wikimedia Commons **T33.1(5)** J. Gall/ Photo Researchers **T33.1(6)** iStockphoto **T33.1(7)** J. K. Lindsey **T33.1(8)** Brand X Pictures/Jupiter Images

Chapter 34 34.2a D. P. Wilson/FLPA/Minden Pictures **34.2b** Juri Vinokurov/Fotolia **34.3b** Jeff Rotman/NPL/ Minden Pictures **34.4a** Sue Scott/OSF/Photolibrary **34.4b** Gary Bell/OceanwideImages.com **34.6** Alan James/ NPL/Minden Pictures **34.7a** Audra Reese Kirkland **34.7b** Jeffrey L. Rotman/CORBIS **34.9** Heather Angel/ Natural Visions **34.10a** Gary Bell/Getty Images **34.10b** Sinclair Stammers/NPL/Minden Pictures **34.15** Rudie Kuiter/OceanwideImages.com **34.17** Pavel Riha **34.21a** Hiroya Minakuchi/Minden Pictures **34.21b** Shutterstock **34.23a** Tom & Theresa Stack/ NHPA/Photoshot **34.23b** James L. Amos/National Geographic Stock **34.24a** Fred McConnaughey/Photo Researchers **34.24b** Paul Souders/Corbis **34.25** Oceans Image/Photoshot **34.26** Arnaz Mehta **34.27a/b** Michael & Patricia Fogden/Minden Pictures **34.28a** Tom McHugh/ Photo Researchers **34.28b** Reg Morrison/Auscape/Minden Pictures **34.29** Frank Lukasseck/Photographer's Choice/ Getty Images **34.30** Anup Shah/NPL/Minden Pictures **34.31** George Grall/National Geographic Stock **34.32** iStockphoto **34.33** iStockphoto **34.34a** Tui De Roy/Minden Pictures **34.34b** Ingo Arndt/Minden Pictures **34.35a** Cyril Ruoso/J. H. Editorial/Minden Pictures **34.35b** Ingo Arndt/Minden Pictures

Chapter 35 Opener Jed Fuhrman, University of Southern California **35.2** SPL/Photo Researchers **35.5** David M. Phillips/Photo Researchers **35.6a** Harold Fisher/Visuals

Unlimited/Getty **35.6b** Biophoto Associates/Photo Researchers **35.6c** Gary Gaugler/Visuals Unlimited/Getty Images **35.6d** Oliver Meckes/E.O.S./Max-Planck-Institut-Tübingen/Photo Researchers **35.12L/R** Abbott Laboratories **35.14a** Hans R. Gelderblom/Eye of Science **35.14b** From R. H. Meints et al., *Virology* 113: 698–703, fig. C (1981). Photo James L. Van Etten, University of Nebraska. **35.16** Jean Roy/CDC **35.17** Philip Leder, Harvard Medical School **35.18** Nigel Cattlin/Holt Studios/Photo Researchers **35.19** Lowell Georgia/Science Source/Photo Researchers **35.20** David Parker/SPL/Photo Researchers

Chapter 36 Opener Thomas Marent/Minden Pictures **36.5a** Matt Meadows/Peter Arnold **36.5bL** Alain Mafart-Renodier/Biosphoto/Peter Arnold **36.5bR** Rukhsana Photography/Fotolia **36.8a** Geoff Dann/Dorling Kindersley **36.8b** Dorling Kindersley **36.8c** Lee W. Wilcox **36.8d** Nigel Cattlin/FLPA/Minden Pictures **36.8e** Will Cook/Charles W. Cook **36.9a/b** Lee W. Wilcox **36.9c/d** David T. Webb **3636.10a/b/c/d** Lee W. Wilcox Victoria Firmston/GAP Photos/Getty Images **36.11** RDF/Visuals Unlimited **36.12a/d** Lee W. Wilcox **36.12b** Ivan Kmit/Alamy **36.12c** Doug Wechsler/Animals Animals–Earth Scenes **36.12e** Torsten Brehm/Nature Picture Library **36.13a** Ed Reschke/Peter Arnold **36.13b** M. I. Walker/Photo Researchers **36.16a** Keith Wheeler/SPL/Photo Researchers **36.16b** ISM/Phototake **36.18L** Ed Reschke/Peter Arnold **36.18R** Ed Reschke/Peter Arnold **36.19** Andrew Syred/SPL/Photo Researchers **36.20a** John Durham/Photo Researchers **36.20b** Lee W. Wilcox **36.21L/R** Lee W. Wilcox **36.22a/c** Lee W. Wilcox **36.22b** G. Shih and R. Kessel/Visuals Unlimited/Getty Images **36.23a** Paul Schulte, University of Nevada **36.23b** Jack Bostrack/Visuals Unlimited **36.24a** John D. Cunningham/Visuals Unlimited/Getty Images **36.24b/c** Richard Kessel & Gene Shih/Visuals Unlimited/Getty Images **36.25L/M/R** Lee W. Wilcox **36.26L** Biodisc/Visuals Unlimited/Alamy **36.26R** Michael Clayton **36.27a** Lee W. Wilcox **36.27b** Biodisc/Visuals Unlimited/Alamy **36.27c** Adam Hart-Davis/SPL/Photo Researchers

Chapter 37 Opener Creatas/Photolibrary **37.4L/R** David Cook/blueshiftstudios/Alamy **37.6L/R** Lee W. Wilcox **37.8** Bruce Peters **37.10T** Ken Wagner/Phototake NYC **37.10M** G. Shih and R. Kessel/Visuals Unlimited **37.10B** Lee W. Wilcox **37.13** Lee W. Wilcox **37.18** Martin H. Zimmerman/Harvard Forest **37.22** Michael R. Sussman/American Society of Plant Biologists

Chapter 38 Opener Angelo Cavalli/Getty Images **38.2a** Photo Researchers **38.2b** Nigel Cattlin/Photo Researchers **38.2c** Photo Researchers **38.3** Emanuel Epstein, University of California at Davis **38.6a** The Institute of Texan Cultures **38.6b** Steve Ringman/ The Seattle Times **38.9** Ed Reschke/Peter Arnold **38.13b** Eduardo Blumwald **38.14T** Hugh Spencer/Photo Researchers **38.14BL** Andrew Syred/SPL/Photo Researchers **38.14BR** E.H. Newcomb & S. R. Tandon/Biological Photo Service **38.16a/b** Ed Reschke/Peter Arnold **38.17L** Frank Greenaway/Dorling Kindersley Media Library **38.17R** Courtesy Carol A. Wilson and Clyde L. Calvin **38.18** Noah Elhardt

Chapter 39 Opener Lee W. Wilcox **39.3** Malcolm B. Wilkins, University of Glasgow **39.4bL/R** John M. Christie/AAAS **39.9T/B** Malcolm B. Wilkins, University of Glasgow **39.11a** American Society of Plant Biologists **39.14** Thomas Bjorkman **39.15** Donald Specker/Animals Animals–Earth Scenes **39.16L** Richard Shiell/Animals Animals–Earth Scenes **39.16R** blickwinkel/Alamy **39.17a/b** Lee W. Wilcox **39.19** Malcolm B. Wilkins, University of Glasgow, Glasgow, Scotland, U.K. **39.23** Adel A. Kader, University of California at Davis **39.29** Nigel Cattlin/Holt Studios International/Photo Researchers

Chapter 40 Opener Walter Siegmund/Wikimedia Commons. GNU Free Documentation License **40.3a** Dan Suzio/Photo Researchers **40.3b/c** Jerome Wexler/Photo Researchers **40.6bT** berniekasper.com **40.6bM** Rod Planck/Photo Researchers **40.6bB** Tom & Therisa Stack/Tom Stack & Associates **40.7a/b** Leonard Lessin/Photo Researchers **40.8a** iStockphoto **40.8bT/B** Lee W. Wilcox **40.11L** Frans Lanting/Corbis **40.11R** Nick Garbutt/NPL/Minden Pictures **40.17aL** Brian Johnston **40.17aR** Pablo Galán Cela/age fotostock **40.17bL** Kerstin Layer/Mauritius/Photolibrary **40.17bR** Laurie Campbell/NHPA/Photoshot **40.17cL** Nicholas and Sherry Lu Aldridge/FLPA/Minden Pictures **40.17cR** James Hardy/PhotoAlto Agency RF Collections/Getty Images **40.18** Lee W. Wilcox **40.19L** Scott Camazine/Alamy **40.19M** Jonathan Watts/OSF/Photolibrary **40.19R** Daniel Heuclin/NHPA/Photoshot

Chapter 41 Opener Joe McDonald/Corbis **41.2** The Royal Society of London **41.3aL** Educational Images/Custom Medical Stock Photo **41.3aR** Nina Zanetti **41.3b** Peter Arnold/Alamy **41.3cL/R** Nina Zanetti **41.3d** Carolina Biological Supply Company/Phototake **41.4a** Deco Images II/Alamy **41.4b** Dennis Kunkel Microscopy/Phototake **41.5a** ISM/Phototake **41.5b** Manfred Kage/Peter Arnold **41.5c** Biophoto Associates/Photo Researchers **41.6** Ed Reschke/Peter Arnold **41.8** Tom Stewart/CORBIS **41.12a** G. Kruitwagen, H.P.M. Geurts and E.S. Pierson/Radboud University Nijmegen (vcbio.science.ru.nl/en) **41.12b** Innerspace Imaging/SPL/Photo Researchers **41.12c** P. M. Motta, A. Caggiati, G. Macchiarelli/SPL/Photo Researchers **41.14** Derrick Hamrick/imagebroker.net/Photolibrary **41.15** Nutscode/T Service/Photo Researchers **41.17** From J. E. Heyning and J. G. Mead, *Science* 278: 1138–1140, fig. 1b (1997). Photo Natural History Museum of Los Angeles.

Chapter 42 Opener Frans Lanting **42.13a** Image by H. Wartenberg, © H. Jastrow, Electron Microscopic Atlas (www.drjastrow.de)

Chapter 43 Opener Jonathan Blair/Corbis **43.2a** Wallace63/Wikimedia. Creative Commons Attribution Share Alike 3.0 License. **43.2b** Kim Taylor/NPL/Minden Pictures **43.4L/M/R** Karel F. Liem **43.12** Biophoto Associates/Photo Researchers

Chapter 44 Opener Tony Freeman/PhotoEdit **44.4a** Walter E. Harvey/Photo Researchers **44.4b** John D. Cunningham/Visuals Unlimited **44.19a** Lennart Nilsson, Albert Bonniers Forlag AB **44.19b** Kessel & Kardon/Tissues & Organs/Visuals Unlimited/Getty Images

Chapter 45 Opener Marcus E. Raichle **45.9a** C. Raines/Visuals Unlimited **45.11** Dennis Kunkel/Visuals Unlimited **45.20a** Genny Anderson **45.22a** Elsevier Science Ltd.

Chapter 46 Opener Neil Hardwick/Alamy **46.1** Dmitry Smirnov **46.3a** Carole M. Hackney **46.7a** David Scharf/Peter Arnold **46.9a** Don Fawcett & T. Kwwabara/Photo Researchers **46.11L/R** David Quillin **46.17T/B** James E. Dennis/Phototake NYC **T46.1L** ISM/Phototake **T46.1M** Manfred Kage/Peter Arnold **T46.1R** Biophoto Associates/Photo Researchers

Chapter 47 Opener Ralph A. Clevenger/Corbis **47.5L** Stephen Dalton/Photo Researchers **47.5M** Duncan McEwan/Nature Picture Library **47.5R** Bernard Castelein/Nature Picture Library **47.6L** BIOS/Heras Joël/Peter Arnold **47.6M** Hans Pfletschinger/Peter Arnold **47.6R** Perennou Nuridsany/Photo Researchers **47.8** The Jackson Laboratory

Chapter 48 Opener Jurgen & Christine Sohns/FLPA/Minden Pictures **48.1a** NHPA/Photoshot **48.1b** Andrew J. Martinez/Photo Researchers **48.1c** Charles J. Cole **48.2** Bruce J. Russell/BioMEDIA Associates **48.5T/B** From C. S. C. Price et al., *Nature* 400:449–452, figs. 2 and 3 (1999). © Macmillan Magazines Ltd. Photo Jerry A. Coyne, University of Chicago. **48.6** From M. Winterbottom, T. Burke, and T. R. Birkhead, *Nature* 399: 28, fig. 1b (May 6 1999). Photo Tim R. Birkhead, The University of Sheffield. **48.7a** Suzanne L. Collins/Photo Researchers **48.8a** Johanna Liljestrand Rönn **48.8a inset** Fleur Champion de Crespigny **48.16a** Geoff Shaw, Zoology, University of Melbourne, Australia/Wikipedia. GNU Free Documentation License. **48.16b** Mitsuaki Iwago/Minden Pictures **48.16c** Mitsuaki Iwago/Minden Pictures **48.17a** Claude Edelmann/Petit Format/Photo Researchers **48.17b** Lennart Nilsson, Albert Bonniers Forlag AB **48.17c** Petit Format/Nestle/Science Source/Photo Researchers **48.19aL/R** AAAS

Chapter 49 Opener Eye of Science/Photo Researchers **49.1** Susumu Nishinaga/Photo Researchers **49.2a** Don W. Fawcett/Photo Researchers **49.2b/c** Steve Gschmeissner/SPL/Photo Researchers **49.5a/c** David M. Phillips/Visuals Unlimited **49.5b** Steve Gschmeissner/SPL/Photo Researchers

Chapter 50 Opener Jeff Rotman/NPL/Minden Pictures **50.1a** Paul Nicklen/National Geographic/Getty Images **50.1b** Sergey Gorshkov/Minden Pictures **50.1c** Werner & Kerstin Layer/Photolibrary **50.1d** Lynn M. Stone/NPL/Minden Pictures **50.6a/b** Gerry Ellis/Minden Pictures **50.6c** Tim Fitzharris/Minden Pictures **50.7T** George Gerster/Photo Researchers **50.7B** Rich Wheater/Aurora/Getty Images **50.8** David Zimmerman/UpperCut Images/Getty Images **50.12** Gerry Ellis/Minden Pictures **50.14** Michael & Patricia Fogden/Minden Pictures **50.16** D. Hartnett, Konza Prairie Biological Station **50.18L/R** Alex Hyde/Minden Pictures **50.20L** Tim Fitzharris/Minden Pictures **50.20R** Stephen J. Krasemann/NHPA/Photoshot **50.22** Pixtal Images/Photolibrary **50.27** Fairfax Photos/Sandy Scheltema/The Age **50.29T** Purestock/Getty Images **50.29B** Comstock/Getty Images **50.30** Bruce Coleman/Alamy **50.31** Michael & Patricia Fogden/Minden Pictures **50.32 inset** Steven Mihok (www.nzitrap.com) **50.33L/R** John M. Randall **50.34L** Shutterstock **50.34R** USGS Canyonlands Research Station (www.soilcrust.org)

Chapter 51 Opener Konrad Wothe/Minden Pictures **51.1** Peter Arnold/Alamy **51.3aL** Peet van Schalkwyk (www.flickr.com/photos/peetvs) **51.3aR** David Dean Peterson **51.4a** iStockphoto **51.6** Gary Woodburn (pbase.com/woody) **51.8** Operation Migration (www.operationmigration.org) **51.11a** Shinji Kusano/Nature Production/Minden Pictures **51.11b** James E. Lloyd/Animals Animals–Earth Scenes **51.12** Bryan D. Neff **51.13** Tom Vezo/Peter Arnold

Chapter 52 Opener Annie Griffiths Belt/NGS Image Collection/Getty Images **52.1** David Kjaer/NPL/Minden Pictures **52.6** Mark Trabue/USDA/NRCS **52.9L** JC Schou/Biopix.dk (www.Biopix.dk) **52.9R** Marko Nieminen, University of Helsinki, Helsinki, Finland **52.13a** John Downer/NPL/Minden Pictures

Chapter 53 Opener Scott Tuason/imagequestmarine.com **53.1a** PREMAPHOTOS/Nature Picture Library **53.1b** David Tipling/Alamy **53.8L/R** From P. R. Grant and B. R. Grant, Evolution of character displacement in Darwin's finches, *Science* 313: 224–226, fig. 1 (2006). **53.9L** blickwinkel/Alamy **53.9M** Chris Newbert/Minden Pictures **53.9R** J. Sneesby/B. Wilkins/Stone/Getty Images **53.11L** Rainer Zenz/Wikimedia Commons.GNU Free Documentation License and Creative Commons Attribution Share Alike. **53.11R** CSIRO Marine and Atmospheric Research (www.scienceimage.csiro.au) **53.13a** Thomas G. Whitham/Northern Arizona University **53.15a/b** Stephen P. Yanoviak **53.15b** Stephen P. Yanoviak **53.16a** Mark Moffett/Minden Pictures **53.16b** David B. Fleetham/OSF/Photolibrary **53.19a** Nancy Sefton/Photo Researchers **53.19b** Thomas Kitchin/Tom Stack & Associates **53.20** Raymond Gehman/Corbis **53.21a** Tony C. Caprio **53.22T** G. Carleton Ray/Photo Researchers **53.22MT** Michael P. Gadomski/Animals Animals–Earth Scenes **53.22MB** James P. Jackson/Photo Researchers **53.22B** Michael P. Gadomski/Photo Researchers **53.23L/R** Glacier Bay National Park **53.23M** Christopher L. Fastie, Middlebury College

Chapter 54 Opener Jan Tove Johansson/The Image Bank/Getty Images **54.8** MODIS/NASA **54.12a** Richard Hartnup/Wikipedia Commons **54.12b** Randall J. Schaetzl, Michigan State University **54.13** John Campbell, U.S. Forest Service, Northern Research Station **54.15** Pornchai Kittiwongsakul/AFP/Getty Images

Chapter 55 Opener Ben Simmons/Photolibrary **55.1a** DLILLC/Corbis **55.1b** Xiaoqiang Wang **55.2** Edward S. Ross, California Academy of Sciences **55.5** Theo Allofs/The Image Bank/Getty Images **55.7** Karl Ammann (KarlAmmann.com) **55.8** Lester Lefkowitz/Corbis **55.9T/B** NASA/USGS (earthobservatory.nasa.gov) **55.12a** F. Mercay, BIOS/Peter Arnold **55.16** Ellen Damschen **55.17** Photo by Ariel Poster, provided by Marlboro Productions, photographed while filming TAKING ROOT: The Vision of Wangari Maathai, A Film by Lisa Merton & Alan Dater (www.takingrootfilm.com), © 2008. **55.18a/b** Daniel H. Janzen, University of Pennsylvania **55.19a/b/c** Charles Curtin **Appendix A BS9.3** From J.P. Ferris et al., *Nature* 381:59–61, fig. 2 (1996). Photo James P. Ferris, Rensselaer Polytechnic Institute. **BS10.1a** Biology Media/Photo Researchers **BS10.1b** Janice Carr/CDC

BS10.2a/b Michael W. Davidson/Florida State University /Molecular Expressions **BS10.3** Rosalind Franklin/Photo Researchers **BS12.1aL** National Cancer Institute **BS12.1aR** E.S. Anderson/Photo Researchers **BS12.1b** Sinclair Stammers/Photo Researchers **BS14.1a** Kwangshin Kim/Photo Researchers **BS14.1b** Mark J. Grimson & Richard L. Blanton **BS14.1c** Holt Studios International/Photo Researchers **BS14.1d** Custom Medical Stock Photo **BS14.1e** Graphic Science/Alamy **BS14.1f** Sinclair Stammers/Photo Researchers **BS14.1g** iStockphoto

Illustration and Table Credits

Chapter 1 **1.3** Adapted by permission from Elsevier from S. P. Moose, J. W. Dudley, and T. R. Rocheford. 2004. Maize selection passes the century mark: A unique resource for 21st century genomics. *Trends in Plant Science* 9: 358–364, Fig 1a. © 2004. **1.7a** Adapted by permission of Wiley-Blackwell from T. P. Young and L. A. Isbell. 1991. Sex differences in giraffe feeding ecology: Energetic and social constraints. *Ethology* 87: 79–80, Figs 5a, 6a. **1.8** Adapted by permission of AAAS and the author from M. Wittlinger, R. Wehner, and H. Wolf. 2006. The ant odometer: Stepping on stilts and stumps. *Science* 312: 1965–1967, Figs 1, 2, 3.

Chapter 2 **Table 2.1** Reproduced by permission of Pearson Education, Inc., from J. McMurry and R. C. Fay, *Chemistry*, 4th ed., Table 8.1. © 2004.

Chapter 3 **Opener** PDB ID: 2DN2. S.Y. Park et al. 2006. 1.25 Å resolution crystal structures of human haemoglobin in the oxy, deoxy and carbonmonoxy forms. *J Mol Biol* 360: 690–701. **3.9a** PDB ID: 1YTB. Y. Kim et al. 1993. Crystal structure of a yeast TBP/TATA-box complex. *Nature* 365: 512–520. **3.9b** PDB ID: 2F1C. G.V. Subbarao and B. van den Berg. 2006. Crystal structure of the monomeric porin OmpG. *J Mol Biol* 360: 750–759. **3.9c** PDB ID: 2PTN. J. Walter et al. 1982. On the disordered activation domain in trypsinogen. chemical labelling and low-temperature crystallography. *Acta Crystallogr, Sect B* 38: 1462. **3.9d** PDB ID: 1CLG. J.M. Chen 1991. An energetic evaluation of a "Smith" collagen microfibril model. *J Protein Chem* 10: 535–552. **3.12bL** PDB ID: 2MHR. S. Sheriff et al. 1987. Structure of myohemerythrin in the azidomet state at 1.7/1.3 Å resolution. *J Mol Biol* 197: 273–296. **3.12bM** PDB ID: 1FTP. N.H. Haunerland et al. 1994. Three-dimensional structure of the muscle fatty-acid-binding protein isolated from the desert locust *Schistocerca gregaria*. *Biochemistry* 33: 12378–12385. **3.12bR** PDB ID: 1IXA. M. Baron et al. 1992. The three-dimensional structure of the first EGF-like module of human factor IX: comparison with EGF and TGF-alpha. *Protein Sci* 1: 81–90. **3.13a** PDB ID: 1D1L. P.B. Rupert et al. 2000. The structural basis for enhanced stability and reduced DNA binding seen in engineered second-generation Cro monomers and dimers. *J Mol Biol* 296: 1079–1090. **3.13b** PDB ID: 2DN2. S. Y. Park et al. 2006. 1.25 resolution crystal structures of human haemoglobin in the oxy, deoxy and carbonmonoxy forms. *J Mol Biol* 360: 690–701. **3.19** PDB IDs: 1Q18, 1SZ2. V.V. Lunin et al. 2004. Crystal structures of *Escherichia coli* ATP-dependent glucokinase and its complex with glucose. *J Bacteriol* 186: 6915–6927. **3.23a/b** Data from T. Hansen, B. Schlichting, and P. Schönheit. 2002. Glucose-6-phosphate dehydrogenase from the hyperthermophilic bacterium *Thermotoga maritima*: Expression of the *g6pd* gene and characterization of an extremely thermophilic enzyme. *FEMS Microbiology Letters*, 216: 249–253, Fig 1; N. N. Nawani and B. P. Kapadnis. 2001. One-step purification of chitinase from *Serratia marcescens* NK1, a soil isolate. *Journal of Applied Microbiology* 90: 803–808, Fig 3; N. N. Nawani et al. 2002. Purification and characterization of a thermophilic and acidophilic chitinase from *Microbispora* sp.V2. *Journal of Applied Microbiology* 93: 965–975, Fig 7.

Chapter 4 **4.11** PDB ID: 1GRZ. B.L. Golden et al. 1998. A preorganized active site in the crystal structure of the *Tetrahymena* ribozyme. *Science* 282: 259–264.

Chapter 6 **6.11** Data fromJ. de Gier, J. G. Mandersloot, and L. L. van Deenen. 1968. Lipid composition and permeability of liposomes. *Biochimica et Biophysica Acta* 150: 666–675. **6.25a** PDB ID: 1J4N. H. Sui et al. Structural basis of water-specific transport through the AQP1 water channel. 2001. *Nature* 414:872–878 **6.25b** PDB IDs: 1ORS, 1ORQ. Y. Jiang et al. 2003. X-ray structure of a voltage-dependent K^+ channel. *Nature* 423: 33–41.

Chapter 8 **8.7b** © 2002. From *Molecular Biology of the Cell*, 4th edition, by B. Alberts et al., Fig 19.5, p. 1069. Reproduced by permission of Garland Science/Taylor & Francis, LLC.

Chapter 9 **9.2L** PDB ID: 1IRK. S. R. Hubbard et al. 1994. Crystal structure of the tyrosine kinase domain of the human insulin receptor. *Nature* 372: 746–754. **9.2R** PDB ID: 1IR3. S. R. Hubbard 1997. Crystal structure of the activated insulin receptor tyrosine kinase in complex with peptide substrate and ATP analog. *EMBO J* 16: 5572–5581. **9.12** PDB ID: 4PFK. P. R. Evans and P. J. Hudson. 1981. Phosphofructokinase: structure and control. *Philos Trans R Soc Lond B Biol Sci* 293: 53–62.

Chapter 14 **14.5** Adapted by permission of Dr. Matthew Meselson after M. Meselson and F. W. Stahl. 1958. The replication of DNA in *Escherichia coli*. *PNAS* 44: 671–682, Fig 6. **14.17a** Data from J. E. Cleaver. 1968. Defective repair replication of DNA in xeroderma pigmentosum. *Nature* 218: 652–656. **14.17b** Adapted by permission of Macmillan Publishers Ltd from J. E. Cleaver. 1968. Defective repair replication of DNA in xeroderma pigmentosum. *Nature* 218: 652–656, Fig 5. © 1968. **14–UN02** Graph adapted by permission of the Radiation Research Society from P. Howard-Flanders and R. P. Boyce. 1966. DNA repair and genetic recombination: Studies on mutants of *Escherichiacoli* defective in these processes. *Radiation Research Supplement* 6: 156–184, Fig 8.

Chapter 16 **16.2** PDB ID: 3IYD. B. P. Hudson et al. 2009. Three-dimensional EM structure of an intact activator-dependent transcription initiation complex. *PNAS* 106:19830–19835. **16.11** PDB ID: 1ZJW. I. Gruic-Sovulj et al. 2005. tRNA-dependent aminoacyl-adenylate hydrolysis by a nonediting class I aminoacyl-tRNA synthetase. *J Biol Chem* 280: 23978–23986. **16.12** Adapted by permission of the publisher from M. B. Hoagland et al. 1958. A soluble ribonucleic acid intermediate in protein synthesis. *Journal of Biological Chemistry* 231: 241–257, Fig 6. © 1958 American Society for Biochemistry and Molecular Biology. **16.14b** PDB IDs: 3FIK, 3FIH. E. Villa et al. 2009. Ribosome-induced changes in elongation factor Tu conformation control GTP hydrolysis. *PNAS* 106: 1063–1068.

Chapter 18 **Opener** PDB ID: 1ZBB. T. Schalch et al. 2005. X-ray structure of a tetranucleosome and its implications for the chromatin fibre. *Nature* 436: 138–141.

Chapter 20 **20.4** Adapted by permission of AAAS and the author from S. J. Giovannoni. 2005. Genome streamlining in a cosmopolitan oceanic bacterium. *Science* 309: 1242–1245, Fig 1. **20.7b** Reproduced by permission of GENDIA from www.paternity.be/information_EN .html#identitytest , Examples 1 and 2. **20.9** Data from www.pantherdb.org . P. D. Thomas et al. 2003. PANTHER (Protein ANalysis Through Evolutionary Relationships): A library of protein families and subfamilies indexed by function.

Chapter 21 **21.10a/b** Adapted by permission of Macmillan Publishers Ltd after S. B. Carroll. 1995. Homeotic genes and the evolution of arthropods and chordates. *Nature* 376: 479–485, Fig 1. © 1999.

Chapter 24 **24.4** After E. B. Daeschler, Neil H. Shubin, and Farish A. Jenkins, Jr. 2006. A Devonian tetrapod-like fish and the evolution of the tetrapod body plan. *Nature* 440:757–763, Fig 6; P. E. Ahlberg and J. A. Clack. 2006. A firm step from water to land. *Nature* 440: 747–749, Fig 1; N. H. Shubin, E. B. Daeschler, and F. A. Jenkins, Jr. 2006. The pectoral fin of *Tiktaalik roseae* and the origin of the tetrapod limb. *Nature* 440: 764–771, Fig 4; M. Hildebrand and G. Goslow. 2001. *Analysis of Vertebrate Structures*, 5th ed. John Wiley and Sons, Inc. **24.11** *Indohyus* reproduced by permission of Macmillan Publishers Ltd after J. G. M. Thewissen et al. 2007. Whales originated from aquatic artiodactyls in the Eocene epoch of India. *Nature* 450: 1190–1194, Fig 5. © 2007. *Rhodocetus* reproduced by permission of AAAS after P. D. Gingerich et al. 2001. Origin of whales from early Artiodactyls: Hands and feet of Eocene Protocetidae from Pakistan. *Science* 293:2239–2242, Fig 3. *Dorudon* reproduced by permission after P. D. Gingerich et al. 2009. New protocetid whale from the middle Eocene of Pakistan: Birth on land, precocial development, and sexual dimorphism. *PLoS ONE* 4(2): e4366, Fig 1B. *Delphinapterus* reproduced by permission of Skulls Unlimited International, Inc. (www.skullsunlimited.com). **24.14** Adapted by permission from D. P. Genereux and C. T. Bergstrom. 2005. Evolution in action: Understanding antibiotic resistance, Fig 3. In J. Cracraft and R. W. Bybee (eds.), *Evolutionary Science and Society: Educating a New Generation*, pp. 145–153. Colorado Springs, CO: BSCS. **24.16** Data from P. T. Boag and P. R. Grant. 1981. Intense natural selection in a population of Darwin's finches (Geospizinae) in the Galápagos. *Science* 214: 82–85, Table 1. **24.17** Body size and beak shape graphs reproduced by permission of AAAS from P. R. Grant and B. R. Grant. 2002. *Science* 296: 707–711, Fig 1. Beak size graph reproduced by permission of AAAS from P. R. Grant and B. R. Grant. 2006. Evolution of Character Displacement in Darwin's Finches. *Science* 313: 224–226, Fig 2.

Chapter 25 **Table 25.2** Data from T. Markow et al. 1993. *HLA* polymorphism in the Havasupai: Evidence for balancing selection. *American Journal of Human Genetics* 53: 943–952, Table 3. **25.3b** Data from C. R. Brown and M. B. Brown. 1998. Intense natural selection on body size and wing and tail asymmetry in cliff swallows during severe weather. *Evolution* 52: 1461–1475. **25.4b** Data from M. N. Karn, H. Lang-Brown, H. MacKenzie, and L. S. Penrose. 1951. Birth weight, gestation time and survival in sibs. *Annals of Eugenics* 15: 306–322. **25.5b** Data from T. B. Smith. 1987. Bill size polymorphism and intraspecific niche utilization in an African finch. *Nature* 329: 717–719. **25.6** Reproduced by permission of Pearson Education, Inc., from S. Freeman and J. Herron, *Evolutionary Analysis*, 3rd ed., Figs 6.15a, 6.15c. © 2004. **25.10b** Adapted by permission of Macmillan Publishers Ltd from E. Postma and A. J. van Noordwijk. 2005. Gene flow maintains a large genetic difference in clutch size at a small spatial scale. *Nature* 433: 65–68, Fig 5b. © 2005. **25.13** Adapted by permission of Wiley-Blackwell from M. O. Johnston. 1992. Effects of cross and self-fertilization on progeny fitness in *Lobelia cardinalis* and *L. siphilitica*. *Evolution* 46: 688–702, Fig 1. **Table 25.4** Reproduced by permission of W. H. Freeman and Company from Curt Stern, *Principles of Human Genetics*, 3rd ed., Table 5.8. © 1973 by W. H. Freeman and Company. **25.14b** Data from J. D. Blount et al. 2003. Carotenoid modulation of immune function and sexual attractiveness in zebra finches. *Science* 300: 125–127, Fig 2. **25.16b, c** Data from B. J. Le Boeuf and R. S. Peterson. 1969. Social status and mating activity in elephant seals. *Science* 163: 91–93.

Chapter 26 **26.3a** Adapted by permission of AAAS and the author after J. C. Avise and W. S. Nelson. 1989. Molecular genetic relationships of the extinct dusky seaside sparrow. *Science* 243: 646–648, Fig 2. **26.3b** J. C. Avise and W. S. Nelson. 1989. Molecular genetic relationships of the extinct dusky seaside sparrow. *Science* 243: 646–648. **26.5b** Data from N. Knowlton et al. 1993. Divergence in proteins, mitochondrial DNA, and reproductive compatibility across the Isthmus of Panama. *Science* 260:1629–1632, Fig 1. **26.7** Adapted by permission of Wiley-Blackwell from H. R. Dambroski et al. 2005. The genetic basis for fruit odor discrimination in *Rhagoletis* flies and its significance for sympatric host shifts. *Evolution* 59: 1953–1964, Figs 1A, 1B. **26.10b** Data from S. E. Rohwer et al. 2001. Plumage and mitochondrial DNA haplotype variation across a moving hybrid zone. *Evolution* 55: 405–422. **26.11** Data from L. H. Rieseberg et al. 1996. Role of gene interactions in hybrid speciation: Evidence from ancient and experimental hybrids. *Science* 272: 741–745.

Chapter 27 **27.5a/b/c** Data adapted by permission of the publisher from M. A. Nikaido, P. Rooney, and N. Okada. 1999. Phylogenetic relationships among cetartiodactyls based on insertions of short and long interspersed elements: Hippopotamuses are the closest extant relatives of whales. *PNAS* 96: 10261–10266, Figs 1, 7. © 1999 National Academy of Sciences, U.S.A. **27.8, 27.9, 27.10** Data from the

International Commission on Stratigraphy, "International Stratigraphic Chart 2009" (www.stratigraphy.org/column.php?id=Chart/Time Scale). The data on this site are modified from F. M. Gradstein and J. C Ogg (eds). 2004. *A Geologic Time Scale 2004.* Cambridge, UK: Cambridge University Press; and J. G. Ogg, G. Ogg, and F. M. Gradstein. 2008. *The Concise Geologic Time Scale.* Cambridge, UK: Cambridge University Press. **27.11** B. G. Baldwin, D. W. Kyhos, and J. Dvorak. 1990. Chloroplast DNA evolution and adaptive radiation in the Hawaiian Silversword Alliance (Asteraceae–Madiinae). *Annals of the Missouri Botanical Garden* 77: 96–109, Fig 2. **27.12c** J. B. Losos, K. I. Warheitt, and T. W. Schoener. 1997. Adaptive differentiation following experimental island colonization in *Anolis* lizards. *Nature* 387: 70–73. **27.15** Data from J. W. Valentine, D. H. Erwin, and D. Jablonski. 1996. Developmental evolution of metazoan body plans: The fossil evidence. *Developmental Biology* 173: 373–381, Fig 3; R. de Rosa et al. 1999. Hox genes in brachiopods and priapulids and protostome evolution. *Nature* 399: 772–776; D. Chourrout et al. 2006. Minimal ProtoHox cluster inferred from bilaterian and cnidarian Hox complements. *Nature* 442: 684–687. **27.16** Data from M. J. Benton. 1995. Diversification and extinction in the history of life. *Science* 268: 52–58. **27.17a** Adapted by permission of AAAS and the author from W. Alvarez, F. Asaro, and A. Montanari. 1990. Iridium profile for 10 million years across the Cretaceous-Tertiary boundary at Gubbio (Italy). *Science* 250: 1700–1702, Fig 1.

Chapter 28 **28.2** Reproduced by permission of the American Medical Association from G. L. Armstrong, L. A. Conn, and R. W. Pinner. 1999. Trends in infectious disease mortality in the United States during the 20th century. *JAMA* 281: 61–66, Fig 1. © 1999 American Medical Association. All rights reserved. **28.6** After J. G. Elkins et al. 2008. A korarchaeal genome reveals insights into the evolution of the Archaea. *PNAS* 105: 8102–8107, Fig 2; F. D. Ciccarelli et al. 2006. Toward automatic reconstruction of a highly resolved tree of life. *Science* 311:1283–1287, Fig 2. **28.11** Adapted by permission of Pearson Education, Inc., from M. T. Madigan and J. M. Martinko. 2006. *Brock Biology of Microorganisms,*12th ed., Fig 5.9. © 2009.

Chapter 29 **29.1, 29.7** S. M. Adl et al. 2005. The new higher level classification of eukaryotes with emphasis on the taxonomy of protists.*Journal of Eukaryotic Microbiology* 52: 399–451; N. Arisue, M. Hasegawa, and T. Hashimoto. 2005. Root of the Eukaryota tree as inferred from combined maximum likelihood analyses of multiple molecular sequence data. *Molecular Biology and Evolution* 22: 409–420; V. Hampl et al. 2009. Phylogenomic analyses support the monophyly of Excavata and resolve relationships among eukaryotic "supergroups." *PNAS* 106: 3859–3864, Figs 1, 2, 3; J. D. Hackett et al. 2007. Phylogenomic analysis supports the monophyly of cryptophytes and haptophytes and the association of Rhizaria with chromalveolates.*Molecular Biology and Evolution* 24: 1702–1713, Fig 1; P. Schaap et al. 2006. Molecular phylogeny and evolution of morphology in the social amoebas. *Science* 314: 661–663.

Chapter 30 **30.3** Adapted by permission of Pearson Education, Inc., from O. Owen et al. 1998. *Natural Resource Conservation: Management for a Sustainable Future* 7th ed., p. 509, Fig 21.2. © 1998.**30.7, 30.10, 30. 24** Y.-L. Qiu et al. 2006. The deepest divergences in land plants inferred from phylogenomic evidence. *PNAS* 103:15511–15516, Fig 1; K. S. Renzaglia et al. 2007. Bryophyte phylogeny: Advancing the molecular and morphological frontiers. *The Bryologist* 110: 179–213; J. F. Pombert et al. 2005. The chloroplast genome sequence of the green alga *Pseudendoclonium akinetum* (Ulvophyceae) reveals unusual structural features and new insights into the branching order of chlorophyte lineages.*Molecular Biology and Evolution* 22: 1903–1918. **30.22** Data from S. D. Johnson and K. E. Steiner. 1997. Long-tongued fly pollination and evolution of floral spur length in the *Disa draconis* complex (Orchidaceae). *Evolution* 51: 45–53. **30.26** P. S. Soltis and D. E. Soltis. 2004. The origin and diversification of angiosperms. *American Journal of Botany* 91: 1614–1626, Figs 1, 2, 3.

Chapter 31 **31.7** S. M. Adl et al. 2005. The new higher level classification of eukaryotes with emphasis on the taxonomy of protists.*Journal of Eukaryotic Microbiology* 52: 399–451; N. Arisue, M. Hasegawa, and T. Hashimoto. 2005. Root of the Eukaryota tree as inferred from combined maximum likelihood analyses of multiple molecular sequence data. *Molecular Biology and Evolution* 22: 409–420; V. Hampl et al. 2009. Phylogenomic analyses support the monophyly of Excavata and resolve relationships among eukaryotic "supergroups." *PNAS* 106: 3859–3864, Figs 1, 2, 3; J. D. Hackett et al. 2007. Phylogenomic analysis supports the monophyly of cryptophytes and haptophytes and the association of Rhizaria with chromalveolates. *Molecular Biology and Evolution* 24: 1702–1713, Fig 1; P. Schaap et al. 2006. Molecular phylogeny and evolution of morphology in the social amoebas. *Science* 314: 661–663. **31.8** T. Y. James et. al. 2006. Reconstructing the early evolution of fungi using a six-gene phylogeny. *Nature* 443: 818–822, Fig 1.

Chapter 32 **32.9** A. M. A. Aguinaldo et al. 1997. Evidence for a clade of nematodes, arthropods, and other moulting animals. *Nature* 387: 489–493, Figs 1, 2, 3; C. W. Dunn et al. 2008. Broad phylogenomic sampling improves resolution of the animal tree of life. *Nature* 452: 745–750, Figs 1, 2.

Chapter 33 **33.2** A. M. A. Aguinaldo et al. 1997. Evidence for a clade of nematodes, arthropods, and other moulting animals. *Nature* 387: 489–493, Figs 1, 2, 3; C. W. Dunn et al. 2008. Broad phylogenomic sampling improves resolution of the animal tree of life. *Nature* 452: 745–750, Figs 1, 2. **33.7** T. H. Struck et al. 2007. Annelid phylogeny and the status of Sipuncula and Echiura. *Evolutionary Biology* 7: 571–11.

Chapter 34 **34.1** After C. W. Dunn et al. 2008. Broad phylogenomic sampling improves resolution of the animal tree of life. *Nature* 452: 745–750, Figs 1, 2; F. Delsuc et al. 2006. Tunicates and not cephalochordates are the closest living relatives of vertebrates. *Nature* 439: 965–968, Fig 1. **34.12, 44.21** J. E. Blair and S. B. Hedges.2005. Molecular phylogeny and divergence times of deuterostome animals. *Molecular Biology and Evolution* 22: 2275–2284, Figs 1, 3, 4. **34.16** Adapted by permission of Macmillan Publishers Ltd after E. B. Daeschler et al. 2006. A Devonian tetrapod-like fish and the evolution of the tetrapod body plan. *Nature* 440:757–763, Fig 6, © 2006; also after N. H. Shubin et al. 2006. The pectoral fin of Tiktaalik roseae and the origin of the tetrapod limb. *Nature* 440:764–771, Fig 4, © 2006. **34.22** J. E. Blair and S. B. Hedges. 2005. Molecular phylogeny and divergence times of deuterostome animals. *Molecular Biology and Evolution* 22: 2275–2284, Figs 1, 3, 4; F. R. Liu et al. 2001. Molecular and morphological supertrees for eutherian (placental) mammals. *Science* 291: 1786–1789; A. B. Prasad et al. 2008. Confirming the phylogeny of mammals by use of large comparative sequence data sets. *Molecular Biology and Evolution* 25:1795–1808. **34.36** A. B. Prasad et al. 2008. Confirming the phylogeny of mammals by use of large comparative sequence data sets. *Molecular Biology and Evolution* 25:1795–1808, Figs 1, 2, 3. **34.39** J. Z. Li et al. 2008. Worldwide human relationships inferred from genome-wide patterns of variation. *Science* 319: 1100–1104, Fig 1. **34.40** After L. L. Cavalli-Sforza and M. W. Feldman. 2003. The application of molecular genetic approaches to the study of human evolution. *Nature Genetics Supplement* 33: 266–275, Fig 3.

Chapter 35 **35.3** Adapted by permission of the publisher from J. G. Bartlett and R. D. Moore. 2003. Improving HIV Therapy. *Scientific American* July 2003: 30–37. Originally published July 1998. © 1998 Scientific American. All rights reserved. **35.4** Data compiled by the United Nations AIDS program. **35.15** F. Gao et al. 1999. Origin of HIV-1 in the chimpanzee *Pan troglodytes troglodytes. Nature* 397: 436–441, Fig 2.

Chapter 36 **36.4** Illustration adapted by permission from "Root Systems of Prairie Plants" by Heidi Natura. © 1995 Conservation Research Institute.

Chapter 37 **37.3** Data from R. G. Cline and W. S. Campbell. 1976. Seasonal and diurnal water relations of selected forest species. *Ecology* 57: 367–373, Fig 1. **37.12** Adapted by permission of the American Society of Plant Biologists from C. M. Wei, M. T. Tyree, and E. Steudle. 1999. Direct measurement of xylem pressure in leaves of intact maize plants. A test of the cohesion-tension theory taking hydraulic architecture into consideration. *Plant Physiology* 121: 1191–1205, Fig 6. © 1999 American Society of Plant Biologists. **37.22** Data from N. D. DeWitt and M. R. Sussman. 1995. Immunocytological localization of an epitope-tagged plasma membrane proton pump (H^+–ATPAse) in phloem companion cells. *Plant Cell* 7: 2053–2067, Fig 7.

Chapter 39 **39.5** After Darwin and Darwin, 1897. *The Power of Movement in Plants* (D. Appleton & Co. New York). **Table 39.1** Data from H. A. Borthwick et al. 1952. A reversible photoreaction controlling seed germination. *PNAS* 38: 662–666, Table 1.

Chapter 40 **40.12** Y.-L. Qiu et al. 2006. The deepest divergences in land plants inferred from phylogenomic evidence. *PNAS* 103:15511–15516, Fig 1; K. S. Renzaglia et al. 2007. Bryophyte phylogeny: Advancing the molecular and morphological frontiers. *The Bryologist* 110: 179–213; J. F. Pombert et al. 2005. The chloroplast genome sequence of the green alga *Pseudendoclonium akinetum* (Ulvophyceae) reveals unusual structural features and new insights into the branching order of chlorophyte lineages. *Molecular Biology and Evolution* 22: 1903–1918.

Chapter 41 **41.10** Data from K. Schmidt-Nielsen. 1984. *Scaling: Why is animal size so important?* Cambridge, UK: Cambridge University Press. **41.11** Adapted by permission of The Company of Biologists from P. R. Wells and A. W. Pinder. 1996. The respiratory development of Atlantic salmon. II. Partitioning of oxygen uptake among gills, yolk sac and body surfaces. *Journal of Experimental Biology* 199: 2737–2744, Figs 1A, 3.

Chapter 42 **42.14a/b** Adapted by permission of Elsevier from K. J. Ullrich, K. Kramer, and J. W. Boyer. 1961. Present knowledge of the counter-current system in the mammalian kidney. *Progress in Cardiovascular Diseases* 3: 395–431, Figs 5, 7. © 1961.

Chapter 43 **43.17a** Graph reproduced with permission from The American Diabetes Association from L. O. Schulz et al. 2006. Effects of traditional and western environments on prevalence of type 2 diabetes in Pima Indians in Mexico and the U.S. *Diabetes Care* 29: 1866–1871, Fig 1. © 2006 American Diabetes Association. **43.17b** Data from L. O. Schulz et al. 2006. Effects of traditional and western environments on prevalence of type 2 diabetes in Pima Indians in Mexico and the U.S. *Diabetes Care* 29: 1866–1871, Table 2.

Chapter 44 **44.7** Adapted by permission from Y. Komai. 1998. Augmented respiration in a flying insect. *Journal of Experimental Biology* 201: 2359–2366, Fig 7. **44.8** Adapted by permission from Y. Komai. 1998. Augmented respiration in a flying insect. *Journal of Experimental Biology* 201: 2359–2366, Fig 1. **44.11** W. Bretz and K. Schmidt Nielsen. 1971. Bird respiration: Flow patterns in the duck lung. *Journal of Experimental Biology* 54:103–118. **44.20** Data from E. H. Starling. 1896. On the absorption of fluids from the connective tissue spaces. *Journal of Physiology* 19: 312–326.

Chapter 45 **45.21** Adapted by permission from Elsevier from K. C. Martin et al. 1997. Synapse-specific, long-term facilitation of Aplysia sensory to motor synapses: A function for local protein synthesis in memory storage. *Cell* 91: 927–938. © 1997. **45.22b** Adapted by permission from Elsevier from K. C. Martin et al. 1997. Synapse-specific, long-term facilitation of Aplysia sensory to motor synapses: A function for local protein synthesis in memory storage. *Cell* 91: 927–938. © 1997.

Chapter 46 **46.2a/b** Data from J. E. Rose et al. 1971. Some effects of stimulus intensity on response of auditory nerve fibers in the squirrel monkey. *Journal of Neurophysiology* 34: 685–699. **46.12** Adapted by permission of AAAS and the author from David M. Hunt et al. 1995. The Chemistry of John Dalton's Color Blindness. *Science* 267: 984–988, Fig 3. **46.19** PDB ID: 1KWO. D. M. Himmel et al. 2002. Crystallographic findings on the internally uncoupled and near-rigor states of myosin: further insights into the mechanics of the motor. *PNAS* 99: 12645–12650.

Chapter 47 **47.10a/b** Data from G. V. Upton et al. 1973. Evidence for the internal feedback phenomenon in human subjects: Effects of ACTH on plasma CRF. *Acta Endocrinologica* 73: 437–443. **47.13** Data from D. Toft, and J. Gorski. 1966. A receptor molecule for estrogens: Isolation from the rat uterus and preliminary characterization. *PNAS* 55: 1574–1581. **47.15b** Data from T. W. Rall et al. 1956. The relationship of epinephrine and glucagon to liver phosphorylase. *Journal of Biological Chemistry* 224: 463–475. **47.16** Reproduced by permission of the Nobel Foundation from E. W. Sutherland. "Studies on the mechanism of hormone action." A lecture delivered in Stockholm, 11 December 1971. © 1971 Nobel Foundation.

Chapter 48 **48.3a** Reproduced by permission of AAAS from R. G. Stross and J. C. Hill. 1965. Diapause induction in Daphnia requires two stimuli. *Science* 150: 1462–1464, Fig 3. **48.3b** Data from O. T. Kleiven, P. Larsson, and A. Hobæk. 1992. Sexual reproduction in Daphnia magna requires three stimuli. *Oikos* 65: 197–206, Table 4. **48.5** Adapted by permission of Macmillan Publishers Ltd from C. S. C. Price, K. A. Dyer, and J. A. Coyne. 1999. Sperm competition between *Drosophila* males involves both displacement and incapacitation. *Nature* 400: 449–452, Fig 3c. © 1999. **48.7b** R. Shine and M. S. Y. Lee. 1999. A reanalysis of the evolution of viviparity and egg-guarding in squamate reptiles. *Herpetologica* 55: 538–549, Figs 1, 2, 3. **48.8b** Reproduced by permission of Elsevier from C. Hotzyand and G. Arnqvist. 2009. Sperm competition favors harmful males in seed beetles. *Current Biology* 19: 404–407. © 2009. **48.14, 48.15** Data from R. Stricker et al. 2006. Establishment of detailed reference values for luteinizing hormone, follicle stimulating hormone, estradiol, and progesterone during different phases of the menstrual cycle on the Abbott ARCHITECT® analyzer. *Clin Chem Lab Med* 44: 883–887, Tables 1A and 1B. **48.19b** Reproduced by permission of AAAS from C. Ikonomidou et al. 2000. Ethanol-induced apoptotic neurodegeneration and fetal alcohol syndrome. *Science* 287: 1056–1060, Fig 4. **48.21** Adapted by permission of the publisher and author from U. Högberg and I. Joelsson. 1985. The decline in maternal mortality in Sweden, 1931–1980. *Acta Obstetricia et Gynecologica Scandinavica* 64: 583–592, Fig 1. Taylor & Francis Group, www.informaworld.com.

Chapter 49 **49.6a** PDB ID: 1IGT. L. J. Harris et al. 1997. Refined structure of an intact IgG2a monoclonal antibody. *Biochemistry* 36: 1581–1597. **49.6b** PDB ID: 1TCR. K. C. Garcia et al. 1996. An αβ cell receptor structure at 2.5Å and its orientation in the TCR-MHC complex. *Science* 274: 209–219.

Chapter 50 **50.4a** Adapted by permission of the University of California Press Journals and the author from M. B. Saffo. 1987. New light on seaweeds. *BioScience* 37: 654–664, Fig 1. © 1987 American Institute of Biological Sciences. **50.28** Graphs adapted by permission of AAAS and the author from A. K. Knapp et al. 2002. Rainfall variability, carbon cycling, and plant species diversity in a mesic grassland. *Science* 298: 2202–2205, Figs 1, 2, 3.

Chapter 51 **51.2** After M. Sokolowski. 2001. *Drosophila*: Genetics meets behavior. *Nature Reviews Genetics* 2: 879–890, Fig 2. **51.3b** Data from R. E. Hegner, S. T. Emlen, and N. J. Demong. 1982. Spatial organization of the white-fronted bee-eater. *Nature* 298: 264–266. **51.4b** Adapted by permission of the author from D. Crews. 1975. Psychobiology of reptilian reproduction. *Science* 189: 1059–1065, Fig 2. **51.16** Data from M. Daly and M. Wilson. 1988. Evolutionary social psychology and family homicide. *Science* 242: 519–524, Fig 1.

Chapter 52

Table 52.1 Data reproduced with kind permission from Springer Science and Business Media from H.Strijbosch and R. C. M. Creemers, 1988. Comparative demography of sympatric populations of *Lacerta vivipara* and *Lacerta agilis*. *Oecologia* 76: 20–26. © 1988 Springer-Verlag. **52.3** Reproduced by permission of AAAS from C. K. Ghalambor and T. E. Martin. 2001. *Science* 292: 494–496, Fig 1c. **52.7b** Data from G. F. Gause. 1934. *The Struggle for Existence*. Baltimore, MD: Williams & Wilkins. **52.8a** Reproduced with kind permission of Springer Science and Business Media and the author from G. E. Forrester. 1995. Strong density-dependant survival and recruitment regulate the abundance of a coral reef fish. *Oecologia* 103: 275–282, Fig 2. © 1995 Springer-Verlag. **52.8b** Data from P. Arcese and J. N. M. Smith. 1988. Effects of population density and supplemental food on reproduction in song sparrows. *Journal of Animal Ecology* 57: 119–136, Fig 2. **52.11** Reproduced by permission of Alberta Sustainable Resource Development. 2002. From www3.gov.ab.ca/srd/fw/watch/rabb_cycles.html. © 2002 Government of Alberta. **52.12** Graphs reproduced by permission of AAAS from C. J. Krebs et al. 1995. Impact of food and predation on the snowshoe hare cycle. *Science* 269: 1112–1115, Fig 3. **52.13b** Data from T. Valverde and J. Silvertown. 1998. Variation in the demography of a woodland understorey herb (*Primula vulgaris*) along the forest regeneration cycle: Projection matrix analysis. *Journal of Ecology* 86: 545–562. **52.14a/b** Data from U.S. Census Bureau, International Data Base (IDB), www.census.gov/ipc/www/idb/.

Chapter 53 **53.2a/b, 53.3, 53.4a/b** G. F. Gause. 1934. *The Struggle for Existence*. Baltimore, MD: Williams & Wilkins. **53.8** Reproduced by permission of AAAS from P. R. Grant and B. R. Grant. 2006. Evolution of character displacement in Darwin's finches. *Science* 313, Fig 2. **53.11** Adapted by permission of Ecological Society of America from G. H. Leonard, M. D. Bertness, and P. O. Yund. 1999. Crab predation, waterborne cues, and inducible defenses in the blue mussel, *Mytilus edulis*. *Ecology* 80: 1–14. © 1999. **53.12** Adapted by permission of Ecological Society of America from G. H. Leonard, M. D. Bertness, and P. O. Yund. 1999. Crab predation, waterborne cues, and inducible defenses in the blue mussel, *Mytilus edulis*. *Ecology* 80: 1–14. © 1999. **53.13b, c** Data from G. D. Martinsen, E. M. Driebe, and T. G. Whitham. 1998. Indirect interactions mediated by changing plant chemistry: Beaver browsing benefits beetles. *Ecology* 79: 192–200. **53.17** Adapted by permission of Ecological Society of America from J. H. Cushman and T. G. Whitham. 1989. Conditional mutualism in a membracid-ant association: Temporal, age-specific, and density-dependant effects. *Ecology* 70: 1040–1047. **53.18** Adapted by permission of Ecological Society of America from D. G. Jenkins and A. L. Buikema, Jr. 1998. Do similar communities develop in similar sites? A test with zooplankton structure and function. *Ecological Monographs* 68: 421–443. **53.19c** Adapted with kind permission from Springer Science and Business Media from R. T. Paine. 1974. Intertidal community structure. *Oecologia* 15: 93–120. © 1974 Springer-Verlag. **53.21b** Data from T. W. Swetnam. 1993. Fire history and climate change in giant sequoia groves. *Science* 262: 885–889. **53.24a** Adapted by permission of Wiley-Blackwell from R. H. MacArthur and E. O. Wilson. 1963. An Equilibrium Theory of Insular Zoogeography. *Evolution*: 17: 373–387, Fig 4. **53.24b, c** Data from R. H. MacArthur and E. O. Wilson. 1963. An equilibrium theory of insular zoogeography. *Evolution*: 17: 373–387, Fig 5. **53.26** Data from W. V. Reid and K. R. Miller. 1989. *Keeping Options Alive: The Scientific Basis for Conserving Biodiversity*. Washington, DC: World Resources Institute.

Chapter 54 **54.3** Data from J. R. Gosz et al. 1978. The flow of energy in a forest ecosystem. *Scientific American* 238: 92–102. **54.7** Reproduced by permission of the publisher from C. A. Mackenzie, A. Lockridge, and M. Keith. 2005. Declining sex ratio in a first nation community. *Environmental Health Perspectives* 113: 12–96, Fig 1. **54.9a/b/c** Data from R. H. Whittaker and G. E. Likens. 1973. Primary production: The biosphere and man. *Human Ecology* 1: 357–369, Table 1; H. Lieth.1973. Primary production: Terrestrial ecosystems. *Human Ecology* 1: 303–332, Tables 2, 3; J. S. Bunt. 1973. Primary production: Marine ecosystems. *Human Ecology* 1: 333–345, Table 1; G. E. Likens. 1973. Primary Production: Freshwater ecosystems. *Human Ecology* 1: 347–356, Table 3. **54.16** Data from W. H. Schlesinger. 1997. *Biogeochemistry: An Analysis of Global Change*, 2nd ed. San Diego, CA: Academic Press. **54.17** Data from P. M. Vitousek et al. 1997. Human alteration of the global nitrogen cycle: Sources and consequences. *Ecological Applications* 7: 737–750. **54.20** Data from BP Statistical Review of World Energy 2007, pp. 6–21 **54.21** Data from NOAA.gov, www.esrl.noaa.gov/gmd/ccgg/trends/. **54.22a** After G. Beaugrand et al. 2002. Reorganization of North Atlantic marine copepod biodiversity and climate. *Science* 296: 1692–1694. **54.22b** Adapted by permission of the publisher fromN. L. Bradley et al. 1999. Phenological changes reflect climate change in Wisconsin. *PNAS* 96: 9701–9704. © 1999 National Academy of Sciences, U.S.A. **54.23** Reproduced by permission of AAAS from R. R. Nemani et al. 2003. Climate-driven increases in global terrestrial net primary production from 1982 to 1999. *Science* 300: 1560–1563, Fig 2. **54.23b** After M. J. Behrenfeld et al. 2006. Climate-driven trends in contemporary ocean productivity. *Nature* 444: 752–755.

Chapter 55 **55.1** F. R. Liu et al. 2001. Molecular and morphological supertrees for eutherian (placental) mammals. *Science* 291: 1786–1789, Fig 1; A. B. Prasad et al. 2008. Confirming the phylogeny of mammals by use of large comparative sequence data sets. *Molecular Biology and Evolution* 25:1795–1808; U. Arnason, A. Gullberg, and A. Janke. 2004. Mitogenomic analyses provide new insights into cetacean origin and evolution. *Gene* 333: 27–34, Fig 1. **55.3a/b** Adapted by permission of Macmillan Publishers Ltd after C. D. L. Orme et al. 2005. Global hotspots of species richness are not congruent with endemism or threat. *Nature* 436: 1016–1019. © 2005. **55.4** Reproduced by permission of Conservation International from *Conservation International 2008 Annual Report*, 12–13. © 2008 Conservation International (www.conservation.org). **55.6** Reproduced by permission of the University of California Press Journals and the author from O.Venter et al. 2006. Threats to endangered species in Canada. *BioScience* 56: 903–910, Fig 2. © 2006 American Institute of Biological Sciences. **55.10b** Adapted by permission of AAAS and the author from W. F. Laurance et al. 1997. Biomass collapse in Amazonian forest fragments. *Science* 278: 1117–1118, Fig 2. **55.11** Adapted by permission of The Royal Society and the author from J. T. Hogg et al. 2006. Genetic rescue of an insular population of large mammals. *Proceedings of the Royal Society B* 273: 1491–1499, Fig 1. **55.12b** Adapted by permission from D. B. Lindenmayer and R. C. Lacy. 1995. Metapopulation viability of Leadbeater's possum, *Gymnobelideus leadbeateri*, in fragmented old-growth forests. *Ecological Applications* 5: 164–182, Fig 5G. **55.13** Adapted by permission of the author fromJ. M. Diamond and E. Mayr. 1976. Species-area relation for birds of the Solomon Archipelago. *PNAS* 73: 262–266, Fig 1. **55.14** Graphs reproduced by permission of AAAS from D. Tilman et al. 1997. The influence of functional diversity and composition on ecosystem processes. *Science* 277: 1300–1302, Fig 1. **55.16** Reproduced by permission of AAAS from E. I. Damschen et al. 2006. Corridors increase plant species richness at large scales. *Science* 313: 1284–1286, Fig 2B. **55-UN01T (England maps)** Reproduced by permission of the author from D. S. Wilcove, C. H. McLellan, and A. P. Dobson. 1986. Habitat fragmentation in the temperate zone. In M. E. Soule (ed.), *Conservation Biology: The Science of Scarcity and Diversity*. Sunderland, MA: Sinauer Associates, pp. 237–256, Fig 1. **55-UN01B (U.S. maps)** After W. B. Greeley, Chief, U.S. Forest Service. 1925. Relation of geography to timber supply. *Economic Geography* 1: 1–11.

Index

Boldface page numbers indicate a glossary entry; page numbers followed by an *f* indicate a figure; page numbers followed by *t* indicate a table.

2' carbon, 66
3' carbon, 61
3-D visualization techniques, B:15–B:16
3' end, DNA, 264, 265, 290
3-phosphoglycerate, 185
3' poly(A) tail, 295
5' cap, DNA, 295
5' carbon, 61
5' end, DNA, 264, 265, 290
−10 box, 291
20-hydroxyecdysone, **936**
30-nanometer fiber, 321
−35 box, 291

A

ABC model, 410
Abdomen, 637, **639**
Abiotic, **995**
Abiotic factors
for density-independent population growth, 1044
for detritus decomposition rate, 1093
for geographic distribution of organisms, 1013, 1015–17
ABO blood group, 245–47
Abomasum, 850
Abortions, spontaneous, 226, 440
Aboveground biomass, **1002**
Abscisic acid (ABA), **769**, 770–75*t*
Abscission, **768**, 774
Abscission zone, **774**
Absolute dating, 422
Absorption, 696–97, 841–42, **845**
Absorption spectrum, **175**–76
Absorptive feeding, 530, 583, 586
Abundance
animal, 624*f*
arthropod, 623
fossil record bias for, 481
prokaryotic, 498
species interactions and, 1059
vertebrate, 655*f*
viral, 675–76
Abuse, human non-kin, 1034
Acanthocephala, 603*t*
Accessory fluids, male reproductive, 958, 959*t*
Accessory pigments, 532*t*
Acclimation (acclimatization), **429**–30, **804**, 806
Acetaldehyde, 34, 167
Acetate, 25
Acetic acid, 25
Acetone, 34
Acetylation, **322**
Acetylcholine, 898*t*, **925**
Acetyl CoA, **157**
catabolic pathways and, 168
cellular respiration and, 153
oxidizing of, to CO_2, 158–61
processing of pyruvate to, 156–58
Achromatopsia, 446
Acid-base reactions, 25–26
Acid-growth hypothesis, **761**–62
Acid hydrolases, 110–11
Acidic habitats, 516
Acidic soils, 743
Acidification, 1100, 1114
Acidity, pH scale and, 26
Acid rain, 490
Acids, **25**–26. *See also* Amino acids; Stomach acid
Acoelomates (Acoelomorpha), 603*t*, **605**, 620
Acorn worms, 646
Acoustic communication, 1029
Acquired characters hypothesis, 231, 234, 415, 429
Acquired immune deficiency syndrome (AIDS), 281, **677**–78, **990**. *See also* HIV (human immunodeficiency virus)
Acquired immune response. *See* Adaptive immune response
Acrosomes, **389**
Actin, 123, **922**
Actin filaments, **123**
image of, 102*f*
in mitosis, 200
movement and, 45
in muscle contraction, 922–25
structure and function of, 123–24*t*
Actinistia, 663
Actinobacteria, 512, 514
Actinopterygii, 656, 662–63
Action potentials, **890**–95, 902–4*f*, 908–9
Action spectrum, **176**
Activation
lymphocyte, 978–79*f*
mitosis-promoting factor, 204
Activation energy, **51**–53. *See also* Enzymes
Active exclusion, plant, 747–48
Active site, **53**–54
Active transport, **97–99**, **732**, **825**–26, 834. *See also* Sodium-potassium pump
Adaptations, **5**, **627**, **804**. *See also* Traits
acclimation vs., 429–30, 804, 806
in animal anatomy and physiology, 804–6
for animal body temperature homeostasis, 818–19
animal mouthparts as, 843–44
animal surface area and, 814
endothermy vs. ectothermy, 817
feeding (*see* Feeding adaptations)
fitness and, 5, 424, 627 (*see also* Fitness)
natural selection and, 435
plant nutritional, 750–52
traits as, 5 (*see also* Traits)
Adaptive immune response, **977**–90
B-cell activation and antibody secretion, 986
B cells and T cells of, 979, 989*t*
characteristics of, 977–78
clonal-selection theory on, 979–83*f*
destruction of viruses by, 987–88
distinguishing self from nonself in, 983–84
immune system rejection of foreign tissues and organs by, 988
immunological memory and, 989–90
killing of bacterial and foreign cells by, 987
lymphocytes and, 978–79*f*
T-cell activation in, 984–86
Adaptive immunity, **974**
Adaptive radiations, **484**, **564**, **844**
angiosperm radiation as, 564–65
Cambrian explosion as, 486–88
cichlids as, 844
ecological opportunity and, 484–85
Hawaiian silverswords as, 700
humans and, 672
morphological innovation and, 485–86
of seed plants, 560–61
Adenine, 60, 63–64, 66–67, 261, 279, 290
Adenosine diphosphate (ADP). *See* ADP (adenosine diphosphate)
Adenosine monophosphate (AMP), 157
Adenosine triphosphate (ATP). *See* ATP (adenosine triphosphate)
Adenoviruses, 679*f*, 690
Adenylyl cyclase, **316**
ADH (antidiuretic hormone), **837**–38, 931, **940**
Adhesion, **22**–23, 136–38, **724**
Adhesion proteins, 131*f*, 136–38
Adipocytes, **938**
Adipose tissue, **807**, 938
ADP (adenosine diphosphate), 124, **148**, 153, 155, 181
Adrenal glands, 837, **933**, 937, 938
Adrenaline, 140, **937**
Adrenocorticotropic hormone (ACTH), **941**
Adults, **616**
Adult testing, genetic, 350
Adventitious roots, **698**, 768
Aerobic respiration, **166**, 487, **510**. *See also* Respiratory systems
Afferent division, **899**
Aflatoxin, 272
Africa, human evolution out of, 672–73*f*
Age, trisomy and maternal, 227
Age classes, **1039**
Age pyramids, 1051*f*
Age-specific fecundity, **1039**, 1115
Age-specific survivorship, 1115
Age structure, **1050**–52
Agglutination, **987**
Aggregate fruits, **797**
Aging, plant, 773–74
Agre, Peter, 95
Agriculture
biotechnology in, 354–56
fungi and, 581
global water cycle and, 1095
habitat destruction by, 1112
plants and, 547, 548
Agrobacterium tumefaciens, 355–56, 515
Aguinaldo, Anna Marie, 607
AIDS (acquired immune deficiency syndrome), 281, **677**–78, **990**. *See also* HIV (human immunodeficiency virus)
Air, animal olfaction and, 919–20. *See also* Atmosphere
Air pressure, animal sensing of, 610
Alanine, 42*t*
Alarm calling, 1031–33
Albumen, **658**
Albumin, **877**
Alcohol fermentation, **167**
Alcohols, 35*f*
Alcohol use, fetal alcohol syndrome and, 969–70
Aldehydes, 34, 35*f*
Aldose, 72
Aldosterone, **837**, **940**
Algae. *See also* Green algae
algal blooms of, 521–22, 1102
brown algae, 535–36, 543
lichens and, 589, 597
Plantae and, 536
red algae, 519*f*, 536, 539
Alimentary canal, 845
Alkaline mucus, 959*t*
Alkaline soil, 743
Alkalinity, pH scale and, 26
Allantois membrane, 658
All-Cells-from-Cells hypothesis, 2–4
Alleles, **212**, **234**
analyzing changes in frequencies of, 436–40
creating mutant, 277
crossing over and, 222
deleterious, 224
effect of inbreeding on frequencies of, 451
female choice for good, 453
foraging, in fruit flies, 1021
frequencies of, in populations, 435
genes and, 212, 221 (*see also* Genes)
genetic bottlenecks and reduction in, 446
identifying human, as recessive or dominant, 250–51
introducing novel, into human cells, 351–52
in Mendelian genetics, 239*t*
mutant, 285
mutations as new, 448–49
particulate inheritance hypothesis and, 234–35
plant pathogen resistance and, 777
recombination of, 244–45
Allergens, **991**
Allergies, 990–**91**
Alligators, 660, 667

Allolactose, 311
Allopatric speciation, 462, **463**, 464
Allopatry, **463**, 468
Allophycocyanin, 532*t*
Alloploidy, 467
Allopolyploidy, **466**, 467–68, 471*t*
Allosteric regulation, **54**–55, **314**
All Taxa Biodiversity Inventory, 1108
α-amylase, 144, **770**
α-glucose, 72, 74
alpha-helix (α-helix), **46**–47
Alpine skypilots, 794
ALS (Lou Gehrig's disease), 376
Alternate leaves, 702*f*
Alternation of generations, **402**, **534**–36, **556**–58, 592, **784**
Alternative splicing, **328**–29, 331, **369**
Altman, Sidney, 68
Altruism, **1031**–34
Alveolata, 524*t*, 537, 540–41
Alveoli (alveolus), 537, **868**
Ambisense viruses, **686**
American elm trees, 581
Amines, 35*f*
Amino acids, **40**
amphipathic proteins and, 92
anabolic synthesis of, 169
as animal essential nutrient, 842
aquaporins and, 95
carbon functional groups and, 34, 35*t*
in central dogma, 280
chemical evolution and formation of, in prebiotic soup, 38–39
experiment on metabolic pathway of, 277–78
genetic code for, 282–83, 297
hormones as derivatives of, 933
major, found in organisms, 41*t*
as neurotransmitters, 898*t*
peptidoglycan and, 76
polymerization of, to form proteins, 42–44 (*see also* Protein synthesis)
side chains of, 40–42 (*see also* Side chains, amino acid)
specification of, by mRNA triplets in translation, 297
structure of, 40
sugars in, 77
transfer of, from transfer RNA to proteins, 297–99
water and solubility of, 42*t*
Aminoacyl tRNA, **298**
Aminoacyl tRNA synthetases, **298**
Amino functional group, 34–35*f*, 40, 43–44, 46–47
Amino-terminus, 43–44, 301
Ammodramus maritimus, 461–62
Ammonia, 19–21*f*, 24*t*, 39–40, 510–12, 748, **829**
Amnion, **968**
Amniotes, **655**
Amniotic eggs, **655**, **658**
Amniotic fluid, 968
Amoebic dysentery, 522*t*
Amoeboid motion, **532**–33
Amoebozoa, 524*t*, 536, 537
AMP (Adenosine monophosphate), 157
Amphibia, 664
Amphibians, **664**. *See also* Animal(s); Deuterostomes
Amphibia lineage and, 664
fungal parasites of, 595
global warming and, 1100–1101
hearts of, 878
hormones in metamorphosis of, 935–36
lateral line system as hearing in, 913
Amphioxus, 651, 652
Amphipathic compounds, **84**, 92
Amplification, signal, 141–42, 908, 947
Amylases, **79**, **847**, 855*t*
Amylopectin, 74
Amyloplasts, **765**
Amylose, 74
Anabolic pathways, **168**, 169
Anadromous, **661**
Anaerobic respiration, **166**, **510**. *See also* Respiratory systems
Anaphase
meiosis, 218
mitosis, **199**–200
Anaphylactic shock, 991
Anatomy, **803**, 810–11*f*. *See also* Animal anatomy and physiology; Plant anatomy and physiology
Ancestral traits, 418–22, 423*f*, 460–61, **475**–79
Anemomes, 619
Aneuploidy, **226**, 227, **286**
Anfinson, Christian, 50
Angiosperms, **550**. *See also* Green plants; Land plants; Plant(s)
adaptive radiations of, 485, 564–65
anatomy and physiology of, 695–96 (*see also* Plant anatomy and physiology)
Anthophyta lineage and, 576–77*f*
in Cenozoic era, 483
development in (*see* Plant development)
diversification of, 551–52
flowers of, 561–62
gametogenesis in, 402–4
life cycle of, 785*f*
parasitic, 751
pollination of, 576–77*f*
reproduction in, 783 (*see also* Plant reproduction)
Anglerfish, 1030
Anhydrite, 490
Animal(s), **601–22**, **958**
biological importance of, 602–3
biological methods for studying, 603–9
bodies (*see* Bodies, animal; Body plans)
Cambrian explosion in, 486–88
cells of (*see* Animal cells)
chemical signals in (*see* Chemical signals, animal)
cloning, 377–78
color vision in nonhuman, 917–18
comparative morphology of, 603–7
deuterostome and protostome, 606–7 (*see also* Deuterostomes; Protostomes)
development of (*see* Animal development)
diseases of (*see* Diseases, animal)
diversification themes of, 610–17
electrical signals in (*see* Electrical signals, animal)
excretory systems (*see* Excretory systems)
feeding strategies and food of, 610–13
form and function of (*see* Animal anatomy and physiology)
gap junctions in tissues of, 139
gas exchange and circulation in (*see* Animal gas exchange and circulation)
immune systems of (*see* Immune systems)
key lineages of non-bilaterian, 617–20
life cycles of, 615–17
major phyla of, 603*t*
meiosis and gametes of, 194, 218
as model organisms, 603
movement of, 613–15 (*see also* Movement, animal)
nervous systems (*see* Nervous systems)
nitrogenous wastes of, 829*t*–30
nutrition for (*see* Animal nutrition)
Paleozoic era and, 483
phylogenetic relationship of fungi to, 6, 584–85
phylogenies of, 607–9
plant pollination by, 562–63, 576–77*f*, 792–94
plant seed dispersal by, 798
relative abundance of lineages of, 624*f*
reproduction of (*see* Animal reproduction)
sensory organs of, 610 (*see also* Sensory systems, animal)
shared traits of, 601
urinary systems (*see* Urinary systems)
water and electolyte balance in (*see* Osmoregulation)
Animal anatomy and physiology, 803–21
adaptations and, 804–6
body size effects on physiology and, 811–14
body temperature regulation and, 816–19
comparative morphology and, 603–7
dry habitats and, 803
fitness trade-offs in, 804, 805*f*
homeostasis in, 814–16
levels of study of, 811*f*
organs and organ systems in, 810–11
structure-function relationships in, 806
tissues in, 806–11
Animal cells. *See also* Cells
cell-cell attachments of, 135–38
cell cultures of, B:17–B:18
chitin as structural polysaccharide of, 76
cytokinesis in, 124, 200
differentiated, as genetically equivalent, 377–78
eukaryotic, 106*f*, 115–16 (*see also* Eukaryotic cells)
extracellular matrix and, 133–34
glycogen as storage polysaccharide in, 168
lysosomes in, 110–11
plant cells vs., 106*f*, 706–7
selective adhesion and adhesion proteins in, 136–38
Animal development, 388–400. *See also* Development
cleavage in, 392–94
fertilization in, 390–92
gamete structure and function in, 389–90
gastrulation in, 394–95
hormones and, 935–37
human embryo, 388*f*
ordered phases of, 389*f*
organogenesis in, 396–99
plant development vs., 401
Animal gas exchange and circulation, 861–84
air and water as respiratory media in, 862–64
blood transport of oxygen and carbon dioxide in, 870–74
circulatory systems in, 874–83
gas exchange organs in, 864–70
respiratory and circulatory systems in, 861–62 (*see also* Circulatory systems; Respiratory systems)
terrestrial adaptations for, 628
Animal models of disease, **350**
Animal nutrition, 841–60
cellulose as human dietary fiber in, 77
digestion and absorption of nutrients in, 845–56
fetal, during pregnancy, 968–70
four steps of, 841–42
glucose and nutritional homeostasis in, 856–58
glycogen as storage polysaccharide for, 74
human digestive tract, 846*f*
ingestion and mouthparts in, 843–45
mammalian digestive enzymes in, 855*t*
nutritional requirements and, 842–43*t*
poor-nutrition plant tissues and, 1066
waste elimination in, 854–55
Animal reproduction, 950–72
asexual, 951
asexual vs. sexual, 950–53
female reproductive systems in, 959–60
fertilization and egg development in, 954–57
human (*see* Human reproduction)
life tables and, 1038–39
male reproductive systems in, 957–59*f*
mammalian pregnancy and birth in, 967–71
mating behaviors and, 1022–25 (*see also* Mating)
pheromones and, 931
protostome, 630
reproductive structures in, 957–60
sex hormones in, 936, 960–66
sexual, and gametogenesis, 952–53
sexual selection and, 452–55
switching modes of, 951–52
variations in, 615
vertebrate adaptations in, 658–59
Anions, **18**–19, 98, 743, 745–46
Annelids, 603*t*, 627, **633**
Annuals, **567**
Annual tree growth rings, 714–15
Anolis lizards, 1023–24
Anoxygenic photosynthesis, **181**, **509**
Antagonistic muscle groups, 921
Antagonistic-pair feedback systems, 818
Antenna complex, **178**
Antennae, **637**, 639, 643
Anterior, **379**
Anterior pituitary, **941**, 943
Antheridia (antheridium), **556**
Anthers, **789**
Anthocerophyta, 571
Anthophyta, 564–65, 576–77*f*. *See also* Angiosperms
Anthrax, 498–99
Anthropoids, **668**–69

Boldface page numbers indicate a glossary entry; page numbers followed by an *f* indicate a figure; page numbers followed by *t* indicate a table.

Antibacterial compounds, 777
Antibiotics, **500**
actinobacteria and, 514
bacterial ribosomes and, 496
evolution of resistance of tuberculosis bacterium to, 424–26
fungi and, 581, 598
in human semen fluids, 959*t*
pathogenic bacteria and, 500
Antibodies, **136**, **324**, **681**, **977**
adaptive immune response and, 977–78
B-cell activation and secretion of, 986
binding of, to epitopes, 981
five classes of immunoglobulins and, 980*t*
humoral response and, 987, 988
humoral response coating of pathogens with, 988
as proteins, 45, 280, 324
selective adhesion of, 136–38
specificity and diversity of, 981–82
viruses and, 681–82
Antibody probes, B:13
Anticancer drugs, 333, 371–72, 575, 579*f*
Anticodons, **299**, 300–301
Antidiuretic hormone (ADH), **837**–38, 931, **940**
Antifungal compounds, 777
Antigen presentation, **980**, 984–85
Antigen-receptor binding, 979
Antigens, **974**, 979
Antihistamines, 991
Antimycin A, 162
Antiparallel DNA strands, **63**–64, 261
Antiporters, **732**, 747–**48**, **826**, 888
Antivirals, 678, **680**
Ants
aphids and, 1068
fungal mutualisms with, 589
Hymenoptera order of, 640*t*
mutualism between treehoppers and, 1069–70
navigation hypothesis experiment with, 10–12
number of chromosomes in, 213*t*
parasite manipulation of, 1067–68
population of, 623
types of, 602
Anus, 606, 845, 846*f*
Aorta, **875**
Aphids, 602, 616, 731, 1068
Aphotic zone, **997**, **1000**–1001
Apical, **404**
Apical-basal axis, **404**
Apical buds, **699**
Apical dominance, **767**–68
Apical meristems, 407, **704**–5, 712
Apical sides, **809**
Apicomplexa, 541
Apomixis, **786**
Apoplastic route, **722**, 723*f*, 747
Apoptosis, **205**, 332–33, 350, **376**, 396. *See also* Programmed cell death
Appendages, animal, 613–15
Appendix, **854**
Apple maggot flies, 465
Aquaporins, **95**–96, 722, **835**, 837–38, **855**
Aquatic ecosystems, 995–1001
animal transition to terrestrial ecosystems from, 627–28
external fertilization in, 954
freshwater, 997–1000
global productivity patterns of, 1090
marine, 1000–1001
nutrient availability in, 995–96
osmoregulation in, 826–28
osmotic stress in, 824
oxygen and carbon dioxide in, 863–64
plant transition to terrestrial ecosystems from, 553–55, 556–64
ventilation in, 867
water depth and flow in, 996–97
Aquatic food chains, 522–23
Arabian oryx, 803, 817
Arabidopsis thaliana
embryogenesis in, 405*f*
embryos, 401*f*
genome of, 362, 369*t*
as model organism, 402, B:20, B:21*f*
overview of development of, 402*f* (*see also* Plant development)
in phloem loading research, 733
Arbuscular mycorrhizal fungi (AMF), **586**, 588*f*, 589, 596, **746**
Archaea, **496**. *See also* Archaea domain
bacteria vs., 496–97 (*see also* Bacteria)
environmental sequencing in, 364
human diseases and, 498
identifying genes in genome sequences of, 362–63
key lineages of, 512, 516
lateral gene transfer in, 364
natural history of genomes of, 363–64
phospholipids in, 84
as prokaryotic, 102–3 (*see also* Prokaryotes)
prokaryotic cell structures and functions of, 102–5 (*see also* Prokaryotic cells)
Archaea domain, 6–7, 102–3, 496–97*t*, 504*f*. *See also* Archaea
Archaeopteryx, 480, 486
Archegonia (archegonium), **556**
Arctic terns, 1027
Arctic toxaphene biomagnification, 1088
Arctic tundra, 1008
Ardipithecus, 670*t*
Area and age hypothesis on species richness, 1080
Area de Conservación Guanacaste, 1122
Area/surface relationships, plant, 696–97
Arginine, 42*t*, 277–78, 280–81, 286
Aristotle, 415
Arms races, coevolutionary, 1059, 1066–67
Arrangement, leaf, 702
Arteries, 832–33, **875**–76, 879
Arterioles, **875**–76, 882
Arthropoda, 639–43
Arthropods, **637**
Arthropoda lineages and, 639–43
body plan of, 626
as ecdysozoans, 603*t*, 637
feeding structures, 628, 629*f*
insects as (*see* Insects)
limbs, 629
as most abundance and diverse protostomes, 623*f*
Articulations, **920**
Artificial membranes, 85–86. *See also* Plasma (cell) membranes
Artificial selection, **4**–5, 231, **548**
Artiodactyla, 477–79
Ascaris, 211
Asci (ascus), **584**
Ascocarp, **598**
Ascomycetes (Ascomycota), 584–85, 593, 597–98
Ascorbic acid, 843*t*
Ascus, 598
Asexual reproduction, **195**, **220**, **786**, **950**
animal, 615, 951
animal switching between sexual and, 951–52
clones and, 221
fungal, 591
mitosis and, 195
plant, 786
prokaryotic vs. eukaryotic, 533
protostome, 630
sexual reproduction vs., 220–21, 223, 533, 950–51 (*see also* Sexual reproduction)
A site, 300
Asparagine, 42*t*
Aspartate, 42*t*
Association studies, 349
Asteroidea, 649
Asteroid impact hypothesis, 490–92. *See also* Meteorites
Asthma, 991
Astragalus, 477
Astrobiologists, 501
Asymmetric competition, **1060**
Athletes, 344, 940
Atmosphere
chemical evolution conditions in, 27, 32, 34*f*, 39–40 (*see also* Chemical evolution; Prebiotic soup)
gas exchange in, 862–63
oxygenic photosynthesis and oxygen in, 181–82
plant nitrogen fixation from nitrogen in, 748–50
water potential in, 721
Atomic level anatomy and physiology, 811*f*
Atomic number, **16**
Atoms
bonding of, in molecules, 17–20 (*see also* Molecules)
energy transformations in, 28–29
found in organisms, B:8*t*
linking carbon, into organic molecules, 34 (*see also* Organic molecules)
radioactive decay of, 417
structure of, 16–17
ATP (adenosine triphosphate), **79**, **148**
in actin-myosin interaction, 124, 924–25
active transport by pumps and, 97–98
cellular respiration and, 153–54, 165–66 (*see also* Cellular respiration)
cell usage of, 116
energy transfer from glucose to, 79
in glycolysis, 155
GTP vs., 158, 297
photosynthesis and, 173–74, 180–81, 183–84 (*see also* Photosynthesis)
prokaryotic metabolic diversity in producing, 506–9
redox reactions and, 151–52
structure and function of, 149–51
synthesis of, 112, 148, 161–66
transfer RNA and, 298
vesicle transport and, 126–27
viruses as unable to produce, 675
ATP synthase, **161**, **164**–65
Atria (atrium), **877**, 879
Atrial natriuretic hormone, **933**
Atrioventricular (AV) node, **881**
Atrioventricular valves, 877
Attenuated virus vaccines, 990
α-tubulin, 125, 126, 201
Australia, 665, 672
Australopithecus, 670*t*, 671
Autocrine signals, **930**
Autoimmune diseases, 983–84. *See also* Immunodeficiency diseases
Autoimmunity, **983**
Autonomic nervous system, **899**, 900*f*
Autophagy, **110**
Autoploidy, 466–67
Autopolyploidy, **466**, 471*t*
Autoradiography, **62**, 196, B:12–B:13
Autosomal inheritance, **242**
Autosomal traits, 250–52, 272–74, 339
Autosomes, **212**, 214*t*, 242
Autotrophs, **172**, **506**, **1083**–84
Auxin, **760**
as blue-light phototropic hormone, 759–62
as gravitropic signal, 765–66
plant body axes and, 406
as plant development master regulator, 380–81
as plant growth regulator, 774, 775*t*
roles of, in apical dominance and plant growth, 767–68
structure and function of, 140*t*
Aves, 668. *See also* Birds
Avian flu virus, 688, 689
Avian malaria, 1015
Avirulence (*avr*) genes, **776**
Axel, Richard, 920
Axes, body, 379–81, 395, 406
Axes, graph, B:2
Axes, leaf, 408
Axillary buds, **699**
Axonemes, **127**–28
Axon hillock, **899**
Axons, **808**, **886**
action potential propagation down, 893–95
diameter of, and action potential propagation speed, 893
measuring membrane potentials of, 890–91
myelination and action potential propagation speed, 894–95
nervous tissue and, 808
neurons and, 886–87
squid, 125–27
Azoneme, **127**

B

Bacillus anthracis, 499
Bacillus thuringiensis, 354, 513
Backbone, DNA, 61, 62, 66, 260, 279. *See also* Sugar-phosphate backbone, nucleic acid
Background extinctions, **489**, 1116. *See also* Extinction; Mass extinctions
BAC libraries, **360**
Bacteria, **496**. *See also* Bacteria domain
antibiotics and, 500
archaea vs., 496–97 (*see also* Archaea)
cell division in, 200
cell theory experiment and, 3–4
chromosome replication in, 265*f*
control of gene expression in (*see* Gene regulation, bacterial)
DNA of, 103–4
environmental sequencing in, 364
evolution of antibiotic resistance in, 424–26
identifying genes in genome sequences of, 362–63

Bacteria, (*continued*)
initiating translation in, 301
introducing recombinant plasmids into cells of, 341
killing of, by adaptive immune response, 987
lateral gene transfer in, 364
model organism (*see Escherichia coli* (*E. coli*))
mRNA synthesis in, 293
natural history of genomes of, 363–64
nitrogenase complex in, 49
nitrogen-fixing, 510–11, 1068, 1093
origin of viruses in symbiotic, 687
pathogenic, 498–500
peptidoglycan as structural polysaccharide in, 76
phospholipids in, 84
photosynthesis in, 173–74, 180, 182
as prokaryotic, 102–3 (*see also* Prokaryotes)
prokaryotic cell structures and functions of, 102–5 (*see also* Prokaryotic cells)
proteins as defense against, 45
quorum sensing as cell-cell communication in, 145–46
RNA polymerase in, 290
symbiotic, in plant nitrogen fixation, 749–50
temperature, pH, and enzyme catalysis in, 55–56
terminating translation in, 303*f*
transcription in, 292–93
transcription in eukaryotes vs. in, 296*t*
transcription promoters in, 291
translation in eukaryotes vs. in, 296–97
Bacteria domain, 6–7, 102–3, 496–97*t*, 504*f*, 512–15. *See also* Bacteria
Bacterial artificial chromosomes (BACs), **360**
Bacteriophages, 49, 679*f*, **680**, 684
Bacteriorhodopsin, 163, 509
Baculum, **958**
Balance, animal, 610
Balancing selection, 442–**43**
Ball-and-stick models, **20**, B:8*f*
Baltimore, David, 690
Baltimore classification, 690
Banting, Frederick, 856
Bar coding, **1106**
Bark, **713**–14
Barnacles, 643, 1061–62
Barn swallows, 1024–25
Barometric pressure, animal sensing of, 610
Baroreceptors, **882**–83
Barriers
to dispersal of organisms, 1014
epithelial tissues as pathogen, 809
human contraceptive, 966*t*
innate immunity and pathogen, 132, 974–75
plant cuticle as pathogen, 707, 775
Bartel, David, 68
Basal, **404**
Basal body, **127**
Basal lamina, **809**
Basal metabolic rate (BMR), **812**–13
Basal transcription complex, **326**–28
Basal transcription factors, 291–**92**, **326**
Bases, **25**–26
Bases, DNA nitrogenous, 260, 279, 280, 282–86
Basic-odor hypothesis, 919–20
Basidia, **584**
Basidiomycetes (Basidiomycota), 584–85, 593, 596
Basilar membrane, **911**–12
Basolateral sides, **809**
Bateman, A. J., 452–53
Bateman-Trivers theory, 452–53
Batesian mimicry, **1064**
Bateson, William, 248
Bats, 912
Bayliss, William, 852
B-cell receptors (BCRs), **980**–82, 986
B cells, **979**
activation of, 986
characteristics of, 989*t*
discovery of, and receptors for, 979–82
as lymphocytes, 978
B-cell tumors, 982
Bdelloids, 615
Beadle, George, 277
Beaks, **636**
Beaumont, William, 848–49
Bedelloids, 631
Beer yeast model organism. *See Saccharomyces cerevisiae*
Bees, 562, 640*t*, 788, 817, 1028–29, 1068
Beeswax, 87*f*
Beetles, 241, 623, 640*t*
Beggiatoa, 507*t*
Behavior, **1019**–21
Behavioral biology, 1019–21
Behavioral contraception methods, 966*t*
Behavioral ecology, 1019–36
altruism in, 1031–34
behavior and behavioral biology in, 1019–21
communication in, 1027–31
conditional strategies and decision making in, 1020–21
foraging in, 1021–22
mating in, 1022–25
migration in, 1026–27
proximate and ultimate causation in, 1020
Behavioral prezygotic isolation, 459*t*
Bendall, Faye, 179
Beneficial alleles, **448**
Beneficial mutations, **286**
Benign tumors, **206**
Benthic, **618**
Benthic zone, **997**, **1000**
Benzene, 24*t*
Benzo[α]pyrene, 272
Beriberi, 54
Best, Charles, 856
β-1, 4-glycosidic linkages, 73–74
β-carotene, 176, 355, 532*t*
Beta-blockers, 140
β-galactosidase, 309–10, 313
β-globin, 247, 322
β-glucose, 72, 76
Beta-pleated sheet (β-pleated sheet), **47**
β-tubulin, 125, 126
Bicoid gene, 379–80
Bikonta, 525
Bilateral symmetry, **605**
Bilateria, 603*t*, **606**–7, 609. *See also* Deuterostomes; Protostomes
Bile, 853–**54**
Bindin, 391
Biodiversity, **1105**–27. *See also* Conservation biology; Diversity; Species richness
benefits of, 1117–20
changes in, through time, 1106–7
current extinction rates as threat to, 1110
current species diversity and, 1107–8
as ethical issue, 1120
exotic and invasive species as threat to, 1110–11
extinction rate prediction methods and, 1114–16
extrapolation techniques for estimating, 1108
habitat destruction as threat to, 1112
habitat fragmentation as threat to, 1113
hotspots of, 1109–10
human population growth and Easter Island loss of, 1105
levels of, 1106–7
phylogenetic tree of life and, 1106
preserving, 1120–23
species endangerment as threat to, 1111
stochastic and genetic problems in metapopulations as threat to, 1114
Biodiversity hot spots, **1109**
Biofilms, **145**
Biogeochemical cycles, **1092**–98
global, 1094–98
global carbon cycle, 1096–97
global nitrogen cycle, 1096
global water cycle, 1095
nutrient cycling, 1092–94
Biogeography, **463**, **1013**. *See also* Geographic distribution of organisms
Bioinformatics, **361**
Biological control agents, 594, 595, 1068
Biological evolution, 15, 39, 59. *See also* Evolution; Natural selection
Biological fitness, **424**. *See also* Fitness; Fitness trade-offs
Biological imaging, B:13–B:16. *See also* Biological methods; Microscopy
cell research and, 116
electron microscopy, B:14–B:15 (*see also* Electron microscopy)
light microscopy, B:13–B:14 (*see also* Light microscopy)
scanning electron microscopy (SEM), B:15 (*see also* Scanning electron microscopy (SEM))
three-dimensional, B:15–B:16
transmission electron microscopy (TEM), B:14–B:15 (*see also* Transmission electron microscopy (TEM))
video microscopy, B:15
Biological methods. *See also* Experiments; Quantitative methods
biological imaging (*see* Biological imaging)
cell and tissue cultures (*see* Cultures, cell and tissue)
direct sequencing (*see* Direct sequencing)
DNA libraries, 342–43
DNA microarrays, 370–72
DNA probes, 342–43, 370–71
enrichment cultures, 501–2
gel electrophoresis, 62, 94, 347
genetic engineering (*see* Genetic engineering)
Gram stains, 505
histograms, 427
molecular phylogenies (*see* Phylogenetic trees; Phylogenies)
replica plating, 310–11
for studying animals, 603–9
for studying fungi, 582–86
for studying green plants, 549–52
for studying membrane proteins, 94
for studying prokaryotes, 501–4
for studying protists, 524–26
for studying viruses, 678–86
Biological species concept, **459**–61
Biology, **954**
astrobiology, 501
behavioral, 1019–21
biotechnology, **343**, 354–56 (*see also* Genetic engineering; Recombinant DNA technology)
cell theory in, 2–4 (*see also* Cells)
characteristics of living organisms in, 1–2 (*see also* Animal(s); Life; Organisms; Plant(s))
conservation (*see* Biodiversity; Conservation biology; Ecology; Ecosystems)
developmental, 374 (*see also* Development)
evo-devo (evolutionary-developmental), 384
forensic, 346
genetics (*see* DNA (deoxyribonucleic acid); Gene expression; Genes; Genetics; Genomics; Meiosis; Mendelian genetics)
geographic distribution of life, 1013–17
microbiology, 498
model organisms of, B:19–B:22 (*see also* Model organisms)
molecular, 276, 279–82, 338 (*see also* Carbohydrates; Chemical evolution; Molecules; Nucleic acids; Plasma (cell) membranes; Proteins)
nanobiology, 678–79
neurobiology, 885
reductionism in, 899
science and experimental methods of, 8–12 (*see also* Biological imaging; Biological methods; BioSkills; Experiments; Quantitative methods)
speciation and phylogenetic tree of life in, 5–8 (*see also* Phylogenies; Speciation; Tree of life)
theory of evolution by natural selection in, 4–5 (*see also* Evolution; Natural selection)
Bioluminescence, **541**
Biomagnification, **1088**–89
Biomass, 738, **1002**, 1005, **1084**–85, **1113**
Biomedical research, 603
Biomes, **1002**, 1090. *See also* Terrestrial ecosystems
Bioprospecting, **1117**
Bioremediation, **500**–501, 598, **1117**
BioSkills, B:1–B:22. *See also* Biological methods; Quantitative methods
biological imaging, B:13–B:16 (*see also* Biological imaging)

Boldface page numbers indicate a glossary entry; page numbers followed by an *f* indicate a figure; page numbers followed by *t* indicate a table.

cell and tissue cultures, B:17–B:18 (*see also* Cultures, cell and tissue)
combining probabilities, B:19
Latin and Greek roots in, B:6
making concept maps, B:10
metric system, B:1–B:2
model organisms, B:19–B:22
reading chemical structures, B:8–B:9
reading graphs, B:2–B:4
reading phylogenetic trees, B:4–B:6
separating and visualizing molecules, B:11–B:13
separating cell components by centrifugation, B:16–B:17
using logorithms, B:9
using statistical tests and interpreting standard error bars, B:7
Biotechnology, **343**, 354–56. *See also* Genetic engineering; Recombinant DNA technology
Biotic, **995**
Biotic factors
density-dependent population growth, 1045–46
of geographic distribution of organisms, 1015–17
Bipedal, **670**
Bipedalism, 671, 672
Bipolar cells, 914, **916**
Birds
Aves lineage and, 668
avian flu virus, 688, 689
avian malaria, 1015
conservation programs for, 1116
crops of, as modified esophagus in, 848
dinosaurs as ancestors of, 486
feathers and flight as vertebrate adaptations, 657–58
female reproductive systems in, 960
gizzards of, as modified stomachs, 851
lungs of, 868–70
mating in, 956
nitrogenous wastes of, 830
optimal foraging in, 1022
parental care in, 659
as reptiles, 660
Birth, mammalian, 970–71
Birthrates, 1037–38, 1040
Bitterness taste receptors, 918–19
Bivalves (Bivalvia), **630**, 634
Black-bellied seedcrackers, 442, 443*f*
Black-tailed prairie dogs, 1031–33
Bladder, **832**
Blade, **701**
Blastocoel, **395**
Blastocysts, **393**
Blastomeres, **393**
Blastopores, **395**
Blastulas, **393**
Bleaching, coral reef, 1100
Blending inheritance hypothesis, 231, 232, 234
Blindness, 355, 364, 514
Blind spot, 914
Blobel, Günter, 120–21
Blood, **807**. *See also* Blood vessels; Circulatory systems
alveolus and, 868
bird lungs and, 870
blood pressure and blood flow patterns, 882–83
carbon dioxide transport in, 873–74
chromatin remodeling in, 322
circulation of, through heart, 879 (*see also* Hearts)
clotting of, 252
glucagon and sugar levels in, 45
hemophilia and human, 252
homeostasis for oxygen in, 870, 940
human blood types, 245–47, 438–39
oxygen transport in hemoglobin of, 870–73 (*see also* Hemoglobin)
patterns in pressure and flow of, 882–83 (*see also* Blood pressure)
regulation of oxygen levels in, 940
sickle-cell disease and cells of human, 46
sickle-cell disease and human, 48
tissues of, 870
transfusions, 988
Blood pressure, 880, 882–83
Blood types, 245–47, 438–39
Blood vessels, 332, 832, 837, 875–76. *See also* Blood; Circulatory systems
Blooms, algal, 521–22, 1102
Blue light
photosynthesis and, 174–75, 176
phototropic responses to, 758–62, 780*t*
Blue mussels, 1064–65*f*
Bodies. *See also* Morphology
animal (*see* Bodies, animal; Body plans)
development of major axes of, 379–81
fungal, 582–84
plant (*see* Bodies, plant)
regulatory genes and specific positional information in development of, 381–83
Bodies, animal
body plans of (*see* Body plans)
defiinition of axes of, in gastrulation, 395
directional selection on size of, 441
evolution of body cavities in, 605–6
gas exchange through diffusion across, 864–65, 874–75
organogenesis and, 396–99
pathogen barriers of, 974–75
physiological effects of size of, 811–14
plant cells vs. cells of, 706–7
symmetry of, 604–5
temperature regulation of, 816–19
Bodies, plant
animal cells vs. cells of, 706–7
apical meristem production of primary, 704–5
brassinosteroids and size of, 773
cells and tissues of, 706–12
components of, 711*t*
definition of axes of, by genes and proteins in embryogenesis, 406
nutrition and, 737
Body mass index, **857**
Body plans, **603**
arthropod, 626
basic features of animal, 603–7, 610
chordate, 650–51
echinoderm, 647–48
mollusk, 626
protostome, 626–27
Bogs, **998**
Bohr shift, **872**, 873*f*
Boluses, 844
Bonds. *See* Carbon-carbon bonds; Chemical bonds; Covalent bonds; Disulfide bonds; Hydrogen bonds; Ionic bonds; Peptide bonds
Bond saturation, lipid, 87–88
Bone, **653**, **807**, **920**
Bone marrow, 352–54, **978**
Bones, animal, 920–21
Bony endoskeletons, 655
Bony exoskeletons, 653
Boreal forests, 1007
Boron, 739*t*
Both-And rule of probability, B:18
Bottom-up limitation hypothesis, 1066
Bouchet Philippe, 1107–8
Boveri, Theodor, 230, 239–41
Bowman's capsule, **833**
Brachiopoda, 603*t*
Braincase, **671**
Brains, **605**. *See also* Nervous systems
anatomy of, 901
antidiuretic hormone (ADH) and, 837
central nervous system and, 886
hominin, 672
Huntington's disease and degeneration of, 233, 251, 348–51
hypothalamus (*see* Hypothalamus)
mapping functional areas in, 901–2
neural tubes and, 396
positron-emission tomography scans of, 885*f*
prions and diseases of, 339
sensory organ transmission of information to, 909
thermoregulation and, 817–18
vertebrate, 653
Brain stem, **901**
Branches, **474**, **699**
Branching adaptations, 814
Brassinosteroids, 140*t*, **773**, 774, 775*t*
Bread mold, 277, 595
Bread yeast model organism. *See* *Saccharomyces cerevisiae*
Breast cancer, 350
Breathing, 866–70. *See also* Respiratory systems
Breeding. *See* Artificial selection; Captive breeding; Inbreeding depression
Brenner, Sydney, 283
Bridled goby, 1045
Briggs, Winslow, 761
Bristlecone pine trees, 401, 575, 695
Britton, Roy, 296
Broca, Paul, 901
Bronchi (bronchus), **867**
Bronchioles, **867**
Bronstein, Judith, 1069
Brown algae, 535–36, 543. *See also* Algae
Brown kiwis, 454
Bryophytes (Bryophyta), **550**, 570
Bryozoa, 603*t*
Bubonic plague, 602
Buchner, Hans and Edward, 155
Buck, Linda, 920
Budding
animal, 619, 620, 652, **951**
flower, 788
viral, 684, 685*f*
Buffers, **25**, **874**
Bugs, 641*t*
Building materials, 548–49
Bulbourethral glands, **958**, 959*t*
Bulk flow, **730**
Bulldog ants, 213*t*
Bumblebees, 817
Bundle-sheath cells, **188**, **728**
Burgess Shale faunas, 486–87
Burials, 480, 671
Bursa, 979
Bursting, viral, 684, 685*f*
Butter, 87*f*
Buttercup root, 71*f*
Butterflies, 640*t*, 1027, 1116
Bypass vessels, 878

C

C_3 photosynthesis, **188**
C_4 photosynthesis, **188**, **728**
Cacti, 189, 701
Caddisflies, 641*t*
Cadherins, 136–**38**
Caecilians, 664
Caenorhabditis elegans
apoptosis in, 376
genome of, 362
as model organism, B:21*f*, B:22
as nematode, 638
number of genes in genome of, 369*t*
as protostome, 623
Caffeine, 775
Calcium
as human nutrient, 843*t*
as plant nutrient, 739*t*
Calcium carbonate, 523, 540, 569
Calcium ions, 143*t*
Callus, **708**
Calmodulin, 428
Calvin, Melvin, 173–74, 185–86
Calvin cycle, **173**, 185–86, 509, 728
Calyx, **788**
CAM (crassulacean acid metabolism), **188**–89, **728**, 1004
Cambium, **712**–13
Cambrian explosion, **486**–88
Camera eyes, **914**
Cancer, **206**–9
anticancer drugs, 333, 371–72, 575, 579*f*
applied genomics and, 371–72
causes of uncontrolled cell growth in, 332
chromosome-level mutations and, 286–87*f*
cytoskeleton-ECM connection failures and, 134
defective Ras proteins and, 144
defects in DNA damage-repair genes and, 274
defects in eukaryotic gene regulation and, 332–33
as family of diseases, 208–9
genetic testing and breast, 350
HIV and, 677
homology and, 421
loss of cell-cycle control and, 207–9
metastasis of, 134
as out-of-control cell division, 206
p53 gene case study on preventing, 332–33
properties of cancer cells, 206–7
Rous sarcoma virus and, 691
severe combined immunodeficiency (SCID) therapy and, 354
sponges and chemotherapy for, 618
telomerases and, 271
Canis familiarus, 369*t*
Cannibalization, 955
Canopy, **1003**
CAP binding site, **315**
CAP-cAMP complex formation, 316
Capillaries, **814**, **875**–76, 882. *See also* Circulatory systems
Capillarity, **723**–24
Capillary action, 722
Capillary beds, **875**

Caps, RNA, **295**
Capsids, 259–60*f*, **679**, 684
Captive breeding, 1122
Carapace, **643**
Carbohydrates, **71**–81
  animal nutrition and, 842
  cell identity role of, 77
  digestion and transportation of, in animal small intestines, 852–53
  energy storage role of, 78–79
  fermentation of, 508
  glycosylation and, 121
  metabolic pathways of, 168–69
  organic molecules and, 36
  photosynthetic production of, 172–74
  polymerization of sugars to form, 42
  polysaccharides, 73–76
  polysaccharide structures, 75*t*
  processing of photosynthetic, 189–90
  roles of, 77–79
  as structural, 77, 113
  sugars as monomers of, 39, 71–73 (*see also* Sugars)
Carbon
  amino acids and, 40
  atomic structure of, 17*f*
  Calvin cycle fixation of, during photosynthesis, 185–86
  as cellular requirement, 168
  electronegativity of, 18
  importance of, in organic molecules, 33–36 (*see also* Organic molecules)
  lipids, hydrocarbons, and, 83
  as plant nutrient, 739*t*
  prokaryotic variation in pathways for fixing, 509
  RNA and, 66
  simple molecules from, 19–20
  six common functional groups attached to atoms of, 35*t*
  Sphagnum moss and, 570
Carbon-carbon bonds, 33–34, 172–73, 506*t*
Carbon cycle, 547, **580**–81, 1096–97
Carbon dioxide (CO_2)
  animal gas exchange and, 861
  behavior of, in air, 862–63
  behavior of, in water, 863–64
  carbon cycle and, 185
  carbonic anhydrase and, 45
  double bonds of, 20
  entry of, into leaves through stomata, 187
  formaldehyde and, 32
  global warming and, 523, 547, 1008–9, 1098–99 (*see also* Global climate change; Global warming)
  iron fertilization and, 1091
  land plants and, 553
  mechanisms for increasing concentration of, in plants, 187–89
  photosynthesis and, 173–74, 185–90
  plant nutrients from, 739*t*
  transport of, in blood, 873–74
  in volcanic gases, 27
  water stress and plant biochemical pathways to increase, 728
Carbon fixation, **185**
Carbonic acid, 27
Carbonic anhydrase, 45, 52, **850**, **873**–74
Carboniferous period, 551
Carbonyl functional group, 34–35*f*, 43, 46–47, 60, 71–73
Carboxyl functional group, 34–35*f*, 40, 43–44, 83–84
Carboxylic acids, 35*f*, **158**
Carboxyl-terminus, 44, 301
Carboxypeptidase, 855*t*
Cardiac cycle, **879**–80. *See also* Hearts
Cardiac muscle, **808**, **921**
Carnauba palms, 707
Carnivores, 547, **612**, 664
Carnivorous plants, 751–52
Carotenes, 176
Carotenoids, **175**–76, 355, 453, 758
Carpels, **403**, **409**, **561**, **789**–90
Carrier proteins (transporters), **96**–97, 731–32, **825**. *See also* Transport proteins
Carriers (disease), **251**
Carrier testing, genetic, 350
Carrion flowers, 562
Carroll, Sean B., 614–15
Carrying capacity, **1042**, 1046, 1065
Cartilage, **653**, **807**, **920**
Casein, 329
Casparian strip, **722**–23
Casts, **480**
Catabolic pathways, **168**–69
Catabolite activator protein (CAP), **315**
Catabolite repression, **314**–17*f*
Catalase, 110
Catalysis
  DNA as poor catalytic molecule, 65–66
  enzymes and, 45, 51–56 (*see also* Enzymes)
  eukaryotic organelles and, 107
  hydrolysis of carbohydrates and, 78–79
  multienzyme complexes and, 49
  nucleic acid polymerization and, 62
  polysaccharides and, 76
  RNA as catalytic molecule, 68
  smooth ER and, 108
  three-step process of, 54*f*
Catalysts, **52**–53. *See also* Catalysis; Enzymes
Catalyze, **45**. *See also* Catalysis
Catastrophic events, 1114
Catecholamines, **943**
Cation exchange, **743**–44*f*
Cations, **18**–19, 98, 743, 745
$CD4^+$ T cells, 984–86, 989*t*
CD4 protein, **681**, **984**
$CD8^+$ T cells, 984–86, 987–88, 989*t*
CD8 protein, **984**
cDNA libraries, **342**–43
Cech, Thomas, 68
Cecum, **854**
Cell body, neuron, 886–**87**
Cell-cell attachments, eukaryotic, 135–38
Cell-cell gaps, 138–39
Cell-cell interactions, 131–47
  carbohydrates, cell-cell recognition, and, 77
  cell surfaces and, 132–34
  communication between distant cells, 139–46
  connection and communication between adjacent cells, 134–39
  in development, 375*t*, 377
  hormone structures and functions for, 140*t*
  second messenger examples, 143*t*
Cell-cell signaling, 139–46
  in animals (*see* Chemical signals, animal)
  auxin as plant master regulator, 380–81
  bacterial quorum sensing and, 145–46
  bicoid gene as animal master regulator, 379–80
  common pathways of, in differing contexts of development, 383–84
  cross-talk, 145
  in development, 375*t*, 377
  in differential gene expression, 379–84 (*see also* Pattern formation)
  evolutionary conservation of signals, 383
  hormones and, in multicellular organisms, 139–40*t*
  in plant sensory systems, 756–57
  proteins and, 45
  reuse of signals in differing developmental contexts, 383–84
  role of carbohydrates in, 77
  signal processing, 140–44
Cell crawling, **124**
Cell cycle, **196**. *See also* Cells
  cancer as out-of-control, 206–9
  cell-cycle checkpoints and regulation of, 204–5
  cell division by meiosis and mitosis and cytokinesis in, 194 (*see also* Meiosis)
  control of, 202–6
  cytokinins and, 768–69*f*
  discovery of cell-cell regulatory molecules, 203–4
  mitosis and, 195–97*f* (*see also* Mitosis)
Cell-cycle checkpoints, 204, **205**, 207–9
Cell determination, 397–98
Cell differentiation, 375*t*, 377
Cell division, **194**
  animal cytokinesis as, 124
  cancer and out-of-control, 206–9
  cell cyle of, 194 (*see also* Cell cycle; Meiosis; Mitosis)
  cytokinins in plant, 768–69*f*
Cell-elongation responses, plant, 761–62
Cell growth, 375*t*, 376–77
Cell identity, 77
Cell-mediated response, **987**–88
Cell movement, 375*t*, 376–77
Cell plate, **200**
Cell proliferation, 375
Cells, **2**, 102–30
  animal vs. plant, 706–7 (*see also* Animal cells; Plant cells)
  blood, 870–71
  cancer, 206–7, 332 (*see also* Cancer)
  carbohydrate functions in, 77–79
  cell cycle of (*see* Cell cycle)
  cell theory, living organisms, and origin of, 2–4
  cellular respiration and fermentation in (*see* Cellular respiration; Fermentation)
  chemical signals and target, 930–31
  chloride, 827–28
  cultures of (*see* Cultures, cell and tissue)
  entry of viruses into, 681–82
  eukaryotic vs. prokaryotic, 7*f*, 105–6*f*, 107*t* (*see also* Eukaryotic cells; Prokaryotic cells)
  first, 91
  genetics and (*see* Genetics)
  hormone target receptors, 943–47
  importance of cell membranes to, 2, 82, 98–99 (*see also* Plasma (cell) membranes)
  interactions between (*see* Cell-cell interactions)
  introducing novel alleles into human, 351–52
  life cycles and diploid vs. haploid, 533–34
  living organisms and, 1
  organogenesis and, 396–99
  photosynthesis in (*see* Photosynthesis)
  programmed death of (*see* Apoptosis)
  prokaryotic cell structures and functions, 102–5
  protein functions in, 45
  protist, 524–25
  separating components of, by centrifugation, B:16–B:17
  studying live, B:15
  two requirements of, 168
  viruses as not, 675
  vocabulary for describing chromosomal makeup of, 214*t*
  water in living, 22
Cell sap, **707**
Cell-suface hormone receptors, 141
Cell-suicide genes, 376
Cell-surface hormone receptors, 945–47
Cell theory, **2**–4, 102. *See also* Cells
Cellular respiration, **154**, **507**
  ATP, ADP, glucose, and metabolism in, 148
  ATP synthesis via electron transport and chemiosis in, 161–66
  catabolic and anabolic pathways and, 168–69
  chemical energy and redox reactions, 149–53
  citric acid cycle in, 158–61
  fermentation and, 166–68
  fermentation vs., 153, 167*f*
  four steps of, 153–54
  gas exchange and, 862
  glycolysis as processing of glucose to pyruvate in, 155–56
  overview of, 154*f*
  photosynthesis vs., 173
  processing of pyruvate to acetyl CoA in, 156–58
  prokaryotic, 506–8
  summary of, 165*f*
Cellular slime mold model organism, 536, B:20, B:21*f*
Cellulases, 590
Cellulose, **76**
  bird digestion of, 848
  in cell walls, 132–33
  fungal digestion of, 579, 581, 590
  as plant structural polysaccharide, 71*f*, 76, 77
  ruminant digestion of, 850–51
  in secondary cell walls, 709
  structure of, 75*t*
Cell walls, **76**, **105**, **707**
  cellulose as structural polysaccharide in, 71*f*, 76, 77
  as characteristic of domains of life, 497*t*

Boldface page numbers indicate a glossary entry; page numbers followed by an *f* indicate a figure; page numbers followed by *t* indicate a table.

cohesion-tension theory and secondary, 726
eukaryotic, 113, 114*t*
in green plants, 549
ingestive feeding and, 530
plant cell-cell attachments and, 135
prokaryotic, 496
prokaryotic diversity in composition of, 504–6
as prokaryotic exoskeleton, 104–5
protist, 528–29
sclerenchyma cells and, 709–10
types of, in plants, 132–33
Cenozoic era, **483**
Centipedes, 639
Central dogma of molecular biology, **280**
Francis Crick's statement of, 280
development of, 279–82
exceptions to, 281
Central nervous system (CNS), **604**, **886**
functional anatomy of, 900–902
hormones, 929
hormone signaling pathways and, 931–32
vertebrate, 653
Centrifugation
in discontinuous replication experiment, 267
in estrogen receptor research, 944
in Hershey-Chase experiment, 259–60*f*
in Meselson-Stahl experiment, 263
overview of, B:16–B:17
Centrioles, **125**, **198**, 202*t*, **389**
Centromeres, **196**, 199, 202*t*, **213**
Centrosomes, **125**, **198**, 202*t*
Cephalization, 603, **605**, 610, 886
Cephalochordates (Cephalachordata), **651**–52
Cephalopods (Cephalopoda), **630**, 636, 914
Cerebellum, **653**, **901**
Cerebral ganglion, 605, 886. *See also* Brains
Cerebrum, **653**, **901**–2
Cervix, **960**
Cetaceans, evolution of, 422
CFTR (cystic fibrosis transmembrane conductance regulator), 94–95, 826–27
Chaetae, **633**
Chaetognatha, 603*t*
Chagas disease, 522*t*
Chambers, heart, 878
Changing-environment hypothesis on sexual reproduction, 224–25
Channel proteins, **95**, **731**, **825**
facilitated diffusion via, 94–96
in phloem loading, 731–32
Chaperone proteins, 330
Character displacement, **1062**
Chargaff, Erwin, 62–63
Charophyceae, 569
Chase, Martha, 119–20, 259–60*f*
Cheatgrass, 1015–17
Checkpoints, cell-cycle, 204, **205**, 207–9
Chelicerae (Chelicerata), **642**
Chemical bonds, **17**. *See also* Carbon-carbon bonds; Covalent bonds; Disulfide bonds; Hydrogen bonds; Ionic bonds; Peptide bonds
Chemical energy, **33**
animal nutrition and (*see* Animal nutrition)
ATP, redox reactions, and, 149–53 (*see also* ATP (adenosine triphosphate))
carbohydrates and storage of, 78–79
cellular respiration, fermentation and (*see* Cellular respiration; Fermentation)
cellular respiration and, in fats and storage carbohydrates, 153
chemical evolution, chemical reactions, and, 27–33 (*see also* Chemical evolution; Chemical reactions)
conversion of light to, 174–75, 178–79 (*see also* Photosynthesis)
plant nutrition and (*see* Plant nutrition)
primary producers and, 1083–84
secondary active transport and, 98
sugars and, 71
Chemical equilibrium, **27**
Chemical evolution, **15**–37. *See also* Evolution
atoms, ions, and molecules as building blocks of, 16–21 (*see also* Molecules)
biological evolution vs., 39, 59
chemical reactions, chemical energy, and, 27–33
environments for, 27
four steps of Oparin-Haldane theory of, 38–39
importance of carbon and organic molecules to, 33–36
monosaccharides and, 71, 73
nucleic acid formation by, 62
nucleotide production by, 60–61
overview of hypothesis of, 34*f*
pattern and process components of, and biological evolution vs., 15
polymerization of amino acids and, 42–43
polysaccharides and monosaccharides in, 76
properties of water and early oceans as environments for, 22–26
RNA as first self-replicating molecule in, 59, 68–69
spark-discharge experiment on, 39–40
Chemical modifications, post-translational, 304
Chemical reactions, **21**
acid-base reactions, 25–26
chemical evolution and, 27–33 (*see also* Chemical evolution)
enzymes and catalysis of, 45, 51–56 (*see also* Enzymes)
eukaryotic organelles and, 107
first, in chemical evolution, 32
plasma membrane and, 82
rate of, in cells, 116
spontaneous, 29–30
temperature and concentration vs, reaction rates of, 30–32
types and transformations of energy in, 27–29
volcanic, 27
water and, 822
Chemical signals, animal, **929**–49
action of hormones on target cells, 943–47
chemical characteristics of hormones as, 933–34
endocrine system components and, 932–33
endocrine system hormones as, 929
experimental identification of hormones as, 934–35
functions of hormones as, 935–40
hormone signaling pathways of, 931–32
major categories of, 930–31
regulation of production of hormones, 940–43
six categories of animal, 930*t*
Chemical structures, reading, B:8–B:9
Chemiosmosis, 153, **162**–63, 181
Chemokines, **976**, 977*t*
Chemolithotrophs, **506**
Chemoorganotrophs, **506**
Chemoreceptors, **908**, 918–20
Chemosenses, 918–20
Chemotherapy, cancer, 371–72, 618
Chestnut blight, 581, 598
Chewing, 846. *See also* Mouthparts; Teeth
Chiasmata (chiasma), **216**
Chicken genome, 369*t*
Child abuse rates, 1034
Childbirth, 970–71
Chimpanzees
ancestors of humans and, 670
HIV and, 688
number of chromosomes in, 213*t*
similarity of genome of, and human genome, 369–70
Chins, 672
Chitin, 75*t*, **76**, 77, 585
Chitinase, 56
Chitons, **630**, 636
Chlamydiae, 512, 514
Chlamydia trachomatis, 364, 514
Chloride cells, 826–28
Chloride ions, 94–95
Chlorine, 739*t*
Chlorophylls, **175**
capture of light energy by, 174–79
in protists, 532*t*
Chloroplasts, **112**, **174**, **707**
Calvin cycle in, 186
in green plants, 549
in moss cells, 172*f*
origin of, in protists, 530–31
photosynthesis in, 112–13, 114*t*, 174 (*see also* Chlorophylls)
phototropins and, 758
pigments in, 176
in plant cells, 116
Choanocytes, **608**
Choanoflagellates, 607–8, 617
Cholecystokinin, **852**, **937**
Cholesterol, 84, 88
Cholodny, N. O., 760–61
Cholodny-Went hypothesis, 760–61
Chondrichthyes, 662
Chordata, 603*t*, 651–52, 660–68
Chordates, **396**, **650**
body plans of, 650–51
Chordata lineages and, 603*t*, 651–52, 660–68
as deuterostome, 646
invertebrates, 651–52
phylogenies of, 654*f*
vertebrates (*see* Vertebrates)
Chorion membrane, 658
Chromalveolata, 525
Chromatids, **195**–96, 202*t*
Chromatin, **197**, 202*t*, **320**–23, 331
Chromatin remodeling, **320**–23, 331*t*
Chromatography, 175–76, 185
Chromosomes, **103**, **195**
circular, as characteristic of domains of life, 497*t*
consequences of meiosis for, 220–23
discovery of, 195–96
eukaryotic, 107–8
genomes and, 364
linkage and genes on same, 243–45, 246*f*
meiosis, heredity, and, 221 (*see also* Chromosome theory of inheritance)
meiosis and, 211–12
in mitosis, 202*t*
movement of, during mitosis, 201–2
mutations in, 286–87*f*
nuclear lamina and, 116
number of, found in some familiar organisms, 197, 213*t*
polyploidy in, and speciation, 465–68
prokaryotic, in nucleoids, 103–4
sister chromatids and, 213–14
trisomy in human births and number of, 227*t*
vocabulary for describing chromosomal makeup of cells, 214*t*
Chromosome theory of inheritance, **241**. *See also* Mendelian genetics
discovery of sex chromosomes and, 241
importance of, 230
meiosis and statement of, 240–41
mutants and, 241
rediscovery of Gregor Mendel's experiments and, 239
sex linkage and testing of, 241–42
Chylomicrons, **854**
Chymotrypsin, 855*t*
Chytridiomycete life cycle, 592
Chytrids, 583, 585, 592, 595
Cichlids, 486, 844–45
Cigarette smoke, 272
Cilia (cilium), **127**–28, 533
Ciliata, 540
Circadian clock, **1026**
Circular chromosomes, 497*t*
Circulatory systems, **862**, 870–83. *See also* Blood
blood pressure and blood flow in, 882–83
closed, 875–77
four steps of gas exchange and, 861–62
hearts in, 877–81
mesoderm and, 604
open, 875
oxygen and carbon dioxide transport in blood, 870–74
pulmonary circulation and, 879*f* (*see also* Respiratory systems)
surface area for gas exchange and, 874–75
types of, 874–77
Cis surface, Golgi apparatus, 109, 121
Cisternae (cisterna), **109**, 121
Citrate, 157
Citric acid, 598, 959*t*
Citric acid cycle, 153, **158**–61
Citrulline, 277
Clades, **460**, 504
Cladistic approach, **475**, 476–79
Clams, 634
Class I and class II MHC proteins, **984**
Clausen, Jens, 700
Clay, 43, 62, 743, 744*t*
Cleaner shrimp, 1068
Cleavage, **392**–94

Cleavage furrow, **200**
Cleaver, James, 273
Clements, Frederick, 1070–72
Clements-Gleason dichotomy, 1070–71
Cliff swallows, 441
Climate, **1002**
biomes and, 1002
effects of global climate change on ecosystems, 1011–13*f*
global patterns in, 1009–11
global warming and, 1008–9, 1098–1102 (*see also* Global climate change; Global warming)
green plant holding of water and moderate, 547
human alteration of, 1098–99
regional effects of mountain ranges and oceans on, 1010–11
seasonality of weather and, 1010
tree growth rings and research on, 715
tropical warmth, polar cold, and, 1010
tropical wetness and, 1009–10
Climax communities, **1071**
Clitoris, **960**
Cloaca, **855**, 954
Clonal-selection theory, **979**–83*f*
Clones, **221**, **378**, 708, **786**, **951**, 979
Cloning, DNA. *See* DNA cloning
Cloning vectors, **340**
Closed circulatory systems, **875**–77
Clostridium acetium, 507*t*, 508
Clotting, blood, 975
Club fungi, 593, 596
Club mosses, **571**, 571
Cnidaria, 601*f*, 603*t*, 616–17, 619
Cnidocyte, **619**
Coal, 548, 551, 571
Cocaine, 775
Cochlea, **910**
Code, genetic. *See* Genetic code
Coding strand, **290**
Codominance, **245**, 250*t*
Codons, **282**–85, 297
Coefficient of relatedness, **1032**
Coelacanths, 663, 917–18
Coelom, **605**, **624**, 626
Coelomates, **605**
Coenocytic, **583**
Coenzyme A (CoA), **157**
Coenzyme Q, **161**–64
Coenzymes, **54**
Coevolution, **793**, **1059**, 1066–67
Coevolutionary arms races, **1059**, 1066–67
Cofactors, enzyme, **54**
Cohesion, **22**–23, **724**
Cohesion-tension theory, 722, **724**–27
Cohorts, **1038**
Cold virus, 692
Coleochaetophyceae (Coleochaetes), 568
Coleoptera, 640*t*
Coleoptiles, **758**
Collagen, 45*f*, **133**–34
Collecting duct, **837**
antidiuretic hormone and, 931, 940
osmoregulation by, 837–38
structure and function of, 832, 833*f*, 838*t*
Collenchyma cells, **708**–9, 711*t*
Colon, **855**
Colonies, **617**
Colonization
allopatric speciation by, 463
founder effect and, 446
Color, flower, 562, 788
Color vision, 446, 915, 916–18
Comb jellies, 620
Commensal, **586**
Commensalism, **1059**, 1070*t*
Common primroses, 1050–51
Communication, **1027**–31
Communities, **994**, **1058**. *See also* Community ecology
biodiversity and stability of, 1119–20
disturbance regimes in, 1073–74
experiments on predictability of structure of, 1070–72
keystone species in, 1072–73
post-disturbance succession in, 1074–77
taxonomic diversity and, 1106, 1107*f*
Community ecology, 994, 1058–82
communities in, 1058 (*see also* Communities)
community dynamics in, 1073–77
community structure in, 1070–73
species interactions in, 1058–70
species richness in, 1077–80
Companion cells, **711**, **729**–30, 733*f*
Comparative morphology, 603–7
Compass orientation, **1026**–27
Competition, **1059**–63*f*
as biotic factor in distribution of species, 1015
competitive exclusion principle of, 1060–61
conservation biology and resistance to invasive species through, 1062
experimenal studies of, 1061–62
fitness and impacts of, 1070*t*
fitness trade-offs in, 1062
giraffe necks and food vs. sexual, 9–10
intraspecific vs. interspecific, 1059
male-male, 454–55
niche differentiation and coexistence to avoid, 1062, 1063*f*
niches and interspecific, 1059–60
Competitive exclusion principle, **1060**–61
Competitive inhibition, enzyme, **54**–55
Complementary base pairing, 63–**64**, 66–67, 76, **261**
Complementary DNA (cDNA), **339**–40, 342, **684**
Complementary strand, DNA, **65**
Complement proteins, 45
Complement system, **987**
Complete digestive tracts, **845**
Complete metamorphosis, **616**
Complex tissues, plant, 710
Compound eyes, **637**, **913**
Compound leaves, 408, 702
Compression, 132
Computer models
of asteroid impact, 490
genetic drift simulations, 444
Concentration
enzyme catalysis and substrate, 55
hormone, 934
increasing carbon dioxide, in plants during photosynthesis, 187–89
role of, in chemical reactions, 30–32
Concentration gradients, **89**–92, 94–95, 380, **823**
Concept maps, B:10
Condensation reactions, **43**, 73, 83–84
Condensed replicated chromosomes, 195–96, 197
Conditional behavior, 1020–21
Condoms, 966*t*
Conduction, 722, **816**
Cones, eye, **915**–17
Cones, gymnosperm, 551, 561, 575
Confocal microscopy, B:15
Conformational homeostasis, 815
Conifers, 1007
Connections, cell. *See* Multicellularity
Connective tissue, **806**–7
Connell, Joseph, 1061–62
Conservation, soil, 742
Conservation biology, **995**, 1105–27. *See also* Biodiversity
bioremediation and, 500–501
competition to resist invasive species in, 1062
designing protected areas, 1120–21
ecology and, 995
ecosystem restoration, 1122
ex situ conservation, 1121–22
Gap Analysis Program (GAP) in, 1120
genetic drift and, 445
Malpai Borderlands case history, 1122–23
phylogenetic species concept and, 461–62
population ecology and endangered species analysis, 1053–55
preserving biodiversity, 1120–23
preserving metapopulations, 1055
preserving phylogenetically distinct species, 1107*f*
stabilizing human population size of and resource use, 1121
sustainability in, 1121
whooping cranes and, 1043–44
wildlife corridors in, 1121
Conservation hotspots, 1110
Conservation International, 1109–10
Conservative, genetic code as, 283–84
Conservative replication, DNA, 261–63
Constant (C) regions, **982**
Constitutive defenses, **1063**–64
Constitutively, **309**
Constitutive mutants, **311**, 313
Constraints, genetic, 431
Constraints, historical, 432
Consumers, 602, 1068, **1084**. *See also* Consumption
Consumption, **1059**, 1063–68
coevolutionary arms races and, 1066–67
constitutive defenses against, 1063–64
consumers as biocontrol agents, 1068
efficiency of predators at reducing prey populations, 1065
fitness and impacts of, 1070*t*
inducible defenses against, 1064–65*f*
limitations on herbivore, 1065–66
parasite manipulation of hosts, 1067–68
types of, 1063
Continental shelf, **1000**
Continents, changes in, 483–84
Continuous population growth, 1043–44
Continuous strand, **266**
Contraception, human, 966
Contractile proteins, 45, 280
Contractile roots, 698
Control. *See* Regulation
Control groups, **12**
Convection, **816**
Convergent evolution, **476**
Conversions, metric system, B:1*f*
Cooperation, 1031–34
Cooperative binding, **871**–72
Coordination, bacterial vs. eukaryotic gene expression, 331–32
Copepods, 643, 1100
Copper, 739*t*, 740, 747
Coprophagy, **854**
Copulation, 630, 954
Copying. *See* Replication
Coral reefs, 539, 649, **1001**, 1058*f*, 1100, 1114
Corals, 619
Co-receptors, **682**
Core enzymes, **291**
Cork cambium, **712**, 713–14
Cork cells, **713**
Corms, **786**
Corn, 4–5, 213*t*, 354, 548
Cornea, **914**
Corolla, **788**
Corona, **631**
Corpus callosum, **901**
Corpus luteum, **963**
Corridors, wildlife, 1055, 1121
Cortex, **706**, **722**, **832**
Cortical granules, **390**, 392
Corticosteroids, 991
Corticotropin-releasing hormone (CRH), **941**
Cortisol, **938**
Costanza, Robert, 1117
Cost-benefit analysis, **1021**
Cotransport, **98**, **732**, **825**
Cotransporters, **732**–33, 745–46, **825**–28, 853, 888
Cotton, 548
Cotyledons, **405**, **564**–65, **796**
Countercurrent exchangers, 818–**19**, 835–37, 866
Covalent bonds, **18**
amino acids and, 40
carbon and, 33
electron sharing and, 17–18
nonpolar and polar, 18
protein tertiary structure and, 48
Cowpox, 973–74
Crabs, 643, 1064–65*f*
Crane, Peter, 554
Cranium, **653**
Crassulacean acid metabolism (CAM), **188**–89, **728**, 1004
Creeks, 999
Crenarchaeota, 512, 516
Creosote bushes, 786
Cretaceous-Paleogene extinction, 490–92
Creutzfeldt-Jakob disease, 51
Crews, David, 1023–24
Crick, Francis
adaptor molecule hypothesis of, 297
articulation of central dogma of molecular biology by, 280
discovery of double helix secondary structure of DNA by, 59*f*, 62–65
genetic code hypothesis of, 279
triplet code reading frame discovery of, 283
wobble hypothesis of, 299–300

Boldface page numbers indicate a glossary entry; page numbers followed by an *f* indicate a figure; page numbers followed by *t* indicate a table.

Crickets, 641*t*
Crime, 366
Cristae, **112**, **157**, 161
Crocodiles, 660, 667, 841*f*
Crocodilia, 667
Cro-Magnons, **671**
Crop, bird, **848**
Cro protein, 48–49
Crops, transgenic, 354–56
Cross-fertilization, **231**
Crossing over, **217**
calculating frequencies of, 246
in meiosis prophase I, 219–20
role of, in meiosis, 222
role of, in Mendelian genetics, 244–45
Cross-pollination, **792–93**
Cross-talk, **145**, **774**
Crustaceans, **643**, 875
Cryoelectron tomography, 157*f*
Cryptic female choice, 955
Cryptic species, **460**
Cryptochromes, 758
Ctenophora, 603*t*, 620
C-terminus, 44, 301
Cultures, cell and tissue, **119–20**
animal cell, B:17–B:18
in anthrax research, 499
in cytokinin research, 768
differentiated plant cells and, 377
discover of gap phases with, 196
in DNA damage research, 273
enrichment cultures in prokaryotic research, 501–2
plant tissue, B:18
in secretory pathway research, 119–20
Cup fungi, **598**
Cushing's disease, **941**
Cuticle, **625**, **707**, **828**
as defense barrier, 775, 974
fossilized, 550
water loss prevention by, 553–54, 727, 828–29
Cuttings, plant, 708
Cuttlefish, 636
Cuvier, Baron George, 417
Cyanobacteria, **510**
characteristics of, 515
lichens and, 566, 589, 597
metabolic diversity and, 507*t*
oxygen revolution and, 509
protist chloroplasts and, 530–31
Cycads (Cycadophyta), 574
Cycles, population. *See* Population cycles
Cyclic adenosine monophosphate (cAMP), 143*t*, **315–16**, **946**
Cyclic guanosine monophosphate (cGMP), 143*t*, 144, 916
Cyclic photophosphorylation, **183–84**
Cyclin-dependent kinase (Cdk), **204**
Cyclins, **204**, 208, 330
Cysteines, 42*t*, 48, 280–81, 296
Cystic fibrosis, 94–95, 350, 826–27
Cytochrome *c*, **163**–64
Cytokines, **930**, 977*t*, 991
Cytokinesis, **124**, **194**–95, 197, 200, **217**, **976**
Cytokinins, **768**–69*f*, 774–75*t*
Cytoplasm, **104**
cleavage of, 393
cytokinesis as division of, 194, 200
cytosol in, 109
ribosomes in, 279
Cytoplasmic determinants, **390**, 393
Cytoplasmic streaming, **124**
Cytosine, 60, 63–64, 66–67, 261, 279
Cytoskeletons, **104**
eukaryotic, 113–14*t*, 123–28, 134
prokaryotic, 104
prokaryotic vs. eukaryotic, 107*t* (*see also* Cytoskeletons, eukaryotic)
protist, 528–29
Cytosol, **109**, 124
Cytotoxic T cells, **985**–89*t*

D

Daddy longlegs, 642
Dalton, **16**
Dalton, John, 916–17
Daly, Martin, 1034
Damaged DNA
cell-cycle checkpoints and, 205
sexual reproduction and, 224
UV radiation study with tumor suppressor gene mutations, 332–33
xeroderma pigmentosum (XP) as failure to repair, 272–74
Damselflies, 641*t*
Dance hypothesis, honeybee language, 1028–29
Danielli, James, 92
Daphnia, 224, 951–52
Darwin, Charles. *See also* Evolution; Natural selection
gravitropic responses research by, 764
orchid pollination hypothesis by, 793
phototropic responses research by, 758–60
sexual selection and, 452
theory of evolution by natural selection by, 4, 414–16, 422–26
Darwin, Francis, 758, 759–60, 764
Data, scientific, 8–9
Data points, graph, B:2
Dating
foraminifera and, 540
radiometric, 417, 481
relative and absolute, 422
Daughter cells
meiosis and, 214–15
mitosis and, 197–200
Davson, Hugh, 92
Day length
Anolis lizard sexual activity and, 1023–24
photoperiodism and, 787
Day-neutral plants, **787**
Deactivation
mitosis-promoting factor, 204
signal, 144
Dead space, **868**, 870
Dead zones, aquatic, 512, 1095, 1102
Death, plant, 773–74
Death rates, human, 206*f*, 1037–40. *See also* Mortality, human
Deception, animal communication and, 1029–31
Deciduous trees, **574**, 1006
Decision making, 1020–21
Decomposer food chains, **1085**
Decomposers, **530**, **1084**
fungi as, 579, 589–90, 595
grazers vs., 1085–86
metabolic rates of, 1093
plasmodial slime molds as, 537
protist, 542
De-differentiation, plant cell, 377, 401
Deep-sea vents, 27, 32, 40, 43, 73, 509. *See also* Prebiotic soup
Defense, cell, 45
Defense responses, plant, 112, 707, 775–81, 1063–66
Definitive hosts, **632**
Deforestation, 742, 1093–94, 1096, 1112
Degeneration hypothesis, 687
Dehydration, human, 803
Dehydration reactions, **43**, 73, 83–84
Deleterious alleles, 224, **448**, 1114
Deleterious mutations, **286**
Demography, **1038**
four processes of, 1037–38
life histories in, 1039–41
life tables in, 1038–40
Denaturation, DNA, 344–45
Denatured, **50**
Dendrites, **808**, **886**–87
Dendritic cells, **984**, 985*f*
Dense connective tissue, **807**
Density, water, 23–24, 26*t*
Density centrifugation, 944
Density dependent population growth, **1042**, 1045–46
Density-gradient centrifugation, 263, B:16–B:17
Density independent population growth, **1042**, 1044
Dental disease, 498
Dental plaque, 145
Deoxyribonucleic acid (DNA). *See* DNA (deoxyribonucleic acid)
Deoxyribonucleoside triphosphates (dNTPs), **264**, 346–47
Deoxyribonucleotides, **60**
anabolic synthesis of, 169
components of, 60
in DNA backbone, 260
DNA composition of, 259
DNA synthesis and, 196
Deoxyribose, 60, 61, 77
Dependent assortment, 236–38
Dephosphorylation, 304
Depolarization, **890–92**, **908–9**
Deposit feeders, **611**, **843**
Derived traits, **475**
Dermal tissue, **707**
Dermal tissue systems, **705**, 707–8, 711*t*
Descending limb, loop of Henle, 835–36, 838*t*
Descent from common ancestors, 418–22, 423*f*
Descent with modification, 4, **416**. *See also* Evolution; Natural selection
Deserts. *See* Subtropical deserts
Design, experimental, 10–12. *See also* Experiments
Desmosomes, **136**
Detergents, **94**
Determination, cell, **397**–98
Detritivores, **612**, **637**, **1084**
Detritus, **530**, **997**, **1084**, 1085, 1092–93
Deuterostomes, **606**, **646**–74. *See also* Animal(s)
characteristics of selected hominins, 670*t*
chordates, 650–52
echinoderms, 647–50
four phyla of, 646–47*f*
major animal phyla and, 603*t*
primates and hominins, 668–73*f*
vertebrates, 653–68
Devegetation study, 1093–94
Development, 374–87
animal (*see* Animal development)
cell-cell signaling of differential gene expression in, 379–84 (*see also* Pattern formation)
cell movement, growth, and differential expansion in, 375*t*, 376–77
differential gene expression in, 377–79 (*see also* Differential gene expression)
evolutionary change and changes in pathways of, 384–85
gene expression regulation and abnormal, 332
plant (*see* Plant development)
reuse of cell-cell signals in differing contexts of, 383–84
shared processes of, 375–77
Developmental biology, 374. *See also* Development
Developmental homology, **420**–21
DeWitt, Natalie, 733
Diabetes insipidus, **837–38**, 857
Diabetes mellitus, 339, **856–58**, 983
Diacylglycerol (DAG), 143*t*
Diamond, Jared, 1105
Diaphragm, human, **868**
Diaphragms, contraceptive, 966*t*
Diarrhea, 522*t*, 691, 854
Diastole, **879**
Diastolic blood pressure, **880**
Diatomaceous earth, 542
Diatoms, 522–23, 542
Dicots, **564**–65
Dictyostelium discoideum, 536, B:20, B:21*f*
Dideoxyribonucleoside triphosphates (ddNTPs), 346–47
Dideoxy sequencing, **346**–47, 349, 353*t*
Diencephalon, **901**
Diet, human, 77, 843*t*. *See also* Animal nutrition; Food
Differential cell growth, 375*t*, 376–77
Differential centrifugation, **116**, 126, B:16–B:17
Differential gene expression, **319**
cell-cell signals that trigger, 379–84 (*see also* Pattern formation)
eukaryotic gene regulation and, 319–20 (*see also* Gene regulation, eukaryotic)
regulatory proteins in, 326
regulatory transcription factors and, 378
role of, in development, 377–79
Differentiation, cell, 375*t*, **377**, 398–99. *See also* Differential gene expression
Diffusion, **89**, **823**
across lipid bilayers, 89–90
facilitated (*see* Facilitated diffusion)
Fick's law of, 864–65
gas exchange and, 862, 864–65, 874–75
in phloem loading, 731
prokaryotic cells and, 105
skin-breathing and, 813
surface area adaptions for, 814
surface area and animal gas exchange by, 874–75
Digestion, **845**
cecum and appendix in, 854
complete human digestive tract, 846*f*
dietary fiber and human, 77
digestive enzymes in, 110–11, 115
digestive enzymes in human, 79
digestive processes in, 846–47
digestive tract in, 845–46
fungal extracellular, 589–90
large intestine in, 854–55
mammalian digestive enzymes, 855*t*

Digestion, (*continued*)
mouth and esophagus in, 847–48
nutrition processes and, 841–42
plant proteinase inhibitors and, 779
small intestine in, 851–54
stomach in, 848–51
Digestive glands, 933
Digestive hormones, 937
Digestive tracts, **845**
animal, 604, 845–46
human, 810*f*, 846*f*
Dihybrid crosses, **236**–38
Dikaryotic, **582**
Dimers, 49, **125**
Dinoflagellata, 521–22*t*, 541
Dinosaurs
birds as, 657–58, 668
extinction of, 483, 486, 490–92
as reptiles, 660
Dioecious, **790**
Diploblasts, **604**
Diploid cells, **213**, 214*t*, 533–34
Diplomadida, 538
Dipnoi, 663
Diptera, 640*t*
Direct counts, 1116
Directed-pollination hypothesis, 562
Direct fitness, 1032
Directional selection, **440**–41
Direct sequencing, **502**–3, **525**, 582, 1107
Disaccharides, **73**
Discontinuous replication hypothesis, 266–67
Discontinuous strand, **267**
Discrete population growth, 1043
Discrete traits, **248**
Diseases, animal
avian flu virus, 688, 689
avian malaria, 1015
herbivore, 1065–66
prions and, 51
simian immunodeficiency viruses (SIVs), 688
viruses and, 681
Diseases, human
achromatopsia, 446
allergies, 990–91
ALS (Lou Gehrig's disease), 376
animals and transmission of, 602–3
anthrax, 498–99
autoimmune, 983–84
bacterial, 513, 514, 515
bacterial quorum sensing and, 145–46
bacteria that cause, 499*t*
beriberi and vitamin deficiencies, 54
cancer (*see* Cancer)
carriers of genetic, 251
caused by protists, 520–22
chlamydia, 364
Cushing's disease, 941
cystic fibrosis, 94–95, 350
dental plaque and tooth decay, 145
diabetes mellitus, 856–58, 983
emerging viruses and emerging, 688–89
food poisoning, 500
fungal, 581, 584, 594
gas exchange and circulation failures and, 861
genetic counselors and inherited, 346
genetic testing and, 350
germ theory of, 499–500
HIV/AIDS, 677–78, 990
Huntington's disease, 233, 251, 348–51
hypertension, 880
immunity, immunization, and, 973–74 (*see also* Immune systems)
immunodeficiency, 990–91
kuru and Creutzfeldt-Jakob disease, 51
malaria, 541, 1066–67
Marfan syndrome, 247
mistakes in meiosis and, 225–27
multiple sclerosis (MS), 894–95, 983
phenylketonuria, 248
pituitary dwarfism, 338–44
prions and, 50–51
prokaryotes and, 498–500
proteins as defense against, 45
protists that cause, 522*t*
red-green color blindness, 916–17
responding to viral outbreaks, 689
rheumatoid arthritis, 983
roundworms and, 638
schistosomiasis, 631–32
severe combined immunodeficiency (SCID), 352–54
sexually transmitted (*see* Sexually transmitted diseases)
sexual reproduction and, 224–25
sickle-cell disease, 46, 251
trisomy in human births, 227*t*
ulcers, 850
viral epidemics, 676–77
vitamin deficiencies, 355
xeroderma pigmentosum (XP), 272–74
Disorder. *See* Entropy
Dispersal, **1013**
allopatric speciation by, 463
barriers to, 1014
of exotic and invasive species, 1014–15
habitat destruction and, 1113
humans as agents of, 1014
physical isolation and, 462–63
seed, 797–98
Dispersive replication, DNA, 262*f*, 263
Disruptive selection, **442**, 443*f*, 465, 471*t*
Distal tubule, 832, **837**–38*t*
Distortion, experimental, 12
Disturbance regimes, **1073**–74, 1080, 1107, 1119–20
Disturbances, **1073**
Disulfide bonds, 35, **48**
Divergence, species, 468, 471*t*. *See also* Speciation
Diversity. *See also* Biodiversity; Ecological diversity; Ecosystem diversity; Genetic diversity; Metabolic diversity; Morphological diversity; Species diversity
adaptive immune response, 978
animal, 610–17
of animal large intestines, 855
animal nitrogenous wastes, 829–30
antibody, 981–82
fungal, 586–94
of genomes, 363
green plant, 553–66
prokaryotic, 498, 504–12
protist, 526–36
protostome, 625–30
viral, 675, 686–89
Dixon, Henry, 724
DNA (deoxyribonucleic acid), **61**, 62–66. *See also* Genes; Nucleic acids
amplification of fossil, with polymerase chain reaction, 345–46
anabolic synthesis of nucleotides for, 169
antiparallel strands of, 63–64
in central dogma, 280–81
as characteristic of domains of life, 497*t*
chloroplast, 113
chromosomes, genes, and, 195–96
cloning of, 340–41, 344–46
closed, as protected from DNase enzyme, 322
comparing *Homo sapiens* and *Homo neanderthalensis*, 345–46
cutting and pasting, 340–41
directionality of strands of, 61
discovery of double helix secondary structure of, 62–65, 260–61
electron micrograph of, 289*f*
as genetic information-containing molecule, 65
mitochondrial, 112
model of condensed, 319*f*
phylogenies based on sequences of, 477–78
physical model of, 59*f*, 63*f*
as poor catalytic molecule, 65–66
primary structure of, 260*f*
prokaryotic, 103–4
prokaryotic vs. eukaryotic, 107*t*
repair of, 271–74
RNA structure vs. structure of, 67*t*
sequencing of (*see* DNA sequencing)
synthesis of (*see* DNA synthesis)
viral, 686, 687*t*
DNA cloning, **340**–41, 344–46
DNA Data Bank of Japan, 360
DNA fingerprinting, 365, **366**, 367
DNA libraries, **342**–43
DNA ligases, **267**–68*t*, 340–**41**, 352*t*
DNA microarrays, **370**–72
DNA polymerases, **263**
in DNA synthesis, 268*t*
DNA synthesis and, 263–66
mutations and, 285–86, 448
in polymerase chain reaction, 344–45
proofreading ability of, 271–72
DNA probes. *See* Probes, DNA
DNA replication. *See* DNA synthesis
DNase enzyme, 322
DNA sequencing. *See also* Whole-genome sequencing
dideoxy DNA sequencing method, 346–47
genetic homology and, 420
phylogenies based on data from, 477–78
pyrosequencing technologies, 347, 360–61
DNA synthesis, 258–75. *See also* Genes; Genetics
comprehensive model of, 263–68
copying mechanism of, 65
correcting mistakes in, 271–72
DNA polymerases and, 263–64
electron micrograph of, 258*f*
Hershey-Chase experiment on DNA as genetic material in genes, 259–60*f*
lagging strand synthesis in, 266–68
leading strand synthesis in, 265–66
Meselson-Stahl experiment on hypotheses about, 261–63
opening and stabiliizing of double helix in, 264–65
proteins required for, 268*t*
repairing damaged DNA and, 272, 273*f*
starting of replication in, 264
structure of DNA and, 260–61
telomere replication in, 269–71
xeroderma pigmentosum (XP) case study of damaged DNA, 272–74
Dogs, 213*t*, 369*t*
Dolly (cloned sheep), 378
Dolphins, 422
Domains, **7**. *See also* Archaea domain; Bacteria domain; Eukarya domain
Domesticated animals, 602
Dominant alleles, 239*t*
Dominant traits, **233**
identifying human traits as, 251
incomplete dominance and codominance vs., 245–47
monohybrid crosses and, 232–33
Dopamine, 898*t*
Dormancy, 570, **714**–15, **769**, 770–71, **798**–99
Dorsal, **379**
Double bonds, 20
Double fertilization, 403–**4**, **562**, **795**
Double helix secondary structure, DNA, **261**. *See also* DNA (deoxyribonucleic acid)
discovery of, 62–65, 260–61
opening and stabiliizing of, in DNA synthesis, 264–65
proteins required for opening, 268*t*
Double-stranded DNA (dsDNA) viruses, 687*t*, 690
Double-stranded RNA (dsRNA) viruses, 687*t*, 691
Doubly compound leaves, 702
Doughnut-shaped proteins, 45*f*
Doushantuo microfossils, 486–87
Down, Langdon, 225
Downstream, **291**
Down syndrome, **225**–27
Dragonflies, 641*t*
Drosophila melanogaster
bicoid gene as master regulator of development in, 379–80
chromosome theory testing with, 241–42
foraging in, 1021
genetic drift in, 445
genetic map of, 246
genome of, 362, 369*t*
linked genes and, 243–45
as model organism, 623, 639, B:21*f*, B:22
mRNA sequences from one gene in, 329
number of chromosomes in, 213*t*
second-male advantage in, 954–55
Drought, 426–27. *See also* Dry habitats
Drugs
animals and testing of, 603
anticancer, 333, 371–72, 575, 579*f*
anti-HIV, 682, 683, 684
antiviral, 678

Boldface page numbers indicate a glossary entry; page numbers followed by an *f* indicate a figure; page numbers followed by *t* indicate a table.

derived from green plants, 549
ephedrine, 576
evolution of resistance to, 424–26, 500
fungal infections and, 584
fungi and, 595
homology and, 421
miRNAs in, 330
Dryer, W. J., 982
Dry habitats
carbon dioxide concentration mechanisms in, 187–89
drought as selective force, 426–27
gas exchange in, 864
plant adaptations to, 720–21, 728
Drying, seed maturation and, 797
Ducts, exocrine, 933
Dusky seaside sparrow, 461–62
Dutch elm disease, 598
Dwarfism, pituitary, 338–44
Dwarf mistletoe, 798
Dwarf plants, 769–70
Dyneins, **128**, 201–2
Dysentery, 522*t*

E

Eardrums, 910–11
Early endosome, **111**
Early prophase I, meiosis, 217
Ear ossicles, **910**–11
Ears, 910–12. *See also* Hearing
Easter Island, 1105
Ebola virus, 688, 689, 692
Ecdysone, **936**
Ecdysozoans, 603*t*, **609**, 625, 637–43
Echidnas, 665
Echinoderms (Echinodermata), 603*t*, 646, **647**–50
Echinoidea, 650
Echolocation, **912**
E. coli. *See Escherichia coli* (*E. coli*)
Ecological disaster, 1105
Ecological diversity, 509–12
Ecological efficiency, 1086–87
Ecological opportunity, 484–85
Ecology, **993**–1018
aquatic ecosystems in, 995–1001
areas of study in, 993–95
behavioral (*see* Behavioral ecology)
biodiversity and (*see* Biodiversity)
climate and climate change in, 1008–13*f*
community, 994 (*see also* Community ecology)
conservation biology and, 995 (*see also* Conservation biology)
ecological importance of protists, 522–23
ecosystem, 995 (*see also* Ecosystems)
geographic distribution of organisms and, 1013–17
organismal, 994
population, 994, 1037–57
terrestrial ecosystems in, 1001–8
Economic impacts
of animals, 602
of biodiversity, 1117
of fungi, 581–82
of sponges, 618
Ecosystem diversity, **1106**
Ecosystem ecology
biogeochemical cycles in, 1092–98
ecosystem energy flow in, 1083–92
ecosystems and, 995, 1083
global warming in, 1098–1102
Ecosystem energy flow, 1083–92
biomagnification in, 1088–89
global productivity patterns in, 1089–90
net primary productivity in, 1084
primary producers in, 1083–84
productivity limiting factors, 1090–92
transformation of solar energy into biomass, 1084–85
trophic cascades and top-down control in, 1087–88
between trophic levels, 1086–87
trophic structure and, 1085–86
Ecosystems, **547**, **995**, **1083**–1104. *See also* Ecology
animals in, 602
aquatic, 995–1001
global climate change and, 1011–13*f*
global productivity patterns of, 1089–90
human impacts on, 1083
most productive, 1089–90
nutrient cycling within, 1092–94
prokaryotes and nitrogen cycle in, 510–11
restoration of, 1122–23
terrestrial, 1001–8
Ecosystem services, **547**, **1117**
Ecotourism, **1117**
Ectoderm, 394, **395**, **604**
Ectomycorrhizal fungi (EMF), **586**–88, 596, 598, **746**
Ectoparasites, **613**
Ectothermic, **660**
Ectotherms, 660, **816**–17, 1086
Edge habitat, 1113
Ediacaran faunas, 486–87
Effectors, **815**, 938
Effector T cells, 986
Efferent division, **899**
Efficiency
asexual reproduction, 224, 786
ecological, 1086–87
Eggs, **211**, **784**, **951**
amniotic, 655, 658
animal, 615, 959–60
cell-cycle regulation and frog, 203–4
evolution of oviparous and viviparous species and, 956–57
fertilization of, 4, 390–92, 954–57
fitness trade-offs for animal, 804, 805*f*
human, 960
land plant retention of, 556
mammalian, 392
meiosis and, 194, 218
monotremes as mammals that lay, 665
plant, 402, 789–90
protostome, 630
sexual selection and, 452–53
vertebrate amniotic, 658
Either-or rule of probability, B:18
Ejaculation, **958**
Ejaculatory duct, **958**
Elastase, 855*t*
Elastic, **868**
Elastin, 134
Electrical activation, heart, 880–81
Electrical energy, lightning and, 39–40
Electrical gradients, 94–95
Electrically charged side chains, 41*t*
Electrical potential, **887**
Electrical signals, animal, 885–906
action potentials, 890–95
calculating equilibrium potentials with Nernst equation, 888
electrical activation of hearts, 880–81
maintenance of resting potentials, 888–90
measuring membrane potentials with microelectrodes, 890
membrane potentials, 887
nervous systems and, 885–86
nervous tissue and, 808
neuron anatomy and, 886–87
principles of electrical signaling and, 885–91
synapses and, 895–99
vertebrate nervous systems and, 899–904*f*
Electrical signals, plant, 767
Electrical stimulation, mapping of cerebrum with, 902
Electric current, **887**
Electric fields, animal sensing of, 610
Electrocardiogram (EKG), **881**
Electrochemical gradients, **94**–98, 745–46, **887**
Electrolytes, **822**, **842**. *See also* Osmoregulation
Electromagnetic spectrum, **174**
Electron acceptors, **151**
Electron carriers, **152**
Electron donors, **151**
Electronegativity, **18**
Electron microscopy. *See also* Scanning electron microscopy (SEM); Transmission electron microscopy (TEM)
of actin and myosin, 924
of bacterial transcription and translation, 296*f*
cell research and, 116
of chromatin nucleosomes, 321*f*
cryoelectron tomography, 157*f*
of DNA, 258*f*, 265*f*
of fossilized plants, 554
freeze-fracture, of membrane proteins, 93
of myofibrils, 923*f*
of noncoding DNA regions, 293–94
of normal and karotypes, 287*f*
overview of, B:14–B:15
of protist cells, 524–25
of skin cells, 131*f*
of transport vesicles, 126*f*
Electrons
atomic structure and, 16–17
chemical bonds and sharing of, 17–20
electron transport chain and (*see* Electron transport chain (ETC))
in photosynthesis, 177–82
potential energy and, 27
prokaryotic variation in donors and acceptors of, 506–8*t*
redox reactions and, 151–52
Electron shells, **17**
Electron tomography, B:15
Electron transport chain (ETC), **161**
chemiosis hypothesis on, 162–63
components of, 161–62
fourth step of cellular respiration and, 153
organization of, 163–64
in photosynthesis, 180–81, 182
prokaryotic cellular respiration and, 507–8
Electrophoresis. *See* Gel electrophoresis
Electroreceptors, **908**
Elements, **16**
Elephantiasis, 638, 973*f*
Elephants, 460–61, 812–13, 912
Elephant seals, 454–55
Elevational gradients
shoot systems and, 700
tree ranges and, 1072
Elimination, waste, 841–42, 854–55
Ellis, Hillary, 376
Elongation, stem, 762–64
Elongation factors, **302**
Elongation phase, **292**
transcription, 292–93
translation, 301–2
Embryogenesis, **392**, **402**, **796**
animal, 615
events in, 404–6
plant seeds and, 796
Embryonic stem cells, 375
Embryonic tissue layers, 394–95, 604
Embryophytes (Embryophyta), **556**
Embryos, **194**, **374**
developmental homology in, 420
human, 388*f*
land plant retention of, 556
plant, 401*f*
Embryo sacs, 402, 403–4, **790**
Emergent vegetation, **998**
Emerging diseases, 224–25, **688**–89
Emerson, Robert, 179
Emigration, 1037–**38**
Emulsification, **853**
Endangered species, **1111**
dusky seaside sparrow case study, 461–62
extinctions and, 1111 (*see also* Extinction)
population ecology and analysis of, 1053–55
preserving metapopulations of, 1055
using life-table data for, 1054–55
Endangered Species Act, 461–62
End-Cretaceous extinction, 490–92
Endemic species, **844**, **1109**
Endergonic reactions, **30**, 62, **149**, 150–51
Endocrine glands, **932**, 935
Endocrine pathway, 931–32
Endocrine signals, **930**
Endocrine systems, **929**, 932–33. *See also* Hormones, animal
Endocytosis, **111**
Endoderm, 394, **395**, **604**
Endodermis, **722**, 747
Endomembrane system, **111**
eukaryotic, 108–12, 114*t*
manufacture and transport of proteins by, 118–23*f*
Endomycorrhizal fungi, **589**
Endoparasites, **612**–13
Endophytic fungi, **586**, 589
Endoplasmic reticulum (ER), **108**, 114*t*, 122–23*f*, 526
Endorphins, 898*t*
Endoskeletons, **647**, 655, **920**–21
Endosperm, 355, **404**, **562**, **795**
Endosymbionts, **514**
Endosymbiosis, **527**, 566
Endosymbiosis theory, **527**–28, 530–31
Endothermic processes, **27**
Endotherms, 660, 668, **816**–19
End-Permian extinction, 489–90
End-point inhibition, 314
Energetic coupling, **150**
Energy, **27**
active transport and, 97–98
animal reserves of, 938–40
capacity of water to absorb, 24–25

Energy, (*continued*)
carbohydrates and storage of chemical, 78–79
as cellular requirement, 168
cellular respiration, fermentation and (*see* Cellular respiration; Fermentation)
chemical (*see* Chemical energy)
chemical evolution and chemical (*see also* Chemical energy; Chemical evolution)
electrical, of lightning, 39–40
enzyme catalysis and activation, 51–53
flow of, through ecosystems, 1083–92
lipid bilayers and, 85
living organisms and, 1
metabolic diversity of prokaryotes and, 506–9
metabolic rates of consumption of, 812–13
photosynthesis and (*see* Photosynthesis)
six general prokaryotic methods for obtaining, 506*t*
specific heats of some liquids, 24*t*
types and transformations of, 27–29
Energy hypothesis on species richness, 1080
Engelmann, T. W., 176
Engineering, genetic. *See* Genetic engineering
Enhancement effect, 179–80, 183
Enhancers, **324**–26
Enkephalins, 898*t*
Enrichment cultures, **501**–2, 508
Entamoeba histolytica, 522*t*
Entoprocta, 603*t*
Entropy, **29**–30
Envelope, **679**
Envelope, nuclear. *See* Nuclear envelope
Enveloped viruses, 679
Environmental genomics, 503
Environmental sequencing, **364**, 1106, 1107
Environments. *See also* Ecosystems
adaptive radiations and ecological opportunities in, 484–85
animal color vision and, 917–18
animals and extreme, 803
animal sensory organs and, 610, 907–8
aquatic (*see* Aquatic ecosystems)
background extinctions vs. mass extinctions and, 489
balancing selection and, 443
cues in, and *Daphnia* reproductive modes, 952
ecology and, 993
effects on phenotypes of physical, 247–48
eukaryotic response to internal, 319–20
gene expression and, 308
genome redundancy and, 363
hormone coordination of responses to changes in, 937–38
limits for net primary productivity in, 1090–91
natural selection and changes in, 426–28
plant sensory systems and, 755
role of, in succession, 1075
seed dormancy and, 798–99
sensing changes in, 907–8 (*see also* Sensory systems, animal)
sexual reproduction and changing, 224–25
succession and, 1075, 1077
terrestrial (*see* Terrestrial ecosystems)
Enzymatic combustion, 590
Enzyme kinetics, 55
Enzyme-linked receptors, 143–44
Enzymes, **45**. *See also* Proteins
catalysis of chemical reactions by, 45, 51–53 (*see also* Catalysis)
cell-cycle regulation and, 203–4
cofactors, coenzymes, and, 54
digestive, in animal lysosomes, 110–11
digestive, in pancreas, 115
DNA synthesis and, 263–68
effects of temperature and pH on function of, 55–56
enzyme kinetics and catalysis rate regulation, 55
eukaryotic organelles and, 107
as first self-replicating molecules, 56
folding in ribonuclease, 50
fungal, 581, 589–90, 746
globular shape of, 45–46
homeostasis and function of, 815
hydrolysis of carbohydrates by, to release glucose, 78–79
lock-and-key model for, 53–54
mammalian digestive, 855*t*
membrane transport proteins and, 97*f*
multienzyme complexes of, 49
nucleic acid polymerization and, 62
one-gene, one-enzyme hypothesis and, 277–78
in peroxisomes, 109–10
post-translational control by, 330
prokaryotic, 104
protein digestion by pancreatic, 851–52
proteins as, in central dogma, 280–81
regulation of, by regulatory molecules, 54–55
regulation of pancreatic, 852
RNA ribozymes as, 68
RNA synthesis and, 279
signal transduction via enzyme-linked receptors, 143–44
smooth ER and, 108
transfer RNA and, 298
Ephedrine, 576
Epicotyls, **796**
Epidemics, **676**
diabetes mellitus, 857–58
fungal, 581
recent viral, in humans, 676–77
Epidemiology, 689
Epidermal cells, plant, 707–8, 711*t*
Epidermis, **406**, **705**, **722**
Epididymis, **958**
Epigenetic inheritance, **323**
Epiithelial tissues, **809**
Epilepsy, 902
Epinephrine, **937**
for anaphylactic shock, 991
binding of, to receptor, 945–47
control of, by sympathetic nerves, 943
fight-or-flight response and, 881, 937–38
as hormone, 933
receptors for, 140
Epiphytes, **572**, **750**, **1003**
Epiphytic plants, 750–51
Epithelia (epithelium, epithelial tissues), **135**, **604**, **809**–10, 815, 834
Epitope, **981**
Equilibrium, 27, 90*f*. *See also* Hardy-Weinberg principle
Equilibrium potentials, 888, **889**
Equisetophyta, 572
Equivalence, genetic, 377–78
Erosion, 547, 741–42
Errors, DNA synthesis, 271–72, 283–84
ER signal sequence, **121**
Erwin, Terry, 1107, 1108
Erythropoietin (EPO), **940**
Escaped-genes hypothesis, 686–87
Escherichia coli (*E. coli*)
BAC libraries and, 360
chromosomes of, 103
discovery of DNA in genes of, 259–60*f*
DNA polymerase proofreading and mutants in, 271–72
food poisoning and, 500
gene expression in, 307–8
growth hormone engineering using, 339–43
lactose metabolism of, as model, 309–10
in Meselson-Stahl experiment, 261–63
metabolic diversity and, 507*t*
as model organism, B:19–B:20, B:21*f*
mutations and fitness in, 449–50
as proteobacteria, 515
scanning electron micrograph of, 307*f*
types of lactose metabolism mutants in, 311*t*
E site, 300
Esophagus, 810*f*, 846*f*, **848**
Essential amino acids, **842**
Essential nutrients, **738**, **842**–43
Ester linkages, 83–**84**
Estradiol, **936**, **960**
binding of, to intracellular receptors, 943–45
biomagnification of estrogen-mimics, 1088–89
identifying receptor for, 944
puberty and, 961–62
sexual activity of *Anolis* lizards and, 1023
Estrogen-mimics, 1088–89
Estrogens, 140*t*, **936**, 943–45, **960**
Estrous cycle, **963**
Estuaries, **1000**
Ethanol, 24*t*, 110, 167, 508
Ethical concerns
over gene therapy, 352–54
over genetic testing, 350–51
over recombinant human growth hormone, 343–44
Ethylene, 140*t*, **773**–**75*t***
Ethylene glycol, 24*t*
Eudicots, **565**, 706*f*, 796
Euglenida, 539
Eukarya domain, 6–7, 84, 497*t*, **519**–20. *See also* Eukaryotes
Eukaryotes, **6**. *See also* Eukarya domain
cell-cell attachments in, 135–38
cell structures and functions of (*see* Eukaryotic cells)
characteristics of, 526
chromosome replication in, 265*f*
control of gene expression in (*see* Gene regulation, eukaryotic)
genomes of, 365–70
identifying genes in genome sequences of, 363
major lineages of protists and, 524*t* (*see also* Protists)
model of condensed DNA of, 319*f*
number of genes in selected, 369*t*
phylogenetic tree of life and, 6–7
plasma membranes of, 92
prokaryotes vs., 519
RNA polymerases of, 290*t*
transcription in bacteria vs. in, 296*t*
transcription promoters in, 291
translation in bacteria vs. in, 296–97
Eukaryotic cells, 105–15. *See also* Cells
animal and plant, 106*f*
bacterial flagella vs. flagella of, 127
benefits of organelles in, 107
cell walls of, 113, 114*t*
chloroplasts of, 112–13, 114*t*
components of, 114*t*
cytoskeletons of, 113, 114*t*
dynamic cytoskeletons of, 123–28
dynamism of whole, 116
endomembrane system and proteins in, 118–23*f*
Golgi apparatus of, 108–9, 114*t*
lysosomes of, 110–11, 114*t*
mitochondria of, 112, 114*t*
nuclear transport in, 116–18
nucleus of, 107–8, 114*t*
peroxisomes of, 109–10, 114*t*
prokaryotic cells vs., 105–6*f*, 107*t*
ribosomes of, 109, 114*t*
rough endoplasmic reticulum of, 108, 114*t*
selective adhesion and adhesion proteins in, 136–38
smooth endoplasmic reticulum of, 108, 114*t*
structure and function of whole, 115–16
vacuoles of, 111–12, 114*t*
views of, 106*f*, 113*f*
Euryarchaeota, 512, 516
Eutherians, 660, **666**, **967**
Evaporation, 717, **816**, 828, 867
Evergreen trees, 1007
Evidence, science and, 528
Evo-devo (evolutionary-developmental biology), **384**, 752
Evolution, **4**
of aerobic respiration, 510
of biodiversity (*see* Biodiversity)
changes in developmental pathways and change in, 384–85
as change through time, 415
chemical vs. biological, 15, 39, 59 (*see also* Chemical evolution)
consumption and arms races in, 1066–67
current examples of, 418
evidence for, 418–22*t*, 423*f*
evolutionary conservation of cell-cell signals and regulatory genes, 383
experimental, 449–50
four processes of (*see* Evolutionary processes)
global warming and evolutionary change within populations, 1101

Boldface page numbers indicate a glossary entry; page numbers followed by an *f* indicate a figure; page numbers followed by *t* indicate a table.

history of life and (*see* Life, history of)
human, 670–73*f*
lateral gene transfer in, 364
living organisms as product of, 2
mapping land plant, on phylogenetic trees, 555–56
of mouthparts, 844
of multichambered hearts with multiple circulations, 878
of neurons and muscle cells, 886
as not goal directed or progressive, 430
of oviparous and viviparous species, 956–57
of pollination, 793–94
of predation, 487–88
prokaryotes and, 509–12
of protostomes, 624–25
religious faith vs. theory of, 8–9
speciation and (*see* Speciation)
theory of, by natural selection, 4–5, 415–16, 422–24 (*see also* Natural selection)
vertebrate, 653–59, 878*f*
of whales, 477–79
Evolutionary-developmental biology (evo-devo), **384**, 752
Evolutionary processes, 435–57. *See also* Evolution
gene flow, 435, 447–48
genetic drift, 435, 443–46
Hardy-Weinberg principle and analyzing change in allele frequencies, 436–40
mutation, 435, 448–50
natural selection, 435, 440–43
nonrandom mating and, 450–55
Excavata, 524*t*, 536, 538–39
Exceptions
central dogma, 281
Mendelian genetics, 250*t*
Excitable membranes, **891**
Excitatory postsynaptic potentials (EPSPs), **897**
Excretory systems
fish osmoregulation and, 824
insect homeostasis and, 830–31
nitrogenous wastes and, 829–30
shark excretion of salt and, 826–27
urinary systems (*see* Urinary systems)
waste elimination, 841–42, 854–55
Exergonic reactions, **30**, **149**
Exhalation, 868–70
Exocrine glands, **933**
Exocytosis, **122**
Exons, **294**, 324
Exonuclease, 272
Exoskeletons, **625**, **920**
animal locomotion and, 920–21
as barriers to pathogens, 974
bony vertebrate, 653
cell walls as prokaryotic, 104–5
Exothermic processes, **27**
Exotic species, **1014**–15, **1110**–11, 1120
Expansins, 133, **762**
Experimental evolution, 449–50
Experiments. *See also* Biological methods; BioSkills
design of, 10–12
as hypothesis testing, 8–12
Gregor Mendel's, on heredity (*see* Mendelian genetics)
origin-of-life (*see* Origin-of-life research)
Exponential population growth, **1042**
Expressed sequence tag (EST), **363**
Ex situ conservation, **1121–22**
Extant species, **416**
Extension, PCR, 344–45
Extensors, **920**
External fertilization, 615, 954
Extinction
background, **489**, 1116
current causes of, 1110–11
as evidence for evolutionary change, 417
genetic variation and, 440
global warming and species, 1100–1101
human population size and species, 1052
mass (*see* Mass extinctions)
metapopulation balance between recolonization and, 1046–47
predicting rates of, 1114–16
rates, 1077–78, 1110, 1114–16
secondary contact between isolated species and, 471*t*
Extinct species, **417**
Extracellular digestion, **589**–90
Extracellular layers, 132–34
Extracellular matrix (ECM), **133**
animal, 601
cell-cell interactions and, 377
connective tissue and, 806–7
functions of, in animals, 133–34
Extrapolation techniques, 1073, 1108
Extraterrestrial life, 501
Extremophiles, **501**
Eyes
arthropod, 637
cancers of human, 208
insect, 913
mollusk, 610
primate, 669
vertebrate, 914–17

F

F_1 generation, **232**
Facilitated diffusion, **96**, **731**, **825**
in phloem loading, 731–32
via carrier proteins, 96–97
via channel proteins, 94–96
Facilitation, **1075**, 1119
Facultative aerobes, **168**
$FADH_2$, **153**, 158–61
Fallopian tubes, **393**, **960**
False penises, 956
Families, gene, 367–68
Family trees. *See* Pedigrees
Farmer ants, 602
Far-red light. *See* Red/far-red light responses, plant
Fats, **83**, 110, 168–69. *See also* Lipids
Fatty-acid binding protein, **854**
Fatty acids, **83**–84, 167, 169, 355. *See also* Lipids
Faunas, **486**–87
Feathers, 486, 657–58, **668**
Features, transitional, 417–18
Feces, **847**, 854
Fecundity, **1039**–40, 1055, 1115
Feedback
in ecosystem responses to global warming, 1099
positive, during action potential depolarization, 892
role of, in homeostasis, 815–16, 818
Feedback inhibition, **156–59*f***, 314, **931**, 941
Feeding adaptations. *See also* Food; Food chains; Metabolic diversity; Metabolism
animal, 610–12, 616, 843–45
echinoderms, 648
fish, 656
protist, 529–30
protostome, 628, 629*f*
Feeding levels, 1085–86
Feet, primate, 669
Females
barn swallow mate selection by, 1024–25
behaviors mimicking, 1030
chromosomes of, 212
eggs as reproductive cells of, 211
embryo development inside animal, 615
gametangia of, 556
oogenesis in mammal, 953
plant flower parts, 789–90
plant reproductive organs, 402
reproductive systems of animal, 959–66
sexual readiness of, and visual cues from male lizards, 1024
sexual selection and female choice, 452–54
status of human, and fertility rates, 1053
Feminization effects, 1088–89
Femmes fatales, 955
Fermentation, 153, **166**–68, **508**, 854. *See also* Cellular respiration
Ferns, 491, 550, 573, 750
Ferredoxin, **182**
Fertility rates, human, 1053, 1121
Fertilization, **211**, **534**
animal (*see* Fertilization, animal)
bioremedial (*see* Fertilization, bioremedial)
fungal, 591
genetic variation and types of, 222–23
meiosis and, 215–16, 220–21
plant (*see* Fertilization, plant)
protist, 534
Fertilization, animal, **389**, **390**, **951**
evolution of egg-bearing and live-bearing species and, 956–57
internal and external, 954–55
multiple, 391–92
protostome, 630
requirements for, 390–91
types of, 615
unusual aspects of mating and, 955–56
Fertilization, bioremedial
of contaminated sites, 500
of ocean with iron to increase net primary productivity, 1091–92
Fertilization, plant, **784**, **792**
angiosperm, 562
double, in plant gametogenesis, 403–4
pollen grains and, 559–60
pollen-stigma interactions and, 402–3
steps in, 794–95
Fertilization envelope, **392**
Fertilizers, 749, 1093, 1095
Fetal alcohol syndrome (FAS), **969**–70
Fetus, **968**–70
Fiber
animal, 602
dietary, 77
extracellular layer, 132, 133*f*
green plants and, 548–49
Fibers, **710**, 711
Fibronectins, **134**
Fick, Adolf, 864–65
Fick's law of diffusion, **864**–65
Fight-or-flight response, 881, **883**, **937**, 943, 946
Filament, **789**
Filamentous algae, 176
Filaments, cytoskeleton, 123–27
Filter feeders, **610**
Filtrates, **830**
Filtration, 832, **833**, 834
Finches, 426–29, 806, 1062, 1063*f*
Finite rate of increase, **1043**
Fins, fish, 656–67
Fire
bark and, 714
cheatgrass and, 1015–16
disturbance regimes and, 1073–74
seed dormancy and, 799
Fireflies, 1030
Firmicutes, 512, 513
Firs, 575
First law of thermodynamics, **29**
Fischer, Emil, **53**–54
Fish
cichlid jaws, 844–45
fertilized eggs and development of, 374*f*
gills of, 865–66
hearts of, 878
lateral line system as hearing in, 913
mutualisms with, 1068
osmoregulation by, 824, 826–28
as vertebrates, 655
Fission, 619, 620, 633, **951**
Fitness, **5**, **1058**. *See also* Fitness trade-offs
competition and, 1059
cost-benefit analysis of, 1021
dominant and recessive phenotypes and, 233
effects of gene flow on, 448
evolution of, 449–50*t*
genetic drift and, 444
hemoglobins and, 873
heterozygous individuals and, 440
inbreeding and human, 452*t*
inbreeding depression in, 451–52
maladaptive traits and, 187
natural selection, adaptation, reproduction, and biological, 424
sexual selection and, 452–55
species interactions and, 1058–59
types of inclusive, 1032–33
Fitness trade-offs, **432**, **1040**, **1062**. *See also* Fitness
in animal anatomy and physiology, 804, 805*f*
in competition, 1062
countervailing directional selection and, 441
endothermy vs. ectothermy, 817
life histories and, 1040, 1041
as limits to natural selection, 432
nitrogenous wastes and, 830
parental care and, 659
Fixed action patterns (FAPs), **1020**
Flaccid, **719**
Flagella (flagellum), **104**, **390**, **524**
bacterial vs. eukaryotic, 127
cell movement by, 127–28
as characteristic of domains of life, 497*t*
fungal, 583–84, 585
prokaryotic, 104
protist, 533
sperm, 390

Flagellin, 127, 526
Flattening adaptations, 814
Flatworms, 631–32
Flavine adenine dinucleotide (FAD), **153**
Flavins, 161–62
Flavonoids, 176
Flemming, Walther, 195
Flexors, **920**
Flies, 629, 640*t*
Flight, 657–58, 668
Flood basalts, 489
Floods, 1074, 1117
Flor, H. H., 775–76
Floral meristems, **409**
Florigen, 787–**88**
Flowers, **561**, **576**, **783**
angiosperm radiation and, 564 (*see also* Angiosperms)
buds and, 699
fading of, 773
female gametophyte production from, 790
floral meristems and, in reproductive development, 409
flowering as production of, 787–88
genetic control of structures of, 409–11
land plants and, 561–62
male gametophyte production from, 790–92
as morphological innovations, 485
pollination of, 562–63
as reproductive structures, 786–92
structure of, 788–90
Fluid connective tissue, **807**
Fluid feeders, **611**, **843**
Fluidity, plasma membrane, 87–89
Fluid-mosaic model, 92, **93**, 94
Flukes, 631–32
Fluorescence, 102*f*, **178**–79, 346–47
Fluorine, 843*t*
Flu virus. *See* Influenza virus
Folate, 843*t*
Folding
adaptations, 814
in animal small intestines, 851
protein, 50–51, 304, 330
Follicles, **961**
Follicle-stimulating hormone (FSH), **943**, **962**, 964–66
Follicular phase, **962**
Foltz, Kathleen, 391
Food, **842**. *See also* Feeding adaptations; Food chains
angiosperms as human, 576
animal nutrition and, 841–42 (*see also* Animal nutrition)
animals as human, 602
animal sources of, 612–13
disruptive selection and, 442
food poisoning and, 500
foraging behaviors and, 1021–22
fungi as human, 581, 596, 598
giraffe necks and competition for, 9–10
green plants as human, 548
mollusks as human, 634
mouthparts and capture of, 843–45
plant reproduction and human, 783
population cycles and, 1048–50
transgenic crops as, 355–56
Food chains, 522–**23**, **1085**
Food webs, **1085**
Foot, **626**, 629
Foraging, **1021**–22
Foraminifera, 540
Forebrain, **653**
Foreign cells, adaptive immune response and, 983–84, 987–88
Forensic biologists, 346, 366
Forestry, 547, 596, 742
Forests. *See* Boreal forests; Temperate forests; Tropical wet forests
Formaldehyde, 29, 32–34, 38, 40, 60
Formulas, molecular and structural, 20–21, B:8*f*
Fossil fuel consumption, 523, 1096, 1098, 1121
Fossilization, 479–80
Fossil record, **416**, **479**. *See also* Fossils
animals in, 617
arthropods in, 637
Cambrian explosion in, 486–87
cetaceans in, 422
DNA and, 66
as evidence for evolutionary change, 416–18
fish in, 656–67
fossilization and fossil formation in, 479–80
hydromedusas in, 601*f*
pollen grains in, 1072
prokaryotes in, 497–98
stromatolites in, 15*f*
studying green plants using, 550–51
as tool for studying history of life, 479–84
vertebrate, 653–55
Fossils, **416**, **479**. *See also* Fossil record
amplification of DNA in, with polymerase chain reaction, 345–46
elongated cells in, 554
feathered dinosaur, 657
fossilization and formation of, 479–80
trilobytes, 474*f*
Founder effect, **446**
Four-o'-clocks, 245
Fovea, **915**
Fox, Arthur, 919
Fragmentation, 570–71, 618–19, 632–33
Frameshift mutations, 286*t*
Franklin, Rosalind, 63
Free-energy change, 30, 42, 52–53, 78–79, 160*f*
Free radicals, **32**, 176, 272, 590
Freeze-fracture electron microscopy, 93
Frequencies
calculating recombination, 246
distribution, 248–49
sound, 911–12
Frequency, **909**
Frequency-dependent selection, **443**
Freshwater ecosystems, 997–1000. *See also* Aquatic ecosystems
estuaries as marine and, 1000
fish osmoregulation in, 827–28
lakes and ponds, 997
lake turnover in, and nutrient availability, 996
osmotic stress in, 824
streams, 999
wetlands, 998
Frogs, 203–4, 394–96, 429–30, 664, 935–36
Fronds, **573**
Frontal lobe, **901**
Fructose, 155–56, 168, 189–90, 959*t*
Fruit fly model organism. *See Drosophila melanogaster*
Fruiting bodies, **515**, 529
Fruits, **563**, **783**
angiosperm radiation and, 564
auxin and, 768
color vision and, 918
development of, and seed dispersal, 797–98
ethylene and ripening of, 773–74
function of, 798
land plants and, 563–64
seeds and, 795
types of, 797
Fucoxanthin, 532*t*
Fuel, green plants and, 548
Functional genomics, **359**, 370–71. *See also* Genomics
Functional groups, **34**–36, 40–42
Function-structure relationships, 806
Fundamental niches, **1061**
Fungi, **579**–600
adaptations of, as decomposers, 589–90
biological methods for studying, 582–86
cell theory experiment and, 3–4
chitinase and, 56
chitin as structural polysaccharide in, 76
economic impacts of, 581–82
four major types of life cycles of, 591–94
importance of relationship between land plants and, 579–80
key lineages of, 585–86, 594–98
lichens and, 566
as model organism (*see Saccharomyces cerevisiae*)
morphological traits of, 582–84
mutualisms of, 586–89
mycorrhizal, and land plant nutrients, 580
phylogenetic relationship of animals to, 6, 584–85
phylogenies of, 584–86
plant nutrient uptake via mycorrhizal, 746
reasons for studying, 580–82
reproductive variation in, 590–91, 592*f*
saprophytic, and carbon cycle on land, 580–81
Silurian-Devonian explosion and, 551
vacuoles in cells of, 111–12
water molds and, 542
Fungicides, **585**
Fusion
cell, 203–4
fungal, 591
gamete, 220–21, 390
population, 471*t*
Fusion inhibitors, 682

G

G_0 state, 202, 271
G_1 phase, **196**, 205, 207–8
G_2 phase, **196**
Galactose, 72, 323–24
Galactoside permease, 313
Galápagos finches, 426–29, 806, 1062, 1063*f*
Galápagos Islands, 419–20
Gallbladder, 846*f*, **854**
Galls, 355–56
Gallus gallus, 369*t*
Gametangia (gametangium), **556**, 560
Gametes, **194**, **211**, **784**. *See also* Eggs; Sperm
egg structure and function, 390
fungal, 591
fusion of, 220–21
land plant production of, 556–64
recognition of each other by same-species, in fertilization, 391
structure and function of, 389–90
Gametic barriers and prezygotic isolation, 459*t*
Gametogenesis, **215**, 402–4, 952–**53**
Gametophytes, **534**, **557**, **784**
land plant evolution away from life cycles dominated by, 558–59
land plant life cycles and, 784–86
production of female, 789–90
production of male, 789, 790–92
Gamma-aminobutyric acid (GABA), 898*t*
Gamow, George, 282
Ganglia, **604**
Ganglion cells, **914**
Gap Analysis Program (GAP), 1120
Gap genes, 381
Gap junctions, **139**, 921
Gap phases, cell cycle and, 196, 205, 207–8
Gaps, cell-cell, 138–39
Garden peas, 213*t*, 230*f*, 231–32. *See also* Mendelian genetics
Gases
chemical evolution and volcanic, 21, 27, 32, 40
water vapor, 24–25
Gas exchange
amniotic egg, 658
amphibian, 664
animal (*see* Animal gas exchange and circulation; Circulatory systems; Respiratory systems)
bark and, 714
diffusion and skin-breathing, 813
fish lungs, 656
gills and, 634, 651, 813
during human pregnancy, 969
insect, 828
osmotic stress and, 824–25
plant stomata and, 708
pneumatophore roots in, 698, 699*f*
in salmon, 813
surface area and animal, 874–75
Gastric juice, 849
Gastrin, **852**
Gastrointestinal (GI) tract, 845
Gastropods (Gastropoda), **630**, 635
Gastrotricha, 603*t*
Gastrovascular cavities, 845
Gastrula, 394
Gastrulation, **376**, **394**–95, 606–7
Gated channels, **96**
Gause, G. F., 1060–61
GDP (guanosine diphosphate), 142–43
Gel electrophoresis, **62**, 94, 347, 853, B:11–B:12
Gemmae, **569**, 571, 573

Boldface page numbers indicate a glossary entry; page numbers followed by an *f* indicate a figure; page numbers followed by *t* indicate a table.

Gene-by-environment interactions, 247–48, 250*t*
Gene-by-gene interactions, 248, 250*t*
Gene duplication, **367**–68, **777**
Gene expression, **276**–88. *See also* Genes; Genetics
  control of (*see* Gene regulation, bacterial; Gene regulation, eukaryotic)
  development of central dogma of molecular biology on, 279–82
  DNA sequencer output, 276*f*
  functional genomics and, 370
  genetic code and, 282–85
  known types of point mutations in, 286*t*
  molecular basis for mutation in, 285–87*f*
  one-gene, one-enzyme hypothesis on, 277–78
  plant cell-cell signals and, 757
  protein synthesis and (*see* Protein synthesis)
  steroid-hormone receptors and, 944–45
  via transcription and translation, 280
Gene families, **367**–68, 476, **777**
Gene flow, **447**
  experimental increase of, 1114
  genetic variation and effect on average fitness of, 448, 450*t*
  Hardy-Weinberg principle and, 438
  in natural populations, 447
  speciation and, 458, 462, 464–65
  wildlife corridors and, 1121
Gene-for-gene hypothesis, 775–**76**
Gene pools, **436**
Generations, **1038**
Gene regulation, bacterial, 307–18. *See also* Gene expression; Genetics
  catabolite repression as positive control mechanism in, 314–16
  eurkaryotic gene regulation vs., 331–32
  identifying genes under regulatory control, 310–12
  information flow and, 307–10
  lactose metabolism as model system for, 309–10
  mechanisms of, 308–9
  overview of positive and negative *lac* operon regulation in, 317*f*
  repressors as negative control mechanisms in, 312–14
Gene regulation, eukaryotic, 319–37. *See also* Gene expression; Genetics
  bacterial gene regulation vs., 331–32
  cancer and defects in, 332–33
  chromatin remodeling in, 320–23
  differential gene expression and, 319–20
  post-transcriptional, 328–30
  regulatory sequences and regulatory proteins in initiating transcription, 323–28
  translational and post-translational, 330
Genes, **103**, **212**, **234**, **326**. *See also* Genetics
  analyzing and engineering (*see* Genetic engineering)
  bicoid gene as master regulator, 379–80
  cancer and, 332
  case study of *HLA*, in humans, 439–40
  case study of *p53*, as cancer preventing, 332–33
  in central dogma, 280
  chemical composition of, 258–61
  chromosomes and, 195–96
  control of flower structures by, 409–11
  discovery of eukaryotic, in pieces, 293–94
  discovery of recombination of, 982–83*f*
  DNA and (*see* DNA (deoxyribonucleic acid))
  dwarf plants and defective, 769–70
  estimating phylogenetic tree of life from rRNA sequences in, 6–7
  finding disease-causing, 348–51
  finding mutant, with replica plating, 310–11
  functions of (*see* Gene expression)
  gene families of, 367–68
  genetic constraints on natural selection, 431
  Hershey-Chase experiment on DNA as genetic material in, 259–60*f*
  identifying, in genome sequences, 362–63 (*see also* Genomics)
  inserting, into plasmids, 341*f*
  lactose metabolism, 312
  leaf shape and, 408
  linked, on same chromosome, 243–45, 246*f*
  living organisms and genetic information in, 1
  in Mendelian genetics, 239*t*
  messenger RNA hypothesis as intermediary between proteins and, 279–80
  natural selection of, in Galápagos finches, 428–29
  number of, in selected eukaryotes, 369*t*
  one-gene, one-enzyme hypothesis on gene expression, 277–78
  particulate inheritance hypothesis and, 234–35
  physical locations (loci) of, 240
  plant body axes and, 406
  plant pathogen resistance, 777
  prokaryotic, 103–4
  redefinition of, 329, 362
  for taste receptors, 919
  and traits, 221
  tumor suppressors, 206
Genes-in-pieces hypothesis, 293–94
Gene therapy, **351**–54
Genetically modified food, 355–56
Genetic bottlenecks, **446**
Genetic code, 279, **282**–85, 297–300
Genetic correlation, **431**
Genetic counselors, 346
Genetic divergence, speciation and, 458. *See also* Speciation
Genetic diversity, 363, 591, **1106**. *See also* Genetic variation
Genetic drift, **443**–46
  allopatric speciation, dispersal, and, 463
  causes of, in natural populations, 445–46
  genetic variation and effect on average fitness of, 448, 450*t*
  Hardy-Weinberg principle and, 438
  metapopulations and, 1114
  simulation studies of, 443–45
Genetic engineering, 338–58. *See also* Genetics
  in agriculture, 548
  amplification of fossil DNA with polymerase chain reaction, 344–46
  common techniques used in, 353*t*
  common tools used in, 352*t*
  development of golden rice with agricultural biotechnology, 354–56
  dideoxy DNA sequencing, 346–48
  efforts to cure pituitary dwarfism with basic recombinant DNA technologies, 338–44
  finding Huntington's disease gene with gene mapping, 348–51
  gene therapy for severe immune disorders, 351–54
Genetic equivalence, 377–78
Genetic homology, **420**–21
Genetic isolation
  allopatry and, 462–64
  autopolyploidy and, 466–67
  speciation and, 458
Genetic maps, **244**, 246, **348**–51, 353*t*
Genetic markers, **348**–49, **445**
Genetic models, **236**
Genetic recombination, **222**, 244–45
Genetics, **230**. *See also* Genes; Nucleic acids
  analyzing and engineering genes (*see* Genetic engineering)
  control of gene expression in (*see* Gene regulation, bacterial; Gene regulation, eukaryotic)
  cystic fibrosis as genetic disease, 94–95
  DNA as genetic information-containing molecule, 65
  DNA sequencing of genomes in (*see* Genomics)
  DNA synthesis and repair in (*see* DNA synthesis)
  estimating phylogenetic tree of life from rRNA sequences in, 6–7
  gene expression in (*see* Gene expression)
  genetic problems in metapopulations, 1114
  human, 436
  meiosis in (*see* Meiosis)
  Mendelian (*see* Mendelian genetics)
  protein synthesis in (*see* Protein synthesis)
Genetic screens, **277**–78, 310
Genetic testing, 350–51
Genetic variation, **440**. *See also* Genetic diversity
  crossing over and, 222
  evolutionary processes and, 448, 450*t*
  evolution of drug resistance and, 425–26
  extinction and, 440
  from independent assortment, 221–22
  lack of, as genetic constraint, 431
  mutations and, 448–50
  reduction of, by directional selection, 440–41
  sexual reproduction and, 533
  switching reproductive modes and, 952
  types of fertilization and, 222–23
Genitalia, **957**
Gennett, J. Claude, 982
Genomes, **263**, **359**. *See also* Genomics
  animals and human, 603
  diversity of viral, 686, 687*t*
  eukaryotic, 365–70
  prokaryotic, 363–65
  reasons for selecting, for sequencing, 361–62
  sequencing of (*see* Whole-genome sequencing)
  sequencing technologies, 347
  smallest eurkaryotic, 594
Genomic libraries, **342**
Genomics, **359**–73. *See also* Genes; Genetics
  bacterial and archaeal genomes, 363–65
  cancer and applied, 371–72
  environmental, 503
  eukaryotic genomes, 365–70
  functional, 370–71
  human genome representation, 359*f*
  number of genes in selected eukaryotes, 369*t*
  proteomics and, 371
  whole-genome sequencing, 359–63
Genotypes, **234**
  body odor and, 439–40
  central dogma and linking phenotypes and, 280–81
  effect of inbreeding on frequencies of, 451
  frequencies of, for human *HLA* gene, 439*t*
  frequencies of, for human MN blood group, 438*t*
  in Mendelian genetics, 239*t*
  mutations and, 285
  particulate inheritance hypothesis and, 234–35
  phenotypes vs., 234
  predicting, with Punnett square, 235–36
Genus (genera), **8**
Geographic areas, similar species in same, 419–20
Geographic distribution of diversity, 1109–10
Geographic distribution of organisms, 1013–17
  dispersal histories in, 1013–15
  global warming and changes in, 1100
  interaction of biotic and abiotic factors in, 1015–17
  mapping current and past, 1072
  protected areas and, 1120
  species interactions and, 1059
Geologic time scale, 182, **416**–17
Geometry. *See* Shape
Geothermal radiation, 509
Germ cells, **409**
Germinal center, 989
Germination, **402**, **794**
  gibberellins and abscisic acid in seed, 770–71
  phases of seed, 799–800
  red/far-red light reception and seed, 762–64
Germ layers, **394**–95, **604**
Germ theory of disease, **499**–500
Gestation, **658**, **967**
Ghosts
  capsids as, 259–60*f*
  plasma membrane, 96–97
Giant axon, squid, 125–27
Giant pandas, 1021
Giant sequoias, 1073–74
Giant sperm, 955
Giardia and giardiasis, 522*t*, 538

Gibberellic acid (GA), 769
Gibberellins, **769**–71, 774–75*t*
Gibbons, Ian, 127–28
Gibbs free-energy change, **30**, 42. *See also* Free-energy change
Gilbert, Walter, 294, 314
Gill arches, **655**–66
Gill filaments, **865**
Gill lamellae, **865**
Gills, **634**, **651**, **813**, **865**
  as countercurrent systems, 866
  developmental changes in gas exchange with, 813
  osmoregulation and, 827–28
  osmotic stress and, 824
  ventilation with, 865
Ginkgoes (Ginkgophyta), 574
Giraffes, 9–10
Gizzards, avian, 851
Glacier Bay succession case history, 1076–77
Glaciers, 1083*f*
Glands, **809**, **930**
Glanville fritillaries (butterflies), 1046–47, 1055
Glaucophyte algae, 536
Gleason, Henry, 1070–72
Glia, **894**
Global air circulation, 1009
Global carbon cycle, **523**, **1096**–97
Global change, prokaryotes and, 509–12
Global climate change
  biological study of, 1011–13*f*
  extinctions and, 1111, 1114
  global warming and, 1008–9, 1098–1102
Global nitrogen cycle, **1096**
Global productivity patterns, 1089–90
Global warming, **1098**–1102
  as biodiversity threat, 1111, 1114
  carbon dioxide concentrations and evidence for, 1098–99
  devegetation and, 547
  ecosystem positive and negative feedback responses to, 1099
  end-Permian extinction and, 489
  global climate change and, 1008–9
  global fossil-fuel consumption and, 1098
  impact of, on organisms, 1099–1101
  iron fertilization and, 1091
  net primary productivity changes due to, 1101–2
  protists and reduction of, 523
  Sphagnum moss and, 570
Global water cycle, **1095**
Globin genes, 368
Globular enzymes, 53
Globular proteins, 45–46, 197
Glomalin, **589**
Glomeromycota, 596
Glomeruli, olfactory, **920**
Glomerulus, renal, **833**
Glucagon, 45, **856**, **930**
Glucocorticoids, **938**, 941–42
Gluconeogenesis, **856**
Glucose, **148**
  animal absorption of, 852–53
  ATP synthesis from, 79, 148*f*, 165–66
  catabolite repression and, 314–17*f*
  in cellular respiration and fermentation, 153–54
  configurations of, 72–73
  cortisol and availability of, 938
  diabetes mellitus and, 856–57
  glycogen and starch as, 74, 168
  glycolysis as processing of, to pyruvate, 155–56
  hydrolysis of carbohydrates by enzymes to release, 78–79
  induced fit and, 53
  lactose metabolism and preference for, 309–10
  membrane protein GLUT-1 and, 97
  nutritional homeostasis and, 856–58
  as organic molecule, 33*f*
  oxidation of, by citric acid cycle, 158–61, 165–66
  processing of photosynthetic, 189–90
  redox reactions and oxidation of, 152
  reduction of carbon dioxide to produce, in photosynthesis, 184–90
  regulation of, 156
  secondary active transport of, 98
  spontaneous chemical reactions and, 30*f*
Glucose-6-phosphate, 55–56, 155
GLUT-1 membrane protein, 97
Glutamate, 42*t*, 46, 898*t*, 919
Glutamic acid, 47
Glutamine, 42*t*, 349–50
Glyceraldehyde-3-phosphate (G3P), **186**, 189–90
Glycerol, **83**–84, 168
Glycine, 42*t*
Glycogen, **74**, 75*t*, 78–79, 168, 585
Glycolipids, **105**
Glycolysis, **153**–56
Glycophosphate, 354
Glycoproteins, **77**, **121**
Glycosidic linkages, **73**–74
Glycosylation, **121**
Glyoxysomes, **110**
Gnathostomes, **653**
Gnathostomulida, 603*t*
Gnetophytes (Gnetophyta), 576
Golden rice, 354–56
Golgi, Camillo, 886–87
Golgi apparatus, **109**
  eukaryotic, 108–9, 114*t*
  extracellular matrix components in, 134
  in mitosis, 200
  pectins and, 132
  post-translational modifications in, 304
  protein transport and sorting by, 122–23*f*
Gonadal hormones, 961–62
Gonadotropin-releasing hormone (GnRH), **962**
Gonads, **936**, **953**
Goose bumps, 418, 419*f*
G proteins, **142**–44
Grades, 552, 567, **655**
Grafting, plant, 787–88
Gram-negative, **505**
Gram-positive, **505**
Gram stains, **505**
Grana (granum), **112**–13, **174**, 184
Grant, Peter and Rosemary, 426–28, 431, 463, 1062, 1063*f*
Graphs, B:2–B:4
Grasshoppers, 212, 641*t*
Grasslands. *See* Temperate grasslands
Graves, 671
Gravitropic responses, plant, 764–66, 780*t*
Gravitropism, **764**
Gravity, hearts and, 878
Gray whales, 819
Grazing food chains, **1085**–86
Great apes, **668–70**
Great chain of being, 415
Great Smoky Mountains National Park, 1108
Great tits, 447–48
Greek word roots, 8, B:6
Green algae, **566**. *See also* Algae
  as green plants, 546, 566–67 (*see also* Green plants)
  lichens and, 597
  lineages, 568–69
  photosystems and, 179
  phylogenetic tree and, 551–52
  Plantae and, 536
  reasons for studying, 546–49
  similarities between land plants and, 549
Green Belt Movement, 1122
Greenhouse gas, **1097**
Green iquanas, 446
Green light, 177
Green plants, 546–78. *See also* Plant(s)
  analysis of morphological traits of, 549–50
  biological methods for studying, 549–52
  diversification themes of land plants, 553–66
  ecosystem services of, 547
  green algae and land plants as, 546, 566–69
  importance of, to humans, 548–49
  key lineages of, 566–77*f*
  non-vascular plants, 567, 569–71
  reasons for studying, 546–49
  seedless vascular plants, 567, 571–73
  seed plants, 567, 574–77*f*
  some drugs derived from land plants, 549*t*
  using fossil record to study, 550–51
  using phylogenetic trees to study, 551–52
Gross photosynthetic efficiency, **1084**
Gross primary productivity, **1084**
Groudine, Mark, 322
Ground meristem, **705**, **796**
Ground pines, 571
Ground tissue, **406**
Ground tissue systems, **705**, 708–10, 711*t*
Groundwater, **1095**
Groups, control, 12
Groups, functional. *See* Functional groups
Growing seasons, plant translocation and, 728
Growth
  cancer as out-of-control, 206
  cell movement and, 376–77
  energy and, 1084
  mammalian, 937
  mitosis and, 195
  by molting, 625
  plant, 554, 766 (*see also* Growth regulators, plant)
  population (*see* Population growth)
  viral, 679–81
Growth factors, **207**–9
Growth hormone (GH), **934**, 935, 937. *See also* Human growth hormone (HGH)
Growth rate, population, 1041–42
Growth regulators, plant, 767–76
  apical dominance and auxin as, 767–68
  cell division and cytokinins as, 768–69*f*
  growth and dormancy and gibberellins and abscisic acid as, 769–73
  overview of, 774, 775*t*
  plant body size and brassinosteroids as, 773
  senescence and ethylene as, 773–74
Growth rings, tree, 714–15, 1073
GTP (guanosine triphosphate), 142–43, **158**
Guanine, 60, 63–64, 66–67, 261, 279
Guanosine diphosphate (GDP), 142–43
Guanosine monophosphate (GMP), 916
Guanosine triphosphate (GTP), 142–43, **158**, 297–98
Guard cells, **187**, **553**–54, **708**, **771**–73
Gustation, **918**–19
Guts, animal, 845
Guttation, **723**
Gymnosperms, **551**
  diversification of, 551–52
  evolution of, 560–61
  key lineages of, 574–76
  in Mesozoic era, 483

H

H^+-ATPases, **733**
Habitat bias, 480
Habitat destruction, **1112**
  human population size and, 1052
  preserving metapopulations from, 1055
  species-area relationships for extinctions due to, 1116
  as threat to biodiversity, 1111, 1112
Habitat fragmentation, **1113**
Habitat prezygotic isolation, 459*t*
Habitats
  carrying capacity of, 1042, 1046
  communication and, 1029
  prokaryotic diversity in, 498
  protecting, 1055
  selection of, 1026
  speciation and preference for, 465
Hadley, George, 1009
Hadley cell, **1009**
Haemophilus influenza, 361–62
Hagfish, 661
Hair cells, **909**–12
Hairpin (RNA secondary structure), **67**, 292–93. *See also* RNA (ribonucleic acid)
Haldane, J. B. S., 38
Halophiles, **503**
Hamilton, William D., 1031–33
Hamilton's rule, 1031–34
Hands, primate, 669
Hanski, Ilkka, 1046–47
Hanson, Jean, 922
Hantaan virus, 689, 692
Hantavirus, 688, 689
Haploid cells, 214*t*, 533–34
Haploid number, **213**, 214*t*

Boldface page numbers indicate a glossary entry; page numbers followed by an *f* indicate a figure; page numbers followed by *t* indicate a table.

Haploid organisms, **213**
Hardy, G. H., 436
Hardy-Weinberg principle, **436**–40
Hartwell, Leland, 205
Hatch, Hal, 188
Haustoria, 751
Havasupai tribe, 439–40
Hawaiian silverswords, 484, 700–701
Hawksworth, David, 582
Hawthorn flies, 465
Hay fever, 991
Head region
  animal, 603, 605, 610
  insect, 639
  sperm, 389
Head-to-tail axes, 381
Health, human. *See* Diseases, human
Hearing, 610, 909–13
Heart murmers, **879**
Hearts, 808, 830, **875**–81. *See also* Blood; Blood vessels; Circulatory systems
Heartwood, **714**
Heat, **28**. *See also* Specific heat
  animal thermoregulation and, 816–19
  denatured proteins and, 50
  DNA copying and, 65
  sterilization using, 3–4
  as thermal energy, 27–29
  water and, 24–25
Heat of vaporization, **24**, 26*t*
Heat-shock proteins, 50, **818**
Heavy chain, **980**
Helicases, **264**, 266, 268*t*
Helicobacter pylori, 850
Heliobacteria, 182, 507*t*
Helper T cells, 677, 681–82, 985–**86**, 989*t*
Hemagglutinin, 981
Heme, **871**
Hemichordata, 603*t*, 646
Hemimetabolous metamorphosis, **616**
Hemiptera, 641*t*
Hemocoel, **626**, 628
Hemoglobin, **871**. *See also* Blood
  β-globin in, 247, 322
  blood oxygen and, 45
  in blood pH buffering, 873–74
  globin genes and, 368
  protein structure of, 49*t*
  sickle-cell disease and, 46
  space-filling model of, 38*f*
  structure and function of, 871–73
Hemolymph, **830**, 875
Hemophilia, **252**
Henle, Jacob, 835
Hepaticophyta, 569
Hepatitis viruses, 692, 990
Herbaceous plants, **697**–98
Herbicides, 176, 354, 589
Herbivores, **612**, **701**, **1063**
  animals as, 612
  limitations on consumption of, 1065–66
  plant defense responses to, 775, 778–81
  plants as food for, 547
  transgenic crops and, 354
Herbivory, **1063**
Heredity, 221, **230**, 280. *See also* Genetics; Mendelian genetics
Heritable traits, **4**
Hermaphroditic, **956**
Hermit warblers, 468–70, 1015
Herpes, 686
Herpesviruses, 690
Hershey, Alfred, 259–60*f*
Hershey-Chase experiment, 259–60*f*
Heterokaryotic, **582**
Heterokonta, 524*t*, 537, 542–43
Heterospory, **559**, 561–62
Heterotherms, **817**
Heterotrophs, **172**, **506**, 601
Heterozygote advantage, **443**, 452
Heterozygous genes, **235**, 239*t*, 245–47
Hexokinase, 53
Hexoses, 60, **72**, 73
Hibernation, **817**
High-productivity hypothesis, 1080
Hill, Robin, 179
Hindbrain, **653**
Hindguts, 830–31
Hippos, 477–79
Histamine, **976**, 977*t*, 991
Histidine, 42*t*, 283
Histograms, 248–49, 427, 441*f*
Histone acetyl transferases (HATs), **322**
Histone code, **323**
Histone deacetylases (HDACs), **322**–23
Histones, **197**, **320**–23, 497*t*
Historical constraints, natural selection and, 432, 914
Historical context, succession and, 1075
History of life. *See* Life, history of
HIV (human immunodeficiency virus), **677**, **990**
  CD4 protein as receptor used by, to enter cells, 681–82
  changing-environment hypothesis and, 224
  current AIDS pandemic and, 677–78
  doorknob hypothesis on, 681–82
  phylogeny of, 688*f*
  as retrovirus, 684, 691
  as RNA virus, 281
  strains of, 688–89
  Toxoplasma and, 541
  two types of, 688
  vaccines and mutations of, 990
Hives, 991
HLA genes, 439–40
Hodgkin, A. L., 890–93
Holoenzymes, **291**, 292, 343
Holometabolous metamorphosis, **616**, 936
Homeobox, 411, 476
Homeosis, **382**
Homeostasis, **25**, **814**–19
  of animal body temperature, 817–19
  animal excretory systems and, 830–31
  of blood oxygen, 870
  of blood pH, 873
  of blood pressure, 882–83
  diabetes mellitus and animal nutritional, 856–58
  general principles of, 814–15
  hormones in, 938–40
  pH buffers and, 25
  regulation and feedback in, 815–16
  ventilation and, 870
  of water and electrolyte balance (*see* Osmoregulation)
Homeotherms, **817**–19
Homeotic genes, **382**–83, 385
Homing pigeons, 1026
Hominids, 668, **669**–70
Hominins, **670**–73*f*
Homo genus
  characteristics of, 670*t*
  DNA comparison of *Homo neanderthalensis* and *Homo sapiens*, 345–46
  fossil record and, 670–72
  Homo erectus, 672
  Homo floresiensis, 671
  Homo neanderthalensis, 345–46, 671, 672
  Homo sapiens, 8, 369*t*, 672–73*f* (*see also* Human(s))
Homologous chromosomes (homologs), **212**, 214–16, 221–22, 225–26
Homology, **363**, **420**, **475**
  animal and human genetic, 603
  animal appendages as, 613–15
  as evidence of descent from common ancestors, 420–22
  genomes and genes, 363
  homoplasy vs., 475–77
Homoplasy, **475**–77
Homosporous, **559**
Homozygous genes, **235**, 239*t*, 451
Honduras, age structure of, 1051
Honeybee language, 1028–29
Hooke, Robert, 2
Hormone-based contraception methods, 966*t*
Hormone binding, plant, 757
Hormone-response elements, **944**
Hormones, **139**–46. *See also* Chemical signals, animal; Hormones, animal; Hormones, plant
Hormones, animal, **380**, **837**, **852**
  central nervous systems, endocrine systems, and, 929
  chemical characteristics and types of, 933–34
  coordination of responses to environmental change by, 937–38
  direction of developmental processes by, 935–37
  discovery of first, 852
  as endocrine signals, 930
  endocrine system components and, 932–33
  experimental identification of, 934–35
  functions of, 935–40
  in homeostasis, 938–40
  human growth hormone (HGH) (*see* Human growth hormone (HGH))
  mammalian sex, 960–66
  regulation of production of, 940–43
  signaling pathways of, 931–32
  urine formation and, 837
Hormones, plant, **756**
  auxin as blue-light phototropic, 759–62
  auxin as development master regulator, 380–81
  auxin as gravitropic signal, 765–66
  auxin in growth responses, 767–68
  brassinosteroids in body size, 773
  cytokinins in cell division, 768–69*f*
  ethylene in senescence, 773–74
  florigen as flowering, 787–88
  gibberellins and abscisic acid in growth reponses, 769–73
  hormone binding and, 757
  hypersensitive response, systemic acquired resistance, and, 777–78
  information processing and, 756
Hornworts, 571
Horowitz, Norman, 277–78
Horseshoe crabs, 642
Horsetails, 572
Horticulture, 547
Horvitz, Robert, 376
Host cells, **675**
Hosts, **1063**, 1067–68
Hotspots
  of biodiversity and endemism, 1109
  conservation, 1110
Howard, Alma, 196
Hox genes, **382**–83, 385, 411, 476, 488
Hozumi, Nobumichi, 982
Hubbard Brook Experimental Forest, 1084, 1085–86, 1093–94
Human(s), **671**
  artificial selection by, 4–5
  blood types, 245–47, 438–39
  brains, 901–2
  digestion, 77, 79, 810*f*, 846*f* (*see also* Digestion)
  diseases (*see* Diseases, human)
  as dispersal agents, 1014
  ears, 910–12
  endocrine systems, 932*f* (*see also* Endocrine systems)
  as endothermic homeotherms, 817
  essential nutrients for, 842, 843*t* (*see also* Animal nutrition)
  evolution of, 345–46, 438–40, 670–73*f*
  excretory systems (*see* Excretory systems)
  gas exchange and circulation (*see* Animal gas exchange and circulation)
  genome of, 328, 369*t* (*see also* Human Genome Project)
  growth hormone, 338–44
  hearts, 879–83 (*see also* Circulatory systems)
  Homo sapiens scientific name for, 8
  hormones in stress responses of, 937–38
  identifying autosomal traits of, 250–52
  immune systems, 975–76, 977*t*, 978 (*see also* Immune systems)
  importance of animals to, 602–3
  importance of fungi to, 581–82
  importance of plants to, 548–49, 783
  importance of protists to, 520–22
  introducing novel alleles into cells of, 351–52
  learning and memory in, 902–4*f*
  life tables for, 1055
  lungs, 868, 869*f* (*see also* Respiratory systems)
  mortality (*see* Mortality, human)
  nervous systems (*see* Nervous systems)
  net primary productivity appropriation by, 1090
  non-kin child abuse rates by, 1034
  number of chromosomes in, 213*t*
  nutrient loss impacts by, 1093
  organs and systems of, parasitized by viruses, 676*f*
  osmoregulation (*see* Osmoregulation)
  population (*see* Human population)
  puberty in, 961–62, 963*t*
  reproduction (*see* Human reproduction)
  sensory systems (*see* Sensory systems, animal)
  urinary systems, 832*f* (*see also* Urinary systems)
  vestigial traits, 418, 419*f*
Human chorionic gonadotropin (hCG), **967**–68
Human Genome Project, **359**
  costs of, 347
  human genome representation, 359*f*
  reasons for similarity of chimpanzee and human genomes, 369–70

Human Genome Project, (*continued*)
  reasons for small number of genes in human genome, 368–69
Human growth hormone (HGH), 338–44
Human immunodeficiency virus (HIV). *See* HIV (human immunodeficiency virus)
Human population
  age structure of, 1051
  growth rate for, 1052–53
  phylogeny of current, 672*f*
  population growth milestones for, 1052*t*
  size of, 1052, 1098, 1121
Human reproduction. *See also* Animal reproduction
  accessory fluids in semen, 959*t*
  childbirth events, 970–71
  contraception in, 966
  embryo development in, 388*f* (*see also* Animal development)
  embryonic tissue layers in, 395
  female reproductive tract in, 960, 961*f*
  fertility rates in, 1053
  fertilization, 211*f*
  gametogenesis in, 953*f*
  inbreeding and reduction in fitness in, 452*t*
  male reproductive tract in, 958–59*f*
  meiotic mistakes in, 226
  ovarian cycle in, 962–66
  pregnancy events in, 967–68
  sex chromosomes and, 212
  sperm cells and, 2*f*
Hummingbirds, 562
Humoral response, **987**
Humus, **741**, **1092**
Hunting, animal, 612
Huntington's disease, 233, **251**, **348**–51
Hutchinson, G. Evelyn, 1060
Hutterites, 440
Huxley, Andrew, 890–93
Huxley, Hugh, 922
Hybridization
  hybrid zones and, 468–70
  new species through, 470–71
  speciation through, 467–68
Hybrids, **232**, 239*t*, 459*t*
Hybrid zones, **468**–70, 471*t*
Hydrocarbons, **83**, 87–88. *See also* Lipids
Hydrochloric acid, 25, 848–50
Hydrogen. *See also* Hydrogen gas
  atomic structure of, 17*f*
  electronegativity of, 18
  lipids, hydrocarbons, and, 83
  as plant nutrient, 739*t*
  proteins and, 46–47
  redox reactions and, 151–52
  simple molecules from, 19–20
Hydrogen bonds, **22**
  amino acids and, 41*t*
  DNA complementary base pairing and, 63–64, 261
  protein structure and, 46–48
  RNA complementary base pairing and, 66–67
  specific heat of liquids and, 24*t*
  water and, 22–25
Hydrogen cyanide, 29, 32, 34, 38, 40, 61
Hydrogen gas, 17–18, 19*f*, 32, 39–40. *See also* Hydrogen
Hydrogen ions, **25**
Hydrogen peroxide, 110
Hydroids, 619
Hydrolysis, **43**
  of ATP, 149
  of carbohydrates by enzymes to release glucose, 78–79
  digestive enzymes and, 110–11
  of polymers to release monomers, 43
Hydrophilic substances, **22**, **42**, 64
Hydrophobic substances, **22**, **42**, 47–48, 64
Hydroponic growth, **740**
Hydrostatic skeletons, **605**, 606, 613, 626, **920**–21
Hydrothermal vents, 509
Hydroxide ions, **25**
Hydroxyl functional group, 35, 60, 71, 72, 83–84, 290
Hydroxyl radicals, 272
Hymenoptera, 640*t*
Hyoid bone, 672
Hyperpolarization, **890**, 893, 897, **908**–9
Hypersensitive reaction, **991**
Hypersensitive response (HR), **775**–78, 780*t*
Hypertension, **880**
Hypertonic solutions, **91**, **824**
Hyphae (hypha), **542**, **582**–83, 590–91, 746
Hypocotyls, **405**, **796**
Hypodermic insemination, 956
Hypothalamic-pituitary axis, **942**–43, 962
Hypothalamus, 817–**18**, 931, **933**, 940–43
Hypotheses, **2**, 8–12
Hypothesis testing, 2–4. *See also* Experiments
Hypotonic solutions, **91**, **718**, **824**

I

Ice, 23–24
Iguanas, 446
Imaging, biological. *See* Biological imaging; Microscopy
Immigration, 1037–**38**, 1077–78
Immune systems, **677**, 973–92
  adaptive immune response (*see* Adaptive immune response)
  antibodies and, 324–26
  evolution of human defenses against malaria, 1066–67
  failures of, 990–91
  HIV and human, 677–78
  HLA genes in human, 439–40
  immunity, immunization, and, 973–74
  immunological memory of, 989–90
  inbreeding and, 452
  innate immune system cells and molecules, 977*t*
  innate immunity and, 974–77
  proteins in, 280
  rejection of foreign tissues and organs by, 988
  severe combined immunodeficiency (SCID) and human, 352–54
  sexual selection, carotenoids, and, 453
  structures of, 978
Immunity, **973**
Immunization, **973**–74, 990
Immunodeficiency diseases, 990. *See also* AIDS (acquired immune deficiency syndrome); HIV (human immunodeficiency virus); Severe combined immunodeficiency (SCID)
Immunoglobulins (Igs), **980**, 982–83*f*
Immunological memory, **989**–90
Impact hypothesis, dinosaur extinction and, **490**–92
Imperfect, **790**
Implantation, 966*t*, **967**–68
Inactivated virus vaccines, 990
Inbreeding, **450**–52
Inbreeding depression, **451**–52, 1114
Inclusive fitness, **1032**–33
Incomplete digestive tracts, **845**
Incomplete dominance, **245**, 250*t*
Incomplete metamorphosis, **616**
Independent assortment, principle of. *See* Principle of independent assortment
Indeterminate growth, **695**
Indicator plates, **311**
Indirect fitness, 1032
Individuals, natural selection and, 414*f*, 429–30
Indole acetic acid (IAA), 760. *See also* Auxin
Induced fit, **53**–54
Inducers, **309**, 314
Inducible defenses, **776**, **1064**–65*f*
Industrial Revolution, 548
Infection thread, **750**
Infidelity, bird, 956
Inflammatory response, **975**–77*t*
Influenza virus, 675, 685–86, 692, 975, 990
Information
  carbohydrates and display of, 77
  gene regulation and flow of, 307–10
  genetic, 1, 65, 67–68 (*see also* Genes; Genetics)
  hormones as carriers of, 139–40
  plant processing of, 756–57
Infrared light, 174–75, 177
Infrasound hearing, elephant, 912
Ingestive feeding
  animal, 601, 841–45
  protist, 529–30
Inhalation, 868–70
Inheritance. *See also* Genetics; Heredity; Mendelian genetics
  of acquired characters, 231, 234, 415, 429
  of chromatin modifications, 323
  chromosome theory of (*see* Chromosome theory of inheritance)
  epigenetic, 323
  evolution by natural selection vs. Lamarkian, 429
  hypotheses on, 231, 232, 234
  particulate, 234–36
  patterns of human, 250–52
  polygenic, 248–50
Inheritance of acquired characters theory, 231, 234, 415, 429
Inhibition, **1075**
Inhibitory postsynaptic potentials (IPSPs), **897**
Initiation, enzyme catalysis, **54**
Initiation factors, **301**
Initiation phase, **291**
  regulatory sequences and proteins in eukaryotic transcription, 323–28
  transcription, 291–92
  translation, 301
Innate behavior, **1020**
Innate immune response, **975**–77*t*. *See also* Immune systems
Innate immunity, **974**–77*t*. *See also* Immune systems
Inner cell mass (ICM), **393**
Inner ear, **910**
Inorganic molecules, 33
Inorganic nutrients
Inorganic phosphate, 189
Inositol triphoshhate (IP_3), 143*t*
Insecta, 639–41
Insects
  adaptive radiations and innovations in, 485
  ants (*see* Ants)
  excretory systems of, 829–31
  eyes and vision in, 913
  Glanville fritillaries (butterflies), 1046–47
  honeybee language, 1028–29
  hormones in metamorphosis of, 929*f*, 936
  Insecta lineage and, **639**–41
  insecticides and, 521, 1088–89
  land plant pollination by, 562–63
  metamorphosis of, 616
  minimization of water loss by, 828–29
  mutualisms of, 1069–70
  open circulatory systems of, 875
  osmoregulation in terrestrial, 828–31
  plant defense responses to, 775, 778–81
  prominent orders of, 640*t*–41*t*
  taxon-specific survey of, 1107
  tracheae of, 866–67
  wings of, 629
In situ hybridization, **380**
  in flower structure research, 410
  in polygenic inheritance research, 428–29
  visualizing mRNAs via, 380
Instantaneous rate of increase, 1044
Insulin, **856**, **930**
  amino acid sequence for, 46
  cortisol and, 938
  discovery of, 856
  as hormone signal, 140*t*
  role of, in homeostasis, 856
  treatment of diabetes mellitus with, 339
Integral membrane proteins, **93**–94
Integrated pest management, **1068**
Integrators, **815**, 938
Integrins, **134**, 135
Interactions, cell-cell. *See* Cell-cell interactions
Interbreeding. *See* Mating
Intercalated discs, **881**
Intergovernmental Panel on Climate Change (IPCC), 1098–99
Interindividual signals, 931
Interleukin 2, 930
Intermediate disturbance hypothesis, **1080**
Intermediate filaments, 124*t*, **125**
Intermediate hosts, **632**
Intermediates, anabolic, 169
Internal environments, 319–20

Boldface page numbers indicate a glossary entry; page numbers followed by an *f* indicate a figure; page numbers followed by *t* indicate a table.

Internal fertilization, animal, 615, 630, 954–55
Internal membranes, 104, 107*t*
International online DNA repositories, 360
International Union for the Conservation of Nature, 1120
Interneurons, **886**
Internodes, **699**
Interphase, **196**
Interspecific competition, **1059**–60
Interstitial fluid, **876**–77
Intertidal zone, **1000**
Intestines. *See* Large intestines; Small intestines
Intraspecific competition, **1059**
Intrauterine device (IUD), 966*t*
Intrinsic rate of increase, **1041**
Introns, **294**
alternative splicing to remove, 328–29
antibodies and regulatory sequences in, 324–26
discovery of, 293–94
RNA splicing to remove, 294–95
Invasive species, **1014**–15, 1062, 1110–11
Inversion, **286**
Invertebrates, **609**, **646**, 651–52, 976–77
In vitro DNA synthesis reaction, 346
Involuntary muscle, **808**
Involuntary responses, 899
Iodine, 843*t*
Ion channels, **94**–95, **888**
Ion concentration gradients, 887–88
Ion currents, action potentials and, 891
Ion exclusion mechanisms, plant, 746–48
Ionic bonds, **18**–19, 48
Ions, **18**, **742**. *See also* Atoms
electrolytes as, 822 (*see also* Osmoregulation)
ionic bonds and, 18–19, 48
lipid bilayers and, 86
plant exclusion mechanism for, 746–48
as plant nutrients in soil, 742–44*f*
plant nutrient uptake mechanisms for, 744–46
Iridium, 490
Iris, **914**
Irish potato famine, 520–21, 542
Iron
as human nutrient, 843*t*
iron-containing heme groups, 161
iron-fertilization experiments, 523, 1091–92
as plant nutrient, 739*t*
Irrigation, 1095
Island biogeography, theory of, 1077–78
Islets of Langerhans, 930
Isolation, reproductive. *See* Reproductive isolation
Isoleucine, 42*t*, 47
Isopods, 643
Isoprene, 83, 84, 177, 496
Isotonic solutions, **91**, **718**, **824**
Isotopes, **16**, 586

J

Jacob, François, 2, 279, 309–14
Jasmonic acid, 779
Jaws
cichlid, 844–45
vertebrate, 655–56
Jeffreys, Alec, 366
Jelly fish, 601*f*, 619
Jelly layer, 390
Jenner, Edward, 973–74, 990
Jet propulsion, mollusk, 630
Johnston, Wendy, 68–69
Jointed limbs, animal, 613, 629, 637
Joints, **920**
Joly, John, 724
Junipers, 575
Jürgens, Gerd, 406
Juvenile hormone (JH), **936**
Juveniles, **616**

K

Kandel, Eric, 903–4*f*
Kangaroos, 967
Karpilov, Y. S., 188
Karyogamy, **591**, 592, 593
Karyotypes, **212**–13, **286**–87*f*
Keel, bird, 658
Kelp forests, 650
Kenrick, Paul, 554
Keratins, 125
Kerr, Warwick, 445
Ketones, 34, 35*f*
Ketose, 72
Keystone species, 1072–**73**, 1113
Kidneys, **832**, 931, **933**, 940
Killer T cells, 985–86
Kilocalories, **149**
Kinesis, **126**–27
Kinetic energy, **27**. *See also* Energy
chemical energy and, 78
energy transformations of, 27–29
enzymes and, 51–53, 55
Kinetochore microtubules, **198**–99, 215, 217
Kinetochores, **199**
chromosome movement and, 201–2
mistakes with, 226
in mitosis, 202*t*
Kingdoms, phylogenetic, 6–7
Kinocilium, **909**
Kinorhyncha, 603*t*
Kin selection, 1031–**33**, 1034
Klinefelter syndrome, **226**
Knock-out alleles, **277**
Koch, Robert, 498–99
Koch's postulates, 498–**99**
Kortschak, Hugo, 188
K-P extinction, 490–92
Krebs, Hans, 158
Krebs cycle. *See* Citric acid cycle
Kudzu, 1015
Kuhn, Werner, 835
Kuru disease, 51

L

Labia majora (labium majus), **960**
Labia minora (labium minus), **960**
Labor, **970**
lac genes, 311–14
lac operon regulation, 311–17*f*
Lactate, 166, 167*f*, **659**
Lactation, 660, **967**
Lacteal, **851**
Lactic acid fermentation, **166**, 167*f*
Lactose, 73–76
Lactose fermentation, 508
Lactose metabolism
identifying genes under regulatory control in, 310–12
as model system for bacterial gene regulation, 309–10 (*see also* Gene regulation, bacterial)
negative control mechanisms in, 312–14
overview of positive and negative *lac* operon regulation in, 317*f*
positive control mechanisms in, 314–16
types of mutants in *E. coli*, 311*t*
Lagging strands, DNA, **266**–68*t*
Lakes, 996, 997
Lamarck, Jean-Baptiste de, 415, 429
Lamellae, **814**
Lampreys, 661
Lancelets, 651, 652
Land. *See* Terrestrial ecosystems
Lander, Eric, 370
Land plants. *See also* Plant(s)
angiosperm radiation of, 564–65
drugs derived from, 549*t*
dry land transition adaptions of, 553–55
key lineages of, 567, 569–77*f*
life cycles of, 784–86
morphological differences among, 549–50
mycorrhizal fungi and nutrients for, 580
origin of, 550
origin of vascular tissue in, 554–55
phylogenetic tree and evolutionary changes of, 555–56
Plantae and, 536
prevention of water loss by cuticles and stomata of, 553–54
reproductive adaptations of, 556–64 (*see also* Plant reproduction)
similarities between green algae and, 549 (*see also* Green plants)
Land transition adaptations. *See* Water-to-land transition adaptations
Language, 672
Language, honeybee, 1028–29
Large intestines, 810*f*, 846*f*, 847, **854**–55
Larvae (larva), **616**, 936
Late endosome, **111**
Latency, **680**
Latent growth, viral, 679–81
Lateral buds, **699**
Lateral gene transfer, **364**
Lateral line system, **913**
Lateral meristem, **712**
Lateral roots, **697**, 706
Latin word roots, 8, B:6
Latitudinal gradient, species richness, 1078–80
Law of succession, 417
Leaching, **743**
Leadbeater's possum, 1115
Leading strands, DNA, 265–**66**, 268*t*
Leak channels, **889**
Learning, **902**–4*f*
Leaves (leaf), **699**
auxin and abscission in, 768
carbon dioxide entry into, through stomata, 187
characteristics of, 701, 702*f*
ethylene and abscission of, 774
genes, proteins, and shape determination for, 408
modified, 703
morphological diversity in, 701–2
phenotypic plasticity in, 702–3
seed, 796
water loss prevention by, 727–28
Leeches, 633
Leghemoglobin, **749**
Legumes, **749**–50
Leishmania and leishmaniasis, 522*t*
Lemmings, 430–31
Lennarz, William, 391
Lens, **914**
Lenski, Richard, 449–50
Lenticells, **714**
Lentiviruses, 688
Lepidoptera, 640*t*
Lepidosauria, 666
Leptin, **933**, 938–**40**
Lesion studies, 901–2
Leucine, 42*t*, 120, 298
Leukocytes (white blood cells), 870, **975**–76, 977*t*
Libraries. *See* cDNA libraries; DNA libraries
Lichens, 515, **566**, **589**, 597, 598
Life
biology, science, and methods for studying, 8–12 (*see also* Biological imaging; Biological methods; Biology; BioSkills; Experiments; Quantitative methods)
cell theory and cells of, 2–4 (*see also* Cells)
characteristics of living organisms and, 1–2, 82 (*see also* Organisms)
extremophiles and extraterrestrial, 501
geographic distribution of, 1013–17
history of (*see* Life, history of)
life tables and, 1038–41, 1053–55
molecules of (*see* Molecules)
origin of viruses in origin of, 687–88
research on origin of (*see* Origin-of-life research)
speciation and phylogenetic tree of, 5–8 (*see also* Phylogenies; Speciation; Tree of life)
theory of chemical evolution of (*see* Chemical evolution)
theory of evolution of, by natural selection, 4–5 (*see also* Evolution; Natural selection)
viruses as not being alive, 675
Life, history of, 474–95. *See also* Evolution
adaptive radiations in, 484–88
fossil record as tool for studying, 479–84
fossil record time line for, 481–84
importance of animals in, 602
mass extinctions in, 488–92
phylogenetic trees and phylogenies as tools for studying, 474–79
Life cycles, **215**
animal, 215–16, 615–17
evolution of land plant, 558–59
fungal, 591–94
land plant, 784–86
plant, 402*f*
protist, 533–36
Life histories, 1039–**41**
Life tables, **1038**–40, 1053–55
Ligand-gated channels, **897**
Ligands, **777**, **897**
Light chain, **980**
Light energy. *See also* Photosynthesis
amino acid production and, 40
animal sensory organs and, 610
carbohydrate storage of, 78
chemical evolution and, 29, 32–33, 34*f*
land plants and, 553, 554
lizard sexual activity and, 1023–24

Light energy, (*continued*)
photosynthetic capture of, by chlorophylls, 172–79
photosynthetic glucose production from, 153
plants and blue (*see* Blue light)
plants and red/far-red (*see* Red/far-red light responses, plant)
prokaryotic photosynthesis and, 509
transformation of, into biomass, 1084–85
vegetative development and, 407
water depth and, 996–97
Light microscopy
cell research and, 116
meiosis and, 215
overview of, B:13–B:14
of protist cells, 524–25
viewing chromosomes with, 196
Lightning energy, 39–40
Light-sensing organs, 913. *See also* Eyes; Vision
Lignin, 113, **133**, 548–49, **555**, 579–81, 590, **709**, 726
Limbs
animal, 613–15
arthropod, 637
endoskeletons and, 920–21
evolutionary loss of, 384–85
loop of Henle, 835–36
protostome, 628, 629
tetrapod, 655, 656–57
Limiting factors
for net primary productivity (NPP), 1090–92
population growth, 1044–46
Limiting nutrients, **738**
Limnetic zone, **997**
Lineages, **460**
animal, 617–20
ecdysozoans, 637–43
echinoderm, 649–50
fungi, 594–98
green plants, 566–77
invertebrate chordates, 651–52
lophotrochozoans, 630–36
monophyletic groups as, 504
prokaryotes, 512–16
protists, 536–43
vertebrates, 660–68
viruses, 689–92
Linear structures, 33*f*, 72
Lingual lipase, 855*t*
Linkage, gene, **243**–45, 246*f*, 250*t*
Linkage maps, **348**
Linnaeus, Carolus, 8
Lipase, **847**
Lipid bilayers, **85**. *See also* Plasma (cell) membranes
artificial membranes as experimental, 85–86
cholesterol and permeability of, 88
diffusion across, 89–90
effect of bond saturation and hydrocarbon chain length on fluidity and permeability of, 87–88
effect of temperature on fluidity and permeability of, 88–89
membrane proteins and, 92–99
micelles vs., 85
osmosis across, 90–92
permeability and selective permeability of, 86
Lipids, **83**
catabolic pathways and, 168–69
digestion of, in animal small intestines, 853–54
glycolipids, 105
hormones and solubility of, 943
lipid-soluble vs. lipid-insoluble hormones, 139–41
nucleic acid polymerization and, 62
plasma membrane, as characteristic of domains of life, 497*t*
smooth ER and, 108
structure of, and cell membrane properties, 87–88
structures of membrane, 84 (*see also* Lipid bilayers; Plasma (cell) membranes)
types of, found in cells, 83–84
Liposomes, 85–86, 88
Liquids
density of water as, 23–24 (*see also* Water)
pinocytosis and, 111
specific heat of some, 24*t*
unsaturated fats as, 88
Littoral zone, **997**
Liver, 810*f*, 846*f*, **854**
Liver cells, 110
Liverworts, 569, 785*f*
Live virus vaccines, 990
Lizards
adaptive radiations of, 484–85
clutch size and egg size in, 804, 805*f*
hearts of, 878
Lepidosauria lineage and, 666
life table case study of *Lacerta vivipara*, 1038–40
as reptiles, 660
sexual activity of *Anolis*, 1023–24
Loams, 742
Lobe-finned fishes, **663**
Lobes, brain, 901
Lobster, 643
Loci (locus), **240**
Lock-and-key model, enzyme catalysis, **53**–54
Locomotion, animal, **920–26**. *See also* Movement
evolution of animal appendages for, as homologous, 613–15
great ape, 670
muscle contraction and, 922–26
muscle tissue and, 808
muscle types and, 921–22
neurons and muscle cells of, 601
protostome, 628–30
sensory systems and, 907
skeletons and, 920–21
Loewi, Otto, 895
Logarithms, 273, 812, B:9
Logistic population growth, **1042**–44, 1046
Long-day plants, **787**
Long interspersed nuclear element (LINE), **365**, 366*f*
Loop of Henle, **835**–37
creation of osmotic gradient by, 835–36
hypothesis on function of, 835
model of functioning of, 836–37
structure and function of, 832, 838*t*
Loose connective tissue, **807**
Lophophore, **624**
Lophotrochozoans, **609**
evolution of, as protostomes, 624–25
lineages of, 630–36
major phyla of animals and, 603*t*
Loss-of-function alleles, **277**, 451
Lou Gehrig's disease (ALS), 376
Love darts, 956
Lubber grasshoppers, 212
Lumber, 548
Lumen, **108**, **174**
Luminescence, 347
Lungfish, 656–57, 663
Lungs, **867**–70, 874
LUREs, 794
Luteal phase, **963**
Luteinizing hormone (LH), **943**, **962**, 964–66
Lycophytes (Lycophyta), 571
Lyme disease, 513
Lymph, **877**, **978**
Lymphatic systems, **877**, **978**
Lymph nodes, **978**
Lymphocytes, 677, **978**–79*f*
Lynx, 1047–50
Lysine, 42*t*, 283
Lysogenic cycle, **680**
Lysosomes, **110**–11, 114*t*
Lysozyme, 681, **975**–76
Lytic cycle, **680**

M

Maathai, Wangari, 1122
MacArthur, Robert, 1077–78
McGinnis, William, 383
Macromolecules, **42**
Macronutrients, **738**, 739*t*
Macrophages, 677, 975*f*, **976**, 977*t*
Mad cow disease, 51
MADS box, 383, 410–**11**
Magnesium, 54, 739*t*, 843*t*
Magnetic fields
animal sensory organs and, 610
navigation, migration and, 1027
Magnetite, 104, 502, 1027
Maidenhair ferns, 466–67
Maintenance, energy and, 1084
Maize. *See* Corn
Major groove, DNA, 64
Major histocompatibility (MHC) protein, **984**
Maladaptive traits, **187**
Malaria, avian, 1015
Malaria, human, 224, 247, **521**, 522*t*, 541, 1066–67
Malate, 157
Males
asexual reproduction and, 224
chromosomes of, 212
female barn swallow mate selection and tail length of, 1024–25
female choice of, 453–54
gametangia of, 556
male-male competition of, 454–55
plant flower parts, 789
plant gametophytes, 790–92
plant reproductive organs, 402
reproductive systems of animal, 957–59*f*
sexual selection and, 452–53
sperm as reproductive cells of, 211
spermatogenesis in mammal, 953
visual cues from, and female lizard sexual readiness, 1024
x-linked inheritance of recessive diseases in, 252
Malignant tumors, **206**–7
Malpai Borderlands group, 1122–23
Malpighian tubules, **830**, 831*f*
Maltose, 73–76, 848
Mammalia, 660, 665–66
Mammals, **660**
amniotic eggs in, 658
in Cenozoic era, 483
cleavage in, 393–94
cloning of, 377–78
digestion in, 846, 855*t*
ears of, 910–12
as endothermic homeotherms, 817
flourishing of, after K-P extinction, 492
fungal mutualisms with, 595
hormones in growth of, 937
Mammalia lineages and, 660, 665–66
mouthparts of, 844
parental care in, 659
primates and humans, 668–73*f*
sex hormones in reproduction of, 960–66
spermatogenesis and oogenesis in, 953
thermoregulation in, 816–19
threatened with extinction, 1116
Mammary glands, **660**
Mandibles, **643**
Manganese, 739*t*
Mangroves, 698, 699*f*
Mantis shrimp, 1030–31
Mantle, **626**
Maps
community, 1072
concept, B:10
genetic (*see* Genetic maps)
Marfan syndrome, **247**
Margulis, Lynn, 527
Marine ecosystems. *See also* Aquatic ecosystems
carbon cycle in, 523
changes in net primary productivity of, 1101–2
estuaries as freshwater and, 1000
global productivity patterns of, 1089–91
metamorphosis in life cycle of animals in, 616–17
oceans, 995–96, 1000–1001
osmoregulation in, 824, 826–27
Mark-recapture studies, 1046–47, 1048
Marshall, Barry, 850
Marshes, **998**
Marsupials (Marsupiala), 660, **665**, **967**
Mass extinctions, 474*f*, **488**–92, 1110. *See also* Extinction
Mass feeders, **611**, **843**
Mass number, **16**
Mast cells, 975*f*, **976**, 977*t*
Master regulators, 379–81
Mate choice, female, 452–53
Maternal age, trisomy and, 227
Maternal chromosomes, **213**
Mathematical models, 436, 994
Mating. *See also* Reproduction
fungal hyphae mating types, 590–91
in genetic experiments, 231
giraffe necks and, 10
Hardy-Weinberg principle and, 438–40

Boldface page numbers indicate a glossary entry; page numbers followed by an *f* indicate a figure; page numbers followed by *t* indicate a table.

inbreeding, 450–52
interbreeding, 468–71
mate selection in, 1023–25
monohybrid crosses, 232–33
nonrandom, 450–55 (*see also* Nonrandom mating)
outcomes of, between populations, 471*t*
sexual selection and in, 452–55
unusual aspects of, 955–56
Mating types, fungal, 590–91, 593
Matthaei, Heinrich, 283
Maturation, seed, 796–97
Measles, 686, 692, 990
Mechanical energy, 28
Mechanical prezygotic isolation, 459*t*
Mechanoreceptors, **908**, 909. *See also* Hearing
Mediator complex, **326**
Medicine. *See* Diseases, human; Drugs
Medium, **311**
Medulla, **832**
Medulla oblongata, **653**
Medullary respiratory center, 870
Medusae (medusa), **617**
Megapascal (MPa), **719**
Megasporangia, **559**, 561, 790
Megaspores, **559**, 561, **790**
Megasporocytes, 790
Meiosis, **194**, **212**, **784**. *See also* Genes; Genetics
chromosome types and, 212
consequences of sexual reproduction by, 220–23
in eukaryotes, 519
events of prophase I, 219–20
meiosis II phases, 218
meiosis I phases, 216–18
mistakes in, 225–27
mitosis vs., 218*t*, 219*f*
number of chromosomes found in some familiar organisms, 213*t*
overview of, 213–16
paradox of sexual reproduction and, 223–25
patterns of mistakes in, 226–27
ploidy concept and, 212–13
and principles of Mendelian genetics, 240–41
as reduction division, 215–16
sexual reproduction and, 211–12
trisomy in human births and mistakes in, 227*t*
as two cell divisions (meiosis I and II), 214–15
types of mistakes in, 225–26
vocabulary for describing chromosomal makeup of cells, 214*t*
Meiosis I, **214**–20
Meiosis II, **214**, 215, 218–19
Meiotic maps, **348**
Melanocortin receptor gene, 280–81, 285–86
Membrane lipids, 84. *See also* Lipid bilayers; Plasma (cell) membranes
Membrane potentials, **745**, **767**, **887**–91, 908–9. *See also* Action potentials
Membrane proteins, 92–99
active transport by pumps, 97–98
amphipathic proteins as, 92
aquaporins (*see* Aquaporins)
biological methods for studying, 94
evolution of fluid-mosaic model of plasma membranes and, 92–94
facilitated diffusion via carrier proteins, 96–97
facilitated diffusion via channel proteins, 94–96
functions of, 45
in osmoregulation, 826–28
in phloem loading, 731–32
plant cell-cell signals and, 757
plasma membranes, intracellular environment, and, 98–99
selective adhesion and, 136–38
in water and electrolyte movement, 825–26
Membranes. *See also* Plasma (cell) membranes
artificial, 85–86
endomembrane system, 118–23*f*
internal, 104, 107
mitochondrial, 112
thylakoid, 112–13
vesicle, 122
Memory, **902**
experiments on human learning and, 902–4*f*
immunological, 978, 989–90
Memory cells, **989**, 990
Mendel, Gregor, 230. *See also* Mendelian genetics
Mendelian genetics, 230–57
applying rules of, to humans, 250–52
chromosome theory of inheritance in, 239–43
determinance of phenotypes in, 247–48
exceptions and extensions to Gregor Mendel's rules in, 250*t*
experimental system in, 230–32
extending rules of, 243–50
garden peas as first model organism of, 231–32
genetic model of, 236*t*
hypotheses of, 231
identifying human alleles as recessive or dominant, 250–51
identifying human traits as autosomal or sex-linked, 251–52
incomplete dominance in, 245
linkage in, 243–45, 246*f*
multiple allelism of genes in, 247
pleiotropic genes in, 247
polygenic inheritance and quantitative traits in, 248–50
quantitative methods in studying linkage, 246*f*
results of monohybrid reciprocal crosses in, 234*t*
single-trait experiments in, 232–36
two-trait experiments in, 236–39
vocabulary of, 239*t*
Meniscus (menisci), 23, **724**–25, 727
Menstrual cycle, mammalian, **962**–66
Menstruation, **962**
Meristems, **375**, **405**, 407–9, **704**–5
Meselson, Matthew, 261–63
Meselson-Stahl experiment, 261–63
Mesoderm, 394, **395**, 396–98, **604**
Mesoglea, **619**
Mesophyll cells, **188**
Mesozoic era, **483**
Messenger RNAs (mRNAs), **117**, **279**
bacterial vs. eukaryotic gene expression regulation and stability of, 331*t*
in central dogma, 280–81
eukaryotic processing of, 293–95, 320
as intermediaries between genes and proteins, 279–80
post-transcriptional alternative splicing of, 328–29
processing, 293–95
ribosome structure and, 300–301
RNA interference and post-transcriptional stability of, 329–30
specification of amino acids by triplets of, 297
synthesis of, through transcription, 289–93
translation and, 295–97
viral, 682–83
visualizing, via in situ hybridization, 380
Meta-analysis, **1065**
Metabolic diversity
of animals, 610–13
deuterostome, 648
fungal, 589–90
prokaryotic, 506–9
prokaryotic electron donors and acceptors and, 508*t*
protist, 529–32
six examples of prokaryotic, 507*t*
Metabolic pathway, **277**
Metabolic rates, **812**–13, 1093
Metabolic water, **803**
Metabolism, **148**, 168
Metagenomics, **364**
Metal ions, 54
Metallothioneins, **747**
Metamorphosis, **616**, **935**
animal, 615–17
arthropod, 637
hormones in amphibian, 935–36
hormones in insect, 929*f*, 936
protostome, 628, 630
Metaphase, **199**, 205
Metaphase I and II, meiosis, 218
Metaphase plate, **199**, **217**
Metapopulations, **1046**–47
balance between extinction and recolonization in, 1046, 1047*f*
habitat destruction and, 1113
mark-recapture experiment on, 1046–47
mark-recapture studies and, 1048
preserving endangered, 1055
stochastic and genetic problems in, 1114
theory of island biogeography and, 1077–78
Metastasis, **134**, **207**, 332. *See also* Cancer
Meteorites, 40, 73, 490–92
Methane, 19–21*f*, 29, 39–40, 509
Methanogens, **503**, 516
Methanotrophs, **509**
Methionine, 42*t*, 47, 283, 296
Methods. *See* Biological methods; Quantitative methods
Methylation, **322**
Methyl salicylate (MeSA), **778**
Metric system, B:1–B:2
Meyerowitz, Elliot, 409–11
Mice, 280–81, 285–86, 369*t*, 812–13
Micelles, 85
Microarrays, DNA. *See* DNA microarrays
Microbes, **498**
Microbiology, **498**
Microcystins, 515
Microelectrodes, 890
Microfibrils, 132
Microfilaments, **123**. *See also* Actin filaments
Micronutrients, 739*t*, **740**
Micropyle, **790**
MicroRNA (miRNA), **329**, 368
Microsatellites, **365**–66
Microscopy, B:13–B:16. *See also* Biological imaging
Anton van Leeuwenhoek and, 2
cell research and, 116
electron (*see* Electron microscopy)
light (*see* Light microscopy)
prokaryotic cell structures and improvements in, 103
video, 165
Microsponangia, 790–91
Microsporangia, **559**, 561
Microspores, **559**, 561, **791**
Microsporidia, 594
Microsporidians, 585
Microsporocytes, 790–91
Microtubule organizing centers, **125**, 202*t*
Microtubules, **125**
in meiosis, 215
mistakes with, 226
in mitosis, 198
in spindle apparatus, 201–2
structure and function of, 124*t*, 125–27
Microvilli (microvillus), **834**, **851**
Midbrain, **653**
Middle ear, **910**, 911
Middle lamella, 135
Mifepristone (RU-486), 966*t*
Migration, **1026**–27
global warming and, 1100
navigation strategies for, 1026–27
Population Viability Analysis (PVA) and, 1115
seasonality of, 1027
Miller, Stanley, 39–40
Millipedes, 639
Millivolt (mV), **887**
Milner, Brenda, 901–2
Mimicry, **1064**
Mineral nutrients, plant, 738
Mineralocorticoids, **940**
Mineral particles, polymerization and, 43, 62
Minisatellites, **365**–66
Minor groove, DNA, 64
miRNA. *See* MicroRNA (miRNA)
Miscarriages, 226, 440
Mismatch repairs, **272**
Missense mutations, **286**
Mistletoe, 751
Mitchell, Peter, 162–63
Mites, 642
Mitochondria (mitochondrion), **112**, **390**
ATP synthesis in, 164–65
eukaryotic, 112, 114*t*
NADH in, 161
origin of, in protists, 527–28
in parietal cells, 850
in proximal tubules, 834
replacement rate for, 116
sperm, 390
Mitochondrial DNA (mtDNA), 469–70, 528
Mitochondrial matrix, **112**, **157**
Mitosis, **194**, 195–202. *See also* Cell cycle
alternation of mitotic (M) phases and interphases in, 196
asexual reproduction and, 195
asexual reproduction vs. sexual reproduction and, 220–21

Mitosis, (*continued*)
cell cycle and, 196–97*f*
chromosomes and, 195–96
cytokinesis and production of daughter cells by, 200
discovery of G_1 and G_2 gap phases in, 196
discovery of synthesis (S) phases in, 196
events in, 197–200
meiosis vs., 218*t*, 219*f*
movement of chromosomes during, 201–2
overview of, 197
plant gametes and, 402
structures involved in, 202*t*
Mitosis-promoting factor (MPF), 203–**4**, 375
Mitotic (M) phase, **196**
Mitotic spindle forces, 201
MN blood types case study, 438–39
Mockingbirds, 419–20
Model organisms, **231**, B:19–B:22
animals as, 603
Arabidopsis thaliana (*see Arabidopsis thaliana*)
Caenorhabditis elegans (*see Caenorhabditis elegans*)
characteristics of, B:19
Dictyostelium discoideum, 536, B:20
Drosophila melanogaster (*see Drosophila melanogaster*)
Escherichia coli (*E. coli*) (*see Escherichia coli* (*E. coli*))
garden peas as, for Mendelian genetics, 231–32
genomes of, 362
liverworts as, 569
Mus musculus, 362, 369*t*, B:21*f*, B:22
pictures of, B:21*f*
protostome, 623
Saccharomyces cerevisiae (*see Saccharomyces cerevisiae*)
Models
animal, of disease, 350
condensed DNA, 319*f*
DNA synthesis, 263–68
mathematical, 436
molecular, 20–21, B:8*f*
population growth, 1043–44
Modified leaves, 703, 751–52
Modified roots, 698–99
Modified shoots, 700–701
Molarity, **21**
Mole, **21**
Molecular biology. *See also* Molecules
beginning of, 276
carbohydrates in (*see* Carbohydrates)
development of central dogma of, 279–82
genetic engineering and, 338
lipids and plasma membranes in (*see* Plasma (cell) membranes)
nucleic acids in (*see* Nucleic acids)
proteins in (*see* Proteins)
water, carbon, and chemical evolution in (*see* Chemical evolution)
Molecular chaperones, **50**, **304**
Molecular formulas, **20**, B:8*f*
Molecular machines, 104
Molecular matching mechanisms, 792
Molecular phylogenies. *See* Phylogenies
Molecular weight, **21**
Molecules, **18**
atomic structure and, 16–17
carbohydrates and sugars (*see* Carbohydrates; Sugars)
carbon and organic, 33–36 (*see also* Organic molecules)
cell-cycle regulatory, 203–4
chemical evolution and (*see* Chemical evolution)
chemical reactions and, 21 (*see also* Chemical reactions)
covalent bonding and, 17–18
diffusion and osmosis of, across lipid bilayers, 89–92
DNA, 65–66
enzyme regulation by regulatory, 54–55
geometry of simple, 20
ionic bonding and, 18–19
lipids and plasma membranes (*see* Lipids; Plasma (cell) membranes)
nucleic acids (*see* Nucleic acids)
phylogenetic tree of life and, 6–7
proteins (*see* Proteins)
proteins as large, 45
representing, 20–21, B:8*f*
RNA, 67–68
self-replicating, as origin of life (*see* Origin-of-life research; Self-replicating molecules)
separating, using gel electrophoresis, B:11–B:12
simple, from carbon, hydrogen, nitrogen, and oxygen, 19–20
transport of, into nucleus, 117–18
visualizing, B:12–B:13
visualizing, with tagging systems, B:12–B:13
Moliason, Henry Gustav, 901–2
Mollusca, 603*t*, 634–36
Mollusks, **610**, 634–36
body plan of, 626
eyes of, 914
foot of, 629
lineages, 630–31
sampling of, 1107–8
Molting, **625**
Molybdenum, 739*t*, 740
Monarch butterflies, 1027
Monkeys, 668–69
Monoamines, 898*t*
Monocots, **564**–65, 706*f*, 796
Monod, Jacques, 279, 309–14
Monoecious, **790**
Monohybrid crosses, **232**–34*t*
Monomers, **42**
amino acids as, 42
monosaccharides as, 71–73
nucleotides as, 59
Monophyletic groups, **460**, **504**
cladistic analysis and, 475
green plants as, 551–52
phylogenies and, 504
synapomorphies and, 524
Monosaccharides, **72**. *See also* Carbohydrates; Sugars
chemical evolution and, 71, 73, 76
differences between, 72–73
as monomer sugars, 71–73
Monosodium glutamate (MSG), 919
Monosomy, **226**
Monotremes (Monotremata), 660, **665**, **967**
Morgan, Thomas Hunt, 216–17, 241–42, 246
Morphogens, **381**
Morphological diversity
adaptive radiations and, 485–86
of animals, 603–7
deuterostome (echinoderm), 647–48
deuterostome chordate, 650–51
land plant, 549–50
in leaves, 701–2
prokaryotic, 504–6
protist, 526–29
protostome, 626–27
in root systems, 697–98
in shoot systems, 699–700
vertebrate, 655–59
Morphological traits
adaptive radiations and innovations in, 485–86
analyzing, in fungi, 582–84
analyzing, in green plants, 549–50
analyzing, in viruses, 679
chordate, 650
phylogeny based on, 477
of protists, 524–25
Morphology, **102**, **420**
comparative, of animals, 603–7 (*see also* Animal anatomy and physiology)
of eukaryotic lineages, 524*t*
of fungi, 582–84
of green plants, 549–50
morphospecies concept and, 460
plant (*see* Plant anatomy and physiology)
structural homology in adult, 420–21
Morphospecies concept, **460**, 461
Mortality, human
bacterial infections and, 500*f*
birth weight and, 442
cancer, 206*f*
childbirth and maternal, 970–71
death rates, 206*f*, 1037–40
from jellyfish stings, 619
Mosquitoes, 521, 602, 616, 640*t*
Mosses, 550, 570
Moths, 640*t*, 907
Motility. *See* Locomotion, animal; Movement
Motor neurons, **886**, 921
Motor proteins, **124**
cell movement and, 45, 128, 280
kinetochores and, 201–2
in meiosis, 215
microtubules, vesicle transport, and, 126–27
Mountain ranges, climate and, 1010, 1011*f*
Mouse mammary tumor virus, 691
Mouse model organism, 362, 369*t*, B:21*f*, B:22
Mouthparts, 843–45
as adaptations, 843–44
cichlid jaws as case study on, 844–45
feeding adaptations and, 610–13
placental mammal, 666
protostome, 628, 629*f*
Mouths
digestion in, 847–48
in digestive tracts, 845
human, 846*f*
protostome vs. deuterostome, 606
Movement
actin filaments and cell, 123–24
animal (*see* Locomotion, animal)
cell, 375*t*, 376–77
cell gastrulation and, 394–95
of chromosomes during mitosis, 201–2
flagella, cilia, and cell, 127–28
flagella and prokaryotic, 104
microtubules and vesicle transport, 125–27
prokaryotic diversity in, 504, 505*f*
proteins and cell, 45
protist diversity in, 532–33
wind/touch responses and plant, 766–67
MPF (mitosis-promoting factor), **204**
M phase. *See* Mitotic (M) phase
mRNA. *See* Messenger RNAs (mRNAs)
Mucigel, **706**
Mucosal-associated lymphoid tissue (MALT), **978**
Mucous cells, **849**
Mucus, **848**, 974–75, 978
Müller-Hill, Benno, 314
Müllerian inhibitory substance, **936**
Müllerian mimicry, **1064**
Multicellularity, **134**, **529**, **806**
Cambrian explosion in, 486–88
cell-cell attachments in eukaryotes, 135–38
cell-cell signaling and, 139–46
as characteristic of domains of life, 497*t*
communications via cell-cell gaps, 138–39
in eukaryotes, 519
eukaryotic, 6, 105, 319
Paleozoic era and, 483
as physical connection of adjacent cells, 134–35
protist, 529
Multienzyme complexes, **49**
Multiple allelism, **247**, 250*t*
Multiple fertilization, 391–92
Multiple fruits, **797**
Multiple sclerosis (MS), **894**–95, 983
Mumps, 692
Münch, Ernst, 729
Murchison meteorite, 73
Murine leukemia virus, 691
Muscle fibers, **808**, **922**
Muscles
alternative splicing and gens for, 328
cells of, 396–98
contraction of, 922–26
mesoderm and, 604
neurons and cells of, 601
organogenesis and tissues of, 396–99
relaxation of, 925
skeletons and, 920–21
types of, 921–22
Muscle tissue, **808**–9*f*
Mushrooms, 581, 583, 593, 596
Mus musculus, 362, 369*t*, B:21*f*, B:22
Mussels, 634, 1064–65*f*
Mustard plant model organism. *See Arabidopsis thaliana*
Mutagens, **310**
Mutants, **241**. *See also* Mutations
chromosome theory of inheritance and, 241
creating alleles as, 277
creating *E. coli*, 310
different classes of lactose metabolism, 311

Boldface page numbers indicate a glossary entry; page numbers followed by an *f* indicate a figure; page numbers followed by *t* indicate a table.

DNA polymerase proofreading and, 271–72
finding mutant genes with replica plating, 310–11
histone, 322
Mutations, **241**, **285**, **448**. *See also* Mutants
animal pollination and plant, 794
antibiotic resistance and bacterial gene, 424–25
Cambrian explosion and, 488
cancer and tumor suppressor gene, 332–33
chromosome-level, 286–87*f*
DNA polymerase proofreading and, 271–72
evolution and, 448–49
experimental studies of, 449–50
gene duplication and, 368
genetic diversity and, 1106–7
genetic variation and effect on average fitness of, 450*t*
and genetic variation in plants, 409
Hardy-Weinberg principle and, 438
known types of point, 286*t*
molecular basis of, 285–87
point, 285–86
polyploidy and speciation by, 465–68
RNA, 69
viral, 686
Mutualisms, **793**, **1059**, 1068–70
diversity of, 1068
dynamism of, 1069–70
fitness and impacts of, 1070*t*
natural selection and, 1069
pollination and, 562
types of fungal, 586–89
Mutualistic, **586**, **746**
Mutualists, **579**
Mycelia (mycelium), **514**, **582**, 583
Mycobacterium tuberculosis, 424–26
Mycorrhizae, **746**
Mycorrhizal fungi, **580**, 746, 1068
Myelination, 894–95
Myelin sheath, **894**
Myelomas, 982
Myoblasts, 398–99
MyoD (myoblast determination), **399**
Myofibrils, **922**
Myosin, **923**
in mitosis, 200
as motor protein, 124
movement and, 45
in muscle contraction, 923–25
Myriapods, **639**
Myxinoidea, 661
Myxogastrida, 537

N

N-acetylglucosamine, 76
NAD⁺
cellular respiration and, 153
fermentation and, 166–67
in glycolysis, 155
photosystem I and, 182
in pyruvate processing, 157–58
NADH, **152**
cellular respiration and, 153
citric acid cycle and, 158–61
fermentation and, 166–67
in pyruvate processing, 157–58
NADPH, **173**–74, 182
Names, scientific, 7–8, 675
Nanobiology, 678–79
Natural experiments, **427**, **1119**
Natural populations, 445–47
Natural selection, **4**, **422**
as acting on individuals while populations evolve, 429–30
allopatric speciation, dispersal, and, 463
artificial selection vs., 4–5
balancing selection, 442–43
biological definitions of fitness and adaptation for, 424
biological evolution and, 15
changes of species through time as pattern component of evolution, 416–22
changing beak size, beak shape, and body size of Galápagos finches as case study for, 426–29
common misconceptions about adaptation and, 429–32
conception in, that organisms do not act for the good of species, 430–31
conditions for, 4
directional selection, 440–41
disruptive selection, 442
evidence for evolution, 422*t*
as evolutionary process, 435
evolution as not goal directed, 430
evolution of antibiotic resistance of *Mycobacterium tuberculosis* as case study for, 424–26
evolution of evolutionary thought and evolution by, 415–16
fitness and adaptations in, 5
fitness trade-offs in, 432
four postulates of Charles Darwin on, 423
genetic constraints on, 431
genetic variation and effect on average fitness of, 448, 450*t*
Hardy-Weinberg principle and, 437–38
historical constraints on, 432
individuals vs. populations and, 414*f*
limitations of, 431–32
mutualisms and, 1069
non-adaptive traits in, 431
as process component of evolution, 422–24
recent research in, 424–29
reproductive success and, 804
sexual reproduction and purifying, 224
species as agents of, 1059
stabilizing selection, 441–42
sympatric speciation by, 465
testing postulates of Charles Darwin on, 425–26
theory of evolution by (*see also* Evolution)
theory of special creation vs. theory of evolution by, 414–15
types of, 440–43
viruses and, 685
Nauplius, **643**
Nautilus, 636
Navigation, migration and, 1026–27
Neanderthals, **671**
Nectar, **562**, **788**
Nectary, **788**
Needlelike leaves, 702
Negative control, **312**–14, 322. *See also* *lac* operon regulation
Negative feedback, **204**, **815**–16, 818, **931**, 1099
Negative pressure ventilation, **868**
Negative-sense viruses, **686**, 687*t*, 692
Neher, Erwin, 892
Nematodes (Nematoda), 211, 603*t*, 605–6, **638**, 973*f*, 1067–68. *See also Caenorhabditis elegans*
Nematomorpha, 603*t*
Nemertea, 603*t*, 627
Nephrons, 832–38*t*
Neritic zone, **1000**
Nernst equation, 888
Nerve cells. *See* Neurons
Nerve cords, **650**
Nerve net, **604**, **886**
Nerves, **886**
Nervous systems
anatomy of neurons in, 886–87
central nervous system in, 900–902
comparative morphology of, 603, 604, 605
control of hormone production by, 940–43
learning, memory, and, 902–4*f*
peripheral nervous system in, 899–900
types of, 885–86
types of neurons in, 886
vertebrate, 899–904*f*
Nervous tissue, **808**
Net primary productivity (NPP), **1002**, **1084**
biodiversity, species richness, and increased, 1118–19
global climate change and, 1012–13*f*
global patterns in, 1089–90
global warming and changes in, 1101–2
human appropriation of global, 1090
limiting factors for, 1090–92
ocean iron-fertilization experiments to increase, 1091–92
pyramid of productivity and, 1086
terrestrial-marine contrast in, 1090–91
Net productive rate, **1040**
Neural signals, **930**–31
Neural tubes, **396**
Neurobiology, 885
Neurodegenerative diseases, 376
Neuroendocrine pathway, 931–32, 943
Neuroendocrine signals, 930*t*, **931**
Neuroendocrine-to-endocrine pathway, 931–32, 943
Neuromuscular junctions, 925–26
Neurons, **601**, **808**, **885**–87, 914. *See also* Synapses
Neurosecretory cells, **942**, 943
Neurospora crassa, 277, 595
Neurotoxins, **892**
Neurotransmitters, **895**
categories of, 898*t*
functions of, 896–97
memory, learning, and, 903–4*f*
as neural signals, 930–31
postsynaptic potentials and, 897–99
synapse structure and release of, 895–96
Neutrality, pH, 26
Neutral mutations, **286**
Neutrons, atomic structure and, 16–17
Neutrophils, 975*f*, **976**, 977*t*
New World monkeys, 668–69
Next-generation sequencing technologies, 347, 360–61
N-formylmethionine (*f*-met), 301
Niacin, 843*t*
Niche differentiation, **1062**, 1063*f*
Niches, **484**, **1059**
Cambrian explosion and, 488
interspecific competition and, 1059–60
types of, 1061
Nickel, 739*t*
Nicolson, Garth, 93
Nicotinamide adenine dinucleotide (NAD⁺), **152**
Nicotine, 775
Niklas, Karl, 546
Nilsson-Ehle, Herman, 249
Ninebarks, 720
Nirenberg, Marshall, 283
Nitrate, 510, 748
Nitrate pollution
fertilizers and, 749
prokaryotes and, 511–12
Nitrogen. *See also* Nitrogen gas
animal nitrogenous wastes, 829–30
atomic structure of, 17*f*
cycads and, 574
cycle of (*see* Nitrogen cycle)
electronegativity of, 18
fertilization with, 500
fixation of (*see* Nitrogen fixation)
hornworts and, 571
mycorrhizal fungi and, 746
nucleotide nitrogenous bases, 60
as plant limiting nutrient, 738
as plant nutrient, 739*t*
simple molecules from, 19–20
sources of human fixed, 1095
Nitrogenase, 49, 749
Nitrogen cycle, 510–11, 1096
Nitrogen fixation, **510**–11, **749**–50, 1068, 1093
Nitrogen gas, 20, 27, 49, 510–11, 748–50. *See also* Nitrogen
Nitrosomonas, 507*t*
Nociceptors, **908**
Node of Ranvier, **894**
Nodes, **474**, **699**
Nod factors, **749**
Nodules, **749**
Noise, experimental, 12
Non-adaptive traits, 431
Non-bilaterian animals, 603*t*, 617–20
Noncoding sequences. *See* Repeated sequences
Noncyclic electron flow, **183**
Nondisjunction, **225**–27, 466
Nonenveloped viruses, 679
Nonesense mutations, 286*t*
Non-kin human child abuse rates, 1034
Nonpolar covalent bonds, **18**, 19*f*
Nonpolar side chains, 41*t*
Nonrandom mating, 450–55
Nonself, adaptive immune response and, 978, 983–84
Nonsense mutations, 286*t*
Non-sex chromosomes. *See* Autosomes
Non-sister chromatids, 214*t*, **216**
Non-template strand, **290**
Non-vascular plants, 550, 551–52, 567, 569–71
Norepinephrine, 898*t*, **943**
Notochords, **396**, 651
N-terminus, 43–44, 301
Nuclear envelope, **107**
as characteristic of domains of life, 497*t*
Eurkarya and, 519
in mitosis, 200
as protist innovation, 526–27
structure and function of, 116–17

Nuclear lamina, **107**, 116–17, 125
Nuclear lamins, **125**
Nuclear localization signal (NLS), **117**–18, 122
Nuclear pore complex, **117**
Nuclear pores, **117**
Nuclear transfer, 378. *See also* Clones
Nucleases, **852**, 855*t*
Nucleic acid probes, B:13
Nucleic acids, **59**–70
chemical evolution and nucleotide production for, 60–61
chemical evolution theory and RNA molecule as first life-form, 68–69
components of, 59–60
DNA structure and function, 62–66 (*see also* DNA (deoxyribonucleic acid))
lipids and, 82
nucleotides and, 39
organic molecules and, 36
polymerization of nucleotides to form, 42, 61–62
ribonuclease enzyme for unfolding, 50
RNA structure and function, 66–68 (*see also* RNA (ribonucleic acid))
RNA world hypothesis on RNA as first self-replicating molecule, 59, 68–69
sugars in, 77
Nucleoids, **103**–4
Nucleolus, **108**, 117
Nucleoplasmin, 117–18
Nucleoside triphosphates, 62, 117
Nucleosomes, **321**
Nucleotide excision repairs, **272**
Nucleotides, **59**
chemical evolution and production of, 60–61
as components of nucleic acids, 59–60
nucleic acids and, 39
polymerization of, to form nucleic acids, 42, 61–62
Nucleus, **107**
atomic structure and, 16–17
cloning by transfer of, 378
DNA in, 279
eukaryotic, 107–8, 114*t*
eukaryotic cells vs. prokaryotic cells and, 6, 7*f*, 102, 107
multinucleate muscle cells, 808
nuclear transport in, 116–18
protist, 527
Nudibranches, 635
Null alleles, **277**
Null hypothesis, **12**, 438–40
Nüsslein-Vohard, Christiane, 379–81
Nutrients, **842**
availability of, in aquatic ecosystems, 995–96
cell-cycle checkpoints and cell, 205
cycling of, 1092–94
essential, for animals and humans, 842–43
essential, for plants, 738–40
plant deficiency in, 740
plant requirements for, 696
plant uptake of, 744–48
Nutrition. *See* Animal nutrition; Plant nutrition
Nutritional homeostasis, 856–58

O

Obesity, diabetes mellitus and, 857–58
Observable traits, 231–32
Observational studies, 1011–12
Occipital lobe, **901**
Oceanic zone, **1000**
Oceans
acidification of, 1100, 1114
changes in net primary productivity of, 1101–2
chemical evolution conditions in, 27, 32, 39–40, 73
deep-sea vents in (*see* Deep-sea vents)
effects of, on climate, 1010–11
global productivity patterns of, 1090
global warming and declining net primary productivity in, 1102
iron-fertilization experiments in, to increase net primary productivity, 1091–92
as marine ecosystems, 1000–1001
polymerization in early, 43
prebiotic soup in early, 38–40
protists in, 519*f*, 520*f*
time line of changes in, 483–84
upwelling of, and nutrient availability, 995–96
water properties and chemical evolution in early, 22–26 (*see also* Water)
Octopus, 636
Odonata, 641*t*
Odor, genotypes and body, 439–40
Oil consumption. *See* Fossil fuel consumption
Oils, **88**. *See also* Lipids
Okazaki, Reiji, 266–67
Okazaki fragments, **267**–68
Old World monkeys, 668–69
Oleanders, 727–28
Olfaction, **918**, 919–20
Olfactory bulb, **919**
Oligodendrocytes, **894**
Oligopeptides, **44**
Oligosaccharides, 71, 77. *See also* Carbohydrates
Omasum, 850
Ommatidia, **913**
Omnivores, 547, **612**
Omnivorous, **663**
Oncogenes, **332**–33
One-gene, one-enzyme hypothesis, **277**–78
On the Origin of Species by Means of Natural Selection (book), 414, 416, 458
Onychophora, 603*t*, 637
Oocytes, **203**, 218
Oogenesis, **953**
Oogonia (oogonium), **953**
Oomycota, 542
Oparin, Alexander I., 38
Oparin-Haldane chemical evolution theory, 38–39
Open circulatory systems, **875**

Boldface page numbers indicate a glossary entry; page numbers followed by an *f* indicate a figure; page numbers followed by *t* indicate a table.

Open reading frames (ORFs), **362**
Operators, **313**–14
Operculum, **865**
Operons, **313**–14, 331
Opisthokonts (Opisthokonta), 524*t*, 536
Opium, 775
Opsins, **915**, 917–18
Optic nerve, **914**
Optimal foraging, **1022**
Oral rehydration therapy, 854
Orbitals, electron, **17**
Organelles, **104**
as characteristic of domains of life, 497*t*
eukaryotic, 107
plant cell, 707
prokaryotic, 104
prokaryotic vs. eukaryotic, 107*t*
Organic matter, 742–43, 744*t*
Organic molecules, **33**. *See also* Carbon
carbon and formation of, 33–34
functional groups and, 34–36
organotrophs and, 506–8
in prebiotic soup, 38
six common functional groups and, 35*t*
Organismal ecology, 994
Organisms, **1**
atoms and molecules of living, 17*f* (*see also* Molecules)
atoms found in, B:8*t*
Cambrian explosion in, 486–88
cells of, 2 (*see also* Cells)
characteristics of living, 1–2, 82
characteristics of viruses vs., 676*t*
ecology and distribution and abundance of, 993
geographic distribution of, 1013–17
impact of global warming on, 1099–1101
major amino acids found in, 41*t*
misconception of higher and lower, in evolution, 430
misconception of self-sacrificing, in evolution, 430–31
model (*see* Model organisms)
multicellular (*see* Multicellularity)
number of chromosomes found in some familiar, 213*t*
organic molecules and, 36
organism level anatomy and physiology, 811*f*
taxonomy and scientific names for, 8
unicellular vs. multicellular, 6
viruses vs., 675, 676*f*
water in living cells of, 22
Organogenesis, **396**–99, 402
Organotrophs, 506–8
Organs, **809**
animal, 810–11
gas exchange, 864–70
immune system rejection of foreign, 988
male animal reproductive, 958–59*f*
organ level anatomy and physiology, 811*f*
organogenesis and, 396–99
plant vegetative, 402
transplants of, 988
viruses and human, 676*f*
Organ systems, 676*f*, **810**–11
Origin-of-life research. *See also* Chemical evolution
extremophiles and, 501
first reactions in chemical evolution, 32
nucleic acid formation in prebiotic soup, 62
origin of viruses in origin of life, 687–88
polysaccharides and, 76
on protein catalysts (enzymes) as first self-replicating molecules, 56
protein polymerization in prebiotic soup, 43
RNA as first self-replicating molecule, 59, 68–69
spark-discharge experiment on chemical evolution in prebiotic soup, 39–40
Origin of replication, **264**
Ornithine, 277
Orthoptera, 641*t*
Oryx, 803, 817
Oryza sativa, 369*t*
Oshima, Yasuji, 323–24
Osmoconformers, **824**
Osmolarity, **823**
Osmoregulation, **823**
in aquatic ecosystems, 826–28
attributes of animal nitrogenous wastes, 829*t*
cell movement of electrolytes and water in, 825–26
diffusion, osmosis, and, 823
hormones of, 940
osmotic stress and, 823–25
overview of, 831
in terrestrial insects, 828–31
in terrestrial vertebrates, 832–38
water and electrolyte balance and, 822
Osmoregulators, **824**
Osmosis, **90**, **718**, **823**
across lipid bilayers, 90–92
across prokaryotic cell walls, 105
in animal kidneys, 832
in animal osmoregulation, 831
Osmotic gradients, 835–37
Osmotic potential, **718**
Osmotic stress. *See also* Osmoregulation
diffusion, osmosis, and osmoregulation of, 823–24
in freshwater, 824
hormonal control of urine formation and, 837
nitrogenous wastes and, 830
in seawater, 824
in terrestrial ecosystems, 825
Ouabain, **826**, 830
Outbreaks, viral, 689
Outcrossing, **222**–23, **792**
Outer coverings, protist, 528–29
Outer ear, **910**
Outgroups, **477**
Out-of-Africa hypothesis, **672**
Ovalbumin genes, 322
Oval window, **910**–11
Ovaries, **393**, **561**, **563**, **789**, **933**, **953**
in female reproductive system, 959
hormones of, 936
ovarian hormones in mammalian menstrual cycle, 964–66
Overexploitation, 1016–17, 1087, 1110–11
Overwatering, 742
Oviducts, **393**, **960**
Oviparous species, **615**, **658**, **956**–57
Ovoviviparous species, **615**, **658**, **956**
Ovulation, **960**
Ovules, **402**, **561**–62, **789**–90
Ovum, **953**
Oxidation, **151**

glucose, 158, 165–66
peroxisomes and eukaryotic, 109–10
of water during photosynthesis, 181
Oxidative phosphorylation, **161**, 166
Oxygen. *See also* Oxygen gas
atomic structure of, 17*f*
deoxyribonucleotide lack of, 60
electronegativity of, 18, 22
proteins and, 46–47
redox reactions and, 151–52
simple molecules from, 19–20
Oxygen gas. *See also* Oxygen
aerobic respiration and, 166
animal gas exchange and, 861
behavior of, in air, 862–63
behavior of, in water, 863–64
betaβ-globin protein and, 247
Cambrian explosion and higher levels of, 487
cyanobacteria and, 509–10, 515
end-Permian extinction and lack of, 489–90
exchange of, between mother and fetus during pregnancy, 969
green plant production of, 547
green plants and atmospheric, 550
hemoglobin and, 45, 46
homeostasis for blood levels of, 940
molecular structure of, 21*f*
photosynthesis and, 173–74
as plant nutrient, 739*t*
Precambrian lack of, 483
prokaryotic oxygen revolution and, 509–10
regulation of blood levels of, 940
in soil, 742
in streams, 999
transport of, in blood hemoglobin, 870–73
Oxygen-hemoglobin equilibrium curve, **871**, **969**
Oxygenic photosynthesis, **181**–82, **509**, 510, 515, 547
Oxytocin, **943**, **970**
Oysters, 634
Ozone, 32

P

p53 gene, **205**, 206, 332–33
P680, 183
P700, 183
Pääbo, Svante, 345–46
Pace, Norman, 498
Pacemaker cells, **880**
Packaging, bacterial vs. eukaryotic gene expression and, 331
Paine, Robert, 1072–73
Pair-rule genes, 382
Palade, George, 119–20
Paleontologists, **481**
Paleozoic era, **483**
Palindromes, 340
Pancreas, **852**, **933**
cells of, 115
as gland, 933
human, 810*f*, 846*f*
identification of pancreatic hormones, 934–35
paracrine signals and, 930
protein digestion by enzymes of, 851–52
pulse-chase experiment on cells of, 119–20
regulation of enzymes of, 852
Pancreatic amylase, 855*t*
Pancreatic lipase, **853**, 855*t*
Pandemics, **677**–78
Papermaking, 548, 575, 590
Parabasalida, 538
Parabiosis, **938**–40
Parabronchi, 870
Paracrine signals, **930**
Paralytic shellfish poisoning, 522
Parameters, 1043
Paramylon, 539
Paranthropus, 670*t*, 671
Paraphyletic groups, **519**–20
Parapodia, 614, **633**
Parasites, **530**, **612**, **751**, **1063**
animal carriers of, 602
animals as, 612–13
ants as, 602
endosymbiont bacteria as, 514
fungal, 581, 594–95, 598
genomes of, 363
human intestinal, 538
leeches as, 633
manipulation of hosts by, 1067–68
nematode as cause of elephantiasis, 973*f*
plants as, 751
protists as, 530, 536
protostomes as, 631
roundworms as, 638
sexual reproduction and, 224–25, 533
viral, 259–60*f*
Parasitic, **586**
Parasitic sequences, 365–67
Parasitism, **1063**. *See also* Parasites
Parasitoids, **780**–81
Parasympathetic nervous system, **900**
Parathyroid glands, **933**
Parenchyma cells, **708**, 709*f*, 711*t*
Parental care, **659**
bird, 668
mammalian, 967
marsupial, 967
reptilian, 667
vertebrate, 659
Parental generation, **232**
Parietal cells, **849**–50
Parietal lobe, **901**
Parker, Geoff, 954
Parkland College, 30–32
Parsimony, **476**–77
Parthenogenesis, **630**, 631, **951**
Partial pressure, **862**–63, 864
Particulate inheritance, **234**–36
Pascal (Pa), **719**
Passive exclusion, plant, 747
Passive transport, **96**–98, **731**, **825**
Pasteur, Louis, 3–4, 498
Patch clamping, **892**
Paternal care, female choice for, 453–54
Paternal chromosomes, **213**
Paternity, 366, 956
Pathogen defense response, plant. *See* Hypersensitive response (HR)
Pathogenic bacteria, **498**
antibiotics and, 500
germ theory of disease and, 499–500
human illnesses caused by, 499*t*
Koch's postulates on, 498–99
medical importance of, 498–500
virulence of, 500
Pathogens, **707**, **776**
barriers to entry of, 974–75
humoral response coating of, with antibodies, 988
plant defense responses to, 775–78, 780*t*
Pattern component
chemical evolution theory, 15, 38
of germ theory of disease, 499
of scientific theories, 2, 414–15
theory of evolution by natural selection, 4, 416
theory of special creation, 415
Pattern formation, **379**–84
Pattern-recognition receptors, **975**
Pearl oysters, 634
Peas. *See* Garden peas
Peat, **570**
Pectins, **132**
Pedigrees, **250**, 251
Pedipalps, 642
Pedometer hypothesis on ant navigation, 10–12
Pelc, Stephen, 196
Penfield, Wilder, 902
Penicilin, 581, 598, 691
Penis, **954**, 957–59*f*
Penis worms, 627
Pentoses, 60, **72**, 73
PEP carboxylase, **188**
Pepsin, **849**, 855*t*
Pepsinogen, **849**
Peptide bonds, **43**
formation of, in protein synthesis, 300, 301–2
polypeptides and, 43–44
protein backbone structure and, 46–48
Peptide proteins, 280
Peptides, **44**, 898*t*
Peptidoglycan, 75*t*, **76**, 77, 105*f*, 496, 505
Perennials, **567**, **698**
Perfect, **790**
Perforations, **710**
Pericarp, **797**
Pericycle, **722**
Periodic table of elements, 16*f*
Periodontitis, 498
Peripheral membrane proteins, **93**–94
Peripheral nervous system (PNS), **886**, 899–900
Peristalsis, **848**
Permafrost, **1008**
Permeability, **86**–89, 92
Permian extinction, 489–90
Permineralized fossils, 480
Peroxisomes, **109**
eukaryotic, 109–10, 114*t*
in green plants, 549
Persistent organic pollutants (POPs), 1088
Petals, **409**, **562**, **788**–89
Petiole, **701**
Petrified wood, 480
Petroleum consumption. *See* Fossil fuel consumption
Petromyzontoidea, 661
pH, **25**
acid-base reactions and, 25–26
of blood, 870
effect of, on hemoglobin, 872, 873*f*
enzyme catalysis and, 55–56
hemoglobin and buffering of blood, 873–74
as measure of acidity or alkalinity of solutions, 26
in soil, 743
Phaeophyta, 543
Phagocytosis, **110–11**, 976, 987
Phanerozoic eon, 482*f*, 483
Pharmaceuticals. *See* Drugs
Pharyngeal gill slits, **650**
Pharyngeal jaws, **656**, 844–45
Pharyngeal pouches, 651
Phelloderm, **713**
Phenetic approach, **475**
Phenology, **1100**
Phenotypes, **231**
central dogma and linking genotypes and, 280–81
codominance in, 245–47
disease genes and improved understanding of, 350
genotypes vs., 234
incomplete dominance in, 245
in Mendelian genetics, 239*t*
mutations and, 285
natural selection and, 440
plasticity in (*see* Phenotypic plasticity)
polygenic inheritance and, 248–50
predicting, with Punnett square, 235–36
ratio of, and segregation principle, 235
Phenotypic plasticity, **698**
in leaves, 702–3
in root systems, 698
in shoot systems, 700
Phenylalanine, 42*t*, 248, 283
Phenylketonuria (PKU), **248**
Pheophytin, **180**–81
Pheromones, animal, **907**, 930*t*, **931**, **954**
Pheromones, plant, **780**–81
Phloem tissue, **710**
anatomy of, 729–30
loading of, in translocation, 731–33
primary plant body, 711*t*
unloading of, in translocation, 733–34
vascular cambium and, 713
vascular tissue system functions of, 711
Phoronida, 603*t*
Phosphatases, 144
Phosphate, 189
Phosphate functional group, 35, 60, 62, 84, 97–98
Phosphodiesterase (PDE), 916
Phosphodiester linkage, **61**, 62, 66, 68, 260
Phosphofructokinase, **156**
Phospholipids, 82*f*, **84**, 496. *See also* Lipid bilayers; Lipids
Phosphorus
in DNA, 196, 259
as human nutrient, 843*t*
as plant limiting nutrient, 738
as plant nutrient, 739*t*
Phosphorylase, **79**, **946**
Phosphorylated molecules, **62**
Phosphorylation, **149**
post-translational control by, 330
as post-translational modification, 304
of proteins by ATP, 149–50
substrate-level, 155–56
Phosphorylation cascades, **144**, **756**
enzyme-linked receptors and, 143–44
hormone signal transduction cascades as, 946–47
Photic zone, **997**, **1000**–1001
Photons, **32**, **175**
Photoperiodism, **787**
Photophosphorylation, **181**
Photoreceptors, **908**
plant, 758
vertebrate, 914, 915, 917–18

Photorespiration, **187**
Photoreversibility, **763**
Photosynthesis, **78**, **172**–93, **509**
autotrophs, heterotrophs, and, 172
capturing of light energy by chlorophyll in, 174–79
carbohydrate storage of energy from, 78
chloroplasts in, 112–13
discovery of photosystems I and II of, 179–84
experiment on enhancement effect of red and far-red light in, 179–80
ground tissue systems and, 708
leaves and, 699
pigments, 532*t*
prokaryotic, 508–9
prokaryotic internal membranes and, 104, 107*t*
protist, 529–32, 536
redox reactions and, 151
reduction of carbion dioxide to produce glucose in, 184–90
regulation of, 189
requirements for, 696
use of sunlight to make carbohydrates in, 172–74
water conservation and, 727
Photosynthesis-transpiration compromise, 727
Photosystem, **178**
Photosystem I and II, **179**–84
Phototrophs, **506**
Phototropins, **758**–59
Phototropism, **758**
blue-light, 758–62
red and far-red light, 762–74
Phycocyanin, 532*t*
Phycoerythrin, 532*t*
Phyla (phylum), **7**, **498**, **603**, 646–47*f*
Phylogenetic species concept, **460**–62
Phylogenetic trees, **474**–79. *See also* Phylogenies
analyzing rRNA to create, 6–7
distinguishing homology from homoplasy with, 475–77
estimating phylogenies with, 475
mapping land plant evolutionary changes on, 555–56
reading, B:4–B:6
studying animals with, 607–9
studying fungi with, 584–86
studying green plants with, 551–52
studying prokaryotes with, 503–4
studying protists with, 525
taxonomy and, 7–8
terminology of, 474–75
tree of life as, 5–7 (*see also* Tree of life)
using, to understand emerging viruses, 688–89
whale evolution as case history for, 477–79
Phylogenies, **6**, **102–3**, **420**, **474**
of animals, 607–9
based on DNA sequence data, 477–78
based on morphological traits, 477
of cetacean whales and dolphins, 422
of chordates, 654*f*
of current human populations, 672*f*
distinguishing homology from homoplasy in, 475–77
domains and, 102–3
endosymbiosis and, 528
of fungi, 584–86
of green plants, 551–52
of HIV, 688*f*
of mockingbirds, 420
of prokaryotes, 503–4
using phylogenetic trees to estimate, 475
vertebrate, 655
of vertebrate circulatory systems, 878*f*
of vertebrates, 654*f*
whale evolution as case history for, 477–79
Physical activity, diabetes mellitus and, 857–58
Physical isolation, 462–64
Physical maps, **348**
Physiology, **803**, 810–11*f*. *See also* Animal anatomy and physiology; Plant anatomy and physiology
Phytoalexins, **777**
Phytochromes, **763**–64, 787
Phytophthora infestans, 520–21, 522*t*, 542
Phytoplankton, 523
Pigments, **174**, **758**
absorption of light by photosynthetic, 175–77
lipids as, 84
in mice, 280–81
in photosynthetic protists, 532*t*
prokaryotic, 509
thylakoid membranes and, 174
in vacuoles, 112
Piloting, **1026**
Pima Indians, 857–58
Pines, 575
Pine trees, 560–61
Pinocytosis, **111**
Pinophyta, 575
Pinworms, 638
Pioneering species, **1075**
Pisum sativum. *See* Garden peas
Pitcher plants, 751–52
Pitches, **909**
Pith, **706**
Pits, **710**
Pituitary dwarfism, 338–44
Pituitary gland, **933**, 937, 940–43, 964–66
Placental mammals, **666**
Placentas, **390**, **393**–94, **658**–59, **968**
Placozoa, 603*t*
Planar bilayers, 86
Plankton, **997**, **1071**
ciliates as, 540
community structure of, 1071
ctenophores as, 620
diatoms as, 542
euglenids as, 539
green algae as, 566
protists as, **522**–23
rotifers as, 631
Plant(s)
bodies (*see* Bodies, plant)
cells of (*see* Plant cells)
deceptive communication in, 1030
defense responses (*see* Defense responses, plant)
development of (*see* Plant development)
form and function of (*see* Plant anatomy and physiology)
fungal mutualisms with, 579–80
global warming and, 1100
green algae and (*see* Green plants)
land plants (*see* Land plants)
meiosis and gametes of, 218
mutualisms with, 1068
nutrition for (*see* Plant nutrition)
oldest known, 786
photosynthesis in (*see* Photosynthesis)
Plantae lineage and, 524*t*, 536
reproduction of (*see* Plant reproduction)
sensory systems, signals, and responses of (*see* Sensory systems, plant)
speciation by polyploidization in, 467–68
sugar transport (translocation) in, 728–34 (*see also* Translocation)
tissue cultures, B:18
transgenic, 354–55 (*see also* Golden rice)
viruses and, 681, 691, 692
water transport in, 717–28 (*see also* Water transport, plant)
Plantae, 524*t*, **536**. *See also* Plant(s)
Plant anatomy and physiology, 695–716
brassinosteroids and plant body size, 773
flowers, 786–92
indeterminate growth and angiosperm, 695–96
leaves in, 701–3
primary growth in, 704–6
primary plant body cells and tissues, 706–12
primary plant body components, 711*t*
root systems in, 697–99
secondary growth components, 713*t*
secondary growth in, 712–15
shoot systems in, 699–701
surface area/volume relationships in, 696–97
Plant cells
animal cells vs., 106*f*, 706–7
cell-cell attachments of, 135
cellulose as structural polysaccharide for, 76
cell walls in, 132–33
cytokinesis in, 200
cytoplasmic streaming in, 124
differentiated, as genetically equivalent, 377
eukaryotic, 106*f*, 115–16 (*see also* Eukaryotic cells)
glyoxysomes and peroxisomes in, 110
middle lamella in cell-cell attachments of, 135
plasmodesmata in, 138–39
starch as storage polysaccharide in, 168
vacuoles in, 111–12
Plant development, 401–13
animal development vs., 401
embryogenesis in, 404–6
gametogenesis, pollination, and fertilization in, 402–4
overview of, 402*f*
reproductive development in, 409–11
vegetative development in, 407–9
Plantlets, **786**
Plant nutrition, 737–54
adaptions for, 750–52
deficiencies in, 740–41
essential nutrient elements and availability, 739*t*
essential nutrients, 738–40
importance of, 717
macronutrients, 738, 739*t*
micronutrients, 739*t*, 740
nitrogen fixation and, 748–50
nutrient uptake and, 744–48
nutritional requirements and, 738–41
soil and, 741–44*f*
soil composition, soil properties, and, 744*t*
starch as storage polysaccharide for, 74
Plant reproduction, 783–802
alternation of generations in, 556–58
asexual reproduction, 786
embryophyte retention of offspring in, 556
evolution from gametophyte-dominant to sporophyte-dominant life cycles in, 558–59
fertilization in, 794–95
flowers as reproductive structures of, 786–92
flowers in, 561–62
fruits in, 563–64
gamete production in protected structures in, 556
heterospory in, 559, 560*f*, 561*f*
importance of, 783
land plant life cycle and, 784–86
land transition adaptations for, 556–64
pollen in, 559–60
pollination in, 562–63, 792–94
seeds in, 560–61, 795–800
sexual, 784–86
Plasma, **870**
Plasma cells, **986**, 988, 989*t*
Plasma (cell) membranes, **82**–101, **102**
of Archaea, 496
cell walls and, 113
crossing of, by hormones, 934
diffusion and osmosis across, 89–92
eukaryotic, 114*t*
excitable, 891
exocytosis and, 122
gamete fusion and, 390
importance of, to cells, 1, 2, 82, 98–99
internal, of photosynthetic prokaryotic cells, 104, 107*t*
lipid bilayers as, 85–89
lipids in, 82*f*, 83–85
lipids in, as characteristic of domains of life, 497*t*
lipid-soluble vs. lipid-insoluble hormones and, 140
membrane proteins and fluid-mosaic model for, 92–99
in mitosis, 200
nuclear envelope and, 526
prokaryotic, 104
proton pumps in, 733
selectively permeable, 823
smooth ER and, 108
transport of solutes across, in plants, 731–32
Plasmids, **103**, **340**
in genomes, 364
introducing recombinant, into bacterial cells for transformation, 341

Boldface page numbers indicate a glossary entry; page numbers followed by an *f* indicate a figure; page numbers followed by *t* indicate a table.

origin of viruses in, 686–87
prokaryotic, 103–4
TI (tumor-inducing), 355–56
uses of, in genetic engineering, 352*t*
using, in DNA cloning, 340
Plasmodesmata (plasmodesma), **138**–39, **568**, **707**
Plasmodial slime molds, 537
Plasmodium falciparum, 369*t*
Plasmodium sp., 521, 522*t*, 541, 1066–67
Plasmogamy, **591**, 592
Plastocyanin, **183**
Plastoquinone (PQ), **180**–81, 183
Platelet-derived growth factor (PDGF), 207
Platelets, **207**, **870**, **975**
Plato, 415
Platyhelminthes, 603*t*, 609, 631–32
Platypuses, 665
Pleiotropic genes, **247**, 250*t*, 431
Ploidy, 212–**13**, 214*t*
Pneumatophores, 698, 699*f*
Pneumonia, 677
Podia, **648**
Point mutations, **285**–86*t*
Poisons
in action potential research, 892
plant, 775
Polar bears, 1100
Polar bodies, **953**
Polar climate, 1010
Polar covalent bonds, **18**, 19*f*
Polarity
amino acid side chain, 42
covalent bond, 18
of water, 22–23
Polar microtubules, **198**
Polar nuclei, **790**
Polar side chains, 41*t*
Polar transport, **768**
Polio virus, 692, 990
Pollen grains, **231**, **402**, **560**, **791**
fossil record and, 480
germination of, 794
land plants and, 559–60
pollen-stigma interactions and, 402–3
seed plants and, 567
stamens and production of, 789, 790–92
Pollen tubes, **403**, **794**–95
Pollination, **562**, **792**–94
of crops, 1117
evolution of, 793–94
flower petals and, 788
in gametogenesis, 402–3
land plants and, 562–63
selfing vs. outcrossing in, 792
speciation and, by animals, 794
syndromes, 792–93
Pollinators, 562
Pollution
bioremediation of, 500–501, 598, 1117
ecosystem ecology and, 995
nitrate, 749
prokaryotes and nitrate, 511–12
soil, 747
stoneworts and water, 569
as threat to biodiversity, 1111
Polygenic inheritance, **249**
exceptions and extensions to Gregor Mendel's rules on, 250*t*
of quantitative traits, 248–50
in situ hybridization and, 428–29
Polymerase chain reaction (PCR), **344**–46
amplification of fossil DNA with, 345–46
direct sequencing with, 502–3
DNA fingerprinting and, 366
extremophiles and, 501
requirements for, 344–45
uses of, in genetic engineering, 353*t*
uses of, in genetic testing, 350
Polymerization, **42**
of amino acids to form proteins, 42–44
chemical evolution and, of monosaccharides, 77
of monosaccharides to form polysaccharides, 73–74
of nucleotides to form nucleic acids, 61–62
of nucleotides to form nucleic acids and lipids, 82
Polymers, **42**, 71, 73–76
Polymorphic, **247**, **348**
Polymorphism, 250*t*
Polypeptides, **43**. *See also* Proteins
as hormones, 933, 936
peptide bonds and formation of, 43–44
translation and extending of, 301–2
Polyphenylalanine, 283
Polyplacophora, 636
Polyploidy, **213**, **286**, **466**
cells, 214*t*
chromosome-level mutations and, 286
sympatric speciation and, 471*t*
sympatric speciation by, 465–68
Polyproteins, 682
Polyps, **617**
Polyribosomes, **297**
Polysaccharides, 71, **73**–**76**, 113, 132–34. *See also* Carbohydrates
Polyspermy, **391**–92
Poly(A) tail, RNA, **295**
Ponds, 997
Poor-nutrition plant tissues, 1066
Population(s), **4**, **416**, **435**, **994**, **1037**. *See also* Population dynamics; Population ecology
bottlenecks in, 446
causes of genetic drift in natural, 445–46
contemporary evolution of, 418
contemporary speciation in, 422
evolutionary processes and evolution of, 435
evolution of, 414*f*, 429–30 (*see also* Evolution; Natural selection)
evolution of, by natural selection, 4, 414
extinction of (*see* Extinction)
gene flow in natural, 447
genetic drift and small, 444
genetic isolation of, 462–64
global warming and evolutionary change within, 1101
growth of (*see* Population growth)
mating between isolated, 468–71
monophyletic, 460–61
outcomes of secondary contract between isolated, 471*t*
population viability analysis for, 1115
secondary contact between isolated, and speciation, 468–71
speciation in (*see* Speciation)
stabilizing size of human, 1121
theory of evolution by natural selection and, 416
Population cycles, 1047–50
Population density, 952, 1021, **1042**
Population dynamics, **1046**–53
human population growth rate changes, 1052–53
mark-recapture studies of, 1048
metapopulation changes, 1046–47
milestones in human population growth, 1052*t*
population age structure and population growth, 1050–52
population cycles, 1047–50
Population ecology, 994, **1037**–57
demography and, 1037–41
endangered species analysis and, 1053–55
human population size and, 1037
population dynamics and, 1046–53
population growth and, 1041–46
preserving metapopulations, 1055
Population growth, 1041–46
age structure and, 1050–52
calculating rates of, using life tables, 1040
changes in rates of, in human populations, 1052–53
developing and applying equations for, 1043–44
exponential, 1042
limiting factors on, 1044–46
logistic, 1042–44
milestones in human, 1052*t*
momentum in, 1051–52
projections of, 1054–55
quantifying rates of, 1041–42
Population momentum, 1051–52
Population thinking, **416**
Population Viability Analysis (PVA), 1115
Pores, 94–95, **554**, **708**
Porifera, 603*t*, 618
Porin, 45
Positive control, **312**, 322. *See also* Catabolite repression
Positive feedback, **892**, 1099
Positive pressure ventilation, **868**
Positive-sense viruses, **686**, 687*t*, 692
Post-anal tails, 650
Posterior, **379**
Posterior pituitary, **941**, 943
Postsynaptic neurons, **896**
Postsynaptic potentials, 897–99
Post-transcriptional control, eukaryotic, 328–30
Post-translational control, **308**
bacterial vs. eukaryotic gene expression regulation and, 308–9, 331*t*
eukaryotic, 330
Postzygotic isolation, **459**
Potassium. *See also* Sodium-potassium pump
action potentials and ions of, 891
as plant limiting nutrient, 738
as plant nutrient, 739*t*
potassium leak channels, 889
Potassium channels, 891, 892
Potatoes, 116, 520–21, 542
Potato famine, 522*t*
Potential energy, **27**. *See also* Energy
aerobic respiration and, 166
of ATP, 148
chemical energy as, 33
energy transformations of, 27–29
phosphate functional group and, 62
spontaneous chemical reactions and, 29–30
water potential and, 718
Pox viruses, 690
Prairie dogs, 1033
Prebiotic soup, **38**. *See also* Chemical evolution; Deep-sea vents; Origin-of-life research; V olcanic gases
chemical evolution in, 38–39
first membranes in, 91
nucleic acid formation in, 62
origin-of-life research on chemical evolution in, 39–40
polymerization in, 43
protein catalysts in, 56
Precambrian eon, 481–**83**
Precipitation
biomes and, 1002
global climate change experiments on, 1012–13*f*
in tropical wet forests, 1009
Predation, **1063**. *See also* Predators
Cambrian explosion and evolution of, 487–88
deceptive communication and, 1030
fungal, 598
of herbivore populations, 1065–66
population cycles and, 1048–50
Predators, **612**. *See also* Predation
animal, 612
deuterostomes as, 646*f*
efficiency of, at reducing prey populations, 1065
mollusk, 636
noxious compounds in vacuoles and, 112
predator-removal programs and, 1065
sharks as, 662
top, 1087
Predictions, **3**, 12, 231, 235–39
Prefixes, metric system, B:1*f*
Pregnancy, 967–71
birth events, 970–71
events in human, 967–68
fetal nourishment by mothers in, 968–70
gestation and early development in marsupials, 967
prenatal genetic testing and human, 350
preventing human, 966
Prenatal testing, 350
Preservation
biodiversity, 1120–23 (*see also* Conservation biology)
fossil, 480
Pressure-flow hypothesis, **730**–31
Pressure potential, **719**–20
Pressure-sensing systems. *See* Hearing
Presynaptic neurons, **897**
Preventive medicine, 685–86
Prey, 1063–65. *See also* Predation
Prezygotic isolation, **459**
Priapula, 603*t*
Priapulids, 627
Pribnow, David, 291
Primary cell walls, **132**–33, **555**
Primary consumers, 523, **1084**
Primary decomposers, **1085**
Primary endosymbiosis, 531–32
Primary growth, plant, **704**–6
Primary immune response, **989**
Primary meristems, 704–5
Primary oocytes, **953**
Primary phloem, 713
Primary plant bodies, 704–5. *See also* Bodies, plant

Primary producers, **522, 1083**
ecosystem energy flow and, 1083–84
green algae, 566, 568
green plants as, 547
protist, 541
Primary root system organization, 705–6
Primary sex determination, 936
Primary shoot system organization, 706
Primary spermatocytes, **953**
Primary structure, **46**
DNA, 260–61
DNA vs. RNA, 67*t*
hemoglobin protein, 49*t*
nucleic acid, 61
protein, 46
RNA, 66
tRNA, 299
Primary succession, **1074**
Primary transcripts, **293**
Primary xylem, 713
Primases, **266**, 268*t*
Primates, **669**
characteristics of, 669
fermentation in digestion by, 854
great apes as hominid, 669–70
humans as, 670–73*f*
lineages of, 668–69
Primer annealing, 344–45
Primers, **266**, 344–45
Primroses, 1050–51
Principle of independent assortment, **238**
dihybrid crosses and, 236–38
genetic variation from, 221–22
linked genes as exception to, 243
meiosis and, 240–41
Principle of segregation, **235**, 240
Prions, 50–**51**, 339
Probabilities, 1032, B:19
Probes, DNA, **342**–43, 353*t*, 370–71
Probes, nucleic acid and antibody, B:13
Proboscis, 563, 611, **627**
Procambium, **705**, 713, **796**
Process component
chemical evolution theory, 15, 38
of germ theory of disease, 499
of scientific theories, 2, 414–15
theory of evolution by natural selection, 4
theory of special creation, 415
Processes, evolutionary. *See* Evolutionary processes
Processing, signal. *See* Signal processing, cell-cell
Producers, primary. *See* Primary producers
Productivity, **996**
global patterns in, 1089–90
ocean, 1001
pyramid of, 1086 (*see also* Net primary productivity (NPP))
species richness and, 1080
Products, **27**
Progesterone, **964**–66
Programmed cell death, 375*t*, **376**. *See also* Apoptosis
Prokaryotes, **6**, 496–516
asexual reproduction in, 223
Bacteria and Archaea domains as, 102–3, 496–97 (*see also* Archaea; Archaea domain; Bacteria; Bacteria domain)
bacteria that cause human illnesses, 499*t*
biological impact of, 497–98
biological methods of studying, 501–4
cell structures and functions of, 102–5 (*see also* Prokaryotic cells)
characteristics of eukarya and, 497*t*
diversification themes of, 504–12
ecological diversity of, and global change, 509–12
environmental sequencing in, 364
eukaryotes vs., 519
as extremophiles, 501
fermentation and, 167
genomes of, 361–65
key lineages of, 512–16
lateral gene transfer in, 364
medical importance of, 498–500
metabolic diversity of, 506–9
morphological diversity of, 504–6
natural history of genomes of, 363–64
phylogenetic tree of life and, 6–7
reasons for studying, 497–501
role of, in bioremediation, 500–501
six examples of metabolic diversity in, 507*t*
six general methods of, for obtaining energy and carbon-carbon bonds, 506*t*
some electron donors and acceptors of, 508*t*
Prokaryotic cells, 102–5. *See also* Cells
bacterial cell-cell communication as quorum sensing in, 145–46
bacterial flagella vs. eukaryotic flagella, 127
cell walls as exoskeletons of, 104–5
chromosomes in nucleoids of, 103–4
cytoskeletons of, 104
eukaryotic cells vs., 105–6*f*, 107*t*
flagella and swimming of, 104
improvements in microscopy and, 103
internal membrane complexes of photosynthetic, 104
organelles in, 104
plasma membranes of, 104
protein manufacture by ribosomes of, 104
quorum sensing as cell-cell communication in, 145–46
views of, 103*f*, 105*f*
Prolactin, **943**
Proliferation, cell, 375
Proline, 42*t*, 47, 283
Prometaphase, 198–**99**
Promoter-proximal elements, 323–**24**
Promoters, **291, 323**
catabolite repression and, 315
eukaryotic, 320
regulatory sequences far from, 324–26
regulatory sequences near, 323–24
transcription and, 291–92
in transcription in bacteria and eurkaryotes, 296*t*
Proofreading, DNA, 271–**72**
Propagation, plant, 708
Prophase, **198**
Prophase, mitosis, 198
Prophase I and II, meiosis, 218–20
Prop roots, 698, 699*f*
Propulsion, seed dispersal by, 798
Prosimians, **668**–69
Prostaglandins, 140*t*, 959*t*
Prostate glands, **958**, 959*t*
Protease inhibitors, 682–83
Proteases, 118, **391, 682, 852**
Proteasomes, 204, **330**
Protected areas, 1055, 1120
Proteinase inhibitors, **779**
Protein filaments, 104, 113
Protein kinases, **142, 204**
Proteins, 38–58
adhesion, 136–38
animal digestion of, 846, 848, 851–52
as animal nutrient, 842
animal stomach digestion of, 849
artificial selection of maize for protein content, 4–5
basal transcription factors, 291–92
betaβ-globin, 247
carbohydrates and glycoproteins, 77
catabolic pathways and, 168–69
catalysis by enzymes as, 45, 51–56 (*see also* Catalysis; Enzymes)
cell-cycle regulation and, 204–5
cell functions of, 45
in central dogma, 280–81
channel (*see* Channel proteins)
chemical evolution in prebiotic soup and formation of, 38–39
cytoskeleton, 104, 113
defined, **42, 44**
destruction of, as regulation mechanism, 204
early origin-of-life experiments on chemical evolution of, 39–40
eukaryotic genomes and genes for coding, 363
filament, 125
as first self-replicating molecules, 56
folding of, 50–51
folding of ribonuclease, 50
four levels of structure of hemoglobin, 49*t*
genes and, 195–96
immunoglobulins (Igs), 980, 982–83*f*
manufacture and transport of, by endomembrane system, 118–23*f*
manufacture of, by prokaryotic ribosomes, 104
membrane (*see* Membrane proteins)
microtubules, vesicle transport, and motor, 126–27
motor, 45, 124, 125–27, 128
movement of, from ER to Golgi to destination, 122–23*f*
one-gene, one-polypeptide hypothesis on, 278
organic molecules and, 36
phosphorylation of, by ATP, 149–50
plant embryogenesis and, that set up body axes, 406
polymerization of amino acids to form, 39
polymerization of amino acids to form polypeptides and, 40–44 (*see also* Amino acids; Polypeptides)
primary structure of, 46
prions as infectious, 50–51
proteomics as study of, 371
proton pump, 733
quaternary structure of, 48–50
regulatory, in differential gene expression, 326
regulatory, in eukaryotic transcription initiation complex, 326–28
repressor, 312–14
required for DNA synthesis, 268*t*
ribosomes and, 109
RNA polymerase for, 290
rough ER and, 108
secondary structure of, 46–47
selective adhesion and adhesion, 131*f*, 136–38
sigma, 291–92
signal hypothesis on entry of, into endomembrane system, 120–21
signal transduction via G proteins, 142–43
structure of, 45–51
synthesis of (*see* Protein synthesis)
targeted destruction of, 330
tertiary structure of, 47–48
in transcription in bacteria and eurkaryotes, 296*t*
transfer of amino acids to, by transfer RNA, 297–99
transport (*see* Transport proteins)
transport of molecules into nucleus and, 117–18
tumor suppressors, 208
in vegetative development that determine leaf shape, 408
viral, 117–18
Protein synthesis, 289–306. *See also* Genetics
central dogma of molecular biology on, 279–82
electron micrograph of transcription, 289*f*
gene expression and, 289 (*see also* Gene expression)
genetic code and, 282–85
ribosomes in, 300–304
RNA processing in eukaryotic, 293–95
transcription in, 289–93
transfer RNA in, 297–300
translation in, 295–97
Proteobacteria, 509, 512, 515
Proteomes, **371**
Proteomics, **371**. *See also* Genomics
Protists, **519**–45
animals and, 607–8, 617
biological methods for studying, 524–26
diversification themes in, 526–36
ecological importance of, 522–23
Eukarya domain and, 519–20 (*see also* Eukaryotes)
feeding and photosynthesis innovations of, 529–32
human health problems caused by, 522*t*
impacts of, on human health and welfare, 520–22
key lineages of, 536–43
life cycles of, 533–36
major lineages of eukaryotes and, 524*t*
morphological innovations of, 526–29
movement of, 532–33
pigments in photosynthetic, 532*t*
reasons for studying, 520–23
reproduction of, 533
Proto-cells, 82
Protoderm, **705, 796**

Boldface page numbers indicate a glossary entry; page numbers followed by an *f* indicate a figure; page numbers followed by *t* indicate a table.

Proton gradients
and ATP production in chemiosmosis, 162–63
nutrient uptake and, 745–46
during photosynthesis, 180–81
Proton-motive force, **162**
Proton pumps, **733**, **745**, **761**
in cell elongation, 761–62
in guard cell closure, 772
location of, in translocation, 733
in photosynthesis, 187
in translocation, 732–33
Protons
acid-base reactions and, 25–26
atomic structure and, 16–17
Proto-oncogenes, **332**
Protostomes, 603*t*, **606**, 623–45. *See also* Animal(s)
abundance of, 623–24
body plan variations of, 626–27
ecdysozoan lineages of, 637–43
ecdysozoans, 625
evolution of, 624–25
feeding adaptations of, 628
lophotrochozoan lineages of, 630–36
lophotrochozoans, 624–25
movement adaptations of, 628–30
prominent orders of insects as, 640*t*–41t
reproduction adaptatoins of, 630
water-to-land transition of, 627–28
Proximal tubule, 832, **834**–35, 838*t*
Proximate causation, **952**, **1020**
Prusiner, Stanley, 50–51
Pseudogenes, **368**, 418, 444
Pseudomonas aeruginosa, 145
Pseudopodia, **530**, 532–33
Psilotopohyta, 572
P site, 300
Pteridophyta, 573
Pterosaurs, 657
Puberty, **936**, **961**–62, 963*t*
Puffballs, 583, 593
Pulmonary arteries, **879**
Pulmonary circulation, **878**
Pulmonary veins, **879**
Pulse-chase experiments, **119**
on Calvin cycle, 185–86
on cell cycle gap phases, 196
on discontinuous replication, 266–67
on photosynthesis, 188
on ribosomes as protein synthesis site, 296
on secretory pathway, 119–20
Pumps, **98**, **732**. *See also* Sodium-potassium pump
active transport by, 97–99
in phloem loading, 732
Punnett, R. C., 248
Punnett squares, **235**
Hardy-Weinberg principle and, 436–37*f*
predicting genotypes and phenotypes with, 235–39
Pupae (pupa), **616**, 936
Pupil, **914**
Pure lines, **231**
crosses between, 234–36
in Mendelian genetics, 239*t*
Purifying selection, 224, **440**
Purines, **60**, 279
chemical evolution and, 61
DNA and, 62–64
nucleotides and, 60
RNA and, 66–67
Purple nonsulfur bacteria, 180
Purple sulfur bacteria, 173, 180
Pyramid of productivity, 1086
Pyrethroids, 775
Pyridoxine, 277
Pyrimidines, **60**
chemical evolution and, 60–61
DNA and, 63–64
nucleotides and, 60
RNA and, 66–67
Pyrosequencing technologies
dideoxy sequencing vs., 347
impact of, on whole-genome sequencing, 360–61
Pyruvate
in cellular respiration, 153
in fermentation, 166
processing glucose to, 155–56
processing of, to acetyl CoA, 156–58
Pyruvate dehydrogenase, **157**

Q

Quadrats, 1048
Quantitative methods. *See also* Biological methods
calculating coefficients of relatedness, 1032
calculating equilibrium potentials with Nernst equation, 888
calculating population growth rates using life tables, 1040
developing and applying population growth equations, 1043–44
extrapolation techniques, 1108
linkage and, 246*f*
mark-recapture studies, 1048
measuring species diversity, 1079
population viability analysis, 1115
Quantitative traits, **248**–50*t*
Quaternary structure, **48**
DNA vs. RNA, 67*t*
hemoglobin protein, 49*t*
protein, 48–50
RNA, 67
Quinones, **161**–62, 163–64
Quorum sensing, bacterial, **145**–46

R

Rabies virus, 692
Racker, Efraim, 164
Radial symmetry, **604**
Radiation, **816**
cancer and, 332–33
creating mutants with, 277
DNA damage and, 272–74
electromagnetic, 174–75
sunlight (*see* Light energy)
Radicles, **796**
Radioactive decay, 417, 481
Radioactive isotopes, B:12
Radiometric dating, 417, 481
Radula, **612**, **635**
Rain forests. *See* Tropical wet forests
Rain shadow, **1010**
Ramón y Cajal, Santiago, 886–87
Rancher ants, 602
Random mutations
DNA and, 283–84
RNA and, 69
Ranges, **1013**, 1072, 1120. *See also* Geographic distribution of organisms
Ras protein, **143**–44, 206
Rats, 369*t*
Ray-finned fishes, 656, 662–63
Rayment, Ivan, 924
Rays, 662, **713**
Rb protein, **208**
Reabsorption mechanisms, 831, 834–35
Reactants, **27**, 30–32
Reaction centers, **178**–79. *See also* Photosystem I and II
Reactions, chemical. *See* Chemical reactions
Reactive oxygen intermediates (ROIs), **777**
Reactivity, amino acid side chain, 40–42
Reading frame, **283**, 362
Realized niches, **1061**
Real-time processes, B:15
Receptacle, plant flower, 788
Receptor-mediated endocytosis, **111**
Receptors
animal B-cell and T-cell, 980–81*f*
animal chemical signal, 930–31
animal hormone, 943–47
animal sensory, 817–18, 886, 908
cell-cell signal, 140, 143–44
plant sensory, 756, 757*f*, 758–59, 763
Receptor tyrosine kinases (RTKs), **143**–44, 207
Recessive alleles
loss-of-function mutations and, 451
in Mendelian genetics, 239*t*
Recessive traits, **233**
autosomal diseases, 272–74
identifying human traits as, 250–51
monohybrid crosses and, 232–33
Reciprocal altruism, **1033**–34
Reciprocal crosses, **234**
in Mendelian genetics, 239*t*
monohybrid crosses and, 233–34
results of, 234*t*
sex linkage and, 241–42
Recolonization, metapopulation, 1046–47
Recombinant, **244**
Recombinant DNA technology, **338**–44. *See also* Biotechnology; Genetic engineering
creating animal models of disease using, 350
discovery of gene recombination, 982–83*f*
ethical concerns over recombinant human growth hormone, 343–44
failure of efforts to treat pituitary dwarfism and, 339
gene therapy and, 351–54
human growth hormone in pituitary dwarfism and, 339
introducing recombinant plasmids into bacterial cells for transformation, 341
mass producing human growth hormone, 343
producing a cDNA library, 342
screening a DNA library, 342–43
steps in engineering human growth hormone, 339–43
transgenic crops and, 354–56
uses of, in genetic engineering, 353*t*
using plasmids in DNA cloning, 340, 341*f*
using restriction endonucleases and DNA ligase to cut and paste DNA, 340–41
using reverse transcriptase to produce cDNAs, 339–40
Recombination frequencies, calculating, 246
Recommended Dietary Allowances (RDAs), 842
Rectal gland, **826**
Rectum, 846*f*, **855**
Red algae, 519*f*, 536, 539. *See also* Algae
Red blood cell ghosts, 96–97
Red blood cells, **870**
Red/far-red light responses, plant, 762–74
germination and, 763*t*
isolation of phytochromes of, 763–64
overview of, 780*t*
photosynthesis and, 762
phytochromes as receptors in, 763
red/far-red switch in, 763
Red-green color blindness, 916–17
Red light, 174–75, 176, 799. *See also* Red/far-red light responses, plant
Redox reactions, **151**
hydrogen, oxygen, and, 151–52
oxidation of glucose and, 152
in photosynthesis, 178–79
photosynthesis and, 151
prokaryotic nitrogen fixation and, 510
Reduction, **151**
Reduction division, meiosis as, 215–16
Reductionism, 899
Reduction-oxidation reactions. *See* Redox reactions
Reduction phase, Calvin cycle, 186
Redundancy
genetic code, 283
genome, 363
Redwoods, 575
Reflexes, **848**
Refractory, **893**
Regeneration phase, Calvin cycle, 186
Regulation. *See also* Hormones
of animal body temperature, 816–19
of animal homeostasis, 814–16
of animal hormone production, 940–43
of animal pancreatic enzymes, 852
of blood pressure and blood flow, 882–83
cancer as loss of cell-cycle, 206–9
of cell cycle, 202–6
of citric acid cycle, 158–59*f*
of enzymes by regulatory molecules, 54–55
of flower structures, 409–11
of gene expression (*see* Gene regulation, bacterial; Gene regulation, eukaryotic)
of glucose in glycolysis, 156
homeostatic, of ventilation, 870
kidney, 832
of mammalian menstrual cycle, 962–66
of mammalian puberty, 961–62
membrane channel, 96
of photosynthesis, 189
plant growth (*see* Growth regulators, plant)
regulatory genes and (*see* Regulatory genes)
regulatory proteins and (*see* Regulatory proteins)
regulatory sequences and (*see* Regulatory sequences)
role of, in homeostasis, 815–16 (*see also* Homeostasis)
of transcription, 291
Regulatory cascades, 382–83, 411

Regulatory genes, **369**
  bicoid gene as master regulator, 379–80
  for cell determination, 398–99
  evolutionary conservation of, 383
  reuse of, in differing developmental contexts, 383–84
  specific developmental positional information and, 381–83
Regulatory molecules
  cell-cycle, 203–4
  enzyme regulation by, 54–55
Regulatory proteins
  in bacterial gene regulation, 308
  in differential gene expression, 326
  in eukaryotic transcription initiation complex, 326–28
  intiation complex for, 326–28
  lac operon model for repressors, 312–14
  in transcription, 291–92
Regulatory sequences, **323**–26, **369**
  transcription initiation and, far from promoters, 324–26
  transcription initiation and, near promoters, 323–24
Regulatory sites, **156**, 314
Regulatory transcription factors, **326**, **378**
  bicoid gene as, 379–80
  cancer and, 332–33
  human and chimpanzee genome similarity and, 369–70
Reinforcement, **468**, 471*t*
Reintroduction programs, 1122
Relative dating, 422
Release factors, **302**–3
Religious faith vs. science, 8–9
Renal arteries, 832, 833
Renal corpuscle, **832**
  filtration by, 832–34
  structure and function of, 838*t*
Renal veins, 832
Reovirus, 691
Repair, DNA, 271–74
Repeated sequences, 365–67
Repetition, scientific experiments and, 12
Replacement mutations, **286**
Replacement rate, **1053**
Replica plating, **310**
Replica plating, finding mutant genes with, 310–11
Replicated chromosomes, 213–14*t*
Replication. *See also* Reproduction; Self-replicating molecules
  chromosome, 195–96
  DNA, 65, 261–63 (*see also* DNA synthesis)
  living organisms and, 2
  RNA, 67–68
  viral, 679–86
Replication bubble, 264
Replication fork, **264**–65
Replicative growth, viral, 679–81
Replisomes, **268**
Repolarization, **890**, 891
Repressors, **312**
  discovery of, as negative control mechanisms, 312–13
  lac operon regulation and, 313–14, 317*f*
Reproduction. *See also* Replication
  animal (*see* Animal reproduction)
  cell cycle and, 194 (*see also* Cell cycle)
  fitness, adaptations, and, 5, 424, 426
  fungal, 583–84, 590–91
  mating and (*see* Mating)
  mitosis and, 195
  natural selection and, 4–5, 423
  plant (*see* Plant reproduction)
  protist, 533
  replication of living organisms as, 2
Reproductive development, plant, 409–11
Reproductive isolation
  allopatric speciation and, 462–64
  mechanisms of, 459
  sympatric speciation and, 464–68
Reptiles (Reptilia), 658, **660**–61, 666–68
Requirements, nutritional, 842–43
Residues, amino acids as, 43–44
Resilience, **1119**
Resistance, **1119**
Resisteance (*R*) genes, **776**
Resonance, 178–79
Resource allocation, life histories based on, 1041
Resource use
  efficiency in, 1119
  human, 1098, 1121
Respiration, cellular. *See* Cellular respiration
Respiratory systems, **862**
  air and water as media in, 862–64
  Fick's law of diffusion and, 864–65
  fish gills in, 865–66
  four steps of gas exchange and, 861–62
  homeostatic control of ventilation in, 870
  insect trachaea in, 866–67
  organs of, 864–70
  pulmonary circulation and, 879*f* (*see also* Circulatory systems)
  vertebrate lungs in, 867–70
Responder cells, plant, 756, 757
Resting potentials, **888**–90
Restoration, ecosystem, 1122–23
Restriction endonucleases, **340**
  recognition sites for, 349
  uses of, in genetic engineering, 352*t*
  using, to cut and paste DNA, 340–41
Restriction sites, 349
Reticulum, 850
Retina, **914**
Retinal, **915**
Retinoblastoma, 208
Retroviruses, **351**–54, **684**, 687*t*, 691
Reverse transcriptase, **281**, **684**
  gene therapy and, 351
  producing cDNAs with, 339–40
  uses of, in genetic engineering, 352*t*
Reverse transcriptase inhibitors, 684
R-groups, amino acid. *See* Side chains, amino acid
Rheumatoid arthritis, 983
Rhizaria, 524*t*, 536–37, 540
Rhizobia, 515, **749**–50
Rhizoids, **567**

Boldface page numbers indicate a glossary entry; page numbers followed by an *f* indicate a figure; page numbers followed by *t* indicate a table.

Rhizomes, **572**, **701**, **786**
Rhizopus stolonifer, 595
Rhodophyta (red algae), 539
Rhodopsin, 364, **915**–16
Rhynie Chert, 554
Rhythm contraceptive method, 966*t*
Ribbon diagrams, 47
Ribbon worms, 627
Ribonuclease, 50
Ribonucleic acid (RNA). *See* RNA (ribonucleic acid)
Ribonucleoside triphosphates, 68
Ribonucleotides, **60**
  anabolic synthesis of, 169
  components of, 60
  diffusion of, across lipid bilayers, 91
  phylogenetic tree of life and, 6–7
Ribonucleotide triphosphate (NTP), 290, 292
Ribose, 60–61, 66, 72, 77
Ribose-5-phosphate, 169
Ribosomal RNAs (rRNAs), **117**, **300**
  analyzing, to create phylogenetic tree of life, 6–7
  phylogenies based on, 503–4
Ribosome binding site, **301**
Ribosomes, **104**, **296**
  elongation of polypeptides during translation and, 301–2
  eukaryotic, 109, 114*t*
  formation of, 117
  initiation of translation and, 301
  mitochondria and, 112
  post-translational modifications and, 304
  prokaryotic, 104
  in protein synthesis, 300–304
  ribosomal RNA and structure of, 300–301
  as ribozymes, 302
  rough ER and, 108
  as site of protein synthesis in translation, 295–96
  termination of translation and, 302–3
  translocation and, 302
Ribozymes, **68**
  lipid bilayers and, 91
  as RNA catalytic molecules, 68
  RNA replicase as, 68–69
Ribulose-1,5-bisphosphate carboxylase/oxygenase. *See* Rubisco
Ribulose bisphosphate (RuBP), **185**–86
Rice
  genetically engineered, 338*f*
  genomes of, 362
  number of genes in genome of, 369*t*
  synthesizing β-carotene in, 355
  as target crop for agricultural biotechnology, 355
  transgenic (*see* Golden rice)
Richness, species. *See* Species richness
Rifampin, 424–25
Rings, tree growth, 714–15
Ring structures, 33*f*, 72
Ripening, fruit, 773–74
Rivers, 999
RNA (ribonucleic acid), **61**. *See also* Nucleic acids
  anabolic synthesis of nucleotides for, 169
  as catalytic molecule, 68
  in central dogma, 280–81
  directionality of strands of, 61
  DNA structure vs. structure of, 67*t*
  DNA synthesis from template of, 270–71
  electron micrograph of, 289*f*
  as first self-replicating molecule, 59, 68–69
  genes and, 195–96
  as genetic information-containing molecule, 67–68
  lipids and, 82
  messenger (*see* Messenger RNAs (mRNAs))
  nucleolus and eukaryotic, 108
  phylogenetic tree of life and, 6–7
  ribonuclease enzyme for unfolding, 50
  structure of, 66–67
  sugar-phosphate backbone of, 61*f*
  synthesis of, 279–80
  synthesis of, using codons, 283–85
  transfer (*see* Transfer RNAs (tRNAs))
  types of, 281
  versatility of, 67
  viral, 686
  viruses and, 682, 686, 687*t*
RNA-induced silencing complex (RISC), 329–30
RNA interference, **329**–30
RNA polymerases, **266**, **279**
  characteristics of, 290
  eukaryotic, 290*t*
  messenger RNA hypothesis and, 279–80
  regulation of, 308
  in transcription in bacteria and eurkaryotes, 296*t*
RNA processing, **295**, **320**. *See also* Protein synthesis
  adding caps and tails to transcripts in, 295
  bacterial vs. eukaryotic gene expression regulation and, 331*t*
  discovery of eukaryotic genes in pieces and, 293–94
  in eukaryotes, 293–95
  RNA splicing in, 294–95
  in transcription in bacteria and eurkaryotes, 296*t*
RNA replicase, **684**
RNA reverse-transcribing viruses. *See* Retroviruses
RNA splicing. *See* Alternative splicing; Splicing, RNA
RNA world hypothesis, 59, 68–69, 302, 687–88
Roberts, Richard, 294
Rods, **915**–16
Root apical meristem (RAM), **405**
Root caps, **705**–6, **764**–66
Root cortex, 722
Root hairs, **706**, **722**, **744**–48
Root pressure, 721, **723**
Roots, **405**, **571**
  auxin and adventitious, 768
  colonization of, by nitrogen-fixing bacteria, 749–50
  modified, 698–99
  plant nutrition and, 737*f*
  seedless vascular plants and, 571
Roots, Latin and Greek word, 8, B:6
Root systems, **696**
  characteristics of, 697
  holding of soil by, 547
  modified roots in, 698–99
  morphological diversity of, 697–98
  nutrient uptake by, 744–48
  organization of primary, 705–6
  phenotypic plasticity in, 698
  water transport in, 721–23

Rotavirus, 679*f*, 691
Rotifera (Rotifers), 603*t*, 631
Rough ER (rough endoplasmic reticulum), **108**
endomembrane system and proteins in, 118–22
eukaryotic, 108, 109, 114*t*
extracellular matrix components in, 134
pectins and, 132
post-translational modifications in, 304
Roundworms, 211, 603*t*, 605–6, **638**, 973*f*, 1067–68. *See also* *Caenorhabditis elegans*
Rous sarcoma virus, 691
Rubisco, **186**, **728**
in C_4 and C_3 photosynthesis, 188
discovery of, 186–87
photosynthesis regulation and, 189
Rudder, 292
Rumen, 167, 850
Ruminants, **850**
fermentation in rumen of, 167
stomachs of, 850–51
Rusts, 596

S

Saccharomyces cerevisiae
alcohol fermentation with, 167
chromatin remodeling in, 322
economic value of, 581, 598
gene expression in, 319
genome of, 362, 369*t*
meaning of scientific name of, 8
as model organism, B:20, B:21*f*, B:22
Sagittal crest, 671
Sago palms, 574
Saharan desert ant, 10–12
St. Martin, Alexis, 849
Sakmann, Bert, 892
Salamanders, 664
Saliva, 846, 848
Salivary amylase, 45, 855*t*
Salivary glands, 810*f*, 846*f*, **847**, 933
Salmon, 813, 827–28, 994–95, 1027
Salmonella, 979
Salps, 652
Salt
ionic bonds of table, 18, 19*f*
osmoregulation of (*see* Osmoregulation)
shark excretion of, 826–27
Saltiness taste reception, 918
Salt marshes, 461–62
Salty habitats
euryarchaeota bacteria and, 516
plant adaptations to, 720–21, 747–48
Sampling effects, 1119
Sampling error, **443**, 445–46
Sand, 744*t*
Sandbox tree, 798
Sand dollars, 650
Sandwich model, plasma membrane, 92
Sanger, Frederick, 46, 346–47
Saprophytes, **580**
carbon cycle on land and, 580–81
fungal, 589–90, 596
Sapwood, **714**
Sarcomeres, **922**–23
Sarcoplasmic reticulum, **926**
SARS virus, 224
Satiation hormone, 939–40
Satin bowerbird, 1019*f*
Saturated lipids, **87**–88, 355
Saturation kinetics, 55
Saxitoxins, 522
Scaffold proteins, 321
Scale of nature, 415
Scales, 662
Scallops, 634
Scanning electron microscopy (SEM). *See also* Electron microscopy
of DNA and RNA, 289*f*
of DNA synthesis, 258*f*
of *Escherichia coli*, 307*f*
of human fertilization, 211*f*
of human immune system cells, 973*f*
overview of, **93**, B:14*f*, B:15
Scarified, **799**
Scent response, fly, 465
Scents, flower, 562
Scheepers, Lue, 9–10
Schistosomiasis, 631–32
Schwann cells, **894**
Science
biology and nature of, 8–9 (*see also* Biology)
evidence in, 528
experiments as hypothesis testing in, 8–12
pattern and process components of theories of, 2, 414–15
reductionism in, 899
taxonomy and, 7–8
Scientific names, **8**, 675
Sclera, 914
Sclereids, **710**, 711
Sclerenchyma cells, **709**–10, 711*t*
Scolex, 613
Scorpions, 642
Scott, J. C., 1107
Scrapie, 51
Screening, DNA library
uses of, in genetic engineering, 353*t*
using DNA probes for, 342–43
Screens, genetic. *See* Genetic screens
Scrotum, **957**
Sea bass, 827–28
Seas. *See* Oceans
Seaside sparrows, 461–62
Seasons
causes of, 1010, 1011*f*
global warming and, 1100
migration and, 1027
phenology and, 1100
plant translocation and, 728
seed dormancy and, 798–99
Sea squirts, 651, 652
Seas slugs, 903–4*f*
Sea stars, 414*f*, 649, 1072–73
Sea urchins, 211, 390–91, 650
Seawater. *See* Marine ecosystems
Secondary active transport (cotransport), **98**, **732**, **825**
Secondary cell walls, **133**, **555**, **709**, 726
Secondary consumers, 523, **1084**
Secondary endosymbiosis, 531–32
Secondary growth, plant, **712**–15
Secondary immune response, **989**
Secondary metabolites, **776**
Secondary phloem, 713
Secondary spermatocytes, **953**
Secondary structure, **46**
DNA, 59*f*, 62–65, 260–61
DNA vs. RNA, 67*t*
hemoglobin protein, 49*t*
protein, 46–47
RNA, 66–67
tRNA, 299
Secondary succession, **1074**
Secondary xylem, 713
Second law of thermodynamics, **29**
Second-male advantage, **954**–55
Second messengers, **142**–43*f*, **756**, **897**, 945–**46**
Secretin, **852**, 933, **934**, 937
Secretions, barrier, 975
Secretory pathway, endomembrane system, 119–20
Secretory vesicles, 120
Sedimentary rocks, **416**
Sediments, fossils and, 480
Seed banks, **1117**
Seed coat, **795**
Seeding, bioremediation with, 500
Seedless vascular plants, 550, 552, 559, 567, 571–73
Seed plants, 550, 552, 559–61, 567, 574–77*f*, 793
Seeds, **402**, **550**, **560**, **783**, 795–800
dormancy of, 798–99
embryogenesis and, 796
fruit development and dispersal of, 797–98
germination of, 799–800
gibberellins and abscisic acid in dormancy and germination of, 770–71
land plants and, 560–61
red/far-red light reception and germination of, 762–64
red/far-red light reception and germination of lettuce, 763*t*
role of drying in maturation of, 797
sclereids and, 710
seed coat and, 795
seed plants and, 567
vacuoles in, 112
Segmentation, **609**
Segmentation genes, **381**–83
Segmented worms, 627, 633
Segment polarity genes, 382
Segments, **381**
Segregation, principle of. *See* Principle of segregation
Selection, directional, 440–41
Selection, natural. *See* Natural selection
Selective adhesion, **136**–38
Selective breeding. *See* Artificial selection
Selectively permeable membranes, **823**
Selective permeability, **86**, 90
Selective reabsorption, 834–35
Selectivity
membrane channel, 95, 96*f*
plasma membrane, 82
Self-fertilization (selfing), **222**, **231**, 451, 468, 620, **792**
Selfish genes, 365, 431
Self molecule, **983**–84
Self-nonself recognition, 978
Self-replicating molecules. *See also* Origin-of-life research
DNA as, 65–66
first cell membrane and, 82
lipid bilayers and, 91
origin-of-life with, 39
protein catalysts (enzymes) as first, 56
RNA world hypothesis on RNA as first, 59, 68–69
Self-sacrificing behavior. *See* Altruism
Semen, human, 959*t*
Semiconservative replication, DNA, 261–63
Seminal vesicles, **958**, 959*t*
Senescence, **773**–74
Sensors, **815**
Sensory neurons, **886**
Sensory receptors, 886, 938
Sensory systems, animal, 907–28
hearing, 909–13
information transmission to brains in, 909
moth pheromones and, 907
movement (locomotion) as response to, 907, 920–26 (*see also* Locomotion, animal)
processes and receptors of, 908
sensory transduction in, 908–9
smell (olfaction), 919–20
taste, 918–19
variations in abilities and structures in sensory organs of, 610
vision, 913–18
Sensory systems, plant, 755–82
defense responses to pathogens and herbivores, 775–81
germination and stem elongation responses to red and far-red light, 762–64
gravitropic responses to gravity, 764–66
growth regulators, 775*t*
growth responses, 767–76
information processing and, 756–57
phototropic responses to blue light, 758–62
responses to wind and touch, 766–67
selected, 780*t*
Sensory transduction, 908–9
Sepals, **409**, **562**, **788**
Septa (septum), **583**
Sequences. *See also* DNA sequencing; Whole-genome sequencing
amino acid, 43–44
DNA, 65
DNA, and genetic homology, 420–21
identifying genes in genome, 362–63
nucleic acid, 61
phylogenetic tree of life and rRNA, 6–7
proteins as amino acid, 46
Sequencing-by-synthesis, 347
Sequencing technology, 346–47, 1106
Serine, 42*t*
Serotonin, 898*t*, **903**–4*t*
Serum, **207**
Sessile, **542**, **607**–8, 611
Set point, **815**, 940
Severe combined immunodeficiency (SCID), **352**–54, 990
Sex chromosomes, **212**, 214*t*, 226–27, 241
Sex hormones, 960–66
contraception methods and, 966*t*
developmental processes and, 935–36
menstrual cycle control by, 962–66
puberty control by, 961–62, 963*t*
sexual activity of *Anolis* lizards and, 1023
testosterone, and estradiol as, 960–61
Sex linkage
chromosome theory of inheritance and, 241–42
exceptions and extensions to Gregor Mendel's rules on, 250*t*
identifying human traits as sex-linked, 251–52
Sex-linked inheritance (sex linkage), **242**, 251–52
Sex ratios, estrogen-mimics and, 1088–89
Sexual activity, lizard, 1023–24
Sexual competition, giraffe necks and, 10

Sexual dimorphism, **455**
Sexually transmitted diseases, 364, 513, 514, 538, 685. *See also* HIV (human immunodeficiency virus)
Sexual reproduction, **220–21, 784, 950**
  animal, 615
  animal switching between asexual and, 951–52
  asexual reproduction vs., 220–21, 223, 533, 950–51
  changing-environment hypothesis on, 224–25
  chromosomes, heredity, and, 221
  effects of fertilization on genetic variation in, 222–23
  in eukaryotes, 519
  fungal, 590–94
  genetic recombination in, 222
  genetic variation from independent assortment in, 221–22
  hormones in vertebrate, 936
  human fertilization micrograph, 211*f*
  mechanisms of, and gametogenesis, 952–53
  meiosis and, 194, 211, 220–25 (*see also* Meiosis)
  mistakes in, 225–27
  paradox of, 223–25
  plant, 784–86
  protist, 533
  protostome, 630
  purifying selection hypothesis on, 224
Sexual selection, **452–55**
Shade, far-red light as, 762
Shade leaves, 703
Shannon index of species diversity, 1079
Shape
  flowers, 562
  genes and proteins in vegetative development that determine leaf, 408
  hormone receptor, 140
  organic molecule, 33*f*
  plant receptor cell, 756
  prokaryotic diversity in, 504, 505*f*
  protein, 45–51, 304
  retinal change in, 915
  shoot system, 699–700
  simple molecule, 20
Sharks, 480, 662, 826–27
Sharp, Phillip, 293–94
Sheep, 377–78
Shells
  eggs, 658
  mollusk, 634
  protist, **528–29**
  turtle, 667
Shine-Dalgarno sequence, **301**
Shoot apical meristem (SAM), **405**, 407, 409
Shoots, **405**
Shoot systems, **696**
  adventitious roots and, 698–99
  characteristics of, 699
  collenchyma cells and shoot support in, 708–9
  formation of, 405
  modified, 700–701
  morphological diversity of, 699–700
  organization of primary, 706
  phenotypic plasticity in, 700
  water transport in, 723–27
  water transport via capillary action in, 723–24
  water transport via cohesion-tension force in, 724–27
Short-day plants, **787**
Short interspersed nuclear elements. *See* SINEs (short interspersed nuclear elements)
Short tandem repeats (STRs), **365**–67
Shotgun sequencing, **360**, 361*f*
Shrimp, 643
Siberian traps, 489
Sickle-cell disease, 46, 251
Side chains, amino acid
  amphipathic proteins and, 92
  enzyme function and, 53–54
  functional groups and reactivity of, 40–42
  major, 41*f*
  peptide bonds and, 43–44
  polarity of, and water solubility, 42
  protein primary structure and, 46
  protein tertiary structure and, 47–48
  of steroids, 84
Sieve plates, **711, 729**–30
Sieve-tube members, **711, 729**–30, 732–33
Sigma proteins, 291–92
Signal amplification, 141–42
Signal deactivation, 144
Signal hypothesis, endomembrane system, 120–21
Signaling, cell-cell. *See* Cell-cell signaling
Signaling pathways, animal hormone, 931–32
Signal processing, cell-cell, 140–44
  lipid-soluble vs. lipid-insoluble signals and signal transduction in, 140–41
  second messengers in signal transduction, 143*t*
  signal amplification in, 141–42
  signal transduction via enzyme-linked receptors in, 143–44
  signal transduction via G proteins in, 142–43
Signal receptors, **140**
Signal recognition particle (SRP), **121**
Signals, animal. *See* Chemical signals, animal; Electrical signals, animal
Signals, communication, **1027**
Signals, plant. *See* Sensory systems, plant
Signals, social, 205
Signal transduction, **141, 756, 945**
  in hair cells, 910
  in hormone binding to cell-surface receptors, 945–47
  of lipid-insoluble signals, 141
  in plant cells, 756, 757*f*
  via enzyme-linked receptors, 143–44
  via G proteins, 142–43
Signal transduction cascades, **946**
Signal transduction pathways, 141, 145, 756, 757*f*
Silencers, **324**–26
Silent mutations, **286**, 431, 444
Silurian-Devonian explosion, green plants in, 550–51
Silverswords. *See* Hawaiian silverswords
Simian immunodeficiency viruses (SIVs), 688
Simmons, Robert, 9–10
Simple eyes, **637**
Simple fruits, **797**
Simple leaves, 408, 701
Simple molecules, 19–20. *See also* Molecules
Simple sequence repeats, **365**–66
Simple sugars. *See* Monosaccharides; Sugars
Simple tissues, plant, 710
Simulation studies
  of genetic drift, 443–45
  of global climate change, 1011
SINEs (short interspersed nuclear elements), **478**–79
Singer, S. Jon, 93
Single bonds, 20
Single nucleotide polymorphisms (SNPs), **349**, 352*t*
Single-strand DNA-binding proteins (SSBPs), **264**, 268*t*
Single-trait experiments, Gregor Mendel's, 232–36
Sinks, plant sugar, **728**
  connections between sources and, 728–29
  low pressure near, 730–31
Sinoatrial (SA) node, **880**–81
Siphons, **630**, 634, 652
Sister chromatids, **196, 213**
  chromosomes and, 213–14*t*
  meiosis mistakes and, 225–26
  in mitosis, 199, 202*t*
Sister species, **464**
Sit-and-wait predators, 612, 627, 664
Site histories, 1073
Size
  cell-cycle checkpoints and cell, 205
  directional selection on cliff swallow body, 441
  disruptive selection on black-bellied seedcracker beak, 442
  eukaryotic genomes, 365
  human population, 1052–53
  island, 1078
  largest fungal mycelium, 582
  male penis, 958
  physiological effects of animal body, 811–14
  of prokaryotic and eukaryotic cells, 105–6*f*, 107*t*
  prokaryotic diversity in, 504, 505*f*
  prokaryotic genomes, 363, 365
  smallest eurkaryotic genome, 594
  testes and ovaries, 1023
  variable, of testes, 955
  virus, 678
Skates, 662
Skeletal muscle, **808, 921**
Skeleton cells, 396–98
Skeletons, animal, 920–21
  cartilaginous, 662
  endoskeleton function, 920–21
  endoskeleton structure, 920
  hydrostatic, 605–6, 613, 626
  notochords and, 396
  types of, 920
Skills, biology. *See* BioSkills
Skin
  amphibian, 664
  as barrier to pathogens, 974
  cancer of, 274
  cells of, 131*f*, 396–98
  ectoderm and, 604
  lymphocytes and, 978
  skin-breathing, 813
  thermoregulation and, 818–19
Skulls, 609, 653, 660, 671
Slack, Roger, 188
Sleeping sickness, 522*t*
Sliding clamps, 266, 268*t*
Sliding-filament model, **922**–26
Slugs, 635
Small interfering RNA (siRNA), 329–30
Small intestines, **851**–54
  carbohydrate digestion and transportation in, 852–53
  digestion in, 846
  fermentation in, 167
  folding and projections that increase surface area of, 851
  human, 810*f*, 846*f*
  lipid digestion and bile transport in, 853–54
  pancreatic enzyme regulation in, 852
  protein processing by pancreatic enzymes in, 851–52
  tight junctions in, 136
  water absorption in, 854
Small nuclear ribonucleoproteins (snRNPs), **294**–95
Smallpox, 686, 690, 973–74, 990
Smell, animal, 610, 919–20
Smith, John Maynard, 223–24
Smoke, seed dormancy and, 799
Smooth ER (smooth endoplasmic reticulum), **108**
  eukaryotic, 108, 114*t*
  in plasmodesmata, 138, 707
Smooth muscle, **808, 922**
Smut, 596
Snails, 224–25, 612, 635
Snakes, 384–85, 660, 666, 844, 848
Snapping shrimp, 464
Snowshoe hare, 1047–50
Social controls, cell-cycle, 205, 207, 208
Social stimulation, lizard sexual activity and, 1023–24
Sockeye salmon, 994–95, 1027
Sodium, action potentials and ions of, 891
Sodium channels, 891, 892
Sodium chloride (table salt), 18, 19*f*
Sodium poisoning, 747
Sodium-potassium pump, **98, 825**
  active transport and, 98–99
  in fish osmoregulation, 827
  resting potentials and, 889–90
  in shark salt excretion, 826–27
Soil, 741–44
  arbuscular mycorrhizal fungi (AMF) and, 589
  conservation of, 742
  effects of composition of, on properties of, 744*t*
  formation of, and textures of, 741–42
  green plant building and holding of, 547
  importance of conservation of, 742
  lichens and, 597
  nutrient availability in, 742–44*f*
  organic matter, **1092**
  plant adaptations to dry and salty, 720–21
  plant nutrients from, 739*t*
  plant nutrition and, 737, 738–40
  Sphagnum moss and, 570
  succession and, 1074
  water potential in, 720
Sokolowski, Maria, 1021

Boldface page numbers indicate a glossary entry; page numbers followed by an *f* indicate a figure; page numbers followed by *t* indicate a table.

Solar energy, transformation of, into biomass, 1084–85
Solid water. *See* Ice
Solubility. *See also* Water
amino acid, 42
of gas in water, 863
of hormones, 934
of lipids, 83
lipid-soluble vs. lipid-insoluble hormones, 139–41
Solute potential, **718**
as negative, 719
water movement with, 719*f*, 720
water potential and, 718
Solutes, **22**, **89**, **718**, **823**
concentration of, and solubility of gases in water, 863
diffusion and, 89
movement of, into roots, 722–23
transport of, across membranes in plants, 731–32
Solutions, **21**, **718**
chemical reactions and, 21
oxygen and carbon dioxide amounts in, 863
pH as indication of proton concentration in, 25
pH scale as measure of acidity and alkalinity of, 26
Solvents, **22**, 26*t*
Soma, **887**
Somatic cells, **194**, **271**
Somatic hypermutation, **989**
Somatic nervous system, **899**
Somatostatin, **930**
Somatotropin, 934
Somites, **396**–98
Song sparrows, 1045–46
Sonication, 360
Sonic hedgehog gene, 385
Soredia, 597
Sori, **573**
Sounds
cochlea detection of frequencies of, 911–12
middle ear amplification of, 911
Sources, plant sugar, **728**
connections between sinks and, 728–29
high pressure near, 730–31
sugar concentration in sieve-tube members at, 732–33
Sourness taste reception, 918
Soybeans, 354
Space-filling models, **21**, B:8*f*
Spanish flu, 677
Spanish moss, 750
Spark-discharge experiment, 39–40, 61
Sparrows, 461–62
Spatial avoidance mechanisms, 792
Spawning, 954
Special creation, theory of, 414–15, 417
Specialization, 616
Speciation, **5**, 458–73. *See also* Evolution; Species
allopatric, 462–64
contemporary, 422
current examples of, 422
definition and identification of species and, 458–62
disruptive selection and, 442
by hybridization, 470–71
mechanisms of reproductive isolation, 459*t*
mechanisms of sympatric, 471*t*
outcomes of secondary contract between isolated populations, 471*t*
pollination by animals and, in plants, 794
secondary contact between isolated populations and, 468–71
secondary contact between isolated species and, 471*t*
species concepts and species identification, 461*t*
species diversity and, 1107
of sunflowers, 458*f*
sympatric, 464–68
tree of life and, 5–6
Species, **459**
biological species concept, 459–60
changes of, through time as pattern component of evolution, 416–22, 423*f*
concepts of, and identification of, 461*t*
deceptive communication within and between, 1020
definition and identification of, 458–62
disruptive selection and formation of new, 442
dusky seaside sparrow as species definition case, 461–62
endangered (*see* Endangered species)
evidence for descent of, from common ancestors, 418–22, 423*f*
evidence for evolution of, 422*t*
evolution and extinct, 417
evolution of, by natural selection, 4–5 (*see also* Evolution; Natural selection; Speciation)
exotic and invasive, 1014–15
ex situ conservation of, 1121–22
extinction of (*see* Extinction)
global warming and loss of, 1098
interactions among (*see* Species interactions)
keystone, 1072–73
life histories and patterns across, 1041
morphospecies concept, 460
phylogenetic species concept, 460–61
protostome, 623–24
reintroduction of, 1087
richness and diversity of (*see* Biodiversity; Species diversity; Species richness)
succession and role of traits in, 1074–76
taxonomy and scientific names for, 8
in theory of evolution by natural selection, 416
traits of (*see* Traits)
Species-area relationships, **1116**
Species diversity, **1077**, **1106**. *See also* Biodiversity
arthropods and, 637
current estimates of, 1107–8
global climate change and, 1012–13*f*
measuring, 1079
of protists, 522
species richness vs., 1077 (*see also* Species richness)
threats to, 1110–14
in tropical wet forests, 1003
Species interactions, 1058–70
competition, 1059–63*f*
consumption, 1063–68
four types of, 1058–59, 1070*t*
mutualisms, 1068–70
role of, in succession, 1075, 1076–77
three themes of, 1059
Species richness, **1077**–80. *See also* Biodiversity
agrea and age hypothesis for, 1080
community stability and, 1119–20
energy hypothesis for, 1080
global patterns in, 1078–80
high-productivity hypothesis for, 1080
increased net primary productivity of, 1118–19
intermediate disturbance hypothesis for, 1080
latitudinal gradient in, 1078–80
measuring, 1079
role of immigration and extinction in, 1077–78
role of island size and isolation in, 1078
species diversity vs. (*see also* Species diversity)
theory of island biogeography on, 1077–78
Specific heat, **24**, **1010–11**
climate and, 1010–11
of some liquids, 24*t*
of water, 24, 26*t*
Specificity
adaptive immune response, 977
antibody, 981–82
Sperm, **211**, **784**, **951**
human, 211*f*
meiosis and, 194, 218
plant, 402, 789, 791 (*see also* Pollen grains)
reasons for fertilization by single, 391–92
sexual selection and, 452–53
structure and function of, 389–90
Spermatids, **953**
Spermatogenesis, **953**, 958
Spermatogonia (spermatogonium), **953**
Spermatophore, **636**, **954**
Sperm cells, 2*f*
Sperm competition, **954**–55, 957–58
Spermicide, 966*t*
Sphagnum moss, 570
S phase. *See* Synthesis (S) phase
Sphenophyta, 572
Sphincters, **848**, **876**
Spicules, **618**
Spiders, 642
Spinal cords, 396, 651, 653, 886, 901
Spindle apparatus, **198**–200, 201–2, **216**
Spine, human, 804
Spines, **701**
Spiracles, **828**–29, **867**
Spirochaetes (spirochetes), 512, 513
Spleen, **978**
Spliceosomes, **294**–95, **328**
Splicing, RNA, **294**
alternative splicing, 328–29, 331
post-transcriptional alternative, of mRNAs, 328–29
in RNA processing, 294–95
Splitting water, 181
Spoken language, 672
Sponge, contraceptive, 966*t*
Sponges, 136, 607–9, 611, 618
Spongiform encephalopathies, 51
Spontaneous abortions, 226, 440
Spontaneous chemical reactions, 29–30, 85. *See also* Chemical reactions
Spontaneous generation hypothesis, 2–4, 415
Sporangia (sporangium), **550**, **592**, **784**
Spores, **583**, **784**
fossilized, 550
as fungal reproductive cells, 590
land plant production of, 556–64
meiosis and, 218
protist sporophytes and, 535
Sporophytes, **535**, **557**, 558–59, **784**, 795
Sporopollenin, **550**, 556, 560, 791–92
Spruces, 575
Spurs, flower, 562–63
Squid, 125–27, 636, 890–92
Srb, Adrian, 277–78
Stability
community, 1119–20
DNA, 66
Stabilizing selection, 441–**42**
Stages, animal life, 616
Stahl, Franklin, 261–63
Stamens, **409**, **561**, **789**, 790–92
Standard error, 224
Standard error bars, B:7
Standing defenses, **1063**
Stapes, **910**
Staphylococcus aureus, 426
Starch, **74**, **189**
hydrolysis of, to release glucose, 78–79
as plant storage polysaccharide, 71*f*, 74, 168
processing of photosynthetic, 189–90
structure of, 75*t*
Starling, Ernest, 852, 876–77
Start codons, **283**, 301
Statistical tests, B:7
Statocyst, **909**
Statolith hypothesis, **764**–65
Statoliths, **765**
Stem-and-loop structure, RNA, 67
Stem cells, 352–54, **375**
Stems, **699**. *See also* Shoot systems
eudicot vs. monocot, 706*f*
modified, 701
Stepparent child abuse rates, 1034
Stereocilia (stereocilium), **909**–10
Sterilization, heat, 3–4
Steroid-hormone receptors, 943, **944**, 945
Steroids, **84**, 595
brassinosteroids as plant, 773
hormone receptors for, 943–45
as hormones, 933, 936
as lipids, 84 (*see also* Lipids)
Steudle, Ernst, 726
Stevens, Nettie, 212, 241
Sticky ends, **340**–41
Stigma, 402–3, **789**
Stolons, **701**
Stomach acid, 848–50
Stomachs, **848**
avian gizzard as modified, 851
discovery of digestion in, 848–49
echinoderm, 648
genes in human, 365
human, 810*f*, 846*f*
hydrochloric acid as stomach acid in, 25
protein digestion in, 846, 849
in ruminants, 850–51
stomach acid production by parietal cells of, 849–50
ulcers as infectious disease of, 850
Stomata (stoma), **187**, **553**, **708**, **771**
abscisic acid role in closing guard cells of, 771–73
carbon dioxide entry into leaves through, 187

Stomata (stoma), (*continued*)
phototropins and, 758
regulation of gas exchange and water loss by, 708
water loss prevention by, 553–54, 717, 727
Stoneworts, 569
Stop codons, **283**, 302
Storage polysaccharides, 71*f*, 74, 75*t*
Stored energy. *See* Potential energy
Strains, viral, **688**–89
Stramenopila, 524*t*, 537, 542–43
Streams, **999**
Streptomyces, 514
Stress
hormones in long-term responses to, 938
hormones in short-term responses to, 937–38
Striated muscle, **808, 921**
Stroma, **113, 174**, 184, 186
Stromatolites, 15*f*
Structural formulas, **20**, B:8*f*
Structural genes, **369**
Structural homology, **420**–21
Structural polysaccharides, 75*t*, 76, 77, 113
Structural proteins, 45, 280
Structure-function relationships, 806
Sturtevant, A. H., 246
Style, **789**
Suberin, **722**
Subpopulations, 1051
Subspecies, **461**
Substance P, 898*t*
Substrate-level phosphorylation, **156**
Substrates, enzyme catalysis, **51**, 53, 55
Subtropical deserts, 1004
Subunit vaccines, 990
Suburbanization, 1096, 1112
Succession, **1074**–77
Glacier Bay case history on, 1076–77
primary and secondary, 1074
role of chance and history in, 1075
role of species interactions in, 1075
role of species traits in, 1074–75
Successional pathways, 1074, 1076–77
Succinate, 157
Succulents, 703
Sucrose, **189**
in glycolysis experiment, 155
processing of photosynthetic, 189–90
transport of, in plants (*see* Translocation)
Sufhydryl functional group, 35
Sugar beets, 733–34
Sugar-phosphate backbone, nucleic acid, 61, 62, 66, 260, 279
Sugars, **60, 71**
carbon functional groups and formation of, 34–36
chemical evolution and nucleotide production from, 60–61
dry seeds and, 797
glucagon and blood levels of, 45
glucose (*see* Glucose)
lactose as (*see* Lactose metabolism)
as monomers of carbohydrates, 39, 42, 71 (*see also* Carbohydrates)
processing of photosynthetic, 189–90
in RNA, 66
table, 148*f*, 155
transport of, in plants (*see* Translocation)
Suicide, cell, 987–88
Sulfate reducers, **503**
Sulfur, 35, 48, 173, 259, 296, 739*t*, 843*t*
Sulfuric acid, 24*t*, 489–90
Summation, **898**–99
Sundews, 752
Sunflowers, 458*f*, 470–71
Sun leaves, 703
Sunlight. *See also* Light energy
biomes and, 1002
primary producers and, 1083–84
transforming, into biomass, 1084–85
tropical, 1010
Sunscreens, flavonoids as natural, 176
Supercoiling, 320
Supporting connective tissue, **807**
Surface area
animal gas exchange and, 874–75
circulatory systems and maximization of, 874–75
in gas exchange organs, 864
oxygen availability in aquatic ecosystems and, 863–64
Surface/area relationships, 696–97
Surface area/volume relationships, 811–14
adaptations that increase surface area, 814
importance of, 811–12
metabolic rates and, 812–13
Surface tension, **23, 724**
water and, 22–23
water transport in plants and, 724–25
Surgery, cancer, 206
Survival, natural selection and, 4–5
Survivorship, **1038**
fitness trade-offs between fecundity and, 1039–40
life tables and, 1038–40
life tables for endangered species and, 1055
Population Viability Analysis (PVA) and, 1115
Survivorship curves, **1038**, 1039*f*
Suspension feeders, **610, 843**
Suspensor, 404
Sussman, Michael, 733
Sustainability, **1121**
Sustainability science, 1121
Sustainable agriculture, **742**
Sustainable development, 1121
Sustainable forestry, 742
Sutton, Walter, 212–13, 215, 230, 239–41
Swamps, **998**
Swans, 1*f*
Sweden
age structure of, 1051
maternal mortality rates in, 970–71
Sweetness taste receptors, 919
Swim bladders, **662**
Swimming
eukaryotic cell, 127–28
fungal gametes and spores, 583
prokaryotic cell, 104
Symbiosis, **527, 850**
Symbiotic organisms, **586**, 687, **746**, 749–50
Symmetric competition, **1060**
Symmetry, animal body, 604–5
Sympathetic nervous system, 883, **900**, 943
Sympatric speciation, **465**
gene flow and, 464–65
isolation and divergence in, 464–68
mechanisms of, 471*t*
by natural selection, 465
by polyploidy, 465–68
Sympatry, **464**, 468
Symplastic route, **722**, 723*f*
Symporters, **732**, 746, **826**
Synapomorphies, **460**–61, 475, **524**
Synapses, **895**–99
memory, learning, and changes in, 903–4*f*
neurotransmitter functions and, 896–97
neurotransmitter release and structure of, 895–96
neurotransmitters and, 895
postsynaptic potentials and, 897–99
Synapsis, **216**
Synaptic cleft, **896**
Synaptic plasticity, **903**
Synaptic vesicles, **896**
Synaptonemal complex, **219**
Synergids, 794
Synthesis, DNA. *See* DNA synthesis
Synthesis, RNA. *See* Transcription
Synthesis (S) phase, **196**
Syphilis, 513
Systemic acquired resistance (SAR), **777**–78
Systemic circulation, **878**
Systemin, 140*t*, **779**
System level anatomy and physiology, 811*f*
Systems, **27**
Systole, **879**
Systolic blood pressure, **880**
Szilard, Leo, 312

T

T2 virus, 259–60*f*
Tabin, Clifford, 428–29
Table salt, 18, 19*f*
Table sugar, 148*f*, 155
Tagging systems, 263, 296, B:12–B:13
Taiga, **1007**
Tails
length of male barn swallow, 1024–25
post-anal, 650
vestigial, 418, 419*f*
Tails, RNA, 295
Talking trees hypothesis, 779
Tannins, 112, 775
Tapeworms, 612–13, 631–32
Taproots, **697**
Taq polymerase, 344–45, 352*t*, 501
Tardigrades, 603*t*, 637
Target cells. *See* Receptors
Target crops, agricultural biotechnology and, 355
Taste, 610, 918–20
Taste buds, **918**–19
TATA-binding protein (TBP), **323**
TATA box, **291, 323**
TATA-box binding protein, 45
Tatum, Edward, 277
Taxol, 575
Taxon (taxa), **7**
Taxonomic diversity, 1106, 1107*f*
Taxonomy, **7**
fossil record and taxonomic bias, 480
tree of life and, 7–8
Taxon-specific surveys, 1107–8
TBP transcription factor, 326
T-cell receptors (TCRs), **980**–82
T cells, **979**
activation and function of, 989*t*
activation of, 984–86
discovery of, 979
discovery of receptors for, 980–81*f*
as lymphocytes, 978
severe combined immunodeficiency (SCID) and, 352–54
T-DNA, 355–56
Tectorial membrane, **912**
Teeth, 480, 612, 666, 667, 844
Teleosts, 663
Telomerases, **270**–71
Telomeres, **269**
replication of, 269–71
Telophase, mitosis, **200**
Telophase I, meiosis, 218
Telophase II, meiosis, 218
Temperate, **1005**
Temperate forests, 1006, 1093–94
Temperate grasslands, 1005, 1012–13*f*, 1122–23
Temperature, **27**
animal regulation of body, 816–19
biomes and, 1002
blood flow and body, 882
effect of, on hemoglobin, 872, 873*f*
enzyme catalysis and, 55–56
global climate change experiments on, 1012
membrane permeability and, 88–89
role of, in chemical reactions, 30–32, 51–52
as thermal energy, 27
tropical species richness and, 1080
of water for gas solubility, 863
Template strand, DNA, **65, 290**
Temporal avoidance mechanisms, 792
Temporal bias, fossil record and, 480–81
Temporal lobe, **901**, 902
Temporal prezygotic isolation, 459*t*
Tendons, **920**
Tendrils, 703
Tenebrio molitor, 241
Tension, 132
Tentacles, 619, 620, **636**
Termination, enzyme catalysis, **54**
Termination phase, **292**
transcription, 292–93
translation, 302–3
Termites, 538
Terrestrial ecosystems, 1001–8
adaptations to (*see* Water-to-land transition adaptations)
animal transition from aquatic ecosystems to, 627–28
arctic tundra, 1008
boreal forests, 1007
changes in net primary productivity of, 1101–2
characteristics of, as biomes, 1001–2
global productivity patterns of, 1089–91
insect osmoregulation in, 828–31
internal fertilization in, 954
limits for net primary productivity in, 1090–91
nutrient cycle of, 1092*f*
nutrient cycling in, 1092
osmotic stress in, 825
plant adaptations to dry, 720–21, 728 (*see also* Land plants)

Boldface page numbers indicate a glossary entry; page numbers followed by an *f* indicate a figure; page numbers followed by *t* indicate a table.

plant transition from aquatic ecosystems to, 553–55, 556–64
subtropical deserts, 1004
temperate forests, 1006
temperate grasslands, 1005
tropical wet forests, 1003
vertebrate osmoregulation in, 831–38
Territory, **454**–55, **1026**
Tertiary consumers, 523, **1084**
Tertiary structure, **47**
DNA vs. RNA, 67*t*
hemoglobin protein, 49*t*
protein, 47–48
RNA, 67
tRNA, 299
Testcrosses, **238**
confirming predictions with, 238–39
in Mendelian genetics, 239*t*
Testes, **933**, **953**
cells of, 115–16
hormones of, 936
male reproductive systems and, 957
variable size of, among species, 955
Testing, drug, 603
Testing, genetic. *See* Genetic testing
Testosterone, **936**, **960**
puberty and, 961–62
sexual activity of *Anolis* lizards and, 1023
testes and, 115–16
Tests, statistical, B:7
Testudinia, 667
Tetrads, 214*t*, **216**
Tetrahydrocannabinol (THC), 775
Tetrahymena, 68, 127–28
Tetramers, 49
Tetrapod limbs, 656–57
Tetrapods, **653**, 655
Texture, **742**
Thalidomide, 969
Theories, **2**, 414–15
Theory of island biogeography, 1077–78
Therapies
gene therapy, 351–54
genetic maps and development of, 350
Thermal energy, **27**, 28
Thermocline, **996**
Thermophiles, **502**, **503**
Thermoreceptors, **908**
Thermoregulation, **816**–19
countercurrent heat exchangers in, 818–19
endotherm homeostasis in, 817–18
endothermy and ectothermy in, 817
mechanisms of heat exchange and, 816
variations in, 816–17
Thermus aquaticus, 345, 501
Thiamine (vitamin B_1), 54, 843*t*
Thiamine pyrophosphate, 54
Thick ascending limb, loop of Henle, 835–36, 838*t*
Thick filaments, **922**
Thigmotropism, **766**
Thin ascending limb, loop of Henle, 835–36, 838*t*
Thin filaments, **922**
Thin layer chromatography, 175–76
Thiols, 35*f*
Thomas, Lewis, 388
Thorax, 637, **639**
Thorns, **701**
Three-dimensional visualization techniques, B:15–B:16
Threonine, 42*t*
Threshold potential, **890**
Throwbacks, evolutionary, 572
Thucydides, 973
Thylakoids, **112**–13, **174**, 549
Thymidine, 196
Thymine, 60, 63–64, 261, 279
Thymine dimer, 272
Thymus, **978**, 979
Thyroid gland, **933**, 934–36
Thyroid-stimulating hormone (TSH), **943**
Thyroxine, 140*t*, **934**
Ticks, 642
Tight junctions, **135**–36
Tilman, David, 1118–19
Time
animal sensory organs and, 610
vastness of geologic, 416–17
TI (tumor-inducing) plasmids, **355–56**
Tips, **474**
Tissue bias, fossil record and, 480
Tissue level anatomy and physiology, 811*f*
Tissues, **135**, **394**, **603**, **704**, **806**. *See also* Tissues, animal; Tissues, plant
Tissues, animal, 806–11
blood, 870–71
connective, 806–7
cultures of (*see* Cultures, cell and tissue)
epithelial, 809–10
immune system rejection of foreign, 988
muscle, 808–9*f*
nervous, 808
organogenesis and, 396–99
organs, organ systems, and, 810–11
origin and diversification of, 603–4
Tissues, plant
cultures of (*see* Cultures, cell and tissue)
dermal tissue systems, 707–8
embryogenesis and, 796
ground tissue systems, 708–10
primary plant body, 706–12
toxic or poor-nutrition, 1066
vascular tissue systems, 710–11
Tobacco mosaic virus, 679*f*, 686
Tolerance, **1075**
Tomatoes, 408, 740
Tonegawa, Susumu, 324–26, 982
Tongues, gray whale, 819
Tonoplast, **734**, **747–48**
Tool use, 671, 672
Tooth decay, 145
Top-down control, **1087**
of herbivore populations, 1065–66
trophic cascades and, 1087–88
Topoisomerases, **265**, 268*t*
Top predators, 1087
Topsoils, 742
Torpor, **817**
Totipotent, **708**
Touch, animal, 610
Touch/wind responses, plant, 766–67, 780*t*
Townsend's warblers, 468–70, 1015
Toxaphene, 1088–**89**
Toxins
animal, 618, 619
biomagnification of, 1088–89
plant, 708
as plant defense responses, 775
plant exclusion of, 747–48
as prey defenses, 1063
Toxoplasma, 522*t*, 541
Toxoplasmosis, 522*t*
Tracers, 586
Tracheae, **828**, 866–**67**
Tracheids, **555**, **710**
Trade-offs, **804**. *See also* Fitness trade-offs
Tragopogon, 467
Traits, **230–31**. *See also* Genotypes; Phenotypes
adaptations as, 5 (*see also* Adaptations)
analyzing morphological, in fungi, 582–84
analyzing morphological, in green plants, 549–50
analyzing morphological, in viruses, 679
complex, 430
dominant vs. recessive, 232–33
evolution and vestigial, 418, 419*f*
genes and, 221
heredity and, in Mendelian genetics, 230–32 (*see also* Mendelian genetics)
heritable, 4
identifying human, as autosomal or sex-linked, 251–52
identifying human, as autosomal recessive or dominant, 250–52
incomplete dominance and codominance in, 245–47
Gregor Mendel's experiments with single, 232–36
Gregor Mendel's experiments with two, 236–39
morphological innovations in, 485–86
natural selection of, 423
non-adaptive, 431
phylogenies based on morphological, 477
polygenic inheritance of quantitative, 248–50
role of, in succession, 1074–77
sexually dimorphic, 455
succession and role of species, 1074–76
synapomorphic, 460–61
transitional, 417–18
vestigial, 418, 431
Transacetylase, 313
Transcription, **280**, 289–93. *See also* Protein synthesis
bacterial vs. eukaryotic gene expression regulation and, 331*t*
in bacteria vs. in eukaryotes, 296*t*
central dogma and role of, 280
characteristics of RNA polymerases, 290
electron micrograph of, 289*f*
elongation and termination phases of, 292–93
eukaryotic RNA polymerases, 290*t*
eurkaryotic post-transcriptional regulation, 328–30
initiation phase of, 291–92
messenger RNA synthesis in, 289–90
regulatory sequences and proteins in initiation of eukaryotic, 323–28
Transcriptional control, **308**–9, 331, 378
Transcriptomes, **371**
Transcripts of unknown function (TUFs), 368, 370
Transduction, **908**–9
Transduction, signal. *See* Signal transduction
Transfer cells, **556**
Transfer RNAs (tRNAs), **298**
anticodons and structure of, 299
experiment on transfer of amino acids from, to proteins, 297–99
ribosome structure and, 300–301
structure and function of, 297–300
wobble hypothesis on types of, 299–300
Transformation, **341**, 364
Transformations, energy, 27–29
Transgenic organisms, **350**
crops as, 354–56
development objectives for, 354–55 (*see also* Golden rice)
target crops for, 355
therapies and, 350
tomatoes, 408
Transitional features, **417**–18
Transition state, enzyme catalysis, **51**–54
Transition state facilitation, enzyme catalysis, **54**
Translation, **280**, 295–97. *See also* Protein synthesis
bacterial vs. eukaryotic gene expression regulation and, 331*t*
in bacteria vs. in eukaryotes, 296–97
central dogma and role of, 280
control of eukaryotic, 330
elongation phase of, 301–2
initiation phase of, 301
post-translational control of eukaryotic, 330
post-translational modifications, 304
ribosomes as site of protein synthesis in, 295–96
specification of amino acids by mRNA triplets in, 297
termination phase of, 302–3
Translational control, **308**
Translocation, **286**, **302**, **728**–34
connectons between sources and sinks in, 728–29
phloem anatomy and, 729–30
phloem loading in, 731–33
phloem unloading in, 733–34
pressure-flow hypothesis on, 730–31
secondary phloem and, 713
sources and sinks in, 728
Transmembrane proteins, **93–94**, 134
Transmembrane route, **722**, 723*f*
Transmission, **685**
animals and human disease, 602–3
sensory signal, 908, 909
viral, 684–86, 689
Transmission electron microscopy (TEM). *See also* Electron microscopy
of bacterial photosynthetic membrane, 104*f*
of cells, 115*f*
overview of, B:14–B:15
of parietal cells, 850
studying viruses with, 679
of synapses, 895–96
Transmission genetics. *See* Mendelian genetics
Transmission vectors, disease, 602–3
Transpiration, **702**, **717**–18, 727
Transport, plant. *See* Translocation; Water transport, plant
Transportation, animals as human, 602
Transporters. *See* Carrier proteins (transporters)

Transport proteins, **94**
carrier proteins as, 96–97
channel proteins as, 94–96
enzymes and, 97*f*
functions of, 45
in osmoregulation, 826–28
plant active exclusion by, 747–48
plant cell-cell signals and, 757
pumps as, 97–98
Transport vesicles, 125–27
Transposable elements, **365**, 366*f*, 478–79, 686–87
Trans surface, Golgi apparatus, 109, 120–21
Treadmilling, 123
Tree-dwelling ants, 1067–68
Tree ferns, 567
Treehoppers, 1069–70
Tree of life, **5**, **503**. *See also* Life; Phylogenetic trees
as molecular phylogeny, 6–7
speciation and, 5–6 (*see also* Speciation)
taxonomy and scientific names for, 7–8
as universal tree, 7*f*
Trees. *See also* Boreal forests; Temperate forests; Tropical wet forests
baobab, 695*f*
bristlecone pine, 401, 575, 695
broad-leaved deciduous, in temperate forests, 1006
fire history of giant sequoias, 1073–74
global warming and, 1114
global warming and, 1100
growth rings of, 714–15, 1073
needle-leaved evergreen, in boreal forests, 1007
trunk structure of, 714–15
Trees, family. *See* Pedigrees
Trees, phylogenetic. *See* Phylogenetic trees
Trends, graph, B:2–B:3
Triacylglycerols/triaglycerides, 83, 88. *See also* Fats
Tricaroxylic acid (TCA) cycle. *See* Citric acid cycle
Trichinosis, 638
Trichomes, **708**
Trichomonas, 522*t*
Trichomoniasis, 522*t*, 538
Trichoptera, 641*t*
Triglycerides, **938**
Triiodothyronine (T_3), 935–**36**
Trilobytes, 474*f*
Trimesters, human pregnancy, 967–68
Trioses, **72**
Triple bonds, 20
Triplet code, **282**–83, 297
Triploblasts, **604**
Trisomies, **225**–27
Trivers, Robert, 452–53
tRNA. *See* Transfer RNAs (tRNAs)
Trochophore, **624**
Trophic cascades, **1087**–88
Trophic levels, **1085**
energy transfer between, 1086–87
trophic structure and, 1085
Trophic structure
energy flow to grazers vs. decomposers in, 1085–86
food chains and food webs in, 1085
trophic levels in, 1085
Trophoblasts, **393**
Tropical climate, 1009–10
Tropical species richness, 1078–80, 1109
Tropical wet forests, 1003
climate and, 1009–10
edge effects in fragmented, 1113
global productivity patterns of, 1090
species richness in, 1109
Tropomyosin, 328, **925**
Troponin, **925**
True navigation, **1026**
Trunks, tree, 714–15
Trypanosomiasis, 522*t*
Trypsin, 45*f*, **852**, 855*t*
Trypsinogen, **852**
Tryptophan, 42*t*
Tsetse flies, 1015
T tubules, **926**
Tube feet, 614, **648**
Tube-like leaves, 703
Tuberculosis, 224, **424**–26
Tubers, **701**
Tube-within-a-tube body design, 607, 626
Tubulin, 127
Tumors, **205**, 206–9
Tumor suppressors, **205**, 206, 208, **332**
Tundra, **1008**
Tunicates, 651, 652
Turgid, **718**
Turgor pressure, **133**, **718**–19, 721
Turner syndrome, **227**
Turnovers, **996**
Turtles, 660, 667, 878, 993*f*
Two-trait experiments, Gregor Mendel's, 236–39
Tympanic membrane, **910**–11
Type 1 diabetes mellitus, 856–57, 983
Type 2 diabetes mellitus, 856–58
Typological thinking, 415
Tyree, Melvin, 726
Tyrosine, 42*t*, 248

U

Ubiquinone, **161**–62, 163–64
Ubiquitins, 204, 330
Ulcers, **850**
Ultimate causation, **952**, **1020**
Ultrasound hearing, bat, 912
Ultraviolet light
cancer and, 332–33
DNA damage and, 272
flower color and, 788, 918
photosynthesis and, 174–75, 176
xeroderma pigmentosum (XP) and, 272–74
Ulvophyceae (Ulvophytes), 568
Umami, **919**
Umbilical cord, **968**, 970
Undersea volcanoes. *See* Deep-sea vents; Volcanic gases
Undifferentiated cells, 375, 405, 407–9
Unequal crossover, **366**–67
Unfolding, protein, 50
Unicellular organisms, 6
as characteristic of domains of life, 497*t*
eurkaryotic, 105, 319
prokaryotes as, 104
prokaryotic (*see also* Prokaryotic cells)
quorum sensing as cell-cell communication in, 145–46
Unikonta, 525
United States
cancer death rates in, 206*f*
diabetes mellitus epidemic in, 857–58
Food and Drug Administration ruling on human growth hormone, 343–44
soil erosion in, 742
Universal tree, **503**
Unjointed limbs, animal, 613
Unreplicated chromosomes, 199, 213–14*t*
Unsaturated lipids, **87**–88, 355
Unstriated muscle, 808
Untranslated regions (UTRs), 295
Upright growth, land plant, 554
Upstream, **291**
Uracil, 60, 66–67, 279, 290
Urbanization, 1096, 1112
Urea, 829*t*, 830, 837–38
Ureter, **832**
Urethra, **832**, **958**, 960
Uric acid, **829**–30
Urinary systems, 832–38
diabetes mellitus and urine formation, 856–57
filtration by renal corpuscle in, 832–34
function of kidneys in, 832
hormonal control of, 837–38
human, 832*f*
osmoregulation by distal tubule and collecting duct in, 837–38
osmotic gradient creation by loop of Henle in, 835–37
osmotic stress and, 825
reabsorption by proximal tublule in, 834–35
structure and function of, 838*t*
structure of kidneys in, 832
Urine
amino group excretion in, 168
animal osmoregulation and, 825
dehydration and, 139
diabetes mellitus and, 856–57
fish osmoregulation and, 824
formation of (*see* Urinary systems)
insect excretory system and, 830–31
Urochordates, **651**, 652
Uterus, **393**, **960**
UV radiation, 176

V

Vaccination, **974**
as immunization, 973–74
immunological memory and, 990
Vaccines, 678, **990**
Vacquier, Victor, 391
Vacuoles, **111**, **707**
eukaryotic, 111–12, 114*t*
plant active exclusion with, 747–48
plant cell, 707
stored starch in, 116
Vagina, **960**
Vagus nerve, 895
Vale, Ronald, 125–27
Valence, **17**
Valence electrons, **17**
Valine, 42*t*, 46, 47
Valves, heart, **876**, 879
Vampire bats, 1033–34
van der Waals interactions, **47**–48
Van Helmont, Jean-Baptiste, 738
Van Leeuwenhoek, Anton, 2, 538
Van Niel, Cornelius, 173
Vapor, water, 721
Variability, temperature and precipitation, 1012–13*f*
Variable number tandem repeats (VNTRs), **365**–66
Variable (V) regions, **982**
Variables, experimental, 12
Variation, genetic. *See* Genetic variation
Vasa recta, **837**
Vascular bundles, **706**
Vascular cambium, **712**
auxin and, 768
secondary growth functions of, 713
Vascular tissue, **188**, **406**, **550**
elaboration of, into tracheids and vessel elements, 555
origin of, in land plants, 554–55
root and shoot systems and, 696
Vascular tissue systems, **705**, 710–11
Vas deferens, **958**
Vectors, **340**
Vegetative development, 407–9
Vegetative organs, **402**
Veins, **875**
blood pressure and, 882
in closed circulatory systems, 875–76
human, 879
renal, 832
Veligers, **634**
Velvet worms, 637
Venae cavae (vena cava), **879**
Ventilation. *See also* Respiratory systems
fish gills and, 865
in gas exchange, 862
homeostatic control of, 870
insect tracheae and, 866–67
vertebrate lungs and, 867–70
Ventral, **379**
Ventricles, **877**, 879
Ventricular systole, 880
Venules, **875**
Venus flytraps, 752, 766–67
Vertebrae, **653**
Vertebrata, 660–68
Vertebrates, **609**, **646**, 653–68
body plan of, 651
as chordate deuterostomes, 646
as chordates, 609
closed circulatory systems of, 875–78
evolution of, 653–59
fishes as, 655
fossil record of, 653–55
hormones in sexual development and activity of, 936
lungs of, 867–70
morphological characteristics of, 653
morphological innovations of, 655–58
nervous systems of, 899–904*f*
osmoregulation in terrestrial, 831–38 (*see also* Urinary systems)
phylogenies of, 655
primates and humans, 668–73*f*
relative abundance of species of, 655*f*
reproductive innovations of, 658–59
Vertebrata lineages and, 660–68
Vervet monkeys, 1034
Vesicles
artificial membrane-bound, 85–86
experiment on microtubules and transport, 125–27

Boldface page numbers indicate a glossary entry; page numbers followed by an *f* indicate a figure; page numbers followed by *t* indicate a table.

in mitosis, 200
prokaryotic, 104
protein transport and sorting by, in Golgi apparatus, 122–23*f*
rough ER in, 108, 109
secretory, 120
synaptic, 896
Vesicular-arbuscular mycorrhizae (VAM), 589
Vessel elements, **555**, **710**
land plants and, 555
Vessels, angiosperm, 564
Vestigial traits, **418**
evolution and, 418, 419*f*
as non-adaptive, 431
vertebrate, 651
in whales, 422
Vicariance, **463**
allopatric speciation by, 463*f*, 464
physical isolation and, 462–63, 464
Video microscopy
of ATP synthase, 165
overview of, B:15
of transport vesicles, 126*f*
Villi (villus), **814**, **851**
Viral proteins, 117–18
Virchow, Rudolf, 2, 194
Virions, **676**
Virulence, 500, **776**
Virulence, bacterial, 500
Virulence genes, 355
Virulent, **677**
Viruses, **675**–94
abundance of, 675–76
analyzing morphological traits of, 679
analyzing phases of replicative cycles of, 681–86
analyzing variation in growth cycles of, 679–81
bacteriophage λ, 49
biological methods for studying, 678–86
characterists of organisms vs., 676*t*
destruction of, by adaptive immune response, 987–88
diversification themes of, 686–89
diversity of viral genomes, 687*t*
emerging diseases and emerging, 688–89
gene therapy using retroviruses, 351–54
Hershey-Chase experiment on DNA in genes of, 259–60*f*
HIV/AIDS as current viral pandemic in humans, 677–78
key lineages of, 689–92
nature of viral genetic material, 686
organisms vs., 676*f*
origins of, 686–88
photomicrograph of, 675*f*
proteins as defense against, 45
reasons for studying, 676–78
recent viral epidemics in humans, 676–77
RNA, 281
viral proteins and, 330
Visceral mass, **626**
Visible light, **174**
Vision, 913–18
color vision in nonhuman, 917–18
defects in human, 446
insect eyes and, 913
light-sensing organs, 913
vertebrate eyes and, 914–17
Visual communication, 1029
Visual cues, lizard sexual activity and, 1024
Vitamin A, 355
Vitamin B_1, 54, 843*t*
Vitamin B_6, 277
Vitamin B_{12}, 843*t*
Vitamin C, 843*t*
Vitamin D, 843*t*
Vitamins, **842**
enzyme cofactors and, 54
important, for humans, 843*t*
lipids as, 84
transgenic crops and deficiencies of, 355
Vitelline envelope, **390**
Viviparous species, **615**, **658**, **956**
Volcanic gases, 21, 27, 32, 40, 43
Volt, **887**
Voltage, **745**, **887**
Voltage clamping, **891**–92
Voltage-gated channels, **891**–92
Voluntary muscle, **808**
Voluntary responses, 899
Von Békésy, Georg, 912
Von Frisch, Karl, 1028–29

W

Waldeyer, Wilhelm, 195
Wallace, Alfred Russel, 4, 414, 415–16, 1014
Wallace line, **1014**
Wall pressure, **718**
Warren, Robin, 850
Wasps, 640*t*, 780–81
Wastes. *See also* Excretory systems; Urinary systems
animal large intestines and, 854–55
animal nutrition and elimination of, 841–42
nitrogenous, 829–30
Water. *See also* Solubility
absorption of, in animal large intestines, 855
acid-base reactions and proton transfer in, 25–26
animal small intestine absorption of, 854
animal water balance (*see* Osmoregulation)
aquaporins and transport of, across plasma membranes, 95–96
behavior of oxygen and carbon dioxide in, 863–64
biomes and precipitation, 1002
cohesion, adhesion, and surface tension of, 22–23
density of, as liquid and as solid, 23–24
depth of, in aquatic ecosystems, 996–97
efficiency of, as solvent, 22
energy absorbing capacity of, 24–25
energy transformations in falling, 28–29
evaporation, 816
flow of, in aquatic ecosystems, 996
green plant holding of, and moderate climate, 547
lipid bilayers and, 86
lipids, hydrocarbons, and, 83
loss of (*see* Water loss)
metabolic, 803
molecular structure of, 18, 19, 20, 21*f*
nitrate pollution of, 511–12
osmosis as diffusion of, across membranes, 90–92
oxidation of, by photosystem II, 181
oxidation of, during photosynthesis, 181
plant biochemical pathway adaptations to water stress, 728
plant nutrients from, 739*t*
plant transport of (*see* Water transport, plant)
polarity of amino acid side chains and solubility in, 42
properties of, 22–25, 26*t*
seed germination and, 799–800
solubility of amino acids in, 42*t*
specific heat of, 24*t*
in volcanic gases, 27
Water balance, animal. *See* Osmoregulation
Water bears, 637
Water breathers, 863–64
Water-conducting cells. *See* Vascular tissue
Water cycle, global, 1095
Water loss
animal terrestrial adaptations to prevent, 628
insect minimization of, 828–29
land plant adaptations to prevent, 553–54
limiting of, in plants, 727–28
as osmotic stress, 825
seed maturation and, 797
stomata and regulation of, 708, 717
Water molds, 542
Water potential, **718**
adaptations of plants in salty or dry habitats, 720–21
calculating, 719–20
creating gradients of, 725–26
potential energy and, 718
pressure potential role in, 718–19
in soils, plants, and atmosphere, 720–21
solute potential role in, 718
transpiration and, 717–18
units of, 719
water movement without pressure and, 719–20
water movement with solute potential and pressure potential and, 720
Water-potential gradient, **721**
Watersheds, **1093**
Water table, **1095**
Water-to-land transition adaptations
of land plants, 553–65
of protostomes, 627–28
Water transport, plant, 717–28. *See also* Translocation
from roots to shoots, 721–27
secondary xylem and, 713
water absorption, water loss, and, 727–28
water potential and, 717–21
Water vascular system, **647**
Watson, James, 59*f*, 62–65
Watson-Crick pairing, **64**. *See also* Complementary base pairing
Wavelength, **174**, 176
Waxes, **87**–88, 133, 707, 828–29
Weather, **1002**, 1010, 1011*f*
Weathering, **741**, 1093
Weeds, 354, **787**, **1075**
Wei, Chunfang, 726
Weinberg, Wilhelm, 436
Weinert, Ted, 205
Weintraub, Harold, 322, 398
Weismann, August, 211, 220
Weissbach, Arthur, 186–87
Welwitschia, 576
Went, Fritz, 760–61
Wet habitats, 793
Wetlands, **998**
Wexler, Nancy, 348, 350
Whales
evolution of, 422
phylogenies of, 477–79
Wheat, 144, 249
Whisk ferns, 572
White blood cells (leukocytes), **870**, 975–76, 977*t*
White pelicans, 1037*f*
Whole-genome sequencing, 359–63. *See also* DNA sequencing; Genomics
cataloging genetic diversity with, 1106
growth in sequenced data of, 359–60
identifying genes in sequences, 362–63
impact of next-generation sequencing strategies on, 360–61
reasons for selecting genomes to sequence, 361–62
shotgun sequencing in, 360, 361*f*
Whooping cranes, 1026, 1043–44, 1055
Whorled leaves, 702*f*
Wieschaus, Eric, 379–81
Wildlife corridors, 1055, **1121**
Wild types, **241**
Wilkins, Maurice, 63
Willow trees, 738
Wilmut, Ian, 378
Wilson, Edward O., 1077–78
Wilson, Margo, 1034
Wilt, **721**
Wind
biomes and, 1002
erosion, 742
nutrient loss and gain by, 1093
pollination by, 576–77*f*, 794
Wind/touch responses, plant, 766–67, 780*t*
Wings
bird, 657–68
insect, 629
as morphological innovation, 486
Wisdom teeth, 844
Withdrawal contraceptive method, 966*t*
Wittlinger, Matthias, 10–12
Wnt (wingless) genes, 384
Wobble hypothesis, **299**–300
Woese, Carl, 6–7, 503
Wolves, 1087
Women, fertility rates and, 1053. *See also* Females
Wood, 548, **555**, 575, 580, **712**, 713
Word roots, Latin and Greek, 8, B:6
World Parks Congress, 1120
World population, 1052–53
Wormlike phyla, 607, 627
Worms, **607**, 620, 627
Wound repair
immune systems and, 973
mitosis and, 195
platelets, growth factors, and, 207
Wounds, plant
attraction of parasitoids with pheromones from, 780–81
parenchyma cells and, 708
systemin as wound-response hormone, 779
Wright, Sewall, 445
Wuchereria bancrofti, 973*f*

X

Xanthophylls, 176, 532*t*
X chromosomes, 212, 241, 246, 352
Xenopus laevis, 203–4
Xenoturbellida, 603*t*, 646
Xeroderma pigmentosum (XP), **272–74**
X-linked immune deficiency, 352–54
X-linked inheritance (X-linkage), **242**, 251–52, 352
X-ray crystallography, **63**, B:15–B:16
 in cancer study, 332–33
 of eukaryotic DNA, 321
 in HIV research, 682–83
 of myosin head, 924
 on tRNA structure, 299
X-ray film, 185, 196
Xu, Xing, 657
Xylem pressure probe, 726–27
Xylem sap, **726**
Xylem tissue, **710**, **804**
 cohesion-tension forces in, 726–27
 primary plant body, 711*t*
 vascular cambium and, 713
 vascular tissue system functions of, 710–11
 water transport into, 722
Xylene, 24*t*

Y

Y chromosomes, 212, 241
Yeast extracts, 155
Yeasts, **582**. *See also Saccharomyces cerevisiae*
Yellowstone National Park, 1087
Yew trees, 575
Y-linked inheritance (Y-linkage), **242**, 251
Yogurt, 513*f*
Yoked hyphae, 583–84, 595
Yolks, **390**, 959
Yucca moths, 1015

Z

Z disk, **922**
Zeatin, 768
Zeaxanthin, 176, 758
Zebra finches, 453
Zebra mussels, 634
Zero population growth (ZPG), **1053**
Zinc, 54, 739*t*
Zinc finger, 944
Zipper, 292
Zona pellucida, **390**
Zone of cellular division, **706**
Zone of cellular elongation, **706**
Zone of cellular maturation, **706**
Zone of maturation, **744**
Zone of nutrient depletion, 744
Z scheme, **182**
 cyclic photophosphorylation, ATP production, and, 183–84
 enhancement effect and, 183
 location of photosystems I and II, 184
 as model for interactions between photosystem I and II, 182–84
 plastocyanin and, 182–83
Zygomycetes, 583–84, 585, 592, 595
Zygosporangia (zygosporangium), 583–**84**, 592
Zygotes, **215**, **390**, 392–94, **784**, 792, 796, **954**

Boldface page numbers indicate a glossary entry; page numbers followed by an *f* indicate a figure; page numbers followed by *t* indicate a table.